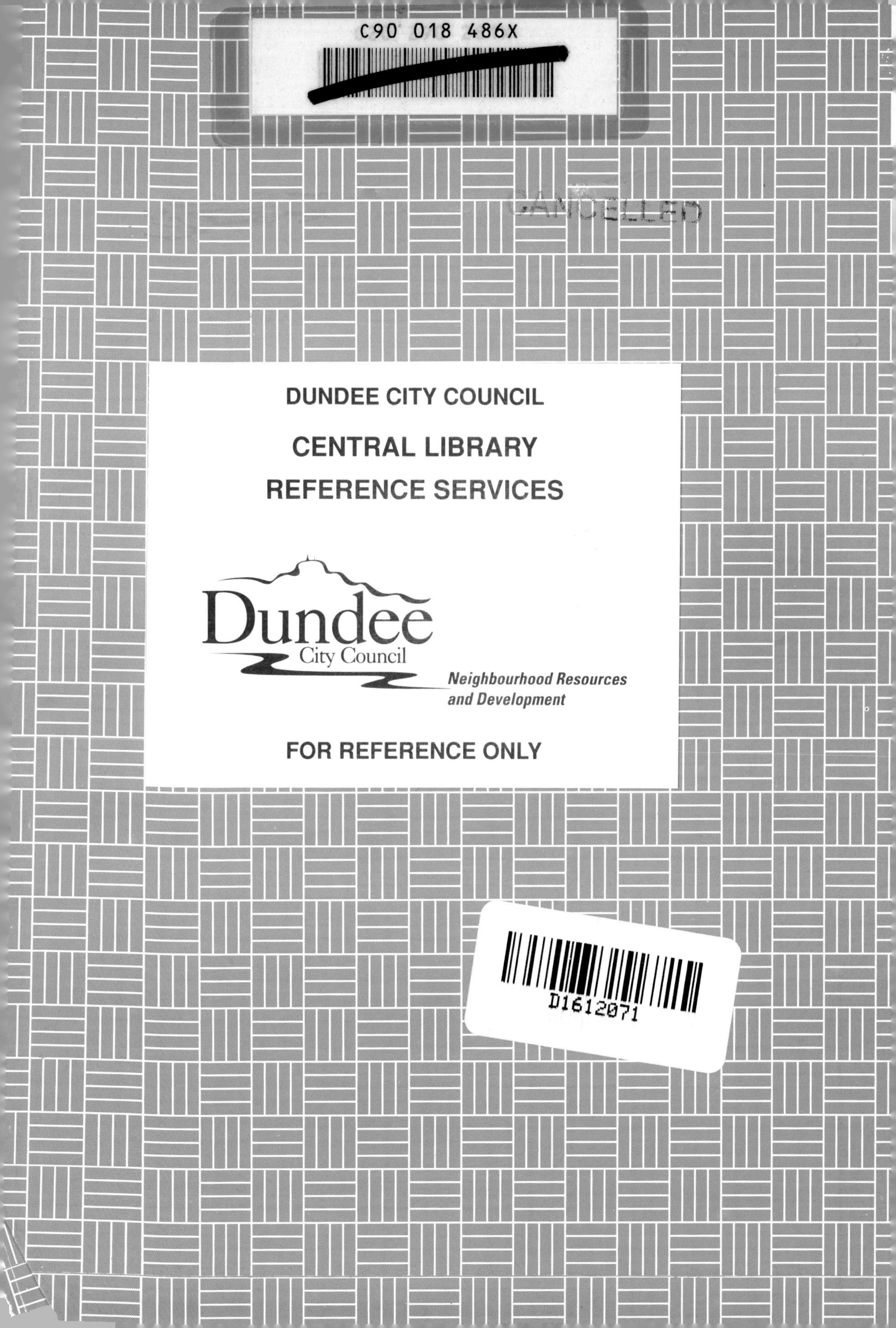
C90 018 486X
CANCELLED
DUNDEE CITY COUNCIL
CENTRAL LIBRARY
REFERENCE SERVICES
Dundee
City Council
Neighbourhood Resources
and Development
FOR REFERENCE ONLY
D1612071

# German Dictionary of Construction

# Wörterbuch Bauwesen

## Deutsch–Englisch/Englisch–Deutsch

# Routledge German Dictionary of Construction

# Wörterbuch Bauwesen

## Deutsch–Englisch/Englisch–Deutsch

First published 1997
by Routledge
11 New Fetter Lane, London EC4P 4EE
Simultaneously published in the USA and Canada
by Routledge
29 West 35th Street, New York, NY 10001


Typeset in Monotype Times, Helvetica Neue and Bauer Bodoni
by Routledge

Printed in England by TJ Press (Padstow) Ltd., Cornwall

Printed on acid-free paper



*British Library Cataloguing-in-Publication Data*
A catalogue record for this book is available from the British Library

*Library of Congress Cataloging-in-Publication Data*
Applied for

ISBN: 0-415-11242-7

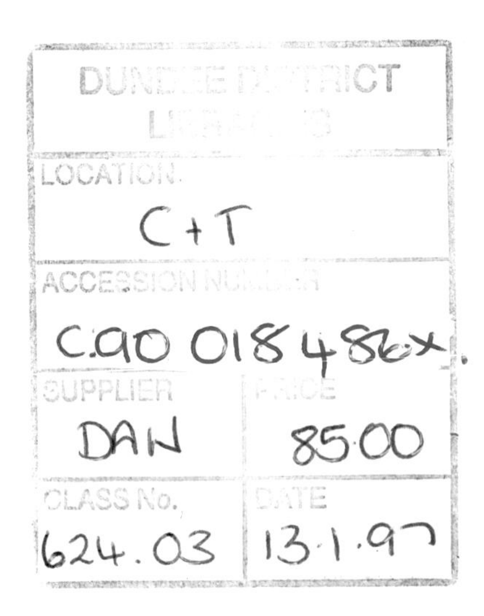

German Dictionary of Construction
Wörterbuch Bauwesen Deutsch–Englisch/Englisch–Deutsch

**Project Manager/Projektmanagement**
Susanne Jordans

**Programme Manager/Programmleitung**
Elizabeth White

**Managing Editor/Redaktionsleitung**
Sinda López

**Editorial/Lektorat**
Martin Barr Lisa Carden Janice McNeillie Robert Timms

**Marketing**
Vanessa Markey Rachel Miller Judith Watts

**Systems/Datenbanksystem**
Omar Raman Simon Thompson

**Administration/Verwaltung**
Amanda Brindley

**Production/Herstellung**
Michelle Draycott Maureen James Nigel Marsh Joanne Tinson

**Terminologists/Terminologie**
Dr Hans-Dieter Junge Christoph Mewes Anthony Kirk Dieter Lukhaup
Ulrike Seeberger Dorothy Thierstein Günter Thierstein

**Consultants/Beratung**
Anna Faherty Madeleine Metcalfe

**Lexicographers/Lexikographie**
Stephen Curtis Anke Kornmüller Karin Schulz

**Proofreaders/Korrekturlesen**
Hazel Curties Martina Dervis Gunhild Prowe Karoline Seel-Gannot
Sabine Schildknecht Martin Stark Vivienne Wattenhofer Jill Williams

**Keyboarders/Datenerfassung**
Kristoffer Blegvad Sara Fenby Antonio Fernández Entrena Rosa Gálvez López
Christiane Grosskopf Ute Krebs Ilona Lehmann Geir Moulson Fabienne Rangeard
Beate Schmitt Debbie Thomas

**Acknowledgements**

We are particularly grateful to Dieter Lukhaup, Bryan Pepper and Heinz R. Schuldes for providing us with the German subject area names.

American terms supplied by Robert Brissot, Dorothy Thierstein and Dr Stephen S. Wilson.

**Danksagung**

Wir bedanken uns besonders bei Dieter Lukhaup, Bryan Pepper und Heinz R. Schuldes, die die deutschen Fachgebietsnamen auswählten.

Einträge in amerikanischem Englisch stammen von Robert Brissot, Dorothy Thierstein und Dr. Stephen S. Wilson.

# Contents/Inhalt

# Foreword/Vorwort

The art of construction practised in the UK and Germany is not a science and only partly academic, yet building and construction cover an ever increasing range of subjects, including the use of rapidly developing new technologies, as well as the enhancement of traditional skills.

The *Routledge German Dictionary of Construction* includes specialized technical words and phrases that are in common use in the industry. The generic elements that are important in many architectural and structural designs are incorporated, covering building physics, building pathology, construction law, project management, specification writing, quality assurance, third-party testing, insurance, liability; and in the UK and in the former British Commonwealth countries, quantity surveying.

Each country within the European Union uses its own unique construction terminology which is in daily use in that country; the *Routledge German Dictionary of Construction* reflects this and presents authenticated terms in a coherent form. Considerable care has been taken to avoid inconsistent or misleading terminology.

Costly confusion can be avoided by using the correct construction term in specifications and contract documents. This dictionary covers the basic vocabulary of the industry and has been enhanced by the inclusion of terms from sixteen commonly recognized subject areas.

The unprecedented current German construction programme means that many nationalities are working in the country and there is a need to respond quickly and accurately to instructions, technical specifications, building contracts, third-party testing, laboratory reports and quality control authorities. Errors in language delay approval of projects, cause work to be stopped, and can ultimately make for bad working relations with both the client and building workforce.

The European Committee for Standardization (CEN) is responsible for establishing

Die in Großbritannien und Deutschland ausgeübte Baukunst ist keine Wissenschaft und nur teilweise eine akademische Disziplin, jedoch umfaßt das Bauwesen eine ständig wachsende Zahl an Fachgebieten, die sowohl den Gebrauch von sich schnell entwickelnden neuen Technologien als auch die Verbesserung traditioneller handwerklicher Fertigkeiten einschließen.

Das *Routledge Wörterbuch Bauwesen Deutsch–Englisch/Englisch–Deutsch* enthält Fachwörter und Formulierungen, die in der Bauwirtschaft allgemein gebräuchlich sind. Aufgenommen wurden Oberbegriffe, die in vielen architektonischen und baulichen Entwürfen wichtig sind, einschließlich Begriffen aus der Bauphysik, Baupathologie, dem Baurecht, der Projektsteuerung, Baubeschreibung, Qualitätssicherung, Prüfung durch eine dritte Person, Versicherung, Haftpflicht und in Großbritannien und in den Ländern des ehemaligen britischen Commonwealth auch der Massenermittlung und Baurechnung.

In der Europäischen Union hat jedes Land seine eigene spezifische Bauterminologie, die in dem jeweiligen Land täglich im Gebrauch ist; das *Routledge Wörterbuch Bauwesen Deutsch–Englisch/Englisch–Deutsch* spiegelt dies wider und stellt gültige Begriffe in verständlicher Form dar. Besondere Sorgfalt wurde darauf verwendet, widersprüchliche oder irreführende Fachvokabeln zu vermeiden.

Durch den Gebrauch der richtigen Fachwörter in Baubeschreibungen und Vertragsunterlagen können kostspielige Mißverständnisse vermieden werden. Das Wörterbuch enthält das Grundvokabular der Bauindustrie und wurde durch Begriffe aus sechzehn allgemein anerkannten Themenkomplexen erweitert.

Aufgrund des bisher einmaligen, gegenwärtigen deutschen Bauprogrammes arbeiten Menschen aus vielen verschiedenen Nationen im Land, und es ist notwendig, schnell und genau auf Arbeitsanweisungen, technische Baubeschrei-

harmonized standards between the member states of the European Union. One of the areas of importance is understanding other countries' standards, particularly those with a well-documented and respected method of working. Germany falls into this category and there are organizations based in Berlin which have a major interest in product assessment and certification: *Bundesanstalt für Materialforschung und Prüfung (BAM),* the Federal German testing establishment; *Deutsches Institut für Normung e. V. (DIN),* the Federal German standards organization and German member of CEN; *Normenausschuß Bauwesen im DIN (NABau im DIN),* the building division of the organization; and *Institut für Bautechnik (IfBt).*

In the UK, quality assurance and product certification have developed around three national bodies: the British Standards Institution (BSI), the National Accreditation Council for Certification Bodies (NACCB) and the National Measurement Accreditation Service Executive (NAMAS), under which the functions of the National Laboratory Testing Accreditation Service (NATLAS) and the British Calibration Service (BCS) are combined at the National Physical Laboratory (NPL).

Although the sixteen states, or *Länder,* in Germany, make their own regulations relating to construction and planning, whilst in the UK a unified system of building regulations has been adopted, European Standards and Codes are common to both. In addition, third-party testing and quality control are on the increase and demanded by both property developers and individual clients. The Building Research Establishment (BRE) is the principal UK organization carrying out research into building and construction, and in recent years has concentrated on promoting the importance of building well. The BRE is an active member of the International Council for Building Research, Studies and Documentation (CIB), as are the German institutes and technical universities.

Concern over environmental issues and the introduction of green building products and techniques has created a number of new terms related to the subject. Sometimes, the complex building physics has been abbreviated and more accessible words have been introduced into the text. The *Routledge German Dictionary of Construction* has included only those terms in common professional usage.

In the UK, the role of the quantity surveyor

bungen, Bauverträge, Prüfungen durch eine dritte Person, Laborberichte und Qualitätssicherungsbehörden zu reagieren. Sprachliche Fehler verzögern Baugenehmigungen, bringen die Arbeit zum Stillstand und können letztendlich schlechte Arbeitsbeziehungen sowohl zum Bauherrn als auch zu den Bauarbeitern verursachen.

Das Europäische Komitee für Normung (CEN) ist für die Harmonisierung der Normen in den Mitgliedsstaaten der Europäischen Union verantwortlich. Ein wichtiger Bereich ist dabei, die Normen anderer Länder zu verstehen, besonders jene aus Ländern mit einer gutdokumentierten und anerkannten Arbeitsmethode. Deutschland zählt zu dieser Kategorie, und es gibt Organisationen in Berlin, die großes Interesse an Produktbeurteilung und -zulassung haben, z.B. die Bundesanstalt für Materialforschung und Prüfung (BAM), das Deutsche Institut für Normung e.V. (DIN) das deutsche Mitglied von CEN, der Normenausschuß für Bauwesen im DIN (NABau im DIN) die Bauabteilung der Organisation, und das Institut für Bautechnik (IfBt).

In Großbritannien haben sich Qualitätssicherung und Produktbescheinigung im Umfeld von drei nationalen Institutionen entwickelt: der *British Standards Institution (BSI),* dem *National Accreditation Council for Certification Bodies (NACCB)* und dem *National Measurement Accreditation Service Executive (NAMAS),* in dem die Aufgaben des *National Laboratory Testing Accreditation Service (NATLAS)* und des *British Calibration Service (BCS)* im *National Physical Laboratory (NPL)* zusammengefaßt sind.

Obwohl die sechzehn Bundesländer in Deutschland jeweils ihre eigenen Vorschriften in bezug auf Bauwesen und Planung aufstellen, während in Großbritannien ein einheitliches System von Bauverordnungen eingeführt wurde, gelten für beide Länder dieselben europäischen Normen und Regeln. Außerdem werden Prüfung durch eine dritte Person sowie Qualitätssicherung in zunehmendem Maße gefordert, und zwar gleichermaßen von Bauträgern und Bauherren. Das *Building Research Establishment (BRE)* ist die wichtigste Organisation in Großbritannien, die im Bereich des Bauwesens Forschung betreibt; es hat sich in den letzten Jahren vor allem für die Bedeutung guten Bauens eingesetzt. Das *BRE* ist ein aktives Mitglied des *Conseil International du Bâtiment pour la Recherche, l'Étude et la Documentation (CIB),* ebenso wie deutsche Institute

has flourished and in many ways has taken over the traditional role of the architect. Elsewhere in Europe this is not the case, and the quantity surveyor function is performed by either the architect or consulting engineer as part of their professional service. Sometimes a technical economist is employed whose work broadly covers work measurement and cost-benefit analysis.

Whilst German and British practice differs in a number of detailed ways, there are many similar features. *The Routledge German Dictionary of Construction* is the common denominator that will enable architects, building physicists, building control officers, civil, service and structural engineers, contractors, lawyers, planners, researchers, surveyors and technicians – in short, the building team – to communicate accurately, effectively and without misunderstanding.

*Anthony Kirk, Dip Arch, RIBA, FBIM, Chartered Architect, London, March 1996*

und technische Universitäten.

Besorgnis über Umweltfragen und die Einbringung "grüner" Bauprodukte und Bautechniken haben eine Anzahl neuer Begriffe in diesem Gebiet hervorgebracht. Manchmal sind komplizierte bauphysikalische Ausdrücke abgekürzt und einfachere Wörter in den Text aufgenommen worden. Im *Routledge Wörterbuch Bauwesen Deutsch–Englisch/Englisch–Deutsch* sind nur diejenigen Fachwörter enthalten, die im beruflichen Umgang gebräuchlich sind.

In Großbritannien hat die Rolle des Massenermittlers und Baukostensachverständigen sehr an Bedeutung gewonnen, und in vielen Fällen hat er die traditionelle Rolle des Architekten übernommen. Anderswo in Europa ist dies nicht der Fall, und der Architekt oder Ingenieur führt die Aufgaben des Baukostensachverständigen als Teil seiner Leistungen aus. Manchmal wird ein technischer Baukostengutachter hinzugezogen, dessen Arbeit gewöhnlich Massenberechnung und Kosten-Nutzen-Analyse einschließt.

Während sich die deutschen und britischen Verfahrensweisen in einigen Detailbereichen unterscheiden, haben sie aber auch vieles gemeinsam. Das *Routledge Wörterbuch Bauwesen Deutsch–Englisch/Englisch–Deutsch* ist der gemeinsame Nenner, der es Architekten, Bauphysikern, Bauaufsichtsbehörden, Ingenieuren, Bauunternehmern, Rechtsanwälten, Planern, Forschern, Landvermessern und Technikern – kurzgesagt dem Bauteam – ermöglicht, genau, effektiv und ohne Mißverständnisse miteinander zu kommunizieren.

*Übersetzung vom Redaktionsteam*

# Introduction/Einleitung

Following on from the programme of general technical and business dictionaries launched in October 1994, the *German Dictionary of Construction* is the second of a new series of specialist dictionaries to be published by Routledge. Given the growing importance of construction in everyday life and the fact that English–German as a language combination has become increasingly important over recent years, this dictionary will be a valuable resource for specialist translators, terminologists and experts working in all fields of construction. Approximately 25,000 entries in each language cover not only terminology used in Architecture & Building Planning and Infrastructure & Design, but also developing areas such as Environmental Issues.

It would not have been possible to compile this dictionary within a realistic timescale, and to the standard achieved, without the use of a custom-designed database. The database used is a relational one: term records for each language are held in separate files, with further files consisting only of link records. Links between terms in different language files represent translations which enable us to handle various types of one-to-many and many-to-one translation equivalences. Links between terms within a single-language file represent cross-references, themselves of a wide variety of types: spelling variants, geographical variants and abbreviations.

The content of the database for this dictionary was compiled and vetted by a team of terminologists (English and German) with current practical experience of a narrowly-defined subject area within construction in order to ensure their currency, the accuracy of explanations, and the adequacy of coverage. Finally, each language file was reviewed by editors to ensure coverage of North American terms and spelling variants; these are clearly labelled and distinguished.

The compilation and editing of the database of terms was, however, only the first stage in the

Das vorliegende *Wörterbuch Bauwesen Deutsch–Englisch/Englisch–Deutsch* ist das zweite im Rahmen einer neuen zweisprachigen Fachwörterbuchreihe von Routledge, die sich den seit Oktober 1994 kontinuierlich erscheinenden Wörterbuchreihen Technik sowie Wirtschaft, Handel und Finanzen anschließt. Das Bauwesen spielt eine wachsende Rolle im Alltagsleben, und gleichzeitig hat die Sprachkombination Englisch–Deutsch in den letzten Jahren an Bedeutung gewonnen. Auf diesem Hintergrund ist das vorliegende Wörterbuch ein wertvolles Mittel für alle Fachübersetzer und Fachübersetzerinnen, Terminologen und Terminologinnen sowie Fachleute im Bereich Bauwesen. Etwa 25 000 Einträge in jeder Sprache decken sowohl die klassischen Gebiete des Bauwesens, wie etwa Architektur & Tragwerksplanung und Infrastruktur & Entwurf ab, als auch Gebiete, die sich erst kürzlich entwickelt haben, wie Umweltthematik.

Es wäre ohne den Einsatz einer für unsere Bedürfnisse speziell zugeschnittenen Datenbank unmöglich gewesen, dieses Wörterbuch innerhalb einer realistischen Zeitspanne und in der vorliegenden Qualität zusammenzustellen. Die Datenbank ist relational: Die Datensätze zu den einzelnen Stichwörtern sind für jede Sprache in einer separaten Datei untergebracht, während gesonderte Dateien nur Datensätze enthalten, die die Verbindungen zwischen Ausgangs- und Zielsprache herstellen. Diese Verbindungen zwischen den Sprachen stellen die Übersetzungen dar, sie schaffen die Möglichkeit, unterschiedliche Verknüpfungen herzustellen: von einem Ausgangswort zu mehreren Übersetzungen oder auch von mehreren Ausgangswörtern zu einer gemeinsamen Übersetzung. Innerhalb der einsprachigen Dateien lassen sich ebenfalls Verknüpfungen schaffen, nämlich Querverweise verschiedener Art wie auf abweichende Schreibweisen, geographische Varianten und Abkürzungen.

making of the dictionary. Within the database, the distinction between source and target languages is not meaningful, but for this printed dictionary it was necessary to format the data to produce separate German–English and English–German sections. The data was processed by a further software module to produce two alphabetic sequences, of German headwords with English translations and vice versa, each displaying the nesting of compounds, ordering of translations, style for cross-references of different types, and other features according to the algorithm.

At this stage the formatted text was edited by a team of experienced English and German lexicographers whose task it was to eliminate duplication or inconsistency; edit the contextual information; and to remove terms that were too general for inclusion in a specialist dictionary.

The editorial team

Der Inhalt der Datenbank für dieses Wörterbuch wurde von einer Gruppe von deutsch- und englischsprachigen Terminologen, die über praktische Erfahrung in eng umrissenen Teilgebieten des Bauwesens verfügt, zusammengestellt und überprüft. So wurde sichergestellt, daß die Wortliste höchste Aktualität hat, daß die Worterklärungen präzise sind und daß wirklich das in Fachkreisen geläufigste Vokabular behandelt wird. Schließlich wurde noch jede Sprachdatei von Redakteuren überprüft, um eine internationale Terminologieabdeckung zu gewährleisten. Varianten in britischem und nordamerikanischem Englisch werden so unterschieden und sind entsprechend gekennzeichnet.

Die Sammlung und Überprüfung der Datenbankgrundlage war jedoch nur der erste Schritt bei der Herstellung dieses Wörterbuches. Innerhalb einer Datenbank erübrigt sich die Unterscheidung zwischen Ausgangs- und Zielsprache im Grunde, aber für das vorliegende Wörterbuch mußten natürlich die Daten so bearbeitet werden, daß separate Teile für Deutsch–Englisch und Englisch–Deutsch entstehen konnten. Mit Hilfe eines weiteren Softwaremoduls wurden zwei alphabetische Listen erstellt, eine deutsche Stichwortliste mit englischen Übersetzungen und umgekehrt. Dabei wurden mit Hilfe eines Algorithmus zusammengesetzte Begriffe in Blöcken aufgeführt, Übersetzungen in sinnvoller Reihenfolge angeordnet und verschiedene Arten von Querverweisen hergestellt.

Dann wurde der formatierte Text von einem Team erfahrener deutscher und englischer Lexikographen redigiert; ihre Aufgabe war es, Doppelnennungen und Unstimmigkeiten auszumerzen, die zum Wortzusammenhang gegebenen Informationen zu bearbeiten und all die Stichwörter zu streichen, die zu allgemeinsprachlich für ein Fachwörterbuch sind.

Das Redaktionsteam

# Features of the dictionary/ Aufbau und Anordnung der Einträge

The main features of the dictionary are highlighted in the text extracts on the following pages. For a more detailed explanation of each of these features and information on how to get the most out of the dictionary, see pages xix–xxi.

Die folgenden Textbeispiele illustrieren Aufbau und Anordnung der Einträge. Weitere Erläuterungen und Hinweise zur Benutzung des Wörterbuches befinden sich auf den Seiten xxiii–xxiv.

*Cross-references from abbreviations to their full forms are shown for both the German and the English translation*

*Sowohl für den deutschen Eintrag wie für die englische Übersetzung werden Querverweise von Abkürzungen auf ihre Vollformen gegeben*

**BSI** *abbr* (*Britisches Institut für Normung*) ARCH & TRAGW, BAURECHT, HOLZ, INFR & ENTW, STAHL, WERKSTOFF BSI (*British Standards Institution*)

**Buche** *f* HOLZ beech; **Buchenholz** *nt* HOLZ, INFR & ENTW, WERKSTOFF beechwood; **Buchenholzschindel** *f* ARCH & TRAGW beech shingle, beechwood shingle; **Buchenparkett** *nt* ARCH & TRAGW, HOLZ beech parquetry

**Buch**: **Buchführung** *f* BAURECHT accounting; **Buchhalter** *m* BAURECHT accountant; **Buchhaltung** *f* BAURECHT accountancy

*Contexts give supplementary information to help locate the right translation*

*Die Angabe von Zusammenhängen ergänzt die gegebenen Informationen und unterstützt die Suche nach dem passenden Übersetzungsäquivalent*

**Buchse** *f* ABWASSER sleeve, ELEKTR socket, HEIZ & BELÜFT sleeve, socket, WERKSTOFF *für Röhre und Kabel* bushing

**Bucht** *f* INFR & ENTW bay

**Buckelquaderverband** *m* NATURSTEIN opus rusticum

**Buckelschweißung** *f* ARCH & TRAGW projection welding

**Budget** *nt* BAURECHT budget

*Genders are indicated at German nouns*

*Bei deutschen Substantiven wird das grammatikalische Geschlecht angegeben*

**Bug** *m* HOLZ, INFR & ENTW, WERKSTOFF angle brace, strut

**Bügel** *m* ARCH & TRAGW binder, fastening, BESCHLÄGE *Schloß* shackle, BETON stirrup, STAHL hoop; **Bügelabstand** *m* BETON pitch of links

*Base form of adjective is replaced by a swung dash in adjective–noun combinations*

*Die Grundform des Adjektivs wird in flektierten Formen durch die Tilde ersetzt*

**bügelbewehrt**: **~e Säule** *f* ARCH & TRAGW, BETON, INFR & ENTW *Stahlbeton* hooped column

***Bügel***: **Bügelschelle** *f* ELEKTR clamp-ring; **Bügelschraube** *f* ARCH & TRAGW strap bolt; **Bügelverlegung** *f* BETON spacing of stirrups

**Buhne** *f* INFR & ENTW groyne

*Subject area labels given in alphabetical order show appropriate translation*

*Sachgebietskürzel in alphabetischer Reihenfolge helfen beim Finden der korrekten Übersetzung*

**Bühne** *f* ARCH & TRAGW attic, platform, stage, BAUMASCHIN platform, INFR & ENTW stage

**Bund-**: **Bundbalken** *m* ARCH & TRAGW, HOLZ binding beam, head runner

**Bündelpfeiler** *m* ARCH & TRAGW multiple rib pillar

**bündig** *adj* ARCH & TRAGW fair-faced, level, *Tür* flush, flush-mounted; **~e Mauerwerksfuge** *f* NATURSTEIN flat joint

**Bündigkeit** *f*: **~ von Stößen** ARCH & TRAGW flush arrangement of joints

*German dummy entries in bold italics indicate the continuation of an earlier nest of compounds*

*Deutscher Blindeintrag in Fettkursivschrift zeigt die Fortsetzung eines vorhergehenden Nests mit zusammengesetzten Formen an*

***Bund-***: **Bundmutter** *f* WERKSTOFF flanged nut; **Bundpfosten** *m* ARCH & TRAGW stud; **Bundsäule** *f* ARCH & TRAGW stud; **Bundständer** *m* ARCH & TRAGW stud; **Bundstiel** *m* ARCH & TRAGW stud

**Bungalow** *m* ARCH & TRAGW bungalow

**Bunker** *m* ARCH & TRAGW, INFR & ENTW bunker

*British English and American English spelling variants are given in full and labelled accordingly*

*Britische und amerikanische Schreibvarianten werden voll ausgeschrieben und entsprechend gekennzeichnet*

**Bunt**: **Buntbartschloß** *nt* BESCHLÄGE, INFR & ENTW warded lock; **Buntglas** *nt* WERKSTOFF colored glass (*AmE*), coloured glass (*BrE*); **Buntmarmor** *m* WERKSTOFF colored marble (*AmE*), coloured marble (*BrE*)

**Bürge** *m* BAURECHT guarantor

**Bürgerhaus** *nt* ARCH & TRAGW public hall

**Bürgersteig** *m* INFR & ENTW pavement (*BrE*), sidewalk (*AmE*)

**Bürgschaft** *f* BAURECHT assurance

**Bürste** *f* OBERFLÄCHE brush

*German terms are ordered in strict letter-by-letter order*

*Deutsche Einträge sind streng alphabetisch geordnet*

**bürsten** *vt* OBERFLÄCHE brush

**Butzenscheibe** *f* ARCH & TRAGW glass roundel, WERKSTOFF bullion

**Bypassventil** *nt* ABWASSER, HEIZ & BELÜFT bypass valve

**Art Nouveau** *n* ARCH & BUILD Jugendstil *m*

*Britische und amerikanische Schreibvarianten werden voll ausgeschrieben und entsprechend gekennzeichnet* — **arts**: **~ center** *AmE*, **~ centre** *BrE* *n* ARCH & BUILD Kulturzentrum *nt* — *British English and American English spelling variants are given in full and are labelled accordingly*

**asbestos** *n* CONST LAW, MAT PROP Asbest *m*; **~ board** *n* MAT PROP Asbestplatte *f*; **~ cement** *n* CONCR Asbestzement *m*, MAT PROP Asbestzement *m*, Eternit® *nt*;

*Zusammengesetzte Begriffe werden alphabetisch hinter dem ersten Element angeordnet* — **~ cement board** *n* CONCR, INFR & DES, MAT PROP Berliner Verbau *m*, Berliner Welle *f*, KW-Platte *f*; **~ cement corrugated roof covering** *n* CONCR, MAT PROP Wellasbestzementdachbelag *m*, Wellfaserzementdachbelag *m*; **~ cement goods** *n pl* CONCR, MAT PROP Asbestzementwaren *f pl*; **~ cement lining** *n* — *Compound forms are nested alphabetically at the first element*

*Die Angabe von Zusammenhängen ergänzt die gegebenen Informationen und unterstützt die Suche nach dem passenden Übersetzungsäquivalent* — CONCR, MAT PROP *fire protection* Asbestzementmörtel *m*, Asbestzementverkleidung *f*; **~ cement panel** *n* CONCR, MAT PROP Asbestzementtafel *f*; **~ cement partition** *n* CONCR, MAT PROP Asbestzementtrennwand *f*; **~ cement product** *n* CONCR, MAT PROP Asbestzementerzeugnis *nt*; **~ cement shingle** *n* CONCR, MAT PROP Asbestzementschindel *f*; **~-containing** *adj* CONST LAW, MAT PROP asbesthaltig; **~ cord** *n* MAT PROP Asbestschnur *f*; **~ dust** *n* CONST LAW, MAT PROP Asbeststaub *m*; **~ fiber** *AmE*, **~ fibre** *BrE* *n* CONST LAW, MAT PROP Asbestfaser *f*; **~ gloves** *n* — *Contexts give supplementary information to help locate the right translation*

*Bei deutschen Substantiven wird das grammatikalische Geschlecht angegeben* — *pl* MAT PROP Asbesthandschuhe *m pl*; **~ insulating board** *n* MAT PROP Asbestdämmplatte *f*; **~ panel** *n* MAT PROP Asbesttafel *f*; **~ pipe** *n* MAT PROP Asbestrohr *nt*; **~ product** *n* MAT PROP Asbesterzeugnis *nt*; **~ PVC floor tile** *n* MAT PROP Asbest-PVC-Bodenplatte *f*; **~ rock** *n* MAT PROP Asbestgestein *nt*; **~ roofing sheet** *n* MAT PROP Asbestdachplatte *f*; **~ sheet** *n* MAT PROP Asbestplatte *f*; **~ wool** *n* MAT PROP Asbestwolle *f* — *Genders are indicated at German noun translations*

**as-built**: **~ drawing** *n* ARCH & BUILD, CONST LAW Bestandszeichnung *f*, Revisionsplan *m*, Revisionszeichnung *f*

**ash** *n* ENVIRON, HEAT & VENT *burnt material* Asche *f*, TIMBER *tree* Esche *f*; **~ and combustion residue** *n* ENVIRON Asche- und Verbrennungsrückstand *m*; **~ content** *n* MAT PROP Aschegehalt *m*; **~-free** *adj* ENVIRON aschefrei

*Sachgebietskürzel in alphabetischer Reihenfolge helfen beim Finden der korrekten Übersetzung* — **ashlar** *n* ARCH & BUILD, INFR & DES, MAT PROP, STONE Naturstein *m*, Naturwerkstein *m*, Quader *m* — *Subject area labels given in alphabetical order show appropriate translation*

**ashlaring** *n* ARCH & BUILD, INFR & DES, STONE Natursteinmauerwerk *nt*, Quadermauerwerk *nt*, Werksteinmauerwerk *nt*

*Englischer Blindeintrag in Fettkursivschrift zeigt die Fortsetzung eines vorhergehenden Nests mit zusammengesetzten Formen an* — ***ashlar***: **~ masonry** *n* ARCH & BUILD, INFR & DES, STONE Natursteinmauerwerk *nt*, Quadermauerwerk *nt*, Werksteinmauerwerk *nt* — *English dummy entries in bold italics indicate the continuation of an earlier nest of compounds*

*Sowohl für den deutschen wie den englischen Eintrag werden Querverweise auf Abkürzungen gegeben* — **ASHRAE** *abbr AmE* (*American Society of Heating, Refrigeration and Air Conditioning Engineers AmE*) HEAT & VENT ASHRAE (*Amerikanische Gesellschaft der Heizungs-, Kühlungs- und Klimatechniker*) — *Cross-references to abbreviations are shown for both the English and the German terms*

***ash***: **~ veneer** *n* TIMBER Eschefurnier *nt*

**aspect** *n* ARCH & BUILD *appearance* Aspekt *m*, Aussehen *nt*, Erscheinung *f*, CONST LAW Aspekt *m*, INFR & DES *appearance* Aussehen *nt*, Erscheinung *f*, *element* Aspekt *m*, *point of view* Gesichtspunkt *m*

*Englische Einträge sind streng alphabetisch geordnet* — **asphalt 1.** *n* ARCH & BUILD Asphalt *m*, Bitumen *nt*, CONCR Asphalt *m*, INFR & DES, MAT PROP, SURFACE Asphalt *m*, Bitumen *nt*; **2.** *vt* INFR & DES, MAT PROP, SURFACE asphaltieren — *English terms are ordered in strict letter-by-letter order*

# Using the dictionary

### Range of coverage

This is a specialist German–English/English–German dictionary that covers the whole range of construction and the scientific knowledge that underlies it. It contains a broad base of terminology drawn from underlying areas of technology such as Architecture & Building Planning and Infrastructure & Design. This dictionary also includes the vocabulary of newly prominent subject areas such as Construction Law, Environmental Issues and Waste Water Treatment.

### Selection of terms

We have aimed to include the essential vocabulary of each specialist area, and the material has been checked by leading subject experts to ensure that both the English and the German terms are accurate and current, that the translations are valid equivalents, and that there are no gaps in coverage.

We have endeavoured to include only genuine specialist terms. Coverage of the subject areas is given proportionally so that an established area, for example, Architecture & Building Planning, has a count of around 7,000 terms, whereas a new area in which terminology is still developing, such as Environmental Issues, will have considerably fewer terms.

### Placement of terms

All terms are ordered alphabetically. This is also the policy for hyphenated compounds.

### Stoplists

Terms in German are not entered under the following elements:

aber, alle, allein, alleine, alleinig, alleinige, alleinigem, alleinigen, alleiniger, alleiniges, allem, allen, aller, alles, allgemein, allgemeine, allgemeinen, allgemeiner, allgemeines, als, am, an, andere, anderem, anderen, anderer, anderes, ans, auch, auf, aufeinander, aufs, aus, außen, außer, äußere, äußerem, äußeren, äußerer, äußeres, bei, beim, das, dasselbe, dem, demselben, den, denselben, der, derselbe, derselben, des, desselben, die, dies, diese, dieselbe, dieselben, diesem, diesen, dieser, dieses, durch, ein, eine, einem, einen, einer, eines, es, etwas, für, im, in, ins, kein, keine, keinem, keinen, keiner, keines, mit, miteinander, nach, neben, neu, nicht, nur, ob, oben, oberem, oberen, oberer, oberes, ohne, per, pro, sehr, sein, seine, seinem, seinen, seiner, seines, selber, selbst, sich, über, um, ums, und, unter, untere, unterem, unteren, unterer, unteres, unterm, vom, von, vor, vors, während, wenig, wenigem, wenigen, weniger, weniges, wenigst, wenigste, wenigstem, wenigsten, wenigster, wenigstes, wieder, zu, zum, zur, zwischen

Terms in English are not entered under the following elements:

a, all, an, any, anybody, anyone, anything, anywhere, are, be, by, during, each, every, everybody, everyone, everything, everywhere, for, from, here, if, in, is, it, no, nobody, no one, nor, not, nothing, nowhere, of, off, on, or, out, over, so, some, somebody, someone, something, somewhere, that, the, then, there, they, thing, this, to, too, under, very, where, while, who, with

Compound terms are listed at their first element. When this first element is itself a headword with a technical sense of its own, compound forms follow the simple form. In the case of noun nests, open forms are listed first with the headword replaced by a swung dash.

If the first element is not itself translated, a colon precedes the compounds. For example:

**accidental: ~ discharge** *n* ENVIRON zufälliger Ausfluß *m*, *of oils, chemicals* havariebedingter Ausfluß *m*

German compounds are spelled out in full and directly follow the nest of open forms. For example:

**Energie** *f* ARCH & TRAGW, INFR & ENTW, UMWELT energy; **~ aus Abfall** UMWELT residue-derived energy; **Energieaufnahme** *f* ARCH & TRAGW, INFR & ENTW energy absorption; **Energieaustausch** *m* HEIZ & BELÜFT, INFR & ENTW, UMWELT energy exchange

Compounds are entered in alphabetical sequence. When a nest is interrupted by other entries, the run of compounds is picked up again later in the correct alphabetical sequence. A dummy headword in bold italics indicates the continuation of a run of compounds. For example:

**Abfuhr** *f* ABWASSER carrying off, UMWELT collection
**Abführbeständigkeit** *f* OBERFLÄCHE chalk resistance
**abführen** *vt* ABWASSER carry off
***Abfuhr*: Abfuhrwagen** *m* UMWELT tipper truck, skip

Adjective–noun combinations are entered at the base form of the adjective. In the nest, the base form is replaced by a swung dash and followed by the relevant inflection. For example:

**dynamisch** *adj* INFR & ENTW dynamic; **~ belasteter Balken** *m* ARCH & TRAGW, INFR & ENTW dynamically loaded beam; **~e Belastung** *f* ARCH & TRAGW, INFR & ENTW dynamic loading; **~e Festigkeit** *f* ARCH & TRAGW, INFR & ENTW dynamic strength; **~es Knicken** *nt* ARCH & TRAGW, INFR & ENTW dynamic buckling; **~e Kraft** *f* INFR & ENTW dynamic force

In German, definite and indefinite articles (*der, die, das, ein, eine*), their inflected forms (*dem, den, des, einem, einen, einer, eines*), the pronoun *es*, demonstrative pronouns (*dasselbe, derselbe, diese, dieser, dieselbe, dieses, selber, selbst*), their inflected forms (*demselben, denselben, derselben, desselben, dieselben, diesem, diesen*), and possessive pronouns (*sein, seine*), their inflected forms (*seinem, seinen, seiner, seines*), the pronoun *etwas*, the adverbs *etwa, nicht, nur, oben, unten*, prepositions with or without articles (*am, an, auf, aufs, aus, bei, beim, durch, für, gegen, im, in, ins, mit, nach, ohne, per, pro, über, übers, um, ums, unter, unterm, vom, von, vor, vors, zu, zum, zur, zwischen*), the adjectives *obere, oberem, oberen, oberer, oberes, untere, unterem, unteren, unterer, unteres*, conjunctions (*als, und, während*), and the reflexive pronoun *sich* are ignored in determining the sequence of nested open forms. For example:

**Gehrung** *f* ARCH & TRAGW bevel, BAUMASCHIN, HOLZ miter (*AmE*), mitre (*BrE*), bevel; **auf ~ geschnitten** *adj* HOLZ mitered (*AmE*), mitred (*BrE*); **auf ~ geschnittener Stahlsturz** *m* NATURSTEIN, STAHL mitered steel lintel (*AmE*), mitred steel lintel (*BrE*); **Gehrungsfuge** *f* BAUMASCHIN, HOLZ miter joint (*AmE*), mitre joint (*BrE*)

In English, definite and indefinite articles (*a, an, the*), prepositions (*of, on, with*) and the conjunction *and* are ignored in determining the sequence of nested open forms. For example:

**end: ~ rafter** *n* ARCH & BUILD, INFR & DES Endpfette *f*; **~ of sleeper** *n* ARCH & BUILD, INFR & DES Schwellenkopf *m*; **~ span** *n* ARCH & BUILD, CONCR, INFR & DES, MAT PROP, STEEL, TIMBER Endfeld *nt*, Endträger *m*

### Homographs

In general, terms are accompanied by a label indicating their part of speech. (For a complete list of these labels and their expansions, please see pages xxvii–xxviii.) When terms beginning with the same element fall into two or more part-of-speech categories, the different categories are distinguished by numbers. The sequence is adjective, adverb, noun and verb, followed by less frequent parts of speech. For example:

**contract 1.** *n* CONST LAW Vertrag *m*; **2.** *vt* CONST LAW *enter into an agreement* schließen; **3.** *vi* MAT PROP *constrict* sich zusammenziehen, *shrink* schrumpfen

### Ordering of translations

Every term is accompanied by one or more labels indicating the technological area in which it is used. (For a complete list of these labels and their expansions, please see pages xxvii–xxviii.)

Where the same term is used in more than one technological area, multiple labels are given as appropriate. These labels appear in alphabetical order. Where a term has the same translation in more than one technological area, this translation is given after the sequence of labels. For example:

**absorb** *vt* INFR & DES, MAT PROP, SURFACE absorbieren

When a term has different translations according to the technological area in which it is used, the appropriate translation is given after each label or set of labels. For example:

**hew** *vt* INFR & DES *tree* fällen, STONE hauen

### Supplementary information

In many cases additional data is given about a term in order to show how it is used. Such contextual information can be:

(a) the typical subject or object of a verb, for example:

**extrudieren** *vt* WERKSTOFF *Metall, Kunststoff* extrude

(b) typical nouns used with an adjective, for example:

**stranggepreßt** *adj* STAHL, WERKSTOFF *Metal, Kunststoff* extruded

(c) words indicating the reference of a noun, for example:

**broom**: **~ finishing** *n* CONCR *concrete surface treatment* Besenabzug *m*

(d) information which supplements the subject area label, for example:

**Einfallen** *nt* INFR & ENTW, UMWELT *Geologie* dip

(e) a paraphrase or broad equivalent, for example:

**agent** *n* CONST LAW *representative* Vertreter *m*

When various different translations apply in the same subject area, contextual information is also used to show which translation is appropriate in different circumstances. For example:

**mounting** *n* ARCH & BUILD *process* Einbau *m*, Einspannung *f*, Montage *f*, *support, base* Untersatz *m*, Sockel *m*

### Cross-references

Geographical variants, both spelling and lexical, are given in full in the German–English section when they are translations:

**gepanzert** *adj* ELEKTR, STAHL armored (*AmE*), armoured (*BrE*)

In the English–German section both British English and North American English terms are covered, and these are differentiated by regional labels. In the case of lexical variants, full information – including translations and cross-references to the other form – is given at each entry. For example:

**earth: ~ fault circuit interrupter** *n BrE* (*cf ground fault circuit interrupter AmE*) ELECTR Fehlerstromschutzschalter *m*

Both abbreviations and their full forms are entered in the main body of the dictionary. Full information – including translations and cross-references to the full form or abbreviation – is given.

Where an abbreviation in the source language translates to an abbreviation in the target language, the full form in each case is given after the abbreviation in brackets and italics. For example:

**PVC** *abbr* (*polyvinyl chloride*) MAT PROP, SURFACE PVC (*Polyvinylchlorid*)

If a source language abbreviation does not translate to a target language abbreviation, the translation of the abbreviation is in roman. For example:

**NWT** *abbr* (*nonwaste technology*) ENVIRON saubere Technologie *f*, umweltfreundliche Technologie *f*

# Hinweise für die Benutzung des Wörterbuches

### Umfang des Wörterbuches

Sie haben ein Fachwörterbuch Deutsch–Englisch/Englisch–Deutsch vor sich, das alle Bereiche des Bauwesens und des zugrundeliegenden technisch-naturwissenschaftlichen Wissens abdeckt. Zusätzlich zu einer breiten Grundlage technischer Ausdrücke aus grundsätzlichen Bereichen wie etwa Architektur & Tragwerksplanung und Infrastruktur & Entwurf enthält dieses Wörterbuch auch Vokabular aus neuen und hochaktuellen Sachgebieten wie etwa Baurecht, Umweltthematik und Abwasserbehandlung.

### Auswahl der Stichwörter

Wir haben uns bemüht, aus jedem Gebiet das wesentliche Fachvokabular aufzunehmen; hierzu wurde die Stichwortliste sowohl im Englischen wie im Deutschen von führenden Experten in den einzelnen Fachgebieten auf Genauigkeit, Richtigkeit und Vollständigkeit überprüft. Die einzelnen Fachgebiete sind proportional zu ihrer Anwendungshäufigkeit vertreten, so daß also ein für die Informationstechnologie klassisches und umfangreiches Gebiet wie Architektur & Tragwerksplanung mit etwa 7 000 Stichwörtern vertreten ist, während ein Gebiet, in dem sich das Vokabular im Augenblick noch stark in der Entwicklung befindet, wie etwa Umweltthematik, dagegen deutlich weniger Stichwörter aufweist.

### Reihenfolge der Anordnung

Alle Einträge sind alphabetisch angeordnet. Zusammengesetzte Einträge erscheinen unter ihrem Basiswort. Dies gilt auch für Komposita mit Bindestrich.

### Stopplisten

Deutsche Mehrwortverbindungen sind nicht unter den folgenden Wörtern aufgeführt:

aber, alle, allein, alleine, alleinig, alleinige, alleinigem, alleinigen, alleiniger, alleiniges, allem, allen, aller, alles, allgemein, allgemeine, allgemeinen, allgemeiner, allgemeines, als, am, an, andere, anderem, anderen, anderer, anderes, anders, ans, auch, auf, aufeinander, aufs, aus, außen, außer, äußere, äußerem, äußeren, äußerer, äußeres, bei, beim, das, dasselbe, dem, demselben, den, denselben, der, derselbe, derselben, des, desselben, die, dies, diese, dieselbe, dieselben, diesem, diesen, dieser, dieses, durch, ein, eine, einem, einen, einer, eines, es, etwas, für, im, in, ins, kein, keine, keinem, keinen, keiner, keines, mit, miteinander, nach, neben, neu, nicht, nur, ob, oben, oberem, oberen, oberer, oberes, ohne, per, pro, sehr, sein, seine, seinem, seinen, seiner, seines, selber, selbst, sich, über, um, ums, und, unter, untere, unterem, unteren, unterer, unteres, unterm, vom, von, vor, vors, während, wenig, wenigem, wenigen, weniger, weniges, wenigst, wenigste, wenigstem, wenigsten, wenigster, wenigstes, wieder, zu, zum, zur, zwischen

Englische zusammengesetzte Einträge sind nicht unter den folgenden Elementen aufgeführt:

a, all, an, any, anybody, anyone, anything, anywhere, are, be, by, during, each, every, everybody, everyone, everything, everywhere, for, from, here, if, in, is, it, no, nobody, no one, nor, not, nothing, nowhere, of, off, on, or, out, over, so, some, somebody, someone, something, somewhere, that, the, then, there, they, thing, this, to, too, under, very, where, while, who, with

Zusammengesetzte Einträge sind in Nestform unter ihrem Basiswort aufgeführt. Wenn das

Basiswort ein eigenes Stichwort mit einer technischen Bedeutung darstellt, folgen alle weiteren Zusammensetzungen in alphabetischer Reihenfolge. Hier ersetzt die Tilde das Stichwort. Zusammengesetzten Einträgen mit nicht übersetztem Basiswort geht ein Doppelpunkt voraus. Zum Beispiel:

**accidental: ~ discharge** *n* ENVIRON zufälliger Ausfluß *m*, *of oils, chemicals* havariebedingter Ausfluß *m*

Deutsche Komposita werden voll ausgeschrieben und folgen auf Einträge mit mehreren Wörtern. Zum Beispiel:

**Energie** *f* ARCH & TRAGW, INFR & ENTW, UMWELT energy; **~ aus Abfall** UMWELT residue-derived energy; **Energieaufnahme** *f* ARCH & TRAGW, INFR & ENTW energy absorption; **Energieaustausch** *m* HEIZ & BELÜFT, INFR & ENTW, UMWELT energy exchange

Deutsche Komposita sowie englische zusammengesetzte Formen sind in alphabetischer Reihenfolge angeordnet. Nester, die aufgrund anderer Stichwörter unterbrochen werden müssen, werden entsprechend alphabetisch wieder aufgenommen. Das Basiswort als Blindeintrag in Fettkursivdruck weist auf die Fortführung des Artikels hin. Zum Beispiel:

**Abfuhr** *f* ABWASSER carrying off, UMWELT collection
**Abführbeständigkeit** *f* OBERFLÄCHE chalk resistance
**abführen** *vt* ABWASSER carry off
***Abfuhr*: Abfuhrwagen** *m* UMWELT tipper truck, skip

Zusammensetzungen aus Adjektiv und Substantiv sind unter der Grundform des Adjektivs aufgeführt. Im sich anschließenden Artikel wird die Grundform des Adjektives durch die Tilde ersetzt. Die entsprechende Adjektivendung schließt sich der Tilde an. Zum Beispiel:

**dynamisch** *adj* INFR & ENTW dynamic; **~ belasteter Balken** *m* ARCH & TRAGW, INFR & ENTW dynamically loaded beam; **~e Belastung** *f* ARCH & TRAGW, INFR & ENTW dynamic loading; **~e Festigkeit** *f* ARCH & TRAGW, INFR & ENTW dynamic strength; **~es Knicken** *nt* ARCH & TRAGW, INFR & ENTW dynamic buckling; **~e Kraft** *f* INFR & ENTW dynamic force

Folgende deutsche Wörter werden bei der alphabetischen Reihenfolge ignoriert: bestimmte und unbestimmte Artikel (*der, die, das, ein, eine*), deren gebeugte Formen (*dem, den, des, einem, einen, einer, eines*), das Pronomen *es*, Demonstrativpronomen (*dasselbe, derselbe, diese, dieser, dieselbe, dieses, selber, selbst*), deren gebeugte Formen (*demselben, denselben, derselben, desselben, dieselben, diesem, diesen*), Possessivpronomen (*sein, seine*), deren gebeugte Formen (*seinem, seinen, seiner, seines*), das unbestimmte Pronomen *etwas*, die Adverbien *etwa, nicht, nur, oben, unten*, Präpositionen mit oder ohne Artikel (*am, an, auf, aufs, aus, bei, beim, durch, für, gegen, im, in, ins, mit, nach, neben, ohne, per, pro, über, übers, um, ums, unter, unterm, vom, von, vor, vors, zu, zum, zur, zwischen*), die Adjektive *obere, oberem, oberen, oberer, oberes, untere, unterem, unteren, unterer, unteres*, Konjunktionen (*als, und, während, weil*) sowie das Reflexivpronomen *sich*.
Zum Beispiel:

**Gehrung** *f* ARCH & TRAGW bevel, BAUMASCHIN, HOLZ miter (*AmE*), mitre (*BrE*), bevel; **auf ~ geschnitten** *adj* HOLZ mitered (*AmE*), mitred (*BrE*); **auf ~ geschnittener Stahlsturz** *m* NATURSTEIN, STAHL mitered steel lintel (*AmE*), mitred steel lintel (*BrE*); **Gehrungsfuge** *f* BAUMASCHIN, HOLZ miter joint (*AmE*), mitre joint (*BrE*)

Folgende englische Wörter werden bei der alphabetischen Reihenfolge ignoriert: bestimmte und unbestimmte Artikel (*a, an, the*) sowie Präpositionen (*of, on, with*) und die Konjunktion *and*. Zum Beispiel:

**end: ~ rafter** *n* ARCH & BUILD, INFR & DES Endpfette *f*; **~ of sleeper** *n* ARCH & BUILD, INFR & DES Schwellenkopf *m*; **~ span** *n* ARCH & BUILD, CONCR, INFR & DES, MAT PROP, STEEL, TIMBER Endfeld *nt*, Endträger *m*;

## Homographen

Im allgemeinen sind Einträge mit einem Label versehen, das die Wortklassenzugehörigkeit angibt. (Eine vollständige Liste dieser Labels befindet sich auf Seiten xxvii–xxviii.) Einträge mit gemeinsamem Basiswort jedoch unterschiedlicher Wortklassenzugehörigkeit erhalten für jede Wortklasse eine eigene Numerierung. Die Einträge sind in der Reihenfolge Adjektiv, Adverb, Substantiv, Verb aufgeführt. Dem schließen sich weniger häufig vorkommende grammatikalische Einheiten an. Zum Beispiel:

**contract 1.** *n* CONST LAW Vertrag *m*; **2.** *vt* CONST LAW *enter into an agreement* schließen; **3.** *vi* MAT PROP *constrict* sich zusammenziehen, *shrink* schrumpfen

## Reihenfolge der Übersetzungen

Auf jedes Stichwort der Ausgangssprache folgen ein oder mehrere Labels, die das Fachgebiet angeben, in dem das Wort benutzt wird. (Eine vollständige Liste dieser Labels mit Erklärungen befindet sich auf den Seiten xxvii–xxviii.)

Wenn dasselbe Stichwort in mehr als einem Fachgebiet benutzt wird, so werden entsprechend mehrere Labels angegeben. Sie werden stets in alphabetischer Reihenfolge aufgeführt. Falls die angegebene Übersetzung in mehr als einem Fachgebiet benutzt wird, so folgt die Übersetzung jeweils im Anschluß an die entsprechenden Labels. Zum Beispiel:

**absorb** *vt* INFR & DES, MAT PROP, SURFACE absorbieren

Hat ein Stichwort verschiedene, von seinen jeweiligen Fachgebieten abhängige Übersetzungen, so ist die zutreffende Übersetzung jeweils nach dem entsprechenden Kürzel zu finden. Zum Beispiel:

**hew** *vt* INFR & DES *tree* fällen, STONE hauen

### Zusatzinformationen

Deutsche Substantive sind mit Geschlechtsangabe versehen. In vielen Fällen wird über das Stichwort noch zusätzliche Information gegeben, die über die Benutzung des Wortes Aufschluß gibt. Solche Informationen zum Benutzungszusammenhang können verschiedene Formen annehmen:

(a) bei einem Verb ein typisches Subjekt oder Objekt:

**extrudieren** *vt* WERKSTOFF *Metall, Kunststoff* extrude

(b) typische Substantive, die mit einem bestimmten Adjektiv verwendet werden:

**stranggepreßt** *adj* STAHL, WERKSTOFF *Metal, Kunststoff* extruded

(c) Wörter, die für Substantive einen typischen Bezug erläutern:

**broom: ~ finishing** *n* CONCR *concrete surface treatment* Besenabzug *m*

(d) Informationen, die das Fachgebiet noch weiter eingrenzen:

**Einfallen** *nt* INFR & ENTW, UMWELT *Geologie* dip

(e) Umschreibungen oder ungefähre Äquivalente:

**agent** *n* CONST LAW *representative* Vertreter *m*

Wenn in einem Fachgebiet verschiedene Übersetzungen für ein Stichwort möglich sind, so soll hier die zusätzliche Information anzeigen, welches im jeweiligen Zusammenhang die korrekte Übersetzung ist. Zum Beispiel:

**mounting** *n* ARCH & BUILD *process* Einbau *m*, Einspannung *f*, Montage *f*, *support, base* Untersatz *m*, Sockel *m*

### Querverweise

Geographische Varianten der Übersetzungen, sowohl orthographischer als auch lexikalischer Art, werden im deutsch–englischen Teil des Wörterbuches immer angegeben und voll ausgeschrieben. Zum Beispiel:

**gepanzert** *adj* ELEKTR, STAHL armored (*AmE*), armoured (*BrE*)

Im englisch–deutschen Teil des Wörterbuches werden Einträge sowohl in britischem als auch in amerikanischem Englisch aufgeführt und sind entsprechend mit geographischen Kürzeln gekennzeichnet. Bei allen lexikalischen Unterscheidungen werden Übersetzungen und Querverweise für beide Englischvarianten gegeben. Zum Beispiel:

**earth: ~ fault circuit interrupter** *n BrE* (*cf ground fault circuit interrupter AmE*) ELECTR Fehlerstromschutzschalter *m*

Im Hauptteil des Wörterbuches sind auch Abkürzungen und ihre ausgeschriebenen Formen angeordnet. Bei jedem Eintrag wird die vollständige Information – inklusive Übersetzungen und Querverweisen zur Vollform respektive Abkürzung in Kursivschrift – aufgeführt. Zum Beispiel:

**PVC** *abbr* (*polyvinyl chloride*) MAT PROP, SURFACE PVC (*Polyvinylchlorid*)

Handelt es sich um Abkürzungen, die in der jeweiligen Zielsprache kein direktes Äquivalent besitzen, bleibt der Übersetzungsvorschlag in Magerschrift.
Zum Beispiel:

**NWT** *abbr* (*nonwaste technology*) ENVIRON saubere Technologie *f*, umweltfreundliche Technologie *f*

# Abbreviations used in this dictionary/ Im Wörterbuch verwendete Abkürzungen

**Parts of speech/Wortarten**

| | | |
|---|---|---|
| *abbr* | abbreviation | Abkürzung |
| *adj* | adjective | Adjektiv |
| *adv* | adverb | Adverb |
| *f* | feminine | Femininum |
| *f pl* | feminine plural | Femininum Plural |
| *in cpds* | base form of compounds | Grundwort bei Komposita |
| *m* | masculine | Maskulinum |
| *m pl* | masculine plural | Maskulinum Plural |
| *n* | noun | Substantiv |
| *n pl* | noun plural | Substantiv Plural |
| *nt* | neuter | Neutrum |
| *nt pl* | neuter plural | Neutrum Plural |
| *phr* | phrase | fachsprachliche Redewendung |
| *pref* | prefix | Präfix |
| *prep* | preposition | Präposition |
| *vi* | intransitive verb | intransitives Verb |
| *v refl* | reflexive verb | reflexives Verb |
| *vt* | transitive verb | transitives Verb |
| *vti* | transitive and intransitive verb | transitives und intransitives Verb |

**Geographic codes/Geographische Kürzel**

| | | |
|---|---|---|
| *AmE* | American English | amerikanisches Englisch |
| *BrE* | British English | britisches Englisch |

**Subject area labels/Fachgebietskürzel**

| | | |
|---|---|---|
| ABWASSER | Abwasserbehandlung | Waste Water Treatment |
| ARCH & BUILD | Architecture & Building Planning | Architektur & Tragwerksplanung |
| ARCH & TRAGW | Architektur & Tragwerksplanung | Architecture & Building Planning |
| BAUMASCHIN | Baumaschinen | Building Machinery |
| BAURECHT | Baurecht | Construction Law |
| BESCHLÄGE | Beschläge & Einrichtung | Building Hardware |
| BETON | Beton | Concrete |
| BUILD HARDW | Building Hardware | Beschläge & Einrichtung |
| BUILD MACH | Building Machinery | Baumaschinen |
| CONCR | Concrete | Beton |
| CONST LAW | Construction Law | Baurecht |
| DÄMMUNG | Dämmung von Schall und Wärme | Sound & Thermal Insulation |
| ELECTR | Electrics | Elektroinstallationen |

| | | |
|---|---|---|
| ELEKTR | Elektroinstallationen | Electrics |
| ENVIRON | Environmental Issues | Umweltthematik |
| HEAT & VENT | Heating, Ventilation & Air Conditioning | Heizung, Be- & Entlüftung, Klimatisierung |
| HEIZ & BELÜFT | Heizung, Be- & Entlüftung, Klimatisierung | Heating, Ventilation & Air Conditioning |
| HOLZ | Holztragwerk | Timber Structures |
| INFR & ENTW | Infrastruktur & Entwurf | Infrastructure & Design |
| INFR & DES | Infrastructure & Design | Infrastruktur & Entwurf |
| MAT PROP | Material Properties | Werkstoffeigenschaften |
| NATURSTEIN | Natursteingebäude | Stone Buildings |
| OBERFLÄCHE | Oberflächenbearbeitung & Korrosionsschutz | Surface Works & Corrosion Protection |
| SOUND & THERMAL | Sound & Thermal Insulation | Dämmung von Schall und Wärme |
| STAHL | Stahl- & Aluminiumkonstruktionen | Steel & Aluminium Structures |
| STEEL | Steel & Aluminium Structures | Stahl- & Aluminiumkonstruktionen |
| STONE | Stone Buildings | Natursteingebäude |
| SURFACE | Surface Works & Corrosion Protection | Oberflächenbearbeitung & Korrosionsschutz |
| TIMBER | Timber Structures | Holztragwerk |
| UMWELT | Umweltthematik | Environmental Issues |
| WASTE WATER | Waste Water Treatment | Abwasserbehandlung |
| WERKSTOFF | Werkstoffeigenschaften | Material Properties |

**Trademarks/Warenzeichen**

Note that neither the presence nor the absence of the symbol ® should be regarded as affecting the legal status of any trademark.

Bitte beachten Sie, daß weder die Anwesenheit noch die Abwesenheit des Symbols ® den gesetzlichen Status irgendeines Warenzeichens beeinflußt.

# Deutsch–Englisch
# German–English

# A

**Aalleiter** *f* INFR & ENTW eel ladder
**Abbau** *m* ARCH & TRAGW dismantling, UMWELT decomposition
**abbaubar** *adj* UMWELT degradable
**abbauen** *vt* ARCH & TRAGW *Gerüst* take down, dismantle
***Abbau*: Abbaustufe** *f* UMWELT stage of decomposition
**abbeizen** *vt* OBERFLÄCHE pickle, strip
**Abbeizen** *nt* OBERFLÄCHE pickling, stripping
**Abbeizer** *m* ARCH & TRAGW *Farbe* remover, BAUMASCHIN, OBERFLÄCHE paint stripper
**Abbeizmittel** *nt* OBERFLÄCHE remover
**Abbildung** *f* ARCH & TRAGW projection
**Abbindebeschleuniger** *m* BETON, NATURSTEIN, WERKSTOFF accelerator
**abbinden** *vi* BETON, WERKSTOFF harden, set
**Abbindzeit** *f* BETON, UMWELT, WERKSTOFF setting time
**abblättern** *vi* ARCH & TRAGW spall
**abböschen** *vt* INFR & ENTW scarp
**abbröckeln** *vi* WERKSTOFF crumble away
**Abbruch** *m* ARCH & TRAGW, BAURECHT, INFR & ENTW, UMWELT clearance, demolition, pulling down, wrecking; **Abbruchabfall** *m* ARCH & TRAGW, UMWELT demolition waste, rubble; **Abbrucharbeit** *f* ARCH & TRAGW, INFR & ENTW, UMWELT demolition work; **Abbrucherlaubnis** *f* ARCH & TRAGW, BAURECHT, UMWELT demolition permit; **Abbruchgenehmigung** *f* ARCH & TRAGW, BAURECHT, UMWELT *offizielle Zustimmung* wrecking permit; **Abbruchmaterial** *nt* ARCH & TRAGW, INFR & ENTW demolition rubbish, UMWELT demolition rubbish, demolition waste, rubble
**abbruchreif** *adj* BAURECHT condemned
**Abdeck-** *in cpds* WERKSTOFF covering; **Abdeckband** *nt* INFR & ENTW, OBERFLÄCHE, WERKSTOFF adhesive masking tape; **Abdeckblech** *nt* STAHL, WERKSTOFF cover plate, covering plate, flashing
**abdecken** *vt* ARCH & TRAGW, NATURSTEIN *Mauerkronen* cap, cope
**Abdecken** *nt* ARCH & TRAGW, NATURSTEIN *Mauerkronen* capping, coping
***Abdeck-*: Abdeckgitter** *nt* STAHL, WERKSTOFF covering grid; **Abdeckleiste** *f* HOLZ, STAHL cover fillet; **Abdeckmaterial** *nt* UMWELT, WERKSTOFF covering material; **Abdeckmatte** *f* BETON curing agent; **Abdeckrost** *m* STAHL cover plate; **Abdeckstein** *m* NATURSTEIN coping stone, capstone
**Abdeckung** *f* ARCH & TRAGW capping, decking, covering, OBERFLÄCHE cover, covering; **Abdeckungsverstärkung** *f* BETON reinforcement of cover
**abdichten** *vt* ABWASSER caulk, DÄMMUNG proof, HEIZ & BELÜFT caulk, INFR & ENTW seal, NATURSTEIN *mit Mörtel* grout, OBERFLÄCHE caulk, proof, seal, UMWELT *Deponie* seal, line
**Abdichten** *nt* ABWASSER caulking, DÄMMUNG proofing, HEIZ & BELÜFT caulking, INFR & ENTW sealing, NATURSTEIN *mit Mörtel* grouting, OBERFLÄCHE, UMWELT sealing
**Abdichtleiste** *f* BESCHLÄGE window bar
**Abdichtung** *f* DÄMMUNG diaphragm, waterproofing, *Dampfsperre* vapor-proofing (*AmE*), vapour-proofing (*BrE*), HEIZ & BELÜFT packing, OBERFLÄCHE sealing, UMWELT lining; **Abdichtungsarbeiten** *f pl* ARCH & TRAGW, INFR & ENTW sealing works; **Abdichtungsband** *nt* ABWASSER, HEIZ & BELÜFT caulking strip; **Abdichtungsmasse** *f* INFR & ENTW, OBERFLÄCHE, WERKSTOFF sealing agent, sealing compound; **Abdichtungsmittel** *nt* OBERFLÄCHE sealant
**abfahren** *vt* ARCH & TRAGW, INFR & ENTW, UMWELT *Erde, Bauschutt, Müll* haul away
**Abfall** *m* ARCH & TRAGW garbage (*AmE*), junk, rubbish (*BrE*), waste, BAUMASCHIN *Leistung* decrease, drop, BAURECHT, INFR & ENTW, UMWELT garbage (*AmE*), junk, rubbish (*BrE*), waste; **als ~ zurückhalten** *phr* INFR & ENTW throw back to waste
**abfallarm**: **~e Technologie** *f* UMWELT clean technology, low-waste technology
***Abfall*: Abfallaufbereitung** *f* UMWELT waste recovery; **Abfallbehälter** *m* UMWELT garbage can (*AmE*), rubbish bin (*BrE*); **Abfallbehandlung** *f* UMWELT waste processing, waste treatment; **Abfallbeseitigung** *f* UMWELT waste disposal, refuse disposal; **Abfallbeseitigungsunternehmen** *nt* UMWELT waste disposal company; **Abfallbörse** *f* UMWELT waste exchange market; **Abfallbrennstoff** *m* UMWELT waste fuel; **Abfallcontainer** *m* UMWELT caster-equipped container; **Abfalldesinfektion** *f* UMWELT waste disinfection; **Abfalleimer** *m* UMWELT garbage can (*AmE*), rubbish bin (*BrE*)
**abfallend**: **~er Bogen** *m* ARCH & TRAGW rising arch, INFR & ENTW sloping arch
***Abfall*: Abfallentsorgung** *f* UMWELT waste disposal; **Abfallerzeuger** *m* UMWELT waste producer, waste generator; **Abfallerzeugung** *f* UMWELT waste formation, waste production; **Abfallfluß** *m* UMWELT flow of waste; **Abfallgärung** *f* UMWELT fermentation of refuse; **Abfallgesetz** *nt* (*AbfG*) UMWELT Waste Avoidance and Management Act, Waste Disposal Act; **Abfallhaufen** *m* INFR & ENTW waste heap; **Abfallagerung** *f* UMWELT wäste storage; **Abfallneutralisation** *f* UMWELT waste neutralization; **Abfallprodukt** *nt* UMWELT *unverwertbar* waste product; **Abfallsäure** *f* UMWELT waste acid; **Abfallsortieranlage** *f* UMWELT refuse separation plant, waste sorting plant; **Abfallsortierung** *f* **am Anfallsort** UMWELT source separation; **Abfallstoff** *m* UMWELT junk; **Abfallstrom** *m* UMWELT flow of waste; **Abfallverbrennungsanlage** *f* UMWELT refuse incinerator; **Abfallverbrennungsofen** *m* UMWELT incinerator; **Abfallvermeidung** *f* UMWELT waste avoidance; **Abfallverursacher** *m* UMWELT waste producer; **Abfallverwertung** *f* UMWELT waste processing, waste treatment; **Abfallverwertungsanlage** *f* UMWELT waste treatment plant, waste utilization plant
**abfallwirtschaftlich**: **~e Planung** *f* UMWELT waste-economical planning

***Abfall***: **Abfallzellstoff** *m* UMWELT waste pulp; **Abfallzerkleinerer** *m* UMWELT waste crusher, waste disintegrator
**Abfangung** *f* ARCH & TRAGW propping, support
**abfärben** *vi* OBERFLÄCHE chalk
**Abfärben** *nt* OBERFLÄCHE chalking
**abfasen** *vt* ARCH & TRAGW, HOLZ bevel, chamfer
**Abfasung** *f* ARCH & TRAGW, HOLZ bevel, chamfer
**AbfG** *abbr* (*Abfallgesetz*) UMWELT Waste Avoidance and Management Act, Waste Disposal Act
**Abflachung** *f* INFR & ENTW *Gelände* flattening
**Abflughalle** *f* ARCH & TRAGW, INFR & ENTW departure lounge
**Abfluß** *m* ABWASSER, UMWELT *Regenwasser* effluent, outlet, runoff; **~ im Oberflächenbereich** UMWELT surface runoff; **Abflußbeiwert** *m* ABWASSER runoff coefficient; **Abflußkanal** *m* ABWASSER, INFR & ENTW *für Abwässer* sewer; **Abflußregler** *m* ABWASSER, UMWELT effluent controller; **Abflußrinne** *f* ABWASSER, INFR & ENTW drainage gutter; **Abflußrohr** *nt* ABWASSER spout, waste pipe, drainpipe
**Abfuhr** *f* ABWASSER carrying off, UMWELT collection
**Abführbeständigkeit** *f* OBERFLÄCHE chalk resistance
**abführen** *vt* ABWASSER carry off
***Abfuhr***: **Abfuhrwagen** *m* UMWELT tipper truck, skip
**Abgas** *nt* HEIZ & BELÜFT, UMWELT exhaust gas, waste gas; **Abgasentschwefelung** *f* UMWELT waste gas desulfurization (*AmE*), waste gas desulphurization (*BrE*); **Abgaskanal** *m* HEIZ & BELÜFT exhaust gas duct; **Abgasreinigung** *f* UMWELT waste gas cleaning
**abgebaut** *adj* ARCH & TRAGW struck
**abgebrannt** *adj* ARCH & TRAGW, BAURECHT destroyed by fire
**abgedeckt** *adj* ARCH & TRAGW, BAURECHT covered, WERKSTOFF capped
**abgedichtet**: **~e Abdeckleiste** *f* ARCH & TRAGW, WERKSTOFF gasketted cover strip
**abgeflacht** *adj* INFR & ENTW *Gelände* flattened
**abgehängt**: **~e Decke** *f* ARCH & TRAGW, BESCHLÄGE counter ceiling, false ceiling, suspended ceiling
**abgekürzt**: **~er Wetterbeständigkeitsversuch** *m* WERKSTOFF accelerated weathering test
**abgerundet** *adj* ARCH & TRAGW rounded; **~e Kantenschutzschiene** *f* BESCHLÄGE nosing; **~es Schließblech** *nt* BESCHLÄGE rounded-off striking plate
**abgeschlossen**: **~e Deponie** *f* UMWELT complete fill
**abgeschrägt** *adj* ARCH & TRAGW *Zimmerhandwerk*, HOLZ beveled (*AmE*), bevelled (*BrE*), canted, splayed; **~e Gehrungsfuge** *f* ARCH & TRAGW splayed miter joint (*AmE*), splayed mitre joint (*BrE*)
**abgesteift**: **~er Schacht** *m* HOLZ timbered shaft
**abgestrahlt** *adj* OBERFLÄCHE blast-cleaned
**abgestuft** *adj* ARCH & TRAGW, BAUMASCHIN *Körnung* screened, stepped; **~e Feuchtigkeitssperre** *f* DÄMMUNG, NATURSTEIN stepped DPC
**abgestumpft**: **~es Dach** *nt* ARCH & TRAGW, HOLZ cut roof
**abgetreppt** *adj* ARCH & TRAGW benched, stepped; **~es Fundament** *nt* ARCH & TRAGW, BETON, INFR & ENTW benched foundation
**abgewalmt**: **~es Mansardendach** *nt* ARCH & TRAGW double pitch roof
**abgezinst**: **~er Cash-flow** *m* BAURECHT discounted cash flow (*DCF*)
**abgezogen**: **~er Estrich** *m* BETON floated screed
**Abgleich** *m* ARCH & TRAGW trim, HEIZ & BELÜFT balance
**abgleichen** *vt* ARCH & TRAGW trim, HEIZ & BELÜFT balance
**abgrenzen** *vt* ARCH & TRAGW, BAURECHT define
**Abgrenzung** *f* ARCH & TRAGW, BAURECHT definition
**Abgrund** *m* INFR & ENTW precipice
**Abhang** *m* ARCH & TRAGW downhill slope, INFR & ENTW downhill slope, slope
**Abhängehöhe** *f* ARCH & TRAGW, INFR & ENTW height of suspension
**abhängen** *vt* ARCH & TRAGW *Decke* suspend
**Abheben** *nt* ARCH & TRAGW, INFR & ENTW lifting-off
**Abhilfemaßnahme** *f* ARCH & TRAGW remedial measure, remedial treatment
**abholen** *vt* UMWELT collect
**Abholen** *nt* UMWELT collection
**Abhorchgerät** *nt* ABWASSER leakage detector
**abkanten** *vt* HOLZ bevel, STAHL *Blech* fold
**Abkanten** *nt* STAHL folding
**Abkantung** *f* STAHL fold
**abkippen** *vt* INFR & ENTW, UMWELT dump
**Abkippförderkorb** *m* BAUMASCHIN dump skip
**Abklärgefäß** *nt* UMWELT decanter
**Abklopfen** *nt* OBERFLÄCHE picking
**Abkommen** *nt* BAURECHT agreement; **Abkommensbescheinigung** *f* BAURECHT agreement certificate
**abkühlen** *vi* HEIZ & BELÜFT cool down
**abkuppeln** *vt* BAUMASCHIN uncouple, HEIZ & BELÜFT disconnect
**abkürzen** *vt* INFR & ENTW abridge
**Abladestation** *f* BAUMASCHIN dumping station
**ablagern** *vt* UMWELT *Müll* deposit, tip
**Ablagerung** *f* INFR & ENTW sediment, UMWELT deposit build-up, *Bergbau* deposition, WERKSTOFF deposit build-up, sediment; **Ablagerungsgeschwindigkeit** *f* UMWELT *radioaktiver Partikel* deposition velocity; **Ablagerungsplatz** *m* UMWELT storage area; **Ablagerungsrate** *f* UMWELT deposition rate; **Ablagerungswert** *m* UMWELT deposition value
**ablängen** *vt* WERKSTOFF cut to length
**Ablaß** *m* ABWASSER *eines Ausgußbeckens* bibcock
**ablassen** *vt* UMWELT *Wasser* discharge
***Ablaß***: **Ablaßhahn** *m* ABWASSER drain cock
**Ablauf** *m* ABWASSER, HEIZ & BELÜFT, UMWELT floor drain, floor gully, floor inlet, outlet, runoff, waste pipe
**ablaufen** *vi* ABWASSER drain, BAURECHT *Garantie* expire
***Ablauf***: **Ablaufkanal** *m* UMWELT flume; **Ablaufplan** *m* ARCH & TRAGW, BAURECHT, INFR & ENTW timetable, work schedule; **Ablauframpe** *f* INFR & ENTW *Eisenbahn* gravity incline
**Ablauge** *f* UMWELT waste lye
**ablaugen** *vt* OBERFLÄCHE pickle, scour
**Ablauger** *m* BAUMASCHIN, OBERFLÄCHE paint stripper
**ableiten** *vt* ABWASSER, HEIZ & BELÜFT *Wärme* carry off
**ableitfähig**: **~er Fußbodenbelag** *m* ELEKTR, WERKSTOFF conductive flooring
**Ableitung** *f* ELEKTR *als Wegbeschreibung beim Blitzschutz* path to earth (*BrE*), path to ground (*AmE*); **Ableitungswiderstand** *m* ELEKTR earthing resistance (*BrE*), grounding resistance (*AmE*)
**Ablenkblech** BESCHLÄGE baffle plate
**Ablenkplatte** *f* HEIZ & BELÜFT baffle plate
**ablesen** *vt* INFR & ENTW take a reading of

**Ablesen** *nt* ELEKTR, HEIZ & BELÜFT, INFR & ENTW reading
**Ablesung** *f* ELEKTR, HEIZ & BELÜFT, INFR & ENTW reading
**ablösen** *vt* BETON strip
**Ablösung** *f* INFR & ENTW separation of flow
**Abluft** *f* HEIZ & BELÜFT, INFR & ENTW exhaust air, exit air, UMWELT cleaned gas, scrubbed gas; **Ablufthaube** *f* BESCHLÄGE, HEIZ & BELÜFT extractor hood; **Abluftklappe** *f* BESCHLÄGE, HEIZ & BELÜFT exhaust air damper; **Abluftventilator** *m* BESCHLÄGE, HEIZ & BELÜFT exhaust fan
**abmessen** *vt* ARCH & TRAGW, INFR & ENTW gauge, measure
**Abmessen** *nt*: ~ **der Baustelle** INFR & ENTW survey of site
**Abmessung** *f* ARCH & TRAGW dimension
**Abnahme** *f* ARCH & TRAGW, BAURECHT acceptance; **Abnahmebescheinigung** *f* ARCH & TRAGW, BAURECHT acceptance certificate; **Abnahmeschein** *m* BAURECHT certificate of acceptance; **Abnahmeunterlagen** *f pl* ARCH & TRAGW, BAURECHT acceptance documents
**abnehmen** *vt* ARCH & TRAGW, BAURECHT *Arbeiten* accept
**Abnutzung** *f* INFR & ENTW wear
**Abort** *m* ABWASSER, BESCHLÄGE, INFR & ENTW toilet
**abplatzen** *vi* ARCH & TRAGW, BETON, INFR & ENTW spall
**Abplatzen** *nt* ARCH & TRAGW, BETON, INFR & ENTW spalling
**Abraumkippe** *f* INFR & ENTW spoil area
**abreiben** *vt* ARCH & TRAGW, BETON, HOLZ, NATURSTEIN, WERKSTOFF *Putz* abrade, float, rub, rub down
**abreißen** *vt* BAURECHT, INFR & ENTW demolish
**Abrieb** *m* ARCH & TRAGW, OBERFLÄCHE, WERKSTOFF abrasion, attrition
**abriebbeständig** *adj* OBERFLÄCHE abrasion-proof
**abriebfest** *adj* OBERFLÄCHE abrasion-proof
***Abrieb***: **Abriebfestigkeit** *f* OBERFLÄCHE abrasion resistance; **Abriebprobe** *f* WERKSTOFF abrasion test; **Abriebverschleiß** *m* WERKSTOFF abrasive wear
**Abriß** *m* BAURECHT, INFR & ENTW demolition; **Abrißerlaß** *m* BAURECHT clearance order
**abrißreif** *adj* ARCH & TRAGW condemned
**abrunden** *vt* ARCH & TRAGW, INFR & ENTW round off
**absacken** *vt* WERKSTOFF bag
**Absackung** *f* WERKSTOFF slump
**Absackwaage** *f* ARCH & TRAGW bagging scale
**absanden** *vt* OBERFLÄCHE sandblast
**Absatz** *m* ABWASSER debris, ARCH & TRAGW ledge, offset, shoulder, HEIZ & BELÜFT debris
**absäuern** *vt* NATURSTEIN, OBERFLÄCHE, WERKSTOFF acidwash
**absaugen** *vt* HEIZ UND BELÜFT *Flüssigkeit, Gas* suck out
**Absaugen** *nt* HEIZ & BELÜFT suction
**Absaugung** *f* ARCH & TRAGW *Staub, Späne in einer Schreinerwerkstatt* exhaustion
**Abschäler** *m* INFR & ENTW scarifier
**abschalten** *vt* ELEKTR disconnect
**abschätzen** *vt* ARCH & TRAGW, BAURECHT, INFR & ENTW *Kosten* estimate, rate, assess
**Abschätzung** *f* ARCH & TRAGW, BAURECHT, INFR & ENTW estmation, assessment
**abscheiden** *vt* UMWELT segregate
**Abscheider** *m* ABWASSER, INFR & ENTW precipitator, separator, trap
**Abscheidung** *f* UMWELT segregation
**abscheren** *vt* INFR & ENTW shear
**Abscherversuch** *m* INFR & ENTW shear test
**Abschirmung** *f* ELEKTR, HEIZ & BELÜFT shielding
**abschlagen** *vt* NATURSTEIN *Fliesen, Putz* strike off
**Abschlagen** *nt* ARCH & TRAGW spalling
**Abschlagzahlung** *f* ARCH & TRAGW, BAURECHT interim payment
**abschleifen** *vt* HOLZ, STAHL abrade
**abschleifend** *adj* HOLZ, STAHL, WERKSTOFF abrasive
**Abschluß** *m* ARCH & TRAGW surround
**Abschnitt** *m* ARCH & TRAGW compartment, stage, section; **Abschnittsfertigstellung** *f* BAURECHT sectional completion
**abschöpfen** *vt* INFR & ENTW, UMWELT skim off
**abschrägen** *vt* ARCH & TRAGW, HOLZ, INFR & ENTW bevel, chamfer, splay, taper
**Abschrägung** *f* ARCH & TRAGW, HOLZ, INFR & ENTW bevel, cant, chamfer
**abschraubbar** *adj* HOLZ, STAHL detachable
**abschrauben** *vt* HOLZ, STAHL detach, screw off, unscrew
**abschrecken** *vt* WERKSTOFF quench
**Abschreibung** *f* BAURECHT depreciation
**abschwemmen** *vt* NATURSTEIN float off
**absenkbar** *adj* ABWASSER submersible
**Absetz-** *in cpds* UMWELT settling; **Absetzbecken** *nt* INFR & ENTW, UMWELT precipitation tank, sedimentation basin, settling basin, settling tank
**Absetzen** *nt* INFR & ENTW deposit
***Absetz-***: **Absetzgefäß** *nt* INFR & ENTW, UMWELT decanter, settling vessel; **Absetzgeschwindigkeit** *f* UMWELT, WERKSTOFF settling velocity; **Absetzglas** *nt* ABWASSER, INFR & ENTW, WERKSTOFF Imhoff cone; **Absetzkammer** *f* UMWELT settling chamber; **Absetzklärung** *f* UMWELT decantation; **Absetzprobe** *f* INFR & ENTW sedimentation test
**absolut**: **~e Dichte** *f* INFR & ENTW, WERKSTOFF absolute density; **~e Feuchte** *f* HEIZ & BELÜFT, INFR & ENTW, WERKSTOFF absolute humidity; **~e Feuchtigkeit** *f* HEIZ & BELÜFT, INFR & ENTW, WERKSTOFF absolute humidity
**Absonderung** *f* ABWASSER separation
**Absorber** *m* OBERFLÄCHE absorber; **Absorberplatte** *f* HEIZ & BELÜFT, INFR & ENTW *Sonnenkollektor* absorber plate
**absorbieren** *vt* INFR & ENTW, OBERFLÄCHE, WERKSTOFF, UMWELT absorb
**absorbierend** *adj* INFR & ENTW, OBERFLÄCHE, UMWELT, WERKSTOFF absorbent; **~es Förderband** *nt* UMWELT *zur Ölaufnahme* absorbent belt skimmer
**absorbiert**: **~es Wasser** *nt* INFR & ENTW, WERKSTOFF absorbed water
**Absorption** *f* INFR & ENTW, OBERFLÄCHE, WERKSTOFF absorbancy, absorption; **Absorptionsgeschwindigkeit** *f* WERKSTOFF rate of absorption; **Absorptionsmaterial** *nt* WERKSTOFF absorbing material; **Absorptionsmittel** *nt* OBERFLÄCHE, UMWELT absorbent; **Absorptionsvermögen** *nt* INFR & ENTW, OBERFLÄCHE, WERKSTOFF absorption capacity
**Abspann-** *in cpds* ARCH & TRAGW stay; **Abspanndraht** *m* ARCH & TRAGW, BETON, STAHL, WERKSTOFF anchoring wire, stay wire; **Abspannpunkt** *m* ARCH & TRAGW

anchor point; **Abspannseil** *nt* ARCH & TRAGW, BESCHLÄGE guy rope
**Abspannung** *f* INFR & ENTW guy, guying
**Absperr-** *in cpds* ARCH & TRAGW cutoff
**absperren** *vt* ABWASSER shut off, ARCH & TRAGW stop, BESCHLÄGE lock, DÄMMUNG isolate, waterproof, HEIZ & BELÜFT shut
**Absperren** *nt* ABWASSER *Rohr* blocking off, shutting off, ARCH & TRAGW stopping, BESCHLÄGE locking, DÄMMUNG isolation
***Absperr-***: **Absperrfurnier** *nt* HOLZ crossband; **Absperrhahn** *m* ABWASSER, HEIZ & BELÜFT cutoff cock, stopcock; **Absperrklappe** *f* ABWASSER, HEIZ & BELÜFT butterfly valve, gate valve, shut-off valve; **Absperrmittel** *nt* OBERFLÄCHE sealer; **Absperrorgan** *nt* ABWASSER, HEIZ & BELÜFT valve; **Absperrschieber** *m* ABWASSER gate valve
**Absperrung** *f* BAURECHT *Sicherheit* barrier, DÄMMUNG barrier, isolation, ELEKTR isolation, HOLZ, STAHL barrier fence
***Absperr-***: **Absperrventil** *nt* BESCHLÄGE, UMWELT check valve, stop valve
**Abspitzen** *nt* NATURSTEIN picking
**absplittern** *vi* BETON, INFR & ENTW, NATURSTEIN, WERKSTOFF flake off
**Absprengung** *f* NATURSTEIN *Stein* spalling
**Abspringen** *nt* OBERFLÄCHE chipping
**Abstand** *m* ARCH & TRAGW, INFR & ENTW distance, interspace, space, spacing; **~ der Ableitung** ELEKTR *Blitzschutz* conductor clearance; **~ zwischen Bewehrungsstäben** BETON bar spacing; **~ der Längseisen** ARCH & TRAGW, BETON longitudinal spacing
**Abstandhalter** *m* ABWASSER, ARCH & TRAGW distance block, spacer, spacer block, spacing block, spacing stay, BETON distance piece, spacer, spacer block, spacing block, spacing stay, ELEKTR, HEIZ & BELÜFT, HOLZ spacer, spacer block, spacer stay, spacing block, spacing stay; **~ für Bewehrungsstahl** BETON bar spacer, spacer
**Absteck-** *in cpds* ARCH & TRAGW pegging
**abstecken** *vt* ARCH & TRAGW peg out, stake out, *Vermessung* set out, BAURECHT peg out
***Absteck-***: **Absteckkette** *f* ARCH & TRAGW surveyor's chain; **Absteckpfahl** *m* ARCH & TRAGW picket, surveyor's staff; **Absteckstange** *f* ARCH & TRAGW ranging pole
**absteifen** *vt* ARCH & TRAGW brace, strut, *mit Stützbalken* needle
**Absteifung** *f* ARCH & TRAGW staying, *Abstützung* bracing, propping, *Bau* reinforcing, stiffening
**Absteller** *m* BESCHLÄGE *Schloß* deadbolt
**Abstellkammer** *f* ARCH & TRAGW closet (*AmE*), cupboard (*BrE*)
**Abstellknopf** *m* ELEKTR stop button
**Abstimmung** *f* ARCH & TRAGW coordination, HEIZ & BELÜFT balance point
**Abstoß** *m* INFR & ENTW repulsion
**Abstrahlblock** *m* HEIZ & BELÜFT heat sink
**Abstrahlen** *nt* OBERFLÄCHE abrasive blasting
**abstrakt** *adj* ARCH & TRAGW abstract
**abstreichen** *vt* INFR & ENTW strike off
**Abstreichplatte** *f* ARCH & TRAGW strickle board
**Abstreifen** *nt* INFR & ENTW *Boden* stripping
**abstumpfend**: **~e Stoffe** *m pl* WERKSTOFF grit
**Absturz** *m* ABWASSER, INFR & ENTW cascade
**Abstütz**: **Abstützbalken** *m* ARCH & TRAGW needle, needle beam; **Abstützbohle** *f* BAUMASCHIN raking shore, shore
**abstützen** *vt* ARCH & TRAGW, INFR & ENTW shore, truss
**Abstützung** *f* ARCH & TRAGW, INFR & ENTW shoring, trussing
**Abszissenachse** *f* ARCH & TRAGW x-axis
**Abteilung** *f* ARCH & TRAGW *Werkstatt* bay
**abteufen** *vt* BAUMASCHIN, INFR & ENTW bore
**abtönen** *vt* OBERFLÄCHE tint
**Abtrag** *m* INFR & ENTW *Erde* cutting, excavation, UMWELT *Erde* excavated area, excavation
**abtragen** *vt* ARCH & TRAGW *Lasten* transfer, *Erdreich* clear out, INFR & ENTW strip, cut, excavate, OBERFLÄCHE skim, UMWELT excavate
**Abtragung** *f* INFR & ENTW stripping, *Erde* cutting, excavation, UMWELT *Erde* excavation
**Abtransport** *m* ARCH & TRAGW removal
**Abtrennung** *f* ARCH & TRAGW *durch Trennwände* partitioning, UMWELT separation
**Abtreppung** *f* ARCH & TRAGW, INFR & ENTW, NATURSTEIN benching, racking, stepback, stepping
**Abwärme** *f* DÄMMUNG, UMWELT thermal discharge, thermal loss, waste heat
**Abwasser** *nt* ABWASSER foul water, sewage, waste water, effluent, INFR & ENTW drain water, sewage, waste water, UMWELT waste water, effluent, drain water, **Abwasserabgabe** *f* BAURECHT sewerage charge; **Abwasseranalyse** *f* ABWASSER, UMWELT waste water analysis; **Abwasseranfall** *m* UMWELT sewage flow, volume of sewage; **Abwasseranlagen** *f pl* INFR & ENTW sewage installation; **Abwasseraufbereitung** *f* UMWELT sewage treatment, waste water treatment; **Abwasserbecken** *nt* UMWELT stabilization pond; **Abwasserbehandlung** *f* INFR & ENTW, UMWELT sewage treatment; **Abwasserbehandlung** *f* **mittels aerober Reinigung** UMWELT aerobic sewage treatment; **Abwasserbehandlungsanlage** *f* UMWELT sewage treatment plant, clarification plant; **Abwasserbehandlungsverfahren** *nt* UMWELT sewage treatment process; **Abwasserbeseitigung** *f* ABWASSER, INFR & ENTW, UMWELT sewage discharge, sewage disposal, sewage water disposal, waste water disposal; **Abwassereinleitung** *f* UMWELT sewage discharge, waste water discharge; **Abwassereinleitung** *f* **ins Meer** UMWELT marine sewage disposal; **Abwassereinleitungsstelle** *f* UMWELT sewage outfall; **Abwasserfaulraum** *m* UMWELT privy tank, hydrolizing tank, septic tank; **Abwasserfischteich** *m* UMWELT waste water fishpond; **Abwasserfluß** *m* INFR & ENTW sewage flow; **Abwasserkanal** *m* ABWASSER, INFR & ENTW sewer; **Abwasserkläranlage** *f* INFR & ENTW, UMWELT clarification plant, sewage purification plant, sewage treatment plant; **Abwasserklärung** *f* UMWELT sewage purification, sewage treatment; **Abwasserkontrolle** *f* UMWELT waste water control; **Abwasserleitung** *f* ABWASSER, INFR & ENTW sewage pipe, sewer, sewer pipe, waste pipe, UMWELT sewer pipe; **Abwassermenge** *f* UMWELT sewage flow, volume of sewage
**abwässern** *vt* WERKSTOFF weather
***Abwasser***: **Abwasserpilz** *m* UMWELT sewage fungus; **Abwasserreinigung** *f* UMWELT sewage treatment, sewage purification; **Abwasserreinigungsanlage** *f* UMWELT sewage treatment plant, sewage treatment

works; **Abwasserrohr** *nt* ABWASSER, INFR & ENTW drainpipe, sewage pipe, sewer pipe, waste pipe, UMWELT sewer pipe; **Abwasserrohrentlüftung** *f* ABWASSER, HEIZ & BELÜFT stack ventilation; **Abwassersammeltank** *m* UMWELT waste water collection tank; **Abwassersammler** *m* UMWELT interceptor sewer; **Abwassersanierung** *f* UMWELT waste water renovation; **Abwasserschlamm** *m* ABWASSER, INFR & ENTW, UMWELT effluent sludge, effluent slurry, sewage sludge; **Abwasserstripper** *m* UMWELT waste water stripper

**Abwässerung** *f* NATURSTEIN cope

***Abwasser***: **Abwasserzusammensetzung** *f* UMWELT sewage composition

**Abweichung** *f* ARCH & TRAGW deviation, UMWELT variance, WERKSTOFF deflection

**abweisend** *adj* WERKSTOFF repellent

**Abweiser** *m* BAUMASCHIN guard

**Abweisstein** *m* NATURSTEIN baffle brick

**Abwicklung** *f* ARCH & TRAGW development

**Abzieh-** *in cpds* ARCH & TRAGW screed; **Abziehbohle** *f* ARCH & TRAGW, NATURSTEIN screed board; **Abziehbrett** *nt* BAUMASCHIN float

**abziehen** *vt* ARCH & TRAGW screed, draw, rub, BETON screed, INFR & ENTW strip, NATURSTEIN *Putz* finish

***Abzieh-***: **Abziehlatte** *f* WERKSTOFF darby

**Abzugshaube** *f* BESCHLÄGE hood

**abzugslos**: **~e Heizung** *f* HEIZ & BELÜFT unflued heater

**Abzugsrohr** *nt* BESCHLÄGE, HEIZ & BELÜFT exhaust pipe, vent pipe, UMWELT exhaust pipe

**Abzweig** *m* ELEKTR tap; **Abzweigdose** *f* BESCHLÄGE joint box, ELEKTR joint; **Abzweigkabel** *nt* ELEKTR stub cable

**Abzweigung** *f* ABWASSER branch, ELEKTR arm, HEIZ & BELÜFT branch

**Achsdruck** *m* BAUMASCHIN axle load

**Achse** *f* ARCH & TRAGW axis, center line (*AmE*), centre line (*BrE*), BAUMASCHIN axle, INFR & ENTW axis, center line (*AmE*), centre line (*BrE*); **~ der Verbindungsmittel** ARCH & TRAGW, HOLZ connector axis; **die ~ verschieben** *phr* INFR & ENTW move the center line (*AmE*), move the centre line (*BrE*)

***Achse***: **Achsenabstand** *m* ARCH & TRAGW center-to-center distance (*AmE*), centre-to-centre distance (*BrE*)

**Achs**: **Achslast** *f* BAUMASCHIN axle load; **Achslinie** *f* ARCH & TRAGW, INFR & ENTW center line (*AmE*), centre line (*BrE*); **Achsmaß** *nt* ARCH & TRAGW center-to-center distance (*AmE*), centre-to-centre distance (*BrE*)

**achsrecht** *adj* ARCH & TRAGW, BETON, INFR & ENTW, STAHL, WERKSTOFF axial

**Achsscheibe** *f* BAUMASCHIN axle pulley

**Achtkantraumfachwerk** *nt* HOLZ octagonal space frame

**Acryl** *nt* INFR & ENTW, OBERFLÄCHE, WERKSTOFF acrylic; **Acrylfarbe** *f* OBERFLÄCHE acrylic paint

**Acrylharz** *nt* WERKSTOFF acrylic resin; **Acrylharzbeton** *m* BETON, WERKSTOFF acrylic concrete; **Acrylharzdispersion** *f* INFR & ENTW, OBERFLÄCHE, WERKSTOFF acrylic resin dispersion; **Acrylharzemulsion** *f* INFR & ENTW, OBERFLÄCHE, WERKSTOFF acrylic resin emulsion

**Adapter** *m* BAUMASCHIN, BETON, ELEKTR, WERKSTOFF adapter

**Additiv** *nt* UMWELT additive

**Ader** *f* INFR & ENTW vein

**Adhäsion** *f* INFR & ENTW, OBERFLÄCHE, WERKSTOFF adhesion

**adiabatisch**: **~e Nachbehandlung** *f* NATURSTEIN adiabatic curing

**adsorbieren** *vt* INFR & ENTW, NATURSTEIN, OBERFLÄCHE, UMWELT, WERKSTOFF adsorb

**Adsorption** *f* INFR & ENTW, NATURSTEIN, OBERFLÄCHE, UMWELT, WERKSTOFF adsorption; **Adsorptionsgeschwindigkeit** *f* INFR & ENTW, NATURSTEIN, OBERFLÄCHE, UMWELT, WERKSTOFF adsorption rate; **Adsorptionsmittel** *nt* INFR & ENTW, NATURSTEIN, OBERFLÄCHE, WERKSTOFF adsorbent; **Adsorptionsphänomen** *nt* INFR & ENTW, UMWELT, WERKSTOFF adsorption phenomenon; **Adsorptionstest** *m* INFR & ENTW, UMWELT, WERKSTOFF adsorption test; **Adsorptionsvermögen** *nt* INFR & ENTW, NATURSTEIN, OBERFLÄCHE, UMWELT, WERKSTOFF adsorption capacity, adsorptivity; **Adsorptionswirkung** *f* INFR & ENTW, NATURSTEIN, OBERFLÄCHE, UMWELT, WERKSTOFF adsorption efficiency

**aerob** *adj* INFR & ENTW, WERKSTOFF aerobic; **~er Abbau** *m* UMWELT aerobic degradation; **~e Bakterien** *f pl* UMWELT aerobic bacteria; **~es Behandlungsverfahren** *nt* UMWELT aerobic treatment process; **~e Gärung** *f* UMWELT aerobic fermentation; **~e Schlammstabilisierung** *f* UMWELT aerobic sludge stabilization; **~ stabilisierter Schlamm** *m* UMWELT aerobically digested sludge; **~e Zersetzung** *f* UMWELT aerobic decomposition

**Aerobier** *m* INFR & ENTW aerobe

**Aero**: **Aerobiologie** *f* UMWELT aerobiology; **Aerobiose** *f* UMWELT aerobiosis

**aerodynamisch**: **~e Kraft** *f* ARCH & TRAGW, INFR & ENTW, UMWELT aerodynamic force, aerodynamic power

***Aero***: **Aeroelastizität** *f* INFR & ENTW aeroelasticity

**Aerosol** *nt* UMWELT aerosol; **Aerosolpackung** *f* UMWELT aerosol dispenser

**Affinität** *f* ARCH & TRAGW, INFR & ENTW, WERKSTOFF affinity

**AG** *abbr* (*Auftraggeber*) ARCH & TRAGW customer

**Agglomerat** *nt* INFR & ENTW, WERKSTOFF agglomerate

**aggressiv** *adj* WERKSTOFF aggressive; **nicht aggressiv** *adj* INFR & ENTW nonaggressive; **~es Wasser** *nt* INFR & ENTW, WERKSTOFF aggressive water

**Agora** *f* ARCH & TRAGW agora

**ähnlich** *adj* INFR & ENTW similar

**Akademie** *f* ARCH & TRAGW, INFR & ENTW academy

**Akanthus** *m* ARCH & TRAGW acanthus; **Akanthusblatt** *nt* ARCH & TRAGW acanthus leaf; **Akanthusfries** *m* ARCH & TRAGW acanthus frieze

**Akkumulation** *f* BAURECHT, INFR & ENTW, WERKSTOFF accumulation

**Akkumulator** *m* BAUMASCHIN, ELEKTR accumulator

**akkumulieren** *vt* BAURECHT, INFR & ENTW, WERKSTOFF accumulate

**Akroterion** *nt* ARCH & TRAGW acroterion

**Aktionsturbine** *f* UMWELT impulse turbine

**aktiv**: **~e Bestandteile** *m pl* **der Konservierungsmittel** WERKSTOFF active ingredients of preservative products; **~er Druck** *m* INFR & ENTW, WERKSTOFF active pressure; **~er Druck** *m* **nach Rankin** INFR & ENTW, WERKSTOFF active Rankine pressure; **~er Erddruck** *m* INFR & ENTW active earth pressure; **~e Solarheizung** *f*

HEIZ & BELÜFT active solar heating; **~es Sonnensystem** *nt* UMWELT active solar system

**Aktivkohle** *f* INFR & ENTW, UMWELT activated carbon, active carbon; **Aktivkohleabsorption** *f* UMWELT active carbon absorption; **Aktivkohlebehandlung** *f* INFR & ENTW, UMWELT activated carbon treatment; **Aktivkohlepatrone** *f* HEIZ & BELÜFT active carbon cartridge

**Akustik** *f* DÄMMUNG acoustics; **Akustikbaustoff** *m* DÄMMUNG, INFR & ENTW, WERKSTOFF acoustic building material; **Akustikdecke** *f* DÄMMUNG *abgehängte Decke zur Schallschluckung*, INFR & ENTW acoustic ceiling; **Akustikmatte** *f* DÄMMUNG, WERKSTOFF acoustic blanket; **Akustikplatte** *f* DÄMMUNG, WERKSTOFF acoustic board; **Akustikverkleidung** *f* DÄMMUNG, INFR & ENTW, WERKSTOFF acoustic lining

**akustisch**: **~e Impedanz** *f* INFR & ENTW, WERKSTOFF acoustic impedance; **~es Verhalten** *nt* DÄMMUNG, WERKSTOFF acoustical behavior (*AmE*), acoustical behaviour (*BrE*)

**akut**: **~e Wirkung** *f* UMWELT acute effect

**AKW** *abbr* (*Atomkraftwerk*) INFR & ENTW nuclear power plant, nuclear power station, UMWELT nuclear power station

**akzeptabel** *adj* BAURECHT acceptable

**Alabaster** *m* WERKSTOFF alabaster

**Alarm** *m* BAURECHT, BESCHLÄGE, ELEKTR alarm, warning; **Alarmanlage** *f* BAURECHT, BESCHLÄGE, ELEKTR alarm system, warning system

**Algen** *f pl* HEIZ & BELÜFT, INFR & ENTW, OBERFLÄCHE, WERKSTOFF algae; **Algenvernichtungsmittel** *nt* OBERFLÄCHE, WERKSTOFF algicide; **Algenwuchs** *m* HEIZ & BELÜFT, INFR & ENTW, OBERFLÄCHE algae growth

**aliphatisch**: **~er Kohlenwasserstoff** *m* WERKSTOFF aliphatic hydrocarbon

**Alkali** *nt* OBERFLÄCHE, WERKSTOFF alkali

***Alkali***: **Alkalibeständigkeit** *f* OBERFLÄCHE, WERKSTOFF alkali resistance; **Alkaliempfindlichkeit** *f* OBERFLÄCHE, WERKSTOFF alkali reactivity

**alkalisch** *adj* INFR & ENTW, WERKSTOFF alkaline; **~es Abbeizmittel** *nt* OBERFLÄCHE, WERKSTOFF alkali paint stripper; **~er Boden** *m* INFR & ENTW, WERKSTOFF alkaline soil

***Alkali***: **Alkali-Silikon-Reaktion** *f* BETON alkali silicon reaction (*ASR*)

**Alkalität** *f* INFR & ENTW, UMWELT, WERKSTOFF *Laugengrad* alkalinity

**Alkoholbeständigkeit** *f* OBERFLÄCHE, WERKSTOFF alcohol resistance

**Alkoven** *m* ARCH & TRAGW alcove

**Alkyd**: **Alkydharzfarbe** *f* WERKSTOFF alkyd paint; **Alkydharzgrundierung** *f* OBERFLÄCHE, WERKSTOFF alkyd paint, alkyd primer

**allgemein**: **~e Bemessungsmethode** *f* ARCH & TRAGW, BETON, INFR & ENTW, STAHL general method of design; **~es Fluchttreppenhaus** *nt* ARCH & TRAGW, BAURECHT common stairs

**Allgemeinbeleuchtung** *f* ELEKTR common lighting

**alluvial** *adj* INFR & ENTW, WERKSTOFF alluvial

**Alt-** *in cpds* UMWELT ancient, end-of-life, old, residual, scrap; **Altablagerung** *f* UMWELT old deposit

**Altar**: **Altarraum** *m* ARCH & TRAGW chancel; **Altarschiff** *nt* ARCH & TRAGW, INFR & ENTW chancel aisle

***Alt-***: **Altblei** *nt* UMWELT scrap lead

**altern** *vi* BETON, OBERFLÄCHE *Farbe, Bitumen, Beton* mature, STAHL, WERKSTOFF age

**Alternativ**: **Alternativeingang** *m* ARCH & TRAGW, BAURECHT alternative entrance; **Alternativfluchtweg** *m* ARCH & TRAGW, BAURECHT alternative escape route, alternative means of escape; **Alternativzugang** *m* ARCH & TRAGW, BAURECHT alternative access

**Alterung** *f* BETON, OBERFLÄCHE maturing, STAHL, WERKSTOFF ageing

**alterungsbeständig** *adj* BETON, STAHL, WERKSTOFF age-proof

***Alt-***: **Altfahrzeug** *nt* UMWELT end-of-life vehicle

**Altglas** *nt* UMWELT waste glass; **Altglasbehälter** *m* UMWELT waste glass container; **Altglascontainer** *m* UMWELT bottle bank; **Altglasverwertung** *f* UMWELT glass recycling

***Alt-***: **Altlast** *f* UMWELT problem site; **Altmetall** *nt* UMWELT scrap metal

**Altöl** *nt* UMWELT residual oil, waste oil; **Altölaufbereitung** *f* UMWELT waste oil preparation; **Altölgesetz** *nt* UMWELT Waste Oil Act; **Altölrückgewinnung** *f* UMWELT waste oil recovery; **Altöltank** *m* UMWELT slop tank; **Altölwiederverwertung** *f* UMWELT waste oil recycling

**Altpapier** *nt* UMWELT waste paper; **Altpapieraufbereitung** *f* UMWELT waste paper preparation; **Altpapierkompressor** *m* UMWELT waste paper compressor; **Altpapierrecycling** *nt* UMWELT waste paper recycling; **Altpapiersammlung** *f* UMWELT paper collection, collection of waste paper

**Aluminium** *nt* ARCH & TRAGW, INFR & ENTW, STAHL, WERKSTOFF aluminium (*BrE*), aluminum (*AmE*); **Aluminiumabdeckung** *f* ARCH & TRAGW, STAHL, WERKSTOFF aluminium coping (*BrE*), aluminum coping (*AmE*); **Aluminiumband** *nt* BESCHLÄGE, STAHL aluminium hinge (*BrE*), aluminum hinge (*AmE*); **Aluminiumblende** *f* BESCHLÄGE aluminium blind (*BrE*), aluminum blind (*AmE*), aluminium trim (*BrE*), aluminum trim (*AmE*), STAHL aluminium trim (*BrE*), aluminum trim (*AmE*); **Aluminiumdach** *nt* ARCH & TRAGW, STAHL aluminium glazing bar (*BrE*), aluminum glazing bar (*AmE*), aluminium roof cladding (*BrE*), aluminum roof cladding (*AmE*); **Aluminiumfassade** *f* ARCH & TRAGW, STAHL aluminium front (*BrE*), aluminum front (*AmE*); **Aluminiumfensterflügel** *m* STAHL aluminium sash (*BrE*), aluminum sash (*AmE*); **Aluminiumfolie** *f* DÄMMUNG, WERKSTOFF aluminium foil (*BrE*), aluminum foil (*AmE*); **Aluminiumgußlegierung** *f* STAHL, WERKSTOFF cast aluminium alloy (*BrE*), cast aluminum alloy (*AmE*); **Aluminiumnagel** *m* BESCHLÄGE, STAHL, WERKSTOFF aluminium nail (*BrE*), aluminum nail (*AmE*); **Aluminiumoberfläche** *f* STAHL aluminium finish (*BrE*), aluminum finish (*AmE*); **Aluminiumschrott** *m* UMWELT aluminium scrap (*BrE*), aluminum scrap (*AmE*); **Aluminiumsonnenblende** *f* BESCHLÄGE aluminium blind (*BrE*), aluminum blind (*AmE*); **Aluminiumsprosse** *f* STAHL *Fenster* aluminium glazing bar (*BrE*), aluminum glazing bar (*AmE*); **Aluminiumstraßenleuchtenmast** *f* ELEKTR, STAHL aluminium street lighting mast (*BrE*), aluminum street lighting mast (*AmE*); **Aluminiumverkleidung** *f* ARCH & T RAGW, INFR & ENTW, WERKSTOFF aluminium facing (*BrE*), aluminum facing (*AmE*);

**Aluminumverwahrung** STAHL aluminium flashing (*BrE*), aluminum flashing (*AmE*)
**aluminothermisch**: **~es Schweißen** *nt* STAHL thermit welding
**Amboß** *m* STAHL anvil
**Amerikanisch**: **~e Gesellschaft** *f* **der Heizungs-, Kühlungs- und Klimatechniker** (*ASHRAE*) HEIZ & BELÜFT American Society of Heating, Refrigeration and Air Conditioning Engineers (*AmE*); **~e Norm** *f* (*AS*) ARCH & TRAGW BAURECHT American Standard (*AS*)
**Amin-Formaldehyd-Harz** *nt* WERKSTOFF amine formaldehyde resin
**Ammoniak** *nt* INFR & ENTW, UMWELT ammonia
**Ampere** *nt* ELEKTR ampere; **Amperezahl** *f* ELEKTR amperage
**Amphibolasbest** *m* WERKSTOFF amphibole asbestos
**Amplifikation** *f* INFR & ENTW, WERKSTOFF amplification
**Amplitude** *f* ELEKTR, STAHL amplitude
**amtlich**: **~e Genehmigung** *f* BAURECHT official approval; **~es Gütezeichen** *nt* BAURECHT seal of approval; **~er Inspektor** *m* ARCH & TRAGW surveyor; **~e Vermessungskarte** *f* INFR & ENTW Ordnance Survey map
**anaerob** *adj* INFR & ENTW, WERKSTOFF anaerob; **~e Faulung** *f* UMWELT anaerobic digestion; **~e Gärung** *f* UMWELT anaerobic digestion, anaerobic fermentation; **~er Teich** *m* UMWELT anaerobic lagoon
**analog** *adj* ARCH & TRAGW, INFR & ENTW analogous
**Analogie** *f* ARCH & TRAGW, INFR & ENTW analogy
**Analyse** *f* ARCH & TRAGW, INFR & ENTW, WERKSTOFF analysis
**analysieren** *vt* ARCH & TRAGW, INFR & ENTW, WERKSTOFF analyse (*BrE*), analyze (*AmE*)
**Anbau** *m* ARCH & TRAGW, INFR & ENTW addition, annex (*AmE*), annexe (*BrE*), extension
**anbauen** *vi* ARCH & TRAGW, INFR & ENTW build an extension
***Anbau***: **Anbaugarage** *f* ARCH & TRAGW attached garage; **Anbaugerät** *nt* BESCHLÄGE attachment; **Anbaugerät** *nt* **für Rad- und Raupenschlepper** BAUMASCHIN attachments for wheeled and crawler tractors
**anbieten** *vt* BAURECHT offer, tender
**Anblattung** *f* HOLZ halved joint, halving
**Anbohrschutz** *m* BAUMASCHIN, BESCHLÄGE *Schloß* antidrill feature
**Anböschung** *f* UMWELT ramp landfill, slope landfill, slope method
**anbringen** *vt* ARCH & TRAGW *installieren*, BETON, HOLZ, INFR & ENTW, STAHL *anfügen* attach, fit, fix, install, mount
**andämmen** *vt* ARCH & TRAGW, INFR & ENTW bank up
**ändern** *vt* ARCH & TRAGW, INFR & ENTW alter, modify
**Änderung** *f* ARCH & TRAGW, INFR & ENTW alteration, modification
**aneinanderfügen** *vt* ARCH & TRAGW, HEIZ & BELÜFT, HOLZ, STAHL *zusammensetzen* assemble, join
**Aneinanderfügung** *f* ARCH & TRAGW, HEIZ & BELÜFT, HOLZ, STAHL assembly
**anerkennen** *vt* BAURECHT acknowledge, admit, recognize
**Anerkennung** *f* BAURECHT acknowledgement, admission, recognition
**Anfall** *m*: **~ von Abfällen** UMWELT waste formation, waste production
**anfänglich**: **~e Vorspannungskraft** *f* BETON initial prestress force
**Anfang**: **Anfangsdruck** *m* ABWASSER *Pumpe* initial pressure; **Anfangsfestigkeit** *f* INFR & ENTW, WERKSTOFF initial strength; **Anfangsporenziffer** *f* INFR & ENTW, WERKSTOFF initial void ratio; **Anfangssetzung** *f* INFR & ENTW initial sediment, WERKSTOFF initial sediment, initial settlement
**anfeuchten** *vt* ARCH & TRAGW dampen, wet, *Sand* temper, DÄMMUNG dampen, STAHL dampen, temper, wet, WERKSTOFF wet, dampen, *Sand* temper
**Anfeuchten** *nt* ARCH & TRAGW damping, tempering, wetting, DÄMMUNG damping, STAHL *Sand* tempering, WERKSTOFF damping, tempering, wetting
**Anforderung** *f* BAURECHT requirement
**anfügen** *vt* INFR & ENTW, WERKSTOFF add
**Angebot** *nt* ARCH & TRAGW, BAURECHT, INFR & ENTW availability, bid, offer, tender, WERKSTOFF availability; **Angebotsabgabetermin** *m* ARCH & TRAGW, BAURECHT, INFR & ENTW tendering date
**Angel** *f* BESCHLÄGE pivot
**angelaufen** *adj* ARCH & TRAGW struck
**angemessen** *adj* ARCH & TRAGW, BAURECHT appropriate, convenient, suitable
**angenähert** *adj* ARCH & TRAGW, INFR & ENTW approximate
**angepaßt** *adj* STAHL cut to fit
**angesäuert**: **~e Bodenfläche** *f* UMWELT acidic area
**angeschuht**: **~e Stange** *f* STAHL shoed bar
**angeschüttet**: **~es Gelände** *nt* INFR & ENTW *Erde* fill, filled ground, filled-up ground
**angeschwemmt** *adj* INFR & ENTW, WERKSTOFF alluvial
**angesetzt**: **~es Holz** *nt* HOLZ pieced wood
**angewandt** *adj* BAURECHT, INFR & ENTW applied; **~e Seismik** *f* INFR & ENTW seismic exploration method
**angleichen** *vt* ARCH & TRAGW, BAUMASCHIN, BAURECHT, BETON, ELEKTR, INFR & ENTW, STAHL adapt, harmonize, match
**Angleichung** *f* BAUMASCHIN, BAURECHT, BETON, ELEKTR, INFR & ENTW, STAHL adaptation, adjustment
**angrenzend** *adj* BAURECHT, INFR & ENTW, WERKSTOFF adjoining; **~er Grundbesitz** *m* ARCH & TRAGW adjacent property; **~e Spannweite** *f* ARCH & TRAGW, BETON, INFR & ENTW, STAHL adjacent span
**anhaften** *vi* INFR & ENTW, STAHL, WERKSTOFF adhere
**Anhalten** *nt* UMWELT cessation
**Anhang** *m* ARCH & TRAGW *Dokument* annex (*AmE*), annexe (*BrE*), appendix, BAURECHT appendix, annex (*AmE*), annexe (*BrE*), INFR & ENTW annex (*AmE*), annexe (*BrE*), appendix
**Anhänge-** *in cpds* BAUMASCHIN towed; **Anhängestraßenhobel** *m* BAUMASCHIN towed grader; **Anhängewalze** *f* BAUMASCHIN tractor-drawn roller
**anhäufen** *vt* INFR & ENTW, WERKSTOFF agglomerate
**Anhäufung** *f* STAHL aggregate
**anheben** *vt* ARCH & TRAGW, BAUMASCHIN, UMWELT elevate, lift
**Anheben** *nt* BAUMASCHIN lifting
**Anhebung** *f* ARCH & TRAGW, UMWELT elevation
**Anhydrit** *nt* WERKSTOFF anhydrite; **Anhydritbinder** *m* NATURSTEIN, WERKSTOFF anhydrite binder; **Anhydritmörtel** *m* NATURSTEIN, WERKSTOFF anhydrite mortar; **Anhydritestrich** *m* WERKSTOFF anhydride screed
**Ankauf** *m* BAURECHT purchase
**Anker** *m* ARCH & TRAGW, BETON, INFR & ENTW, STAHL

anchor; **Ankerausziehversuch** *m* INFR & ENTW, WERKSTOFF anchor pulling test; **Ankerblock** *m* ARCH & TRAGW, INFR & ENTW deadman; **Ankerbolzen** *m* BESCHLÄGE lag screw, BETON, INFR & ENTW, STAHL, WERKSTOFF anchor bolt, anchoring bolt, rockbolt; **Ankereisen** *nt* ARCH & TRAGW tie bar; **Ankerkopf** *m* INFR & ENTW anchor head

**ankern** *vt* INFR & ENTW, STAHL, WERKSTOFF anchor

***Anker***: **Ankernagel** *m* INFR & ENTW, STAHL, WERKSTOFF anchor nail; **Ankerplatte** *f* INFR & ENTW, STAHL, WERKSTOFF anchor plate; **Ankerschiene** *f* INFR & ENTW, STAHL, WERKSTOFF anchor channel, anchoring rail; **Ankerschraube** *f* BESCHLÄGE lag screw; **Ankerstab** *m* ARCH & TRAGW, INFR & ENTW, STAHL anchor rod, stay rod, tie bar; **Ankerwand** *f* ARCH & TRAGW, INFR & ENTW anchor wall

**ankleben** *vi* INFR & ENTW, STAHL, WERKSTOFF adhere

**Ankunft** *f* ARCH & TRAGW arrival; **Ankunftsbahnsteig** *m* ARCH & TRAGW arrival platform; **Ankunftshalle** *f* ARCH & TRAGW arrival lounge

**Anlage** *f* ARCH & TRAGW *Fabrik* facility, annexe (*BrE*), system, plant, BAURECHT annex (*AmE*), annexe (*BrE*), *Geld* investment, ELEKTR, HEIZ & BELÜFT system, INFR & ENTW annex (*AmE*), annexe (*BrE*), facility, plant, system; **~ zur Schwefelrückgewinnung** UMWELT sulfur recovery plant (*AmE*), sulphur recovery plant (*BrE*); **Anlagekosten** *pl* ARCH & TRAGW, BAURECHT, INFR & ENTW initial cost; **Anlagenfahrer** *m* BAUMASCHIN operator; **Anlagen- und Friedhofsbau** *m* INFR & ENTW construction of parks and cemeteries

**Anlagerung** *f* INFR & ENTW, NATURSTEIN, OBERFLÄCHE, UMWELT, WERKSTOFF adsorption

**Anlauf** *m* ARCH & TRAGW, INFR & ENTW inverted cavetto, short ramp

**Anlaufen** *nt* OBERFLÄCHE blooming

***Anlauf***: **Anlaufschiene** *f* ARCH & TRAGW facing rail

**Anlegen** *nt*: **~ von Terrassen** INFR & ENTW terracing

**Anleitung** *f* BAURECHT instruction

**Anlieferung** *f* ARCH & TRAGW, INFR & ENTW, WERKSTOFF delivery; **Anlieferungsrampe** *f* ARCH & TRAGW, INFR & ENTW delivery ramp

**Anlieger** *m* BAURECHT resident

**anmachen** *vt* ARCH & TRAGW temper, BETON, NATURSTEIN, WERKSTOFF *Mörtel* gage (*AmE*), gauge (*BrE*), mix, temper

**Anmachen** *nt* ARCH & TRAGW tempering, BETON, NATURSTEIN, WERKSTOFF *Mörtel* gaging (*AmE*), gauging (*BrE*), mixing, tempering

**Anmach**: **Anmachflüssigkeit** *f* BETON, NATURSTEIN mixing liquid; **Anmachwasser** *nt* BETON, NATURSTEIN *Mörtel* gaging water (*AmE*), gauging water (*BrE*), mixing water

**anmelden** *vt* BAURECHT apply for

**Anmeldung** *f* BAURECHT application

**annageln** *vt* ARCH & TRAGW, HOLZ nail

**Annahme** *f* ARCH & TRAGW, BAURECHT, INFR & ENTW *Lasten* assumption

**annässen** *vt* ARCH & TRAGW wet

**annehmbar** *adj* BAURECHT acceptable

**annehmen** *vt* ARCH & TRAGW *Lasten*, BAURECHT assume

**anodisch**: **~e Oxidierung** *f* OBERFLÄCHE anodization

**anordnen** *vt* ARCH & TRAGW arrange

**Anordnung** *f* ARCH & TRAGW arrangement, layout, system, BAURECHT system, INFR & ENTW arrangement, system

**anorganisch** *adj* INFR & ENTW, WERKSTOFF inorganic; **~er Boden** *m* INFR & ENTW, WERKSTOFF inorganic soil; **~er Schluff** *m* INFR & ENTW, WERKSTOFF inorganic silt

**anpassen** *vt* ARCH & TRAGW harmonize, BAUMASCHIN, BETON, ELEKTR, HEIZ & BELÜFT, INFR & ENTW, STAHL adapt

**Anpassung** *f* BAURECHT adjustment

**Anpflanzung** *f* INFR & ENTW cultivation

**Anprall** *m* ARCH & TRAGW, INFR & ENTW impact; **Anprallast** *f* INFR & ENTW impact load; **Anprallkraft** *f* ARCH & TRAGW, INFR & ENTW impact force

**Anpreßdruck** *m*: **~ im Glas** WERKSTOFF rebate pressure

**Anreicherungsbecken** *nt* UMWELT infiltration basin

**anreißen** *vi* ARCH & TRAGW, BAUMASCHIN, BETON, NATURSTEIN, WERKSTOFF line out, scribe

**Anreißen** *nt* ARCH & TRAGW marking off, plotting

**Anreißer** *m* HOLZ scribe

**Anreißtisch** *m* ARCH & TRAGW marking table

**anrühren** *vt* NATURSTEIN *Mörtel* mix

**ansammeln** *vt* BAURECHT, INFR & ENTW, WERKSTOFF accumulate

**Ansammlung** *f* BAURECHT, INFR & ENTW, WERKSTOFF accumulation

**Ansatzsäge** *f* ARCH & TRAGW tenon saw

**ansäuern** *vt* UMWELT acidify

**ansäuernd** *adj* UMWELT acidifying

**Ansäuerung** *f* UMWELT acidification

**ansaugen** *vt* HEIZ & BELÜFT induct, take in

**Ansauggitter** *nt* HEIZ & BELÜFT intake grille, intake louver (*AmE*), intake louvre (*BrE*)

**Ansaugrohr** *nt* HEIZ & BELÜFT induction pipe

**Anschlag** *m* ARCH & TRAGW *Fenster, Tür* rabbet, *Tür* stop, BAUMASCHIN *Kran* limit stop, rabbet, BESCHLÄGE rabbet; **Anschlaganleitung** *f* BESCHLÄGE fitting instructions; **Anschlagdämpfung** *f* ARCH & TRAGW *Tür* stop cushion; **Anschlagfalz** *m* ARCH & TRAGW rebate; **Anschlagmaße** *nt pl* ARCH & TRAGW fitting dimensions; **Anschlagschiene** *f* ARCH & TRAGW stop rail, striker bar; **Anschlagtafel** *f* BESCHLÄGE bulletin board

**anschließen** *vt* ARCH & TRAGW, ELEKTR connect

**Anschluß** *m* ARCH & TRAGW junction, ELEKTR connection; **~ mit Versatz** ARCH & TRAGW, HOLZ housed joint; **Anschlußbewehrungsstab** *m* ARCH & TRAGW starter bar; **Anschlußblech** *nt* STAHL joint plate; **Anschlußfahne** *f* ELEKTR connection lug; **Anschlußfuge** *f* ARCH & TRAGW connection joint; **Anschlußgleis** *nt* INFR & ENTW siding; **Anschlußhahn** *m* ABWASSER union cock; **Anschlußkasten** *m* ELEKTR terminal box; **Anschlußklemme** *f pl* ELEKTR, HEIZ & BELÜFT connector; **Anschlußkraft** *f* ARCH & TRAGW, INFR & ENTW bearing reaction; **Anschlußleitung** *f* ABWASSER connecting line, connection line, ELEKTR junction line, connecting corridor, HEIZ & BELÜFT connection line; **Anschlußmaß** *nt* ARCH & TRAGW connection dimension; **Anschlußstück** *nt* ARCH & TRAGW union, BAUMASCHIN nipple; **Anschlußstutzen** *m* ABWASSER, HEIZ & BELÜFT pipe union; **Anschlußwert** ELEKTR connected load

**anschrauben** *vt* BETON, HOLZ, STAHL attach

**Anschraubplatte** *f* BETON, HOLZ, STAHL attaching plate

**Anschüttung** *f* INFR & ENTW fill
**anschwellen** *vi* INFR & ENTW, WERKSTOFF belly out, bulge
**Anschwellen** *nt* INFR & ENTW, WERKSTOFF intumescence
**anschwellend**: **~e Dichtung** *f* INFR & ENTW, WERKSTOFF intumescent strip
**Anschwemm-** *in cpds* INFR & ENTW, WERKSTOFF alluvial; **Anschwemmboden** *m* INFR & ENTW, WERKSTOFF alluvial soil; **Anschwemmsand** *m* INFR & ENTW, WERKSTOFF alluvial sand
**anspritzen** *vt* NATURSTEIN *Mörtel* spray on
**Anspruch** *m* BAURECHT claim; **~ bei Zeitverlängerung** BAURECHT extension of time claim
**ansteigend** *adj* INFR & ENTW uphill
**Anstieg** *m* ARCH & TRAGW, UMWELT *Wasser* elevation, rise
**Anstreichen** *nt* OBERFLÄCHE brushing, coating, painting, painting work
**Anstrich** *m* OBERFLÄCHE coat, coat of paint, painting, painting work; **Anstrichfarbe** *f* OBERFLÄCHE, WERKSTOFF paint; **Anstrichfehler** *m* OBERFLÄCHE coating defect
**Anteil** *m* INFR & ENTW, WERKSTOFF fraction
**Antenne** *f* BAUMASCHIN, ELEKTR aerial, antenna; **Antennenhalterung** *f* BESCHLÄGE, ELEKTR antenna pipe mount; **Antennenmast** *m* ELEKTR, STAHL aerial mast, antenna mast; **Antennensteckdose** *f* ELEKTR aerial socket, antenna socket
**anthropogen**: **~ bedingte Übersäuerung** *f* UMWELT anthropogenic acidification
**Antidröhn-** *in cpds* DÄMMUNG sound-deadening; **Antidröhnbeschichtung** *f* DÄMMUNG, OBERFLÄCHE sound-deadening coating
**Antikglas** *nt* WERKSTOFF antique glass
**Antiklopfmittel** *nt* BAUMASCHIN, UMWELT antiknock additive
**Antioxidans** *nt* WERKSTOFF antioxidant
**Antrag** *m* BAURECHT application
**Antritt** *m* ARCH & TRAGW starting step; **Antrittspfosten** *m* ARCH & TRAGW newel post; **Antrittsstufe** *f* ARCH & TRAGW bottom step
**Anvisieren** *nt* ARCH & TRAGW sighting
**anvisiert**: **~e Dioptrie** *f* ARCH & TRAGW sighted alidade; **~e Höhe** *f* ARCH & TRAGW sighted level
**anwenden** *vt* ARCH & TRAGW employ
**Anwendung** *f* ARCH & TRAGW, INFR & ENTW use, OBERFLÄCHE application; **Anwendungsbereich** *m* ARCH & TRAGW, INFR & ENTW area of application; **Anwendungsmethode** *f* ARCH & TRAGW, INFR & ENTW applied method
**Anwerfen** *nt* NATURSTEIN *Putz* throwing on
**Anwurf** *m* NATURSTEIN rendering, roughcast, *Putz* throwing on
**Anzahl** *f* ARCH & TRAGW number
**anzeichnen** *vt* BAUMASCHIN scribe
**Anzeige** *f* BAUMASCHIN indication; **Anzeigebereich** *m* BAUMASCHIN indication range; **Anzeigefehler** *m* BAUMASCHIN indication error; **Anzeigen** *f pl* BAURECHT notices
**Appartement** *nt* ARCH & TRAGW apartment
**Apsidenbogen** *m* ARCH & TRAGW apse arch; **Apsidenbogenkämpfer** *m* ARCH & TRAGW apse arch impost
**Aquädukt** *nt* INFR & ENTW, UMWELT aqueduct
**Aquifer** *m* UMWELT aquifer
**äquivalent** *adj* ARCH & TRAGW, INFR & ENTW equivalent
**Arabeske** *f* ARCH & TRAGW arabesque
**Aragonit** *m* WERKSTOFF aragonite
**Arbeit** *f* ARCH & TRAGW, INFR&ENTW work; **Arbeitsablauf** *m* ARCH & TRAGW, INFR & ENTW sequence of operations, working schedule; **Arbeitsablaufplan** *m* ARCH & TRAGW, INFR & ENTW working schedule; **Arbeitsbrücke** *f* ARCH & TRAGW staging; **Arbeitsbühne** *f* ARCH & TRAGW platform, *Gerüst* working platform; **Arbeitsdruck** *m* BAUMASCHIN working pressure; **Arbeitsfläche** *f* ARCH & TRAGW working surface; **Arbeitsfortgang** *m* ARCH & TRAGW work progress; **Arbeitsfuge** *f* ARCH & TRAGW, BETON construction joint; **Arbeitsgemeinschaft** *f* BAURECHT joint venture; **Arbeitsgrube** *f* ARCH & TRAGW, INFR & ENTW *für Wartungsarbeiten* inspection pit; **Arbeitskosten** *pl* ARCH & TRAGW, BAURECHT, INFR & ENTW labor charges (*AmE*), labor costs (*AmE*), labour charges (*BrE*), labour costs (*BrE*); **Arbeitskraft** *f* ARCH & TRAGW manpower; **Arbeitskraftplanung** *f* BAURECHT manpower planning; **Arbeitslärm** *m* DÄMMUNG work noise; **Arbeitslohn** *m* ARCH & TRAGW, BAURECHT wages; **Arbeitsplatte** *f* ARCH & TRAGW *Küche* tabletop; **Arbeitsraum** *m* ARCH & TRAGW, INFR & ENTW workspace; **Arbeitsstunden** *f pl* BAURECHT hours of work
**arbeitstäglich**: **~e Abdeckung** *f* UMWELT *einer Deponie* daily cover
***Arbeit***: **Arbeitstemperatur** *f* BAUMASCHIN working temperature; **Arbeitsumfang** *m* ARCH & TRAGW, INFR & ENTW scope of work; **Arbeitsvermögen** *nt* UMWELT working capacity; **Arbeitszeit** *f* ARCH & TRAGW, BAURECHT working hours
**Architekt** *m* ARCH & TRAGW architect; **Architektenleistung** *f* ARCH & TRAGW architectural work; **Architektenzeichnung** *f* ARCH & TRAGW architectural drawing
**architektonisch** *adj* ARCH & TRAGW, BAURECHT architectural
**Architektur** *f* ARCH & TRAGW architecture; **Architekturausbildung** *f* ARCH & TRAGW architectural education; **Architekturpreis** *m* ARCH & TRAGW architectural award
**Architrav** *m* ARCH & TRAGW, HOLZ architrave
**Argillit** *m* WERKSTOFF argillite
**Arkade** *f* ARCH & TRAGW arcade, colonnade; **Arkadenhof** *m* ARCH & TRAGW arcaded court
**Arkatur** *f* ARCH & TRAGW arcade, arcature
**Arm** *m* ARCH & TRAGW arm
**Armatur** *f* HEIZ & BELÜFT fitting
***Arm***: **Armauflage** *f* BESCHLÄGE elbow rail
**armieren** *vt* BETON reinforce
**Armierung** *f* ARCH & TRAGW, BETON reinforcement; **Armierungsgewebe** *nt* NATURSTEIN *Putz* reinforcing tape; **Armierungsmatte** *f* BETON reinforcement mat, reinforcement steel mesh, steel wire mesh
**aromatisch**: **~e Verbindung** *f* WERKSTOFF aromatic compound
**Arretierschraube** *f* BESCHLÄGE stop screw
**Arretierung** *f* BESCHLÄGE, WERKSTOFF *Schloß* catch, lock
**artesisch** *adj* INFR & ENTW artesian; **~er Brunnen** *m* INFR & ENTW, UMWELT artesian well; **~er Druck** *m* INFR & ENTW artesian pressure
**AS** *abbr* (*Amerikanische Norm*) ARCH & TRAGW BAURECHT AS (*American Standard*)

**Asbest** *m* BAURECHT, WERKSTOFF asbestos; **Asbestdachplatte** *f* WERKSTOFF asbestos roofing sheet; **Asbestdämmplatte** *f* WERKSTOFF asbestos insulating board; **Asbesterzeugnis** *nt* WERKSTOFF asbestos product; **Asbestfaser** *f* BAURECHT, WERKSTOFF asbestos fiber (*AmE*), asbestos fibre (*BrE*)

**asbestfaserhaltig** *adj* WERKSTOFF containing asbestos fibers (*AmE*), containing asbestos fibres (*BrE*)

***Asbest***: **Asbestgestein** *nt* WERKSTOFF asbestos rock

**asbesthaltig** *adj* BAURECHT, WERKSTOFF asbestos-containing

***Asbest***: **Asbesthandschuhe** *m pl* WERKSTOFF asbestos gloves; **Asbestplatte** *f* WERKSTOFF asbestos board, asbestos sheet; **Asbest-PVC-Bodenplatte** *f* WERKSTOFF asbestos PVC floor tile; **Asbestrohr** *nt* WERKSTOFF asbestos pipe; **Asbestschnur** *f* WERKSTOFF asbestos cord; **Asbeststaub** *m* BAURECHT, WERKSTOFF asbestos dust; **Asbesttafel** *f* WERKSTOFF asbestos panel; **Asbestwolle** *f* WERKSTOFF asbestos wool

**Asbestzement** *m* BETON, WERKSTOFF asbestos cement; **Asbestzementerzeugnis** *nt* BETON, WERKSTOFF asbestos cement product; **Asbestzementmörtel** *m* BETON, WERKSTOFF *Feuerschutz* asbestos cement lining; **Asbestzementschindel** *f* BETON, WERKSTOFF asbestos cement shingle; **Asbestzementtafel** *f* BETON, WERKSTOFF asbestos cement panel; **Asbestzementtrennwand** *f* ARCH & TRAGW, BETON asbestos cement partition; **Asbestzementverkleidung** *f* BETON, WERKSTOFF asbestos cement lining; **Asbestzementwaren** *f pl* BETON, WERKSTOFF asbestos cement goods; **Asbestzementwellplatte** *f* BETON, WERKSTOFF corrugated asbestos cement sheet

**Asche** *f* HEIZ & BELÜFT, UMWELT ash

**aschefrei** *adj* UMWELT ash-free

***Asche***: **Aschegehalt** *m* WERKSTOFF ash content; **Asche- und Verbrennungsrückstand** *m* UMWELT ash and combustion residue

**ASHRAE** *abbr* (*Amerikanische Gesellschaft der Heizungs-, Kühlungs- und Klimatechniker*) HEIZ & BELÜFT ASHRAE (*AmE*) (*American Society of Heating, Refrigeration and Air Conditioning Engineers*)

**Aspekt** *m* ARCH & TRAGW, BAURECHT, INFR & ENTW *Seite, Element* aspect

**Asphalt** *m* ARCH & TRAGW, BETON, INFR & ENTW, OBERFLÄCHE, WERKSTOFF asphalt, bitumen; **Asphaltbeton** *m* BETON, INFR & ENTW, WERKSTOFF asphalt concrete, bitumen concrete, bituminous concrete; **Asphaltbetonfertiger** *m* BAUMASCHIN, BETON, INFR & ENTW asphaltic concrete finisher; **Asphaltbinder** *m* INFR & ENTW, WERKSTOFF asphaltic binder; **Asphaltbinderschicht** *f* NATURSTEIN binder course; **Asphaltbrecher** *m* INFR & ENTW asphalt crusher; **Asphaltdecke** *f* INFR & ENTW, WERKSTOFF asphalt pavement, asphalt surfacing, asphaltic pavement, bituminous pavement; **Asphaltfußboden** *m* ARCH & TRAGW, WERKSTOFF asphalt floor, asphalt flooring, asphaltic floor covering, mastic flooring; **Asphaltfußbodenbelag** *m* ARCH & TRAGW, WERKSTOFF asphalt floor, asphalt flooring, asphaltic floor covering, mastic flooring; **Asphaltgestein** *nt* NATURSTEIN bituminous rock

**asphaltieren** *vt* INFR & ENTW, OBERFLÄCHE, WERKSTOFF asphalt, bituminize, tarmac

**Asphaltieren** *nt* INFR & ENTW, OBERFLÄCHE, WERKSTOFF asphalting, bituminization, tarmacking

**asphaltiert** *adj* INFR & ENTW, OBERFLÄCHE, WERKSTOFF bituminized, tarmacked

**asphaltisch** *adj* ARCH & TRAGW, INFR & ENTW, OBERFLÄCHE, WERKSTOFF asphaltic

***Asphalt***: **Asphaltmastix** *f* INFR & ENTW, WERKSTOFF asphalt mastic, mastic asphalt; **Asphaltmörtel** *m* BETON, WERKSTOFF bitumen mortar; **Asphaltmühle** *f* INFR & ENTW, WERKSTOFF asphalt mastic; **Asphaltplatte** *f* INFR & ENTW asphaltic tile; **Asphaltstrich** *m* ARCH & TRAGW, WERKSTOFF asphalt floor, asphalt flooring, asphaltic floor covering, mastic flooring

**Ast** *m* HOLZ knot; **Astbohrer** *m* BAUMASCHIN, HOLZ knot borer

**astfrei** *adj* HOLZ clean, knotless, UMWELT clean

**ästhetisch** *adj* ARCH & TRAGW, OBERFLÄCHE aesthetic; **~ korrekt** *adj* ARCH & TRAGW, OBERFLÄCHE aesthetically correct

***Ast***: **Astloch** *nt* HOLZ knot hole

**Atem** *m* UMWELT breath; **Atemgrenzwert** *m* UMWELT breathing capacity; **Atemschutzgerät** *nt* BAURECHT breathing apparatus, UMWELT breathing apparatus, respiratory protection equipment; **Atemschutzsystem** *nt* UMWELT breathing protection system

**atmen** *vt* UMWELT respire

**Atmosphäre** *f* INFR & ENTW, UMWELT atmosphere

**atmosphärisch**: **~er Auswaschvorgang** *m* HEIZ & BELÜFT, UMWELT atmospheric scrubbing; **~e Erscheinung** *f* UMWELT atmospheric phenomenon; **~e Luftbelastung** *f* UMWELT atmospheric loading; **~e Masse** *f* UMWELT air mass; **~er Niederschlag** *m* UMWELT precipitation, atmospheric precipitation, *radioaktiv* atmospheric fallout; **~e Säurekapazität** *f* UMWELT atmospheric acidity; **~er Schwefel** *m* UMWELT atmospheric sulfur (*AmE*), atmospheric sulphur (*BrE*); **~e Verdunklung** *f* UMWELT atmospheric obscurity

**Atmung** *f* UMWELT breathing, respiration; **Atmungsgerät** *nt* INFR & ENTW respirator; **Atmungsvermögen** *nt* UMWELT, WERKSTOFF breathing capability

**Atom** *nt* WERKSTOFF atom; **Atomgewicht** *nt* WERKSTOFF atomic weight; **Atomkraftwerk** *nt* (*AKW*) INFR & ENTW, UMWELT nuclear power plant, nuclear power station; **Atommüll** *m* UMWELT radioactive waste

**Atrium** *nt* ARCH & TRAGW atrium, central courtyard, INFR & ENTW atrium; **~ in ganzer Höhe** ARCH & TRAGW full-height atrium

**Atterberg**: **~sche Konsistenzgrenzen** *f pl* INFR & ENTW Atterberg limits

**Attika** *f* ARCH & TRAGW roof parapet

**ätzen** *vt* OBERFLÄCHE, WERKSTOFF corrode

**ätzend** *adj* OBERFLÄCHE, WERKSTOFF corrosive

**Ätzung** *f* OBERFLÄCHE, WERKSTOFF corrosion

**Audiometrie** *f* ARCH & TRAGW, DÄMMUNG audibility range, audiometry

**Aufarbeitung** *f* ARCH & TRAGW, INFR & ENTW reconditioning, refurbishment, UMWELT *Müll* processing, reprocessing, WERKSTOFF processing; **Aufarbeitungsvorrichtung** *f* UMWELT recovery device; **Aufarbeitungszentrum** *nt* **für wiederverwertbare feste Abfallmaterialien** UMWELT processing center for recyclable solid waste materials (*AmE*), processing centre for recyclable solid waste materials (*BrE*)

**Aufbau** *m* ARCH & TRAGW arrangement, superstructure, INFR & ENTW arrangement, WERKSTOFF composition

**aufbauen** *vt* ARCH & TRAGW rig, build up, set up
***Aufbau***: **Aufbauleuchte** *f* ELEKTR surface-mounted light fixture
**aufbereiten** *vt* UMWELT condition, treat, WERKSTOFF prepare
**Aufbereitung** *f* BAUMASCHIN, INFR & ENTW, UMWELT preparation, treatment; **Aufbereitungsanlage** *f* INFR & ENTW preparation plant, treatment plant, UMWELT preparation plant, treatment plant, processing plant; **Aufbereitungsmaschinen** *f pl* **für Baustoffe** BAUMASCHIN machinery for recycling of building materials; **Aufbereitungsverfahren** *nt* UMWELT treatment process, preparation process
**Aufbeton** *m* BETON concrete topping
**aufbewahren** *vt* ARCH & TRAGW keep
**aufbiegen** *vt* WERKSTOFF *Bewehrungsstahl* bend up
**aufblähen** ARCH & TRAGW, BAUMASCHIN, HEIZ & BELÜFT, INFR & ENTW, WERKSTOFF expand
**Aufblähen** *nt* ARCH & TRAGW, BAUMASCHIN, HEIZ & BELÜFT, INFR & ENTW, WERKSTOFF expansion
**Aufbohrschutz** *m* BAUMASCHIN, BESCHLÄGE *Schloß* antidrill feature
**aufbrechen** *vi* INFR & ENTW *Straßenbelag* break open
**Aufbringen** *nt* OBERFLÄCHE application; **~ einer neuen Schotterschicht** INFR & ENTW *Straßenbau* remetaling (*AmE*), remetalling (*BrE*)
**aufdämmen** *vt* ARCH & TRAGW bank up
**Aufdoppelung** *f* HOLZ *Kante* false edge
**aufeinanderfolgend** *adj* ARCH & TRAGW back-to-back
**Aufenthaltsraum** *m* ARCH & TRAGW, INFR & ENTW dayroom, recreation room
**Auffang**: **Auffangbecken** *nt* INFR & ENTW catch-basin; **Auffanggefäß** *nt* ABWASSER trap; **Auffangring** *m* ELEKTR *Blitzschutz* conductor loop; **Auffangschale** *f* ABWASSER, HEIZ & BELÜFT collecting reservoir; **Auffangtrichter** *m* HEIZ & BELÜFT collecting funnel, collecting hopper; **Auffangvorrichtung** *f* ABWASSER trap; **Auffangwanne** *f* ABWASSER, HEIZ & BELÜFT collecting reservoir
**auffrischen** *vt* ARCH & TRAGW refurbish
**Auffüllbeton** *m* ARCH & TRAGW, BETON backfill concrete
**auffüllen** *vt* ARCH & TRAGW *Graben* fill up, BETON backfill, DÄMMUNG deaden, INFR & ENTW, NATURSTEIN backfill
**Auffüllung** *f* ARCH & TRAGW refilling, BETON backfilling, backfill, DÄMMUNG *zur Schalldämmung* pugging, INFR & ENTW *Erdreich* made ground, backfill, backfilling, refilling, NATURSTEIN backfill, backfilling
**Aufgabe** *f* BAURECHT abandonment
**aufgeben** *vt* BAURECHT abandon
**aufgebogen**: **~er Stab** *m* BETON inclined bar
**aufgegliedert**: **~e Fassade** *f* ARCH & TRAGW articulated elevation
**aufgehängt** *adj* ARCH & TRAGW suspended
**aufgehend**: **~e Wand** *f* ARCH & TRAGW rising wall
**aufgelagert** *adj* ARCH & TRAGW supported
**aufgelegt**: **~er Gitterrost** *m* STAHL lay-on grating
**aufgenietet** *adj* BAUMASCHIN, STAHL riveted
**aufgenommen**: **~e Last** *f* ARCH & TRAGW accepted load
**aufgesattelt**: **~e Treppe** *f* ARCH & TRAGW open string stairs
**aufgeschäumt** *adj* DÄMMUNG, WERKSTOFF expanded, *Kunststoff, Gummi* foamed
**aufgeschrumpft** *adj* WERKSTOFF shrunk-on
**aufgeschüttet** *adj* INFR & ENTW *Erdreich* made-up
**Aufhängeeisen** *nt* BESCHLÄGE suspension bracket
**Aufhängevorrichtung** *f* WERKSTOFF hanger
**Aufhäufen** *nt* ARCH & TRAGW piling up
**aufheben** *vt* BAURECHT abolish
**Aufhebung** *f* BAURECHT abolition
**aufhellen** *vt* OBERFLÄCHE tint
**Aufhöhung** *f* ARCH & TRAGW *einer Mauer* raising
**Aufkantung** *f* ARCH & TRAGW upstand; **Aufkantungsrinne** *f* ABWASSER upstand gutter
**Aufkippbauweise** *f* ARCH & TRAGW tilt-up method
**aufkleben** *vt* HOLZ, OBERFLÄCHE glue on
**Aufkleber** *m* WERKSTOFF *auf technischen Anlagen, Typenschildern* adhesive label
**aufkratzen** *vt* OBERFLÄCHE scratch
**aufladen** *vt* BAUMASCHIN recharge
**Aufladezeit** *f* BAUMASCHIN, HEIZ & BELÜFT charging period
**Aufladung** *f* ELEKTR charge
**Auflage** *f* ARCH & TRAGW impost, support, *Rohr, Kabel* bearing surface, BAURECHT condition, INFR & ENTW bearing surface, WERKSTOFF packing rubber; **Auflageflanschstab** *m* ARCH & TRAGW bearing bar; **Auflageholz** *nt* HOLZ pole plate
**Auflager** *nt* INFR & ENTW footing; **Auflagerbedingungen** *f pl* ARCH & TRAGW support conditions; **Auflagerfläche** *f* ARCH & TRAGW, BETON, NATURSTEIN, STAHL area of support, bearing area, seating
***Auflage***: **Auflagering** *m* ABWASSER, INFR & ENTW manhole top ring
***Auflager***: **Auflagerkraft** *f* ARCH & TRAGW, INFR & ENTW bearing reaction; **Auflagerrahmen** *m* ARCH & TRAGW supporting frame; **Auflagerschräge** *f* BETON tapered haunch; **Auflagerstein** *m* ARCH & TRAGW padstone; **Auflagerverschiebung** *f* INFR & ENTW displacement of support
**Auflast** *f* ARCH & TRAGW, INFR & ENTW superimposed load
**aufliegend** *adj* ARCH & TRAGW supported; **~es Schloß** *nt* BESCHLÄGE surface-mounted lock
**auflockern** *vt* INFR & ENTW, WERKSTOFF loosen
**Aufmaß** *nt* ARCH & TRAGW, INFR & ENTW measured work, quantity survey, site measuring, survey
**aufmauern** *vt* ARCH & TRAGW, NATURSTEIN brick up, cope
**Aufmauern** *nt* ARCH & TRAGW, NATURSTEIN *Mauerkronen* bricking up, coping
**Aufnahme** *f* ARCH & TRAGW mapping out, acceptance, *Gebäude, Bausubstanz* survey, BAURECHT acceptance, admission, INFR & ENTW survey; **Aufnahmebunker** *m* UMWELT receiving bunker
**aufnehmen** *vt* ARCH & TRAGW *Last*, BAURECHT accept
**aufpolieren** *vt* ARCH & TRAGW refurbish
**Aufputzausführung** *f* ABWASSER, ELEKTR, HEIZ & BELÜFT surface type
**Aufputzinstallation** *f* ELEKTR, HEIZ & BELÜFT exposed wiring
**aufrauhen** *vt* BESCHLÄGE, INFR & ENTW, NATURSTEIN, OBERFLÄCHE hack off, roughen, scarify, score
**Aufrauhen** *nt* BESCHLÄGE, INFR & ENTW, NATURSTEIN, OBERFLÄCHE hacking off, roughening, scarification
**aufrecht** *adj* ARCH & TRAGW standing
**aufrechtstehend** *adj* ARCH & TRAGW upright; **~e Gesimsziegelschicht** *f* NATURSTEIN soldier string course; **~er Ziegel** *m* NATURSTEIN soldier
**Aufreibdorn** *m* BAUMASCHIN reaming iron
**aufreißen** *vt* INFR & ENTW rip, scarify

**Aufreißen** *nt* INFR & ENTW *Straßen* scarification
**Aufreißer** *m* INFR & ENTW scarifier, *Straßenbau* ripper
**Aufriß** *m* ARCH & TRAGW vertical plane, *eines Gebäudes* elevation, UMWELT elevation
**aufrunden** *vt* ARCH & TRAGW, INFR & ENTW round up
**Aufsattelung** *f* HOLZ bolster, head tree, corbel piece
**Aufsatz** *m* ABWASSER *Straßen, Abläufen* frame, ARCH & TRAGW crown; **Aufsatzschloß** *nt* BESCHLÄGE rim lock
**aufsaugend** *adj* OBERFLÄCHE absorbent
**aufschäumen** *vt* WERKSTOFF expand, foam
**aufschichten** *vt* INFR & ENTW pile
**Aufschichtung** *f* ARCH & TRAGW piling
**Aufschiebling** *m* HOLZ *Traufbrett* cant strip
**Aufschlag** *m* BAURECHT additional charge, UMWELT impaction
**Aufschluß** *m* INFR & ENTW exposure; **Aufschlußbohrung** *f* INFR & ENTW trial boring
**aufschraubbar** *adj* BAUMASCHIN threaded
**Aufschraubschloß** *nt* BESCHLÄGE rim lock
**aufschrumpfen** *vt* WERKSTOFF shrink on
**aufschütten** *vt* ARCH & TRAGW bank up, raise
**Aufschüttung** *f* INFR & ENTW *Erde* fill, filled ground, filled-up ground, filling, made-up ground
**aufsetzen** *vt* ARCH & TRAGW, BETON, HOLZ, STAHL attach
**Aufsetzkranz** *m* INFR & ENTW *Schacht* bearing frame
**Aufsicht** *f* ARCH & TRAGW, BAURECHT supervision; **mit Aufsicht** *phr* INFR & ENTW *für ältere Bewohner* warden-assisted; **Aufsichtsbehörde** *f* BAURECHT, INFR & ENTW supervising authority
**aufspachteln** *vt* NATURSTEIN *Putz* float
**Aufspaltung** *f* ARCH & TRAGW splitting
**Aufsperrsicherung** *f* BAURECHT, BESCHLÄGE *Schloß* intruder protection
**aufspitzen** *vt* BAUMASCHIN, BETON, NATURSTEIN *Beton- oder Steinoberflächen* granulate, scabble
**aufspritzen** *vt* OBERFLÄCHE spray, *Farbe* spray on
**Aufsteckhülse** *f* HEIZ & BELÜFT extension tube
**aufsteigend**: **~e Feuchtigkeit** *f* ARCH & TRAGW, DÄMMUNG, NATURSTEIN rising damp, rising humidity, rising moisture
**aufstellen** *vt* ARCH & TRAGW, BAUMASCHIN, INFR & ENTW position, rig
**Aufstellung** *f* ARCH & TRAGW, BAUMASCHIN, INFR & ENTW erection, installation; **Aufstellungsplan** *m* ARCH & TRAGW installation plan, BAURECHT space assignment plan, INFR & ENTW installation plan, space assignment plan
**aufstocken** *vt* ARCH & TRAGW add a storey to (*BrE*), add a story to (*AmE*), *Gebäude* heighten, raise, BETON scabble, pick, NATURSTEIN scabble, *Bruchsteine* ax (*AmE*), pick, axe (*BrE*)
**Aufstockung** *f* ARCH & TRAGW *Gebäude* heightening
**Aufstreichen** *nt* OBERFLÄCHE application by brushing
**aufteilen** *vt* ARCH & TRAGW, INFR & ENTW partition
**Aufteilung** *f* ARCH & TRAGW division
**Auftrag** *m* BAURECHT order, INFR & ENTW, WERKSTOFF *Erde* filling
**auftragen** *vt* ARCH & TRAGW plot, OBERFLÄCHE apply, fill
***Auftrag***: **Auftraggeber** *m* (*AG*) ARCH & TRAGW, BAURECHT client, customer; **Auftragmenge** *f* OBERFLÄCHE application rate; **Auftragnehmer** *m* BAURECHT contractor; **Auftragserteilung** *f* BAURECHT award of the contract; **Auftragsschreiben** *nt* ARCH & TRAGW notice of award; **Auftragsverfahren** *nt* OBERFLÄCHE application method
**Auftrieb** *m* ARCH & TRAGW, INFR & ENTW, UMWELT, WERKSTOFF *Aerodynamik* buoyancy, lift; **Auftriebsbeiwert** *m* (*CL*) UMWELT lift coefficient (*CL*); **Auftriebssicherung** *f* ARCH & TRAGW, INFR & ENTW buoyancy protection; **Auftriebszahl** *f* (*CL*) UMWELT lift coefficient (*CL*)
**Auftritt** *m* ARCH & TRAGW *Treppe* tread; **Auftrittbreite** *f* ARCH & TRAGW run; **Auftrittsbreite** *f* ARCH & TRAGW foothold
**aufwärts** *adv* INFR & ENTW upwards
**Aufwärtsströmung** *f* INFR & ENTW upward flow
**Aufweichen** *nt* INFR & ENTW, UMWELT, WERKSTOFF maceration
**aufweiten** *vt* OBERFLÄCHE bulge, planish, WERKSTOFF bulge
**aufwendig** *adj* ARCH & TRAGW costly
**aufwerten** *vt* ARCH & TRAGW, WERKSTOFF *Gebäude* grade up
**Aufwertung** *f* ARCH & TRAGW upgrading
**Aufwölbung** *f* ARCH & TRAGW, BETON, INFR & ENTW camber; **Aufwölbungsmoment** *nt* ARCH & TRAGW, BETON, INFR & ENTW, STAHL, WERKSTOFF hogging moment; **Aufwölbungsquerbiegung** *f* ARCH & TRAGW, BETON, INFR & ENTW, STAHL *Verbundträger* hogging transverse bending
**Aufwuchs** *m* INFR & ENTW growth
**aufzeichnen** *vt* ARCH & TRAGW record, plot
**Aufzeichnung** *f* ARCH & TRAGW record; **~ der Untersuchungsergebnisse** BAURECHT recording inspection findings
**Aufzug** *m* ARCH & TRAGW, BESCHLÄGE, ELEKTR elevator (*AmE*), lift (*BrE*); **~ zur Feuerbekämpfung** BAURECHT, INFR & ENTW firefighting lift; **Aufzugsschacht** *m* ARCH & TRAGW well; **Aufzugstür** *f* ARCH & TRAGW, BESCHLÄGE elevator door (*AmE*), lift door (*BrE*)
**Auge**: **Augendusche** *f* UMWELT eye shower; **Augenhöhe** *f* ARCH & TRAGW eye level; **Augenscheinkontrolle** *f* ARCH & TRAGW, INFR & ENTW visual check, visual examination; **Augenschutzausrüstung** *f* BAURECHT, BESCHLÄGE eye protection equipment; **Augenstab** *m* STAHL *Brücken, Dächer* eyebar
**Augitporphyr** *m* WERKSTOFF augite porphyry
**Aula** *f* ARCH & TRAGW, INFR & ENTW aula
**ausbaggern** *vt* INFR & ENTW, UMWELT excavate
**Ausbau** *m* ARCH & TRAGW extension; **Ausbauarbeiten** *pl* ARCH & TRAGW finishing work, interior work, *Innenausbau* completion work
**Ausbauchung** *f* WERKSTOFF belly, bulge
**ausbauen** *vt* ARCH & TRAGW extend, remove
**Ausbauen** *nt* ARCH & TRAGW extending, removing
**ausbaufähig** *adj* ARCH & TRAGW extendable, extensible
***Ausbau***: **Ausbaugewerbe** *nt pl* ARCH & TRAGW finishing trades; **Ausbaugröße** *f* ARCH & TRAGW design capacity; **Ausbaumaß** *nt* ARCH & TRAGW size when completed; **Ausbauverhältnis** *nt* ARCH & TRAGW interior work ratio
**ausbessern** *vt* BAUMASCHIN, NATURSTEIN, OBERFLÄCHE adjust, make good
**Ausbessern** *nt* BAUMASCHIN, NATURSTEIN, OBERFLÄCHE adjusting, making good
**Ausbesserungsarbeiten** *f pl* ARCH & TRAGW, INFR & ENTW repair work

**ausbeulen** *vt* OBERFLÄCHE, WERKSTOFF bulge, planish
**Ausbeulen** *nt* OBERFLÄCHE, WERKSTOFF planishing
**ausbeuten** *vt* INFR & ENTW, UMWELT exploit
**Ausbeutung** *f* INFR & ENTW, UMWELT exploitation; **~ der geothermalen Energie** UMWELT exploitation of geothermal energy
**Ausbildungsgebäude** *nt* ARCH & TRAGW educational building
**ausblasen** *vt* HEIZ & BELÜFT, INFR & ENTW blow out
**Ausblasjalousie** *f* HEIZ & BELÜFT louver vent (*AmE*), louvre vent (*BrE*)
**Ausblühung** *f* WERKSTOFF efflorescence
**ausbreiten**: **sich ausbreiten** *v refl* DÄMMUNG *Wellen* travel
**Ausbreitmaß** *nt* BETON, INFR & ENTW, WERKSTOFF *Beton* slump; **Ausbreitmaßprüfung** *f* BETON, WERKSTOFF slump test
**Ausbreitung** *f* BAURECHT, INFR & ENTW spreading, UMWELT diffusion, propagation, WERKSTOFF propagation; **Ausbreitungsgeschwindigkeit** *f* ARCH & TRAGW rate of spread; **Ausbreitungsmaß** *nt* BETON *Betonprüfung*, INFR & ENTW, WERKSTOFF *Betonprüfung* slump
**Ausbruch** *m* INFR & ENTW *Tunnel*, UMWELT excavation
**ausdehnbar** *adj* WERKSTOFF expandable
**ausdehnen**: **sich ausdehnen** *v refl* WERKSTOFF expand
**Ausdehnung** *f* ARCH & TRAGW, BAUMASCHIN, HEIZ & BELÜFT expansion, INFR & ENTW, WERKSTOFF expansion, extension; **Ausdehnungsfuge** *f* ARCH & TRAGW, INFR & ENTW expansion joint; **Ausdehnungsgefäß** *nt* HEIZ & BELÜFT expansion tank, expansion vessel; **Ausdehnungskraft** *f* ARCH & TRAGW, INFR & ENTW, WERKSTOFF expansion force; **Ausdehnungsmodul** *m* ARCH & TRAGW, BETON, INFR & ENTW, STAHL, WERKSTOFF modulus of longitudinal deformation
**auseinandergezogen-isometrisch** *adj* ARCH & TRAGW, INFR & ENTW exploded isometric
**Ausfachung** *f* ARCH & TRAGW filler wall, nogging, NATURSTEIN infilling, infilling wall
**ausfahren** *vi* INFR & ENTW ride
**Ausfall** *m* BAUMASCHIN, INFR & ENTW failure
**ausfällen** *vt* INFR & ENTW, UMWELT precipitate
***Ausfall***: **Ausfallkörnung** *f* WERKSTOFF gap grading, omitted-size fraction
**Ausfällung** *f* UMWELT precipitation, coagulation
***Ausfall***: **Ausfallwarnleuchte** *f* BAURECHT, ELEKTR, INFR & ENTW, WERKSTOFF failure warning light
**Ausfaulgrube** *f* INFR & ENTW sludge digestion tank
**ausfließen** *vi* ABWASSER, UMWELT discharge, flow out, leak out
**Ausflockung** *f* UMWELT, WERKSTOFF flocculation; **Ausflockungsmittel** *nt* INFR & ENTW, WERKSTOFF flocculating agent
**ausfluchten** *vt* ABWASSER *Linie* line out, ARCH & TRAGW align
**Ausflußzahl** *f* UMWELT discharge coefficient
**ausfugen** *vt* NATURSTEIN join, point, *Mauerwerk, Fliesen* fill in
**Ausfugen** *nt* NATURSTEIN filling in, joining, pointing
***ausfugen***: **wieder ausfugen** *vt* NATURSTEIN repoint
**Ausfugmasse** *f* WERKSTOFF jointing compound
**ausführen** *vt* ARCH & TRAGW, BAUMASCHIN, INFR & ENTW, WERKSTOFF *Arbeiten* execute, perform
**ausführend**: **~er Ingenieur** *m* ARCH & TRAGW, INFR & ENTW project engineer
**Ausführung** *f* ARCH & TRAGW workmanship, type, *Arbeiten* execution *n*; **Ausführungsbestimmungen** *f pl* BAURECHT code of practice; **Ausführungshöhe** *f* ARCH & TRAGW mounting height; **Ausführungsplan** *m* ARCH & TRAGW final plan; **Ausführungsunterlagen** *f pl* ARCH & TRAGW, INFR & ENTW final design, final planning documents; **Ausführungszeichnung** *f* ARCH & TRAGW, INFR & ENTW final drawing, working drawing
**Ausfüllung** *f* NATURSTEIN hearting, infilling wall
**Ausgang** *m* ARCH & TRAGW, HEIZ & BELÜFT, INFR & ENTW exit; **Ausgangsöffnungen** *f pl* ELEKTR, HEIZ & BELÜFT exit holes; **Ausgangstransparent** *nt* ELEKTR illuminated exit sign
**ausgefüllt** *adj* ARCH & TRAGW filled
**ausgeglichen**: **~e Trittstufenfläche** *f* INFR & ENTW balance step
**ausgeglüht**: **~er Weichdraht** *m* INFR & ENTW, WERKSTOFF annealed soft wire
**ausgekleidet** *adj* ARCH & TRAGW, INFR & ENTW lined
**ausgelaugt** *adj* INFR & ENTW, WERKSTOFF leached
**ausgespachtelt** *adj* BETON, NATURSTEIN filled
**ausgespart** *adj* ARCH & TRAGW recessed
**ausgeweitet** *adj* WERKSTOFF expanded
**ausgewuchtet** *adj* HEIZ & BELÜFT *Ventilator* balanced
**ausgießen** *vt* BETON, NATURSTEIN pour out
**Ausgleich** *m* ARCH & TRAGW equilibrium, BAURECHT adjustment, INFR & ENTW adjustment, equilibrium; **Ausgleichbecken** *nt* UMWELT surge tank; **Ausgleichbehälter** *m* BAUMASCHIN make-up tank; **Ausgleichbeton** *m* BETON blinding concrete, leveling concrete (*AmE*), levelling concrete (*BrE*)
**ausgleichen** *vt* ARCH & TRAGW level, average out, INFR & ENTW, WERKSTOFF average out
***Ausgleich***: **Ausgleichsbogen** *m* HEIZ & BELÜFT, INFR & ENTW *Rohr* expansion bend; **Ausgleichsfuge** *f* ARCH & TRAGW compensation joint; **Ausgleichsgewicht** *nt* BAUMASCHIN *Kran, Tor* counterweight; **Ausgleichsring** *m* BETON equalizing ring; **Ausgleichsschicht** *f* INFR & ENTW leveling course (*AmE*), levelling course (*BrE*), leveling layer (*AmE*), levelling layer (*BrE*); **Ausgleichsverfahren** *nt* ARCH & TRAGW balance method
**ausgraben** *vt* INFR & ENTW, UMWELT excavate
**Ausguß** *m* ABWASSER sink, ARCH & TRAGW lip
**aushärten** *vt* BETON *Mörtel, Beton* mature
**Aushebe-** *in cpds* ARCH & TRAGW lift-off, lifting
**ausheben** *vt* ABWASSER, ARCH & TRAGW lift, INFR & ENTW excavate
***Aushebe-***: **Aushebeschlüssel** *m* ABWASSER, ARCH & TRAGW *Schachtdeckel* lifting key; **Aushebesicherung** *f* ABWASSER, ARCH & TRAGW *Schachtdeckel* lift-off guard
**aushöhlen** *vt* ARCH & TRAGW, INFR & ENTW hollow out
**Aushöhlung** *f* ARCH & TRAGW, INFR & ENTW cavity, hollow, NATURSTEIN *in einem Ziegel* frog
**Aushub** *m* INFR & ENTW *Tiefbau* excavation, excavated material, UMWELT excavated material, excavation; **Aushubfläche** *f* INFR & ENTW, UMWELT excavated area; **Aushubgebiet** *nt* INFR & ENTW excavation area; **Aushubmaterial** *nt* INFR & ENTW, UMWELT excavated material; **Aushubsprengung** *f* INFR & ENTW excavation blasting; **Aushubsverstrebung** *f* ARCH & TRAGW excavation support; **Aushubtiefe** *f* INFR & ENTW excavation depth
**Auskalkung** *f* NATURSTEIN, WERKSTOFF lime leaching

**auskehlen** *vt* ARCH & TRAGW, HOLZ channel, hollow
**Auskehlung** *f* ARCH & TRAGW, HOLZ channeling (*AmE*), channelling (*BrE*), hollowing
**auskipppen** *vt* INFR & ENTW, UMWELT dump
**Auskippen** *nt* INFR & ENTW, UMWELT dumping
**auskitten** *vt* ARCH & TRAGW stop with putty
**auskleiden** *vt* ARCH & TRAGW, INFR & ENTW, WEKSTOFF line
**Auskleidung** *f* ARCH & TRAGW, INFR & ENTW, WERKSTOFF liner, lining
**ausklinken** *vt* ARCH & TRAGW, HOLZ notch
**Ausklinkhaken** *m* BESCHLÄGE releasing hook
**Ausklinkung** *f* ARCH & TRAGW jog, HOLZ cutout
**ausknicken** *vi* INFR & ENTW, WERKSTOFF buckle out
**auskoffern** *vt* INFR & ENTW *Straßenbau*, UMWELT excavate
**Auskofferung** *f* INFR & ENTW road bed excavation
**auskolken** *vt* INFR & ENTW scour
**Auskolkungstiefe** *f* INFR & ENTW scour depth
**auskragen 1.** *vt* ARCH & TRAGW corbel, corbel out; **2.** *vi* ARCH & TRAGW project
**auskragend** *adj* ARCH & TRAGW, BAUMASCHIN, BESCHLÄGE, NATURSTEIN, WERKSTOFF cantilevering, corbelling, overhanging, projecting; **~e Schicht** *f* NATURSTEIN corbelling
**Auskragung** *f* ARCH & TRAGW, BETON, STAHL cantilever, overhang, projection
**auskreiden** *vt* OBERFLÄCHE chalk
**Auskreiden** *nt* OBERFLÄCHE chalking
**auskreidungsbeständig** *adj* WERKSTOFF chalkproof
**ausladend** *adj* ARCH & TRAGW overhanging; **~e Plattform** *f* ARCH & TRAGW projecting platform, protruding platform
**Ausladung** *f* ARCH & TRAGW cantilever, overhang
**Auslaß** *m* ABWASSER, BAUMASCHIN, HEIZ & BELÜFT exhaust, outlet, UMWELT outlet
**Auslastung** *f* ARCH & TRAGW utilization
**Auslauf** *m* ABWASSER outlet; **Auslaufbogen** *m* ABWASSER outlet elbow
**auslaufen** *vi* ABWASSER, UMWELT leak
**auslaufsicher** *adj* ARCH & TRAGW, HEIZ & BELÜFT *Tank* leakproof
***Auslauf***: **Auslaufventil** *nt* ABWASSER tap
**Auslaug-** *in cpds* UMWELT leaching
**auslaugen** *vt* INFR & ENTW, UMWELT, WERKSTOFF leach, lixiviate
**Auslaugen** *nt* INFR & ENTW, UMWELT, WERKSTOFF leaching
***Auslaug-***: **Auslaugkontrollschicht** *f* INFR & ENTW leachate detection layer; **Auslaugtest** *m* UMWELT leaching test
**Auslaugung** *f* UMWELT elutriation, lixiviation
***Auslaug-***: **Auslaugverfahren** *nt* UMWELT leaching property
**Ausleger** *m* ARCH & TRAGW cantilever, BAUMASCHIN *Kran* jib, flange, boom, STAHL cantilever; **Auslegerbrücke** *f* INFR & ENTW cantilever bridge; **Auslegergerüst** *nt* BAUMASCHIN projecting scaffolding; **Auslegerkran** *m* ARCH & TRAGW jib crane, BAUMASCHIN boom crane, jib crane; **Auslegerstellung** *f* ARCH & TRAGW boom position, BAUMASCHIN jib position
**Auslegung** *f* ARCH & TRAGW *einer Anlage oder Konstruktion* layout, plant layout
**Auslöse-** *in cpds* BESCHLÄGE release; **Auslösehebel** *m* BAUMASCHIN trip lever; **Auslösenadel** *f* BESCHLÄGE *Profilzylinder* release pin
**Auslösung** *f* BAURECHT *bei entferntem Wohnsitz* living allowance, ELEKTR *Alarm* release
**ausloten** *vt* STAHL lead
**Ausmaß** *nt* ARCH & TRAGW, INFR & ENTW *der Arbeit* extent, scheme, scope
**Ausmauerung** *f* ARCH & TRAGW, NATURSTEIN *Fachwerk* filling masonry, infill brickwork, infiller masonry, nogging
**Ausnahme** *f* ARCH & TRAGW exception
**Ausnehmung** *f* HOLZ *in Holz gefräste Aussparung zur Aufnahme von Beschlägen* clearance space
**Ausnutung** *f* ARCH & TRAGW groove
**Ausnutzung** *f* ARCH & TRAGW utilization; **Ausnutzungsgrad** *m* ARCH & TRAGW, INFR & ENTW utilization factor; **Ausnutzungskurve** *f* UMWELT utilization curve
**auspflanzen** *vt* ARCH & TRAGW bed out
**auspressen** *vt* NATURSTEIN *mit Mörtel* pressure-grout
**Auspreßverfahren** *nt* NATURSTEIN pressure grouting
**Auspuff** *m* BAUMASCHIN exhaust
**Ausputzdeckel** *m* ABWASSER cleanout cover
**ausreichend** *adj* ARCH & TRAGW *Qualität* satisfactory
**ausrichten** *vt* ARCH & TRAGW align, arrange, *Bewehrung* adjust, BETON align, adjust, arrange; **wieder ausrichten** *phr* ARCH & TRAGW *Mauer* re-align, throw back into alignment
**Ausrollgrenze** *f* INFR & ENTW plastic limit
**Ausrollversuch** *m* WERKSTOFF plastic limit test
**Ausrundungshalbmesser** *m* ARCH & TRAGW vertical curve radius
**ausrüsten** *vt* ARCH & TRAGW take down; **neu ausrüsten** *vt* BAUMASCHIN re-equip
**Ausrüstung** *f* ARCH & TRAGW plant, rig, BAUMASCHIN equipment, INFR & ENTW plant, rig
**Ausschachtung** *f* INFR & ENTW, UMWELT excavation
**ausschalen** *vt* ARCH & TRAGW strip framework, BETON strike, strip
**Ausschalen** *nt* ARCH & TRAGW framework stripping, BETON form removal, form stripping, formwork removal, shuttering removal
**Ausschalfrist** *f* BETON stripping time
**Ausschalter** *m* ELEKTR circuit breaker
**ausschöpfen** *vt* INFR & ENTW bail
**ausschreiben** *vt* BAURECHT put out to tender
**Ausschreibung** *f* ARCH & TRAGW, BAURECHT, INFR & ENTW invitation to tender, tendering; **Ausschreibungsdauer** *f* BAURECHT bidding period
**Ausschuß** *m* BAURECHT spoilage
**Ausschütten** *nt* INFR & ENTW, UMWELT dumping
**Ausschüttung** *f* INFR & ENTW *Erde* fill, filled ground, filled-up ground
**Ausschwitzung** *f* BETON, NATURSTEIN exudation
**Aussehen** *nt* ARCH & TRAGW appearance, *Äußeres* aspect, INFR & ENTW *Äußeres* aspect
**Außen-** *in cpds* ARCH & TRAGW, ELEKTR, HEIZ & BELÜFT, INFR & ENTW exterior, external, outdoor; **Außenanlagen** *f pl* ARCH & TRAGW, INFR & ENTW outdoor facilities, external features; **Außenanstrich** *m* OBERFLÄCHE weathercoat; **Außenarbeiten** *f pl* ARCH & TRAGW, INFR & ENTW external works; **Außenbeleuchtung** *f* ELEKTR exterior lighting; **Außendruck** *m* ABWASSER outside pressure, ARCH & TRAGW, INFR & ENTW external pressure; **Außendurchmesser** *m* ABWASSER, ARCH & TRAGW, INFR &

ENTW outside diameter; **Außeneck** *nt* ARCH & TRAGW outer corner; **Außenfenster** *nt* ARCH & TRAGW, HOLZ outer window, outside window; **Außenfühler** *m* HEIZ & BELÜFT outdoor sensor; **Außenfurnier** *nt* HOLZ face veneer, outer veneer; **Außengewinde** *nt* WERKSTOFF external thread; **Außenhydrant** *m* BAURECHT *Brandbekämpfung* external hydrant; **Außenkante** *f* ARCH & TRAGW outer edge; **Außenkorrosion** *f* INFR & ENTW, WERKSTOFF exterior corrosion; **Außenluft** *f* HEIZ & BELÜFT outside air; **Außenlufttemperatur** *f* HEIZ & BELÜFT outside air temperature; **Außenmauer** *f* ARCH & TRAGW, NATURSTEIN external wall; **Außenmauerisolierung** *f* DÄMMUNG, WERKSTOFF external wall insulation; **Außenputz** *m* ARCH & TRAGW, NATURSTEIN, WERKSTOFF exterior plaster, exterior render, external rendering; **Außenrüttler** *m* BAUMASCHIN external vibrator; **Außenschale** *f* ARCH & TRAGW outer leaf; **Außenseite** *f* INFR & ENTW outside; **Außenspannung** *f* ARCH & TRAGW, INFR & ENTW external stress; **Außenstütze** *f* ARCH & TRAGW perimeter column; **Außenthermostat** *m* HEIZ & BELÜFT outdoor thermostat; **Außentreppe** *f* ARCH & TRAGW perron, outdoor staircase, outdoor stairs, flyers; **Außentür** *f* ARCH & TRAGW exterior door; **Außenverputz** *m* ARCH & TRAGW external rendering, NATURSTEIN exterior plaster, render, WERKSTOFF exterior render; **Außenwand** *f* ARCH & TRAGW, NATURSTEIN enclosing wall, external wall; **Außenwandisolierung** *f* DÄMMUNG external wall insulation, WERKSTOFF external wall insulation, perimeter insulation; **Außenwandplatte** *f* **zwischen Geschoßfenstern** ARCH & TRAGW, HOLZ spandrel panel; **Außenwange** *f* ARCH & TRAGW *Treppe* external string; **Außenzylinder** *m* BESCHLÄGE outside cylinder

**äußere**: ~ **Abmessung** *f* ARCH & TRAGW, INFR & ENTW external dimension; ~ **Kraft** *f* ARCH & TRAGW, INFR & ENTW external force; ~ **Last** *f* ARCH & TRAGW, INFR & ENTW external load; ~ **s Moment** *nt* ARCH & TRAGW, INFR & ENTW external moment; ~ **Oberfläche** *f* DÄMMUNG external surface; ~ **Querkraft** *f* ARCH & TRAGW, INFR & ENTW external transverse force; ~ **Reibung** *f* ARCH & TRAGW external friction; **~er Ringerder** *m* ELEKTR *Blitzschutz* exterior ring earth connection (*BrE*), exterior ring ground connection (*AmE*); ~ **Schale** *f* NATURSTEIN *Hohlwand* external cavity wall, outer leaf; ~ **Torsion** *f* ARCH & TRAGW, INFR & ENTW external torsion

**außerhalb**: ~ **der Baustelle** *phr* ARCH & TRAGW off-site

**außermittig**: **~e Verbindung** *f* ARCH & TRAGW, BETON, HOLZ, INFR & ENTW, STAHL, WERKSTOFF eccentric connection

**Außermittigkeit** *f* ARCH & TRAGW, HOLZ arm of eccentricity, eccentricity

**aussetzen** *vt* WERKSTOFF *Feuereinwirkung* expose

**Aussetzung** *f* WERKSTOFF *Feuereinwirkung, Bewitterung, Strahlung* exposure; **Aussetzungsgrad** *m* HEIZ & BELÜFT, WERKSTOFF degree of exposure

**ausspachteln** *vt* BETON, NATURSTEIN, OBERFLÄCHE fill, trowel off

**aussparen** *vt* ARCH & TRAGW, HOLZ leave open, notch, recess, spare

**Aussparen** *nt* ARCH & TRAGW recessing, HOLZ notching

**Aussparung** *f* ARCH & TRAGW recess, notch

**ausspülen** *vt* INFR & ENTW underwash, UMWELT flush out

**ausstatten**: **neu ausstatten** *vt* BAUMASCHIN re-equip

**Ausstattung** *f* BAUMASCHIN equipment

**aussteifen** *vt* ARCH & TRAGW brace, stiffen

**aussteifend**: **~e Trennwand** *f* ARCH & TRAGW tie wall

**Aussteifung** *f* ARCH & TRAGW bracing, reinforcing, stiffening; **Aussteifungsrahmen** *m* ARCH & TRAGW reinforcing frame; **Aussteifungsriegel** *m* BETON reinforcing cross member

**ausstellen** *vt* ARCH & TRAGW, INFR & ENTW exhibit

**Ausstellung** *f* ARCH & TRAGW, INFR & ENTW exhibition, exposition

**ausstemmen** *vt* HOLZ, NATURSTEIN chisel, mortise

**ausstrahlen** *vi* HEIZ & BELÜFT, INFR & ENTW radiate

**Ausstreichen** *nt* NATURSTEIN jointing, *Fugen* pointing

**Ausstrich** *m* INFR & ENTW outcrop

**Ausstrippen** *nt*: ~ **mit Luft** UMWELT air stripping

**Ausströmöffnung** *f* BESCHLÄGE, HEIZ & BELÜFT *Halon* discharge valve, outlet opening

**Austauscher** *m* HEIZ & BELÜFT exchanger

**austenitisch**: **~er Edelstahl** *m* STAHL austenitic stainless steel

**Austritt** *m* ARCH & TRAGW, HEIZ & BELÜFT, INFR & ENTW exit; **Austrittspfosten** *m* ARCH & TRAGW, HOLZ newel; **Austrittstemperatur** *f* HEIZ & BELÜFT, UMWELT outlet temperature

**Austrocknen** *nt* ARCH & TRAGW, BETON drying out, HOLZ, WERKSTOFF drying out, seasoning

**Austrocknung** *f* WERKSTOFF exsiccation

**Auswahl** *f* ARCH & TRAGW selection

**auswalzen** *vt* ARCH & TRAGW *Tiefbau* sheet out

**auswaschbar** *adj* OBERFLÄCHE washable, WERKSTOFF removable by washing

**auswaschen** *vt* INFR & ENTW underwash

**Auswaschung** *f* INFR & ENTW, UMWELT washout

**auswechselbar**: **~es Kation** *nt* UMWELT exchangeable cation

**Auswechselung** *f* HOLZ, STAHL trimmer, trimming

**auswerten** *vt* ARCH & TRAGW, BAURECHT, INFR & ENTW, WERKSTOFF analyse (*BrE*), analyze (*AmE*), evaluate

**Auswertung** *f* ARCH & TRAGW, BAURECHT, INFR & ENTW, WERKSTOFF analysis, evaluation

**auswittern** *vt* WERKSTOFF weather

**ausziehbar** *adj* ARCH & TRAGW extendable

**Auszieher** *m* BAUMASCHIN, HEIZ & BELÜFT, UMWELT extractor

**Ausziehwiderstand** *m* INFR & ENTW, WERKSTOFF extraction resistance

**Autobahn** *f* INFR & ENTW freeway (*AmE*), motorway (*BrE*); **Autobahnzubringer** *m* ARCH & TRAGW, INFR & ENTW slip road

**Autofriedhof** *m* UMWELT used car dump

**Autogen-** *in cpds* STAHL oxyacetylene

**autogen**: **~es Schweißen** *nt* STAHL autogenous welding

***Autogen-***: **Autogenbrenner** *m* BAUMASCHIN, STAHL oxyacetylene blowpipe; **Autogenschweißen** *nt* STAHL oxyacetylene welding

**Autoklav-Schaumbeton-Erzeugnis** *nt* BETON autoclaved aerated concrete product

**Autokran** *m* BAUMASCHIN mobile crane

**Automatik**: **Automatikmeißel** *m* BAUMASCHIN *Holzbau* self-coring chisel; **Automatiktür** *f* BESCHLÄGE self-closing door

**Automation** *f* BAUMASCHIN, ELEKTR, HEIZ & BELÜFT automation

**automatisch** *adj* ARCH & TRAGW, BAUMASCHIN automatic; **~er Auslösemechanismus** *m* BAURECHT automatic release mechanism; **~er Betrieb** *m* BAUMASCHIN automatic operation; **~e Brandmeldeanlage** *f* BESCHLÄGE, ELEKTR automatic fire alarm; **~e Brandschutztür** *f* BAURECHT, BESCHLÄGE automatic fire door; **~er Brandschutz-türschließer** *m* BAURECHT, BESCHLÄGE automatic fire door closer; **~e Müllsortierung** *f* UMWELT mechanical separation; **~es Rohrventil** *nt* ABWASSER self-closing cock; **~es Schließen** *nt* BAUMASCHIN automatic closing; **~e Steuerung** *f* BAUMASCHIN automatic control; **~er Türschließer** *m* BAURECHT, BESCHLÄGE automatic door closer, automatic self-closing device; **~e Waage** *f* ARCH & TRAGW automatic scales

**automatisieren** *vt* ARCH & TRAGW, BAUMASCHIN automate, automatize

**automatisiert** *adj* ARCH & TRAGW, BAUMASCHIN automated

**Automatisierung** *f* BAUMASCHIN automation

**Auto**: **Autoplätze** *m pl* INFR & ENTW car spaces; **Autoverschrottungsanlage** *f* UMWELT car fragmentation plant; **Autowrack** *nt* UMWELT scrap motor car

**axial** *adj* ARCH & TRAGW, BETON, INFR & ENTW, STAHL, WERKSTOFF axial; **~e Druckkraft** *f* ARCH & TRAGW, BETON, INFR & ENTW, STAHL, WERKSTOFF axial compressive force

**Axial**: **Axialdrucklager** *nt* ARCH & TRAGW thrust bearing; **Axialgeschwindigkeit** *f* UMWELT axial velocity; **Axialkraft** *f* ARCH & TRAGW, INFR & ENTW, WERKSTOFF axial force; **Axiallager** *nt* UMWELT axial thrust bearing; **Axiallast** *f* INFR & ENTW axial load; **Axiallüfter** *m* BESCHLÄGE, HEIZ & BELÜFT axial fan; **Axialschub** *m* INFR & ENTW, WERKSTOFF axial shear; **Axialventilator** *m* BESCHLÄGE, HEIZ & BELÜFT axial fan

**axonometrisch**: **~er Schnitt** *m* ARCH & TRAGW, HEIZ & BELÜFT, HOLZ axonometric cut-away section; **~e Schnittzeichnung** *f* ARCH & TRAGW, INFR & ENTW sectional axonometric drawing; **~e Zeichnung** *f* ARCH & TRAGW axonometric drawing

**Axt** *f* BAUMASCHIN, HOLZ ax (*AmE*), axe (*BrE*)

**Azetylen** *nt* STAHL, WERKSTOFF acetylene; **Azetylenentwickler** *m* BAUMASCHIN, STAHL acetylene generator; **Azetylensauerstoffbrenner** *m* STAHL oxyacetylene blowpipe; **Azetylenschweißbrenner** *m* BAUMASCHIN, STAHL acetylene blowpipe

**Azimut** *m* ARCH & TRAGW, INFR & ENTW azimuth

**azurblau**: **~ gefärbt** *adj* WERKSTOFF azure-colored (*AmE*), azure-coloured (*BrE*)

# B

**Backen**: **Backenbrecher** *m* BAUMASCHIN jaw crusher; **Backenfänger** *m* BAUMASCHIN casing spears; **Backenquetsche** *f* BAUMASCHIN jaw crusher

**Backstein** *m* ARCH & TRAGW, NATURSTEIN brick; **Backsteinarchitektur** *f* ARCH & TRAGW brick architecture; **Backsteinbau** *m* ARCH & TRAGW, NATURSTEIN brick building, clay brick building; **Backsteinverband** *m* ARCH & TRAGW, NATURSTEIN brickwork

**Bad** *nt* ARCH & TRAGW bathroom; **Badezimmer** *nt* ARCH & TRAGW bathroom

**Badüberlauf** *m* ABWASSER bath overflow

**Bagger** *m* BAUMASCHIN excavator; **Baggereimer** *m* BAUMASCHIN excavator bucket; **Baggerkette** *f* BAUMASCHIN bucket chain; **Baggerkorb** *m* BAUMASCHIN grab; **Baggerpumpe** *f* BAUMASCHIN *Naßbogen* dredging pump; **Baggerschaufel** *f* BAUMASCHIN bucket

**Bahnen** *f pl* WERKSTOFF sheeting

**Bake** *f* ARCH & TRAGW ranging pole

**Bakelitfarbe** *f* OBERFLÄCHE bakelite paint

**bakteriell**: **~e Verseuchung** *f* INFR & ENTW bacterial contamination

**Bakterien-** *in cpds* UMWELT bacterial; **Bakterienverseuchung** *f* UMWELT bacterial contamination

**bakteriologisch**: **~e Reinigung** *f* UMWELT bacteriological treatment, bacteriological purification

**Balken** *m* ARCH & TRAGW, BETON, HOLZ, STAHL balk (*AmE*), baulk (*BrE*), beam, girder, joist; **aus ~ gebaut** *phr* ARCH & TRAGW trabeated; **Balkenabstand** *m* HOLZ beam distance; **Balkenachse** *f* HOLZ beam mid-line; **Balkenanker** *m* BETON, HOLZ, INFR & ENTW, STAHL, WERKSTOFF beam anchor, beam tie; **Balkenauflagerplatte** *f* ARCH & TRAGW wall plate; **Balkenaussparung** *f* ELEKTR wall box; **Balkenbemessung** *f* ARCH & TRAGW, BETON, HOLZ, STAHL beam design; **Balkenbiegung** *f* ARCH & TRAGW beam bending; **Balkenbruchprüfung** *f* ARCH & TRAGW, INFR & ENTW, WERKSTOFF beam test; **Balkendach** *nt* HOLZ beam roof; **Balkendecke** *f* ARCH & TRAGW, HOLZ beam floor, joist ceiling, single floor, span ceiling

**balkenfrei** *adj* ARCH & TRAGW, BETON, HOLZ, STAHL beamless

***Balken***: **Balkenkopf** *m* ARCH & TRAGW, HOLZ beam end, beam head, joist end; **Balkenkreuzung** *f* BETON, HOLZ, INFR & ENTW, STAHL beam crossing; **Balkenlage** *f* HOLZ binders and joists, frame of joists, timberwork; **Balkenprofil** *nt* HOLZ beam profile; **Balkenprüfung** *f* ARCH & TRAGW, INFR & ENTW, WERKSTOFF beam test; **Balkenquerschnitt** *f* BETON, INFR & ENTW, STAHL, WERKSTOFF beam cross section; **Balkenschuh** *m* HOLZ, NATURSTEIN *Mauerwerk* joist hanger; **Balkenstoß** *m* HOLZ beam joint; **Balkenträgerdecke** *f* ARCH & TRAGW, HOLZ joist ceiling, single floor; **Balkenuntersicht** *f* ARCH & TRAGW, BETON, STAHL beam bottom

**Balkon** *m* ARCH & TRAGW balcony; **Balkonträger** *m* ARCH & TRAGW, BETON, HOLZ, STAHL balcony girder, balcony support; **Balkontür** *f* ARCH & TRAGW, BESCHLÄGE balcony door

**ballistisch**: **~e Auslesevorrichtung** *f* UMWELT ballistic sorter; **~e Sichtung** *f* UMWELT ballistic separation; **~e Sortierung** *f* UMWELT *von Müll* ballistic sorting

**Baluster** *m* BESCHLÄGE baluster

**Balustrade** *f* ARCH & TRAGW, HOLZ, STAHL balustrade

**Band** *nt* ARCH & TRAGW strip, *Bauschmuck* band, BAUMASCHIN band, belt, BESCHLÄGE *Baubeschlag* hinge, WERKSTOFF band; **Bandanlage** *f* **für Langstreckenförderung** BAUMASCHIN *Großförderbandanlagen* long-distance conveyor belt; **Bandantrieb** *m* BAUMASCHIN belt drive; **Banddosiereinrichtung** *f* BAUMASCHIN belt-type proportioner; **Bandeisen** *nt* STAHL strap steel, strip iron; **Banderder** *m* ELEKTR earthing strip (*BrE*), grounding strip (*AmE*); **Bandförderer** *m* BAUMASCHIN belt conveyor; **Bandholz** *nt* ARCH & TRAGW strut, HOLZ, INFR & ENTW, WERKSTOFF angle brace; **Bandmaß** *nt* ARCH & TRAGW measuring tape; **Bandraster** *nt* ARCH & TRAGW modular grid; **Bandrasterleuchte** *f* ELEKTR modular grid light fixture; **Bandrastersystem** *nt* ARCH & TRAGW modular grid system; **Bandsäge** *f* ARCH & TRAGW, HOLZ band saw; **Bandscharnier** *nt* ARCH & TRAGW strap hinge; **Band- und Schloßfräsungen** *f pl* BESCHLÄGE hardware cutouts; **Bandstahl** *m* STAHL, WERKSTOFF flat steel, hoop, rolled steel, strip steel; **Bandtasche** *f* BESCHLÄGE, HOLZ hinge pocket

**bandverzinkt** *adj* OBERFLÄCHE continuously galvanized

***Band***: **Bandverzinkung** *f* OBERFLÄCHE continuous galvanizing

**Bank** *f* ARCH & TRAGW, BAUMASCHIN bench, BAURECHT *Geldinstitut* bank, INFR & ENTW bank, bench; **Bankakzept** *nt* BAURECHT bank approval; **Bankbürgschaft** *f* BAURECHT bank guarantee

**Bankeisen** *nt* WERKSTOFF *zur Verbindung von Steinen, Holz* cramp iron

**Bankett** *nt* ARCH & TRAGW, INFR & ENTW banquette, bench, berm, continuous footing, continuous foundation, flank

**bankfinanziert** *adj* BAURECHT bank-funded

***Bank***: **Bankgenehmigung** *f* BAURECHT bank approval; **Bankkredit** *m* BAURECHT bank credit

**Bankrott** *m* BAURECHT bankruptcy

**Bankschraube** *f* BAUMASCHIN bench screw

**Bankschraubstock** *m* BAUMASCHIN bench vice (*BrE*), bench vise (*AmE*)

***Bank***: **Bankwechsel** *m* BAURECHT bank draft

**Bär** *m* BAUMASCHIN hammer

**barock** *adj* ARCH & TRAGW baroque

**Barock**: **Barockkirche** *f* ARCH & TRAGW, INFR & ENTW baroque church; **Barockschloß** *nt* ARCH & TRAGW, INFR & ENTW baroque castle

**barometrisch**: **~e Höhenmessung** *f* ARCH & TRAGW barometrical height measurement

**Barrikade** *f* INFR & ENTW barricade

**Barytzement** *m* BETON, WERKSTOFF baritic cement

**Basalt** *m* UMWELT, WERKSTOFF basalt; **Basaltedelsplitt** *m* WERKSTOFF finely crushed basalt chippings; **Basaltpflasterstein** *m* WERKSTOFF basalt paver; **Basaltsplitt** *m* WERKSTOFF basalt chippings

**Basenkation** *nt* UMWELT base cation

**Basis** *f* ARCH & TRAGW, INFR & ENTW *einer Wand, Säule* base, basis; **an der Basis** *phr* INFR & ENTW basal; **Basisabdichtung** *f* UMWELT bottom sealing; *einer Deponie* base sealing

**basisch** *adj* INFR & ENTW, WERKSTOFF alkaline

**Baskülverschluß** *m* BESCHLÄGE *Dreifachriegelverschluß* bascule-bolt

**Bastion** *f* ARCH & TRAGW, INFR & ENTW, NATURSTEIN bastion

**Batterie** *f* ABWASSER *Mischventil* battery; **Batteriefertigung** *f* BETON *Betonfertigteile* vertical multimolding (*AmE*), vertical multimoulding (*BrE*)

**Bau** *m* ARCH & TRAGW action of building, *Gebäude* building, construction, edifice, structure, INFR & ENTW *Gebäude* building, construction, edifice, structure; **im Bau** *phr* ARCH & TRAGW under construction; **Bauablaufgeschwindigkeit** *f* ARCH & TRAGW rate of progress; **Bauablaufplan** *m* ARCH & TRAGW, INFR & ENTW progress chart, work schedule, working schedule; **Bauabmessung** *f* ARCH & TRAGW, INFR & ENTW building dimension; **Bauabnahme** *f* BAURECHT acceptance of work; **Bauabort** *m* ABWASSER site toilet; **Bauabschlagszahlung** *f* ARCH & TRAGW progress payment; **Bauabwicklung** *f* ARCH & TRAGW, INFR & ENTW execution of construction work; **Bauakustik** *f* DÄMMUNG architectural acoustics

**bauakustisch**: **~e Prüfung** *f* DÄMMUNG, WERKSTOFF acoustic testing of buildings

***Bau***: **Bauangebot** *nt* BAURECHT bid; **Bauantrag** *m* ARCH & TRAGW, BAURECHT, INFR & ENTW building proposal; **Bauarbeit** *f* ARCH & TRAGW construction work; **Bauarbeiten** *f pl* ARCH & TRAGW, BAURECHT, INFR & ENTW building works, construction work; **Bauarbeiter** *m* ARCH & TRAGW building worker; **Bauart** *f* ARCH & TRAGW style

**bauartgeprüft** *adj* ARCH & TRAGW, BAURECHT approved

**bauartzugelassen** *adj* BAURECHT with design certification

***Bau***: **Bauaufseher** *m* ARCH & TRAGW, BAURECHT, INFR & ENTW inspector; **Bauaufsicht** *f* ARCH & TRAGW, BAURECHT construction supervision

**bauaufsichtlich**: **~ zugelassen** *adj* BAURECHT admitted for use by the construction supervising authority, building regulations approved

***Bau***: **Bauaufsichtsbeamter** *m* BAURECHT building inspector; **Bauaufsichtsbeamtin** *f* BAURECHT building inspector; **Bauaufsichtsbeauftragte** *f* BAURECHT building inspector; **Bauaufsichtsbeauftragter** *m* BAURECHT building inspector; **Bauaufsichtsbehörde** *f* BAURECHT building inspection; **Bauausführung** *f* ARCH & TRAGW building construction, construction; **Baubarkeit** *f* ARCH & TRAGW, HEIZ & BELÜFT, HOLZ, INFR & ENTW buildability; **Baubeginn** *m* ARCH & TRAGW start of work; **Baubehelf** *m* ARCH & TRAGW, INFR & ENTW temporary installations; **Baubehörde** *f* BAURECHT building authorities, building authority, construction authority; **Baubeschlag** *m* BESCHLÄGE hardware; **Baubeschränkung** *f* BAURECHT building restriction; **Baubeschreibung** *f* ARCH & TRAGW, BAURECHT description of work, specifications; **Baubesprechungsprotokoll** *nt* ARCH & TRAGW memorandum; **Baubestandszeichnungen** *f pl* BAURECHT record drawings; **Baubestimmungen** *f pl* BAURECHT, INFR & ENTW construction regulations; **Baubüro** *nt* ARCH & TRAGW, INFR & ENTW building office, construction office, contractor's site office, job site office (*AmE*), site office (*BrE*); **Bauchemie** *f* WERKSTOFF construction chemistry; **Baudämmfolie** *f* DÄMMUNG, INFR & ENTW, WERKSTOFF building insulating foil; **Baueingabeplan** *m* ARCH & TRAGW preconstruction drawing; **Baueinheit** *f* HEIZ & BELÜFT, HOLZ, STAHL assembly; **Baueisen** *nt* STAHL structural iron; **Bauelement** *nt* ARCH & TRAGW component, structural element, WERKSTOFF constructional unit

**bauen** *vt* ARCH & TRAGW, INFR & ENTW build

***Bau***: **Bauentwurf** *m* ARCH & TRAGW, BAURECHT, INFR & ENTW construction plan, structural design; **Bauerfahrung** *f* ARCH & TRAGW, INFR & ENTW building experience; **Baufachmann** *m* ARCH & TRAGW builder, construction expert

**baufällig** *adj* ARCH & TRAGW dilapidated, out-of-repair, *Decke, Gewölbe* unsound

***Bau***: **Baufeuchte** *f* ARCH & TRAGW trapped humidity, BETON, NATURSTEIN, WERKSTOFF building moisture; **Baufirma** *f* BAURECHT construction firm; **Baufluchtlinie** *f* ARCH & TRAGW, INFR & ENTW alignment; **Baufolge** *f* ARCH & TRAGW, INFR & ENTW sequence of construction work; **Baufolie** *f* WERKSTOFF construction sheeting; **Bauform** *f* ARCH & TRAGW type of construction; **Bauforschung** *f* ARCH & TRAGW, BAURECHT, INFR & ENTW building research; **Baufortschritt** *m* ARCH & TRAGW progress of construction work; **Baufortschrittsbericht** *m* ARCH & TRAGW progress report; **Baufristenplan** *m* ARCH & TRAGW progress chart; **Baugelände** *nt* ARCH & TRAGW, BAURECHT, INFR & ENTW building ground, construction ground; **Baugeländebodenprobenentnahme** *f* ARCH & TRAGW, INFR & ENTW site sampling; **Baugenehmigung** *f* BAURECHT building permission, building permit, planning permission; **Baugesetze** *nt pl* BAURECHT Building Acts; **Baugesuch** *nt* ARCH & TRAGW, BAURECHT, INFR & ENTW building proposal; **Baugewerbe** *nt* INFR & ENTW building trade; **Baugips** *m* WERKSTOFF building plaster; **Bauglas** *nt* WERKSTOFF construction glass

**Baugrube** *f* ARCH & TRAGW trench, INFR & ENTW building pit; **Baugrubensprengung** *f* INFR & ENTW building pit blasting; **Baugrubenverbauarbeiten** *f pl* ARCH & TRAGW, INFR & ENTW building pit sheeting work; **Baugrubenverfüllung** *f* INFR & ENTW ditch refilling; **Baugrubenverkleidung** *f* ARCH & TRAGW, INFR & ENTW building pit lining; **Baugrubenverkleidungsarbeiten** *f pl* ARCH & TRAGW, INFR & ENTW building pit sheeting work

***Bau***: **Baugrund** *m* ARCH & TRAGW, INFR & ENTW foundation soil, subsoil; **Baugrundsätze** *m pl* ARCH & TRAGW, BAURECHT, INFR & ENTW building principles; **Baugrundstück** *nt* ARCH & TRAGW, BAURECHT, INFR & ENTW building estate, building site (*BrE*); **Baugrunduntersuchung** *f* INFR & ENTW soil examination, subsurface investigation, UMWELT soil examination; **Baugrundverbesserung** *f* ARCH & TRAGW, INFR & ENTW, WERKSTOFF artificial soil stabilization, earth improvement (*BrE*), ground improvement (*AmE*); **Baugruppe** *f* ARCH & TRAGW structural component; **Baugutachten** *nt* ARCH & TRAGW survey; **Baugüte** *f* BAURECHT building materials quality; **Baugüteüberwachung** *f* BAURECHT building materials quality control; **Bauherrenmodell** *nt* BAURECHT house-builders' scheme; **Bauhochkonjunktur** *f* ARCH & TRAGW, INFR & ENTW building boom; **Bauhof** *m* HOLZ timber yard; **Bauhöhe** *f* ARCH & TRAGW overall height, total height; **Bauholz** *nt* HOLZ construction lumber (*AmE*), constructional timber (*BrE*), lumber (*AmE*), structural timber, timber (*BrE*); **Bauhütte** *f* ARCH & TRAGW site hut; **Bauhygiene** *f* ARCH & TRAGW, BAURECHT building hygiene; **Bauindustrie** *f* ARCH & TRAGW, BAURECHT, INFR & ENTW building industry, construction industry; **Bauingenieur** *m* ARCH & TRAGW, INFR & ENTW building engineer, civil engineer, construction engineer; **Bauingenieurwesen** *nt* ARCH & TRAGW, INFR & ENTW structural engineering; **Baujahr** *nt* ARCH & TRAGW year of construction; **Baukalk** *m* NATURSTEIN, WERKSTOFF building lime; **Baukastenbauweise** *f* ARCH & TRAGW modular building system

**Baukastenprinzip** *nt* ARCH & TRAGW modular building system; **nach dem ~ gebaut** *phr* ARCH & TRAGW modular

***Bau***: **Baukastensystem** *nt* ARCH & TRAGW modular design; **Baukeramik** *f* NATURSTEIN structural ceramics; **Baukies** *m* WERKSTOFF construction gravel; **Bauklammer** *f* BAUMASCHIN clamping iron, clamp; **Bauklempner** *m* ABWASSER, HEIZ & BELÜFT building plumber; **Baukonstruktion** *f* ARCH & TRAGW building construction; **Baukonstruktionslehre** *f* ARCH & TRAGW, INFR & ENTW structural theory; **Baukörper** *m* ARCH & TRAGW structure; **Baukosten** *f pl* ARCH & TRAGW, BAURECHT, INFR & ENTW building costs, construction costs; **Baukostenkalkulator** *m* ARCH & TRAGW quantity surveyor; **Baukostenkalkulatorin** *f* ARCH & TRAGW quantity surveyor; **Baukostenvoranschlag** *m* ARCH & TRAGW contractor's estimate; **Baukran** *m* BAUMASCHIN building crane; **Baukunst** *f* ARCH & TRAGW architecture; **Bauland** *nt* BAURECHT construction ground; **Baulänge** *f* ARCH & TRAGW overall length; **Baulärm** *m* BAURECHT construction noise; **Bauleistung** *f* ARCH & TRAGW building work

**bauleitend**: **~er Architekt** *m* ARCH & TRAGW, INFR & ENTW architect in charge, architect on site

***Bau***: **Bauleiter** *m* ARCH & TRAGW site engineer; **Bauleitung** *f* ARCH & TRAGW, BAURECHT construction supervision, supervision; **Bauleitungsbüro** *nt* **des Auftragnehmers** ARCH & TRAGW contractor's site office

**baulich** *adj* ARCH & TRAGW structural; **~e Abhilfemaßnahmen** *f pl* ARCH & TRAGW remedial building work; **~e Änderung** *f* ARCH & TRAGW, BAURECHT structural alteration; **~e Anlagen** *f pl* ARCH & TRAGW, INFR & ENTW buildings and structures; **~e Aufnahme** *f* ARCH & TRAGW survey of buildings and site; **~e Veränderung** *f* ARCH & TRAGW conversion

***Bau***: **Baulinie** *f* ARCH & TRAGW, BAURECHT, INFR & ENTW building line; **Baulücke** *f* ARCH & TRAGW gap; **Baumangel** *m* ARCH & TRAGW constructional defect; **Baumaschinen** *f pl* BAUMASCHIN building machinery, construction equipment; **Baumaß** *nt* ARCH & TRAGW structural dimension; **Baumaßnahme** *f* ARCH & TRAGW construction work; **Baumaterial** *nt* ARCH & TRAGW, WERKSTOFF building material, building ware, construction material; **Baumeister** *m* ARCH & TRAGW master builder; **Baumerkmal** *f* ARCH & TRAGW architectural feature

**baumkantig** *adj* HOLZ rough-hewn

**Baumstumpf** *m* ARCH & TRAGW stub, stump

**Baumwoll**: **Baumwollgewebe** *nt* WERKSTOFF cotton cloth; **Baumwollisolierung** *f* DÄMMUNG cotton insulation

***Bau***: **Baunorm** *f* ARCH & TRAGW, BAURECHT, INFR & ENTW construction standard, constructional standard; **Bauöffnungsmaße** *nt pl* ARCH & TRAGW *Zarge* structural opening dimensions; **Bauordnung** *f* ARCH & TRAGW, BAURECHT, INFR & ENTW building regulations; **Baupappe** *f* DÄMMUNG, WERKSTOFF building paper; **Bauphase** *f* ARCH & TRAGW, BAURECHT, INFR & ENTW building phase; **Bauplan** *m* ARCH & TRAGW working plan, construction plan; **Bauplanung** *f* BAURECHT, INFR & ENTW building planning; **Bauplatte** *f* ARCH & TRAGW building block module, BETON building slab, WERKSTOFF building board; **Bauplatz** *m* ARCH & TRAGW, BAURECHT, INFR & ENTW building ground, building lot (*AmE*), building plot, building site (*BrE*), job site (*AmE*), site (*BrE*); **Baupreis** *m* ARCH & TRAGW construction price; **Baupreisindex** *m* ARCH & TRAGW, BAURECHT, INFR & ENTW building price index, construction price index; **Bauprüfung** *f* ARCH & TRAGW, BAURECHT, INFR & ENTW building inspection; **Bauraster** *m* ARCH & TRAGW structural module; **Baurichtmaß** *nt* ARCH & TRAGW basic dimension, controlling dimension; **Bausachverständiger** *m* ARCH & TRAGW, BAURECHT, INFR & ENTW building expert; **Bausand** *m* WERKSTOFF building sand, construction sand; **Bauschaden** *m* ARCH & TRAGW building deficiency; **Bauschäden- und Verlustversicherung** *f* BAURECHT insurance against loss or damage to works; **Bauschaum** *m* DÄMMUNG, INFR & ENTW, WERKSTOFF expanding foam; **Bauschild** *nt* BAURECHT construction sign; **Bauschreinerei** *f* HOLZ joinery; **Bauschutt** *m* ARCH & TRAGW, INFR & ENTW, UMWELT construction waste, demolition waste, garbage (*AmE*), rubbish (*BrE*), WERKSTOFF construction waste, waste building material; **Bausektor** *m* ARCH & TRAGW, INFR & ENTW construction industry; **Bausparkasse** *f* BAURECHT building society; **Baustadium** *nt* ARCH & TRAGW, BAURECHT building phase; **Baustahl** *m* ARCH & TRAGW, INFR & ENTW, STAHL, WERKSTOFF construction steel, constructional steel, structural steel

**baustahlarmiert** *adj* WERKSTOFF reinforced

***Bau***: **Baustahlgewebe** *nt* BETON reinforcing steel mesh, welded wire mesh; **Baustahlmatte** *f* BETON reinforcement mat, reinforcement steel mesh, reinforcing steel mesh, steel wire mesh; **Baustahlprofil** *nt* STAHL structural steel section; **Baustatik** *f* ARCH & TRAGW, INFR & ENTW structural analysis, structural engineering, theory of structures; **Baustein** *m* ARCH & TRAGW module, NATURSTEIN building block

**Baustelle** *f* ARCH & TRAGW, BAURECHT, INFR & ENTW building site (*BrE*), job site (*AmE*); **auf der Baustelle** *phr* ARCH & TRAGW on site; **Baustellenabfall** *m* UMWELT construction waste; **Baustellenaufzug** *m* ARCH & TRAGW, BAUMASCHIN, INFR & ENTW service elevator (*AmE*), service lift (*BrE*); **Baustellenbaracke** *f* ARCH & TRAGW site hut; **Baustellenbegrenzung** *f* ARCH & TRAGW, BAURECHT boundary of the site; **Baustellenbesprechung** *f* ARCH & TRAGW site meeting; **Baustellenbüro** *nt* ARCH & TRAGW, INFR & ENTW job site office (*AmE*), site office (*BrE*); **Baustelleneinrichtung** *f* ARCH & TRAGW job site installations (*AmE*), job site mobilization (*AmE*), site facilities, site installations (*BrE*), site mobilization (*BrE*); **Baustelleneinrichtungsplan** *m* ARCH & TRAGW job site mobilization plan (*AmE*), site mobilization plan (*BrE*); **Baustellenfortschrittsfotografien** *f pl* BAURECHT site progress photographs

**baustellengeschweißt** *adj* STAHL site-welded

**baustellenmontiert** *adj* ARCH & TRAGW, INFR & ENTW, STAHL site-assembled

***Baustelle***: **Baustellenstraße** *f* INFR & ENTW site road; **Baustellenzufahrt** *f* ARCH & TRAGW, INFR & ENTW access to site

***Bau***: **Baustil** *m* ARCH & TRAGW architectural style, style; **Baustoff** *m* ARCH & TRAGW, WERKSTOFF building material, construction material; **Baustoffindustrie** *f* WERKSTOFF building materials industry

**baustoffschädlich** *adj* BAURECHT, OBERFLÄCHE, WERKSTOFF aggressive to building material

***Bau***: **Bausystem** *nt* ARCH & TRAGW construction system; **Bautätigkeit** *f* ARCH & TRAGW building activity; **Bautechnik** *f* ARCH & TRAGW, INFR & ENTW building engineering, construction engineering, structural engineering; **Bauteil** *nt* ARCH & TRAGW member, unit, WERKSTOFF constructional element

**Bautenanstrichfarbe** *f* OBERFLÄCHE, WERKSTOFF house paint

**Bautenschutzmittel** *nt* OBERFLÄCHE preservation of structures, building protecting agents

***Bau***: **Bautischlerei** *f* HOLZ joinery; **Bautoleranz** *f* ARCH & TRAGW constructional tolerance; **Bauübergabe** *f* ARCH & TRAGW handing over, handover; **Bauüberwachung** *f* ARCH & TRAGW, BAURECHT building supervision, construction supervision; **Bauunterkunft** *f* ARCH & TRAGW site accommodation; **Bauunternehmen** *nt* ARCH & TRAGW building firm; **Bauunternehmer** *m* ARCH & TRAGW, BAURECHT builder, contractor; **Bauunternehmung** *f* BAURECHT construction firm; **Bauvertrag** *m* ARCH & TRAGW, BAURECHT, INFR & ENTW building contract; **Bauverwaltung** *f* BAURECHT building administration; **Bauvorbesprechung** *f* ARCH & TRAGW preconstruction conference; **Bauvorhaben** *nt* ARCH & TRAGW, BAURECHT, INFR & ENTW building project, project; **Bauvorschrift** *f* ARCH & TRAGW, BAURECHT, INFR & ENTW building regulation, planning regulation; **Bauvorschriften** *f pl* BAURECHT, INFR & ENTW building regulations, construction regulations; **Bauweise** *f* ARCH & TRAGW, BAURECHT, INFR & ENTW building method, construction method; **Bauwerk** *nt* ARCH & TRAGW, INFR & ENTW *Gebäude* building, construction, edifice, structure; **Bauwerksabdichtung** *f* DÄMMUNG waterproofing; **Bauwerksbeton** *m* ARCH & TRAGW, BETON structural concrete; **Bauwesen** *nt* ARCH & TRAGW *Geschäft* building; **Bauwirtschaft** *f* ARCH & TRAGW, INFR & ENTW construction industry

**Bauxit** *m* WERKSTOFF bauxite; **Bauxitzement** *m* BETON, WERKSTOFF bauxitic cement

***Bau***: **Bauzaun** *m* INFR & ENTW hoarding, site fence; **Bauzeichnung** *f* ARCH & TRAGW constructional drawing, construction drawing, building drawing; **Bauzeit** *f* ARCH & TRAGW construction time; **Bauzeitenplan** *m* ARCH & TRAGW, BAURECHT, INFR & ENTW time schedule

**BCS** *abbr* (*Britische Anstalt für Kalibrierung*) BAURECHT BCS (*British Calibration Service*)

**beanspruchen** *vt* ARCH & TRAGW, BAUMASCHIN load, strain, stress, INFR & ENTW load, stress

**beansprucht** *adj* ARCH & TRAGW, BAUMASCHIN, INFR & ENTW loaded

**Beanspruchung** *f* ARCH & TRAGW, BAUMASCHIN, INFR & ENTW load, strain, stress; **Beanspruchungsgruppe** *f* ARCH & TRAGW, INFR & ENTW load group

**bearbeitbar** *adj* BETON workable

**bearbeiten** *vt* INFR & ENTW *Außenanlagen* cultivate, WERKSTOFF process

**bearbeitet** *adj* BAUMASCHIN, WERKSTOFF machined; **~e Betonoberfläche** *f* BETON tooled concrete finish

**Bearbeitung** *f* INFR & ENTW *gärtnerisch* cultivation

**Beaufsichtigung** *f* BAURECHT inspection

**bebaut**: **~e Fläche** *f* INFR & ENTW building area

**Bebauung** *f* BAURECHT, INFR & ENTW development; **Bebauungsplan** *m* BAURECHT master plan, zoning map

**beben** *vt* ARCH & TRAGW, INFR & ENTW pulsate

**Becherleiter** *f* BAUMASCHIN bucket ladder

**Becherwerk** *nt* ARCH & TRAGW, BAUMASCHIN bucket conveyor, bucket elevator

**Becken** *nt* ABWASSER, INFR & ENTW basin; **~ für Rückstände** UMWELT tailings pond

**Bedachung** *f* ARCH & TRAGW roof cladding, roof covering, roofing

**bedecken** *vt* NATURSTEIN cope, OBERFLÄCHE *cover* top, WERKSTOFF put under cover

**Bedienung** *f* ARCH & TRAGW attendance, ELEKTR, HEIZ & BELÜFT servicing; **Bedienungsanleitung** *f* BAUMASCHIN operating manual; **Bedienungsgang** *m* ARCH & TRAGW, INFR & ENTW service gangway; **Bedienungsgriff** *m* BESCHLÄGE operating handle; **Bedienungskette** *f* BESCHLÄGE *Jalousie* operating chain; **Bedienungsstange** *f* BESCHLÄGE operating rod; **Bedienungssteg** *m* ARCH & TRAGW, INFR & ENTW service gangway; **Bedienungstrupp** *m* ARCH & TRAGW service staff

**beeinträchtigt** *adj* ARCH & TRAGW, BAURECHT impaired

**befahrbar** *adj* ARCH & TRAGW, INFR & ENTW fit for traffic

**Befall** *m* BETON, HOLZ, OBERFLÄCHE, WERKSTOFF *Pilz* attack; **~ von Insekten** HOLZ insect attack

**Befensterung** *f* ARCH & TRAGW arrangement of windows, fenestration

**befestigen** *vt* ARCH & TRAGW *festmachen* fix, attach, BESCHLÄGE fix, BETON attach, HOLZ attach, *mit Latten* batten, INFR & ENTW *Böschung* revet, pave, STAHL attach

**Befestigung** *f* ARCH & TRAGW fastening, BESCHLÄGE fixing, BETON, HOLZ attachment, INFR & ENTW *Böschungen* revetment, NATURSTEIN *von Wegen oder Straßen* paving, STAHL attachment, WERKSTOFF mounting device; **Befestigungsklemme** *f* WERKSTOFF fastening clamp; **Befestigungskonstruktion** *f* ARCH & TRAGW fastening structure; **Befestigungslasche** *f* WERKSTOFF mounting strap; **Befestigungsloch** *nt* ARCH & TRAGW fastener hole; **Befestigungsmaterial** *nt* WERKSTOFF mounting material; **Befestigungspratze** *f* WERKSTOFF fixing clamp; **Befestigungsrahmen** *m* ARCH & TRAGW fixing frame; **Befestigungsschiene** *f* WERKSTOFF fastening rail; **Befestigungsschraube** *f* WERKSTOFF fixing bolt, retaining bolt, retaining screw, fixing screw; **Befestigungsstelle** *f* ARCH & TRAGW fixing point; **Befestigungsvorrichtung** *f* WERKSTOFF fixing device, mounting device

**befeuchten** *vt* ABWASSER water, BETON moisten, HEIZ & BELÜFT humidify, NATURSTEIN moisten

**Befeuchter** *m* HEIZ & BELÜFT humidifier

**Befeuchtung** *f* HEIZ & BELÜFT humidification

**befolgen** *vt* BAURECHT adhere to

**befördern** *vt* ARCH & TRAGW, BAUMASCHIN, WERKSTOFF carry, convey, transport

**Beförderung** *f* BAUMASCHIN conveying, transport

**Befüllung** *f* UMWELT *einer Kaverne* waste injection

**begehbar** *adj* ABWASSER *Kanal* man-sized

**Begehung** *f*: **~ einer Baustelle** ARCH & TRAGW site inspection

**begradigen** *vt* ARCH & TRAGW straighten, rectify

**begrenzen** *vt* BAURECHT restrict

**Begrenzung** *f* ARCH & TRAGW, BAURECHT, INFR & ENTW boundary, limitation; **Begrenzungsbedingung** *f* BAURECHT, UMWELT boundary condition; **Begrenzungslinie** *f* ARCH & TRAGW boundary line

**Begriffsbestimmung** *f* ARCH & TRAGW, BAURECHT definition

**Behälter** *m* ARCH & TRAGW container, BESCHLÄGE vessel, HEIZ & BELÜFT vessel, tank, INFR & ENTW, UMWELT tank

**behandeln** *vt* WERKSTOFF treat

**Behandlung** *f* WERKSTOFF treatment; **~ von Klärschlamm** UMWELT treatment of sewage sludge

**Beharrungsvermögen** *nt* ARCH & TRAGW, INFR & ENTW, UMWELT inertia force

**behauen** *vt* ARCH & TRAGW *Stein* square, BAUMASCHIN ax (*AmE*), axe (*BrE*), NATURSTEIN mill

**Behauen** *nt* NATURSTEIN dressing

**behauen**: **~er Naturstein** *m* NATURSTEIN dressed stone

**Behelf**: **Behelfsarbeiten** *f pl* BAURECHT, INFR & ENTW provisional works; **Behelfsbau** *m* ARCH & TRAGW, BAURECHT, INFR & ENTW temporary structure; **Behelfsbrücke** *f* INFR & ENTW temporary bridge

**behindert** *adj* ARCH & TRAGW, BAURECHT, INFR & ENTW disabled; **nicht behindert** *adj* BAURECHT able-bodied

**Behinderung** *f* ARCH & TRAGW impediment

**Behörde** *f* BAURECHT, INFR & ENTW authority, council

**beidseitig**: **~ beschichtet** *adj* OBERFLÄCHE coated on both sides; **~ eingespannter Balken** *m* ARCH & TRAGW fixed end beam, fully restrained beam

**beifügen** *vt* INFR & ENTW, WERKSTOFF add

**Beil** *nt* BAUMASCHIN ax (*AmE*), axe (*BrE*)

**Beilagescheibe** *f* BESCHLÄGE washer

**Beilegung** *f* BAURECHT adjustment

**beimauern** *vt* NATURSTEIN fill up with masonry

**Beimengung** *f* INFR & ENTW, WERKSTOFF addition

**Beimischung** *f* BETON, WERKSTOFF *in Beton, Farben- und Beschichtungsmitteln* admixture

**Beiputzarbeiten** *f pl* NATURSTEIN additional plasterwork, auxiliary plastering work

**Beiputzen** *nt* NATURSTEIN additional plasterwork, auxiliary plastering work, plastering-in

**beißend**: **~er Geruch** *m* UMWELT acrid odor (*AmE*), acrid odour (*BrE*)

**beistellen** *vt* ARCH & TRAGW provide

**Beiwert** *m* WERKSTOFF coefficient

**Beize** *f* OBERFLÄCHE mordant, remover, stain

**beizen** *vt* OBERFLÄCHE pickle, scour, *Holz* stain

**Beizmittel** *nt* OBERFLÄCHE mordant, remover

**Bekiesung** *f* ARCH & TRAGW graveling (*AmE*), gravelling (*BrE*)

**Belade- und Entladerampe** *f* ARCH & TRAGW, INFR & ENTW loading ramp

**beladen 1.** *adj* ARCH & TRAGW, INFR & ENTW loaded; **2.** *vt* ARCH & TRAGW, INFR & ENTW load

**Belag** *m* ARCH & TRAGW *Schutzschicht* covering, INFR & ENTW overlaying, *Straßenbau* paving, overlay, pavement, OBERFLÄCHE *Schutzschicht* cover, WERKSTOFF *Straßenbau* paving

**belagsbündig** *adj* ARCH & TRAGW *Straße* level with the pavement

**belastbar** *adj* BAURECHT chargeable

**belasten** *vt* ARCH & TRAGW *Gewicht* charge, load, stress, INFR & ENTW *Gewicht* charge, load, strain

**belastet** *adj* ARCH & TRAGW, INFR & ENTW loaded

**Belastung** *f* ARCH & TRAGW, INFR & ENTW *Gewicht* charge, load, strain, stress; **Belastungsausgleich** *m* ARCH & TRAGW, INFR & ENTW load compensation; **Belastungsfall** *m* ARCH & TRAGW, INFR & ENTW load scheme, loading case; **Belastungsgeschichte** *f* ARCH & TRAGW, BETON, INFR & ENTW, WERKSTOFF loading history; **Belastungsgewicht** *nt* ARCH & TRAGW, INFR & ENTW *Aufzug* loading weight; **Belastungsgrad** *m* ARCH & TRAGW, INFR & ENTW utilization factor; **Belastungskontrolle** *f* INFR & ENTW loading control; **Belastungsprobe** *f* ARCH & TRAGW, INFR & ENTW load test; **Belastungsschlupfverhalten** *nt* ARCH & TRAGW, BETON, WERKSTOFF *Verbundträger* load; **Belastungsschwankung** *f* ARCH & TRAGW fluctuation of load; **Belastungsversuch** *m* ARCH & TRAGW, INFR & ENTW load test

**Belebtschlamm** *m* INFR & ENTW, UMWELT activated sludge; **Belebtschlammanlage** *f* INFR & ENTW, UMWELT activated sludge plant; **Belebtschlammbecken** *nt* INFR & ENTW, UMWELT activated sludge tank, aeration tank; **Belebtschlammverfahren** *nt* INFR & ENTW, UMWELT activated sludge process

**belegen** *vt* BAURECHT *Gebäude* occupy

**Belegschaft** *f* ARCH & TRAGW, BAURECHT workforce

**Belegungsplan** *m* INFR & ENTW *Energiehaushaltsanalyse* schedule of building occupancy

**Beleuchtung** *f* ELEKTR illumination, lighting; **Beleuchtungsanlage** *f* ELEKTR lighting installation, lighting system; **Beleuchtungskörper** *m* ELEKTR light fixture; **Beleuchtungsstärke** *f* ARCH & TRAGW intensity of illumination

**belüften** *vt* ABWASSER, ARCH & TRAGW, BAUMASCHIN aerate, BESCHLÄGE ventilate, vent, BETON aerate, HEIZ & BELÜFT aerate, ventilate, INFR & ENTW aerate

**Belüfter** *m* BAUMASCHIN, INFR & ENTW, UMWELT, WERKSTOFF aerator

**belüftet** *adj* HEIZ & BELÜFT ventilated

**Belüftung** *f* ABWASSER, BAUMASCHIN aeration, HEIZ & BELÜFT aeration, airing, bioaeration, ventilation, INFR & ENTW, STAHL aeration; **Belüftungsanlage** *f* INFR & ENTW aeration plant; **Belüftungsbecken** *nt* BETON, INFR & ENTW, UMWELT activated sludge tank, aeration tank; **Belüftungsmöglichkeiten** *f pl* ARCH & TRAGW, HEIZ & BELÜFT means of ventilation; **Belüftungsöffnung** *f* HEIZ & BELÜFT vent, vent opening, ventilation opening; **Belüftungsschlitz** *m* HEIZ & BELÜFT ventilation slot; **Belüftungsverfahren** *nt* ABWASSER, HEIZ & BELÜFT aeration method

**bemessen** *vt* ARCH & TRAGW rate

**Bemessung** *f* INFR & ENTW structural design; **Bemessungsannahme** *f* ARCH & TRAGW, BETON, HOLZ, INFR & ENTW, STAHL design assumption; **Bemessungsdruck** *m* HEIZ & BELÜFT design pressure; **Bemessungsfehler** *m* ARCH & TRAGW design error; **Bemessungskriterium** *nt* ARCH & TRAGW design criterion; **Bemessungslast** *f* ARCH & TRAGW assumed load; **Bemessungstabelle** *f* INFR & ENTW design table; **Bemessungstafel** *f* INFR & ENTW design table; **Bemessungstemperatur** *f* HEIZ & BELÜFT design temperature; **Bemessungswindlast** *f* ARCH & TRAGW, INFR & ENTW design wind load

**Bemusterung** *f* ARCH & TRAGW sampling; **Bemusterungsunterlagen** *f pl* ARCH & TRAGW samples

**benachbart** *adj* ARCH & TRAGW neighboring (*AmE*), neighbouring (*BrE*), BAURECHT, INFR & ENTW, WERKSTOFF adjoining

**benannt**: **~er Subunternehmer** *m* BAURECHT nominated sub-contractor

**benetzen** *vt* BETON, NATURSTEIN, OBERFLÄCHE moisten, wet

**Bentonit** *m* BETON bentonite; **Bentonitzahl** *f* INFR & ENTW, WERKSTOFF ACC test

**benutzen** *vt* ARCH & TRAGW use

**Benutzer** *m* BAURECHT user

**Benutzung** *f* ARCH & TRAGW, INFR & ENTW use

**Benzin**: **Benzinabscheider** *m* ABWASSER gasoline separator (*AmE*), petrol separator (*BrE*); **Benzindampfrückgewinnungsanlage** *f* UMWELT gas vapor recovery plant (*AmE*), petrol vapour recovery plant (*BrE*); **Benzinmotor** *m* BAUMASCHIN, UMWELT gasoline engine (*AmE*), petrol engine (*BrE*)

**Beobachtung** *f* ABWASSER, ARCH & TRAGW, INFR & ENTW, UMWELT observation; **Beobachtungsbrunnen** *m* INFR & ENTW, UMWELT monitoring well, observation well; **Beobachtungspunkt** *m* INFR & ENTW observation point; **Beobachtungsrohr** *nt* ABWASSER observation tube

**beplankt** *adj* HOLZ planked

**Beplankung** *f* HOLZ paneling (*AmE*), panelling (*BrE*), OBERFLÄCHE veneering

**bequem** *adj* ARCH & TRAGW, BAURECHT convenient

**beraten 1.** *vt* ARCH & TRAGW advise; **2. sich ~ mit** *v refl* ARCH & TRAGW consult

**beratend**: **~er Ingenieur** *m* ARCH & TRAGW, INFR & ENTW consulting engineer

**Beratung** *f* ARCH & TRAGW consultation

**berechnen** *vt* ARCH & TRAGW, BAURECHT, INFR & ENTW calculate

**berechnet** *adj* ARCH & TRAGW calculated

**Berechnung** *f* ARCH & TRAGW, BAURECHT, INFR & ENTW calculation, computation; **~ und Bemessung von Baugliedern** ARCH & TRAGW, INFR & ENTW structural analysis

**Beregnungsfläche** *f* ABWASSER sprinkler area

**Bereich** *m* ARCH & TRAGW, BAURECHT, INFR & ENTW range, scope, zone

**bereitstellen** *vt* ARCH & TRAGW provide

**Berg** *m* INFR & ENTW mountain

**bergauf** *adv* INFR & ENTW uphill

**bergmännisch**: **~ hergestellter Hohlraum** *m* UMWELT mined space

**Bericht** *m* ARCH & TRAGW report

**berichten** *vt* ARCH & TRAGW report

**berichtigen** *vt* BAURECHT adjust

**Berichtigung** *f* BAURECHT adjustment

**Berlin**: **~er Lattendecke** *f* NATURSTEIN counterlathing; **~er Schwarz** *nt* OBERFLÄCHE Berlin black; **~er Verbau** *m* BETON, INFR & ENTW, WERKSTOFF asbestos cement board; **~er Welle** *f* BETON, WERKSTOFF asbestos cement board

**Berme** *f* ARCH & TRAGW bench, INFR & ENTW *waagerechter Böschungsabsatz* bench, berm, UMWELT *waagerechter Böschungsabsatz* segregation berm

**Bernoulli**: **Bernoulli-Gleichung** *f* ARCH & TRAGW Bernoulli equation

**Berstfestigkeit** *f* INFR & ENTW, WERKSTOFF bursting strength

**berücksichtigen** *vt* ARCH & TRAGW take into account

**beruflich**: **~e Haftpflichtversicherung** *f* BAURECHT professional indemnity insurance; **~e Sorgfalt** *f* BAURECHT good practice

**Berufsschule** *f* BAURECHT technical college

**Beruhigungskammer** *f* UMWELT settling chamber

**Berührung** *f* ARCH & TRAGW, BESCHLÄGE, INFR & ENTW, WERKSTOFF contact; **Berührungsfläche** *f* ARCH & TRAGW contact area; **Berührungsschutz** *m* ELEKTR protection against accidental contact

**berührungssicher**: **~e Verdrahtung** *f* ELEKTR fully insulated wiring

**besandet** *adj* NATURSTEIN sand-surfaced

**besäumen** *vt* HOLZ square

**besäumt** *adj* ARCH & TRAGW *Holz* square-edged

**beschädigen** *vt* BAURECHT damage

**beschädigt** *adj* BAURECHT damaged

**Beschaffenheit** *f* ARCH & TRAGW state

**beschäftigen** *vt* BAURECHT *Personal* employ

**Bescheinigung** *f* BAURECHT certification

**beschichten** *vt* ARCH & TRAGW, OBERFLÄCHE *Material* overcoat, laminate, surface, coat, *mit Metall* plate

**Beschichten** *nt* OBERFLÄCHE coating, *mit Metall* plating

**beschichtet** *adj* INFR & ENTW laminated, OBERFLÄCHE coated, WERKSTOFF laminated, *mit Metall* plated

**Beschichtung** *f* OBERFLÄCHE, WERKSTOFF coating, plating; **Beschichtungsdefekt** *m* WERKSTOFF coating defect; **Beschichtungsharz** *nt* OBERFLÄCHE coating resin; **Beschichtungsmasse** *f* OBERFLÄCHE coating compound

**Beschickung**: **Beschickungseinrichtung** *f* UMWELT loading mechanism

**beschiefert**: **~e Dachpappe** *f* WERKSTOFF roofing felt with granulated slate surface

**Beschlag** *m* BESCHLÄGE fitting, hardware; **Beschlagsarbeit** *f* ARCH & TRAGW hardware work; **Beschlagsatz** *m* BESCHLÄGE hardware kit

**beschleunigen** *vt* BAUMASCHIN, STAHL, UMWELT, WEKSTOFF accelerate

**Beschleuniger** *m* BETON activator, accelerating admixture, NATURSTEIN accelerator, WERKSTOFF accelerating additive

**beschleunigt** *adj* BAUMASCHIN, STAHL, UMWELT, WERKSTOFF accelerated; **~es Betonabbinden** *nt* BETON concrete accelerated curing; **~e Kompostierung** *f* UMWELT accelerated composting, mechanical composting, rapid fermentation

**Beschleunigung** *f* BAUMASCHIN, BETON, STAHL, WERKSTOFF acceleration

**beschränken** *vt* BAURECHT restrict

**Beschränkung** *f* BAURECHT limitation, restriction

**beschriften** *vt* ARCH & TRAGW letter

**Beschriftung** *f* ARCH & TRAGW lettering
**Beschwerungsschiene** *f* BESCHLÄGE *Jalousie* weighting rail
**Beseitigung** *f* ARCH & TRAGW removal, UMWELT *von organischen Bestandteilen* disposal, *von Schwebstoffen durch Sedimentablagerung* removal
**Besen** *m* BETON, NATURSTEIN broom; **Besenabzug** *m* BETON *Fertigbehandlung von Betondecken* broom finishing
**besenrein** *adj* ARCH & TRAGW, BAURECHT well-swept
***Besen***: **Besenstrich** *m* BETON, NATURSTEIN *Fertigbehandlung von Betondecken* broom finish, brooming; **Besenwurf** *m* NATURSTEIN *Putz* regrating skin
**Besichtigung** *f* ARCH & TRAGW inspection, surveying
**Besitz** *m* BAURECHT ownership, possession; **wieder in ~ nehmen** *phr* BAURECHT repossess; **Besitzergreifung** *f* BAURECHT occupancy; **Besitzübergabe** *f* BAURECHT completion
**Besonderheit** *f* ARCH & TRAGW feature
**besonders**: **~ korrosionsfördernde Einflüsse** *m pl* ARCH & TRAGW, INFR & ENTW severe exposure conditions
**Besprechung** *f* ARCH & TRAGW *Baubesprechung vor Ort* conference; **Besprechungszimmer** *nt* INFR & ENTW conference room
**Bessemerstahl** *m* WERKSTOFF Bessemer steel
**beständig** *adj* ELEKTR, WERKSTOFF proof, resistant
**Beständigkeit** *f* ARCH & TRAGW, INFR & ENTW stability, WERKSTOFF resistance
**Bestand**: **Bestandsaufnahme** *f* ARCH & TRAGW inventory; **Bestandsaufnahme** *f* **der Mängel** ARCH & TRAGW investigation of deficiencies; **Bestandszeichnung** *f* ARCH & TRAGW, BAURECHT as-built drawing; **Bestandteil** *m* INFR & ENTW, WERKSTOFF component, constituent
**Besteck** *nt* ARCH & TRAGW, BAUMASCHIN set of instruments
**bestellen** *vt* ARCH & TRAGW, BAURECHT order
**Bestellungsabweichungen** *f pl* BAURECHT order variations
**besteuern** *vt* BAURECHT tax
**bestimmbar** *adj* ARCH & TRAGW, BAURECHT definable
**bestimmen** *vt* ARCH & TRAGW, BAURECHT, INFR & ENTW define, determine
**bestimmt** *adj* ARCH & TRAGW defined
**Bestimmung** *f* ARCH & TRAGW detailed plans, BAURECHT detailed plans, determination; **Bestimmungsbescheid** *m* BAURECHT notice of determination
**Bestrahlung** *f* UMWELT irradiation; **Bestrahlungsdichte** *f* UMWELT irradiance
**Beton** *m* BETON concrete; **~ einbringen** *phr* BETON pour concrete; **~ mit Kalksteinzuschlag** BETON limestone-filled concrete; **Betonabstellplatz** *m* BETON, INFR & ENTW hardstand; **Betonabwasserleitung** *f* ABWASSER, INFR & ENTW concrete sewer; **Betonalter** *nt* BETON, WERKSTOFF age; **Betonarbeiten** *f pl* BETON concrete work; **Betonaufbruchhammer** *m* BETON concrete breaker; **Betonausbesserung** *f* BETON, INFR & ENTW concrete patching; **Betonausblühung** *f* BETON concrete efflorescence; **Betonbindemittel** *nt* BETON concrete binder; **Betonblockstein** *m* BETON concrete block; **Betonbogen** *m* BETON concrete arch; **Betonbordstein** *m* BETON, INFR & ENTW concrete curb (*AmE*), concrete kerb (*BrE*); **Betonbrecher** *m* BAUMASCHIN concrete breaker; **Betondachpfanne** *f* BETON concrete pantile; **Betondachstein** *m* BETON concrete roofing tile; **Betondecke** *f* ARCH & TRAGW, BETON, INFR & ENTW concrete slab, *Tiefbau* concrete pavement; **Betondeckenherstellung** *f* **im Schnellverfahren** BETON concrete fast track paving; **Betondeckung** *f* ARCH & TRAGW, BETON concrete cover, coverage; **Betondichte** *f* BETON concrete density; **Betondicke** *f* BETON concrete thickness; **Betondöbel** *m* BETON concrete insert; **Betondosieranlage** *f* BETON concrete-batching plant; **Betondruckplatte** *f* BETON concrete topping; **Betondübel** *m* BESCHLÄGE, BETON concrete dowel; **Betoneinbringung** *f* BETON concreting, placing of concrete; **Betoneisen** *nt* BAUMASCHIN reinforcing bar; **Betonentwässerungsrohr** *nt* ABWASSER, BETON, INFR & ENTW concrete drain pipe; **Betonfahrbahn** *f* BETON, INFR & ENTW concrete carriageway; **Betonfertigteil** *nt* ARCH & TRAGW, BETON, INFR & ENTW, WERKSTOFF concrete component, prefabricated concrete unit; **Betonfertigteilwerk** *nt* BETON precasting plant; **Betonfestigkeit** *f* BETON, WERKSTOFF concrete strength; **Betonfläche** *f* BETON concrete area; **Betonflansch** *m* ARCH & TRAGW, BETON *Verbundträger* concrete flange, concrete slab; **Betonform** *f* BETON, HOLZ concrete mold (*AmE*), concrete mould (*BrE*); **Betonformstück** *nt* WERKSTOFF concrete fitting, purpose-made concrete element; **Betonfundament** *nt* ARCH & TRAGW concrete foundation; **Betonfußboden** *m* BETON concrete floor; **Betonfußbodenplatte** *f* BETON concrete floor slab; **Betongebäude** *nt* ARCH & TRAGW, BETON concrete building
**betongefüllt** *adj* BETON concrete-filled
***Beton***: **Betongelenk** *nt* ARCH & TRAGW, BETON, INFR & ENTW concrete joint; **Betongratstein** *m* BETON concrete hip tile; **Betongurt** *m* BETON *Verbundträger* concrete chord, concrete flange, concrete slab; **Betongüte** *f* BETON concrete quality, grade of concrete; **Betonhaltbarkeit** *f* BETON concrete durability; **Betonhärter** *m* BETON concrete hardener; **Betonherstellung** *f* BETON concrete manufacture, concrete production; **Betonhinterfüllung** *f* BETON, INFR & ENTW concrete backing; **Betonhohlblockstein** *m* WERKSTOFF hollow concrete block; **Betonhydratation** *f* BETON concrete hydration
**Betonier**: **Betonieranlage** *f* INFR & ENTW concreting plant; **Betoniereindrücke** *m pl* ARCH & TRAGW, BETON casting marks
**betonieren** *vt* BETON cast, pour, concrete, pour concrete
**Betonieren** *nt* BETON pouring of concrete, casting concrete, concreting
***Betonier***: **Betonierfuge** *f* ARCH & TRAGW, BETON construction joint; **Betoniertrichter** *m* BETON tremie
**Betonierung** *f* BETON concreting, placing of concrete, pouring of concrete
***Beton***: **Betonimprägniermittel** *nt* BETON concrete-impregnating agent; **Betonkabelabdeckstein** *m* BETON concrete cable cover; **Betonkamin** *m* BETON concrete chimney; **Betonkonsistenz** *f* BETON concrete consistency; **Betonkragarm** *m* BETON concrete cantilever; **Betonkranz** *m* BETON *Aufsatz* concrete crown, *Einfassung* concrete collar; **Betonkriechen** *nt* ARCH & TRAGW, BETON creep of concrete; **Betonmantel** *m* BETON concrete encasement;

**Betonmauerdeckung** *f* BETON concrete capping; **Betonmauerwerk** *nt* BETON concrete masonry; **Betonmischanlage** *f* BETON concrete mixing plant; **Betonmischer** *m* BAUMASCHIN, BETON concrete mixer; **Betonmischung** *f* **mit niedrigem Wasser-Zement-Faktor** WERKSTOFF low w/c mix (*low water/cement mix*); **Betonnest** *nt* ARCH & TRAGW, BETON cavity in concrete, concrete pocket, honeycomb; **Betonoberfläche** *f* BETON, OBERFLÄCHE concrete finish; **Betonpfahl** *m* INFR & ENTW concrete pile; **Betonpfosten** *m* BETON concrete post; **Betonplastizität** *f* BETON plasticity of concrete; **Betonplatte** *f* ARCH & TRAGW, BETON concrete slab, mattress; **Betonprobewürfel** *m* BETON concrete sample, test cube; **Betonprüfung** *f* BETON, WERKSTOFF concrete test; **Betonpumpe** *f* BAUMASCHIN concrete pump; **Betonreparatur** *f* BETON, INFR & ENTW concrete patching; **Betonrippendecke** *f* ARCH & TRAGW ribbed concrete floor; **Betonrohr** *nt* BETON concrete pipe; **Betonsand** *m* WERKSTOFF natural sand; **Betonsanierung** *f* BETON, INFR & ENTW concrete patching; **Betonsäule** *f* BETON, INFR & ENTW concrete column; **Betonschacht** *m* BETON, INFR & ENTW concrete manhole; **Betonschale** *f* ARCH & TRAGW, BETON, INFR & ENTW concrete shell; **Betonschalung** *f* BETON formwork, HOLZ, STAHL concrete formwork; **Betonschaum** *m* BETON concrete foam; **Betonschienenschwelle** *f* BETON, INFR & ENTW concrete railroad tie (*AmE*), concrete railway sleeper (*BrE*); **Betonschleuderverfahren** *nt* BETON concrete centrifugal casting; **Betonschornstein** *m* BETON concrete chimney; **Betonschutz** *m* OBERFLÄCHE concrete protection; **Betonschutzmittel** *nt* BETON, OBERFLÄCHE concrete preservative; **Betonschwinden** *nt* BETON concrete shrinkage; **Betonsichtfläche** *f* BETON visible concrete surface; **Betonskelettkonstruktion** *f* ARCH & TRAGW, BETON, INFR & ENTW concrete skeleton construction; **Betonsockel** *m* BETON concrete base; **Betonspannung** *f* BETON, WERKSTOFF concrete stress; **Betonstahl** *m* BETON reinforcement steel, reinforcing steel; **Beton- und Stahlbetonarbeiten** *f pl* BETON concrete and reinforced concrete work; **Betonstahlgewebe** *nt* BETON steel web; **Betonstahlmatte** *f* ARCH & TRAGW, BETON, INFR & ENTW, STAHL, WERKSTOFF bar mat, reinforcing steel mesh, welded wire mesh; **Betonstahlschere** *f* BAUMASCHIN iron bar cutter; **Betonsteife** *f* WERKSTOFF concrete consistency; **Betonstein** *m* BETON precast concrete block; **Betonstraßenfertiger** *m* BAUMASCHIN concrete lane finisher; **Betonsturz** *m* BETON concrete lintel; **Betonstütze** *f* BETON, INFR & ENTW concrete column; **Betontabelle** *f* BETON concrete table; **Betontechnik** *f* BETON concrete engineering; **Betontemperatur** *f* BETON concrete temperature; **Betonträger** *m* ARCH & TRAGW, BETON concrete girder; **Betontragwerk** *nt* ARCH & TRAGW, BETON, INFR & ENTW concrete supporting framework; **Betontransportfahrzeug** *nt* BAUMASCHIN mixer conveyor, BETON mixer truck; **Betonüberdeckung** *f* ARCH & TRAGW, BETON concrete cover

**betonumhüllt** *adj* ARCH & TRAGW, BETON, INFR & ENTW, STAHL, WERKSTOFF concrete-encased

**betonummantelt** *adj* ARCH & TRAGW, BETON, INFR & ENTW, STAHL, WERKSTOFF concrete-encased, haunched; **~er Träger** *m* ARCH & TRAGW, BETON, STAHL *Verbundkonstruktion* cased beam

***Beton***: **Betonummantelung** *f* BETON concrete encasement; **Betonverarbeitbarkeit** *f* BETON concrete workability; **Betonverflüssiger** *m* BETON plastifying admixture; **Betonverschleißoberfläche** *f* OBERFLÄCHE concrete wearing surface; **Betonvorschrift** *f* BETON concrete specification; **Betonwand** *f* BETON concrete wall; **Betonware** *f* BETON concrete goods; **Betonwerk** *nt* BETON, INFR & ENTW concrete factory; **Betonwerkstein** *m* BESCHLÄGE, INFR & ENTW artificial stone, NATURSTEIN, WERKSTOFF cast stone; **Betonzersetzung** *f* BETON concrete disintegration; **Betonzug** *m* BETON concrete flue; **Betonzugfestigkeit** *f* ARCH & TRAGW, BETON tensile strength of concrete; **Betonzusammensetzung** *f* BETON concrete composition; **Betonzusatz** *m* BETON additive, concrete admixture; **Betonzuschlagstoff** *m* BETON concrete aggregate

**Betrag** *m* ARCH & TRAGW amount, sum, total, BESCHLÄGE amount

**Betrieb** *m* ARCH & TRAGW, BAUMASCHIN operation, HEIZ & BELÜFT service, INFR & ENTW operation, UMWELT existing plant; **außer ~ nehmen** *phr* BAUMASCHIN take out of service; **außer ~ setzen** *phr* BAURECHT, ELEKTR put out of service

***Betrieb***: **Betriebsablauf** *m* UMWELT industrial process

**betriebsbedingt**: **~er Fehler** *m* UMWELT operational error

***Betrieb***: **Betriebsdruck** *m* ABWASSER, HEIZ & BELÜFT operating pressure, working pressure; **Betriebserfahrung** *f* ARCH & TRAGW, INFR & ENTW practical experience; **Betriebserlaubnis** *f* BAURECHT type approval; **Betriebsgenehmigung** *f* BAURECHT type approval; **Betriebsgewicht** *nt* BAUMASCHIN operating weight, ELEKTR service weight; **Betriebskennwerte** *m pl* BAUMASCHIN operating characteristics; **Betriebskosten** *pl* ARCH & TRAGW, BAUMASCHIN, BAURECHT operating costs; **Betriebskrankheiten** *f pl* BAURECHT, UMWELT occupational diseases; **Betriebslast** *f* INFR & ENTW rolling load; **Betriebsprüfung** *f* BAURECHT operating test

**betriebssicher** *adj* BAUMASCHIN, BAURECHT reliable

***Betrieb***: **Betriebssicherheit** *f* BAUMASCHIN, BAURECHT operational dependability, operational reliability, working safety; **Betriebsstrom** *m* ELEKTR working current; **Betriebsstunde** *f* BAUMASCHIN operating hour; **Betriebs- und Wartungshandbücher** *nt pl* ARCH & TRAGW operating and maintenance manuals

**Bett** *nt* ARCH & TRAGW, INFR & ENTW bed

**Bettung** *f* ARCH & TRAGW *Eisenbahn*, BETON bedding, INFR & ENTW ballast, underlay, bedding, WERKSTOFF ballast, bedding; **Bettungsbeton** *m* BETON bedding concrete; **Bettungsmodul** *m* ARCH & TRAGW, INFR & ENTW, WERKSTOFF bedding module; **Bettungsmörtelschicht** *f* BETON bedding course; **Bettungsplatte** *f* ARCH & TRAGW bed plate; **Bettungsschicht** *f* INFR & ENTW underlay

**Bevölkerung** *f* ABWASSER, BAURECHT, UMWELT population; **Bevölkerungsteildosis** *f* UMWELT subpopulation collective dose

**Bewässerung** *f* INFR & ENTW irrigation, watering, UMWELT irrigation; **Bewässerungsfläche** *f* INFR & ENTW irrigated area; **Bewässerungsgraben** *m* INFR & ENTW, UMWELT irrigation ditch; **Bewässerungsleitung** *f* INFR & ENTW irrigation line; **Bewässerungstechnik** *f* INFR & ENTW, UMWELT irrigation engineering

**beweglich** *adj* BAUMASCHIN mobile, moveable; **~er Abschluß** *m* ARCH & TRAGW shutter; **~e Arbeitsbühne** *f* INFR & ENTW moving platform; **~es Auflager** *nt* ARCH & TRAGW, INFR & ENTW expansion bearing; **~e Last** *f* INFR & ENTW moving load

**Bewegung** *f* ARCH & TRAGW, WERKSTOFF movement; **Bewegungsfuge** *f* ARCH & TRAGW, INFR & ENTW expansion joint, movement joint; **Bewegungsmelder** *m* ELEKTR motion detector, motion sensor; **Bewegungsübertragung** *f* ARCH & TRAGW transmission of motion

**Bewehren** *nt* ARCH & TRAGW reinforcing

**bewehrt** *adj* BETON reinforced, ELEKTR, STAHL armored (*AmE*), armoured (*BrE*); **nicht bewehrt** *adj* BETON, WERKSTOFF nonreinforced, unreinforced; **~er Beton** *m* BETON reinforced concrete; **~e Erde** *f* INFR & ENTW reinforced soil; **~es Kabel** *nt* ELEKTR, INFR & ENTW, WERKSTOFF armoured cable (*BrE*); **~es Mauerwerk** *nt* NATURSTEIN reinforced brickwork

**Bewehrung** *f* ARCH & TRAGW, BETON reinforcement, ELEKTR *Kabel* armoring (*AmE*), armouring (*BrE*); **~ im Bereich negativer Momente** BETON negative reinforcement; **Bewehrungsarbeiten** *f pl* STAHL steel fixing; **Bewehrungseisen** *nt* ARCH & TRAGW, BETON, WERKSTOFF reinforcing bar; **Bewehrungskorb** *m* BETON reinforcing cage; **Bewehrungsliste** *f* BETON bar schedule; **Bewehrungsmatte** *f* ARCH & TRAGW, BETON, HOLZ, INFR & ENTW, STAHL, WERKSTOFF bar mat, reinforcement mat, reinforcement steel mesh, reinforcing steel mesh, spiling, steel fabric; **Bewehrungsnetz** *nt* BETON, INFR & ENTW mat reinforcement; **Bewehrungsstab** *m* BETON, INFR & ENTW, STAHL, WERKSTOFF bar, bar iron; **Bewehrungsstahl** *m* STAHL reinforcing bar; **Bewehrungsverteilung** *f* BETON *Stahlbau* lacing

**Beweis** *m* BAURECHT evidence, proof

**beweisen** *vt* BAURECHT prove

**bewerfen** *vt* NATURSTEIN *mit Putz* daub

**bewerten** *vt* ARCH & TRAGW, BAURECHT, INFR & ENTW, WERKSTOFF evaluate, value

**Bewertung** *f* ARCH & TRAGW, BAURECHT, INFR & ENTW, WERKSTOFF evaluation

**Bewilligung** *f* BAURECHT permit

**bewohnbar** *adj* ARCH & TRAGW, BAURECHT habitable

**Bewohner** *m* BAURECHT inhabitant, occupant, resident

**Bewuchs** *m* INFR & ENTW vegetation

**Bewurf** *m* NATURSTEIN *Putz* daub

**Bezeichnung** *f* ARCH & TRAGW, BAURECHT designation; **Bezeichnungsschild** *nt* BESCHLÄGE label

**beziehen**: **sich ~ auf** *v refl* BAURECHT apply to

**beziffert**: **~er Schadenersatz** *m* BAURECHT liquidated damages

**Bezug**: **Bezugsebene** *f* ARCH & TRAGW, INFR & ENTW datum plane; **Bezugsgröße** *f* ARCH & TRAGW, INFR & ENTW datum, reference value; **Bezugshöhe** *f* ARCH & TRAGW, INFR & ENTW datum, datum level, fixed datum; **Bezugslinie** *f* ARCH & TRAGW, INFR & ENTW datum line; **Bezugspunkt** *m* ARCH & TRAGW datum, datum point, reference point, *Vermessung* reference mark, INFR & ENTW reference mark, datum point, reference point, datum; **Bezugsschalldruck** *m* UMWELT reference sound pressure; **Bezugswert** *m* INFR & ENTW fixed datum

**Biberschwanz** *m* HOLZ plain tile; **Biberschwanzziegel** *m* WERKSTOFF plain tile

**Biegeangaben** *f pl* ARCH & TRAGW bending specifications

**biegebeansprucht** *adj* ARCH & TRAGW, BETON, STAHL, WERKSTOFF subject to bending

**Biege-**: **Biegebruch** *m* BETON, STAHL, WERKSTOFF bending failure; **Biegedrillknicken** *nt* ARCH & TRAGW, BETON, STAHL, WERKSTOFF torsional-flexural buckling; **Biegedrillknicknachweis** *m* ARCH & TRAGW, BETON, STAHL, WERKSTOFF torsional-flexural buckling analysis

**biegefähig** *adj* NATURSTEIN flexural

***Biege-***: **Biegefestigkeit** *f* ARCH & TRAGW, INFR & ENTW, STAHL, WERKSTOFF flexural strength; **Biegelinie** *f* WERKSTOFF *Balken, Träger* deflection curve; **Biegeliste** *f* WERKSTOFF *Bewehrung* bending list; **Biegemaß** *nt* ARCH & TRAGW bending dimension; **Biegemoment** *nt* ARCH & TRAGW, INFR & ENTW, STAHL, WERKSTOFF bending moment, moment of flexion; **Biegemomentdiagramm** *nt* ARCH & TRAGW, HOLZ bending moment diagram

**biegen** *vt* ARCH & TRAGW, BETON bend

**Biegen** *nt* ARCH & TRAGW, BETON bending

**biegend**: **nach oben biegend** *phr* ARCH & TRAGW, BETON, INFR & ENTW, WERKSTOFF bending

***Biege-***: **Biegeplatz** *m* BETON bending yard; **Biegepresse** *f* ARCH & TRAGW bending press; **Biegepunkt** *m* WERKSTOFF bend point; **Biegeradius** *m* ARCH & TRAGW, BETON, INFR & ENTW, STAHL, WERKSTOFF bending radius; **Biegeriß** *m* BETON, STAHL bending crack; **Biegespannung** *f* ARCH & TRAGW, BETON, HOLZ, INFR & ENTW, STAHL, WERKSTOFF bending stress

**biegesteif** *adj* STAHL, WERKSTOFF bend-resistant, bendproof, flexurally rigid, stiff; **mit biegesteifen Knoten** *phr* ARCH & TRAGW stiff-jointed

***Biege-***: **Biegesteifigkeit** *f* ARCH & TRAGW, BETON, INFR & ENTW, STAHL, WERKSTOFF bending rigidity, flexural stiffness; **Biegesteifigkeit** *f* **nach Rißbildung** ARCH & TRAGW, BETON, INFR & ENTW, WERKSTOFF cracked flexural stiffness; **Biegesteifigkeit** *f* **ohne Rißbildung** ARCH & TRAGW, BETON, INFR & ENTW, WERKSTOFF uncracked flexural stiffness; **Biegestelle** *f* STAHL bend point; **Biegetisch** *m* BETON *Betonbewehrung* bender; **Biegetoleranz** *f* WERKSTOFF bending allowance; **Biegeversuch** *m* WERKSTOFF bending test; **Biegewechselzahl** *f* WERKSTOFF bending cycles; **Biegewinkel** *m* BETON, WERKSTOFF bending angle

**biegsam** *adj* WERKSTOFF flexible

**Biegsamkeit** *f* WERKSTOFF flexibility

**Biegung** *f* ARCH & TRAGW, INFR & ENTW bend, curve, flexion, turn

**Bieter** *m* BAURECHT tenderer

**bilden** *vt* ARCH & TRAGW, INFR & ENTW form

**Bildsamkeit** *f* WERKSTOFF coefficient of plasticity

**bilinear**: **~es Druck-Dehnungs-Verhältnis** *nt* BETON, INFR & ENTW, WERKSTOFF bilinear stress-strain relationship

**Bimetallstreifen** *m* STAHL, WERKSTOFF bimetal strip

**Bims** *m* BETON, NATURSTEIN pumice; **Bimsbeton** *m* BETON, WERKSTOFF expanded cinder concrete, pumice concrete; **Bimsbetondiele** *f* BETON pumice concrete panel; **Bimsmauerwerk** *nt* NATURSTEIN pumice masonry; **Bimsplatte** *f* NATURSTEIN pumice slab; **Bimsstein** *m* BETON, DÄMMUNG, NATURSTEIN pumice

**Binde**: **Bindebeginn** *m* BETON commencement of setting; **Bindeblech** *nt* ARCH & TRAGW, HOLZ, INFR & ENTW, STAHL *Gleise* batten plate, brace, tie plate; **Bindedraht** *m* BETON binding wire; **Bindemittel** *nt* BETON cement, cementing material, UMWELT binder, WERKSTOFF binder, binding agent, binding material; **Bindemittelanteil** *m* WERKSTOFF binder content

**binden** *vt* BETON bind

**Binder** *m* ARCH & TRAGW binder, roof frame, roof framing, *Tragwerk* truss, HOLZ roof framing, roof frame, NATURSTEIN header, WERKSTOFF binder; **Binderfarbe** *f* ABWASSER water-based paint; **Binderlage** *f* ARCH & TRAGW header course

**binderlos**: **~e Dachkonstruktion** *f* ARCH & TRAGW untrussed roof

***Binder***: **Binderschicht** *f* NATURSTEIN base course, bonding course; **Bindersparren** *m* ARCH & TRAGW, HOLZ binding rafter, common rafter; **Binderstein** *m* ARCH & TRAGW header, *in Wandstärke* through stone, NATURSTEIN *Mauerwerk* binding stone; **Binderverband** *m* NATURSTEIN header bond

***Binde***: **Bindeschicht** *f* BETON knitting layer, OBERFLÄCHE *Straßenbau* tack coat; **Bindeton** *m* ABWASSER pipe clay

**bindig**: **~es Erdmaterial** *nt* INFR & ENTW, WERKSTOFF binder soil

**Bindigkeit** *f* INFR & ENTW, WERKSTOFF *Boden* coherence

**Bio-** *in cpds* UMWELT biological; **Bioabfall** *m* UMWELT biological waste

**biochemisch**: **~er Indikator** *m* UMWELT biochemical tracer

***Bio-***: **Biofilter** *nt* UMWELT biofilter; **Biogas** *nt* UMWELT digester gas, fermentation gas, biogas; **Bioindikator** *m* UMWELT bioindicator, biological indicator

**biologisch** *adj* UMWELT biological; **~er Abbau** *m* UMWELT biodegradation, biological degradation; **~ abbaubar** *adj* INFR & ENTW, UMWELT, WERKSTOFF biodegradable; **~ abbaubarer Abfall** *m* INFR & ENTW, UMWELT, WERKSTOFF biodegradable waste; **~ nicht abbaubarer Abfall** *m* UMWELT nonbiodegradable waste; **~ abbaubare Substanz** *f* UMWELT biodegradable substance; **~e Abbaubarkeit** *f* UMWELT biodegradability; **~es Agens** *nt* UMWELT biological agent; **~e Behandlung** *f* UMWELT biological treatment; **~es Filter** *nt* UMWELT biological filter; **~es Gleichgewicht** *nt* UMWELT biological equilibrium; **~e Kläranlage** *f* INFR & ENTW, UMWELT biological clarification plant; **~e Korrosion** *f* INFR & ENTW, OBERFLÄCHE, WERKSTOFF bacterial corrosion; **~e Nachklärung** *f* INFR & ENTW secondary sewage treatment; **~e Nachreinigung** *f* UMWELT secondary sewage treatment; **~er Sauerstoffbedarf** *m* (*BSB*) INFR & ENTW, UMWELT, WERKSTOFF biological oxygen demand (*BOD*); **~e Tropfkörperanlage** *f* INFR & ENTW biological filter; **~e Umwandlung** *f* UMWELT biological energy conversion; **~er Verfall** *m* UMWELT, WERKSTOFF biodeterioration

**Biomasse** *f* INFR & ENTW biomass

***Bio-***: **Biomüllkompost** *m* UMWELT biowaste compost; **Biomüllkompostierung** *f* UMWELT biological waste composting; **Biostabilisator** *m* UMWELT biostabilizer

**Biotop** *nt* UMWELT biota

**Biozid** *nt* ABWASSER, BAURECHT, HEIZ & BELÜFT, INFR & ENTW biocide

**Biozönose** *f* UMWELT biocenosis (*AmE*), biocoenosis (*BrE*)

**Birkenholz** *nt* ARCH & TRAGW, HOLZ, WERKSTOFF birch wood

**Bitukies** *m* INFR & ENTW, WERKSTOFF asphalt-coated gravel, bitumen-coated gravel

**Bitumen** *nt* ARCH & TRAGW asphalt, BETON bitumen, INFR & ENTW, OBERFLÄCHE asphalt, WERKSTOFF asphalt, bitumen; **Bitumenanstrich** *m* INFR & ENTW, OBERFLÄCHE, WERKSTOFF bituminous coat; **Bitumenaufstrich** *m* INFR & ENTW, OBERFLÄCHE, WERKSTOFF bituminous coat; **Bitumenbahn** *f* DÄMMUNG welded asphalt sheeting, OBERFLÄCHE *Abdichtung gegen Wasser* bituminous sheeting; **Bitumenband** *nt* WERKSTOFF asphalt tape; **Bitumenbelag** *m* INFR & ENTW asphalt pavement, bitumen pavement, bituminous pavement

**bitumenbeständig** *adj* WERKSTOFF asphalt-resistant

***Bitumen***: **Bitumenbeton** *m* INFR & ENTW asphalt concrete; **Bitumenbinde** *f* WERKSTOFF asphalt tape; **Bitumendachbahn** *f* WERKSTOFF asphalt roofing felt; **Bitumendachpappe** *f* DÄMMUNG, WERKSTOFF bitumen sheet roofing, bituminous roofing felt; **Bitumenemulsion** *f* BETON, WERKSTOFF bitumen emulsion; **Bitumenfarbe** *f* OBERFLÄCHE, WERKSTOFF bituminous paint

**bitumengetränkt** *adj* WERKSTOFF asphalt-impregnated, asphalt-saturated, bitumen-impregnated, bitumen-saturated

**bitumenhaltig** *adj* INFR & ENTW, OBERFLÄCHE, WERKSTOFF asphalt-based, bituminous

***Bitumen***: **Bitumenheißklebemasse** *f* WERKSTOFF asphaltic hot adhesive compound; **Bitumenholzfaserplatte** *f* HOLZ asphalt-treated chipboard; **Bitumenkaltaufstrich** *m* OBERFLÄCHE cold asphaltic coating; **Bitumenkies** *m* INFR & ENTW, WERKSTOFF asphalt-coated gravel, bitumen-coated gravel; **Bitumenmasse** *f* WERKSTOFF bituminous compound; **Bitumenmastix** *m* BETON, WERKSTOFF bitumen mastic; **Bitumenmörtel** *m* BETON, WERKSTOFF bitumen mortar; **Bitumenpapier** *nt* WERKSTOFF asphalt-saturated paper; **Bitumenpappe** *f* INFR & ENTW, OBERFLÄCHE, WERKSTOFF bituminized felt, bituminized paper; **Bitumenschweißbahn** *f* WERKSTOFF asphalt sheet, asphalt sheeting; **Bitumensperrpappe** *f* DÄMMUNG, WERKSTOFF insulating asphalt felt; **Bitumenspritzgerät** *nt* INFR & ENTW *für die Straßenausbesserung* bitumen spraying machine

**bitumenverträglich** *adj* WERKSTOFF asphalt-compatible

**bituminieren** *vt* INFR & ENTW, OBERFLÄCHE, WERKSTOFF bituminize

**Bituminieren** *nt* INFR & ENTW, OBERFLÄCHE, WERKSTOFF bituminization

**bituminisiert** *adj* INFR & ENTW, OBERFLÄCHE, WERKSTOFF bituminized

**bituminös** *adj* ARCH & TRAGW, INFR & ENTW, OBERFLÄCHE, WERKSTOFF asphaltic, bituminous; **~es Bautenschutzmittel** *nt* ARCH & TRAGW, INFR & ENTW, WERKSTOFF bituminous buildings preservative; **~e Beschichtungsmasse** *f* INFR & ENTW, OBERFLÄCHE, WERKSTOFF bituminous coating compound; **~er Beton** *m* BETON, INFR & ENTW bituminous concrete; ~es **Bindemittel** *nt* INFR & ENTW, WERKSTOFF asphaltic road binder, bituminous binder; ~e **Fahrbahndecke** *f* INFR & ENTW, WERKSTOFF asphaltic street pavement, bituminous carriageway pavement (*BrE*), bituminous street pavement (*AmE*); ~es **Mischgut** *nt* INFR & ENTW, WERKSTOFF asphaltic mixture; ~er **Schutzanstrich** *m* INFR & ENTW, OBERFLÄCHE, WERKSTOFF bituminous coat; ~er **Sperranstrich** *m* OBERFLÄCHE bituminous waterproofing coat; ~es **Sperrmittel** *nt* OBERFLÄCHE bituminous damp-proofing agent; ~er **Unterbau** *m* INFR & ENTW, WERKSTOFF bituminous base course

**Bläh-** *in cpds* BETON, INFR & ENTW, WERKSTOFF bloated, swelling; **Blähbeton** *m* BETON, WERKSTOFF expanded concrete, gas concrete; **Blähglimmer** *m* WERKSTOFF vermiculite; **Blähperlit** *m* WERKSTOFF bloated pearlite; **Blähton** *m* INFR & ENTW swelling clay, WERKSTOFF expanded clay

**blank** *adj* ELEKTR, WERKSTOFF bare; ~e **Ableitung** *f* **aus Kupfer** ELEKTR *Blitzschutz* bare copper down conductor; ~**geglühter Draht** *m* STAHL bright annealed wire

**Blase** *f* BETON *Materialfehler* blister, bubble, void, OBERFLÄCHE bubble, void, *Farbe* blister; **Blasebalg** *m* BESCHLÄGE bellows; **Blasenbildung** *f* BETON, OBERFLÄCHE *Farbe* blistering

**Blatt** *nt* ARCH & TRAGW leaf

**Blattung** *f* ARCH & TRAGW splice joint, *Holzbau* halved joint

***Blatt***: **Blattwerk** *nt* ARCH & TRAGW leafage; **Blattzapfen** *m* HOLZ scarf tenon

**Blaufäule** *f* HOLZ blueing, blue stain

**Blaupause** *f* ARCH & TRAGW blueprint

**Blaupausen** *nt* ARCH & TRAGW blueprinting

**Bläuung** *f* OBERFLÄCHE blueing

**Blech** *nt* BAUMASCHIN, BESCHLÄGE, INFR & ENTW, STAHL plate, WERKSTOFF plate, sheet; **Blechanschluß** *m* STAHL sheet metal flashing; **Blechanschlußstreifen** *m* STAHL sheet metal flashing piece; **Blechauskleidung** *f* STAHL sheet lining; **Blechdicke** *f* STAHL plate thickness, WERKSTOFF sheet thickness; **Blechmantel** *m* STAHL sheet metal jacket

**Blechner** *m* ABWASSER, ARCH & TRAGW, STAHL plumber; **Blechnerarbeiten** *f pl* ABWASSER, ARCH & TRAGW plumber's work, STAHL plumber's work, sheet metal work

***Blech***: **Blechrahmen** *m* STAHL plate frame; **Blechrohr** *nt* STAHL sheet iron pipe; **Blechschere** *f* BAUMASCHIN plate shears, sheet shears; **Blechschornstein** *m* STAHL steel chimney; **Blechstegträger** *m* ARCH & TRAGW plate web girder; **Blechträger** *m* STAHL plate girder; **Blechtreibschraube** *f* WERKSTOFF parker screw; **Blechummantelung** *f* STAHL sheet metal jacket; **Blechverkleidung** *f* STAHL sheet lining

**Blei** *nt* WERKSTOFF lead; **Bleiauflager** *nt* WERKSTOFF lead saddle

**bleibend**: ~e **Senkung** *f* ARCH & TRAGW permanent set; ~e **Verformung** *f* ARCH & TRAGW, INFR & ENTW permanent deformation

***Blei***: **Bleiblech** *nt* WERKSTOFF lead sheet, sheet lead

**bleichen** *vt* OBERFLÄCHE, WERKSTOFF bleach

**Bleichen** *nt* OBERFLÄCHE, WERKSTOFF bleaching

**Bleichstoff** *m* OBERFLÄCHE, WERKSTOFF bleach, bleaching agent

***Blei***: **Bleidichtung** *f* WERKSTOFF caulking, lead joint, lead packing; **Bleieinlage** *f* WERKSTOFF lead filler; **Bleifeile** *f* BESCHLÄGE shave hook; **Bleiglas** *nt* WERKSTOFF lead glass

**bleihaltig**: ~e **Pigmente** *nt pl* OBERFLÄCHE, WERKSTOFF lead-containing pigments

***Blei***: **Bleikeil** *m* WERKSTOFF lead wedge; **Bleimennige** *f* OBERFLÄCHE, WERKSTOFF mineral orange, mineral red, red lead; **Bleischweißen** *nt* WERKSTOFF lead welding; **Bleiverglasung** *f* WERKSTOFF lead glazing; **Bleizusatzstoffe** *m pl* UMWELT lead additives

**Blend-**: **Blendboden** *m* ARCH & TRAGW dead floor; **Blendbogen** *m* ARCH & TRAGW blind arch

**Blende** *f* HOLZ edging board, INFR & ENTW *Flachdach* gravel stop

**blendfrei** *adj* OBERFLÄCHE antidazzle

***Blend-***: **Blendgiebel** *m* ARCH & TRAGW *Kirchenbau* false gable; **Blendmauer** *f* BAUMASCHIN screen wall, NATURSTEIN facing wall; **Blendmauerwerk** *nt* NATURSTEIN facing masonry; **Blendrahmen** *m* ARCH & TRAGW window frame; **Blendstein** *m* NATURSTEIN facing brick

**Blendung** *f* ARCH & TRAGW, INFR & ENTW glare

***Blend-***: **Blendziegel** *m* NATURSTEIN face brick

**Blickwinkel** *m* INFR & ENTW angle of view

**blind**: ~e **Mauer** *f* ARCH & TRAGW dead wall

**Blind**: **Blindboden** *m* ARCH & TRAGW false floor, counterfloor; **Blindbohrung** *f* ARCH & TRAGW blind hole; **Blinddeckel** *m* ABWASSER, BESCHLÄGE, HEIZ & BELÜFT blank cover; **Blindfenster** *nt* ARCH & TRAGW blind window; **Blindflansch** *m* HEIZ & BELÜFT blank flange, blind flange; **Blindfurnier** *nt* HOLZ crossband; **Blindniete** *f* STAHL, WERKSTOFF blind rivet; **Blindschacht** *m* INFR & ENTW *Bergbau* blind pit, blind shaft, WERKSTOFF *Bergbau* blind pit, blind shaft; **Blindsparren** *m* HOLZ dummy rafter; **Blindtür** *f* ARCH & TRAGW blind door

**Blitz** *m* ELEKTR lightning; **Blitzauffangstange** *f* ELEKTR lightning arrester rod; **Blitzeinschlag** *m* ELEKTR lightning strike

**blitzgefährdet**: ~e **Gebiete** *nt pl* UMWELT areas particularly exposed to lightning

***Blitz***: **Blitzkugel** *f* ELEKTR *Blitzableiter* lightning globe; **Blitzschutzableitung** *f* ELEKTR lightning down conductor

**Block** *m* ARCH & TRAGW, HOLZ, INFR & ENTW, NATURSTEIN block; **Blockfundament** *nt* ARCH & TRAGW, INFR & ENTW foundation block, single footing; **Blockhaus** *nt* HOLZ cabin; **Blocklast** *f* INFR & ENTW, WERKSTOFF block load; **Blockrahmen** *m* ARCH & TRAGW window frame; **Blockstufe** *f* ARCH & TRAGW solid rectangular step, square, flyer; **Blockverband** *m* NATURSTEIN block bond

**Bluten** *nt* BETON, WERKSTOFF bleeding

**Bock**: **Bockbrücke** *f* INFR & ENTW trestle bridge; **Bockschere** *f* BAUMASCHIN bench shears; **Bockstütze** *f* INFR & ENTW trestle shore

**Boden** *m* ARCH & TRAGW *von Behälter* bottom, INFR & ENTW soil, earth, *Grundlagen* subsoil, ground, *von Behälter* bottom; ~ **profilgerecht lösen** *phr* INFR & ENTW excavate soil true to profile; **Bodenablauf** *m* ABWASSER floor drain, floor gully, floor inlet; **Bodenabtrag** *m* INFR & ENTW soil erosion; **Bodenanalyse** *f* INFR & ENTW soil analysis; **Bodenart** *f* INFR & ENTW nature of soil, soil type; **Bodenaufschüttung** *f* INFR & ENTW, WERKSTOFF filling; **Bodenaushub** *m* INFR & ENTW, UMWELT excavated soil, excavation; **Boden-**

**aussparung** *f* ARCH & TRAGW floor recess; **Bodenbelag** *m* ARCH & TRAGW floor covering, floor decking; **Bodenbelastung** *f* UMWELT impact of soil; **Bodenbelüftung** *f* INFR & ENTW soil aeration; **Bodenbeschaffenheit** *f* INFR & ENTW nature of the ground, UMWELT soil composition; **Bodenbewässerung** *f* INFR & ENTW soil irrigation; **Bodenbewegung** *f* ARCH & TRAGW earth displacement, earthwork, INFR & ENTW earth displacement, earthwork; **Bodenchemie** *f* WERKSTOFF soil chemistry; **Bodendämmung** *f* DÄMMUNG floor insulation; **Bodendeckung** *f* INFR & ENTW soil covering plants; **Bodendruck** *m* ARCH & TRAGW bearing load, INFR & ENTW soil pressure, earth pressure; **Bodeneigenschaften** *f pl* WERKSTOFF soil character, soil characteristics; **Bodeneinbau** *m* INFR & ENTW soil placement; **Bodeneinstufung** *f* WERKSTOFF soil classification; **Bodenfeuchtigkeit** *f* WERKSTOFF soil humidity; **Bodenfilter** *nt* UMWELT soil filter; **Bodenfliese** *f* ARCH & TRAGW, BESCHLÄGE, INFR & ENTW, WERKSTOFF floor tile, flooring tile

**bodenfremd**: **~e Substanz** *f* UMWELT allochthonous matter

***Boden***: **Bodenfuge** *f* ARCH & TRAGW flooring joint

**bodengleich** *adj* OBERFLÄCHE on grade

***Boden***: **Bodengutachten** *nt* BAURECHT, INFR & ENTW geological survey, subsoil expertise; **Bodenhebung** *f* INFR & ENTW land uplift; **Bodenkanal** *m* ARCH & TRAGW, HEIZ & BELÜFT *Erdreich* buried duct, *Fußboden* floor channel; **Bodenkippe** *f* UMWELT tip (*BrE*); **Bodenklasse** *f* INFR & ENTW soil class; **Bodenluftkonzentration** *f* UMWELT soil atmosphere concentration; **Bodenluke** *f* ARCH & TRAGW trap door; **Bodenmechanik** *f* INFR & ENTW soil mechanics; **Bodenparameter** *m* INFR & ENTW soil parameter; **Bodenplatte** *f* BETON floor slab; **Bodenpressung** *f* ARCH & TRAGW ground pressure, bearing load, bearing pressure, INFR & ENTW soil pressure, ground pressure, subgrade reaction; **Bodenprobe** *f* INFR & ENTW, UMWELT soil sample; **Bodenreform** *f* BAURECHT, INFR & ENTW land reform; **Bodenrost** *m* STAHL floor grating; **Bodensatz** *m* INFR & ENTW, WERKSTOFF deposit, sediment; **Bodenschätze** *m pl* INFR & ENTW natural resources; **Bodenschicht** *f* INFR & ENTW soil stratum; **Bodenschiene** *f* BESCHLÄGE floor track; **Bodenschürfung** *f* INFR & ENTW exploratory dig; **Bodenschwelle** *f* ARCH & TRAGW ground plate; **Bodensenkung** *f* INFR & ENTW settlement, settling, subsidence of ground; **Bodensetzung** *f* INFR & ENTW, UMWELT settlement, settling; **Bodenstabilisierung** *f* INFR & ENTW soil cementation; **Bodenstruktur** *f* WERKSTOFF soil structure; **Bodenteilchen** *nt* WERKSTOFF soil particle; **Bodentürschließer** *m* BESCHLÄGE floor closer; **Bodenuntersuchung** *f* INFR & ENTW, UMWELT soil examination; **Bodenventil** *nt* ABWASSER foot valve; **Bodenverankerung** *f* ARCH & TRAGW, INFR & ENTW deadman; **Bodenverbesserung** *f* ARCH & TRAGW, BAURECHT, INFR & ENTW amelioration; **Bodenverbundfliese** *f* ARCH & TRAGW, WERKSTOFF composite flooring finish, composite flooring tile; **Bodenverdichtung** *f* INFR & ENTW consolidation, soil compaction, soil densification, solidification; **Bodenverdichtungsgerät** *nt* INFR & ENTW earth compacting equipment; **Bodenverdrängung** *f* INFR & ENTW soil displacement; **Bodenverfestigung** *f* INFR & ENTW consolidation, soil compaction, soil densification, WERKSTOFF solidification; **Bodenverfestigungsgerät** *nt* BAUMASCHIN compactor; **Bodenverhalten** *nt* WERKSTOFF soil behavior (*AmE*), soil behaviour (*BrE*); **Bodenverhältnisse** *nt pl* INFR & ENTW earth conditions, ground conditions, soil conditions; **Bodenvermörtelung** *f* INFR & ENTW, WERKSTOFF soil stabilization

**bodenverschmutzend**: **~er Stoff** *m* UMWELT land pollutant, soil pollutant

***Boden***: **Bodenverschmutzung** *f* BAURECHT, INFR & ENTW, UMWELT land pollution, soil pollution; **Bodenvertiefung** *f* ARCH & TRAGW *Fußboden* floor recess

**bodenverunreinigend**: **~er Stoff** *m* UMWELT land pollutant, soil pollutant

***Boden***: **Bodenverunreinigung** *f* UMWELT land pollution, soil pollution; **Bodenzusammensetzung** *f* UMWELT soil composition

**Bogen** *m* ARCH & TRAGW, INFR & ENTW arch; **~ mit Zugband** ARCH & TRAGW tied arch; **Bogenabschluß** *m* ARCH & TRAGW *bei sakralen Bauwerken*, INFR & ENTW arched head; **Bogenachse** *f* ARCH & TRAGW arch center line (*AmE*), arch centre line (*BrE*), flank of an arch; **Bogenanfang** *m* INFR & ENTW springing; **Bogenauflager** *nt* ARCH & TRAGW, NATURSTEIN arch bearing; **Bogenbrücke** *f* **mit Zugband** INFR & ENTW bowstring bridge; **Bogendach** *nt* ARCH & TRAGW, INFR & ENTW arched roof; **Bogendickenmesser** *m* BAUMASCHIN bow calipers (*AmE*), bow callipers (*BrE*); **Bogendreieck** *nt* ARCH & TRAGW *Kirchenbau* curved sided triangle; **Bogenfederzirkel** *m* BAUMASCHIN bow spring compasses; **Bogenfeld** *nt* ARCH & TRAGW arch bay

**bogenförmig** *adj* ARCH & TRAGW, HOLZ, INFR & ENTW, NATURSTEIN arched

***Bogen***: **Bogengang** *m* ARCH & TRAGW arcade, archway; **Bogengewichtsmauer** *f* ARCH & TRAGW, INFR & ENTW arch-gravity dam; **Bogengewölbe** *nt* ARCH & TRAGW, INFR & ENTW arched roof, arched vault; **Bogenhälfte** *f* ARCH & TRAGW flank of an arch; **Bogenkämpferstein** *m* ARCH & TRAGW, NATURSTEIN arch springer; **Bogenleibung** *f* ARCH & TRAGW intrados; **Bogenmauerwerk** *nt* NATURSTEIN curved brickwork; **Bogennische** *f* ARCH & TRAGW *bei sakralen Bauwerken*, INFR & ENTW arched recess; **Bogenpfeiler** *m* ARCH & TRAGW, BETON, INFR & ENTW, NATURSTEIN buttress; **Bogenpfeilermauer** *f* ARCH & TRAGW, INFR & ENTW multiple-arch dam; **Bogenradius** *m* ARCH & TRAGW, INFR & ENTW arch radius; **Bogenrohr** *nt* ABWASSER quadrant pipe; **Bogensäge** *f* BAUMASCHIN backsaw; **Bogenschenkel** *m* ARCH & TRAGW flank of an arch, haunch; **Bogenspannung** *f* ARCH & TRAGW arch stress; **Bogenstaumauer** *f* ARCH & TRAGW, INFR & ENTW arched dam; **Bogenstichhöhe** *f* ARCH & TRAGW, INFR & ENTW arch rise; **Bogenträger** *m* ARCH & TRAGW, BETON, HOLZ, INFR & ENTW, NATURSTEIN arched beam, arched girder, curved girder; **Bogenträger** *m* **mit Zugband** ARCH & TRAGW bowstring girder; **Bogenverankerung** *f* ARCH & TRAGW grappling of an arch; **Bogenziegel** *m* ARCH & TRAGW voussoir, NATURSTEIN gage brick (*AmE*), gauge brick (*BrE*); **Bogenzirkel** *m* BAUMASCHIN bow compass

**Bohle** *f* HOLZ board, deal, INFR & ENTW plank, WERKSTOFF board, plank; **Bohlen hochkant verlegen** *phr* ARCH & TRAGW set boards edgewise;

**Bohlenbelag** *m* ARCH & TRAGW, HOLZ, INFR & ENTW decking, plank covering, planking; **Bohlenrost** *m* HOLZ plank grating

**bohnern** *vt* BAUMASCHIN, OBERFLÄCHE, WERKSTOFF polish

**Bohr-** *in cpds* ARCH & TRAGW, INFR & ENTW, HOLZ, STAHL boring, drilling; **Bohransatzpunkt** *m* INFR & ENTW boring site; **Bohrausrüstung** *f* **für Pfahlgründungen** ARCH & TRAGW boring equipment for pile foundations; **Bohrdiagramm** *nt* ARCH & TRAGW drilling chart; **Bohrdurchmesser** *m* HOLZ, STAHL size of bore

**bohren** *vt* BAUMASCHIN, INFR & ENTW bore, drill

**Bohren** *nt* BAUMASCHIN, INFR & ENTW boring, drilling

**Bohrer** *m* BAUMASCHIN, INFR & ENTW borer

***Bohr-***: **Bohrgestell** *nt* BAUMASCHIN, UMWELT drilling rig; **Bohrhammer** *m* BAUMASCHIN rock drill; **Bohrinsel** *f* INFR & ENTW oil rig; **Bohrkern** *m* INFR & ENTW *Bergbau* test core; **Bohrloch** *nt* BAUMASCHIN, HEIZ & BELÜFT, UMWELT borehole; **Bohrlochpumpe** *f* BAUMASCHIN borehole pump; **Bohrlöffel** *m* INFR & ENTW drilling spoon; **Bohrmeißel** *m* BAUMASCHIN bore bit, jumper; **Bohrplattform** *f* INFR & ENTW oil rig; **Bohrrohr** *nt* BAUMASCHIN well casing; **Bohrstange** *f* BAUMASCHIN bore rod, boring rod; **Bohrstelle** *f* INFR & ENTW boring site; **Bohrstift** *m* BAUMASCHIN drill pin; **Bohrturm** *m* BAUMASCHIN, UMWELT derrick, drilling rig

**Bohrung** *f* UMWELT bore

***Bohr-***: **Bohrunternehmer** *m* ARCH & TRAGW boring contractor; **Bohrwerkzeug** *nt* BAUMASCHIN drilling tool

**Boiler** *m* ABWASSER, ELEKTR, HEIZ & BELÜFT boiler, hot water tank, storage water heater

**Bolzen** *m* ARCH & TRAGW bolt, stay, BAUMASCHIN screw, BESCHLÄGE fang bolt, pin, pintle, STAHL pin, WERKSTOFF bolt, pin; **~ mit versenktem Kopf** ARCH & TRAGW countersunk head-bolt; **Bolzenschlaggerät** *nt* BAUMASCHIN bolt driving tool; **Bolzenschweißen** *nt* STAHL stud welding; **Bolzenschweißpistole** *f* STAHL stud welding gun; **Bolzensetzwerkzeug** *nt* BAUMASCHIN, BETON cartridge-firing studder; **Bolzenverbindung** *f* BESCHLÄGE bolt connection, HEIZ & BELÜFT *für Rohre* nose

**bombensicher** *adj* ARCH & TRAGW bombproof

**Bommerband** *nt* BESCHLÄGE *Beschlag* Bommer hinge

**Borax** *m* WERKSTOFF borax; **Boraxperle** *f* WERKSTOFF borax bead

**Bordeinfassung** *f* INFR & ENTW *entlang einer Straße* curb (*AmE*), kerb (*BrE*)

**Bördelblech** *nt* BESCHLÄGE flanged plate

**Bordkante** *f* INFR & ENTW curb (*AmE*), kerb (*BrE*)

**Bordstein** *m* INFR & ENTW, WERKSTOFF curb (*AmE*), curbstone (*AmE*), kerb (*BrE*), kerbstone (*BrE*)

**Böschung** *f* INFR & ENTW slope; **Böschungsbruch** *m* INFR & ENTW slope failure; **Böschungskreis** *m* INFR & ENTW slope circle; **Böschungsmähgerät** *nt* BAUMASCHIN slope mower; **Böschungsmauer** *f* ARCH & TRAGW toe wall; **Böschungsneigung** *f* INFR & ENTW inclination of slope; **Böschungsschutz** *m* BAURECHT, INFR & ENTW slope protection; **Böschungssicherung** *f* INFR & ENTW slope protection; **Böschungsstabilität** *f* INFR & ENTW slope stability; **Böschungsverdichtung** *f* INFR & ENTW slope compaction; **Böschungsverkleidung** *f* INFR & ENTW revetment of slopes; **Böschungswinkel** *m* INFR & ENTW slope angle

**Bosse** *f* ARCH & TRAGW, WERKSTOFF boss; **Bossenmauerwerk** *nt* NATURSTEIN quarry-faced masonry; **Bossenwerk** *nt* HOLZ bossage, NATURSTEIN bossage, opus rusticum, rustication

**bossieren** *vt* NATURSTEIN ax (*AmE*), axe (*BrE*), rough-hew

**Brand** *m* ARCH & TRAGW, BAURECHT fire; **Brandabschluß** *m* BAURECHT fire-resistant shutter; **Brandabschnitt** *m* ARCH & TRAGW, BAURECHT fire compartment, fire lobby; **Brandabschnittswand** *f* BAURECHT fire-resistant shutter; **Brandgefahr** *f* BAURECHT fire risk, fire hazard; **Brandklasse** *f* BAURECHT, WERKSTOFF fire classification, fire rating; **Brandlast** *f* BAURECHT fire load; **Brandmauer** *f* ARCH & TRAGW, BAURECHT compartment wall, fire partition, fire wall, fireproof wall, party wall; **Brandmeldeanlage** *f* BAURECHT, ELEKTR fire alarm system; **Brandmelder** *m* BAURECHT, ELEKTR fire alarm; **Brandmeldezentrale** *f* ELEKTR, INFR & ENTW central fire alarm station; **Brandrisiko** *nt* BAURECHT fire risk; **Brandschott** *nt* ARCH & TRAGW, BAURECHT fire partition, fire wall, fireproof wall; **Brandschutz** *m* BAURECHT fire protection; **Brandschutzbeschichtung** *f* OBERFLÄCHE fireproofing coat; **Brandschutzdecke** *f* ARCH & TRAGW, BAURECHT, BESCHLÄGE fire-resistant ceiling; **Brandschutzmaßnahmen** *f pl* BAURECHT fire precautions; **Brandschutztür** *f* ARCH & TRAGW, BAURECHT, BESCHLÄGE fire door, fire-resistant door, fireproof door; **Brandschutzüberzug** *m* OBERFLÄCHE fireproofing coat; **Brandsicherheitsmaßnahmen** *f pl* BAURECHT fire precautions; **Brandversuch** *m* BAURECHT, BESCHLÄGE fire test; **Brandverzug** *m* INFR & ENTW, WERKSTOFF deformation during burning; **Brandwand** *f* ARCH & TRAGW, BAURECHT compartment wall

**brauchbar** *adj* ARCH & TRAGW, INFR & ENTW suitable, useful

**Brauchbarkeit** *f* ARCH & TRAGW, INFR & ENTW suitability

**Brauchwasser** *nt* ABWASSER, BAURECHT, INFR & ENTW, UMWELT industrial water, nonpotable water, process water, raw water, water for domestic use, water for industrial use

**Brause** *f* ABWASSER shower; **Brauseauslauf** *m* ABWASSER shower outlet

**BRE** *abbr* (*Institution für Forschung im Bauwesen*) BAURECHT BRE (*Building Research Establishment*)

**Brecheisen** *nt* BAUMASCHIN crowbar, pinch bar

**Brech-** *in cpds* BAUMASCHIN, WERKSTOFF breaking, crushing, broken, crushed

**brechen** *vt* INFR & ENTW refract

**Brecher** *m* BAUMASCHIN crusher, crushing machine

***Brech-***: **Brechgut** *nt* WERKSTOFF broken material, crushed material; **Brechsand** *m* WERKSTOFF crushed rock fine aggregate, crushed stone sand; **Brechstange** *f* BAUMASCHIN pinch bar, BESCHLÄGE *mit Finne* claw bar; **Brechwerk** *nt* BAUMASCHIN, INFR & ENTW, WERKSTOFF crushing equipment

**Brei** *m* UMWELT pulp, slurry

**breit** *adj* ARCH & TRAGW, BAUMASCHIN, WERKSTOFF broad, wide

**Breit-** *in cpds* ARCH & TRAGW, BAUMASCHIN, WERKSTOFF broad, wide

**Breite** *f* ARCH & TRAGW breadth, width, HEIZ & BELÜFT latitude, INFR & ENTW, WERKSTOFF breadth

***Breit-***: **Breitende** *nt* BAUMASCHIN *eines Hammers* poll

**breitflanschig** *adj* STAHL, WERKSTOFF broad-flanged

***Breit-***: **Breitflanschprofil** *nt* WERKSTOFF H-section; **Breitflanschträger** *m* ARCH & TRAGW, BETON, STAHL broad-flanged beam (*BFB*), H-beam, wide-flanged beam, wide-flanged girder; **Breithacke** *f* BAUMASCHIN mattock; **Breitkopfstift** *m* WERKSTOFF clout nail

**Brekzie** *f* INFR & ENTW breccia

**Brennanlage** *f* INFR & ENTW furnace

**brennbar** *adj* BAURECHT, DÄMMUNG, WERKSTOFF combustible; **nicht brennbar** *adj* BAURECHT, INFR & ENTW, UMWELT, WERKSTOFF incombustible, noncombustible; **~er Abfall** *m* UMWELT combustible waste; **~es Material** *nt* UMWELT combustible material

**Brennbarkeit** *f* BAURECHT, DÄMMUNG, WERKSTOFF combustibility

**Brennbohren** *nt* BAUMASCHIN oxygen lancing

**brennen 1.** *vt* ARCH & TRAGW, HEIZ & BELÜFT, UMWELT, WERKSTOFF fire; **2. ~** *vi* ARCH & TRAGW, HEIZ & BELÜFT, UMWELT, WERKSTOFF burn

**Brennen** *nt* NATURSTEIN firing

**Brenner** *m* STAHL torch; **Brennerleistung** *f* HEIZ & BELÜFT, UMWELT burner capacity; **Brennerlöten** *nt* BAUMASCHIN torch brazing; **Brennermundstück** *nt* BAUMASCHIN tip; **Brennerspitze** *f* ARCH & TRAGW tip

**Brenngut** *nt* UMWELT incinerator charge

**Brennkammer** *f* HEIZ & BELÜFT combustion chamber

**Brennofen** *m* WERKSTOFF burning kiln

**Brennstoff** *m* WERKSTOFF fuel; **Brennstoffverbrauch** *m* HEIZ & BELÜFT fuel consumption

**Brennwert** *m* HEIZ & BELÜFT, UMWELT calorific value

**Brett** *nt* HOLZ, WERKSTOFF board; **Brettbinder** *m* ARCH & TRAGW, HOLZ plank truss, sandwiched truss

**Bretter**: **Bretterdach** *nt* HOLZ board roof; **Bretterschalung** *f* HOLZ planking; **Bretterverkleidung** *f* ARCH & TRAGW, HOLZ boarding, plank revetment; **Bretterwand** *f* HOLZ batten wall

***Brett***: **Brettlasche** *f* HOLZ wooden butt strap joint

**brettverleimt** *adj* HOLZ glue-laminated (*glulam*)

**Brinellhärte** *f* STAHL, WERKSTOFF Brinell hardness

**Britisch**: **~e Anstalt** *f* **für Kalibrierung** (*BCS*) BAURECHT British Calibration Service (*BCS*); **~es Institut** *nt* **für Normung** (*BSI*) ARCH & TRAGW, BAURECHT, HOLZ, INFR & ENTW, STAHL, WERKSTOFF British Standards Institution (*BSI*)

**britisch**: **~e Norm** *f* (*BS*) ARCH & TRAGW, BETON, HOLZ, INFR & ENTW, STAHL British Standard (*BS*)

**bröckelig**: **~es Gestein** *nt* INFR & ENTW, WERKSTOFF friable rock

**Bronze** *f* WERKSTOFF bronze; **Bronzeband** *nt* BESCHLÄGE bronze hinge

**Bruch** *m* INFR & ENTW failure, WERKSTOFF failure, *Trennbruch* fracture, rupture; **zu ~ gehen** *phr* ARCH & TRAGW fail; **Bruchbedingung** *f* INFR & ENTW, WERKSTOFF failure condition; **Bruchbelastung** *f* ARCH & TRAGW, INFR & ENTW, WERKSTOFF breaking load, failure load, ultimate loading; **Bruchdehnung** *f* ARCH & TRAGW, INFR & ENTW breaking elongation, WERKSTOFF breaking elongation, ultimate elongation; **Bruchdruckfestigkeit** *f* WERKSTOFF ultimate compressive strength

**bruchfest** *adj* WERKSTOFF breakproof, unbreakable

***Bruch***: **Bruchfestigkeit** *f* ARCH & TRAGW ultimate strength; **Bruchfläche** *f* INFR & ENTW, WERKSTOFF failure plane; **Bruchgrenze** *f* INFR & ENTW, WERKSTOFF failure limit; **Bruchhypothese** *f* INFR & ENTW, WERKSTOFF failure hypothesis; **Bruchkreis** *m* INFR & ENTW failure circle, slip circle, WERKSTOFF failure circle, failure load; **Bruchkriterium** *nt* INFR & ENTW, WERKSTOFF failure criterium; **Bruchkurve** *f* INFR & ENTW, WERKSTOFF failure curve; **Bruchlast** *f* ARCH & TRAGW, INFR & ENTW breaking load, failure load, load at rupture, ultimate loading, WERKSTOFF breaking load, failure load, load at rupture; **Bruchlinie** *f* WERKSTOFF rupture line, yield line; **Bruchlinientheorie** *f* INFR & ENTW yield line method; **Bruchmoment** *nt* INFR & ENTW, WERKSTOFF failure moment

**bruchrauh** *adj* NATURSTEIN natural cleft, quarry-faced, WERKSTOFF *Stein* naturally split

**bruchsicher** *adj* WERKSTOFF breakproof, fracture-proof

***Bruch***: **Bruchspannung** *f* ARCH & TRAGW, INFR & ENTW, WERKSTOFF breaking stress, failure stress, ultimate stress; **Bruchstaub** *m* ARCH & TRAGW, BAURECHT crusher dust; **Bruchstein** *m* ARCH & TRAGW, NATURSTEIN broken stone, quarry stone, rubble, WERKSTOFF rubblestone; **Bruchsteinmauerwerk** *nt* NATURSTEIN, WERKSTOFF coursed masonry, quarry stone work, rubble masonry; **Bruchsteinschüttung** *f* INFR & ENTW random rubble fill; **Bruchstreckgrenze** *f* STAHL, WERKSTOFF ultimate yield strength; **Bruchtest** *m* INFR & ENTW, WERKSTOFF failure test; **Bruchversuch** *m* WERKSTOFF breaking test; **Bruchwahrscheinlichkeit** *f* INFR & ENTW probability of failure; **Bruchzustand** *m* WERKSTOFF state of failure

**Brücke** *f* ELEKTR, INFR & ENTW bridge; **~ mit obenliegender Fahrbahn** INFR & ENTW deck bridge; **Brückenbaustahl** *m* STAHL bridge steel; **Brückenblech** *nt* STAHL bridge plate; **Brückenbogen** *m* ARCH & TRAGW, INFR & ENTW bridge arch; **Brückenbuch** *nt* BAURECHT, INFR & ENTW bridge record book; **Brückendrehkran** *m* ARCH & TRAGW, BAUMASCHIN jib crane; **Brückenfachwerkträger** *m* ARCH & TRAGW, INFR & ENTW bridge truss; **Brückengeländer** *nt* BESCHLÄGE bridge railing; **Brückenkran** *m* BAUMASCHIN, INFR & ENTW bridge crane, overhead crane; **Brückenpfeiler** *m* ARCH & TRAGW pylon, BETON bridge pier, INFR & ENTW pylon, NATURSTEIN, STAHL bridge pier; **Brückenrost** *m* INFR & ENTW floor system; **Brückensteinpflaster** *nt* BAUMASCHIN, NATURSTEIN bridge paver; **Brückenwaage** *f* BAUMASCHIN platform weighing machine

**Brüden**: **Brüdendampf** *m* HEIZ & BELÜFT exhaust vapor (*AmE*), exhaust vapour (*BrE*); **Brüdenkühler** *m* HEIZ & BELÜFT exhaust vapor cooler (*AmE*), exhaust vapour cooler (*BrE*), vapor cooler (*AmE*), vapour cooler (*BrE*)

**Brunnen** *m* ABWASSER, BESCHLÄGE, INFR & ENTW fountain, spring, well; **Brunnenbau** *m* INFR & ENTW, UMWELT well sinking; **Brunnengründung** *f* ARCH & TRAGW, INFR & ENTW well foundation; **Brunnenkopf** *m* UMWELT wellhead; **Brunnenkopfdruck** *m* UMWELT wellhead pressure; **Brunnenkopftemperatur** *f* UMWELT wellhead temperature; **Brunnenkopfventil** *nt* UMWELT wellhead valve; **Brunnenleistung** *f* INFR & ENTW well capacity; **Brunnenring** *m* INFR & ENTW, UMWELT well casing; **Brunnenstube** *f* ABWASSER well house

**Brust** *f* ARCH & TRAGW, INFR & ENTW breast, front, front portion
**brusthoch** *adj* ARCH & TRAGW chest high
***Brust*: Brustholz** *nt* BETON, HOLZ, INFR & ENTW, WERKSTOFF *Grabenverbau* ranger, scantling, waler, waling
**Brüstung** *f* ARCH & TRAGW, BESCHLÄGE, BETON, HOLZ, STAHL balustrade, breast, parapet, railing; **Brüstungshöhe** *f* ARCH & TRAGW parapet height; **Brüstungsholm** *m* ARCH & TRAGW parapet cross beam; **Brüstungsmauer** *f* ARCH & TRAGW, INFR & ENTW breast wall, parapet, parapet wall; **Brüstungsplatte** *f* ARCH & TRAGW parapet slab
***Brust*: Brustzapfen** *m* HOLZ shouldered tenon; **Brustzapfenaufwölbung** *f* ARCH & TRAGW tusk; **Brustzapfenverbindung** *f* ARCH & TRAGW tusk tenon joint
**Bruttogeschoßfläche** *f* ARCH & TRAGW gross storey area (*BrE*), gross story area (*AmE*)
**BS** *abbr* (*Britische Norm*) ARCH & TRAGW, BETON, HOLZ, INFR & ENTW, STAHL BS (*British Standard*)
**BSB** *abbr* (*biologischer Sauerstoffbedarf*) INFR & ENTW, UMWELT, WERKSTOFF BOD (*biological oxygen demand*)
**BSI** *abbr* (*Britisches Institut für Normung*) ARCH & TRAGW, BAURECHT, HOLZ, INFR & ENTW, STAHL, WERKSTOFF BSI (*British Standards Institution*)
**Buche** *f* HOLZ beech; **Buchenholz** *nt* HOLZ, INFR & ENTW, WERKSTOFF beechwood; **Buchenholzschindel** *f* ARCH & TRAGW beech shingle, beechwood shingle; **Buchenparkett** *nt* ARCH & TRAGW, HOLZ beech parquetry
**Buch**: **Buchführung** *f* BAURECHT accounting; **Buchhalter** *m* BAURECHT accountant; **Buchhaltung** *f* BAURECHT accountancy
**Buchse** *f* ABWASSER sleeve, ELEKTR socket, HEIZ & BELÜFT sleeve, socket, WERKSTOFF *für Röhre und Kabel* bushing
**Bucht** *f* INFR & ENTW bay
**Buckelquaderverband** *m* NATURSTEIN opus rusticum
**Buckelschweißung** *f* ARCH & TRAGW projection welding
**Budget** *nt* BAURECHT budget
**Bug** *m* HOLZ, INFR & ENTW, WERKSTOFF angle brace, strut
**Bügel** *m* ARCH & TRAGW binder, fastening, BESCHLÄGE *Schloß* shackle, BETON stirrup, STAHL hoop; **Bügelabstand** *m* BETON pitch of links
**bügelbewehrt**: **~e Säule** *f* ARCH & TRAGW, BETON, INFR & ENTW *Stahlbeton* hooped column
***Bügel*: Bügelschelle** *f* ELEKTR clamp-ring; **Bügelschraube** *f* ARCH & TRAGW strap bolt; **Bügelverlegung** *f* BETON spacing of stirrups
**Buhne** *f* INFR & ENTW groyne
**Bühne** *f* ARCH & TRAGW attic, platform, stage, BAUMASCHIN platform, INFR & ENTW stage
**Bund-**: **Bundbalken** *m* ARCH & TRAGW, HOLZ binding beam, head runner
**Bündelpfeiler** *m* ARCH & TRAGW multiple rib pillar
**bündig** *adj* ARCH & TRAGW fair-faced, level, *Tür* flush, flush-mounted; **~e Mauerwerksfuge** *f* NATURSTEIN flat joint
**Bündigkeit** *f*: **~ von Stößen** ARCH & TRAGW flush arrangement of joints
***Bund-*: Bundmutter** *f* WERKSTOFF flanged nut; **Bundpfosten** *m* ARCH & TRAGW stud; **Bundsäule** *f* ARCH & TRAGW stud; **Bundständer** *m* ARCH & TRAGW stud; **Bundstiel** *m* ARCH & TRAGW stud
**Bungalow** *m* ARCH & TRAGW bungalow
**Bunker** *m* ARCH & TRAGW, INFR & ENTW bunker
**Bunt**: **Buntbartschloß** *nt* BESCHLÄGE, INFR & ENTW warded lock; **Buntglas** *nt* WERKSTOFF colored glass (*AmE*), coloured glass (*BrE*); **Buntmarmor** *m* WERKSTOFF colored marble (*AmE*), coloured marble (*BrE*)
**Bürge** *m* BAURECHT guarantor
**Bürgerhaus** *nt* ARCH & TRAGW public hall
**Bürgersteig** *m* INFR & ENTW pavement (*BrE*), sidewalk (*AmE*)
**Bürgschaft** *f* BAURECHT assurance
**Bürste** *f* OBERFLÄCHE brush
**bürsten** *vt* OBERFLÄCHE brush
**Butzenscheibe** *f* ARCH & TRAGW glass roundel, WERKSTOFF bullion
**Bypassventil** *nt* ABWASSER, HEIZ & BELÜFT bypass valve
**byzantinisch**: **~er Bogen** *m* ARCH & TRAGW stilted arch

# C

**c** *abbr* (*Konzentration*) INFR & ENTW, UMWELT, WERKSTOFF c (*concentration*)
**Cafeteria** *f* ARCH & TRAGW, INFR & ENTW cafeteria
**Caisson** *m* INFR & ENTW caisson
**Carbonatnester** *nt pl* BETON carbonate pockets
**Carbonylsulfid** *nt* UMWELT carbonyl sulfide (*AmE*), carbonyl sulphide (*BrE*)
**Carport** *m* INFR & ENTW carport
**Carraramarmor** *m* WERKSTOFF Carrara marble
**Catnic-Stahlsturz**® *m* STAHL Catnic steel lintel®
**Cella** *f* ARCH & TRAGW cella
**CEN** *abbr* (*Europäisches Komitee für Normung*) ARCH & TRAGW, BAURECHT, INFR & ENTW CEN (*European Committee for Standardisation*)
**CENELEC** *abbr* (*Europäischer Ausschuß für Elektrizitätsnormung*) BAURECHT, ELEKTR CENELEC (*European Body of Electrical Standards*)
**Ceresit** *nt* WERKSTOFF ceresit
**Charge** *f* WERKSTOFF batch
**Chemikalien** *f pl* WERKSTOFF chemicals; **Chemikalienangriff** *m* OBERFLÄCHE chemical attack
**chemikalienbeständig** *adj* WERKSTOFF resistant to chemicals
**chemisch**: **~e Abfälle** *m pl* BAURECHT, INFR & ENTW chemical waste; **~e Ausfällung** *f* INFR & ENTW, WERKSTOFF chemical precipitation; **~e Bodenverfestigung** *f* INFR & ENTW chemical stabilization; **~e Eigenschaft** *f* WERKSTOFF chemical property; **~e Feuchtigkeitssperre** *f* NATURSTEIN, WERKSTOFF chemical dampproof course (*chemical dpc*); **~e Formel** *f* WERKSTOFF chemical formula; **~er Kampfstoff** *m* UMWELT *Militär* toxic agent; **~e Oberflächenbehandlung** *f* OBERFLÄCHE chemical surface treatment; **~es Polieren** *nt* OBERFLÄCHE chemical hardening; **~er Prozeß** *m* **in der Atmosphäre** UMWELT atmospheric chemical process; **~er Sauerstoffbedarf** *m* (*CSB*) INFR & ENTW, UMWELT chemical oxygen demand (*COD*); **~e Toilette** *f* BESCHLÄGE chemical closet, chemical toilet; **~e Wasseruntersuchung** *f* ABWASSER, INFR & ENTW, WERKSTOFF chemical water analysis; **~er Wirkstoff** *m* UMWELT chemical agent; **~e Zusammensetzung** *f* WERKSTOFF chemical composition
**Chicago**: **~er Fenster** *nt* ARCH & TRAGW Chicago window
**Chlor** *nt* ABWASSER, UMWELT, WERKSTOFF chlorine
**chloren** *vt* ABWASSER, UMWELT, WERKSTOFF chlorinate
**Chlorierung** *f* ABWASSER, UMWELT, WERKSTOFF chlorination
***Chlor***: **Chlorkalk** *m* WERKSTOFF chlorinated lime; **Chlorkautschuk** *m* WERKSTOFF chlorinated rubber; **Chlorkautschukanstrich** *m* OBERFLÄCHE chlorinated rubber coat; **Chlorkautschukfarbe** *f* OBERFLÄCHE chlorinated rubber paint; **Chlorkautschuklack** *m* OBERFLÄCHE chlorinated rubber lacquer; **Chlorschwefelisocyanat** *nt* (*CSI*) UMWELT chlorosulfonyl isocyanate (*AmE*) (*CSI*), chlorosulphonyl isocyanate (*BrE*) (*CSI*)
**Chlorung** *f* INFR & ENTW chlorination
**Chor** *m* ARCH & TRAGW *Kirchenbau* choir; **Chorbogen** *m* ARCH & TRAGW choir arch; **Chorgewölbe** *nt* ARCH & TRAGW choir termination; **Chorhals** *m* ARCH & TRAGW chancel neck; **Chorschluß** *m* ARCH & TRAGW termination of choir; **Chorseitenschiff** *nt* ARCH & TRAGW choir aisle; **Chorumgang** *m* ARCH & TRAGW deambulatory
**Chromat** *nt* OBERFLÄCHE chromate; **Chromatierverfahren** *nt* OBERFLÄCHE, STAHL, WERKSTOFF chromating process
**Chrom**: **Chromnickel** *nt* WERKSTOFF chromium-nickel, chrome-nickel; **Chrom-Nickel-Stahl** *m* (*Cr-Ni-Stahl*) STAHL, WERKSTOFF chrome-nickel steel, chromium-nickel steel; **Chrompigment** *nt* OBERFLÄCHE, STAHL, WERKSTOFF chrome pigment; **Chromrot** *nt* OBERFLÄCHE chrome red, chromium red; **Chromstahl** *m* STAHL, WERKSTOFF chrome steel, chromium steel; **Chromüberzug** *m* OBERFLÄCHE, STAHL, WERKSTOFF chrome coat, chrome finish, chromium coat, chromium finish
**chronisch**: **~e Wirkung** *f* UMWELT chronic effect
**Chrysotil** *m* WERKSTOFF *Asbest* chrysotile
**CIB** *abbr* (*Internationaler Rat für Forschung, Studium und Dokumentation des Bauwesens*) BAURECHT CIB (*International Council for Building Research and Documentation*)
**CL** *abbr* (*Auftriebszahl*) UMWELT CL (*lift coefficient*)
**CO** *abbr* (*Kohlenmonoxid*) UMWELT CO (*carbon monoxide*)
**College** *nt* ARCH & TRAGW, INFR & ENTW college
**Computer** *m* ARCH & TRAGW, INFR & ENTW computer; **Computerberechnung** *f* ARCH & TRAGW, INFR & ENTW computer calculation; **Computermodellvalidation** *f* ARCH & TRAGW, INFR & ENTW computer model validation; **Computersimulation** *f* ARCH & TRAGW, INFR & ENTW computer simulation
**computerunterstützt**: **~es Entwerfen** *nt* ARCH & TRAGW, INFR & ENTW computer-aided design (*CAD*)
***Computer***: **Computervoraussage** *f* ARCH & TRAGW, INFR & ENTW computer prediction
**Copolymer** *nt* WERKSTOFF copolymer
***CO***: **CO-Warnanlage** *f* BAURECHT, ELEKTR, HEIZ & BELÜFT CO-warning system
**CO2** *abbr* (*Kohlendioxid*) UMWELT CO2 (*carbon dioxide*)
***CO2***: **CO2-Emissionsindex** *m* HEIZ & BELÜFT, UMWELT CO2 emission index
**CPL** *abbr* (*Zementputz*) NATURSTEIN CEM PLAS (*cement plaster*)
**Cremona-Plan** *m* ARCH & TRAGW *Stab- und Lagerkräftebestimmung* Cremona diagram
**Cr-Ni-Stahl** *m* (*Chrom-Nickel-Stahl*) STAHL, WERKSTOFF chrome-nickel steel, chromium-nickel steel
**Cross**: **Cross-Verfahren** *nt* ARCH & TRAGW, INFR & ENTW Cross method, method of moment distribution
**CSB** *abbr* (*chemischer Sauerstoffbedarf*) INFR & ENTW, UMWELT COD (*chemical oxygen demand*)

**CSI** *abbr* (*Chlorschwefelisocyanat*) UMWELT CSI (*chlorosulfonyl isocyanate (AmE), chlorosulphonyl isocyanate (BrE)*)

**Cullmann**: **~sche E-Linie** *f* INFR & ENTW Cullmann's method

# D

**Dach** *nt* ARCH & TRAGW, BESCHLÄGE, DÄMMUNG, ELEKTR, HEIZ & BELÜFT, HOLZ, INFR & ENT, STAHL, WERKSTOFF roof; **Dachabschlußblende** *f* HOLZ, INFR & ENTW edging board, gravel stop; **Dachanschluß** *m* STAHL roof flashing; **Dachanschlußstreifen** *m* STAHL roof flashing; **Dachantenne** *f* ELEKTR roof antenna; **Dacharbeiten** *f pl* ARCH & TRAGW roofing work; **Dachaufbau** *m* ARCH & TRAGW turret; **Dachaufbauten** *m pl* ARCH & TRAGW roof structures; **Dachaufsatz** *m* ARCH & TRAGW lantern tower, skylight; **Dachbelastung** *f* INFR & ENTW roof loading; **Dachbinder** *m* ARCH & TRAGW, HOLZ, INFR & ENTW, WERKSTOFF roof frame, roof framing, roof truss; **Dachboden** *m* ARCH & TRAGW attic, garret; **Dachbodeneinstiesgluke** *f* ARCH & TRAGW attic access hatch; **Dachbrüstung** *f* ARCH & TRAGW roof parapet; **Dachdämmung** *f* DÄMMUNG roof insulation; **Dachdecker** *m* ARCH & TRAGW roofer, slater

**Dachdeckung** *f* ARCH & TRAGW, WERKSTOFF roof cladding, roof covering, roof sheathing, roofing, *mit Pfannen* pantiling; **Dachdeckungsdiele** *f* WERKSTOFF roofing plank

***Dach*: Dachdichtungsbahn** *f* DÄMMUNG, WERKSTOFF moisture-proof roofing sheet; **Dachdurchbruch** *m* ARCH & TRAGW roof opening, roof penetration; **Dacheindeckung** *f* ARCH & TRAGW roof cladding, roof covering, roofing; **Dacheinfassung** *f* ARCH & TRAGW roof surround; **Dacherker** *m* ARCH & TRAGW gabled dormer, gabled dormer window; **Dachfenster** *nt* ARCH & TRAGW, BESCHLÄGE gabled dormer, gabled dormer window, rooflight, skylight; **Dachfirst** *m* ARCH & TRAGW roof ridge; **Dachfläche** *f* ARCH & TRAGW roof area, *als Maß* roofage; **Dachform** *f* ARCH & TRAGW roof profile, roof shape, roof type; **Dachgarten** *m* ARCH & TRAGW roof garden, rooftop terrace garden, terrace garden; **Dachgefälle** *nt* ARCH & TRAGW roof pitch; **Dachgeschoß** *nt* ARCH & TRAGW attic, loft; **Dachgesims** *nt* ARCH & TRAGW principal molding (*AmE*), principal moulding (*BrE*); **Dachgully** *m* ABWASSER roof gully, roof inlet; **Dachhaken** *m* BESCHLÄGE roof hook; **Dachhammer** *m* BAUMASCHIN slate ax (*AmE*), slate axe (*BrE*), slate hammer; **Dachhaube** *f* ARCH & TRAGW, HEIZ & BELÜFT extractor hood; **Dachhaut** *f* ARCH & TRAGW, DÄMMUNG, INFR & ENTW roofing skin, roof covering, roof membrane, *im Flachdach* vapor barrier (*AmE*), vapour barrier (*BrE*); **Dachhubschrauberlandeplatz** *m* ARCH & TRAGW, INFR & ENTW roof heliport, roof-top heliport; **Dachkante** *f* ARCH & TRAGW roof edge; **Dachkehle** *f* ARCH & TRAGW valley, valley gutter; **Dachkies** *m* WERKSTOFF roofing gravel; **Dachkollektor** *m* ELEKTR, HEIZ & BELÜFT *Sonnenwärme* roof collector; **Dachkonsole** *f* ARCH & TRAGW roof-mounted bracket; **Dachkonstruktion** *f* ARCH & TRAGW, HOLZ, INFR & ENTW roof structure; **Dachkuppel** *f* ARCH & TRAGW light cupola; **Dachlast** *f* ARCH & TRAGW, INFR & ENTW roof load; **Dachlaterne** *f* ARCH & TRAGW lantern light, rooflight; **Dachlatte** *f* HOLZ, WERKSTOFF batten, counter batten, roof batten, roof lath, roofing batten; **Dachleitungshalter** *m* ELEKTR *Blitzschutz* lightning conductor holder, roof conductor holder; **Dachlüfter** *m* HEIZ & BELÜFT roof fan, roof ventilator; **Dachluke** *f* ARCH & TRAGW skylight; **Dachlukenaufsatz** *m* ARCH & TRAGW skylight turret; **Dachneigung** *f* ARCH & TRAGW, WERKSTOFF roof pitch

**Dachpappe** *f* WERKSTOFF roofing felt; **Dachpappennagel** *m* WERKSTOFF roofing felt nail; **Dachpappenrandstreifen** *m* WERKSTOFF selvage; **Dachpappenunterlage** *f* WERKSTOFF sarking felt, underlining felt

***Dach*: Dachpfanne** *f* NATURSTEIN, WERKSTOFF pantile, roofing tile; **Dachpfette** *f* ARCH & TRAGW, BESCHLÄGE, HOLZ, STAHL purlin; **Dachpfettenstoß** *m* ARCH & TRAGW, HOLZ, STAHL purlin joint; **Dachplatte** *f* ARCH & TRAGW roof panel; **Dachrahmen** *m* ARCH & TRAGW, BESCHLÄGE purlin; **Dachrandprofil** *nt* HOLZ, INFR & ENTW edging board, gravel stop

**Dachrinne** *f* ABWASSER, ARCH & TRAGW eaves gutter, eaves trough, gutter, *hinter einer Brüstungsmauer* parapet gutter; **Dachrinnenendstück** *nt* ABWASSER gutter stop end; **Dachrinnenhalter** *m* BESCHLÄGE gutter bracket, gutter clamp, gutter hanger

***Dach*: Dachschale** *f* ARCH & TRAGW roof sheathing; **Dachschalung** *f* HOLZ, WERKSTOFF roof boarding, roofing; **Dachschiefer** *m* NATURSTEIN roofing slate; **Dachschräge** *f* ARCH & TRAGW, WERKSTOFF pitch of roof, roof pitch, roof slope; **Dachsilhouette** *f* ARCH & TRAGW roofline; **Dachstein** *m* NATURSTEIN roofing tile, saddle stone; **Dachstuhl** *m* ARCH & TRAGW, HOLZ, INFR & ENTW, WERKSTOFF roof framework, roof structure, roof truss; **Dachstuhl-Auflageplatte** *f* WERKSTOFF roof plate; **Dachterrasse** *f* ARCH & TRAGW roof terrace; **Dachträger** *m* ARCH & TRAGW, HOLZ, STAHL roof girder; **Dachtraufe** *f* ARCH & TRAGW, HOLZ eaves; **Dachventilator** *m* HEIZ & BELÜFT roof fan, roof ventilator; **Dachverglasung** *f* ARCH & TRAGW roof glazing; **Dachverwahrung** *f* STAHL roof flashing; **Dachwasser** *nt* ABWASSER roof water, stormwater; **Dachziegel** *m* HOLZ pantile, roof tile, roofing tile, INFR & ENTW, NATURSTEIN, WERKSTOFF pantile, roofing tile, tile

**Damm** *m* INFR & ENTW dam, bank

**Dämm-** *in cpds* DÄMMUNG, ELEKTR, OBERFLÄCHE, WERKSTOFF insulating; **Dämmatte** *f* DÄMMUNG, WERKSTOFF insulating slab

**dämmen** *vt* DÄMMUNG deaden, insulate, ELEKTR, OBERFLÄCHE, WERKSTOFF insulate

**dämmend** *adj* DÄMMUNG, ELEKTR, OBERFLÄCHE, WERKSTOFF insulating

***Dämm-*: Dämmfilz** *m* WERKSTOFF insulating felt; **Dämmplatte** *f* DÄMMUNG, WERKSTOFF insulating slab; **Dämmschicht** *f* DÄMMUNG insulating course, insulating layer, WERKSTOFF insulating layer; **Dämmstoff** *m* DÄMMUNG, ELEKTR, WERKSTOFF insulant, insulating material

***Damm***: **Dammstraße** *f* INFR & ENTW causeway
**Dämmung** *f* DÄMMUNG insulation
***Dämm-***: **Dämmwert** *m* DÄMMUNG insulation value
**Dampf** *m* DÄMMUNG, HEIZ & BELÜFT, UMWELT vapor (*AmE*), vapour (*BrE*); **Dampfbehandlung** *f* BETON steam curing
**dampfbeständig** *adj* DÄMMUNG vapor-resistant (*AmE*), vapour-resistant (*BrE*)
**dampfdicht** *adj* DÄMMUNG vapor-proof (*AmE*), vapor-tight (*AmE*), vapour-proof (*BrE*), vapour-tight (*BrE*)
***Dampf***: **Dampfdruck** *m* DÄMMUNG, HEIZ & BELÜFT vapor pressure (*AmE*), vapour pressure (*BrE*); **Dampfdruckausgleichsschicht** *f* DÄMMUNG vapor pressure equalizing layer (*AmE*), vapour pressure equalizing layer (*BrE*)
**dampfdurchlässig** *adj* DÄMMUNG, WERKSTOFF pervious to steam, pervious to vapor (*AmE*), pervious to vapour (*BrE*)
**dampfen** *vi* DÄMMUNG, HEIZ & BELÜFT, UMWELT steam
**dämpfen** *vt* DÄMMUNG damp, deaden, muffle
**Dämpfen** *nt* DÄMMUNG deadening
**dämpfend** *adj* DÄMMUNG deadening
***Dampf***: **Dampffahne** *f* UMWELT vapor plume (*AmE*), vapour plume (*BrE*); **Dampffernheizung** *f* HEIZ & BELÜFT, INFR & ENTW district steam heating, long-distance steam heating; **Dampfhärten** *nt* BETON steam curing; **Dampfhärtungsprozeß** *m* BETON steam-hardening process; **Dampfheizung** *f* HEIZ & BELÜFT steam heating; **Dampfkessel** *m* ELEKTR, HEIZ & BELÜFT boiler; **Dampfleitung** *f* HEIZ & BELÜFT steam pipe; **Dampfspannung** *f* DÄMMUNG, HEIZ & BELÜFT vapor pressure (*AmE*), vapour pressure (*BrE*); **Dampfsperre** *f* DÄMMUNG vapor-proof barrier (*AmE*), vapour-proof barrier (*BrE*), vapor seal (*AmE*), vapour seal (*BrE*); **Dampfstrahl** *m* UMWELT steam jet
**Dämpfung** *f* ARCH & TRAGW attenuation, DÄMMUNG *Schall* damping, deadening, attenuation, *von Stößen* absorption, ELEKTR attenuation; **Dämpfungsvermögen** *nt* DÄMMUNG damping capacity
**darstellen** *vt* ARCH & TRAGW represent, plot
**darstellend** *adj* ARCH & TRAGW representing
**Darstellung** *f* ARCH & TRAGW representation
**Datenblatt** *nt* ARCH & TRAGW, BAURECHT, INFR & ENTW data sheet
**Datum** *nt* ARCH & TRAGW date
**Dauer** *f* ARCH & TRAGW, BAURECHT, WERKSTOFF term; **Dauerbelastung** *f* ARCH & TRAGW, INFR & ENTW sustained loading; **Dauerdehngrenze** *f* ARCH & TRAGW, BAURECHT, WERKSTOFF fatigue yield limit
**dauerelastisch** *adj* WERKSTOFF permanently elastic
***Dauer***: **Dauerelektrode** *f* ELEKTR permanent electrode; **Dauerfestigkeit** *f* INFR & ENTW, WERKSTOFF fatigue limit; **Dauerfestigkeit** *f* WERKSTOFF durability
**dauerhaft** *adj* ARCH & TRAGW lasting, permanent, BAURECHT, DÄMMUNG permanent, INFR & ENTW lasting, permanent, WERKSTOFF permanent
***Dauer***: **Dauerlast** *f* ARCH & TRAGW, INFR & ENTW, WERKSTOFF continuous load; **Dauerlüftung** *f* BETON, HEIZ & BELÜFT permanent shuttering
**dauerplastisch** *adj* WERKSTOFF permanently plastic, nonhardening
***Dauer***: **Dauerschalung** *f* BETON permanent formwork; **Dauerschwingfestigkeit** *f* INFR & ENTW, WERKSTOFF fatigue strength; **Dauerversuch** *m* WERKSTOFF extended time test
**dB** *abbr* (*Dezibel*) DÄMMUNG, INFR & ENTW, UMWELT, WERKSTOFF dB (*decibel*)
**DD** *abbr* (*Widerstandsbeiwert*) UMWELT coefficient of drag
**Deck** *nt* INFR & ENTW deck; **Deckanstrich** *m* OBERFLÄCHE finish coat, opaque coat; **Deckasphaltschicht** *f* INFR & ENTW asphalt surfacing
**Decke** *f* ARCH & TRAGW ceiling, INFR & ENTW roof, *Straßenbau* pavement, paving, WERKSTOFF *Straßenbau* paving; **Deckenabhängung** *f* ARCH & TRAGW, STAHL ceiling suspension, suspended suspension; **Deckenauslegung** *f* ARCH & TRAGW flooring layout; **Deckenbalken** *m* ARCH & TRAGW, HOLZ ceiling joist, floor beam, floor joist, joist; **Deckenbeleuchtung** *f* BESCHLÄGE, ELEKTR ceiling illumination, ceiling lighting; **Deckenbemessung** *f* INFR & ENTW *Straßenbau* pavement design (*BrE*), sidewalk design (*AmE*); **Deckenfeld** *nt* BESCHLÄGE ceiling panel; **Deckenfertiger** *m* BAUMASCHIN finisher, finishing machine; **Deckenfüllkörper** *m* ARCH & TRAGW filler tile; **Deckengitter** *nt* BESCHLÄGE, HEIZ & BELÜFT ceiling grille; **Deckenhaken** *m* BESCHLÄGE ceiling hook; **Deckenheizung** *f* HEIZ & BELÜFT ceiling heating; **Deckenhöhe** *f* ARCH & TRAGW ceiling height; **Deckenhohlkörper** *m* WERKSTOFF hollow filler block; **Deckenkuppel** *f* ARCH & TRAGW, INFR & ENTW spherical vault; **Deckenlüfter** *m* HEIZ & BELÜFT ceiling ventilator; **Deckenlufterhitzer** *m* HEIZ & BELÜFT ceiling-mounted air heater; **Deckenluftverteiler** *m* HEIZ & BELÜFT ceiling diffuser; **Deckenniveau** *nt* ARCH & TRAGW ceiling level; **Deckenoberflächenbehandlung** *f* WERKSTOFF ceiling finish; **Deckenplatte** *f* WERKSTOFF ceiling tile; **Deckenputz** *m* NATURSTEIN ceiling plaster; **Deckenschalung** *f* BETON, HOLZ, INFR & ENTW ceiling boarding, formwork, shuttering; **Deckenstein** *m* WERKSTOFF ceiling stone; **Deckenstrahlungsheizung** *f* HEIZ & BELÜFT radiant ceiling heating; **Deckentafel** *f* BESCHLÄGE, HOLZ ceiling panel, panel; **Deckenträger** *m* ARCH & TRAGW filler joist, floor beam, floor joist; **Deckenunterzug** *m* ARCH & TRAGW, BESCHLÄGE floor joist; **Deckenventilator** *m* HEIZ & BELÜFT ceiling-mounted fan, ceiling fan; **Deckenverkleidung** *f* HOLZ ceiling lining
***Deck***: **Deckfuge** *f* ARCH & TRAGW covering joint; **Deckfurnier** *nt* HOLZ face veneer, outer veneer; **Deckleiste** *f* BESCHLÄGE, HOLZ cover fillet, cover strip, fillet, trim, tringle; **Deckplatte** *f* ARCH & TRAGW crown, top panel, BESCHLÄGE flange tile, STAHL cover plate; **Deckputz** *m* NATURSTEIN finish coat, finishing coat, final rendering, finishing coat plaster; **Deckschicht** *f* ARCH & TRAGW, INFR & ENTW *Straße* finishing layer, covering layer, topping, wearing course, OBERFLÄCHE covering; **Deckstein** *m* NATURSTEIN coping stone
**defekt** *adj* BAUMASCHIN, BAURECHT, INFR & ENTW defective
**Defekt** *m* BAUMASCHIN failure, BAURECHT damage, defect, INFR & ENTW failure; **Defektdatum** *nt* BAURECHT defects date
**definieren** *vt* ARCH & TRAGW, BAURECHT define
**Definition** *f* ARCH & TRAGW, BAURECHT definition
**Deflektor** *m* UMWELT deflector; **Deflektorhaube** *f* HEIZ & BELÜFT deflector cap

**dehnen** *vt* ARCH & TRAGW, BAUMASCHIN, HEIZ & BELÜFT, INFR & ENTW WERKSTOFF expand, strain, stretch
**Dehnfugenabdichtung** *f* ARCH & TRAGW, INFR & ENTW expansion joint sealing
**Dehnung** *f* ARCH & TRAGW, BAUMASCHIN, HEIZ & BELÜFT, INFR & ENTW, WERKSTOFF expansion, extension, stretching; **Dehnungsausgleicher** *m* ABWASSER, HEIZ & BELÜFT tension-compensating member, *Rohr* extension bellows, extension-compensating member; **Dehnungsbogen** *m* HEIZ & BELÜFT *Rohr* expansion bend; **Dehnungsfuge** *f* ARCH & TRAGW, BETON, INFR & ENTW *Straßenbau* construction joint, expansion joint, running joint; **Dehnungskoeffizient** *m* WERKSTOFF strain modulus; **Dehnungskraft** *f* ARCH & TRAGW, WERKSTOFF racking force; **Dehnungsmesser** *m* BAUMASCHIN, INFR & ENTW, WERKSTOFF extensometer, strain gage (*AmE*), strain gauge (*BrE*), strain meter; **Dehnungsmeßstreifen** *m* (*DMS*) WERKSTOFF wire strain gage (*AmE*), wire strain gauge (*BrE*); **Dehnungsriß** *m* INFR & ENTW, WERKSTOFF expansion crack; **Dehnungsschlaufe** *f* HEIZ & BELÜFT expansion bend; **Dehnungsschleife** *f* HEIZ & BELÜFT *Rohr* expansion bend; **Dehnungsverteilung** *f* WERKSTOFF strain distribution; **Dehnungszone** *f* HEIZ & BELÜFT expansion zone
**Dehydration** *f* HEIZ & BELÜFT, INFR & ENTW dehydration
**Deich** *m* INFR & ENTW dike
**Deklination** *f* INFR & ENTW declination
**Dekontaminierung** *f* BAURECHT, INFR & ENTW, UMWELT, WERKSTOFF decontamination
**Dekor** *nt* ARCH & TRAGW décor, decoration
**dekorativ** *adj* ARCH & TRAGW decorative
***Dekor***: **Dekorbeschichtung** *f* OBERFLÄCHE decorative coat, decorative finish
**dekoriert**: **~er Stil** *m* ARCH & TRAGW, INFR & ENTW decorated style
***Dekor***: **Dekortür** *f* ARCH & TRAGW, BESCHLÄGE decorated door
**Demontage** *f* INFR & ENTW disassembly, dismantling
**demontierbar** *adj* WERKSTOFF collapsible; **~e Konstruktion** *f* ARCH & TRAGW, BESCHLÄGE demountable structure
**demontieren** *vt* ABWASSER, ARCH & TRAGW, ELEKTR, INFR & ENTW disassemble, dismount
**Demulgator** *m* UMWELT demulsifier
**Denitrierung** *f*: **~ des Abfalls** UMWELT waste denitrification
**Denitrifikation** *f* ABWASSER, INFR & ENTW, UMWELT denitrification
**denkmalgeschützt**: **~es Bauwerk** *nt* BAURECHT, INFR & ENTW listed building (*BrE*)
**Deponie** *f* INFR & ENTW, UMWELT dumping ground, landfill, repository, storage site, waste dump, waste tip (*BrE*); **Deponiebetrieb** *m* UMWELT waste site operation; **Deponiegelände** *nt* INFR & ENTW dumping site, UMWELT dumping site, tipping site (*BrE*); **Deponiegut** *nt* UMWELT waste mass, fill mass; **Deponieoberfläche** *f* UMWELT operating face, working face; **Deponieschluß** *m* UMWELT waste site closure; **Deponiestandort** *m* INFR & ENTW, UMWELT landfill site; **Deponietyp** *m* INFR & ENTW, UMWELT landfill design, landfill type
**Derrick** *m* BAUMASCHIN derrick; **Derrickkran** *m* BAUMASCHIN derrick crane; **Derrickmast** *m* BAUMASCHIN derrick post
**Design** *nt* ARCH & TRAGW, INFR & ENTW, WERKSTOFF design; **Designziel** *nt* ARCH & TRAGW, INFR & ENTW design objective
**Dessin** *nt* ARCH & TRAGW, INFR & ENTW, OBERFLÄCHE, WERKSTOFF design, pattern; **Dessinblech** *nt* STAHL fancy sheet metal; **Dessindraht** *m* STAHL profiled wire
**Destillation** *f* UMWELT distillation; **~ mittels Sonnenenergie** UMWELT solar distillation
**Detail** *nt* ARCH & TRAGW, INFR & ENTW detail
**detailliert** **1.** *adj* ARCH & TRAGW, BAURECHT, INFR & ENTW detailed; **2. ~ beschreiben** *vt* ARCH & TRAGW, BAURECHT specify
***detailliert***: **~er Kostenvoranschlag** *m* ARCH & TRAGW, BAURECHT, INFR & ENTW detailed estimate; **~e Pläne** *m pl* ARCH & TRAGW, BAURECHT, INFR & ENTW detailed plans
***Detail***: **Detailzeichnung** *f* ARCH & TRAGW, INFR & ENTW detail drawing
**Detergentien** *f pl* ABWASSER detergents
**Deutsch**: **~e Industrienorm** *f* (*DIN*) ARCH & TRAGW, BAURECHT, INFR & ENTW German Industrial Standard (*DIN*); **~es Institut** *nt* **für Normung** (*DIN-Institut*) ARCH & TRAGW, BAURECHT, INFR & ENTW German Industrial Standards Institute
**Devastierung** *f* INFR & ENTW, UMWELT *von Land* land degradation, land disturbance
**Dezibel** *nt* (*dB*) DÄMMUNG, INFR & ENTW, UMWELT, WERKSTOFF decibel (*dB*)
**Dezimetersystem** *nt* ARCH & TRAGW *Maßordnung* decimetric system
**Diagenese** *f* UMWELT diagenesis
**diagenetisch** *adj* UMWELT diagenetic
**diagonal** *adj* ARCH & TRAGW, HOLZ diagonal
**Diagonal**: **Diagonalaussteifung** *f* STAHL *von Deckenträgern* diagonal bracing; **Diagonaldruck** *m* ARCH & TRAGW diagonal compression
**Diagonale** *f* ARCH & TRAGW, STAHL diagonal member
***Diagonal***: **Diagonalstab** *m* WERKSTOFF diagonal bar; **Diagonalstrebe** *f* ARCH & TRAGW, HOLZ cross stay, diagonal rod; **Diagonalverband** *m* ARCH & TRAGW, STAHL diagonal bracing, raker pile
**Diagramm** *nt* ARCH & TRAGW, BAURECHT chart
**Diamant** *m* BAUMASCHIN, BESCHLÄGE diamond; **Diamantbohrer** *m* BAUMASCHIN diamond drill; **Diamantschneider** *m* BAUMASCHIN cutting diamond; **Diamantspitze** *f* BESCHLÄGE diamond point
**dicht** *adj* DÄMMUNG tight, WERKSTOFF impermeable
**Dichte** *f* WERKSTOFF specific gravity
**dichten** *vt* ARCH & TRAGW, OBERFLÄCHE stop
**Dichten** *nt* ARCH & TRAGW, OBERFLÄCHE stopping
**dichtgelagert**: **~er Kies** *m* WERKSTOFF tight gravel
**Dichtheit** *f* DÄMMUNG tightness, INFR & ENTW, WERKSTOFF impermeability
**Dichtigkeitsprüfung** *f* WERKSTOFF permeability test
**Dichtleiste** *f* DÄMMUNG sealing strip
**Dichtring** *m* ABWASSER sealing ring
**Dichtung** *f* ABWASSER, DÄMMUNG gasket, HEIZ & BELÜFT gasket, *zwischen beweglichen Teilen* packing, INFR & ENTW, WERKSTOFF gasket; **Dichtungsbahn** *f* DÄMMUNG seal sheeting, *wasserfest* waterproof sheeting, UMWELT liner sheet; **Dichtungsband** *nt* ABWASSER, HEIZ & BELÜFT caulking strip; **Dichtungshaut** *f* DÄMMUNG *Bauwerksabdichtung*, WERKSTOFF dampproof membrane (*DPM*); **Dichtungskern** *f* INFR & ENTW core, sealing core;

**Dichtungsmanschette** *f* OBERFLÄCHE sealing sheet; **Dichtungsmasse** *f* OBERFLÄCHE, WERKSTOFF sealing compound; **Dichtungsmaterial** *nt* ABWASSER sealing material, OBERFLÄCHE sealing end; **Dichtungsmittel** *nt* BETON *Betonoberfläche* membrane curing compound, DÄMMUNG water repeller, *Mörtel, Beton* waterproofing agent, OBERFLÄCHE sealing material, sealer, WERKSTOFF *Betonoberfläche* curing compound; **Dichtungspackung** *f* ABWASSER gland packing; **Dichtungsschicht** *f* DÄMMUNG, OBERFLÄCHE sealing layer; **Dichtungsschlämme** *f* NATURSTEIN grouting compound; **Dichtungsschleier** *m* DÄMMUNG, INFR & ENTW curtain wall; **Dichtungsstreifen** *m* DÄMMUNG sealing strip; **Dichtungswand** *f* DÄMMUNG, HOLZ, INFR & ENTW, UMWELT diaphragm wall, slurry wall

**dick** *adj* ARCH & TRAGW thick

**Dicke** *f* ARCH & TRAGW thickness, OBERFLÄCHE *Beschichtungen* build

**Dickformat** *nt* NATURSTEIN *Mauerstein* large format

**dickwandig** *adj* ARCH & TRAGW thick-walled

**diebstahlsicher** *adj* ARCH & TRAGW burglarproof

**Diele** *f* ARCH & TRAGW lobby, HOLZ deal, INFR & ENTW, WERKSTOFF plank

**dielen** *vt* HOLZ floor

**Dielung** *f* BESCHLÄGE flooring, INFR & ENTW planking

**Differential**: **Differentialdruck** *m* HEIZ & BELÜFT differential force; **Differentialgleichung** *f* ARCH & TRAGW differential equation; **Differentialkraft** *f* HEIZ & BELÜFT differential force; **Differentialthermoanalyse** *f* (*DTA*) UMWELT differential thermal analysis (*DTA*); **Differentialthermostat** *m* DÄMMUNG, ELEKTR, HEIZ & BELÜFT differential thermostat

**Differenz** *f* ARCH & TRAGW, WERKSTOFF difference; **Differenzdruck** *m* UMWELT differential pressure; **Differenzdruckhöhe** *f* ABWASSER, INFR & ENTW, UMWELT differential head

**diffus**: **~es Schallfeld** *nt* DÄMMUNG diffuse sound field

**Diffusion** *f* DÄMMUNG diffusion; **Diffusionsbeiwert** *m* DÄMMUNG diffusion coefficient

**diffusionsdicht** *adj* DÄMMUNG diffusion-tight

**Dimensionierung** *f* ARCH & TRAGW dimensional analysis

**Dimmer** *m* ELEKTR dimmer

**DIN** *abbr* (*Deutsche Industrienorm*) ARCH & TRAGW, BAURECHT, INFR & ENTW DIN (*German Industrial Standard*)

***DIN***: **DIN-Institut** *nt* (*Deutsches Institut für Normung*) ARCH & TRAGW, BAURECHT, INFR & ENTW German Industrial Standards Institute

**direkt** *adj* ARCH & TRAGW straight; **~e Beleuchtung** *f* ELEKTR direct lighting; **~e Kupplung** *f* UMWELT direct coupling; **~e Lagerung** *f* ARCH & TRAGW, BETON direct support; **~es Licht** *nt* ELEKTR direct light; **~er Solargewinn** *m* HEIZ & BELÜFT, UMWELT direct solar gain; **~e Strahlung** *f* UMWELT direct radiation; **~e Übertragung** *f* DÄMMUNG direct transmission

**Direkt**: **Direktheizung** *f* HEIZ & BELÜFT direct heating; **Direktumwandlung** *f* UMWELT direct conversion

**diskontiert**: **~er Cash-flow** *m* BAURECHT discounted cash flow (*DCF*)

**Dispergens** *nt* OBERFLÄCHE, UMWELT dispersant

**dispergieren** *vt* OBERFLÄCHE, UMWELT disperse

**dispergiert**: **~es Harz** *nt* OBERFLÄCHE dispersed resin

**Dispersant** *nt* UMWELT dispersant

**Dispersionsfarbe** *f* ABWASSER, WERKSTOFF dispersion paint, water-based paint, OBERFLÄCHE dispersion paint

**Disput** *m* BAURECHT dispute

**Distanz** *f* INFR & ENTW distance; **Distanzhalter** *m* ABWASSER, ARCH & TRAGW, BETON, ELEKTR, HEIZ & BELÜFT, HOLZ spacer; **Distanzklemme** *f* ELEKTR spacer clamp

**DMS** *abbr* (*Dehnungsmeßstreifen*) WERKSTOFF wire strain gage (*AmE*), wire strain gauge (*BrE*)

**Dole** *f* ABWASSER drain

**Dollen** *m* HOLZ peg, plug

**Dolomit**: **Dolomitkalk** *m* WERKSTOFF dolomitic lime; **Dolomitmarmor** *m* NATURSTEIN dolomite marble; **Dolomitstein** *m* NATURSTEIN dolomite brick

**Domikalgewölbe** *nt* ARCH & TRAGW covered vault, domical vault

**Doppel-** *in cpds* ABWASSER, ARCH & TRAGW, BESCHLÄGE, INFR & ENTW, UMWELT, WERKSTOFF double; **Doppelabzweig** *m* ABWASSER double bay; **Doppelbartschlüssel** *m* BESCHLÄGE two-way key; **Doppelboden** *m* ARCH & TRAGW false floor, double floor; **Doppelbrücke** *f* INFR & ENTW twin bridge; **Doppeldurchfahrt** *f* INFR & ENTW double archway; **Doppelfenster** *nt* ARCH & TRAGW double window

**doppelflügelig**: **~e Tür** *f* ARCH & TRAGW two-leaf door

***Doppel-***: **Doppelfußboden** *m* ARCH & TRAGW framed floor; **Doppelgewindeschraube** *f* WERKSTOFF double-threaded screw; **Doppelhaus** *nt* ARCH & TRAGW, INFR & ENTW two semidetached houses; **Doppelhaushälfte** *f* ARCH & TRAGW, INFR & ENTW semidetached house; **Doppelkopfniete** *f* BESCHLÄGE bullhead rivet; **Doppelläufer** *m* NATURSTEIN double stretcher; **Doppelrippe** *f* ARCH & TRAGW, WERKSTOFF double rib; **Doppelsäule** *f* ARCH & TRAGW twin column; **Doppelschaufelabscheider** *m* UMWELT double bucket collector; **Doppelschneckenextruder** *m* UMWELT twin screw extruder; **Doppelstehfalz** *m* ARCH & TRAGW double-skin partition; **Doppelsturzriegel** *m* HOLZ double cross timber; **Doppelsymmetrie** *f* ARCH & TRAGW, BETON, INFR & ENTW, STAHL, WERKSTOFF double symmetry

**doppelsymmetrisch** *adj* ARCH & TRAGW, BETON, INFR & ENTW, STAHL, WERKSTOFF double symmetric

**doppelt**: **~es Rohrgewebe** *nt* INFR & ENTW, WERKSTOFF double reed lathing; **~e Zwischendecke** *f* ARCH & TRAGW double false ceiling

***Doppel-***: **Doppel-T-Träger** *m* ARCH & TRAGW, BETON, STAHL H-beam

**doppelüberlappend**: **~er, glatter Ziegel** *m* ARCH & TRAGW, INFR & ENTW double lap plain tile

***Doppel-***: **Doppelüberlappung** *f* ARCH & TRAGW, INFR & ENTW double lap; **Doppel-U-Profil** *nt* STAHL, WERKSTOFF double channel section; **Doppelverglasung** *f* ARCH & TRAGW double glazing; **Doppelwand** *f* NATURSTEIN double wall

**doppelwandig** *adj* ARCH & TRAGW double-walled

***Doppel-***: **Doppelzirkel** *m* BAUMASCHIN double calipers (*AmE*), double callipers (*BrE*)

**Döpper** *m* BAUMASCHIN *Nieten* rivet set, rivet snap, riveting set

**dorisch** *adj* ARCH & TRAGW Doric; **~es Kapitell** *nt* ARCH & TRAGW Doric capital; **~e Säule** *f* ARCH & TRAGW Doric column; **~er Stil** *m* ARCH & TRAGW Doric style

**Dorn** *m* ARCH & TRAGW spur, bolt, pin drift, *Schleifscheibe* arbor, BAUMASCHIN *Schleifscheibe* arbor, spur, BESCHLÄGE pin drift, bolt, spur, *Schleifscheibe* arbor, *Schloß* pin, STAHL, WERKSTOFF *Schloß* pin

**Dose** *f* ELEKTR jack

**Dosieranlage** *f* ARCH & TRAGW batcher, BAUMASCHIN proportioning equipment

**dosieren** *vt* ABWASSER, BETON, WERKSTOFF batch

**Dosierer** *m* BAUMASCHIN proportioner

**Dosiergerät** *nt* BAUMASCHIN proportioning device

**Dosierpumpe** *f* ABWASSER dosage pump

**Dosiswirkung** *f* UMWELT dose response

**Draht** *m* BESCHLÄGE wire; **Drahtbewehrung** *f* ELEKTR wire armoring (*AmE*), wire armouring (*BrE*); **Drahtbürste** *f* BAUMASCHIN scratch brush; **Drahteinlage** *f* WERKSTOFF wire core, wire reinforcement; **Drahtgeflecht** *nt* BESCHLÄGE, WERKSTOFF mesh wire, wire fabric, wire netting; **Drahtgewebe** *nt* STAHL metal cloth; **Drahtgitter** *nt* WERKSTOFF mesh wire; **Drahtglas** *nt* BESCHLÄGE, WERKSTOFF armoured glass (*BrE*), wire glass, wired glass; **Drahtornamentglas** *nt* ARCH & TRAGW wired pattern glass; **Drahtputzdecke** *f* NATURSTEIN wire plaster ceiling; **Drahtputzwand** *f* NATURSTEIN wire plaster wall; **Drahtschere** *f* BAUMASCHIN wire cutters; **Drahtseilbahn** *f* INFR & ENTW, STAHL aerial cableway, aerial railway, ropeway, wire ropeway; **Drahtseilsägemaschine** *f* BAUMASCHIN cable cutting machine; **Drahtverbundsicherheitsglas** *nt* WERKSTOFF laminated wired glass; **Drahtzaun** *m* INFR & ENTW wire fence; **Drahtziegelgewebe** *nt* WERKSTOFF clay lathing, clayed wire mesh

**Drainage** *f* INFR & ENTW drainage; **Drainagegraben** *m* INFR & ENTW drainage trench

**Drängraben** *m* INFR & ENTW drainage trench

**Dränschicht** *f* INFR & ENTW filter layer

**draußen** *adv* ARCH & TRAGW outside

**Dreh-** *in cpds* BAUMASCHIN, UMWELT, WERKSTOFF rotary; **Drehachse** *f* INFR & ENTW axis of rotation; **Drehbank** *f* BAUMASCHIN lathe; **Drehbohren** *nt* BAUMASCHIN rotary drilling; **Drehbohrer** *m* BAUMASCHIN, HOLZ rotary drill, twist gimlet; **Drehbrücke** *f* INFR & ENTW pivot bridge, swing bridge, turn bridge, turning bridge; **Drehdeckel** *m* ABWASSER screw cap

**drehen** *vt* BAUMASCHIN turn

**Drehen** *nt* BAUMASCHIN slewing (*BrE*), sluing (*AmE*), *eines Krans* turning

***Dreh-***: **Drehfenster** *nt* ARCH & TRAGW, BESCHLÄGE pivot-hung window, side-hung window; **Drehfensterflügel** *m* BESCHLÄGE pivot-hung sash; **Drehflügel** *m* BESCHLÄGE casement; **Drehflügelfenster** *nt* ARCH & TRAGW, BESCHLÄGE casement window, pivoted sash, side-hung window; **Drehkippfenster** *nt* ARCH & TRAGW pivot window, *horizontal oder vertikal* center-hung window (*AmE*), centre-hung window (*BrE*); **Drehkippflügel** *m* ARCH & TRAGW, WERKSTOFF tilt and turn window, tip-and-turn sash; **Drehkippverschluß** *m* BESCHLÄGE tip-and-turn hardware; **Drehknopf** *m* BESCHLÄGE turn knob; **Drehkran** *m* BAUMASCHIN slewing crane (*BrE*), sluing crane (*AmE*), rotary crane; **Drehmomentskoeffizient** *m* ARCH & TRAGW coefficient of torque; **Drehpunkt** *m* ARCH & TRAGW moment pole; **Drehregelventil** *nt* ABWASSER plug cock, plug tap; **Drehriegel** *m* BESCHLÄGE *Schloß* rotary lever; **Drehriegelverschluß** *m* BESCHLÄGE rotary lever lock; **Drehrohrofen** *m* BAUMASCHIN, UMWELT rotary furnace; **Drehschalter** *m* ELEKTR rotary switch; **Drehschieber** *m* UMWELT rotary valve; **Drehspülbohren** *nt* BAUMASCHIN rotary drilling; **Drehstift** *m* BESCHLÄGE hinge pin, turn pin; **Drehstromlichtmaschine** *f* ELEKTR, UMWELT alternator; **Drehtor** *nt* ARCH & TRAGW swing gate; **Drehtrommel** *f* BAUMASCHIN, UMWELT revolving drum, tumbler; **Drehtür** *f* ARCH & TRAGW revolving door

**Drehung** *f* ARCH & TRAGW rotation, turn

***Dreh-***: **Drehverschluß** *m* BESCHLÄGE *halbe Drehung* turnlock fastener; **Drehwinkel** *m* INFR & ENTW angle of rotation; **Drehzahl** *f* ABWASSER, BAUMASCHIN, HEIZ & BELÜFT *Pumpe, Ventilator* revolutions per minute (*rpm*), speed; **Drehzahlregler** *m* UMWELT speed control device

**drehzapfengelagert** *adj* BESCHLÄGE pivoted

**Drei-** *in cpds* ARCH & TRAGW triple; **Dreiaxialdruckgerät** *nt* BAUMASCHIN triaxial compression cell; **Dreiaxialprüfung** *f* INFR & ENTW triaxial test; **Dreibein** *nt* BAUMASCHIN gin; **Dreibeinkran** *m* INFR & ENTW shear leg

**dreidimensional** *adj* ARCH & TRAGW, INFR & ENTW three-dimensional; **~er Raster** *m* ARCH & TRAGW modular space grid

**Dreieck** *nt* ARCH & TRAGW triangle; **Dreieckbogen** *m* ARCH & TRAGW triangular arch; **Dreiecksbinder** *m* ARCH & TRAGW triangular truss; **Dreiecksrahmen** *m* ARCH & TRAGW, HOLZ triangulated frame; **Dreieckstufe** *f* ARCH & TRAGW spandrel step; **Dreiecksverband** *m* STAHL diagonal bracing; **Dreiecksvermessung** *f* INFR & ENTW triangulation

**dreifeldrig** *adj* ARCH & TRAGW threebay

**Dreigelenk-** *in cpds* ARCH & TRAGW three-hinged; **Dreigelenkbogen** *m* ARCH & TRAGW three-hinged arch

***Drei-***: **Dreikammerstein** *m* WERKSTOFF three-core block

**Dreikant-** *in cpds* ARCH & TRAGW, HOLZ triangular; **Dreikantleiste** *f* ARCH & TRAGW, HOLZ triangular cleat, triangular fillet

**Dreilagen-** *in cpds* HOLZ, WERKSTOFF three-ply; **Dreilagenholz** *nt* HOLZ three-ply wood

**dreilagig** *adj* NATURSTEIN three-layer; **~er Putz** *m* NATURSTEIN float and set, three-coat plaster, three-coat work, render, float and set (*RFS*)

**dreipolig** *adj* ELEKTR three-pole

***Drei-***: **Dreiradwalze** *f* BAUMASCHIN three-wheeled roller

**dreischiffig** *adj* ARCH & TRAGW threebay

**Dreiviertel-** *in cpds* ARCH & TRAGW, NATURSTEIN three-quarter; **Dreiviertelquartier** *nt* NATURSTEIN king closer; **Dreiviertelziegel** *m* ARCH & TRAGW, NATURSTEIN king closer, three-quarter brick

**Dreiwege-** *in cpds* ABWASSER, ARCH & TRAGW three-way; **Dreiwegehahn** *m* ABWASSER three-way tap, three-way cock; **Dreiwegemischer** *m* ABWASSER three-way mixer; **Dreiwegeventil** *nt* ABWASSER three-way valve

***Drei-***: **Dreizentrenbogen** *m* ARCH & TRAGW three-centered arch (*AmE*), three-centred arch (*BrE*)

**Drempel** *m* ARCH & TRAGW, BAUMASCHIN miter sill

(*AmE*), mitre sill (*BrE*); **Drempelwand** *f* ARCH & TRAGW, BETON, NATURSTEIN jamb wall

**Drillstem-Test** *m* UMWELT drill stem test

**dritt**: **~e Putzlage** *f* NATURSTEIN set, setting coat; **~e Reinigungsstufe** *f* UMWELT tertiary sewage treatment

**Drossel** *f* BAUMASCHIN throttle, ELEKTR *Leuchtstoffröhre* ballast; **Drosselklappe** *f* BAUMASCHIN throttle, HEIZ & BELÜFT damper, throttle valve; **Drosselklappenventil** *nt* ABWASSER, HEIZ & BELÜFT butterfly valve

**drosseln** *vt* BAUMASCHIN, HEIZ & BELÜFT throttle

***Drossel***: **Drosselschieber** *m* HEIZ & BELÜFT throttle slide valve; **Drosselventil** *nt* HEIZ & BELÜFT throttle valve, damper

**Druck** *m* ABWASSER, ARCH & TRAGW, BAUMASCHIN, HEIZ & BELÜFT, INFR & ENTW, UMWELT pressure; **Druckabfall** *m* ABWASSER, BAUMASCHIN, HEIZ & BELÜFT decrease in pressure, drop in pressure, pressure drop; **Druckabnahme** *f* ABWASSER, BAUMASCHIN, HEIZ & BELÜFT decompression, pressure drop; **Druckausdehnungsgefäß** *nt* HEIZ & BELÜFT pressure-compensating tank; **Druckausgleich** *m* ARCH & TRAGW, HEIZ & BELÜFT, UMWELT pressure compensation, pressure equalization; **Druckausgleichsbehälter** *m* HEIZ & BELÜFT, INFR & ENTW, UMWELT pressure-compensating tank, surge tank; **Druckausgleichsschicht** *f* ARCH & TRAGW relieving layer, INFR & ENTW pressure compensation layer

**druckbeansprucht** *adj* ARCH & TRAGW, BETON, STAHL, WERKSTOFF subject to compression

***Druck***: **Druckbehälter** *m* ABWASSER, HEIZ & BELÜFT pressure tank; **Druckbelastung** *f* ARCH & TRAGW pressure load

**druckbeständig** *adj* ARCH & TRAGW pressure-resistant

***Druck***: **Druckbeton** *m* BETON compressed concrete, tamped concrete; **Druckbewehrung** *f* ARCH & TRAGW compressive reinforcement; **Druckbiegespannung** *f* ARCH & TRAGW, BETON, HOLZ, INFR & ENTW, STAHL, WERKSTOFF compression-bending stress; **Druckblock** *m* ARCH & TRAGW, BETON, WERKSTOFF stress-block; **Druck-Dehnungs-Beziehung** *f* ARCH & TRAGW, BETON, INFR & ENTW, STAHL, WERKSTOFF stress-strain relationship; **Druck-Dehnungs-Diagramm** *nt* ARCH & TRAGW, BETON, INFR & ENTW, WERKSTOFF stress-strain diagram

**druckdicht** *adj* ABWASSER, HEIZ & BELÜFT pressure-tight

***Druck***: **Druckdifferenz** *f* ABWASSER, HEIZ & BELÜFT pressure difference

**drucken** *vt* BAUMASCHIN, INFR & ENTW press

**Drücker** *m* BESCHLÄGE lever handle, *Tür* door handle; **Drückergarnitur** *f* BESCHLÄGE set of door handles, set of handles

***Druck***: **Druckerhöhung** *f* ABWASSER, HEIZ & BELÜFT pressure rise; **Druckerhöhungsanlage** *f* ABWASSER, ARCH & TRAGW, INFR & ENTW booster pump system; **Druckerhöhungspumpe** *f* ABWASSER, ARCH & TRAGW, BAUMASCHIN booster pump

***Drücker***: **Drücker-Klappring-Garnitur** *f* BESCHLÄGE flush handle set

***Druck***: **Druckfeld** *nt* INFR & ENTW pressure field

**druckfest** *adj* BETON, WERKSTOFF compression-resistant; **~e Ausführung** *f* ARCH & TRAGW pressure-proof design

***Druck***: **Druckfestigkeit** *f* ARCH & TRAGW, BETON, HOLZ, STAHL, WERKSTOFF compression strength, crushing strength; **Druckfilter** *nt* UMWELT pressure filter; **Druckfühler** *m* ABWASSER, HEIZ & BELÜFT pressure pickup, pressure transducer; **Druckgasflasche** *f* ARCH & TRAGW *zum Schweißen* compressed gas cylinder; **Druckgefäß** *nt* ABWASSER, HEIZ & BELÜFT pressure vessel; **Druckglied** *nt* ARCH & TRAGW strut; **Druckgurt** *m* ARCH & TRAGW, HOLZ, STAHL, WERKSTOFF compression boom, compression flange; **Druckgußwerkstoff** *m* STAHL die-casting metal; **Druckhöhe** *f* INFR & ENTW *hydraulisch* head, UMWELT effective head; **Druckhöhenverlust** *m* ABWASSER, HEIZ & BELÜFT loss of head; **Druckkoeffizient** *m* UMWELT pressure coefficient; **Druckkraft** *f* ARCH & TRAGW, BETON, HOLZ, NATURSTEIN, STAHL compression force, compressive force; **Druckkurve** *f* INFR & ENTW pressure curve; **Drucklast** *f* ARCH & TRAGW compressive force; **Druckleitung** *f* ABWASSER, HEIZ & BELÜFT, UMWELT delivery pipe, pressure pipe; **Drucklinie** *f* ARCH & TRAGW *Bogen* pressure line

**drucklos** *adj* ABWASSER unpressurized

**Druckluft** *f* WERKSTOFF compressed air; **Druckluftantrieb** *m* BAUMASCHIN pneumatic drive; **Druckluftbohrer** *m* BAUMASCHIN pneumatic drill; **Druckluftgründung** *f* INFR & ENTW caisson foundation; **Druckluftkammer** *f* INFR & ENTW caisson; **Druckluftkompressor** *m* BAUMASCHIN, HEIZ & BELÜFT air compressor; **Druckluftnietung** *f* STAHL pneumatic riveting; **Drucklufttramme** *f* BAUMASCHIN air hammer; **Druckluftschlagbohrer** *m* BAUMASCHIN pneumatic hammer drill; **Druckluftstampfer** *m* BAUMASCHIN air rammer

***Druck***: **Druckmesser** *m* BAUMASCHIN, HEIZ & BELÜFT pressure gage (*AmE*), pressure gauge (*BrE*); **Druckminderer** *m* ABWASSER, HEIZ & BELÜFT pressure reducer, pressure-reducing valve; **Druckpfahl** *m* INFR & ENTW compression pile; **Druckplatte** *f* ARCH & TRAGW topping slab; **Druckprobe** *f* ABWASSER, HEIZ & BELÜFT, WERKSTOFF pressure test; **Druckpumpe** *f* ABWASSER forcing pump; **Druckregelung** *f* ABWASSER, HEIZ & BELÜFT pressure control; **Druckregler** *m* ABWASSER, HEIZ & BELÜFT pressure controller, pressure regulator; **Druckring** *m* ARCH & TRAGW compression ring; **Druckrohr** *nt* ABWASSER, HEIZ & BELÜFT pressure pipe; **Druckrohrleitung** *f* UMWELT pen trough; **Druckschicht** *f* INFR & ENTW topping; **Druckschwankung** *f* BAUMASCHIN pressure fluctuation; **Druckspeicher** *m* HEIZ & BELÜFT pressurized hot-water tank; **Druckspüler** *m* ABWASSER flushing valve, flush valve; **Druckstab** *m* HOLZ diagonal strut

**druckstark** *adj* HEIZ & BELÜFT *Flügelrad* high-pressure

***Druck***: **Druckstrebe** *f* HOLZ diagonal strut; **Druckstufe** *f* ABWASSER, HEIZ & BELÜFT pressure stage; **Druckstutzen** *m* ABWASSER, HEIZ & BELÜFT pressure joint; **Drucktastenschloß** *nt* BESCHLÄGE cipher lock; **Drucktragfähigkeit** *f* ARCH & TRAGW, BETON, HOLZ, STAHL, WERKSTOFF compression resistance; **Druckveränderung** *f* ARCH & TRAGW, BAUMASCHIN, HEIZ & BELÜFT, INFR & ENTW change in pressure, pressure change; **Druckverhältnis** *nt* ABWASSER, HEIZ & BELÜFT pressure ratio; **Druckverlust** *m* ABWASSER, BAUMASCHIN, HEIZ & BELÜFT loss of head, pressure drop, pressure loss, UMWELT pressure drop; **Druckverteilung** *f* ARCH & TRAGW, INFR & ENTW pressure distribution;

**Druckverteilungsblock** *m* ARCH & TRAGW, BETON, WERKSTOFF stress-block; **Druckwiderstand** *m* ARCH & TRAGW, BETON, HOLZ, INFR & ENTW, STAHL compression resistance; **Druckzwiebel** *f* INFR & ENTW bulb of pressure

**DTA** *abbr* (*Differentialthermoanalyse*) UMWELT DTA (*differential thermal analysis*)

**Dübel** *m* ABWASSER plug, ARCH & TRAGW joggle, BESCHLÄGE dowel, dowel pin, HOLZ dowel, WERKSTOFF plug, connector; **Dübelbalken** *m* HOLZ flitched beam; **Dübelloch** *nt* BESCHLÄGE dowel hole

**dübeln** *vt* HOLZ plug

**Dübeln** *nt* HOLZ plugging

***Dübel***: **Dübelstein** *m* ARCH & TRAGW fixing block; **Dübelverbindung** *f* NATURSTEIN dowelled connection

**Düker** *m* ABWASSER, INFR & ENTW *unter einer Straße, einem Gehweg* culvert

**Düne** *f* INFR & ENTW dune

**Dunkelkammer** *f* ARCH & TRAGW dark room

**dünn** *adj* ARCH & TRAGW, OBERFLÄCHE, WERKSTOFF thin, *Beschichtung* fine; **~e Verblendung** *f* OBERFLÄCHE veneer

**Dünn-** *in cpds* ARCH & TRAGW, OBERFLÄCHE, WERKSTOFF thin; **Dünnbettkleber** *m* BETON cement adhesive; **Dünnbettverfahren** *nt* NATURSTEIN *Fliesen* thin-bed method; **Dünnglas** *nt* ARCH & TRAGW thin window glass

**dünnwandig** *adj* ARCH & TRAGW thin-walled; **~e Schale** *f* ARCH & TRAGW, INFR & ENTW thin shell

**Dunst** *m* DÄMMUNG, HEIZ & BELÜFT, UMWELT vapor (*AmE*), vapour (*BrE*); **Dunstabzugshaube** *f* BESCHLÄGE, HEIZ & BELÜFT *über einem Herd* exhaust hood; **Dunstrohr** *nt* ABWASSER, HEIZ & BELÜFT outlet vent, vent pipe; **Dunstrohrziegel** *m* ABWASSER, NATURSTEIN outlet vent tile, vent pipe tile

**Duplikat** *nt* BAURECHT duplicate

**durchbiegen 1.** *vt* ARCH & TRAGW, WERKSTOFF deflect; **2. sich durchbiegen** *v refl* ARCH & TRAGW deflect

**Durchbiegen** *nt* ARCH & TRAGW, BETON, WERKSTOFF sagging bend

**Durchbiegung** *f* ARCH & TRAGW, INFR & ENTW flexure, WERKSTOFF deflection; **Durchbiegungsmesser** *m* BAUMASCHIN deflectometer; **Durchbiegungsmeßwertaufnehmer** *m* ARCH & TRAGW, INFR & ENTW, STAHL, WERKSTOFF deflection transducer

**Durchbinder** *m* NATURSTEIN perpend

**durchbohren** *vt* ARCH & TRAGW pierce

**durchbrechen** *vt* ARCH & TRAGW pierce

**durchbrochen**: **~es Maßwerk** *nt* ARCH & TRAGW openwork tracery; **~es Mauerwerk** *nt* NATURSTEIN honeycomb masonry, pigeon-holed masonry, trellis work; **~e Rippe** *f* ARCH & TRAGW honeycomb rib; **~er Schornsteinaufsatz** *m* ARCH & TRAGW openwork chimney top; **~e Turmspitze** *f* ARCH & TRAGW openwork spire

**Durchbruch** *m* ARCH & TRAGW, BAUMASCHIN, INFR & ENTW, OBERFLÄCHE, WERKSTOFF breakthrough, hole, opening, penetration

**durchdringen** *vt* ARCH & TRAGW, INFR & ENTW penetrate

**Durchfahrt** *f* ARCH & TRAGW thoroughfare (*BrE*), thruway (*AmE*); **Durchfahrtshöhe** *f* ARCH & TRAGW, INFR & ENTW headroom, headway, overhead clearance; **Durchfahrtshöhe und -breite** *f* ARCH & TRAGW clearance height and width

**Durchfeuchtungsschaden** *m* BAURECHT, INFR & ENTW damage due to penetration of moisture

**Durchfluß** *m* ABWASSER, HEIZ & BELÜFT, INFR & ENTW, UMWELT, WERKSTOFF flow; **Durchflußkoeffizient** *m* INFR & ENTW flow coefficient

**durchführbar** *adj* ARCH & TRAGW, INFR & ENTW feasible

**Durchführbarkeit** *f* ARCH & TRAGW, INFR & ENTW feasibility; **Durchführbarkeitsstudie** *f* ARCH & TRAGW, INFR & ENTW feasibility study

**durchführen** *vt* ARCH & TRAGW, INFR & ENTW, WERKSTOFF *Arbeiten* execute

**Durchführung** *f* ARCH & TRAGW, INFR & ENTW *Arbeiten* execution, WERKSTOFF *für Rohre und Kabel* bushing

**Durchgang** *m* ARCH & TRAGW connecting passage, passage; **Durchgangshöhe** *f* ARCH & TRAGW, INFR & ENTW headroom; **Durchgangshöhe und -breite** *f* ARCH & TRAGW clearance height and width

**durchgebogen** *adj* ARCH & TRAGW sagged

**durchgefärbt** *adj* WERKSTOFF integrally colored (*AmE*), integrally coloured (*BrE*), pigmented

**durchgehend** *adj* ARCH & TRAGW continuous; **~er Anschluß** *m* STAHL through-wall flashing; **~er Anschlußstreifen** *m* STAHL through-wall flashing piece; **~e Decke** *f* ARCH & TRAGW continuous floor; **~es Fachwerk** *nt* ARCH & TRAGW continuous truss; **~er Handlauf** *m* ARCH & TRAGW, BESCHLÄGE continuous handrail; **~es Mauerwerk** *nt* ARCH & TRAGW blind wall; **~er Riß** *m* NATURSTEIN through crack; **~e Säule** *f* ARCH & TRAGW *durch mehrere Geschosse* passing column, through column; **~er Ständer** *m* ARCH & TRAGW, BAUMASCHIN passing pillar, passing post, through pillar; **~er Träger** *m* ARCH & TRAGW continuous girder; **~ verglaste Fläche** *f* WERKSTOFF continuous glass surface

**durchgerostet** *adj* STAHL rusted through

**durchgesickert** *adj* ABWASSER, ARCH & TRAGW, INFR & ENTW percolated

**Durchhang** *m* HEIZ & BELÜFT *von Leitungen* dip

**durchhängen** *vi* ARCH & TRAGW sag

**durchlässig** *adj* BETON pervious, porous, HOLZ permeable, pervious, INFR & ENTW pervious, porous, NATURSTEIN permeable, pervious, UMWELT, WERKSTOFF pervious, porous; **~er Boden** *m* INFR & ENTW, WERKSTOFF permeable ground; **~e Schicht** *f* INFR & ENTW, WERKSTOFF permeable layer; **~er Untergrund** *m* INFR & ENTW permeable subsoil

**Durchlässigkeit** *f* BETON, INFR & ENTW porosity, UMWELT, WERKSTOFF permeability, porosity; **Durchlässigkeitsbeiwert** *m* UMWELT coefficient of permeability, hydraulic conductivity; **Durchlässigkeitskoeffizient** *m* UMWELT, WERKSTOFF coefficient of permeability, permeability coefficient

**Durchlaßkanal** *m* ABWASSER, INFR & ENTW *unter einer Straße, einem Gehweg* culvert

**Durchlauf** *m* ABWASSER, HEIZ & BELÜFT, INFR & ENTW, UMWELT, WERKSTOFF flow

**durchlaufen** *vi* ARCH & TRAGW, BETON, HEIZ & BELÜFT flow through, run through, INFR & ENTW run through

**durchlaufend** *adj* ARCH & TRAGW continuous; **~e Bewehrung** *f* BETON continuous reinforcement; **~er Spannbetonträger** *m* ARCH & TRAGW, BETON, INFR & ENTW continuous prestressed concrete girder, continuous prestressed girder; **~er Sturzbalken** *m* INFR & ENTW lintel beam; **~er Unterzug** *m* ARCH & TRAGW continuous girder

***Durchlauf***: **Durchlauferhitzer** *m* ABWASSER, HEIZ &

BELÜFT flow heater, flow-type calorifier, instantaneous heater; **Durchlaufmischer** *m* BAUMASCHIN continuous mixer; **Durchlaufpfette** *f* HOLZ continuous purlin; **Durchlaufplatte** *f* BETON continuous slab; **Durchlaufrahmen** *m* ARCH & TRAGW continuous frame; **Durchlaufstabarmierung** *f* BETON continuous rod reinforcement; **Durchlaufträger** *m* ARCH & TRAGW continuous beam

**Durchreiche** *f* ARCH & TRAGW hatch

**durchrosten** *vt* STAHL rust through

**Durchschnitt** *m* ARCH & TRAGW, INFR & ENTW, WERKSTOFF average, mean value

**durchschnittlich** *adj* ARCH & TRAGW, INFR & ENTW, WERKSTOFF average; **~e Festigkeit** *f* BETON, WERKSTOFF average strength; **~er Korndurchmesser** *m* BETON average grain diameter; **~e Tagesleistung** *f* UMWELT average daily output; **~e Windgeschwindigkeit** *f* UMWELT average wind speed

***Durchschnitt***: **Durchschnittskosten** *pl* ARCH & TRAGW, INFR & ENTW average cost; **Durchschnittsstrahlungstemperatur** *f* BAURECHT mean radiant temperature; **Durchschnittsverbrauch** *m* ARCH & TRAGW, HEIZ & BELÜFT, INFR & ENTW average consumption; **Durchschnittswert** *m* ARCH & TRAGW mean value; **Durchschnittswindgeschwindigkeit** *f* INFR & ENTW average wind speed

**Durchsicht** *f* ARCH & TRAGW, BAURECHT inspection

**durchsichtig** *adj* ARCH & TRAGW, OBERFLÄCHE, WERKSTOFF transparent

**durchsickern** *vi* ABWASSER, ARCH & TRAGW percolate; ABWASSER leak, INFR & ENTW percolate, pass through, UMWELT percolate, leak

**durchspülen** *vt* ABWASSER flush

**durchstecken** *vt* BAUMASCHIN, BESCHLÄGE, HOLZ, WERKSTOFF perforate

**Durchsteckschraube** *f* BESCHLÄGE bolt and nut

**Durchstich** *m* INFR & ENTW pilot cutting

**Durchstoß** *m* UMWELT puncture; **Durchstoßfestigkeit** *f* UMWELT puncturability, puncture resistance

**durchströmen** *vt* HEIZ & BELÜFT, INFR & ENTW pass through

**Durchtreiber** *m* BAUMASCHIN pin punch

**durchzeichnen** *vt* ARCH & TRAGW caulk

**Duroplast** *m* OBERFLÄCHE thermosetting plastic

**Duschanlage** *f* ABWASSER shower installation

**Dusche** *f* ABWASSER shower

**Duschkopf** *m* ABWASSER shower head

**Duschwanne** *f* ABWASSER shower receptor, shower tray, shower tub

**Düse** *f* HEIZ & BELÜFT, UMWELT nozzle; **Düsenflansch** *m* HEIZ & BELÜFT nozzle flange; **Düsenventil** *nt* HEIZ & BELÜFT jet valve

**dynamisch** *adj* INFR & ENTW dynamic; **~ belasteter Balken** *m* ARCH & TRAGW, INFR & ENTW dynamically loaded beam; **~e Belastung** *f* ARCH & TRAGW, INFR & ENTW dynamic loading; **~e Festigkeit** *f* ARCH & TRAGW, INFR & ENTW dynamic strength; **~es Knicken** *nt* ARCH & TRAGW, INFR & ENTW dynamic buckling; **~e Kraft** *f* INFR & ENTW dynamic force; **~e Last** *f* INFR & ENTW dynamic load; **~e Untersuchung** *f* WERKSTOFF dynamic test; **~e Zähigkeit** *f* WERKSTOFF dynamic viscosity

# E

**E** *abbr* (*Evaporation*) ARCH & TRAGW, HEIZ & BELÜFT, INFR & ENTW, UMWELT, WERKSTOFF E (*evaporation*)

**Ebbe** *f* UMWELT falling tide, low tide; **Ebbekrafterzeugung** *f* UMWELT ebb generation; **Ebbeströmung** *f* UMWELT ebb tide

**eben** *adj* ARCH & TRAGW, INFR & ENTW, WERKSTOFF even, flat, level, plane, smooth; **~es Lager** *nt* ARCH & TRAGW surface bearing

**Ebene** *f* ARCH & TRAGW level, plane, INFR & ENTW level, plain

**ebenerdig** *adj* ARCH & TRAGW, BETON at grade, at ground level, even with the ground

**Ebenheit** *f*: **~ der Straßendecke** INFR & ENTW pavement surface evenness (*BrE*), sidewalk surface evenness (*AmE*)

**EC** *abbr* (*Eurocode*) ARCH & TRAGW, BAURECHT, INFR & ENTW, STAHL, WERKSTOFF EC (*Eurocode*)

**Echinus** *m* BAUMASCHIN ovolo

**echt**: **~e Kohäsion** *f* OBERFLÄCHE, WERKSTOFF true cohesion

**Eck-** *in cpds* ARCH & TRAGW, HEIZ & BELÜFT, INFR & ENTW corner; **Eckabsperrventil** *nt* HEIZ & BELÜFT angle-body valve, angle shut-off valve, right-angled valve; **Eckbalkon** *m* ARCH & TRAGW corner balcony; **Eckbereich** *m* INFR & ENTW corner region

**Ecke** *f* ARCH & TRAGW corner; **an der Ecke** *phr* ARCH & TRAGW at the corner; **Eckenschutzschiene** *f* NATURSTEIN corner bead

***Eck-***: **Eckhaus** *nt* ARCH & TRAGW corner house; **Eckkamin** *m* ARCH & TRAGW corner fireplace; **Eckklemme** *f* ARCH & TRAGW, BESCHLÄGE corner cleat; **Eck- und Knotenpunkte** *m pl* ARCH & TRAGW corner and nodal points; **Eckpfeiler** *m* ARCH & TRAGW corner pillar, jamb stone; **Eckpfosten** *m* ARCH & TRAGW corner post; **Eckprofil** *nt* BESCHLÄGE, NATURSTEIN corner bead; **Eckpunkte** *m pl* **des Raumes** ARCH & TRAGW corners of the room; **Eckschiene** *f* BETON, INFR & ENTW, STAHL, WERKSTOFF angle bead; **Eckschutzleiste** *f* NATURSTEIN corner bead; **Eckstein** *m* BETON, INFR & ENTW, NATURSTEIN, WERKSTOFF *eines Gebäudes* angle-quoin, pillar stone, quoin, cornerstone; **Eckstütze** *f* ARCH & TRAGW corner column; **Eckülberblattung** *f* **mit schrägem Schnitt** HOLZ bevelled halving; **Eckventil** *nt* HEIZ & BELÜFT angle shut-off valve, right-angled valve; **Eckverband** *m* HOLZ corner bench

**eckverschweißt** *adj* STAHL corner welded

***Eck-***: **Eckversteifung** *f* ARCH & TRAGW corner bracing; **Eckwinkel** *m* ARCH & TRAGW corner angle

***EC***: **EC-Übereinstimmungszeichen** *nt* BAURECHT EC conformity mark

**Edel**: **Edelfeuerton** *m* WERKSTOFF high-grade fireclay; **Edelmetall** *nt* UMWELT noble metal; **Edelstahl** *m* WERKSTOFF stainless steel

**effektiv**: **~e Biegesteifigkeit** *f* ARCH & TRAGW, BETON, INFR & ENTW, WERKSTOFF effective lateral stiffness; **~e Knicksteifigkeit** *f* ARCH & TRAGW, BETON, INFR & ENTW, WERKSTOFF effective lateral stiffness; **~er Modul** *m* BETON, INFR & ENTW, WERKSTOFF effective modulus; **~er Schalldruck** *m* DÄMMUNG, UMWELT effective sound pressure

**EGW** *abbr* (*Einwohnergleichwert*) ABWASSER, BAURECHT, UMWELT *Abwasser* population equivalence

**Eiche** *f* HOLZ oak

**eichen** *vt* ARCH & TRAGW, BAUMASCHIN, BAURECHT, BESCHLÄGE, INFR & ENTW, WERKSTOFF adjust, calibrate

***Eiche***: **Eichenholz** *nt* HOLZ oak, oak timber

**Eichung** *f* ARCH & TRAGW, BAURECHT, BAUMASCHIN, BESCHLÄGE, INFR & ENTW, WERKSTOFF adjustment, calibration

**eidlich**: **~e Erklärung** *f* BAURECHT affidavit

**Eigen-** *in cpds* ARCH & TRAGW, BAURECHT, BETON, DÄMMUNG, HEIZ & BELÜFT, INFR & ENTW, UMWELT, WERKSTOFF natural, permanent; **Eigenbau** *m* BAURECHT self build; **Eigenfestigkeit** *f* WERKSTOFF inherent strength; **Eigenfrequenz** *f* DÄMMUNG, HEIZ & BELÜFT, INFR & ENTW natural frequency; **Eigengewicht** *nt* ARCH & TRAGW, INFR & ENTW, WERKSTOFF dead load, dead weight; **Eigenlast** *f* ARCH & TRAGW, INFR & ENTW, WERKSTOFF permanent load, permanent weight; **Eigenlastdurchbiegung** *f* ARCH & TRAGW dead-load deflection; **Eigenmasse** *f* ARCH & TRAGW, INFR & ENTW, WERKSTOFF dead weight, permanent load, permanent weight

**Eigenschaft** *f* WERKSTOFF nature, property

**eigensicher** *adj* ELEKTR inherently safe

***Eigen-***: **Eigenspannung** *f* ARCH & TRAGW, INFR & ENTW internal stress

**Eigentum** *nt* BAURECHT property

**Eigentümer** *m* BAURECHT owner

***Eigentum***: **Eigentumserwerb** *m* ARCH & TRAGW, BAURECHT acquisition of property; **Eigentumsübertragung** *f* BAURECHT conveyancing

**Eignung** *f* INFR & ENTW suitability; **Eignungsnachweis** *m* INFR & ENTW, WERKSTOFF indication of suitability; **Eignungsprüfung** *f* ARCH & TRAGW performance test, *Personal* qualification examination, BAURECHT qualification examination, aptitude test, INFR & ENTW, WERKSTOFF aptitude test; **Eignungstest** *m* ARCH & TRAGW performance test, BAURECHT qualification examination

**Eimer** *m* BESCHLÄGE, INFR & ENTW bucket; **Eimerkettenaufzug** *m* BAUMASCHIN bucket elevator; **Eimerkettenbagger** *m* BAUMASCHIN bucket excavator; **Eimertragring** *m* INFR & ENTW *Schacht* bucket-bearing ring

**einachsig** *adj* ARCH & TRAGW, BAUMASCHIN, ELEKTR uniaxial

**Einarmzapfverbindung** *f* HOLZ housed joint

**Einbau** *m* ARCH & TRAGW, BAUMASCHIN, HEIZ & BELÜFT, HOLZ, INFR & ENTW, STAHL assembly, installation, mounting

**einbauen** *vt* ARCH & TRAGW incorporate, install, mount, build in, embed, fit in, HEIZ & BELÜFT, HOLZ *installieren* assemble, INFR & ENTW embed, NATURSTEIN embed, tail, STAHL *installieren* assemble

**Einbauen** *nt* HEIZ & BELÜFT, HOLZ assembling, NATURSTEIN tailing, STAHL assembling
***Einbau***: **Einbaugarnitur** *f* BESCHLÄGE mounting accessories; **Einbauküche** *f* HOLZ built-in kitchen; **Einbauküchenfittings** *nt pl* ABWASSER, ARCH & TRAGW built-in kitchen fittings; **Einbauleuchte** *f* ELEKTR recessed light fixture
**Einbauort**: **am ~ betoniert** *adj* BETON poured-in-place
***Einbau***: **Einbaurahmen** *m* WERKSTOFF mounting frame; **Einbauschrank** *m* HOLZ built-in cupboard (*BrE*), closet (*AmE*)
**Einbehaltung** *f*: **~ des Rechtsanspruches** BAURECHT retention of title; **Einbehaltungsbetrag** *m* BAURECHT retention fee amount; **Einbehaltungssumme** *f* BAURECHT retention money
**einbetonieren** *vt* BETON cast in, set in concrete
**einbetoniert** *adj* ARCH & TRAGW, BETON, INFR & ENTW, STAHL, WERKSTOFF *Verbundstütze* concrete-encased; **~e Befestigungsschienen** *f pl* BETON, STAHL cast-in fixing rails; **~es Gehäuse** *nt* ARCH & TRAGW, BETON cast-in casings; **~e Rinne** *f* BETON cast-in gutter
**einbetten** *vt* ARCH & TRAGW, INFR & ENTW, NATURSTEIN bed in, embed
**Einbettung** *f* ARCH & TRAGW, INFR & ENTW, NATURSTEIN embedding; **Einbettungsmasse** *f* WERKSTOFF matrix
**Einbindelänge** *f* ARCH & TRAGW, BETON *Spannbeton* bond length
**einbinden** *vt* ARCH & TRAGW incorporate, NATURSTEIN tail
**Einbinden** *nt* ARCH & TRAGW, NATURSTEIN tailing
**Einbindung** *f* ARCH & TRAGW, INFR & ENTW *einer Stütze* embedment, UMWELT *Sondermüll* grain encapsulation
**Einbrenn-** *in cpds* OBERFLÄCHE stoved, stoving
**einbrennen** *vt* OBERFLÄCHE bake
**Einbrennen** *nt* OBERFLÄCHE *Farbe* baking
***Einbrenn-***: **Einbrennlack** *m* OBERFLÄCHE baking varnish
**einbrennlackiert** *adj* OBERFLÄCHE stove-enameled (*AmE*), stove-enamelled (*BrE*)
***Einbrenn-***: **Einbrennlackierung** *f* OBERFLÄCHE stoved enamel finish; **Einbrenntemperatur** *f* OBERFLÄCHE stoving temperature
**Einbringen** *nt* BETON, NATURSTEIN pouring
**Einbringung** *f*: **~ des Betons** BETON placing of concrete
**Einbruchmeldeanlage** *f* BAURECHT, ELEKTR burglar alarm system, intrusion detection system
**einbruchsicher** *adj* ARCH & TRAGW, BAURECHT burglar-proof, intruder-proof
**Eindampfen** *nt*: **~ einer Säure** UMWELT *Aktivität* acid concentration
**Eindeckrahmen** *m* BESCHLÄGE *Dachfenster* covering frame
**Eindring-** *in cpds* OBERFLÄCHE penetration
**eindringen** *vi* ARCH & TRAGW, BAUMASCHIN, INFR & ENTW, OBERFLÄCHE, WERKSTOFF penetrate
**Eindringen** *nt* ARCH & TRAGW, BAUMASCHIN, INFR & ENTW, OBERFLÄCHE, WERKSTOFF penetration
***Eindring-***: **Eindringlastkurve** *f* INFR & ENTW, WERKSTOFF penetration load curve; **Eindringtiefe** *f* INFR & ENTW, WERKSTOFF penetration depth; **Eindringtiefenmesser** *m* BAUMASCHIN penetrometer
**Eindringung**: **Eindringungstiefe** *f* INFR & ENTW, OBERFLÄCHE, WERKSTOFF depth of penetration
***Eindring-***: **Eindringwiderstand** *m* INFR & ENTW, WERKSTOFF penetration resistance
**einebnen** *vt* Arch & TRAGW, INFR & ENTW grade, level, plane, *Boden* level out
**Einebnen** *nt* ARCH & TRAGW, INFR & ENTW leveling (*AmE*), levelling (*BrE*)
**einfach**: **~es Auflager** *nt* ARCH & TRAGW simple support; **~e Auflagerung** *f* ARCH & TRAGW simple support; **~e bituminöse Oberflächenbehandlung** *f* INFR & ENTW single bituminous surface treatment; **~es Hängewerk** *nt* ARCH & TRAGW king post truss; **~es Sparrendach** *nt* ARCH & TRAGW, HOLZ couple roof, single rafter roof; **~ symmetrisch** *adj* ARCH & TRAGW, BETON, INFR & ENTW, STAHL, WERKSTOFF mono-symmetric
**Einfachabzweig** *m* ABWASSER *Rohr* single-branch pipe
**Einfahrt** *f* ARCH & TRAGW, INFR & ENTW access way
**Einfallen** *nt* INFR & ENTW, UMWELT *Geologie* dip; **Einfallenschloß** *nt* BESCHLÄGE single-latch bolt lock
**Einfallwinkel** *m* ARCH & TRAGW, INFR & ENTW, UMWELT angle of incidence
**einfassen** *vt* ARCH & TRAGW, HOLZ border, hem
**Einfaßprofil** *nt* ARCH & TRAGW sectional surround
**Einfassung** *f* ARCH & TRAGW, HOLZ skirting (*BrE*), surround
**Einfeldbrücke** *f* INFR & ENTW simple bridge
**Einfeldträger** *m* ARCH & TRAGW, BETON, STAHL simple beam, simply-supported beam, single-span beam
**einfluchten** *vt* ARCH & TRAGW range in
**Einfluchten** *nt* ARCH & TRAGW running
**einflügelig** *adj* ARCH & TRAGW *Wand* single-leaf; **~e Tür** *f* ARCH & TRAGW single-leaf door, single-wing door
**Einflußlinie** *f* ARCH & TRAGW line of influence
**einflutig**: **~e Pumpe** *f* ABWASSER single-flow pump
**Einfriedung** *f* ARCH & TRAGW, BAURECHT, INFR & ENTW boundary fence, enclosure; **Einfriedungsmauer** *f* ARCH & TRAGW, BAURECHT, INFR & ENTW enclosure wall, fence wall
**einfügen** *vt* INFR & ENTW, WERKSTOFF add
**Einführung** *f* ELEKTR lead-in
**Einfüll-** *in cpds* ABWASSER, BAUMASCHIN, INFR & ENTW inlet; **Einfüllstutzen** *m* ABWASSER, INFR & ENTW inlet connector; **Einfülltrichter** *m* BAUMASCHIN hopper
**Eingang** *m* ARCH & TRAGW, INFR & ENTW entrance; **Eingangmischer** *m* BAUMASCHIN single-pass stabilizer; **Eingangshalle** *f* ARCH & TRAGW porch, vestibule; **Eingangskontrolle** *f* UMWELT *einer Deponie* weigh office
**eingebaut** *adj* ARCH & TRAGW, INFR & ENTW built-in, encastré, fitted; **~e Dachrinne** *f* ABWASSER secret gutter
**eingebettet** *adj* ARCH & TRAGW, INFR & ENTW, NATURSTEIN, WERKSTOFF bedded, embedded
**eingebrannt** *adj* OBERFLÄCHE baked
**eingebunden** *adj* ARCH & TRAGW, NATURSTEIN, UMWELT incorporated
**eingefluchtet** *adj* ARCH & TRAGW ranged in
**eingefügt** *adj* INFR & ENTW, WERKSTOFF added
**eingegossen** *adj* BETON poured in
**eingelassen** *adj* ARCH & TRAGW recessed, INFR & ENTW *Randstein* flush
**eingelegt** *adj* ARCH & TRAGW inserted
**eingepaßt** *adj* ARCH & TRAGW fitted
**eingeschalt** *adj* HOLZ timbered
**eingeschäumt** *adj* ARCH & TRAGW *Fenster* foamed in place

**eingeschlämmt**: **~e Erdmassen** *f pl* INFR & ENTW hydraulic fill
**eingeschnitten**: **~es Walmdach** *nt* ARCH & TRAGW hip and valley roof
**eingeschossig** *adj* ARCH & TRAGW single-floor, single-storey (*BrE*), single-story (*AmE*)
**eingesickert** *adj* ABWASSER, ARCH & TRAGW, INFR & ENTW *Boden* percolated
**eingespannt** *adj* ARCH & TRAGW, BAUMASCHIN, BETON, INFR & ENTW, STAHL restrained; **~er Bogen** *m* ARCH & TRAGW fixed arch, rigid arch; **~e Stufen** *f pl* ARCH & TRAGW *einseitig* hanging steps; **~er Träger** *m* ARCH & TRAGW, HOLZ, INFR & ENTW encastré beam; **~es Trägerende** *nt* ARCH & TRAGW fixed end
**eingespült**: **~e Erdmassen** *f pl* INFR & ENTW hydraulic fill
**eingezapft**: **~er Mauerstein** *m* ARCH & TRAGW tusk
**eingezäunt**: **nicht eingezäunt** *adj* INFR & ENTW unfenced
**eingießen** *vt* BETON, NATURSTEIN pour in
**einhalten** *vt* BAURECHT adhere to
**Einhängefeld** *nt* ARCH & TRAGW *Brücke* suspended span
**Einhängeträger** *m* ARCH & TRAGW suspended beam
**einhauptig**: **~e Schalung** *f* BETON single-side formwork
**Einhausung** *f* ARCH & TRAGW housing
**Einhebelbatterie** *f* ABWASSER single-lever mixer, single-lever mixing valve
**einheimisch**: **~es Holz** *nt* HOLZ domestic timber, domestic wood; **~e Materialien** *nt pl* WERKSTOFF vernacular materials
**Einheit** *f* ARCH & TRAGW unit
**einheitlich** *adj* ARCH & TRAGW, INFR & ENTW uniform
***Einheit***: **Einheitsmasse** *f* ARCH & TRAGW, BETON, WERKSTOFF unit mass; **Einheitspreis** *m* ARCH & TRAGW unit price
**einholen** *vt* BAURECHT catch up, *erbitten* request
**Einholung** *f*: **~ von Angeboten** ARCH & TRAGW request for bids
**einkapseln** *vt* UMWELT seal
**Einkapselung** *f* UMWELT *Deponie* sealing
**Einkaufszentrum** *nt* INFR & ENTW shopping center (*AmE*), shopping centre (*BrE*), shopping complex, shopping mall (*AmE*)
**einkerben** *vt* ARCH & TRAGW, HOLZ score
**Einkerbung** *f* ARCH & TRAGW, HOLZ indentation
**Einklemm-** *in cpds* ABWASSER wafer-type; **Einklemmarmatur** *f* ABWASSER wafer-type valve; **Einklemmventil** *nt* ABWASSER wafer-type valve
**einknicken** *vti* ARCH & TRAGW, INFR & ENTW, WERKSTOFF buckle
**Einkorn-** *in cpds* BETON, WERKSTOFF short-range, single-grained; **Einkornbeton** *m* BETON like-grained concrete, short-range aggregate concrete, single-sized concrete; **Einkornstruktur** *f* WERKSTOFF single-grained structure
**Einlage** *f* ARCH & TRAGW insert, DÄMMUNG sealing strip
**einlagig** *adj* ARCH & TRAGW, INFR & ENTW single-layer, NATURSTEIN single-coat, *Putz* one-coat, OBERFLÄCHE single-coat
**Einlaß** *m* ABWASSER, HEIZ & BELÜFT, INFR & ENTW inlet, intake; **Einlaßdübel** *m* BESCHLÄGE, HOLZ split ring connector
**einlassen** *vt* ARCH & TRAGW recess
***Einlaß***: **Einlaßgrund** *m* OBERFLÄCHE impregnating primer, penetrating stopper, sealer; **Einlaßmöbelschloß** *nt* BESCHLÄGE half-mortice cabinet lock, half-mortise cabinet lock; **Einlaßrohr** *nt* ABWASSER, HEIZ & BELÜFT, INFR & ENTW inlet pipe; **Einlaßzapfen** *m* HOLZ tusk tenon
**Einlauf** *m* ARCH & TRAGW inlet, *Regenrohr* hopper head; **Einlaufrost** *m* BESCHLÄGE, INFR & ENTW intake screen; **Einlaufsieb** *nt* ABWASSER, INFR & ENTW inlet strainer
**einlegen** *vt* ARCH & TRAGW *Dichtungsband* insert, lay in
**Einleimer** *m* HOLZ concealed edge band
**Einleitung** *f* UMWELT *von Abwässern* discharge
**Einleitungen** *f pl* BAURECHT preambles
***Einleitung***: **Einleitungsanlage** *f* UMWELT discharge system
**Einloch-** *in cpds* ABWASSER one-hole; **Einlochbatterie** *f* ABWASSER mixer tap (*BrE*), mixing faucet (*AmE*), one-hole mixer; **Einlochdüsenbrausekopf** *m* ABWASSER one-hole shower head
**Einmannbedienung** *f* BAUMASCHIN one-man operation
**einmauern** *vt* NATURSTEIN fix, wall in, wall up, *Kessel* brick in
**Einmessen** *nt* INFR & ENTW survey
**einölen** *vt* BAUMASCHIN, BESCHLÄGE, OBERFLÄCHE oil
**einpassen** *vt* ARCH & TRAGW, BESCHLÄGE, HEIZ & BELÜFT adjust, fit in
**einpeilen** *vt* ARCH & TRAGW locate
**Einphasen-** *in cpds* ELEKTR single-phase; **Einphasenstrom** *m* ELEKTR single-phase electric current
**einphasig** *adj* ELEKTR single-phase
**Einpreß-** *in cpds* BAUMASCHIN, BETON, INFR & ENTW, NATURSTEIN injection
**einpressen** *vt* BAUMASCHIN, BETON, INFR & ENTW, NATURSTEIN inject
**Einpressen** *nt* BAUMASCHIN grout injection, grouting, injection, BETON injection, INFR & ENTW, NATURSTEIN *von Mörtel* grout injection, grouting, injection
***Einpreß-***: **Einpreßgut** *nt* INFR & ENTW, WERKSTOFF injection material; **Einpreßmörtel** *m* BETON, NATURSTEIN intrusion mortar; **Einpreßpumpe** *f* BAUMASCHIN injection pump; **Einpreßverfahren** *nt* BETON, INFR & ENTW, NATURSTEIN injection method
**Einputzrahmen** *m* ARCH & TRAGW flush-mounting frame
**Einradwalze** *f* BAUMASCHIN one-wheel roller, single roller
**Einrastknopf** *m* BESCHLÄGE lock knob
**einregulieren** *vt* HEIZ & BELÜFT regulate
**Einreiber** *m* BESCHLÄGE casement fastener
**einreihig** *adj* ARCH & TRAGW single-row; **~e Nietüberlappung** *f* STAHL single-riveted lap joint; **~e Nietverbindung** *f* STAHL single-riveted joint
**Einreißfestigkeit** *f* WERKSTOFF initial tearing resistance, tear resistance, tear strength
**einrichten** *vt* ARCH & TRAGW align
**Einrichtung** *f* ARCH & TRAGW, BAUMASCHIN, INFR & ENTW appliance, facility; **Einrichtungen** *f pl* **für Behinderte** ARCH & TRAGW, BAURECHT, INFR & ENTW disabled facilities; **Einrichtungsplan** *m* ARCH & TRAGW *Baustelle* site facilities program (*AmE*), site facilities programme (*BrE*)
**Einrohrsystem** *nt* ABWASSER one-pipe system
**einrüsten** *vt* ARCH & TRAGW scaffold
**Einrüsten** *nt* ARCH & TRAGW scaffolding

**Einsackwaage** *f* WERKSTOFF bagging scale
**einsäen** *vt* INFR & ENTW seed
**Einsammeln** *nt* UMWELT collection; **~ von Altöl** UMWELT collection of waste oil; **~ von Müllsäcken** UMWELT refuse sack collection
**einschalen** *vt* BETON form, shutter
**einschalig** *adj* ARCH & TRAGW *Wand* single-leaf, WERKSTOFF *Dach* nonventilated
**Einschalung** *f* BETON formwork, HOLZ timbering
**einschäumen** *vt* ARCH & TRAGW *Fenster* foam in
**Einscheibensicherheitsglas** *nt* ARCH & TRAGW tempered glass, toughened safety glass, INFR & ENTW prestressed glass, WERKSTOFF heat-treated glass, prestressed glass
**einschichtig** *adj* ARCH & TRAGW, INFR & ENTW single-layer
**einschiebbar** *adj* ARCH & TRAGW extendable, BESCHLÄGE retractable
**Einschienenbahn** *f* INFR & ENTW monorail
**Einschienenhängebahn** *f* INFR & ENTW monorail
**einschlägig** *adj* BAURECHT *Vorschriften* applicable, pertinent
**Einschlagtiefe** *f* ARCH & TRAGW, INFR & ENTW driving depth
**Einschließung** *f* ARCH & TRAGW housing
**einschneiden** *vt* ARCH & TRAGW notch
**Einschnitt** *m* ARCH & TRAGW, HOLZ indentation
**einschnittig**: **~es Gelenk** *nt* ARCH & TRAGW, INFR & ENTW single shear joint; **~e Verbindung** *f* ARCH & TRAGW, INFR & ENTW single shear connection
**einschränken** *vt* BAURECHT restrict
**Einschränkung** *f* BAURECHT limitation, restriction
**einschraubbar** *adj* BAUMASCHIN threaded
**einschrauben** *vt* BESCHLÄGE drive
**Einschraubstutzen** *m* ABWASSER, HEIZ & BELÜFT screw union, threaded bushing, threaded socket
**Einschub** *m* ELEKTR slide-in unit; **Einschubdecke** *f* ARCH & TRAGW, DÄMMUNG inserted ceiling, sound boarding, sound-boarded floor; **Einschubtreppe** *f* ARCH & TRAGW folding ladder, BESCHLÄGE folding ladder, folding staircase
**einschwalben** *vt* HOLZ dovetail
**einschweißbar** *adj* STAHL weldable
**einsehen** *vt* ARCH & TRAGW *Unterlagen* consult, inspect
**einsickern** *vi* ABWASSER, ARCH & TRAGW penetrate, percolate, seep in, INFR & ENTW penetrate, seep in
**Einsickern** *nt* ABWASSER, ARCH & TRAGW, INFR & ENTW percolation, seepage
**Einsinken** *nt* UMWELT subsidence
**Einspann-** *in cpds* ARCH & TRAGW, BAUMASCHIN, INFR & ENTW fixing, mounting, STAHL mounting
**einspannen** *vt* ARCH & TRAGW *Balken, Träger*, BAUMASCHIN, BETON, INFR & ENTW, STAHL fix, restrain
***Einspann-***: **Einspannmoment** *nt* ARCH & TRAGW, INFR & ENTW end moment; **Einspannstelle** *f* ARCH & TRAGW fixing point
**Einspannung** *f* ARCH & TRAGW, BAUMASCHIN, BETON, HEIZ & BELÜFT, HOLZ, INFR & ENTW, STAHL mounting, restraining; **Einspannungsbügel** *m pl* BAUMASCHIN, STAHL restraining straps
***Einspann-***: **Einspannvorrichtung** *f* BAUMASCHIN mounting device
**Einsparung** *f* ARCH & TRAGW saving
**einspeisen** *vt* ABWASSER, ELEKTR feed
**einspeisend**: **~es Netz** *nt* ELEKTR, INFR & ENTW supply mains
**Einspeisung** *f* ABWASSER feed-in, feeder, ELEKTR feed-in, line entry
**Einspritzpumpe** *f* BAUMASCHIN injection pump
**einstampfbar**: **~er Klebstoff** *m* UMWELT repulpable adhesive
**Einsteck-** *in cpds* ABWASSER plug-in; **Einsteckende** *nt* ARCH & TRAGW *einer Muffenrohrverbindung* spigot; **Einsteckfallenschloß** *nt* BESCHLÄGE mortise latch; **Einsteckfeder** *f* HOLZ loose tongue; **Einsteckschloß** *nt* BESCHLÄGE, HOLZ mortise deadlock, mortise lock
**einstellbar** *adj* BAUMASCHIN, BESCHLÄGE, HEIZ & BELÜFT adjustable; **~es Schneidmesser** *nt* BAUMASCHIN *für Furniere* cutting gage (*AmE*), cutting gauge (*BrE*)
**einstellen** *vt* ARCH & TRAGW, BESCHLÄGE, ELEKTR, HEIZ & BELÜFT, INFR & ENTW, WERKSTOFF adjust, calibrate, set up
**Einstellung** *f* ARCH & TRAGW, BESCHLÄGE, ELEKTR, HEIZ & BELÜFT, INFR & ENTW, WERKSTOFF adjustment, calibration, setup
**Einstellvorrichtung** *f* BAUMASCHIN adjustment
**Einstemmband** *nt* BESCHLÄGE butt hinge
**Einstemmen** *nt* HOLZ morticing, mortising
**Einstieg** *m* ARCH & TRAGW, BAURECHT, INFR & ENTW access; **Einstiegsleiter** *f* INFR & ENTW, STAHL, WERKSTOFF access ladder; **Einstiegsluke** *f* ARCH & TRAGW access hatch; **Einstiegsöffnung** *f* ARCH & TRAGW, BAUMASCHIN access opening; **Einstiegsschacht** *m* ABWASSER, INFR & ENTW manhole
**einstielig**: **~es strebenloses Pfettendach** *nt* ARCH & TRAGW, HOLZ king post truss, single-post purlin roof
**Einstufung** *f* INFR & ENTW grading, WERKSTOFF classification, grading
**Einsturz** *m* ARCH & TRAGW, INFR & ENTW, WERKSTOFF collapse
**einstürzen** *vi* ARCH & TRAGW, INFR & ENTW, WERKSTOFF collapse, fall in
***Einsturz***: **Einsturzlast** *f* ARCH & TRAGW, INFR & ENTW, WERKSTOFF collapse load, failure load; **Einsturzschutt** *m* ARCH & TRAGW subsidence rubbish
**Eintauchen** *nt* ABWASSER plunging
**eintourig** *adj* BESCHLÄGE *Schloß* single-turn
**Einträgerlaufkran** *m* BAUMASCHIN single-beam travel crane, *Brückenkran* single-beam bridge crane
**Eintragungslänge** *f* BETON *Spannbeton* transmission length
**eintreiben** *vt* INFR & ENTW *von Pfählen* pile
**Eintritt** *m* ARCH & TRAGW inlet; **Eintrittsöffnung** *f* ARCH & TRAGW throat; **Eintrittstemperatur** *f* UMWELT inlet temperature
**Einwalzen** *nt* INFR & ENTW rolling
**einwandfrei** *adj* ARCH & TRAGW flawless, perfect
**Einweichanlage** *f* BESCHLÄGE, WERKSTOFF macerator
**einweichen** *vt* INFR & ENTW, UMWELT, WERKSTOFF macerate
**Einweisung** *f*: **~ von Personal** BAURECHT instruction of staff
**Einwohner** *m* BAURECHT inhabitant; **Einwohnergleichwert** *m* (*EGW*) ABWASSER, BAURECHT, UMWELT *Abwasser* population equivalence; **Einwohnertreppenhaus** *nt* ARCH & TRAGW, INFR & ENTW accommodation stair
**einzapfen** *vt* HOLZ notch
**Einzäunung** *f* INFR & ENTW fencing
**Einzel-** *in cpds* ARCH & TRAGW, BETON single; **Einzelfundament** *nt* ARCH & TRAGW *Gebäude*, BETON, INFR

& ENTW independent footing, foundation pad, individual footing, foundation block, isolated footing
**Einzelheit** *f* ARCH & TRAGW item
***Einzel-***: **Einzellast** *f* ARCH & TRAGW single load
**einzeln**: **~e Rohrleitung** *f* ABWASSER single duct
**Einziehen** *nt* ARCH & TRAGW, INFR & ENTW tapering
**Einzugsgebiet** *nt* INFR & ENTW, UMWELT catchment area
**Eiprofil** *nt* HEIZ & BELÜFT, INFR & ENTW, WERKSTOFF oval-shaped sewer pipe
**Eis** *nt* INFR & ENTW ice; **Eisbrecher** *m* INFR & ENTW *vor einer Brücke* ice apron
**Eisen** *nt* WERKSTOFF iron
**Eisenbahn** *f* INFR & ENTW railroad (*AmE*), railway (*BrE*); **Eisenbahnbau** *m* INFR & ENTW railroad construction (*AmE*), railway construction (*BrE*); **Eisenbahnbrücke** *f* INFR & ENTW railroad bridge (*AmE*), railway bridge (*BrE*); **Eisenbahntunnel** *m* INFR & ENTW railroad tunnel (*AmE*), railway tunnel (*BrE*); **Eisenbahnunterführung** *f* INFR & ENTW railroad underpass (*AmE*), railway underbridge (*BrE*)
***Eisen***: **Eisenbeschläge** *m pl* BESCHLÄGE ironwork, WERKSTOFF ironware; **Eisenbieger** *m* BETON bar bender; **Eisenblech** *nt* WERKSTOFF sheet iron; **Eisenbrücke** *f* STAHL iron bridge; **Eisenportlandzement** *m* (*EPZ*) WERKSTOFF iron Portland cement, Portland blast-furnace cement; **Eisenquerschnitt** *m* WERKSTOFF iron cross section; **Eisenrohr** *nt* STAHL iron pipe; **Eisenschrott** *m* STAHL junk iron, UMWELT ferrous scrap, scrap iron; **Eisenschrottpresse** *f* UMWELT scrap-baling press; **Eisenträger** *m* ARCH & TRAGW, STAHL iron girder
***Eis***: **Eiskristall** *nt* INFR & ENTW, WERKSTOFF ice crystal; **Eisschicht** *f* INFR & ENTW, WERKSTOFF ice layer; **Eisschutz** *m* INFR & ENTW, WERKSTOFF ice protection
**elastisch** *adj* WERKSTOFF flexible, resilient; **~e Berechnung** *f* ARCH & TRAGW, BETON, INFR & ENTW, STAHL, WERKSTOFF elastic analysis; **~e Eigenschaften** *f pl* INFR & ENTW, WERKSTOFF elastic properties; **~es Equilibrium** *nt* INFR & ENTW, WERKSTOFF elastic equilibrium; **~e Konstante** *f* INFR & ENTW elastic constant; **~e Lagerung** *f* ARCH & TRAGW, BETON, INFR & ENTW, STAHL, WERKSTOFF elastic support; **~e Momententragfähigkeit** *f* ARCH & TRAGW, BETON, INFR & ENTW, STAHL, WERKSTOFF elastic resistance to bending; **~e Querschnitteigenschaften** *f pl* ARCH & TRAGW, BETON, INFR & ENTW, WERKSTOFF elastic section properties; **~e Rückbildung** *f* INFR & ENTW elastic rebound; **~e Theorie** *f* ARCH & TRAGW, BETON, INFR & ENTW, STAHL, WERKSTOFF elastic theory; **~e Tragfähigkeit** *f* ARCH & TRAGW, BETON, INFR & ENTW, STAHL, WERKSTOFF elastic resistance; **~e Unterstützung** *f* ARCH & TRAGW, BETON, INFR & ENTW, STAHL, WERKSTOFF elastic support; **~e Zusammendrückung** *f* INFR & ENTW, WERKSTOFF elastic compression; **elastisch-perfekt-plastisch** *adj* ARCH & TRAGW elastic-perfectly-plastic; **elastisch-plastische Methode** *f* ARCH & TRAGW, BETON, STAHL elastic-plastic method
**Elastizität** *f* ARCH & TRAGW, WERKSTOFF elasticity; **Elastizitätsgesetz** *nt* ARCH & TRAGW elastic law; **Elastizitätsgrenze** *f* ARCH & TRAGW, BETON, STAHL, WERKSTOFF elastic limit, limit of elasticity, yield point; **Elastizitätsmodul** *m* (*E-Modul*) BETON, STAHL, WERKSTOFF coefficient of elasticity, elastic modulus, modulus of elasticity, Young's modulus; **Elastizitätstheorie** *f* ARCH & TRAGW, BETON, INFR & ENTW, STAHL, WERKSTOFF elastic theory
**Elastomer** *nt* ARCH & TRAGW, INFR & ENTW, WERKSTOFF elastomer; **Elastomerlager** *nt* ARCH & TRAGW, INFR & ENTW, WERKSTOFF elastomer support
**Elefantenrüssel** *m* BETON tremie
**elektrisch** *adj* ELEKTR electrical; **~e Bohrmaschine** *f* BAUMASCHIN power drill; **~e Sondierung** *f* ELEKTR, INFR & ENTW electrical sounding; **~er Strom** *m* ELEKTR current
**Elektroinstallationsrohr** *nt* ELEKTR cable conduit
**Elektron** *nt* UMWELT electron; **Elektroneneinfangdetektor** *m* UMWELT electron capture detector
**elektronisch**: **~es Überwachungssystem** *nt* ELEKTR electronic control system
**Elektro-Osmose-Verfahren** *nt* ABWASSER, UMWELT electro-osmotic-method
**elektrostatisch**: **~er Staubabscheider** *m* (*ESA*) UMWELT electrostatic precipitator (*ESP*)
**Element** *nt* ARCH & TRAGW member, INFR & ENTW element; **Elementbauweise** *f* ARCH & TRAGW prefabricated construction method
**Ellipse** *f* ARCH & TRAGW, STAHL ellipse; **Ellipsenbogen** *m* ARCH & TRAGW, STAHL elliptical arch
**elliptisch**: **~es Gewölbe** *nt* ARCH & TRAGW three-centered arch (*AmE*), three-centred arch (*BrE*)
**eloxiert** *adj* OBERFLÄCHE anodized
**Eloxierung** *f* OBERFLÄCHE anodic oxidation, anodization
**Elutionsversuch** *m* UMWELT leaching test
**Emaille** *nt* WERKSTOFF enamel; **Emaillelack** *m* OBERFLÄCHE baking varnish, WERKSTOFF enamel
**Emission** *f* UMWELT emission; **Emissionsdaten** *pl* UMWELT emission data; **Emissionsort** *m* UMWELT emission point; **Emissionsquelle** *f* UMWELT emission source, pollution emitter, *in einem Gebiet* area emission source; **Emissionsstandard** *m* UMWELT emission standard; **Emissionsvermeidung** *f* UMWELT avoidance of emissions; **Emissionsvermögen** *nt* UMWELT emissivity; **Emissionsverzeichnis** *nt* UMWELT emission inventory
**Emittanz** *f* UMWELT emittance
**E-Modul** *m* (*Elastizitätsmodul*) BETON, STAHL, WERKSTOFF coefficient of elasticity, elastic modulus, modulus of elasticity, Young's modulus
**Empfänger** *m* UMWELT receptor; **Empfängerbereich** *m* UMWELT receptor region
**Empfangsgebäude** *nt* ARCH & TRAGW, INFR & ENTW terminal building
**Empfehlung** *f* BAURECHT recommendation
**Empfindlichkeit** *f* BAUMASCHIN sensitivity
**empirisch** *adj* ARCH & TRAGW, INFR & ENTW empirical; **~es Ermittlungsverfahren** *nt* INFR & ENTW trial and error method
**Empore** *f* ARCH & TRAGW gallery
**Emulsion** *f* OBERFLÄCHE emulsion; **Emulsionsfarbe** *f* OBERFLÄCHE emulsion paint; **Emulsionsspalter** *m* UMWELT demulsifier
**End-** *in cpds* ARCH & TRAGW, INFR & ENTW, NATURSTEIN, UMWELT final, ultimate; **Endabdeckung** *f* UMWELT *einer Deponie* final cover; **Endabnahme** *f* ARCH & TRAGW, BAURECHT, INFR & ENTW final inspection; **Endabstand** *m* HOLZ *parallel zu Faser* end distance; **Endauflager** *nt* ARCH & TRAGW, INFR & ENTW end bearing; **Endauflagerung** *f* ARCH & TRAGW, INFR &

ENTW end bearing; **Endausgang** *m* BAURECHT, INFR & ENTW final exit; **Enddruck** *m* ABWASSER final pressure; **Endfeld** *nt* ARCH & TRAGW, BETON, HOLZ, INFR & ENTW, STAHL, WERKSTOFF end span; **Endfestigkeit** *f* BETON ultimate strength

**endgültig**: **~er Entwurf** *m* INFR & ENTW final design

***End-***: **Endlager** *nt* ARCH & TRAGW, INFR & ENTW end bearing support, UMWELT disposal zone; **Endlagerstätte** *f* UMWELT *für radioaktiven Abfall* repository

**Endlagerung** *f* UMWELT *von Müll* final storage, ultimate storage; **~ von Abfällen** UMWELT permanent waste storage; **Endlagerungsstätte** *f* UMWELT disposal zone

***End-***: **Endmoment** *nt* ARCH & TRAGW, INFR & ENTW end moment

**Endosmose** *f* INFR & ENTW end osmosis

**endothermisch**: **~er Effekt** *m* INFR & ENTW, UMWELT endothermic effect

***End-***: **Endpfette** *f* ARCH & TRAGW, INFR & ENTW end rafter; **Endspreize** *f* INFR & ENTW *Grabenverbau* face piece; **Endstandsicherheit** *f* INFR & ENTW final stability; **Endträger** *m* ARCH & TRAGW, BETON, HOLZ, INFR & ENTW, STAHL, WERKSTOFF end span; **Endverankerung** *f* ARCH & TRAGW, INFR & ENTW end anchoring; **Endverankerung** *f* **der Bewehrung** ARCH & TRAGW, INFR & ENTW end-anchored reinforcement; **Endverankerungsbereich** *m* ARCH & TRAGW, INFR & ENTW *Bewehrung* end zone; **Endverschluß** *m* ABWASSER, BESCHLÄGE blank cap, OBERFLÄCHE sealing end; **Endvorspannkraft** *f* BETON *Spannbeton* final pre-stress; **Endvorspannung** *f* BETON *Spannbeton* final pre-stress; **Endwert** *m* ARCH & TRAGW, INFR & ENTW ultimate value; **Endzahlung** *f* BAURECHT, INFR & ENTW final payment

**energetisch**: **~e Verwertung** *f* HEIZ & BELÜFT, UMWELT energy recovery

**Energie** *f* ARCH & TRAGW, INFR & ENTW, UMWELT energy; **~ aus Abfall** UMWELT residue-derived energy; **Energieaufnahme** *f* ARCH & TRAGW, INFR & ENTW energy absorption; **Energieaustausch** *m* HEIZ & BELÜFT, INFR & ENTW, UMWELT energy exchange; **Energiebedarf** *m* BAURECHT, ELEKTR, INFR & ENTW power requirement; **Energiebilanz** *f* ARCH & TRAGW, INFR & ENTW energy balance; **Energieerhaltung** *f* ARCH & TRAGW, HEIZ & BELÜFT, INFR & ENTW, UMWELT energy conservation; **Energiegewinnung** *f* UMWELT energy extraction; **Energiehaushalt** *m* ARCH & TRAGW, INFR & ENTW, UMWELT energy budget; **Energiemusterfaktor** *m* UMWELT energy pattern factor; **Energieressourcen** *f pl* UMWELT energy resources; **Energierückgewinnung** *f* HEIZ & BELÜFT, UMWELT energy recovery; **Energierückgewinnungsfaktor** *m* UMWELT energy recovery factor

**energiesparend**: **~e Technologie** *f* UMWELT energy-saving technology

***Energie***: **Energietechnik** *f* UMWELT energy technology; **Energieverbrauch** *m* ARCH & TRAGW, ELEKTR, HEIZ & BELÜFT, INFR & ENTW, UMWELT energy consumption, power consumption; **Energieverlust** *m* HEIZ & BELÜFT energy loss; **Energievernichtungspfeiler** *m* BETON, INFR & ENTW, STAHL baffle pier; **Energieverschwendung** *f* UMWELT waste of energy; **Energieversorgung** *f* UMWELT energy supply; **Energiewirtschaftlichkeit** *f* HEIZ & BELÜFT, INFR & ENTW, UMWELT energy efficiency

**eng** *adj* ARCH & TRAGW narrow

**Engobeton** *m* BETON coating clay

**engstehend** *adj* ARCH & TRAGW closely spaced

**entbinden** *vt* ARCH & TRAGW, BAURECHT, INFR & ENTW acquit

**Entdröhnungsmittel** *nt* DÄMMUNG, WERKSTOFF antidrumming agent

**enteignen** *vt* BAURECHT expropriate

**Enteignung** *f* BAURECHT expropriation; **Enteignungsbeschluß** *m* BAURECHT expropriating order; **Enteignungsverfahren** *nt* BAURECHT expropriating proceedings

**Enteisenung** *f* UMWELT deferrization

**Entfärbung** *f* OBERFLÄCHE decolorization (*AmE*), decolourization (*BrE*)

**entfeint**: **~er Beton** *m* BETON no-fines concrete

**Entfernung** *f* ARCH & TRAGW *Abbruch* removal, INFR & ENTW distance; **Entfernungsmessen** *nt* **mit einer Kette** INFR & ENTW chaining; **Entfernungsmesser** *m* BAUMASCHIN, INFR & ENTW, WERKSTOFF mileometer, odometer

**Entfettung** *f* ABWASSER degreasing

**entfeuchten** *vt* HEIZ & BELÜFT, WERKSTOFF dehumidify

**Entfeuchter** *m* HEIZ & BELÜFT dehumidifier, air dehumidifier

**entflammbar** *adj* BAURECHT, WERKSTOFF inflammable

**Entgasung** *f* UMWELT degassing, *einer Deponie* degasification

**entgraten** *vt* STAHL debur

**Enthalpie** *f* UMWELT enthalpy

**Enthärteranlage** *f* ABWASSER, INFR & ENTW softening plant

**enthärtet**: **~es Wasser** *nt* ABWASSER, INFR & ENTW softened water

**Enthärtungsanlage** *f* ABWASSER softening unit, INFR & ENTW softening plant

**entionisieren** *vt* ABWASSER, WERKSTOFF de-ionize

**entkalken** *vt* ABWASSER, HEIZ & BELÜFT descale

**Entkalken** *nt* ABWASSER, HEIZ & BELÜFT descaling

**Entkeimer** *m* ABWASSER, HEIZ & BELÜFT disinfectant

**Entlade-** *in cpds* ARCH & TRAGW, BAUMASCHIN, UMWELT unloading; **Entladebunker** *m* UMWELT unloading hopper; **Entladefläche** *f* ARCH & TRAGW unloading area

**entladen** **1.** *adj* BAUMASCHIN unloaded; **2.** *vt* BAUMASCHIN unload

***Entlade-***: **Entladepumpe** *f* BAUMASCHIN unloading pump

**Entladung** *f* ELEKTR discharge; **Entladungslampe** *f* ELEKTR discharge lamp

**entlanggleiten** *vi* INFR & ENTW ride

**entlasten** *vt* ARCH & TRAGW, BAURECHT, INFR & ENTW acquit, unload

**Entlasten** *nt* ARCH & TRAGW, BAURECHT, INFR & ENTW acquittal, removing, unloading

**entlastet** *adj* ARCH & TRAGW, BAURECHT, INFR & ENTW acquitted, unloaded

**Entlastung** *f* ARCH & TRAGW, BAURECHT, INFR & ENTW, WERKSTOFF discharge, relief; **Entlastungsbogen** *m* ARCH & TRAGW *Mauerwerk* discharging arch, BAURECHT safety arch; **Entlastungskurve** *f* WERKSTOFF unloading curve; **Entlastungsstraße** *f* ARCH & TRAGW, BAURECHT, INFR & ENTW auxiliary lane; **Entlastungsventil** *nt* UMWELT relief valve

**entleeren** *vt* ABWASSER, HEIZ & BELÜFT drain, exhaust

**entleert** *adj* ABWASSER, HEIZ & BELÜFT drained, exhausted
**entlüften** *vt* ABWASSER, BESCHLÄGE, HEIZ & BELÜFT *Rohre* vent
**Entlüften** *nt* ABWASSER, BESCHLÄGE, HEIZ & BELÜFT venting
**Entlüfter** *m* ABWASSER, BESCHLÄGE, HEIZ & BELÜFT exhaust fan, exhauster, extract ventilation unit, vent; **Entlüfterstutzen** *m* BESCHLÄGE vent plug
**entlüftet** *adj* ARCH & TRAGW, BESCHLÄGE, HEIZ & BELÜFT air-free
**Entlüftung** *f* HEIZ & BELÜFT airing, ventilation; **Entlüftungsfirstkappe** *f* HEIZ & BELÜFT ridge ventilation cap; **Entlüftungsklappe** *f* BESCHLÄGE vent cap; **Entlüftungsloch** *nt* HEIZ & BELÜFT vent hole; **Entlüftungsrohr** *nt* BESCHLÄGE vent pipe, HEIZ & BELÜFT air chimney, air flue, *Lüftungstechnik* exhaust pipe, UMWELT exhaust pipe; **Entlüftungsschacht** *m* HEIZ & BELÜFT extract ventilation shaft, uptake; **Entlüftungsstutzen** *m* HEIZ & BELÜFT vent connection; **Entlüftungsventil** *nt* HEIZ & BELÜFT vent valve; **Entlüftungsziegel** *m* NATURSTEIN outlet vent tile
**Entnahmegerät** *nt* BAUMASCHIN, HEIZ & BELÜFT, UMWELT extractor
**Entnahmestelle** *f* ABWASSER tapping point, tap
**entnehmen** *vt* ABWASSER, INFR & ENTW, WERKSTOFF extract
**entnommen** *adj* ABWASSER, INFR & ENTW, WERKSTOFF extracted
**Entölen** *nt* UMWELT oil removal, oil separation
**Entropie** *f* UMWELT entropy
**entrosten** *vt* OBERFLÄCHE derust
**entrostet** *adj* OBERFLÄCHE derusted
**Entsalzung** *f* ABWASSER, UMWELT demineralization; **Entsalzungsgerät** *nt* ABWASSER, UMWELT demineralizer; **Entsalzungsverfahren** *nt* UMWELT *Abfall* demineralization technique
**entschalen** *vt* BETON strike, strip
**Entschalen** *nt* BETON form removal, form stripping, shuttering removal
**entschalt** *adj* BETON stripped
**Entschrottung** *f* UMWELT scrap metal separation
**Entschuppungshammer** *m* BAUMASCHIN scaling hammer
**entschwefeln** *vt* UMWELT desulfurize (*AmE*), desulphurize (*BrE*)
**entschwefelt** *adj* UMWELT desulfurized (*AmE*), desulphurized (*BrE*)
**Entschwefelung** *f* UMWELT desulfurization (*AmE*), desulphurization (*BrE*)
**entseuchen** *vt* BAURECHT, INFR & ENTW, UMWELT, WERKSTOFF *Boden* decontaminate
**entseucht** *adj* BAURECHT, INFR & ENTW, UMWELT, WERKSTOFF *Boden* decontaminated
**Entsorgung** *f* BAURECHT, INFR & ENTW, UMWELT disposal; **Entsorgungsleitung** *f* INFR & ENTW disposal line; **Entsorgungslogistik** *f* UMWELT logistics of disposal; **Entsorgungsweg** *m* UMWELT disposal route
**Entspanner** *m* OBERFLÄCHE antistress agent
**Entspannung** *f* BETON relaxation; **Entspannungsmittel** *nt* OBERFLÄCHE antistress agent
**entwässern** *vt* ABWASSER, INFR & ENTW drain
**entwässert**: **~er Abfall** *m* UMWELT dewatered waste; **~er Schlamm** *m* UMWELT dewatered sludge
**Entwässerung** *f* HEIZ & BELÜFT dehydration, INFR & ENTW dehydration, drainage, dewatering, UMWELT dewatering; **Entwässerungsgraben** *m* ABWASSER drain, INFR & ENTW drainage trench, open drain; **Entwässerungshahn** *m* ABWASSER drip cock; **Entwässerungskanal** *m* INFR & ENTW drainage duct, UMWELT dike; **Entwässerungskanäle** *m pl* INFR & ENTW drainage ways; **Entwässerungsleitung** *f* INFR & ENTW drainage line; **Entwässerungsleitungen** *f pl* INFR & ENTW drainage ways; **Entwässerungs- und Lüftungsrohr** *nt* ABWASSER soil and vent pipe; **Entwässerungsöffnung** *f* NATURSTEIN weep hole; **Entwässerungsrinne** *f* INFR & ENTW drainage channel, drainage gutter; **Entwässerungsrohr** *nt* ABWASSER drain, NATURSTEIN weep hole; **Entwässerungsschicht** *f* INFR & ENTW, UMWELT *einer Deponie* drainage layer
**entwerfen** *vt* ARCH & TRAGW, INFR & ENTW design, plan
**Entwurf** *m* ARCH & TRAGW model, *Plan* design, INFR & ENTW *Plan*, WERKSTOFF *Entwerfen* design; **Entwurfsdaten** *nt pl* ARCH & TRAGW design data; **Entwurfstragfähigkeit** *f* ARCH & TRAGW, BETON, INFR & ENTW, WERKSTOFF design resistance; **Entwurfswiderstand** *m* ARCH & TRAGW, BETON, INFR & ENTW, WERKSTOFF design resistance; **Entwurfszeichnung** *f* ARCH & TRAGW design drawing
**entziehen** *vt* INFR & ENTW, WERKSTOFF extract
**Entziehung** *f* INFR & ENTW, WERKSTOFF extraction
**entzinnen** *vt* UMWELT detin
**Entzinnung** *f* UMWELT detinning
**entzündbar** *adj* BAURECHT, WERKSTOFF inflammable
**entzundern** *vt* OBERFLÄCHE descale
**Entzundern** *nt* OBERFLÄCHE descaling
**Entzunderung** *f* OBERFLÄCHE scaling off
**entzündlich** *adj* BAURECHT, WERKSTOFF inflammable
**ENV** *abbr* (*Europäische Vornorm*) ARCH & TRAGW, INFR & ENTW, WERKSTOFF ENV (*European initial standard, European tentative standard*)
**EOTA** *abbr* (*Europäische Organisation für Technische Zulassung*) BAURECHT EOTA (*European Organisation for Technical Approvals*)
**EOTC** *abbr* (*Europäische Organisation für Prüfung und Normung*) BAURECHT EOTC (*European Organisation for Testing and Certification*)
**Epizentrum** *nt* INFR & ENTW epicenter (*AmE*), epicentre (*BrE*)
**EPM** *abbr* (*Äquivalent je Million*) UMWELT EPM (*equivalent per million*)
**EPZ** *abbr* (*Eisenportlandzement*) WERKSTOFF iron Portland cement, Portland blast-furnace cement
**Erbau** *m* BAURECHT building, constructing
**Erbauer** *m* ARCH & TRAGW builder
***Erbbau***: **Erbbauberechtigter** *m* BAURECHT lessee; **Erbbaurecht** *nt* BAURECHT lease
**Erd-** *in cpds* ARCH & TRAGW earth, ground, soil, ELEKTR earth (*BrE*), ground (*AmE*), INFR & ENTW, STAHL earth, ground, soil; **Erdableitwiderstand** *m* ELEKTR earthing resistance (*BrE*), grounding resistance (*AmE*); **Erdandeckung** *f* INFR & ENTW soil cover; **Erdanker** *m* ARCH & TRAGW, INFR & ENTW, STAHL deadman, ground anchor; **Erdanziehungskraft** *f* INFR & ENTW gravitation; **Erdaufschüttung** *f* INFR & ENTW earth fill
**Erdbau** *m* ARCH & TRAGW, INFR & ENTW earthwork, soil engineering; **Erdbauarbeiten** *f pl* ARCH & TRAGW, INFR & ENTW earthwork; **Erdbaumaschinen** *f pl* BAUMASCHIN earthworking machinery

**Erdbeben** *nt* INFR & ENTW earthquake, seism; **Erdbebengebiet** *nt* INFR & ENTW seismic area; **Erdbebenherd** *m* INFR & ENTW seismic focus; **Erdbebenmesser** *m* BAUMASCHIN seismometer, INFR & ENTW seismograph; **Erdbebenmessung** *f* INFR & ENTW earthquake measurement

**erdbebensicher** *adj* INFR & ENTW earthquake-proof; **~es Bauen** *nt* ARCH & TRAGW, INFR & ENTW seismic construction; **~e Bemessung** *f* INFR & ENTW seismic design

***Erd-***: **Erdbecken** *nt* INFR & ENTW earth reservoir

**Erdbewegung** *f* ARCH & TRAGW, INFR & ENTW earth movement; **Erdbewegungsanlage** *f* BAUMASCHIN earth moving plant

**Erdboden** *m* INFR & ENTW earth, ground, soil; **auf dem Erdboden** *phr* ARCH & TRAGW on grade

***Erd-***: **Erdbohrer** *m* BAUMASCHIN, INFR & ENTW auger, earth auger, ground auger; **Erdböschung** *f* INFR & ENTW earth slope; **Erddamm** *m* INFR & ENTW earth dam; **Erddraht** *m* ELEKTR earth lead (*BrE*), ground lead (*AmE*)

**Erddruck** *m* ARCH & TRAGW, INFR & ENTW foundation pressure; **Erddruckdiagramm** *nt* INFR & ENTW earth pressure diagram; **Erddruckkoeffizient** *m* INFR & ENTW earth pressure coefficient; **Erddruckmeßdose** *f* INFR & ENTW earth pressure cell

**Erde** *f* ARCH & TRAGW earth, ground, soil, ELEKTR earth (*BrE*), ground (*AmE*), INFR & ENTW, STAHL earth, ground, soil; **unter der Erde** *phr* ABWASSER, ARCH & TRAGW, ELEKTR, HEIZ & BELÜFT, INFR & ENTW below ground, underground

***Erd-***: **Erdeinsturz** *m* INFR & ENTW fall of earth

**erden** *vt* ELEKTR earth (*BrE*), ground (*AmE*)

***Erd-***: **Erdgas** *nt* INFR & ENTW, UMWELT natural gas

**erdgelagert**: **~e Betonbodenplatten** *f pl* BETON ground-supported concrete floor slabs; **~e Ortbetonböden** *m pl* BETON ground-supported in-situ concrete floors

**Erdgeschoß** *nt* ARCH & TRAGW first floor (*AmE*), ground floor (*BrE*)

**erdgeschossig**: **~er Wohnungsbau** *m* ARCH & TRAGW, INFR & ENTW low-rise housing

***Erdgeschoß***: **Erdgeschoßplan** *m* ARCH & TRAGW first-floor plan (*AmE*), ground-floor plan (*BrE*)

***Erd-***: **Erdkabel** *nt* INFR & ENTW underground cable; **Erdkern** *m* UMWELT earth's core; **Erdklumpen** *m* WERKSTOFF clod; **Erdkonstruktion** *f* INFR & ENTW earth construction; **Erdkörper** *m* INFR & ENTW earth body; **Erdkruste** *f* UMWELT earth's crust

**Erdöl** *nt* INFR & ENTW, UMWELT petroleum; **Erdölerzeugnis** *nt* INFR & ENTW, UMWELT petroleum product

***Erd-***: **Erdpech** *nt* INFR & ENTW, NATURSTEIN mineral pitch; **Erdplanum** *nt* INFR & ENTW finished grade; **Erdreich** *nt* INFR & ENTW earth, ground, soil; **Erdruhedruck** *m* INFR & ENTW earth pressure at rest; **Erdstampfer** *m* BAUMASCHIN earth rammer; **Erdstützmauer** *f* INFR & ENTW earth-retaining wall; **Erdtank** *m* ABWASSER, HEIZ & BELÜFT buried tank, underground storage tank, underground tank; **Erdteer** *m* INFR & ENTW mineral tar

**Erdung** *f* ELEKTR earth (*BrE*), earthing (*BrE*), ground (*AmE*), grounding (*AmE*); **Erdungsanlage** *f* ELEKTR earthing system (*BrE*), grounding system (*AmE*); **Erdungsbox** *f* ELEKTR earth box (*BrE*), ground box (*AmE*); **Erdungsbuchse** *f* ELEKTR threaded earthing sleeve (*BrE*), threaded grounding sleeve (*AmE*); **Erdungsfeld** *nt* ELEKTR earthing field (*BrE*), grounding field (*AmE*); **Erdungsleitung** *f* ELEKTR earth conductor (*BrE*), ground conductor (*AmE*); **Erdungsmeßgerät** *nt* ELEKTR earth tester (*BrE*), ground tester (*AmE*); **Erdungsmeßschacht** *m* ELEKTR earthing well (*BrE*), grounding well (*AmE*); **Erdungsplatte** *f* ELEKTR earthing plate (*BrE*), grounding plate (*AmE*), earth plate (*BrE*), ground plate (*AmE*); **Erdungsrohrschelle** *f* ELEKTR earthing pipe clamp (*BrE*), grounding pipe clamp (*AmE*); **Erdungsschiene** *f* ELEKTR earth busbar (*BrE*), ground busbar (*AmE*); **Erdungswiderstand** *m* ELEKTR earthing resistance (*BrE*), grounding resistance (*AmE*)

***Erd-***: **Erdverkabelung** *f* INFR & ENTW underground cabling

**erdverlegt** *adj* ABWASSER, ELEKTR, HEIZ & BELÜFT, INFR & ENTW buried, underground

**Erfahrungswert** *m* INFR & ENTW empirical value

**erforderlich**: **~e Gelenkrotation** *f* ARCH & TRAGW, BETON, INFR & ENTW required hinge rotation

**erfüllen** *vt* ARCH & TRAGW *Anforderungen* meet

**ergänzen** *vt* ARCH & TRAGW, BAURECHT, INFR & ENTW, WERKSTOFF add to

**ergänzend** *adj* ARCH & TRAGW, BAURECHT, INFR & ENTW additional

**Ergänzung** *f* BAURECHT addition; **Ergänzungsbauten** *m pl* ARCH & TRAGW supplementary structures; **Ergänzungsbestimmung** *f* BAURECHT, INFR & ENTW supplementary regulation; **Ergänzungswinkel** *m* ARCH & TRAGW complementary angle

**ergeben** *vt* ARCH & TRAGW, BESCHLÄGE *Summe* amount to, yield

**Ergebnis** *nt* ARCH & TRAGW, BESCHLÄGE amount, yield

**Ergiebigkeit** *f* INFR & ENTW, WERKSTOFF coverage

**ergonomisch**: **~es Design** *nt* ARCH & TRAGW, HEIZ & BELÜFT, INFR & ENTW ergonomic design

**Ergußgestein** *nt* INFR & ENTW, UMWELT igneous rocks

**erhaben** *adj* ARCH & TRAGW, OBERFLÄCHE raised

**erhältlich** *adj* ARCH & TRAGW, INFR & ENTW, WERKSTOFF available

**Erhaltung** *f* HOLZ preservation

**erhärten** *vt* BETON, WERKSTOFF *Mörtel, Gips* harden, set

**Erhärten** *nt* BETON, NATURSTEIN, WERKSTOFF hardening, setting

**Erhärtungsbeschleuniger** *m* BETON, WERKSTOFF hardening accelerator

**erheben** *adj* INFR & ENTW raised

**Erhebungswinkel** *m*: **~ der Sonne** UMWELT solar altitude angle

**erhitzen** *vt* ABWASSER heat

**erhöhen** *vt* ARCH & TRAGW heighten, raise

**Erhöhen** *nt* ARCH & TRAGW heightening, raising

**Erhöhung** *f* ARCH & TRAGW prominence, *Kosten* increase

**Erholungszentrum** *nt* ARCH & TRAGW, INFR & ENTW recreation center (*AmE*), recreation centre (*BrE*)

**Erker** *m* ARCH & TRAGW oriel; **Erkerfenster** *nt* ARCH & TRAGW bay window; **Erkerturm** *m* ARCH & TRAGW turret

**Erkundigung** *f* BAURECHT inquiry

**Erlaubnis** *f* ARCH & TRAGW, BAURECHT approval

**Ermüdung** *f* WERKSTOFF fatigue; **Ermüdungserscheinungen** *f pl* WERKSTOFF fatigue;

**Ermüdungsgrenze** *f* INFR & ENTW *Wechselbeanspruchung*, WERKSTOFF fatigue limit; **Ermüdungsverhalten** *nt* INFR & ENTW, WERKSTOFF fatigue behavior (*AmE*), fatigue behaviour (*BrE*); **Ermüdungsversagen** *nt* INFR & ENTW, WERKSTOFF fatigue failure

**erneuern** *vt* ARCH & TRAGW, OBERFLÄCHE renew, renovate

**Erneuerung** *f* ARCH & TRAGW, OBERFLÄCHE renewal, renovation; **Erneuerungsanstrich** *m* OBERFLÄCHE repainting

**Erodierbarkeit** *f* ARCH & TRAGW, INFR & ENTW, UMWELT, WERKSTOFF erodibility

**erodieren** *vt* ARCH & TRAGW, INFR & ENTW, UMWELT, WERKSTOFF erode

**Erosion** *f* ARCH & TRAGW, INFR & ENTW, UMWELT erosion; **Erosionsgefahr** *f* INFR & ENTW, UMWELT risk of erosion; **Erosionsgeschwindigkeit** *f* INFR & ENTW, WERKSTOFF rate of erosion; **Erosionsschutz** *m* INFR & ENTW erosion protection; **Erosionsvermögen** *nt* INFR & ENTW, UMWELT, WERKSTOFF erosive capacity

**erratisch**: **~e Bodenverhältnisse** *nt pl* INFR & ENTW erratic subsoil; **~e Struktur** *f* INFR & ENTW erratic structure

**errechnen** *vt* ARCH & TRAGW calculate

**Erreger** *m* UMWELT exciter

**errichten** *vt* ARCH & TRAGW, INFR & ENTW build, construct, put up

**Errichtung** *f* ARCH & TRAGW, INFR & ENTW erection

**Ersatz** *m* INFR & ENTW replacement; **Ersatzbalkenmethode** *f* ARCH & TRAGW, INFR & ENTW equivalent-beam method; **Ersatzheizungsanlage** *f* HEIZ & BELÜFT backup heating system; **Ersatzkraft** *f* ARCH & TRAGW, INFR & ENTW equivalent force

**Erschaffung** *f* ARCH & TRAGW creation

**Erscheinung** *f* ARCH & TRAGW, INFR & ENTW *Äußeres* aspect

**Erschließung** *f* INFR & ENTW site development; **~ von Baugelände** BAURECHT, INFR & ENTW land development; **Erschließungskosten** *pl* BAURECHT land development fees

**erschlossen**: **~es Bauland** *nt* ARCH & TRAGW, INFR & ENTW developed building land

**Erschwernis** *f* ARCH & TRAGW impediment; **Erschwerniszulage** *f* ARCH & TRAGW, BAURECHT, BESCHLÄGE allowance for extra-heavy work

**erstarren** *vi* UMWELT set, WERKSTOFF *Mörtel, Beton, Gips* set, solidify

**Erstarren** *nt* WERKSTOFF solidification

**Erstarrung** *f* UMWELT setting; **Erstarrungsbeginn** *m* BETON initial set; **Erstarrungsgeschwindigkeit** *f* WERKSTOFF rate of curing; **Erstarrungsgestein** *nt* INFR & ENTW igneous rock

**erste**: **~ Etage** *f* ARCH & TRAGW *über dem Erdgeschoß* first floor (*BrE*), second floor (*AmE*)

**Ersteinregulierung** *f* HEIZ & BELÜFT first adjustment

**Erstellungskosten** *pl* ARCH & TRAGW, BAURECHT, INFR & ENTW building costs

**erste**: **~r Spatenstich** *m* ARCH & TRAGW groundbreaking; **~s Stockwerk** *nt* ARCH & TRAGW *über dem Erdgeschoß* first floor (*BrE*), second floor (*AmE*); **~ und zweite Oberflächenbehandlungen** *f pl* OBERFLÄCHE, WERKSTOFF first and second finishes

**erstrecken**: **sich erstrecken** *v refl* ARCH & TRAGW extend

**Ertrag** *m* UMWELT yield

**erwärmen** *vt* ABWASSER *Wasser* heat

**erwartet**: **~e Baukosten** *f pl* ARCH & TRAGW expected construction costs

**erweiterbar** *adj* ARCH & TRAGW extendable

**erweitern** *vt* ARCH & TRAGW extend

**Erweiterung** *f* ARCH & TRAGW addition, annex (*AmE*), annexe (*BrE*), extension, BAURECHT addition, INFR & ENTW addition, annex (*AmE*), annexe (*BrE*), extension; **Erweiterungsarbeiten** *f pl* ARCH & TRAGW expansion work; **Erweiterungsbau** *m* ARCH & TRAGW, INFR & ENTW annex (*AmE*), annexe (*BrE*), extension

**Erwerb** *m* ARCH & TRAGW, BAURECHT acquisition

**erzeugen** *vt* BETON produce, ELEKTR, HEIZ & BELÜFT generate, INFR & ENTW, STAHL produce, WERKSTOFF manufacture, prepare

**Erzeuger** *m* BAUMASCHIN manufacturer

**Erzeugnis** *nt* WERKSTOFF product

**Erzeugung** *f* ELEKTR, HEIZ & BELÜFT generation, WERKSTOFF manufacture

**ESA** *abbr* (*elektrostatischer Staubabscheider*) UMWELT ESP (*electrostatic precipitator*)

**Eschefurnier** *nt* HOLZ ash veneer

**Esse** *f* ARCH & TRAGW chimney

**Essenkopf** *m* ARCH & TRAGW chimney head

**Estrich** *m* NATURSTEIN screed topping; **Estrichförderer** *m* BAUMASCHIN floor screed conveyor; **Estrichstärke** *f* ARCH & TRAGW screed height

**E-Stück** *nt* HEIZ & BELÜFT flanged socket

**Etage** *f* ARCH & TRAGW floor, storey (*BrE*), story (*AmE*); **Etagenbogen** *m* ABWASSER swan neck; **Etagenofen** *m* UMWELT multiple-hearth incinerator

**Eternit**® *nt* WERKSTOFF asbestos cement, eternit®

**Euler**: **~ Fälle** *m pl* ARCH & TRAGW, BETON, HOLZ, INFR & ENTW, STAHL, WERKSTOFF Euler's cases; **~sche Knickformel** *f* ARCH & TRAGW, BETON, HOLZ, INFR & ENTW, STAHL, WERKSTOFF Euler's formula; **~sche Knickspannung** *f* ARCH & TRAGW, BETON, HOLZ, INFR & ENTW, STAHL, WERKSTOFF Euler's critical tension

**Eurocode** *m* (*EC*) ARCH & TRAGW, BAURECHT, INFR & ENTW, STAHL, WERKSTOFF Eurocode (*EC*)

**Euronorm** *f* ARCH & TRAGW, BAURECHT, INFR & ENTW Euronorm

**Europäisch**: **~er Ausschuß** *m* **für Elektrizitätsnormung** (*CENELEC*) BAURECHT, ELEKTR European Body of Electrical Standards (*CENELEC*); **~es Komitee** *nt* **für Normung** (*CEN*) ARCH & TRAGW, BAURECHT, INFR & ENTW European Committee for Standardisation (*CEN*); **~e Organisation** *f* **für Prüfung und Normung** (*EOTC*) BAURECHT European Organisation for Testing and Certification (*EOTC*); **~e Organisation** *f* **für Technische Zulassung** (*EOTA*) BAURECHT European Organisation for Technical Approvals (*EOTA*); **~e Technische Vorschriften** *f pl* ARCH & TRAGW, BAURECHT, INFR & ENTW European Technical Specifications; **~e Vornorm** *f* (*ENV*) ARCH & TRAGW, BAURECHT, INFR & ENTW, WERKSTOFF European initial standard (*ENV*), European tentative standard (*ENV*); **~e Vorschrift** *f* ARCH & TRAGW, BAURECHT, INFR & ENTW European standard

**Eurostar**® *m* INFR & ENTW Eurostar®

**eustatisch**: **~e Bewegung** *f* INFR & ENTW eustatic movement

**Eutrophierung** *f* UMWELT eutrophication

**evakuieren** *vt* BAURECHT evacuate
**evakuiert**: **~er Rohrkollektor** *m* HEIZ & BELÜFT evacuated tube collector
**Evakuierungsaufzug** *m* BAURECHT, INFR & ENTW evacuation elevator (*AmE*), evacuation lift (*BrE*)
**Evaporation** *f* (*E*) ARCH & TRAGW, HEIZ & BELÜFT, INFR & ENTW, UMWELT, WERKSTOFF evaporation (*E*)
**evaporieren** *vi* ARCH & TRAGW, HEIZ & BELÜFT, INFR & ENTW, UMWELT, WERKSTOFF evaporate
**Eventualposition** *f* ARCH & TRAGW contingency item
**exakt** *adj* ARCH & TRAGW, BAUMASCHIN, BAURECHT, BETON, INFR & ENTW, STAHL, WERKSTOFF accurate, exact
**ex-geschützt** *adj* (*explosionsgeschützt*) ARCH & TRAGW, ELEKTR, INFR & ENTW, WERKSTOFF explosion-proof
**expandieren** *vi* WERKSTOFF expand
**expandiert** *adj* WERKSTOFF expanded; **~er Kork** *m* WERKSTOFF expanded cork
**Experiment** *nt* ARCH & TRAGW, INFR & ENTW, WERKSTOFF experiment
**Experimentalbau** *m* ARCH & TRAGW, INFR & ENTW experimental building
**experimentell** *adj* ARCH & TRAGW, INFR & ENTW, WERKSTOFF experimental; **~e Methode** *f* ARCH & TRAGW, INFR & ENTW, WERKSTOFF experimental method; **~e Untersuchung** *f* ARCH & TRAGW, INFR & ENTW, WERKSTOFF experimental investigation
**experimentieren** *vi* ARCH & TRAGW, INFR & ENTW, WERKSTOFF experiment
**Exploitation** *f* INFR & ENTW, UMWELT exploitation
**Explosion** *f* INFR & ENTW, WERKSTOFF explosion; **Explosionsdruck** *m* ARCH & TRAGW, HEIZ & BELÜFT, INFR & ENTW, WERKSTOFF explosion pressure
**explosionsgefährdet**: **~er Betriebsraum** *m* BAURECHT hazardous location
**explosionsgeschützt** *adj* (*ex-geschützt*) ARCH & TRAGW, ELEKTR, INFR & ENTW, WERKSTOFF explosion-proof
***Explosion***: **Explosionsramme** *f* BAUMASCHIN combustion rammer, internal combustion rammer
**explosionssicher** *adj* ARCH & TRAGW, ELEKTR, INFR & ENTW, WERKSTOFF explosion-proof; **~e Ausführung** *f* ARCH & TRAGW, ELEKTR, INFR & ENTW, WERKSTOFF explosion-proof design
***Explosion***: **Explosionsstampfer** *m* BAUMASCHIN frog rammer; **Explosionswelle** *f* INFR & ENTW, WERKSTOFF explosion wave
**Extrakosten** *pl* ARCH & TRAGW, BAURECHT surcharge
**extrudieren** *vt* WERKSTOFF *Metall, Kunststoff* extrude
**extrudiert** *adj* INFR & ENTW, STAHL, WERKSTOFF *Metall, Kunststoff* extruded; **~er Hartschaum** *m* WERKSTOFF extruded rigid foam
**Extrusion** *f* INFR & ENTW, STAHL, WERKSTOFF extrusion; **~ im Sekundenbruchteil** WERKSTOFF split-section extrusion
**exzentrisch**: **~e Last** *f* ARCH & TRAGW, HOLZ eccentric load

# F

**Fabrik** *f* INFR & ENTW factory
**fabrikfertig** *adj* WERKSTOFF factory-built
**fabrizieren** *vt* WERKSTOFF fabricate
**fabriziert** *adj* WERKSTOFF fabricated
**Fach** *nt* ARCH & TRAGW bay, compartment, *Decke* coffer, panel; **Facharbeiter** *m* ARCH & TRAGW skilled worker; **Fachboden** *m* HOLZ *Einlegeboden für Schrank* shelf
**Fächer** *m* ARCH & TRAGW, BESCHLÄGE, HEIZ & BELÜFT fan; **Fächergewölbe** *nt* ARCH & TRAGW *Kirchenbau* fan vault; **Fächermethode** *f* UMWELT cell method
**fachgerecht** *adj* BAURECHT workmanlike
***Fach***: **Fachliteratur** *f* ARCH & TRAGW, INFR & ENTW specialist literature; **Fachmann** *m* ARCH & TRAGW specialist
**fachmännisch**: **~e Dienstleistung** *f* BAURECHT specialist service
***Fach***: **Fachpersonal** *nt* ARCH & TRAGW technical staff
**Fachwerk** *nt* ARCH & TRAGW, HOLZ, INFR & ENTW, WERKSTOFF framework, lattice, roof truss, truss; **mit ~ versehen** *phr* HOLZ timber
***Fachwerk***: **Fachwerkbinder** *m* ARCH & TRAGW, HOLZ open-web girder, roof frame, trussed girder; **Fachwerkbinderdach** *nt* HOLZ trussed roof; **Fachwerkbinderdachstuhl** *m* NATURSTEIN trussed rafter roof; **Fachwerkbrücke** *f* HOLZ truss bridge, INFR & ENTW frame bridge; **Fachwerkdach** *nt* ARCH & TRAGW framed roof; **Fachwerkhaus** *nt* ARCH & TRAGW frame house, half-timbered house; **Fachwerkholzträger** *m* ARCH & TRAGW, HOLZ trussed wooden beam; **Fachwerkschwelle** *f* ARCH & TRAGW sill of framework; **Fachwerkstab** *m* ARCH & TRAGW frame member; **Fachwerkstrebe** *f* ARCH & TRAGW, HOLZ brace; **Fachwerkträger** *m* ARCH & TRAGW, HOLZ trussed beam, trussed girder, lattice girder; **Fachwerkwand** *f* ARCH & TRAGW framed wall, stud wall
***Fach***: **Fachzeitschrift** *f* ARCH & TRAGW, INFR & ENTW specialist journal, trade journal
**Fahrbahn** *f* INFR & ENTW carriageway (*BrE*), highway (*AmE*), road bed, roadway, *Brücke* deck; **Fahrbahnmarkierung** *f* BAUMASCHIN, INFR & ENTW road marking, road painting; **Fahrbahnmarkierungsgerät** *nt* BAUMASCHIN road-marking machine; **Fahrbahnplatte** *f* INFR & ENTW carriageway slab (*BrE*), highway slab (*AmE*), *Brücke* deck
**fahrbar** *adj* BAUMASCHIN mobile; **~es Gerüst** *nt* BAUMASCHIN jumbo; **~e Hubbühne** *f* BAUMASCHIN portable hoisting platform; **~er Schraubstock** *m* BAUMASCHIN portable vice (*BrE*), portable vise (*AmE*); **~er Verdichter** *m* BAUMASCHIN, UMWELT packer lorry
**Fahrenheit** *nt* HEIZ & BELÜFT Fahrenheit; **Fahrenheitskala** *f* HEIZ & BELÜFT Fahrenheit scale
**Fahr**: **Fahrkran** *m* BAUMASCHIN portable crane; **Fahrspur** *f* INFR & ENTW lane
**Fahrzeug** *nt* BAUMASCHIN vehicle; **Fahrzeugzufahrt** *f* INFR & ENTW vehicular access
**Faktor** *m* ARCH & TRAGW, INFR & ENTW, WERKSTOFF factor
**Fallbär** *m* BAUMASCHIN tup
**Falle** *f* BESCHLÄGE *Schloß* latch bolt
**Falleitung** *f* ABWASSER fall pipe
**fällen** *vt* INFR & ENTW *Bäume* hew, cut down
***Falle***: **Fallenfeststeller** *m* BESCHLÄGE *Schloß* latch bolt stop; **Fallenriegel** *m* BESCHLÄGE *Schloß* drop latches
**Fall**: **Fallgewicht** *nt* BAUMASCHIN tup; **Fallhammer** *m* BAUMASCHIN pile driver; **Fallhöhe** *f* ARCH & TRAGW head; **Fallinie** *f* INFR & ENTW slope line; **Falluft** *f* HEIZ & BELÜFT downdraught; **Fallmischer** *m* BAUMASCHIN tumbler
**Fällmittel** *nt* UMWELT coagulant
**Fallout** *m* UMWELT fallout
**Fallrecht** *nt* BAURECHT case law
***Fall***: **Fallrohr** *nt* ABWASSER *Dachrinne* rainwater pipe, downspout, downpipe, fall pipe, ARCH & TRAGW *Regenrinne* gutter pipe, BAUMASCHIN *Kaminschacht* band, BESCHLÄGE *Regenrinne* gutter pipe, HEIZ & BELÜFT downpipe; **Fallrohrauslauf** *m* BAUMASCHIN shoe; **Fallrohre** *nt pl* HEIZ & BELÜFT falls; **Fallschloß** *nt* BESCHLÄGE spring lock; **Fallstufe** *f* ABWASSER cascade; **Falltür** *f* ARCH & TRAGW flap door, trap door
**Fällung** *f* INFR & ENTW, WERKSTOFF coagulation; **Fällungsbecken** *nt* ABWASSER, INFR & ENTW coagulation tank; **Fällungsmittel** *nt* WERKSTOFF coagulant
**falsch** *adj* ARCH & TRAGW improper; **~e Kohäsion** *f* INFR & ENTW, WERKSTOFF false cohesion
**Falt**: **Faltbohrmast** *m* BAUMASCHIN jackknife drilling mast
**Falte** *f* ARCH & TRAGW V-unit
**Falt**: **Falttor** *nt* ARCH & TRAGW folding door; **Falttür** *f* ARCH & TRAGW, BESCHLÄGE folding door; **Faltwand** *f* ARCH & TRAGW, BESCHLÄGE folding partition, accordion partition; **Faltwerk** *nt* ARCH & TRAGW folded-plate structure, V-unit
**Falz** *m* ARCH & TRAGW rebate, joggle, lap, rabbet, BAUMASCHIN plough (*BrE*), plow (*AmE*), rabbet, BESCHLÄGE rabbet, HOLZ mortise, STAHL fold
**falzen** *vt* STAHL seam
**Falzen** *nt* STAHL folding, lockseaming
***Falz***: **Falzfuge** *f* ARCH & TRAGW rebated joint; **Falzhobel** *m* BAUMASCHIN rabbet plane, rebate plane, BESCHLÄGE fillister; **Falztür** *f* BESCHLÄGE rebated door; **Falzziegel** *m* WERKSTOFF interlocking tile
**Fang**: **Fangdamm** *m* ARCH & TRAGW, INFR & ENTW batardeau; **Fangeinrichtung** *f* BAURECHT, ELEKTR *Blitzschutz* arrester device, lightning arrester; **Fangleitung** *f* BAURECHT, ELEKTR *Blitzschutz* arrester line; **Fangstange** *f* BAURECHT, ELEKTR *Blitzschutz* arrester rod; **Fangvorrichtung** *f* BESCHLÄGE *Aufzug* safety catch
**Farb**: **Farbabbrennlampe** *f* OBERFLÄCHE paint-burning lamp; **Farbanstrich** *m* OBERFLÄCHE coat of paint; **Farbanwendungsspezifikation** *f* WERKSTOFF painting specification; **Farbaufrollen** *nt* OBERFLÄCHE roller painting

**Farbe** *f* OBERFLÄCHE color (*AmE*), colour (*BrE*), paint, WERKSTOFF paint

**farbecht** *adj* WERKSTOFF colorfast (*AmE*), colourfast (*BrE*)

***Farbe-***: **Farbenkarte** *f* OBERFLÄCHE color chart (*AmE*), colour chart (*BrE*); **Farbentemperatur** *f* OBERFLÄCHE color temperature (*AmE*), colour temperature (*BrE*)

***Farb***: **Farbhaltung** *f* OBERFLÄCHE color retention (*AmE*), colour retention (*BrE*)

**farbig**: **~ gekennzeichnet** *adj* ELEKTR, HEIZ & BELÜFT color-coded (*AmE*), colour-coded (*BrE*); **~es Glas** *nt* WERKSTOFF colored glass (*AmE*), coloured glass (*BrE*)

**farblos** *adj* OBERFLÄCHE, WERKSTOFF clear, colorless (*AmE*), colourless (*BrE*)

***Farb***: **Farbmarkierung** *f* ELEKTR *Kabel* color coding (*AmE*), colour coding (*BrE*); **Farbmittel** *nt* WERKSTOFF coloring matter (*AmE*), colouring matter (*BrE*); **Farbspritzpistole** *f* BAUMASCHIN, OBERFLÄCHE paint sprayer; **Farbtonänderung** *f* OBERFLÄCHE change in color (*AmE*), change in colour (*BrE*); **Farbwahl** *f* OBERFLÄCHE color selection (*AmE*), colour selection (*BrE*)

**Faschine** *f* INFR & ENTW, WERKSTOFF fagot, fascine; **Faschinenbau** *m* INFR & ENTW fascine work, *Deichbau* fagotting; **Faschinenpackwerk** *nt* INFR & ENTW fascine work, *Deichbau* fagotting; **Faschinenverbauung** *f* INFR & ENTW fascine work, *Deichbau* fagotting

**Fase** *f* ARCH & TRAGW bevel, HOLZ bevel, chamfer, chamfered edge

**Faser** *f* WERKSTOFF fiber (*AmE*), fibre (*BrE*); **Faserdämmstoff** *m* DÄMMUNG, INFR & ENTW, WERKSTOFF fiber insulating material (*AmE*), fiber insulation material (*AmE*), fibre insulating material (*BrE*), fibre insulation material (*BrE*)

**faserhaltig**: **~er Torf** *m* INFR & ENTW fibrous peat

***Faser***: **Faserholz** *nt* UMWELT pulpwood; **Faserplatte** *f* DÄMMUNG, WERKSTOFF fiberboard (*AmE*), fibreboard (*BrE*); **Faservlies** *m* WERKSTOFF nonwoven fabric

**Faß** *nt* BESCHLÄGE barrel

**Fassade** *f* ARCH & TRAGW facade, frontage; **die ~ erneuern** *phr* ARCH & TRAGW reface

***Fassade***: **Fassadenbogen** *m* ARCH & TRAGW facing arch; **Fassadenfarbe** *f* OBERFLÄCHE, WERKSTOFF house paint; **Fassadenfuge** *f* NATURSTEIN facing joint; **Fassadengerüst** *nt* ARCH & TRAGW facade scaffolding; **Fassadenlift** *m* ARCH & TRAGW window-cleaning cradle; **Fassadenstein** *m* NATURSTEIN facing stone

**Fassungsvermögen** *nt* ARCH & TRAGW capacity

**Faszie** *f* ARCH & TRAGW fascia

**Faul-** *in cpds* UMWELT digesting; **Faulanlagen** *f pl* ABWASSER digestion plants, UMWELT digestion plant; **Faulbehälter** *m* UMWELT digester, digestion tank, digestion sump

**faulfähig**: **~er Schlamm** *m* UMWELT putrescible sludge

***Faul-***: **Faulfähigkeit** *f* UMWELT putrescibility; **Faulgas** *nt* INFR & ENTW, UMWELT, WERKSTOFF biogas, digester gas, fermentation gas

**Fäulnis** *f* INFR & ENTW putrefaction, WERKSTOFF rot

**fäulnisbeständig** *adj* HOLZ, OBERFLÄCHE antirot, imputrescible, rotproof, UMWELT antirot, WERKSTOFF antirot, imputrescible, rotproof

**fäulnisfähig**: **~er Stoff** *m* UMWELT putrescible matter

**fäulnisfest** *adj* HOLZ, OBERFLÄCHE antirot, imputrescible, rotproof, UMWELT antirot, WERKSTOFF antirot, imputrescible, rotproof

***Fäulnis***: **Fäulnisgärung** *f* INFR & ENTW putrefactive fermentation; **Fäulnisschutz** *m* WERKSTOFF rot protection

***Faul-***: **Faulraum** *m* UMWELT digestion tank, digestion sump; **Faulschlamm** *m* UMWELT digested sludge; **Faulteich** *m* UMWELT anaerobic lagoon; **Faulturm** *m* ABWASSER, INFR & ENTW digestion tower

**Faulung** *f* UMWELT fouling

**FCKW** *abbr* (*Fluorchlorkohlenwasserstoff*) UMWELT CFC (*chlorofluorocarbon*)

**Feder** *f* ARCH & TRAGW tongue, BAUMASCHIN, BESCHLÄGE spring, HOLZ feather; **Federband** *nt* BESCHLÄGE helical hinge; **Federbolzen** *m* BESCHLÄGE spring bolt, spring hanger pin; **Federkeil** *m* HOLZ feather, feather tongue; **Federschloß** *nt* BESCHLÄGE spring lock; **Federverbindung** *f* ARCH & TRAGW ploughed-and-feathered joint (*BrE*), plowed-and-feathered joint (*AmE*), slip tongue joint, HOLZ tongue-and-groove joint

**Fehl**: **Fehlboden** *m* ARCH & TRAGW dead floor, DÄMMUNG sound-boarded ceiling

**Fehler** *m* BAURECHT fault

**fehlerhaft** *adj* ARCH & TRAGW defective, BAURECHT defective, unsound, INFR & ENTW defective; **~ geplant** *adj* ARCH & TRAGW, BAURECHT, INFR & ENTW defectively designed

***Fehler***: **Fehlerquelle** *f* ARCH & TRAGW source of errors; **Fehlerstromschutzschalter** *m* ELEKTR earth fault circuit interrupter (*BrE*), ground fault circuit interrupter (*AmE*); **Fehlertheorie** *f* ARCH & TRAGW error theory

**Fehl**: **Fehlkante** *f* HOLZ wane

**Feilkolben** *m* BAUMASCHIN pin vice (*BrE*), pin vise (*AmE*)

**fein 1.** *adj* WERKSTOFF *Sand, Kies* fine; **2. ~ verteilen** *vt* OBERFLÄCHE disperse

***fein***: **~es Glaspapier** *nt* WERKSTOFF glass paper

**Fein-** *in cpds* WERKSTOFF fine; **Feinblech** *nt* WERKSTOFF thin sheet

**feingezahnt** *adj* ARCH & TRAGW denticulated

***Fein-***: **Feinkies** *m* WERKSTOFF fine gravel; **Feinkiesbett** *nt* INFR & ENTW fine gravel bed

**feinkörnig** *adj* WERKSTOFF fine-grained; **~er Bruch** *m* WERKSTOFF fine-grained fracture; **~es Pulver** *nt* WERKSTOFF fine-grained powder; **~er Sand** *m* WERKSTOFF fine sand; **~er Zuschlag** *m* BETON fine aggregate

**feinmaschig** *adj* WERKSTOFF fine-mesh

***Fein-***: **Feinmühle** *f* BAUMASCHIN pulverizer; **Feinplanierung** *f* INFR & ENTW final grading, finegrading; **Feinplanum** *nt* INFR & ENTW final grade; **Feinplanumshöhe** *f* INFR & ENTW formation level; **Feinputz** *m* NATURSTEIN finish coat, finishing coat, final rendering, finish plaster; **Feinrechen** *m* UMWELT fine screen; **Feinregelung** *f* BAUMASCHIN fine adjustment; **Feinsplitt** *m* WERKSTOFF fine chippings; **Feinzuschlag** *m* BETON fine aggregate

**Feld** *nt* ARCH & TRAGW bay, zone, *Träger* span, HOLZ *Balkenwerk* panel, INFR & ENTW field, zone; **Felddaten** *pl* INFR & ENTW field data; **Feldmeßkette** *f* INFR & ENTW land measuring chain; **Feldmessung** *f* INFR & ENTW land measuring; **Feldmitte** *f* ARCH &

TRAGW midspan; **Feldmoment** *nt* ARCH & TRAGW, INFR & ENTW field moment; **Feldspat** *m* INFR & ENTW, WERKSTOFF feldspar

**feldspathaltig**: **~er Sandstein** *m* WERKSTOFF feldspathic sandstone

**Fels** *m* INFR & ENTW, NATURSTEIN rock; **mit ~ durchsetzter Boden** *m* INFR & ENTW rocky soil; **Felsanker** *m* INFR & ENTW rock anchor; **Felsformation** *f* INFR & ENTW rock formation; **Felsgestein** *nt* NATURSTEIN rock material; **Felslage** *f* INFR & ENTW rock layer; **Felsmechanik** *f* INFR & ENTW rock mechanics; **Felsoberfläche** *f* INFR & ENTW rock surface

**Fenn** *nt* INFR & ENTW fen

**Fenster** *nt* ARCH & TRAGW window; **Fensteranordnung** *f* ARCH & TRAGW arrangement of windows, fenestration; **Fensteranschlag** *m* ARCH & TRAGW window rabbet; **Fensteraufteilung** *f* ARCH & TRAGW fenestration; **Fensteraussparung** *f* ARCH & TRAGW window opening; **Fensterband** *nt* BESCHLÄGE window hinge; **Fensterbank** *f* ARCH & TRAGW cill (*BrE*), window sill, sill; **Fensterbankkanal** *m* ARCH & TRAGW window sill duct; **Fensterbankplatte** *f* ARCH & TRAGW window sill slab; **Fensterbau** *m* ARCH & TRAGW construction of windows; **Fensterbeschläge** *m pl* BESCHLÄGE window fittings, window hardware; **Fensterblendrahmen** *m* ARCH & TRAGW, BESCHLÄGE window frame; **Fensterbrett** *nt* ARCH & TRAGW window board; **Fensterbrüstung** *f* ARCH & TRAGW apron wall, window parapet; **Fensterdichtleiste** *f* ARCH & TRAGW fillet, DÄMMUNG window sealing fillet; **Fensterdichtungsprofil** *nt* DÄMMUNG window gasket; **Fensterfernbedienungsgeräte** *nt pl* BESCHLÄGE remote window controls; **Fensterfläche** *f* ARCH & TRAGW window area; **Fensterflügel** *m* ARCH & TRAGW, BESCHLÄGE casement; **Fensterflügelrahmen** *m* ARCH & TRAGW, BESCHLÄGE sash; **Fensterführungsschiene** *f* ARCH & TRAGW window guide rail; **Fenstergewände** *nt* ARCH & TRAGW, HOLZ jamb, window jamb, window surround; **Fenstergitter** *nt* BESCHLÄGE, STAHL window bar, window grille; **Fensterglas** *nt* ARCH & TRAGW, BESCHLÄGE window glass; **Fensterglasflügel** *m* ARCH & TRAGW, BESCHLÄGE glazed sash; **Fensterkämpfer** *m* ARCH & TRAGW window transom; **Fensterklimagerät** *nt* HEIZ & BELÜFT window air-conditioning unit; **Fensterladen** *m* ARCH & TRAGW folding shutter, shutter, BESCHLÄGE folding shutter

**fensterlos** *adj* ARCH & TRAGW windowless

***Fenster***: **Fensternischenbogen** *m* ARCH & TRAGW window recess arch; **Fensteröffnung** *f* ARCH & TRAGW, BESCHLÄGE window opening; **Fensterpfeiler** *m* ARCH & TRAGW window column; **Fensterpfosten** *m* ARCH & TRAGW monial, window post; **Fensterputzwagen** *m* ARCH & TRAGW window-cleaning cradle; **Fensterrahmen** *m* ARCH & TRAGW window frame; **Fensterreiber** *m* BESCHLÄGE casement fastener; **Fensterriegel** *m* ARCH & TRAGW sash rail, BESCHLÄGE window catch; **Fensterrosette** *f* ARCH & TRAGW wheel window; **Fensterscheibe** *f* ARCH & TRAGW, BESCHLÄGE window pane; **Fensterschließer** *m* BESCHLÄGE window fastener; **Fenstersprosse** *f* ARCH & TRAGW glazing bar, munnion, BAUMASCHIN sash bar, BESCHLÄGE window bar; **Fensterstab** *m* BAUMASCHIN sash bar; **Fenstersturz** *m* ARCH & TRAGW window lintel; **Fenstertür** *f* ARCH & TRAGW casement door, French door, French window, glazed door, HOLZ casement door, French door, French window; **Fensterventilator** *m* HEIZ & BELÜFT window fan; **Fensterverglasung** *f* ARCH & TRAGW window glazing

**fern** *adj* INFR & ENTW long-distance

**Fern**: **Fernbedienung** *f* ELEKTR remote control; **Fernfühler** *m* ELEKTR remote sensor; **Fernheizleitung** *f* HEIZ & BELÜFT, INFR & ENTW district heating line, long-distance heating line; **Fernheizung** *f* HEIZ & BELÜFT, INFR & ENTW district heating; **Fernheizwerk** *nt* HEIZ & BELÜFT, INFR & ENTW district heating plant; **Fernmeldekabel** *nt* ELEKTR, INFR & ENTW telecommunication cable; **Fernsprechnetz** *nt* ELEKTR, INFR & ENTW telephone network; **Fernsteuerung** *f* ELEKTR remote control; **Fernverkehrsstraße** *f* INFR & ENTW highway; **Fernwärme** *f* INFR & ENTW district heat; **Fernwärmeversorgung** *f* INFR & ENTW district heat supply

**fertig** *adj* BETON, STAHL precast

**Fertig-** *in cpds* ARCH & TRAGW prefabricated, BETON ready-mixed, WERKSTOFF prefabricated; **Fertiganstrich** *m* OBERFLÄCHE finish coat; **Fertigbauteil** *nt* WERKSTOFF prefabricated building unit; **Fertigbauweise** *f* ARCH & TRAGW prefabricated construction method; **Fertigbeton** *m* BETON ready-mixed concrete, transit-mixed concrete

**fertigen** *vt* WERKSTOFF manufacture

***Fertig-***: **Fertigfußboden** *m* ARCH & TRAGW *als Höhenangabe* finished floor

**fertiggestellt** *adj* ARCH & TRAGW *Gebäude* completed

***Fertig-***: **Fertighaus** *nt* ARCH & TRAGW prefabricated house

**Fertigkeit** *f* BAURECHT skill

***Fertig-***: **Fertigpackung** *f* UMWELT prepackaging

**fertigstellen** *vt* ARCH & TRAGW complete

**Fertigstellung** *f* ARCH & TRAGW, BAURECHT completion; **Fertigstellungstermin** *m* ARCH & TRAGW, BAURECHT, INFR & ENTW completion date, time limit

***Fertig-***: **Fertigsturz** *m* ARCH & TRAGW prefabricated lintel, WERKSTOFF precast unit, prefabricated lintel

**Fertigteil** *nt* BETON, WERKSTOFF precast unit, prefabricated building unit, prefabricated element; **Fertigteilbau** *m* ARCH & TRAGW system building construction; **Fertigteilbauweise** *f* **mit Skelett und Ausfachungsplatten** ARCH & TRAGW, INFR & ENTW frame and panel construction; **Fertigteilbeton** *m* BETON precast concrete; **Fertigteilbetonbalken** *m* ARCH & TRAGW, BETON cast beam

**Fertigung** *f* WERKSTOFF manufacture

**fest** *adj* ARCH & TRAGW fixed, rigid, *Untergrund* solid, BETON, DÄMMUNG, INFR & ENTW rigid, WERKSTOFF firm, solid, rigid; **~er Abfall** *m* UMWELT solid waste; **~e Abfälle** *m pl* BAURECHT, INFR & ENTW solid waste; **~er Abfallstoff** *m* UMWELT solid waste; **~ angepaßt** *adj* HOLZ, WERKSTOFF tight fit; **~e Einbauten** *m pl* BESCHLÄGE built-in units; **~es Gehrungsdreieck** *nt* BAUMASCHIN miter square (*AmE*), mitre square (*BrE*); **~er Grund** *m* ARCH & TRAGW, INFR & ENTW firm ground; **~e Phase** *f* WERKSTOFF solid phase; **~er Siedlungsabfall** *m* UMWELT municipal solid waste (*MSW*); **~es Stadium** *nt* WERKSTOFF solid state; **~er Ton** *m* INFR & ENTW, WERKSTOFF firm clay; **~ verlegt** *adj* ARCH & TRAGW, BAURECHT, DÄMMUNG, INFR & ENTW, WERKSTOFF permanent

**festeingebaut**: **~e Wohnungsgegenstände** *m pl* BESCHLÄGE fixtures and fittings

**festgelegt** *adj* ARCH & TRAGW, BAURECHT defined
**Festigkeit** *f* UMWELT stability, WERKSTOFF solidity, strength; **Festigkeitslehre** *f* ARCH & TRAGW, INFR & ENTW, WERKSTOFF mechanics of materials; **Festigkeitsprüfung** *f* INFR & ENTW, WERKSTOFF strength test
**Fest**: **Festkörper** *m* ARCH & TRAGW, INFR & ENTW solid; **Festmaß** *nt* ARCH & TRAGW *Festmeter* solid measure
**Festpunkt** *m* ARCH & TRAGW, INFR & ENTW bench mark, fixed datum, fixed point; **Festpunktnetz** *nt* INFR & ENTW observation grid
**Festsaal** *m* ARCH & TRAGW, INFR & ENTW ballroom
**festschrauben** *vt* BESCHLÄGE bolt
**festsetzen** *vt* ARCH & TRAGW, BAURECHT peg out
***Fest***: **Festspannen** *nt* ARCH & TRAGW stretching
**feststehend** *adj* ARCH & TRAGW fixed, static, stationary
**Feststoff** *m* WERKSTOFF solid; **Feststoffabfall** *m* UMWELT solid waste; **Feststoffgehalt** *m* UMWELT solids content; **Feststoffteilchen** *nt* INFR & ENTW, UMWELT, WERKSTOFF solid particle
***Fest***: **Festverglasung** *f* ARCH & TRAGW fixed glazing
**Fett** *nt* WERKSTOFF grease
**fett**: **~er Kalkputzmörtel** *m* NATURSTEIN lime putty; **~er Ton** *m* INFR & ENTW, WERKSTOFF fat clay
***Fett***: **Fettabscheider** *m* ABWASSER, UMWELT grease separator, skimming tank
**fettbeständig** *adj* WERKSTOFF grease-resistant
***Fett***: **Fettfilter** *nt* BESCHLÄGE *Dunstabzugshaube* grease filter
**Fettheit** *f*: **~ des Betons** BETON fatness of concrete
***Fett***: **Fettkalk** *m* WERKSTOFF fat lime
**feucht** *adj* ARCH & TRAGW *Räume* damp, HEIZ & BELÜFT moist, WERKSTOFF *Mörtel, Beton* green, damp; **~er Untergrund** *m* INFR & ENTW moist subsoil
**Feuchte** *f* ARCH & TRAGW, DÄMMUNG, HEIZ & BELÜFT, INFR & ENTW, WERKSTOFF dampness, humidity, moisture
**Feuchtigkeit** *f* ARCH & TRAGW, DÄMMUNG, HEIZ & BELÜFT, INFR & ENTW, WERKSTOFF dampness, humidity, moisture; **Feuchtigkeitsabdichtung** *f* DÄMMUNG vapor barrier (*AmE*), vapour barrier (*BrE*), weathering, weatherproofing
**feuchtigkeitsbeständig** *adj* DÄMMUNG, WERKSTOFF dampproof, moisture-proof
***Feuchtigkeit***: **Feuchtigkeitsgehalt** *m* HEIZ & BELÜFT, HOLZ, INFR & ENTW, WERKSTOFF humidity content, moisture content; **Feuchtigkeitsgehaltkontrolle** *f* INFR & ENTW, WERKSTOFF moisture content control; **Feuchtigkeitsindex** *m* INFR & ENTW, WERKSTOFF moisture index; **Feuchtigkeitsisolierung** *f* DÄMMUNG, WERKSTOFF dampproofing; **Feuchtigkeitskorrosion** *f* OBERFLÄCHE aqueous corrosion; **Feuchtigkeitsmesser** *m* HEIZ & BELÜFT, INFR & ENTW, WERKSTOFF hygrometer; **Feuchtigkeitsschaden** *m* DÄMMUNG, INFR & ENTW damage due to humidity; **Feuchtigkeitssperrschicht** *f* DÄMMUNG dampproof course (*DPC*), dampproofing course, humidity barrier, vapor barrier (*AmE*), vapour barrier (*BrE*), NATURSTEIN, WERKSTOFF dampproof course (*DPC*), dampproofing course; **Feuchtigkeitsverlust** *m* INFR & ENTW, WERKSTOFF moisture loss
**Feuchtraum** *m* (*FR*) ARCH & TRAGW humid room, damp room, DÄMMUNG damp room; **Feuchtraumausstattung** *f* ABWASSER dampproof equipment; **Feuchtraumgewicht** *nt* WERKSTOFF moist unit weight; **Feuchtraumleuchte** *f* (*FR-Leuchte*) ELEKTR dampproof lighting fixture
**Feuer** *nt* ARCH & TRAGW, BAURECHT fire; **Feueranweisung** *f* ARCH & TRAGW, BAURECHT fire notice; **Feueranzeiger** *m* BAURECHT, ELEKTR fire detector
**feuerbeschädigt**: **~es Gebäude** *nt* ARCH & TRAGW, BAURECHT fire-damaged building
**feuerbeständig 1.** *adj* BAURECHT, WERKSTOFF fire-resistant, fireproof; **2. ~ machen** *vt* WERKSTOFF fireproof
***feuerbeständig***: **~er Boden** *m* ARCH & TRAGW, BAURECHT fire-resisting floor
***Feuer***: **Feuerbeständigkeit** *f* BAURECHT, DÄMMUNG, WERKSTOFF fire resistance; **Feuerdecke** *f* BAURECHT, BESCHLÄGE fire blanket; **Feuereimer** *m* BAURECHT, BESCHLÄGE fire bucket
**feuerfest** *adj* WERKSTOFF fireproof; **~er Baustoff** *m* BAURECHT, DÄMMUNG, INFR & ENTW, WERKSTOFF fireproof building material; **~er Ton** *m* WERKSTOFF fire clay; **~er Ziegel** *m* NATURSTEIN fire brick; **~ machen** *phr* WERKSTOFF fireproof
***Feuer***: **Feuergefahr** *f* BAURECHT fire risk
**feuergefährlich** *adj* BAURECHT, WERKSTOFF highly inflammable
**feuergeschützt** *adj* BAURECHT, ELEKTR, INFR & ENTW fire-protected, protected; **~er Aufzug** *m* BAURECHT, INFR & ENTW firefighting lift; **~er Schacht** *m* BAURECHT, INFR & ENTW firefighting shaft; **~er Stromkreis** *m* BAURECHT fire-protected circuit; **~es Treppenhaus** *nt* BAURECHT, INFR & ENTW firefighting stair; **~es Vestibül** *nt* BAURECHT, INFR & ENTW firefighting lobby
***Feuer***: **Feuerhahn** *m* BESCHLÄGE fire hydrant
**feuerhemmend** *adj* WERKSTOFF fire-retardant
***Feuer***: **Feuerleiter** *f* BAURECHT fire ladder, firefighters' ladder; **Feuerlöscher** *m* BESCHLÄGE fire extinguisher; **Feuerlöschkasten** *m* ARCH & TRAGW, BESCHLÄGE fire cabinet, fire extinguisher box; **Feuermeldeanlage** *f* BAURECHT, ELEKTR fire alarm system; **Feuermelder** *m* BAURECHT, ELEKTR fire alarm; **Feuerrichtlinie** *f* BAURECHT fire policy; **Feuerrisiko** *nt* BAURECHT fire risk; **Feuerschaden** *m* BAURECHT, WERKSTOFF fire damage; **Feuerschutzklappe** *f* HEIZ & BELÜFT fire damper; **Feuerschutztür** *f* ARCH & TRAGW fire door; **Feuerstellenboden** *m* ARCH & TRAGW, HEIZ & BELÜFT hearth; **Feuerton** *m* WERKSTOFF fire clay; **Feuerungsanlage** *f* HEIZ & BELÜFT furnace; **Feuerversiegelung** *f* BAURECHT fire stop; **Feuerverzinken** *nt* OBERFLÄCHE, STAHL hot-dip galvanizing
**feuerverzinkt** *adj* OBERFLÄCHE hot-dip galvanized; **~er Stahl** *m* STAHL hot-dipped galvanized steel
***Feuer***: **Feuerwehrzufahrt** *f* ARCH & TRAGW, BAURECHT access for fire services; **Feuerwiderstandsfähigkeit** *f* BAURECHT, DÄMMUNG fire resistance, WERKSTOFF fire endurance, fire resistance; **Feuerwiderstandsklasse** *f* BAURECHT, WERKSTOFF fire classification, fire rating
**Fiale** *f* ARCH & TRAGW pinnacle, *Kirchenbau* buttress pinnacle
**Fiber** *f* WERKSTOFF fiber (*AmE*), fibre (*BrE*); **Fiberglas** *nt* WERKSTOFF fiberglass (*AmE*), fibreglass (*BrE*)
**fiktiv** *adj* ARCH & TRAGW notional
**Filter** *m* ABWASSER, HEIZ & BELÜFT, INFR & ENTW, WERKSTOFF filter; **Filteranlage** *f* INFR & ENTW filter plant, UMWELT filtering unit; **Filterbett** *nt* INFR & ENTW filter bed; **Filtergeschwindigkeit** *f* ABWASSER filter velocity, filtration rate, INFR & ENTW, WERK-

STOFF filter velocity; **Filtergewebe** *nt* WERKSTOFF filter fabric; **Filterkies** *m* INFR & ENTW, WERKSTOFF filter gravel; **Filterlaufzeit** *f* DÄMMUNG, HEIZ & BELÜFT filter run; **Filtermaterial** *nt* INFR & ENTW filter material; **Filtermatte** *f* WERKSTOFF woven filter medium; **Filterrohr** *nt* ABWASSER, INFR & ENTW filter drain, filter pipe; **Filtersand** *m* INFR & ENTW, WERKSTOFF filter sand; **Filterschicht** *f* INFR & ENTW filter layer; **Filtersediment** *nt* ABWASSER, INFR & ENTW filter sediment; **Filterstandzeit** *f* DÄMMUNG, HEIZ & BELÜFT filter run; **Filterströmung** *f* INFR & ENTW filter flow; **Filterventilatoreinheit** *f* HEIZ & BELÜFT filter-fan unit

**Filtrat** *nt* HEIZ & BELÜFT filtrate

**Filtration** *f* INFR & ENTW, UMWELT filtration; **Filtrationsanlage** *f* INFR & ENTW filtration plant

**Filtrieren** *nt* HEIZ & BELÜFT filtering

**Filtrierung** *f* INFR & ENTW, UMWELT filtration

**Filz** *m* WERKSTOFF felt

**filzen** *vt* NATURSTEIN *Putz* felt

**finit**: **~e Differenzmethode** *f* ARCH & TRAGW, INFR & ENTW finite-difference method; **~es Element** *nt* ARCH & TRAGW, INFR & ENTW finite element; **~e Elementmethode** *f* ARCH & TRAGW, INFR & ENTW finite-element method

**First** *m* ARCH & TRAGW ridge; **Firstabdeckung** *f* ARCH & TRAGW ridge capping, ridge covering; **Firstanschlußziegel** *m* ARCH & TRAGW ridge starting tile; **Firstbalken** *m* ARCH & TRAGW ridge beam; **Firstbrett** *nt* ARCH & TRAGW ridge piece; **Firsthöhe** *f* ARCH & TRAGW ridge height; **Firstkappe** *f* ARCH & TRAGW ridge capping tile, ridge covering tile; **Firstlinie** *f* ARCH & TRAGW ridge line; **Firstpfette** *f* ARCH & TRAGW, HOLZ arris rail; **Firstpfosten** *m* ARCH & TRAGW *Dach* crown post; **Firstpunkt** *m* ARCH & TRAGW ridge joint; **Firststein** *m* ARCH & TRAGW, WERKSTOFF crown tile, ridge tile; **Firststück** *nt* ARCH & TRAGW ridge piece; **Firstziegel** *m* ARCH & TRAGW, INFR & ENTW, NATURSTEIN, WERKSTOFF crest tile, head, hip tile, ridge tile

**Fisch**: **Fischband** *nt* BESCHLÄGE butt hinge

**fischbauchig** *adj* ARCH & TRAGW fish-bellied

***Fisch***: **Fischbauchklappe** *f* ARCH & TRAGW fish-bellied flap; **Fischbauchträger** *m* ARCH & TRAGW fish beam, fish-bellied girder; **Fischblasenmotiv** *nt* ARCH & TRAGW vesica piscis; **Fischgerinne** *nt* INFR & ENTW, UMWELT fish pass; **Fischgraben** *m* INFR & ENTW, UMWELT fish pass; **Fischgrätenform** *f* HOLZ *Parkett* herringbone pattern; **Fischgrätenmuster** *nt* HOLZ *Parkett* herringbone pattern; **Fischgrätenverband** *m* NATURSTEIN raking bond; **Fischleiter** *f* INFR & ENTW, UMWELT fish pass; **Fischschwanzrohrverlegung** *f* HEIZ & BELÜFT fishtail ducting

**Fitschband** *nt* BESCHLÄGE butt hinge

**Fitsche** *f* BESCHLÄGE butt hinge

**Fitting** *nt* HEIZ & BELÜFT fitting

**fixiert**: **~es Walzenwehr** *nt* UMWELT fixed roller sluice gate

**Fixpunkt** *m* ARCH & TRAGW, INFR & ENTW fixed datum, fixed end

**flach** *adj* ABWASSER shallow, ARCH & TRAGW, INFR & ENTW even, flat, level

**Flach-** *in cpds* ARCH & TRAGW even, flat, level, INFR & ENTW even

***flach***: **~es Sicherheitsglas** *nt* INFR & ENTW, WERKSTOFF flat safety glass

***Flach-***: **Flachaushub** *m* INFR & ENTW shallow excavation; **Flachbaggerung** *f* INFR & ENTW shallow excavation, surface digging; **Flachbaugrube** *f* INFR & ENTW shallow building pit, shallow excavation; **Flachboden** *m* ARCH & TRAGW flat top; **Flachbogen** *m* ARCH & TRAGW flat arch, jack arch, scheme arch, WERKSTOFF scheme arch; **Flachdach** *nt* ARCH & TRAGW flat roof, cut roof, HOLZ cut roof; **Flachdachablauf** *m* ABWASSER flat roof outlet; **Flachdachaufbau** *m* ARCH & TRAGW flat roof system; **Flachdachpfanne** *f* WERKSTOFF flat roofing tile, flat clay roofing tile

**Fläche** *f* ARCH & TRAGW, INFR & ENTW area, surface

***Flach-***: **Flacheisen** *nt* STAHL, WERKSTOFF flat steel, rolled steel; **Flacheisenzapfen** *nt* INFR & ENTW, STAHL, WERKSTOFF lug bolt

***Fläche***: **Flächenaushub** *m* INFR & ENTW areal excavation; **Flächenbefestigung** *f* INFR & ENTW paved area, pavement; **Flächenberechnung** *f* ARCH & TRAGW, INFR & ENTW calculation of area, quadrature

**flächenbündig** *adj* ARCH & TRAGW flush with the adjacent areas

***Fläche***: **Flächeneinheit** *f* ARCH & TRAGW unit of area; **Flächengerüst** *nt* INFR & ENTW areal scaffolding; **Flächengründung** *f* ARCH & TRAGW pad foundation, spread footing; **Flächenheizung** *f* HEIZ & BELÜFT radiant panel heating; **Flächeninhaltsmesser** *m* ARCH & TRAGW, BAUMASCHIN planimeter; **Flächenlast** *f* ARCH & TRAGW area load, INFR & ENTW area load, distributed load; **Flächenmuster** *nt* ARCH & TRAGW modular grid; **Flächennutzungsplan** *m* BAURECHT master plan, zoning map; **Flächenrüttler** *m* BAUMASCHIN surface-vibrating machine, surface vibrator

***flach***: **~ geneigt** *adj* INFR & ENTW low-gradient

***Flach-***: **Flachglas** *nt* WERKSTOFF rolled glass, plate glass; **Flachgründung** *f* ARCH & TRAGW, INFR & ENTW spread foundation; **Flachrelief** *nt* ARCH & TRAGW, INFR & ENTW bas-relief; **Flachrundkopfniet** *m* BESCHLÄGE cup head rivet; **Flachschweißnaht** *f* STAHL flush weld; **Flachsolarkollektor** *m* HEIZ & BELÜFT flat plate solar collector; **Flachspülklosett** *nt* (*Flachspül-WC*) ABWASSER, BESCHLÄGE shallow bowl water closet; **Flachspül-WC** *nt* (*Flachspülklosett*) ABWASSER, BESCHLÄGE shallow bowl WC; **Flachstahl** *m* STAHL rolled steel, WERKSTOFF flat steel; **Flachstütze** *f* WERKSTOFF flush bracket; **Flachwulstprofil** *nt* STAHL, WERKSTOFF flat-beaded profile; **Flachziegel** *m* HOLZ, WERKSTOFF flat clay roofing tile, flat roofing tile, plain tile

**flammenhemmend** *adj* BAURECHT, HOLZ flame-retardant; **~e Behandlung** *f* ARCH & TRAGW, BAURECHT, DÄMMUNG, HOLZ flame-retardant treatment

**Flamm**: **Flammpunkt** *m* WERKSTOFF point of ignition; **Flammrohr** *nt* HEIZ & BELÜFT flame tube

**flammsicher** *adj* INFR & ENTW, WERKSTOFF flameproof

***Flamm***: **Flammspritzverzinkung** *f* OBERFLÄCHE flamesprayed zinc coating

**flammwidrig** *adj* WERKSTOFF flame-resistant

**Flanke** *f* ARCH & TRAGW flank; **Flankenübertragung** *f* DÄMMUNG flanking transmission

**Flansch** *m* BAUMASCHIN boom, HEIZ & BELÜFT *beim Rohrende*, STAHL *I-Träger* flange; **Flanschgußrohr** *nt* BESCHLÄGE flanged cast-iron pipe; **Flanschkrümmer** *m* HEIZ & BELÜFT flange bend; **Flanschmuffe** *f* HEIZ & BELÜFT flanged socket; **Flanschpaßstück** *nt* HEIZ &

BELÜFT flanged adapter; **Flanschrohr** *nt* BESCHLÄGE, HEIZ & BELÜFT flange pipe, flange tube; **Flanschstutzen** *m* HEIZ & BELÜFT flanged connection; **Flanschverbindung** *f* HEIZ & BELÜFT, STAHL flanged connection

**Flasche** *f* ABWASSER, BAUMASCHIN bottle; **Flaschenrüttler** *m* BAUMASCHIN internal vibrator; **Flaschensiphon** *m* ABWASSER bottle trap; **Flaschenverschluß** *m* ABWASSER bottle trap; **Flaschenzug** *m* BAUMASCHIN tackle

**Flaschner** *m* ABWASSER, ARCH & TRAGW, STAHL plumber; **Flaschnerarbeiten** *f pl* ABWASSER, ARCH & TRAGW plumber's work, STAHL plumber's work, sheet metal work

**Flattern** *nt* ARCH & TRAGW, BAUMASCHIN, WERKSTOFF vibration

**Flechtwerk** *nt* HOLZ trellis, trellis work, WERKSTOFF plaiting

**Fleck** *m* OBERFLÄCHE stain

**Fletton** *nt* NATURSTEIN fletton

**flexibel** *adj* WERKSTOFF flexible; **flexible Abzugsverkleidung** *f* HEIZ & BELÜFT, INFR & ENTW, WERKSTOFF flexible flue liner; **flexibles Fundament** *nt* ARCH & TRAGW, INFR & ENTW flexible foundation; **flexible Rauchdichtung** *f* BAURECHT, INFR & ENTW, WERKSTOFF flexible smoke seal

**Flexibilität** *f* WERKSTOFF flexibility

**fliegend**: **~es Betonwerk** *nt* BAUMASCHIN mobile concrete factory; **~e Brücke** *f* INFR & ENTW flying bridge

**Fliegengitter** *nt* BESCHLÄGE fly screen

**Fliehkraft** *f* UMWELT centrifugal force; **Fliehkraftregler** *m* BAUMASCHIN, HEIZ & BELÜFT centrifugal governor

**Fliese** *f* BESCHLÄGE, INFR & ENTW tile; **Fliesenarbeiten** *f pl* ARCH & TRAGW tiling work; **Fliesenbelag** *m* ARCH & TRAGW tiling; **Fliesenbild** *nt* ARCH & TRAGW tile pattern; **Fliesenfußboden** *m* ARCH & TRAGW, HOLZ tile flooring; **Fliesenleger** *m* ARCH & TRAGW, NATURSTEIN tiler; **Fliesenmörtel** *m* WERKSTOFF tiling mortar; **Fliesenmuster** *nt* ARCH & TRAGW tile pattern; **Fliesenraster** *nt* ARCH & TRAGW tile pattern

**Fließ-** *in cpds* ABWASSER, HEIZ & BELÜFT, INFR & ENTW, UMWELT flow, WERKSTOFF flow, yield; **Fließdiagramm** *nt* ABWASSER, INFR & ENTW flow chart

**fließen** *vi* UMWELT flow, WERKSTOFF yield

**Fließen** *nt* UMWELT flow WERKSTOFF yield

**fließend**: **~er Verkehr** *m* INFR & ENTW moving traffic

***Fließ-***: **Fließfertigung** *f* ARCH & TRAGW flow-line construction; **Fließformel** *f* INFR & ENTW, WERKSTOFF flow formula; **Fließgeschwindigkeit** *f* ABWASSER, INFR & ENTW flow velocity; **Fließgesetz** *nt* WERKSTOFF yield law; **Fließgrenzversuch** *m* INFR & ENTW, WERKSTOFF liquid limit test; **Fließkriterium** *nt* WERKSTOFF yield criterium

**Flintkonglomerat** *nt* WERKSTOFF puddingstone

**Floatglas** *nt* WERKSTOFF float glass

**Flockung** *f* UMWELT flocculation

**Fluatierung** *f* OBERFLÄCHE fluosilicate treatment

**Flucht** *f* ARCH & TRAGW, BAURECHT, INFR & ENTW building line

**fluchteben** *adj* ARCH & TRAGW *Fassade* flush

**fluchtgerecht** *adj* ARCH & TRAGW in true alignment

**flüchtig**: **nicht flüchtiger Inhalt** *m* OBERFLÄCHE *Farbe* nonvolatile contents

***Flucht***: **Fluchtlinie** *f* ARCH & TRAGW, INFR & ENTW alignment; **Fluchtmöglichkeiten** *f pl* ARCH & TRAGW, BAURECHT, INFR & ENTW means of escape; **Fluchtstab** *m* INFR & ENTW range pole, range rod; **Fluchtstange** *f* ARCH & TRAGW ranging pole, *Vermessung* target; **Fluchttreppe** *f* BAURECHT, INFR & ENTW emergency stairs, escape staircase; **Fluchtweg** *m* BAURECHT escape route, escape way, fire escape; **Fluchtwege** *m pl* **bei Feuer** BAURECHT means of escape in case of fire

**Flugasche** *f* UMWELT fly ash, WERKSTOFF fly ash, pulverized fuel ash

**Flugdach** *nt* ARCH & TRAGW shed roof

**Flügel** *m* ARCH & TRAGW leaf; **Flügelfenster** *nt* BESCHLÄGE casement window; **Flügelhahn** *m* UMWELT butterfly cock, butterfly valve; **Flügelmauer** *f* ARCH & TRAGW, BESCHLÄGE, INFR & ENTW wing wall; **Flügelmutter** *f* BESCHLÄGE butterfly nut, wing nut; **Flügelpumpe** *f* BAUMASCHIN vane pump; **Flügelrahmen** *m* ARCH & TRAGW *Fenster* sash, HOLZ casement frame; **Flügelschraube** *f* BESCHLÄGE butterfly screw, thumb screw, wing bolt, wing screw; **Flügelsondenversuch** *m* STAHL, WERKSTOFF vane shear test; **Flügeltür** *f* ARCH & TRAGW, BESCHLÄGE folding door, leaf door; **Flügelwand** *f* ARCH & TRAGW return wall

**Flughafenabfertigungsgebäude** *nt* ARCH & TRAGW air terminal

**Flugstaub** *m* UMWELT, WERKSTOFF airborne dust

**Flugzeug**: **Flugzeugabgas** *nt* UMWELT aircraft waste gas; **Flugzeughalle** *f* ARCH & TRAGW, INFR & ENTW aircraft hangar, hangar

**Fluorchlorkohlenwasserstoff** *m* (*FCKW*) UMWELT chlorofluorocarbon (*CFC*)

**Fluoridierung** *f* ABWASSER, UMWELT fluoridation

**Flur** *m* ARCH & TRAGW corridor, hall, hallway, INFR & ENTW corridor

**Flurbereinigung** *f* BAURECHT, INFR & ENTW land consolidation

**Fluß** *m* ABWASSER, HEIZ & BELÜFT flow, INFR & ENTW, UMWELT, WERKSTOFF flow, river, stream; **Flußablagerung** *f* INFR & ENTW, UMWELT fluvial deposit, fluvial sediment; **Flußbett** *nt* INFR & ENTW river bed; **Flußeisen** *nt* STAHL structural iron; **Flußerosion** *f* INFR & ENTW stream erosion

**flüssig** *adj* INFR & ENTW, WERKSTOFF liquid

**Flüssig-** *in cpds* INFR & ENTW liquid, UMWELT fluid, WERKSTOFF liquid

***flüssig***: **~er Abfall** *m* UMWELT liquid waste

***Flüssig-***: **Flüssigbett** *nt* UMWELT fluidized bed; **Flüssigchromatographie** *f* UMWELT fluid chromatography

**Flüssigkeit** *f* INFR & ENTW, WERKSTOFF fluid, liquid; **Flüssigkeitsheber** *m* ABWASSER siphon

***Flüssig-***: **Flüssigmist** *m* UMWELT *Abfall* slurry; **Flüssigschlamm** *m* INFR & ENTW, UMWELT liquid sludge, slurry

***Fluß***: **Flußkies** *m* NATURSTEIN pebble stone, WERKSTOFF river gravel; **Flußprojektierung** *f* UMWELT run-of-river scheme; **Flußsand** *m* INFR & ENTW, UMWELT fluvial sand; **Flußspat** *m* INFR & ENTW fluorspar

**Flußstahl** *m* STAHL mild steel, WERKSTOFF ingot steel; **Flußstahlrohr** *nt* STAHL mild steel pipe

***Fluß***: **Flußton** *m* INFR & ENTW, UMWELT, WERKSTOFF fluvial clay

**Flut** *f* HEIZ & BELÜFT flood, UMWELT high tide; **Flutbecken** *nt* UMWELT tidal basin; **Flutkraftwerk** *nt* UMWELT tidal power station; **Flutlichtmast** *m* INFR

& ENTW floodlight mast; **Flutschleuse** *f* INFR & ENTW tide lock; **Flutschreiber** *m* BAUMASCHIN marigraph; **Flutströmung** *f* UMWELT flowing tide; **Flutverlust** *m* UMWELT flood loss

**fluvial** *adj* INFR & ENTW fluvial

**Foamglas** *nt* WERKSTOFF foam glass

**Folgenutzung** *f* UMWELT after use

**Folie** *f* WERKSTOFF *Kunststoff* film, sheet, *Metall* foil; **Folien** *f pl* WERKSTOFF *Kunststoff* sheeting

**Förder-** *in cpds* BAUMASCHIN conveying, hauling, hoisting; **Förderanlage** *f* **für die grobkeramische Industrie** BAUMASCHIN conveying equipment for heavy clay industry

**Förderband** *nt* BAUMASCHIN band conveyor, belt conveyor, conveying belt; **Förderbandskimmer** *m* UMWELT conveyor belt skimmer

**Förderer** *m* BAUMASCHIN conveyor

***Förder-***: **Fördergüter** *nt pl* BAUMASCHIN handling materials; **Förderhöhe** *f* ARCH & TRAGW *Pumpe* head, BAUMASCHIN hoisting height; **Fördermaschine** *f* BAUMASCHIN hauling machine; **Förderrohr** *nt* BAUMASCHIN screw elevator; **Förderschnecke** *f* BAUMASCHIN, INFR & ENTW charging screw; **Förderseil** *nt* BAUMASCHIN haulage rope

**Forderung** *f* BAURECHT requirement

**Förderung** *f* BAUMASCHIN haulage, transport

**Form** *f* ARCH & TRAGW form, shape, BAUMASCHIN *Betonprobewürfel* mold (*AmE*), mould (*BrE*), INFR & ENTW shape; **~ und Funktion** *f* ARCH & TRAGW, INFR & ENTW form and function

**Format** *nt* ARCH & TRAGW format, size

**Formbarkeit** *f* BETON, WERKSTOFF plasticity

***Form***: **Formblech** *nt* STAHL profiled sheet; **Formdraht** *m* STAHL, WERKSTOFF profiled wire

**formen** *vt* ARCH & TRAGW, INFR & ENTW form, shape, WERKSTOFF mold (*AmE*), mould (*BrE*)

***Form***: **Formfliese** *f* WERKSTOFF precast tile; **Formgebung** *f* ARCH & TRAGW, BETON, INFR & ENTW forming, shaping; **Formgenauigkeit** *f* ARCH & TRAGW, INFR & ENTW, WERKSTOFF accuracy of shape; **Formkachel** *f* ARCH & TRAGW, NATURSTEIN trimmer; **Formmaschine** *f* **für Betonfertigbauteile** BAUMASCHIN molding machine for prefabricated concrete elements (*AmE*), moulding machine for prefabricated concrete elements (*BrE*)

**formschön** *adj* ARCH & TRAGW, OBERFLÄCHE aesthetically pleasing

***Form***: **Formstahl** *m* ARCH & TRAGW, STAHL steel section; **Formstück** *nt* WERKSTOFF pipe fitting

**Formular** *nt* BAURECHT form

**Forschung** *f* BAURECHT, INFR & ENTW research; **Forschungsanstalt** *f* BAURECHT, INFR & ENTW research institute; **Forschungsprogramm** *nt* BAURECHT, INFR & ENTW research program (*AmE*), research programme (*BrE*)

**Forstwirtschaftsforschung** *f* UMWELT forestry research

**Fortgang** *m* ARCH & TRAGW progress

**Fortluft** *f* HEIZ & BELÜFT, INFR & ENTW exhaust air, exit air; **Fortluftklappe** *f* BESCHLÄGE, HEIZ & BELÜFT exhaust air damper

**fortschreitend**: **~e Erosion** *f* INFR & ENTW progressive erosion; **~e Setzung** *f* ARCH & TRAGW, INFR & ENTW progressive settlement, progressive subsidence

**Fortschritt** *m* ARCH & TRAGW progress

**Fotogrammetrie** *f* INFR & ENTW photogrammetry

**FR** *abbr* (*Feuchtraum*) ARCH & TRAGW, DÄMMUNG damp room

**Francis-Turbine** *f* BAUMASCHIN Francis turbine

**Franki-Bohrpfahl** *m* INFR & ENTW Franki-pile

**französisch**: **~es Fenster** *nt* ARCH & TRAGW French door, glazed door, HOLZ French door; **~er Nußbaum** *m* HOLZ *Farbton* French walnut

**fräsen** *vt* NATURSTEIN mill

**Fräswerkzeug** *nt*: **~ für Vielkeilverzahnung** HOLZ spline cutting tool

**Frei-** *in cpds* ARCH & TRAGW, HEIZ & BELÜFT, INFR & ENTW free, open, open-air, outdoor

**frei**: **~ aufgelagert** *adj* ARCH & TRAGW simply supported; **~es Auflager** *nt* ARCH & TRAGW free bearing, free support; **~ aufliegend** *adj* ARCH & TRAGW simply supported, BESCHLÄGE freely supported; **~es Ende** *nt* ARCH & TRAGW, BETON, INFR & ENTW, STAHL, WERKSTOFF free end; **~e Oberfläche** *f* ARCH & TRAGW, INFR & ENTW free surface; **~es Porenwasser** *nt* INFR & ENTW, WERKSTOFF free pore-water; **~er Raum** *m* ARCH & TRAGW, INFR & ENTW free space; **~e Schwingung** *f* DÄMMUNG free vibration; **~er Träger** *m* ARCH & TRAGW free beam; **~es Walzenwehr** *nt* UMWELT free roller sluice gate

***Frei-***: **Freibad** *nt* INFR & ENTW open-air swimming pool

**Freien**: **im Freien** *phr* ARCH & TRAGW, HEIZ & BELÜFT, INFR & ENTW open-air

***Frei-***: **Freifallbohrung** *f* BAUMASCHIN free-fall boring, free-fall drilling; **Freifallmischer** *m* BAUMASCHIN reversing-drum mixer, rotary-drum mixer; **Freifallramme** *f* BAUMASCHIN drop pile hammer; **Freifallstanze** *f* BAUMASCHIN free-falling stamp; **Freifläche** *f* ARCH & TRAGW concourse, open area, BAURECHT *nicht zu bebauende Fläche* open area, open space, INFR & ENTW concourse; **Freigabevermerk** *m* ARCH & TRAGW, BAURECHT approval note

**freigelegt**: **~e Betonoberfläche** *f* **mit Zuschlagstoffen** BETON, INFR & ENTW, WERKSTOFF exposed aggregate finish; **~es Entwässerungsrohr** *nt* ABWASSER, HEIZ & BELÜFT exposed drainpipe

**Freiheitsgrad** *m* ARCH & TRAGW, INFR & ENTW degree of freedom

***Frei-***: **Freilagerfläche** *f* INFR & ENTW outdoor storage area

**freilegen** *vt* ARCH & TRAGW *Kabel, Rohre* lay bare, INFR & ENTW *Baugelände* clear

***Frei-***: **Freileitung** *f* ELEKTR, INFR & ENTW overhead line

**freiliegend** *adj* ARCH & TRAGW exposed

***Frei-***: **Freiliegen** *nt* **der Zuschlagstoffe** BETON aggregate exposure; **Freiluft** *f* ARCH & TRAGW open air; **Freiluftanlage** *f* INFR & ENTW open-air plant, outdoor facilities; **Freiluftbauwerk** *nt* ARCH & TRAGW open-air structure; **Freipfeiler** *m* ARCH & TRAGW, INFR & ENTW pillar; **Freisäule** *f* ARCH & TRAGW free-standing column

**Freisetzung** *f* UMWELT release

***Frei-***: **Freispannsystem** *nt* ARCH & TRAGW, INFR & ENTW free-span system; **Freistadium** *nt* INFR & ENTW open-air stadium

**freistehend** *adj* ARCH & TRAGW detached, self-supporting, BESCHLÄGE free-standing, INFR & ENTW detached; **~es Gebäude** *nt* ARCH & TRAGW, INFR & ENTW detached building, isolated building; **~er Kran** *m* BAUMASCHIN independent crane; **~er Pfeiler** *m* ARCH & TRAGW, INFR & ENTW pillar; **~e Zwischenwand** *f* ARCH & TRAGW self-supporting partition

**freistrahlend**: **~e Leuchte** *f* ELEKTR general diffused light

**freitragend** *adj* ARCH & TRAGW cantilevered, self-supporting, overhanging, WERKSTOFF self-contained; **~es Pultdach** *nt* ARCH & TRAGW single pitch roof; **~e Treppe** *f* ARCH & TRAGW flyers

***Frei-***: **Freiträger** *m* ARCH & TRAGW, BETON, HOLZ cantilever beam, semibeam; **Freiträgertreppe** *f* ARCH & TRAGW hanging stairs; **Freitreppe** *f* ARCH & TRAGW perron; **Freiwange** *f* ARCH & TRAGW outer string

**Freizeit** *f* INFR & ENTW leisure time; **Freizeitanlagen** *f pl* INFR & ENTW leisure facilities; **Freizeiteinrichtung** *f* ARCH & TRAGW, INFR & ENTW recreation center (*AmE*), recreation centre (*BrE*); **Freizeitpark** *m* INFR & ENTW theme park; **Freizeitzentrum** *nt* INFR & ENTW leisure center (*AmE*), leisure centre (*BrE*)

**Fremd-** *in cpds* UMWELT external, foreign; **Fremdemissionen** *f pl* UMWELT foreign emissions

**Fremdenheim** *nt* ARCH & TRAGW, INFR & ENTW hospice

***Fremd-***: **Fremdstoffniederschlag** *m* UMWELT contamination fallout; **Fremdzufuhr** *f* UMWELT external input

**Frequenz** *f* DÄMMUNG, ELEKTR frequency; **Frequenzkurve** *f* DÄMMUNG frequency curve; **Frequenzmesser** *m* BAUMASCHIN, DÄMMUNG frequency meter

**Frettbohrer** *m* BAUMASCHIN auger gimlet

**Fries** *m* ARCH & TRAGW frieze, molding (*AmE*), moulding (*BrE*)

**frisch** *adj* BETON, WERKSTOFF *Mörtel* green, fresh

**Frisch-** *in cpds* BETON, HEIZ & BELÜFT, INFR & ENTW, UMWELT fresh

**Frisch**: **Frischbeton** *m* BETON, WERKSTOFF freshly mixed concrete, green concrete, wet concrete; **Frischluft** *f* HEIZ & BELÜFT fresh air; **Frischschlamm** *m* INFR & ENTW, UMWELT fresh sludge; **Frischwasser** *nt* INFR & ENTW, UMWELT fresh water; **Frischwasserversorgung** *f* ABWASSER, INFR & ENTW freshwater supply

**Frist** *f* BAURECHT deadline, term

**FR-Leuchte** *f* (*Feuchtraumleuchte*) ELEKTR dampproof lighting fixture

**Front** *f* ARCH & TRAGW front; **Frontbogen** *m* ARCH & TRAGW facing arch

**Frontispiz** *m* ARCH & TRAGW frontispiece

***Front***: **Frontlader** *m* BAUMASCHIN front loader, head end loader; **Frontschürze** *f* ARCH & TRAGW apron; **Frontwand** *f* NATURSTEIN face wall

**Frost** *m* ARCH & TRAGW, BETON, HEIZ & BELÜFT, INFR & ENTW, WERKSTOFF frost; **Frostaufbruch** *m* INFR & ENTW frost heave

**frostbeständig** *adj* BETON, INFR & ENTW, WERKSTOFF frost-resistant

***Frost***: **Frostbeständigkeit** *f* BETON, INFR & ENTW, WERKSTOFF frost resistance; **Frosteindringtiefe** *f* ARCH & TRAGW, INFR & ENTW frost penetration depth; **Frosteinwirkung** *f* ARCH & TRAGW, BETON, HEIZ & BELÜFT, INFR & ENTW, OBERFLÄCHE action of frost, effect of frost

**frostempfindlich** *adj* INFR & ENTW, WERKSTOFF frost-susceptible

**frostfrei** *adj* ARCH & TRAGW *Fundament, Rohr*, DÄMMUNG, INFR & ENTW, WERKSTOFF *Fundament, Rohr* frost-protected

***Frost***: **Frostgrenze** *f* ARCH & TRAGW, INFR & ENTW depth of frost penetration, frost line; **Frostmauer** *f* ARCH & TRAGW ice wall; **Frostschaden** *m* ARCH & TRAGW frost damage; **Frostschürze** *f* ARCH & TRAGW ice wall; **Frostschutz** *m* ARCH & TRAGW frost protection, BAUMASCHIN antifreeze, DÄMMUNG frost protection, HEIZ & BELÜFT antifreeze, INFR & ENTW, WERKSTOFF antifreeze, frost protection; **Frostschutzkies** *m* WERKSTOFF frost blanket gravel; **Frostschutzmittel** *nt* BAUMASCHIN, HEIZ & BELÜFT, INFR & ENTW, WERKSTOFF antifreeze agent, frost protection agent; **Frostschutzschicht** *f* HEIZ & BELÜFT antifreeze layer, frost layer, INFR & ENTW *Tiefbau* subbase, antifreeze layer, frost layer, *Straßenbau* frost blanket, WERKSTOFF antifreeze layer, frost layer

**frostsicher** *adj* ARCH & TRAGW, DÄMMUNG, INFR & ENTW, WERKSTOFF *Fundament, Rohr* frost-protected

***Frost***: **Frost-Tau-Versuch** *m* WERKSTOFF freezing and thawing test; **Frosttiefe** *f* ARCH & TRAGW, INFR & ENTW depth of frost penetration, frost line

**frostunempfindlich** *adj* INFR & ENTW, WERKSTOFF frost-proof

**Frühfestigkeit** *f* INFR & ENTW, WERKSTOFF early strength

**frühhochfest** *adj* NATURSTEIN *Zement* high-early-strength; **~er Zement** *m* BETON, WERKSTOFF rapid-hardening cement

**Frühlingsmaximum** *nt* UMWELT spring maximum of fallout

**Fuchs** *m* ARCH & TRAGW flue; **Fuchsauskleidung** *f* DÄMMUNG, NATURSTEIN flue lining; **Fuchskanal** *m* ARCH & TRAGW smoke flue, *Schornstein* uptake; **Fuchsschwanz** *m* ARCH & TRAGW tenon saw, BAUMASCHIN handsaw

**Fuge** *f* ARCH & TRAGW joint, meeting, HOLZ mortise, INFR & ENTW, NATURSTEIN joint

**Fugeisen** *nt* ARCH & TRAGW, NATURSTEIN jointer

**Fugen**: **~ ausfüllen** *phr* NATURSTEIN *vergießen* repoint

***Fuge***: **Fugenband** *nt* DÄMMUNG waterstop, WERKSTOFF joint tape, preformed gasket; **Fugenbeton** *m* BETON joint concrete; **Fugendeckblech** *nt* ARCH & TRAGW, WERKSTOFF apron flashing; **Fugendichtung** *f* NATURSTEIN joint packing, joint sealing, WERKSTOFF caulking; **Fugendurchlaßkoeffizient** *m* INFR & ENTW, NATURSTEIN, WERKSTOFF joint permeability coefficient; **Fugeneinlage** *f* ARCH & TRAGW, DÄMMUNG , INFR & ENTW, WERKSTOFF expansion joint filler, joint filler, joint lining, joint sealing strip, sealing strip; **Fugeneinteilung** *f* NATURSTEIN joint pattern; **Fugenfüller** *m* WERKSTOFF jointing compound; **Fugengips** *m* NATURSTEIN joint plaster; **Fugenkelle** *f* BAUMASCHIN filling trowel; **Fugenkitt** *m* WERKSTOFF joint cement; **Fugenleiste** *f* HOLZ cover strip

**fugenlos** *adj* ARCH & TRAGW jointless; **~er Fußboden** *m* ARCH & TRAGW composition flooring, magnesite flooring

***Fuge***: **Fugenmasse** *f* WERKSTOFF joint sealer; **Fugenmesser** *nt* BAUMASCHIN joint cutter; **Fugenmörtel** *m* NATURSTEIN, WERKSTOFF joint mortar; **Fugenreißen** *nt* NATURSTEIN joint raking; **Fugenschneider** *m* BAUMASCHIN joint cutter; **Fugenschnitt** *m* BETON joint cutting; **Fugenvergußmasse** *f* OBERFLÄCHE, WERKSTOFF joint sealing compound, sealing compound; **Fugenwerkzeug** *nt* NATURSTEIN jointing and pointing tools

**Fuhrpark** *m* BAUMASCHIN, INFR & ENTW rolling stock

**Führung** *f* BESCHLÄGE *Schloß* forcing; **Führungsleiste** *f* ARCH & TRAGW guide fillet; **Führungsnut** *f* HOLZ *Schiebetür eines Schrankes* guide groove; **Führungsrolle** *f* BAUMASCHIN guide pulley; **Führungsschiene** *f* BESCHLÄGE *Tür, Fenster* guide rail; **Führungsstange** *f* BESCHLÄGE *Ramme* guide pole

**Füll-**: **Füllart** *f* **der Deponie** UMWELT tip-filling method; **Füllelement** *nt* WERKSTOFF infill panel

**füllen** *vt* OBERFLÄCHE fill

**Füller-** *in cpds* WERKSTOFF filler, filling, infill; **Fülleranteil** *m* WERKSTOFF filler content

**Fullererde** *f* WERKSTOFF Fuller's earth

***Füller-***: **Füllergehalt** *m* WERKSTOFF filler content

**Fullerkreide** *f* WERKSTOFF Fuller's chalk

***Füll-***: **Füllholz** *nt* HOLZ filler, packing piece; **Füllkörper** *m* WERKSTOFF *Decke* infill block; **Füllkörperdecke** *f* ARCH & TRAGW hollow filler block floor, filler block floor; **Füllmasse** *f* WERKSTOFF filling compound, filling material; **Füllmaterial** *nt* WERKSTOFF filler, filling, infill; **Füllschicht** *f* ARCH & TRAGW, INFR & ENTW back-up; **Füllstoff** *m* WERKSTOFF filler, filling, infill

**Füllung** *f* ARCH & TRAGW, BESCHLÄGE, INFR & ENTW lining, stuffing

**Fundament** *nt* ARCH & TRAGW, INFR & ENTW footing, groundwork, *eines Gebäudes* foundation; **Fundamentarbeit** *f* ARCH & TRAGW, INFR & ENTW foundation work; **Fundamentaushub** *m* INFR & ENTW excavation for foundation; **Fundamentbalken** *m* ARCH & TRAGW grade beam; **Fundamentblock** *m* ARCH & TRAGW footing block, foundation block, INFR & ENTW foundation block; **Fundamentbreite** *f* ARCH & TRAGW, INFR & ENTW foundation width; **Fundamenterder** *m* ELEKTR foundation earthing (*BrE*), foundation grounding (*AmE*); **Fundamentgewölbe** *nt* ARCH & TRAGW inverted arch; **Fundamentgraben** *m* ARCH & TRAGW, INFR & ENTW footing trench, foundation trench; **Fundamentklotz** *m* ARCH & TRAGW, INFR & ENTW foundation block; **Fundamentplan** *m* ARCH & TRAGW, INFR & ENTW foundation drawing; **Fundamentplatte** *f* ARCH & TRAGW, BESCHLÄGE, INFR & ENTW bottom plate, foundation plate, foundation raft, foundation slab; **Fundamentschraube** *f* BESCHLÄGE foundation bolt, lag screw; **Fundamentsohle** *f* ARCH & TRAGW foundation base, INFR & ENTW final grade, bottom of foundation, foundation base; **Fundamentstreifen** *m* ARCH & TRAGW foundation strip; **Fundamenttiefe** *f* ARCH & TRAGW, INFR & ENTW depth of foundation, foundation depth

**Fungizid** *nt* BAURECHT, WERKSTOFF fungicide

**Funken** *m* ELEKTR spark; **Funkenfang** *m* ELEKTR spark arrester; **Funkenregen** *m* INFR & ENTW shower of sparks

**Funktion** *f* ARCH & TRAGW function

**funktionsfähig**: **~e Windgeschwindigkeit** *f* UMWELT survival wind speed

**Furnier** *nt* HOLZ, OBERFLÄCHE veneer

**furnieren** *vt* HOLZ, OBERFLÄCHE veneer

**Furnieren** *nt* HOLZ, OBERFLÄCHE veneering

***Furnier***: **Furnierholz** *nt* OBERFLÄCHE veneer; **Furnierplatte** *f* HOLZ scale board, veneer plywood, *dreilagig* three-ply wood

**Fuß** *m* ARCH & TRAGW, INFR & ENTW *einer Wand, Säule* base, foot; **~ des Turmes** ARCH & TRAGW tower base; **Fußabstreifmatte** *f* BESCHLÄGE foot scraper mat

**Fußboden** *m* WERKSTOFF *eines Zimmers* floor; **Fußbodenbelag** *m* ARCH & TRAGW, BESCHLÄGE, NATURSTEIN, WERKSTOFF floor covering, floor decking, floor pavement, flooring; **Fußbodenbretter** *nt pl* HOLZ floor boards

**fußbodenbündig** *adj* ARCH & TRAGW flush with the floor, level with the floor

***Fußboden***: **Fußbodenfliese** *f* ARCH & TRAGW, BESCHLÄGE, INFR & ENTW, WERKSTOFF floor tile, flooring tile; **Fußbodenführung** *f* BESCHLÄGE *Schiebetür* floor guide; **Fußbodenheizung** *f* ARCH & TRAGW, HEIZ & BELÜFT floor heating, screed heating, underfloor heating; **Fußbodenisolierung** *f* DÄMMUNG floor insulation; **Fußbodennagel** *m* BESCHLÄGE flooring nail; **Fußbodenplatte** *f* ARCH & TRAGW, BESCHLÄGE, BETON, INFR & ENTW, WERKSTOFF floor slab, floor tile, flooring tile

**Fußgänger** *m* ARCH & TRAGW, INFR & ENTW pedestrian; **Fußgängerbrücke** *f* INFR & ENTW footbridge, pedestrian bridge; **Fußgängerübergang** *m* INFR & ENTW crosswalk (*AmE*), pedestrian crossing (*BrE*); **Fußgängerweg** *m* INFR & ENTW pavement (*BrE*), sidewalk (*AmE*), walkway; **Fußgängerzugang** *m* INFR & ENTW pedestrian access

***Fuß***: **Fußholz** *nt* ARCH & TRAGW groundsill, sole piece, HOLZ *Dachkonstruktion* pole plate; **Fußleiste** *f* ARCH & TRAGW, BESCHLÄGE, HOLZ baseboard (*AmE*), skirting (*BrE*), skirting board (*BrE*); **Fußpfette** *f* ARCH & TRAGW wall plate, HOLZ inferior purlin; **Fußplatte** *f* ARCH & TRAGW base plate; **Fußpunkt** *m* ARCH & TRAGW base point; **Fußschwelle** *f* HOLZ inferior purlin; **Fußventil** *nt* ABWASSER foot valve; **Fußverbreiterung** *f* ARCH & TRAGW, INFR & ENTW *Pfahlgründung* underream

**Futter** *nt* ARCH & TRAGW lining, HOLZ packing piece, INFR & ENTW lining; **Futterblech** *nt* STAHL filler; **Futterbrett** *nt* ARCH & TRAGW *Stufe* riser; **Futterholz** *nt* HOLZ firring, furring, packing piece, furring piece; **Futterleiste** *f* HOLZ filling rod; **Futtermauer** *f* ARCH & TRAGW, NATURSTEIN breast wall, lining wall; **Futterstück** *nt* HOLZ filler

**Fütterungsstab** *m* ARCH & TRAGW web member

# G

**g** *abbr* (*Gramm*) ARCH & TRAGW, INFR & ENTW, WERKSTOFF g (*gram*)

**Gabbro** *m* WERKSTOFF gabbro

**Gabel-** *in cpds* ARCH & TRAGW, HOLZ forked; **Gabelanker** *m* ARCH & TRAGW forked tie; **Gabelholz** *nt* HOLZ forked wood; **Gabelkopf** *m* STAHL clevis; **Gabelrohr** *nt* ABWASSER forked pipe; **Gabelstapler** *m* BAUMASCHIN fork-lift truck

**Gabelung** *f* INFR & ENTW bifurcation, Y-junction

**Galerie** *f* ARCH & TRAGW gallery

**galvanisieren** *vt* OBERFLÄCHE, WERKSTOFF galvanize

**galvanisiert** *adj* OBERFLÄCHE, WERKSTOFF galvanized

**Galvanisierung** *f* OBERFLÄCHE, WERKSTOFF galvanizing

**Gammasonde** *f* INFR & ENTW, WERKSTOFF gamma probe

**Gang** *m* ARCH & TRAGW *Gebäude, Zug* corridor, gangway, *Kirche, Kino* aisle; **Ganghöhe** *f* BETON *Spiralbewehrung* pitch of spiral; **Ganglinie** *f* ARCH & TRAGW *Treppe* walking line; **Gang-Nail**[R] *nt* INFR & ENTW, STAHL, WERKSTOFF Gang-Nail[R]; **Gang-Nail verzinkte Nagelplatte**[R] *f* INFR & ENTW, STAHL, WERKSTOFF Gang-Nail galvanized connector plate[R]

**ganz** *adj* ARCH & TRAGW, BESCHLÄGE, STAHL all

**ganzflächig**: **~ aufkleben** *vt* HOLZ, OBERFLÄCHE glue on throughout

***Ganz-***: **Ganzglasfassade** *f* STAHL fully glazed facade; **Ganzglastür** *f* BESCHLÄGE fully glazed door; **Ganzglaswand** *f* ARCH & TRAGW, STAHL all-glass wall; **Ganzholztür** *f* ARCH & TRAGW, BESCHLÄGE, HOLZ all-wood door; **Ganzstahlkonstruktion** *f* STAHL all-steel design

**Garage** *f* ARCH & TRAGW garage

**Garantie** *f* BAURECHT assurance, guarantee, warranty

**garantieren** *vt* BAURECHT guarantee

**Gardine** *f* BESCHLÄGE curtain

**Garnitur** *f* BESCHLÄGE fittings, set

**Gas** *nt* HEIZ & BELÜFT, UMWELT, WERKSTOFF gas; **Gasabzug** *m* HEIZ & BELÜFT gas vent; **Gasarmatur** *f* HEIZ & BELÜFT gas fittings; **Gasbeton** *m* BETON, WERKSTOFF *Trennwände* aerated concrete, cellular concrete, foam concrete, gas concrete, porous concrete; **Gasbrenner** *m* HEIZ & BELÜFT gas burner

**gasdicht** *adj* HEIZ & BELÜFT, INFR & ENTW, WERKSTOFF gasproof, gastight

***Gas***: **Gasdruck** *m* HEIZ & BELÜFT, INFR & ENTW gas pressure; **Gasentschwefelung** *f* UMWELT gas desulfurization (*AmE*), gas desulphurization (*BrE*); **Gasfeuerung** *f* HEIZ & BELÜFT gas firing; **Gasflasche** *f* INFR & ENTW, STAHL, WERKSTOFF gas bottle, gas cylinder

**gasförmig**: **~es Medium** *nt* UMWELT gaseous medium

**gasgefeuert** *adj* HEIZ & BELÜFT gas-fired

***Gas***: **Gashahn** *m* BESCHLÄGE gas tap; **Gasheizung** *f* HEIZ & BELÜFT gas heating; **Gasheizwert** *m* HEIZ & BELÜFT, INFR & ENTW, WERKSTOFF gas calorific value; **Gasinstallation** *f* HEIZ & BELÜFT gas installation; **Gasleitung** *f* HEIZ & BELÜFT gas conduit, gas pipe; **Gaslöten** *nt* BAUMASCHIN torch brazing; **Gasmesser** *m* HEIZ & BELÜFT gas meter; **Gasrohr** *nt* HEIZ & BELÜFT gas pipe; **Gasrohrleitung** *f* INFR & ENTW gas pipeline; **Gasrohrzange** *f* BAUMASCHIN gas pliers; **Gasrückgewinnung** *f* UMWELT gas recovery; **Gasrückstand** *m* UMWELT residual gas; **Gasverbrauch** *m* HEIZ & BELÜFT gas consumption; **Gasversorgungsanlage** *f* HEIZ & BELÜFT, INFR & ENTW gas supply system; **Gasversorgungsgesellschaft** *f* BAURECHT, HEIZ & BELÜFT gas supply company; **Gasversorgungsleitung** *f* HEIZ & BELÜFT gas connection; **Gaszentralheizung** *f* HEIZ & BELÜFT gas-fired central heating

**Gebälk** *nt* ARCH & TRAGW, HOLZ binders and joists, frame of joists, timberwork, *klassische Säulenverbindung* entablature

**Gebäude** *nt* ARCH & TRAGW, INFR & ENTW building, edifice; **~ in Hallenbauweise** ARCH & TRAGW, INFR & ENTW hall-type building; **~ ohne Umfassungswände** ARCH & TRAGW open-sided building; **Gebäudeanschlußleitung** *f* ABWASSER building sewer; **Gebäudebelegung** *f* ARCH & TRAGW, INFR & ENTW building occupants; **Gebäudeerweiterung** *f* ARCH & TRAGW addition; **Gebäudeflucht** *f* ARCH & TRAGW, BAURECHT, INFR & ENTW building line, frontage; **Gebäudeflügel** *m* ARCH & TRAGW wing; **Gebäudeform** *f* ARCH & TRAGW, HOLZ, INFR & ENTW building form; **Gebäudefundament** *nt* ARCH & TRAGW building foundation; **Gebäudegerippe** *nt* ARCH & TRAGW building skeleton; **Gebäudeinstallation** *f* ABWASSER, ELEKTR, HEIZ & BELÜFT indoor installations; **Gebäudekern** *m* ARCH & TRAGW core; **Gebäudekomplex** *m* ARCH & TRAGW, INFR & ENTW block of buildings, building complex; **Gebäudenummer** *f* ARCH & TRAGW, INFR & ENTW building number; **Gebäudenutzung** *f* ARCH & TRAGW, BAURECHT, INFR & ENTW use of building; **Gebäudeschnittzeichnung** *f* ARCH & TRAGW, INFR & ENTW sectional drawing; **Gebäudeskelett** *nt* ARCH & TRAGW skeleton; **Gebäudestütze** *f* ARCH & TRAGW, BETON, HOLZ, STAHL building column; **Gebäudeteil** *m* ARCH & TRAGW building part; **Gebäudeumbau** *m* ARCH & TRAGW building alteration; **Gebäudeunterhaltung** *f* ARCH & TRAGW, BAURECHT, INFR & ENTW building maintenance; **Gebäudewartung** *f* ARCH & TRAGW, BAURECHT, INFR & ENTW building maintenance

**gebeizt** *adj* OBERFLÄCHE stained

**Gebiet** *nt* BAURECHT, INFR & ENTW *Bereich* area, district, territory

**Gebinde** *nt* ARCH & TRAGW *Dach* truss, HOLZ couple

**Gebläse** *nt* ARCH & TRAGW, BESCHLÄGE, HEIZ & BELÜFT fan, ventilating fan, ventilator

**gebogen** *adj* ARCH & TRAGW, INFR & ENTW, STAHL, WERKSTOFF bent; **~e Antrittsstufe** *f* ARCH & TRAGW commode step; **~es Glas** *nt* WERKSTOFF cast glass; **~er Sparren** *m* ARCH & TRAGW, HOLZ curved roof

**gebohnert** *adj* OBERFLÄCHE, WERKSTOFF polished

**geböscht** *adj* INFR & ENTW sloped

**gebrannt** *adj* WERKSTOFF fired; **~er Kalk** *m* NATUR-

STEIN quicklime, WERKSTOFF calcium oxide; **~er Ziegel** *m* NATURSTEIN burnt brick

**Gebrauch** *m* ARCH & TRAGW, INFR & ENTW use

**gebrauchen** *vt* ARCH & TRAGW employ, use, INFR & ENTW use

***Gebrauch***: **Gebrauchsspannung** *f* INFR & ENTW working stress

**gebrochen** *adj* WERKSTOFF crushed; **~es Gestein** *nt* WERKSTOFF crushed rock; **~es Mansardendach** *nt* ARCH & TRAGW curb roof, double-pitched roof; **~es Material** *nt* WERKSTOFF broken material, crushed material

**Gebühr** *f* ARCH & TRAGW, BAURECHT fee

**gedämmt** *adj* DÄMMUNG, ELEKTR insulated

**gedämpft**: **~e Buche** *f* HOLZ steamed beech

**gedeckt**: **~er Fußgängerweg** *m* ARCH & TRAGW covered walkway; **~er Weg** *m* ARCH & TRAGW covered way

**gedehnt** *adj* ARCH & TRAGW, WERKSTOFF stretched

**gedrosselt** *adj* ABWASSER, BAUMASCHIN, HEIZ & BELÜFT throttled

**gedrückt**: **~er Spitzbogen** *m* ARCH & TRAGW four-centered arch (*AmE*), four-centred arch (*BrE*)

**geeicht**: **~es Wasserrückhaltebecken** *nt* UMWELT *Landwirtschaft* calibrated watershed

**geeignet** *adj* ARCH & TRAGW, BAURECHT appropriate, convenient, suitable

**Gefahr** *f* BAURECHT, INFR & ENTW danger, hazard; **~ von Holzfäule** HOLZ risk of rot; **Gefahrenfunktion** *f* BAURECHT emergency function; **Gefahrenklasse** *f* BAURECHT danger class; **Gefahrensituation** *f* UMWELT episode; **Gefahrgutbeauftragter** *m* UMWELT official responsible for hazardous goods; **Gefahrgutumschlag** *m* UMWELT transshipment of hazardous goods

**gefährlich** *adj* ARCH & TRAGW, BAURECHT, INFR & ENTW dangerous, hazardous; **~e Abfälle** *m pl* BAURECHT, INFR & ENTW, UMWELT hazardous waste; **~e Konstruktion** *f* BAURECHT, INFR & ENTW dangerous structure; **~es Schmutzwasser** *nt* BAURECHT, INFR & ENTW, UMWELT hazardous sewage; **~e Stoffe** *m pl* UMWELT hazardous material

***Gefahr***: **Gefahrstoffkataster** *m* UMWELT register of hazardous substances

**Gefälle** *nt* ARCH & TRAGW *Neigung eines Bauwerks* slant, gradient, falling gradient, slope, INFR & ENTW falling gradient, *Druck, Temperatur* fall, *Neigung eines Bauwerks* gradient, slope, grade, downward gradient; **im ~ verlegt** *phr* DÄMMUNG laid to falls; **Gefällebeton** *m* BETON sloping concrete; **Gefälleestrich** *m* NATURSTEIN sloping screed; **Gefälleverlust** *m* INFR & ENTW loss of head

**gefalzt** *adj* ARCH & TRAGW rebated; **~e Verbindung** *f* ARCH & TRAGW rabbeted joint

**Gefäßreinigungsanlage** *f* INFR & ENTW tank-cleaning plant

**gefedert**: **~er Dielenfußboden** *m* HOLZ tongued flooring; **~e Verbindung** *f* HOLZ feather joint

**gefeuert** *adj* WERKSTOFF fired

**gefiltert** *adj* HEIZ & BELÜFT filtered

**gefilzt** *adj* NATURSTEIN *Putz* felted

**Geflecht** *nt* STAHL netting, WERKSTOFF plaiting, netting

**gefleckt** *adj* WERKSTOFF dappled

**geformt** *adj* WERKSTOFF molded (*AmE*), moulded (*BrE*)

**Gefrier-** *in cpds* ARCH & TRAGW, INFR & ENTW, WERKSTOFF freezing

**gefrieren** *vi* ABWASSER *Wasser in einem Rohr*, HEIZ & BELÜFT freeze

**Gefrieren** *nt* ABWASSER, HEIZ & BELÜFT freezing

***Gefrier-***: **Gefrierpunkt** *m* INFR & ENTW, WERKSTOFF freezing point; **Gefrierraum** *m* ARCH & TRAGW freezer room; **Gefrierverfahren** *nt* INFR & ENTW, WERKSTOFF freezing process

**gefroren** *adj* INFR & ENTW, WERKSTOFF frozen; **~er Boden** *m* ARCH & TRAGW, INFR & ENTW, WERKSTOFF frozen soil

**Gefüge** *nt* INFR & ENTW structure

**gefüllt** *adj* ARCH & TRAGW filled

**Gegen-** *in cpds* ARCH & TRAGW, BAURECHT, INFR & ENTW adverse

**gegen**: **~ die Faser arbeiten** *phr* HOLZ work against the grain

***Gegen-***: **Gegenbogen** *m* ARCH & TRAGW inflected arch, reversed arch; **Gegenböschung** *f* INFR & ENTW counterslope

**Gegend** *f* INFR & ENTW *Bereich* area

***Gegen-***: **Gegendruck** *m* ABWASSER backpressure; **Gegenflansch** *m* HEIZ & BELÜFT companion flange, counterflange; **Gegengefälle** *nt* INFR & ENTW reverse gradient; **Gegengewicht** *nt* ARCH & TRAGW, BAUMASCHIN balancing weight; **Gegengewölbe** *nt* ARCH & TRAGW inverted arch; **Gegenhalter** *m* BAUMASCHIN rivet dolly

**gegenläufig**: **~e Treppe** *f* ARCH & TRAGW dogleg stairs

***Gegen-***: **Gegenneigung** *f* ARCH & TRAGW, INFR & ENTW adverse slope; **Gegensprechanlage** *f* ELEKTR intercom system

**Gegenstand** *m* ARCH & TRAGW, BAURECHT item

***Gegen-***: **Gegenstromfilter** *nt* UMWELT reverse flow filter; **Gegenstrommischer** *m* BAUMASCHIN *zur Betonwarenherstellung* countercurrent mixer; **Gegenstromprinzip** *nt* INFR & ENTW counter current method; **Gegenverkehr** *m* INFR & ENTW opposing traffic

**gegliedert** *adj* ARCH & TRAGW jointed

**Gehänge** *nt* ARCH & TRAGW *Ornament* festoon

**gehärtet** *adj* STAHL tempered

**Gehäuse** *nt* ARCH & TRAGW, UMWELT housing

**Geheimtür** *f* ARCH & TRAGW jib door

**Gehfläche** *f* ARCH & TRAGW, INFR & ENTW pedestrian concourse

**Gehflügel** *m* BESCHLÄGE active leaf

**gehobelt**: **~e Schalung** *f* BETON, HOLZ planed formwork, planed shuttering

**Gehrstoß** *m* BAUMASCHIN, HOLZ miter joint (*AmE*), mitre joint (*BrE*)

**Gehrung** *f* ARCH & TRAGW bevel, BAUMASCHIN, HOLZ miter (*AmE*), mitre (*BrE*), bevel; **auf ~ geschnitten** *adj* HOLZ mitered (*AmE*), mitred (*BrE*); **auf ~ geschnittener Stahlsturz** *m* NATURSTEIN, STAHL mitered steel lintel (*AmE*), mitred steel lintel (*BrE*); **Gehrungsfuge** *f* BAUMASCHIN, HOLZ miter joint (*AmE*), mitre joint (*BrE*); **Gehrungsschnittlehre** *f* BAUMASCHIN miter board (*AmE*), miter box (*AmE*), mitre board (*BrE*), mitre box (*BrE*); **Gehrungsstanzmaschine** *f* BAUMASCHIN miter-cutting machine (*AmE*), mitre-cutting machine (*BrE*); **Gehrungswinkel** *m* BAUMASCHIN miter square (*AmE*), mitre square (*BrE*)

**Gehweg** *m* INFR & ENTW banquette, pavement (*BrE*),

sidewalk (*AmE*), walkway; **Gehwegplatte** *f* INFR & ENTW flag, flagstone, pavement expansion joint (*BrE*), pavement paving flag (*BrE*), paving stone (*BrE*), sidewalk paving flag (*AmE*), NATURSTEIN flag, flagstone, paving stone, pavement expansion joint (*BrE*), WERKSTOFF pavement paving flag (*BrE*), sidewalk paving flag (*AmE*); **Gehwegplattenbelag** *m* INFR & ENTW flagstone pavement (*BrE*), flagstone sidewalk (*AmE*)

**Geigerzähler** *m* UMWELT Geiger counter

**gekerbt** *adj* WERKSTOFF notched

**gekrümmt** *adj* ARCH & TRAGW bent, curved, *Straße* cambered, BETON cambered, curved, INFR & ENTW bent, cambered, STAHL, WERKSTOFF bent; **~er Balken** *m* ARCH & TRAGW, BETON curved beam; **~e Gurtung** *f* ARCH & TRAGW, INFR & ENTW arched boom; **~e Platte** *f* ARCH & TRAGW curved slab; **~er Sparren** *m* HOLZ compass rafter; **~er Träger** *m* ARCH & TRAGW, BETON curved girder

**Gelände** *nt* INFR & ENTW ground, site; **Geländeabschnitt** *m* ARCH & TRAGW *Areal* area; **Geländeaufnahmeplan** *m* INFR & ENTW site survey plan; **Geländebruch** *m* INFR & ENTW sliding; **Geländeerkundung** *f* ARCH & TRAGW, INFR & ENTW site exploration

**geländegängig**: **~es Fahrzeug** *nt* **für den Erdbau** BAUMASCHIN off-road truck for earthwork

***Gelände***: **Geländehöhe** *f* ARCH & TRAGW, INFR & ENTW grade, ground level; **Geländeneigung** *f* INFR & ENTW fall of ground; **Geländeoberfläche** *f* ARCH & TRAGW ground level, INFR & ENTW grade, ground level; **Geländeplanierung** *f* BAURECHT, INFR & ENTW land leveling (*AmE*), land levelling (*BrE*)

**Geländer** *nt* ARCH & TRAGW, BESCHLÄGE, HOLZ, STAHL balustrade, guard rail, railing; **Geländerausfachung** *f* ARCH & TRAGW, INFR & ENTW paling; **Geländerpfosten** *m* ARCH & TRAGW railing, BESCHLÄGE baluster, HOLZ banister

***Gelände***: **Geländesetzung** *f* INFR & ENTW ground settlement

**Geldeinbehaltung** *f* BAURECHT retention of money, retention money

**Gelenk** *nt* ARCH & TRAGW, BESCHLÄGE, STAHL articulation, hinge; **Gelenkbolzen** *m* BESCHLÄGE hinge bolt, joint bolt

**gelenkig** *adj* ARCH & TRAGW jointed

***Gelenk***: **Gelenkträger** *m* ARCH & TRAGW articulated beam; **Gelenkverbindung** *f* BESCHLÄGE hinge joint

**gelocht** *adj* WERKSTOFF cored, perforated; **~er Backstein** *m* NATURSTEIN, WERKSTOFF perforated brick

**gelöst**: **~er Sauerstoff** *m* WERKSTOFF dissolved oxygen; **~e Stoffe** *m pl* WERKSTOFF dissolved matter

**gelötet** *adj* ELEKTR, HEIZ & BELÜFT, STAHL soldered; **~e Blechverbindung** *f* ARCH & TRAGW plumb joint

**gelten**: **~ für** *vi* BAURECHT apply to

**Geltungsdauer** *f* BAURECHT validity period

**gemasert** *adj* WERKSTOFF grained

**Gemäuer** *nt* ARCH & TRAGW, NATURSTEIN masonry, walling

**gemauert** *adj* ARCH & TRAGW, NATURSTEIN brick-built, bricked up; **~es Bauteil** *nt* ARCH & TRAGW, NATURSTEIN brick structure; **~er Damm** *m* INFR & ENTW, NATURSTEIN masonry dam

**Gemeinde** *f* BAURECHT community; **Gemeindeordnung** *f* BAURECHT by-law; **Gemeindewasserversorgung** *f* ABWASSER, BAURECHT, INFR & ENTW public water supply; **Gemeindezentrum** *nt* INFR & ENTW community center (*AmE*), community centre (*BrE*)

**Gemeinkosten** *pl* ARCH & TRAGW, BAURECHT overhead costs, overheads

**gemeinsam**: **~es Fundament** *nt* ARCH & TRAGW *mehrerer Stützen* combined footing

**gemeinschaftlich**: **~e Teile** *m pl* BAURECHT, INFR & ENTW common parts

**Gemeinschaftswaschküche** *f* ARCH & TRAGW communal laundry

**Gemeinwesen** *nt* BAURECHT community

**Gemisch** *nt* WERKSTOFF mixture

**gemischt**: **~e Ablagerung** *f* UMWELT codeposition, codisposal; **~er hydraulischer Zement** *m* BETON blended hydraulic cement

**gemischtkörnig** *adj* WERKSTOFF mixed-grained

**genagelt** *adj* ARCH & TRAGW, HOLZ, STAHL nailed; **~es Sperrholzknotenblech** *nt* HOLZ, STAHL nailed plywood gusset; **~er Träger** *m* ARCH & TRAGW, HOLZ nailed beam

**genarbt** *adj* OBERFLÄCHE grained

**genau 1.** *adj* ARCH & TRAGW, BAUMASCHIN, BAURECHT, BETON, INFR & ENTW, STAHL, WERKSTOFF accurate, exact, precise; **2. ~ beschreiben** *vt* ARCH & TRAGW, BAURECHT specify

***genau***: **~ fluchtend** *adj* ARCH & TRAGW in true alignment; **~e Überprüfung** *f* ARCH & TRAGW meticulous inspection

**Genauigkeit** *f* ARCH & TRAGW, INFR & ENTW, WERKSTOFF accuracy; **Genauigkeitsgrad** *m* INFR & ENTW, WERKSTOFF degree of accuracy

**genehmigt**: **~e Liste** *f* ARCH & TRAGW, BAURECHT approved list

**Genehmigung** *f* ARCH & TRAGW, BAURECHT approval, authorization, permit

**genehmigungspflichtig** *adj* BAURECHT subject to authorization

**geneigt** *adj* ARCH & TRAGW *Fläche* inclined, sloped, sloping, *Gebäude* tilted; **~es Dach** *nt* ARCH & TRAGW pitched roof; **~e Pfette** *f* HOLZ, WERKSTOFF angled purlin; **~er Sparren** *m* ARCH & TRAGW, HOLZ angled rafter

**General-** *in cpds* ARCH & TRAGW, BAURECHT master, main; **Generalauftragnehmer** *m* ARCH & TRAGW main contractor; **Generalbebauungsplan** *m* BAURECHT master plan; **Generalschlüssel** *m* BAURECHT, BESCHLÄGE master key; **Generalunternehmer** *m* ARCH & TRAGW master builder

**Generator** *m* ELEKTR generator

**genietet** *adj* BAUMASCHIN, STAHL riveted

**genutet** *adj* WERKSTOFF grooved

**Geodäsie** *f* INFR & ENTW geodesy

**geodätisch**: **~e Tragwerke** *nt pl* INFR & ENTW geodesics

**geographisch**: **~e Abweichung** *f* UMWELT geographic variation

**Geologie** *f* INFR & ENTW geology

**geologisch**: **~e Aufnahme** *f* INFR & ENTW geological survey; **~er Plan** *m* INFR & ENTW geological map; **~es Profil** *nt* INFR & ENTW geological profile

**Geomembrane** *f* INFR & ENTW, WERKSTOFF geomembrane

**Geomembranverkleidung** *f* INFR & ENTW, WERKSTOFF geomembrane liner

**Geometrie** *f* INFR & ENTW shape
**geordnet**: **~es Ablagern** *nt* UMWELT landfilling; **~e Ablagerung** *f* UMWELT sanitary landfilling, proper disposal; **~e Beseitigung** *f* UMWELT safe disposal; **~e Deponie** *f* UMWELT controlled dumping, sanitary landfill
**geosynthetisch**: **~e Verkleidung** *f* INFR & ENTW, WERKSTOFF geosynthetic liner
**Geotextilien** *f pl* INFR & ENTW, WERKSTOFF geotextiles
**geothermisch** *adj* UMWELT geothermal; **~e Anlage** *f* UMWELT geothermal plant; **~e Bohrausrüstung** *f* UMWELT geothermal drilling equipment; **~e Energie** *f* HEIZ & BELÜFT, UMWELT geothermal energy, geothermal power; **~es Feld** *nt* UMWELT geothermal field; **~er Kreislauf** *m* UMWELT geothermal cycle; **~e Quellen** *f pl* UMWELT geothermal resources; **~e Tiefenstufe** *f* UMWELT geothermal gradient, geothermal step
**Geovlies** *nt* INFR & ENTW, WERKSTOFF geofabric
**Gepäck** *nt* ARCH & TRAGW, INFR & ENTW baggage, luggage; **Gepäckaufzug** *m* ARCH & TRAGW baggage elevator (*AmE*), baggage lift (*BrE*), luggage elevator (*AmE*), luggage lift (*BrE*); **Gepäckausgabebereich** *m* ARCH & TRAGW, INFR & ENTW baggage reclaim area; **Gepäckschalter** *m* ARCH & TRAGW, INFR & ENTW baggage-handling counter
**gepanzert** *adj* ELEKTR, STAHL armored (*AmE*), armoured (*BrE*); **~es Kabel** *nt* ELEKTR, INFR & ENTW, WERKSTOFF armored cable (*AmE*), armoured cable (*BrE*)
**gepfuscht** *adj* BAURECHT jerry-built
**geplant** *adj* ARCH & TRAGW, BAURECHT, INFR & ENTW planned
**gerade** *adj* ARCH & TRAGW straight
**Gerade** *f* ARCH & TRAGW straight line
***gerade***: **~r Bogen** *m* ARCH & TRAGW flat arch, jack arch, straight arch; **~ Kante** *f* ARCH & TRAGW straight edge
**geradlinig** *adj* ARCH & TRAGW, INFR & ENTW linear
**Gerät** *nt* BAUMASCHIN appliance, utensil; **Gerätschuppen** *m* ARCH & TRAGW tool shed
**geräumig** *adj* INFR & ENTW, WERKSTOFF ample
**Geräusch** *nt* BAURECHT, DÄMMUNG, INFR & ENTW, UMWELT noise
**geräuscharm** *adj* BAUMASCHIN low-noise
**geräuschdämpfend** *adj* DÄMMUNG noise-absorbent
***Geräusch***: **Geräuschmessung** *f* BAURECHT noise measurement; **Geräuschminderung** *f* DÄMMUNG noise reduction; **Geräuschpegelanzeiger** *m* UMWELT weighted noise level indicator; **Geräuschstärke** *f* DÄMMUNG noise intensity
**Gerber**: **Gerber-Fachwerkbinder** *m* ARCH & TRAGW articulated beam; **Gerbergelenk** *nt* ARCH & TRAGW, NATURSTEIN bucket handle joint; **Gerberpfette** *f* HOLZ hinged ridge purlin
**gereinigt**: **~es Wasser** *nt* ABWASSER, UMWELT depolluted water
**Gericht** *nt* BAURECHT court
**gerichtet** *adj* STAHL *Blech* flattened; **~es Blech** *nt* STAHL straightened sheet
***Gericht***: **Gerichtsgebäude** *nt* INFR & ENTW courthouse; **Gerichtsverhandlungen** *f pl* BAURECHT tribunal proceedings
**geriefelt** *adj* ARCH & TRAGW, BETON, HEIZ & BELÜFT, STAHL, WERKSTOFF ribbed
**gering-kontinuierlich**: **~e Belüftung** *f* DÄMMUNG, HEIZ & BELÜFT trickle ventilation
**Gerinne** *nt* ABWASSER chute, raceway, ARCH & TRAGW launder, INFR & ENTW channel, flume, UMWELT flume
**Gerippe** *nt* ARCH & TRAGW, BETON carcass, skeleton, *Fachwerk* stud; **Gerippetrennwand** *f* ARCH & TRAGW framed partition, stud wall; **Gerippewand** *f* ARCH & TRAGW *Innenwand* framed partition, *Innenwand* stud wall
**gerippt** *adj* ARCH & TRAGW, BETON, HEIZ & BELÜFT, STAHL, WERKSTOFF ribbed; **~e Absorberplatte** *f* HEIZ & BELÜFT finned absorber plate; **~es Bewehrungseisen** *nt* BETON, STAHL ribbed bar
**Geröll** *nt* NATURSTEIN pebble stone; **Geröllblock** *m* INFR & ENTW boulder; **Geröllstein** *m* NATURSTEIN pebble stone
**gerollt** *adj* STAHL coiled
**Geruch** *m* ABWASSER, UMWELT odor (*AmE*), odour (*BrE*); **Geruchsbekämpfung** *f* UMWELT odor control (*AmE*), odour control (*BrE*); **Geruchsbelästigung** *f* UMWELT odor nuisance (*AmE*), odour nuisance (*BrE*); **Geruchsbeseitigung** *f* ABWASSER deodorization; **Geruchsemission** *f* UMWELT odor emission (*AmE*), odour emission (*BrE*); **Geruchsverschluß** *m* ABWASSER drain trap, odor trap (*AmE*), odour trap (*BrE*), stench trap
**Gerümpel** *nt* UMWELT litter
**Gerüst** *nt* ARCH & TRAGW scaffold, frame, scaffolding, BAUMASCHIN gantry, BESCHLÄGE scaffold, INFR & ENTW, STAHL frame; **ein ~ aufstellen** *phr* ARCH & TRAGW scaffold; **Gerüstbau** *m* ARCH & TRAGW staging; **ein ~ bauen** *phr* ARCH & TRAGW scaffold; **Gerüstbock** *m* ARCH & TRAGW, INFR & ENTW trestle; **Gerüstbohle** *f* ARCH & TRAGW scaffold board; **Gerüstloch** *nt* NATURSTEIN putlog hole; **Gerüststange** *f* ARCH & TRAGW scaffold pole
**Gesamt-** *in cpds* ARCH & TRAGW, BAURECHT, INFR & ENTW, UMWELT total; **Gesamtablagerung** *f* UMWELT total deposition; **Gesamtbaukosten** *pl* ARCH & TRAGW, BAURECHT, INFR & ENTW total building costs; **Gesamtbetrag** *m* ARCH & TRAGW total; **Gesamtdruck** *m* ARCH & TRAGW, INFR & ENTW total pressure; **Gesamterdungswiderstand** *m* ELEKTR total earth resistance (*BrE*), total ground resistance (*AmE*); **Gesamtevakuierung** *f* BAURECHT, INFR & ENTW total evacuation; **Gesamtfläche** *f* ARCH & TRAGW total area, UMWELT *eines Kollektors* gross area; **Gesamtgewicht** *nt* ARCH & TRAGW, INFR & ENTW total weight; **Gesamthärte** *f* WERKSTOFF total hardness; **Gesamtinhalt** *m* ARCH & TRAGW total capacity; **Gesamtkosten** *pl* BAURECHT total cost; **Gesamtlast** *f* ARCH & TRAGW, INFR & ENTW total design load; **Gesamtquerschnitt** *m* ARCH & TRAGW, INFR & ENTW total cross-section; **Gesamtreibung** *f* ARCH & TRAGW, INFR & ENTW total friction; **Gesamtverlust** *m* BAURECHT, WERKSTOFF total loss; **Gesamtwirkung** *f* INFR & ENTW total efficiency; **Gesamtwirkungsgrad** *m* HEIZ & BELÜFT, UMWELT overall efficiency
**Geschäftsbuch** *nt* ARCH & TRAGW, BAURECHT account book
**geschichtet** *adj* WERKSTOFF stratified; **~er Boden** *m* INFR & ENTW stratified soil; **~es Bruchsteinmauerwerk** *nt* NATURSTEIN coursed masonry
**Geschiebe** *nt* INFR & ENTW bed load
**geschleudert** *adj* WERKSTOFF *Rohr* centrifugally cast

**geschliffen** *adj* WERKSTOFF ground; **~es Glas** *nt* ARCH & TRAGW cut glass; **~e Kante** *f* ARCH & TRAGW earth edge (*BrE*), ground edge (*AmE*)
**geschlitzt** *adj* ARCH & TRAGW slotted; **~es Filterrohr** *nt* INFR & ENTW perforated pipe
**geschlossen**: **~e Ecke** *f* ARCH & TRAGW tight corner; **~e Kompostierung** *f* UMWELT mechanical composting, rapid fermentation; **~er Polygonzug** *m* INFR & ENTW closed traverse; **~e Rundzelle** *f* ARCH & TRAGW covered vault
**geschlossenzellig** *adj* WERKSTOFF *Kunststoff* with closed cells
**geschmeidig** *adj* WERKSTOFF ductile
**Geschoß** *nt* ARCH & TRAGW storey (*BrE*), story (*AmE*); **~ mit dem Haupteingang** ARCH & TRAGW, INFR & ENTW principal entrance storey (*BrE*), principal entrance story (*AmE*); **Geschoßdecke** *f* ARCH & TRAGW floor; **Geschoßhöhe** *f* ARCH & TRAGW floor-to-floor height; **Geschoßquerbalken** *m* HOLZ summer tree
**geschützt** *adj* BAUMASCHIN screened
**geschweift**: **~es Dach** *nt* ARCH & TRAGW curved roof
**geschweißt** *adj* STAHL welded; **~er Abschnitt** *m* ARCH & TRAGW, STAHL, WERKSTOFF welded section; **~e Maschen** *f pl* ARCH & TRAGW, BETON, STAHL, WERKSTOFF welded mesh; **~es Stahlprofil** *nt* ARCH & TRAGW, STAHL, WERKSTOFF welded structural steel
**Geschwindigkeit** *f* ABWASSER, BAUMASCHIN, HEIZ & BELÜFT, INFR & ENTW speed, velocity; **Geschwindigkeitsdiagramm** *nt* UMWELT velocity diagram; **Geschwindigkeitsdruck** *m* ABWASSER, HEIZ & BELÜFT, INFR & ENTW velocity pressure; **Geschwindigkeitshöhe** *f* ABWASSER, HEIZ & BELÜFT, INFR & ENTW velocity head; **Geschwindigkeitskoeffizient** *m* UMWELT velocity coefficient; **Geschwindigkeitspotential** *nt* UMWELT velocity potential; **Geschwindigkeitsverteilung** *f* ABWASSER, HEIZ & BELÜFT, INFR & ENTW velocity distribution; **Geschwindigkeitszahl** *f* WERKSTOFF velocity index
**geschwungen**: **~e Antrittsstufe** *f* ARCH & TRAGW curtail step, curtail
**Gesellschaft** *f* BAURECHT association, society
**Gesetz** *nt* BAURECHT law, Act; **~ zur Reinhaltung der Luft** UMWELT Clean Air Act; **Gesetzesänderungen** *f pl* BAURECHT changes in the law; **Gesetzgebung** *f* BAURECHT legislation
**Gesichtspunkt** *m* ARCH & TRAGW criterion, INFR & ENTW *Ansicht, Betrachtungsweise* aspect
**gesiebt** *adj* BAUMASCHIN screened
**Gesims** *nt* ARCH & TRAGW cornice, molding (*AmE*), moulding (*BrE*); **Gesimsband** *nt* ARCH & TRAGW string
**gespalten** *adj* ARCH & TRAGW split
**gespannt**: **~es Grundwasser** *nt* INFR & ENTW confined ground water
**gespitzt**: **~e Fläche** *f* BETON, NATURSTEIN scabbled area
**gespleißt**: **~e Stelle** *f* HOLZ splice
**gespundet** *adj* HOLZ tongued-and-grooved
**gestaffelt** *adj* ARCH & TRAGW, BETON staggered
**Gestalt** *f* ARCH & TRAGW form
**gestatten** *vt* BAURECHT permit
**Gestein** *nt* ARCH & TRAGW, NATURSTEIN rock, stone; **Gesteinsanker** *m* INFR & ENTW rockbolt; **Gesteinsbohrer** *m* BAUMASCHIN rock borer, rock drill; **Gesteinskunde** *f* WERKSTOFF mineralogy
**Gestell** *nt* ARCH & TRAGW frame, trestle, BESCHLÄGE rack, HOLZ shelf, INFR & ENTW frame, trestle, STAHL frame
**gestrichen** *adj* OBERFLÄCHE, WERKSTOFF painted
**gesund** *adj* BAURECHT able-bodied
**Gesundheit** *f*: **~ und Komfort** HEIZ & BELÜFT health and comfort; **Gesundheitstechnik** *f* ABWASSER, BAURECHT, INFR & ENTW public health engineering
**getäfelt**: **~e Decke** *f* HOLZ paneled ceiling (*AmE*), panelled ceiling (*BrE*)
**geteilt** *adj* ARCH & TRAGW split, louvered (*AmE*), louvred (*BrE*)
**getönt** *adj* OBERFLÄCHE tinted; **~e Plexiglasscheibe** *f* ARCH & TRAGW tinted plexiglass pane
**getränkt** *adj* DÄMMUNG impregnated, WERKSTOFF saturated
**getrennt**: **~e Abfallagerung** *f* UMWELT waste segregation; **~e Müllabfuhr** *f* UMWELT selective collection; **~e Müllsammlung** *f* UMWELT selective collection
**getrocknet** *adj* HOLZ baked
**gewachsen**: **~er Boden** *m* UMWELT unspoilt land; **~er Fels** *m* INFR & ENTW bed rock
**Gewächshaus** *nt* ARCH & TRAGW greenhouse, hothouse
**gewähren** *vt* BAURECHT allow
**gewährleisten** *vt* BAURECHT guarantee
**Gewährleistung** *f* BAURECHT guarantee, warranty; **Gewährleistungsabnahme** *f* BAURECHT warranty inspection; **Gewährleistungszeit** *f* BAURECHT warranty period
**Gewändepfosten** *m* ARCH & TRAGW, HOLZ jamb
**Gewässer** *nt* BAURECHT, UMWELT water, waters; **Gewässerbelastung** *f* UMWELT water pollution; **Gewässerkunde** *f* INFR & ENTW, UMWELT hydrography; **Gewässerschutz** *m* BAURECHT, UMWELT river and lake protection, water protection
**Gewebe** *nt* WERKSTOFF fabric, tissue; **Gewebeeinlage** *f* WERKSTOFF cloth insert; **Gewebefilter** *m* INFR & ENTW, UMWELT fabric filter; **Gewebeträgermaterial** *nt* WERKSTOFF cloth backing, cloth base
**gewebeverstärkt** *adj* WERKSTOFF cloth-reinforced
***Gewebe***: **Gewebeverstärkung** *f* WERKSTOFF scrim
**gewellt** *adj* WERKSTOFF corrugated
**Gewerbe** *nt* ARCH & TRAGW, BAURECHT, INFR & ENTW, UMWELT business, trade; **Gewerbeabfall** *m* UMWELT industrial waste; **Gewerbesteuer** *f* BAURECHT, INFR & ENTW business tax
**gewerblich**: **~er Abfall** *m* BAURECHT, UMWELT commercial waste, industrial waste, process waste, trade waste; **~es Abwasser** *nt* ABWASSER, INFR & ENTW, UMWELT industrial effluent, industrial sewage; **~es Gebäude** *nt* ARCH & TRAGW, BAURECHT, INFR & ENTW commercial building; **~e Nutzfläche** *f* BAURECHT, INFR & ENTW commercial area
**Gewerk** *nt* ARCH & TRAGW trade
**Gewicht** *nt* WERKSTOFF weight; **Gewichtsanteil** *m* INFR & ENTW, WERKSTOFF fraction weight; **Gewichtsausgleich** *m* BAUMASCHIN equilibration; **Gewichtseinheit** *f* WERKSTOFF unit weight; **Gewichtsteil** *nt* INFR & ENTW, WERKSTOFF part by weight; **Gewichtsverlust** *m* INFR & ENTW, WERKSTOFF loss of weight; **Gewichtswiderlager** *nt* BETON, INFR & ENTW gravity abutment
**Gewinde** *nt* ABWASSER, BAUMASCHIN, HEIZ & BELÜFT, STAHL, WERKSTOFF thread; **Gewindeanschluß** *m* ABWASSER, HEIZ & BELÜFT threaded connection;

**Gewindebohrer** *m* BAUMASCHIN screw tap; **Gewindebolzen** *m* BESCHLÄGE threaded bolt; **Gewindedreher** *m* ABWASSER pipe threader; **Gewindemuffe** *f* ABWASSER, HEIZ & BELÜFT screw socket; **Gewindenachbohrer** *m* ABWASSER, BAUMASCHIN plug tap; **Gewinderohr** *nt* ABWASSER threaded pipe; **Gewindeschneidöl** *nt* STAHL thread-cutting oil; **Gewindestab** *m* BESCHLÄGE threaded rod; **Gewindestange** *f* BESCHLÄGE threaded rod; **Gewindestift** *m* ARCH & TRAGW stud bolt; **Gewindestopfen** *m* BAUMASCHIN screw plug; **Gewindetiefe** *f* BAUMASCHIN thread depth

**Gewinn** *m* BAURECHT, INFR & ENTW profit, UMWELT yield

**gewinnen** *vt* INFR & ENTW recover

***Gewinn***: **Gewinnspanne** *f* BAURECHT, INFR & ENTW profit margin

**gewöhnlich**: **~er Portlandzement** *m* BETON, WERKSTOFF ordinary Portland cement

**Gewölbe** *nt* ARCH & TRAGW, NATURSTEIN vault; **Gewölbeanker** *m* ARCH & TRAGW tie anchor; **Gewölbeleibung** *f* ARCH & TRAGW intrados; **Gewölbepfeilermauer** *f* INFR & ENTW *Staumauer* multiple-arch dam; **Gewölberippe** *f* ARCH & TRAGW vaulting rib; **Gewölborücken** *m* ARCH & TRAGW extrados; **Gewölbeschenkel** *m* ARCH & TRAGW haunch; **Gewölbeschub** *m* ARCH & TRAGW arch thrust

**gewölbt** *adj* ARCH & TRAGW, BETON, HOLZ, INFR & ENTW, NATURSTEIN arched, cambered; **~e Decke** *f* ARCH & TRAGW vaulted ceiling

**gewunden** *adj* INFR & ENTW meandering

**gezackt**: **~e Schaufel** *f* BAUMASCHIN pronged shovel

**Gezeit** *f* UMWELT tide; **Gezeitenbereich** *m* UMWELT tidal range; **Gezeitenbewegung** *f* UMWELT tidal movement; **Gezeitenenergie** *f* INFR & ENTW, UMWELT tidal power; **Gezeitenkraft** *f* INFR & ENTW, UMWELT tidal power; **Gezeitenkraftwerk** *nt* INFR & ENTW, UMWELT tidal power plant, tidal power station; **Gezeitenmesser** *m* UMWELT tide gage (*AmE*), tide gauge (*BrE*); **Gezeitenmühle** *f* UMWELT tide mill; **Gezeitenprisma** *nt* UMWELT tidal prism; **Gezeitenstrom** *m* UMWELT tidal current

**Giebel** *m* ARCH & TRAGW gable; **Giebelbogen** *m* ARCH & TRAGW triangular arch; **Giebeldach** *nt* ARCH & TRAGW gable roof, ridge roof; **Giebelfenster** *nt* ARCH & TRAGW gable window; **Giebelfußstein** *m* ARCH & TRAGW skew; **Giebelgaube** *f* ARCH & TRAGW gabled dormer window, gabled dormer; **Giebelkante** *f* ARCH & TRAGW verge; **Giebelmauer** *f* NATURSTEIN flank wall; **Giebelschlußstein** *m* ARCH & TRAGW, NATURSTEIN apex stone; **Giebelseite** *f* ARCH & TRAGW gable wall; **Giebelspitze** *f* ARCH & TRAGW gable peak; **Giebelstufe** *f* ARCH & TRAGW corbel step, corbel vault, crowstep; **Giebelstütze** *f* ARCH & TRAGW gable column; **Giebel- und Traufkanten** *f pl* ARCH & TRAGW gable and eaves edges; **Giebeltürmchen** *nt* ARCH & TRAGW bell cote; **Giebelwand** *f* ARCH & TRAGW, NATURSTEIN flank wall, gable end

**Gieß-** *in cpds* BAUMASCHIN, BETON, WERKSTOFF casting

**gießen** *vt* BETON cast, pour, STAHL, WERKSTOFF cast

**Gießen** *nt* BETON, NATURSTEIN pouring

***Gieß-***: **Gießharz** *nt* WERKSTOFF casting resin; **Gießlöffel** *m* BAUMASCHIN *Maurerwerkzeug* ladle; **Gießmaschine** *f* BAUMASCHIN casting machine; **Gießmörtel** *m* WERKSTOFF casting mortar

**Gift** *nt* BAURECHT, INFR & ENTW, UMWELT poison, toxin

**giftig** *adj* BAURECHT, INFR & ENTW, UMWELT toxic; **~er Abfall** *m* BAURECHT, INFR & ENTW toxic waste; **~es Nebenprodukt** *nt* BAURECHT, INFR & ENTW toxic by-product

***Gift***: **Giftmüll** *m* UMWELT poisonous waste, toxic waste; **Giftmüllentsorgungsanlage** *f* UMWELT toxic waste disposal plant; **Giftstoff** *m* BAURECHT, INFR & ENTW, UMWELT toxic agent, toxic matter, toxicant

**Gipfel** *m* ARCH & TRAGW summit, vertex

**Gips** *m* WERKSTOFF gypsum; **Gipsdecke** *f* NATURSTEIN stucco ceiling; **Gipsestrich** *m* NATURSTEIN plaster floor; **Gipsgestein** *nt* NATURSTEIN plaster rock, plaster stone; **Gipskartonplatte** *f* WERKSTOFF gypsum plaster, plasterboard; **Gipsmörtel** *m* BETON staff, WERKSTOFF plaster; **Gipsputz** *m* NATURSTEIN stucco; **Gipsputzunterlage** *f* STAHL *aus Streckmetall* furring; **Gipsschlackenzement** *m* BETON, WERKSTOFF gypsum slag cement, supersulfated cement (*AmE*), supersulphated cement (*BrE*); **Gipsunterputz** *m* WERKSTOFF gypsum plaster

**Gitter** *nt* HOLZ lattice, trellis, STAHL grating, grid, grille; **Gitterbalken** *m* HOLZ lattice truss; **Gitterbrücke** *f* HOLZ lattice bridge; **Gittergewebe** *nt* NATURSTEIN, WERKSTOFF *Putz* cloth lathing, mesh lath; **Gittergewebematte** *f* NATURSTEIN *Putz* mesh lath; **Gittermast** *m* ARCH & TRAGW, BESCHLÄGE, INFR & ENTW pylon, tower; **Gittermauer** *f* BAUMASCHIN screen wall; **Gitterpfosten** *m* HOLZ trellis post; **Gitterrost** *m* ARCH & TRAGW bar grate, BESCHLÄGE grate, STAHL grid iron, *Bodenrost* grating, WERKSTOFF grillage; **Gitterschale** *f* HOLZ lattice shell; **Gitterstab** *m* BAUMASCHIN screen bar, STAHL grate bar; **Gitterstütze** *f* HOLZ latticed column; **Gitterträger** *m* ARCH & TRAGW lattice girder, HOLZ lattice beam, lattice girder; **Gitterwerk** *nt* HOLZ latticework; **Gitterzerkleinerer** *m* UMWELT grid crusher; **Gitterziegel** *m* NATURSTEIN, WERKSTOFF honeycomb brick, perforated brick

**Glanz** *m* OBERFLÄCHE gloss, luster (*AmE*), lustre (*BrE*)

**glanzverzinkt** *adj* OBERFLÄCHE bright galvanized

**Glas** *nt* WERKSTOFF glass; **Glasart** *f* WERKSTOFF glass type, type of glass; **Glasbaustein** *m* WERKSTOFF glass block; **Glasbehälter** *m* UMWELT bottle bank; **Glasbruch** *m* UMWELT cullet; **Glascontainer** *m* UMWELT bottle bank

**Glaser** *m* ARCH & TRAGW glazier; **Glaserdiamant** *m* ARCH & TRAGW, BESCHLÄGE diamond pencil, glazier's diamond; **Glaserkitt** *m* ARCH & TRAGW sash putty

***Glas***: **Glasfalz** *m* ARCH & TRAGW rebate for putty

**Glasfaser** *f* WERKSTOFF fiberglass (*AmE*), fibreglass (*BrE*), glass fiber (*AmE*), glass fibre (*BrE*)

**glasfaserbewehrt**: **~er Zement** *m* BETON, WERKSTOFF glass fiber reinforced cement (*AmE*) (*GRC*), glass fibre reinforced cement (*BrE*) (*GRC*); **~e Zementverkleidung** *f* UMWELT glass fiber reinforced cement cladding (*AmE*), glass fibre reinforced cement cladding (*BrE*), GRC cladding

***Glasfaser***: **Glasfaserdachrinne** *f* ABWASSER glass fiber gutter (*AmE*), glass fibre gutter (*BrE*); **Glasfasergewebe** *nt* WERKSTOFF glass fabric; **Glasfasermatte** *f* DÄMMUNG, WERKSTOFF glass fiber quilt (*AmE*), glass fibre quilt (*BrE*); **Glasfaserschalung** *f* ARCH & TRAGW glass fiber formwork (*AmE*), glass fibre formwork (*BrE*)

**glasfaserverstärkt** *adj* WERKSTOFF *Kunststoff* glass

fiber reinforced (*AmE*), glass fibre reinforced (*BrE*); **~e Kunststoffe** *m pl* WERKSTOFF fiberglass-reinforced plastics (*AmE*), fibreglass-reinforced plastics (*BrE*), glass fiber reinforced plastics (*AmE*), glass fibre reinforced plastics (*BrE*)

***Glas***: **Glasfassade** *f* ARCH & TRAGW glass front; **Glasfüllungstür** *f* BAUMASCHIN sash door

**glasieren** *vt* OBERFLÄCHE glaze

**glasiert** *adj* OBERFLÄCHE glazed

**glasig**: **~es Gestein** *nt* NATURSTEIN vitreous rock

***Glas***: **Glasleiste** *f* BESCHLÄGE window bar; **Glaspapier** *nt* WERKSTOFF glass paper; **Glasscheibe** *f* ARCH & TRAGW glass pane; **Glasschneider** *m* BAUMASCHIN glass cutter; **Glassorte** *f* WERKSTOFF glass type, type of glass; **Glastür** *f* ARCH & TRAGW French door, glazed door, HOLZ French door

**Glasur** *f* OBERFLÄCHE glaze, WERKSTOFF enamel

***Glas***: **Glasvlies** *nt* WERKSTOFF glass fiber mat (*AmE*), glass fibre mat (*BrE*); **Glaswatte** *f* WERKSTOFF glass wadding; **Glaswolle** *f* DÄMMUNG, WERKSTOFF glass wool, spun glass; **Glasziegel** *m* ARCH & TRAGW, WERKSTOFF *Dachdeckung* glass tile

**glatt** *adj* ARCH & TRAGW, WERKSTOFF fair-faced, smooth; **~e Bewehrungseisen** *nt pl* ARCH & TRAGW, BETON, STAHL plain bars; **~e Fuge** *f* BETON, NATURSTEIN straight joint; **~e Schalung** *f* BETON smooth formwork

**Glättbohle** *f* NATURSTEIN screed board, smoothing board; **~ für Oberbeton oder Deckenschluß** BETON final float for road surfacing

**glätten** *vt* BAUMASCHIN plane, BETON, NATURSTEIN float, OBERFLÄCHE skim

**Glätten** *nt* BAUMASCHIN planing, BETON, NATURSTEIN floating, OBERFLÄCHE skimming

**Glätter** *m* BAUMASCHIN finishing trowel

**glatthobeln** *vt* INFR & ENTW *Holzkanten* shoot

**Glättkelle** *f* BAUMASCHIN, NATURSTEIN float, smoothing trowel, trowel

**Glattmantelwalze** *f* BAUMASCHIN, INFR & ENTW smooth roller

**glattstreichen** *vt* NATURSTEIN trowel

**Glattstrich** *m* NATURSTEIN float coat, trowel finish, trowel-finished layer

**Glaubersalz** *nt* HEIZ & BELÜFT, WERKSTOFF Glauber's salt

**Glaukonit** *nt* WERKSTOFF glauconite

**Gleichdruckturbine** *f* UMWELT impulse turbine

**gleichförmig** *adj* ARCH & TRAGW, INFR & ENTW uniform

**Gleichgewicht** *nt* ARCH & TRAGW, INFR & ENTW equilibrium, UMWELT balance, equilibrium; **Gleichgewichtsbedingungen** *f pl* ARCH & TRAGW, INFR & ENTW equilibrium conditions; **Gleichgewichtsgleichung** *f* ARCH & TRAGW, INFR & ENTW equilibrium equation; **Gleichgewichtsgrenze** *f* ARCH & TRAGW, INFR & ENTW equilibrium limit; **Gleichgewichtsmodell** *nt* ARCH & TRAGW, INFR & ENTW equilibrium model; **Gleichgewichtszustand** *m* ARCH & TRAGW, INFR & ENTW equilibrium position, state of equilibrium; **Gleichgewichtszustand** *m* **des Gezeitenwechsels** UMWELT equilibrium tide

**gleichkörnig**: **~er Kieszuschlagstoff** *m* WERKSTOFF single-size gravel aggregate

**Gleichlast** *f* INFR & ENTW uniformly distributed load

**gleichmäßig**: **~ gekörnt** *adj* WERKSTOFF evenly graded; **~e Korngröße** *f* WERKSTOFF uniform grain size; **~er Querschnitt** *m* ARCH & TRAGW, BETON, HOLZ, INFR & ENTW, STAHL *Träger* uniform depth; **~e Setzung** *f* ARCH & TRAGW, INFR & ENTW uniform settlement; **~ verteilte Last** *f* INFR & ENTW uniformly distributed load

**gleichschenklig** *adj* INFR & ENTW isosceles; **~es Winkeleisen** *nt* BESCHLÄGE equal-sided angle

**gleichseitig**: **~es Giebeldach** *nt* ARCH & TRAGW span roof

**gleichwertig** *adj* ARCH & TRAGW, INFR & ENTW equivalent

**gleichzeitig** *adj* BAUMASCHIN simultaneous

**Gleis** *nt* INFR & ENTW track; **Gleisanschluß** *m* INFR & ENTW siding; **Gleisbett** *nt* INFR & ENTW track bedding; **Gleiskette** *f* BAUMASCHIN caterpillar chain; **Gleisschotter** *m* INFR & ENTW track ballast; **Gleistrageplatten** *f pl* INFR & ENTW track base plates; **Gleistrasse** *f* INFR & ENTW track line; **Gleisverbindung** *f* INFR & ENTW track connection; **Gleisverlegung** *f* INFR & ENTW track laying

**Gleit-** *in cpds* INFR & ENTW, WERKSTOFF sliding; **Gleitbewegung** *f* INFR & ENTW sliding motion

**gleiten** *vi* INFR & ENTW slip

**gleitfest**: **~e Schraube** *f* STAHL *hochfest vorgespannt* friction-grip bolt; **~e Schraubenverbindung** *f* ARCH & TRAGW friction-grip bolt connection, high-tensile bolted structural joint, STAHL friction-grip bolting, high-tensile bolted structural joint

***Gleit-***: **Gleitfläche** *f* ARCH & TRAGW sliding surface, INFR & ENTW slip plane; **Gleitfuge** *f* ABWASSER *Rohr* sliding joint, INFR & ENTW slip joint; **Gleitkreis** *m* INFR & ENTW *Bodenmechanik* slip circle; **Gleitlager** *nt* BAUMASCHIN sliding bearing; **Gleitpassung** *f* BESCHLÄGE easy fit; **Gleitreibung** *f* WERKSTOFF sliding friction, sliding resistance; **Gleitringdichtung** *f* ABWASSER *Pumpe* slide ring packing, slide ring sealing, HEIZ & BELÜFT *Pumpe* slide ring sealing; **Gleitschalung** *f* ARCH & TRAGW sliding formwork, sliding shuttering, BETON traveling form (*AmE*), travelling form (*BrE*), INFR & ENTW sliding formwork; **Gleitschiene** *f* STAHL slideway

**gleitsicher** *adj* WERKSTOFF nonskid, skidproof

***Gleit-***: **Gleitsicherheit** *f* INFR & ENTW stability against gliding; **Gleitstange** *f* BETON jack rod; **Gleitwiderstand** *m* INFR & ENTW sliding resistance

**Gletscher** *m* INFR & ENTW glacier

**Gliederung** *f* ARCH & TRAGW division

**Glimmer** *m* INFR & ENTW, WERKSTOFF mica; **Glimmerschiefer** *m* WERKSTOFF mica slate; **Glimmerton** *m* WERKSTOFF micaceous clay

**Glocke** *f* ARCH & TRAGW bell, *eines Dachfirstes* cap; **Glockenmuffenrohr** *nt* ABWASSER, WERKSTOFF bell and spigot pipe; **Glockenmuffenverbindung** *f* ABWASSER, WERKSTOFF bell and spigot joint; **Glockenstuhl** *m* ARCH & TRAGW bell frame; **Glockenturm** *m* INFR & ENTW campanile

**glühen** *vi* STAHL anneal

**Glühschmelzer** *m* BAUMASCHIN *Tiefbau* heating melter

**Glühverlust** *m* UMWELT ignition loss

**Gneis** *m* INFR & ENTW, WERKSTOFF gneiss

**Göpel** *m* BAUMASCHIN *Bohrarbeiten* gin

**gotisch**: **~er Bogen** *m* ARCH & TRAGW lancet arch; **~e Rose** *f* ARCH & TRAGW wheel window

**Graben** *m* ARCH & TRAGW *Entwässerung*, INFR & ENTW, UMWELT trench, ditch, dike; **Grabenarbeiten** *f pl* ARCH & TRAGW, INFR & ENTW trenchwork; **Grabenaushub** *m* ARCH & TRAGW trenching; **Grabenbagger**

*m* ARCH & TRAGW, BAUMASCHIN trench excavator; **Grabenfräser** *m* BAUMASCHIN trench excavator; **Grabenherstellung** *f* ARCH & TRAGW trenching; **Grabenmethode** *f* UMWELT trench method; **Grabenverbau** *m* ARCH & TRAGW trench sheeting
**Gradientenzug** *m* INFR & ENTW line of levels
**Grad** *m*: **~ der statischen Unbestimmtheit** ARCH & TRAGW redundancy, INFR & ENTW degree of statical indeterminacy
**Gradtag** *m* BAURECHT, HEIZ & BELÜFT degree day
**Gramm** *nt* (*g*) ARCH & TRAGW, INFR & ENTW, WERKSTOFF gram (*g*), gramme (*g*)
**Granit** *m* INFR & ENTW, WERKSTOFF granite
**granuliert**: **~e Hochofenschlacke** *f* WERKSTOFF granulated cinder
**graphisch**: **~e Analyse** *f* BAURECHT graphical analysis
**Graphit** *m* WERKSTOFF graphite
**Gras** *nt* INFR & ENTW grass; **Grasboden** *m* INFR & ENTW grassed area; **Grasböschung** *f* INFR & ENTW grass slope
**Grat** *m* ARCH & TRAGW ridge, INFR & ENTW *Architektur* groin; **Gratbalken** *m* ARCH & TRAGW angle rafter, angled rafter, HOLZ *Dach* hip rafter, angle rafter, angled rafter, WERKSTOFF angle rafter; **Gratlinie** *f* ARCH & TRAGW arris, piend; **Gratsparren** *m* ARCH & TRAGW angle rafter, angled rafter, HOLZ *Dach* hip rafter, angle rafter, angled rafter, WERKSTOFF angle rafter; **Gratsparrendach** *nt* ARCH & TRAGW, HOLZ hip and ridge roof; **Gratziegel** *m* ARCH & TRAGW, INFR & ENTW, NATURSTEIN, WERKSTOFF hip tile
**Grauguß** *m* WERKSTOFF gray cast iron (*AmE*), grey cast iron (*BrE*)
**Gravitation** *f* INFR & ENTW, UMWELT gravity
**Greifbagger** *m* BAUMASCHIN grab crane
**Greifer** *m* BAUMASCHIN grab, nippers; **Greiferbagger** *m* BAUMASCHIN *Schwimmbagger* grab dredger; **Greiferkorb** *m* BAUMASCHIN grab bucket
**Grenz-** *in cpds* ARCH & TRAGW, BAURECHT, INFR & ENTW limit; **Grenzbelastung** *f* ARCH & TRAGW, INFR & ENTW, WERKSTOFF breaking load, failure load, limit load, limit of load stress
**Grenze** *f* ARCH & TRAGW margin, limit, boundary, BAURECHT, INFR & ENTW boundary, limit; **Grenzenbestimmung** *f* INFR & ENTW boundary determination
***Grenz-***: **Grenzkonzentration** *f* UMWELT limiting concentration; **Grenzlast** *f* ARCH & TRAGW, INFR & ENTW, WERKSTOFF breaking load, failure load, limit load; **Grenzmarkierung** *f* ARCH & TRAGW boundary mark; **Grenzpfosten** *m* ARCH & TRAGW boundary post; **Grenzschicht** *f* UMWELT boundary layer; **Grenzstein** *m* BAURECHT, INFR & ENTW boundary stone, landmark
**grenzüberschreitend**: **~e Abfallverbringung** *f* UMWELT transboundary movement of waste
***Grenz-***: **Grenzwert** *m* ARCH & TRAGW critical value; **Grenzwertgeber** *m* HEIZ & BELÜFT limit indicator, limit value indicator; **Grenzzustand** *m* **der Tragfähigkeit** ARCH & TRAGW, BETON, INFR & ENTW, STAHL, WERKSTOFF ultimate limit state
**Griff** *m* BESCHLÄGE *Tür* door handle, handle
**Griffigkeit** *f* OBERFLÄCHE grip
***Griff***: **Griffstange** *f* BESCHLÄGE *neben einer Badewanne oder Dusche* grab bar; **Griffverschluß** *m* BESCHLÄGE locking handle; **Griffzwinge** *f* BAUMASCHIN handle collar
**Grob-** *in cpds* HOLZ rough, INFR & ENTW, WERKSTOFF coarse
**grob**: **~er Kies** *m* NATURSTEIN pebble stone, WERKSTOFF coarse gravel; **~ zugehauen** *phr* NATURSTEIN rough-axed; **~er Zuschlag** *m* WERKSTOFF coarse aggregate; **~e Zuschlagstoffe** *m pl* WERKSTOFF coarse aggregate
***Grob-***: **Grobblech** *nt* WERKSTOFF checkered plate (*AmE*), chequered plate (*BrE*); **Grobfraktion** *f* WERKSTOFF coarse fraction
**grobgängig**: **~e Schraube** *f* BESCHLÄGE coarse-pitch screw
***Grob-***: **Grobholz** *nt* HOLZ rough wood; **Grobkies** *m* WERKSTOFF coarse gravel, rubble; **Grobkiesschüttung** *f* INFR & ENTW coarse gravel filling; **Grobkorn** *nt* WERKSTOFF coarse fraction
**grobkörnig** *adj* HOLZ rough-grained, WERKSTOFF *Boden, Sand* coarse-grained
***Grob-***: **Grobmörtel** *m* NATURSTEIN coarse mortar; **Grobputzschicht** *f* NATURSTEIN pricking-up coat; **Grobrechen** *m* UMWELT coarse screen; **Grobsand** *m* WERKSTOFF coarse sand; **Grobzuschlagstoffe** *m pl* WERKSTOFF ballasting material
**groß** *adj* ARCH & TRAGW tall, NATURSTEIN, WERKSTOFF large
**Groß-** *in cpds* NATURSTEIN, WERKSTOFF large
***groß***: **~es Grundstück** *nt* INFR & ENTW estate
***Groß-***: **Großblock** *m* NATURSTEIN large block
**Größe** *f* ARCH & TRAGW size; **Größenordnung** *f* INFR & ENTW magnitude
***groß***: **mit großer Spannweite** *phr* ARCH & TRAGW long-span
**großflächig**: **~es Fenster** *nt* ARCH & TRAGW picture window
***Groß-***: **Großformat** *nt* NATURSTEIN *Mauerstein* large format
**großformatig**: **~er Ziegelstein** *m* WERKSTOFF jumbo brick
***Groß-***: **Großpflaster** *nt* NATURSTEIN large sett; **Großplatte** *f* WERKSTOFF large panel; **Großplattenbauweise** *f* ARCH & TRAGW large-panel construction method
**Großraum-** *in cpds* ARCH & TRAGW, INFR & ENTW open-plan; **Großraumplanung** *f* ARCH & TRAGW, INFR & ENTW open spatial planning
**größte** *adj* ARCH & TRAGW, ELEKTR, INFR & ENTW maximum; **~ Tiefe** *f* INFR & ENTW maximum depth
**Grube** *f* INFR & ENTW pit, cesspool, UMWELT trench landfill; **Grubenarbeit** *f* INFR & ENTW pit work; **Gruben- und Kanalreinigungsfahrzeug** *nt* BAUMASCHIN, INFR & ENTW cesspool and sewer cleaning vehicle; **Grubenkies** *m* NATURSTEIN pit gravel; **Grubenwand** *f* UMWELT *einer Deponie* embankment
**Gruftgewölbe** *nt* ARCH & TRAGW undercroft
**grün** *adj* OBERFLÄCHE, UMWELT green
**Grün-** *in cpds* INFR & ENTW, OBERFLÄCHE, UMWELT, WERKSTOFF green
***grün***: **~es Gebäude** *nt* UMWELT green building
**Grund** *m* ARCH & TRAGW, INFR & ENTW, STAHL earth, ground, soil; **Grundabmessungen** *f pl* ARCH & TRAGW basic assumption; **Grundanstrich** *m* OBERFLÄCHE first coat, priming; **Grundausbildung** *f* BAURECHT basic training; **Grundausführung** *f* ARCH & TRAGW basic design; **Grundbalken** *m* ARCH & TRAGW ground beam; **Grundbau** *m* INFR & ENTW soil engineering; **Grundbaustein** *m* ARCH & TRAGW

basic module; **Grundbesitz** *m* BAURECHT estate, land ownership; **Grundbesitzer** *m* BAURECHT land owner, lessor; **Grundbestandteil** *m* ARCH & TRAGW basic ingredient; **Grundbogen** *m* ARCH & TRAGW reversed arch; **Grundbruch** *m* INFR & ENTW shear failure; **Grundbuch** *nt* BAURECHT land register (*BrE*), register of real estates (*AmE*); **Grundbuchamt** *nt* BAURECHT land registry (*BrE*); **Grundbucheintragung** *f* BAURECHT land registry entry (*BrE*); **Grundbuchüberprüfung** *f* BAURECHT land registry search (*BrE*); **Grundeigenschaften** *f pl* WERKSTOFF fundamental properties; **Grundeigentum** *nt* BAURECHT land ownership, real estate; **Grundeigentümer** *m* BAURECHT land owner, landholder, lessor; **Grunderwerb** *m* BAURECHT land purchase; **Grundfläche** *f* ARCH & TRAGW floor space, floor area, *Oberfläche* area, *eines Körpers* base, INFR & ENTW *eines Körpers* base, floor space; **Grundgefüge** *nt* INFR & ENTW, NATURSTEIN backing, WERKSTOFF matrix; **Grundgesetz** *nt* BAURECHT basic law

**Grundieranstrich** *m* OBERFLÄCHE prime coat

**grundieren** *vt* OBERFLÄCHE *Farbe* precoat, prime coat

**Grundierschicht** *f* OBERFLÄCHE base coat

**grundiert** *adj* OBERFLÄCHE precoated

**Grundierung** *f* OBERFLÄCHE primer

***Grund*: Grundisolierung** *f* DÄMMUNG basic insulation; **Grundkonzentration** *f* UMWELT background concentration, instantaneous concentration; **Grundlage** *f* ARCH & TRAGW, INFR & ENTW *Fundament* base, basis; **Grundlast** *f* ARCH & TRAGW basic load

**grundlegend**: **~e Verschmutzung** *f* UMWELT background pollution

***Grund*: Grundleitung** *f* ABWASSER ground line, ELEKTR earth line (*BrE*), ground line (*AmE*); **Grundlinie** *f* ARCH & TRAGW base line; **Grundlohn** *m* BAURECHT basic wage; **Grundmaß** *nt* ARCH & TRAGW basic size; **Grundmasse** *f* WERKSTOFF matrix; **Grundmauer** *f* ARCH & TRAGW base wall; **Grundplatte** *f* ARCH & TRAGW base plate, INFR & ENTW bed plate; **Grundriß** *m* ARCH & TRAGW first-floor plan (*AmE*), ground-floor plan (*BrE*); **Grundschubspannung** *f* BETON, ARCH & TRAGW, WERKSTOFF basic shear strength; **Grundschwelle** *f* ARCH & TRAGW groundsill, sill, sill plate; **Grundstein** *m* ARCH & TRAGW foundation stone; **Grundsteinlegung** *f* BAURECHT cornerstone ceremony, laying of foundation stone; **Grundsteuer** *f* BAURECHT property tax (*BrE*), real estate tax (*AmE*)

**Grundstück** *nt* BAURECHT plot, premises; **Grundstücksbegrenzungsmauer** *f* ARCH & TRAGW party wall; **Grundstücksentwässerung** *f* ABWASSER estate drainage, sewerage system, INFR & ENTW, UMWELT sewerage system; **Grundstücksgrenze** *f* ARCH & TRAGW, BAURECHT, INFR & ENTW land boundary, property line; **Grundstücksmakler** *m* BAURECHT broker, estate agent (*BrE*), real estate agent (*AmE*); **Grundstücksmauer** *f* BAURECHT, INFR & ENTW boundary wall; **Grundstücksräumung** *f* BAURECHT, INFR & ENTW land clearing

**Gründung** *f* ARCH & TRAGW, INFR & ENTW foundation, groundwork, *Gebäude* footing; **Gründungsdruck** *m* ARCH & TRAGW, INFR & ENTW foundation pressure; **Gründungsfläche** *f* ARCH & TRAGW bearing area; **Gründungspfahl** *m* ARCH & TRAGW, INFR & ENTW foundation pile, pier; **Gründungsplatte** *f* ARCH & TRAGW, INFR & ENTW foundation slab; **Gründungsrost** *m* ARCH & TRAGW, INFR & ENTW grillage foundation; **Gründungsschwelle** *f* ARCH & TRAGW grade beam; **Gründungssohle** *f* ARCH & TRAGW, INFR & ENTW base level, foundation level; **Gründungstiefe** *f* ARCH & TRAGW depth of foundation, foundation depth, INFR & ENTW foundation depth

**Grundwasser** *nt* INFR & ENTW groundwater, UMWELT underground water; **Grundwasserabpumpen** *nt* BAURECHT, INFR & ENTW groundwater pumping; **Grundwasserabsenkung** *f* ARCH & TRAGW, INFR & ENTW lowering of the groundwater level; **Grundwasserbelastung** *f* UMWELT impact on groundwater; **Grundwasserdichtungsschicht** *f* ARCH & TRAGW tanking, INFR & ENTW basement waterproofing; **Grundwasserdrainage** *f* INFR & ENTW groundwater drainage; **Grundwassergüte** *f* BAURECHT, INFR & ENTW, WERKSTOFF groundwater quality; **Grundwasserhaltung** *f* ARCH & TRAGW unwatering, INFR & ENTW groundwater lowering; **Grundwasserhöhe** *f* ARCH & TRAGW, INFR & ENTW groundwater level; **Grundwasserhydraulik** *f* INFR & ENTW groundwater hydraulics; **Grundwasserlinse** *f* INFR & ENTW groundwater lens; **Grundwasserschicht** *f* INFR & ENTW groundwater layer; **Grundwasserschutz** *m* ARCH & TRAGW, BAURECHT, INFR & ENTW, UMWELT basement waterproofing, groundwater protection; **Grundwasserstockwerke** *nt pl* INFR & ENTW water-bearing layers; **Grundwasserstrom** *m* INFR & ENTW groundwater current, UMWELT underground water flow; **Grundwasserströmung** *f* INFR & ENTW groundwater current; **Grundwasserverschmutzung** *f* BAURECHT, INFR & ENTW groundwater pollution, UMWELT groundwater contamination, groundwater pollution; **Grundwasserwanne** *f* ARCH & TRAGW basement waterproofing, tank, INFR & ENTW basement waterproofing

**Grünfläche** *f* INFR & ENTW grassed area

**Grünling** *m* NATURSTEIN sun-dried brick

**Grünspan** *m* OBERFLÄCHE verdigris

**Grünstein** *m* WERKSTOFF green stone

**Gruppen-** *in cpds* ELEKTR, WERKSTOFF group; **Gruppenschaltung** *f* ELEKTR multiple series; **Gruppenschlüssel** *m* BESCHLÄGE floor master key, submaster key

**Gülle** *f* UMWELT slurry

**Gully** *m* ABWASSER, INFR & ENTW gully; **Gullydeckel** *m* ABWASSER, INFR & ENTW gully cover

**Gummi** *nt* BAUMASCHIN, BESCHLÄGE, OBERFLÄCHE, WERKSTOFF rubber; **Gummibelag** *m* OBERFLÄCHE rubber coating; **Gummieinlage** *f* WERKSTOFF rubber core, rubber ply; **Gummifußbodenbelag** *m* WERKSTOFF rubber flooring; **Gummilippendichtung** *f* BESCHLÄGE rubber gasket, rubber lip sealing; **Gummimanschette** *f* BESCHLÄGE rubber collar; **Gummiprofil** *nt* WERKSTOFF rubber profile; **Gummiradwalze** *f* BAUMASCHIN *Straßenbau* pneumatic-tired roller (*AmE*), pneumatic-tyred roller (*BrE*), *Straßenbau* rubber-wheel roller, rubber-tired roller (*AmE*), rubber-tyred roller (*BrE*)

**Gurt** *m* ARCH & TRAGW chord, BAUMASCHIN boom, BETON waler, waling, INFR & ENTW waler, waling, *Spundwände* ranger, STAHL flange, WERKSTOFF *Spundwände* ranger; **Gurtblech** *nt* BAUMASCHIN boom plate; **Gurtgesims** *nt* ARCH & TRAGW cornice, belt course, string course; **Gurtholz** *nt* ARCH & TRAGW runner, BETON waler, waling, HOLZ waler, INFR & ENTW ranger, waler, waling, WERKSTOFF ranger; **Gurtplatte** *f* BESCHLÄGE *eines Trägers* flange plate; **Gurtsims** *m* ARCH & TRAGW fascia

**Guß** *m* STAHL, WERKSTOFF casting; **Gußasphalt** *m* INFR & ENTW, WERKSTOFF asphaltic mastic; **Gußauslauf** *m* ABWASSER, STAHL cast-iron outlet; **Gußbronze** *f* WERKSTOFF cast bronze; **Gußdrahtglas** *nt* WERKSTOFF roughcast wired glass

**Gußeisen** *nt* WERKSTOFF cast iron (*CI*); **Gußeisenabflußrohr** *nt* ABWASSER, STAHL cast-iron drainpipe; **Gußeisendeckel** *m* INFR & ENTW *Schacht*, STAHL cast-iron cover; **Gußeisenflansch** *m* STAHL cast-iron flange; **Gußeisengelenk** *nt* BESCHLÄGE cast-iron joint; **Gußeisenkessel** *m* HEIZ & BELÜFT, STAHL cast-iron boiler; **Gußeisenrohr** *nt* STAHL, WERKSTOFF cast-iron pipe (*CIP*), CIP (*cast-iron pipe*); **Gußeisenträger** *m* ARCH & TRAGW, STAHL cast-iron girder

**gußeisern**: **~er Heizkörper** *m* HEIZ & BELÜFT, STAHL cast-iron radiator; **~er Krümmer** *m* BESCHLÄGE cast-iron elbow; **~es Rohr** *nt* STAHL, WERKSTOFF cast-iron pipe (*CIP*), CIP (*cast-iron pipe*)

***Guß***: **Gußform** *f* BAUMASCHIN *Betonprobewürfel* mold (*AmE*), mould (*BrE*); **Gußglas** *nt* WERKSTOFF cast glass, roughcast glass; **Gußmessing** *nt* WERKSTOFF cast brass (*CB*), CB (*cast brass*); **Gußstück** *nt* STAHL, WERKSTOFF casting; **Gußteil** *nt* STAHL, WERKSTOFF casting

**gut**: **~ haften** *vi* STAHL adhere firmly; **in ~em Zustand** *phr* ARCH & TRAGW in good repair

**Güte** *f* WERKSTOFF quality; **~ des Zuschlagstoffs** WERKSTOFF quality of aggregate; **Gütebeschreibung** *f* BAURECHT, WERKSTOFF quality description; **Gütebestimmungen** *f pl* BAURECHT quality requirements, WERKSTOFF quality; **Gütefaktor** *m* (*Q-Faktor*) UMWELT quality factor (*Q factor*); **Gütegemeinschaft** *f* BAURECHT, WERKSTOFF quality control association; **Güteklasse** *f* HOLZ quality class, WERKSTOFF grade; **Gütesicherung** *f* BAURECHT, WERKSTOFF quality control; **Gütezeichen** *nt* BAURECHT, WERKSTOFF quality control mark

# H

**Haarriß** *m* INFR & ENTW, STAHL hair crack, WERKSTOFF hair crack, craze
**Haarrisse** *m pl* WERKSTOFF crazing
**Hafen** *m* INFR & ENTW harbor (*AmE*), harbour (*BrE*); **Hafenbahn** *f* INFR & ENTW harbor railroad (*AmE*), harbour railway (*BrE*); **Hafenbau** *m* INFR & ENTW wharf construction; **Hafenbauerschließung** *f* INFR & ENTW waterfront development, wharf development; **Hafendamm** *m* ARCH & TRAGW jetty, INFR & ENTW jetty, wharf; **Hafengebiet** *nt* INFR & ENTW waterfront; **Hafengleis** *nt* INFR & ENTW harbor railroad (*AmE*), harbour railway (*BrE*)
**Haft-** *in cpds* INFR & ENTW, OBERFLÄCHE, WERKSTOFF adhesion; **Haftbrücke** *f* NATURSTEIN *Estrich* bonding course, WERKSTOFF bonding area
**haftend** *adj* OBERFLÄCHE, WERKSTOFF adhesive
***Haft-***: **Haftfähigkeit** *f* INFR & ENTW adhesion, adhesion power, OBERFLÄCHE adhesion, adhesion power, grip, WERKSTOFF adhesion, adhesion power; **Haftfestigkeit** *f* ARCH & TRAGW, INFR & ENTW, OBERFLÄCHE, WERKSTOFF bond strength, bonding strength; **Haftgipsputz** *m* NATURSTEIN bonding plaster; **Haftgrenze** *f* WERKSTOFF sticky limit; **Haftgrund** *m* OBERFLÄCHE wash primer; **Haftgrundierung** *f* OBERFLÄCHE wash primer; **Haftlänge** *f* ARCH & TRAGW, INFR & ENTW bond length; **Haftmittel** *nt* WERKSTOFF adhesion-promoting agent
**Haftpflicht** *f* BAURECHT liability; **Haftpflichtgrenze** *f* BAURECHT limit of liability
***Haft-***: **Haftputz** *m* NATURSTEIN bonding finish
**Haftung** *f* INFR & ENTW, NATURSTEIN, OBERFLÄCHE, WERKSTOFF bonding; **Haftungsbegrenzung** *f* **des Bauunternehmers** BAURECHT limitation of contractor's liabilities
**haftverbessernd** *adj* INFR & ENTW, OBERFLÄCHE, WERKSTOFF bond-improving
***Haft-***: **Haftvermittler** *m* OBERFLÄCHE, WERKSTOFF adhesion promoter; **Haftvermögen** *nt* INFR & ENTW adhesion, adhesion power, OBERFLÄCHE adhesion, adhesion power, grip, WERKSTOFF adhesion, adhesion power
**Hahn** *m* ABWASSER cock, faucet (*AmE*), tap (*BrE*)
**Haken** *m* BESCHLÄGE, WERKSTOFF hook; **Hakenfalle** *f* BESCHLÄGE *Schloß* hook bolt; **Hakennagel** *m* ARCH & TRAGW spike; **Hakenschraube** *f* BAUMASCHIN screw hook; **Hakenweg** *m* BAUMASCHIN *Kran* hook track
**Halb-** *in cpds* ARCH & TRAGW, INFR & ENTW half-, semi-
**halb**: **~e Balkenhöhe** *f* HOLZ beam mid-line
***Halb-***: **Halbbalken** *m* ARCH & TRAGW half-beam, HOLZ half-timber; **Halbbinder** *m* ARCH & TRAGW half truss; **Halbdach** *nt* ARCH & TRAGW pent roof, shed roof
**halbfest** *adj* UMWELT, WERKSTOFF semisolid
***Halb-***: **Halbfeuchttrennung** *f* UMWELT semiwet sorting
**halbgedreht**: **~e Treppe** *f* ARCH & TRAGW half-turn stairs
***Halb-***: **Halbgeschoß** *nt* ARCH & TRAGW mezzanine; **Halbglanz** *m* ARCH & TRAGW, OBERFLÄCHE semigloss; **Halbholzbalken** *m* ARCH & TRAGW half-beam, HOLZ half-timber; **Halbkreis** *m* ARCH & TRAGW hemicycle, semicircle; **Halbkreisbogen** *m* ARCH & TRAGW semicircular arch
**halbkreisförmig** *adj* ARCH & TRAGW semicircular
***Halb-***: **Halbparabelbrücke** *f* INFR & ENTW hogbacked bridge; **Halbportalkran** *m* INFR & ENTW semigantry crane
**halbrund**: **~e Dachrinne** *f* ABWASSER half-round gutter; **~er Ziegel** *m* ARCH & TRAGW, NATURSTEIN, WERKSTOFF half-round tile
***Halb-***: **Halbrundstahl** *m* STAHL half-round bar; **Halbrundstange** *f* STAHL half-round rod; **Halbsäule** *f* ARCH & TRAGW pilaster, BESCHLÄGE demi-column; **Halbschale** *f* INFR & ENTW open channel
**halbstarr** *adj* WERKSTOFF *Kunststoff* semirigid
**halbsteif** *adj* ARCH & TRAGW *Rahmen*, INFR & ENTW, WERKSTOFF semirigid
**halbsteindick** *adj* NATURSTEIN half-brick
***Halb-***: **Halbziegel** *m* NATURSTEIN snap header
**Halde** *f* INFR & ENTW spoil heap
**Halle** *f* ARCH & TRAGW hall; **Hallenbad** *nt* ARCH & TRAGW, INFR & ENTW indoor swimming pool
**Hals** *m* ARCH & TRAGW, BESCHLÄGE, HEIZ & BELÜFT neck; **Halsstück** *nt* ARCH & TRAGW throat
**Haltbarmachung** *f* HOLZ preservation
**Haltebolzen** *m* BAUMASCHIN, BESCHLÄGE holding bolt
**Haltelasche** *f* BESCHLÄGE mounting bracket
**Halter** *m* BESCHLÄGE, ELEKTR clip
**Halterung** *f* ARCH & TRAGW fastener
**Haltespur** *f* INFR & ENTW lay-by
**Haltestock** *m* BAUMASCHIN *Zimmermannsarbeiten* bench stop
**Haltewinkel** *m* WERKSTOFF holding angle, mounting angle
**Hammer** *m* BAUMASCHIN hammer; **~ mit gerader Finne** BAUMASCHIN straight-pane hammer, straight-peen hammer; **Hammermühle** *f* UMWELT hammer mill
**hämmern** *vt* BAUMASCHIN hammer, sledge, OBERFLÄCHE hammer
**Hämmern** *nt* BAUMASCHIN sledging, OBERFLÄCHE hammering
***Hammer***: **Hammerschlaglack** *m* OBERFLÄCHE hammer-finish enamel, hammertone enamel; **Hammerzerkleinerer** *m* BAUMASCHIN, UMWELT hammer crusher
**Hand** *f* BAUMASCHIN hand
**Hand-** *in cpds* BAUMASCHIN manual
***Hand***: **Handabsperrklappe** *f* HEIZ & BELÜFT hand-operated shut-off valve; **Handarbeit** *f* ARCH & TRAGW manual labor (*AmE*), manual labour (*BrE*); **Handbetrieb** *m* BAUMASCHIN hand operation; **Handbohrer** *m* BAUMASCHIN gimlet; **Handbuch** *nt* ARCH & TRAGW manual; **Handdurchschläger** *m* BAUMASCHIN nail set
**Handel** *m* ARCH & TRAGW, BAURECHT, INFR & ENTW trade
**handelsüblich** *adj* WERKSTOFF commercial; **~e Länge** *f* WERKSTOFF commercial length
***Hand***: **Handfeuerlöscher** *m* BAURECHT hand-operated

fire extinguisher; **Handgriff** *m* BESCHLÄGE handle; **Handkette** *f* BAUMASCHIN handchain; **Handkurbel** *f* BESCHLÄGE winch; **Handlauf** *m* ARCH & TRAGW, BESCHLÄGE handrail; **Handleuchte** *f* BESCHLÄGE portable light; **Handrad** *nt* HEIZ & BELÜFT hand wheel; **Handsäge** *f* BAUMASCHIN handsaw; **Handschachten** *nt* INFR & ENTW shovel work; **Handschrapper** *m* BAUMASCHIN hand scraper; **Handschutzschild** *m* STAHL welding handshield; **Handsondiergerät** *nt* BAUMASCHIN manual penetrometer; **Handsortierung** *f* UMWELT *von Müll* manual separation, manual sorting; **Handstampfer** *m* BAUMASCHIN hand tamper; **Handstrichziegel** *m* NATURSTEIN hand-formed brick, hand-made brick, hand-molded brick (*AmE*), hand-moulded brick (*BrE*); **Handwerkszeug** *nt* BAUMASCHIN set of tools

**Hang** *m* INFR & ENTW slope

**Hangar** *m* ARCH & TRAGW, INFR & ENTW hangar

**Hänge-** *in cpds* ARCH & TRAGW hanging, suspended; **Hängebahn** *f* INFR & ENTW suspension railroad (*AmE*), suspension railway (*BrE*); **Hängebaugerüst** *nt* BAUMASCHIN boat scaffold; **Hängebrücke** *f* INFR & ENTW hanging bridge, suspension bridge; **Hängebrücke** *f* **mit steifer Fahrbahntafel** INFR & ENTW stiffened suspension bridge; **Hängebühne** *f* ARCH & TRAGW, BAUMASCHIN cradle, hanging scaffold; **Hängedach** *nt* ARCH & TRAGW cable roof, suspended roof; **Hängedecke** *f* ARCH & TRAGW suspended ceiling; **Hängegerüst** *nt* ARCH & TRAGW, BAUMASCHIN, INFR & ENTW flying scaffold, hanging scaffold, hanging stage, suspended scaffold, traveling cradle (*AmE*), travelling cradle (*BrE*); **Hängegurtung** *f* ARCH & TRAGW suspension boom

***Hang***: **Hangeinschnitt** *m* INFR & ENTW sidehill cut

***Hänge-***: **Hängelaufkatze** *f* BAUMASCHIN suspended trolley

**hängend** *adj* ARCH & TRAGW suspended; **~es Gerüst** *nt* ARCH & TRAGW, BAUMASCHIN, INFR & ENTW cradle, flying scaffold, suspended scaffold; **~er Torpfosten** *m* ARCH & TRAGW, BESCHLÄGE hanging post, hinge post

***Hänge-***: **Hängeregal** *nt* BESCHLÄGE hanging shelf; **Hängerinne** *f* ABWASSER, ARCH & TRAGW hanging gutter, hanging stormwater gutter; **Hängerüstung** *f* ARCH & TRAGW, BAUMASCHIN, INFR & ENTW cradle, suspended scaffold; **Hängesäule** *f* ARCH & TRAGW king post, HOLZ suspender, STAHL kingbolt; **Hängeschiene** *f* BESCHLÄGE suspended rail; **Hängestange** *f* HOLZ suspender; **Hängewerk** *nt* ARCH & TRAGW hanging truss, suspension truss; **Hängewerksbrücke** *f* INFR & ENTW pendant-framed bridge; **Hängezwickel** *m* ARCH & TRAGW pendentive

***Hang***: **Hangkanal** *m* UMWELT headrace canal; **Hangschutt** *m* INFR & ENTW talus

**Harmonika**: **Harmonikatrennwand** *f* ARCH & TRAGW slip partition; **Harmonikatür** *f* ARCH & TRAGW, BESCHLÄGE accordion door, concertina door, folding door, sliding folding door

**harmonisch**: **~e Schwingung** *f* DÄMMUNG harmonic vibration; **~ anordnen** *vt* ARCH & TRAGW harmonize

**hart 1.** *adj* WERKSTOFF hard; **2. ~ werden** *vi* BETON, WERKSTOFF *Gips, Mörtel* harden, set

***hart***: **~er Boden** *m* INFR & ENTW, WERKSTOFF hard soil; **~er Ton** *m* INFR & ENTW, WERKSTOFF hard clay; **~es Wasser** *nt* ABWASSER, INFR & ENTW, UMWELT hard water

**Hart-** *in cpds* WERKSTOFF hard; **Hartbeton** *m* WERKSTOFF granolithic concrete; **Hartbrandstein** *m* NATURSTEIN, WERKSTOFF engineering brick; **Hartbrandziegel** *m* INFR & ENTW, NATURSTEIN, WERKSTOFF hard-burnt brick, well-burnt brick

**Härte** *f* INFR & ENTW, WERKSTOFF hardness; **Härtegrad** *m* WERKSTOFF degree of hardness; **Härtemittel** *nt* OBERFLÄCHE *Lacke und Farben* hardening agent

**härten 1.** *vt* WERKSTOFF *Gußeisen* chill, set; **2.** *vi* BETON, WERKSTOFF harden

**Härten** *nt* BETON hardening, WERKSTOFF hardening, *Gußeisen* chilling

**Härter** *m* OBERFLÄCHE *Lacke und Farben* hardening agent

***Härte***: **Härteskala** *f* WERKSTOFF hardness scale; **Härtetest** *m* WERKSTOFF hardness test

**Hart**: **Hartfaserplatte** *f* HOLZ, WERKSTOFF hardboard, molded fiber board (*AmE*), moulded fibre board (*BrE*)

**hartgebrannt** *adj* WERKSTOFF hard-burnt

**hartgelötet** *adj* STAHL, WERKSTOFF brazed

***Hart***: **Hartgestein** *nt* INFR & ENTW, WERKSTOFF hard rock; **Hartglas** *nt* BAURECHT toughened glass; **Hartgummileitung** *f* WERKSTOFF hard rubber pipe; **Hartgummistab** *m* WERKSTOFF hard rubber bar; **Hartholz** *nt* HOLZ, WERKSTOFF hardwood; **Hartlot** *nt* STAHL, WERKSTOFF brazing solder; **Hartlöten** *nt* OBERFLÄCHE, STAHL, WERKSTOFF brazing; **Hartschaum** *m* WERKSTOFF rigid foam

**Härtung**: **die ~ fördern** *phr* BETON *Beton oder Mörtel durch Feuchthalten* cure

**Härtungskurve** *f* INFR & ENTW, WERKSTOFF hardening graph

**Harz** *nt* HOLZ, WERKSTOFF resin; **Harzeinschluß** *m* HOLZ resin pocket

**harzhaltig** *adj* WERKSTOFF resinous

**harzreich** *adj* HOLZ highly resinous; **~es Holz** *nt* HOLZ highly resinous wood

**Haspe** *f* BESCHLÄGE *Schloß* hasp, staple

**Haubenscharnier** *nt* BESCHLÄGE bonnet hinge

**Haueisen** *nt* STAHL mattock

**hauen** *vt* NATURSTEIN hew

**Hauer** *m* BAUMASCHIN, HOLZ, NATURSTEIN hewer

**Haufen** *m* ARCH & TRAGW heap

**Häufigkeit** *f* ARCH & TRAGW, INFR & ENTW frequency; **Häufigkeitskurve** *f* DÄMMUNG frequency curve; **Häufigkeitsmesser** *m* DÄMMUNG frequency meter

**haufwerksporig**: **~er Beton** *m* BETON no-fines concrete

**Haupt** *nt* ARCH & TRAGW crown

**Haupt-** *in cpds* ABWASSER, ARCH & TRAGW, BAUMASCHIN, BETON, ELEKTR, HEIZ & BELÜFT, INFR & ENTW, UMWELT main, principal

***Haupt***: **Hauptabsperrung** *f* ABWASSER main shut-off valve; **Hauptbalken** *m* ARCH & TRAGW, HOLZ main beam, principal beam; **Hauptbauwerk** *nt* ARCH & TRAGW main structure; **Hauptbebauungsplan** *m* BAURECHT master plan; **Hauptbewehrung** *f* BETON, INFR & ENTW main reinforcement, principal reinforcement; **Haupteingang** *m* ARCH & TRAGW, INFR & ENTW main entrance, principal entrance; **Haupterdungskasten** *m* ELEKTR main earth box (*BrE*), main ground box (*AmE*); **Hauptgasleitung** *f* HEIZ & BELÜFT, INFR & ENTW gas mains; **Haupthub** *m* BAUMASCHIN *Kran* main lift; **Hauptlagerplatz** *m* ARCH & TRAGW main storage yard; **Hauptleitung** *f* ABWASSER, ELEKTR, HEIZ & BELÜFT, INFR & ENTW main line,

*Strom, Wasser* main; **Hauptleitungsrohr** *nt* ABWASSER, INFR & ENTW main pipe; **Hauptmoment** *nt* ARCH & TRAGW principal moment; **Hauptpfeiler** *m* ARCH & TRAGW, INFR & ENTW main pier; **Hauptquelle** *f* UMWELT major source; **Hauptrohrleitung** *f* UMWELT pipeline; **Hauptsammelkanal** *m* UMWELT interceptor sewer; **Hauptsammler** *m* ABWASSER, INFR & ENTW, UMWELT interceptor sewer, main collector, main drain, main sewer; **Hauptschalter** *m* ELEKTR main switch, master switch; **Hauptschlüssel** *m* BAURECHT, BESCHLÄGE master key; **Hauptschlüsselanlage** *f* BAURECHT, BESCHLÄGE master key system, *Schließanlage* grandmaster keyed system; **Hauptspannung** *f* ARCH & TRAGW *mechanisch* principal tension; **Hauptsparren** *m* ARCH & TRAGW, HOLZ principal rafter; **Hauptstütze** *f* ARCH & TRAGW, INFR & ENTW main column; **Haupttragbalken** *m* ARCH & TRAGW bearer; **Hauptträger** *m* ARCH & TRAGW, INFR & ENTW main girder, principal girder; **Haupttragglied** *nt* INFR & ENTW main bearing member; **Hauptträgheitsmoment** *nt* ARCH & TRAGW principal moment of inertia; **Hauptunternehmer** *m* BAURECHT, INFR & ENTW main contractor; **Hauptverkehrsader** *f* INFR & ENTW thoroughfare (*BrE*), thruway (*AmE*); **Hauptverteiler** *m* ELEKTR main distribution panel; **Hauptverteilleitung** *f* ELEKTR main supply line; **Hauptwasserleitung** *f* HEIZ & BELÜFT, INFR & ENTW water main; **Hauptzugang** *m* INFR & ENTW main access

**Haus** *nt* ARCH & TRAGW house; **~ mit niedrigem Energieverbrauch** HEIZ & BELÜFT low-energy consumption house; **Hausanschluß** *m* ABWASSER, ELEKTR, HEIZ & BELÜFT house connection; **Hausanschlußkasten** *m* ELEKTR house connection box, service switch cabinet; **Hausanschlußleitung** *f* ABWASSER branch line, service line, ELEKTR house service connection, service line, HEIZ & BELÜFT service line, INFR & ENTW branch line; **Hausbesitzer** *m* BAURECHT landlord; **Hausbockkäfer** *m* HOLZ, UMWELT, WERKSTOFF wood-boring beetle; **Hauseigentümer** *m* BAURECHT landlord; **Hausentwässerung** *f* ABWASSER, ARCH & TRAGW building drainage, house drainage

**Haushalt** *m* INFR & ENTW, UMWELT household; **Haushaltsabwässer** *nt pl* UMWELT household waste water; **Haushaltsgerät** *nt* ELEKTR domestic appliance

***Haus***: **Hauslaube** *f* ARCH & TRAGW loggia

**häuslich**: **~es Abwasser** *nt* ABWASSER, INFR & ENTW domestic waste water; **~e Steuer** *f* BAURECHT, INFR & ENTW council tax (*BrE*); **~es Trinkwasser** *nt* ABWASSER, INFR & ENTW domestic drinking water; **~e Wasserheizung** *f* HEIZ & BELÜFT domestic water heating

**Hausmüll** *m* INFR & ENTW, UMWELT consumer waste, domestic waste, household refuse, municipal waste; **Hausmülldeponie** *f* UMWELT municipal waste landfill; **Hausmüllzusammensetzung** *f* UMWELT waste composition

***Haus***: **Hausnummer** *f* ARCH & TRAGW, INFR & ENTW building number; **Haustechnik** *f* ABWASSER, ELEKTR, HEIZ & BELÜFT *Gesamtheit der haustechnischen Anlagen* mechanical services

**haustechnisch**: **~e Anlagen** *f pl* ABWASSER, ELEKTR, HEIZ & BELÜFT domestic service facilities

**Haustein** *m* NATURSTEIN cut stone

***Haus***: **Hausverbrauch** *m* INFR & ENTW domestic consumption; **Hauswand** *f* ARCH & TRAGW building wall

**Haut** *f* DÄMMUNG skin

**Häutchenwasser** *nt* INFR & ENTW pellicular water

***Haut***: **Hautverhinderungsmittel** *nt* WERKSTOFF antiskinning agent

**havariebedingt**: **~er Ausfluß** *m* UMWELT *Chemikalien, Öl* accidental discharge

**Hebe-** *in cpds* BAUMASCHIN, INFR & ENTW lifting; **Hebeanlage** *f* ABWASSER, INFR & ENTW pump station; **Hebebock** *m* BAUMASCHIN derrick; **Hebebrücke** *f* INFR & ENTW lift bridge; **Hebekraft** *f* BAUMASCHIN, INFR & ENTW lifting force

**Hebel** *m* ARCH & TRAGW lever; **Hebelarm** *m* ARCH & TRAGW, BAUMASCHIN, INFR & ENTW lever arm; **Hebelträger** *m* ARCH & TRAGW fulcrum bracket

**heben** *vt* ARCH & TRAGW raise, BAUMASCHIN hoist, lift

**Heben** *nt* ARCH & TRAGW raising, BAUMASCHIN hoisting, lifting

**Heber** *m* BAUMASCHIN lifter

**hebern** *vt* UMWELT, WERKSTOFF siphon, syphon

***Hebe-***: **Hebetür** *f* ARCH & TRAGW lifting door; **Hebetürsicherung** *f* BAURECHT security lock for lifting doors; **Hebevorrichtung** *f* BAUMASCHIN jack, lifting device; **Hebezeug** *nt* BAUMASCHIN elevating equipment, gin, hoist, lifting equipment, tackle

**Hebung** *f* ARCH & TRAGW uplift

**Heim** *nt* ARCH & TRAGW, BAURECHT home

**heiß** *adj* WERKSTOFF hot

**Heiß-** *in cpds* WERKSTOFF hot; **Heißbitumen** *nt* WERKSTOFF hot asphalt, penetration-grade asphalt

***heiß***: **~ geteert** *adj* WERKSTOFF hot-tarred

**Heiß**: **Heißklebemasse** *f* WERKSTOFF hot adhesive compound; **Heißwasserspeicher** *m* ABWASSER, ELEKTR, HEIZ & BELÜFT boiler, hot water reservoir, hot water tank, storage water heater

**Heiz-** *in cpds* HEIZ & BELÜFT heating

**heizen** *vti* DÄMMUNG, HEIZ & BELÜFT, UMWELT heat

***Heiz-***: **Heizfläche** *f* HEIZ & BELÜFT heating surface; **Heizgerät** *nt* HEIZ & BELÜFT heater; **Heizkörper** *m* HEIZ & BELÜFT heating element, radiator; **Heizkörperglied** *nt* HEIZ & BELÜFT radiator element; **Heizkreis** *m* HEIZ & BELÜFT heating circuit; **Heizlast** *f* HEIZ & BELÜFT heating load; **Heizleistung** *f* HEIZ & BELÜFT calorific value; **Heizmaterial** *nt* WERKSTOFF fuel

**Heizöl** *nt* WERKSTOFF fuel; **Heizölabscheider** *m* ABWASSER fuel separator; **Heizöltank** *m* HEIZ & BELÜFT fuel tank

**Heizung** *f* HEIZ & BELÜFT heating; **~ für Wohnhäuser** HEIZ & BELÜFT domestic heating; **Heizungsanlage** *f* HEIZ & BELÜFT heating system; **Heizungskeller** *m* ARCH & TRAGW, HEIZ & BELÜFT basement boiler room; **Heizungsrohr** *nt* HEIZ & BELÜFT heating pipe, heating tube

***Heiz-***: **Heizwert** *m* HEIZ & BELÜFT, UMWELT calorific value

**Heliport** *m* ARCH & TRAGW, INFR & ENTW *allgemein* heliport, helistop, *auf dem Dach eines Gebäudes* helipad

**hell** *adj* ARCH & TRAGW light

**Helmdach** *nt* ARCH & TRAGW helm roof

**Hemmnis** *nt* ARCH & TRAGW, BAURECHT obstruction, INFR & ENTW obstacle

**Hemmschiene** *f* INFR & ENTW skid track

**Heraklithplatte** *f* INFR & ENTW, WERKSTOFF Heraklith insulating board
**herausheben** *vt* BAUMASCHIN, HEIZ & BELÜFT, UMWELT extract
**Herausheber** *m* BAUMASCHIN, HEIZ & BELÜFT, UMWELT extractor
**heraussägen** *vt* HOLZ saw out
**herausziehen** *vt* ARCH & TRAGW *Nagel* pull out
**herkömmlich** *adj* ARCH & TRAGW conventional; **~e Bauweise** *f* ARCH & TRAGW, INFR & ENTW conventional construction method
**herstellen** *vt* WERKSTOFF manufacture
**Hersteller** *m* WERKSTOFF manufacturer
**Herstellung** *f* WERKSTOFF manufacture
**herunterhängen**: **~ lassen** *vt* ARCH & TRAGW suspend
**Herzblattpolierschaufel** *f* ARCH & TRAGW, BAUMASCHIN heart trowel
**HFCK** *abbr* (*Hydrofluorchlorkohlenwasserstoff*) UMWELT HCFC (*hydrochlorofluorocarbon*)
**Hilfe** *f* BAURECHT, INFR & ENTW, WERKSTOFF aid
**Hilfs-** *in cpds* ARCH & TRAGW, BAURECHT, ELEKTR, HEIZ & BELÜFT, INFR & ENTW auxiliary; **Hilfsarbeiter** *m* ARCH & TRAGW helper; **Hilfsauflager** *nt* ARCH & TRAGW, INFR & ENTW auxiliary bearing; **Hilfsbewehrung** *f* ARCH & TRAGW, BETON auxiliary reinforcement; **Hilfserder** *m* ELEKTR *Blitzschutz* auxiliary ground connection; **Hilfsgerüst** *nt* ARCH & TRAGW shoring; **Hilfskessel** *m* HEIZ & BELÜFT supplementary boiler; **Hilfskonstruktion** *f* ARCH & TRAGW, BAURECHT, INFR & ENTW auxiliary construction; **Hilfskraft** *f* ARCH & TRAGW *Statik* auxiliary force; **Hilfsphasenextrusion** *f* WERKSTOFF split-section extrusion; **Hilfsplatten** *f pl* BESCHLÄGE accessory plates; **Hilfsquelle** *f* INFR & ENTW, UMWELT resource; **Hilfsrüstung** *f* ARCH & TRAGW shoring; **Hilfssäule** *f* ARCH & TRAGW subsidiary shaft; **Hilfsstütze** *f* ARCH & TRAGW flying shore
**Hindernis** *nt* ARCH & TRAGW, BAURECHT obstruction, INFR & ENTW obstacle, barricade
**hinten**: **nach ~ versetzen** *phr* ARCH & TRAGW set back
**hinter** *adj* ARCH & TRAGW back, rear
**Hinter-** *in cpds* ARCH & TRAGW back, rear; **Hinterfüllbeton** *m* ARCH & TRAGW, BETON backfill concrete
**hinterfüllen** *vt* BETON, INFR & ENTW, NATURSTEIN backfill
***Hinter-***: **Hinterfüllung** *f* BETON, INFR & ENTW, NATURSTEIN backfill, backfilling; **Hinterhof** *m* ARCH & TRAGW backyard
**hinterlüften** *vt* HEIZ & BELÜFT ventilate
***Hinter-***: **Hinterlüftung** *f* ARCH & TRAGW ventilation
**hintermauert**: **~es Mauerwerk** *nt* NATURSTEIN backed-up masonry
***Hinter-***: **Hintermauerung** *f* ARCH & TRAGW, NATURSTEIN backing masonry, backup masonry, masonry backup; **Hinterseite** *f* ARCH & TRAGW back, rear side; **Hintertreppe** *f* ARCH & TRAGW, INFR & ENTW back stairs
**Hinweisschild** *nt* ARCH & TRAGW guide board, BESCHLÄGE indicating label, INFR & ENTW guide board
**hinzufügen** *vt* ARCH & TRAGW, BAURECHT, BETON, INFR & ENTW, WERKSTOFF *Zusatzstoffe, Zuschlagstoffe* add
**Hinzufügung** *f* BAURECHT addition
**hinzukommend** *adj* ARCH & TRAGW, BAURECHT additional
**Hirnholz** *nt* HOLZ cross-grained wood, grain-cut timber
**Hirnschnittfläche** *f* HOLZ top surface
**Hitze** *f* ABWASSER, DÄMMUNG, HEIZ & BELÜFT, UMWELT heat
**hitzebeständig** *adj* WERKSTOFF heatproof
**hitzefest** *adj* WERKSTOFF heatproof
**HLZ** *abbr* (*Hochlochziegel*) NATURSTEIN, WERKSTOFF honeycomb brick, vertically perforated brick
**Hobel** *m* BAUMASCHIN plane; **Hobelbank** *f* BAUMASCHIN joiner's bench, workbench; **Hobeleisen** *nt* BAUMASCHIN plane iron; **Hobelkasten** *m* BAUMASCHIN plane stock; **Hobelmaschine** *f* BAUMASCHIN planer, planing machine; **Hobelmeißel** *m* BAUMASCHIN paring chisel; **Hobelmesser** *nt* BAUMASCHIN plane iron
**hobeln** *vt* BAUMASCHIN, HOLZ plane
**Hobeln** *nt* BAUMASCHIN, HOLZ planing
***Hobel***: **Hobelspan** *m* ARCH & TRAGW, HOLZ shaving; **Hobelwerkzeug** *nt* BAUMASCHIN planing tool
**hoch** *adj* ARCH & TRAGW, INFR & ENTW high, tall; **hoher Balkenträger** *m* ARCH & TRAGW deep beam; **hoher Schornsteinaufsatz** *m* ARCH & TRAGW tallboy; **hohe Wassersäule** *f* UMWELT high head
**Hoch-** *in cpds* ARCH & TRAGW, INFR & ENTW high, raised; **Hochbahn** *f* INFR & ENTW high-level railroad (*AmE*), high-level railway (*BrE*); **Hochbau** *m* ARCH & TRAGW, INFR & ENTW building construction, structural engineering
**hochbeanspruchbar** *adj* INFR & ENTW, WERKSTOFF heavy-duty; **~er Sturz** *m* INFR & ENTW, WERKSTOFF heavy-duty lintel
**hochbelastbar**: **~er Stahl** *m* ARCH & TRAGW, WERKSTOFF high-yield stress steel
***Hoch-***: **Hochbiegen** *nt* ARCH & TRAGW, BETON, INFR & ENTW, WERKSTOFF hogging bend; **Hochbordstein** *m* INFR & ENTW raised curb (*AmE*), raised kerb (*BrE*); **Hochbrücke** *f* INFR & ENTW *auf Pfeilern* viaduct; **Hochdruckbehälter** *m* BAUMASCHIN high-pressure tank; **Hochdruckdampf** *m* HEIZ & BELÜFT, INFR & ENTW high-pressure steam; **Hochdruckspülfahrzeug** *nt* **für die Kanalreinigung** ABWASSER, BAUMASCHIN, INFR & ENTW sewer jetting truck
**hochfest** *adj* WERKSTOFF high-strength; **~er Bolzen** *m* ARCH & TRAGW, WERKSTOFF high-strength bolt; **~ verschraubte Verbindung** *f* (*HV-Verbindung*) ARCH & TRAGW, STAHL high-strength friction grip, high-tensile bolted structural joint; **~ vorgespannte Schraube** *f* (*HV-Schraube*) STAHL friction-grip bolt
***Hoch-***: **Hochfrequenzfertiger** *m* BAUMASCHIN high-frequency finisher; **Hochgeschwindigkeitsspur** *f* INFR & ENTW *Eisenbahn* high-speed track; **Hochgeschwindigkeitszug** *m* INFR & ENTW high-speed train
**hochglanzpoliert** *adj* OBERFLÄCHE extra-bright
***Hoch-***: **Hochhaus** *nt* ARCH & TRAGW, INFR & ENTW high-rise building (*BrE*), skyscraper; **Hochheben** *nt* ARCH & TRAGW raising, BAUMASCHIN lifting
**hochintensiv**: **~er Wohnungsbau** *m* ARCH & TRAGW high-density housing
**hochkantbiegen** *vt* WERKSTOFF bend on edge
**hochkanten** *vt* ARCH & TRAGW raise on edge
**hochkorrosionsfest** *adj* WERKSTOFF highly corrosion-proof
***Hoch-***: **Hochleistungseinsteckschloß** *nt* BESCHLÄGE heavy-duty mortise lock; **Hochleistungsfilterung** *f* UMWELT high-rate filtration; **Hochlochziegel** *m* (*HLZ*) NATURSTEIN, WERKSTOFF honeycomb brick, vertically perforated brick

**hochmodern** *adj* ARCH & TRAGW state-of-the-art; **~e Technik** *f* ARCH & TRAGW state-of-the-art technique

***Hoch-***: **Hochofen** *m* BETON, STAHL blastfurnace; **Hochofenschlacke** *f* STAHL blastfurnace slag; **Hochofenzement** *m* BETON, WERKSTOFF blastfurnace cement, slag cement; **Hochreservoir** *nt* INFR & ENTW elevated tank

**hochschlagfest** *adj* WERKSTOFF highly shock-resistant, *Kunststoff* highly impact-resistant

***Hoch-***: **Hochspannungsleitung** *f* ELEKTR, INFR & ENTW transmission line

**Höchst-** *in cpds* ARCH & TRAGW, ELEKTR, INFR & ENTW maximum

**höchste**: **~r Hochwasserstand** *m* INFR & ENTW highest high water level

***Höchst-***: **Höchstleistung** *f* **bei Nennwindgeschwindigkeit** UMWELT maximum power at rated wind speed

***Hoch-***: **Hochstraße** *f* INFR & ENTW flyover

***Höchst-***: **Höchstschweißstrom** *m* ELEKTR maximum welding current

**höchstzulässig** *adj* ARCH & TRAGW maximum permissible; **~e Konzentration** *f* (*HZK*) UMWELT threshold limit value (*TLV*); **~es Niveau** *nt* BAURECHT, INFR & ENTW, UMWELT maximum permissible level (*MPL*), MPL (*maximum permissible level*)

**hochverdichtet** *adj* WERKSTOFF highly compressed

**Hochwasser** *nt* INFR & ENTW high tide; **Hochwassermarke** *f* INFR & ENTW high tide mark; **Hochwasserschaden** *m* ARCH & TRAGW, INFR & ENTW flood damage; **Hochwasserschutz** *m* ARCH & TRAGW, INFR & ENTW, UMWELT flood control, flood protection; **Hochwasserspiegel** *m* INFR & ENTW high water level; **Hochwasserstand** *m* INFR & ENTW high flood, high water level

**hochwertig** *adj* WERKSTOFF high-quality; **~e Energie** *f* HEIZ & BELÜFT high-grade energy

**Hof** *m* INFR & ENTW court, courtyard, yard; **Hofablauf** *m* ABWASSER yard gully, yard inlet; **Hofeingang** *m* INFR & ENTW court entrance; **Hofeinlauf** *m* ABWASSER yard gully, *Rohr* yard inlet; **Hofpumpe** *f* ABWASSER yard pump; **Hofsinkkasten** *m* ABWASSER yard gully hole, yard inlet; **Hoftopf** *m* ABWASSER yard inlet

**Höhe** *f* ARCH & TRAGW *Niveau* level, *eines Bauwerks* height, *Zahnradzahn* depth, elevation, BAUMASCHIN *Zahnradzahn*, INFR & ENTW depth, UMWELT elevation, WERKSTOFF depth; **~ über dem Meeresspiegel** ARCH & TRAGW height above sea level; **Höhenaufnahme** *f* ARCH & TRAGW leveling (*AmE*), levelling (*BrE*); **Höhenfestpunkt** *m* INFR & ENTW bench mark; **Höhenfries** *m* ARCH & TRAGW *einer Tür* stile; **Höhenkote** *f* ARCH & TRAGW, INFR & ENTW height indication; **Höhenlage** *f* ARCH & TRAGW, INFR & ENTW altitude; **Höhenlinie** *f* ARCH & TRAGW contour; **Höhenmesser** *m* BAUMASCHIN altimeter; **Höhenplan** *m* ARCH & TRAGW, INFR & ENTW *Landvermessung* contour map; **Höhenschichtlinie** *f* ARCH & TRAGW, INFR & ENTW contour line

**höhenverstellbar** *adj* ARCH & TRAGW vertically adjustable, BESCHLÄGE height adjustable

***Höhe***: **Höhenwinkel** *m* ARCH & TRAGW, UMWELT elevation angle

**höher**: **~e Energieleistung** *f* UMWELT upper calorific value (*UCV*)

**hohl** *adj* ARCH & TRAGW, WERKSTOFF hollow

**Hohl-** *in cpds* ARCH & TRAGW, WERKSTOFF hollow

***hohl***: **~e Rundstahlstütze** *f* ARCH & TRAGW, BETON, INFR & ENTW, STAHL, WERKSTOFF circular hollow steel column, round hollow steel column

***Hohl-***: **Hohlblockstein** *m* NATURSTEIN, WERKSTOFF cavity block, hollow-core block; **Hohldiele** *f* WERKSTOFF hollow-core plank; **Hohleisen** *nt* STAHL gouge

**Hohlheit** *f* WERKSTOFF hollowness

***Hohl-***: **Hohlkastenträger** *m* ARCH & TRAGW, BETON, INFR & ENTW, STAHL box girder; **Hohlkehle** *f* ARCH & TRAGW concave molding (*AmE*), concave moulding (*BrE*), cove, gorge, scotia, BESCHLÄGE fillet, quirk; **Hohlkehlenverfugung** *f* NATURSTEIN keyed pointing; **Hohlkörper** *m* ARCH & TRAGW hollow, BESCHLÄGE vessel, WERKSTOFF hollow block; **Hohlkörperdecke** *f* ARCH & TRAGW hollow brick ceiling, hollow filler block floor; **Hohlmeißel** *m* STAHL *Holzbearbeitung* gouge; **Hohlpfahl** *m* ARCH & TRAGW, BAUMASCHIN, INFR & ENTW tubular pile; **Hohlplatte** *f* WERKSTOFF core slab, cored panel; **Hohlprofil** *nt* STAHL hollow section, hollow shape

**Hohlraum** *m* ARCH & TRAGW cavity, gap, interstice, BETON pocket, void; **Hohlraumbildung** *f* BETON, UMWELT cavitation; **Hohlraumfeuchtigkeitssperre** *f* DÄMMUNG, NATURSTEIN cavity wall dampproof course, cavity wall DPC

**hohlraumfrei** *adj* BETON free of cavities

***Hohlraum***: **Hohlraumfüllung** *f* ARCH & TRAGW, DÄMMUNG, WERKSTOFF cavity fill; **Hohlraumgehalt** *m* BETON, INFR & ENTW, UMWELT, WERKSTOFF porosity; **Hohlraumisolierung** *f* ARCH & TRAGW cavity insulation, DÄMMUNG cavity wall insulation, NATURSTEIN cavity insulation, WERKSTOFF cavity wall insulation; **Hohlraumverschluß** *m* ARCH & TRAGW, BAURECHT cavity barrier

***Hohl-***: **Hohlstein** *m* WERKSTOFF hollow brick; **Hohlträger** *m* INFR & ENTW box girder; **Hohlwanddose** *f* ELEKTR hollow wall box; **Hohlziegel** *m* NATURSTEIN hollow brick

**Holm** *m* ARCH & TRAGW *Leiter* runner

**Holz** *nt* HOLZ wood; **aus Holz** *phr* HOLZ wooden; **mit Holzverschalung versehen** *phr* BETON board up

***Holz***: **Holzabfall** *m* UMWELT wood waste; **Holzarbeiten** *f pl* HOLZ woodwork; **Holzasbest** *m* STAHL ligneous asbestos; **Holzbalkenboden** *m* HOLZ wooden joist floor; **Holzbalkendecke** *f* HOLZ wooden joist ceiling; **Holzbalkenzugankerverbindung** *f* ARCH & TRAGW, HOLZ haunched mortice and tenon joint, haunched mortise and tenon joint; **Holzbau** *m* HOLZ wood construction, woodwork; **Holzbauteile** *m pl* HOLZ woodwork; **Holzbauwerk** *nt* HOLZ timber structure; **Holzbekleidung** *f* HOLZ wooden lining; **Holzbock** *m* HOLZ timber jack; **Holzbogen** *m* HOLZ wooden arch; **Holzbohrer** *m* BAUMASCHIN, HOLZ auger; **Holzbrücke** *f* HOLZ timber bridge; **Holzdrehbank** *f* HOLZ wood-turning lathe; **Holzdübel** *m* HOLZ nog, nogging

**hölzern** *adj* HOLZ wooden

***Holz***: **Holzfachwerk** *nt* HOLZ timber framing; **Holzfachwerkträger** *m* HOLZ timber truss; **Holzfachwerkwand** *f* HOLZ timber frame wall; **Holzfaserbeton** *m* BETON wood fiber concrete (*AmE*), wood fibre concrete (*BrE*); **Holzfaserplatte** *f* HOLZ hard fiber board (*AmE*), hard fibre board (*BrE*); **Holzfäule** *f* HOLZ, WERKSTOFF rot; **Holzfestigkeit** *f* HOLZ wood strength; **Holzfeuchte** *f* HOLZ, WERKSTOFF timber moisture, wood moisture;

**Holzfeuchtigkeit** *f* HOLZ, WERKSTOFF timber moisture, wood moisture; **Holzfloß** *nt* HOLZ timber raft
**holzfressend** *adj* HOLZ, UMWELT *Schädling* lignivorous, xylophagous
***Holz*: Holzfußboden** *m* HOLZ wooden floor; **Holzgebälk** *nt* HOLZ binders and joists, timberwork; **Holzgrundierung** *f* OBERFLÄCHE wood primer, wood sealer; **Holzgurtgesims** *nt* ARCH & TRAGW stringer; **Holzgüte** *f* BAURECHT, HOLZ timber quality; **Holzhammer** *m* BAUMASCHIN *Pflasterarbeiten* beetle; **Holzimprägnieranlage** *f* HOLZ wood-impregnating plant; **Holzimprägnierung** *f* HOLZ impregnation of wood; **Holzkitt** *m* HOLZ crack filler; **Holzklammer** *f* HOLZ timber dog; **Holzkonditionierung** *f* HOLZ wood conditioning; **Holzkonstruktion** *f* HOLZ wooden structure; **Holzlack** *m* HOLZ, OBERFLÄCHE wood lacquer; **Holzlager** *nt* HOLZ, INFR & ENTW wood yard; **Holzlasche** *f* HOLZ wooden cleat; **Holzleimbau** *m* ARCH & TRAGW, HOLZ bonded wood construction; **Holzmetallverbindung** *f* HOLZ, STAHL metal timber connector; **Holzmontagebau** *m* ARCH & TRAGW, HOLZ wooden prefabricated construction; **Holznagel** *m* HOLZ tree nail; **Holzöl** *nt* HOLZ wood oil, OBERFLÄCHE tung oil; **Holzpfahl** *m* HOLZ spile, timber pile; **Holzpflaster** *nt* HOLZ timber pavement, timber paving, wood-block pavement, wood-block paving; **Holzpflasterklotz** *m* HOLZ wooden paver, wooden paving block, wood paving block; **Holzpflock** *m* HOLZ nog, spile; **Holzplatz** *m* HOLZ timber yard; **Holzrahmen** *m* HOLZ timber frame; **Holzrahmentafel** *f* HOLZ timber frame panel; **Holzraspel** *f* BAUMASCHIN wood rasp; **Holzriegel** *m* ARCH & TRAGW nogging piece; **Holzriemenboden** *m* HOLZ matchboarding floor; **Holzrohbau** *m* HOLZ carpentry; **Holzsäge** *f* BAUMASCHIN wood saw; **Holzschalung** *f* BETON, HOLZ planking, roof boarding, timber formwork; **Holzschindel** *f* HOLZ wood shingle; **Holzschlaghammer** *m* BAUMASCHIN bossing mallet; **Holzschraube** *f* HOLZ woodscrew; **Holzschutzmittel** *nt* HOLZ, OBERFLÄCHE, WERKSTOFF timber preservative, wood preservative; **Holzschwelle** *f* HOLZ wooden sleeper; **Holzskelettbau** *m* ARCH & TRAGW, HOLZ wood skeleton structure, wooden skeleton construction; **Holzsockelleiste** *f* HOLZ, WERKSTOFF timber baseboard (*AmE*), timber skirting (*BrE*); **Holzspaltkeil** *m* HOLZ timber splitting wedge; **Holzspanplatte** *f* HOLZ, WERKSTOFF chipboard, particle board; **Holzsparren** *m* ARCH & TRAGW, HOLZ timber rafter; **Holzspundwand** *f* INFR & ENTW timber sheet piling; **Holzständerkonstruktion** *f* ARCH & TRAGW, HOLZ wooden post-and-beam structure; **Holzstift** *m* HOLZ nog; **Holzsturz** *m* HOLZ timber lintel; **Holztafel** *f* HOLZ wooden panel; **Holztafelbauweise** *f* HOLZ wood panel construction; **Holztäfelung** *f* HOLZ paneling (*AmE*), panelling (*BrE*), wooden paneling (*AmE*), wooden panelling (*BrE*); **Holzträger** *m* ARCH & TRAGW, HOLZ timber girder, wood girder, wooden girder; **Holzverbundstoff** *m* HOLZ, WERKSTOFF wood composite; **Holzverfall** *m* HOLZ timber decay; **Holzverkleidung** *f* HOLZ timbering; **Holzverlattung** *f* HOLZ battening, lathing; **Holzverschalung** *f* ARCH & TRAGW, HOLZ weather boarding; **Holzverschlag** *m* HOLZ wooden crate; **Holzwerk** *nt* HOLZ timbering; **Holzwolle** *f* ARCH & TRAGW, DÄMMUNG wood wool; **Holzwollebauplatte** *f* WERKSTOFF excelsior building slab, wood wool building slab; **Holzwollebeton** *m* WERKSTOFF excelsior concrete; **Holzwolleleichtbauplatte** *f* BESCHLÄGE wood wool lightweight building board; **Holzwurm** *m* HOLZ, UMWELT woodworm; **Holzzementdach** *nt* ARCH & TRAGW Häusler roofing; **Holzziegel** *m* HOLZ wood brick
**homogen** *adj* INFR & ENTW, WERKSTOFF homogenous
**Hooke**: **~sches Gesetz** *nt* ARCH & TRAGW, INFR & ENTW, WERKSTOFF Hooke's law
**Hörbereich** *m* BAURECHT, DÄMMUNG, INFR & ENTW audibility range
**horizontal** *adj* ARCH & TRAGW, INFR & ENTW horizontal
**Horizontalschub** *m* ARCH & TRAGW horizontal shear
**Hosenrohr** *nt* ABWASSER wye, Y-tube
**Hospiz** *nt* ARCH & TRAGW, INFR & ENTW hospice
**Hotel** *nt* ARCH & TRAGW, INFR & ENTW hotel
**Hourdi** *m* WERKSTOFF clay pot, hollow clay block, hollow-gaged brick (*AmE*), hollow-gauged brick (*BrE*)
**H-Profil** *nt* WERKSTOFF H-section
**HP-Schale** *f* (*hyperbolische Paraboloidschale*) ARCH & TRAGW, INFR & ENTW hyperbolic paraboloid shell
**H-Träger** *m* ARCH & TRAGW, BETON, STAHL H-beam, H-girder
**Hub** *m* ABWASSER *Pumpe*, BAUMASCHIN, INFR & ENTW lift; **Hubbrücke** *f* INFR & ENTW lifting bridge; **Hubfenster** *nt* ARCH & TRAGW sash window; **Hubgeschwindigkeit** *f* BAUMASCHIN hoisting speed; **Hubhöhe** *f* ABWASSER *Pumpe* lift, BAUMASCHIN, INFR & ENTW hoisting height, lift, lifting height; **Hubkette** *f* BAUMASCHIN lifting chain; **Hublader** *m* BAUMASCHIN *Tiefbau* loading shovel; **Hubplatte** *f* ARCH & TRAGW lift slab; **Hubplattenverfahren** *nt* ARCH & TRAGW lift-slab method; **Hubpresse** *f* BAUMASCHIN lifting press; **Hubrohr** *nt* HEIZ & BELÜFT lift pipe
**Hubschrauberlandeplatz** *m* ARCH & TRAGW, INFR & ENTW *allgemein* heliport, helistop, *auf dem Dach eines Gebäudes* helipad
***Hub*: Hubstapler** *m* BAUMASCHIN fork-lift truck; **Hubtor** *nt* ARCH & TRAGW lift gate
**Hufeisengewölbe** *nt* ARCH & TRAGW horseshoe arch
**Hülle** *f* ARCH & TRAGW, BAUMASCHIN shell
**Hüllrohr** *nt* BETON sheath
**Hülse** *f* ABWASSER sleeve, BESCHLÄGE tube, *eines Schlosses* barrel, HEIZ & BELÜFT sleeve
**Humus** *m* DÄMMUNG, INFR & ENTW humus; **Humusgehalt** *m* INFR & ENTW, WERKSTOFF humus content; **Humusverfüllung** *f* **der Hohlräume** INFR & ENTW *Gartenarbeit* filling of openings with humus soil
**Hundezahnornament** *nt* ARCH & TRAGW dogtooth
**Hutabdeckung** *f* WERKSTOFF top-hat cover piece
**Hutquerschnitt** *m* WERKSTOFF top-hat section
**Hütte** *f* STAHL, WERKSTOFF foundry, smelter, smeltery, steelworks; **Hüttensand** *m* WERKSTOFF granulated cinder; **Hüttenwolle** *f* WERKSTOFF slag wool; **Hüttenzement** *m* BETON, WERKSTOFF blastfurnace cement
**HV-Schraube** *f* (*hochfest vorgespannte Schraube*) STAHL friction-grip bolt; **HV-Schraubenverbindung** *f* ARCH & TRAGW, STAHL friction-grip bolt connection, friction-grip bolting, high-strength friction grip, high-tensile bolted structural joint
**HV-Verbindung** *f* (*hochfest verschraubte Verbindung*) ARCH & TRAGW, STAHL high-strength friction grip, high-tensile bolted structural joint

**Hybrid-Betonkonstruktion** *f* ARCH & TRAGW, WERKSTOFF hybrid concrete construction
**Hydrant** *m* ABWASSER , BAURECHT *Brandbekämpfung* hydrant (*BrE*), fireplug (*AmE*), INFR & ENTW standpipe
**Hydratation** *f* INFR & ENTW, WERKSTOFF hydration; **Hydratationswärme** *f* BETON, NATURSTEIN heat of hydration, hydration heat; **Hydratationswasser** *nt* BETON, INFR & ENTW, WERKSTOFF hydration water
**Hydraulik** *f* ABWASSER, INFR & ENTW hydraulics; **Hydraulikpresse** *f* BAUMASCHIN hydraulic press
**hydraulisch**: **~es Bindemittel** *nt* WERKSTOFF hydraulic binder; **~er Druck** *m* ARCH & TRAGW hydraulic pressure; **~e Druckhöhe** *f* INFR & ENTW hydraulic head; **~er Lift** *m* ARCH & TRAGW, INFR & ENTW hydraulic lift; **~es Modell** *nt* INFR & ENTW, UMWELT, WERKSTOFF hydraulic model; **~er Modellversuch** *m* INFR & ENTW, WERKSTOFF hydraulic model test; **~e Rißbildung** *f* UMWELT hydraulic fracturing; **~er Widder** *m* UMWELT hydraulic ram; **~er Wirkungsgrad** *m* UMWELT hydraulic efficiency; **~er Zement** *m* BETON, WERKSTOFF hydraulic cement
**hydrodynamisch** *adj* INFR & ENTW hydrodynamic; **~es Modell** *nt* INFR & ENTW hydrodynamic model
**Hydrofluorchlorkohlenwasserstoff** *m* (*HFCK*) UMWELT hydrochlorofluorocarbon (*HCFC*)
**Hydrographie** *f* UMWELT hydrography
**Hydropulsor** *m* UMWELT hydraulic ram
**hydrostatisch**: **~er Druck** *m* UMWELT hydraulic thrust
**hydrothermal**: **~er Prozeß** *m* UMWELT hydrothermal process; **~es Verfahren** *nt* UMWELT hydrothermal process
**hygienisch** *adj* BAURECHT sanitary
**hyperbolisch**: **~e Paraboloidschale** *f* (*HP-Schale*) ARCH & TRAGW, INFR & ENTW hyperbolic paraboloid shell
**Hypothek** *f* BAURECHT mortgage; **Hypothekengläubiger** *m* BAURECHT mortgagee; **Hypothekenschuldner** *m* BAURECHT mortgagor
**HZK** *abbr* (*höchstzulässige Konzentration*) UMWELT TLV (*threshold limit value*)
***HZK*** *f*: **~ am Arbeitsplatz** UMWELT TLV in the workplace; **~ in der Umwelt** UMWELT TLV in the free environment

# I

**Idealgeschwindigkeit** *f* UMWELT ideal velocity

**identifiziert**: **~e Quellen** *f pl* UMWELT identified resources

**IfBt** *abbr* (*Institut für Bautechnik*) BAURECHT Institute for Building Technology

**Imhoffbrunnen** *m* ABWASSER, INFR & ENTW, UMWELT Imhoff tank

**Imhoff-Trichter** *m* ABWASSER, INFR & ENTW, WERKSTOFF Imhoff cone

**Immissionsgrenzwert** *m*: **~ der Luft** UMWELT ambient air emission standard, ambient air quality standard

**Immobilien**: **Immobilienhändler** *m* BAURECHT broker, estate agent (*BrE*), real estate agent (*AmE*); **Immobilienmakler** *m* BAURECHT broker, estate agent (*BrE*), real estate agent (*AmE*)

**Impaktion** *f* UMWELT impaction

**Imperfektion** *f* ARCH & TRAGW, INFR & ENTW, WERKSTOFF imperfection

**Impermeabilität** *f* INFR & ENTW, WERKSTOFF impermeability

**Impfkompostrückführung** *f* UMWELT recycling of inoculated compost

**impliziert**: **~e Übereinkunft** *f* BAURECHT implied agreement

**imprägnieren** *vt* DÄMMUNG impregnate, waterproof, proof, HOLZ impregnate, OBERFLÄCHE proof

**imprägniert** *adj* DÄMMUNG impregnated, proofed, waterproof, HOLZ impregnated, OBERFLÄCHE proofed

**Imprägnierung** *f* DÄMMUNG impregnation, proofing, HOLZ impregnation, OBERFLÄCHE preservation, proofing

**indirekt**: **~e Beleuchtung** *f* ELEKTR concealed lighting; **~er Gewinn** *m* HEIZ & BELÜFT indirect gain; **~e künstliche Beleuchtung** *f* ELEKTR indirect artificial lighting; **~e Lagerung** *f* BETON indirect support

**Industrie** *f* INFR & ENTW, UMWELT industry; **Industrieabfall** *m* BAURECHT commercial waste; **Industrieabwasser** *nt* INFR & ENTW, UMWELT industrial waste water; **Industrieanlage** *f* INFR & ENTW industrial plant; **Industriebau** *m* INFR & ENTW industrial building; **Industriegebäude** *nt* INFR & ENTW industrial building; **Industriegebiet** *nt* INFR & ENTW industrial estate; **Industriegleis** *nt* INFR & ENTW industrial railroad (*AmE*), industrial railway (*BrE*)

**industriell**: **~e Bauweise** *f* ARCH & TRAGW, INFR & ENTW industrial building method, industrial construction method, industrialized construction method; **~e Heizungsanlage** *f* ABWASSER, INFR & ENTW industrial heating; **~er Klärschlamm** *m* INFR & ENTW, UMWELT industrial sewage sludge; **~er Teppich** *m* INFR & ENTW, WERKSTOFF heavy-duty carpet

***Industrie***: **Industriemüll** *m* UMWELT industrial waste; **Industriemülldeponie** *f* UMWELT industrial landfill

**induzierend**: **~er Durchfluß** *m* ELEKTR *Brunnen* inducing flow

**inert** *adj* INFR & ENTW, UMWELT inert; **~er Abfall** *m* UMWELT inert waste; **~es Material** *nt* UMWELT inert material

**inertisiert**: **~er Rückstand** *m* UMWELT noncombustible residue

**infektiös**: **~er Abfall** *m* UMWELT pathological waste, infectious waste, anatomical waste

**Infiltration** *f* ABWASSER, INFR & ENTW, UMWELT *Schadstoffe* infiltration; **Infiltrationsbeiwert** *m* INFR & ENTW, WERKSTOFF infiltration coefficient; **Infiltrationswasser** *nt* INFR & ENTW, WERKSTOFF infiltration water

**infrarot**: **~e Thermographie** *f* WERKSTOFF infrared thermography

**Ingenieur** *m* INFR & ENTW engineer; **Ingenieurbüro** *nt* ARCH & TRAGW, INFR & ENTW consulting engineers; **Ingenieurhochbau** *m* ARCH & TRAGW structural engineering; **Ingenieurwesen** *nt* INFR & ENTW engineering

**Inhalt** *m*: **~ der Kanalhaltung** UMWELT pondage

**Injektion** *f* BAUMASCHIN, BETON, INFR & ENTW, NATURSTEIN injection; **Injektionsanker** *m* INFR & ENTW grout anchor; **Injektionsverfahren** *nt* NATURSTEIN pressure grouting

**inkohärent** *adj* INFR & ENTW, WERKSTOFF incoherent, noncohesive

**inkompressibel** *adj* INFR & ENTW, WERKSTOFF incompressible

**Inlandsquelle** *f* UMWELT internal source

**Innen-** *in cpds* ABWASSER, ARCH & TRAGW, BAUMASCHIN, BESCHLÄGE, BETON, ELEKTR, INFR & ENTW, NATURSTEIN, OBERFLÄCHE inner, interior, internal; **Innenarchitektur** *f* ARCH & TRAGW interior decoration, interior design; **Innenausbau** *m* ARCH & TRAGW interior work, internal finishing work; **Innenausstattung** *f* BESCHLÄGE interior furnishings; **Innendeckanstrich** *m* OBERFLÄCHE interior finishing coat; **Innendruck** *m* ARCH & TRAGW, INFR & ENTW internal pressure; **Innendruckprüfung** *f* ABWASSER *Wasser*, INFR & ENTW water pressure test; **Innendurchmesser** *m* ABWASSER, INFR & ENTW internal diameter; **Inneneck** *nt* ARCH & TRAGW inner corner; **Innenfeld** *nt* ARCH & TRAGW interior span; **Innengerippe** *nt* ARCH & TRAGW internal carcass; **Innengewinde** *nt* STAHL female thread, WERKSTOFF internal thread; **Innenhof** *m* ARCH & TRAGW patio; **Innenperspektive** *f* ARCH & TRAGW, INFR & ENTW interior perspective; **Innenputz** *m* NATURSTEIN interior plaster; **Innenrüttler** *m* BAUMASCHIN internal vibrator, poker vibrator; **Innenrüttlung** *f* BETON internal vibration; **Innenspannung** *f* ARCH & TRAGW, INFR & ENTW internal stress; **Innentür** *f* ARCH & TRAGW interior door; **Innenwand** *f* ARCH & TRAGW interior wall; **Innenwandputz** *m* NATURSTEIN interior wall plaster, internal wall plaster; **Innenweite** *f* ARCH & TRAGW inner span; **Innenzylinder** *m* BESCHLÄGE inside cylinder

**innere** *adj* ARCH & TRAGW, INFR & ENTW inner, internal; **~ Abmessung** *f* ARCH & TRAGW, INFR & ENTW internal dimension; **~r Blitzschutz** *m* ELEKTR interior lightning protection; ~ **Kraft** *f* ARCH & TRAGW, INFR & ENTW internal force; ~ **Last** *f* ARCH & TRAGW, INFR & ENTW internal load; **~s Moment** *nt* ARCH & TRAGW, INFR & ENTW internal moment; ~ **Oberfläche** *f* DÄMMUNG internal surface; ~ **Querkraft** *f* ARCH & TRAGW, INFR & ENTW internal transverse force; ~ **Reibung** *f* ARCH & TRAGW, INFR & ENTW internal friction; **~r Reibungswinkel** *m* INFR & ENTW angle of internal friction; **~r Ringerder** *m* ELEKTR interior ring earth (*BrE*), interior ring ground (*AmE*); ~ **Schale** *f* NATURSTEIN *Hohlwand* inner leaf, internal cavity wall; ~ **Torsion** *f* ARCH & TRAGW, INFR & ENTW internal torsion; ~ **Verkleidung** *f* ARCH & TRAGW, DÄMMUNG internal lining

**Insektenschutzgitter** *nt* BESCHLÄGE insect screen

**inspizieren** *vt* BAURECHT inspect

**instabil** *adj* INFR & ENTW unstable

**Installateur** *m* ABWASSER, ARCH & TRAGW, STAHL plumber

**Installation** *f* ABWASSER plumbing, ARCH & TRAGW installation, plumbing, HEIZ & BELÜFT plumbing, INFR & ENTW installation, STAHL plumbing; **Installationsarbeiten** *f pl* ABWASSER, ARCH & TRAGW, HEIZ & BELÜFT, INFR & ENTW installation work; **Installationsblock** *m* ABWASSER modular building unit, HEIZ & BELÜFT plumbing unit; **Installationsgeschoß** *nt* ARCH & TRAGW mechanical floor; **Installationskern** *m* ABWASSER, ELEKTR utility core, HEIZ & BELÜFT plumbing core, utility core; **Installationsobjekte** *nt pl* BESCHLÄGE fixtures and fittings; **Installationsrohr** *nt* ELEKTR conduit; **Installationsschlitz** *m* NATURSTEIN installation chase; **Installationswand** *f* ARCH & TRAGW *für Rohre* plumbing wall; **Installationszelle** *f* ABWASSER modular building unit, ARCH & TRAGW building block module, HEIZ & BELÜFT plumbing unit, INFR & ENTW building block module

**installieren** *vt* ABWASSER, ELEKTR, HEIZ & BELÜFT install, plumb

**installiert**: **~e Leistung** *f* ELEKTR installed capacity

**instand halten** *vt* ARCH & TRAGW, BAURECHT, INFR & ENTW maintain

**Instandhaltung** *f* ARCH & TRAGW, BAURECHT, INFR & ENTW maintenance; **Instandhaltungshängebühne** *f* BAUMASCHIN maintenance cradle

**Instandsetzungsarbeiten** *f pl* ARCH & TRAGW repairs

**Institut** *nt*: ~ **für Bauingenieure** BAURECHT, INFR & ENTW Institution of Civil Engineers (*ICE*); ~ **für Bautechnik** (*IfBt*) BAURECHT Institute for Building Technology

**Institution** *f* INFR & ENTW institution; ~ **für Forschung in Bauwesen** (*BRE*) BAURECHT Building Research Establishment (*BRE*)

**Instruktion** *f* BAURECHT instruction

**Instrumentenhöhe** *f* ARCH & TRAGW, INFR & ENTW *Vermessung* height of instrument

**intelligent**: **~e Baustelle** *f* ARCH & TRAGW, INFR & ENTW intelligent site; **~es Gebäude** *nt* ARCH & TRAGW, INFR & ENTW intelligent building

**Interaktion** *f* INFR & ENTW, WERKSTOFF interaction; **Interaktionsdiagramm** *nt* ARCH & TRAGW, BETON, INFR & ENTW, STAHL, WERKSTOFF interaction diagram

**Interferenz** *f* INFR & ENTW interference

**Interkostalträger** *m* ARCH & TRAGW intercostal girder

**International**: **~e Organisation** *f* **für Standardisierung** (*ISO*) BAURECHT International Standards Organization (*ISO*); **~er Rat** *m* **für Forschung, Studium und Dokumentation des Bauwesens** (*CIB*) BAURECHT International Council for Building Research and Documentation (*CIB*)

**interstitiell**: **~e Kondensation** *f* HEIZ & BELÜFT, WERKSTOFF interstitial condensation

**Intervall** *nt* ARCH & TRAGW interval

**Inversionsschicht** *f* UMWELT atmospheric inversion

**ionisch**: **~e Säulenanordung** *f* ARCH & TRAGW Ionian order

**irreversibel** *adj* ARCH & TRAGW, INFR & ENTW irreversible

**irrtümlich** *adj* BAURECHT erroneous

**ISO** *abbr* (*Internationale Organisation für Standardisierung*) BAURECHT ISO (*International Standards Organization*)

**Isolation** *f* DÄMMUNG insulation; **Isolationsplatte** *f* DÄMMUNG, WERKSTOFF insulating slab

**Isolier-** *in cpds* DÄMMUNG, ELEKTR, OBERFLÄCHE, WERKSTOFF insulating; **Isolieranstrich** *m* DÄMMUNG, OBERFLÄCHE impenetrable paint coat, insulating paint coat, sealing coat; **Isolierbeton** *m* BETON, WERKSTOFF insulating concrete; **Isolierdicke** *f* DÄMMUNG insulation thickness

**isolieren** *vt* DÄMMUNG deaden, insulate, ELEKTR, OBERFLÄCHE, WERKSTOFF insulate

**isolierend** *adj* DÄMMUNG, ELEKTR, OBERFLÄCHE, WERKSTOFF insulating

***Isolier-***: **Isolierglas** *nt* DÄMMUNG, WERKSTOFF insulating glass, thermopane glazing; **Isoliermantel** *m* DÄMMUNG, WERKSTOFF insulating jacket; **Isoliermasse** *f* DÄMMUNG, WERKSTOFF insulating compound; **Isoliermaterial** *nt* ELEKTR, WERKSTOFF insulant; **Isolierputz** *m* NATURSTEIN, UMWELT heat-insulating plaster; **Isolierrohr** *nt* UMWELT conduit; **Isolierschicht** *f* DÄMMUNG, WERKSTOFF insulating layer

**isoliert** *adj* DÄMMUNG, ELEKTR insulated; **~e Jalousien** *f pl* DÄMMUNG insulated blinds

**Isolierung** *f* DÄMMUNG deadening, insulation ELEKTR, OBERFLÄCHE, WERKSTOFF insulation

**isolierverglast** *adj* DÄMMUNG thermopane-glazed

***Isolier-***: **Isolierverglasung** *f* DÄMMUNG thermopane glazing; **Isolierwert** *m* DÄMMUNG insulation value, ELEKTR insulance, insulation resistance, WERKSTOFF insulance

**isometrisch**: **~e Ansicht** *f* INFR & ENTW isometric view

**Isostasie** *f* UMWELT isostasy

**Isotherme** *f* HEIZ & BELÜFT, INFR & ENTW, WERKSTOFF isotherm

**isotrop** *adj* INFR & ENTW, WERKSTOFF isotropic

**Istmaß** *nt* ARCH & TRAGW, INFR & ENTW actual dimension

**Istwert** *m* HEIZ & BELÜFT actual value

**Iteration** *f* ARCH & TRAGW iteration

# J

**Jahresverbrauch** *m* ELEKTR, HEIZ & BELÜFT, INFR & ENTW annual consumption
**Jalousie** *f* BESCHLÄGE blind; **Jalousiefenster** *nt* ARCH & TRAGW louver window (*AmE*), louvre window (*BrE*)
**Jauche** *f* INFR & ENTW liquid manure, UMWELT slurry
**JCB®** *f* BAUMASCHIN *Bagger* JCB® (*BrE*)
**Joch** *nt* ARCH & TRAGW cross plate; **Jochbalken** *m* BAUMASCHIN straining piece
**Jugendstil** *m* ARCH & TRAGW Art Nouveau
**justieren** *vt* BAUMASCHIN, BESCHLÄGE adjust
**Justierung** *f* BAUMASCHIN, BESCHLÄGE adjustment
**Jutetuch** *nt* WERKSTOFF gunny cloth

# K

**Kabel** *nt* ELEKTR cable; **Kabelabdeckhaube** *f* ELEKTR cable cover; **Kabelabdeckstein** *m* ELEKTR cable cover; **Kabelboden** *m* ARCH & TRAGW raised floor; **Kabelcode** *m* ELEKTR *Liste* cable code; **Kabeleinführung** *f* ELEKTR cable entry; **Kabelformstein** *m* ELEKTR conduit tile; **Kabelführung** *f* ELEKTR cable run, cable routing, cable line; **Kabelgehäuse** *nt* ELEKTR cable housing; **Kabelgraben** *m* ELEKTR, INFR & ENTW cable trench; **Kabelhauptleitung** *f* ELEKTR, HEIZ & BELÜFT cable trunking; **Kabelkanal** *m* ELEKTR cable duct; **Kabelkanalstein** *m* ELEKTR conduit tile; **Kabelkeller** *m* ARCH & TRAGW underground distribution chamber, *unter dem Erdgeschoß* cable store; **Kabelkran** *m* BAUMASCHIN cable crane; **Kabellager** *nt* BESCHLÄGE, ELEKTR cable store; **Kabelleiter** *f* BESCHLÄGE, ELEKTR cable ladder; **Kabelmerkstein** *m* BESCHLÄGE, ELEKTR cable marker; **Kabelmesser** *nt* BAUMASCHIN hacking knife; **Kabelmontage** *f* BESCHLÄGE, ELEKTR cable assembly; **Kabelmuffe** *f* ELEKTR cable sleeve; **Kabelpritsche** *f* ELEKTR cable tray; **Kabelrinne** *f* ELEKTR cable channel; **Kabelschacht** *m* ARCH & TRAGW cable shaft; **Kabelschelle** *f* ELEKTR cable clip, cable clamp; **Kabelschutzrohr** *nt* ELEKTR cable conduit; **Kabelsteigtrasse** *f* ELEKTR cable ladder; **Kabelsuchgerät** *nt* BAUMASCHIN cable detector; **Kabelsystem** *nt* ELEKTR, HEIZ & BELÜFT cabling system; **Kabelträger** *m* ELEKTR cable tray; **Kabeltrasse** *f* ELEKTR location line of the cable; **Kabeltrommel** *f* ELEKTR cable drum; **Kabeltrommeltransport- und -verlegewagen** *m* BAUMASCHIN cable-laying truck with drum; **Kabelunterstützung** *f* ELEKTR, HEIZ & BELÜFT cable support; **Kabelverbindung** *f* ELEKTR cable connection; **Kabelverlegung** *f* ELEKTR cable laying; **Kabelwarnband** *nt* ELEKTR cable warning tape

**Kabine** *f* ARCH & TRAGW *kleiner Raum* cubicle

**Kachel** *f* BESCHLÄGE, INFR & ENTW, NATURSTEIN tile; **Kachelofen** *m* HEIZ & BELÜFT tiled stove

**Kadmiumbeschichtung** *f* OBERFLÄCHE cadmium coating, cadmium finish

**Kai** *m* BETON, INFR & ENTW, NATURSTEIN quay; **Kaimauer** *f* BETON, INFR & ENTW, NATURSTEIN quay wall

**Kaiserdach** *nt* ARCH & TRAGW imperial roof

**Kalibrierung** *f* ELEKTR, INFR & ENTW calibration

**Kalidarium** *nt* ARCH & TRAGW calidarium

**Kalk** *m* NATURSTEIN lime

**kalken** *vt* OBERFLÄCHE limewash, limewhite, whiten, whitewash

**Kalken** *nt* OBERFLÄCHE limewashing, liming, whitewashing

***Kalk***: **Kalkestrich** *m* NATURSTEIN, WERKSTOFF lime floor; **Kalkgehalt** *m* NATURSTEIN, WERKSTOFF lime content

**kalkhaltig** *adj* ABWASSER, WERKSTOFF calcareous

**kalkig** *adj* ABWASSER, WERKSTOFF calcareous

***Kalk***: **Kalkmörtel** *m* BETON, NATURSTEIN lime mortar; **Kalkputz** *m* NATURSTEIN lime plaster; **Kalksandlochstein** *m* NATURSTEIN perforated sandlime brick; **Kalksandmörtel** *m* BETON, NATURSTEIN, WERKSTOFF coarse stuff; **Kalksandstein** *m* NATURSTEIN sandlime brick, solid sandlime brick; **Kalkschlämme** *m pl* BETON neat lime; **Kalkstein** *m* NATURSTEIN limestone; **Kalktuff** *m* WERKSTOFF calcareous tuff; **Kalktünche** *f* OBERFLÄCHE limewash

**Kalkulation** *f* ARCH & TRAGW, INFR & ENTW calculation

***Kalk***: **Kalkzementmörtel** *m* BETON, NATURSTEIN cement-lime mortar; **Kalkzementputz** *m* NATURSTEIN lime-cement plaster

**Kalotte** *f* ARCH & TRAGW, INFR & ENTW calotte, spherical cap

**kalt** *adj* HEIZ & BELÜFT cold; **~er Punkt** *m* DÄMMUNG, HEIZ & BELÜFT cold spot

**Kalt**: **Kaltanstrich** *m* OBERFLÄCHE cold coating; **Kaltbiegen** *nt* STAHL cold bending; **Kaltdach** *nt* ARCH & TRAGW roof with air circulation, ventilated roof

**Kälte** *f* HEIZ & BELÜFT cold; **Kältebedarf** *m* HEIZ & BELÜFT cold requirement; **Kältebeständigkeit** *f* WERKSTOFF low-temperature stability; **Kältebrücke** *f* ARCH & TRAGW, DÄMMUNG thermal bridge; **Kältemaschine** *f* HEIZ & BELÜFT refrigerating machine

**kaltgewalzt** *adj* STAHL, WERKSTOFF cold-rolled; **~er Balken** *m* WERKSTOFF cold-rolled beam; **~er Stahl** *m* WERKSTOFF cold-rolled steel

**kalthärtend** *adj* WERKSTOFF cold-hardening, cold-setting

***Kalt***: **Kalthärtung** *f* OBERFLÄCHE cold-curing; **Kaltleim** *m* WERKSTOFF cold glue; **Kaltluft** *f* HEIZ & BELÜFT cold air; **Kaltluftkanal** *m* HEIZ & BELÜFT cold air duct; **Kaltverarbeitbarkeit** *f* WERKSTOFF cold-workability

**kaltvergießbar** *adj* WERKSTOFF cold-pourable

**kaltvergossen** *adj* WERKSTOFF cold-poured

**kaltverzinken** *vt* OBERFLÄCHE cold-galvanize

***Kalt***: **Kaltwalzen** *nt* WERKSTOFF cold rolling

**Kaltwasser** *nt* ABWASSER cold water; **Kaltwasserhahn** *m* ABWASSER, BESCHLÄGE cold water tap; **Kaltwasserleitung** *f* ABWASSER cold water conduit, cold water pipe; **Kaltwasserversorgung** *f* ABWASSER cold water supply

***Kalt***: **Kaltwetterbetonieren** *nt* BETON cold weather concreting

**Kamin** *m* ARCH & TRAGW fireplace, stack, *Schornstein* chimney; **Kaminformstein** *m* NATURSTEIN, WERKSTOFF chimney block; **Kaminhut** *m* ARCH & TRAGW chimney head; **Kaminkopf** *m* ARCH & TRAGW chimney head; **Kaminrohr** *nt* ARCH & TRAGW, STAHL chimney pipe; **Kaminverwahrung** *f* ARCH & TRAGW chimney flashing; **Kaminvorsprung** *m* ARCH & TRAGW chimney breast; **Kaminzug** *m* ARCH & TRAGW, HEIZ & BELÜFT draft (*AmE*), draught (*BrE*)

**Kämmen** *nt* NATURSTEIN combing

**Kammer** *f* ARCH & TRAGW, UMWELT cell, chamber, compartment

**Kammputz** *m* NATURSTEIN comb rendering

**Kämpfer** *m* ARCH & TRAGW *Fenster* crossbar, springer, impost, INFR & ENTW *eines Bogens* springer, *Fenster* crossbar, impost; **Kämpferholz** *nt* HOLZ *Fenster, Tür* wooden transom; **Kämpferlinie** *f* ARCH & TRAGW,

INFR & ENTW spring line, springing line; **Kämpferriegel** *m* ARCH & TRAGW transom; **Kämpferschicht** *f* INFR & ENTW springing course; **Kämpferstein** *m* ARCH & TRAGW *Bogen, Gewölbe* springer, NATURSTEIN springer stone

**Kanal** *m* ABWASSER canal, channel, ELEKTR conduit, HEIZ & BELÜFT, INFR & ENTW canal, channel; **Kanalbildung** *f* ARCH & TRAGW, INFR & ENTW, UMWELT channeling (*AmE*), channelling (*BrE*); **Kanaldiele** *f* INFR & ENTW trench sheeting; **Kanaldüker** *m* ABWASSER, INFR & ENTW sewer culvert; **Kanalgas** *nt* INFR & ENTW sewerage gas

**Kanalisation** *f* ABWASSER, INFR & ENTW, UMWELT drain system, public sewers, sewer system, sewerage system; **Kanalisationsrohr** *nt* ABWASSER, INFR & ENTW, UMWELT sewage pipe, sewer pipe

**Kanalisierung** *f* INFR & ENTW canalization

***Kanal***: **Kanallänge** *f* ABWASSER, INFR & ENTW, UMWELT length of channel; **Kanalreinigung** *f* ABWASSER, INFR & ENTW sewer cleaning; **Kanalschlammabsauggerät** *nt* BAUMASCHIN sewer mud extractor; **Kanalschleuse** *f* INFR & ENTW canal lock; **Kanalsystem** *nt* INFR & ENTW ductwork; **Kanaltunnel** *m* INFR & ENTW Channel Tunnel; **Kanalwartung** *f* ABWASSER, INFR & ENTW sewer maintenance; **Kanalziegel** *m* ABWASSER, NATURSTEIN sewer brick

**Kannelieren** *nt* ARCH & TRAGW, INFR & ENTW channeling (*AmE*), channelling (*BrE*)

**kanneliert**: **~e Säule** *f* ARCH & TRAGW fluted column

**Kante** *f* ARCH & TRAGW verge, HOLZ skirt

**kanten** *vt* INFR & ENTW tip

***Kante***: **Kantenpressung** *f* NATURSTEIN edge pressure; **Kantenschloß** *nt* BESCHLÄGE flush lock; **Kantenschutz** *m* NATURSTEIN corner bead; **Kantenschutzleiste** *f* NATURSTEIN corner bead; **Kantenschutzprofil** *nt* NATURSTEIN corner bead; **Kantenschutzschiene** *f* BESCHLÄGE nosing; **Kantenschutzwinkel** *m* BESCHLÄGE, STAHL metal angle bead

**Kantholz** *nt* HOLZ scantling, square timber

**Kapazität** *f* ARCH & TRAGW capacity

**Kapelle** *f* ARCH & TRAGW, INFR & ENTW chapel

**kapillarbrechend**: **~e Schicht** *f* ARCH & TRAGW, INFR & ENTW *Straßenbau* anticapillary course, dry area

**Kapillar**: **Kapillarfitting** *f* HEIZ & BELÜFT, INFR & ENTW capillary fitting; **Kapillarwirkung** *f* INFR & ENTW *Boden* capillary action

**Kapital** *nt* BAURECHT capital

**Kapitalkosten** *pl* BAURECHT capital costs

**Kapitel** *nt* BAURECHT chapter

**Kapitell** *nt* ARCH & TRAGW capital

**Kapo** *m* ARCH & TRAGW head mason

**Kappe** *f* ARCH & TRAGW dome

**Kappstreifen** *m* ARCH & TRAGW cap flashing, cap piece

**Karbonitrierung** *f* WERKSTOFF carbonitriding

**Karnies** *nt* ABWASSER, ARCH & TRAGW ogee; **Karnieshobel** *m* BAUMASCHIN ogee plane; **Karniesrinne** *f* ABWASSER, ARCH & TRAGW ogee gutter

**Karrbohle** *f* ARCH & TRAGW, HOLZ barrow way, runway plank

**Karte** *f* ARCH & TRAGW, BAURECHT, INFR & ENTW chart, map

**Kartierer** *m* ARCH & TRAGW, BAURECHT, INFR & ENTW *Vermessung* mapper

**Kartierung** *f* ARCH & TRAGW, BAURECHT, INFR & ENTW mapping

**Karyatide** *f* ARCH & TRAGW caryatid

**kaschieren** *vt* INFR & ENTW laminate, OBERFLÄCHE coat, WERKSTOFF laminate

**Kaschieren** *nt* INFR & ENTW laminating, OBERFLÄCHE coating, WERKSTOFF laminating

**Kasein** *nt* HOLZ, OBERFLÄCHE, WERKSTOFF casein; **Kaseinbindemittel** *nt* OBERFLÄCHE casein binder; **Kaseinbinder** *m* OBERFLÄCHE casein binder; **Kaseinfarbe** *f* OBERFLÄCHE casein distemper; **Kaseinleim** *m* WERKSTOFF casein glue

**Kasematte** *f* ARCH & TRAGW casemate

**Kaserne** *f* ARCH & TRAGW, INFR & ENTW barracks, caserne

**Kasino** *nt* ARCH & TRAGW, INFR & ENTW casino

**Kaskade** *f* ABWASSER, INFR & ENTW cascade

**Kassenraum** *m* ARCH & TRAGW customer service area

**Kassette** *f* ARCH & TRAGW coffer, *Decke* waffle, UMWELT landfill cell, refuse cell; **in Kassetten teilen** *phr* ARCH & TRAGW *Decke* coffer; **Kassettendecke** *f* ARCH & TRAGW, BETON, HOLZ cassette ceiling, coffered ceiling, pan ceiling, paneled ceiling (*AmE*), panelled ceiling (*BrE*), waffle slab ceiling; **Kassettenfeld** *nt* ARCH & TRAGW coffer, *Decke* waffle; **Kassettenplatte** *f* ARCH & TRAGW waffle panel, waffle plate

**kassettiert** *adj* ARCH & TRAGW coffered

**Kasten** *m* ARCH & TRAGW, HOLZ, INFR & ENTW box, crate; **Kastenblechträger** *m* ARCH & TRAGW, STAHL box plate girder; **Kastendüker** *m* ARCH & TRAGW box culvert; **Kastendurchlaß** *m* ARCH & TRAGW box culvert; **Kastenfangdamm** *m* INFR & ENTW coffer dam; **Kastenfundament** *nt* BETON, INFR & ENTW box foundation; **Kastenprofil** *nt* ARCH & TRAGW, STAHL box profile; **Kastenquerschnitt** *m* ARCH & TRAGW, STAHL box profile; **Kastenrinne** *f* ABWASSER, ARCH & TRAGW *Gebäude* parallel gutter, rectangular section gutter, INFR & ENTW box gutter, *Dach* trough gutter, WERKSTOFF box gutter; **Kastenschloß** *nt* BESCHLÄGE box lock; **Kastenträger** *m* ARCH & TRAGW, BETON, INFR & ENTW box girder

**Kataster** *m* BAURECHT land register (*BrE*)

**kathodisch**: **~e Korrosionsschutzanlage** *f* BETON, ELEKTR, OBERFLÄCHE, STAHL cathodic corrosion protection system

**kationisch**: **~es Netzmittel** *nt* OBERFLÄCHE cationic wetting agent

**Katzfahren** *nt* BAUMASCHIN *Kran* cross traversing

**Käufer** *m* WERKSTOFF buyer; **Käuferin** *f* WERKSTOFF buyer

**Kautschuk** *m* BAUMASCHIN, BESCHLÄGE, OBERFLÄCHE, WERKSTOFF rubber

**Kavitation** *f* BETON, INFR & ENTW, WERKSTOFF cavitation; **Kavitationsschaden** *m* BETON, INFR & ENTW, WERKSTOFF cavitation damage

**kcal** *abbr* (*Kilokalorie*) HEIZ & BELÜFT kcal (*kilocalorie*)

**KEG** *abbr* (*Kommission der Europäischen Gemeinschaft*) BAURECHT EC (*European Commission*)

**Kegel** *m* ARCH & TRAGW cone; **Kegeldach** *nt* ARCH & TRAGW conical roof; **Kegeleindringungsversuch** *m* WERKSTOFF cone penetration test; **Kegelgewölbe** *nt* ARCH & TRAGW cone vault; **Kegelkopfniet** *m* BESCHLÄGE cone head rivet; **Kegelschnitt** *m* ARCH & TRAGW conic section

**Kehl-** *in cpds* ARCH & TRAGW valley; **Kehlbalken** *m* ARCH & TRAGW valley beam, valley girder, *Sparren* collar beam, span piece, top beam, HOLZ *Sparren* collar beam, span piece; **Kehlbalkenbinder** *m* ARCH

& TRAGW, HOLZ collar beam truss; **Kehlbalkendach** *nt* ARCH & TRAGW, HOLZ collar beam roof, collar roof; **Kehlbalkenstütze** *f* ARCH & TRAGW side post; **Kehlbeitel** *m* BAUMASCHIN *Holzverarbeitung* gouge; **Kehlblech** *nt* STAHL valley flashing; **Kehlbrett** *nt* ARCH & TRAGW valley board

**Kehle** *f* ARCH & TRAGW groove, valley

***Kehl-***: **Kehlgratbalken** *m* ARCH & TRAGW valley rafter; **Kehlhobel** *m* BAUMASCHIN plough (*BrE*), plow (*AmE*); **Kehlleiste** *f* ARCH & TRAGW fillet; **Kehlnaht** *f* HOLZ fillet joint; **Kehlrinne** *f* ABWASSER valley gutter; **Kehlrinnenweg** *m* ARCH & TRAGW valley gutter walkway; **Kehlsparren** *m* ARCH & TRAGW, HOLZ valley rafter; **Kehlstein** *m* ARCH & TRAGW valley tile

**Kehlung** *f* ARCH & TRAGW molding (*AmE*), moulding (*BrE*), coving, valley, weather check, INFR & ENTW, WERKSTOFF coving

***Kehl-***: **Kehlziegel** *m* ARCH & TRAGW *am Dach* valley tile

**kehren** *vt* ARCH & TRAGW *Kamin* sweep

**Keil** *m* ARCH & TRAGW wedge; **mit einem ~ spalten** *phr* ARCH & TRAGW split with wedges; **Keilklaue** *f* BAUMASCHIN lewis; **Keilnut** *f* BAUMASCHIN keyway; **Keilriemen** *m* BAUMASCHIN V-belt; **Keilstein** *m* ARCH & TRAGW, NATURSTEIN voussoir, wedge; **Keilzapfenverbindung** *f* ARCH & TRAGW wedged mortice and tenon joint; **Keilziegel** *m* ARCH & TRAGW feather-edged brick, NATURSTEIN compass brick, gage brick (*AmE*), gauge brick (*BrE*); **Keilzinkung** *f* HOLZ wedge finger jointing

**keimfrei** *adj* UMWELT, WERKSTOFF germ-free

**Kelle** *f* BAUMASCHIN trowel; **Kellenputz** *m* NATURSTEIN trowel plaster; **Kellenwurf** *m* NATURSTEIN trowel plaster

**Keller** *m* ARCH & TRAGW basement, cellar; **Kellereingang** *m* ARCH & TRAGW basement entrance, cellar entrance; **Kellergarage** *f* INFR & ENTW underground parking; **Kellergeschoß** *nt* ARCH & TRAGW basement, cellar; **Kellergeschoßwand** *f* ARCH & TRAGW basement wall, cellar wall; **Kellergewölbe** *nt* ARCH & TRAGW basement vault, cellar vault, NATURSTEIN cellar vault; **Kellergitter** *nt* ARCH & TRAGW, BESCHLÄGE, STAHL basement grating, cellar grating; **Kellertank** *m* HEIZ & BELÜFT basement tank; **Kellertreppe** *f* ARCH & TRAGW basement stairs, cellar stairs; **Kellertür** *f* ARCH & TRAGW, BESCHLÄGE basement door, cellar door

**Kenndaten** *nt pl* ARCH & TRAGW, ELEKTR, HEIZ & BELÜFT, INFR & ENTW, WERKSTOFF characteristic data

**Kennzeichnung** *f* ARCH & TRAGW marking, WERKSTOFF notation

**Keramik** *f* WERKSTOFF ceramics; **Keramikbodenfliese** *f* WERKSTOFF ceramic floor tile; **Keramikduschwanne** *f* ABWASSER, BESCHLÄGE ceramic shower tray; **Keramikfliese** *f* NATURSTEIN, WERKSTOFF ceramic tile; **Keramikverkleidung** *f* NATURSTEIN ceramic facing; **Keramikwandfliese** *f* NATURSTEIN, WERKSTOFF ceramic wall tile; **Keramikwaschbecken** *nt* ABWASSER, BESCHLÄGE ceramic washbasin; **Keramikwerkstoff** *m* WERKSTOFF ceramic material

**keramisch**: **~er Zuschlagstoff** *m* WERKSTOFF ceramic aggregate

**Kerbe** *f* ARCH & TRAGW groove, notch, slot

**Kerbempfindlichkeit** *f* WERKSTOFF notch sensitivity

**Kerbverbindung** *f* ARCH & TRAGW notch joint

**Kern** *m* ARCH & TRAGW mid-third, core; **Kernbeton** *m* BETON, INFR & ENTW mass concrete; **Kernbohrmaschine** *f* BAUMASCHIN core driller; **Kernbohrung** *f* INFR & ENTW core drilling; **Kerndämmung** *f* DÄMMUNG core insulation; **Kernfüllung** *f* NATURSTEIN hearting; **Kernholz** *nt* HOLZ, WERKSTOFF heartwood; **Kernnagel** *m* BESCHLÄGE core nail; **Kernquerschnitt** *m* ARCH & TRAGW core cross section

**Kessel** *m* ELEKTR, HEIZ & BELÜFT boiler; **Kesselbauer** *m* HEIZ & BELÜFT boiler manufacturer; **Kesselblech** *nt* HEIZ & BELÜFT boiler plate; **Kesselhaus** *nt* ARCH & TRAGW boiler house; **Kesselregelung** *f* ELEKTR, HEIZ & BELÜFT boiler control; **Kesselspeisewasser** *nt* HEIZ & BELÜFT boiler feed water (*BFW*); **Kesselstein** *m* ABWASSER, HEIZ & BELÜFT, WERKSTOFF scale; **Kesselthermostat** *nt* BESCHLÄGE, HEIZ & BELÜFT boiler thermostat

**Kette** *f* ARCH & TRAGW, BAUMASCHIN, BESCHLÄGE, INFR & ENTW, WERKSTOFF chain; **Kettenabsperrung** *f* INFR & ENTW chain barrier; **Kettenbrücke** *f* INFR & ENTW chain bridge; **Kettenbügel** *m* BESCHLÄGE *Schloß* chain shackle; **Kettenzugmaschine** *f* BAUMASCHIN tracked tractor

**Keupermergel** *m* INFR & ENTW, WERKSTOFF saliferous marl

**Kf-Wert** *m* UMWELT coefficient of permeability

**kg** *abbr* (*Kilogramm*) ARCH & TRAGW, INFR & ENTW, WERKSTOFF kg (*kilogram, kilogramme*)

**K-Glas** *nt* WERKSTOFF K glass

**Kiefer** *f* HOLZ pine; **Kiefernholz** *nt* HOLZ pine, pine wood

**Kielbogen** *m* ARCH & TRAGW keel arch

**Kies** *m* INFR & ENTW, WERKSTOFF gravel; **Kiesablagerung** *f* INFR & ENTW, WERKSTOFF gravel deposit; **Kiesbank** *f* INFR & ENTW, WERKSTOFF gravel bank; **Kiesbeton** *m* BETON gravel concrete; **Kiesboden** *m* INFR & ENTW, WERKSTOFF gravelly soil; **Kiesdrain** *m* INFR & ENTW, WERKSTOFF gravel drain

**Kiesel** *m pl* NATURSTEIN, WERKSTOFF pebbles; **Kieselerde** *f* INFR & ENTW, WERKSTOFF silica; **Kieselfilterschicht** *f* UMWELT *Deponie* gravel filter layer; **Kieselstein** *m* NATURSTEIN pebble

***Kies***: **Kiesfang** *m* INFR & ENTW gravel stop; **Kiesfilter** *m* INFR & ENTW, WERKSTOFF gravel filter; **Kiesfilterschicht** *f* ARCH & TRAGW, INFR & ENTW gravel filter layer; **Kiesfraktion** *f* INFR & ENTW, WERKSTOFF gravel fraction; **Kiesfüllung** *f* ARCH & TRAGW, INFR & ENTW gravel packing; **Kiesgrube** *f* INFR & ENTW gravel pit

**kieshaltig**: **~er Boden** *m* INFR & ENTW, WERKSTOFF gravelly soil; **~er Lehm** *m* INFR & ENTW, WERKSTOFF gravelly loam

***Kies***: **Kiesnest** *nt* ARCH & TRAGW, BETON gravel pocket, pocket of loose gravel; **Kiespacklage** *f* ARCH & TRAGW, INFR & ENTW gravel packing; **Kiespreßdach** *nt* ARCH & TRAGW, INFR & ENTW felt-and-gravel roof; **Kiesprobe** *f* INFR & ENTW, WERKSTOFF gravel sample; **Kiesschicht** *f* INFR & ENTW, WERKSTOFF gravel bed; **Kiesschotter** *m* INFR & ENTW, WERKSTOFF ballast; **Kiessohle** *f* INFR & ENTW, WERKSTOFF gravel bottom; **Kiessplitt** *m* WERKSTOFF gravel chippings; **Kiesunterbau** *m* INFR & ENTW gravel substructure; **Kiesverfüllung** *f* INFR & ENTW gravel fill

**Kilo** *nt* (*Kilogramm*) ARCH & TRAGW, INFR & ENTW, WERKSTOFF kilo (*kilogram, kilogramme*); **Kilogramm**

*nt* ARCH & TRAGW, INFR & ENTW, WERKSTOFF kilogram (*kg, kilo*), kilogramme (*kg, kilo*); **Kilokalorie** *f* (*kcal*) HEIZ & BELÜFT kilocalorie (*kcal*); **Kilonewton** *nt* (*kN*) ARCH & TRAGW, INFR & ENTW, WERKSTOFF kilonewton (*kN*); **Kilowatt** *nt* (*kW*) ELEKTR, HEIZ & BELÜFT kilowatt (*kW*); **Kilowattstunde** *f* (*kWh*) ELEKTR, HEIZ & BELÜFT kilowatt-hour (*kWh*)

**kinetisch**: **~e Energie** *f* INFR & ENTW kinetic energy; **~e Reibung** *f* INFR & ENTW kinetic friction

**Kinogebäude** *nt* ARCH & TRAGW cinema building

**Kipp-** *in cpds* ARCH & TRAGW tilting

**kippbar**: **~er Turm** *m* UMWELT tiltable tower

**kippen** *vt* BAUMASCHIN, INFR & ENTW tip

**Kippen** *nt* BAUMASCHIN, INFR & ENTW tipping

**Kipper** *m* BAUMASCHIN, UMWELT skip, tipper, tipper truck

***Kipp-***: **Kippfenster** *nt* ARCH & TRAGW, HOLZ bottom-hinged window, hopper window, pivot-hung window; **Kippfensterflügel** *m* BESCHLÄGE pivot-hung sash; **Kippflügel** *m* BESCHLÄGE *Fenster*, HOLZ bottom-hinged sash; **Kippkübel** *m* BAUMASCHIN dumping bucket, tipping bucket; **Kippmoment** *nt* ARCH & TRAGW, HOLZ, INFR & ENTW, UMWELT overturning moment, pitching moment; **Kippoberlicht** *nt* ARCH & TRAGW, HOLZ *Außenfenster* bottom-hinged skylight, *halbrundes Fenster mit Streben über einer Tür* bottom-hinged fanlight, bottom-hinged transom light; **Kippregel** *f* BAUMASCHIN, INFR & ENTW *Vermessung* leveling alidade (*AmE*), levelling alidade (*BrE*); **Kippsicherheit** *f* ARCH & TRAGW, BAURECHT stability against tilting, INFR & ENTW stability against gliding; **Kipptrommelmischer** *m* BAUMASCHIN tilting drum mixer; **Kippvorrichtung** *f* BAUMASCHIN tipping device

**Kitt** *m* ARCH & TRAGW, BESCHLÄGE, WERKSTOFF putty; **Kittentfernungsmesser** *nt* BAUMASCHIN, BESCHLÄGE hacking knife; **Kittfalz** *m* ARCH & TRAGW rebate, putty rebate, BESCHLÄGE *Fenster* fillister; **Kittfuge** *f* ARCH & TRAGW putty joint

**kittlos**: **~e Verglasung** *f* ARCH & TRAGW patent glazing

***Kitt***: **Kittmesser** *nt* BAUMASCHIN, BESCHLÄGE putty knife

**Klammer** *f* BESCHLÄGE, WERKSTOFF staple, *zur Verbindung von Steinen, Holz* cramp

**Klapp** *in cpds* ARCH & TRAGW, BESCHLÄGE, HOLZ, STAHL hinged, folding; **Klappbohrmast** *m* BAUMASCHIN jackknife drilling mast; **Klappbrücke** *f* INFR & ENTW balance bridge, counterpoise bridge

**Klappe** *f* ARCH & TRAGW, HEIZ & BELÜFT flap, register; **Klappenscharnier** *nt* BESCHLÄGE flap hinge

**Klapp-**: **Klappflügelfenster** *nt* ARCH & TRAGW top-hinged sash window, top-hung window; **Klappladen** *m* BESCHLÄGE, HOLZ box shutter, folding shutter; **Klapprost** *m* STAHL hinged grating; **Klappschütz** *nt* UMWELT tilting gate; **Klapptor** *nt* UMWELT flap gate

**Klar-** *in cpds* ARCH & TRAGW, OBERFLÄCHE, WERKSTOFF clear, transparent

***Klär-*** *in cpds*, ABWASSER, UMWELT clarification, purification, sewage; **Kläranlage** *f* ABWASSER, INFR & ENTW, UMWELT clarification plant, purification plant, sewage treatment plant, sewage works, waste water purification plant, waste water treatment plant; **Klärbecken** *nt* ABWASSER, INFR & ENTW, UMWELT clarification basin, clarification tank, sediment tank

**klären** *vt* ABWASSER purify, treat, ARCH & TRAGW clarify

***Klar-***: **Klarglas** *nt* ARCH & TRAGW, WERKSTOFF clear glass, transparent glass; **Klarlack** *m* OBERFLÄCHE clear lacquer, *chemisch trocknend* clear varnish, transparent varnish

**Klär-**: **Klärschlamm** *m* INFR & ENTW, UMWELT sewage sludge, waste water sludge

**Klärung** *f* ABWASSER, UMWELT purification

***Klär-***: **Klärwerk** *nt* ABWASSER sewage works, INFR & ENTW, UMWELT clarification plant, purification plant, sewage treatment plant, sewage works, waste water treatment plant, waste water treatment works

**klassieren** *vt* WERKSTOFF grade

**klassiert** *adj* WERKSTOFF graded

**Klassierung** *f* BAUMASCHIN screening, INFR & ENTW, WERKSTOFF *nach Korngrößen* grading

**Klassier-**: **~ und Waschmaschine** *f* BAUMASCHIN *für Baustoffe* sizing and washing machine

**Klausel** *f* BAURECHT clause

**Klavierband** *nt* BESCHLÄGE angle hinge, continuous hinge, piano hinge

**Klebeanker** *m* WERKSTOFF adhesive anchor

**Klebeband** *nt* DÄMMUNG, WERKSTOFF adhesive tape

**kleben** *vt* OBERFLÄCHE glue

**Kleber** *m* BETON cementing material, HOLZ, OBERFLÄCHE adhesive, WERKSTOFF adhesive, glue

**klebrig** *adj* OBERFLÄCHE *Lack* adhesive, tacky, WERKSTOFF *Lack* adhesive; **~er Ton** *m* INFR & ENTW sticky clay

**Klebstoff** *m* OBERFLÄCHE adhesive

**Klei** *m* INFR & ENTW clay

**klein**: **~er Holzpfropfen** *m* HOLZ spile

**Klein**: **Kleineisenzeug** *nt* BESCHLÄGE, WERKSTOFF hardware, ironware; **Kleinmaterial** *nt* WERKSTOFF incidentals; **Kleinpflaster** *nt* NATURSTEIN pebble pavement; **Kleinpflasterstein** *m* NATURSTEIN, WERKSTOFF small cobble

**Klemmdose** *f* ELEKTR connection box

**Klemmkasten** *m* ELEKTR terminal box

**Klempner** *m* ABWASSER, ARCH & TRAGW, STAHL plumber; **Klempnerarbeiten** *f pl* ABWASSER, ARCH & TRAGW, HEIZ & BELÜFT, STAHL plumbing

**Klettereisen** *nt* BAURECHT, INFR & ENTW access hook

**Kletterstange** *f* BETON jack rod

**Klienteninteresse** *nt* BAURECHT client's interest

**Klima** *nt* HEIZ & BELÜFT, UMWELT climate; **Klimaanlage** *f* ARCH & TRAGW, BAUMASCHIN, HEIZ & BELÜFT air conditioning, air-conditioning equipment; **Klimafestigkeit** *f* OBERFLÄCHE weathering resistance; **Klimagerät** *nt* HEIZ & BELÜFT air conditioner, air-conditioning unit, conditioning unit; **Klimakanal** *m* ARCH & TRAGW, BAUMASCHIN, HEIZ & BELÜFT air-conditioning duct, conditioning duct; **Klimakarte** *f* HEIZ & BELÜFT, INFR & ENTW design temperature map; **Klimatechnik** *f* BAUMASCHIN, HEIZ & BELÜFT air-conditioning duct, conditioning duct; **Klimatgerät** *nt* HEIZ & BELÜFT conditioner

**klimatisieren** *vt* ARCH & TRAGW air-condition, HEIZ & BELÜFT air-condition, condition

**klimatisiert** *adj* ARCH & TRAGW, HEIZ & BELÜFT air-conditioned, conditioned; **~e Luft** *f* HEIZ & BELÜFT conditioned air

**Klimatisierung** *f* ARCH & TRAGW, HEIZ & BELÜFT air conditioning

**Klingel** *f* BESCHLÄGE bell

**Klinke** *f* BESCHLÄGE lever handle, *Tür* door handle

**klinkenlos**: **~es Schloß** *nt* BESCHLÄGE deadlock

**Klinker** *m* NATURSTEIN, WERKSTOFF clinker, engineer-

ing brick; **Klinkerplatte** *f* NATURSTEIN, WERKSTOFF clinker slab; **Klinkerstein** *m* NATURSTEIN, WERKSTOFF clinker, engineering brick; **Klinkerziegel** *m* NATURSTEIN, WERKSTOFF clinker
**Klosett** *nt* ABWASSER, BESCHLÄGE, INFR & ENTW toilet
**Klothoide** *f* ARCH & TRAGW, INFR & ENTW clothoid
**klumpenfrei** *adj* BETON, INFR & ENTW, WERKSTOFF lump-free
**klumpig** *adj* INFR & ENTW, WERKSTOFF lumpy; **~e Erde** *f* INFR & ENTW lumpy soil
**kN** *abbr* (*Kilonewton*) ARCH & TRAGW, INFR & ENTW, WERKSTOFF kN (*kilonewton*)
**Knagge** *f* ARCH & TRAGW, HOLZ cleat; **Knaggenanschluß** *m* ARCH & TRAGW, HOLZ cleated joint
**Knebel** *m* BESCHLÄGE toggle, *am Schraubstock* handle; **Knebelgriff** *m* BESCHLÄGE locking handle; **Knebelschraube** *f* BESCHLÄGE capstan-headed screw; **Knebelverschluß** *m* BESCHLÄGE slotted head locking device
**Kneifzange** *f* BAUMASCHIN nippers, nipper pliers, pincers
**knetbar** *adj* INFR & ENTW, WERKSTOFF kneadable
**kneten** *vt* INFR & ENTW, WERKSTOFF knead
**Knettest** *m* INFR & ENTW, WERKSTOFF kneading test
**Knick** *m* ARCH & TRAGW bend
**knickbeansprucht** *adj* ARCH & TRAGW, BETON, STAHL, WERKSTOFF subject to buckling
***Knick***: **Knickbereich** *m* ARCH & TRAGW bend; **Knickfestigkeit** *f* ARCH & TRAGW, BETON, INFR & ENTW, STAHL, WERKSTOFF buckling resistance, resistance to buckling; **Knickgefahr** *f* ARCH & TRAGW, BETON, HOLZ, INFR & ENTW, STAHL, WERKSTOFF buckling risk; **Knicklänge** *f* ARCH & TRAGW, BETON, HOLZ, STAHL effective length; **Knicklast** *f* ARCH & TRAGW, INFR & ENTW, WERKSTOFF buckling load
**Knickung** *f* BETON, HOLZ, INFR & ENTW, STAHL, WERKSTOFF buckling
**Knie** *nt* ABWASSER knee; **Knierohr** *nt* ABWASSER pipe bend; **Kniestock** *m* ARCH & TRAGW, BETON, NATURSTEIN jamb wall; **Kniestück** *nt* ABWASSER angle, elbow joint, knee, BESCHLÄGE *Rohr* elbow joint, HEIZ & BELÜFT angle
**Knolle** *f* ARCH & TRAGW *mittelalterliche Verzierung* crocket; **Knollenfußpfahl** *m* ARCH & TRAGW, INFR & ENTW underreamed pile
**Knopf** *m* BESCHLÄGE, ELEKTR button
**Knospenkapitell** *nt* ARCH & TRAGW crocket capital
**Knoten** *m* ARCH & TRAGW point of junction, nodal point, HEIZ & BELÜFT, HOLZ, STAHL assemblage point; **Knotenblech** *nt* HOLZ gusset plate, junction plate, STAHL gusset plate; **Knotenpunkt** *m* ARCH & TRAGW intersection, junction, nodal point, ELEKTR intersection, INFR & ENTW intersection, junction, nodal point; **Knotenpunktverbindung** *f* BESCHLÄGE knee bracket plate, HOLZ *Fachwerk* nodal joint; **Knotenpunktverfahren** *nt* ARCH & TRAGW method of joints
**Knüppeldamm** *m* INFR & ENTW log road
**Koagulation** *f* INFR & ENTW, WERKSTOFF coagulation
**Koaxialkabel** *nt* ELEKTR coaxial cable
**Kobalt** *nt* WERKSTOFF cobalt; **Kobaltblau** *nt* OBERFLÄCHE *Pigment* cobalt blue; **Kobaltgrün** *nt* OBERFLÄCHE *Pigment* cobalt green; **Kobaltviolett** *nt* OBERFLÄCHE *Pigment* cobalt violet
**Köcherfundament** *nt* ARCH & TRAGW, BETON bucket foundation
**Kodifizierung** *f* ARCH & TRAGW, BAURECHT, INFR & ENTW, WERKSTOFF codification
**Koeffizient** *m* BETON, WERKSTOFF coefficient
**Kohäsion** *f* WERKSTOFF cohesion
**kohäsionslos** *adj* INFR & ENTW, WERKSTOFF *Boden* friable, noncohesive
**kohäsiv**: **nicht kohäsiv** *adj* INFR & ENTW, WERKSTOFF noncohesive
**Kohlen-** *in cpds* BESCHLÄGE coal, UMWELT carbon; **Kohlendioxid** *nt* ($CO_2$) UMWELT carbon dioxide ($CO_2$); **Kohlendioxidtreibhauseffekt** *m* UMWELT carbon dioxide greenhouse effect; **Kohlenkessel** *m* HEIZ & BELÜFT, UMWELT coal-fired boiler; **Kohlenmonoxid** *nt* (*CO*) UMWELT carbon monoxide (*CO*); **Kohlenrutsche** *f* BESCHLÄGE coal chute; **Kohlenschachtabdeckung** *f* BESCHLÄGE coal manhole cover
**Kohlenstoff** *m* WERKSTOFF carbon; **Kohlenstoffanteil** *m* WERKSTOFF carbon content; **Kohlenstoffgehalt** *m* WERKSTOFF carbon content
***Kohlen-***: **Kohlenwasserstoff** *m* UMWELT hydrocarbon
**Kokosfaser** *f* WERKSTOFF coir; **Kokosfaserdämmstoff** *m* DÄMMUNG, WERKSTOFF coir insulating material; **Kokosfaserdiele** *f* DÄMMUNG, WERKSTOFF coir-reinforced gypsum plank; **Kokosfasermatte** *f* DÄMMUNG, WERKSTOFF coir mat
**koksgefeuert** *adj* HEIZ & BELÜFT coke-fired
**Kokskessel** *m* HEIZ & BELÜFT coke boiler
**Kolben** *m* BAUMASCHIN piston; **Kolbenentlastungskanal** *m* BAUMASCHIN piston relief duct; **Kolbenhub** *m* BAUMASCHIN piston stroke; **Kolbenhublänge** *f* BAUMASCHIN length of piston stroke
**kollateral**: **~e Gewährleistung** *f* BAURECHT collateral warranty
**Kollektor** *m* UMWELT collector; **Kollektorleistungsvermögen** *nt* UMWELT collector efficiency
**Kolloid-** *in cpds* BETON, OBERFLÄCHE, WERKSTOFF colloidal; **Kolloidbeton** *m* BETON, OBERFLÄCHE, WERKSTOFF colloidal concrete; **Kolloidmörtel** *m* BETON, OBERFLÄCHE, WERKSTOFF colloid mortar, colloidal mortar
**Kolonnade** *f* ARCH & TRAGW colonnade
**Kombinationsschloß** *nt* ARCH & TRAGW, BESCHLÄGE combination lock, puzzle lock
**Kombinationssicherheitsschloß** *nt* BESCHLÄGE combination safety lock
**kombiniert**: **~e Hauptschlüsselanlage** *f* ARCH & TRAGW *Schloß* combined master key system; **~e Raumheizung** *f* **und Warmwasserversorgung** HEIZ & BELÜFT combined heating and hot water supply system; **~e Wärme** *f* **und Energie** HEIZ & BELÜFT combined heat and power (*CHP*)
**Kommission** *f*: **~ der Europäischen Gemeinschaft** (*KEG*) BAURECHT European Commission (*EC*)
**kommunal** *adj* BAURECHT communal; **~er Abfall** *m* UMWELT municipal waste, urban waste; **~es Abwasser** *nt* UMWELT municipal sewage; **~e Kläranlage** *f* UMWELT municipal sewage works
**Kommunalbehörde** *f* BAURECHT local authority
**kompaktiert** *adj* INFR & ENTW, UMWELT compacted
**Kompaktierung** *f* INFR & ENTW, UMWELT compaction
**Kompaktklimagerät** *nt* HEIZ & BELÜFT compact air-conditioning unit
**Kompaktor** *m* INFR & ENTW, UMWELT landfill compactor, packer unit

**Kompaß** *m* BAUMASCHIN leveling compass (*AmE*), levelling compass (*BrE*), INFR & ENTW compass; **Kompaßdiopter** *m* ARCH & TRAGW, INFR & ENTW sight vane

**Kompensator** *m* ABWASSER *Rohr* bellow expansion joint

**Komplementärwinkel** *m* ARCH & TRAGW complementary angle

**Komplex** *m* ARCH & TRAGW, BAURECHT, INFR & ENTW complex

**kompostierbar**: **nicht kompostierbarer Abfall** *m* UMWELT noncompostable waste

**Kompression** *f* INFR & ENTW, UMWELT compaction, WERKSTOFF compression; **Kompressionsversuch** *m* WERKSTOFF compression test

**Kompressor** *m* BAUMASCHIN, HEIZ & BELÜFT compressor

**Kondensat** *nt* DÄMMUNG, HEIZ & BELÜFT condensate, condensation water

**Kondensation** *f* DÄMMUNG, HEIZ & BELÜFT, UMWELT condensation; **Kondensationskern** *m* UMWELT condensation nucleus; **Kondensationskernzähler** *m* UMWELT condensation nucleus counter

***Kondensat***. **Kondensatleitung** *f* HEIZ & BELÜFT condensate piping

**Kondensator** *m* ELEKTR, HEIZ & BELÜFT condenser

**Kondenswasser** *nt* DÄMMUNG, HEIZ & BELÜFT condensate, condensation water; **Kondenswasserentstehung** *f* ABWASSER, HEIZ & BELÜFT formation of condensate; **Kondenswasserregelung** *f* HEIZ & BELÜFT condensation control; **Kondenswasserrinne** *f* HEIZ & BELÜFT condensation gutter

**Konferenzzimmer** *nt* ARCH & TRAGW conference room

**Konglomerat** *nt* INFR & ENTW conglomerate

**konisch**: ~ **zulaufender Treppentritt** *m* ARCH & TRAGW, HOLZ, STAHL tapered tread

**Konkurs** *m* BAURECHT bankruptcy

**Konservatorium** *nt*: ~ **als thermischer Puffer** DÄMMUNG, UMWELT conservatory as thermal buffer

**konservieren** *vt* BAURECHT, HOLZ, OBERFLÄCHE preserve

**Konservierung** *f* BAURECHT, HOLZ preservation; **Konservierungsmittel** *nt* OBERFLÄCHE preservative

**Konsistenz** *f* BETON, INFR & ENTW, WERKSTOFF consistence, consistency

**Konsolbalken** *m* ARCH & TRAGW semigirder

**Konsole** *f* ARCH & TRAGW bracket, BETON bracket, console, HOLZ console, NATURSTEIN corbel, STAHL, WERKSTOFF console

**Konsolidation** *f* INFR & ENTW consolidation

**konstruieren** *vt* ARCH & TRAGW construct

**Konstruktion** *f* ARCH & TRAGW structure, construction, *Bau* design, INFR & ENTW *Entwerfen*, WERKSTOFF *Bau* design; **Konstruktionsbaustahl** *m* STAHL structural grade steel; **Konstruktionsleichtbeton** *m* BETON structural light concrete, structural lightweight concrete; **Konstruktionsrippe** *f* ARCH & TRAGW structural fin; **Konstruktionsteil** *nt* ARCH & TRAGW member

**konstruktiv** *adj* ARCH & TRAGW, BETON, INFR & ENTW, STAHL, WERKSTOFF structural; ~**e Ausbildung** *f* ARCH & TRAGW structural design; ~**er Beton** *m* ARCH & TRAGW, BETON structural concrete; ~**e Bewehrung** *f* ARCH & TRAGW, BETON, INFR & ENTW structural reinforcement; ~**e Details** *nt pl* ARCH & TRAGW, INFR & ENTW structural details; ~**e Form** *f* ARCH & TRAGW structural form; ~**er Leichtbeton** *m* BETON structural light concrete, structural lightweight concrete; ~ **verlegter Baustahl** *m* ARCH & TRAGW, INFR & ENTW, STAHL, WERKSTOFF structural steel

**Kontakt** *m* ARCH & TRAGW contact; **Kontaktfläche** *f* ARCH & TRAGW contact area; **Kontaktkleber** *m* ARCH & TRAGW, WERKSTOFF contact adhesive; **Kontaktschlammverfahren** *nt* UMWELT sludge contact process; **Kontakttermin** *m* BAURECHT contact date; **Kontaktunterbrecher** *m* ELEKTR contact breaker

**Kontaminationsüberwachung** *f* BAURECHT, INFR & ENTW, UMWELT contamination control

**kontaminiert**: ~**er Standort** *m* UMWELT contaminated site, problem site

**Kontenabstimmung** *f* BAURECHT adjustment

**Konterlattung** *f* HOLZ counterlathing, cross lathing

**Kontraktion** *f* WERKSTOFF contraction; **Kontraktionsfuge** *f* ARCH & TRAGW contraction joint, shrinkage joint

**Kontroll-** *in cpds* UMWELT inspection, monitoring, observation; **Kontrollbrunnen** *m* UMWELT monitoring well, observation well

**Kontrolle** *f* ARCH & TRAGW, BAURECHT control, inspection

**kontrollieren** *vt* UMWELT monitor

**kontrolliert**: ~**es Abladen** *nt* **von Schutt** UMWELT controlled dumping; ~**e Müllablagerung** *f* UMWELT controlled dumping, sanitary landfill

***Kontroll-***: **Kontrollmodi** *m pl* BAUMASCHIN, ELEKTR, HEIZ & BELÜFT modes of control; **Kontrollöffnung** *f* ABWASSER, ARCH & TRAGW, INFR & ENTW inspection opening; **Kontrollschacht** *m* ABWASSER, INFR & ENTW inspection manhole

**Konvektionsheizung** *f* HEIZ & BELÜFT convection heating

**Konvektionsheizungsgerät** *nt* HEIZ & BELÜFT convector heating element

**Konvektor** *m* HEIZ & BELÜFT finned radiator, heating convector, convector, ribbed radiator; **Konvektorheizkörper** *m* HEIZ & BELÜFT convector heating element

**Konventionalstrafe**: ~ **zahlen** *phr* BAURECHT pay stipulated fine

**konvex**: ~**er Stab** *m* BAUMASCHIN ovolo

**Konzentration** *f* (*c*) INFR & ENTW, UMWELT, WERKSTOFF concentration (*c*); ~ **in der Atmosphäre** UMWELT atmospheric concentration; ~ **der Umweltschadstoffe** UMWELT ambient pollutant concentration

**konzentriert**: ~**e Last** *f* ARCH & TRAGW, BETON, WERKSTOFF concentrated load

**Kopal** *m* OBERFLÄCHE copal; **Kopalharz** *m* OBERFLÄCHE copal resin

**Kopf** *m* ARCH & TRAGW head; **Kopfanker** *m* ARCH & TRAGW iron tie; **Kopfband** *nt* ARCH & TRAGW strut, BAUMASCHIN raker, HOLZ knee brace, strut; **Kopfbandpfette** *f* HOLZ strutted ridge purlin; **Kopfbolzendübel** *m* BETON *Verbundträger* shear connector; **Kopffläche** *f* HOLZ top surface; **Kopfhöhe** *f* ARCH & TRAGW, INFR & ENTW headroom; **Kopfkipper** *m* BAUMASCHIN front dumper; **Kopfmontage** *f* BESCHLÄGE *Türschließer* transom mounting; **Kopfstein** *m* ARCH & TRAGW *Mauerwerk*, NATURSTEIN header, WERKSTOFF *Pflaster* cobble; **Kopfsteinschicht** *f* ARCH & TRAGW header course;

**Kopfverband** *m* ARCH & TRAGW, NATURSTEIN header bond

**Korbbogen** *m* ARCH & TRAGW basket arch, basket handle arch, three-centered arch (*AmE*), three-centred arch (*BrE*)

**Kork** *m* WERKSTOFF cork; **Korkbelag** *m* DÄMMUNG cork covering; **Korkbodenbelag** *m* WERKSTOFF cork flooring; **Korkdämmung** *f* DÄMMUNG cork insulation; **Korkfliese** *f* WERKSTOFF cork tile

**korkhaltig** *adj* WERKSTOFF cork-based

***Kork*: Korkplatte** *f* WERKSTOFF corkboard; **Korkunterlage** *f* WERKSTOFF cork backing

**Korn** *nt* INFR & ENTW, WERKSTOFF grain; **Kornanordnung** *f* INFR & ENTW, WERKSTOFF grain arrangement; **Kornbeschaffenheit** *f* INFR & ENTW, WERKSTOFF grain character; **Korndurchmesser** *m* INFR & ENTW, WERKSTOFF grain diameter; **Kornform** *f* INFR & ENTW, WERKSTOFF grain shape; **Kornfraktion** *f* WERKSTOFF grain-size fraction, grain-size range, sieve fraction, screening fraction; **Korngerüst** *nt* WERKSTOFF grain skeleton; **Korngröße** *f* INFR & ENTW, WERKSTOFF grain size; **Korngrößeneinteilung** *f* INFR & ENTW, WERKSTOFF grain-size classification; **Korngrößenverteilung** *f* WERKSTOFF grain-size distribution, particle-size distribution; **Korngruppe** *f* WERKSTOFF size range

**körnig** *adj* INFR & ENTW, WERKSTOFF grainy, granular; **~er Anteil** *m* INFR & ENTW, WERKSTOFF granular fraction; **~er Boden** *m* INFR & ENTW, WERKSTOFF granular soil

**Körnigkeit** *f* INFR & ENTW, WERKSTOFF granularity

***Korn*: Kornoberfläche** *f* WERKSTOFF grain surface; **Kornpackung** *f* WERKSTOFF packing; **Kornstruktur** *f* WERKSTOFF grain structure

**Körnung** *f* INFR & ENTW, WERKSTOFF grading, size range; **Körnungsgemisch** *nt* INFR & ENTW, WERKSTOFF grain mixture

***Korn*: Kornverteilungskurve** *f* WERKSTOFF *Kies, Sand* grading curve

**Körperschall** *m* DÄMMUNG impact sound, solid-borne sound, structure-borne sound; **Körperschalldämmung** *f* DÄMMUNG impact sound insulation; **Körperschalldämpfer** *m* DÄMMUNG structure-borne sound absorber

**Korridor** *m* ARCH & TRAGW, INFR & ENTW corridor, hall, hallway

**korrodieren** *vti* ARCH & TRAGW, OBERFLÄCHE, STAHL, WERKSTOFF corrode

**korrodierend** *adj* ARCH & TRAGW, OBERFLÄCHE, STAHL, WERKSTOFF corrosive

**Korrosion** *f* OBERFLÄCHE, STAHL, WERKSTOFF corrosion; **Korrosionsbekämpfung** *f* OBERFLÄCHE, STAHL, WERKSTOFF corrosion control

**korrosionsbeständig** *adj* OBERFLÄCHE, WERKSTOFF corrosion-resistant, stainless

***Korrosion*: Korrosionsbeständigkeit** *f* OBERFLÄCHE, WERKSTOFF corrosion resistance

**korrosionsgeschützt** *adj* OBERFLÄCHE, WERKSTOFF corrosion-proof

***Korrosion*: Korrosionsschaden** *m* OBERFLÄCHE, WERKSTOFF corrosion damage; **Korrosionsschutz** *m* OBERFLÄCHE, WERKSTOFF corrosion protection, corrosion-proofer; **Korrosionsschutzanstrich** *m* OBERFLÄCHE, WERKSTOFF corrosion protection coat; **Korrosionsschutzbinde** *f* OBERFLÄCHE, WERKSTOFF corrosion protection tape; **Korrosionsschutzmittel** *nt* INFR & ENTW, WERKSTOFF anticorrosive agent

**korrosiv** *adj* INFR & ENTW, OBERFLÄCHE, UMWELT, WERKSTOFF corrosive

**Korundstein** *m* WERKSTOFF corundum stone

**Kosten** *pl* ARCH & TRAGW, BAURECHT, INFR & ENTW charge, cost; **~ für unvorhergesehene Arbeiten** ARCH & TRAGW, BAURECHT, INFR & ENTW contingency cost; **Kostenberechnung** *f* ARCH & TRAGW, BAURECHT cost calculation

**kosteneffektiv** *adj* ARCH & TRAGW, INFR & ENTW cost-effective

**kostengünstig**: **~e Montage** *f* ARCH & TRAGW low-cost assembly

***Kosten*: Kostenkontrolle** *f* ARCH & TRAGW, INFR & ENTW cost control; **Kostennutzungsprogramm** *nt* ARCH & TRAGW, INFR & ENTW cost benefit program (*AmE*), cost benefit programme (*BrE*); **Kostenplaner** *m* ARCH & TRAGW, INFR & ENTW quantity surveyor; **Kostenschätzung** *f* ARCH & TRAGW, BAURECHT, INFR & ENTW approximate cost, cost estimate, estimate of costs; **Kostenüberprüfung** *f* ARCH & TRAGW, INFR & ENTW cost check; **Kostenvergleich** *m* ARCH & TRAGW, BAURECHT, INFR & ENTW cost benefit comparison, cost comparison; **Kostenvoranschlag** *m* ARCH & TRAGW, BAURECHT, INFR & ENTW estimate of costs

**Kote** *f* ARCH & TRAGW, INFR & ENTW height notation, level; **~ über NN** (*Kote über Normalnull*) INFR & ENTW height above zero level; **~ über Normalnull** (*Kote über NN*) INFR & ENTW height above zero level

**Krabbe** *f* ARCH & TRAGW *mittelalterliche Verzierung* crocket

**Kraft** *f* ARCH & TRAGW force, BAUMASCHIN, ELEKTR power, INFR & ENTW force, power, UMWELT power; **Kraftanlage** *f* UMWELT powerhouse

**Kräfte-** *in cpds* ARCH & TRAGW, INFR & ENTW of forces; **Kräftedreieck** *nt* ARCH & TRAGW, INFR & ENTW triangle of forces; **Kräftegleichgewicht** *nt* ARCH & TRAGW, INFR & ENTW equilibrium of forces; **Kräftemaßstab** *m* ARCH & TRAGW, INFR & ENTW scale factor; **Kräfteplan** *m* ARCH & TRAGW, INFR & ENTW force diagram; **Kräftepolygon** *m* ARCH & TRAGW, INFR & ENTW polygon of forces; **Kräftevieleck** *nt* ARCH & TRAGW, INFR & ENTW polygon of forces; **Kräftezerlegung** *f* ARCH & TRAGW, INFR & ENTW decomposition of forces, resolution of forces

***Kraft*: Kraftschluß** *m* ARCH & TRAGW, BAUMASCHIN, ELEKTR traction; **Kraftturm** *m* UMWELT power tower; **Kraftwerk** *nt* ARCH & TRAGW, ELEKTR, INFR & ENTW, UMWELT power station

**Krag-** *in cpds* ARCH & TRAGW cantilever; **Kragarm** *m* ARCH & TRAGW cantilever; **Kragarmlast** *f* ARCH & TRAGW cantilever load; **Kragbalken** *m* ARCH & TRAGW, HOLZ semibeam; **Kragdach** *nt* ARCH & TRAGW, HOLZ cantilever roof; **Kraggewölbe** *nt* ARCH & TRAGW corbel vault; **Kraglänge** *f* ARCH & TRAGW cantilevering length, protruding length; **Kraglast** *f* ARCH & TRAGW, INFR & ENTW cantilever load; **Kragplatte** *f* ARCH & TRAGW, BETON cantilever plate, cantilever slab; **Kragstein** *m* ARCH & TRAGW perch, summer stone, HOLZ bracket, NATURSTEIN bracket, console, corbel; **Kragträger** *m* ARCH & TRAGW, BETON, HOLZ cantilever beam, overhanging beam, semibeam; **Kragtreppe** *f* ARCH & TRAGW hanging steps, hanging stairs; **Kragtreppenpodest** *nt* ARCH & TRAGW cantilever landing

**Krählarm** *m* UMWELT rabble arm

**Krampe** *f* BESCHLÄGE staple, WERKSTOFF *zur Verbindung von Steinen, Holz* cramp

**Kran** *m* BAUMASCHIN crane; **Kranbahn** *f* BAUMASCHIN crane runway, craneway; **Kranbahnträger** *m* BAUMASCHIN crane girder, runner; **Kranbrücke** *f* BAUMASCHIN crane bridge; **Kranfahren** *nt* BAUMASCHIN crane driving; **Kranführer** *m* BAUMASCHIN crane operator; **Krangerüst** *nt* BAUMASCHIN crane gantry

**krank**: **~es Gebäudesyndrom** *nt* BAURECHT, INFR & ENTW, UMWELT sick building syndrome

***Kran***: **Kranlasthaken** *m* BAUMASCHIN crane hook; **Kranportal** *nt* BAUMASCHIN gantry; **Kranschiene** *f* BAUMASCHIN crane rail

**kranversetzbar**: **~e Verschalung** *f* BETON crane-displaceable formwork

**Kranz** *m* ARCH & TRAGW wreath; **Kranzbogen** *m* ARCH & TRAGW, NATURSTEIN annular arch; **Kranzgesims** *nt* ARCH & TRAGW cornice; **Kranzleiste** *f* BESCHLÄGE platband

**Kratz-** *in cpds* BAUMASCHIN, NATURSTEIN scraped, scratched; **Kratzbagger** *m* BAUMASCHIN chain dredger; **Kratzband** *nt* BAUMASCHIN scraper conveyor; **Kratzeisen** *nt* BAUMASCHIN raker; **Kratzputz** *m* NATURSTEIN roughcast, scraped rendering, scraped stucco

**Kreis** *m* ARCH & TRAGW, INFR & ENTW circle; **Kreisbecken** *nt* UMWELT clarification basin; **Kreisbogen** *m* ARCH & TRAGW circular arc, *Geometrie* arc of a circle

**Kreiselpumpe** *f* BAUMASCHIN, HEIZ & BELÜFT centrifugal pump

**kreisförmig**: **~er Grundriß** *m* ARCH & TRAGW circular ground plan; **~e Kammer** *f* ARCH & TRAGW circular chamber

***Kreis***: **Kreisfrequenz** *f* UMWELT angular velocity; **Kreisgewölbe** *nt* ARCH & TRAGW circular vault; **Kreislast** *f* ARCH & TRAGW, INFR & ENTW circular load; **Kreislauf** *m* ABWASSER, HEIZ & BELÜFT circuit, INFR & ENTW circulation; **Kreislaufwirtschaft** *f* UMWELT recycling economy; **Kreissäge** *f* BAUMASCHIN circular saw; **Kreissektor** *m* INFR & ENTW sector; **Kreisverkehr** *m* INFR & ENTW rotary (*AmE*), rotary traffic, roundabout (*BrE*)

**Kremper** *m* ARCH & TRAGW, WERKSTOFF *Dachdeckung* flap tile

**Krempziegel** *m* ARCH & TRAGW, WERKSTOFF *Dachdeckung* flap tile

**kreneliert** *adj* ARCH & TRAGW crenelated (*AmE*), crenellated (*BrE*)

**Kreosot** *nt* HOLZ creosote

**kreosotieren** *vt* HOLZ creosote

**Kreuz** *nt* HEIZ & BELÜFT cross; **Kreuzband** *nt* ARCH & TRAGW, INFR & ENTW X-brace; **Kreuzfachwerkbinder** *m* ARCH & TRAGW, HOLZ, INFR & ENTW lattice truss

**kreuzförmig**: **~er Grundriß** *m* ARCH & TRAGW cruciform ground plan

***Kreuz***: **Kreuzfuge** *f* NATURSTEIN cross joint; **Kreuzgewölbe** *nt* ARCH & TRAGW, HOLZ, INFR & ENTW cross-vault; **Kreuzholz** *nt* HOLZ scantling, quarter timber; **Kreuzlochschraube** *f* BESCHLÄGE capstan screw; **Kreuzrippe** *f* ARCH & TRAGW cross rib; **Kreuzschaltung** *f* ELEKTR cross mounting; **Kreuzscheibe** *f* ARCH & TRAGW cross-staff head; **Kreuzstoß** *m* NATURSTEIN cross joint; **Kreuzstrebe** *f* ARCH & TRAGW, HOLZ, STAHL cross stud, diagonal strut; **Kreuzstück** *nt* HEIZ & BELÜFT cross, *Rohr* cross joint

**Kreuzung** *f* ARCH & TRAGW *Stahlträger* crossing area, INFR & ENTW crossing; **Kreuzungsbereich** *m* ARCH & TRAGW, INFR & ENTW crossing area; **Kreuzungspunkt** *m* ARCH & TRAGW, INFR & ENTW crossing point

***Kreuz***: **Kreuzverband** *m* HOLZ cross bracing, NATURSTEIN *Mauerwerk* cross bond, STAHL cross bracing; **Kreuzverbinder** *m* BESCHLÄGE, HOLZ, STAHL cross connector; **Kreuzverbindung** *f* ARCH & TRAGW spider connection, HOLZ, STAHL cross connection; **Kreuzverstrebung** *f* ARCH & TRAGW, INFR & ENTW X-bracing; **Kreuzverzapfung** *f* HOLZ cross joint

**kreuzweise**: **~ bewehrte Stahlbetonplatte** *f* BETON two-way reinforcement concrete slab

**Kriech-** *in cpds* ARCH & TRAGW, BETON, INFR & ENTW, WERKSTOFF creep; **Kriechblume** *f* ARCH & TRAGW *mittelalterliche Verzierung* crocket; **Kriechblumenkapitell** *nt* ARCH & TRAGW crocket capital; **Kriecheinfluß** *m* ARCH & TRAGW, BETON, INFR & ENTW, STAHL, WERKSTOFF creep effect; **Kriecheinwirkung** *f* BETON, INFR & ENTW effect of creep

**Kriechen** *nt* ARCH & TRAGW, BETON, INFR & ENTW, WERKSTOFF creep, plastic flow

***Kriech-***: **Kriechfestigkeit** *f* ARCH & TRAGW, BETON, STAHL, WERKSTOFF creep resistance, resistance to creep; **Kriechgeschwindigkeit** *f* WERKSTOFF creep rate, rate of creep; **Kriechkeller** *m* ARCH & TRAGW crawlway; **Kriechmodul** *m* BETON creep modulus

**Kristall** *m* WERKSTOFF crystal; **Kristallspiegelglas** *nt* WERKSTOFF polished plate glass

**Kriterium** *nt* ARCH & TRAGW, BAURECHT, INFR & ENTW criterion

**kritisch** *adj* ARCH & TRAGW *Last* critical; **~e Länge** *f* ARCH & TRAGW, BETON, INFR & ENTW, WERKSTOFF critical length; **~e Last** *f* ARCH & TRAGW, BETON, HOLZ, STAHL critical load; **~er Querschnitt** *m* ARCH & TRAGW, BETON, INFR & ENTW, WERKSTOFF critical cross-section; **~er Standort** *m* ARCH & TRAGW critical location; **~e Windgeschwindigkeit** *f* ARCH & TRAGW, INFR & ENTW critical wind velocity

**Kritische-Pfad-Analyse** *f* BAURECHT critical path analysis

**Kritische-Pfad-Methode** *f* BAURECHT critical path method

**Krone** *f* ARCH & TRAGW crest, crown, NATURSTEIN cope; **Kronendach** *nt* ARCH & TRAGW crown-tile roof; **Kronenmutter** *f* BESCHLÄGE castellated nut

**kröpfen** *vt* ARCH & TRAGW joggle, HEIZ & BELÜFT bend at right angles, crank

**Kröpfling** *m* ARCH & TRAGW *Holztreppe* wreath piece, wreath

**Kropfstück** *nt* ARCH & TRAGW *Handlauf* string wreath, *Holztreppe* wreath, wreath piece

**Kröseleisen** *nt* BAUMASCHIN, BESCHLÄGE grooving iron, rabbet iron

**krümmen** **1.** *vt* ARCH & TRAGW, INFR & ENTW, WERKSTOFF arch, curve, buckle; **2. sich krümmen** *v refl* ARCH & TRAGW, INFR & ENTW, WERKSTOFF arch, buckle, curve

**Krümmer** *m* ABWASSER *Rohr*, BESCHLÄGE bend; **Krümmerüberwurf** *m* ABWASSER elbow union

**Krümmling** *m* ARCH & TRAGW *Handlauf* string wreath, *Holztreppe* wreath, wreath piece
**Krümmung** *f* ARCH & TRAGW bend, curve, flexion, INFR & ENTW *Straßen, Wege, Eisenbahnlinie* curve, flexion
**Krüppelwalm** *m* ARCH & TRAGW, HOLZ false hip, partial hip; **Krüppelwalmdach** *nt* ARCH & TRAGW, HOLZ false hip roof
**Kübel** *m* BAUMASCHIN bucket, pail; **Kübelfördergerät** *nt* BAUMASCHIN bucket conveyor
**Kubikwindgeschwindigkeit** *f* UMWELT wind velocity cubed
**Kugel** *f* ABWASSER ball, ARCH & TRAGW sphere
**kugelförmig** *adj* ARCH & TRAGW spherical
***Kugel***: **Kugelgelenk** *nt* BESCHLÄGE ball hinge, STAHL, WERKSTOFF ball and socket joint; **Kugelgewölbe** *nt* ARCH & TRAGW, INFR & ENTW spherical vault; **Kugelgraphitguß** *m* WERKSTOFF nodular cast iron; **Kugelhahn** *m* ABWASSER ball valve, ball cock; **Kugelmühle** *f* BAUMASCHIN ball mill; **Kugelprobe** *f* **nach Brinell** STAHL Brinell hardness test; **Kugelschale** *f* ARCH & TRAGW, INFR & ENTW, WERKSTOFF spherical shell; **Kugelschlagprobe** *f* WERKSTOFF ball impact test; **Kugelsegment** *nt* ARCH & TRAGW calotte
**kugelsicher** *adj* WERKSTOFF bulletproof
***Kugel***: **Kugelventil** *nt* ABWASSER ball cock, ball valve; **Kugelventilschwimmer** *m* ABWASSER ball valve float
**Kuhfuß** *m* BAUMASCHIN claw bar
**Kühl-** *in cpds* HEIZ & BELÜFT cooling
**kühlen** *vt* HEIZ & BELÜFT refrigerate
**Kühler** *m* HEIZ & BELÜFT cooler
***Kühl-***: **Kühllast** *f* HEIZ & BELÜFT cooling load; **Kühlluft** *f* HEIZ & BELÜFT cooling air; **Kühlmittel** *nt* HEIZ & BELÜFT, UMWELT, WERKSTOFF coolant; **Kühlraum** *m* ARCH & TRAGW, HEIZ & BELÜFT cold room, cold store; **Kühlraumtür** *f* ARCH & TRAGW cold-storage door
**Kühlung** *f* HEIZ & BELÜFT cooling, refrigeration
***Kühl-***: **Kühlwasser** *nt* HEIZ & BELÜFT cooling water; **Kühlwasserkreislauf** *m* HEIZ & BELÜFT cooling load
**Kulturzentrum** *nt* ARCH & TRAGW arts center (*AmE*), arts centre (*BrE*)
**kumulativ**: **~er Ausfluß** *m* UMWELT cumulative discharge
**Kunde** *m* ARCH & TRAGW customer; **Kundenschalter** *m* HOLZ counter
**Kunst-** *in cpds* WERKSTOFF artificial; **Kunstdenkmal** *nt* INFR & ENTW artistic monument; **Kunstfertigkeit** *f* ARCH & TRAGW, BAURECHT workmanship; **Kunstharz** *nt* OBERFLÄCHE synthetic resin
**kunstharzgebunden** *adj* OBERFLÄCHE synthetic-resin-bound, WERKSTOFF resinoid-bonded
***Kunst-***: **Kunstharzlack** *m* OBERFLÄCHE synthetic enamel, synthetic resin varnish; **Kunstharzmörtel** *m* WERKSTOFF resin-based mortar; **Kunstharzputz** *m* OBERFLÄCHE synthetic resin plaster
**künstlerisch**: **~er Ausdruck** *m* ARCH & TRAGW, INFR & ENTW artistic expression; **~e Gestaltung** *f* ARCH & TRAGW, INFR & ENTW artistic design
**künstlich** *adj* WERKSTOFF artificial, synthetic; **~e Abdichtung** *f* UMWELT synthetic lining; **~es Altern** *nt* WERKSTOFF artificial ageing; **~e Alterung** *f* UMWELT artificial ageing; **~e Beleuchtung** *f* ELEKTR artificial lighting; **~e Bewitterung** *f* WERKSTOFF artificial weathering; **~e Lüftung** *f* HEIZ & BELÜFT artificial ventilation; **~e Übersäuerung** *f* UMWELT artificial acidification
***Kunst-***: **Kunstmarmor** *m* NATURSTEIN, WERKSTOFF artificial marble; **Kunstmarmorgießmaschine** *f* BAUMASCHIN casting machine for artificial marble; **Kunststein** *m* NATURSTEIN, WERKSTOFF cast stone
**Kunststoff** *m* ARCH & TRAGW, UMWELT, WERKSTOFF plastic; **Kunststoffabfall** *m* UMWELT plastic waste; **Kunststoffdispersion** *f* OBERFLÄCHE plastic dispersion; **Kunststoffolie** *f* BESCHLÄGE plastic foil; **Kunststoffgebinde** *nt* UMWELT plastic container; **Kunststoffklebeband** *nt* WERKSTOFF plastic adhesive tape; **Kunststoffmantelleitung** *f* ELEKTR plastic-sheathed cable; **Kunststoffmembran** *f* WERKSTOFF synthetic membrane; **Kunststoffrecycling** *nt* UMWELT plastics recycling; **Kunststoffrohr** *nt* WERKSTOFF plastic conduit; **Kunststoffverpackung** *f* UMWELT plastic packing material, plastic container; **Kunststoffverwertung** *f* UMWELT plastics recycling
***Kunst-***: **Kunsttischlerei** *f* HOLZ cabinet-making
**Kupfer** *nt*, WERKSTOFF copper; **Kupferbedachung** *f* ARCH & TRAGW copper roofing; **Kupferblech** *nt* WERKSTOFF copper sheet; **Kupferbogen** *m* WERKSTOFF copper elbow; **Kupferdach** *nt* ARCH & TRAGW copper roof, copper roofing, copper-clad roof; **Kupferdachrinne** *f* ABWASSER copper roof gutter; **Kupferdraht** *m* WERKSTOFF copper sheet, copper wire; **Kupferfitting** *nt* ABWASSER, WERKSTOFF copper fitting; **Kupferkapillarfitting** *nt* ELEKTR, HEIZ & BELÜFT, WERKSTOFF copper capillary fitting; **Kupferkreuzstück** *nt* HEIZ & BELÜFT copper cross; **Kupferkrümmer** *m* HEIZ & BELÜFT copper bend; **Kupferlamellenrohr** *nt* HEIZ & BELÜFT copper fin pipe; **Kupferlegierung** *f* WERKSTOFF copper alloy; **Kupfermantel** *m* ELEKTR copper cladding; **Kupfermantelkabel** *nt* ELEKTR copper-sheathed cable; **Kupfermatte** *f* WERKSTOFF copper wire mesh; **Kupfernagel** *m* WERKSTOFF copper nail; **Kupferniet** *m* WERKSTOFF copper rivet; **Kupferplattierung** *f* OBERFLÄCHE copper plating; **Kupferrohr** *nt* ABWASSER, HEIZ & BELÜFT copper pipe; **Kupfersammelschiene** *f* ELEKTR copper bus bar; **Kupferschiene** *f* ELEKTR, WERKSTOFF copper bar
**kupferüberzogen**: **~er Staberder** *m* ELEKTR *Blitzschutz* copper-clad earth rod (*BrE*), copper-clad ground rod (*AmE*)
**kupferummantelt**: **~er Staberder** *m* ELEKTR *Blitzschutz* copper-clad earth rod (*BrE*), copper-clad ground rod (*AmE*)
***Kupfer***: **Kupferverbindungsstück** *nt* ABWASSER, WERKSTOFF copper fitting; **Kupfervitriol** *nt* WERKSTOFF blue vitriol
**Kuppe** *f* ARCH & TRAGW summit, INFR & ENTW knoll
**Kuppel** *f* ARCH & TRAGW cupola, dome; **Kuppelbauwerk** *nt* ARCH & TRAGW domed structure; **Kuppeldach** *nt* ARCH & TRAGW domed roof, dome roof
**kuppelförmig** *adj* ARCH & TRAGW domed
***Kuppel***: **Kuppeloberfläche** *f* ARCH & TRAGW cupola surface; **Kuppelring** *m* ARCH & TRAGW dome ring
**Kupplung** *f* BAUMASCHIN clutch, HEIZ & BELÜFT coupling
**kurz**: **~es Verbindungsstück** *nt* ABWASSER *von Rohren* faucet joint; **~er Zapfen** *m* ARCH & TRAGW stub tenon; **~es Zapfenloch** *nt* ARCH & TRAGW stub mortice, stub mortise
**Kurzzapfverbindung** *f* ARCH & TRAGW spur tenon joint
**Kurzzeitbelastung** *f* BETON, WERKSTOFF short-term loading

**Küstenschutz** *m* INFR & ENTW coastal defence; **Küstenschutzkonstruktion** *f* INFR & ENTW sea defence construction

**kW** *abbr* (*Kilowatt*) ELEKTR, HEIZ & BELÜFT kW (*kilowatt*)

**k-Wert** *m* INFR & ENTW, WERKSTOFF heat transition coefficient, k-value

**kWh** *abbr* (*Kilowattstunde*) ELEKTR, HEIZ & BELÜFT kWh (*kilowatt-hour*)

**KW-Platte** *f* BETON, WERKSTOFF asbestos cement board

# L

**labil**: **~es Gleichgewicht** *nt* INFR & ENTW unstable equilibrium; **~er Zustand** *m* INFR & ENTW unstable state

**Labor** *nt* ABWASSER, ARCH & TRAGW, ELEKTR, HEIZ & BELÜFT, INFR & ENTW laboratory; **~ für Geotextilien** INFR & ENTW, WERKSTOFF geotextile laboratory; **Laborresultat** *nt* INFR & ENTW, WERKSTOFF laboratory result; **Labortest** *m* INFR & ENTW, WERKSTOFF laboratory test; **Laboruntersuchung** *f* INFR & ENTW, WERKSTOFF laboratory investigation; **Laborversuch** *m* WERKSTOFF bench test

**Lack** *m* OBERFLÄCHE, WERKSTOFF lacquer, varnish; **Lackfarbe** *f* OBERFLÄCHE, WERKSTOFF enamel, lacquer, shellac, varnish; **Lackfirnis** *m* OBERFLÄCHE, WERKSTOFF varnish

**lackieren** *vt* OBERFLÄCHE, WERKSTOFF paint, varnish

**Lackieren** *nt* OBERFLÄCHE, WERKSTOFF painting, varnishing

**lackiert** *adj* OBERFLÄCHE, WERKSTOFF painted, varnished

**Lackierung** *f* OBERFLÄCHE painting, varnishing

***Lack***: **Lackschlamm** *m* UMWELT paint sludge

**Lade-** *in cpds* ARCH & TRAGW, BAUMASCHIN, INFR & ENTW loading; **Ladebagger** *m* BAUMASCHIN excavator-loader; **Ladebrücke** *f* INFR & ENTW loading bridge; **Ladebühne** *f* ARCH & TRAGW handling platform; **Ladegerät** *nt* BAUMASCHIN loader

**laden** *vt* BAUMASCHIN, INFR & ENTW load

**Laden** *m* ARCH & TRAGW shop; **Ladenfenster** *nt* ARCH & TRAGW display window; **Ladentisch** *m* HOLZ counter

***Lade-***: **Ladeplatz** *m* ARCH & TRAGW, INFR & ENTW loading bay; **Laderampe** *f* ARCH & TRAGW, INFR & ENTW loading platform, loading ramp; **Ladeschaufel** *f* BAUMASCHIN loading shovel; **Ladetank** *m* UMWELT cargo tank

**Ladung** *f* ARCH & TRAGW, ELEKTR charge

**LAGA** *abbr* (*Länderarbeitsgemeinschaft für Abfall*) UMWELT *in Deutschland* Federal States Working Group on Waste

**Lage** *f* ARCH & TRAGW location, position, site, INFR & ENTW layer, course, lie, NATURSTEIN course; **Lagenschüttung** *f* INFR & ENTW layer-construction; **Lageplan** *m* ARCH & TRAGW, BAURECHT, INFR & ENTW location plan, site plan

**Lager** *nt* ARCH & TRAGW bearing, *Raum, Gebäude* store, warehouse; **nicht auf Lager** *phr* INFR & ENTW, WERKSTOFF nonstandard; **Lagerbett** *nt* ARCH & TRAGW, INFR & ENTW bed; **Lagerbuchse** *f* ARCH & TRAGW bearing bush; **Lagerbunker** *m* UMWELT receiving bin, refuse bunker; **Lagerdruck** *m* ARCH & TRAGW bearing pressure; **Lagerfläche** *f* INFR & ENTW stacking ground; **Lagerfuge** *f* ARCH & TRAGW, NATURSTEIN bed joint, horizontal joint; **Lagergut** *nt* UMWELT fill mass; **Lagerhaus** *nt* ARCH & TRAGW warehouse; **Lagerhof** *m* INFR & ENTW stacking yard; **Lagerholz** *nt* HOLZ batten, floor joist; **Lagerkammer** *f* UMWELT *für radioaktive Abfälle* storage chamber; **Lagermöglichkeit** *f* UMWELT storage facility

**lagern** *vt* ARCH & TRAGW bear, carry, keep, support, HOLZ bear, INFR & ENTW bear, store, STAHL bear, WERKSTOFF bear, store

***Lager***: **Lagerschuppen** *m* ARCH & TRAGW storage shed; **Lagerstätte** *f* UMWELT repository, storage facility

**Lagerung** *f* ARCH & TRAGW bearing; **Lagerungsdichte** *f* INFR & ENTW *des Untergrundes* compactness of the ground

**Lamelle** *f* HEIZ & BELÜFT fin, rib; **Lamellengraphit** *m* WERKSTOFF flake graphite; **Lamellenheizkörper** *m* HEIZ & BELÜFT finned radiator, ribbed radiator; **Lamellentür** *f* ARCH & TRAGW louver door (*AmE*), louvre door (*BrE*); **Lamellenverfahren** *nt* INFR & ENTW slice method; **Lamellenvorhang** *m* BESCHLÄGE slatted curtain

**lamelliert**: **~es Holz** *nt* HOLZ, WERKSTOFF laminated wood

**laminar** *adj* INFR & ENTW, WERKSTOFF laminar

**Laminarströmung** *f* HEIZ & BELÜFT, INFR & ENTW, WERKSTOFF laminar flow

**laminiert**: **~er Boden** *m* INFR & ENTW, WERKSTOFF laminated soil

**Lammfellrolle** *f* OBERFLÄCHE lamb's wool roll

**Lampe** *f* ELEKTR lamp

**Land** *nt* BAURECHT, INFR & ENTW, UMWELT land; **Landauffüllung** *f* UMWELT landfill; **Landbesitz** *m* BAURECHT real estate

**Landebrücke** *f* ARCH & TRAGW, INFR & ENTW pier

***Land***: **Landentwässerung** *f* BAURECHT, INFR & ENTW land drainage

**Landeplatz** *m* INFR & ENTW landing area

**Länderarbeitsgemeinschaft** *f*: **~ für Abfall** (*LAGA*) UMWELT *in Deutschland* Federal States Working Group on Waste

***Land***: **Landerschließung** *f* BAURECHT, INFR & ENTW, UMWELT reclamation of land; **Landesvermessungsnetz** *nt* ARCH & TRAGW triangulation network; **Landgewinnung** *f* BAURECHT, INFR & ENTW, UMWELT reclamation of land

**ländlich** *adj* BAURECHT, INFR & ENTW rural

***Land***: **Landmesser** *m* ARCH & TRAGW surveyor

**Landschaft** *f* INFR & ENTW landscape; **Landschaftsgärtner** *m* INFR & ENTW landscape gardener, landscaper; **Landschaftsgärtnerei** *f* INFR & ENTW landscape architecture

**landschaftsgärtnerisch**: **~e Arbeiten** *f pl* INFR & ENTW landscaping work, landscaping

***Land***: **Landstück** *nt* INFR & ENTW lot; **Landvermessung** *f* INFR & ENTW geodetic surveying, land surveying; **Landwiedergewinnung** *f* BAURECHT, INFR & ENTW land reclamation

**landwirtschaftlich** *adj* ARCH & TRAGW, INFR & ENTW agricultural; **~e Drainage** *f* INFR & ENTW agricultural drainage; **~es Gebäude** *nt* ARCH & TRAGW agricultural building

**lang**: **~er Sturzbalken** *m* ARCH & TRAGW breastsummer

**Länge** *f* ARCH & TRAGW length, *in Yard* yardage; **Längenänderung** *f* ARCH & TRAGW change in length; **Längenmaß** *nt* ARCH & TRAGW measure of length

**Langhobel** *m* BAUMASCHIN, HOLZ trying plane

**Langloch** *nt* STAHL oblong hole; **Langlochplatte** *f* WERKSTOFF horizontal core slab; **Langlochstein** *m* NATURSTEIN, WERKSTOFF horizontal coring block; **Langlochziegel** *m* NATURSTEIN, WERKSTOFF horizontal coring brick

**Längs-** *in cpds* ARCH & TRAGW, BETON, INFR & ENTW, WERKSTOFF longitudinal; **Längsabscheren** *nt* ARCH & TRAGW, BETON, INFR & ENTW, WERKSTOFF longitudinal shear; **Längsachse** *f* INFR & ENTW longitudinal axis; **Längsbalken** *m* ARCH & TRAGW stringer; **Längsbetonierfuge** *f* ARCH & TRAGW longitudinal construction joint; **Längseisen** *nt* ARCH & TRAGW, BETON, INFR & ENTW, WERKSTOFF longitudinal bar; **Längsfalz** *m* STAHL longitudinal fold; **Längsfuge** *f* ARCH & TRAGW longitudinal joint; **Längsnaht** *f* STAHL longitudinal seam; **Längspreßfuge** *f* ARCH & TRAGW longitudinal compression joint; **Längsschlupf** *m* ARCH & TRAGW, BETON, INFR & ENTW, WERKSTOFF longitudinal slip; **Längsschnitt** *m* ARCH & TRAGW longitudinal section; **Längsschnittsäge** *f* BAUMASCHIN rip saw; **Längsstange** *f* HOLZ *Gerüst* ledger; **Längsträger** *m* ARCH & TRAGW longitudinal beam, longitudinal girder

**Langstreckenfördergerät** *nt* BAUMASCHIN, INFR & ENTW long-distance conveyor

***Längs-***: **Längsverband** *m* ARCH & TRAGW, HOLZ, STAHL horizontal bracing, longitudinal bracing; **Längsverbindung** *f* HOLZ splice; **Längsverformung** *f* BETON, INFR & ENTW, STAHL, WERKSTOFF longitudinal deformation

**Langzeit-** *in cpds* BAURECHT long-term; **Langzeitbelastung** *f* ARCH & TRAGW, INFR & ENTW sustained loading; **Langzeitversuch** *m* WERKSTOFF extended time test; **Langzeitwirkung** *f* BAURECHT long-term effect

**Lärm** *m* BAURECHT, DÄMMUNG, INFR & ENTW, UMWELT noise; **Lärmbekämpfung** *f* DÄMMUNG noise control, noise suppression; **Lärmbelästigung** *f* BAURECHT, UMWELT noise nuisance, noise pollution, sound pollution; **Lärmpegel** *m* DÄMMUNG noise level; **Lärmquelle** *f* UMWELT noise source; **Lärmschutz** *m* DÄMMUNG noise insulation, noise protection, noise abatement; **Lärmschutzfenster** *nt* ARCH & TRAGW, DÄMMUNG noise protection window; **Lärmschutzsperre** *f* DÄMMUNG noise barrier; **Lärmschutzwall** *m* INFR & ENTW noise protection embankment; **Lärmschutzzaun** *m* DÄMMUNG, INFR & ENTW acoustic fencing; **Lärmstärke** *f* BAURECHT, DÄMMUNG intensity of noise; **Lärmtesthammer** *m* DÄMMUNG sounding hammer

**Larssenbohle** *f* INFR & ENTW Larssen sheet pile

**Lasche** *f* ARCH & TRAGW butt strap, tongue, HOLZ fish, fishplate, gusset plate, STAHL gusset plate; **Laschenanschluß** *m* ARCH & TRAGW, HOLZ cleated joint; **Laschenbolzen** *m* WERKSTOFF fish bolt; **Laschenschraube** *f* WERKSTOFF fish bolt; **Laschenstoß** *m* HOLZ fished joint; **Laschenverbindung** *f* ARCH & TRAGW splice joint, HOLZ fished joint, scarf

**Laschung** *f* HOLZ scarf

**Lasergerät** *nt* BAUMASCHIN laser

**Last** *f* ARCH & TRAGW charge, load; **Lastangriffspunkte** *m pl* ARCH & TRAGW, HOLZ loading points; **Lastannahme** *f* ARCH & TRAGW loading assumption, assumed load, design load, BETON design load, INFR & ENTW design load, loading assumption, STAHL design load; **Lastansatz** *m* ARCH & TRAGW, BETON, INFR & ENTW, STAHL design load; **Lastdauer** *f* INFR & ENTW load duration; **Lastenaufzug** *m* ARCH & TRAGW, BAUMASCHIN, INFR & ENTW goods elevator (*AmE*), goods lift (*BrE*), service elevator (*AmE*), service lift (*BrE*); **Lastenfreiheit** *f* ARCH & TRAGW freedom from load; **Lastfall** *m* ARCH & TRAGW loading condition, load scheme; **Lasthaken** *m* BAUMASCHIN load hook; **Lastkombination** *f* ARCH & TRAGW, INFR & ENTW load combination; **Lastkraftwagen** *m* (*LKW*) BAUMASCHIN lorry (*BrE*), truck (*AmE*)

**Lastöse** *f* BESCHLÄGE *Schloß* shackle

***Last***: **Lastplattendruckversuch** *m* INFR & ENTW plate load test; **Lastplattenversuch** *m* INFR & ENTW *Boden* plate load test; **Lastschwellungsdiagramm** *nt* INFR & ENTW, WERKSTOFF load-swelling diagram; **Lastsetzungsdiagramm** *nt* INFR & ENTW, WERKSTOFF load settlement diagram; **Laststufe** *f* ARCH & TRAGW, INFR & ENTW load category; **Lastübertragung** *f* ARCH & TRAGW, INFR & ENTW load transmission, transmission of load; **Lastverformungsdiagramm** *nt* INFR & ENTW load-deflection curve; **Lastverteilung** *f* INFR & ENTW load distribution; **Lastverteilungsplatte** *f* HOLZ wall plate, INFR & ENTW load distribution plate; **Lastverteilungsprinzip** *nt* ARCH & TRAGW load-sharing concept

**Lasur** *f* OBERFLÄCHE glaze, scumble

**latent** *adj* INFR & ENTW, WERKSTOFF latent; **~e hydraulische Bindemittel** *nt pl* HEIZ & BELÜFT latent hydraulic binders

**Laterne** *f* ARCH & TRAGW lantern, lantern light, skylight turret; **Laternenoberlicht** *nt* ARCH & TRAGW skylight

**Latexfarbe** *f* OBERFLÄCHE latex paint

**Latte** *f* HOLZ lath; **Lattenholz** *nt* HOLZ lathwood; **Lattenstift** *m* BESCHLÄGE lath nail; **Lattenverschlag** *m* HOLZ lath partition; **Lattenzaun** *m* HOLZ lattice fence, paling

**Lattung** *f* ARCH & TRAGW roof battening

**Laube** *f* INFR & ENTW gazebo; **Laubengang** *m* ARCH & TRAGW exterior corridor, covered walk, INFR & ENTW access gallery

**Laubfang** *m* ARCH & TRAGW leaf trap

**Lauf** *m* BAUMASCHIN *eines Kolbens* travel, INFR & ENTW course, track; **Laufbohle** *f* ARCH & TRAGW barrow way, plank, WERKSTOFF runway plank; **Laufbreite** *f* ARCH & TRAGW *Treppe* flight width; **Laufbrett** *nt* ARCH & TRAGW barrow way

**Läufer** *m* NATURSTEIN outbond brick, stretcher; **Läuferlage** *f* NATURSTEIN stretcher course; **Läuferschicht** *f* ARCH & TRAGW, NATURSTEIN stretcher course, stretching course; **Läuferstein** *m* NATURSTEIN stretcher; **Läuferverband** *m* ARCH & TRAGW, NATURSTEIN *Mauerwerk* running bond, stretcher bond, stretching bond

***Lauf***: **Laufgang** *m* ARCH & TRAGW gallery; **Laufkatze** *f* BAUMASCHIN *Kran* traveling crab (*AmE*), traveling trolley (*AmE*), travelling crab (*BrE*), travelling trolley (*BrE*), INFR & ENTW traveling winch (*AmE*), travelling winch (*BrE*); **Laufkraftwerk** *nt* UMWELT run-of-river station; **Laufkran** *m* BAUMASCHIN traveling crane (*AmE*), travelling crane (*BrE*), overhead crane; **Lauflinie** *f* ARCH & TRAGW run, *Treppe* walking line; **Laufrad** *nt* BAUMASCHIN *Pumpe* rotor, HEIZ & BELÜFT

*Ventilator*, INFR & ENTW impeller; **Laufschritt** *m* DÄMMUNG footstep
**laugenhaft** *adj* INFR & ENTW, WERKSTOFF alkaline
**laut** *adj* DÄMMUNG noisy
**Lautstärke** *f* DÄMMUNG loudness; **Lautstärkepegel** *m* DÄMMUNG volume level
**Lava** *f* INFR & ENTW lava
**Lawine** *f* ARCH & TRAGW, INFR & ENTW avalanche; **Lawinengalerie** *f* ARCH & TRAGW, INFR & ENTW avalanche gallery
**lawinengeschützt** *adj* ARCH & TRAGW, INFR & ENTW avalanche-proof
***Lawine***: **Lawinenschutz** *m* ARCH & TRAGW avalanche protector; **Lawinenwehr** *f* ARCH & TRAGW, INFR & ENTW avalanche screen
**Leben**: **Lebensdauer** *f* ABWASSER, BAUMASCHIN, ELEKTR, HEIZ & BELÜFT, WERKSTOFF service expectancy, service life; **Lebensdauer** *f* **der Atmosphäre** UMWELT atmospheric lifetime; **Lebenszykluskostenberechnung** *f* BAURECHT life cycle costing
**Leck** *nt* ABWASSER leak, UMWELT leak, leakage; **Leckage** *f* UMWELT leakage; **Leckanzeigegerät** *nt* ABWASSER, HEIZ & BELÜFT leak detector; **Leckbestimmung** *f* UMWELT leak detection; **Leckloch** *nt* NATURSTEIN weep hole
**lecksicher** *adj* ARCH & TRAGW, HEIZ & BELÜFT leakproof
***Leck***: **Leckstrom** *m* ELEKTR leakage current, leakage; **Lecksucher** *m* ABWASSER, HEIZ & BELÜFT leak detector; **Leckverlust** *m* INFR & ENTW leakage loss; **Leckwasser** *nt* INFR & ENTW leakage water
**Leder** *nt* WERKSTOFF leather
**lederfarben**: **~er Ziegelstein** *m* NATURSTEIN buff-colored brick (*AmE*), buff-coloured brick (*BrE*)
***Leder***: **Lederhobel** *m* BAUMASCHIN spokeshave; **Ledermanschette** *f* WERKSTOFF leathering
**Leergewicht** *nt* WERKSTOFF empty weight
**Leerlauf** *m* BAURECHT idle running
**Leerrohr** *nt* ELEKTR conduit
**legen** *vt* ARCH & TRAGW, HOLZ *Parkett* lay
**Legende** *f* ARCH & TRAGW coding legend
**legiert** *adj* STAHL, WERKSTOFF alloyed; **~er Stahl** *m* STAHL, WERKSTOFF alloy steel
**Legierung** *f* STAHL, WERKSTOFF alloy; **Legierungsstoff** *m* STAHL, WERKSTOFF alloying material
**Legionärskrankheit** *f* HEIZ & BELÜFT, UMWELT legionnaire's disease
**Lehm** *m* WERKSTOFF clay, loam; **Lehmbaustein** *m* NATURSTEIN sun-dried brick; **Lehmestrich** *m* ARCH & TRAGW clay floor; **Lehmfußboden** *m* ARCH & TRAGW clay floor; **Lehmgehalt** *m* INFR & ENTW, WERKSTOFF loam content; **Lehmgrube** *f* INFR & ENTW loam pit
**lehmig** *adj* INFR & ENTW, WERKSTOFF loamy; **~er Boden** *m* INFR & ENTW, WERKSTOFF loamy soil; **~er Sand** *m* INFR & ENTW, WERKSTOFF loamy sand
***Lehm***: **Lehmkern** *m* INFR & ENTW loam core; **Lehmquader** *m* NATURSTEIN clay block; **Lehmschicht** *f* INFR & ENTW loam layer; **Lehmwall** *m* UMWELT clay barrier; **Lehmziegel** *m* ARCH & TRAGW adobe
**Lehrbogen** *m* ARCH & TRAGW, HOLZ, INFR & ENTW, STAHL cradling
**Lehrbrief** *m* BAURECHT certificate of apprenticeship
**Lehre** *f* BAUMASCHIN caliper (*AmE*), calliper (*BrE*), gage (*AmE*), gauge (*BrE*), *Bohrvorrichtung* setting jig
**Lehrgerüst** *nt* ARCH & TRAGW, INFR & ENTW falsework, NATURSTEIN centring
**Lehrjahre** *nt pl* BAURECHT apprenticeship
**Lehrling** *m* BAURECHT apprentice
**Lehrmodell** *nt* INFR & ENTW mock-up
**Lehrvertrag** *m* BAURECHT articles of apprenticeship
**Leibung** *f* ARCH & TRAGW *Bogen, Gewölbe* soffit, *Fenster, Tür* reveal
**leicht** *adj* WERKSTOFF light, lightweight
**Leicht-** *in cpds* WERKSTOFF light, lightweight
***leicht***: **~er Holzbinder** *m* ARCH & TRAGW, HOLZ lightweight timber truss
***Leicht-***: **Leichtbau** *m* ARCH & TRAGW lightweight construction; **Leichtbauplatte** *f* WERKSTOFF light building board, lightweight building board; **Leichtbauweise** *f* ARCH & TRAGW light construction method; **Leichtbeton** *m* ARCH & TRAGW, BETON lightweight concrete
***leicht***: **~ entflammbar** *adj* BAURECHT, WERKSTOFF highly inflammable; **~ entzündbar** *adj* BAURECHT, WERKSTOFF highly inflammable; **~ entzündlich** *adj* BAURECHT, WERKSTOFF highly inflammable
***Leicht-***: **Leichtgewichtsbeton** *m* ARCH & TRAGW, BETON lightweight concrete; **Leichtverputz** *m* NATURSTEIN lightweight plaster; **Leichtziegel** *m* NATURSTEIN, WERKSTOFF lightweight clay brick
***leicht***: **~ zugänglich** *adj* BAUMASCHIN easily accessible
***Leicht-***: **Leichtzuschlagstoff** *m* WERKSTOFF light aggregate, lightweight aggregate
**Leim** *m* WERKSTOFF glue; **Leimbürste** *f* BAUMASCHIN glue brush
**leimen** *vt* HOLZ glue
***Leim***: **Leimfarbe** *f* OBERFLÄCHE, WERKSTOFF distemper, nonwashable distemper; **Leimgrundierung** *f* HOLZ, OBERFLÄCHE glue priming; **Leimholz** *nt* HOLZ glulam timber; **Leim-Nagel-Verbindung** *f* HOLZ glue-nail joint
**Leistchen** *nt* ARCH & TRAGW, BESCHLÄGE listel
**Leiste** *f* BESCHLÄGE, HOLZ, WERKSTOFF batten, fascia, molding (*AmE*), moulding (*BrE*)
**leisten** *vt* ARCH & TRAGW perform
***Leiste***: **Leistenhobelmaschine** *f* BAUMASCHIN, HOLZ molding machine (*AmE*), moulding machine (*BrE*)
**Leistung** *f* ARCH & TRAGW work, BAUMASCHIN power, ELEKTR power, output, HEIZ & BELÜFT output, INFR & ENTW, UMWELT power; **Leistungsabgabe** *f* UMWELT power output; **Leistungsbeschreibung** *f* ARCH & TRAGW, BAURECHT specifications; **Leistungsdichte** *f* UMWELT power density; **Leistungsfähigkeit** *f* ARCH & TRAGW, BAUMASCHIN efficiency; **Leistungskoeffizient** *m* UMWELT power coefficient; **Leistungskurve** *f* UMWELT power curve; **Leistungsumfang** *m* ARCH & TRAGW, INFR & ENTW scope of work; **Leistungsverzeichnis** *nt* (*LV*) ARCH & TRAGW, BAURECHT specifications
**Leit-** *in cpds* ELEKTR conductive, INFR & ENTW guiding; **Leitblech** *nt* BESCHLÄGE baffle plate
**Leiter** *m* ARCH & TRAGW manager, *f* BAUMASCHIN ladder, ELEKTR conductor; **Leitergerüst** *nt* BAUMASCHIN ladder scaffolding; **Leitersprosse** *f* ARCH & TRAGW rung
**leitfähig** *adj* ELEKTR, WERKSTOFF conductive; **~es Material** *nt* ELEKTR, WERKSTOFF conductive material
***Leit-***: **Leitfähigkeit** *f* ELEKTR, WERKSTOFF conductivity; **Leitkegel** *m* INFR & ENTW *Verkehr* traffic cone; **Leitpfosten** *m* INFR & ENTW *Verkehr* guide post;

**Leitplanke** *f* INFR & ENTW *Verkehr* side rail; **Leitschiene** *f* BESCHLÄGE *Tür, Fenster* guide rail
**Leitung** *f* ABWASSER conduit, line, ARCH & TRAGW management, ELEKTR line, HEIZ & BELÜFT line, conduit, WERKSTOFF conduit; **Leitungsgraben** *m* ARCH & TRAGW utility trench, INFR & ENTW trench for pipes and cables, utility trench; **Leitungsrohr** *nt* ABWASSER, UMWELT conduit pipe; **Leitungswasser** *nt* INFR & ENTW tap water
***Leit-***: **Leitzentrale** *f* ELEKTR control room
**lenken** *vt* UMWELT channel
**letzte**: **~ r Anstrich** *m* OBERFLÄCHE finish coat, finishing coat
**Leucht-** *in cpds* ARCH & TRAGW, ELEKTR fluorescent, luminous; **Leuchtdecke** *f* ARCH & TRAGW, ELEKTR luminous ceiling
**Leuchte** *f* ELEKTR light fixture, luminaire
**leuchtend** *adj* ELEKTR luminous
***Leucht-***: **Leuchtmittel** *nt* ELEKTR lamp; **Leuchtstoff** *m* ELEKTR fluorescent; **Leuchtstoffbeleuchtung** *f* ARCH & TRAGW, ELEKTR fluorescent lighting; **Leuchtstofflampe** *f* ELEKTR fluorescent lamp; **Leuchtstoffröhre** *f* ELEKTR fluorescent lamp
**licht** *adj* ARCH & TRAGW clear
**Licht** *nt* ELEKTR light
***licht***: **~e Breite** *f* ARCH & TRAGW clear width; **~e Höhe** *f* ARCH & TRAGW, INFR & ENTW clear height, clearance, headroom; **~es Maß** *nt* ARCH & TRAGW, INFR & ENTW clear dimension, clearance; **~e Öffnung** *f* ARCH & TRAGW clear opening; **~e Öffnungsbreite** *f* ARCH & TRAGW, INFR & ENTW clear opening width; **~er Strebepfeilerabstand** *m* ARCH & TRAGW, BETON, INFR & ENTW, NATURSTEIN buttress clearance; **~e Weite** *f* ARCH & TRAGW, INFR & ENTW *Spannweite* bearing distance, clear span
***Licht***: **Lichtband** *nt* ARCH & TRAGW *Fenster* row of windows, ELEKTR lighting row, striplight; **Lichtbogensauerstoffschweißen** *nt* STAHL oxygen arc welding; **Lichtbogenschneiden** *nt* STAHL arc cutting; **Lichtbogenschweißen** *nt* STAHL arc welding, electric arc welding, arc weld, WERKSTOFF arc welding; **Lichtdach** *nt* ARCH & TRAGW transparent roof; **Lichtdecke** *f* ARCH & TRAGW luminous ceiling
**lichtdurchlässig** *adj* ARCH & TRAGW, OBERFLÄCHE, WERKSTOFF transparent
***Licht***: **Lichtechtheit** *f* OBERFLÄCHE light-fastness; **Lichtkuppel** *f* ARCH & TRAGW light cupola, saucer dome; **Lichtmaß** *nt* ARCH & TRAGW, INFR & ENTW clear dimension; **Lichtpause** *f* ARCH & TRAGW blueprint; **Lichtschacht** *m* ARCH & TRAGW light shaft, BAUMASCHIN funnel; **Lichtwange** *f* ARCH & TRAGW *Treppe* face string
**Lieferant** *m* BAURECHT supplier; **Lieferantenraum** *m* ARCH & TRAGW, INFR & ENTW delivery room; **Lieferantentür** *f* ARCH & TRAGW, INFR & ENTW delivery door
**lieferbar**: **~ ab Lager** *adj* ARCH & TRAGW, WERKSTOFF available from stock
**Lieferbedingungen** *f pl* BAURECHT terms of delivery
**Lieferbeton** *m* BETON truck-mixed concrete
**liefern** *vt* ARCH & TRAGW supply, INFR & ENTW, WERKSTOFF deliver; **~ und montieren** *phr* ARCH & TRAGW furnish and install
**Lieferung** *f* ARCH & TRAGW delivery, supply, INFR & ENTW, WERKSTOFF delivery; **~ frei Verwendungsstelle** BAURECHT free delivery
**Lignit** *nt* WERKSTOFF lignite
**Limba** *nt* HOLZ *afrikanische Holzart* limba
**Lineal** *nt* ARCH & TRAGW rule (*BrE*), ruler
**linear** *adj* ARCH & TRAGW, INFR & ENTW linear; **~er Ausdehnungsbeiwert** *m* WERKSTOFF coefficient of linear expansion; **~e Wärmeausdehnung** *f* BETON, INFR & ENTW, STAHL, WERKSTOFF linear thermal expansion
**Linie** *f* ARCH & TRAGW, INFR & ENTW line; **in eine ~ bringen** *phr* ARCH & TRAGW align; **Linienführung** *f* INFR & ENTW route mapping; **Linienlast** *f* ARCH & TRAGW, INFR & ENTW line load; **Liniennetz** *nt* ARCH & TRAGW, INFR & ENTW traction system
**linkswirkend**: **~er Türschließer** *m* BESCHLÄGE left-hand door closer
**Linoleum** *nt* WERKSTOFF linoleum
**Linsenkopfschraube** *f* BESCHLÄGE cheese-head screw
**Linsenschraube** *f* BESCHLÄGE slotted fillister head screw
**Lippendichtung** *f* BESCHLÄGE lip sealing
**Liste** *f* ARCH & TRAGW, BAURECHT list, schedule; **Listenpreis** *m* INFR & ENTW, WERKSTOFF list price
**Lithosphäre** *f* UMWELT earth's crust, lithosphere
**Litze** *f* WERKSTOFF strand
**LKW** *m* (*Lastkraftwagen*) BAUMASCHIN lorry (*BrE*), truck (*AmE*)
**Lobby** *f* ARCH & TRAGW lobby
**Loch 1.** *nt* ARCH & TRAGW hole; **2.** *in cpds* WERKSTOFF cored
***Loch***: **Lochband** *nt* WERKSTOFF perforated steel strip; **Lochblech** *nt* STAHL punched plate, WERKSTOFF perforated plate; **Lochbohrung** *f* INFR & ENTW boreholing
**lochen** *vt* BAUMASCHIN, BESCHLÄGE, HOLZ, STAHL, WERKSTOFF perforate
**Lochen** *nt* BAUMASCHIN boring
***Loch***: **Lochlehre** *f* ABWASSER *Meßtechnik* boring rod, plug gage (*AmE*), plug gauge (*BrE*); **Lochpappe** *f* WERKSTOFF cored paper; **Lochplatte** *f* WERKSTOFF core slab, cored panel; **Lochreihe** *f* WERKSTOFF *Ziegel* coring; **Lochsäge** *f* BAUMASCHIN, BESCHLÄGE lock saw; **Lochsteinmauerwerk** *nt* NATURSTEIN cored block masonry, perforated brick masonry
**Lochung** *f* WERKSTOFF *Ziegel* coring
***Loch***: **Lochziegel** *m* NATURSTEIN, WERKSTOFF cored brick, perforated brick
**locker** *adj* INFR & ENTW loose, loosened; **~e Erde** *f* INFR & ENTW loosened soil; **~ gelagert** *adj* INFR & ENTW, WERKSTOFF loose-packed; **~ gelagerter Sand** *m* INFR & ENTW, WERKSTOFF loose sand
**Löffel** *m* BAUMASCHIN *Bagger* shovel; **Löffelbagger** *m* BAUMASCHIN power shovel; **Löffelbohrer** *m* BAUMASCHIN, INFR & ENTW *Tiefbau* shell auger, spoon auger
**Loggia** *f* ARCH & TRAGW loggia
**Lore** *f* BAUMASCHIN car (*AmE*), tipper, truck (*BrE*), wagon (*BrE*)
**Los** *nt* BAURECHT lot
**lösbar** *adj* WERKSTOFF *in Flüssigkeit* soluble
**Löscheinrichtung** *f* BAUMASCHIN, BAURECHT fire extinguishing equipment
**Löscher** *m* BAURECHT extinguisher
**Löschmittel** *nt* BAURECHT extinguishing agent
**lose** *adj* INFR & ENTW, WERKSTOFF loose; **~ Ablagerung** *f* UMWELT bulk deposition; **~ Dämmung** *f* DÄMMUNG fill insulation

**Lösen** *nt*: **~ und Fördern von Boden** INFR & ENTW soil displacement

*lose*: **loses Material** *nt* INFR & ENTW, WERKSTOFF loose material

**löslich** *adj* WERKSTOFF soluble; **~es Holzschutzmittel** *nt* WERKSTOFF solvent-based wood preservative

**Losnummer** *f* WERKSTOFF *Fliesen, Farbe* batch number

**Löß** *m* INFR & ENTW loess; **Lößlehm** *m* NATURSTEIN loess clay

**Lösungsmittel** *nt* UMWELT, WERKSTOFF solvent

**lösungsmittelfrei** *adj* OBERFLÄCHE *Farbe* solvent-free

**Lösungsweg** *m* ARCH & TRAGW approach

**Lot** *nt* ARCH & TRAGW plummet, BAUMASCHIN bob

**Löt-** *in cpds* WERKSTOFF soldering

**löten** *vt* WERKSTOFF solder

**Löten** *nt* ABWASSER, ARCH & TRAGW, HEIZ & BELÜFT, STAHL plumbing

**lotgerecht**: **nicht lotgerecht** *adj* ARCH & TRAGW off plumb

*Löt-*: **Lötkolben** *m* BAUMASCHIN, ELEKTR copper bit, STAHL soldering iron; **Lötlampe** *f* STAHL blowlamp, blowtorch; **Lötnaht** *f* WERKSTOFF brazed joint

**lotrecht** *adj* ARCH & TRAGW perpendicular, plumb, vertical, INFR & ENTW perpendicular

*Löt-*: **Lötrohr** *nt* WERKSTOFF soldering blowpipe

*Lot*: **Lotschnur** *f* ARCH & TRAGW plumb line

*Löt-*: **Lötverbindung** *f* ELEKTR, HEIZ & BELÜFT, WERKSTOFF soldered joint; **Lötzinn** *nt* WERKSTOFF plumber's solder

**Lücke** *f* ARCH & TRAGW gap, interstice

**Luft** *f* HEIZ & BELÜFT air

**luftabbindend** *adj* NATURSTEIN, WERKSTOFF air-setting

*Luft*: **Luftabscheider** *m* UMWELT air separator; **Luftausfällung** *f* UMWELT air shed; **Luftauslaß** *m* ARCH & TRAGW air outlet; **Luftblasenflotation** *f* UMWELT air flotation; **Luftchemie** *f* UMWELT atmospheric chemistry

**luftdicht** *adj* ARCH & TRAGW, DÄMMUNG, HEIZ & BELÜFT, UMWELT, WERKSTOFF airtight

*Luft*: **Luftdichtigkeit** *f* ARCH & TRAGW, DÄMMUNG, HEIZ & BELÜFT, UMWELT, WERKSTOFF airtightness; **Luftdruck** *m* HEIZ & BELÜFT air pressure, INFR & ENTW atmospheric pressure; **Luftdurchgängigkeit** *f* ARCH & TRAGW air permeability; **Lufteinlaß** *m* ARCH & TRAGW, HEIZ & BELÜFT air inlet

**lüften** *vt* ARCH & TRAGW, BESCHLÄGE, HEIZ & BELÜFT vent, ventilate

**Lüfter** *m* ARCH & TRAGW, BESCHLÄGE fan, HEIZ & BELÜFT fan, ventilator; **Lüfterhaube** *f* HEIZ & BELÜFT fan guard

*Luft*: **Lufterhitzer** *m* HEIZ & BELÜFT air heater

*Lüfter*: **Lüfterstein** *m* ARCH & TRAGW, NATURSTEIN ventilator block, ventilator tile

*Luft*: **Luftfeuchtigkeit** *f* ARCH & TRAGW, BETON, HEIZ & BELÜFT, OBERFLÄCHE air humidity, atmospheric moisture; **Luftgehalt** *m* INFR & ENTW, WERKSTOFF air content

**luftgehärtet** *adj* WERKSTOFF air-cured

*Luft*: **Luftgeschwindigkeit** *f* ARCH & TRAGW, HEIZ & BELÜFT air velocity; **Luftgeschwindigkeitsmesser** *m* BAUMASCHIN anemometer

**luftgetrocknet** *adj* WERKSTOFF air-dried

**lufthärtend** *adj* WERKSTOFF *Mörtel, Beton* air-hardening

*Luft*: **Luftheizung** *f* HEIZ & BELÜFT air heating; **Lufthohlraum** *m* DÄMMUNG, HEIZ & BELÜFT air space; **Lufthygiene** *f* UMWELT air pollution control; **Luftkalk** *m* NATURSTEIN air-hardening lime; **Luftkanal** *m* HEIZ & BELÜFT air duct; **Luftklappe** *f* BESCHLÄGE, HEIZ & BELÜFT air shutter; **Luftklassierer** *m* UMWELT air classifier, air separation plant; **Luftkompressor** *m* BAUMASCHIN, HEIZ & BELÜFT air compressor; **Luftkühler** *m* BAUMASCHIN, HEIZ & BELÜFT air cooler; **Luftkühlung** *f* HEIZ & BELÜFT air cooling; **Luftleistung** *f* HEIZ & BELÜFT *Ventilator* air-handling capacity; **Luftlenkblech** *nt* HEIZ & BELÜFT air deflector; **Luftloch** *nt* HEIZ & BELÜFT vent hole

**luftlos** *adj* OBERFLÄCHE, WERKSTOFF airless; **~es Farbspritzen** *nt* OBERFLÄCHE airless paint spraying

*Luft*: **Luft-Luft-Wärmepumpe** *f* HEIZ & BELÜFT air to air heat pump; **Luftnest** *nt* BETON air pocket

**Luftporen** *f pl* BETON, INFR & ENTW, NATURSTEIN, WERKSTOFF air voids; **Luftporenanteil** *m* BETON, INFR & ENTW, WERKSTOFF air voids ratio; **Luftporenbeton** *m* BETON air-entrained concrete; **Luftporenbildner** *m* BETON, INFR & ENTW, WERKSTOFF air-entraining admixture, air-entraining agent; **Luftporengehalt** *m* BETON, DÄMMUNG, INFR & ENTW, WERKSTOFF air voids content; **Luftporenminderer** *m* BETON, INFR & ENTW, WERKSTOFF air-reducing agent; **Luftporenmörtel** *m* NATURSTEIN, WERKSTOFF air mortar, air-entrained mortar

*Luft*: **Luftqualitätsdaten** *nt pl* UMWELT air quality data; **Luftqualitätsnorm** *f* UMWELT air quality standard; **Luftreinheit** *f* HEIZ & BELÜFT, UMWELT air purity; **Luftschadstoff** *m* UMWELT air pollutant; **Luftschall** *m* BAURECHT, DÄMMUNG airborne sound; **Luftschalldämmung** *f* DÄMMUNG airborne sound insulation; **Luftschallpegel** *m* BAURECHT, DÄMMUNG, INFR & ENTW, WERKSTOFF airborne sound level; **Luftschlauch** *m* HEIZ & BELÜFT *Druckluft* air hose; **Luftschleiertür** *f* HEIZ & BELÜFT air curtain door; **Luftschleuse** *f* ARCH & TRAGW, HEIZ & BELÜFT air lock

**Luftschlitzen mit ~** *phr* ARCH & TRAGW louvered (*AmE*), louvred (*BrE*)

**Luft: Luftschütz** *m* ELEKTR air-break contactor; **Luftschutzbunker** *m* ARCH & TRAGW air-raid shelter; **Luftschutzraum** *m* INFR & ENTW bunker; **Luftsortierer** *m* UMWELT air separator, air classifier, air separation plant; **Luftstrom** *m* HEIZ & BELÜFT air flow; **Luftstrommessung** *f* HEIZ & BELÜFT air flow measurement; **Lufttemperatur** *f* HEIZ & BELÜFT, INFR & ENTW air temperature; **Lufttrocknung** *f* WERKSTOFF air drying; **Luftumwälzung** *f* HEIZ & BELÜFT air circulation

**Lüftung** *f* HEIZ & BELÜFT ventilation; **Lüftungsanlage** *f* HEIZ & BELÜFT ventilation system; **Lüftungsflügel** *m* ARCH & TRAGW, BESCHLÄGE projected window, vent window; **Lüftungsflügelfenster** *nt* ARCH & TRAGW projected window; **Lüftungsgitter** *nt* HEIZ & BELÜFT louver (*AmE*), louvre (*BrE*), ventilation grille; **Lüftungsjalousie** *f* HEIZ & BELÜFT ventilation louver (*AmE*), ventilation louvre (*BrE*); **Lüftungskanal** *m* BESCHLÄGE ventiduct, ventilation duct, HEIZ & BELÜFT air duct; **Lüftungsklappe** *f* HEIZ & BELÜFT ventilation flap; **Lüftungs- und Klimakanäle** *m pl* HEIZ & BELÜFT ventilation and air-conditioning ducts; **Lüftungsleitung** *f* HEIZ & BELÜFT ventilation line; **Lüftungsöffnung** *f* HEIZ & BELÜFT air opening, ventilation opening, vent opening; **Lüftungsrohr** *nt*

HEIZ & BELÜFT ventilation pipe; **Lüftungsschacht** *m* BAUMASCHIN funnel, HEIZ & BELÜFT ventilation shaft; **Lüftungsschraube** *f* HEIZ & BELÜFT propeller fan; **Lüftungsstein** *m* ARCH & TRAGW, HEIZ & BELÜFT, NATURSTEIN ventilation block, ventilation brick; **Lüftungstür** *f* BESCHLÄGE ventilating door

***Luft***: **Luftventil** *nt* UMWELT air valve; **Luftvergiftung** *f* UMWELT air poisoning; **Luftverschmutzung** *f* BAURECHT, INFR & ENTW, UMWELT air contamination, air pollution, atmospheric pollution; **Luftverschmutzungsvorhersage** *f* UMWELT air pollution forecast; **Luftverteilung** *f* HEIZ & BELÜFT air distribution

**luftverunreinigend** *adj* UMWELT air-polluting; **~er Stoff** *m* UMWELT air-polluting substance

***Luft***: **Luftverunreinigung** *f* UMWELT air pollution; **Luftverunreinigungsemission** *f* UMWELT air pollution emission; **Luftverunreinigungsereignis** *nt* UMWELT air pollution incident; **Luftverunreinigungsgefahrensituation** *f* UMWELT air pollution episode; **Luftvorwärmer** *m* HEIZ & BELÜFT air preheater; **Luftwechsel** *m* HEIZ & BELÜFT air change; **Luftwechselzahl** *f* HEIZ & BELÜFT number of air changes; **Luftwiderstandsbeiwert** *m* (*DD*) UMWELT coefficient of drag; **Luftzufuhr** *f* STAHL aeration; **Luftzug** *m* HEIZ & BELÜFT draft (*AmE*), draught (*BrE*); **Luftzwischenraum** *m* (*LZR*) BESCHLÄGE, HEIZ & BELÜFT air gap

**Luke** *f* ARCH & TRAGW hatch

**Luttenrohre** *f pl* HEIZ & BELÜFT air pipes

**Lux** *nt* ELEKTR, INFR & ENTW, WERKSTOFF lux

**LV** *abbr* (*Leistungsverzeichnis*) ARCH & TRAGW, BAURECHT specifications

**LZR** *abbr* (*Luftzwischenraum*) BESCHLÄGE, HEIZ & BELÜFT air gap

# M

**m** *abbr* (*Meter*) ARCH & TRAGW m (*meter AmE, metre BrE*)
**mager** *adj* BETON, WERKSTOFF lean
**Mager-** *in cpds* BETON, WERKSTOFF lean
***mager***: **~er Ton** *m* INFR & ENTW, WERKSTOFF lean clay, sandy clay
**Mager-**: **Magerbeton** *m* BETON, WERKSTOFF lean concrete; **Magerkalk** *m* WERKSTOFF poor lime; **Magermischung** *f* BETON, WERKSTOFF lean mix
**Magnesit** *m* INFR & ENTW, WERKSTOFF magnesite; **Magnesitstein** *m* INFR & ENTW magnesite brick; **Magnesitziegel** *m* INFR & ENTW, NATURSTEIN magnesite brick
**Magnesiumchlorid** *nt* INFR & ENTW, WERKSTOFF magnesium chloride
**Magnet** *m* UMWELT magnet; **Magnetabscheider** *m* UMWELT magnetic separator; **Magnetabscheidung** *f* UMWELT magnetic separation; **Magnetsortierung** *f* UMWELT *von Müll* magnetic separation
**mahlen** *vt* BAUMASCHIN grind
**Mahlen** *nt* BAUMASCHIN grinding
**Maisonette** *f* ARCH & TRAGW maisonette
**Makadam** *m* WERKSTOFF macadam
**makadamisieren** *vt* INFR & ENTW macadamize
**Makadamisierung** *f* INFR & ENTW macadamization
**Makler** *m* BAURECHT *Immobilien* agent, broker
**Maler** *m* ARCH & TRAGW, OBERFLÄCHE decorator, painter; **Malerarbeiten** *f pl* OBERFLÄCHE painting work, painting
**malerfertig** *adj* OBERFLÄCHE ready-to-paint
**Mammutpumpe** *f* BAUMASCHIN mammoth pump
**Managementvertrag** *m* BAURECHT management contract
**Manager** *m* ARCH & TRAGW manager
**Mangel** *m* BAURECHT defect, fault, WERKSTOFF flaw
**Mängel** *m pl* ARCH & TRAGW, BAURECHT deficiencies; **Mängelbehebung** *f* ARCH & TRAGW, BAURECHT correction of deficiencies; **Mängelbeseitigung** *f* ARCH & TRAGW, BAURECHT correction of deficiencies
**mangelhaft** *adj* ARCH & TRAGW, BAURECHT defective, faulty, unsound, INFR & ENTW defective, WERKSTOFF flawed
***Mangel***: **Mangelhaftigkeit** *f* BAURECHT unsoundness
***Mängel***: **Mängelliste** *f* ARCH & TRAGW, BAURECHT list of defects, list of deficiencies
**Mannloch** *nt* ABWASSER, INFR & ENTW manhole
**Manometer** *nt* BAUMASCHIN, HEIZ & BELÜFT manometer, pressure gage (*AmE*), pressure gauge (*BrE*)
**manövrieren** *vt* ARCH & TRAGW maneuver (*AmE*), manoeuvre (*BrE*)
**Mansardendach** *nt* ARCH & TRAGW, HOLZ gambrel roof, mansard roof
**Manschette** *f* ABWASSER, HEIZ & BELÜFT *Rohr* collar
**Mantel** *m* ARCH & TRAGW jacket, shell, BAUMASCHIN shell, ELEKTR *Kabel* sheath, HEIZ & BELÜFT jacket; **Mantelkasten** *m* BETON, INFR & ENTW open caisson; **Mantelleitung** *f* ELEKTR plastic-sheathed cable; **Mantelpfahl** *m* ARCH & TRAGW, INFR & ENTW cased pile; **Mantelreibungspfahl** *m* ARCH & TRAGW, INFR & ENTW suspended pile; **Mantelrohr** *nt* HEIZ & BELÜFT jacket pipe
**Marina** *f* INFR & ENTW *Bootshafen* marina
**marine**: **~ Ablagerung** *f* INFR & ENTW marine deposit, marine sediment
**Marke** *f* ARCH & TRAGW datum level
**Markenname** *m* BAURECHT, WERKSTOFF brand name
**markieren** *vt* BAURECHT beacon
**Markiernadel** *f* BAUMASCHIN scratch awl
**Markierung** *f* ARCH & TRAGW marking, BESCHLÄGE label; **Markierungsanstrich** *m* INFR & ENTW marking coat; **Markierungseisen** *nt* BAUMASCHIN scribing iron; **Markierungsfarbe** *f* INFR & ENTW *Straßenmarkierung* road marking paint, OBERFLÄCHE marking-paint; **Markierungsstoff** *m* WERKSTOFF marking material
**märkisch**: **~er Verband** *m* NATURSTEIN flying bond
**Markise** *f* ARCH & TRAGW awning
**Marmor** *m* NATURSTEIN marble; **Marmorgips** *m* WERKSTOFF Keene's cement®; **Marmorstaub** *m* NATURSTEIN, WERKSTOFF marble dust
**Marsch** *f* INFR & ENTW fen, marsh; **Marschland** *nt* INFR & ENTW marshland
**Masche** *f* WERKSTOFF mesh; **Maschendrahtgewebe** *nt* WERKSTOFF mesh wire; **Maschendrahtzaun** *m* INFR & ENTW mesh wire fence; **Maschengitter** *nt* WERKSTOFF mesh; **Maschenweite** *f* WERKSTOFF mesh size
**maschinell**: **~ gerichtetes Blech** *nt* STAHL, WERKSTOFF machine-straightened sheet
**Maschine** *f* BAUMASCHIN, BETON, INFR & ENTW machine; **Maschinen** *f pl* **und Geräte** *nt pl* **für die Straßeninstandhaltung** BAUMASCHIN machinery and equipment for road maintenance; **Maschinen** *f pl* BAUMASCHIN, INFR & ENTW machinery; **Maschinenanlage** *f* BAUMASCHIN, INFR & ENTW machinery; **Maschinenfundament** *nt* BETON, INFR & ENTW machine foundation; **Maschinenputz** *m* NATURSTEIN machine-applied plaster; **Maschinenraum** *m* ARCH & TRAGW, BAUMASCHIN engine room, machine room
**Maschinist** *m* BAURECHT operator
**Masern** *nt* HOLZ graining
**Maserung** *f* HOLZ graining
**Maß** *nt* ARCH & TRAGW measure, size; **Maßabweichung** *f* ARCH & TRAGW dimensional variation; **Maßaufnahme** *f* INFR & ENTW survey; **Maßband** *nt* INFR & ENTW tape
**Masse** *f* ARCH & TRAGW mass, quantity, WERKSTOFF bulk; **~ der Luft** HEIZ & BELÜFT air mass
***Maß***: **Maßeinheit** *f* ARCH & TRAGW unit
***Masse***: **Massenausgleichung** *m* INFR & ENTW balancing of masses; **Massenberechner** *m* BAURECHT quantity surveyor; **Massenberechnung** *f* ARCH & TRAGW quantity survey, INFR & ENTW *Erdbewegung* mensuration; **Massenbeton** *m* BETON, INFR & ENTW mass concrete; **Massenermittlung** *f* ARCH & TRAGW quantity survey; **Massenermittlungsdienstleistungen** *f pl* BAURECHT quantity surveying services;

**Massengut** *nt* WERKSTOFF bulk material; **Massentransport** *m* BAUMASCHIN mass transportation; **Massenverzeichnis** *nt* ARCH & TRAGW, BAURECHT bill of quantity

**maßgerecht** *adj* ARCH & TRAGW true to dimensions

**maßgleich**: **~e Darstellung** *f* ARCH & TRAGW, INFR & ENTW isometric projection, isometric representation

**massiv** *adj* ARCH & TRAGW solid, sturdy; **~e Treppenspindel** *f* HOLZ newel post

**Massiv-** *in cpds* ARCH & TRAGW, BETON, INFR & ENTW mass, solid; **Massivbau** *m* ARCH & TRAGW solid construction; **Massivbeton** *m* BETON, INFR & ENTW mass concrete; **Massivboden** *m* ARCH & TRAGW solid floor; **Massivdecke** *f* ARCH & TRAGW solid ceiling; **Massivstufe** *f* ARCH & TRAGW solid step

***Maß***: **Maßkasten** *m* BAUMASCHIN gage box (*AmE*), gauge box (*BrE*); **Maßmauerwerk** *nt* NATURSTEIN gaged brickwork (*AmE*), gauged brickwork (*BrE*)

**Maßnahme** *f* ARCH & TRAGW, BAURECHT measure; **Maßnahmen** *f pl* **gegen Holzwurmbefall** BAURECHT, HOLZ woodworm treatment

***Maß***: **Maßsprung** *m* ARCH & TRAGW, INFR & ENTW increment

**Maßstab** *m* ARCH & TRAGW, INFR & ENTW scale, scale factor

**maßstäblich** *adj* ARCH & TRAGW according to scale, **~ gezeichnet** *adj* ARCH & TRAGW drawn to scale; **~ zeichnen** *vt* ARCH & TRAGW draw to scale

**maßstabsgerecht** *adj* ARCH & TRAGW, INFR & ENTW according to scale, to scale; **~e Bauaufnahmezeichnungen** *f pl* ARCH & TRAGW, INFR & ENTW measured drawings; **~e Zeichnung** *f* ARCH & TRAGW drawing to scale

***Maßstab***: **Maßstabsmodell** *nt* ARCH & TRAGW, INFR & ENTW scale model

***Maß***: **Maßtoleranz** *f* ARCH & TRAGW dimensional tolerance; **Maßwerk** *nt* ARCH & TRAGW tracery; **Maßwerkgiebel** *m* ARCH & TRAGW traceried gable

**Mast** *m* HOLZ pole, INFR & ENTW mast, pole; **Mastabstand** *m* INFR & ENTW pole spacing; **Mastenkran** *m* BAUMASCHIN derrick

**Mastixharz** *nt* WERKSTOFF mastic

**Material** *nt* WERKSTOFF material, substance; **Materialeigenschaften** *f pl* INFR & ENTW, WERKSTOFF material properties

**Materialien** *nt pl*: **~ mit begrenzter Brennbarkeit** BAURECHT, INFR & ENTW, WERKSTOFF material of limited combustibility; **Materialientauglichkeit** *f* BAURECHT, WERKSTOFF suitability of materials

***Material***: **Materialkennwerte** *m pl* INFR & ENTW, WERKSTOFF material characteristics

**Materialprüfung** *f* **gemäß Spezifikationen** BAURECHT, INFR & ENTW specification test

***Material***: **Materialrückgewinnung** *f* UMWELT material recovery; **Materialveränderung** *f* WERKSTOFF change of material; **Materialverhalten** *nt* INFR & ENTW, WERKSTOFF material behavior (*AmE*), material behaviour (*BrE*)

**Materie** *f* ARCH & TRAGW, INFR & ENTW, WERKSTOFF matter

**mathematisch** *adj* ARCH & TRAGW, INFR & ENTW mathematical; **~e Gleichung** *f* ARCH & TRAGW, INFR & ENTW mathematical equation

**Matrix** *f* ARCH & TRAGW matrix

**matt** *adj* ARCH & TRAGW, INFR & ENTW, OBERFLÄCHE flat

**Matt-** *in cpds* ARCH & TRAGW, INFR & ENTW, OBERFLÄCHE matt; **Mattätzen** *nt* OBERFLÄCHE, STAHL acid embossing

**Matte** *f* WERKSTOFF mat; **Mattenbewehrung** *f* ARCH & TRAGW wire mesh reinforcement

***Matt-***: **Mattglanz** *m* OBERFLÄCHE low luster (*AmE*), low lustre (*BrE*); **Mattglas** *nt* WERKSTOFF earth glass (*BrE*), ground glass (*AmE*), opal glass

**Mauer** *f* ARCH & TRAGW wall, NATURSTEIN masonry wall; **Mauerabdeckung** *f* ARCH & TRAGW wall coping; **Mauerabsatz** *m* ARCH & TRAGW wall offset; **Maueranker** *m* ARCH & TRAGW, BESCHLÄGE, NATURSTEIN wall hook, wall tie; **Maueranstrichfarbe** *f* INFR & ENTW, OBERFLÄCHE, WERKSTOFF masonry paint; **Mauerarbeiten** *f pl* ARCH & TRAGW, NATURSTEIN masonry work; **Mauerband** *nt* ARCH & TRAGW, NATURSTEIN cordon, string; **Mauerblock** *m* INFR & ENTW, NATURSTEIN block; **Mauerbohrer** *m* BAUMASCHIN masonry drill; **Mauerecke** *f* NATURSTEIN quoin; **Mauergründung** *f* ARCH & TRAGW, BETON, INFR & ENTW wall footing, wall foundation; **Mauerhaken** *m* ARCH & TRAGW spike; **Mauerhammer** *m* BAUMASCHIN axhammer; **Mauerkante** *f* ARCH & TRAGW wall edge; **Mauerkappe** *f* BESCHLÄGE hood; **Mauerkrone** *f* ARCH & TRAGW crest, wall crown, wall top; **Mauermantel** *m* NATURSTEIN lining; **Mauermörtelzement** *m* BETON, INFR & ENTW, WERKSTOFF masonry cement

**mauern** *vt* ARCH & TRAGW, NATURSTEIN brick up, lay bricks

**Mauern** *nt* ARCH & TRAGW, NATURSTEIN walling; **~ von der Außenseite** NATURSTEIN overhand bricklaying

***Mauer***: **Mauerpfeiler** *m* ARCH & TRAGW, NATURSTEIN counterfort; **Mauerschlitz** *m* NATURSTEIN keyway, masonry wall slot; **Mauerstein** *m* INFR & ENTW block, NATURSTEIN block, brick

**Mauerung** *f* ARCH & TRAGW, NATURSTEIN walling

***Mauer***: **Mauerverband** *m* NATURSTEIN wall bond; **Mauervorsprung** *m* ARCH & TRAGW spur; **Mauerwerk** *nt* ARCH & TRAGW, BAUMASCHIN walling, NATURSTEIN masonry, walling; **Mauerwerksarbeiten** *f pl* ARCH & TRAGW, NATURSTEIN masonry; **Mauerwerkverankerung** *f* NATURSTEIN tying-in of brickwork; **Mauerzacke** *f* ARCH & TRAGW, NATURSTEIN battlement; **Mauerziegel** *m* NATURSTEIN brick

**Maurer** *m* NATURSTEIN mason; **Maurerarbeit** *f* ARCH & TRAGW, NATURSTEIN walling; **Maurerkelle** *f* BAUMASCHIN brick trowel, BESCHLÄGE, NATURSTEIN bricklayer's trowel; **Maurermeister** *m* ARCH & TRAGW, NATURSTEIN master mason

**Mautbrücke** *f* INFR & ENTW toll bridge

**maximal**: **~es Aufwölbungsmoment** *nt* ARCH & TRAGW, BETON, INFR & ENTW, STAHL, WERKSTOFF maximum hogging moment; **~er Axialdruck** *m* ARCH & TRAGW, UMWELT maximum axial thrust; **~es Biegemoment** *nt* (*max. M*) ARCH & TRAGW, BETON maximum bending moment (*max. M*); **~er Druck** *m* INFR & ENTW maximum pressure; **~e Gesamtbelastung** *f* ARCH & TRAGW, INFR & ENTW maximum total load; **~e Wellengeschwindigkeit** *f* UMWELT maximum shaft speed

**max.**: **~ M** *abbr* (*maximales Biegemoment*) ARCH & TRAGW, BETON max. M (*maximum bending moment*)

**Maxwell**: **~scher Kräfteplan** *m* ARCH & TRAGW Maxwell diagram

**mechanisch**: **~e Abnutzung** *f* BAUMASCHIN mechanical wear; **~er Abscheider** *m* UMWELT mechanical collector; **~e Abwasserreinigung** *f* UMWELT primary sewage treatment; **~ belüftet** *adj* HEIZ & BELÜFT mechanically ventilated; **~ betriebener Greifer** *m* BAUMASCHIN mechanical grab; **~-chemisch** *adj* INFR & ENTW, WERKSTOFF physico-chemical; **~-chemische Bodenstabilisierung** *f* INFR & ENTW physico-chemical stabilization; **~es Drucktastenschloß** *nt* BESCHLÄGE mechanical cipher lock; **~e Eigenschaften** *f pl* INFR & ENTW, WERKSTOFF mechanical properties; **~er Einschluß** *m* UMWELT *von Schadstoffen* physical stabilization; **~e Kraftübertragung** *f* BAUMASCHIN, UMWELT mechanical transmission; **~e Sortierung** *f* UMWELT mechanical separation; **~e Trennung** *f* UMWELT automatic sorting, mechanical separation; **~er Verbund** *m* ARCH & TRAGW, BETON, INFR & ENTW, STAHL, WERKSTOFF mechanical connection; **~es Verhalten** *nt* WERKSTOFF mechanical behavior (*AmE*), mechanical behaviour (*BrE*); **~es Versagen** *nt* INFR & ENTW, WERKSTOFF mechanical failure; **~er Wirkungsgrad** *m* UMWELT mechanical efficiency

**Mechanismus** *m* BAUMASCHIN, BESCHLÄGE mechanism

**Meer**: **Meeresboden** *m* INFR & ENTW ocean floor, UMWELT seabed; **Meeresgrund** *m* INFR & ENTW, UMWELT seabed; **Meereshöhe** *f* ARCH & TRAGW sea level; **Meeressand** *m* INFR & ENTW marine sand; **Meeresspiegel** *m* INFR & ENTW, UMWELT sea level; **Meeresspiegelbezugsebene** *f* INFR & ENTW *Normalnull* sea level datum plane; **Meerestiefe** *f* INFR & ENTW sea depth; **Meeresverschmutzung** *f* UMWELT marine pollution; **Meersalz** *nt* INFR & ENTW, WERKSTOFF marine salt; **Meerton** *m* INFR & ENTW, NATURSTEIN, WERKSTOFF marine clay

**Meganewton** *nt* (*MN*) ARCH & TRAGW, INFR & ENTW, WERKSTOFF meganewton (*MN*)

**Megapond** *nt* (*Mp*) ARCH & TRAGW, INFR & ENTW, WERKSTOFF megapond (*Mp*)

**Mehr-** *in cpds* ARCH & TRAGW, INFR & ENTW additional, extra; **Mehraufwand** *m* ARCH & TRAGW, BAURECHT, INFR & ENTW extra work

**mehrbalkig** *adj* ARCH & TRAGW, HOLZ multibeam

**mehrere**: **~ Währungen** *f pl* BAURECHT multiple currencies

***Mehr-***: **Mehrfeldplatte** *f* ARCH & TRAGW, BETON continuous slab; **Mehrfeldrahmen** *m* ARCH & TRAGW continuous frame

**mehrfeldrig** *adj* ARCH & TRAGW multispan

***Mehr-***: **Mehrfeldträger** *m* ARCH & TRAGW continuous beam

**mehrgeschossig** *adj* ARCH & TRAGW multistorey (*BrE*), multistory (*AmE*)

***Mehr-***: **Mehrkomponentenepoxidharz** *nt* WERKSTOFF multicomponent epoxy resin; **Mehrkosten** *pl* BAURECHT additional charges; **Mehrpunkterdung** *f* ELEKTR multiple-point earthing (*BrE*), multiple-point grounding (*AmE*); **Mehrscheibenisolierglas** *nt* WERKSTOFF laminated insulating glass; **Mehrschichtenplatte** *f* WERKSTOFF laminated board

**mehrschichtig** *adj* INFR & ENTW, WERKSTOFF laminated

***Mehr-***: **Mehrschichtprinzip** *nt* UMWELT *Deponie* multibarrier principle; **Mehrschlüssel** *m* BESCHLÄGE additional change key

**mehrschnittig**: **~es Gelenk** *nt* ARCH & TRAGW, INFR & ENTW multiple shear joint; **~e Verbindung** *f* ARCH & TRAGW, INFR & ENTW multiple shear connection

**mehrstöckig** *adj* ARCH & TRAGW multistorey (*BrE*), multistory (*AmE*); **~es Gebäude** *nt* ARCH & TRAGW, INFR & ENTW multistorey building (*BrE*), multistory building (*AmE*)

**mehrteilig**: **~e Säule** *f* ARCH & TRAGW spaced column

**Mehrwertsteuer** *f* (*MWSt*) BAURECHT value-added tax (*VAT*)

**Meißelhammer** *m* BAUMASCHIN chipper

**Meister** *m* ARCH & TRAGW master

**Melamin** *nt* HOLZ, INFR & ENTW, OBERFLÄCHE, WERKSTOFF melamine; **Melamin-Formaldehydharz** *nt* HOLZ, INFR & ENTW, OBERFLÄCHE, WERKSTOFF melamine formaldehyde resin

**melaminharzbeschichtet** *adj* OBERFLÄCHE melamine resin coated

***Melamin***: **Melaminleim** *m* HOLZ, INFR & ENTW, WERKSTOFF melamine adhesive; **Melamin/Urea-Formaldehyd** *nt* (*MF/UF*) HOLZ, UMWELT, WERKSTOFF melamine/urea formaldehyde (*MF/UF*)

**Melder** *m* BAURECHT, BESCHLÄGE, ELEKTR *Feuer* alarm

**meliert** *adj* OBERFLÄCHE, WERKSTOFF *Teppichboden, Textilien* blended, mottled

**Melioration** *f* ARCH & TRAGW, BAURECHT, INFR & ENTW amelioration

**Membranschutz** *m* INFR & ENTW membrane protection

**Menge** *f* ARCH & TRAGW amount, mass, quantity, BESCHLÄGE amount, STAHL aggregate

**Menhir** *m* ARCH & TRAGW monolith

**Menschenmenge**: **durch ~ verursacht** *phr* ARCH & TRAGW crowd-induced

**menschlich**: **~ bedingte Übersäuerung** *f* UMWELT anthropogenic acidification

**Mergel** *m* INFR & ENTW marl; **Mergelboden** *m* INFR & ENTW marly soil; **Mergelsandstein** *m* INFR & ENTW marly sandstone

**Merkmal** *nt* ARCH & TRAGW feature

**Meß-** *in cpds* ARCH & TRAGW, BAUMASCHIN, ELEKTR, HEIZ & BELÜFT, INFR & ENTW, WERKSTOFF measuring; **Meßapparat** *m* BAUMASCHIN measuring apparatus; **Meßband** *nt* ARCH & TRAGW surveyor's tape; **Meßbereich** *m* BAUMASCHIN measuring range; **Meßbildverfahren** *nt* INFR & ENTW photogrammetry; **Meßelektrode** *f* ARCH & TRAGW, INFR & ENTW, WERKSTOFF measuring electrode

**messen** *vt* ARCH & TRAGW, INFR & ENTW, WERKSTOFF measure

**Messerfurnier** *nt* HOLZ knife-cut veneer, sliced veneer

***Meß-***: **Meßfehler** *m* ARCH & TRAGW, INFR & ENTW, WERKSTOFF measurement error; **Meßfühler** *m* ARCH & TRAGW, HEIZ & BELÜFT gage (*AmE*), gauge (*BrE*), probe, sensor; **Meßgefäß** *nt* BAUMASCHIN *Beton* measuring tank; **Meßgenauigkeit** *f* ARCH & TRAGW accuracy of measurement, measuring accuracy, BAUMASCHIN measuring accuracy, INFR & ENTW, WERKSTOFF accuracy of measurement, measuring accuracy; **Meßgerät** *nt* HEIZ & BELÜFT meter

**Messing** *nt* WERKSTOFF brass; **Messingband** *nt* BESCHLÄGE brass hinge; **Messinglot** *nt* WERKSTOFF brass solder; **Messingrundkopfschraube** *f* BESCHLÄGE brass round-head wood screw; **Messingstab** *m* BESCHLÄGE brass bar

***Meß-***: **Meßkette** *f* ARCH & TRAGW *Vermessung* chain, BAUMASCHIN measuring chain, *Vermessung* band chain, INFR & ENTW chain; **Meßkunde** *f* ARCH &

TRAGW, INFR & ENTW surveying; **Meßlatte** *f* ARCH & TRAGW gage (*AmE*), gauge (*BrE*), *mit Ablesemarkierungen* speaking rod, *Vermessung* staff, INFR & ENTW *mit Ablesemarkierungen* speaking rod, gage (*AmE*), gauge (*BrE*) ; **Meßlattenträger** *m* ARCH & TRAGW, INFR & ENTW staff holder; **Meßlehre** *f* BAUMASCHIN gage (*AmE*), gauge (*BrE*); **Meß-, Steuerungs- und Regeleinrichtungen** *f pl* (*MSR-Einrichtungen*) ELEKTR, HEIZ & BELÜFT measuring and control equipment; **Meßtisch** *m* INFR & ENTW *Vermessung* plane table

**Messung** *f* ARCH & TRAGW, ELEKTR measurement, INFR & ENTW mensuration, WERKSTOFF measuring; **~ der Luftqualität** UMWELT air quality measurement

***Meß-***: **Meßverfahren** *nt* ARCH & TRAGW, INFR & ENTW method of measurement; **Meßzeitraum** *m* ARCH & TRAGW, INFR & ENTW, WERKSTOFF measurement period

**Metall** *nt* WERKSTOFF metal; **Metallabdeckung** *f* STAHL, WERKSTOFF metal decking; **Metallabfall** *m* UMWELT metal waste; **Metallanker** *m* WERKSTOFF metal anchor, metal tie

**metallausgekleidet**: **~e Tür** *f* ARCH & TRAGW, BAURECHT metal-clad door

***Metall***: **Metallauskleidung** *f* STAHL metal cladding, metal lining; **Metallbarometer** *nt* ARCH & TRAGW *Vermessung* surveying aneroid barometer; **Metallbau** *m* STAHL metal construction, metal structure; **Metallbauarbeiten** *f pl* ARCH & TRAGW, STAHL metal construction work; **Metallbaumaterial** *nt* WERKSTOFF metal building material; **Metallbauwerk** *nt* ARCH & TRAGW, STAHL metallic structure; **Metallbelag** *m* OBERFLÄCHE, STAHL, WERKSTOFF metal coat, metalization (*AmE*), metallization (*BrE*); **Metallbeschichtung** *f* OBERFLÄCHE, STAHL metal coating; **Metallbolzen** *m* BESCHLÄGE gate hook; **Metalldachkehle** *f* STAHL metal valley; **Metallleichtbau** *m* ARCH & TRAGW, STAHL lightweight metal construction; **Metalleinlage** *f* WERKSTOFF metal insertion; **Metallermüdung** *f* WERKSTOFF metal fatigue; **Metallfassade** *f* ARCH & TRAGW, STAHL metal façade; **Metallfenster** *nt* STAHL metal window; **Metallfolie** *f* WERKSTOFF metal foil; **Metallfolieneinlage** *f* WERKSTOFF metal foil insertion

**metallgekapselt** *adj* STAHL metal-enclosed

***Metall***: **Metallgrundierung** *f* OBERFLÄCHE metal primer; **Metallgrundierungsmittel** *nt* OBERFLÄCHE metal primer; **Metallinertgasschweißen** *nt* STAHL metal inert gas welding; **Metallkante** *f* BESCHLÄGE pressed metal edging; **Metallkern** *m* WERKSTOFF metal core; **Metallklammer** *f* STAHL, WERKSTOFF metal cramp; **Metallmaueranker** *m* STAHL, WERKSTOFF metal wall anchor; **Metallprobe** *f* WERKSTOFF assay; **Metallputzträger** *m* STAHL, WERKSTOFF metal lathing, steel lathing; **Metallschiene** *f* WERKSTOFF metal rail; **Metallschrott** *m* UMWELT metal waste; **Metallschuh** *m* STAHL, WERKSTOFF metal hanger; **Metallskelettwand** *f* ARCH & TRAGW, STAHL metal stud partition, metal stud wall; **Metallsortieranlage** *f* UMWELT *magnetische Abtrennung* metal separator; **Metallsprosse** *f* BAUMASCHIN *Leiter* metal rung; **Metallstütze** *f* ARCH & TRAGW, STAHL metal column; **Metallüberzug** *m* STAHL metal coat

**Meter** *m* (*m*) ARCH & TRAGW meter (*AmE*) (*m*), metre (*BrE*) (*m*)

**Methan** *nt* UMWELT methane gas; **Methangärung** *f* UMWELT alkaline fermentation, methane digestion, methane fermentation

**Methode** *f* ARCH & TRAGW method; **~ wiederholter Momentenverteilung** ARCH & TRAGW, INFR & ENTW method of moment distribution

**Methylenblautest** *m* UMWELT methylene blue test

**metrisch**: **~es System** *nt* ARCH & TRAGW, BAURECHT, INFR & ENTW metric system

**Mezzanin** *nt* ARCH & TRAGW mezzanine

**MF/UF** *abbr* (*Melamin/Urea-Formaldehyd*) HOLZ, UMWELT, WERKSTOFF MF/UF (*melamine/urea formaldehyde*)

**Mietkosten** *pl* ARCH & TRAGW, BAURECHT rental charge

**MIG-MAG-Schweißbrenner** *m* BAUMASCHIN torch for MIG-MAG welding

**Mikro-** *pref* INFR & ENTW, UMWELT, WERKSTOFF micro; **Mikroklima** *nt* UMWELT microclimate

**Mikron** *nt* WERKSTOFF micron

***Mikro-***: **Mikroorganismus** *m* INFR & ENTW, UMWELT, WERKSTOFF microorganism; **Mikroriß** *m* INFR & ENTW, STAHL, WERKSTOFF hair crack

**Mikroskop** *nt* INFR & ENTW microscope

***Mikro-***: **Mikrostruktur** *f* WERKSTOFF microstructure; **Mikroverschmutzer** *m* INFR & ENTW, UMWELT micropollutant; **Mikroverschmutzung** *f* BAURECHT micropollution

**Milchglas** *nt* WERKSTOFF milk glass

**Milieu** *nt* ARCH & TRAGW, BAURECHT, INFR & ENTW environment

**Millimeter** *m* (*mm*) ARCH & TRAGW millimeter (*AmE*) (*mm*), millimetre (*BrE*) (*mm*); **Millimeterpapier** *nt* ARCH & TRAGW millimeter paper (*AmE*), millimitre paper (*BrE*), plotting paper

**minderwertig** *adj* ARCH & TRAGW substandard; **~e Energie** *f* HEIZ & BELÜFT low-grade energy

**Mindest-** *in cpds* ARCH & TRAGW, ELEKTR, INFR & ENTW, WERKSTOFF minimum; **Mindestgüte** *f* WERKSTOFF minimum quality; **Mindestquerschnitt** *m* ARCH & TRAGW, ELEKTR, HEIZ & BELÜFT minimum cross section; **Mindestschweißstrom** *m* ELEKTR minimum welding current; **Mindesttiefe** *f* ARCH & TRAGW, INFR & ENTW minimum depth

**Mineral** *nt* WERKSTOFF mineral; **Mineralbeton** *m* BETON scalpings; **Mineralfaser** *f* WERKSTOFF mineral fiber (*AmE*), mineral fibre (*BrE*); **Mineralfasereinlage** *f* DÄMMUNG mineral fiber filling (*AmE*), mineral fibre filling (*BrE*); **Mineralfasermatte** *f* DÄMMUNG mineral fiber mat (*AmE*), mineral fibre mat (*BrE*); **Mineralfaserplatte** *f* DÄMMUNG, INFR & ENTW, WERKSTOFF mineral fiber slab (*AmE*), mineral fibre slab (*BrE*)

**mineralisch** *adj* WERKSTOFF mineral

**Mineralogie** *f* INFR & ENTW, WERKSTOFF mineralogy

***Mineral***: **Mineralwolle** *f* WERKSTOFF mineral wool

**Minimum**: **auf ein ~ reduzieren** *phr* ARCH & TRAGW minimize

**Misch-** *in cpds* ABWASSER, BETON, INFR & ENTW, UMWELT mixed, combined; **Mischbatterie** *f* ABWASSER blending valve, *Bad* mixer tap (*BrE*), mixing faucet (*AmE*); **Mischboden** *m* INFR & ENTW mixed soil; **Mischbrett** *nt* NATURSTEIN gaging board (*AmE*), gauging board (*BrE*); **Mischdauer** *f* BETON mixing time; **Mischdeponie** *f* UMWELT codisposal landfill

**mischen** *vt* BETON mix, WERKSTOFF prepare

**Mischen** *nt* BETON, WERKSTOFF mixing

***Misch-***: **Mischentwässerung** *f* ABWASSER, INFR & ENTW combined drainage

**Mischer** *m* BAUMASCHIN mixer; **Mischerbühne** *f* BAUMASCHIN mixer platform

***Misch-***: **Mischgutart** *f* WERKSTOFF type of mixture; **Mischkanalisation** *f* ABWASSER, INFR & ENTW, UMWELT combined drainage system, combined sewerage system; **Mischmaschine** *f* BETON mixer; **Mischpodest** *nt* NATURSTEIN gaging board (*AmE*), gauging board (*BrE*); **Mischpolymerisat** *nt* WERKSTOFF copolymer; **Mischsystem** *nt* ABWASSER, INFR & ENTW, UMWELT *Entwässerung* combined drainage system, combined sewerage system; **Mischtrommel** *f* BAUMASCHIN mixing drum

**Mischung** *f* WERKSTOFF mixing, mixture; **Mischungsverhältnis** *nt* BETON, WERKSTOFF mix proportions, mixture ratio

***Misch-***: **Mischwasser** *nt* ABWASSER, INFR & ENTW combined sewage; **Mischwassersammler** *m* ABWASSER, INFR & ENTW combined sewer; **Mischzeit** *f* BETON mixing time

**mitnageln** *vt* ARCH & TRAGW, HOLZ nail

**Mitnehmer** *m* BESCHLÄGE *eines Türriegels* nosing

**Mitte** *f* ARCH & TRAGW middle, mid-depth, BETON, INFR & ENTW, WERKSTOFF mid-depth

**Mittel** *nt* ARCH & TRAGW, INFR & ENTW average, WERKSTOFF agent, average; **Mittelachse** *f* ARCH & TRAGW central axis; **Mittelbau** *m* ARCH & TRAGW central structure; **Mittelbogen** *m* ARCH & TRAGW central arch

**mittel**: **mittleres Drittel** *nt* ARCH & TRAGW middle third

***Mittel***: **Mitteldruck** *m* HEIZ & BELÜFT medium pressure; **Mitteldruckdampfkessel** *m* HEIZ & BELÜFT medium pressure boiler; **Mittelfeld** *nt* ARCH & TRAGW, INFR & ENTW center span (*AmE*), centre span (*BrE*), interior span, middle span; **Mittelgang** *m* ARCH & TRAGW central corridor; **Mittelgewölbe** *nt* ARCH & TRAGW, NATURSTEIN central vault; **Mittelgiebel** *m* ARCH & TRAGW, NATURSTEIN central gable, main gable

**mittelhoch** *adj* ARCH & TRAGW medium-rise

***Mittel***: **Mittelhof** *m* ARCH & TRAGW central courtyard; **Mitteljoch** *nt* ARCH & TRAGW central bay; **Mittelkies** *m* INFR & ENTW, WERKSTOFF medium gravel

**mittelkörnig** *adj* INFR & ENTW, WERKSTOFF medium-grained

***Mittel***: **Mittellage** *f* ARCH & TRAGW central position; **Mittellängswand** *f* ARCH & TRAGW spine wall; **Mittellinie** *f* ARCH & TRAGW, INFR & ENTW axis

***mittel***: **mittlere Monatsdosis** *f* UMWELT average monthly dose

***Mittel***: **Mittelmosaik** *nt* NATURSTEIN medium-sized mosaic; **Mittelpfette** *f* ARCH & TRAGW, HOLZ center purlin (*AmE*), central purlin, centre purlin (*BrE*), middle purlin; **Mittelpunkt** *m* ARCH & TRAGW, INFR & ENTW center (*AmE*), centre (*BrE*); **Mittelsäule** *f* ARCH & TRAGW, BETON, NATURSTEIN, STAHL central column; **Mittelschiene** *f* ARCH & TRAGW middle rail; **Mittelschiff** *nt* ARCH & TRAGW center span (*AmE*), centre span (*BrE*); **Mittelschluff** *m* INFR & ENTW, WERKSTOFF medium silt

**mittelschwer** *adj* WERKSTOFF medium-heavy

***Mittel***: **Mittelsprosse** *f* ARCH & TRAGW *Fenster* middle muntin

***mittel***: **mittlerer Tidehub** *m* UMWELT mean tidal range

***Mittel***: **Mittelwasser** *nt* INFR & ENTW mean water

***mittel***: **mittlerer Wasserstand** *m* INFR & ENTW mean water level

***Mittel***: **Mittelwert** *m* ARCH & TRAGW, INFR & ENTW, WERKSTOFF average, mean value

***mittel***: **mittlere Windgeschwindigkeit** *f* UMWELT mean wind speed

**mittig** *adj* ARCH & TRAGW BETON, INFR & ENTW, STAHL, WERKSTOFF axial, central; **~ belastet** *adj* ARCH & TRAGW, INFR & ENTW, STAHL axially loaded; **~e Belastung** *f* ARCH & TRAGW central loading; **~er Druck** *m* INFR & ENTW, WERKSTOFF axial pressure

**mitwirkend**: **~e Gurtweite** *f* ARCH & TRAGW, BETON, INFR & ENTW, WERKSTOFF effective width

**mm** *abbr* (*Millimeter*) ARCH & TRAGW mm (*millimeter AmE, millimetre BrE*)

**MN** *abbr* (*Meganewton*) ARCH & TRAGW, INFR & ENTW, WERKSTOFF MN (*meganewton*)

**Möbel** *nt pl* HOLZ furniture; **Möbelschloß** *nt* BESCHLÄGE cabinet lock; **Möbelschreiner** *m* HOLZ cabinet-maker; **Möbeltischlerei** *f* HOLZ cabinet-making

**Modell** *nt* ARCH & TRAGW model, type; **Modellmaßstab** *m* INFR & ENTW, WERKSTOFF model scale; **Modellstatik** *f* ARCH & TRAGW model analysis; **Modellstudie** *f* INFR & ENTW, WERKSTOFF model study; **Modelltest** *m* INFR & ENTW, WERKSTOFF model test

**modern** *adj* ARCH & TRAGW modern

**modernisieren** *vt* ARCH & TRAGW modernize

**Modernisierung** *f* ARCH & TRAGW modernization

**Modifikation** *f* ARCH & TRAGW modification

**Modul** *m* ARCH & TRAGW modulus

**modular** *adj* ARCH & TRAGW modular

***Modul***: **Modulbauweise** *f* ARCH & TRAGW modular building system; **in Modulbauweise** *phr* ARCH & TRAGW modular; **Modulordnung** *f* ARCH & TRAGW modular coordination; **Modulrasterebene** *f* ARCH & TRAGW modular plane; **Modulspiel** *nt* ARCH & TRAGW modular gap

**Mohr**: **~scher Kreis** *m* INFR & ENTW Mohr's circle

**Mole** *f* ARCH & TRAGW jetty, INFR & ENTW jetty, pier

**Molekular-** *in cpds* INFR & ENTW, WERKSTOFF molecular; **Molekularanalyse** *f* WERKSTOFF molecular analysis; **Molekularkräfte** *f pl* INFR & ENTW, WERKSTOFF molecular forces; **Molekularstruktur** *f* INFR & ENTW, WERKSTOFF molecular structure

**Moment** *nt* ARCH & TRAGW moment; **Momentenausgleichsverfahren** *nt* ARCH & TRAGW, INFR & ENTW Cross method, method of moment distribution, moment distribution method; **Momentenfläche** *f* ARCH & TRAGW, INFR & ENTW moment area; **Momentengleichgewicht** *nt* ARCH & TRAGW, INFR & ENTW moment equilibrium; **Momentennullpunkt** *m* ARCH & TRAGW, INFR & ENTW zero point of moment; **Momentenpunkt** *m* ARCH & TRAGW moment pole

**Mönch** *m* NATURSTEIN convex tile, overtile, WERKSTOFF convex tile, *Deckziegel* overtile; **Mönch-Nonne-Ziegeldeckung** *f* ARCH & TRAGW, NATURSTEIN mission tiling

**Monolit** *m* ARCH & TRAGW monolith

**monolithisch**: **~e Bauweise** *f* ARCH & TRAGW monolithic construction method; **~e Verbindung** *f* ARCH & TRAGW, BETON, INFR & ENTW, WERKSTOFF monolithic connection

**Montage** *f* ARCH & TRAGW installation, mounting, BAUMASCHIN mounting, HEIZ & BELÜFT assembly, mounting, HOLZ mounting, INFR & ENTW erection, installation, mounting, STAHL mounting; **Montage-**

**ablauf** *m* BAURECHT sequence of assembly; **Montageanker** *m* WERKSTOFF mounting anchor; **Montagearbeiten** *f pl* ARCH & TRAGW, INFR & ENTW installation work, mounting work; **Montagebetonbalken** *m* ARCH & TRAGW, BETON precast concrete beam; **Montageelement** *nt* ARCH & TRAGW structural element, unit, INFR & ENTW unit, WERKSTOFF constructional unit, prefabricated building member; **Montagegrube** *f* ARCH & TRAGW maintenance pit; **Montageplan** *m* BAURECHT, BETON, STAHL assembly schedule; **Montageplatte** *f* WERKSTOFF mounting plate; **Montageschablone** *f* BESCHLÄGE mounting template; **Montagespannung** *f* ARCH & TRAGW, BETON, INFR & ENTW, STAHL temporary stress

**Monteur** *m* ARCH & TRAGW, HEIZ & BELÜFT, HOLZ, STAHL fitter

**montieren** *vt* ARCH & TRAGW fit, install, mount, set up, HEIZ & BELÜFT, HOLZ *aus Einzelteilen*, STAHL assemble

**Moor** *nt* INFR & ENTW fen, marsh; **Moorboden** *m* INFR & ENTW marshland, moor

**Moosgummiabdichtung** *f* DÄMMUNG, WERKSTOFF cellular rubber sealing, sponge rubber sealing, sponge sealing

**Moräne** *f* INFR & ENTW moraine; **Moränenfilterschicht** *f* UMWELT *Deponie* morainic filter layer

**Morast** *m* INFR & ENTW fen, marsh

**Morphologie** *f* INFR & ENTW geomorphology

**Mörtel** *m* NATURSTEIN, WERKSTOFF mortar, plaster; ~ **anmachen** *phr* NATURSTEIN prepare mortar; **mit ~ ausgießen** *phr* NATURSTEIN seal with mortar; **Mörtelabfälle** *m pl* NATURSTEIN mortar droppings; **Mörtelanalyse** *f* NATURSTEIN mortar analysis; **Mörtelbett** *nt* NATURSTEIN mortar bed, underbed; **Mörteleinpressung** *f* NATURSTEIN mortar intrusion; **Mörtelfuge** *f* BETON, NATURSTEIN abreuvoir; **Mörtelgruppe** *f* WERKSTOFF mortar class

**mörtellos**: **~e Fuge** *f* HEIZ & BELÜFT, NATURSTEIN dry joint

***Mörtel***: **Mörtelmischen** *nt* NATURSTEIN mortar mixing; **Mörtelmischmaschine** *f* BAUMASCHIN mortar mixer; **Mörtelpigment** *nt* NATURSTEIN mortar pigment; **Mörtelverhältnisse** *nt pl* NATURSTEIN mortar ratios; **Mörtelzusatz** *m* NATURSTEIN, WERKSTOFF mortar additive

**Mosaik** *nt* ARCH & TRAGW, NATURSTEIN mosaic; **Mosaikbelag** *m* ARCH & TRAGW, NATURSTEIN mosaic flooring; **Mosaikfußboden** *m* ARCH & TRAGW, NATURSTEIN mosaic flooring; **Mosaikpflaster** *nt* INFR & ENTW, NATURSTEIN mosaic paving; **Mosaikstein** *m* NATURSTEIN tessera

**Motor** *m* BAUMASCHIN motor; **Motorschürfkübel** *m* BAUMASCHIN tractor scraper; **Motorstraßenhobel** *m* BAUMASCHIN motor grader; **Motorwalze** *f* BAUMASCHIN motor roller

**Mp** *abbr* (*Megapond*) ARCH & TRAGW, INFR & ENTW, WERKSTOFF Mp (*megapond*)

**MSR-Einrichtungen** *f pl* (*Meß-, Steuerungs- und Regeleinrichtungen*) ELEKTR, HEIZ & BELÜFT measuring and control equipment

**Muffe** *f* ABWASSER bell, sleeve, socket, ELEKTR socket, HEIZ & BELÜFT *pipe* coupler, sleeve, socket, WERKSTOFF bell

**muffenlos** *adj* HEIZ & BELÜFT sleeveless

***Muffe***: **Muffenrohr** *nt* ABWASSER, HEIZ & BELÜFT socket pipe; **Muffenrohrverbindung** *f* ABWASSER spigot joint; **Muffenverbindung** *f* ABWASSER spigot and socket joint, ARCH & TRAGW tailpiece, WERKSTOFF bell and spigot joint; **Muffenverbindungsrohre** *nt pl* ABWASSER spigot and socket joint pipes; **Muffenverschraubung** *f* ABWASSER, HEIZ & BELÜFT threaded sleeve joint

**Mulde** *f* NATURSTEIN *in einem Ziegel* frog

**muldenförmig** *adj* ARCH & TRAGW, INFR & ENTW shallow

***Mulde***: **Muldenkipper** *m* BAUMASCHIN skip lorry (*BrE*), skip truck (*AmE*), dumper; **Muldentransporter** *m* BAUMASCHIN skip lorry (*BrE*), skip truck (*AmE*)

**Müll** *m* BAURECHT, INFR & ENTW, UMWELT garbage (*AmE*), rubbish (*BrE*), waste; **Müllabfuhr** *f* BAURECHT, UMWELT garbage collection (*AmE*), garbage disposal (*AmE*), refuse collection service, rubbish collection (*BrE*), rubbish disposal (*BrE*), waste collection, waste disposal; **Müllabfuhrwagen** *m* BAUMASCHIN, UMWELT garbage truck (*AmE*), refuse collection vehicle; **Müllabladen** *nt* INFR & ENTW, UMWELT dumping; **Müllabladeplatz** *m* INFR & ENTW, UMWELT dump ground, dump site, dumping ground; **Müllabwurfschacht** *m* ARCH & TRAGW, UMWELT garbage chute (*AmE*), rubbish chute (*BrE*); **Müllanfall** *m* UMWELT waste formation, waste production; **Müllaufbereitung** *f* BAURECHT, UMWELT conditioning of waste, waste treatment; **Müllaufbereitungsanlage** *f* BAURECHT, INFR & ENTW waste treatment plant; **Müllbehälter** *m* UMWELT garbage container (*AmE*), rubbish container (*BrE*); **Müllbeseitigung** *f* BAURECHT, UMWELT waste disposal; **Müllbrennstoff** *m* UMWELT refuse-derived fuel (*RDF*); **Müllbunker** *m* UMWELT receiving bunker; **Müllcontainer** *m* UMWELT roll-out container; **Mülldeponie** *f* INFR & ENTW, UMWELT dump ground, dump site, dumping ground, dumping site, landfill, sanitary landfill, waste tip (*BrE*); **Mülleimer** *m* UMWELT garbage can (*AmE*), rubbish bin (*BrE*); **Mülleinfülltrichter** *m* UMWELT loading hopper; **Müllfahrzeug** *nt* BAUMASCHIN, UMWELT refuse collection vehicle; **Müllhalde** *f* INFR & ENTW, UMWELT dump ground, dump site, dumping ground, tip (*BrE*), waste dump; **Müllkippe** *f* UMWELT waste dump, waste tip (*BrE*), refuse disposal site; **Müllpresse** *f* BAUMASCHIN, UMWELT garbage press (*AmE*), rubbish press (*BrE*); **Müllsack** *m* UMWELT refuse sack; **Müllsammelfahrzeug** *nt* BAUMASCHIN, UMWELT garbage truck (*AmE*), refuse collection vehicle; **Müllsammlung** *f* UMWELT garbage collection (*AmE*), garbage disposal (*AmE*), rubbish collection (*BrE*), rubbish disposal (*BrE*); **Müllschacht** *m* ARCH & TRAGW, INFR & ENTW waste chute; **Müllschlacke** *f* UMWELT, WERKSTOFF clinker, slag; **Müllschlucker** *m* ARCH & TRAGW, UMWELT garbage chute (*AmE*), rubbish chute (*BrE*); **Müllsortierungsanlage** *f* UMWELT refuse separation plant; **Mülltonne** *f* UMWELT dustbin (*BrE*), garbage can (*AmE*), garbage container (*AmE*), rubbish container (*BrE*); **Müllverbrennung** *f* UMWELT refuse incineration; **Müllverbrennungsanlage** *f* (*MVA*) UMWELT refuse incineration plant, waste incineration plant; **Müllverdichter** *m* UMWELT landfill compactor, packer unit; **Müllverdichtung** *f* INFR & ENTW, UMWELT compaction; **Müllwagen** *m* BAUMASCHIN, UMWELT refuse collection vehicle; **Müllwiedergewinnungsanlage** *f* BAURECHT, INFR & ENTW waste recycling plant; **Müllzerkleinerer** *m* UMWELT refuse grinder

**multidiziplinarisch**: **~es Team** *nt* ARCH & TRAGW, INFR & ENTW multidisciplinary team

**Mündung** *f* INFR & ENTW mouth

**Muster** *nt* ARCH & TRAGW master, WERKSTOFF pattern, specimen, sample; **Musterbau** *m* ARCH & TRAGW prototype building; **Musterküche** *f* ARCH & TRAGW demonstration kitchen; **Musterschlüssel** *m* BESCHLÄGE sample key

**Mutter** *f* BESCHLÄGE nut; **~ und Bolzen** *m* BESCHLÄGE bolt and nut;**Mutterboden** *m* ARCH & TRAGW, INFR & ENTW humus, topsoil; **Mutterbodenabtrag** *m* ARCH & TRAGW, INFR & ENTW topsoil stripping; **Muttergestein** *nt* INFR & ENTW parent rock; **Mutterpause** *f* ARCH & TRAGW blueprint; **Mutterschraube** *f* BESCHLÄGE bolt and nut

**MVA** *abbr* (*Müllverbrennungsanlage*) UMWELT garbage incineration plant (*AmE*), refuse incineration plant, waste incineration plant

**MWSt** *abbr* (*Mehrwertsteuer*) BAURECHT VAT (*value-added tax*)

# N

**N** *abbr* (*Newton*) ARCH & TRAGW, INFR & ENTW, WERKSTOFF N (*newton*)
**NACCS** *abbr* (*Nationaler Akkreditationsrat für Normungsausschüsse*) BAURECHT NACCB (*BrE*) (*National Accreditation Council for Certification Bodies*)
**Nacharbeiten** *nt* BETON touching-up; **~ der Kanten** BETON touching-up of edges
**Nachbarschaft** *f* ARCH & TRAGW neighborhood (*AmE*), neighbourhood (*BrE*)
**Nachbearbeitung** *f* OBERFLÄCHE finishing
**nachbehandeln** *vt* BETON cure
**Nachbehandeln** *nt* BETON curing
**Nachbehandlung** *f* BAURECHT aftertreatment, BETON aftertreatment, curing, HOLZ aftertreatment, INFR & ENTW retreatment, OBERFLÄCHE, STAHL aftertreatment
**Nachbrennkammer** *f* (*SCC*) UMWELT afterburner chamber, secondary combustion chamber (*SCC*)
**Nachfrage** *f* BAURECHT inquiry
**nachgeben** *vi* ARCH & TRAGW, WERKSTOFF give way
**Nachinstallation** *f* ARCH & TRAGW, HEIZ & BELÜFT subsequent installation
**Nachklärbecken** *nt* INFR & ENTW final settling tank, UMWELT secondary sedimentation basin, secondary settling tank
**Nachmischer** *m* BAUMASCHIN *Beton* truck agitator
**nachrüsten** *vt* UMWELT retrofit
**Nachrüsten** *nt* UMWELT retrofit
**Nachschlagtabelle** *f* ARCH & TRAGW reference table
**Nachsorge** *f* UMWELT *einer Deponie* monitoring after site closure
**Nachspannen** *nt* BETON post tensioning
**Nacht**: **Nachtabsenkung** *f* HEIZ & BELÜFT *Heizung* night economy, night economy feature; **Nachtisolierung** *f* DÄMMUNG night insulation
**Nachtrag** *m* ARCH & TRAGW annex (*AmE*), annexe (*BrE*), BAURECHT *Dokument* addition, annexe (*BrE*), annex (*AmE*), INFR & ENTW annex (*AmE*), annexe (*BrE*)
**nachträglich** *adj* ARCH & TRAGW, BAURECHT additional; **~ verstellbar** *adj* ELEKTR, HEIZ & BELÜFT subsequently adjustable
**Nachtragsangebot** *nt* BAURECHT revised tender
***Nacht***: **Nachtspeicherheizkörper** *m* ELEKTR, HEIZ & BELÜFT off-peak electricity heaters; **Nachtspeicherheizung** *f* ELEKTR off-peak electricity heating, HEIZ & BELÜFT night storage heating, off-peak electricity heating; **Nachtstrom** *m* ELEKTR night current; **Nachtstromspeicherheizung** *f* ELEKTR off-peak electricity heating, HEIZ & BELÜFT night storage heating, off-peak electricity heating; **Nachtstromversorgung** *f* ELEKTR, HEIZ & BELÜFT off-peak electricity supply; **Nachtstromzähler** *m* ELEKTR, HEIZ & BELÜFT off-peak electricity supply meter
**Nachunternehmer** *m* BAURECHT subcontractor
**nachverdichten** *vt* INFR & ENTW *Boden* recompact
**Nachvermessung** *f* ARCH & TRAGW resurvey
**Nachweis** *m* BAURECHT certificate, evidence, proof
**nachweisen** *vt* BAURECHT certify, prove
**Nadel** *f* BESCHLÄGE pin, STAHL needle, pin, WERKSTOFF pin; **Nadelfilz** *m* ARCH & TRAGW, WERKSTOFF needle felt; **Nadelholz** *nt* HOLZ, WERKSTOFF coniferous wood, softwood; **Nadellochkorrosion** *f* WERKSTOFF pinhole corrosion; **Nadelplatte** *f* HOLZ, STAHL punched metal plate fastener; **Nadelvliesteppichboden** *m* ARCH & TRAGW, BESCHLÄGE tufted floor covering
**Nagel** *m* BAUMASCHIN, HOLZ, WERKSTOFF nail; **Nagelabstand** *m* HOLZ, WERKSTOFF nail distance; **Nagelbinder** *m* ARCH & TRAGW, HOLZ nailed framework, nailed truss, plank truss; **Nagelbohrer** *m* BAUMASCHIN gimlet; **Nageldachbinder** *m* ARCH & TRAGW nail roof truss; **Nageldurchmesser** *m* HOLZ, WERKSTOFF nail diameter; **Nagelhaftlänge** *f* ARCH & TRAGW grip length of nail; **Nagelklaue** *f* BESCHLÄGE, HOLZ nail claw; **Nagelkopf** *m* WERKSTOFF nail head; **Nagellänge** *f* WERKSTOFF nail length
**nageln** *vt* BAUMASCHIN, HOLZ, WERKSTOFF nail
***Nagel***: **Nagelspitze** *f* WERKSTOFF nail point; **Nageltreiber** *m* BAUMASCHIN nail punch; **Nagelverbindung** *f* HOLZ nailed connection; **Nagelzieheisen** *nt* BESCHLÄGE, HOLZ nail claw; **Nagelzieher** *m* BAUMASCHIN nail puller, nail extractor
**Nahansicht** *f* ARCH & TRAGW close-up
**Näherung**: **Näherungsannahme** *f* ARCH & TRAGW, INFR & ENTW approximate assumption; **Näherungsverfahren** *nt* ARCH & TRAGW, INFR & ENTW approximation method
**näherungsweise** *adj* ARCH & TRAGW, INFR & ENTW approximate
**Naht** *f* STAHL seam
**nahtlos**: **~es Rohr** *nt* HEIZ & BELÜFT seamless pipe
**NAMAS** *abbr* (*Nationaler Messungs-Akkreditationsservice*) BAURECHT NAMAS (*BrE*) (*National Measurement Accreditation Service*)
**Nase** *f* ARCH & TRAGW curved slab; **Nasenbogen** *m* ARCH & TRAGW cusped arch; **Nasenverzierung** *f* ARCH & TRAGW *gothische Architektur* cuspidation, cusping
**Naß-**: **Naß-auf-Naßbeschichtung** *f* OBERFLÄCHE wet-on-wet coating; **Naß-auf-Naß-Methode** *f* BETON, OBERFLÄCHE wet-on-wet method
**naß**: **nasser Boden** *m* INFR & ENTW wet soil
***Naß-***: **Naßdach** *nt* ARCH & TRAGW wet roof; **Naßfäule** *f* ARCH & TRAGW *Holz*, HOLZ, WERKSTOFF wet rot; **Naßpresse** *f* BAUMASCHIN stream machine
***naß***: **nasser Sand** *m* INFR & ENTW wet sand
***Naß-***: **Naßschlamm** *m* UMWELT slurry, liquid sludge; **Naßsieben** *nt* INFR & ENTW wet screening; **Naßsiebung** *f* INFR & ENTW wet screening; **Naßzelle** *f* ABWASSER, ARCH & TRAGW plumbing unit
**National**: **~er Akkreditationsrat** *m* **für Normungsausschüsse** (*NACCS*) BAURECHT National Accreditation Council for Certification Bodies (*BrE*) (*NACCB*); **~er Messungs-Akkredita-**

tionsservice *m* (*NAMAS*) BAURECHT National Measurement Accreditation Service (*BrE*) (*NAMAS*)
**NATLAS** *abbr* (*Staatliche Anstalt für die Beglaubigung von Laboruntersuchungen*) BAURECHT NALTLAS (*BrE*) (*National Laboratory Testing Accreditation Service*)
**NATM®** *abbr* (*neue österreichische Tunnelbauweise*) INFR & ENTW NATM® (*new Austrian tunnelling method*)
**Natur** *f* INFR & ENTW nature; **Naturasphalt** *m* INFR & ENTW, NATURSTEIN mineral pitch; **Naturbaustein** *m* NATURSTEIN building stone
**naturbelassen** *adj* HOLZ natural finish
***Natur***: **Naturboden** *m* INFR & ENTW natural soil; **Naturbordstein** *m* WERKSTOFF natural curb (*BrE*), natural kerb (*AmE*), natural curbstone (*BrE*), natural kerbstone (*AmE*); **Natureckstein** *m* NATURSTEIN *beidseitig sichtbar* perpend stone
**naturfarben** *adj* WERKSTOFF natural color (*AmE*), natural colour (*BrE*)
***Natur***: **Naturholzfarbe** *f* OBERFLÄCHE oleoresinous paint
**natürlich** *adj* ARCH & TRAGW, BESCHLÄGE, BETON, HEIZ & BELÜFT, UMWELT, WERKSTOFF natural; **~e Abdichtung** *f* UMWELT *einer Deponie* natural lining; **~e Belüftung** *f* INFR & ENTW natural ventilation; **~e Böschung** *f* ARCH & TRAGW natural slope; **~er Böschungswinkel** *m* ARCH & TRAGW angle of repose, INFR & ENTW natural slope; **~e Farbe** *f* WERKSTOFF natural color (*AmE*), natural colour (*BrE*); **~e Filtrierung** *f* INFR & ENTW natural filtration; **~es Gefälle** *nt* INFR & ENTW natural fall; **~e Konvektion** *f* HEIZ & BELÜFT natural convection; **~e Konvektionsschleife** *f* HEIZ & BELÜFT natural convective loop; **~e Puzzolanerde** *f* BETON, WERKSTOFF natural pozzolana; **~e Reinigung** *f* INFR & ENTW natural purification; **~e Umwelt** *f* UMWELT natural environment; **~er Wasserinhalt** *m* INFR & ENTW, WERKSTOFF natural water content
***Natur***: **Naturpflaster** *nt* NATURSTEIN natural pavement; **Natursand** *m* WERKSTOFF natural sand
**natursauer**: **natursaurer See** *m* UMWELT naturally acid lake
***Natur***: **Naturschiefer** *m* WERKSTOFF natural slate; **Naturstein** *m* INFR & ENTW, NATURSTEIN, WERKSTOFF ashlar, natural stone; **Naturstein** *m* **zur Wandverkleidung** NATURSTEIN facing stone; **Natursteinmauerwerk** *nt* ARCH & TRAGW, NATURSTEIN ashlar masonry, ashlaring, natural stone masonry; **Natursteinplatte** *f* INFR & ENTW, NATURSTEIN flag, flagstone, quarry tile; **Natursteinsturz** *m* NATURSTEIN architrave; **Naturwerkstein** *m* INFR & ENTW, NATURSTEIN ashlar, hewn stone; **Naturzement** *m* INFR & ENTW, WERKSTOFF natural cement
**Neben-** *in cpds* ARCH & TRAGW auxiliary, BAURECHT supplementary; **Nebengebäude** *nt* ARCH & TRAGW, INFR & ENTW annex (*AmE*), annexe (*BrE*), auxiliary building; **Nebengleis** *nt* INFR & ENTW spur track; **Nebenleistung** *f* BAURECHT supplementary work; **Nebenluft** *f* HEIZ & BELÜFT secondary air; **Nebenprodukt** *nt* UMWELT by-product; **Nebenraum** *m* ARCH & TRAGW side room; **Nebensäule** *f* ARCH & TRAGW subsidiary shaft; **Nebenstraße** *f* ARCH & TRAGW, BAURECHT, INFR & ENTW auxiliary lane, byroad, byway; **Nebenzugang** *m* ARCH & TRAGW, INFR & ENTW secondary access
**negativ**: **~es Biegemoment** *nt* HOLZ, WERKSTOFF negative bending moment; **~er Druck** *m* ARCH & TRAGW negative pressure; **~er Immobilienwert** *m* BAURECHT negative equity
**neigen**: **sich neigen** *v refl* ABWASSER lean
**Neigung** *f* ARCH & TRAGW, INFR & ENTW falling gradient, inclination, incline, slant, slope, tilt, UMWELT pitch; **Neigungsmesser** *m* ARCH & TRAGW batter level, INFR & ENTW clinometer, slope level; **Neigungswinkel** *m* ARCH & TRAGW, INFR & ENTW angle of inclination, angle of slope, pitch
**NE-Metall** *nt* WERKSTOFF nonferrous metal
**Nenn-** *in cpds* ARCH & TRAGW, ELEKTR, HEIZ & BELÜFT nominal, basic; **Nenndruck** *m* ABWASSER, HEIZ & BELÜFT nominal pressure; **Nennmaß** *nt* ARCH & TRAGW basic dimension, nominal dimension, nominal size; **Nennquerschnitt** *m* ARCH & TRAGW nominal cross section; **Nennschweißstrom** *m* ELEKTR rated welding current; **Nennspannung** *f* ARCH & TRAGW, HOLZ nominal stress; **Nennstrom** *m* ELEKTR nominal current; **Nennweite** *f* ABWASSER bore, ARCH & TRAGW nominal width, HEIZ & BELÜFT, INFR & ENTW bore
**Neopren** *nt* ARCH & TRAGW, WERKSTOFF neoprene; **Neoprendichtungsband** *nt* WERKSTOFF neoprene gasket, neoprene sealing; **Neoprendichtungsprofil** *nt* WERKSTOFF neoprene gasket, neoprene sealing; **Neoprenlagerkissen** *nt* ARCH & TRAGW neoprene bearing pad
**Nest** *nt* BETON void
**netto** *adj* ARCH & TRAGW, INFR & ENTW, WERKSTOFF net
**Netto-** *in cpds* ARCH & TRAGW, INFR & ENTW, WERKSTOFF net; **Nettofläche** *f* ARCH & TRAGW, INFR & ENTW net area
**Netz** *nt* ABWASSER network, ARCH & TRAGW *von Rohren, Kabeln* system, BAURECHT network, ELEKTR network, mains, HEIZ & BELÜFT network, INFR & ENTW network, mains, WERKSTOFF *Stromverteilung, Rohrleitungen* network; **Netzbetrieb** *nt* ELEKTR mains power supply; **Netzbewehrung** *f* BETON, INFR & ENTW mat reinforcement; **Netzflechtwerk** *nt* STAHL *Draht* netting; **Netzgerät** *nt* ELEKTR power supply unit; **Netzgewölbe** *nt* ARCH & TRAGW reticulated vault; **Netzmittel** *nt* OBERFLÄCHE surface-active agent, WERKSTOFF wetting agent; **Netzplantechnik** *f* **nach CPM** BAURECHT critical path network; **Netzwerk** *nt* ABWASSER network, ARCH & TRAGW tracery, BAURECHT, ELEKTR, HEIZ & BELÜFT, INFR & ENTW, WERKSTOFF network; **Netzwerkdiagramm** *nt* BAURECHT network diagram; **Netzwerkfenster** *nt* ARCH & TRAGW reticulated window; **Netzwerkplanung** *f* INFR & ENTW network planning
**neu** *adj* ARCH & TRAGW new
**Neu-** *in cpds* ARCH & TRAGW new
***neu***: **~e Betriebsanlage** *f* UMWELT new plant; **~e österreichische Tunnelbauweise** *f* (*NATM®*) INFR & ENTW new Austrian tunnelling method (*NATM*)
***Neu-***: **Neuaufnahme** *f* INFR & ENTW *Vermessung* releveling (*AmE*), relevelling (*BrE*); **Neubau** *m* ARCH & TRAGW new building
**neugestalten** *vt* ARCH & TRAGW redevelop
***Neu-***: **Neuordnung** *f* ARCH & TRAGW rearrangement; **Neuprofilieren** *nt* ARCH & TRAGW reprofiling; **Neusilber** *nt* WERKSTOFF nickel silver
**neutral** *adj* ARCH & TRAGW, WERKSTOFF neutral; **~e**

**Achse** *f* ARCH & TRAGW neutral axis; **~e Oberfläche** *f* WERKSTOFF neutral surface
**Neutralisationsmittel** *nt* UMWELT neutralizer, neutralizing agent
**neutralisieren** *vt* UMWELT neutralize
**Neutralisierung** *f* UMWELT neutralization
**Newton** *nt* (*N*) ARCH & TRAGW, INFR & ENTW, WERKSTOFF newton (*N*)
**NF** *abbr* (*Normalformat*) WERKSTOFF standard size
**nichtbindig** *adj* INFR & ENTW *Boden* friable
**nichtdrückend**: **~es Wasser** *nt* ARCH & TRAGW, INFR & ENTW moisture
**Nichteisenmetall** *nt* WERKSTOFF nonferrous metal
**nichtkohäsiv** *adj* INFR & ENTW *Boden* friable
**nichtleitend** *adj* DÄMMUNG, ELEKTR, OBERFLÄCHE insulating, WERKSTOFF insulating, nonconductive
**nichtrostend** *adj* STAHL rustproofed; **~er Stahl** *m* WERKSTOFF stainless steel
**nichttragend** *adj* ARCH & TRAGW nonstructural
**nichtunterkellert**: **~e Fußbodenplatte** *f* ARCH & TRAGW slab-on-grade
**Nickel** *nt* OBERFLÄCHE, WERKSTOFF nickel; **Nickeloxydbeschichtung** *f* WERKSTOFF nickel oxide coating
**Niederdruck** *m* HEIZ & BELÜFT low pressure
**niederdrücken** *vt* ARCH & TRAGW weigh down
**niedergebrannt** *adj* ARCH & TRAGW, BAURECHT destroyed by fire
**niederreißen** *vt* BAURECHT, INFR & ENTW demolish
**Niederschlag** *m* ABWASSER rainfall, stormwater, INFR & ENTW precipitation, rainfall, stormwater, UMWELT *radioaktiv* fallout; **Niederschlagssammler** *m* UMWELT precipitation collector; **Niederschlagselektrode** *f* UMWELT collecting electrode; **Niederschlagsmesser** *m* BAUMASCHIN, UMWELT rain gage (*AmE*), rain gauge (*BrE*); **Niederschlagsmeßgerät** *nt* BAUMASCHIN rainwater collector; **Niederschlagsvorfall** *m* UMWELT precipitation event; **Niederschlagswasserkanal** *m* ABWASSER, INFR & ENTW storm sewer, stormwater sewer
**Niederschraub-** *in cpds* BAUMASCHIN screw-down; **Niederschraubabsperrventil** *nt* BAUMASCHIN screw-down stop valve; **Niederschraubhahn** *m* BAUMASCHIN screw-down cock; **Niederschraubventil** *nt* BAUMASCHIN screw-down valve
**niedrig** *adj* ARCH & TRAGW shallow, *Gebäude* low-rise; **~er Lärmpegel** *m* DÄMMUNG low-level noise
**Niedrigwasserstand** *m* INFR & ENTW low water level
**Nießbrauch** *m* BAURECHT usufruct
**Nießbraucher** *m* BAURECHT usufructuary
**Niet** *m* BAUMASCHIN rivet, BESCHLÄGE, STAHL rivet, riveting pressure; **Nietabstand** *m* STAHL spacing of rivets; **Nietbolzenkette** *f* BESCHLÄGE pin chain; **Nietdruck** *m* STAHL riveting pressure; **Nietdurchmesser** *m* STAHL rivet diameter
**Niete** *f* BAUMASCHIN, BESCHLÄGE, STAHL rivet
**nieten** *vt* BAUMASCHIN, STAHL rivet
**Nieten** *nt* BAUMASCHIN, STAHL riveting
**Nieter** *m* BAUMASCHIN, STAHL riveter
***Niet***: **Niethammer** *m* BAUMASCHIN rivet hammer, riveting hammer; **Nietkopf** *m* BESCHLÄGE, STAHL rivet head; **Nietkopfsetzer** *m* BAUMASCHIN rivet set; **Nietmaschine** *f* BAUMASCHIN riveter, riveting machine, STAHL riveter; **Nietplatte** *f* BESCHLÄGE, STAHL riveted plate; **Nietteilung** *f* STAHL spacing of rivets
**Nietung** *f* BAUMASCHIN, STAHL riveting
***Niet***: **Nietverbindung** *f* BAUMASCHIN, STAHL rivet joint, riveted joint
**Nippel** *m* BAUMASCHIN nipple
**Nipptide** *f* UMWELT neap tide
**Nische** *f* ARCH & TRAGW housing, niche, recess
**Nitrat** *nt* UMWELT, WERKSTOFF nitrate
**Nitrolack** *m* OBERFLÄCHE nitrocellulose lacquer
**Niveau** *nt* ARCH & TRAGW level
**Nivellier-** *in cpds* ARCH & TRAGW, BAUMASCHIN, BESCHLÄGE, INFR & ENTW leveling (*AmE*), levelling (*BrE*)
**Nivellieren** *nt* ARCH & TRAGW, INFR & ENTW leveling (*AmE*), levelling (*BrE*)
***Nivellier-***: **Nivelliergerät** *nt* ARCH & TRAGW, BAUMASCHIN surveyor's level; **Nivellierinstrument** *nt* ARCH & TRAGW, BAUMASCHIN leveling instrument (*AmE*), levelling instrument (*BrE*); **Nivellierkreuz** *nt* INFR & ENTW leveling rod (*AmE*), levelling rod (*BrE*); **Nivellierlatte** *f* BAUMASCHIN, BESCHLÄGE, INFR & ENTW leveling pole (*AmE*), leveling staff (*AmE*), levelling pole (*BrE*), levelling staff (*BrE*), sighting rod; **Nivellierlatte** *f* **mit Anzeige** INFR & ENTW *Vermessung* target leveling staff (*AmE*), target levelling staff (*BrE*), target leveling rod (*AmE*), target levelling rod (*BrE*); **Nivellierpunkt** *m* ARCH & TRAGW, BAUMASCHIN leveling point (*AmE*), levelling point (*BrE*)
**Nivellierung** *f* ARCH & TRAGW leveling (*AmE*), levelling (*BrE*)
***Nivellier-***: **Nivellierwaage** *f* BAUMASCHIN spirit level
**NN** *abbr* (*Normalnull*) INFR & ENTW zero level
**Nonius** *m* BAUMASCHIN vernier
**Noppe** *f* WERKSTOFF nap
**Norm** *f* BAURECHT standard; **unter der Norm** *phr* ARCH & TRAGW *minderwertig* substandard
**normal** *adj* GEN normal, standard
**Normal-** *in cpds* ABWASSER, BAURECHT, BESCHLÄGE, BETON, HEIZ & BELÜFT, INFR & ENTW, UMWELT, WERKSTOFF normal, standard
***normal***: **~e Springzeitflut** *f* UMWELT high-water ordinary spring tide (*HWOST*), HWOST (*high-water ordinary spring tide*)
***Normal-***: **Normalanschlag** *m* BESCHLÄGE standard mounting; **Normalbeton** *m* BETON normal concrete; **Normalformat** *nt* (*NF*) ARCH & TRAGW, WERKSTOFF normal format, standard size; **Normalgewichtsbeton** *m* BETON, INFR & ENTW, WERKSTOFF normal-weight concrete; **Normalluft** *f* HEIZ & BELÜFT, UMWELT standard air; **Normalnull** *nt* (*NN*) INFR & ENTW mean sea level, zero level; **Normalzement** *m* BETON ordinary cement
***Norm***: **Normfarbtafel** *f* ARCH & TRAGW, INFR & ENTW International Commission on Illumination diagram (*CIE diagram*)
**Normung** *f* BAURECHT standardization; **Normungsvorschrift** *f* BAURECHT standard specification
***Norm***: **Normversuch** *m* BAURECHT, INFR & ENTW standard test
**Not**: **Notausgang** *m* BAURECHT, INFR & ENTW emergency exit, fire exit; **Notbeleuchtung** *f* BAURECHT, ELEKTR emergency lighting, escape lighting, safety lighting; **Notfall** *m* BAURECHT emergency; **Not-**

**generator** *m* ELEKTR emergency generator; **Notschlüssel** *m* BAURECHT emergency key; **Notschlüsselfunktion** *f* BAURECHT emergency key function; **Notwasserversorgung** *f* INFR & ENTW emergency water supply

**NPL** *abbr* (*Staatliches Physikalisches Labor*) BAURECHT NPL (*BrE*) (*National Physical Laboratory*)

**Nuklearschutzbunker** *m* ARCH & TRAGW, BETON fallout shelter

**Null-** *in cpds* ARCH & TRAGW neutral, zero, WERKSTOFF neutral; **Nullachse** *f* ARCH & TRAGW neutral axis; **Nulleiter** *m* ELEKTR neutral wire; **Nullinie** *f* ARCH & TRAGW, INFR & ENTW zero line; **Nullstab** *m* ARCH & TRAGW *Fachwerk* unstrained member

**numerisch**: **~e Analyse** *f* ARCH & TRAGW, INFR & ENTW numerical analysis; **~e Methode** *f* ARCH & TRAGW, INFR & ENTW numerical method; **~es Modell** *nt* ARCH & TRAGW, INFR & ENTW numerical model

**Nut** *f* ARCH & TRAGW groove, slot, housing, notch, rabbet, BAUMASCHIN rabbet, BESCHLÄGE rabbet, flute, quirk

**Nuten** *nt* WERKSTOFF fluting

***Nut***: **Nut- und Federbrett** *nt* HOLZ matchboard; **Nut- und Federverbindung** *f* HOLZ tongue-and-groove joint; **Nut- und Federverspundung** *f* HOLZ grooved and tongued joint; **Nuthobel** *m* BAUMASCHIN grooving plane, plough (*BrE*), plow (*AmE*), plough plane (*BrE*), plow plane (*AmE*), rabbet plane, rebate plane; **Nutverbindung** *f* HOLZ matched joint

**nutzbar**: **~es Gefälle** *nt* UMWELT effective head

**nutzen** *vt* BAURECHT *Gebäude* occupy

**Nutzer** *m* BAURECHT user

**Nutz-** *in cpds* ARCH & TRAGW, INFR & DES effective; **Nutzfläche** *f* ARCH & TRAGW, INFR & ENTW floor space, useful area; **Nutzhöhe** *f* ARCH & TRAGW, INFR & ENTW effective depth; **Nutzholz** *nt* HOLZ lumber (*AmE*), timber (*BrE*); **Nutzlast** *f* ARCH & TRAGW, INFR & ENTW imposed load, live load, superimposed load; **Nutzleistung** *f* ELEKTR net capacity

**Nutzungsänderung** *f* BAURECHT change of use

***Nutz-***: **Nutzwasser** *nt* BAURECHT, UMWELT process water

# O

**oben** *adj* INFR & ENTW overhead; **~ beleuchtet** *adj* INFR & ENTW top-lit

**obenliegend**: **~er Türschließer** *m* BESCHLÄGE overhead door closer

**Ober-**: **Oberbau** *m* ARCH & TRAGW, INFR & ENTW superstructure; **Oberbaumaterial** *nt* INFR & ENTW *Gleisbau* permanent-way equipment; **Oberbekleidung** *f* ARCH & TRAGW overcoat; **Oberboden** *m* INFR & ENTW topsoil; **Oberbodenauftrag** *m* INFR & ENTW topsoil filling

**obere**: **~r Anschluß** *m* STAHL cap flashing; **~s Ende** *nt* ARCH & TRAGW head, top, BETON, INFR & ENTW, OBERFLÄCHE, STAHL, WERKSTOFF top; **~ Etage** *f* ARCH & TRAGW upper floor, upper storey (*BrE*), upper story (*AmE*); **~ Grenze** *f* INFR & ENTW upper limit; **~ Schicht** *f* INFR & ENTW upper layer; **~s Schleusentor** *nt* UMWELT head gate, sluicegate; **~s Speicherbecken** *nt* UMWELT upper storage basin; **~ Tragschicht** *f* INFR & ENTW pavement; **~r Wasserspiegel** *m* ARCH & TRAGW, INFR & ENTW upper water level

**Oberfläche** *f* ARCH & TRAGW, INFR & ENTW, OBERFLÄCHE area, finish, surface; **die ~ bearbeiten** *phr* OBERFLÄCHE finish; **die ~ erneuern** *phr* ARCH & TRAGW reface; **Oberflächenabfluß** *m* ARCH & TRAGW runoff; **Oberflächenabschluß** *m* WERKSTOFF blinding

**oberflächenaktiv**: **~er Stoff** *m* OBERFLÄCHE, WERKSTOFF surface-acting agent

***Oberfläche***: **Oberflächenbearbeitung** *f* OBERFLÄCHE finishing work, surface finishing; **Oberflächenbehandlung** *f* INFR & ENTW *Straße* surface dressing, OBERFLÄCHE surface treatment; **Oberflächenbelag** *m* OBERFLÄCHE finish; **Oberflächenbereich** *m* UMWELT *Erde, Wasser* surface area; **Oberflächenbeschaffenheit** *f* OBERFLÄCHE finish; **Oberflächeneigenschaften** *f pl* ARCH & TRAGW, BETON, OBERFLÄCHE surface characteristics; **Oberflächenentwässerung** *f* INFR & ENTW surface drainage; **Oberflächenerder** *m* ELEKTR surface earthing electrode (*BrE*), surface grounding electrode (*AmE*); **Oberflächenerdung** *f* ELEKTR surface earthing (*BrE*), surface grounding (*AmE*); **Oberflächenerosion** *f* INFR & ENTW surface erosion; **Oberflächengestaltung** *f* OBERFLÄCHE finishing work; **Oberflächenmethode** *f* UMWELT *Ablagerungstechnik* surface method; **Oberflächenrauheit** *f* OBERFLÄCHE surface roughness; **Oberflächenreibung** *f* INFR & ENTW, WERKSTOFF surface friction; **Oberflächenrost** *m* OBERFLÄCHE surface rust; **Oberflächenschutz** *m* OBERFLÄCHE surface protection; **Oberflächenspannung** *f* OBERFLÄCHE, WERKSTOFF surface tension; **Oberflächentemperatur** *f* HEIZ & BELÜFT, OBERFLÄCHE surface temperature; **Oberflächenvermarkung** *f* OBERFLÄCHE *Vermessung* surface demarcation; **Oberflächenversiegelung** *f* UMWELT *einer Deponie* final cover; **Oberflächenvorbehandlung** *f* OBERFLÄCHE surface preparation; **Oberflächenwasser** *nt* ABWASSER, INFR & ENTW, OBERFLÄCHE, UMWELT stormwater, surface water; **Oberflächenwasserkanal** *m* ABWASSER, INFR & ENTW storm sewer, stormwater sewer; **Oberflächenwiderstand** *m* DÄMMUNG surface resistance

**oberflächig**: **~e Faserrichtung** *f* WERKSTOFF face grain direction

***Ober-***: **Obergeschoß** *nt* ARCH & TRAGW upper floor, upper storey (*BrE*), upper story (*AmE*); **Obergurt** *m* ARCH & TRAGW, BETON, INFR & ENTW, STAHL, WERKSTOFF top boom, top chord, top flange, upper boom

**oberhalb** *adv* INFR & ENTW upstream

**Oberholm** *m* ARCH & TRAGW head beam

**oberirdisch** *adj* ARCH & TRAGW, BETON above-grade, above-ground

***Ober-***: **Oberkante** *f* ARCH & TRAGW top edge, upper edge; **Oberkuppel** *f* ARCH & TRAGW light cupola; **Oberleitung** *f* INFR & ENTW overhead line; **Oberlicht** *nt* ARCH & TRAGW transom light, overhead light, skylight, *halbrundes Fenster über einer Tür* fanlight; **Oberputz** *m* NATURSTEIN final coat, final rendering; **Oberschwelle** *f* ARCH & TRAGW head piece, head plate, runner; **Oberteil** *nt* ARCH & TRAGW, BETON, INFR & ENTW, OBERFLÄCHE, STAHL, WERKSTOFF top; **Obertor** *nt* UMWELT head gate; **Oberwasser** *nt* UMWELT upstream head; **Oberwasserstand** *m* INFR & ENTW upstream water level

**Ochsenauge** *nt* WERKSTOFF bull's eye glass

**Odometer** *nt* BAUMASCHIN, INFR & ENTW, WERKSTOFF mileometer, odometer

**Ofen** *m* HEIZ & BELÜFT furnace, stove; **Ofenrohr** *nt* HEIZ & BELÜFT stove pipe

**offen** *adj* ARCH & TRAGW open; **~er Abzugsgraben** *m* UMWELT open drain, open sewer; **~e Feuerstelle** *f* ARCH & TRAGW fireplace; **~es Gerinne** *nt* ABWASSER, INFR & ENTW open channel; **~e Gründung** *f* INFR & ENTW open foundation method; **~er Senkkasten** *m* BETON, INFR & ENTW open caisson; **~e Spindeltreppe** *f* ARCH & TRAGW open newel stairs; **~er Wasserkanal** *m* INFR & ENTW water channel

**öffentlich**: **~es Ausschreiben** *nt* ARCH & TRAGW, BAURECHT, INFR & ENTW open tender; **~e Bauarbeiten** *f pl* ARCH & TRAGW, BAURECHT, INFR & ENTW public works; **~e Bedürfnisanstalt** *f* ARCH & TRAGW, INFR & ENTW public convenience; **~e Fläche** *f* BAURECHT, INFR & ENTW public area; **~es Gebäude** *nt* ARCH & TRAGW, BAURECHT, INFR & ENTW public building; **~e Straße** *f* INFR & ENTW public road; **~e Wasserversorgung** *f* ABWASSER, BAURECHT, INFR & ENTW public water supply; **~er Weg** *m* BAURECHT, INFR & ENTW public way

**Öffentlichkeit** *f* BAURECHT general public

**öffentlich-rechtlich** *adj* BAURECHT under public law

**Offerte** *f* ARCH & TRAGW, BAURECHT bid, offer, tender, INFR & ENTW tender

**Öffnung** *f* ARCH & TRAGW aperture, opening; **Öffnungsbreite** *f* ARCH & TRAGW opening width; **Öffnungsmaß** *nt* BESCHLÄGE *Tür* opening angle; **Öffnungstemperatur** *f* ELEKTR *Ansprechtemperatur*

*eines Sprinklers* response temperature; **Öffnungswange** *f* ARCH & TRAGW *Treppe* external string, face string; **Öffnungswinkel** *m* ARCH & TRAGW *Tür* opening angle

**Off-shore-** *in cpds* INFR & ENTW offshore; **Off-shore-Bohren** *nt* INFR & ENTW offshore drilling; **Off-shore-Konstruktion** *f* BETON, INFR & ENTW, STAHL offshore structure; **Off-shore-Rohrleitung** *f* INFR & ENTW, STAHL offshore pipeline

**Ohm** *nt* ELEKTR ohm

**Ökologie** *f* UMWELT ecology

**ökologisch**: **~es Gleichgewicht** *nt* UMWELT ecological balance; **~er Zusammenbruch** *m* UMWELT ecological disaster

**Ökosystem** *nt* UMWELT ecosystem

**Öl** *nt* ABWASSER, BAUMASCHIN, BESCHLÄGE, HEIZ & BELÜFT, INFR & ENTW, OBERFLÄCHE, UMWELT, WERKSTOFF oil; **mit ~ tränken** *phr* WERKSTOFF oil; **Ölabfall** *m* UMWELT residual oil, oil waste; **Ölabscheider** *m* ABWASSER, UMWELT oil separator; **Ölabscheidung** *f* UMWELT oil removal, oil separation; **Ölaufbereitungsanlage** *f* UMWELT oil regeneration plant

**ölbeheizt** *adj* HEIZ & BELÜFT fuel-heated, oil-heated

**ölbeständig** *adj* WERKSTOFF oil-resistant, *Heizöl* fuel-resistant

**ölen** *vt* BAUMASCHIN, BESCHLÄGE, OBERFLÄCHE oil

***Öl***: **Ölfarbe** *f* OBERFLÄCHE, WERKSTOFF oil paint, oil-based paint; **Ölfeuerung** *f* HEIZ & BELÜFT oil furnace, oil firing; **Ölfilm** *m* WERKSTOFF oil film

**ölgekühlt** *adj* BAUMASCHIN oil-cooled

**ölhaltig**: **~es Abwasser** *nt* UMWELT oleiferous waste water, oil-containing waste water

***Öl***: **Ölheizung** *f* HEIZ & BELÜFT fuel heating, oil furnace, oil heating

**ölig**: **~es Wasser** *nt* UMWELT black water

***Öl***: **Ölregenerat** *nt* UMWELT recovered oil; **Ölrückstände** *m pl* UMWELT oil waste, residual oil

**ölverschmutzt**: **~es Abwasser** *nt* INFR & ENTW, UMWELT oil-polluted waste water

***Öl***: **Ölverschmutzung des Meeres** *f* UMWELT oil spill; **Ölverschmutzungsnotfall** *m* UMWELT oil pollution emergency

**ölverseucht**: **~e Gewässer** *nt pl* UMWELT oil-contaminated waters

**Omega-Verfahren** *nt* ARCH & TRAGW, BETON, HOLZ, INFR & ENTW, STAHL Omega method

**Opakglas** *nt* WERKSTOFF opaque glass

**opal**: **~e Abdeckhaube** *f* ELEKTR *Leuchte* opal louver (*AmE*), opal louvre (*BrE*)

**Opalglas** *nt* WERKSTOFF opal glass

**optimal**: **~e Verdichtung** *f* INFR & ENTW, WERKSTOFF optimum compaction, optimum compression; **~er Wassergehalt** *m* BETON, WERKSTOFF optimum moisture content

**Optimierung** *f* ARCH & TRAGW, BAURECHT, INFR & ENTW optimization

**optisch**: **~er Abscheider** *m* UMWELT optical sorter; **~er Fühler** *m* WERKSTOFF optical probe

**Ordinatenachse** *f* ARCH & TRAGW y-axis

**Ordnung**: **in ~ bringen** *phr* BAURECHT adjust

**organisch** *adj* INFR & ENTW, UMWELT, WERKSTOFF organic; **~er Abfall** *m* UMWELT organic waste; **~er Boden** *m* INFR & ENTW, WERKSTOFF organic soil; **~e Chemie** *f* WERKSTOFF organic chemistry; **~es Lösungsmittel** *nt* OBERFLÄCHE organic solvent; **~es Material** *nt* INFR & ENTW, WERKSTOFF organic matter; **~e Salze** *nt pl* WERKSTOFF organic salts; **~er Stoff** *m* UMWELT organic matter; **~er Ton** *m* INFR & ENTW organic clay; **~e Verbindung** *f* INFR & ENTW, UMWELT, WERKSTOFF organic compound

**Orientierung** *f* ARCH & TRAGW, INFR & ENTW orientation

**Original** *nt* ARCH & TRAGW master, BAURECHT master, original document

**Ornament** *nt* ARCH & TRAGW ornament; **Ornamentglas** *nt* WERKSTOFF figured glass

**Ort** *m* ARCH & TRAGW location; **vor ~** *phr* ARCH & TRAGW on the spot; **Ortbeton** *m* ARCH & TRAGW, BETON cast-in-place concrete, in-situ concrete, site concrete

**ortbetoniert** *adj* BETON poured-in-place

***Ort***: **Ortblech** *nt* STAHL verge flashing; **Ortgang** *m* ARCH & TRAGW verge; **Ortgangbrett** *nt* HOLZ barge board; **Ortgangrinne** *f* ABWASSER verge gutter; **Ortgangverwahrung** *f* STAHL verge flashing

**orthotrop** *adj* ARCH & TRAGW orthotropic

**örtlich**: **~e Bauleitung** *f* INFR & ENTW site supervision; **~e behördliche Überprüfung** *f* BAURECHT local search; **~e Betonprobenentnahme** *f* BETON concrete site sampling; **~e Fernheizung** *f* HEIZ & BELÜFT local district heating; **~e Gegebenheiten** *f pl* ARCH & TRAGW local conditions

***Ort***: **Ortpfahl** *m* INFR & ENTW cast-in-place concrete pile; **Ortsbesichtigung** *f* ARCH & TRAGW, BAURECHT, INFR & ENTW site survey; **Ortschaum** *m* DÄMMUNG in-situ foam

**ortsfest** *adj* ARCH & TRAGW stationary; **~e Emissionsquelle** *f* UMWELT stationary emission source

***Ort***: **Ortsstatut** *nt* BAURECHT by-law; **Ortstein** *m* ARCH & TRAGW margin tile

**ortsveränderlich** *adj* BAUMASCHIN mobile

**Öse** *f* ARCH & TRAGW eye, ELEKTR lug

**Osmose** *f* INFR & ENTW, UMWELT, WERKSTOFF osmosis

**osmotisch**: **~er Druck** *m* INFR & ENTW, UMWELT osmotic pressure

**Ovaltürknopf** *m* BESCHLÄGE oval knob

**Oxidation** *f* OBERFLÄCHE, UMWELT oxidation; **Oxidationsgraben** *m* UMWELT oxidation ditch; **Oxidationsmittel** *nt* UMWELT, WERKSTOFF oxidizing agent; **Oxidationsteich** *m* UMWELT aerated lagoon, oxidation pond, sewage oxidation pond

**oxidierend**: **~e Flamme** *f* STAHL oxidizing flame

**Ozon** *nt* UMWELT ozone; **Ozonloch** *nt* UMWELT hole in the ozone layer; **Ozonschicht** *f* UMWELT ozone layer

# P

**Packlage** *f* BETON hardcore
**Packstoff** *m* UMWELT packaging material
**Packung** *f* BESCHLÄGE stuffing; **Packungsstopfbuchse** *f* ABWASSER gland packing
**Packwerk** *nt* INFR & ENTW, WERKSTOFF enrockment, stone filling
**Palast** *m* ARCH & TRAGW palace
**Palisade** *f* INFR & ENTW stockade; **Palisadenzaun** *m* HOLZ, INFR & ENTW palisade
**Paneel** *nt* HOLZ *Täfelung* panel; **Paneeldecke** *f* ARCH & TRAGW, HOLZ ceiling paneling (*AmE*), ceiling panelling (*BrE*)
**Panik**: **Panikbeleuchtung** *f* BAURECHT, ELEKTR safety lighting; **Panikdrückergarnitur** *f* BAURECHT, BESCHLÄGE panic handle set; **Panikriegelfallenschloß** *nt* BAURECHT, BESCHLÄGE antipanic bolt lock; **Panikverschluß** *m* BAURECHT, BESCHLÄGE panic bolt
**Panorama**: **Panoramafenster** *nt* ARCH & TRAGW picture window
**Panzer-** *in cpds* ELEKTR, STAHL armored (*AmE*), armoured (*BrE*); **Panzerblech** *nt* INFR & ENTW, STAHL, WERKSTOFF armor plate (*AmE*); **Panzerglas** *nt* ARCH & TRAGW, WERKSTOFF bullet-proof glass; **Panzerkabel** *nt* ELEKTR, INFR & ENTW, WERKSTOFF armored cable (*AmE*), armoured cable (*BrE*); **Panzerrohr** *nt* ELEKTR *zum Schutz von Elektrokabeln* armored conduit (*AmE*), armoured conduit (*BrE*)
**Papier** *nt* UMWELT, WERKSTOFF paper; **Papierbahn** *f* WERKSTOFF paper web; **Papierhandtuchspender** *m* ABWASSER paper towel dispenser; **Papierholz** *nt* UMWELT pulpwood
**Pappe** *f* WERKSTOFF paperboard
**parabolisch** *adj* ARCH & TRAGW, INFR & ENTW parabolic
**paraboloid** *adj* ARCH & TRAGW, INFR & ENTW paraboloid
**Paragraph** *m* BAURECHT paragraph
**parallel** *adj* ARCH & TRAGW, HOLZ parallel
**Parallelfachwerk** *nt* ARCH & TRAGW flat truss, HOLZ parallel chord truss
**parallelgurtig**: **~es Fachwerk** *nt* ARCH & TRAGW flat truss
**Parallelträger** *m* ARCH & TRAGW parallel girder
**Parameter** *m* ARCH & TRAGW, INFR & ENTW, WERKSTOFF parameter
**Park**: **Park-and-Ride-Parkplätze** *m pl* (*PR-Parkplätze*) INFR & ENTW park-and-ride lots (*AmE*), park-and-ride sites (*BrE*); **Parkanlage** *f* INFR & ENTW park; **Parkbucht** *f* INFR & ENTW parking bay
**parken** *vti* INFR & ENTW park
**Parkett** *nt* HOLZ parquetry; **Parkettbodenbelag** *m* HOLZ parquet flooring; **Parkettfußboden** *m* HOLZ parquetry
***Park***: **Parkfläche** *f* INFR & ENTW parking area; **Parkhaus** *nt* INFR & ENTW multistorey car park (*BrE*), multistory car park (*AmE*); **Parkplatz** *m* INFR & ENTW car park (*BrE*), parking area, parking lot (*AmE*); **Parkstreifen** *m* INFR & ENTW lay-by
**Partikel** *f* HEIZ & BELÜFT, UMWELT particulate material
**Partner** *m* ARCH & TRAGW, BAURECHT, INFR & ENTW partner
**Parzelle** *f* BAURECHT, INFR & ENTW allotment, lot
**Paß**: **Paßeinsatz** *m* ELEKTR *für Sicherungen* gage piece (*AmE*), gauge piece (*BrE*)
**passend** *adj* ARCH & TRAGW, BAURECHT appropriate, convenient, suitable
**passiv**: **~er Druck** *m* INFR & ENTW passive pressure; **~er Erddruck** *m* INFR & ENTW passive earth pressure (*BrE*), passive ground pressure (*AmE*); **~er Rankindruck** *m* INFR & ENTW passive Rankin pressure; **~es System** *nt* UMWELT passive system
***Paß***: **Paßleiste** *f* BESCHLÄGE, HOLZ cover fillet; **Paßstift** *m* BESCHLÄGE dowel pin; **Paßstück** *nt* BAUMASCHIN, BESCHLÄGE, HEIZ & BELÜFT, STAHL adapter
**Passung** *f* ARCH & TRAGW fit
**Paste** *f* WERKSTOFF paste
**pastös**: **~er Abfall** *m* UMWELT pasty waste
**Patent** *nt* BAURECHT patent
**patentieren**: **~ lassen** *vt* BAURECHT patent
**patentiert** *adj* BAURECHT patented
**pathogen**: **~er Abfall** *m* UMWELT pathological waste, infectious waste, anatomical waste
**Patio** *m* ARCH & TRAGW patio
**Pauschalabfindung** *f* BAURECHT lump sum settlement
**Pauschale** *f* ARCH & TRAGW, BAURECHT lump sum
**Pauschalvertrag** *m* ARCH & TRAGW, BAURECHT lump sum contract
**pausen** *vt* ARCH & TRAGW caulk
**Pavillon** *m* ARCH & TRAGW pavilion
**PE** *abbr* (*Polyethylen*) WERKSTOFF PE (*polyethylene*)
**Pegel** *m* ARCH & TRAGW, INFR & ENTW gage (*AmE*), gauge (*BrE*); **Pegelstand** *m* UMWELT water depth
**PE**: **~ hart** *abbr* (*Polyethylen hoher Dichte*) WERKSTOFF HDPE (*high-density polyethylene*)
**Peillinie** *f* INFR & ENTW collimation line
**Peilrohr** *nt* INFR & ENTW sounding pipe
**Pelletisierung** *f* UMWELT pelletization
**Pendel** *nt* ARCH & TRAGW swing; **Pendelaufhängung** *f* ARCH & TRAGW pendulum suspension; **Pendelstütze** *f* ARCH & TRAGW hinged column, rocking pier; **Pendeltor** *nt* ARCH & TRAGW swing gate; **Pendeltür** *f* ARCH & TRAGW swing door, swinging door
**Pendentif** *nt* ARCH & TRAGW pendentive
**Pendler** *m* INFR & ENTW commuter
**Penetration** *f* ARCH & TRAGW, BAUMASCHIN, INFR & ENTW, OBERFLÄCHE, WERKSTOFF penetration
**Penetrometer** *nt* BAUMASCHIN penetrometer
**Penthouse** *nt* ARCH & TRAGW penthouse
**perforieren** *vt* BAUMASCHIN, BESCHLÄGE, HOLZ, WERKSTOFF perforate
**perforiert**: **~es Papier** *nt* WERKSTOFF perforated paper
**Pergola** *f* ARCH & TRAGW pergola
**Periode** *f* ARCH & TRAGW, BAURECHT, INFR & ENTW period
**periodisch** *adj* ARCH & TRAGW, INFR & ENTW periodic
**Peripherie** *f* INFR & ENTW periphery
**Perlit** *m* DÄMMUNG, WERKSTOFF perlite; **Perlitisolierung** *f* DÄMMUNG perlite insulation

**permeabel** *adj* HOLZ, INFR & ENTW, NATURSTEIN, WERKSTOFF permeable
**persisch**: **~er Bogen** *m* ARCH & TRAGW keel arch
**persistent**: **~es Öl** *nt* UMWELT persistent oil; **~e Packstoffe** *m pl* UMWELT nonbiodegradable packaging
**Personal** *nt* ARCH & TRAGW, BAURECHT personnel, staff, workforce
**Personenaufzug** *m* ARCH & TRAGW, BAUMASCHIN, INFR & ENTW passenger elevator (*AmE*), passenger lift (*BrE*)
**Perspektivzeichnung** *f* ARCH & TRAGW, INFR & ENTW perspective drawing
**PETRIFIX-Verfahren** *nt* (*Verfestigungsverfahren für Sonderabfälle*) UMWELT PETRIFIX process
**PE**: **~ weich** *abbr* (*Polyethylen niedriger Dichte*) WERKSTOFF LDPE (*low-density polyethylene*)
**PF** *abbr* (*Phenol-Formaldehyd*) HOLZ, WERKSTOFF PF (*phenol-formaldehyde*)
**Pfahl** *m* ARCH & TRAGW pile, standard, stake, BAUMASCHIN pile, HOLZ picket, pole, INFR & ENTW pile, pole; **~ mit angeschnittenem Fuß** ARCH & TRAGW underreamed pile; **Pfahlabschnitthöhe** *f* INFR & ENTW pile cut-off level; **Pfahlabstand** *m* INFR & ENTW pile spacing; **Pfahlabweichung** *f* INFR & ENTW pile deflection; **Pfahlbau** *m* ARCH & TRAGW pilework; **Pfahlbewehrung** *f* BETON, INFR & ENTW pile reinforcement; **Pfahldurchmesser** *m* INFR & ENTW pile diameter; **Pfahlfuß** *m* STAHL, UMWELT pile toe; **Pfahlgründung** *f* ARCH & TRAGW, INFR & ENTW pile foundation, pile group, piled foundation; **mit Pfahlgründung** *phr* INFR & ENTW pile-supported; **Pfahljoch** *nt* ARCH & TRAGW trestle; **Pfahlkappe** *f* INFR & ENTW pile cap; **Pfahlkopf** *m* INFR & ENTW pile head; **Pfahlkopfplatte** *f* INFR & ENTW pile cap; **Pfahllänge** *f* INFR & ENTW pile length; **Pfahllast** *f* INFR & ENTW pile load; **Pfahllasttest** *m* INFR & ENTW pile load test; **Pfahlplan** *m* INFR & ENTW pile plan; **Pfahlramme** *f* ARCH & TRAGW, BAUMASCHIN pile driver, piling frame; **Pfahlrammung** *f* ARCH & TRAGW, INFR & ENTW pile driving; **Pfahlrammversuch** *m* INFR & ENTW pile-driving test; **Pfahlring** *m* INFR & ENTW pile ferrule; **Pfahlrost** *m* ARCH & TRAGW, INFR & ENTW pile foundation grille; **Pfahlschuh** *m* ARCH & TRAGW, INFR & ENTW, STAHL pile shoe; **Pfahl- und Spundwandramme** *f* BAUMASCHIN pile and sheet-pile driver; **Pfahltragfähigkeit** *f* INFR & ENTW pile resistance; **Pfahltreiben** *nt* ARCH & TRAGW piling; **Pfahlwand** *f* ARCH & TRAGW, INFR & ENTW pile wall; **Pfahlzieher** *m* BAUMASCHIN, INFR & ENTW *pile drawer, pile extractor;* **Pfahlzwinge** *f* INFR & ENTW *pile ferrule*
**Pfanne** *f* WERKSTOFF pantile; **Pfannenziegel** *m* NATURSTEIN, WERKSTOFF clay pantile
**Pfeiler** *m* ARCH & TRAGW *freistehend* pile, pillar, pier, column, BAUMASCHIN pile, BETON column, INFR & ENTW pillar, pile, NATURSTEIN, STAHL column; **Pfeilerbasilika** *f* ARCH & TRAGW pier basilica; **Pfeilergründung** *f* INFR & ENTW pier foundation; **Pfeilerkopf** *m* ARCH & TRAGW pier head, INFR & ENTW *Brücke* cutwater; **Pfeilerverband** *m* NATURSTEIN pier bond; **Pfeilervorlage** *f* ARCH & TRAGW pilaster strip
**Pfeilhöhe** *f* ARCH & TRAGW pitch, rise, INFR & ENTW pitch
**Pfette** *f* ARCH & TRAGW, BESCHLÄGE, HOLZ binding rafter, purlin; **Pfettennagel** *m* BESCHLÄGE purlin nail; **Pfettenstützholz** *nt* ARCH & TRAGW, HOLZ purlin post
**Pflanzenschutzgesetz** *nt* UMWELT Plant Protection Act
**Pflaster** *nt* NATURSTEIN *Straßenbelag* paving; **Pflasterbelag** *m* NATURSTEIN pavement (*BrE*), WERKSTOFF pavement paving flag (*BrE*), sidewalk paving flag (*AmE*)
**Pflasterer** *m* ARCH & TRAGW pavior (*AmE*), paviour (*BrE*)
***Pflaster***: **Pflasterhammer** *m* BAUMASCHIN sledgehammer
**pflastern** *vt* INFR & ENTW pave, NATURSTEIN floor
***Pflaster***: **Pflasterstein** *m* INFR & ENTW road stone, NATURSTEIN pavior (*AmE*), *klein* sett, pitcher; **Pflasterweiche** *f* INFR & ENTW tramway switch
**Pflege** *f* ARCH & TRAGW, BAURECHT, INFR & ENTW maintenance
**Pflicht** *f* BAURECHT duty; **Pflichtenheft** *nt* BAURECHT specification
**Pflock** *m* HOLZ picket
**Pforte** *f* ARCH & TRAGW, BETON, HOLZ portal
**Pfosten** *m* ARCH & TRAGW stake, stanchion, stay, post, upright, BAUMASCHIN, BETON post; **Pfostenfundament** *nt* ARCH & TRAGW *Zaun* post foundation
**PF/RF** *abbr* (*Phenol/Resorcin-Formaldehyd*) HOLZ, UMWELT, WERKSTOFF PF/RF (*phenol/resorcinol-formaldehyde*)
**Pfund** *nt*: **~ pro Quadratzoll** ABWASSER, HEIZ & BELÜFT pounds per square inch (*psi*)
**Pfusch** *m* BAURECHT bungled work; **Pfuscharbeit** *f* BAURECHT bungled work
**Pfuscher** *m* BAURECHT jerry builder
**pH-Abnahme** *f* UMWELT pH drop
**Phasentrennung** *f* UMWELT phase separation
**Phenol** *nt* HOLZ, WERKSTOFF phenol; **Phenol-Formaldehyd** *nt* HOLZ (*PF*), UMWELT (*PH*), WERKSTOFF (*PF*) phenol-formaldehyde (*PF, PH*); **Phenolharz** *nt* HOLZ phenol resin; **Phenol/Resorcin-Formaldehyd** *nt* (*PF/RF*) HOLZ, UMWELT, WERKSTOFF phenol/resorcinol-formaldehyde (*PF/RF*); **Phenolschaumstoff** *m* DÄMMUNG, WERKSTOFF phenolic foam
**pH-Wert** *m* UMWELT, WERKSTOFF pH-value
**physikalisch** *adj* INFR & ENTW, UMWELT, WERKSTOFF physical; **~e Eigenschaften** *f pl* INFR & ENTW, WERKSTOFF physical properties; **~es Modell** *nt* INFR & ENTW, WERKSTOFF physical model; **~e Verwitterung** *f* INFR & ENTW, WERKSTOFF physical weathering
**Pickel** *m* BAUMASCHIN pick
**Piezometer** *nt* BAUMASCHIN, INFR & ENTW, WERKSTOFF piezometer
**Pigment** *nt* WERKSTOFF pigment; **Pigmentschlamm** *m* UMWELT pigment sludge
**Pilaster** *m* ARCH & TRAGW pilaster
**Pilz-** *in cpds* ARCH & TRAGW flared, umbrella, UMWELT, WERKSTOFF fungal; **Pilzbefall** *m* WERKSTOFF fungal attack; **Pilzkopf** *m* ARCH & TRAGW flared head; **Pilzschale** *f* ARCH & TRAGW umbrella shell
**pilztötend**: **~es Mittel** *nt* BAURECHT, WERKSTOFF fungicide
***Pilz-***: **Pilzventil** *nt* HEIZ & BELÜFT mushroom valve
**Pinselauftrag** *m* OBERFLÄCHE brush painting
**Pipeline** *f* ABWASSER, INFR & ENTW, UMWELT pipeline

**Plan** *m* ARCH & TRAGW design, *Zeichnung* layout, *Zeittafel* project, *Entwurf* plan, drawing, schedule, INFR & ENTW *Entwurf* design, project, WERKSTOFF *Entwurf* design

**Planebenheit** *f* ARCH & TRAGW, INFR & ENTW, WERKSTOFF accuracy of level, evenness

**planen** *vt* ARCH & TRAGW plan, project

**Planer** *m* ARCH & TRAGW, INFR & ENTW designer, planner

**Planier-** *in cpds* BAUMASCHIN, INFR & ENTW, WERKSTOFF grading, leveling (*AmE*), levelling (*BrE*); **Planierarbeiten** *f pl* INFR & ENTW, WERKSTOFF grading

**planieren** *vt* INFR & ENTW level, plane

**Planieren** *nt* BAUMASCHIN planing, INFR & ENTW leveling (*AmE*), levelling (*BrE*)

***Planier-***: **Planiergerät** *nt* BAUMASCHIN grader; **Planiermaschine** *f* BAUMASCHIN leveling machine (*AmE*), levelling machine (*BrE*); **Planierraupe** *f* BAUMASCHIN crawler dozer, crawler tractor, crawler, bulldozer; **Planierstange** *f* INFR & ENTW *Vermessung* leveling rod (*AmE*), levelling rod (*BrE*)

**Planimeter** *nt* ARCH & TRAGW, BAUMASCHIN, INFR & ENTW *Vermessung* planimeter

**Planimetrie** *f* ARCH & TRAGW, BAUMASCHIN, INFR & ENTW planimetry

**Planke** *f* HOLZ deal, INFR & ENTW, WERKSTOFF plank

**Plankton** *nt* INFR & ENTW plankton

**Planum** *nt* INFR & ENTW formation level

**Planung** *f* ARCH & TRAGW planning; **Planungsfehler** *m* ARCH & TRAGW design error; **Planungsraster** *m* ARCH & TRAGW, INFR & ENTW planning grid

**Plasma** *nt* BAUMASCHIN, STAHL plasma; **Plasmalichtbogenschneiden** *nt* STAHL plasma arc cutting; **Plasmaschneiden** *nt* BAUMASCHIN plasma cutting; **Plasmaschweißbrenner** *m* BAUMASCHIN torch for plasma welding; **Plasmatrennen** *nt* BAUMASCHIN plasma cutting

**Plastifizieren** *nt* BETON *Schweißen*, WERKSTOFF plastification

**Plastik** *nt* ARCH & TRAGW, UMWELT, WERKSTOFF plastic

**plastisch** *adj* ARCH & TRAGW, UMWELT, WERKSTOFF plastic; **nicht plastisch** *adj* WERKSTOFF nonplastic; **~er Betonsetzungsriß** *m* BETON concrete plastic settlement crack; **~e Dehnung** *f* WERKSTOFF plastic elongation; **~er Fluß** *m* WERKSTOFF plastic flow; **~es Gelenk** *nt* ARCH & TRAGW, BETON, STAHL plastic hinge; **~e Momententragfähigkeit** *f* ARCH & TRAGW, BETON, WERKSTOFF plastic resistance moment; **~e Nullinie** *f* ARCH & TRAGW, BETON, WERKSTOFF plastic neutral axis; **~e Schubtragfähigkeit** *f* ARCH & TRAGW, BETON, STAHL, WERKSTOFF plastic shear resistance; **~e Verformung** *f* ARCH & TRAGW, BETON, INFR & ENTW inelastic deformation, WERKSTOFF plastic deformation, inelastic deformation, yield; **~e Verzierung** *f* ARCH & TRAGW carved ornament

**Plastizität** *f* BETON, WERKSTOFF plasticity; **Plastizitätsgrenze** *f* INFR & ENTW, WERKSTOFF *Boden* plastic limit; **Plastizitätsindex** *m* WERKSTOFF plasticity index; **Plastizitätsmodul** *m* ARCH & TRAGW, INFR & ENTW, WERKSTOFF modulus of plasticity; **Plastizitätstheorie** *f* STAHL, WERKSTOFF plastic theory

**Platte** *f* BAUMASCHIN, BESCHLÄGE, INFR & ENTW, STAHL plate, WERKSTOFF panel, sheet, plate, slab; **Plattenbalkendecke** *f* ARCH & TRAGW slab-and-beam ceiling; **Plattenbefestigung** *f* BESCHLÄGE, WERKSTOFF plate fixing; **Plattenbelag** *m* INFR & ENTW, NATURSTEIN flagging; **Plattenbrücke** *f* INFR & ENTW slab bridge; **Plattendämmung** *f* DÄMMUNG board insulation; **Plattendruckversuch** *m* INFR & ENTW *Boden* plate load test; **Plattenfundament** *nt* ARCH & TRAGW, BETON, INFR & ENTW foundation raft, slab footing, slab foundation; **Plattengründung** *f* BETON, INFR & ENTW raft foundation; **Plattenguß** *m* STAHL plate casting; **Plattenkern** *m* WERKSTOFF board core; **Plattenleger** *m* ARCH & TRAGW, NATURSTEIN tiler; **Plattenrüttler** *m* BAUMASCHIN plate vibrator, vibrating plate compactor; **Plattenschleifautomat** *m* ARCH & TRAGW automatic slab grinder, BAUMASCHIN automatic tile grinder; **Plattenträger** *m* ARCH & TRAGW plate girder; **Plattentrennwand** *f* ARCH & TRAGW slab partition, slab wall; **Plattenverkleidung** *f* ARCH & TRAGW slab lining, NATURSTEIN stone facing; **Plattenverlegen** *nt* INFR & ENTW flagging; **Plattenwand** *f* ARCH & TRAGW slab partition, slab wall; **Plattenwandverkleidung** *f* NATURSTEIN tile hanging

**Plattform** *f* ARCH & TRAGW platform

**plattieren** *vt* OBERFLÄCHE clad

**plattiert** *adj* OBERFLÄCHE cladded; **~es Stahlblech** *nt* WERKSTOFF cladded steel plate

**Plattierung** *f* OBERFLÄCHE *Metall* cladding

**platzen** *vi* INFR & ENTW, WERKSTOFF burst

**Plexiglas** *nt* WERKSTOFF plexiglass

**Plombe** *f* BESCHLÄGE seal, OBERFLÄCHE sealing

**plombiert** *adj* BAURECHT sealed

**Pluviometer** *nt* BAUMASCHIN rain gage (*AmE*), rain gauge (*BrE*)

**pneumatisch**: **~e Sortieranlage** *f* UMWELT pneumatic sorter; **~es Sortieren** *nt* UMWELT pneumatic classification

**Podest** *nt* ARCH & TRAGW platform; **Podeststufe** *f* ARCH & TRAGW landing step; **Podestträger** *m* ARCH & TRAGW bearer; **Podestwechselbalken** *m* BAUMASCHIN landing trimmer

**Poisson**: **~sche Querdehnzahl** *f* BETON, INFR & ENTW, WERKSTOFF Poisson's ratio

**Polder** *m* INFR & ENTW, UMWELT landfill cell, refuse cell

**Polier** *m* ARCH & TRAGW foreman, head mason

**Polierasche** *f* WERKSTOFF putty powder

**Polierbarkeit** *f* OBERFLÄCHE buffability

**polieren** *vt* ARCH & TRAGW rub, BAUMASCHIN plane, polish, OBERFLÄCHE, WERKSTOFF polish

**Polieren** *nt* ARCH & TRAGW, BAUMASCHIN planing, polish, OBERFLÄCHE, WERKSTOFF polishing

**Poliermaschine** *f* **mit biegsamer Welle** BAUMASCHIN polishing machine with flexible shaft

**poliert** *adj* OBERFLÄCHE, WERKSTOFF polished; **~es Messing** *nt* WERKSTOFF polished brass; **~e Steinoberfläche** *f* OBERFLÄCHE polished stone finish

**Polonceauträger** *m* ARCH & TRAGW French truss

**Polster** *nt* WERKSTOFF pad

**Polychloroprenkautschuk** *m* WERKSTOFF polychloroprene rubber

**Polyester** *nt* OBERFLÄCHE, UMWELT, WERKSTOFF polyester; **Polyesterfarbe** *f* OBERFLÄCHE polyester paint; **Polyesterharz** *nt* WERKSTOFF polyester resin; **Polyesterplatte** *f* WERKSTOFF polyester board; **Polyesterschaumstoff** *m* UMWELT polyester foam; **Polyesterwellplatte** *f* WERKSTOFF corrugated polyester board

**Polyethylen** *nt* (*PE*) WERKSTOFF polyethylene (*PE*);

~ **hoher Dichte** (*PE hart*) WERKSTOFF high-density polyethylene (*HDPE*); ~ **niedriger Dichte** (*PE weich*) WERKSTOFF low-density polyethylene (*LDPE*); **Polyethylenfolie** *f* WERKSTOFF polyethylene foil;

**polygonal** *adj* ARCH & TRAGW polygonal

**Polygonmauerwerk** *nt* NATURSTEIN random rubble masonry

**Polymer** *nt* WERKSTOFF polymer; **Polymerbetongießmaschine** *f* BAUMASCHIN casting machine for polymer concrete

**Polystyrol** *nt* (*PS*) WERKSTOFF polystyrene (*PS*)

**Polyurethan** *nt* (*PUR*) DÄMMUNG, WERKSTOFF polyurethane (*PUR*)

**Polyvinylchlorid** *nt* (*PVC*) OBERFLÄCHE, WERKSTOFF polyvinyl chloride (*PVC*)

**polyzyklisch**: **~er aromatischer Kohlenwasserstoff** *m* UMWELT polycyclic aromatic hydrocarbon

**Ponton** *m* BAUMASCHIN, BETON, STAHL pontoon

**Pool** *m* ARCH & TRAGW pool

**Pore** *f* BETON pore, void, WERKSTOFF pore; **Porenbeton** *m* BETON, WERKSTOFF cellular concrete, foam concrete, porous concrete, *Trennwände* aerated concrete; **Porenbildung** *f* BETON, INFR & ENTW, WERKSTOFF air entrainment; **Porendruck** *m* BETON pore pressure; **Porendurchmesser** *m* BETON pore diameter; **Porenestrich** *m* BETON, OBERFLÄCHE aerated cement floor

**porenfrei** *adj* WERKSTOFF nonporous

***Pore***: **Porenfüller** *m* OBERFLÄCHE sealer; **Porengehalt** *m* WERKSTOFF void ratio; **Porengröße** *f* WERKSTOFF pore size; **Porensaugwirkung** *f* ARCH & TRAGW, INFR & ENTW capillary attraction; **Porenwassergehalt** *m* BETON pore-water content; **Porenzahl** *f* BETON void ratio; **Porenziffer** *f* WERKSTOFF void ratio

**porig** *adj* WERKSTOFF foamed

**porig-zellig** *adj* WERKSTOFF *Gummi, Kunststoffe* expanded

**porös** *adj* BETON, INFR & ENTW, UMWELT, WERKSTOFF porous; **~e Abdeckung** *f* UMWELT *einer Deponie* porous cover; **~er Boden** *m* INFR & ENTW porous soil; **~es Gestein** *nt* INFR & ENTW porous rock

**Porosität** *f* BETON, INFR & ENTW, UMWELT, WERKSTOFF porosity; **Porositätstest** *m* WERKSTOFF porosity test

**Porphyr** *m* WERKSTOFF porphyry

**Portal** *f* ARCH & TRAGW, BETON, HOLZ portal; **Portalkran** *m* BAUMASCHIN gantry crane, portal crane; **Portalrahmen** *m* BETON, HOLZ portal frame

**Portikus** *m* ARCH & TRAGW portal, portico, BETON, HOLZ portal

**Portlandzement** *m* BETON slag cement

**Position** *f* ARCH & TRAGW position

**Potential** *nt* ELEKTR, INFR & ENTW, WERKSTOFF potential; ~ **der Schallschnelle** UMWELT velocity potential; **Potentialausgleich** *m* ELEKTR equipotential bonding; **Potentialausgleichsschiene** *f* ELEKTR equipotential busbar; **Potentialmethode** *f* INFR & ENTW potential method; **Potentialströmung** *f* ELEKTR potential flow

**potentiell**: **~e Energie** *f* WERKSTOFF potential energy

**praktisch**: **~e Anwendung** *f* ARCH & TRAGW, INFR & ENTW practical application

**Prallblech** *nt* HEIZ & BELÜFT baffle plate

**Prämie** *f*: ~ **für vorzeitige Fertigstellung** BAURECHT bonus for early completion

**präzis** *adj* ARCH & TRAGW, BAUMASCHIN, BAURECHT, BETON, INFR & DES, STAHL accurate

**Preis** *m* ARCH & TRAGW, BAURECHT, INFR & ENTW cost, price; **Preisindex** *m* **im Bauwesen** ARCH & TRAGW, BAURECHT, INFR & ENTW building index

**Preiße** *f* NATURSTEIN, WERKSTOFF *Deckziegel* convex tile

***Preis***: **Preistabelle** *f* ARCH & TRAGW, BAURECHT table of prices

**Prell-** *in cpds* ARCH & TRAGW fender, BESCHLÄGE baffle; **Prellbalken** *m* ARCH & TRAGW fender beam; **Prellpfahl** *m* ARCH & TRAGW fender pile; **Prellplatte** *f* BESCHLÄGE baffle plate

**Preß-** *in cpds* BETON, STAHL, WERKSTOFF pressed; **Preßblechdachrinne** *f* STAHL, WERKSTOFF pressed steel gutter

**Presse** *f* BAUMASCHIN press; ~ **zur Betonrohrherstellung** BAUMASCHIN press for the manufacture of concrete pipes

***Preß-***: **Preßfuge** *f* ARCH & TRAGW compression joint; **Preßglas** *nt* WERKSTOFF molded glass (*AmE*), moulded glass (*BrE*), pressed glass

**Pressiometer** *nt* INFR & ENTW pressiometer

***Preß-***: **Preßkork** *m* WERKSTOFF compressed cork; **Preßlufthammer** *m* BAUMASCHIN pneumatic hammer; **Preßmüllwagen** *m* BAUMASCHIN, UMWELT packer lorry; **Preßschweißen** *nt* STAHL pressure welding; **Preßsperrholz** *nt* HOLZ densified plywood; **Preßstrohplatte** *f* DÄMMUNG strawboard; **Preßtorf** *m* WERKSTOFF compressed peat

**Primär-** *in cpds* ARCH & TRAGW, BETON, HEIZ & BELÜFT, HOLZ, INFR & ENTW, UMWELT primary; **Primärbinder** *m* BETON, HOLZ primary truss; **Primärkonstruktion** *f* ARCH & TRAGW primary structure; **Primärluft** *f* HEIZ & BELÜFT primary air; **Primärschlamm** *m* INFR & ENTW, UMWELT primary sludge

**Prismenwanne** *f* ELEKTR *Leuchte* prismatic diffuser

**Privaträume** *m pl* ARCH & TRAGW private rooms

**Probe** *f* WERKSTOFF specimen; **eine ~ entnehmen** *phr* INFR & ENTW sample; **Probebohrung** *f* INFR & ENTW trial boring; **Probedruck** *m* ABWASSER, HEIZ & BELÜFT test pressure; **Probeentnahme** *f* INFR & ENTW sampling; **Probeentnahmen** *f pl* **gleicher Menge** UMWELT constant-volume sampling; **Probeflasche** *f* BAUMASCHIN sample bottle; **Probelauf** *m* BAUMASCHIN, HEIZ & BELÜFT trial run; **Probelöffel** *m* BAUMASCHIN spoon auger; **Probenahme** *f* UMWELT sampling; **Probenentnahmestelle** *f* INFR & ENTW sampling point; **Proberammung** *f* INFR & ENTW trial pile driving; **Probesondierung** *f* INFR & ENTW, WERKSTOFF exploratory sounding, sounding test; **Probewürfel** *m* BETON test cube, WERKSTOFF compression cube; **Probezylinder** *m* BETON test cylinder

**Proctor-** *in cpds* INFR & ENTW, WERKSTOFF Proctor; **Proctordichte** *f* INFR & ENTW, WERKSTOFF Proctor density; **Proctortest** *m* INFR & ENTW, WERKSTOFF Proctor test; **Proctorversuch** *m* INFR & ENTW compaction test

**Produkt** *nt* WERKSTOFF product

**Produktion** *f* ARCH & TRAGW, INFR & ENTW production, UMWELT yield, WERKSTOFF manufacture; **Produktionsabfall** *m* BAURECHT, UMWELT process waste

**produktionsspezifisch**: **~er Abfall** *m* BAURECHT, UMWELT process waste

**Produktivität** *f* ARCH & TRAGW, INFR & ENTW productivity

**produzieren** *vt* WERKSTOFF manufacture

**Profil** *nt* ARCH & TRAGW, BESCHLÄGE profile, INFR & ENTW shape, profile, STAHL section, WERKSTOFF shape, profile; **~e setzen** *phr* NATURSTEIN set up profiles; **Profilblech** *nt* STAHL sectional sheet; **Profilbleche** *nt pl* ARCH & TRAGW, STAHL, WERKSTOFF profile steel sheeting; **Profilbrett** *nt* HOLZ matchboard; **Profildraht** *m* STAHL, WERKSTOFF profiled wire

**profilgemäß** *adj* INFR & ENTW *Bodenaushub* true to profile

***Profil***: **Profilglas** *nt* WERKSTOFF figured glass

**profilieren** *vt* ARCH & TRAGW, BESCHLÄGE, INFR & ENTW profile, NATURSTEIN set up profiles, WERKSTOFF profile

**profiliert** *adj* ARCH & TRAGW, BESCHLÄGE, INFR & ENTW, WERKSTOFF profiled; **~es Leichtgewicht-PVC-Dach** *nt* ARCH & TRAGW, WERKSTOFF lightweight-profiled PVC roof

***Profil***: **Profilmessung** *f* ARCH & TRAGW, BESCHLÄGE, INFR & ENTW, WERKSTOFF profiling; **Profilschere** *f* BAUMASCHIN section cutter; **Profilschneider** *m* BAUMASCHIN section cutter; **Profilstahl** *m* ARCH & TRAGW, STAHL sectional steel, steel section; **Profilstahldeck** *nt* BETON, STAHL profiled steel deck; **Profilstahlkonstruktion** *f* ARCH & TRAGW, STAHL sectional steel construction; **Profilstange** *f* BETON, STAHL, WERKSTOFF profiled rod; **Profilstein** *m* WERKSTOFF purpose-made block; **Profilzylinder** *m* BESCHLÄGE profile cylinder

**Programm** *nt* ARCH & TRAGW schedule

**Projekt** *nt* ARCH & TRAGW, BAURECHT, INFR & ENTW project; **Projektänderung** *f* BAURECHT variation order; **Projektfinanzierung** *f* BAURECHT funding of projects

**projektiert**: **~er Treppenauftritt** *m* ARCH & TRAGW going

**Projektion** *f* ARCH & TRAGW projection; **Projektionsfläche** *f* ARCH & TRAGW projected area; **Projektionslänge** *f* ARCH & TRAGW projection length

***Projekt***: **Projektleiter** *m* BAURECHT project manager; **Projektleitung** *f* ARCH & TRAGW, BAURECHT project management; **Projektüberwachung** *f* ARCH & TRAGW, BAURECHT project monitoring

**Propellergebläse** *nt* HEIZ & BELÜFT propeller fan

**Proportionalitätsgrenze** *f* WERKSTOFF proportionality limit

**Prospektion** *f* INFR & ENTW prospection

**provisorisch** *adj* ARCH & TRAGW temporary; **~er Wasseranschluß** *m* INFR & ENTW temporary water connection

**Prozentsatz** *m*: **~ der Einbehaltung** BAURECHT retention percentage

**Prozeß** *m* ARCH & TRAGW, BAURECHT, INFR & ENTW process

**PR-Parkplätze** *m pl* (*Park-and-Ride-Parkplätze*) INFR & ENTW park-and-ride lots (*AmE*), park-and-ride sites (*BrE*)

**Prüf-** *in cpds* ARCH & TRAGW, BAUMASCHIN, BAURECHT, INFR & ENTW inspection, test, testing; **Prüfausrüstung** *f* INFR & ENTW testing kit; **Prüfbericht** *m* BAURECHT test report; **Prüfbescheinigung** *f* BAURECHT test certificate; **Prüfbestimmungen** *f pl* BAURECHT test specifications

**prüfen** *vt* ARCH & TRAGW examine, BAURECHT, ELEKTR, HEIZ & BELÜFT, INFR & ENTW test

***Prüf-***: **Prüflast** *f* INFR & ENTW test load; **Prüfnorm** *f* BAURECHT test standard; **Prüfprotokoll** *nt* ARCH & TRAGW, BAURECHT, INFR & ENTW inspection sheet

**Prüfung** *f* ARCH & TRAGW examination, inspection, BAURECHT check, checking, examination, ELEKTR, HEIZ & BELÜFT test, INFR & ENTW examination, test

***Prüf-***: **Prüfverfahren** *nt* BAURECHT, INFR & ENTW test method; **Prüfvorrichtung** *f* BAUMASCHIN test apparatus; **Prüfvorschriften** *f pl* BAURECHT test specifications; **Prüfwaage** *f* BAUMASCHIN assay balance; **Prüfzeichen** *nt* BAURECHT test symbol, test mark; **Prüfzeugnis** *nt* BAURECHT test certificate

**PS** *abbr* (*Polystyrol*) WERKSTOFF PS (*polystyrene*)

**Puder** *m* WERKSTOFF powder

**Puffer** *m* ARCH & TRAGW *Kran*, BAURECHT, STAHL buffer, WERKSTOFF pad

**pulsieren** *vi* ARCH & TRAGW, INFR & ENTW pulsate

**pulsierend** *adj* ELEKTR pulsating; **~e Last** *f* ARCH & TRAGW, INFR & ENTW pulsating load

**Pultanbau** *m* ARCH & TRAGW lean-to

**Pultdach** *nt* ARCH & TRAGW, HOLZ lean-to roof, monopitch roof, pitch roof, shed roof

**Pulver** *nt* WERKSTOFF powder

**pulverbeschichtet** *adj* OBERFLÄCHE powder-coated

**pulverisieren** *vt* WERKSTOFF pulverize

***Pulver***: **Pulverlackbeschichtung** *f* OBERFLÄCHE powder coating

**Pump-** *in cpds* BETON pumping; **Pumpbeton** *m* BETON pumpcrete, pumped concrete

**Pumpe** *f* ABWASSER, BAUMASCHIN, HEIZ & BELÜFT pump

**Pumpen** *nt* BETON pumping

***Pumpe***: **Pumpenbagger** *m* BAUMASCHIN pump dredger

***Pump-***: **Pumpversuch** *m* INFR & ENTW pumping test, test pumping

**Punkt** *m* ARCH & TRAGW, ELEKTR point, spot; **Punkte** *m pl* **der Gegenbiegung** ARCH & TRAGW points of contraflexure; **Punkterdung** *f* ELEKTR single-point earthing (*BrE*), single-point grounding (*AmE*); **Punktschweißen** *nt* STAHL spot welding; **Punktschweißung** *f* STAHL spot welding

**PUR** *abbr* (*Polyurethan*) DÄMMUNG, WERKSTOFF PUR (*polyurethane*)

***PUR***: **PUR-Hartschaum** *m* DÄMMUNG, WERKSTOFF polyurethane rigid foam

**Putz** *m* NATURSTEIN/parget, plaster, stucco,WERKSTOFF *Innenputz* plaster; **auf ~** *phr* ABWASSER, ELEKTR, HEIZ & BELÜFT surface-mounted; **unter ~** *phr* ABWASSER, ELEKTR concealed, *Schalter, Steckdosen* flush, HEIZ & BELÜFT *Kabel, Rohre* concealed; **mit ~ bewerfen** *phr* NATURSTEIN daub; **unter ~ montiert** *phr* ELEKTR *Schalter, Steckdose* flush-mounted; **auf ~ montiert sein** *phr* ABWASSER surface-mounted; **Putzabstandshalter** *m* BAURECHT, NATURSTEIN furring; **Putzarbeiten** *f pl*/NATURSTEIN plaster and stucco work, plaster work, WERKSTOFF plastering; **Putzbewehrung** *f* NATURSTEIN plaster reinforcement; **Putzbund** *m* NATURSTEIN plaster base; **Putzdeckel** *m* ABWASSER cleanout cover; **Putzeckleiste** *f* NATURSTEIN corner bead

**putzfähig** *adj* NATURSTEIN *Untergrund* ready for plastering

***Putz***: **Putzhobel** *m* BAUMASCHIN fine plane; **Putzkelle**

*f* BAUMASCHIN plastering trowel; **Putzlage** *f* BETON, NATURSTEIN facing work; **Putzleiste** *f* ARCH & TRAGW, BESCHLÄGE, NATURSTEIN counterlath; **Putzmaschine** *f* BAUMASCHIN plastering machine, plaster-throwing machine; **Putzmasse** *f* ARCH & TRAGW stuff; **Putzmörtel** *m* ARCH & TRAGW, NATURSTEIN stuff; **Putzstück** *nt* ABWASSER cleanout piece; **Putzträger** *m* NATURSTEIN lathwork; **Putzträger** *m* **aus Leisten** NATURSTEIN counterlathing; **Putzwinkel** *m* BESCHLÄGE, NATURSTEIN, WERKSTOFF plaster angle

**Puzzolanzement** *m* BETON, WERKSTOFF pozzolanic cement

**PVC** *abbr* (*Polyvinylchlorid*) OBERFLÄCHE, WERKSTOFF PVC (*polyvinyl chloride*)

**P-Verschluß** *m* ABWASSER p-trap

**Pylon** *m* ARCH & TRAGW, INFR & ENTW pylon

**Pyramidenturmdach** *nt* ARCH & TRAGW spire roof

**Pyrit** *m* INFR & ENTW pyrite

**Pyrolyse** *f* UMWELT pyrolysis

**Q-Faktor** *m* (*Qualitätsfaktor*) UMWELT Q factor (*quality factor*)

**Quader** *m* INFR & ENTW ashlar, NATURSTEIN ashlar, cut stone; **Quadermauerwerk** *nt* ARCH & TRAGW, NATURSTEIN *Liniensteinblötzen* ashlar masonry, ashlaring, freestone masonry, natural stone masonry, regular coursed ashlar stone work

**Quadrat** *nt* ARCH & TRAGW square

**quadratisch** *adj* ARCH & TRAGW square; **~er Wasserstandmittelwert** *m* INFR & ENTW root line mean square water level

**Qualität** *f* WERKSTOFF grade, quality; **Qualitätsbewertung** *f* BAURECHT quality assessment; **Qualitätsfaktor** *m* (*Q-Faktor*) UMWELT quality factor (*Q factor*); **Qualitätskontrolle** *f* BAURECHT, WERKSTOFF quality control; **Qualitätssicherung** *f* BAURECHT quality assurance; **Qualitätssicherungssystem** *nt* BAURECHT quality assurance system

**Quantität** *f* ARCH & TRAGW quantity

**Quartier** *nt* ARCH & TRAGW quarter

**Quarz** *m* INFR & ENTW, NATURSTEIN, WERKSTOFF quartz

**quarzhaltig** *adj* INFR & ENTW, WERKSTOFF quartz-containing

**Quarzit** *m* INFR & ENTW, WERKSTOFF quartzite

***Quarz*:** **Quarzkorn** *nt* WERKSTOFF quartz grain; **Quarzsand** *m* DÄMMUNG, INFR & ENTW silica, WERKSTOFF siliceous sand, quartz sand; **Quarzton** *m* INFR & ENTW, WERKSTOFF quartz clay

**Quecksilber** *nt* BAURECHT, INFR & ENTW, UMWELT, WERKSTOFF mercury; **Quecksilberdampf** *m* WERKSTOFF mercury vapor (*AmE*), mercury vapour (*BrE*)

**Quell-** *in cpds* BESCHLÄGE, DÄMMUNG, INFR & ENTW, UMWELT source, spring; **Quellbereich** *m* INFR & ENTW, UMWELT source area

**Quelle** *f* BESCHLÄGE spring, DÄMMUNG *Schall* source, INFR & ENTW spring, UMWELT source

***Quell-*:** **Quellvermögen** *nt* INFR & ENTW, WERKSTOFF ACC test; **Quellvolumen** *nt* WERKSTOFF bulking; **Quellwasser** *nt* INFR & ENTW spring water

**quer** *adj* ARCH & TRAGW, BETON, INFR & ENTW, STAHL, WERKSTOFF lateral, transverse

**Quer-** *in cpds* ARCH & TRAGW, BETON, INFR & ENTW, STAHL, WERKSTOFF crossways, diagonally, transverse; **Queraussteifung** *f* ARCH & TRAGW, BETON, HOLZ, STAHL, WERKSTOFF cross bracing, transverse stiffening; **Querbalken** *m* ARCH & TRAGW, HOLZ, INFR & ENTW, STAHL crossbar, cross beam, cross girder, intertie beam; **Querbewehrung** *f* ARCH & TRAGW, BETON cross reinforcement, transverse reinforcement; **Querdehnung** *f* WERKSTOFF lateral expansion, lateral extension; **Querdehnzahl** *f* BETON, INFR & ENTW, WERKSTOFF Poisson's ratio; **Querfries** *m* ARCH & TRAGW *Senkrechtschiebefenster* meeting rail; **Querfuge** *f* ARCH & TRAGW transverse joint

**quergelüftet** *adj* HEIZ & BELÜFT cross-ventilated

***Quer-*:** **Querholz** *nt* ARCH & TRAGW *Fenster* transom, HOLZ ledger; **Querkraft** *f* ARCH & TRAGW, BETON, INFR & ENTW, STAHL, WERKSTOFF lateral force; **Querkrafttragfähigkeit** *f* ARCH & TRAGW, BETON, INFR & ENTW, STAHL, WERKSTOFF lateral force resistance; **Querneigung** *f* ARCH & TRAGW lateral inclination, INFR & ENTW slope; **Querneigungsgefälle** *nt* INFR & ENTW transverse slope; **Querprofil** *nt* ARCH & TRAGW cross profile; **Querreibung** *f* INFR & ENTW, WERKSTOFF lateral friction; **Querriegel** *m* ARCH & TRAGW, BETON, INFR & ENTW, STAHL beam, crossbar, strap; **Querrippe** *f* ARCH & TRAGW transverse rib; **Querschnitt** *m* ARCH & TRAGW, HOLZ, STAHL, WERKSTOFF cross section; **Querschnittsfläche** *f* ARCH & TRAGW cross-section; **Quersparren** *m* HOLZ cross rafter; **Querstab** *m* ARCH & TRAGW, INFR & ENTW crossbar; **Querstange** *f* ARCH & TRAGW, INFR & ENTW crossbar; **Querstraße** *f* INFR & ENTW crossway, crossroad; **Querstrebe** *f* ARCH & TRAGW, HOLZ, STAHL cross stud; **Querstück** *nt* ARCH & TRAGW, INFR & ENTW crossbar; **Querträger** *m* ARCH & TRAGW filler joist, cross girder, secondary beam, transverse girder, HOLZ cross girder, INFR & ENTW wind brace, STAHL cross girder; **Querverband** *m* ARCH & TRAGW sway bracing, transverse bracing; **Querverformung** *f* BETON, INFR & ENTW, STAHL, WERKSTOFF lateral strain; **Quersteifung** *f* ARCH & TRAGW, BETON, STAHL, WERKSTOFF cross stay, sway bracing, transverse stiffener; **Querverstrebung** *f* HOLZ, STAHL cross bracing; **Querwand** *f* ARCH & TRAGW crosswall; **Querzusammendrückung** *f* INFR & ENTW, WERKSTOFF lateral compression

**Quetschhahn** *m* BAUMASCHIN pinchcock

# R

**Rabatte** *f* INFR & ENTW border

**Rabitz-** *in cpds* ARCH & TRAGW, NATURSTEIN wire plaster; **Rabitzdecke** *f* ARCH & TRAGW, NATURSTEIN wire plaster ceiling; **Rabitzgewebe** *nt* WERKSTOFF *Putz* cloth lathing; **Rabitzwand** *f* ARCH & TRAGW, NATURSTEIN wire plaster wall

**Rad** *nt* ARCH & TRAGW, BAUMASCHIN, INFR & ENTW wheel; **Radabweiser** *m* ARCH & TRAGW spur post, INFR & ENTW spur stone; **Raddruck** *m* BAUMASCHIN, INFR & ENTW wheel pressure

**Rädeldraht** *m* BETON binding wire

***Rad***: **Radfahrzeug** *nt* BAUMASCHIN wheeled vehicle

**Radial-** *in cpds* WERKSTOFF radial; **Radialdruck** *m* WERKSTOFF radial pressure; **Radialnetz** *nt* ABWASSER radial system; **Radialspannung** *f* WERKSTOFF radial stress; **Radialstrecker** *m* NATURSTEIN compass header; **Radialventilator** *m* BESCHLÄGE, HEIZ & BELÜFT centrifugal fan, radial fan; **Radialziegel** *m* NATURSTEIN compass brick

**Radiator** *m* ELEKTR, HEIZ & BELÜFT radiator

**radioaktiv** *adj* BAURECHT, UMWELT, WERKSTOFF radioactive; **~er Abfall** *m* BAURECHT, UMWELT nuclear waste, radioactive waste; **~ kontaminiertes Wasser** *nt* UMWELT contaminated water; **~er Niederschlag** *m* UMWELT radioactive fallout; **~er Stoff** *m* UMWELT radioactive substance; **~e Verschmutzung** *f* UMWELT radioactive pollution

**radiometrisch**: **~e Bohrlochvermessung** *f* UMWELT well logging

**Radius** *m* ARCH & TRAGW radius

***Rad***: **Radlast** *f* INFR & ENTW wheel load

**Radon** *nt* HEIZ & BELÜFT, WERKSTOFF radon; **Radonausschluß** *m* **in Gebäuden** BAURECHT, HEIZ & BELÜFT, INFR & ENTW, WERKSTOFF radon exclusion in buildings; **Radonrisiken** *nt pl* BAURECHT risks from radon

***Rad***: **Radweg** *m* INFR & ENTW bicycle track

**Raffinerierückstände** *m pl* UMWELT refinery waste

**Rähm** *m* HOLZ head runner

**Rahmen** *m* ARCH & TRAGW, INFR & ENTW, STAHL frame; **Rahmenberechnung** *f* ARCH & TRAGW, INFR & ENTW frame analysis

**rahmenfrei** *adj* ARCH & TRAGW frameless

**rahmenlos** *adj* ARCH & TRAGW frameless

***Rahmen***: **Rahmenriegel** *m* ARCH & TRAGW, BETON, INFR & ENTW, STAHL horizontal member; **Rahmenschenkel** *m* HOLZ frame leg, frame piece; **Rahmenstatik** *f* ARCH & TRAGW, INFR & ENTW frame analysis; **Rahmenträgerbrücke** *f* ARCH & TRAGW, BETON, INFR & ENTW portal frame bridge, rigid frame bridge; **Rahmentür** *f* ARCH & TRAGW framed door

**Rahmholz** *nt* ARCH & TRAGW top rail, HOLZ head runner

**Rähmstück** *nt* ARCH & TRAGW summer beam, breastsummer

**Ramm-** *in cpds* ARCH & TRAGW, BAUMASCHIN, INFR & ENTW pile, piling; **Rammarbeiten** *f pl* ARCH & TRAGW, INFR & ENTW pile driving; **Rammbär** *m* ARCH & TRAGW, BAUMASCHIN pile driver, piling hammer; **Rammbrunnen** *m* INFR & ENTW Abyssinian well, driven well

**Ramme** *f* BAUMASCHIN hammer, pile driver, ram

**rammen** *vt* INFR & ENTW pile

**Rammen** *nt* ARCH & TRAGW piling, HOLZ *von Pfählen* spiling

***Ramm-***: **Rammgerüst** *nt* BAUMASCHIN piling frame; **Rammhammer** *m* BAUMASCHIN pile driver; **Rammhaube** *f* BAUMASCHIN head packing, helmet, *Pfahlgründung* cushion head, INFR & ENTW pile cap; **Rammprotokoll** *nt* INFR & ENTW penetration record; **Rammsondierung** *f* INFR & ENTW Standard Penetration Test (*STP*)

**Rampe** *f* ARCH & TRAGW inverted cavetto, ramp, short ramp, INFR & ENTW ramp; **Rampenaufgang** *m* ARCH & TRAGW ramp incline; **Rampenschräge** *f* ARCH & TRAGW ramp incline

**Rand** *m* ARCH & TRAGW border, margin, surround, *eines Abhangs* brow, verge, HOLZ skirt; **Randabstand** *m* HOLZ *senkrecht zur Faser* edge distance; **Randbalken** *m* ARCH & TRAGW, HOLZ, INFR & ENTW boundary beam; **Randbereich** *m* ARCH & TRAGW, INFR & ENTW edge region; **Randbogen** *m* ARCH & TRAGW, INFR & ENTW boundary arch; **Randeinfassung** *f* ARCH & TRAGW surround; **Randfliese** *f* ABWASSER, ARCH & TRAGW, INFR & ENTW, WERKSTOFF border tile; **Randstein** *m* INFR & ENTW, WERKSTOFF curbstone (*AmE*), kerbstone (*BrE*); **Randstreifen** *m* INFR & ENTW verge, *Straßenbau* marginal strip; **Randträger** *m* ARCH & TRAGW rim beam; **Randverstärkung** *f* WERKSTOFF reinforced border; **Randziegel** *m* ARCH & TRAGW, NATURSTEIN margin tile

**Rangierlok** *f* BAUMASCHIN shunting engine (*BrE*), switching engine (*AmE*)

**Rangierwinde** *f* BAUMASCHIN shunting winch (*BrE*), switching winch (*AmE*)

**Rasen**: **Rasengitterstein** *m* INFR & ENTW, WERKSTOFF grass paver; **Rasenstein** *m* INFR & ENTW, WERKSTOFF grass paver; **Rasenziegel** *m* INFR & ENTW sod square

**Raspe** *f* BAUMASCHIN rasp

**Rastbolzen** *m* BESCHLÄGE index bolt

**Raster** *m* ARCH & TRAGW structural module, INFR & ENTW grid; **Rastergrundmaß** *nt* ARCH & TRAGW module; **Rasterlinie** *f* ARCH & TRAGW, INFR & ENTW grid line; **Rastermaß** *nt* ARCH & TRAGW modular dimensions; **Rasterpunkt** *m* ARCH & TRAGW modular point; **Rastersystem** *nt* ARCH & TRAGW bay system

**Raststift** *m* BESCHLÄGE latch pin

**Rathaus** *nt* ARCH & TRAGW, INFR & ENTW city hall, town hall

**Ratsche** *f* BAUMASCHIN ratchet

**Rauch** *m* BAURECHT, HEIZ & BELÜFT smoke, UMWELT fumes; **Rauchabzug** *m* ARCH & TRAGW, BAURECHT smoke funnel, smoke outlet, HEIZ & BELÜFT vent

**rauchdicht** *adj* BAURECHT smoke-tight; **~e Tür** *f* ARCH & TRAGW, BAURECHT, BESCHLÄGE smoke-tight door

***Rauch***: **Rauchgas** *nt* UMWELT fumes

**rauchgasdicht** *adj* ARCH & TRAGW flue-gas-tight

***Rauch***: **Rauchgase** *nt pl* UMWELT fumes; **Rauch-**

gasentschwefelung *f* UMWELT flue gas desulfurization (*AmE*), flue gas desulphurization (*BrE*); **Rauchgasentstaubung** *f* UMWELT particulate collection; **Rauchmelder** *m* BAURECHT, BESCHLÄGE smoke alarm, smoke detector; **Rauchöffnung** *f* ARCH & TRAGW, BAURECHT smoke outlet; **Rauchrohr** *nt* ARCH & TRAGW smoke pipe, smoke tube, BAURECHT smoke pipe, HEIZ & BELÜFT vent; **Rauchtest** *m* BAURECHT smoke test

**rauh** *adj* WERKSTOFF rough; **~er Putz** *m* NATURSTEIN rendering, roughcast

**Rauheit** *f* WERKSTOFF roughness

**Rauh-** *in cpds* WERKSTOFF in grain, rough; **Rauhfasertapete** *f* WERKSTOFF ingrain wallpaper; **Rauhhobel** *m* BAUMASCHIN jack plane

**Rauhigkeit** *f* WERKSTOFF roughness

***Rauh-***: **Rauhmaß** *nt* HOLZ nominal measure; **Rauhputz** *m* NATURSTEIN roughcast

**rauhschleifen** *vt* ARCH & TRAGW rough down

**Raum** *m* ARCH & TRAGW chamber, room, space, BESCHLÄGE, HEIZ & BELÜFT, WERKSTOFF room; **Raumausdehnungskoeffizient** *m* WERKSTOFF coefficient of volume expansion; **Raumbeständigkeit** *f* WERKSTOFF volume stability; **Raumdecke** *f* ARCH & TRAGW ceiling

**räumen** *vt* ARCH & TRAGW vacate

**Räumer** *m* BAUMASCHIN reamer

***Raum***: **Raumfachwerk** *nt* ARCH & TRAGW, INFR & ENTW space truss; **Raumfachwerkträger** *m* ARCH & TRAGW, INFR & ENTW spatial lattice girder; **Raumfuge** *f* ARCH & TRAGW running joint; **Raumgestaltung** *f* ARCH & TRAGW interior decoration, interior design; **Raumgewicht** *nt* WERKSTOFF density, specific gravity

**raumhoch** *adj* ARCH & TRAGW room-high

***Raum***: **Rauminhalt** *m* ARCH & TRAGW capacity, cubage, cubical contents, volume, content; **Raumkonstanz** *f* WERKSTOFF volume stability

**räumlich**: **~es Fachwerk** *nt* ARCH & TRAGW, INFR & ENTW space truss; **~es Gittertragwerk** *nt* ARCH & TRAGW, INFR & ENTW space lattice, space truss; **~e Struktur** *f* UMWELT spatial pattern; **~e Tendenz** *f* UMWELT spatial trend; **~es Tragwerk** *nt* ARCH & TRAGW, INFR & ENTW space frame, space framework, space structure, spatial structure; **~e Veränderlichkeit** *f* UMWELT spatial variability; **~e Verteilung** *f* UMWELT spatial distribution

**Räumlöffel** *m* BAUMASCHIN raker

***Raum***: **Raumluftqualität** *f* HEIZ & BELÜFT indoor air quality

**raumlufttechnisch**: **~e Anlage** *f* HEIZ & BELÜFT ventilation and air-conditioning system

***Raum***: **Raumnutzung** *f* ARCH & TRAGW, BAURECHT, INFR & ENTW space utilization; **Raumordnungsplanung** *f* BAURECHT, INFR & ENTW rural planning; **Raumraster** *m* ARCH & TRAGW modular space grid

**raumsparend** *adj* ARCH & TRAGW, INFR & ENTW space-saving

***Raum***: **Raumteil** *nt* INFR & ENTW, WERKSTOFF part by volume; **Raumteiler** *m* ARCH & TRAGW partition, room divider; **Raumteilung** *f* ARCH & TRAGW partitioning; **Raumtemperatur** *f* HEIZ & BELÜFT, WERKSTOFF room temperature; **Raumthermostat** *m* BESCHLÄGE, HEIZ & BELÜFT room thermostat; **Raumtragwerk** *nt* ARCH & TRAGW, INFR & ENTW space framework, space structure, spatial structure

**Räumung** *f* INFR & ENTW clearing

**Raupe** *f* BAUMASCHIN caterpillar tractor

**Raupenbagger** *m* BAUMASCHIN crawler excavator

**Raupenkette** *f* BAUMASCHIN caterpillar chain

**Raute** *nt* ARCH & TRAGW diamond; **Rautendach** *nt* ARCH & TRAGW helm roof; **Rautendrahtgitter** *nt* BESCHLÄGE diamond wire lattice

**rautenförmig** *adj* ARCH & TRAGW diamond-shaped

***Raute***: **Rautenmuster** *nt* ARCH & TRAGW diamond pattern

**Re** *abbr* (*Reynoldszahl*) INFR & ENTW, UMWELT, WERKSTOFF Re (*Reynolds number*)

**reagieren** *vi* WERKSTOFF react

**Reaktion** *f* WERKSTOFF reaction; **Reaktionsharz** *nt* WERKSTOFF cold-curing resin; **Reaktionsklebestoff** *m* WERKSTOFF cold-setting adhesive; **Reaktionsmittel** *nt* UMWELT solidifying agent; **Reaktionsturbine** *f* UMWELT reaction turbine; **Reaktionsverzögerer** *m* UMWELT retarder, retarding agent

**Rechen** *m* ABWASSER trash rack, ARCH & TRAGW calculation, UMWELT trash rack

***Rechen***: **Rechenformel** *f* ARCH & TRAGW, INFR & ENTW calculating formula; **Rechenmodell** *nt* ARCH & TRAGW, INFR & ENTW mathematical model

**rechenschaftspflichtig** *adj* ARCH & TRAGW, BAURECHT accountable

***Rechen***: **Rechentechnik** *f* ARCH & TRAGW computation

**rechnerisch** *adj* ARCH & TRAGW, BAURECHT, INFR & ENTW, WERKSTOFF designed; **~e Spannungsdehnungslinie** *f* ARCH & TRAGW, BETON, WERKSTOFF designed stress-strain diagram

**Rechnung** *f* ARCH & TRAGW, BAURECHT account, invoice; **Rechnungsprüfer** *m* BAURECHT auditor; **Rechnungswesen** *nt* BAURECHT accounting

**rechte** *adj* ARCH & TRAGW right; **rechter Winkel** *m* ARCH & TRAGW right angle

**Rechteck** *nt* ARCH & TRAGW triangular; **Rechteckdeckleiste** *f* ARCH & TRAGW square staff

**rechteckig** *adj* ARCH & TRAGW rectangular; **~er Querschnitt** *m* ARCH & TRAGW rectangular cross-section; **~e Verblattung** *f* ARCH & TRAGW square splice

***Rechteck***: **Rechteckkanal** *m* ELEKTR, HEIZ & BELÜFT rectangular conduit

**rechts** *adv* ARCH & TRAGW right

**Rechtsanwalt** *m* BAURECHT solicitor; **Rechtsanwaltshonorar** *nt* BAURECHT solicitor's fee

**rechtskräftig** *adj* BAURECHT legally valid

**rechtswirkend**: **~er Türschließer** *m* BESCHLÄGE right-hand door closer

**Rechtswirksamkeit** *f* BAURECHT *eines Vertrages* validity

**rechtwinklig** *adj* ARCH & TRAGW rectangular; **~es Dreieck** *nt* ARCH & TRAGW rectangular triangle; **~er Falzhobel** *m* BAUMASCHIN square rabbet plane; **~e Verbindung** *f* ARCH & TRAGW square joint

**Rechtwinkligschneiden** *nt* ARCH & TRAGW squaring

**recyceln** *vt* UMWELT recycle

**Recycling** *nt* UMWELT recycling; **Recyclinganlage** *f* UMWELT recycling plant

**recyclingfähig** *adj* UMWELT recyclable

***Recycling***: **Recyclingpapier** *nt* UMWELT recycled paper; **Recyclingprozeß** *m* UMWELT recycling process; **Recyclingquote** *f* UMWELT recycling rate

**Reduktionsstück** *nt* ABWASSER, HEIZ & BELÜFT decreaser, reducing fitting, transition piece
**Reduzierstück** *nt* ABWASSER reducer, reducing pipe fitting, ARCH & TRAGW reducer
**Reet** *nt* ARCH & TRAGW thatch; **mit ~ decken** *phr* ARCH & TRAGW thatch; **Reetdach** *nt* ARCH & TRAGW reed roof
**Reflektor** *m* ELEKTR *Leuchte* reflector
**Reflexionsseismik** *f* INFR & ENTW seismic reflections
**Refraktion** *f* INFR & ENTW refraction
**Regel** *f* BAURECHT rule; **Regeleinrichtung** *f* ELEKTR, HEIZ & BELÜFT controller, controlling equipment, control system; **Regelgerät** *nt* ELEKTR, HEIZ & BELÜFT control instrument
**regelmäßig** *adj* INFR & ENTW regular; **nicht ~ verteilte Kräfte** *f pl* ARCH & TRAGW noncomplanar forces; **~e Wartung** *f* ARCH & TRAGW, BAUMASCHIN, BAURECHT scheduled service
**regeln** *vt* BAURECHT adjust
***Regel***: **Regelquerschnitt** *m* ABWASSER, ARCH & TRAGW, HEIZ & BELÜFT, INFR & ENTW typical cross section
**Regelung** *f* BAURECHT, BESCHLÄGE adjustment, ELEKTR, HEIZ & BELÜFT controlling equipment
**Regen** *m* UMWELT rain
**regenabweisend** *adj* OBERFLÄCHE, WERKSTOFF rain-repellent
***Regen***: **Regenabweiser** *m* ARCH & TRAGW rain repeller; **Regenbeständigkeit** *f* INFR & ENTW, WERKSTOFF rain resistance; **Regendichte** *f* INFR & ENTW intensity of rainfall
**Regeneration** *f* UMWELT regeneration
**regenerativ**: **~e Energie** *f* UMWELT renewable energy
***Regen***: **Regenfallrohr** *nt* ABWASSER stack pipe; **Regenleiste** *f* ARCH & TRAGW stormwater deflector; **Regenmeßgerät** *nt* BAUMASCHIN rain gage (*AmE*), rain gauge (*BrE*); **Regenrinne** *f* ABWASSER stormwater gutter; **Regenrohrschelle** *f* BESCHLÄGE stormwater pipe clamp; **Regenrückhaltebecken** *nt* ABWASSER, INFR & ENTW rainwater retention basin, stormwater retention basin; **Regenschaden** *m* WERKSTOFF rain damage; **Regenschauer** *m* ABWASSER shower; **Regenschirmschale** *f* ARCH & TRAGW umbrella shell; **Regenschutz** *m* INFR & ENTW rain protection; **Regenschutzschiene** *f* ARCH & TRAGW stormwater deflector; **Regenspende** *f* INFR & ENTW, UMWELT rainfall per second per area
**regenundurchlässig** *adj* OBERFLÄCHE, WERKSTOFF rainproof
***Regen***: **Regenwasser** *nt* ABWASSER, INFR & ENTW rainwater, stormwater; **Regenwasserablauf** *m* ABWASSER rainwater inlet, stormwater inlet, gully, INFR & ENTW gully; **Regenwasserfallrohr** *nt* ABWASSER rainwater downpipe; **Regenwasserkanal** *m* ABWASSER, INFR & ENTW rain-drainage channel, storm sewer; **Regenwasserleitung** *f* INFR & ENTW storm drain; **Regenwassernutzung** *f* UMWELT utilization of rainwater
**Regler** *m* ELEKTR, HEIZ & BELÜFT controller, regulator
**regulierbar**: **~es Grundwehr** *nt* UMWELT adjustable submersion weir
**regulieren** *vt* BAUMASCHIN, BESCHLÄGE adjust, HEIZ & BELÜFT adjust, regulate
**Regulierung** *f* BAUMASCHIN, BESCHLÄGE, HEIZ & BELÜFT adjustment, regulation
**Regulierventil** *nt* HEIZ & BELÜFT control valve, regulating valve
**Reibahle** *f* BAUMASCHIN reamer
**Reibeisen** *nt* BAUMASCHIN rasp
**Reibeputz** *m* NATURSTEIN float and set
**Reibung** *f* WERKSTOFF friction; **äußere Reibung** *f* INFR & ENTW external friction; **Reibungsbeiwert** *m* WERKSTOFF coefficient of friction; **Reibungselektrizität** *f* ELEKTR static electricity; **Reibungskoeffizient** *m* ARCH & TRAGW, INFR & ENTW, WERKSTOFF friction coefficient; **Reibungskraft** *f* ARCH & TRAGW, INFR & ENTW, WERKSTOFF friction force; **Reibungskreis** *m* INFR & ENTW, WERKSTOFF friction circle
**reibungslos** *adj* INFR & ENTW, WERKSTOFF frictionless
***Reibung***: **Reibungspfahl** *m* INFR & ENTW, WERKSTOFF friction pile; **Reibungsspannung** *f* ARCH & TRAGW, INFR & ENTW, WERKSTOFF frictional stress; **Reibungstest** *m* INFR & ENTW, WERKSTOFF friction test; **Reibungsverbund** *m* ARCH & TRAGW, BETON, INFR & ENTW, STAHL, WERKSTOFF friction connection; **Reibungsverlust** *m* INFR & ENTW, UMWELT, WERKSTOFF friction loss, frictional loss; **Reibungswiderstand** *m* ARCH & TRAGW, INFR & ENTW, WERKSTOFF frictional resistance; **Reibungswinkel** *m* INFR & ENTW, WERKSTOFF angle of friction; **Reibungszugkraft** *f* ARCH & TRAGW, INFR & ENTW *Windkraft* frictional drag
**Reif** *m* INFR & ENTW frost
**Reihe** *f* ARCH & TRAGW row; **Reihenfolge** *f* BAURECHT order; **Reihenschaltung** *f* ELEKTR series mounting
**Rein-** *in cpds* HOLZ clean, UMWELT clean, cleaned
**rein**: **~er Kalkmörtel** *m* BETON, NATURSTEIN, WERKSTOFF neat lime
***Rein-***: **Reingas** *nt* HEIZ & BELÜFT, INFR & ENTW, UMWELT natural gas; **Reinhaltung** *f* UMWELT *Luft, Gewässer* pollution control
**reinigen** *vt* BAURECHT, INFR & ENTW purify
**Reiniger** *m* OBERFLÄCHE, WERKSTOFF cleaner, cleaning agent
**Reinigung** *f* INFR & ENTW clarification, OBERFLÄCHE cleaning; **Reinigungsdeckel** *m* ABWASSER cleanout cover; **Reinigungsfällung** *f* UMWELT below-cloud scavenging; **Reinigungsformstück** *nt* ABWASSER cleanout fitting; **Reinigungsgrad** *m* INFR & ENTW, WERKSTOFF degree of clarification; **Reinigungsmittel** *nt* WERKSTOFF cleaner, cleaning agent; **Reinigungsöffnung** *f* ABWASSER cleanout opening; **Reinigungsstopfen** *m* ABWASSER cleanout plug; **Reinigungsvermögen** *nt* UMWELT purification capacity; **Reinigungsverschluß** *m* ABWASSER cleanout opening; **Reinigungswagen** *m* BAUMASCHIN cradle machine
***Rein-***: **Reinluft** *f* HEIZ & BELÜFT, INFR & ENTW clean air; **Reinluftgebiet** *nt* UMWELT area of pure air; **Reinraum** *m* ARCH & TRAGW, HEIZ & BELÜFT clean room; **Reinregen** *m* UMWELT clean rain
**Reiß-** *in cpds* BAUMASCHIN, WERKSTOFF scratch, scribing, tear
**reißen 1.** *vt* ARCH & TRAGW, INFR & ENTW tear; **2.** *vi* BETON, NATURSTEIN, WERKSTOFF crack
***Reiß-***: **Reißfestigkeit** *f* WERKSTOFF tear strength; **Reißlehre** *f* BAUMASCHIN scribing gage (*AmE*), scribing gauge (*BrE*); **Reißnadel** *f* BAUMASCHIN scratch awl; **Reißspitze** *f* BAUMASCHIN scribing awl
**rekonstruieren** *vt* ARCH & TRAGW reconstruct
**Rekonstruktion** *f* ARCH & TRAGW redevelopment, reconstruction

**Rekontamination** *f* BAURECHT, UMWELT recontamination
**Rekultivieren** *nt* BAURECHT, INFR & ENTW land restoration
**Rekultivierung** *f* INFR & ENTW, UMWELT revegetation
**relativ**: **~e Feuchtigkeit** *f* DÄMMUNG relative humidity, HEIZ & BELÜFT relative water velocity, WERKSTOFF relative humidity; **~e Wassergeschwindigkeit** *f* UMWELT relative water velocity
**Relaxation** *f* BETON relaxation
**renovieren** *vt* ARCH & TRAGW facelift, refurbish, renovate
**Renovierung** *f* ARCH & TRAGW, OBERFLÄCHE renovation; **Renovierungsanstrich** *m* OBERFLÄCHE renovation coat
**Reparatur** *f* ARCH & TRAGW, BAUMASCHIN, OBERFLÄCHE repair; **Reparaturhaken** *m* ARCH & TRAGW roof hook
**reparieren** *vt* ARCH & TRAGW, BAUMASCHIN, OBERFLÄCHE repair
**Reservoir** *nt* INFR & ENTW container
**Resonanz** *f* DÄMMUNG, UMWELT, WERKSTOFF resonance; **Resonanzsieb** *nt* UMWELT resonance screen
**Resorcin-Formaldehyd** *nt* (*RF*) HOLZ, UMWELT, WERKSTOFF resorcinol-formaldehyde (*RF*)
**respirieren** *vt* UMWELT respire
**Rest** *m* BAURECHT, WERKSTOFF, UMWELT remainder, residue, remnant; **Restarbeitenliste** *f* BAURECHT snagging list
**restaurieren** *vt* ARCH & TRAGW restore
***Rest***: **Restspannung** *f* WERKSTOFF residual stress; **Reststoff** *m* UMWELT remainder; **Reststoffdeponie** *f* UMWELT residue landfill
**Resultat** *nt* INFR & ENTW result
**resultierend**: **~e Kraft** *f* ARCH & TRAGW, INFR & ENTW resultant force
**Resultierende** *f* ARCH & TRAGW, INFR & ENTW resultant
**Revision** *f* ABWASSER, ARCH & TRAGW, BAURECHT, INFR & ENTW revision, inspection; **Revisionsplan** *m* ARCH & TRAGW, BAURECHT as-built drawing; **Revisionsschacht** *m* ABWASSER, INFR & ENTW inspection manhole; **Revisionsstück** *nt* ABWASSER, INFR & ENTW inspection fitting, inspection piece; **Revisionszeichnung** *f* ARCH & TRAGW, BAURECHT as-built drawing, revised drawing
**Reynoldszahl** *f* (*Re*) INFR & ENTW, UMWELT, WERKSTOFF Reynolds number (*Re*)
**RF** *abbr* (*Resorcin-Formaldehyd*) HOLZ, UMWELT, WERKSTOFF RF (*resorcinol-formaldehyde*)
**Rheinsand** *m* WERKSTOFF Rhine sand
**Rhombendach** *nt* ARCH & TRAGW helm roof
**Ribbungsfaktor** *m* BETON, WERKSTOFF rib factor
**Richt-** *in cpds* ARCH & TRAGW tilt-up; **Richtaufbauweise** *f* ARCH & TRAGW tilt-up method; **Richtblei** *nt* ARCH & TRAGW, BAUMASCHIN plummet
**richten** *vt* ARCH & TRAGW straighten
**Richten** *nt* ARCH & TRAGW straightening, STAHL *Blech* flattening
**Richterskala** *f* INFR & ENTW Richter scale
***Richt-***: **Richtfest** *nt* ARCH & TRAGW topping-out ceremony
**richtig** *adj* BAURECHT correct
***Richt-***: **Richtlatte** *f* ARCH & TRAGW level, ruler; **Richtlinie** *f* BAURECHT code of practice, guideline; **Richtprofil** *nt* ARCH & TRAGW, INFR & ENTW gage (*AmE*), gauge (*BrE*); **Richtscheit** *m* ARCH & TRAGW level; **Richtstellung** *f* ARCH & TRAGW, OBERFLÄCHE aiming position; **Richtstollen** *m* INFR & ENTW pilot drift
**Richtungswinkel** *m* ARCH & TRAGW, INFR & ENTW, WERKSTOFF azimuth, bearing
***Richt-***: **Richtwaage** *f* ARCH & TRAGW, BAUMASCHIN level; **Richtwert** *m* ARCH & TRAGW, INFR & ENTW, WERKSTOFF approximate value
**Rieddach** *nt* ARCH & TRAGW reed roof
**Riegel** *m* ARCH & TRAGW beam, cross-frame member, BESCHLÄGE lock bolt, shutter, BETON beam, waler, waling, HOLZ ledger, cross-frame member, INFR & ENTW ranger, waler, waling, STAHL beam, cross-frame member, WERKSTOFF ranger; **Riegelbolzen** *m* BESCHLÄGE, WERKSTOFF bolt; **Riegelwand** *f* ARCH & TRAGW framed wall, HOLZ timber framework, NATURSTEIN framework wall
**Riemchen** *nt* ARCH & TRAGW listel, NATURSTEIN closer, queen closer, WERKSTOFF quarter brick, *Fliesen* fillet
**Riemen** *m* BAUMASCHIN belt, HOLZ *Fußboden* matchboarding; **Riemenspanner** *m* BAUMASCHIN belt stretcher; **Riemenstück** *nt* NATURSTEIN quarter brick; **Riementrieb** *m* BAUMASCHIN belt drive
**Riesel-** *in cpds* INFR & ENTW, UMWELT irrigation; **Rieselfeld** *nt* INFR & ENTW, UMWELT irrigation field; **Rieselfurche** *f* INFR & ENTW irrigation furrow
**Riffelblech** *nt* STAHL, WERKSTOFF checkered plate (*AmE*), chequered plate (*BrE*)
**Riffelglas** *nt* WERKSTOFF ribbed glass
**Riffeln** *nt* ARCH & TRAGW channeling (*AmE*), channelling (*BrE*), WERKSTOFF fluting
**Riffelung** *f* BESCHLÄGE flute
**Rigipsplatte**® *f* WERKSTOFF plasterboard
**Rigole** *f* ABWASSER gravel-filled drain trench
**Rille** *f* ARCH & TRAGW furrow, *Baustil* groove
**Ring** *m* ARCH & TRAGW, BAUMASCHIN, ELEKTR, INFR & ENTW ring; **Ringanker** *m* ARCH & TRAGW ring beam, ring girder
**ringarmiert**: **~e Säule** *f* ARCH & TRAGW *Stahlbeton*, BETON, INFR & ENTW hooped column
***Ring***: **Ringbalken** *m* ARCH & TRAGW, BETON circular beam, ring beam, ring girder
**ringbewehrt**: **~e Säule** *f* ARCH & TRAGW *Stahlbeton*, BETON, INFR & ENTW hooped column
***Ring***: **Ringerder** *m* ELEKTR ring earth (*BrE*), ring ground (*AmE*); **Ringerdersystem** *nt* ELEKTR ring earth system (*BrE*), ring ground system (*AmE*); **Ringgewölbe** *nt* ARCH & TRAGW, NATURSTEIN annular vault; **Ringkanalsystem** *nt* INFR & ENTW perimeter system; **Ringkluft** *f* HOLZ annular shake; **Ringleitung** *f* ABWASSER, HEIZ & BELÜFT, INFR & ENTW closed circular pipeline
**Ringscher**: **Ringschertest** *m* BAUMASCHIN, INFR & ENTW, WERKSTOFF ring-shear test; **Ringschertester** *m* BAUMASCHIN, WERKSTOFF ring-shear tester; **Ringschertestgerät** *nt* BAUMASCHIN, INFR & ENTW ring-shear tester
***Ring***: **Ringträger** *m* ARCH & TRAGW ring girder, circular girder; **Ringziegel** *m* NATURSTEIN compass brick
**Rinne** *f* ABWASSER chute, ARCH & TRAGW trench, INFR & ENTW gutter; **Rinnenbalkenträger** *m* ARCH & TRAGW valley beam, valley girder; **Rinnendeckel** *m* ABWASSER, INFR & ENTW gully cover; **Rinnenfirst** *m* ABWASSER gutter ridge; **Rinnenhaken** *m* BESCHLÄGE gutter bracket; **Rinnenhalter** *m* BESCHLÄGE gutter bracket, gutter clamp, gutter hanger; **Rinnenkasten**

*m* ABWASSER rainwater head; **Rinnenstutzen** *m* ARCH & TRAGW, BESCHLÄGE gutter outlet

**Rinn**: **Rinnhaken** *m* BESCHLÄGE gutter bracket; **Rinnleiste** *f* ARCH & TRAGW weather molding (*AmE*), weather moulding (*BrE*); **Rinnstein** *m* INFR & ENTW, WERKSTOFF curbstone (*AmE*), kerbstone (*BrE*)

**Rippe** *f* ARCH & TRAGW rib; **Rippendecke** *f* ARCH & TRAGW ribbed floor, ribbed ceiling; **Rippenheizkörper** *m* HEIZ & BELÜFT finned radiator, ribbed radiator; **Rippenplatte** *f* ARCH & TRAGW ribbed concrete floor; **Rippenstreckmetall** *nt* NATURSTEIN rib mesh

**Rispe** *f* ARCH & TRAGW cocking piece

**Riß** *m* ARCH & TRAGW *durch Witterungseinflüsse* weather check, *in Wand* fissure, BETON crack, INFR & ENTW fissure, NATURSTEIN, WERKSTOFF crack; **Rißbewehrung** *f* BETON anticrack reinforcement; **Rißbildung** *f* BETON, NATURSTEIN, WERKSTOFF crack formation, cracking; **Rißbreite** *f* BETON, NATURSTEIN crack width

**risseüberbrückend**: **~es Armierungsgewebe** *nt* NATURSTEIN crack-covering reinforcement tape

***Riß***: **Rißgefahr** *f* BETON, NATURSTEIN cracking risk

**rissig**: **~ werden** *vi* WERKSTOFF fissure

**rißsicher** *adj* BETON, INFR & ENTW, WERKSTOFF crackproof

***Riß***: **Rißspannung** *f* BETON, INFR & ENTW, NATURSTEIN, WERKSTOFF cracking stress; **Rißüberbrückung** *f* NATURSTEIN crack bridging

**Ritterdach** *nt* ARCH & TRAGW crown-tile roof

**robust**: **~e Ausführung** *f* ARCH & TRAGW sturdy design

**Rödeldraht** *m* BETON binding wire, tie wire

**Rodung** *f* INFR & ENTW uprooting

**roh** *adj* BETON, OBERFLÄCHE, UMWELT, WERKSTOFF untreated

**Roh-** *in cpds* UMWELT, WERKSTOFF crude, raw, rough

***roh***: **~e Schraube** *f* STAHL black bolt

***Roh-***: **Rohabwasser** *nt* ABWASSER, BAURECHT, INFR & ENTW raw sewage

**Rohbau** *m* ARCH & TRAGW preliminary building works, carcass, main works; **Rohbauarbeiten** *f pl* ARCH & TRAGW, HOLZ carcassing, rough work, shell work

**rohbaufertig** *adj* ARCH & TRAGW topped-out

***Rohbau***: **Rohbaulichtmaß** *nt* ARCH & TRAGW clear opening dimensions, rough opening dimensions; **Rohbauskelett** *nt* ARCH & TRAGW skeleton

***Roh-***: **Rohbehauen** *nt* ARCH & TRAGW, NATURSTEIN *Stein* rough dressing

***roh***: **~ behauenes Quaderwerk** *nt* NATURSTEIN rustication

***Roh-***: **Rohdichte** *f* INFR & ENTW, WERKSTOFF apparent density, bulk density; **Rohfußboden** *m* ARCH & TRAGW unfinished floor; **Rohgips** *m* NATURSTEIN plaster rock, plaster stone

**Rohling** *m* WERKSTOFF blank

***Roh-***: **Rohmauerung** *f* NATURSTEIN rough walling; **Rohmüll** *m* UMWELT raw refuse; **Rohputz** *m* NATURSTEIN coarse plaster

**Rohr** *nt* ABWASSER, BESCHLÄGE, HEIZ & BELÜFT, UMWELT, WERKSTOFF conduit, pipe, tube; **in ~e leiten** *phr* ABWASSER, INFR & ENTW, UMWELT pipe; **~e** *nt pl* **und Armaturen** BESCHLÄGE pipes and fittings; **~e verlegen** *phr* ABWASSER pipe; **Rohranschluß** *m* ABWASSER pipe connection; **Rohraufweitepresse** *f* BAUMASCHIN tube-expanding press; **Rohraufweiter** *m* BESCHLÄGE tube expander; **Rohrbau** *m* ABWASSER, BESCHLÄGE pipework; **Rohrbegleitheizung** *f* ABWASSER, ELEKTR pipeline heating; **Rohrbelüfter** *m* ABWASSER antisiphoning device, antivacuum device, HEIZ & BELÜFT antivacuum device; **Rohrbogen** *m* ABWASSER bend, pipe bend, BESCHLÄGE bend, INFR & ENTW ell; **Rohrbrunnen** *m* INFR & ENTW tubular well; **Rohrbündel** *nt* BESCHLÄGE tube nest; **Rohrdichter** *m* BAUMASCHIN casing expander; **Rohrdichtung** *f* ABWASSER pipe gasket; **Rohrdurchlaß** *m* ABWASSER, INFR & ENTW *Straßen* pipe culvert

**Röhre** *f* ABWASSER, BESCHLÄGE, HEIZ & BELÜFT, WERKSTOFF pipe, tube

***Rohr***: **Rohrelevator** *m* BAUMASCHIN casing elevator

**röhrenförmig** *adj* ARCH & TRAGW tubular

***Röhre***: **Röhrenheizkörper** *m* HEIZ & BELÜFT column radiator, tubular heating element, tubular radiator; **Röhrensiphon** *m* ABWASSER stink trap

***Rohr***: **Rohrentlüfter** *m* ABWASSER antisiphoning device, antivacuum device, HEIZ & BELÜFT antivacuum device; **Rohrgefälle** *nt* ABWASSER, ARCH & TRAGW, INFR & ENTW pipe gradient; **Rohrgerüst** *nt* ARCH & TRAGW tubular scaffolding; **Rohrgewebe** *nt* NATURSTEIN reed lathing; **Rohrgewinde** *nt* ABWASSER pipe thread; **Rohrgewindebohrer** *m* BAUMASCHIN pipe tap; **Rohrgraben** *m* ABWASSER pipe top; **Rohrhaken** *m* ABWASSER pipe hook; **Rohrhülse** *f* ABWASSER, BESCHLÄGE, HEIZ & BELÜFT tube sleeve, tube socket, tube-jointing sleeve; **Rohrkopf** *m* BAUMASCHIN casing head; **Rohrkrümmer** *m* ABWASSER quadrant pipe, bend, pipe bend, pipe knee, BESCHLÄGE bend

**Rohrleitung** *f* ABWASSER, BESCHLÄGE, UMWELT conduit, pipeline, tubing; **Rohrleitungsbau** *m* INFR & ENTW pipeline construction; **Rohrleitungsnetz** *nt* ARCH & TRAGW, INFR & ENTW pipework; **Rohrleitungsplan** *m* ARCH & TRAGW piping plan; **Rohrleitungsverlegung** *f* ABWASSER pipe laying

***Rohr***: **Rohrmanschette** *f* ABWASSER pipe collar; **Rohrmatte** *f* NATURSTEIN reed lathing; **Rohrmuffe** *f* ABWASSER pipe joint; **Rohrnetz** *nt* ABWASSER, ARCH & TRAGW, INFR & ENTW pipe system, system of pipes, UMWELT tubing; **Rohrpfahl** *m* ARCH & TRAGW tubular pile; **Rohrquerschnitt** *m* ABWASSER, ARCH & TRAGW pipe cross section; **Rohrreduzierstück** *nt* ABWASSER pipe reducer; **Rohrscheitel** *m* ABWASSER pipe top; **Rohrschelle** *f* ABWASSER pipe strap, BESCHLÄGE tube clip, WERKSTOFF pipe clamp, pipe hanger; **Rohrschlosser** *m* ABWASSER pipe fitter; **Rohrschlüssel** *m* ABWASSER pipe wrench; **Rohrschneider** *m* ABWASSER pipe cutter, BAUMASCHIN casing cutter; **Rohrschraubstock** *m* BAUMASCHIN pipe vice (*BrE*), pipe vise (*AmE*); **Rohrschweißung** *f* STAHL tube welding; **Rohrstahl** *m* WERKSTOFF pipe steel; **Rohrsteckverbindung** *f* ABWASSER spigot joint; **Rohrstrang** *m* ABWASSER pipeline; **Rohrstutzen** *m* WERKSTOFF pipe socket; **Rohrverbindung** *f* ABWASSER pipe connection, pipe coupling, pipe joint; **Rohrverbindungsstück** *nt* ABWASSER pipe union, HEIZ & BELÜFT fitting; **Rohrverlegung** *f* ABWASSER pipe laying; **Rohrverschraubung** *f* ABWASSER screwed pipe joint, threaded pipe connection, compression fitting, pipe screwing, BESCHLÄGE tube fitting, HEIZ & BELÜFT screwed pipe joint, compression fitting, union; **Rohrverstopfung** *f* ABWASSER pipe blockage; **Rohrweite** *f*

WERKSTOFF pipe diameter; **Rohrwickler** *m* ABWASSER pipe twister; **Rohrwiderstand** *m* INFR & ENTW pipe resistance; **Rohrzange** *f* ABWASSER pipe wrench, BAUMASCHIN pipe tongs

***Roh-***: **Rohschlamm** *m* UMWELT raw sludge, WERKSTOFF *Zementherstellung* slurry; **Rohstoff** *m* WERKSTOFF raw material; **Rohstoffrückgewinnung** *f* UMWELT resource recovery; **Rohwasser** *nt* INFR & ENTW untreated water

**Roll-** *in cpds* ARCH & TRAGW, BESCHLÄGE, INFR & ENTW, OBERFLÄCHE roller

**Rolladen** *m* ARCH & TRAGW, BESCHLÄGE *an Fenster, Tür* roller shutter, shutter; **Rolladenkasten** *m* ARCH & TRAGW roller shutter housing; **Rolladentor** *nt* ARCH & TRAGW rolling shutter

***Roll-***: **Rollbrücke** *f* INFR & ENTW roller bridge, rolling bridge

**Rolle** *f* BAUMASCHIN pulley, BESCHLÄGE *einer Walze* barrel, WERKSTOFF cylinder; **Rollenlager** *nt* ARCH & TRAGW roller bearing, BAUMASCHIN rolling bearing

***Roll-***: **Rollfalle** *f* BESCHLÄGE *Schloß* roller bolt; **Rollgrenze** *f* INFR & ENTW, WERKSTOFF plastic limit

**rollig** *adj* INFR & ENTW *Boden* friable, WERKSTOFF nonplastic

***Roll-***: **Rollkies** *m* WERKSTOFF small cobbles; **Rollschicht** *f* INFR & ENTW *Mauerwerk* upright course, NATURSTEIN brick-on-edge course, rolock, rowlock; **Rollschrank** *m* HOLZ shutter cabinet; **Rollsplitt** *m* INFR & ENTW, WERKSTOFF loose chippings

**Rollstuhl** *m* BAURECHT wheelchair; **Rollstuhlanlagen** *f pl* ARCH & TRAGW, BAURECHT, INFR & ENTW wheelchair facilities

**rollstuhlgeeignet** *adj* WERKSTOFF suitable for wheelchairs

***Rollstuhl***: **Rollstuhlplatz** *m* ARCH & TRAGW, BAURECHT, INFR & ENTW wheelchair space; **Rollstuhlrampe** *f* ARCH & TRAGW, BAURECHT, INFR & ENTW wheelchair ramp; **Rollstuhlzugang** *m* ARCH & TRAGW, BAURECHT access for wheelchairs

***Roll-***: **Rolltor** *nt* ARCH & TRAGW *senkrecht rollend* rolling shutter door, *waagrecht rollend* sliding gate; **Rolltreppe** *f* INFR & ENTW escalator, moving staircase, moving stairway

**Romanzement** *m* WERKSTOFF Roman cement

**römisch** *adj* ARCH & TRAGW Roman

**Röntgenstrahl** *m* INFR & ENTW, UMWELT, WERKSTOFF *Prüfung* X-ray

**Rosette** *f* ARCH & TRAGW rosette, BESCHLÄGE *Tür* rose

**Rost** *m* OBERFLÄCHE, STAHL rust, *Gitter* grid iron

**rosten** *vi* OBERFLÄCHE, STAHL rust

***Rost***: **Rostentferner** *m* OBERFLÄCHE, WERKSTOFF derusting agent

**rostfest** *adj* OBERFLÄCHE, STAHL, WERKSTOFF rustproofed, stainless

***Rost***: **Rostfilm** *m* STAHL rust film; **Rostfläche** *f* ARCH & TRAGW grate area

**rostfrei** *adj* OBERFLÄCHE, STAHL, WERKSTOFF rustproofed; **~er Chromstahl** *m* WERKSTOFF stainless chromium steel; **~es Spülbecken** *nt* ABWASSER stainless steel sink; **~er Stahlsturz** *m* ARCH & TRAGW, STAHL stainless steel lintel

**rostgeschützt** *adj* STAHL rustproofed

**Rostschutz** *m* OBERFLÄCHE rustproofing; **Rostschutzfarbe** *f* OBERFLÄCHE rustproofing paint; **Rostschutzgrundierung** *f* OBERFLÄCHE, STAHL, WERKSTOFF antirust primer, rustproofing primer; **Rostschutzmittel** *nt* OBERFLÄCHE rustproofing agent

**rot**: **~er Stab** *m* UMWELT red rod

**Rotation** *f* ARCH & TRAGW rotation; **Rotationsachse** *f* ARCH & TRAGW, INFR & ENTW axis of rotation; **Rotationsbohren** *nt* INFR & ENTW rotary drilling; **Rotationskapazität** *f* ARCH & TRAGW, BETON, WERKSTOFF rotation capacity; **Rotationsofen** *m* UMWELT rotary furnace

**Rot**: **Rotguß** *m* WERKSTOFF red brass; **Rotocker** *m* INFR & ENTW red ocher (*AmE*), red ochre (*BrE*); **Rotschlamm** *m* UMWELT red mud

**Rottedeponie** *f* UMWELT digestion deposit

**Rück-** *in cpds* ARCH & TRAGW back; **Rückansicht** *f* ARCH & TRAGW, INFR & ENTW back view

**Rücken**: **~ an Rücken** *phr* ARCH & TRAGW back-to-back; **Rückenschutz** *m* ARCH & TRAGW *Leiter an Schornstein, Kran* safety cage

***Rück-***: **Rückfließen** *nt* HEIZ & BELÜFT reverse flow; **Rückfluß** *m* ABWASSER backflow; **Rückflußverhinderer** *m* ABWASSER backflow preventer; **Rückführung** *f* UMWELT recycling

**rückgewinnbar**: **~er Abfall** *m* UMWELT recoverable waste; **nicht ~er Abfall** *m* UMWELT nonrecoverable waste

**Rückgewinnung** *f* UMWELT recycling; **Rückgewinnungsanlage** *f* UMWELT reclamation plant; **Rückgewinnungskessel** *m* UMWELT recovery boiler

**rückgewonnen**: **~e Pulpe** *f* UMWELT recovered pulp; **~e Wärme** *f* HEIZ & BELÜFT, UMWELT recovered heat

***Rück-***: **Rückhaltebecken** *nt* INFR & ENTW retention basin; **Rücklagenwand** *f* INFR & ENTW retention basin

**Rücklauf** *m* HEIZ & BELÜFT *Leitung* return flow; **Rücklaufschlamm** *m* UMWELT return sludge, recycle sludge; **Rücklauftemperatur** *f* HEIZ & BELÜFT return temperature; **Rücklaufwasser** *nt* BAUMASCHIN, HEIZ & BELÜFT return water

***Rück-***: **Rückschlagklappe** *f* ABWASSER flap trap; **Rückschlagventil** *nt* ABWASSER, BAUMASCHIN, UMWELT check valve, nonreturn valve; **Rückseite** *f* ARCH & TRAGW back, rear side; **Rücksprung** *m* ARCH & TRAGW offset, recess, set-off, INFR & ENTW retreat, NATURSTEIN *Mauerwerk* offset; **Rückstand** *m* UMWELT residue, WERKSTOFF residue, *Sieb* screen residue; **Rückstandsdeponie** *f* UMWELT residue landfill

**Rückstau** *m* ABWASSER backpressure; **Rückstaudruck** *m* ABWASSER backpressure; **Rückstauebene** *f* ABWASSER backpressure level; **Rückstauklappe** *f* ABWASSER check valve, ARCH & TRAGW trap; **Rückstauverhinderer** *m* ABWASSER backpressure preventer; **Rückstauverschluß** *m* ABWASSER antiflooding valve, backwash shut-off valve; **Rückstauwirkung** *f* INFR & ENTW, UMWELT backwater effect

***Rück-***: **Rückstrahlungsvermögen** *nt* UMWELT reflectance; **Rückvergütungsvertrag** *m* BAURECHT cost reimbursement contract; **Rückwand** *f* ARCH & TRAGW rear wall

**rückwärtsschreitend**: **~e Korrosion** *f* INFR & ENTW, OBERFLÄCHE, WERKSTOFF retrogressive erosion

**Ruflampe** *f* BAURECHT, ELEKTR emergency call lamp

**Ruhe** *f* ARCH & TRAGW, INFR & ENTW repose, rest; **Ruhelage** *f* ARCH & TRAGW, INFR & ENTW equilibrium

**ruhend**: **~e Last** *f* ARCH & TRAGW, INFR & ENTW permanent load

***Ruhe***: **Ruheplattform** *f* ARCH & TRAGW, BAUMASCHIN *Stahlgitterturmaufgang* resting platform; **Ruhewinkel** *m* ARCH & TRAGW angle of repose; **Ruhezustand** *m* ARCH & TRAGW, BAUMASCHIN, INFR & ENTW state of rest

**Rühr-** *in cpds* BAUMASCHIN, BETON, INFR & ENTW, OBERFLÄCHE agitating, UMWELT rabble, WERKSTOFF agitating; **Rührarm** *m* UMWELT rabble arm; **Rühreinrichtung** *f* BAUMASCHIN, BETON, INFR & ENTW, OBERFLÄCHE, WERKSTOFF agitating equipment

**rühren** *vt* BAUMASCHIN, BETON, INFR & ENTW, OBERFLÄCHE, WERKSTOFF agitate

**Rühren** *nt* BAUMASCHIN, BETON, INFR & ENTW, OBERFLÄCHE, WERKSTOFF agitation

**Rührer** *m* BAUMASCHIN, BETON, OBERFLÄCHE agitator

**Rührwerk** *nt* BAUMASCHIN, BETON, OBERFLÄCHE agitator; **Rührwerkszwangsmischer** *m* BAUMASCHIN, BETON *zur Betonwarenherstellung* agitator-type compulsory mixer

**Ruine** *f* ARCH & TRAGW ruin

**rund** *adj* ARCH & TRAGW round

**Rund-** *in cpds* ARCH & TRAGW round

***rund***: **~es Fenster** *nt* ARCH & TRAGW roundel, wheel window; **~e Nische** *f* ARCH & TRAGW roundel

***Rund-***: **Rundbogen** *m* ARCH & TRAGW round arch, semicircular arch

**rundbogig** *adj* ARCH & TRAGW round-arched

***Rund-***: **Rundbolzenriegel** *m* BESCHLÄGE *Schloß* round bolt; **Runddach** *nt* ARCH & TRAGW compass roof; **Rundfenster** *nt* ARCH & TRAGW roundel, wheel window, WERKSTOFF bull's eye glass; **Rundgang** *m* ARCH & TRAGW circular corridor; **Rundgehweg** *m* ARCH & TRAGW circular walkway; **Rundhobel** *m* BAUMASCHIN compass plane; **Rundkopfbolzen** *m* BESCHLÄGE button-head bolt

**Rundlingszaun** *m* ARCH & TRAGW, INFR & ENTW paling

***Rund-***: **Rundrohr** *nt* WERKSTOFF round pipe; **Rundsäule** *f* ARCH & TRAGW circular column; **Rundschacht** *m* ABWASSER, INFR & ENTW circular manhole, circular manway, cylindrical manhole; **Rundschnittverfahren** *nt* ARCH & TRAGW method of joints; **Rundschornstein** *m* ARCH & TRAGW, BETON, STAHL circular chimney, cylindrical chimney; **Rundstab** *m* WERKSTOFF round bar; **Rundstahl** *m* WERKSTOFF round steel; **Rundstahlstütze** *f* STAHL cylindrical steel column; **Rundstütze** *f* ARCH & TRAGW cylindrical column; **Rundumsicht** *f* ARCH & TRAGW all-round visibility; **Rundwulstabdeckung** *f* BAUMASCHIN torus roll flashing

**Ruß** *m* BAURECHT, INFR & ENTW soot

**Rüst-** *in cpds* BAUMASCHIN, NATURSTEIN putlog

**rüsten** *vt* ARCH & TRAGW rig

***Rüst-***: **Rüstholz** *nt* ARCH & TRAGW, BAUMASCHIN, NATURSTEIN putlog

**rustikal** *adj* ARCH & TRAGW rustic

**Rustikaverband** *m* NATURSTEIN opus rusticum

***Rüst-***: **Rüstloch** *nt* NATURSTEIN putlog hole; **Rüststange** *f* BAUMASCHIN, NATURSTEIN putlog

**Rutschen** *nt* ARCH & TRAGW slippage

**rutschfest** *adj* OBERFLÄCHE, WERKSTOFF antiskid, nonskid, skidproof

**rutschig**: **~er Boden** *m* BAURECHT slippery flooring

**Rüttel-** *in cpds* BAUMASCHIN vibrating, BETON jolted, vibrated; **Rüttelbeton** *m* BETON jolted concrete, vibrated concrete

**rütteln** *vt* BAUMASCHIN agitate

***Rüttel-***: **Rüttelplatte** *f* BAUMASCHIN plate vibrator; **Rüttelschaffußwalze** *f* BAUMASCHIN vibrating sheepsfoot roller; **Rüttelschotter** *m* INFR & ENTW vibrated crushed rock; **Rüttelsieb** *nt* BAUMASCHIN vibration sieve; **Rüttelstampfer** *m* BAUMASCHIN vibrating tamper; **Rütteltisch** *m* BAUMASCHIN shaking chute

**rüttelverdichten** *vt* BAUMASCHIN *Beton* vibrate

***Rüttel-***: **Rüttelverdichter** *m* BAUMASCHIN vibrating compactor

**Rüttler** *m* BAUMASCHIN vibrator

# S

**Saal** *m* ARCH & TRAGW hall

**Sachkenntnis** *f* ARCH & TRAGW, BAURECHT expertise, special experience

**Sack** *m* WERKSTOFF bag; **in Säcke verpacken** *phr* WERKSTOFF bag; **Sackbahnhof** *m* INFR & ENTW dead-end railroad station (*AmE*), dead-end railway station (*BrE*)

**sacken** *vi* ARCH & TRAGW *Untergrund* subside, INFR & ENTW settle

***Sack***: **Sackgasse** *f* INFR & ENTW cul-de-sac, dead-end; **Sackleinen** *nt* WERKSTOFF burlap; **Sackloch** *nt* ARCH & TRAGW *Bohren* blind hole

**Sackung** *f* UMWELT settlement, settling

***Sack***: **Sackzement** *m* BETON bagged cement

**Säge** *f* BAUMASCHIN saw; **Sägebank** *f* BAUMASCHIN saw bench; **Sägeblock** *m* BAUMASCHIN saw log; **Sägebock** *m* BAUMASCHIN, HOLZ buck (*AmE*), sawbuck (*AmE*), sawhorse (*BrE*); **Sägemaschine** *f* BAUMASCHIN sawing machine; **Sägemehl** *nt* BAUMASCHIN, HOLZ sawdust

**Sägen** *nt* BAUMASCHIN sawing

***Säge***: **Sägeschnitt** *m* BAUMASCHIN, HOLZ saw cut; **Sägewerk** *nt* BAUMASCHIN, INFR & ENTW sawmill; **Sägezahndach** *nt* ARCH & TRAGW northlight roof

**Saline** *f* INFR & ENTW saline

**Salpeter** *m* UMWELT niter (*AmE*), nitre (*BrE*), potassium nitrate, saltpeter (*AmE*), saltpetre (*BrE*); **Salpeterfraß** *m* BETON damage by efflorescence, NATURSTEIN exudation; **Salpetersäure** *f* UMWELT nitric acid

**Salz** *nt* WERKSTOFF salt; **Salzbadlöten** *nt* STAHL salt bath brazing

**salzhaltig** *adj* INFR & ENTW saliferous

**Salzhaltigkeit** *f* UMWELT salinity

***Salz***: **Salzsole** *f* INFR & ENTW, WERKSTOFF brine; **Salzton** *m* INFR & ENTW, WERKSTOFF saliferous clay

**salzwasserbeständig** *adj* WERKSTOFF saltwater-proof

***Salz***: **Salzwerk** *nt* INFR & ENTW saline plant

**Sammel**: **Sammelbecken** *nt* ABWASSER, INFR & ENTW, UMWELT reservoir, storage basin; **Sammelbehälter** *m* INFR & ENTW sump pan, UMWELT storage tank; **Sammeldrän** *m* ABWASSER French drain; **Sammelgrube** *f* INFR & ENTW catch-basin, collecting pit; **Sammelkammer** *f* ABWASSER, INFR & ENTW well chamber; **Sammelkanal** *m* ABWASSER catch drain, collector, intercepting drain, intercepting sewer, HEIZ & BELÜFT catch drain, INFR & ENTW catch drain, collector, intercepting sewer, UMWELT main collector, main sewer; **Sammelkläranlage** *f* ABWASSER, UMWELT joint sewage treatment plant; **Sammelrohrleitung** *f* HEIZ & BELÜFT collecting pipeline; **Sammeltank** *m* UMWELT collection tank; **Sammelzug** *m* ARCH & TRAGW *Schornstein*, HEIZ & BELÜFT common flue

**Sammler** *m* ABWASSER catch drain, collector, intercepting drain, HEIZ & BELÜFT accumulator, catch drain, INFR & ENTW accumulator, catch drain, collector, container

**Sammlung** *f*: **~ von Hausmüll** UMWELT garbage collection (*AmE*), garbage disposal (*AmE*), rubbish disposal (*BrE*)

**Sanatorium** *nt* INFR & ENTW sanatorium (*BrE*), sanitarium (*AmE*)

**Sand** *m* NATURSTEIN sand; **mit ~ abdecken** *phr* WERKSTOFF sand; **Sandablagerung** *f* INFR & ENTW sand deposit; **Sandabscheider** *m* ABWASSER, INFR & ENTW sand trap; **Sandanteil** *m* NATURSTEIN sand content; **Sandäquivalent** *nt* WERKSTOFF sand equivalent; **Sandasphalt** *m* INFR & ENTW, NATURSTEIN sand asphalt; **Sandbank** *f* INFR & ENTW sand bank; **Sandbett** *nt* INFR & ENTW sand bedding; **Sandboden** *m* INFR & ENTW sandy ground; **Sandböschung** *f* INFR & ENTW sand slope; **Sanddamm** *m* INFR & ENTW sand dam; **Sanddüne** *f* INFR & ENTW sand dune; **Sandeinschluß** *m* INFR & ENTW, WERKSTOFF sand inclusion; **Sandfang** *m* ABWASSER, INFR & ENTW sand trap, UMWELT sand filter; **Sandfanganlage** *f* UMWELT grit chamber, sand trap; **Sandfangbecken** *nt* INFR & ENTW catch-basin; **Sandfänger** *m* UMWELT grit chamber, sand trap; **Sandfilter** *nt* UMWELT sand filter; **Sandformation** *f* INFR & ENTW sand formation; **Sandfraktion** *f* NATURSTEIN sand fraction

**sandgestrahlt** *adj* OBERFLÄCHE sandblasted

**sandhaltig** *adj* UMWELT, WERKSTOFF arenaceous

**sandig** *adj* UMWELT, WERKSTOFF arenaceous; **~er Lehm** *m* INFR & ENTW, WERKSTOFF sandy loam

***Sand***: **Sand- und Kiesaufbereitungsmaschine** *f* BAUMASCHIN sand and gravel processing equipment; **Sandpapier** *nt* INFR & ENTW, WERKSTOFF abrasive paper; **Sandpolstergründung** *f* BETON sand cushion foundation; **Sandprobe** *f* NATURSTEIN sand sample; **Sandschicht** *f* INFR & ENTW sand layer; **Sandschüttung** *f* ARCH & TRAGW sand filling; **Sandstein** *m* INFR & ENTW, NATURSTEIN sandstone; **Sandstrahlen** *nt* NATURSTEIN, OBERFLÄCHE sandblasting; **Sandton** *m* INFR & ENTW, WERKSTOFF sandy clay

**Sandwichplatte** *f* ARCH & TRAGW sandwich board, sandwich panel, BESCHLÄGE flitch plate, DÄMMUNG composite board, WERKSTOFF sandwich board, sandwich panel

**sanieren** *vt* ARCH & TRAGW redevelop, *verbessern* renovate

**Sanierung** *f* ARCH & TRAGW rehabilitation, INFR & ENTW urban renewal; **Sanierungsmaßnahme** *f* UMWELT measure of redevelopment

**sanitär** *adj* ABWASSER, BAURECHT, INFR & ENTW sanitary

**Sanitär-** *in cpds* ABWASSER, BAURECHT, INFR & ENTW sanitary

***sanitär***: **~e Einrichtungen** *f pl* ABWASSER, ARCH & TRAGW sanitary equipment, sanitation, INFR & ENTW sanitary equipment

***Sanitär-***: **Sanitäranlagen** *f pl* ARCH & TRAGW, INFR & ENTW sanitary facilities; **Sanitärinstallation** *f* ABWASSER sanitary installations; **Sanitärkeramik** *f* WERKSTOFF sanitary china; **Sanitärporzellan** *nt*

WERKSTOFF sanitary china; **Sanitärtechnik** *f* ABWASSER, BAURECHT, INFR & ENTW public health engineering, sanitary engineering; **Sanitärzelle** *f* ABWASSER sanitary module

**satiniert**: **~es Glas** *nt* ARCH & TRAGW, WERKSTOFF satin-finish glass, velvet finish glass

**satt**: **~ anliegend** *adj* ARCH & TRAGW faying; **nicht ~ aufliegend** *adj* ARCH & TRAGW false-bearing; **~ aufliegend** *adj* HOLZ well set; **~e Farben** *f pl* OBERFLÄCHE saturated colors (*AmE*), saturated colours (*BrE*);

**Sattel** *m* ARCH & TRAGW saddle; **Sattelblech** *nt* ARCH & TRAGW ridge plate; **Satteldach** *nt* ARCH & TRAGW gable roof, couple close roof, ridge roof; **Sattelholz** *nt* HOLZ bolster, head tree, corbel piece, saddle; **Sattelschiene** *f* ARCH & TRAGW bearing rail; **Sattelschlepper** *m* BAUMASCHIN articulated lorry (*BrE*), semitrailer truck (*AmE*); **Sattelwange** *f* ARCH & TRAGW *Treppe* open wall string

**Sättigung** *f* DÄMMUNG, WERKSTOFF saturation; **Sättigungsdruck** *m* DÄMMUNG, INFR & ENTW, WERKSTOFF saturated vapor pressure (*AmE*), saturated vapour pressure (*BrE*); **Sättigungsgrad** *m* WERKSTOFF saturation ratio; **Sättigungspunkt** *m* WERKSTOFF saturation point; **Sättigungszone** *f* INFR & ENTW zone of saturation

**sauber**: **~e Technologie** *f* UMWELT clean technology, nonwaste technology (*NWT*)

**Sauberkeitsschicht** *f* ARCH & TRAGW, BETON, INFR & ENTW subbase

**sauer** *adj* INFR & ENTW, OBERFLÄCHE, STAHL, UMWELT, WERKSTOFF acid; **saurer Abfall** *m* INFR & ENTW, WERKSTOFF acid waste; **saurer Abfluß** *m* UMWELT acid runoff; **saures Abgas** *nt* UMWELT acid waste gas; **saure Ablagerung** *f* UMWELT acid deposit; **saures Aerosol** *nt* UMWELT acid aerosol; **saurer Boden** *m* INFR & ENTW, UMWELT, WERKSTOFF acid soil; **saure Erde** *f* UMWELT acid earth

**säuern** *vt* UMWELT acidify

**Säuern** *nt* OBERFLÄCHE, STAHL acid treatment

***sauer***: **saurer Niederschlag** *m* ARCH & TRAGW, INFR & ENTW acid rain, UMWELT acid fallout, acidic rain, acid rain; **saurer Regen** *m* ARCH & TRAGW, INFR & ENTW, UMWELT acid fallout, acid rain, acidic rain; **saurer Schnee** *m* UMWELT acid snow; **saurer See** *m* UMWELT acid lake, acidified lake; **saures Teilchen** *nt* UMWELT acid particle, acidic particle; **saure Umweltverschmutzung** *f* UMWELT acid pollution; **saures Wasser** *nt* UMWELT acid water

**Sauerstoff** *m* BAUMASCHIN, INFR & ENTW, STAHL, WERKSTOFF oxygen; **Sauerstoffbedarf** *m* INFR & ENTW oxygen requirement; **Sauerstoffbohren** *nt* BAUMASCHIN oxygen lancing; **Sauerstofferzeuger** *m* BAUMASCHIN oxygen generator; **Sauerstoffgehalt** *m* INFR & ENTW oxygen content; **Sauerstofflanze** *f* BAUMASCHIN oxygen lance; **Sauerstofflichtbogenschneiden** *nt* STAHL oxyarc cutting, oxygen arc cutting

**Saug-** *in cpds* HEIZ & BELÜFT suction; **Saugbagger** *m* BAUMASCHIN pump dredger

**Saugen** *nt* ELEKTR, HEIZ & BELÜFT suction

**saugfähig** *adj* OBERFLÄCHE absorbent; **~er Untergrund** *m* OBERFLÄCHE absorbent base

***Saug-***: **Saugfilter** *nt* ABWASSER, INFR & ENTW, UMWELT suction filter; **Saugheber** *m* BAUMASCHIN plunger elevator; **Saughöhe** *f* INFR & ENTW suction head; **Saugkorb** *m* BAUMASCHIN pump strainer; **Saugmund** *m* BAUMASCHIN, UMWELT *am Kehrfahrzeug* suction port; **Saugpumpe** *f* ABWASSER, BAUMASCHIN suction pump; **Saugrohr** *nt* ABWASSER, HEIZ & BELÜFT suction pipe; **Saugseite** *f* HEIZ & BELÜFT *Pumpe, Ventilator* suction side; **Saugwasserspiegel** *m* ABWASSER suction water level; **Saugzug** *m* ARCH & TRAGW forced draft (*AmE*), forced draught (*BrE*)

**Säule** *f* ARCH & TRAGW, BETON, INFR & ENTW, NATURSTEIN, STAHL column, head, pillar; **Säulenabstand** *m* ARCH & TRAGW intercolumniation; **Säulenbündel** *nt* ARCH & TRAGW clustered column; **Säulendurchmesser** *m* ARCH & TRAGW column diameter; **Säulenform** *f* ARCH & TRAGW column shape; **Säulenfuß** *m* ARCH & TRAGW patten; **Säulengebälk** *nt* HOLZ entablature; **Säulenhalle** *f* ARCH & TRAGW columned hall; **Säulenhals** *m* ARCH & TRAGW neck molding (*AmE*), neck moulding (*BrE*); **Säulenkapitell** *nt* ARCH & TRAGW column capital; **Säulenkonsollager** *nt* ARCH & TRAGW post bracket; **Säulenkopf** *m* ARCH & TRAGW column head; **Säulenkran** *m* BAUMASCHIN post crane; **Säulenordnung** *f* ARCH & TRAGW order; **Säulenplatte** *f* ARCH & TRAGW plinth; **Säulenportal** *nt* ARCH & TRAGW columnar portal; **Säulenquerschnitt** *m* ARCH & TRAGW, BETON, HOLZ, STAHL column cross section; **Säulenrumpf** *m* ARCH & TRAGW shaft, shank; **Säulensockel** *m* ARCH & TRAGW column pedestal

**Saum** *m* ARCH & TRAGW verge

**säumen** *vt* ARCH & TRAGW, HOLZ border, hem

***Saum***: **Saumleiste** *f* ARCH & TRAGW fillet

**Säure** *f* INFR & ENTW, OBERFLÄCHE, STAHL, UMWELT, WERKSTOFF acid; **durch ~ belastet** *adj* UMWELT acid-stressed; **Säureangriff** *m* INFR & ENTW, OBERFLÄCHE, STAHL acid attack; **Säurebad** *nt* OBERFLÄCHE, STAHL acid bath; **Säurebelastung** *f* UMWELT acid loading, acid stress

**säurebeständig** *adj* OBERFLÄCHE, STAHL, UMWELT, WERKSTOFF acid-resistant, acidproof

***Säure***: **Säurebeständigkeit** *f* OBERFLÄCHE, STAHL, UMWELT, WERKSTOFF acid resistance; **Säurefällung** *f* UMWELT acid precipitation

**säurefest** *adj* OBERFLÄCHE, STAHL, UMWELT, WERKSTOFF acid-resistant; **~e Farbe** *f* BESCHLÄGE, OBERFLÄCHE, STAHL, UMWELT acidproof paint, WERKSTOFF acid-resisting paint; **~er Klinker** *m* INFR & ENTW, NATURSTEIN, WERKSTOFF acid-resistant brick

***Säure***: **Säuregehalt** *m* INFR & ENTW, UMWELT, WERKSTOFF acid content, acidity; **Säuregrad** *m* UMWELT acidity level; **Säurekonzentration** *f* UMWELT acid concentration; **Säurenebel** *m* UMWELT acid fog; **Säureneutralisation** *f* UMWELT acid neutralizing; **Säureneutralisierungsvermögen** *nt* UMWELT acid-neutralizing capacity; **Säurenvorstufe** *f* UMWELT acidic precursor; **Säureschock** *m* UMWELT acid shock; **Säureschutzfarbe** *f* BESCHLÄGE, OBERFLÄCHE, STAHL, UMWELT acidproof paint; **Säureverträglichkeit** *f* UMWELT *des Bodens* acid tolerance; **Säurewiderstandsfähigkeit** *f* UMWELT acid tolerance

**SCC** *abbr* HEIZ & BELÜFT (*zweiter Brennraum*), UMWELT (*Nachbrennkammer, zweiter Brennraum*) SCC (*secondary combustion chamber*)

**Schab** *nt* BESCHLÄGE shave hook

**Schaber** *m* BAUMASCHIN scraper

**Schablone** *f* ARCH & TRAGW template, strickle board

**Schabputz** *m* NATURSTEIN scraped stucco
**Schachbrettmuster** *nt* NATURSTEIN checkerwork (*AmE*), chequerwork (*BrE*)
**Schachbrettverband** *m* NATURSTEIN checkerboard bond (*AmE*), chequerboard bond (*BrE*)
**Schacht** *m* ABWASSER, ARCH & TRAGW, INFR & ENTW manhole, shaft; **Schachtabdeckung** *f* ABWASSER manhole covering, BESCHLÄGE cowl, INFR & ENTW manhole covering; **Schachtdeckel** *m* ABWASSER, INFR & ENTW manhole cover; **Schachtring** *m* ABWASSER, INFR & ENTW, UMWELT manhole ring, well casing; **Schachtsohle** *f* ABWASSER, INFR & ENTW manhole bottom; **Schachttür** *f* ARCH & TRAGW manhole door
**Schaden** *m* BAURECHT damage; **Schadenersatz** *m* BAURECHT award of damages; **Schadenersatzanspruch** *m* BAURECHT claim for compensation
**schadhaft** *adj* ARCH & TRAGW defective, BAURECHT *Gebäude* unsound, *Elektrotechnik* faulty, *fehlerhaft* defective, *schlecht behandelt* damaged, INFR & ENTW defective
**Schädigung** *f* BAURECHT damage
**Schädlichkeit** *f* UMWELT harmfulness
**Schädlingsbekämpfungsmittel** *nt* ABWASSER, BAURECHT, HEIZ & BELÜFT, INFR & ENTW biocide
**Schadstoff** *m* BAURECHT deleterious substance, INFR & ENTW contaminant, harmful substance, pollutant, OBERFLÄCHE aggressive substance, UMWELT harmful substance, *Luft, Wasser* pollutant, *Industrie, Nebenprodukte* toxic substance, aggressive substance, contaminant, WERKSTOFF harmful substance, pollutant, contaminant, *Säure* aggressive substance, deleterious substance; **Schadstoffablagerung** *f* UMWELT pollutant deposition; **Schadstoffausbreitung** *f* UMWELT propagation of pollutant
**schadstoffbelastet**: **~es Erdreich** *nt* UMWELT pollutant-impacted ground
***Schadstoff***: **Schadstoffbelastung** *f* UMWELT pollution burden
**Schaffung** *f*: **~ der Baufreiheit** ARCH & TRAGW clearing operations
**Schaffußwalze** *f* ARCH & TRAGW, INFR & ENTW sheepsfoot roller
**Schaft** *m* ARCH & TRAGW shaft, shank
**Schaftverbindung** *f* HOLZ scarf joint
**Schalbrett** *nt* ARCH & TRAGW, BETON formboard, formwork board
**Schale** *f* ARCH & TRAGW preformed section, shell, BAUMASCHIN shell
**Schäleisen** *nt* BAUMASCHIN paring chisel
**schalen** *vt* ARCH & TRAGW, HOLZ, INFR & ENTW board, line, plank
**schälen** *vt* HOLZ peel
***Schale***: **Schalenbauweise** *f* ARCH & TRAGW shell construction; **Schalenbeton** *m* BETON shell concrete; **Schalendach** *nt* ARCH & TRAGW shell roof; **Schalenkonstruktion** *f* ARCH & TRAGW, INFR & ENTW shell structure; **Schalentheorie** *f* ARCH & TRAGW, INFR & ENTW theory of thin shells
**Schalgerüst** *nt* ARCH & TRAGW, INFR & ENTW falsework
**Schälholz** *nt* HOLZ barked timber
**Schall** *m* DÄMMUNG sound; **Schallabsorption** *f* DÄMMUNG sound absorption; **Schalldämmstoff** *m* DÄMMUNG sound-absorbent material, sound insulation material; **Schalldämpfer** *m* DÄMMUNG muffler (*AmE*), silencer (*BrE*); **Schalldämpfung** *f* DÄMMUNG sound damping, sound deadening
**schalldicht** *adj* DÄMMUNG soundproof; **~e Zwischenschicht** *f* ABWASSER plugging
***Schall***: **Schalldruckpegel** *m* DÄMMUNG, UMWELT sound pressure level; **Schalldruckspektrum** *nt* UMWELT sound pressure spectrum; **Schallgeschwindigkeit** *f* ARCH & TRAGW, DÄMMUNG, WERKSTOFF sound velocity; **Schallgeschwindigkeitsprofil** *nt* DÄMMUNG, WERKSTOFF acoustic velocity log; **Schallimpedanz** *f* DÄMMUNG characteristic impedance; **Schallisolationsmaterial** *nt* DÄMMUNG, UMWELT acoustic insulating materials
**schallisolierend**: **~e Schicht** *f* DÄMMUNG resilient layer
***Schall***: **Schallpegel** *m* DÄMMUNG sound level; **Schallquelle** *f* UMWELT sound source; **Schallschirm** *m* UMWELT baffle collector
**schallschluckend**: **~er Boden** *m* ARCH & TRAGW, DÄMMUNG dead floor
***Schall***: **Schallschluckmaterial** *nt* DÄMMUNG, WERKSTOFF acoustic material; **Schallschutz** *m* DÄMMUNG sound insulation, soundproofing; **Schallschutzfenster** *nt* ARCH & TRAGW, DÄMMUNG noise protection window; **Schallstärke** *f* BAURECHT, DÄMMUNG sound intensity; **Schallübertragung** *f* ARCH & TRAGW, BAURECHT, DÄMMUNG sound transmittance; **Schallwelle** *f* DÄMMUNG sound wave
**Schälmaschine** *f* BAUMASCHIN paring machine
**Schalöl** *nt* BETON formwork lube (*AmE*), formwork lubricant, formwork oil (*BrE*), WERKSTOFF separating oil; **Schalölflecken** *m pl* BETON formwork lube stains (*AmE*), formwork lubricant stains, formwork oil stains (*BrE*)
**Schalrohr** *nt* ARCH & TRAGW, BESCHLÄGE casing pipe
**Schalt-** *in cpds* ELEKTR switching; **Schaltbild** *nt* ELEKTR wiring diagram; **Schaltbolzen** *m* WERKSTOFF index bolt
**schalten** *vt* ELEKTR switch
**Schalter** *m* ELEKTR switch
**Schalterhalle** *f* ARCH & TRAGW, INFR & ENTW customer service area
***Schalt-***: **Schalthahn** *m* ABWASSER switch cock; **Schaltplan** *m* ELEKTR circuit diagram, wiring diagram; **Schaltschrank** *m* ELEKTR switch cabinet; **Schaltverbindung** *f* ELEKTR circuit connection; **Schaltwarte** *f* ELEKTR control room
**Schalung** *f* ARCH & TRAGW, BETON boarding, form, formwork, HEIZ & BELÜFT shuttering, HOLZ boarding, concrete formwork, timbering, STAHL concrete formwork; **Schalungsfrist** *f* BETON stripping time; **Schalungsöl** *nt* BETON form oil, mold oil (*AmE*), mould oil (*BrE*), release lube; **Schalungspaste** *f* BETON form paste, release paste
**schalungsrauh** *adj* ARCH & TRAGW, BETON board-marked, rough-shuttered, HOLZ natural
***Schalung***: **Schalungsrüttler** *m* BAUMASCHIN external vibrator; **Schalungsspanndraht** *m* BETON tie wire; **Schalungstafel** *f* BETON fit-up, formwork panel, STAHL shuttering panel
**Schalwand** *f* ARCH & TRAGW plank partition
**Schamotte** *f* WERKSTOFF chamotte; **Schamotteauskleidung** *f* HEIZ & BELÜFT, WERKSTOFF chamotte lining; **Schamotterohr** *nt* WERKSTOFF chamotte pipe; **Schamottestein** *m* NATURSTEIN fire brick

**scharf** *adj* BAUMASCHIN *Kante* keen; **~e Kante** *f* HOLZ quoin
**Schärfe** *f* BAUMASCHIN *Werkzeugschneide* keenness
**scharfkantig**: **~er Sand** *m* WERKSTOFF grit
**Scharnier** *nt* BESCHLÄGE hinge, hinge joint; **~ mit lösbaren Bolzen** BESCHLÄGE loose-pin hinge; **Scharnierband** *nt* BESCHLÄGE flap hinge, hinge, strap hinge, piano hinge; **Scharniergelenk** *nt* BESCHLÄGE knuckle; **Scharnierstift** *m* BESCHLÄGE hinge pin
**Scharrieren** *nt* ARCH & TRAGW, BETON, WERKSTOFF charring
**Schaufel** *f* BAUMASCHIN *Bagger* shovel; **Schaufelbagger** *m* BAUMASCHIN bucket dredger; **Schaufelentnahmegerät** *nt* UMWELT scraper extractor; **Schaufellader** *m* BAUMASCHIN shovel loader
**Schaufeln** *nt* INFR & ENTW shovel work
***Schaufel***: **Schaufelrad** *nt* BAUMASCHIN bucket wheel; **Schaufelradbagger** *m* ARCH & TRAGW bucket-wheel excavator; **Schaufelreihe** *f* BAUMASCHIN *Förderband* line of buckets
**Schaufenster** *nt* ARCH & TRAGW display window
**Schaukelförderer** *m* BAUMASCHIN jigging conveyor
**Schaum** *m* INFR & ENTW scum, WERKSTOFF foam; **Schaumbeton** *m* BETON *Trennwände* aerated concrete
**schäumen** *vi* ABWASSER, ARCH & TRAGW, BAUMASCHIN, BETON, HEIZ & BELÜFT, WERKSTOFF foam
***Schaum***: **Schaumglas** *nt* DÄMMUNG foam glass, WERKSTOFF foam glass, multicellular glass; **Schaumgummiband** *nt* DÄMMUNG, WERKSTOFF expanded rubber band, foam rubber tape; **Schaumstoff** *m* WERKSTOFF expanded plastic, plastic foam
**Scheibenabschöpfer** *m* UMWELT disc skimmer (*BrE*), disk skimmer (*AmE*)
**Scheibenbauweise** *f* ARCH & TRAGW, BETON frameless construction
**Scheibenwand** *f* BESCHLÄGE shear wall
**scheinbar**: **~e Dichte** *f* INFR & ENTW, WERKSTOFF apparent density
**Scheingewölbe** *nt* ARCH & TRAGW blind vault
**Scheinkohäsion** *f* INFR & ENTW, WERKSTOFF apparent cohesion
**Scheitel** *m* ARCH & TRAGW apex, crown, crest, vertex, NATURSTEIN apex; **Scheitelbruchlast** *f* ARCH & TRAGW, INFR & ENTW, WERKSTOFF crushing load; **Scheitelfuge** *f* ARCH & TRAGW, NATURSTEIN apex joint; **Scheitelpunkt** *m* ARCH & TRAGW vertex; **Scheitelrippe** *f* ARCH & TRAGW axial rib, ridge rib
**scheitrecht**: **~er Bogen** *m* ARCH & TRAGW floor arch, straight arch
**Schellack** *m* WERKSTOFF shellac
**Schelle** *f* BESCHLÄGE clip, strap, ELEKTR clip, clamp, *Klingel* bell
**Schema** *nt* ARCH & TRAGW, INFR & ENTW scheme; **Schemabild** *nt* BAUMASCHIN, INFR & ENTW schematic drawing
**Schematisierung** *f* ARCH & TRAGW, INFR & ENTW schematization
**Schenkel** *m* ARCH & TRAGW side
**Scher-** *in cpds* ARCH & TRAGW shearing
**Scherengitter** *nt* BAUMASCHIN slidable lattice grate, INFR & ENTW worm fence
**Scherentreppe** *f* ARCH & TRAGW folding ladder, BESCHLÄGE folding ladder, folding staircase
***Scher-***: **Scherkraft** *f* ARCH & TRAGW shearing force; **Schermoment** *nt* ARCH & TRAGW shear moment; **Scherprobe** *f* ARCH & TRAGW, INFR & ENTW, WERKSTOFF shearing test; **Scherspannung** *f* ARCH & TRAGW, INFR & ENTW shear stress, shearing stress, transverse strain
**Scherung** *f* INFR & ENTW shear
***Scher-***: **Scherversuch** *m* ARCH & TRAGW, INFR & ENTW, WERKSTOFF shearing test; **Scherwinkel** *m* INFR & ENTW, WERKSTOFF angle of shear; **Scherzone** *f* INFR & ENTW shear zone
**scheuerbeständig** *adj* OBERFLÄCHE scrubbable
**scheuerfest** *adj* OBERFLÄCHE scrubbable
**Scheuerleiste** *f* ARCH & TRAGW, BESCHLÄGE, HOLZ baseboard (*AmE*), skirting board (*BrE*)
**Schicht** *f* INFR & ENTW course, stratum, subsurface, NATURSTEIN course, OBERFLÄCHE coat, layer; **Schichtanordnung** *f* NATURSTEIN *Mauerwerk* coursing; **Schichtdicke** *f* INFR & ENTW layer thickness, OBERFLÄCHE coating thickness; **Schichtenfolge** *f* INFR & ENTW, NATURSTEIN, WERKSTOFF sequence of strata; **Schichtengrenze** *f* INFR & ENTW strata boundary; **Schichtenkarte** *f* ARCH & TRAGW *Vermessung* contour map; **Schichtenmauerwerk** *nt* NATURSTEIN coursed masonry; **Schichtenpappe** *f* WERKSTOFF pasteboard; **Schichtpreßholz** *nt* HOLZ compregnated wood, glulam timber, laminated timber, laminated wood, stacked wood; **Schichtpreßstoff** *m* WERKSTOFF molded laminated plastic (*AmE*), moulded laminated plastic (*BrE*); **Schichtpreßstoffplatte** *f* WERKSTOFF laminated plastic board; **Schichttiefe** *f* INFR & ENTW layer depth
**schichtverleimt** *adj* HOLZ glue-laminated (*glulam*)
***Schicht***: **Schichtwasser** *nt* INFR & ENTW foreign water
**Schiebe-** *in cpds* ARCH & TRAGW sliding; **Schiebebühne** *f* ARCH & TRAGW transfer table, INFR & ENTW traveling platform (*AmE*), travelling platform (*BrE*); **Schiebefenster** *nt* ARCH & TRAGW sliding hatch, sliding sash, sash window, *halböffnend* sliding window; **Schiebefensterbeschläge** *m pl* BESCHLÄGE sash hardware; **Schiebefensterfeststeller** *m* BESCHLÄGE sash fastener; **Schiebefensterrahmen** *m* BESCHLÄGE sash; **Schiebeleiter** *f* INFR & ENTW traveling ladder (*AmE*), travelling ladder (*BrE*); **Schiebeluke** *f* ARCH & TRAGW sliding hatch
**Schieber** *m* ABWASSER slide valve
***Schiebe-***: **Schieberahmen** *m* BAUMASCHIN sash frame, BESCHLÄGE window sash; **Schiebetor** *nt* ARCH & TRAGW slide gate, sliding gate; **Schiebetür** *f* ARCH & TRAGW sliding door; **Schiebetürführung** *f* BESCHLÄGE floor guide
**Schiedsverfahren** *nt* BAURECHT arbitration
**schief** *adj* ARCH & TRAGW slanting, *Turm* leaning; **~e Brücke** *f* INFR & ENTW skew bridge; **~e Ebene** *f* ARCH & TRAGW inclined plane
**Schiefbogen** *m* ARCH & TRAGW oblique arch
**Schiefer** *m* BAUMASCHIN, INFR & ENTW, WERKSTOFF schist, shale, slate; **Schieferbedachung** *f* BAUMASCHIN slate roof cladding; **Schiefergestein** *nt* INFR & ENTW schistose rock; **Schiefernagel** *m* BAUMASCHIN slate nail; **Schieferplatte** *f* BAUMASCHIN slate; **Schieferton** *m* WERKSTOFF slate clay; **Schieferunterlegschicht** *f* WERKSTOFF sarking
**schiefwinklig**: **~e Brücke** *f* INFR & ENTW oblique bridge; **~e Koordinaten** *f pl* ARCH & TRAGW, INFR & ENTW oblique coordinates; **~e Platte** *f* ARCH & TRAGW, BETON, INFR & ENTW oblique-angled slab
**Schiene** *f* INFR & ENTW rail, track; **Schienengleis** *nt*

INFR & ENTW tram track; **Schienenheber** *m* BAUMASCHIN rail jack; **Schienenkopf** *m* INFR & ENTW rail head; **Schienenstoß** *m* INFR & ENTW rail connection, rail joint

**Schiff** *nt* ARCH & TRAGW, INFR & ENTW *Kirche* aisle, bay

**Schiffscontainer** *m* INFR & ENTW shipping container

**Schiffstransport** *m* INFR & ENTW water carriage

**Schifter** *m* BAUMASCHIN jack rafter

**Schiftsparren** *m* ARCH & TRAGW rafter trimmer, BAUMASCHIN jack rafter, HOLZ jack rafter, creeping rafter

**Schild** *m* ARCH & TRAGW shield, BESCHLÄGE label, sign; **Schildbauweise** *f* INFR & ENTW *Tunnelbau* shield tunneling (*AmE*), shield tunnelling (*BrE*); **Schildvortrieb** *m* INFR & ENTW *Tunnelbau* shield driving method, shield tunneling (*AmE*), shield tunnelling (*BrE*)

**Schimmel** *m* HEIZ & BELÜFT, OBERFLÄCHE mold (*AmE*), mould (*BrE*)

**schimmelbeständig** *adj* WERKSTOFF mold-resistant (*AmE*), mould-resistant (*BrE*)

***Schimmel***: **Schimmelgeruch** *m* HEIZ & BELÜFT moldy odor (*AmE*), mouldy odour (*BrE*); **Schimmelwuchs** *m* HEIZ & BELÜFT mold growth (*AmE*), mould growth (*BrE*)

**Schindel** *f* HOLZ clapboard, WERKSTOFF *Dach* shingle; **Schindeldach** *nt* ARCH & TRAGW shingle roof; **Schindeltür** *f* ARCH & TRAGW louver door (*AmE*), louvre door (*BrE*)

**Schlacke** *f* UMWELT clinker, WERKSTOFF cinder, *Lava* scoria; **Schlackenbeton** *m* BETON clinker concrete, slag concrete, cinder concrete; **Schlackenbetonstein** *m* WERKSTOFF cinder block; **Schlackenstein** *m* NATURSTEIN scoria brick; **Schlackenwolle** *f* WERKSTOFF slag wool; **Schlackenzement** *m* WERKSTOFF slag cement; **Schlackenzuschlagstoff** *m* WERKSTOFF cinder aggregate

**schlaff** *adj* ARCH & TRAGW *Mechanik* unstressed; **~ armiert** *adj* BETON untensioned; **~ bewehrt** *adj* BETON nonprestressed, untensioned

**Schlafstadt** *f* INFR & ENTW dormitory town

**Schlag-** *in cpds* WERKSTOFF impact, shock; **Schlagbohren** *nt* BAUMASCHIN boring by percussion; **Schlagbohrer** *m* BAUMASCHIN hammer drill

**schlagfest** *adj* WERKSTOFF impact-resistant, shock-resistant

***Schlag-***: **Schlagfestigkeit** *f* ARCH & TRAGW, BETON, STAHL, WERKSTOFF impact resistance, resistance to impact, shock resistance; **Schlaghaube** *f* BAUMASCHIN helmet; **Schlagleiste** *f* ARCH & TRAGW baffle plate, *Fenster* rabbet ledge, HOLZ, INFR & ENTW, STAHL, WERKSTOFF baffle plate; **Schlagloch** *nt* INFR & ENTW pothole; **Schlagprobe** *f* WERKSTOFF impact test

**schlagregensicher** *adj* ELEKTR resistant to heavy rain

***Schlag-***: **Schlagregensicherheit** *f* ARCH & TRAGW, ELEKTR resistance to heavy rain; **Schlagseite** *f* ARCH & TRAGW *zweiflügeliges Fenster* rebate ledge; **Schlagversuch** *m* WERKSTOFF impact test

**Schlamm** *m* ABWASSER sludge, HEIZ & BELÜFT, INFR & ENTW mud, sludge, UMWELT sludge, slurry; **Schlammablagerung** *f* INFR & ENTW mud deposit; **Schlammablagerungen** *f pl* UMWELT alluvial deposits; **Schlammaufbereitung** *f* UMWELT sludge processing; **Schlammbank** *f* INFR & ENTW mud bank; **Schlammbehandlung** *f* INFR & ENTW, UMWELT sludge processing; **Schlammbelebung** *f* UMWELT activation of sludge; **Schlammbeseitigung** *f* UMWELT sludge removal; **Schlammbett** *nt* INFR & ENTW sludge bed; **Schlammeimer** *m* ABWASSER, INFR & ENTW grit box, silt box; **Schlammeindickung** *f* INFR & ENTW, UMWELT sludge thickening

**schlämmen** *vt* OBERFLÄCHE limewash, limewhite

**Schlämmen** *nt* OBERFLÄCHE limewashing, liming

***Schlamm***: **Schlammentwässerung** *f* INFR & ENTW, UMWELT dehydration of sludge, sludge dewatering; **Schlammfang** *m* ABWASSER silt box, INFR & ENTW grit chamber, UMWELT sludge sump; **Schlammfaulbehälter** *m* UMWELT sludge digestion tank, digestion sump; **Schlammfaulraum** *m* ABWASSER, INFR & ENTW digestion chamber, sludge digestion chamber; **Schlammfaulung** *f* UMWELT sludge digestion

**Schlämmgerät** *nt* BAUMASCHIN *Materialprüfung* sedimentation machine

**schlammig**: **~er Boden** *m* INFR & ENTW, WERKSTOFF muddy soil; **~er Sand** *m* INFR & ENTW, WERKSTOFF muddy sand

***Schlamm***: **Schlammkompostierung** *f* INFR & ENTW, UMWELT sludge composting; **Schlammkuchen** *m* UMWELT sludge cake; **Schlammprobe** *f* INFR & ENTW, WERKSTOFF mud sample; **Schlammräumer** *m* UMWELT sludge rake; **Schlammsammelbehälter** *m* UMWELT silt container, sludge sump; **Schlammschicht** *f* INFR & ENTW mud layer; **Schlammschwelle** *f* ARCH & TRAGW mudsill; **Schlammstabilisierung** *f* INFR & ENTW, UMWELT sludge stabilization; **Schlammtrockenbett** *nt* UMWELT sludge drying bed; **Schlammtrocknung** *f* UMWELT sludge drying; **Schlammverdickung** *f* UMWELT dehydration of sludge, sludge dewatering; **Schlammversiegelung** *f* INFR & ENTW slurry seal

**Schlämmversuch** *m* STAHL, UMWELT elutriation test

***Schlamm***: **Schlammverwertung** *f* UMWELT recycling of sludge

**schlangenförmig** *adj* INFR & ENTW meandering

**Schlankheit** *f* ARCH & TRAGW, INFR & ENTW slenderness; **Schlankheitsgrad** *m* ARCH & TRAGW, INFR & ENTW slenderness ratio

**Schlauch** *m* ABWASSER, BESCHLÄGE hose, UMWELT tube; **Schlauchanschluß** *m* BESCHLÄGE hose connection; **Schlauchhalter** *m* ABWASSER, BESCHLÄGE hose clip; **Schlauchklemme** *f* ABWASSER hose clip; **Schlauchschelle** *f* ABWASSER hose clip; **Schlauchtrommel** *f* ABWASSER, BESCHLÄGE hose reel; **Schlauchtülle** *f* ABWASSER, BESCHLÄGE hose nozzle; **Schlauchverschraubung** *f* ABWASSER, BESCHLÄGE threaded hose connection, threaded hose joint; **Schlauchwaage** *f* BAUMASCHIN hydrostatic tube balance

**Schlauder** *m* ARCH & TRAGW iron tie

**schlecht**: **~er Boden** *m* INFR & ENTW, UMWELT poor soil; **~ fluchtend** *adj* ARCH & TRAGW misaligned

**Schleif-** *in cpds* BAUMASCHIN grinding, HOLZ, STAHL, WERKSTOFF abrasive

**schleifen** *vt* BAUMASCHIN grind, HOLZ, STAHL abrade

**Schleifen** *nt* BAUMASCHIN grinding

***Schleif-***: **Schleiffläche** *f* WERKSTOFF abrasive surface; **Schleifmaschine** *f* BAUMASCHIN grinder; **Schleif- und Poliermaschine** *f* **mit biegsamer Welle** BAUMASCHIN grinding and polishing machine with flexible shaft; **Schleif- und Polierstraße** *f* BAUMASCHIN *Marmor* grinding and polishing line;

**Schleifsand** *m* WERKSTOFF abrasive sand; **Schleifscheibe** *f* BAUMASCHIN grinding wheel; **Schleifstein** *m* BAUMASCHIN grindstone; **Schleifwirkung** *f* HOLZ, STAHL, WERKSTOFF abrasion
**Schleppdach** *nt* ARCH & TRAGW monopitch roof, penthouse roof
**Schlepper** *m* BAUMASCHIN tractor
**Schleppgaube** *f* ARCH & TRAGW shed dormer
**Schleppzug** *m* BAUMASCHIN truck trailer, trailer truck
**Schleuder** *f* BAUMASCHIN centrifuge; **Schleuderbeton** *m* BETON spun concrete; **Schleuderbetonrohr** *nt* BETON, INFR & ENTW centrifugally cast concrete pipe; **Schleuderguß** *m* WERKSTOFF centrifugal casting; **Schleudermaschine** *f* **für Betonrohre** BAUMASCHIN centrifugal machine for concrete pipes; **Schleudertrennung** *f* UMWELT ballistic separation
**Schleuse** *f* UMWELT sluice; **Schleusentor** *nt* BESCHLÄGE gate
**schlichten** *vt* BAURECHT adjust, settle
**Schlichter** *m* BAURECHT arbitrator
**Schlick** *m* ABWASSER, HEIZ & BELÜFT, INFR & ENTW silt, sludge; **Schlickablagerung** *f* UMWELT deposition of silt
**Schließ-** *in cpds* BESCHLÄGE locking; **Schließanlage** *f* BESCHLÄGE locking system, master-keyed system; **Schließbart** *m* BESCHLÄGE key bit; **Schließbeschlag** *m* BESCHLÄGE lock fitting; **Schließblech** *nt* ARCH & TRAGW keeper, BESCHLÄGE lock plate, striking plate; **Schließbügel** *m* BESCHLÄGE strike
**schließen** *vt* BAURECHT *Vertrag* contract, HEIZ & BELÜFT shut
***Schließ-***: **Schließhaken** *m* BESCHLÄGE locking hook; **Schließhebel** *m* BESCHLÄGE key bit; **Schließkasten** *m* BESCHLÄGE lock case, lock casing, *Schloß* strike; **Schließklappe** *f* ARCH & TRAGW box staple; **Schließkraft** *f* ARCH & TRAGW *Türschließer* closing power; **Schließpfosten** *m* INFR & ENTW shutting post; **Schließwinkel** *m* BESCHLÄGE *Schloß* strike; **Schließzylinder** *m* BESCHLÄGE lock cylinder
**Schlinge** *f* BESCHLÄGE loop; **Schlingengewebe** *nt* WERKSTOFF looped fabric, *Teppich* terry cloth; **Schlingenware** *f* WERKSTOFF looped fabric
**Schlitz** *m* ARCH & TRAGW slot, NATURSTEIN chase
**schlitzartig**: **~e Aussparung** *f* ARCH & TRAGW keyway
***Schlitz***: **Schlitzdränung** *f* INFR & ENTW mole drainage; **Schlitzniet** *m* BESCHLÄGE slotted rivet; **Schlitzrohr** *nt* ABWASSER slot pipe; **Schlitzwandverfahren** *nt* INFR & ENTW, UMWELT slurry trenching
**Schloß** *nt* ARCH & TRAGW palace, BESCHLÄGE lock; **~ DIN rechts** BESCHLÄGE right-hand lock; **~ mit einseitigem Stulp** BESCHLÄGE lock for rebated door; **~ mit Stulp auf Mitte** BESCHLÄGE lock for flush door; **Schloßgarnitur** *f* BESCHLÄGE lockset; **Schloßkasten** *m* ARCH & TRAGW box staple, BESCHLÄGE case, lock casing, lock case; **Schloßriegel** *m* BESCHLÄGE deadbolt; **Schloßschutzblech** *nt* BESCHLÄGE finger plate; **Schloßstulp** *m* BESCHLÄGE lock foreend
**Schlot** *m* ARCH & TRAGW chimney, BAUMASCHIN funnel, INFR & ENTW chimney
**Schlucken** *nt* DÄMMUNG absorption
**Schluffgehalt** *m* WERKSTOFF silt content
**schluffig**: **~er Boden** *m* INFR & ENTW silty soil
**Schluffschicht** *f* INFR & ENTW silt layer
**Schlupf** *m* ARCH & TRAGW slippage
**Schlußanstrich** *m* OBERFLÄCHE finish coat
**Schlüssel** *m* BESCHLÄGE key; **Schlüsselbart** *m* BESCHLÄGE bit, keybit
**schlüsselfertig** *adj* ARCH & TRAGW turnkey
***Schlüssel***: **Schlüsselformblech** *nt* INFR & ENTW ward; **Schlüsselkanal** *m* BESCHLÄGE keyway; **Schlüssellinie** *f* ARCH & TRAGW keyline; **Schlüsselloch** *nt* BESCHLÄGE keyhole; **Schlüssellochabdeckung** *f* BESCHLÄGE key drop; **Schlüsselschalter** *m* ELEKTR key-operated switch; **Schlüsselschild** *nt* ARCH & TRAGW scutcheon, BESCHLÄGE keyhole plate, escutcheon, key plate
**Schlußstein** *m* ARCH & TRAGW arch stone, trap, NATURSTEIN arch stone, *eines Gewölbes* keystone, capstone
**Schlußzahlung** *f* BAURECHT, INFR & ENTW final payment
**schmal** *adj* ARCH & TRAGW narrow
**Schmelz-** *in cpds* BESCHLÄGE, WERKSTOFF fusible; **Schmelzbarkeit** *f* WERKSTOFF fusibility; **Schmelzmittel** *nt* WERKSTOFF *Schweißen* flux; **Schmelzpatrone** *f* BESCHLÄGE *zum Brandschutz bei Türfeststellern* fusible link; **Schmelztauchverfahren** *nt* OBERFLÄCHE hot-dipping process
**Schmiedbarkeit** *f* STAHL, WERKSTOFF malleability
**Schmiede** *f* STAHL smithy; **Schmiedearbeit** *f* BESCHLÄGE ironwork, STAHL smithery; **Schmiedeeisen** *nt* ARCH & TRAGW wrought iron; **Schmiedehandwerk** *nt* STAHL smithery; **Schmiedezange** *f* BAUMASCHIN smith's pliers
**Schmier-** *in cpds* WERKSTOFF lubricating; **Schmiermittel** *nt* WERKSTOFF lubricant; **Schmiervorrichtung** *f* BAUMASCHIN lubricating device
**Schmirgelpapier** *nt* INFR & ENTW, WERKSTOFF abrasive paper
**Schmuckbeton** *m* BETON ornamental concrete
**Schmutz** *m* UMWELT dirt
**schmutzabweisend** *adj* OBERFLÄCHE dirt-repellent
***Schmutz***: **Schmutzfang** *m* INFR & ENTW gully trap; **Schmutzfilter** *m* ABWASSER, UMWELT dirt filter; **Schmutzsieb** *nt* ABWASSER, INFR & ENTW strainer; **Schmutzstoff** *m* INFR & ENTW, UMWELT, WERKSTOFF contaminant, pollutant; **Schmutzwasser** *nt* ABWASSER, INFR & ENTW, UMWELT drain water, foul water, sewage, waste water; **Schmutzwasserkanal** *m* ABWASSER, INFR & ENTW foul water sewer; **Schmutzwasserleitung** *f* ABWASSER sanitary sewer
**Schnäpperschloß** *nt* BESCHLÄGE spring bolt lock
**Schnappschloß** *nt* BESCHLÄGE latch lock
**Schnecke** *f* ARCH & TRAGW volute, BAUMASCHIN auger; **Schneckenauge** *nt* ARCH & TRAGW well hole; **Schneckenbeschickung** *f* BAUMASCHIN worm feed; **Schneckenbohrer** *m* BAUMASCHIN gimlet; **Schneckenförderer** *m* BAUMASCHIN screw conveyor; **Schneckenhandbohrer** *m* HOLZ shell gimlet; **Schneckentreppe** *f* ARCH & TRAGW spiral stairs
**Schnee** *m* ABWASSER snow; **Schneedecke** *f* INFR & ENTW snow cover; **Schneefang** *m* ARCH & TRAGW *Dach* snow guard; **Schneelast** *f* ARCH & TRAGW snowload; **Schneepflug** *m* BAUMASCHIN snow plough (*BrE*), snow plow (*AmE*)
**Schneid-** *in cpds* BAUMASCHIN cutting; **Schneidbrenner** *m* BAUMASCHIN flame cutter
**Schneide** *f* BAUMASCHIN cutting edge
**Schneideholz** *nt* BAUMASCHIN saw timber
**schneiden** *vt* WERKSTOFF cut

**Schneiden** *nt* WERKSTOFF cutting
**schneidend** *adj* BAUMASCHIN cutting
***Schneide***: **Schneidescheibe** *f* BAUMASCHIN cutter disc (*BrE*), cutter disk (*AmE*); **Schneidewerkzeug** *nt* BAUMASCHIN cutter
***Schneid-***: **Schneidfuß** *m* BAUMASCHIN *Pfahlgründung* cutting foot; **Schneidkante** *f* BAUMASCHIN cutting edge; **Schneidzange** *f* BAUMASCHIN cutting pliers
**schnell**: **~e Beton-Tilt-up-Fabrikherstellung** *f* BETON concrete tilt-up fast factory construction; **~ erstellbares Rahmensystem** *nt* ARCH & TRAGW, INFR & ENTW rapid frame system
**Schnell-**: **Schnellfilter** *m* ABWASSER rapid filter; **Schnellkompostierung** *f* UMWELT mechanical composting, rapid fermentation; **Schnellstraße** *f* INFR & ENTW motorway, dual carriageway, thoroughfare (*BrE*), thruway (*AmE*)
**Schnitt** *m* ARCH & TRAGW cut, sectional drawing, section, INFR & ENTW cut, sectional drawing; **Schnittfuge** *f* NATURSTEIN flat joint; **Schnittholz** *nt* HOLZ lumber (*AmE*), timber (*BrE*); **Schnittkante** *f* WERKSTOFF cutting edge; **Schnittkräfte** *f pl* ARCH & TRAGW, BETON, HOLZ, INFR & ENTW, STAHL, WERKSTOFF internal forces, static forces; **Schnittliste** *f* HOLZ cutting list; **Schnittnagel** *m* WERKSTOFF cut nail; **Schnittpunkt** *m* ARCH & TRAGW, ELEKTR, INFR & ENTW intersection; **Schnittverfahren** *nt* ARCH & TRAGW method of sections; **Schnittzeichnung** *f* ARCH & TRAGW, INFR & ENTW sectional drawing
**schnitzen** *vt* HOLZ carve
**Schnitzerei** *f* HOLZ carving
**Schnitzornament** *nt* ARCH & TRAGW carved ornament
**Schnur** *f* BESCHLÄGE cord, string; **Schnurbrett** *nt* ARCH & TRAGW, INFR & ENTW batter board; **Schnurbretter** *nt pl* ARCH & TRAGW, INFR & ENTW sight rail; **Schnurgerüst** *nt* ARCH & TRAGW, INFR & ENTW batter board, sight rail; **Schnurlot** *nt* ARCH & TRAGW, BAUMASCHIN plumb bob, plumb line; **Schnurnagel** *m* BESCHLÄGE line pin; **Schnurzug** *m* BESCHLÄGE *Jalousie* pull-cord
**schöpferisch** *adj* ARCH & TRAGW creative
**Schopfwalm** *m* ARCH & TRAGW, HOLZ partial hip
**Schöpfwerk** *nt* ABWASSER pumping station
**Schornstein** *m* ARCH & TRAGW chimney, stack, BAUMASCHIN funnel, INFR & ENTW, NATURSTEIN chimney; **Schornsteinanschluß** *m* ARCH & TRAGW chimney junction; **Schornsteinaufsatz** *m* ARCH & TRAGW chimney top, BESCHLÄGE cowl, STAHL chimney top; **Schornsteinauskleidung** *f* NATURSTEIN, STAHL chimney lining; **Schornsteinbauer** *m* ARCH & TRAGW chimney builder; **Schornsteinbemessung** *f* ARCH & TRAGW chimney design; **Schornsteinblechrinne** *f* BESCHLÄGE fillet gutter; **Schornsteindachrinne** *f* ABWASSER back gutter; **Schornsteineffekt** *m* HEIZ & BELÜFT stack effect; **Schornsteinelement** *nt* WERKSTOFF chimney component; **Schornsteinfuß** *m* NATURSTEIN chimney base; **Schornsteinhaube** *f* ARCH & TRAGW, STAHL chimney top; **Schornsteinkappe** *f* ARCH & TRAGW, STAHL chimney top; **Schornsteinleiter** *f* ARCH & TRAGW chimney ladder; **Schornsteinmantel** *m* ARCH & TRAGW chimney mantle; **Schornsteinöffnung** *f* ARCH & TRAGW, STAHL chimney outlet; **Schornsteinquerschnitt** *m* ARCH & TRAGW chimney cross section; **Schornsteinrohr** *nt* ARCH & TRAGW, STAHL chimney pipe; **Schornsteinschieber** *m* HEIZ & BELÜFT damper; **Schornsteinventilator** *m* HEIZ & BELÜFT chimney fan; **Schornsteinziegel** *m* WERKSTOFF chimney brick; **Schornsteinzug** *m* ARCH & TRAGW, NATURSTEIN chimney flue; **Schornsteinzunge** *f* ARCH & TRAGW midfeather, withe
**Schottenbauweise** *f* ARCH & TRAGW crosswall construction
**Schotter** *m* ARCH & TRAGW broken stone, BETON ballast, INFR & ENTW *Straßen* macadam, metal (*BrE*), WERKSTOFF broken stone, crushed rock, crushed stone, broken rock; **Schotterbrecher** *m* BAUMASCHIN stone breaker; **Schotterdecke** *f* INFR & ENTW macadam
**schottern** *vt* INFR & ENTW macadamize
***Schotter***: **Schottersand** *m* WERKSTOFF coarse sand; **Schotterstraße** *f* INFR & ENTW metaled road (*AmE*), metalled road (*BrE*); **Schottertragschicht** *f* INFR & ENTW crushed stone base course, crushed stone subbase
**schraffieren** *vt* ARCH & TRAGW hatch
**Schraffur** *f* ARCH & TRAGW hatching
**schräg** *adj* ARCH & TRAGW slanting, HOLZ beveled (*AmE*), bevelled (*BrE*)
**Schräg-** *in cpds* ARCH & TRAGW slanted
***schräg***: **~es Blatt** *nt* ARCH & TRAGW splayed scarf; **~er Kamm** *m* INFR & ENTW bevelled cogging; **~er Stoß** *m* HOLZ bevelled joint; **~e Widerlager** *nt pl* INFR & ENTW, NATURSTEIN raking abutments
***Schräg-***: **Schrägbalken** *m* BAUMASCHIN raker; **Schrägdach** *nt* ARCH & TRAGW pitched roof
**Schräge** *f* ARCH & TRAGW, INFR & ENTW haunch, inclination, incline, pitch
**schrägen** *vt* ARCH & TRAGW chamfer
***Schräg-***: **Schrägfuge** *f* ARCH & TRAGW chamfered joint; **Schrägkopfriegel** *m* BESCHLÄGE bevel-headed bolt
**schrägliegend** *adj* ARCH & TRAGW inclined
***Schräg-***: **Schrägpfahl** *m* ARCH & TRAGW, INFR & ENTW raker pile; **Schrägstab** *m* BETON inclined bar; **Schrägstellung** *f* INFR & ENTW skewing; **Schrägtrommelmischer** *m* BAUMASCHIN tilting drum mixer
**Schrägung** *f* INFR & ENTW *Aushubarbeiten* gain
***Schräg-***: **Schrägverband** *m* NATURSTEIN raking bond, STAHL diagonal bracing; **Schrägverblattung** *f* ARCH & TRAGW splayed joint; **Schrägwand** *f* ARCH & TRAGW batter wall
**Schramme** *f* BAUMASCHIN scratch, INFR & ENTW curbstone (*AmE*), WERKSTOFF scratch
**Schrank** *m* HOLZ built-in cupboard (*BrE*), cabinet, closet (*AmE*)
**Schränkverband** *m* NATURSTEIN diagonal masonry bond
**Schrapper** *m* BAUMASCHIN scraper
**Schraub-** *in cpds* BAUMASCHIN screw; **Schraubdeckel** *m* ABWASSER screw cap
**Schraube** *f* ARCH & TRAGW bolt, BAUMASCHIN, BESCHLÄGE screw, WERKSTOFF bolt; **Schraubendreher** *m* BAUMASCHIN screwdriver; **Schraubennagel** *m* BESCHLÄGE screw nail; **Schraubenpumpe** *f* BAUMASCHIN screw pump; **Schraubenschlüssel** *m* BAUMASCHIN spanner (*BrE*), wrench; **Schraubenwinde** *f* BAUMASCHIN screw jack; **Schraubenzieher** *m* BAUMASCHIN screwdriver
***Schraub-***: **Schraubloch** *nt* BAUMASCHIN screw hole; **Schraubmuffe** *f* BESCHLÄGE union; **Schraubstempel** *m* BAUMASCHIN ratchet brace; **Schraubstock** *m* BAUMASCHIN vice (*BrE*), vise

(*AmE*); **Schraubverbindung** *f* NATURSTEIN, STAHL, WERKSTOFF screwed connection; **Schraubzwinge** *f* BAUMASCHIN joiner's clamp
**Schrebergarten** *m* INFR & ENTW allotment
**Schreinerei** *f* HOLZ joinery
**Schreitschürfbagger** *m* BAUMASCHIN walking dragline
**Schritt** *m* ARCH & TRAGW step; **Schrittweite** *f* ARCH & TRAGW, INFR & ENTW increment
**Schrotbohren** *nt* BAUMASCHIN boring by shot drills
**Schrott** *m* BAURECHT, INFR & ENTW scrap; **Schrottplatz** *m* UMWELT scrapyard; **Schrottpresse** *f* BAUMASCHIN, UMWELT scrap-baling press; **Schrottsortierung** *f* UMWELT scrap sorting; **Schrottverwertung** *f* INFR & ENTW, UMWELT scrap processing, scrap recovery, scrap recycling
**Schrumpf-** *in cpds* WERKSTOFF shrinking; **Schrumpfabschottung** *f* ABWASSER shrink-on bushing
**schrumpfen** *vi* WERKSTOFF contract, shrink
**Schrumpfen** *nt* BETON shrinkage, shrinking
***Schrumpf-***: **Schrumpfmanschette** *f* ABWASSER shrink-on collar, shrunk-on sleeve; **Schrumpfmuffe** *f* ABWASSER shrink-on sleeve; **Schrumpfriß** *m* ARCH & TRAGW, BETON, ELEKTR, INFR & ENTW contraction crack
**Schrumpfung** *f* WERKSTOFF contraction
**Schub** *m* ARCH & TRAGW, INFR & ENTW shear, thrust; **auf ~ beanspruchen** *phr* INFR & ENTW shear; **Schubbewehrung** *f* ARCH & TRAGW, BETON web reinforcement
**schubfest** *adj* ARCH & TRAGW shear-resistant
***Schub***: **Schubfläche** *f* ARCH & TRAGW, BETON, STAHL, WERKSTOFF shear area; **Schubkarre** *f* BAUMASCHIN barrow, wheelbarrow; **Schubkarren** *m* BAUMASCHIN barrow, wheelbarrow; **Schubknicken** *nt* ARCH & TRAGW, BETON, STAHL, WERKSTOFF shear buckling; **Schubknicktragfähigkeit** *f* ARCH & TRAGW, BETON, STAHL, WERKSTOFF shear buckling resistance; **Schubkraft** *f* ARCH & TRAGW, BAUMASCHIN horizontal shear; **Schubkrafttragfähigkeit** *f* ARCH & TRAGW, BETON, STAHL, WERKSTOFF shear force resistance
**schubkraftübertragend**: **~es Panel** *nt* ARCH & TRAGW shear panel; **~e Täfelung** *nt* HOLZ shear panel
***Schub***: **Schubrichtung** *f* ARCH & TRAGW, BETON, INFR & ENTW, WERKSTOFF direction of thrust; **Schubriegel** *m* INFR & ENTW tower bolt; **Schubspannung** *f* ARCH & TRAGW shear, transverse strain; **Schubtragfähigkeit** *f* ARCH & TRAGW, BETON, HOLZ, STAHL, WERKSTOFF resistance to shear; **Schubverbindung** *f* BETON shear connector
**Schuppen** *m* ARCH & TRAGW shed
**Schürfgrube** *f* INFR & ENTW test pit, trial pit
**Schürfloch** *nt* INFR & ENTW trial hole
**Schürze** *f* STAHL flashing
**schußsicher** *adj* WERKSTOFF bulletproof
**Schutt** *m* ARCH & TRAGW, INFR & ENTW, UMWELT, WERKSTOFF debris, demolition rubbish, demolition waste, rubble
**Schütt-** *in cpds* WERKSTOFF bulk, fill
***Schutt***: **Schuttabladegebühren** *f pl* BAURECHT, INFR & ENTW dumping fees; **Schuttabladeplatz** *m* INFR & ENTW, UMWELT dumping site
***Schütt-***: **Schüttdichte** *f* INFR & ENTW, WERKSTOFF apparent density
**Schüttelförderer** *m* BAUMASCHIN jigging conveyor
**Schüttelrinne** *f* BAUMASCHIN shaking chute
**schütten** *vt* BETON pour
***Schütt-***: **Schüttgut** *nt* WERKSTOFF bulk material; **Schüttgutbehälter** *m* BAUMASCHIN hopper
***Schutt***: **Schuttkegel** *m* INFR & ENTW debris cone
***Schütt-***: **Schüttmaterial** *nt* INFR & ENTW, WERKSTOFF fill material; **Schüttrohr** *nt* BETON tremie; **Schüttsteine** *m pl* INFR & ENTW rip-rap
**Schüttung** *f* BETON pouring, INFR & ENTW *Erde* filled ground, filled-up ground, NATURSTEIN pouring
**Schutz** *m* BAURECHT protection, OBERFLÄCHE coat of paint
**Schütz** *nt* ELEKTR contactor
***Schutz***: **Schutzanstrich** *m* OBERFLÄCHE finish coat, protective paint coat, protective coating, finishing coat; **Schutzart** *f* ELEKTR protective system; **Schutzband** *nt* WERKSTOFF protective tape; **Schutzbehandlung** *f* HOLZ, OBERFLÄCHE preservative treatment, UMWELT, WERKSTOFF suspended matter; **Schutzbinde** *f* WERKSTOFF protective tape; **Schutzbrett** *nt* WERKSTOFF baffle board; **Schutzdach** *nt* ARCH & TRAGW shelter
**schützen** *vt* ARCH & TRAGW, BAURECHT guard
**schützend** *adj* ARCH & TRAGW, ELEKTR, NATURSTEIN, OBERFLÄCHE, WERKSTOFF protective
***Schutz***: **Schutzerdung** *f* ELEKTR earthing (*BrE*), grounding (*AmE*), protective earth (*BrE*), protective ground (*AmE*); **Schutzgitter** *nt* STAHL protective grating; **Schutzgurt** *m* INFR & ENTW safety belt; **Schutzkappe** *f* HEIZ & BELÜFT *am unteren Ende eines Fallrohres* boot; **Schutzkeller** *m* ARCH & TRAGW, INFR & ENTW cellar shelter; **Schutzkorb** *m* ARCH & TRAGW *Schornsteinleiter* protective cage, ELEKTR *Leuchte* basket guard; **Schutzraum** *m* ARCH & TRAGW shelter; **Schutzrohr** *nt* ABWASSER conduit, ELEKTR protective conduit; **Schutzrohr** *nt* **für Erdeinführungen von Leitungen** ELEKTR protective conduit for underground cabling; **Schutzschicht** *f* NATURSTEIN protective layer, OBERFLÄCHE preventive coating; **Schutzvorrichtung** *f* ARCH & TRAGW, BAUMASCHIN, BAURECHT, INFR & ENTW guard, guarding, safety device; **Schutzwerke** *nt pl* BAURECHT, INFR & ENTW protection works
**schwach** *adj* WERKSTOFF weak
**Schwächerwerden** *nt* ARCH & TRAGW, INFR & ENTW tapering
**Schwalbenschwanz** *m* HOLZ dovetail; **Schwalbenschwanzmaueranker** *m* NATURSTEIN dovetail tie; **Schwalbenschwanzverbindung** *f* HOLZ dovetail joint
**schwammartig** *adj* WERKSTOFF spongy; **~er Boden** *m* INFR & ENTW spongy soil
**Schwanenhals** *m* ABWASSER, ARCH & TRAGW offset, swan neck
**Schwankung** *f* ARCH & TRAGW, BAURECHT, INFR & ENTW variation
**Schwarz-** *in cpds* WERKSTOFF black
**schwarz**: **~es Brett** *nt* ARCH & TRAGW, BESCHLÄGE blackboard, bulletin board
***Schwarz-***: **Schwarzbeton** *m* BETON, INFR & ENTW, WERKSTOFF bituminous concrete; **Schwarzchrom** *nt* WERKSTOFF black chrome; **Schwarzdecke** *f* INFR & ENTW asphalt pavement, bituminous pavement, black top, WERKSTOFF black top; **Schwarzdeckenfertiger** *m* BAUMASCHIN bituminous surface finisher; **Schwarzkalk** *m* WERKSTOFF semihydraulic lime; **Schwarzkalkputz** *m* NATURSTEIN graystone lime

plaster (*AmE*), greystone lime plaster (*BrE*), semihydraulic lime plaster; **Schwarzmaterial** *nt* INFR & ENTW, WERKSTOFF asphaltic mixture

**Schweb-** *in cpds* INFR & ENTW suspension; **Schwebebahn** *f* INFR & ENTW suspension railroad (*AmE*), suspension railway (*BrE*)

**schwebend**: **~e Mikroorganismen** *m pl* HEIZ & BELÜFT, UMWELT airborne microorganisms; **~er Pfahl** *m* ARCH & TRAGW, INFR & ENTW suspended pile; **~er Schienenstoß** *m* ARCH & TRAGW suspended joint

**schwebestoffhaltig**: **~e Luft** *f* UMWELT aerosol

***Schweb-***: **Schwebeträger** *m* ARCH & TRAGW suspended beam; **Schwebstoffteilchen** *nt* UMWELT particulate matter; **Schwebteilchen** *nt* WERKSTOFF suspended particle

**Schwefel** *m* UMWELT, WERKSTOFF sulfur (*AmE*), sulphur (*BrE*); **Schwefeldioxid** *nt* UMWELT sulfur dioxide (*AmE*), sulphur dioxide (*BrE*); **Schwefeldioxidreduktion** *f* UMWELT sulfur dioxide reduction (*AmE*), sulphur dioxide reduction (*BrE*)

**schwefelig** *adj* WERKSTOFF sulfurous (*AmE*), sulphurous (*BrE*)

***Schwefel***: **Schwefeloxid** *nt* UMWELT sulfur oxide (*AmE*), sulphur oxide (*BrE*); **Schwefelsäure** *f* UMWELT, WERKSTOFF sulfuric acid (*AmE*), sulphuric acid (*BrE*); **Schwefelsäurenanhydrid** *nt* UMWELT sulfuric anhydride (*AmE*), sulphuric anhydride (*BrE*); **Schwefelzement** *m* WERKSTOFF sulfur cement (*AmE*), sulphur cement (*BrE*)

**schweflig**: **~e Säure** *f* UMWELT sulfurous acid (*AmE*), sulphurous acid (*BrE*)

**Schweiß-** *in cpds* BAUMASCHIN, STAHL welded, welding; **Schweißapparat** *m* BAUMASCHIN welding equipment; **Schweißbahn** *f* DÄMMUNG welded asphalt sheeting, OBERFLÄCHE *Abdichtung gegen Wasser* bituminous sheeting; **Schweißbrenner** *m* BAUMASCHIN welding blowpipe; **Schweißdraht** *m* STAHL welding wire

**schweißen** *vt* STAHL weld

**Schweißen** *nt* STAHL sheet metal work, welding

**Schweißer** *m* STAHL welder; **Schweißerhandschirm** *m* BAUMASCHIN welding handshield

***Schweiß-***: **Schweißfolge** *f* STAHL welding sequence; **Schweißhelm** *m* BAUMASCHIN welding helmet; **Schweißmethode** *f* STAHL, WERKSTOFF welding technique; **Schweißnaht** *f* STAHL welding seam; **Schweißprogramm** *nt* STAHL welding program (*AmE*), welding programme (*BrE*); **Schweißprozeß** *m* STAHL welding process; **Schweißstromkreis** *m* STAHL welding circuit; **Schweißtakt** *m* STAHL welding cycle

**Schweißung** *f* STAHL welding

***Schweiß-***: **Schweißverfahren** *nt* STAHL welding procedure; **Schweißvorrichtung** *f* BAUMASCHIN welding equipment; **Schweißzyklus** *m* STAHL welding cycle

**Schwell-** *in cpds* WERKSTOFF swelling; **Schwellboden** *m* INFR & ENTW swelling ground; **Schwelldruck** *m* WERKSTOFF swelling pressure

**Schwelle** *f* ARCH & TRAGW ground plate, sole piece, threshold, sill, sleeper, *Fachwerk* cross-sill, BETON sleeper, HOLZ sole piece, sole plate, sleeper; **Schwellenkopf** *m* ARCH & TRAGW, INFR & ENTW end of sleeper; **Schwellenträger** *m* ARCH & TRAGW sleeper-carrying girder

***Schwell-***: **Schwellholz** *nt* ARCH & TRAGW sill plate; **Schwellkurve** *f* WERKSTOFF swelling curve; **Schwellmittel** *nt* WERKSTOFF swelling agent; **Schwelltest** *m* INFR & ENTW, WERKSTOFF swell test

**Schwellung** *f* WERKSTOFF swelling; **Schwellungsvermögen** *nt* WERKSTOFF swelling capacity

***Schwell-***: **Schwellzone** *f* INFR & ENTW swelling zone

**Schwemmland** *nt* INFR & ENTW alluvium

**Schwenk-** *in cpds* ABWASSER, INFR & ENTW swiveling (*AmE*), swivelling (*BrE*); **Schwenkbatterie** *f* ABWASSER swivel mixer tap, swiveling mixer faucet (*AmE*), swivelling mixer tap (*BrE*); **Schwenkbrücke** *f* INFR & ENTW swing bridge, swivel bridge; **Schwenkbühne** *f* ARCH & TRAGW, BAUMASCHIN swinging platform

**schwenken** *vi* INFR & ENTW pivot

**Schwenken** *nt* BAUMASCHIN slewing (*BrE*), sluing (*AmE*)

***Schwenk-***: **Schwenkkran** *m* ARCH & TRAGW, BAUMASCHIN slewing crane (*BrE*), sluing crane (*AmE*); **Schwenkradius** *m* ARCH & TRAGW, BAUMASCHIN *eines Krans* swinging round; **Schwenkrinne** *f* BAUMASCHIN swinging chute

**schwer** *adj* INFR & ENTW heavy, WERKSTOFF weighty; **~es Baugerüst** *nt* ARCH & TRAGW, BAUMASCHIN gantry; **~er Boden** *m* INFR & ENTW heavy soil; **~er Holzhammer** *m* BAUMASCHIN maul; **~er Ton** *m* INFR & ENTW, WERKSTOFF heavy clay

**Schwer-** *in cpds* ARCH & TRAGW, BETON, heavy, weighty; **Schwerbeton** *m* ARCH & TRAGW, BETON barite concrete

***schwer***: **~ entflammbar** *adj* WERKSTOFF hardly inflammable; **~ entzündlich** *adj* WERKSTOFF hardly inflammable

**Schwergewicht** *nt* INFR & ENTW, UMWELT gravity; **Schwergewichtsdamm** *m* INFR & ENTW gravity dam; **Schwergewichtsmauer** *f* INFR & ENTW gravity dam; **Schwergewichtsstützmauer** *f* BETON, INFR & ENTW, NATURSTEIN gravity retaining wall

**Schwerkraft** *f* INFR & ENTW, WERKSTOFF gravitational force, gravity force; **Schwerkraftabfluß** *m* ABWASSER, INFR & ENTW gravity drainage; **Schwerkraftheizung** *f* HEIZ & BELÜFT gravity heating; **Schwerkraftströmung** *f* ABWASSER, INFR & ENTW gravity flow; **Schwerkraftsystem** *nt* ABWASSER, HEIZ & BELÜFT, INFR & ENTW gravity system

***Schwer-***: **Schwerlastkran** *m* BAUMASCHIN goliath crane; **Schwermetall** *nt* BAURECHT, UMWELT, WERKSTOFF heavy metal; **Schwerpunkt** *m* ARCH & TRAGW, INFR & ENTW, WERKSTOFF center of gravity (*AmE*), centre of gravity (*BrE*), centroid; **Schwerspat** *m* WERKSTOFF barite

**Schwerstbeton** *m* WERKSTOFF loaded concrete

**Schwimm-** *in cpds* BETON, BAUMASCHIN, INFR & ENTW floating, swimming; **Schwimmbad** *nt* INFR & ENTW swimming pool, swimming bath; **Schwimmbagger** *m* BAUMASCHIN grab dredger; **Schwimmbühne** *f* BAUMASCHIN, INFR & ENTW floating support

**schwimmend** *adj* ARCH & TRAGW, BAUMASCHIN, INFR & ENTW floating; **~es Bauwerk** *nt* INFR & ENTW floating structure; **~er Derrick** *m* BAUMASCHIN floating derrick; **~er Estrich** *m* ARCH & TRAGW, DÄMMUNG, NATURSTEIN floating flooring, floating layer; **~er Kran** *m* BAUMASCHIN floating crane; **~e Pfahlgründung** *f* INFR & ENTW, STAHL floating pile foundation; **~e Rohrleitung** *f* INFR & ENTW, STAHL floating pipeline

***Schwimm-***: **Schwimmgründung** *f* BETON, INFR &

ENTW floating foundation; **Schwimmkastengründung** *f* BETON floating caisson foundation; **Schwimmkran** *m* BAUMASCHIN pontoon crane; **Schwimmschirm** *m* UMWELT floating boom
**Schwind-** *in cpds* ARCH & TRAGW, BETON shrinkage
**schwindarm** *adj* WERKSTOFF low-shrink
**schwindbeständig** *adj* WERKSTOFF shrinkproof
***Schwind-***: **Schwindeinfluß** *m* ARCH & TRAGW, BETON, STAHL, WERKSTOFF effect of shrinkage, shrinkage effect
**Schwinden** *nt* ARCH & TRAGW, BETON, WERKSTOFF contraction, shrinkage, shrinking
***Schwind-***: **Schwindfuge** *f* ARCH & TRAGW, BETON, INFR & ENTW contraction joint, shrinkage joint; **Schwindriß** *m* ARCH & TRAGW, BETON shrinkage crack; **Schwindschutzzusatz** *m* WERKSTOFF antishrinkage admixture; **Schwindverformung** *f* ARCH & TRAGW, BETON, STAHL, WERKSTOFF shrinkage deformation
**Schwing-** *in cpds* ARCH & TRAGW sway, BAUMASCHIN jigging, INFR & ENTW sway
**schwingen** *vti* ARCH & TRAGW swing
***Schwing-***: **Schwingfenster** *nt* ARCH & TRAGW horizontally pivoted sash window; **Schwingflügel** *m* ARCH & TRAGW horizontally pivoted sash, BESCHLÄGE horizontally pivoted sach window; **Schwingstärke** *f* UMWELT amplitude of vibration; **Schwingsteifigkeit** *f* ARCH & TRAGW, INFR & ENTW, STAHL sway stiffness
**Schwingung** *f* ARCH & TRAGW, DÄMMUNG, WERKSTOFF oscillation, vibration; **Schwingungsamplitude** *f* DÄMMUNG oscillation amplitude
**schwingungsdämpfend** *adj* ARCH & TRAGW, DÄMMUNG vibration-absorbing
***Schwingung***: **Schwingungserreger** *m* BAUMASCHIN vibrator; **Schwingungsmessung** *f* WERKSTOFF vibration measurement; **Schwingungsperiode** *f* WERKSTOFF vibration period; **Schwingungsversuch** *m* INFR & ENTW, WERKSTOFF vibration test; **Schwingungsweite** *f* UMWELT amplitude of vibration
***Schwing-***: **Schwing- und Vibrationssieb** *nt* BAUMASCHIN jigging and vibrating screen; **Schwingwascher** *m* BAUMASCHIN jigging washer
**Schwitzwasser** *nt* DÄMMUNG, HEIZ & BELÜFT condensate, condensation water, perspiration
**Sediliennische** *f* ARCH & TRAGW sedilia niche
**Sedimentablagerung** *f* INFR & ENTW sedimentation, UMWELT *Geologie* deposition, sedimentation
**Sedimentationsanalyse** *f* INFR & ENTW sedimentation analysis
**Sedimentierglas** *nt*: ~ **nach Imhoff** ABWASSER, INFR & ENTW, WERKSTOFF Imhoff cone
**See**: **Seebuhne** *f* INFR & ENTW sea groin (*AmE*), sea groyne (*BrE*); **Seeverklappung** *f* UMWELT ocean dumping
**Segeltuch** *nt* ARCH & TRAGW canvas; **Segeltuchplane** *f* BESCHLÄGE tarpaulin
**Segment** *nt* ARCH & TRAGW segment; **Segmentbogen** *m* ARCH & TRAGW arched lintel, segmental arch, INFR & ENTW, NATURSTEIN arched lintel; **Segmentstahl** *m* STAHL half-round bar
**Sehlinie** *f* INFR & ENTW collimation line, line of sight
**seidenmatt**: **~es Glas** *nt* WERKSTOFF satin-finish glass
**Seifenschüssel** *f* BESCHLÄGE soap dish
**Seifenspender** *m* ABWASSER, BESCHLÄGE soap dispenser
**Seil** *nt* BAUMASCHIN cable, WERKSTOFF rope; **Seilbahn** *f* BAUMASCHIN, INFR & ENTW funicular railway; **Seilverspannung** *f* ARCH & TRAGW guying, INFR & ENTW guy, guying; **Seilwinde** *f* BAUMASCHIN aerial railway
**seismisch** *adj* INFR & ENTW seismic; **~e Aktivität** *f* INFR & ENTW seismic activity; **~er Aufschluß** *m* INFR & ENTW seismic survey; **~e Auswirkungen** *f pl* WERKSTOFF seismic effects; **~e Bodenforschung** *f* INFR & ENTW seismic inspection; **~e Bodenuntersuchung** *f* INFR & ENTW seismic exploration method
**Seismograph** *m* BAUMASCHIN, INFR & ENTW seismograph
**seismographisch**: **~e Beobachtung** *f* INFR & ENTW seismographic observation
**Seismologie** *f* INFR & ENTW seismology
**Seismometer** *nt* BAUMASCHIN seismometer
**Seite** *f* ARCH & TRAGW side; **Seitenablagerung** *f* INFR & ENTW *von Bodenmaterial* spoil area; **Seitenansicht** *f* ARCH & TRAGW side face; **Seitenbewegung** *f* ARCH & TRAGW, HOLZ, INFR & ENTW lateral movement; **Seitenblech** *nt* WERKSTOFF side plate; **Seitendruck** *m* ARCH & TRAGW, INFR & ENTW, WERKSTOFF lateral pressure; **Seitenfläche** *f* ARCH & TRAGW cheek, side; **Seitenführung** *f* BESCHLÄGE *Jalousie* side track; **Seitenführungsschiene** *f* BESCHLÄGE *Jalousie* side track; **Seitenkanal** *m* HEIZ & BELÜFT bypass channel; **Seitenlast** *f* ARCH & TRAGW, INFR & ENTW lateral load; **Seitenrippe** *f* ARCH & TRAGW *Gewölbe* nervure; **Seitenschiff** *nt* ARCH & TRAGW, INFR & ENTW *Kirchenbau* aisle; **Seitenstraße** *f* INFR & ENTW side street, byroad, byway; **Seitenstreifen** *m* INFR & ENTW verge; **Seitenversteifung** *f* ARCH & TRAGW, BETON, INFR & ENTW, STAHL, WERKSTOFF lateral reinforcing structure; **Seitenwinde** *f* ARCH & TRAGW, BAUMASCHIN side pulley
**seitlich** *adj* ARCH & TRAGW, INFR & ENTW lateral; **~e Abstützung** *f* ARCH & TRAGW, INFR & ENTW lateral support; ~ **festgehalten** *adj* ARCH & TRAGW, BETON, INFR & ENTW, STAHL laterally restrained; **~e Halterung** *f* ARCH & TRAGW, BETON, INFR & ENTW, WERKSTOFF lateral restraint; ~ **stabil** *adj* ARCH & TRAGW, BETON, INFR & ENTW, STAHL, WERKSTOFF laterally stable; **~e Stabilität** *f* ARCH & TRAGW, BETON, INFR & ENTW, STAHL, WERKSTOFF lateral stability; **~es Torsionsausknicken** *nt* ARCH & TRAGW, INFR & ENTW, STAHL, WERKSTOFF lateral-torsional buckling; ~ **versetzt** *adj* ARCH & TRAGW, BETON, INFR & ENTW, STAHL, WERKSTOFF off-center (*AmE*), off-centre (*BrE*)
**Sekundärrohstoff** *m* UMWELT *wiederverwertbar* waste product
**Sekundärträger** *m* ARCH & TRAGW intercostal girder
**Selbst**: **Selbstentladeeimer** *m* BAUMASCHIN self-dumping bucket; **Selbstentwässerungssystem** *nt* ARCH & TRAGW, INFR & ENTW self-draining system
**selbstfahrend**: **~e Walze** *f* BAUMASCHIN self-propelled roller
**selbstklebend** *adj* WERKSTOFF self-adhesive
**selbstlöschend** *adj* WERKSTOFF self-quenching
***Selbst***: **Selbstreinigung** *f* UMWELT natural purification; **Selbstreinigungskraft** *f* UMWELT assimilative capacity
**selbstschließend**: **~e Tür** *f* BESCHLÄGE self-closing door
***Selbst***: **Selbstschlußbatterie** *f* BESCHLÄGE *Wasserhahn* self-closing faucet (*AmE*), self-closing tap (*BrE*)
**selbstspannend** *adj* BETON self-tensioning

**selbsttragend** *adj* ARCH & TRAGW self-supporting
**selbstverlöschend** *adj* WERKSTOFF self-extinguishing
**selektiv**: **~e Oberfläche** *f* WERKSTOFF selective surface
**Senk**: **Senkblei** *nt* ARCH & TRAGW, BAUMASCHIN bob, plumb, plumb bob; **Senkbrunnen** *m* ABWASSER, INFR & ENTW well; **Senkbrunnengründung** *f* ARCH & TRAGW, INFR & ENTW well foundation
**Senke** *f* INFR & ENTW sink
**senken**: **sich senken** *v refl* ARCH & TRAGW subside
***Senk***: **Senkgrube** *f* ABWASSER cloaca, cesspit, cesspool, INFR & ENTW cesspit; **Senkkasten** *m* INFR & ENTW caisson; **Senkkastengründung** *f* INFR & ENTW caisson foundation; **Senkkastenkrankheit** *f* BAURECHT, INFR & ENTW caisson disease, decompression sickness; **Senkkopfniet** *m* BESCHLÄGE countersunk button-head rivet; **Senkkopfvernietung** *f* STAHL countersunk riveting; **Senklot** *nt* BAUMASCHIN plumb
**senkrecht 1.** *adv* ARCH & TRAGW plumb; **2.** *adj* ARCH & TRAGW, INFR & ENTW perpendicular, upright, vertical, plumb
**Senkrecht-** *in cpds* ARCH & TRAGW perpendicular, vertical
***senkrecht***: **nicht ~** *adj* ARCH & TRAGW off plumb; **~e Absteifung** *f* ARCH & TRAGW vertical bracing; **~e Abwasserleitung** *f* ABWASSER stack; **~e Stange** *f* ARCH & TRAGW vertical member; **~e Verkleidungsschiene** *f* BESCHLÄGE, STAHL vertical cladding rail
***Senkrecht-***: **Senkrechtförderschnecke** *f* BAUMASCHIN screw elevator; **Senkrechtschnitt** *m* ARCH & TRAGW vertical section; **Senkrechtstab** *m* ARCH & TRAGW vertical member
**Senkung** *f* ARCH & TRAGW sagging, INFR & ENTW settlement, settling, sagging, *Boden* sinking, WERKSTOFF slump
**Sensor** *m* BAURECHT, ELEKTR sensor
**Serienschaltung** *f* ELEKTR series mounting
**Serpentinisierung** *f* UMWELT serpentinization
**Servomotor** *m* BAUMASCHIN, UMWELT servomotor
**Setz-** *in cpds* ARCH & TRAGW, INFR & ENTW settlement, WERKSTOFF slump; **Setzbecher** *m* WERKSTOFF *Ausbreitversuch* slump cone
**setzen** *vt* BAURECHT settle, INFR & ENTW set
**Setzen** *nt* INFR & ENTW settlement
***Setz-***: **Setzfuge** *f* ARCH & TRAGW, INFR & ENTW settlement joint; **Setzholz** *nt* ARCH & TRAGW window post; **Setzmaß** *nt* WERKSTOFF slump; **Setzprobe** *f* WERKSTOFF slump test; **Setzriß** *m* ARCH & TRAGW, INFR & ENTW settlement crack; **Setzstufe** *f* ARCH & TRAGW *Holztreppe* riser
**Setzung** *f* ARCH & TRAGW, INFR & ENTW sagging, settlement, subsidence; **Setzungsbeobachtung** *f* ARCH & TRAGW, INFR & ENTW settlement observation; **Setzungsdauer** *f* BETON, INFR & ENTW, WERKSTOFF settlement duration; **Setzungskurve** *f* BETON, WERKSTOFF settlement curve; **Setzungsmeßgerät** *nt* BAUMASCHIN settlement gage (*AmE*), settlement gauge (*BrE*); **Setzungsspannung** *f* INFR & ENTW settlement stress
***Setz-***: **Setzversuch** *m* INFR & ENTW settlement test
**S-Farbe** *f* OBERFLÄCHE silicate paint
**Sheddach** *nt* ARCH & TRAGW northlight roof, sawtooth roof
**Shorehärte** *f* WERKSTOFF Shore hardness
**Shredderabfälle-Deponie** *f* UMWELT shredded refuse landfill
**SHZ** *abbr* (*Sulfathüttenzement*) BETON supersulfated cement (*AmE*), supersulphated cement (*BrE*)
**SI** *abbr* (*Système International*) ARCH & TRAGW, INFR & ENTW, WERKSTOFF SI (*Système International*)
**sicher** *adj* BAURECHT safe; **~e Arbeitsübereinkünfte** *f pl* BAURECHT safe working arrangements; **~e Belastung** *f* ARCH & TRAGW, BAURECHT, INFR & ENTW safe working load (*SWL*); **~es Zerbrechen** *nt* WERKSTOFF safe breakage
**Sicherheit** *f* BAURECHT safety, assurance, security; **Sicherheitsbeiwert** *m* BAURECHT safety factor; **Sicherheitsbeleuchtung** *f* BAURECHT, ELEKTR emergency illumination, emergency lighting, safety lighting; **Sicherheitsberater** *m* BAURECHT safety adviser; **Sicherheitsfaktor** *m* ARCH & TRAGW, BAURECHT, INFR & ENTW safety factor; **Sicherheitsglas** *nt* BAURECHT, WERKSTOFF safety glass, tempered safety glass; **Sicherheitsgrenze** *f* BAURECHT safety limit; **Sicherheitsprotokoll** *nt* BAURECHT safety record; **Sicherheitsschloß** *nt* BAURECHT, BESCHLÄGE security lock, safety lock; **Sicherheitsspielraum** *m* BAURECHT safety margin; **Sicherheitsstreifen** *m* INFR & ENTW safety strip; **Sicherheitstor** *nt* BAURECHT safety door; **Sicherheitstür** *f* BAURECHT safety door; **Sicherheitsventil** *nt* HEIZ & BELÜFT, UMWELT relief valve, safety valve; **Sicherheitsverbundglas** *nt* BAURECHT, WERKSTOFF laminated safety glass
**sichern** *vt* BAURECHT secure
**sicherstellen** *vt* BAURECHT secure
**Sicherstellung** *f* BAURECHT indemnity
**Sicherung** *f* ELEKTR cutout; **Sicherungseinsatz** *m* ELEKTR fuse link; **Sicherungspatrone** *f* ELEKTR fuse cartridge
**Sicht** *f* ARCH & TRAGW visibility
**sichtbar** *adj* ARCH & TRAGW visible; **~e Fuge** *f* NATURSTEIN face joint
***Sicht***: **Sichtbeton** *m* BETON, INFR & ENTW, WERKSTOFF exposed concrete, exposed concrete finish, facing concrete; **Sichtbetonschalung** *f* BETON, HOLZ exposed concrete formwork; **Sichtfenster** *nt* ARCH & TRAGW window; **Sichtfläche** *f* ARCH & TRAGW visible area, NATURSTEIN facing; **Sichtfuge** *f* NATURSTEIN facing joint; **Sichtkontrolle** *f* ARCH & TRAGW, INFR & ENTW sight check, visual check, visual examination; **Sichtmauerstein** *m* INFR & ENTW, NATURSTEIN, WERKSTOFF fair-faced brick; **Sichtmauerwerk** *nt* INFR & ENTW, NATURSTEIN exposed masonry, fair-faced mansory; **Sichtprüfung** *f* ARCH & TRAGW, BAURECHT, INFR & ENTW visual check, visual examination; **Sichtschalung** *f* BETON, HOLZ exposed concrete formwork; **Sichtschutz** *m* UMWELT *einer Deponie* screen; **Sichtschutzwand** *f* ARCH & TRAGW, INFR & ENTW screen wall; **Sichtseite** *f* ARCH & TRAGW visible face; **Sichtweite** *f* ARCH & TRAGW sight distance
**Sicker-** *in cpds* UMWELT seeping; **Sickerbecken** *nt* ABWASSER oozing basin, UMWELT infiltration basin; **Sickerdränage** *f* ABWASSER weep drain; **Sickergraben** *m* ABWASSER, INFR & ENTW rubble drain, seepage trench; **Sickergrube** *f* ABWASSER trickle pool, INFR & ENTW soakaway; **Sickerleitung** *f* ABWASSER filter drain; **Sickerlinie** *f* INFR & ENTW seepage line
**sickern** *vi* ABWASSER trickle, INFR & ENTW seep
***Sicker-***: **Sickerrohr** *nt* ABWASSER filter drain;

**Sickerschacht** *m* ABWASSER, ARCH & TRAGW, INFR & ENTW absorbing well, seepage shaft, soakaway, *Senkbrunnen* well drain; **Sickerströmungsdruck** *m* INFR & ENTW seepage pressure

**Sickerung** *f* ABWASSER percolation

***Sicker-***: **Sickerverlust** *m* INFR & ENTW seepage loss; **Sickerwasser** *nt* ABWASSER, INFR & ENTW, UMWELT leakage water, percolating water, seepage water; **Sickerweg** *m* INFR & ENTW seepage path

**Sieb** *nt* ABWASSER strainer, BAUMASCHIN screen, BESCHLÄGE sieve, UMWELT screening equipment; **Siebanalyse** *f* WERKSTOFF sieve analysis; **Siebblech** *nt* BAUMASCHIN screening sheet, punched-plate screen; **Siebboden** *m* BAUMASCHIN sieve bottom

**sieben** *vt* BAUMASCHIN screen

**Sieben** *nt* BAUMASCHIN screening, BESCHLÄGE sieving, UMWELT screening

***Sieb***: **Siebkurve** *f* WERKSTOFF *Kies, Sand* grading curve; **Sieblinie** *f* BETON, WERKSTOFF *Beton* particle-size distribution curve; **Siebmasche** *f* BAUMASCHIN mesh; **Siebrest** *m* UMWELT screenings; **Siebrückstand** *m* UMWELT screenings; **Siebschutt** *m* WERKSTOFF sieve refuse; **Siebstein** *m* ARCH & TRAGW ventilation tile; **Siebtrommel** *f* BAUMASCHIN rotary screen

**Siebung** *f* UMWELT sieving

**Siedebereich** *m* WERKSTOFF boiling range

**Siederohr** *nt* HEIZ & BELÜFT boiler pipe

**Siedlung** *f* INFR & ENTW settlement; **Siedlungsabfall** *m* UMWELT municipal waste, urban waste, urban solid waste; **Siedlungsplanung** *f* INFR & ENTW, UMWELT community planning

**Siegel** *nt* BESCHLÄGE seal, OBERFLÄCHE sealing

**SI-Einheit** *f* ARCH & TRAGW, INFR & ENTW, WERKSTOFF SI unit

**Siel** *nt* ABWASSER sewer

**Sifonbogen** *m* ABWASSER trap elbow

**Signal** *nt* ARCH & TRAGW beacon

**SI-Grundeinheit** *f* ARCH & TRAGW, INFR & ENTW, WERKSTOFF basic SI unit

**Silikat** *nt* WERKSTOFF silicate; **Silikatfarbe** *f* OBERFLÄCHE silicate paint; **Silikatglas** *nt* WERKSTOFF silicate glass

**Silikon** *nt* WERKSTOFF silicone; **Silikonzelle** *f* UMWELT silicon cell

**Siliziumoxid** *nt* WERKSTOFF silica

**Silo** *m* BAUMASCHIN silo

**Sims** *m* ARCH & TRAGW cornice, ledge; **Simsbrett** *nt* ARCH & TRAGW fascia board, window board; **Simshobel** *m* BAUMASCHIN side rabbet plane

**Sink-** *in cpds* INFR & ENTW, WERKSTOFF decantation

**sinken** *vi* INFR & ENTW sink

***Sink-***: **Sinkgeschwindigkeit** *f* INFR & ENTW, UMWELT, WERKSTOFF decantation rate; **Sinkgut** *nt* UMWELT deposited matter; **Sinkstoff** *m* UMWELT deposited matter

**Sinterbeton** *m* BETON hooped concrete

**Siphon** *m* ABWASSER, UMWELT siphon, stink trap; **Siphonhöhe** *f* UMWELT siphon crest; **Siphonüberlauf** *m* UMWELT siphon spillway

**Sitzungsraum** *m* ARCH & TRAGW conference room

**Skelett** *nt* ARCH & TRAGW, INFR & ENTW, STAHL *Fachwerk* framework, skeleton framing, *Gerippe eines Gebäudes* frame, *Gebäude* skeleton; **Skelettbau** *m* ARCH & TRAGW, INFR & ENTW framed structure; **Skelettbauweise** *f* ARCH & TRAGW, INFR & ENTW frame construction, skeleton construction; **Skelettplattenbauweise** *f* ARCH & TRAGW panel-frame construction; **Skelettträger** *m* ARCH & TRAGW skeleton girder

**Skizze** *f* ARCH & TRAGW sketch

**Smog** *m* UMWELT smog

**Sockel** *m* ARCH & TRAGW mounting, *eines Denkmals, einer Statue* socle, *einer Wand, Säule* base, foot, footing, *eines Gebäudes* foundation, pedestal, *einer Raumwand* dado, plinth, BAUMASCHIN mounting, BETON *einer Wand, Säule* base, HEIZ & BELÜFT, HOLZ mounting, INFR & ENTW foundation, *einer Wand* footing, mounting, foot, STAHL mounting; **~ mit Hohlkehle** ARCH & TRAGW, WERKSTOFF coved skirting; **Sockelfliese** *f* ABWASSER, NATURSTEIN dado tile; **Sockelfuß** *m* ARCH & TRAGW dado base; **Sockelleiste** *f* ARCH & TRAGW, BESCHLÄGE, HOLZ baseboard (*AmE*), skirting board (*BrE*); **Sockelleiste** *f* **mit Hohlkehle** ARCH & TRAGW, WERKSTOFF coved skirting; **Sockelmauer** *f* ARCH & TRAGW plinth wall; **Sockelschaft** *m* ARCH & TRAGW die

**Soglast** *f* ARCH & TRAGW *Statik* suction load

**Sohlbank** *f* ARCH & TRAGW sill, window sill

**Sohlbreite** *f* INFR & ENTW *Graben* bottom width

**Sohle** *f* ABWASSER *Rohr* invert, *Rohr, Graben* bottom, ARCH & TRAGW *Boden* bottom, *Baugrund* foot, INFR & ENTW *Tunnelbau* foot wall, *Boden* bottom, foot, WERKSTOFF *Becken* floor; **Sohlenabdichtung** *f* INFR & ENTW bottom sealing; **Sohlendrainageschicht** *f* INFR & ENTW basal drainage blanket; **Sohlengewölbe** *nt* ARCH & TRAGW floor arch

**Solar-** *in cpds* UMWELT solar; **Solaranlage** *f* ELEKTR, HEIZ & BELÜFT, INFR & ENTW, UMWELT solar collector, solar energy plant, solar plant; **Solardynamik** *f* solar dynamics; **Solarenergie** *f* UMWELT solar energy; **Solargewinn** *m* HEIZ & BELÜFT solar gain; **Solarheizungssystem** *nt* UMWELT solar heating system

**Solarimeter** *nt* UMWELT solarimeter

***Solar-***: **Solarkollektor** *m* UMWELT solar energy plant; **Solarkonstante** *f* UMWELT solar constant; **Solarkontrolle** *f* ARCH & TRAGW, HEIZ & BELÜFT solar control; **Solarpond** *nt* UMWELT solar pond; **Solartechnik** *f* UMWELT solar engineering, solar technology; **Solarzelle** *f* UMWELT solar cell; **Solarzellenlaken** *nt* UMWELT array blanket; **Solarzellenplatte** *f* UMWELT solar panel

**Solnhofer**: **~ Platte** *f* WERKSTOFF Solnhofer stone

**Sonde** *f* INFR & ENTW test probe; **Sondenspitze** *f* INFR & ENTW sounding cone

**Sonder-** *in cpds* ARCH & TRAGW, INFR & ENTW, UMWELT special; **Sonderabfall** *m* UMWELT special waste; **Sonderabfallzwischenlager** *nt* UMWELT temporary deposit for hazardous waste; **Sonderanfertigung** *f* ARCH & TRAGW, INFR & ENTW special design; **Sonderbronze** *f* WERKSTOFF high-strength bronze; **Sondererzeugnisse** *nt pl* ARCH & TRAGW, NATURSTEIN specials

**Sondermüll** *m* UMWELT special waste, hazardous waste; **Sondermülldeponie** *f* UMWELT hazardous waste landfill; **Sondermülleinsammlung** *f* UMWELT hazardous waste collection

***Sonder-***: **Sonderstahl** *m* STAHL, WERKSTOFF special steel

**Sondiergestänge** *nt* INFR & ENTW sounding rod

**Sondiermethode** *f* INFR & ENTW sounding method

**Sondierung** *f* INFR & ENTW, WERKSTOFF sounding

**Sonnen-** *in cpds* INFR & ENTW, UMWELT solar; **Sonnenabsorptionskoeffizient** *m* UMWELT solar absorption coefficient; **Sonnenabsorptionsvermögen** *nt* UMWELT solar absorptivity; **Sonnenazimut** *m* UMWELT solar azimuth; **Sonnenbatterie** *f* UMWELT solar battery; **Sonnenblende** *f* ARCH & TRAGW awning, BESCHLÄGE blind; **Sonnenenergie** *f* UMWELT solar energy; **mit Sonnenenergie betrieben** *phr* UMWELT solar-powered; **Sonnenfarm** *f* UMWELT solar farm; **Sonnenhöhe** *f* UMWELT solar altitude; **Sonnenkollektor** *m* ELEKTR, INFR & ENTW, HEIZ & BELÜFT solar collector, solar energy plant; **Sonnenofen** *m* UMWELT solar furnace; **Sonnenschutzglas** *nt* WERKSTOFF antisun glass; **Sonnenschutzvorrichtung** *f* ARCH & TRAGW shading device; **Sonnenstand** *m* UMWELT solar altitude; **Sonnenstrahlung** *f* INFR & ENTW, UMWELT solar radiation; **Sonnenturm** *m* UMWELT solar tower; **Sonnenwärme** *f* UMWELT solar heat; **Sonnenwärmekollektor** *m* HEIZ & BELÜFT, UMWELT solar collector; **Sonnenwärmekonzentrator** *m* UMWELT solar concentrator; **Sonnenwege** *m pl* **im jeweiligen Monat** HEIZ & BELÜFT, UMWELT monthly sun paths

**Sorgepflicht** *f* BAURECHT duty of care

**sortieren** *vt* WERKSTOFF grade, *Recycling* assort

**Sortiergerät** *nt* BAUMASCHIN grader

**sortiert** *adj* WERKSTOFF graded

**Sortierung** *f*: **~ von Abfällen** UMWELT waste sorting

**Souterrain** *nt* ARCH & TRAGW semibasement

**sozial**: **~er Wohnungsbau** *m* BAURECHT publicly financed housing, INFR & ENTW public housing units (*AmE*), council housing (*BrE*)

**Sozialgebäude** *nt* BAURECHT welfare building

**Spachtel** *f* BAUMASCHIN trowel, flat trowel, spatula; **Spachtelmasse** *f* ARCH & TRAGW, BESCHLÄGE putty, OBERFLÄCHE stopper, WERKSTOFF filling compound, knifing filler, putty, filling material; **Spachtelmesser** *nt* BAUMASCHIN stopping knife; **Spachtelverbindung** *f* BESCHLÄGE putty joint

**Spalt** *m* ARCH & TRAGW *in einer Wand* fissure, slot, split, BETON crack, INFR & ENTW fissure, NATURSTEIN, WERKSTOFF crack; **Spaltaxt** *f* BAUMASCHIN splitting ax (*AmE*), splitting axe (*BrE*)

**Spalten** *nt* BESCHLÄGE cleaving, INFR & ENTW, WERKSTOFF ripping

***Spalt***: **Spaltfuge** *f* ARCH & TRAGW *Dachsparren* split; **Spalthammer** *m* BAUMASCHIN cleaver; **Spaltsäge** *f* BAUMASCHIN cleaving saw

**Spaltung** *f* ARCH & TRAGW splitting

**Span** *m* HOLZ splinter

**Spange** *f* ARCH & TRAGW stay bolt

**Spann-** *in cpds* ARCH & TRAGW bracing, stretching, tensioning, BETON, INFR & ENTW jacking; **Spannbalken** *m* ARCH & TRAGW tie beam; **Spannbeton** *m* BETON, INFR & ENTW, WERKSTOFF prestressed concrete; **Spannbett** *nt* BETON stressbed; **Spannblock** *m* BETON, INFR & ENTW jacking block; **Spannbohle** *f* ARCH & TRAGW strutting board; **Spanndraht** *m* ARCH & TRAGW bracing wire, tensioning wire, stretching wire, BETON, INFR & ENTW, WERKSTOFF prestressed concrete wire; **Spanndrahtbündel** *nt* BETON, STAHL strand

**spannen** *vt* ARCH & TRAGW stretch, tighten, BETON stress, tension, WERKSTOFF stretch

**Spannen** *nt* ARCH & TRAGW tensioning

***Spann-***: **Spannglied** *nt* BETON tendon; **Spannschloß** *nt* BESCHLÄGE *Zaun* turnbuckle; **Spannschlüssel** *m* BAUMASCHIN ratchet wrench; **Spannseil** *nt* ARCH & TRAGW stay; **Spannstab** *m* ARCH & TRAGW, BETON, STAHL stressing bar; **Spanntraverse** *f* ARCH & TRAGW, BETON *Spannbeton* crosshead

**Spannung** *f* ARCH & TRAGW stress, tension, ELEKTR voltage; **Spannungsabbau** *m* BETON loss of prestress; **Spannungsabfall** *m* BETON loss of prestress, ELEKTR loss of voltage; **Spannungsarten** *f pl* ARCH & TRAGW, BETON, HOLZ forms of stress; **Spannungserhöhung** *f* ARCH & TRAGW increase of tension; **Spannungsermittlung** *f* INFR & ENTW stress analysis

**spannungsfrei** *adj* ARCH & TRAGW, WERKSTOFF stress-free, unstressed

***Spannung***: **Spannungsgleichgewicht** *nt* ARCH & TRAGW, INFR & ENTW equilibrium of stresses; **Spannungskorrosion** *f* STAHL, WERKSTOFF stress corrosion; **Spannungskurve** *f* WERKSTOFF tension curve; **Spannungslehre** *f* ARCH & TRAGW, INFR & ENTW stress theory; **Spannungsnachweis** *m* ARCH & TRAGW stress analysis; **Spannungsriß** *m* STAHL, WERKSTOFF stress crack; **Spannungstrajektorie** *f* ARCH & TRAGW, WERKSTOFF trajectory of stress; **Spannungsverteilung** *f* ARCH & TRAGW, BETON, INFR & ENTW, STAHL, WERKSTOFF stress distribution; **Spannungsweg** *m* ARCH & TRAGW, WERKSTOFF trajectory of stress

***Spann-***: **Spannverankerung** *f* BETON, INFR & ENTW jacking anchorage; **Spannweite** *f* ARCH & TRAGW, INFR & ENTW bearing distance, span; **Spannweite** *f* **des Bogens** ARCH & TRAGW, INFR & ENTW buttress span

***Span***: **Spanplatte** *f* HOLZ, WERKSTOFF chipboard, particle board

**Sparren** *m* ARCH & TRAGW, HOLZ rafter; **Sparrenabstand** *m* ARCH & TRAGW, HOLZ distance between rafters; **Sparrendach** *nt* ARCH & TRAGW, HOLZ rafter roof; **Sparrengebinde** *nt* HOLZ couple; **Sparrenkopf** *m* ARCH & TRAGW, HOLZ rafter head; **Sparrenlage** *f* ARCH & TRAGW, HOLZ rafter system; **Sparrennagel** *m* BESCHLÄGE purlin nail; **Sparrenpaar** *nt* HOLZ couple; **Sparrenpfettenanker** *m* ARCH & TRAGW, INFR & ENTW, STAHL rafter-to-purlin connector; **Sparrenschwelle** *f* HOLZ inferior purlin

**Sparschalung** *f* BETON open formwork

**Spaten** *m* BAUMASCHIN spade

**Speicher** *m* ARCH & TRAGW attic, loft, silo, INFR & ENTW storage space; **Speicherbecken** *nt* UMWELT storage basin; **Speicherheizung** *f* HEIZ & BELÜFT night storage heating, storage heating; **Speicherschema** *nt* UMWELT storage scheme

**Speirohr** *nt* ABWASSER spout

**Speise-** *in cpds* ABWASSER, ELEKTR feeder; **Speisebecken** *nt* ABWASSER feeder basin; **Speisekammer** *f* ARCH & TRAGW larder, pantry; **Speiseleitung** *f* ELEKTR feeder, *Straße, Eisenbahn* feeder line

**Spengler** *m* ABWASSER, ARCH & TRAGW, STAHL plumber; **Spenglerarbeiten** *f pl* ABWASSER, ARCH & TRAGW, STAHL plumber's work

**Sperr-** *in cpds* BETON water-repellent, waterproof, OBERFLÄCHE, WERKSTOFF dampproofing; **Sperranstrichmittel** *nt* DÄMMUNG, OBERFLÄCHE dampproofing paint; **Sperrbeton** *m* BETON waterproof concrete, water-repellent concrete

**Sperre** *f* INFR & ENTW barricade, WERKSTOFF *Schloß* catch

**sperren** *vt* OBERFLÄCHE *Feuchtigkeit* stop, waterproof
***Sperr-***: **Sperrflüssigkeit** *f* DÄMMUNG liquid waterproofing agent; **Sperrfolie** *f* ARCH & TRAGW, DÄMMUNG *Abdichtung* barrier membrane; **Sperrfurnier** *nt* HOLZ crossband; **Sperrhaken** *m* BAUMASCHIN ratchet; **Sperrholz** *nt* HOLZ plywood
**Sperrigkeit** *f* WERKSTOFF bulkiness
***Sperr-***: **Sperrklinke** *f* BESCHLÄGE ratchet; **Sperrleiste** *f* HOLZ ledger; **Sperrmauer** *f* INFR & ENTW barrage; **Sperrmörtel** *m* BETON, NATURSTEIN water-repellent mortar, waterproof mortar; **Sperrmüllabfuhr** *f* UMWELT bulk collection; **Sperrpappe** *f* WERKSTOFF insulating felt; **Sperrputz** *m* BETON, NATURSTEIN water-repellent finish, waterproofing finish; **Sperrschicht** *f* DÄMMUNG barrier layer, impervious course, UMWELT barrier layer; **Sperrstoff** *m* ARCH & TRAGW, DÄMMUNG, ELEKTR, WERKSTOFF barrier material, *gegen Feuchtigkeit* dampproofing material, *Abdichtung gegen Wasser* insulant, waterproofer
**Sperrung** *f* DÄMMUNG insulation, *Abdichtung* waterproofing
***Sperr-***: **Sperrwand** *f* UMWELT slurry wall; **Sperrzusatz** *m* DÄMMUNG, WERKSTOFF dampproofing additive
**Spezialfliese** *f* WERKSTOFF purpose-made tile
**spezialisiert**: **~er Subunternehmer** *m* BAURECHT specialist sub-contractor
**Spezifikation** *f* ARCH & TRAGW, BAURECHT spec, specification (*spec*)
**spezifisch**: **~e Brunnenkapazität** *f* UMWELT specific capacity of a well; **~e Drehzahl** *f* UMWELT specific speed; **~er elektrischer Widerstand** *m* ELEKTR specific electrical resistivity; **~es Gewicht** *nt* **je cbm umbauten Raumes** WERKSTOFF specific weight of building volume; **~e Masse** *f* WERKSTOFF specific mass; **~es Volumen** *nt* WERKSTOFF specific volume; **~e Wärme** *f* HEIZ & BELÜFT, WERKSTOFF specific heat; **~e Wärmekapazität** *f* HEIZ & BELÜFT, WERKSTOFF specific heat capacity; **~e Wärmeleitfähigkeit** *f* DÄMMUNG, WERKSTOFF coefficient of thermal conductivity, specific thermal conductivity
**spezifizieren** *vt* ARCH & TRAGW, BAURECHT specify
**sphärisch** *adj* INFR & ENTW spherical; **~es Korn** *nt* WERKSTOFF spherical grain
**Sphäroguß** *m* WERKSTOFF nodular cast iron, *sphärolitisches Gußeisen* spheroidal cast iron
**Spiegel** *m* WERKSTOFF mirror; **Spiegelglas** *nt* WERKSTOFF mirror glass, plate glass
**spiegeloptisch**: **~es Verfahren** *nt* ARCH & TRAGW mirror method
***Spiegel***: **Spiegelreflektor** *m* ELEKTR *Leuchte* mirror reflector
**Spielplatz** *m* INFR & ENTW playground; **Spielplatzbau** *m* INFR & ENTW playground construction
**Spindel** *f* ABWASSER *Pumpe* spindle, ARCH & TRAGW stem, newel, BAUMASCHIN *Drehbank* spindle, HOLZ *Treppe* newel, WERKSTOFF *Pumpe* spindle; **Spindelfräsmaschine** *f* BAUMASCHIN *Holzbau* spindle molding machine (*AmE*), spindle moulding machine (*BrE*); **Spindelkappe** *f* ARCH & TRAGW *Treppe* newel cap; **Spindelstab** *m* BAUMASCHIN spindle; **Spindeltreppe** *f* ARCH & TRAGW helical staircase, solid newel stair
**Spinnwebenmuster** *nt* ARCH & TRAGW cobweb pattern
**Spirale** *f* ARCH & TRAGW spiral, volute
**spiralenförmig** *adj* ARCH & TRAGW spiral
***Spirale***: **Spiralenzirkel** *m* ARCH & TRAGW volute compass
**Spirituslack** *m* OBERFLÄCHE spirit lacquer
**spitz**: **~er Maurerhammer** *m* BAUMASCHIN mattock
**Spitz**: **Spitzboden** *m* ARCH & TRAGW cock loft; **Spitzbogen** *m* ARCH & TRAGW lancet arch, pointed arch, ogive
**Spitze** *f* ARCH & TRAGW, BETON, HEIZ & BELÜFT, INFR & ENTW, OBERFLÄCHE, STAHL, UMWELT, WERKSTOFF peak, summit, *Höhepunkt* top, *Mathematik* vertex; **Spitzenbedarf** *m* ELEKTR peak demand
**Spitzende** *nt* ARCH & TRAGW spigot
***Spitze***: **Spitzendruckdiagramm** *nt* INFR & ENTW sounding graph; **Spitzendruckpfahl** *m* ARCH & TRAGW, INFR & ENTW end-bearing pile; **Spitzenkonzentration** *f* UMWELT peak concentration; **Spitzenleistung** *f* HEIZ & BELÜFT peak capacity; **Spitzenstärke** *f* INFR & ENTW, WERKSTOFF peak strength; **Spitzentechnik** *f* ARCH & TRAGW state-of-the-art technique; **Spitzentechnologie** *f* ARCH & TRAGW state-of-the-art technology; **Spitzentragfähigkeit** *f* ARCH & TRAGW, INFR & ENTW end-bearing capacity
***Spitz***: **Spitzkopfniet** *m* BESCHLÄGE, STAHL steeple head rivet; **Spitzturm** *m* ARCH & TRAGW flèche; **Spitztürmchen** *nt* ARCH & TRAGW pinnacle
***spitz***: **~ zulaufend** *adj* ARCH & TRAGW, INFR & ENTW tapered
**Spließdach** *nt* ARCH & TRAGW, HOLZ split-tiled roof
**Splintholz** *nt* HOLZ sapwood
**Splintverbindung** *f* ARCH & TRAGW cottered joint
**Splitt** *m* NATURSTEIN stone chippings, stone chips, WERKSTOFF chippings, *Stein* grit; **Splittbeton** *m* BETON chipping concrete
**Splitter** *m* ARCH & TRAGW *Holz* splinter; **Splitterschutzwand** *f* ARCH & TRAGW revetment wall
**splittreich** *adj* NATURSTEIN stone-filled; **~e Asphaltbetondeckschicht** *f* INFR & ENTW stone-filled asphaltic concrete pavement
***Splitt***: **Splittstreuer** *m* BAUMASCHIN *für die Straßenausbesserung* grit spreader
**Sport**: **Sportanlagen** *f pl* ARCH & TRAGW, INFR & ENTW sports facilities; **Sportplatz** *m* INFR & ENTW sportsground; **Sportplatzpflegegerät** *nt* BAUMASCHIN sportsground maintenance equipment; **Sportstadion** *nt* INFR & ENTW sports stadium
**Spotlicht** *nt* ARCH & TRAGW, ELEKTR downlighter
**Spray** *nt* OBERFLÄCHE spray
**sprayen** *vt* OBERFLÄCHE spray
**Spreizdübel** *m* WERKSTOFF expansion bolt
**Spreizen** *nt*: **~ zwischen Querbalken** ARCH & TRAGW bridging
**Spreng-** *in cpds* ARCH & TRAGW, BAUMASCHIN straining, INFR & ENTW explosion, seismic
**sprengen** *vt* INFR & ENTW shoot, *Gestein* blow up
**Sprengen** *nt* INFR & ENTW *Gestein* blasting
***Spreng-***: **Sprengladung** *f* INFR & ENTW explosion charge; **Sprengseismik** *f* INFR & ENTW seismic exploration method; **Sprengstrebe** *f* ARCH & TRAGW, BAUMASCHIN, HOLZ straining beam, strut
**Sprengung** *f* INFR & ENTW blasting
***Spreng-***: **Sprengwerk** *nt* ARCH & TRAGW strutted frame, truss, truss frame, HOLZ strutted frame, strutted roof; **Sprengwirkung** *f* INFR & ENTW explosive effect
**Spring-** *in cpds* UMWELT spring; **Springnipptide-**

**Zyklus** *m* UMWELT spring neap cycle; **Springtide** *f* UMWELT spring tide; **Springtide-Marke** *f* INFR & ENTW, UMWELT *Gezeiten* spring tide mark
**Sprinkler** *m* ABWASSER sprinkler; **Sprinkleranlage** *f* ABWASSER sprinkler system
**Spritz-** *in cpds* WERKSTOFF spray, spraying; **Spritzbeton** *m* BETON jetcrete®, shotcrete, WERKSTOFF gunite; **Spritzbewurf** *m* NATURSTEIN machine-applied plaster
**Spritze** *f* BAUMASCHIN gun, spraying device
**spritzen** *vt* BAUMASCHIN, OBERFLÄCHE, WERKSTOFF spray
**Spritzer** *m* STAHL *Schweißen* spatter
***Spritz-***: **Spritzlackierung** *f* OBERFLÄCHE spray painting; **Spritzmörtel** *m* NATURSTEIN air-blown mortar; **Spritzpistole** *f* BAUMASCHIN spraying gun; **Spritzputz** *m* NATURSTEIN, WERKSTOFF machine-applied plaster, sprayed mortar; **Spritzturm** *m* ARCH & TRAGW broach; **Spritzwand** *f* BESCHLÄGE splashback
**spritzwassergeschützt** *adj* BAUMASCHIN, WERKSTOFF splashproof
**Sprödbruch** *m* INFR & ENTW, STAHL, WERKSTOFF brittle fracture
**Sprödigkeit** *f* STAHL, WERKSTOFF brittleness
**Sprosse** *f* ARCH & TRAGW rung, muntin, BESCHLÄGE *Fenster* rail, *Leiter* rung; **Sprosseneisen** *nt* BAUMASCHIN sash bar
**sprühen** *vt* OBERFLÄCHE spray
**Sprühmittel** *nt* OBERFLÄCHE spray
**Sprung** *m* BETON, NATURSTEIN, WERKSTOFF crack; **Sprungrohr** *nt* ABWASSER, ARCH & TRAGW offset, swan neck
**Spül-** *in cpds* ABWASSER, HEIZ & BELÜFT flushing; **Spülbohren** *nt* INFR & ENTW wash boring, wash drilling
**Spüle** *f* ABWASSER sink; **~ mit Mittelbecken** BESCHLÄGE center bowl sink (*AmE*), centre bowl sink (*BrE*)
**spulen** *vt* INFR & ENTW wind
**spülen** *vti* ABWASSER flush
**Spülen** *nt* ABWASSER, HEIZ & BELÜFT flushing
***Spül-***: **Spülhilfe** *f* ABWASSER, HEIZ & BELÜFT flushing aid; **Spülkanal** *m* ABWASSER water course; **Spülkasten** *m* ABWASSER flushing cistern, flushing tank; **Spülkastenschwimmer** *m* ABWASSER cistern float; **Spülrohr** *nt* ABWASSER flushing pipe; **Spültisch** *m* ABWASSER sink unit
**Spülung** *f* ABWASSER, HEIZ & BELÜFT, UMWELT flushing
**Spund** *m* ABWASSER plug, HOLZ spile; **Spundbohle** *f* ARCH & TRAGW, INFR & ENTW, STAHL pile plank, sheet pile; **Spundbrett** *nt* HOLZ matchboard; **Spundeisen** *nt* STAHL tonguing iron; **Spundhobel** *m* ARCH & TRAGW tongue plane, BAUMASCHIN grooving plane, matching plane, plough plane (*BrE*), plow plane (*AmE*), STAHL tonguing plane; **Spundholzlage** *f* HOLZ matching; **Spundmaschine** *f* HOLZ tonguing-and-grooving machine; **Spund- und Nutmaschine** *f* BAUMASCHIN matching machine; **Spundpfahl** *m* ARCH & TRAGW, INFR & ENTW needle beam, sheet pile; **Spundschalung** *f* ARCH & TRAGW, INFR & ENTW tight sheathing
**Spundung** *f* HOLZ tonguing-and-grooving
***Spund***: **Spundverbindung** *f* ARCH & TRAGW ploughed-and-tongued-joint (*BrE*), plowed-and-tongued-joint (*AmE*), HOLZ tongue-and-groove joint; **Spundwand** *f* ARCH & TRAGW, INFR & ENTW sheet piling, STAHL steel piling; **Spundwandramme** *f* BAUMASCHIN sheet-piling driver; **Spundwandverankerung** *f* INFR & ENTW sheet pile anchorage; **Spundwerk** *nt* ARCH & TRAGW, INFR & ENTW sheet piling
**Spurenelement** *nt* UMWELT, WERKSTOFF trace element
**Spurzapfen** *m* BESCHLÄGE pintle
**S-Rohr** *nt* ABWASSER swan neck
**Staatlich**: **~e Anstalt** *f* **für die Beglaubigung von Laboruntersuchungen** (*NATLAS*) BAURECHT National Laboratory Testing Accreditation Service (*BrE*) (*NATLAS*); **~es Physikalisches Labor** *nt* (*NPL*) BAURECHT National Physics Laboratory (*BrE*) (*NPL*)
**Stab** *m* STAHL, WERKSTOFF bar, rod
**Stäbchenparkett** *nt* HOLZ strip flooring
**Stäbchenpalette** *f* HOLZ blockboard
***Stab***: **Stabeisen** *nt* DÄMMUNG rock wool; **Staberder** *m* ELEKTR earth rod (*BrE*), ground rod (*AmE*); **Stabfußboden** *m* HOLZ strip flooring; **Stabgitter** *nt* ARCH & TRAGW bar grate
**stabil** *adj* ARCH & TRAGW, INFR & ENTW stable
**Stabilisator** *m* WERKSTOFF stabilizer
**stabilisieren** *vt* INFR & ENTW stabilize
**Stabilisierung** *f* INFR & ENTW stabilization
**Stabilität** *f* ARCH & TRAGW, INFR & ENTW stability; **Stabilitätsanalyse** *f* INFR & ENTW stability analysis; **Stabilitätsbedingungen** *f pl* ARCH & TRAGW, INFR & ENTW stability conditions; **Stabilitätsfaktor** *m* INFR & ENTW stability factor; **Stabilitätsstudie** *f* INFR & ENTW stability study; **Stabilitätstheorie** *f* ARCH & TRAGW, INFR & ENTW theory of stability; **Stabilitätsversuch** *m* INFR & ENTW stability test
***Stab***: **Stabparkettfußboden** *m* HOLZ strip flooring; **Stabvertauschung** *f* ARCH & TRAGW exchange of members; **Stabwerk** *nt* ARCH & TRAGW framework
**Stacheldraht** *m* STAHL, WERKSTOFF barbed wire; **Stacheldrahtzaun** *m* INFR & ENTW barbed wire fence
**Stadiometer** *nt* INFR & ENTW *Vermessung* stadiometer
**Stadium** *nt* ARCH & TRAGW, INFR & ENTW stage
**Stadt** *f* INFR & ENTW town; **Stadtentwässerung** *f* ABWASSER, INFR & ENTW, UMWELT sewerage system
**Städteplanung** *f* INFR & ENTW town planning
***Stadt***: **Stadtgebiet** *nt* INFR & ENTW urban area
**städtisch** *adj* INFR & ENTW urban; **~es Kanalnetz** *nt* ABWASSER municipal sewer system
***Stadt***: **Stadtstraße** *f* INFR & ENTW urban road; **Stadtverkehr** *m* INFR & ENTW urban traffic; **Stadtversorgung** *f* ABWASSER, ELEKTR, HEIZ & BELÜFT city supply
**Staffelgiebel** *m* ARCH & TRAGW stepped gable
**Stahl** *m* STAHL steel; **Stahlarbeiten** *f pl* STAHL, WERKSTOFF steelwork; **Stahlart** *f* BETON, STAHL, WERKSTOFF steel grade; **Stahlbandkette** *f* BAUMASCHIN, STAHL *Vermessung* steel band chain; **Stahlbau** *m* ARCH & TRAGW, STAHL steel construction, structural steel work; **Stahlbautechnik** *f* ARCH & TRAGW, STAHL structural steel engineering; **Stahlbeton** *m* BETON reinforced concrete; **Stahlbetondecke** *f* ARCH & TRAGW, BETON reinforced concrete floor; **Stahlbetongurt** *m* ARCH & TRAGW, BETON, INFR & ENTW, STAHL *Verbundträger* composite slab; **Stahlbetonrippendecke** *f* ARCH & TRAGW ribbed concrete floor; **Stahlblech** *nt* STAHL steel sheet; **Stahldrahtseil** *nt* STAHL steel wire rope; **Stahleinsatz** *m* STAHL steel insert; **Stahlflechter** *m* STAHL *Stahlbeton* steelfixer; **Stahlgeflecht** *nt* STAHL steel mesh, steel netting; **Stahlgittermast** *m* STAHL

lattice tower; **Stahlgitterturm** *m* STAHL steel lattice tower; **Stahlgurt** *m* ARCH & TRAGW, STAHL steel flange; **Stahlkabel** *nt* STAHL steel cable; **Stahlkelle** *f* BAUMASCHIN steel trowel; **Stahlkern** *m* STAHL steel core; **Stahlkonstruktion** *f* STAHL steel construction; **Stahlleichtbau** *m* ARCH & TRAGW, STAHL light steel construction, lightweight steel construction; **Stahlliste** *f* ARCH & TRAGW, BETON, STAHL bar iron list, bar schedule, steel schedule; **Stahlmantelwalze** *f* BAUMASCHIN, INFR & ENTW smooth roller; **Stahlnagel** *m* BESCHLÄGE, STAHL masonry nail; **Stahlpfahl** *m* STAHL steel pile; **Stahlpfette** *f* ARCH & TRAGW steel purlin; **Stahlrohrgerüst** *nt* ARCH & TRAGW tubular steel scaffolding, BAUMASCHIN tubular steel scaffolfing; **Stahlrohrgitter** *nt* STAHL tubular steel grating; **Stahlschalung** *f* BETON, STAHL steel formwork, steel forms; **Stahlschlüssel** *m* BESCHLÄGE steel key; **Stahlschrott** *m* UMWELT steel scrap; **Stahlschubbolzen** *m pl* ARCH & TRAGW steel shear studs; **Stahlseele** *f* STAHL steel core; **Stahlskelettbau** *m* ARCH & TRAGW, STAHL steel skeleton structure; **Stahlsorte** *f* BETON, STAHL, WERKSTOFF steel grade; **Stahlspundwand** *f* INFR & ENTW, STAHL steel sheet piling; **Stahlträger** *m* ARCH & TRAGW steel girder, STAHL steel beam; **Stahlüberkreuzung** *f* STAHL bar intersection

**Stamm** *m* INFR & ENTW *Baum* trunk; **Stammholz** *nt* HOLZ standing timber

**Stampf-** *in cpds* BETON tamped, INFR & ENTW compressed; **Stampfasphalt** *m* INFR & ENTW compressed asphalt; **Stampfbeton** *m* BETON compressed concrete, tamped concrete; **Stampfbohle** *f* BAUMASCHIN, INFR & ENTW tamper

**stampfen** *vt* BAUMASCHIN, INFR & ENTW tamp

**Stampfen** *nt* BAUMASCHIN, INFR & ENTW tamping

**Stampfer** *m* BAUMASCHIN compactor, rammer, tamper, INFR & ENTW tamper

***Stampf-***: **Stampfmaschine** *f* **und Rüttelstampfmaschine zur Betonsteinherstellung** BAUMASCHIN compactor and vibrating compactor for the manufacture of concrete; **Stampfstange** *f* BAUMASCHIN, INFR & ENTW tamping rod

**Standard** *m* ARCH & TRAGW, BAUMASCHIN, BAURECHT, BESCHLÄGE, BETON, HEIZ & BELÜFT, INFR & ENTW, UMWELT, WERKSTOFF standard; **Standardabweichung** *f* ARCH & TRAGW standard deviation; **Standardleistungsbuch** *nt* **für das Bauwesen** ARCH & TRAGW Standard Construction Services Manual

**Ständer** *m* ARCH & TRAGW standard, stay, post, prop, stud, support, upright, pillar, vertical member, BAUMASCHIN, BETON post, HOLZ timber pillar, INFR & ENTW pillar; **Ständerbau** *m* ARCH & TRAGW post-and-beam structure; **Ständertrennwand** *f* HOLZ, STAHL stud wall; **Ständerverbindung** *f* ARCH & TRAGW stud union; **Ständerwand** *f* ARCH & TRAGW stud wall, stud partition; **Ständerwerk** *nt* ARCH & TRAGW post-and-beam structure

**standfest** *adj* ARCH & TRAGW, INFR & ENTW stable; **~er Boden** *m* INFR & ENTW stable soil; **~e Böschung** *f* INFR & ENTW stable slope

**Standfestigkeit** *f* ARCH & TRAGW, INFR & ENTW stability under load

**Standmast** *m* ARCH & TRAGW fixed post

**Standort** *m* ARCH & TRAGW location, INFR & ENTW site; **~ der Deponie** UMWELT tip (*BrE*); **~ festlegen** ARCH & TRAGW locate; **Standortbestimmung** *f* ARCH & TRAGW, INFR & ENTW siting; **Standortkriterien** *nt pl* UMWELT site criteria; **Standortvermessung** *f* ARCH & TRAGW, INFR & ENTW field survey; **Standortwahl** *f* ARCH & TRAGW, INFR & ENTW siting

**Standrohr** *nt* ABWASSER vertical pipe

**standsicher** *adj* ARCH & TRAGW, INFR & ENTW stable; **nicht ~e Böschung** *f* INFR & ENTW unstable slope

**Standsicherheit** *f* INFR & ENTW stability

**Stange** *f* ARCH & TRAGW, STAHL bar, WERKSTOFF rod; **Stangengriff** *m* BESCHLÄGE *neben Badewanne oder Dusche* grab bar; **Stangenverschluß** *m* BESCHLÄGE *Stahlfensterbeschlag* espagnolette, WERKSTOFF bolt with handle

**stanzen** *vt* BAUMASCHIN, STAHL punch

**Stapel** *m* ARCH & TRAGW stack

**stapeln** *vt* INFR & ENTW pile

**Stapeln** *nt* ARCH & TRAGW piling

***Stapel***: **Stapelplatz** *m* INFR & ENTW stacking ground

**stark** *adj* ARCH & TRAGW, BETON, DÄMMUNG, INFR & ENTW, WERKSTOFF rigid, thick; **~e Verbindung** *f* BETON, INFR & ENTW, WERKSTOFF high-bond action; **~er Weichmacher** *m* WERKSTOFF *Beton* superplastisizer

**Stärke** *f* ARCH & TRAGW thickness

***stark***: **~ saugender Putzgrund** *m* NATURSTEIN highly absorbent plaster base

**starr**: **~es Bauwerk** *nt* ARCH & TRAGW, INFR & ENTW rigid structure; **~es Fundament** *nt* ARCH & TRAGW, INFR & ENTW rigid foundation; **~e Verbindung** *f* ARCH & TRAGW, INFR & ENTW rigid connection

**Starrahmen** *m* WERKSTOFF rigid frame

**Starrpunkt** *m* ARCH & TRAGW, INFR & ENTW breaking point, WERKSTOFF brittle point

**Starthilfekabel** *nt* ELEKTR jumper cable

**Statik** *f* ARCH & TRAGW, INFR & ENTW structural analysis, theory of structures

**Statiker** *m* ARCH & TRAGW, INFR & ENTW structural designer, structural engineer

**stationär** *adj* ARCH & TRAGW stationary

**statisch** *adj* ARCH & TRAGW static; **~e Beanspruchung** *f* ARCH & TRAGW static stress; **~e Belastungstestanlage** *f* ARCH & TRAGW, BAUMASCHIN, HOLZ, INFR & ENTW static proof-loading machine

***statisch***: **~ berechnen** *vt* ARCH & TRAGW, INFR & ENTW, WERKSTOFF analyse (*BrE*), analyze (*AmE*), determine statically

***statisch***: **~e Berechnung** *f* ARCH & TRAGW, INFR & ENTW static calculation; **~ bestimmbar** *adj* ARCH & TRAGW statically definable, statically determinable; **~ bestimmt** *adj* ARCH & TRAGW isostatic, statically defined; **~er Druck** *m* ARCH & TRAGW static pressure; **~e Elektrizität** *f* ELEKTR static electricity; **~e Erfordernisse** *nt pl* ARCH & TRAGW static requirements; **~e Festigkeit** *f* ARCH & TRAGW static strength; **~es Gleichgewicht** *nt* ARCH & TRAGW, INFR & ENTW static equilibrium; **~e Gleichung** *f* ARCH & TRAGW, INFR & ENTW static equation; **~e Last** *f* INFR & ENTW static load; **~ mitwirkender Ziegel** *m* ARCH & TRAGW load-bearing brick; **~es Moment** *nt* ARCH & TRAGW, INFR & ENTW static moment; **~er Nachweis** *m* ARCH & TRAGW, INFR & ENTW structural analysis; **~e Prüfung** *f* ARCH & TRAGW static test; **~ überbestimmt** *adj* ARCH & TRAGW statically overdefined, statically overdetermined; **~ unbestimmbar** *adj* ARCH & TRAGW statically indeterminable; **~ unbestimmt** *adj* ARCH &

TRAGW, INFR & ENTW hyperstatic, statically indeterminate; **~e Unbestimmtheit** *f* ARCH & TRAGW statical indeterminacy; **~e Untersuchung** *f* ARCH & TRAGW, INFR & ENTW static investigation; **~ zulässig** *adj* ARCH & TRAGW statically admissible

**Statistik** *f* ARCH & TRAGW, BAURECHT, INFR & ENTW, UMWELT statistics

**statistisch** *adv* ARCH & TRAGW, BAURECHT, INFR & ENTW, UMWELT statistically

**Stativkompaß** *m* ARCH & TRAGW, BAUMASCHIN surveyor's compass

**Stau-** *in cpds* ABWASSER build-up, retaining, ARCH & TRAGW, BETON retaining, INFR & ENTW build-up, retaining, NATURSTEIN retaining, UMWELT build-up, retaining, WERKSTOFF retaining

**Staub** *m* UMWELT dust; **Staubabscheidung** *f* UMWELT particulate collection

***Stau-***: **Staubalken** *m* BESCHLÄGE stop log

***Staub***: **Staubanteil** *m* UMWELT dust content

***Stau-***: **Staubecken** *nt* ELEKTR *Elektro* power basin, INFR & ENTW retaining basin

***Staub***: **Staubfilter** *m* UMWELT fabric filter; **Staub- und Lärmschutz** *m* BAURECHT, DÄMMUNG, INFR & ENTW dust and noise control; **Staubschutz** *m* BAURECHT, INFR & ENTW dust protection; **Staubschutzmittel** *nt* OBERFLÄCHE, WERKSTOFF dust preventer; **Staubteilchen** *nt* UMWELT dust particle

***Stau-***: **Staudamm** *m* INFR & ENTW dam, retaining dam

**stauen** *vt* ABWASSER, ARCH & TRAGW, BETON, INFR & ENTW, NATURSTEIN, UMWELT,WERKSTOFF retain

***Stau-***: **Staufläche** *f* UMWELT surface area; **Staukörper** *m* BESCHLÄGE gate; **Staumauer** *f* INFR & ENTW dam, multiple-arch dam; **Staumauer** *f* **aus übergroßen Schottersteinen** INFR & ENTW tailings dam; **Stauraumbemessung** *f* ARCH & TRAGW, INFR & ENTW calculation of storage capacity; **Stauraumsedimentierung** *f* INFR & ENTW reservoir sedimentation

**Stauung** *f* UMWELT dike

***Stau-***: **Stauwasser** *nt* ABWASSER, INFR & ENTW backwater, tail water; **Stauwasserhöhe** *f* INFR & ENTW backwater level; **Stauwasserzone** *f* INFR & ENTW backwater zone; **Stauwehr** *nt* INFR & ENTW barrage

**Stechfase** *f* ARCH & TRAGW, HOLZ *Tischlerarbeiten* chamfer stop

**Steckbuchse** *f* BESCHLÄGE lock bush

**Steckdorn** *m* BAUMASCHIN pin spanner, pin wrench

**Steckdose** *f* ELEKTR, HEIZ & BELÜFTUNG outlet, receptacle, socket

**Stecker** *m* ELEKTR plug

**Steg** *m* ARCH & TRAGW web, stud link, *Träger* stem; **Stegbewehrung** *f* BETON web reinforcement; **Stegblechstoß** *m* HOLZ web splice; **Stegdiele** *f* WERKSTOFF hollow-core plank; **Stegkette** *f* ARCH & TRAGW, STAHL studded link cable chain; **Stegzementdiele** *f* WERKSTOFF hollow-core plank

**Stehbolzen** *m* ARCH & TRAGW, BESCHLÄGE stay bolt

**stehend** *adj* ARCH & TRAGW, INFR & ENTW standing, upright; **~es Wasser** *nt* INFR & ENTW stagnant water

**Stehfalz** *m* STAHL standing seam

**Stehlampe** *f* ELEKTR floor lamp

**steif** *adj* ARCH & TRAGW, BETON, DÄMMUNG, INFR & ENTW, WERKSTOFF rigid, stiff; **~er Beton** *m* BETON low-slump concrete; **~er Boden** *m* INFR & ENTW stiff soil; **~es Fiberglas** *nt* ARCH & TRAGW, DÄMMUNG, WERKSTOFF rigid fiberglass (*AmE*), rigid fibreglass (*BrE*); **~e Konstruktion** *f* ARCH & TRAGW rigid construction; **~e Plastizitätstheorie** *f* BETON, WERKSTOFF rigid-plastic theory; **~er Ton** *m* INFR & ENTW stiff clay

**Steife** *f* ARCH & TRAGW strut, brace, HOLZ brace

**Steifigkeit** *f* ARCH & TRAGW, DÄMMUNG, INFR & ENTW, WERKSTOFF rigidity, stiffness

**steifknotig** *adj* ARCH & TRAGW stiff-jointed

**steif-plastisch**: **~e Theorie** *f* BETON, WERKSTOFF rigid-plastic theory

**Steig-** *in cpds* ABWASSER, ARCH & TRAGW step; **Steigeisen** *nt* ABWASSER *Schacht* step iron, ARCH & TRAGW step iron, spur, BAURECHT, INFR & ENTW access hook

**steigend**: **~er Bogen** *m* ARCH & TRAGW rampant arch, rising arch

**Steigerung** *f* ARCH & TRAGW, BAURECHT *Kosten* increase

***Steig-***: **Steigleitung** *f* ABWASSER rising main; **Steigmaß** *nt* ARCH & TRAGW rise; **Steigrohr** *nt* ABWASSER standpipe, riser pipe

**Steigung** *f* ARCH & TRAGW pitch, rising gradient, upgrade, INFR & ENTW slope, pitch; **Steigungsverhältnis** *nt* ARCH & TRAGW rise-run ratio; **Steigungswinkel** *m* ARCH & TRAGW, INFR & ENTW pitch angle

**steil** *adj* ARCH & TRAGW, INFR & ENTW steep; **~er Abhang** *m* ARCH & TRAGW steep gradient; **~e Neigung** *f* INFR & ENTW steep slope; **~e Straße** *f* ARCH & TRAGW steep road

**Steil-** *in cpds* ARCH & TRAGW, INFR & ENTW steep; **Steilböschung** *f* ARCH & TRAGW steep slope; **Steildach** *nt* ARCH & TRAGW high-pitched roof, steep roof; **Steilhang** *m* INFR & ENTW scarp

**Steilheit** *f* ARCH & TRAGW, INFR & ENTW steepness

***Steil-***: **Steilkantenvorbereitung** *f* ARCH & TRAGW *Schweißen* square edge preparation

**Stein** *m* ARCH & TRAGW, NATURSTEIN stone; **Steinauflage** *f* INFR & ENTW, WERKSTOFF enrockment; **Steinbandsägemaschine** *f* BAUMASCHIN stone band saw; **Steinbau** *m* NATURSTEIN stone construction, stone structure; **Steinbauwerk** *nt* NATURSTEIN stone structure; **Steinbearbeitung** *f* INFR & ENTW, NATURSTEIN stoneworking; **Steinbildhauerei** *f* NATURSTEIN stone carving; **Steinblume** *f* ARCH & TRAGW *mittelalterliche Verzierung* crocket; **Steinblumenkapitell** *nt* ARCH & TRAGW crocket capital; **Steinbrecher** *m* BAUMASCHIN stone crusher, rock breaker; **Steinbruch** *m* INFR & ENTW, NATURSTEIN rock quarry, stone pit, stone quarry; **Steindamm** *m* INFR & ENTW stone dam; **Steindübel** *m* BAUMASCHIN rock dowel; **Steinfräse** *f* INFR & ENTW stone mill; **Steingreifer** *m* BAUMASCHIN nippers

**Steingut** *nt* NATURSTEIN stoneware, faience; **Steingutrohr** *nt* ABWASSER earthenware pipe

**Steinholz** *nt* WERKSTOFF magnesite composition; **Steinholzfußboden** *m* ARCH & TRAGW magnesite flooring, stone wood floor; **Steinholzplatte** *f* NATURSTEIN xylolite slab

**steinig** *adj* INFR & ENTW stony; **~er Untergrund** *m* INFR & ENTW stony ground

**Steinmetz** *m* ARCH & TRAGW, INFR & ENTW stone dresser, stonemason; **Steinmetzarbeit** *f* NATURSTEIN carving

***Stein***: **Steinpackung** *f* INFR & ENTW rubble bedding, rip-rap, stone packing; **Steinpaketierungsanlage** *f* ARCH & TRAGW, BAUMASCHIN brick packaging machine; **Steinpflaster** *nt* INFR & ENTW block pavement, stone pavement; **Steinputz** *m* NATURSTEIN stone plaster, *Edelputz mit Kiesel* pebble dash, *Edelputz mit Kieseln gemischt* roughcast; **Steinramme** *f* BAUMASCHIN beetle; **Steinsalz** *nt* INFR & ENTW rock salt; **Steinschraube** *f* BESCHLÄGE, BETON barb bolt, rag bolt; **Steinschüttdamm** *m* INFR & ENTW rock fill dam; **Steinschüttung** *f* INFR & ENTW rip-rap, enrockment, rock fill, WERKSTOFF enrockment, rubble; **Steinsetzer** *m* ARCH & TRAGW pavior (*AmE*), paviour (*BrE*); **Steinskulptur** *f* NATURSTEIN stone carving; **Steinspaltmaschine** *f* BAUMASCHIN stone-splitting machine; **Steinverband** *m* NATURSTEIN stone bond; **Steinverblendung** *f* NATURSTEIN hewn stone facing; **Steinverklammerung** *f* NATURSTEIN joggle joint; **Steinwolf** *m* BAUMASCHIN lewis; **Steinwolle** *f* DÄMMUNG rock wool; **Steinwurf** *m* INFR & ENTW rubble layer

**Steinzeug** *nt* NATURSTEIN stoneware; **Steinzeugabzweig** *m* ABWASSER stoneware branch; **Steinzeugbogen** *m* ABWASSER stoneware bend; **Steinzeugfliese** *f* NATURSTEIN stoneware tile; **Steinzeugplatte** *f* NATURSTEIN stoneware tile; **Steinzeugrohr** *nt* ABWASSER, ARCH & TRAGW, NATURSTEIN, WERKSTOFF ceramic pipe, stoneware pipe, vitrified clay pipe

**Stellelement** *nt* HEIZ & BELÜFT actuator

**Stelle** *f*: **~ des plastischen Gelenkes** ARCH & TRAGW, BETON, STAHL plastic hinge location

**Stellung** *f* ARCH & TRAGW position; **Stellungslinie** *f* INFR & ENTW earth pressure line

**Stellvertretung** *f* BAURECHT agency

**Stelzbogen** *m* ARCH & TRAGW stilted arch

**Stelze** *f* ARCH & TRAGW stilt

**Stemm-** *in cpds* HOLZ morticing, mortise, mortising; **Stemmaschine** *f* BAUMASCHIN, HOLZ morticing machine, mortising machine; **Stemmeisen** *nt* BAUMASCHIN, HOLZ mortise chisel; **Stemmeißel** *m* BAUMASCHIN, HOLZ mortise chisel

**stemmen** *vt* HOLZ mortise

***Stemm-***: **Stemmloch** *nt* HOLZ mortise

**Stempel** *m* ARCH & TRAGW *Absteifung* dead shore, BAUMASCHIN *Schaffußwalze* tamping foot

**Stempelgebühr** *f* BAURECHT stamp duty

**steril** *adj* ABWASSER sterile

**Sternbild** *nt* ARCH & TRAGW *Muster* stellar pattern

**Sternriß** *m* STAHL *Holz* star shake

**Stetigmischer** *m* BAUMASCHIN continuous mixer

**Steuer** *f* BAURECHT tax, taxation; **2.** *in cpds* HEIZ & BELÜFT control

**steuerlich**: **~e Einstufung** *f* BAURECHT tax band

***Steuer***: **Steuersatz** *m* BAURECHT tax rate; **Steuerschieber** *m* HEIZ & BELÜFT control slide valve

**Steuerung** *f* ELEKTR, HEIZ & BELÜFT control

***Steuer***: **Steuerventil** *nt* HEIZ & BELÜFT control valve

**Stich**: **Stichbalken** *m* ARCH & TRAGW hammer beam, tail beam; **Stichbogen** *m* ARCH & TRAGW segmental arch; **Stichfallrohr** *nt* ABWASSER stub stack

**stichfest** *adj* UMWELT, WERKSTOFF semisolid

***Stich***: **Stichhöhe** *f* ARCH & TRAGW, INFR & ENTW pitch, rise; **Stichkabel** *nt* HEIZ & BELÜFT branch cable; **Stichleitung** *f* ABWASSER branch line, ELEKTR *Abzweigkabel* stub cable, INFR & ENTW branch line; **Stichprobe** *f* BAURECHT *Probematerial* random sample, *Prüfung* random test; **Stichsäge** *f* BAUMASCHIN compass saw, keyhole saw

**Stickoxid** *nt* UMWELT nitrogen oxide, nitric oxide

**Stickstoff** *m* UMWELT, WERKSTOFF nitrogen; **Stickstoffdioxid** *nt* UMWELT nitrogen dioxide, nitrogen peroxide; **Stickstoffpentoxid** *nt* UMWELT nitrogen pentoxide

**Stiel** *m* ARCH & TRAGW upright, post, standard, stud, BAUMASCHIN, BETON post

**Stift** *m* ABWASSER plug, BESCHLÄGE dowel, pin, HOLZ dowel, STAHL, WERKSTOFF pin; **Stiftfeder** *f* BESCHLÄGE *Schloß* lever spring; **Stiftkolben** *m* BAUMASCHIN pin vice (*BrE*), pin vise (*AmE*); **Stiftschlüssel** *m* BAUMASCHIN pin spanner, pin wrench; **Stiftventil** *nt* BAUMASCHIN pin valve; **Stiftzuhaltung** *f* BESCHLÄGE *Schloß* tumbler

**Stil** *m* ARCH & TRAGW style

**stillegen** *vt* BAURECHT shut down, put out of service, INFR & ENTW put out of service, shut down

**Stillegung** *f* BAURECHT, INFR & ENTW shutdown

**stillstehend**: **~e Luft** *f* UMWELT still air

**Stirn-** *in cpds* ARCH & TRAGW, BETON, INFR & ENTW, NATURSTEIN, WERKSTOFF face; **Stirnbogen** *m* ARCH & TRAGW face arch; **Stirnbrett** *nt* ARCH & TRAGW side board, HOLZ barge board, eaves board, fascia board; **Stirnfläche** *f* BESCHLÄGE butt end; **Stirnseite** *f* ARCH & TRAGW front; **Stirnseite** *f* **eines Bogens** ARCH & TRAGW face of arch; **Stirnseite** *f* **der Tür** BESCHLÄGE door front; **Stirnspreize** *f* INFR & ENTW *Grabenverbau* face piece

**stochastisch** *adj* WERKSTOFF stochastic; **~es Modell** *nt* INFR & ENTW stochastic model

**stochern** *vt* BETON *Betonverdichtung* puddle

**stocken** *vi* BAUMASCHIN *Beton- oder Steinoberflächen*, BETON granulate

**Stocken** *nt* BETON, NATURSTEIN bush hammering

**Stockhammer** *m* BAUMASCHIN bush hammering

**Stockputz** *m* NATURSTEIN scraped stucco

**Stockwerk** *nt* ARCH & TRAGW floor, storey (*BrE*), story (*AmE*)

**Stoff** *m* WERKSTOFF fabric, material, matter, substance; **Stoffdach** *nt* ARCH & TRAGW, INFR & ENTW, WERKSTOFF fabric roof; **Stoffilter** *nt* UMWELT fabric filter; **Stofftrennprozeß** *m* UMWELT material separation operation; **Stoffwechselschlacken** *f pl* UMWELT metabolic waste

**Stopfbuchsenpackung** *f* ABWASSER gland packing

**stopfen** *vt* ABWASSER plug, INFR & ENTW, NATURSTEIN *Schotter* pack

**Stopfen** *m* ABWASSER plug

**stören** *vt* INFR & ENTW, WERKSTOFF disturb

**Störung** *f* ABWASSER, HEIZ & BELÜFT, INFR & ENTW defect, trouble

**Stoß** *m* ARCH & TRAGW joint, stack, meeting, BESCHLÄGE butt, INFR & ENTW joint; **Stoßblech** *nt* BESCHLÄGE *Tür* kick plate, HOLZ splice plate; **Stoßdämpfer** *m* BAUMASCHIN shock absorber

**stoßen** *vt* ARCH & TRAGW join, abut

**stoßfest** *adj* WERKSTOFF shock-proof, shock-resistant

***Stoß***: **Stoßfestigkeit** *f* WERKSTOFF shock resistance; **Stoßfuge** *f* ARCH & TRAGW butt joint, HOLZ straight joint, NATURSTEIN vertical joint; **Stoßgriff** *m* BESCHLÄGE grip handle, push handle

**Strahl-** *in cpds* INFR & ENTW shot

**Strahlenschutz** *m* UMWELT radiation protection; **Strahlenschutzbeton** *m* WERKSTOFF loaded concrete

***Strahl-***: **Strahlreinigen** *nt* INFR & ENTW shot blasting;

**Strahlrohrmundstück** *nt* BAUMASCHIN jet pipe nozzle; **Strahlsand** *m* WERKSTOFF *Strahlreinigen* grit
**Strahlung** *f* HEIZ & BELÜFT, UMWELT radiation; **Strahlungsbündler** *m* UMWELT solar concentrator; **Strahlungsintensität** *f* UMWELT irradiance; **Strahlungswärme** *f* HEIZ & BELÜFT radiant heat
**Strandkies** *m* WERKSTOFF beach gravel
**Strandmauer** *f* INFR & ENTW sea wall
**Strang-** *in cpds* STAHL, WERKSTOFF extruded; **Strangdachziegel** *m* INFR & ENTW, NATURSTEIN, WERKSTOFF extruded clay roof tile; **Strangfalzziegel** *m* INFR & ENTW, NATURSTEIN, WERKSTOFF extruded interlocking tile
**stranggepreßt** *adj* STAHL, WERKSTOFF *Metall, Kunststoff* extruded
***Strang-***: **Strangpreßprofil** *nt* WERKSTOFF extruded section, extruded profile
**strapazierfähig** *adj* WERKSTOFF hardwearing
**Straße** *f* BAUMASCHIN, INFR & ENTW road, street; **~ in Tieflage** INFR & ENTW sunk freeway (*AmE*), sunk motorway (*BrE*), sunken road; **Straßenablauf** *m* ARCH & TRAGW *Regenwasser* storm drain, INFR & ENTW grit trap, road inlet, street inlet; **Straßenabsteckung** *f* INFR & ENTW road setting out; **Straßenabwasser** *nt* ABWASSER, INFR & ENTW surface water; **Straßenaufreißer** *m* BAUMASCHIN road ripper; **Straßenbahn** *f* INFR & ENTW streetcar (*AmE*), tramcar (*BrE*); **Straßenbahnweiche** *f* INFR & ENTW tramway switch; **Straßenbau** *m* INFR & ENTW road construction; **Straßenbauarbeiten** *f pl* INFR & ENTW road construction work, roadworks; **Straßenbaumaschine** *f* BAUMASCHIN road construction machinery; **Straßenbefestigung** *f* INFR & ENTW pavement; **Straßenbeleuchtung** *f* BAURECHT, ELEKTR, INFR & ENTW street lighting; **Straßenbett** *nt* INFR & ENTW subgrade; **Straßenbrücke** *f* BAUMASCHIN road bridge; **Straßendamm** *m* INFR & ENTW embankment, road embankment; **Straßendecke** *f* INFR & ENTW topping; **Straßendeckenbeton** *m* BETON pavement-quality concrete (*BrE*), sidewalk-quality concrete (*AmE*); **Straßenentwässerung** *f* INFR & ENTW road drainage; **Straßenfertiger** *m* BAUMASCHIN finisher, finishing machine; **Straßenfräsmaschine** *f* BAUMASCHIN road groover; **Straßenfront** *f* ARCH & TRAGW frontage; **Straßengabelung** *f* INFR & ENTW Y-junction; **Straßengründung** *f* INFR & ENTW road foundation; **Straßenkappe** *f* INFR & ENTW valve box; **Straßenkehricht** *m* UMWELT litter; **Straßenkreuzung** *f* ARCH & TRAGW intersection, INFR & ENTW crossing, intersection, junction, road crossing; **Straßenkreuzungspunkt** *m* INFR & ENTW road crossing; **Straßenmarkierung** *f* INFR & ENTW road marking, street line; **Straßenmarkierungsgerät** *nt* BAUMASCHIN road marking machine; **Straßenpflaster** *nt* NATURSTEIN pavement (*BrE*), WERKSTOFF pavement paving flag (*BrE*), sidewalk paving flag (*AmE*); **Straßenplatte** *f* NATURSTEIN pad; **Straßenrauhmaschine** *f* BAUMASCHIN road grader; **Straßenreinigung** *f* INFR & ENTW, UMWELT street cleaning, street sweeping; **Straßenrinne** *f* INFR & ENTW gutter; **Straßenschotter** *m* BAUMASCHIN road metal; **Straßentunnel** *m* INFR & ENTW road tunnel; **Straßenunterbau** *m* INFR & ENTW road bed; **Straßenunterführung** *f* INFR & ENTW underbridge, undergrade crossing, underpass; **Straßenwalze** *f* BAUMASCHIN road roller; **Straßenzubehör** *nt* INFR & ENTW street furniture
**Strebe** *f* ARCH & TRAGW, HOLZ diagonal rod, strut shore, spur, stay; **Strebenkopf** *m* ARCH & TRAGW strutting head
**strebenlos**: **~er Balken** *m* ARCH & TRAGW unsupported beam
***Strebe***: **Strebepfeiler** *m* ARCH & TRAGW, BETON, INFR & ENTW, NATURSTEIN, STAHL abutment pier, buttress, counterfort, flying buttress
**Streck-** *in cpds* STAHL, WERKSTOFF yield; **Streckbalken** *m* ARCH & TRAGW string piece
**streckbar** *adj* WERKSTOFF tensile
**Strecke** *f* BAUMASCHIN *Kabel, Rohr* route, INFR & ENTW distance
**strecken** *vt* ARCH & TRAGW, WERKSTOFF stretch
***Strecke***: **Streckenführung** *f* INFR & ENTW route mapping; **Streckenlast** *f* ARCH & TRAGW, INFR & ENTW distributed load, line load, linear load
**Strecker** *m* HOLZ bonder, NATURSTEIN binder, bonder, header, outbond brick; **Streckerschicht** *f* NATURSTEIN bonding course; **Streckerverband** *m* NATURSTEIN header bond
***Streck-***: **Streckgrenze** *f* ARCH & TRAGW, STAHL, WERKSTOFF yield point, yield strength; **Streckmetall** *nt* STAHL, WERKSTOFF expanded metal, expanded mesh
**Streich-** *in cpds* ARCH & TRAGW head, trimmer, BAUMASCHIN spreader; **Streichbalken** *m* ARCH & TRAGW head piece, head plate, trimmer, trimmer beam
**streichen** *vt* OBERFLÄCHE, WERKSTOFF paint; **neu ~** *vt* OBERFLÄCHE repaint
**Streichen** *nt* OBERFLÄCHE brush application
**streichfähig** *adj* OBERFLÄCHE brushable
***Streich-***: **Streichfähigkeit** *f* OBERFLÄCHE brushability; **Streichgerät** *nt* BAUMASCHIN spreader; **Streichmaß** *nt* BAUMASCHIN scratch gage (*AmE*), scratch gauge (*BrE*); **Streichpfahl** *m* ARCH & TRAGW fender pile
**Streifen** *m* ARCH & TRAGW strip; **Streifenbelastung** *f* ARCH & TRAGW, INFR & ENTW strip load; **Streifenfundament** *nt* ARCH & TRAGW, BETON continuous footing, continuous foundation, strip foundation
**Streu- und Bruchverlust** *m* INFR & ENTW wastage, WERKSTOFF spoilage
**streuen** *vt* WERKSTOFF scatter
**Stroh** *nt* ARCH & TRAGW thatch; **mit ~ decken** *phr* ARCH & TRAGW thatch; **Strohbauplatte** *f* DÄMMUNG strawboard; **Strohdach** *nt* ARCH & TRAGW thatched roof; **Strohplatte** *f* WERKSTOFF thatchboard
**Strom** *m* ABWASSER flow, ELEKTR current, HEIZ & BELÜFT flow, INFR & ENTW flow, stream, UMWELT, WERKSTOFF flow
**stromabwärts** *adj* INFR & ENTW downstream
**stromaufwärts** *adv* INFR & ENTW, UMWELT upstream; **~ gelegen** *adj* INFR & ENTW, UMWELT upstream
**strömen** *vi* UMWELT flow
**stromführend**: **~er Leiter** *m* ELEKTR live wire
***Strom***: **Stromführung** *f* ELEKTR, INFR & ENTW power supply; **Stromkreis** *m* ELEKTR circuit, power circuit; **Stromlaufplan** *m* ELEKTR circuit diagram; **Stromleitung** *f* ELEKTR, INFR & ENTW transmission line; **Strommesser** *m* UMWELT current meter; **Stromschnelleüberlauf** *m* UMWELT chute spillway
**Strömung** *f* ABWASSER, INFR & ENTW, UMWELT, WERKSTOFF current, flow; **Strömungsgeschwindigkeit** *f* ABWASSER, HEIZ & BELÜFT, INFR & ENTW velocity of flow; **Strömungslehre** *f* INFR & ENTW fluid mech-

anics; **Strömungsrichtung** *f* HEIZ & BELÜFT direction of circulation
***Strom***: **Stromversorgungsnetz** *nt* ELEKTR, INFR & ENTW mains
**Struktur** *f* INFR & ENTW structure, WERKSTOFF texture
**strukturell**: **~e Hierarchie** *f* ARCH & TRAGW structural hierarchy
**strukturiert** *adj* STAHL *Blech* structured; **~e und profilierte Oberflächen** *phr* BETON, OBERFLÄCHE textured and profiled finishes
**Stubben** *m* INFR & ENTW stump
**Stuck** *m* NATURSTEIN stucco, *gemustert* parget; **Stuckarbeit** *f* NATURSTEIN stucco work; **Stuckarbeit ausführen** *phr* NATURSTEIN parget; **Stuckdecke** *f* NATURSTEIN stucco ceiling; **Stuckgips** *m* NATURSTEIN plaster of Paris
**Stückliste** *f* ARCH & TRAGW piece list
***Stuck***: **Stuckmörtel** *m* OBERFLÄCHE stucco mortar
**Stückzahl** *f* ARCH & TRAGW number
**Studie** *f* ARCH & TRAGW, BAURECHT, INFR & ENTW study
**Stufe** *f* ARCH & TRAGW step, tread; **Stufenbreite** *f* ARCH & TRAGW tread, length of step; **Stufenkantenlinie** *f* ARCH & TRAGW nosing line
**Stukkatur** *f* NATURSTEIN stucco work
**Stulpbreite** *f* BESCHLÄGE *Schloß* lockplate width
**Stülpschalung** *f* ARCH & TRAGW weather boarding
**Stumpf** *m* ARCH & TRAGW stub
**stumpf**: **~ eingeschlagene Tür** *f* ARCH & TRAGW flush door; **~es Ende** *nt* BESCHLÄGE butt
***Stumpf***: **Stumpffuge** *f* HOLZ *Holzbau* header joint
**stumpf**: **~ gestoßen** *adj* ARCH & TRAGW, HOLZ, NATURSTEIN, STAHL butt jointed
***Stumpf***: **Stumpfschweißung** *f* ARCH & TRAGW, HEIZ & BELÜFT, STAHL, WERKSTOFF butt weld; **Stumpfstoß** *m* ARCH & TRAGW butt joint, HOLZ butt joint, scarf joint, NATURSTEIN, STAHL butt joint
**stumpfstoßen** *vt* BESCHLÄGE butt-joint
**Stumpfstoßen** *nt* HOLZ scarf jointing
***Stumpf***: **Stumpfwinkel** *m* ARCH & TRAGW, INFR & ENTW obtuse angle
**stündlich**: **~e Solarstrahlung** *f* HEIZ & BELÜFT hourly solar radiation
**Sturm** *m* ARCH & TRAGW sprocket; **Sturmhaken** *m* BESCHLÄGE *Fenster* casement stay; **Sturmlatte** *f* ARCH & TRAGW sprocket piece, HOLZ cross lath
**Sturz** *m* ARCH & TRAGW lintel, head, INFR & ENTW, NATURSTEIN lintel; **Sturzbalken** *m* ARCH & TRAGW lintel, bressumer, summer beam, INFR & ENTW bressumer, lintel, summer beam; **Sturzfutterwinkel** *m* BESCHLÄGE angular soffit bracket; **Sturzriegel** *m* ARCH & TRAGW, INFR & ENTW intertie beam, lintel; **Sturzträger** *m* ARCH & TRAGW, INFR & ENTW lintel
**Stütz-** *in cpds* ARCH & TRAGW supporting, INFR & ENTW sustaining; **Stützbalken** *m* ARCH & TRAGW stringer, supported beam; **Stützblech** *nt* HOLZ, STAHL gusset plate
**Stütze** *f* ARCH & TRAGW bearer, pillar, support, column, spur, standard, stay, BETON column, INFR & ENTW pillar, NATURSTEIN, STAHL column
**stützen 1.** *vt* ARCH & TRAGW bear, carry, support, INFR & ENTW, HOLZ, STAHL, WERKSTOFF bear; **2. sich stützen** *v refl* INFR & ENTW, WERKSTOFF lean
**Stutzen** *m* ABWASSER connection piece, ARCH & TRAGW stub, HEIZ & BELÜFT connection piece
**Stützen** *nt* ARCH & TRAGW bearing
***Stützen***: **Stützenabstand** *m* ARCH & TRAGW column spacing; **Stützenauflage** *f* ARCH & TRAGW column bearing
**stützend** *adj* ARCH & TRAGW carrying, supporting
**stützenfrei** *adj* ARCH & TRAGW column-free
***Stützen***: **Stützenfundament** *nt* ARCH & TRAGW, BETON column footing; **Stützenfuß** *m* ARCH & TRAGW, BETON, STAHL column base; **Stützenkopf** *m* ARCH & TRAGW column head; **Stützenraster** *m* ARCH & TRAGW column grid; **Stützensenkung** *f* ARCH & TRAGW, INFR & ENTW settlement of support; **Stützenverkleidung** *f* ARCH & TRAGW, DÄMMUNG, NATURSTEIN column lining
***Stütz-***: **Stützflüssigkeit** *f* INFR & ENTW, WERKSTOFF sustaining fluid; **Stützgerüst** *nt* ARCH & TRAGW supporting frame; **Stützholz** *nt* HOLZ propwood; **Stützkonstruktion** *f* ARCH & TRAGW load-bearing structure, supporting structure; **Stützlänge** *f* ARCH & TRAGW span; **Stützmauer** *f* ARCH & TRAGW retaining wall, supporting wall; **Stützmoment** *nt* ARCH & TRAGW, INFR & ENTW moment at support; **Stützpfeiler** *m* ARCH & TRAGW, BETON, INFR & ENTW buttress, pillar, NATURSTEIN buttress; **Stützplatte** *f* ARCH & TRAGW bearing plate; **Stützschicht** *f* INFR & ENTW supporting layer; **Stützwand** *f* BETON, INFR & ENTW, NATURSTEIN retaining wall; **Stützweite** *f* ARCH & TRAGW, INFR & ENTW bearing distance
**Substanz** *f* WERKSTOFF substance
**Subunternehmer** *m* BAURECHT subcontractor
**Sulfat** *nt* UMWELT, WERKSTOFF sulfate (*AmE*), sulphate (*BrE*); **Sulfatangriff** *m* WERKSTOFF *Beton* sulfate attack (*AmE*), sulphate attack (*BrE*)
**sulfatbeständig**: **~er Portlandzement** *m* BETON, WERKSTOFF sulfate-resisting Portland cement (*AmE*), sulphate-resisting Portland cement (*BrE*) **~er Zement** *m* BETON, WERKSTOFF sulfate-resistant cement (*AmE*), sulphate-resistant cement (*BrE*)
***Sulfat***: **Sulfathüttenzement** *m* (*SHZ*) BETON, WERKSTOFF supersulfated cement (*AmE*), supersulphated cement (*BrE*)
**Summe** *f* ARCH & TRAGW sum, total; **Summenganglinie** *f* INFR & ENTW summation curve
**Summierung** *f* INFR & ENTW summation
**Sumpf** *m* INFR & ENTW fen, marsh, swamp; **Sumpfboden** *m* INFR & ENTW marshland, swampy soil; **Sumpfgebiet** *nt* INFR & ENTW swamp
**sumpfig** *adj* INFR & ENTW swampy; **~er Boden** *m* INFR & ENTW swampy ground
***Sumpf***: **Sumpfland** *nt* INFR & ENTW wetland
**superhoch**: **superhohe Frequenz** *f* ELEKTR superhigh frequency
**Suspension** *f* WERKSTOFF suspension
**Süßwasser** *nt* INFR & ENTW, UMWELT fresh water; **Süßwasserlinse** *f* INFR & ENTW freshwater lens
**S-Verschluß** *m* ABWASSER S-trap
**Symmetrie** *f* ARCH & TRAGW symmetry; **Symmetrieachse** *f* ARCH & TRAGW central axis; **Symmetrieebene** *f* ARCH & TRAGW, INFR & ENTW plane of symmetry
**symmetrisch** *adj* ARCH & TRAGW symmetric
**Synchrongeschwindigkeit** *f* UMWELT synchronous speed
**synthetisch** *adj* WERKSTOFF synthetic
**System** *nt* ARCH & TRAGW, BAURECHT, INFR & ENTW system
**systematisch**: **~e Fehler** *m pl* ARCH & TRAGW systematic errors

***System***: **Systembau** *m* ARCH & TRAGW systems building

**Système** *nt*: ~ **International** (*SI*) ARCH & TRAGW, INFR & ENTW, WERKSTOFF Système International (*SI*)

***System***: **Systemlinie** *f* ARCH & TRAGW modular line; **Systemliniengitter** *nt* ARCH & TRAGW modular grid; **Systemliniennetz** *nt* ARCH & TRAGW modular grid; **Systemnetz** *nt* ARCH & TRAGW modular grid

# T

**TAA** *abbr* (*Technische Anleitung Abfall*) UMWELT Technical Instruction on Waste Management
**Tabelle** *f* ARCH & TRAGW table
**tachometrisch**: **~e Vermessung** *f* ARCH & TRAGW stadia surveying
**Tachymeter** *nt* ARCH & TRAGW, BAUMASCHIN *Vermessung* tacheometer
**Tachymetrie** *f* ARCH & TRAGW *Vermessung* tacheometry
**Tafel** *f* ARCH & TRAGW chart, *einer Wand* plate, BAUMASCHIN plate, BAURECHT chart, BESCHLÄGE, INFR & ENTW, STAHL plate, WERKSTOFF panel, plate; **Tafelbauweise** *f* ARCH & TRAGW panel construction; **Tafelblei** *nt* STAHL sheet lead; **Tafelglas** *nt* WERKSTOFF plate glass, sheet frame
**täfeln** *vt* ARCH & TRAGW, BESCHLÄGE, HOLZ panel
**Täfelung** *f* ARCH & TRAGW wainscoting, HOLZ paneling (*AmE*), panelling (*BrE*), panel
***Tafel***: **Tafelwand** *f* ARCH & TRAGW filler wall, panel wall
**Tag**: **Tagesarbeitsblätter** *nt pl* BAURECHT daywork sheets; **Tagesarbeitsbücher** *nt pl* BAURECHT daywork accounts; **Tageslicht** *nt* ARCH & TRAGW daylight; **Tageslichtbeleuchtung** *f* ARCH & TRAGW daylight illumination; **Tagesordnung** *f* ARCH & TRAGW, BAURECHT agenda; **Tagesverbrauchsmenge** *f* ARCH & TRAGW, INFR & ENTW daily consumption
**täglich**: **~er Sonnenweg** *m* ARCH & TRAGW, HEIZ & BELÜFT, INFR & ENTW, UMWELT daily sun path
***Tag***: **Tagwasser** *nt* ABWASSER, INFR & ENTW stormwater
**talkumiert** *adj* OBERFLÄCHE talcumed
**Talsperre** *f* INFR & ENTW dam
**Tandem** *nt* BAUMASCHIN, INFR & ENTW tandem; **Tandemvibrationswalze** *f* BAUMASCHIN *Straßenbau* tandem vibrating roller; **Tandemwalze** *f* BAUMASCHIN tandem roller
**Tangente** *f* ARCH & TRAGW, INFR & ENTW tangent
**Tangentialpunkt** *m* ARCH & TRAGW, INFR & ENTW tangent point
**Tank** *m* HEIZ & BELÜFT, INFR & ENTW, UMWELT tank
**Tanne** *f* HOLZ fir; **Tannenholzbohle** *f* HOLZ fir plank
**Tapete** *f* OBERFLÄCHE wall covering, wallpaper, WERKSTOFF wallpaper
**Tapezierarbeiten** *f pl* OBERFLÄCHE wallpapering
**tapezieren** *vt* ARCH & TRAGW, OBERFLÄCHE paper, wallpaper
**Tarif** *m* ARCH & TRAGW, BAURECHT, INFR & ENTW tariff
**Tasche** *f* WERKSTOFF *für Beschläge* pocket
**Taster** *m* ELEKTR push button
**Tätigkeit** *f* ARCH & TRAGW function
**tatsächlich**: **~e Belastung** *f* INFR & ENTW actual load
**Tau 1.** *m* HEIZ & BELÜFT, UMWELT dew; **2.** WERKSTOFF rope
**Tauch-** *in cpds* OBERFLÄCHE dip, dipping
**tauchgrundiert** *adj* OBERFLÄCHE dip-primed
***Tauch-***: **Tauchkolben** *m* BAUMASCHIN plunger; **Tauchlack** *m* OBERFLÄCHE dipping varnish; **Tauchlackierung** *f* OBERFLÄCHE dip coating; **Tauchrüttler** *m* BAUMASCHIN internal vibrator
***Tau***: **Taupunkt** *m* DÄMMUNG, HEIZ & BELÜFT dew point, thawing point; **Tautropfenglas** *nt* WERKSTOFF dewdrop glass; **Tauwasser** *nt* DÄMMUNG, HEIZ & BELÜFT condensate, condensation water
**T-Band** *nt* BESCHLÄGE cross garnet, T-hinge
**Technik** *f* ARCH & TRAGW, INFR & ENTW engineering, technique
**Techniker** *m* ARCH & TRAGW technician, INFR & ENTW engineer
***Technik***: **Technikraum** *m* ARCH & TRAGW, INFR & ENTW engineering room
**technisch** *adj* ABWASSER, ARCH & TRAGW, BAUMASCHIN, BAURECHT, ELEKTR, HEIZ & BELÜFT, INFR & ENTW technical; **~e Angaben** *f pl* ARCH & TRAGW, BAUMASCHIN, BAURECHT, INFR & ENTW technical specifications
**Technisch**: **~e Anleitung** *f* **Abfall** (*TAA*) UMWELT Technical Instruction on Waste Management
***technisch***: **~e Beratung** *f* ARCH & TRAGW, INFR & ENTW technical advisory service; **~e Einrichtung** *f* ABWASSER, ELEKTR, HEIZ & BELÜFT technical equipment; **~e Gebäudeausrüstung** *f* ABWASSER, ELEKTR, HEIZ & BELÜFT mechanical services, building services; **~e Gebäudeausstattung** *f* ABWASSER, ELEKTR, HEIZ & BELÜFT technical equipment; **~e Lösung** *f* ARCH & TRAGW technical solution; **~e Planung** *f* INFR & ENTW engineering
**Technologie** *f* ARCH & TRAGW, INFR & ENTW technology
**Teer** *m* DÄMMUNG, INFR & ENTW, OBERFLÄCHE tar, WERKSTOFF tar, pitch; **Teerbitumenpappe** *f* DÄMMUNG, WERKSTOFF tar-saturated paper; **Teerkessel** *m* BAUMASCHIN, INFR & ENTW tar boiler; **Teerleinwand** *f* BESCHLÄGE tarpaulin; **Teermakadam** *m* INFR & ENTW *Tiefbau* tarmac, tarmacadam; **Teerpappe** *f* WERKSTOFF asphaltic felt, tarred felt; **Teerspritzgerät** *nt* BAUMASCHIN, INFR & ENTW tar sprinkler; **Teerspritzmaschine** *f* BAUMASCHIN, INFR & ENTW tar sprayer
**Teich** *m* INFR & ENTW pond, pool
**Teil** *m* ARCH & TRAGW section, member; **Teilannahme** *f* BAURECHT, INFR & ENTW partial acceptance
**Teilchen** *nt* UMWELT particulate material, WERKSTOFF particle; **Teilchendichte** *f* WERKSTOFF particle density; **Teilchengröße** *f* WERKSTOFF particle size
**teilen** *vt* ARCH & TRAGW, INFR & ENTW partition
***Teil***: **Teilsicherheitsfaktor** *m* ARCH & TRAGW, BAURECHT, INFR & ENTW partial safety factor
**Teilungsfläche** *f* BAUMASCHIN jointing plane
**teilweise**: **~ einbetoniertes Profil** *nt* ARCH & TRAGW, BETON, INFR & ENTW, STAHL partially cased section; **~r Schubanschluß** *m* ARCH & TRAGW, BETON, INFR & ENTW, WERKSTOFF *Verbundträger* partial shear connection; **~ Verdübelung** *f* ARCH & TRAGW, BETON, INFR & ENTW, WERKSTOFF *Verbundträger* partial shear connection
***Teil***: **Teilzahlung** *f* BAURECHT part-payment
**T-Eisen** *nt* STAHL, WERKSTOFF T-iron
**Tellerventil** *nt* HEIZ & BELÜFT mushroom valve
**Tellurmesser** *m* BAUMASCHIN tellurometer

**Tempel** *m* ARCH & TRAGW, INFR & ENTW temple
**Temperafarbe** *f* OBERFLÄCHE, WERKSTOFF distemper
**Temperatur** *f* ARCH & TRAGW, BETON, DÄMMUNG, HEIZ & BELÜFT, UMWELT, WERKSTOFF temperature; **Temperaturabfall** *m* HEIZ & BELÜFT, UMWELT drop in temperature, temperature drop; **Temperaturanstieg** *m* HEIZ & BELÜFT, UMWELT temperature rise; **Temperaturbewehrung** *f* BETON temperature reinforcement; **Temperatureinfluß** *m* ARCH & TRAGW, BETON, WERKSTOFF temperature effect; **Temperaturgradient** *m* WERKSTOFF temperature gradient; **Temperaturleitzahl** *f* DÄMMUNG, HEIZ & BELÜFT thermal diffusivity; **Temperaturspannung** *f* WERKSTOFF thermal stress
**temperaturwechselbeständig** *adj* WERKSTOFF resistant to temperature changes
**Temperguß** *m* STAHL, WERKSTOFF malleable cast iron
**tempern** *vt* STAHL temper
**Teppichboden** *m* ARCH & TRAGW wall-to-wall carpeting, BESCHLÄGE fitted carpet
**Termin** *m* ARCH & TRAGW, BAURECHT date
**termingemäß** *adj* ARCH & TRAGW, BAURECHT on schedule
***Termin***: **Terminplan** *m* ARCH & TRAGW, BAURECHT, INFR & ENTW construction schedule
**Terrainaufnahme** *f* INFR & ENTW land surveying
**Terrakotta** *f* NATURSTEIN terracotta
**Terrasse** *f* ARCH & TRAGW terrace
**Terrassieren** *nt* INFR & ENTW terracing
**Terrazzo** *m* NATURSTEIN terrazzo, Venetian mosaic
**Tesafilm**® *m* ARCH & TRAGW Scotch tape® (*AmE*), Sellotape® (*BrE*)
**Test** *m* INFR & ENTW test
**Testen** *nt*: ~ **durch Dritte** BAURECHT, WERKSTOFF third party testing
**T-förmig**: ~**e Verbindung** *f* ARCH & TRAGW t-piece union
**Theodolit** *m* INFR & ENTW, WERKSTOFF theodolite, transit
**theoretisch** *adj* ARCH & TRAGW, INFR & ENTW theoretical; ~**es Gebäude** *nt* ARCH & TRAGW, INFR & ENTW notional building
**Theorie** *f* ARCH & TRAGW, BAURECHT, INFR & ENTW theory; ~ **I. Ordnung** ARCH & TRAGW, BETON, INFR & ENTW, STAHL, WERKSTOFF first order theory; ~ **II. Ordnung** ARCH & TRAGW, BETON, INFR & ENTW, STAHL, WERKSTOFF second order theory
**thermisch**: ~**e Eigenschaften** *f pl* WERKSTOFF thermal properties; ~ **erzeugter Auftrieb** *m* UMWELT thermally induced buoyancy; ~**e Konvektion** *f* HEIZ & BELÜFT thermal convection; ~**er Puffer** *m* DÄMMUNG thermal buffer; ~**e Reaktion** *f* WERKSTOFF thermal reaction; ~**er Überlastungsschutz** *m* ELEKTR, HEIZ & BELÜFT thermal overload protection; ~**es Verbrennungsverfahren** *nt* UMWELT incineration train; ~**e Zersetzung** *f* UMWELT pyrolysis; ~ **wirksame Masse** *f* DÄMMUNG, NATURSTEIN, UMWELT thermal mass
**thermoelektrisch**: ~**e Sonnenenergieumwandlung** *f* UMWELT solar thermoelectric conversion
**thermofixiert**: ~**e Polyesterpulverbeschichtung** *f* OBERFLÄCHE thermosetting polyester powder coating
**Thermoplaste** *m pl* WERKSTOFF thermoplastics
**thermoplastisch**: ~**e Kunststoffe** *m pl* WERKSTOFF thermoplastic materials
**Thermostat** *m* HEIZ & BELÜFT thermostat; **Thermostatbatterie** *f* ABWASSER thermostatic blending valve, thermostatic mixer, thermostatic valve
**thermostatisch** *adj* HEIZ & BELÜFT thermostatic; ~**es Heizkörperventil** *nt* HEIZ & BELÜFT thermostatic radiator valve
***Thermostat***: **Thermostatventil** *nt* ABWASSER, HEIZ & BELÜFT thermostatic valve
**Thermosiphon** *m* HEIZ & BELÜFT thermosiphon
**Thiokol**® *nt* INFR & ENTW, WERKSTOFF Thiokol®; **auf Thiokolbasis**® *phr* INFR & ENTW, WERKSTOFF Thiokol-based®
**Tide** *f* INFR & ENTW *Gezeiten* tide; **Tideablauf** *m* UMWELT ebb tide
**Tief-** *in cpds* ARCH & TRAGW underground
**tief**: ~**e Wassersäule** *f* UMWELT low head
***Tief-***: **Tiefbau** *m* INFR & ENTW deep workings; **Tiefbautechnik** *f* INFR & ENTW civil engineering; **Tiefbordstein** *m* INFR & ENTW flush curb (*AmE*), flush kerb (*BrE*); **Tiefbrunnen** *m* INFR & ENTW deep well
**Tiefe** *f* ARCH & TRAGW, INFR & ENTW, WERKSTOFF depth; **Tiefenerder** *m* ELEKTR earth rod (*BrE*), ground rod (*AmE*); **Tiefenfaktor** *m* ARCH & TRAGW, HOLZ depth factor
***Tief-***: **Tiefgarage** *f* ARCH & TRAGW, INFR & ENTW underground garage; **Tiefgeschoßkonstruktion** *f* ARCH & TRAGW, INFR & ENTW deep basement construction
**tiefgezogen** *adj* WERKSTOFF deep-drawn (*DD*)
***Tief-***: **Tiefkühlanlage** *f* HEIZ & BELÜFT freezer plant
**tiefkühlen** *vt* HEIZ & BELÜFT refrigerate
***Tief-***: **Tiefkühlen** *nt* HEIZ & BELÜFT freezing; **Tiefkühlraum** *m* ARCH & TRAGW freezer room; **Tiefspülklosett** *nt* ABWASSER flush-down water closet
**Tierleim** *m* WERKSTOFF animal glue
**TIG-Schweißbrenner** *m* BAUMASCHIN torch for TIG welding
**Tilt-up-Bauweise** *f* ARCH & TRAGW tilt-up method
**Tisch** *m* HOLZ table; **Tischdrehbank** *f* BAUMASCHIN bench lathe; **Tischhobel** *m* BAUMASCHIN bench plane
**Tischler** *m* HOLZ joiner; **Tischlerarbeit** *f* HOLZ joiner's work
**Tischlerei** *f* HOLZ joinery
***Tischler***: **Tischlerleim** *m* HOLZ, WERKSTOFF joiner's glue; **Tischlerplatte** *f* HOLZ coreboard, blockboard
***Tisch***: **Tischrechner** *m* ARCH & TRAGW calculator, laptop computer
**tödlich**: ~**e Konzentration** *f* UMWELT lethal concentration; ~**e Wirkung** *f* UMWELT lethal effect
**Toilette** *f* ABWASSER, BESCHLÄGE, INFR & ENTW toilet; **Toilettenanlage** *f* ABWASSER, INFR & ENTW toilet installations
**TOK-Band** *nt* ARCH & TRAGW TOK-joint ribbon
**Toleranz** *f* ARCH & TRAGW tolerance, BAURECHT allowance, INFR & ENTW tolerance
**Ton** *m* DÄMMUNG sound, OBERFLÄCHE tint, WERKSTOFF clay
**tonartig** *adj* UMWELT, WERKSTOFF argillaceous, clayey
**tönen** *vt* OBERFLÄCHE tint
***Ton***: **Tonerde** *f* WERKSTOFF clay; **Tonerdemörtel** *m* NATURSTEIN alumina mortar; **Tongestein** *nt* INFR & ENTW, NATURSTEIN, WERKSTOFF argillaceous rock, argillite; **Tongrube** *f* WERKSTOFF clay pit
**tonhaltig** *adj* INFR & ENTW argillaceous, UMWELT, WERKSTOFF argillaceous, clayey

***Ton***: **Tonhohlplatte** *f* NATURSTEIN, WERKSTOFF clay pot, hollow clay pot, hollow-gaged brick (*AmE*), hollow-gauged brick (*BrE*)
**tonig** *adj* UMWELT, WERKSTOFF argillaceous, clayey
***Ton***: **Tonmörtel** *m* WERKSTOFF clay mortar
**Tonnendach** *nt* ARCH & TRAGW compass roof
**Tonnengewölbe** *nt* ARCH & TRAGW cradle vault, *Kirchenbau* barrel vault, INFR & ENTW wagon vault, tunnel vault, NATURSTEIN barrel vault, cradle vault
***Ton***: **Tonrohr** *nt* ABWASSER clay pipe, earthenware pipe; **Tonsandstein** *m* INFR & ENTW clayey sandstone
**Topographie** *f* ARCH & TRAGW, INFR & ENTW topography
**topographisch**: **~e Landaufnahme** *f* BAURECHT, INFR & ENTW land survey
**Tor** *nt* ARCH & TRAGW gate, portal; **Tordurchfahrt** *f* ARCH & TRAGW gateway
**Torf** *m* INFR & ENTW, WERKSTOFF peat, turf; **Torfablagerung** *f* INFR & ENTW peat deposit
**torfhaltig** *adj* INFR & ENTW, WERKSTOFF peaty; **~e Erde** *f* INFR & ENTW, WERKSTOFF peaty soil
***Torf***: **Torfmoor** *nt* INFR & ENTW peat bog; **Torfschicht** *f* INFR & ENTW peat layer
**Torkretbeton** *m* WERKSTOFF gunite
**torkretieren** *vt* BETON gun
**Torkretieren** *nt* BETON gunning
***Tor***: **Torpfosten** *m* ARCH & TRAGW swinging post; **Torriegel** *m* BESCHLÄGE gate latch; **Torschließbolzen** *m* BESCHLÄGE barrel bolt
**Torsion** *f* ARCH & TRAGW torsion; **Torsionsspannung** *f* ARCH & TRAGW torsional stress, torsional rigid
**TOR-Stahl** *m* STAHL tor steel
***Tor***: **Torweg** *m* ARCH & TRAGW gateway
**Tosbecken** *nt* ABWASSER, INFR & ENTW stilling basin, whirlpool basin
**tot**: **~e Leitung** *f* ELEKTR dead wire
**Totzeit** *f* ARCH & TRAGW, INFR & ENTW idle time
**toxisch**: **~e Abfälle** *m pl* UMWELT toxic waste; **~es Abfallprodukt** *nt* UMWELT toxic degradation product, toxic waste product; **~e Wirkung** *f* UMWELT toxic effect
**Traforaum** *abbr* ELEKTR transformer room
**Trag-** *in cpds* ARCH & TRAGW, INFR & ENTW bearing; **Tragbalken** *m* ARCH & TRAGW beam, girder
**Trage** *f* BAUMASCHIN handbarrow; **Tragekonstruktion** *f* ARCH & TRAGW, BETON, STAHL bearing structure
**tragen** *vt* ARCH & TRAGW bear, carry, support, HOLZ, INFR & ENTW, STAHL, WERKSTOFF bear
**Tragen** *nt* ARCH & TRAGW bearing
**tragend** *adj* ARCH & TRAGW load-bearing, supporting, INFR & ENTW load-bearing; **~e Bundsäulen** *f pl* ARCH & TRAGW, WERKSTOFF load-bearing studs; **~e Konstruktion** *f* ARCH & TRAGW load-bearing structure, supporting structure; **~er Leichtbeton** *m* BETON structural light concrete, structural lightweight concrete; **nicht ~e Mauerverkleidung** *f* NATURSTEIN, OBERFLÄCHE veneer; **~e Wand** *f* ARCH & TRAGW bearing wall, load-bearing wall, main wall;
**Träger** *m* ARCH & TRAGW arm, beam, bearer, girder, joist, stay, BETON beam, HOLZ joist, INFR & ENTW arm, STAHL beam; **~ auf zwei Stützen** ARCH & TRAGW, BETON, STAHL simple beam, simply-supported beam; **Trägerabstand** *m* ARCH & TRAGW, INFR & ENTW distance between girders, girder spacing; **Trägeranordnung** *f* ARCH & TRAGW, INFR & ENTW arrangement of girders; **Trägerbrücke** *f* ARCH & TRAGW, INFR & ENTW girder bridge; **Trägerflansch** *m* BAUMASCHIN flange; **Trägergewebe** *nt* WERKSTOFF *Putz* base fabric; **Trägermitte** *f* ARCH & TRAGW, BETON midspan; **Trägerschalung** *f* BETON girder forms; **Trägerschicht** *f* INFR & ENTW backing, underlay, NATURSTEIN backing, WERKSTOFF substrate
**tragfähig**: **~er Boden** *m* ARCH & TRAGW natural foundation, INFR & ENTW good bearing soil; **~er Naturgrund** *m* ARCH & TRAGW natural foundation, INFR & ENTW good bearing soil
***Trag-***: **Tragfähigkeit** *f* ARCH & TRAGW bearing capacity, load-bearing capacity, INFR & ENTW load-bearing capacity, stability, OBERFLÄCHE *Farbuntergrund* stability, UMWELT carrying capacity; **Tragfläche** *f* UMWELT airfoil (*AmE*)
**Trägheit**: **Trägheitskraft** *f* ARCH & TRAGW, INFR & ENTW, UMWELT inertia force; **Trägheitsmoment** *nt* ARCH & TRAGW moment of inertia; **Trägheitsradius** *m* ARCH & TRAGW radius of gyration
***Trag-***: **Tragkonstruktion** *f* ARCH & TRAGW, INFR & ENTW load-bearing structure, supporting structure; **Tragkraft** *f* ARCH & TRAGW, INFR & ENTW load-bearing capacity; **Traglast** *f* ARCH & TRAGW, INFR & ENTW, WERKSTOFF limit load; **Traglastverfahren** *nt* ARCH & TRAGW, BETON, INFR & ENTW, WERKSTOFF limit design, plastic analysis, plastic theory; **Tragmauer** *f* INFR & ENTW load-bearing wall; **Tragplatte** *f* ARCH & TRAGW bearing plate, WERKSTOFF mounting base; **Tragrahmen** *m* ARCH & TRAGW, INFR & ENTW bearing frame; **Tragsäule** *f* ARCH & TRAGW pillar, supporting column, INFR & ENTW pillar; **Tragschicht** *f* ARCH & TRAGW bearing course, *Geologie* bearing bed, BETON base course, INFR & ENTW substratum, base course, bearing course, WERKSTOFF bearing course; **Tragschiene** *f* ARCH & TRAGW bearing rail; **Tragseil** *nt* ARCH & TRAGW supporting cable; **Tragvermögen** *nt* ARCH & TRAGW, INFR & ENTW load-bearing capacity, UMWELT carrying capacity; **Tragwerk** *nt* ARCH & TRAGW, INFR & ENTW load-bearing structure; **Tragwerksabbau** *m* ARCH & TRAGW striking framework
**Traktor** *m* BAUMASCHIN tractor
**tränken** *vt* DÄMMUNG, HOLZ impregnate, OBERFLÄCHE *Wasser* saturate
**Tränkung** *f* DÄMMUNG, HOLZ, OBERFLÄCHE impregnation
**Transformator** *m* ELEKTR transformer; **Transformatorraum** *m* ELEKTR transformer room
**Transparentpause** *f* ARCH & TRAGW *Zeichnung* transparent copy
**Transport** *m* BAUMASCHIN conveying, haulage, transport; **Transportbelastung** *f* HOLZ handling load; **Transportbeton** *m* BETON ready-mixed concrete
**transportieren** *vt* BAUMASCHIN haul, transport, WERKSTOFF convey
**transportiert** *adj* BAUMASCHIN transported
***Transport***: **Transportmischer** *m* BAUMASCHIN, BETON mixer conveyor, mixer truck; **Transportsystem** *nt* INFR & ENTW transportation system
**Trapezblech** *nt* STAHL *Feinblech* trapezoidal sheet metal
**Traß** *m* WERKSTOFF tuff
**Trasse** *f* ARCH & TRAGW *Kabel, Rohr* route; **Trassenband** *nt* ELEKTR *zur Kabelmarkierung* cable marking tape

**trassieren** *vt* ARCH & TRAGW locate, map, ELEKTR locate
**Trassierung** *f* ARCH & TRAGW location, ELEKTR location of the line
**Traßzement** *m* BETON trass cement
**Trauf**: **Traufbohle** *f* ARCH & TRAGW, HOLZ, INFR & ENTW, WERKSTOFF eaves board, gutter board; **Traufbrett** *nt* ARCH & TRAGW gutter board, fascia board, HOLZ eaves board, INFR & ENTW, WERKSTOFF gutter board
**Traufe** *f* ARCH & TRAGW, HOLZ eaves
***Trauf***: **Traufhöhe** *f* ARCH & TRAGW eaves height; **Traufkante** *f* ARCH & TRAGW, HOLZ eaves; **Trauflatte** *f* NATURSTEIN tilting fillet; **Traufrinne** *f* ARCH & TRAGW eaves gutter; **Traufstein** *m* ARCH & TRAGW margin tile
**Traverse** *f* ARCH & TRAGW crossbar, strong-back, traverse, BESCHLÄGE suspension bracket, INFR & ENTW crossbar
**Travertin** *m* NATURSTEIN travertin
**Treib**: **Treibhaus** *nt* ARCH & TRAGW greenhouse, hothouse; **Treibhauseffekt** *m* UMWELT greenhouse effect; **Treibkessel** *m* BAUMASCHIN pressure cylinder; **Treibriß** *m* INFR & ENTW expansion crack; **Treibsand** *m* INFR & ENTW quicksand
**Trenn-** *in cpds* ABWASSER, INFR & ENTW separation
**trennen** *vt* ABWASSER separate, ARCH & TRAGW partition, INFR & ENTW separate, partition
**Trennen** *nt* ABWASSER, INFR & ENTW separation
***Trenn-***: **Trennentwässerung** *f* ABWASSER, INFR & ENTW two-pipe system
**Trenner** *m* ABWASSER, INFR & ENTW separator
***Trenn-***: **Trennfuge** *f* ARCH & TRAGW, INFR & ENTW expansion joint, separation joint; **Trennkanalisation** *f* ABWASSER, INFR & ENTW two-pipe system; **Trennmittel** *nt* WERKSTOFF separating agent; **Trennschicht** *f* ARCH & TRAGW separation layer, INFR & ENTW parting; **Trennsystem** *nt* ABWASSER, INFR & ENTW two-pipe system
**Trennung** *f* ABWASSER, UMWELT separation; **~ durch Schwerkraft** INFR & ENTW gravity separation
***Trenn-***: **Trennverfahren** *nt* ABWASSER separate sewage system; **Trennwand** *f* ARCH & TRAGW partition, separation wall, UMWELT slurry wall
**Treppe** *f* ARCH & TRAGW staircase, stairs, stairway; **Treppenabsatz** *m* ARCH & TRAGW landing; **Treppenauge** *nt* ARCH & TRAGW open well, round stair well, well hole
**treppenförmig**: **~ anlegen** *vt* OBERFLÄCHE terrace
***Treppe***: **Treppenhaus** *nt* ARCH & TRAGW staircase, stairway; **Treppenlauf** *m* ARCH & TRAGW flight of stairs; **Treppenlift** *m* ARCH & TRAGW, BAUMASCHIN stairlift; **Treppenlochwange** *f* ARCH & TRAGW external string, face string, outer string; **Treppenpfosten** *m* ARCH & TRAGW, HOLZ newel; **Treppenpodest** *nt* ARCH & TRAGW landing; **Treppenpodestträger** *m* ARCH & TRAGW bearer; **Treppenstufe** *f* ARCH & TRAGW stair, step; **Treppenwange** *f* ARCH & TRAGW stair stringer, string, stringer
**Triangulation** *f* ARCH & TRAGW triangulation; **Triangulationspunkt** *m* ARCH & TRAGW triangulation point
**Trichter** *m* BAUMASCHIN funnel
**trigonometrisch**: **~e Funktion** *f* ARCH & TRAGW trigonometrical function
**Trink-** *in cpds* ABWASSER, INFR & ENTW drinking
**trinkbar** *adj* ABWASSER potable
***Trink-***: **Trinkbrunnen** *m* INFR & ENTW drinking fountain (*DF*), water fountain; **Trinkwasser** *nt* INFR & ENTW drinking water; **Trinkwasserversorgung** *f* INFR & ENTW drinking water supply
**Tritt** *m* ARCH & TRAGW stair; **Trittfläche** *f* ARCH & TRAGW tread; **Tritthöhe** *f* ARCH & TRAGW rise-run ratio; **Trittschall** *m* DÄMMUNG footfall sound, footstep sound; **Trittschalldämmung** *f* DÄMMUNG footfall sound insulation, footstep sound insulation
**trittsicher** *adj* OBERFLÄCHE antiskid, WERKSTOFF skidproof
**Triumphbogen** *m* ARCH & TRAGW triumphal arch
**Trocken-** *in cpds* ARCH & TRAGW, HEIZ & BELÜFT, HOLZ, INFR & ENTW, NATURSTEIN, UMWELT, WERKSTOFF dry, drying; **Trockenbagger** *m* BAUMASCHIN excavator; **Trockenbett** *nt* INFR & ENTW drying bed; **Trockendampf** *m* UMWELT dry steam; **Trockenentschwefelungsprozeß** *m* UMWELT dry desulfurization process (*AmE*), dry desulphurization process (*BrE*); **Trockenfäule** *f* ARCH & TRAGW, HOLZ, WERKSTOFF dry rot
**trockengemauert** *adj* NATURSTEIN laid-dry
***Trocken-***: **Trockenputz** *m* NATURSTEIN, WERKSTOFF premixed stuff, ready-mixed stuff; **Trockenputzsystem** *nt* ARCH & TRAGW, INFR & ENTW, NATURSTEIN, WERKSTOFF dry lining system; **Trockenschlammdeponie** *f* UMWELT dry-sludge disposal site; **Trockenschwinden** *nt* BETON, INFR & ENTW, WERKSTOFF initial shrinkage; **Trockensprinklersystem** *nt* ARCH & TRAGW dry sprinkler system; **Trockenverglasung** *f* ARCH & TRAGW patent glazing; **Trockenzeiten** *f pl* WERKSTOFF drying times
**trocknen** *vti* HOLZ bake
**Trogbrücke** *f* INFR & ENTW trough bridge
**T-Rohr** *nt* ABWASSER, HEIZ & BELÜFT T-pipe
**Trombe-Wand** *f* DÄMMUNG, NATURSTEIN Trombe wall
**Trommel** *f* WERKSTOFF cylinder; **Trommelwascher** *m* BAUMASCHIN drum washer
**Tropf** *m* UMWELT drip; **Tropfbecher** *m* BESCHLÄGE drip cup; **Tropfenabscheider** *m* UMWELT mist eliminator; **Tropfkörper** *m* UMWELT percolating filter, sprinkling filter; **Tropfkörperanlage** *f* INFR & ENTW percolating filter, trickling filter; **Tropfnase** *f* ARCH & TRAGW weather drip, weather groove, window drip; **Tropfzylinder** *m* BESCHLÄGE drip cup
**tropisch**: **~es Hartholz** *nt* HOLZ tropical hardwood
**Trübglas** *nt* WERKSTOFF obscured glass, opal glass
**Trübheit** *f* BAURECHT, UMWELT, WERKSTOFF turbidity
**Trübung** *f* BAURECHT, UMWELT, WERKSTOFF turbidity; **Trübungskoeffizient** *m* UMWELT turbidity coefficient
**Trümmer** *pl* ARCH & TRAGW debris, rubble, INFR & ENTW ruins; **Trümmerverwertung** *f* INFR & ENTW debris utilization
**T-Stück** *nt* ABWASSER, HEIZ & BELÜFT *Rohr* T-piece, union-T
**T-Träger** *m* ARCH & TRAGW, STAHL T-beam, T-girder
**Tubularzylinder** *m* BESCHLÄGE *Schloß* tubular cylinder
**Tuchfilter** *m* UMWELT fabric filter
**Tudorbogen** *m* ARCH & TRAGW Tudor arch
**Tuff** *m* INFR & ENTW, NATURSTEIN tuff
**tünchen** *vt* OBERFLÄCHE limewash, whiten, whitewash, limewhite
**Tünchen** *nt* OBERFLÄCHE limewashing, liming
**Tünchsandputz** *m* NATURSTEIN fine stuff
**Tünchscheibe** *f* NATURSTEIN *Putz* hawk

**Tunnel** *m* INFR & ENTW, UMWELT tunnel; **Tunnelabschnitt** *m* INFR & ENTW tunnel section; **Tunnelachse** *f* INFR & ENTW tunnel axis; **Tunnelauskleidung** *f* INFR & ENTW tunnel lining; **Tunnelbau** *m* INFR & ENTW tunneling (*AmE*), tunnelling (*BrE*); **Tunnelbaumaschine** *f* BAUMASCHIN tunneling machine (*AmE*), tunnelling machine (*BrE*); **Tunnelbauverfahren** *nt* INFR & ENTW tunneling technique (*AmE*), tunnelling technique (*BrE*); **Tunnelbohrmaschine** *f* BAUMASCHIN tunnel-boring machine; **Tunneldecke** *f* INFR & ENTW tunnel soffit; **Tunneleinfahrt** *f* INFR & ENTW tunnel portal; **Tunneleingang** *m* INFR & ENTW tunnel portal, portal; **Tunnelmund** *m* INFR & ENTW tunnel portal; **Tunnelsohle** *f* INFR & ENTW tunnel floor, tunnel invert; **Tunnelvortrieb** *m* INFR & ENTW tunneling (*AmE*), tunnelling (*BrE*); **Tunnelwand** *f* INFR & ENTW tunnel wall; **Tunnelwandung** *f* INFR & ENTW tunnel wall; **Tunnelzimmerung** *f* HOLZ, INFR & ENTW tunnel timbering

**Tür** *f* BESCHLÄGE door; **~ mit Luftschlitzfüllung** ARCH & TRAGW louver door (*AmE*), louvre door (*BrE*); **Türangel** *f* BESCHLÄGE garnet hinge, hinge; **Türanschlag** *m* ARCH & TRAGW *am Fußboden*, BESCHLÄGE floor stop; **Türband** *nt* BESCHLÄGE door hinge; **Türbeschläge** *m pl* BESCHLÄGE door fittings, door furniture

**Turbine** *f* ELEKTR, UMWELT turbine; **Turbinenleistung** *f* ELEKTR, UMWELT turbine output; **Turbinenleistungsvermögen** *nt* UMWELT turbine efficiency; **Turbinenschaufel** *f* UMWELT turbine blade

***Tür***: **Türblatt** *nt* BESCHLÄGE door leaf

**turbulent**: **~e Strömung** *f* ABWASSER, HEIZ & BELÜFT, INFR & ENTW turbulent flow

**Turbulenz** *f* ABWASSER, ARCH & TRAGW, HEIZ & BELÜFT, INFR & ENTW, UMWELT turbulence

***Tür***: **Türdrücker** *m* BESCHLÄGE door opener; **Türflügel** *m* BESCHLÄGE active leaf, wing; **Türfüllung** *f* BESCHLÄGE *Täfelung* door panel, panel; **Türfutter** *nt* BESCHLÄGE door case, jamb lining; **Türgröße** *f* ARCH & TRAGW door size; **Türgucker** *m* BESCHLÄGE judas, judas hole, viewer; **Türkantenschoner** *m* BESCHLÄGE edge plate; **Türkette** *f* BESCHLÄGE door chain; **Türklopfer** *m* BESCHLÄGE door knocker; **Türknauf** *m* BESCHLÄGE door knob; **Türleibung** *f* ARCH & TRAGW, BESCHLÄGE door reveal

**Turm** *m* ARCH & TRAGW tower; **Turmansatz** *m* ARCH & TRAGW stump; **Turmbau** *m* ARCH & TRAGW tower construction

**Türmchen** *nt* ARCH & TRAGW turret

***Turm***: **Turmdrehkran** *m* BAUMASCHIN, INFR & ENTW tower crane; **Turmfuß** *m* ARCH & TRAGW tower base; **Turmhelm** *m* ARCH & TRAGW spire; **Turmkran** *m* BAUMASCHIN, INFR & ENTW tower crane; **Turmpfeiler** *m* INFR & ENTW tower pier; **Turmringerder** *m* ELEKTR tower ring earth (*BrE*), tower ring ground (*AmE*); **Turmschaft** *m* ARCH & TRAGW tower shaft; **Turmspitze** *f* ARCH & TRAGW spire; **Turmverbindung** *f* BAUMASCHIN derrick girt

**Turnhalle** *f* ARCH & TRAGW, INFR & ENTW gymnasium

***Tür***: **Türoberlicht** *nt* ARCH & TRAGW fanlight, transom light; **Türöffner** *m* BESCHLÄGE door opener; **Türöffnung** *f* ARCH & TRAGW doorway; **Türpfosten** *m* ARCH & TRAGW jamb, BESCHLÄGE door post, HOLZ jamb; **Türpuffer** *m* BESCHLÄGE door bumper; **Türquerriegel** *m* BESCHLÄGE *in Schloßhöhe* lock rail; **Türrahmen** *m* BESCHLÄGE, HOLZ door frame; **Türriegel** *m* BESCHLÄGE bolt, door bolt; **Türschließer** *m* BESCHLÄGE *Brandschutz* door closer; **Türschwelle** *f* ARCH & TRAGW doorsill, sill; **Türstange** *f* BESCHLÄGE door bar; **Türstopper** *m* ARCH & TRAGW, BESCHLÄGE *am Fußboden* floor stop; **Türsturz** *m* ARCH & TRAGW, WERKSTOFF browpiece, door lintel; **Türverkleidung** *f* BESCHLÄGE door panel, DÄMMUNG, INFR & ENTW, WERKSTOFF door lining; **Türzarge** *f* BESCHLÄGE, HOLZ buck, door buck, door case, door casing, door frame; **Türzubehör** *nt* BESCHLÄGE door accessories

**Typ** *m* ARCH & TRAGW type

**Tyton-Muffe** *f* ABWASSER Tyton joint

# U

**U-Bahn** *f* INFR & ENTW underground railroad (*AmE*), underground railway (*BrE*), subway (*AmE*), underground (*BrE*); **U-Bahnhof** *m* INFR & ENTW subway station (*AmE*), underground station (*BrE*)

**Überbau** *m* ARCH & TRAGW, INFR & ENTW superstructure

**Überbeanspruchung** *f* ARCH & TRAGW overstress

**Überbelastung** *f* INFR & ENTW surcharge

**Überbewertung** *f* ARCH & TRAGW, BAURECHT, INFR & ENTW overestimate

**Überblattung** *f* HOLZ halved joint, halving, INFR & ENTW overleap joint

**Überblick** *m* ARCH & TRAGW, INFR & ENTW general view, survey

**Überbrückung** *f* ARCH & TRAGW, ELEKTR *von Rissen* bridging, spanning,

**überdacht**: **nicht ~** *adj* ARCH & TRAGW open

**Überdachung** *f* ARCH & TRAGW roofing

**Überdeckung** *f* ARCH & TRAGW overlap, shelter, INFR & ENTW cover, overlap

**Überdimensionierung** *f* ARCH & TRAGW, INFR & ENTW overdimensioning, oversizing

**Überdrehzahlkontrolle** *f* UMWELT overspeed control

**Überdruck** *m* ABWASSER, HEIZ & BELÜFT superpressure, INFR & ENTW excess pressure; **Überdruckturbine** *f* UMWELT reaction turbine

**Überfalle** *f* BESCHLÄGE *Schloß* padlock hasp

**Überfallfischgerinne** *nt* UMWELT overfall-type fish pass

**Überfallwehr** *nt* INFR & ENTW, UMWELT spillway

**Überfangglas** *nt* WERKSTOFF flashed glass

**Überflurhydrant** *m* ABWASSER, INFR & ENTW aboveground hydrant, pillar hydrant

**Überführung** *f* INFR & ENTW overbridge

**Übergabe** *f* ARCH & TRAGW, BAURECHT *Schlüssel* handing over

**Übergang** *m* ABWASSER, BAURECHT, HEIZ & BELÜFT, INFR & ENTW transition; **Übergangsbogen** *m* INFR & ENTW transition curve; **Übergangsbogenlänge** *f* INFR & ENTW transition curve length; **Übergangsheizperiode** *f* HEIZ & BELÜFT heating period between seasons; **Übergangsrohr** *nt* ABWASSER reducer, tapered pipe; **Übergangsschicht** *f* INFR & ENTW transition layer; **Übergangsstück** *nt* ABWASSER, ARCH & TRAGW, HEIZ & BELÜFT reducer; **Übergangszeit** *f* BAURECHT transition time; **Übergangszone** *f* INFR & ENTW transition zone; **Übergangszustand** *m* BAURECHT, INFR & ENTW state of transition

**Überhang** *m* ARCH & TRAGW overhang, projection; **Überhangwand** *f* ARCH & TRAGW overhanging wall

**Überheizung** *f* HEIZ & BELÜFT overheating

**Überhitzen** *nt* HEIZ & BELÜFT superheating

**Überhitzung** *f* HEIZ & BELÜFT superheating

**Überhöhung** *f* ARCH & TRAGW camber, BETON banking, camber, INFR & ENTW banking, camber, *Schienen, Straßen* superelevation

**Überholspur** *f* INFR & ENTW passing lane, *Straße* acceleration lane

**Überladung** *f* ARCH & TRAGW overload

**Überlagerung** *f* INFR & ENTW superposition

**Überlandleitung** *f* ELEKTR, INFR & ENTW land line

**Überlappung** *f* ARCH & TRAGW overlap, lap, step joint, INFR & ENTW overlap; **Überlappungsfuge** *f* HOLZ, STAHL lap joint; **Überlappungsstoß** *m* HOLZ lap joint, INFR & ENTW lap joint, overlapping joint, STAHL lap joint

**Überlast** *f* ARCH & TRAGW, ELEKTR overload, INFR & ENTW excess load

**Überlastung** *f* ABWASSER, HEIZ & BELÜFT overcharge

**Überlauf** *m* ABWASSER, HEIZ & BELÜFT overflow, INFR & ENTW overflow, spillway, WERKSTOFF screen residue; **Überlaufgut** *nt* INFR & ENTW *Siebreste* tailings; **Überlaufkanal** *m* ABWASSER, INFR & ENTW spillway channel; **Überlaufkante** *f* ARCH & TRAGW lip, INFR & ENTW overflow edge; **Überlaufrinne** *f* INFR & ENTW overflow gutter; **Überlaufrohr** *nt* ABWASSER, HEIZ & BELÜFT, INFR & ENTW overflow pipe, spillway pipe

**Übernahme** *f* ARCH & TRAGW, BAURECHT acceptance

**Überprüfung** *f* ARCH & TRAGW, BAURECHT, INFR & ENTW check, examination, inspection

**Übersäuerung**: **~ des Wassers** *f* ABWASSER, UMWELT aquatic acidification

**Überschattung** *f* ARCH & TRAGW, HEIZ & BELÜFT overshadowing

**Überschiebmuffe** *f* ABWASSER, HEIZ & BELÜFT *Rohr* collar, coupler

**Überschlagsrechnung** *f* ARCH & TRAGW rough calculation

**Überschwemmung** *f* HEIZ & BELÜFT, UMWELT flooding; **Überschwemmungsgebiet** *nt* INFR & ENTW, UMWELT flood plane

**Übersicht** *f* ARCH & TRAGW, INFR & ENTW survey; **Übersichtskarte** *f* ARCH & TRAGW, INFR & ENTW outline map; **Übersichtsplan** *m* ARCH & TRAGW, INFR & ENTW block plan, general plan, key plan

**Überstand** *m* ARCH & TRAGW projecting end, projection

**Übersteckring** *m* WERKSTOFF retaining ring

**Überströmungshöhe** *f* INFR & ENTW head of water over weir

**Überstromvorrichtung** *f* ELEKTR overload device

**Überstunden** *f pl* BAURECHT overtime

**Übertragung** *f* BAURECHT conveyance, conveyancing, ELEKTR *Alarm* transmission

**Überwachung** *f* ABWASSER observation, ARCH & TRAGW observation, supervision, BAURECHT checking, control, monitoring, supervision, surveillance, INFR & ENTW monitoring, observation, UMWELT observation; **Überwachungsanlage** *f* BAURECHT, INFR & ENTW monitoring equipment; **Überwachungsgerät** *nt* BAUMASCHIN, BAURECHT monitor; **Überwachungsnetz** *nt* INFR & ENTW monitoring network

**Überweisung** *f* BAURECHT transfer

**Überwurf** *m* BESCHLÄGE lock bush; **Überwurfkrümmer** *m* ABWASSER union elbow

**Überzug** *m* ARCH & TRAGW upstand beam, INFR & ENTW overlay, overlaying; **Überzugslack** *m* OBERFLÄCHE top coat

**üblich**: **allgemein ~ sein** *phr* ARCH & TRAGW to be in common practice
**U-Eisen** *nt* STAHL channel iron
**UF** *abbr* (*Urea-Formaldehyd*) HOLZ, UMWELT UF (*urea formaldehyde*)
**Ufer** *nt* INFR & ENTW shore; **Uferbefestigung** *f* INFR & ENTW revetment of slopes; **Uferschutz** *m* INFR & ENTW revetment of slopes
**Uhrenanlage** *f* ELEKTR clock system
**Ulme** *f* HOLZ elm
**Ultraschallbewegungsmelder** *m* ELEKTR ultrasonic motion detector
**Ultraschallsonde** *f* ELEKTR, INFR & ENTW, WERKSTOFF ultrasonic probe
**Ultraviolett**: **Ultraviolettbestrahlung** *f* UMWELT, WERKSTOFF ultraviolet radiation; **Ultraviolettstrahlung** *f* UMWELT ultraviolet radiation
**Umänderung** *f* BAURECHT alteration
**umarbeiten** *vt* ARCH & TRAGW redesign
**Umbau** *m* ARCH & TRAGW reconstruction, alteration, conversion
**umbauen** *vt* ARCH & TRAGW alter, convert, rebuild, renovate
**umbaut**: **~er Raum** *m* ARCH & TRAGW cubical content, cubic yardage, enclosed space
**Umbemessung** *f* ARCH & TRAGW, INFR & ENTW redesign
**umbördeln** *vt* HOLZ bead over
**Umdrehung** *f* BAUMASCHIN rotation; **Umdrehungen** *f pl* **pro Minute** (*UpM*) ABWASSER, BAUMASCHIN, HEIZ & BELÜFT *Pumpe, Ventilator* revolutions per minute (*rpm*)
**Umfang** *m* ARCH & TRAGW scope, perimeter, circumference, *der Arbeiten* extent, INFR & ENTW extent, scope, periphery
**Umfassung** *f* ARCH & TRAGW, BAURECHT, INFR & ENTW span, *eines Standorts* enclosure, perimeter; **Umfassungsmauer** *f* ARCH & TRAGW fence wall, enclosing wall; **Umfassungswand** *f* ARCH & TRAGW, BAURECHT enclosure wall, INFR & ENTW enclosure wall, perimeter wall; **Umfassungszaun** *m* ARCH & TRAGW, INFR & ENTW perimeter fence
**Umgangsgewölbe** *nt* ARCH & TRAGW deambulatory
**umgebaut** *adj* ARCH & TRAGW altered
**umgebend** *adj* DÄMMUNG, HEIZ & BELÜFT, INFR & ENTW, UMWELT ambient, surrounding; **~e Luft** *f* UMWELT ambient air
**Umgebung** *f* ARCH & TRAGW, BAURECHT, INFR & ENTW environment, surroundings; **Umgebungslärm** *m* DÄMMUNG, INFR & ENTW ambient noise; **Umgebungstemperatur** *f* HEIZ & BELÜFT, INFR & ENTW ambient temperature
**Umgehung** *f* HEIZ & BELÜFT bypass; **Umgehungskanal** *m* HEIZ & BELÜFT bypass channel; **Umgehungsrohr** *nt* ABWASSER, HEIZ & BELÜFT bypass pipe; **Umgehungsstraße** *f* INFR & ENTW bypass
**umhüllen** *vt* ARCH & TRAGW sheathe
**Umhüllen** *nt* OBERFLÄCHE coating
**Umhüllung** *f* ARCH & TRAGW jacket, sheathing, BESCHLÄGE, BETON casing, HEIZ & BELÜFT jacket, OBERFLÄCHE coating, STAHL casing
**Umkehrosmose** *f* UMWELT, WERKSTOFF reverse osmosis
**Umkehrtrommel** *f* BAUMASCHIN *Mischer* reverse drum
**umkippen 1.** *vt* ARCH & TRAGW tip up; **2.** *vi* ARCH & TRAGW tilt
**Umkippen** *nt* BAUMASCHIN tipping
**Umkleideeinrichtung** *f* ARCH & TRAGW, INFR & ENTW changing facility
**Umkleideraum** *m* ARCH & TRAGW, INFR & ENTW changing room
**Umkreis** *m* ARCH & TRAGW perimeter, INFR & ENTW periphery
**Umlage** *f*: **~ von Gemeinkosten** BAURECHT allocation of expenses
**Umlaufsrichtung** *f* BAUMASCHIN direction of rotation
**umlegbar**: **~e Falle** *f* BESCHLÄGE *Schloß* reversible latch bolt
**Umleitung** *f* INFR & ENTW bypass
**Umlenkblech** *nt* BESCHLÄGE baffle plate
**Umlenkplatte** *f* BESCHLÄGE baffle plate
**Umluft** *f* HEIZ & BELÜFT recirculated air
**ummanteln** *vt* ARCH & TRAGW sheathe, jacket, HEIZ & BELÜFT jacket
**ummantelt** *adj* ARCH & TRAGW cased, sheathed, BESCHLÄGE cased; **~e Stütze** *f* ARCH & TRAGW cased column
**Ummantelung** *f* ARCH & TRAGW sheathing, jacket, BESCHLÄGE, BETON casing, HEIZ & BELÜFT jacket, STAHL casing
**ummauert**: **~er Raum** *m* ARCH & TRAGW walled enclosure
**umplanen** *vt* ARCH & TRAGW redesign
**umrahmen** *vt* ARCH & TRAGW frame
**Umrahmung** *f* BESCHLÄGE *Trennwand* framing
**Umrandung** *f* ARCH & TRAGW border
**umrechnen** *vt* ARCH & TRAGW convert
**Umrechnungsfaktor** *m* ARCH & TRAGW conversion factor
**Umriß** *m* ARCH & TRAGW contour, outline; **Umrißplanung** *f* INFR & ENTW outline planning
**Umschalter** *m* ELEKTR changeover switch
**Umschaltventil** *nt* HEIZ & BELÜFT changeover valve
**Umschlagbühne** *f* ARCH & TRAGW handling platform
**Umsetzungsprodukt** *nt* UMWELT solidified waste, solidified product, solidified material
**umstellbar**: **~er Zylinder** *m* BESCHLÄGE *Schloß* reversible cylinder
**umstellen** *vt* ARCH & TRAGW relocate
**Umverpackung** *f* UMWELT overpackaging
**Umwälz-** *in cpds* HEIZ & BELÜFT circulation
**umwälzen** *vt* HEIZ & BELÜFT circulate
***Umwälz-***: **Umwälzleitung** *f* HEIZ & BELÜFT circulation line; **Umwälzpumpe** *f* HEIZ & BELÜFT circulation pump
**Umwälzung** *f* HEIZ & BELÜFT circulation
***Umwälz-***: **Umwälzverlust** *m* HEIZ & BELÜFT circulation loss
**Umwandlung** *f* ELEKTR transformation; **~ von Abfallstoffen** UMWELT waste processing; **Umwandlungsrate** *f* UMWELT transformation rate
**Umwehrung** *f* ARCH & TRAGW protection device, *bis Brusthöhe* breastwork
**Umwelt** *f* ARCH & TRAGW, BAURECHT, INFR & ENTW, UMWELT environment; **Umweltbedingungen** *f pl* UMWELT environmental conditions; **Umweltbelastung** *f* UMWELT environmental impact; **Umweltbewußtsein** *nt* UMWELT ecological awareness; **Umwelteinschätzung** *f* UMWELT environmental assessment; **Umwelterhaltungsbescheid** *m* BAURECHT, UMWELT conservation order; **Umweltfaktor** *m* UMWELT ecological factor
**umweltfreundlich** *adj* UMWELT environmentally

friendly; **~e Technologie** *f* UMWELT clean technology, nonwaste technology (*NWT*)
***Umwelt***: **Umweltführer** *m* BAURECHT, UMWELT environmental guide; **Umweltgefahr** *f* UMWELT ecological menace; **Umweltgeologie** *f* INFR & ENTW, UMWELT environmental geology; **Umweltgeotechnik** *f* INFR & ENTW, UMWELT environmental geotechnology; **Umweltgesetz** *nt* BAURECHT, UMWELT environmental law; **Umweltkatastrophe** *f* UMWELT ecological disaster, environmental disaster; **Umweltplanung** *f* UMWELT environmental planning, planned environment
**umweltschädlich** *adj* UMWELT harmful to the environment
***Umwelt***: **Umweltschutz** *m* ARCH & TRAGW, BAURECHT, INFR & ENTW, UMWELT conservation, environmental protection, environmentalism; **Umweltschutzbehörde** *f* BAURECHT, UMWELT environmental protection agency; **Umweltschutz-ministerium** *nt* BAURECHT Department of the Environment (*BrE*); **Umweltverschmutzung** *f* UMWELT pollution; **Umweltverschmutzungsforschung** *f* UMWELT pollution research; **Umweltverschmutzung** *f* **durch Wärme** UMWELT heat pollution; **Umweltvertraglichkeit** *f* UMWELT environmental compatibility; **Umweltverwaltungsprojekt** *nt* BAURECHT, UMWELT environmental management scheme; **Umweltziel** *nt* UMWELT environmental target
**Umzäunung** *f* ARCH & TRAGW, BAURECHT, INFR & ENTW boundary fence, enclosure; **Umzäunungspfahl** *m* INFR & ENTW, WERKSTOFF fence post
**Umziehgerüst** *nt* ARCH & TRAGW, BAUMASCHIN traveling cradle (*AmE*), travelling cradle (*BrE*)
**unangemessen** *adj* ARCH & TRAGW improper, unsuitable
**unarmiert** *adj* BETON, WERKSTOFF nonreinforced
**unbedenklich**: **~es Lösungsmittel** *nt* BAURECHT, OBERFLÄCHE safety solvent
**unbefestigt** *adj* INFR & ENTW unfixed
**unbehandelt** *adj* UMWELT untreated, WERKSTOFF nontreated
**unbelastet**: **~er Zustand** *m* ARCH & TRAGW, INFR & ENTW unloaded state
**unbelüftet** *adj* HEIZ & BELÜFT unvented
**unbesandet**: **~e Pappe** *f* DÄMMUNG, WERKSTOFF *Dach* smooth-surface roofing felt, smooth-surface roofing paper
**unbestimmbar** *adj* ARCH & TRAGW indeterminable, undeterminable
**unbestimmt** *adj* ARCH & TRAGW indefinite, indeterminate, indetermined, redundant
**Unbestimmtheit** *f* ARCH & TRAGW indeterminacy; **Unbestimmtheitsgrad** *m* ARCH & TRAGW, INFR & ENTW degree of indeterminacy
**unbewehrt** *adj* BETON, WERKSTOFF nonreinforced, unreinforced; **~er Beton** *m* BETON nonreinforced concrete, unreinforced concrete, plain concrete, mass concrete, INFR & ENTW mass concrete
**Undichtigkeit** *f* ABWASSER, HEIZ & BELÜFT, INFR & ENTW leakage
**undurchdringbar** *adj* ARCH & TRAGW impenetrable
**undurchlässig** *adj* INFR & ENTW, WERKSTOFF impermeable, impervious; **~er Boden** *m* INFR & ENTW, WERKSTOFF impermeable soil
**Undurchlässigkeit** *f* INFR & ENTW, WERKSTOFF impermeability
**uneben** *adj* ARCH & TRAGW uneven
**Unebenheit** *f* ARCH & TRAGW unevenness
**unelastisch** *adj* ARCH & TRAGW, BETON, INFR & ENTW, WERKSTOFF inelastic
**unerwünscht**: **~e Wärmezuwächse** *m pl* HEIZ & BELÜFT unwanted heat gains
**Unfall** *m* BAURECHT accident; **Unfallverhütung** *f* BAURECHT accident prevention; **Unfallverhütungsausschuß** *m* BAURECHT Health and Safety Executive (*BrE*) (*HSE*); **Unfallverhütungsvorschrift** *f* **an der Arbeitsstelle** BAURECHT *Health and Safety at Work Act*
**unfertig** *adj* ARCH & TRAGW unfinished
**ungebrannt** *adj* NATURSTEIN, WERKSTOFF *Ziegel* green
**ungeeignet** *adj* ARCH & TRAGW improper, unsuitable
**ungefähr** *adj* ARCH & TRAGW, INFR & ENTW approximate
**ungefüllt**: **~es Bitumen** *nt* WERKSTOFF bitumen without fillers
**ungehärtet** *adj* STAHL unhardened
**ungehobelt** *adj* HOLZ rough, unplaned
**ungenehmigt**: **~e Deponie** *f* UMWELT phantom dump
**ungenutzt**: **~e Stauung** *f* UMWELT dead dike
**ungeordnet**: **~e Ablagerung** *f* UMWELT uncontrolled dumping; **~e Deponie** *f* UMWELT open dump, uncontrolled tipping (*BrE*)
**ungerissen** *adj* ARCH & TRAGW, BETON, OBERFLÄCHE, WERKSTOFF uncracked
**ungerollt** *adj* STAHL decoiled
**ungesättigt** *adj* INFR & ENTW unsaturated; **~er Bereich** *m* INFR & ENTW unsaturated zone
**ungeschützt** *adj* ARCH & TRAGW exposed, open; **~e Lage** *f* ARCH & TRAGW, INFR & ENTW exposed position
**ungespannt**: **~es Grundwasser** *nt* INFR & ENTW free ground water; **~er Stab** *m* ARCH & TRAGW *Fachwerk* unstrained member
**ungestört**: **~er Boden** *m* INFR & ENTW undisturbed soil; **~e Bodenprobe** *f* INFR & ENTW undisturbed soil sample
**ungezieferbeständig** *adj* INFR & ENTW, WERKSTOFF vermin-proof, vermin-resistant
**ungezieferfest** *adj* INFR & ENTW, WERKSTOFF vermin-proof, vermin-resistant
**ungeziefersicher** *adj* INFR & ENTW, WERKSTOFF vermin-proof, vermin-resistant
**ungiftig** *adj* WERKSTOFF nonpoisonous, nontoxic
**ungleichförmig** *adj* ARCH & TRAGW, BETON, INFR & ENTW, STAHL, WERKSTOFF nonuniform
**ungültig**: **~ werden** *vi* BAURECHT *Garantie, Angebot* expire
**unisex** *adj* BAURECHT unisex
**Universal**: **Universalgrundierung** *f* OBERFLÄCHE all-purpose primer; **Universalmobilbagger** *m* BAUMASCHIN all-purpose mobile excavator
**unklassifiziert**: **~e Zuschlagstoffe** *m pl* WERKSTOFF all-in ballast
**unlöslich** *adj* BETON, OBERFLÄCHE, WERKSTOFF colloidal
**unproduktiv**: **~e Zeit** *f* ARCH & TRAGW, BAURECHT, INFR & ENTW idle time
**unregelmäßig** *adj* ARCH & TRAGW, INFR & ENTW irregular
**unsortiert**: **~er Bruchstein** *m* ARCH & TRAGW, INFR & ENTW, NATURSTEIN random rubble
**unsymmetrisch** *adj* ARCH & TRAGW, BETON, INFR &

ENTW, STAHL, WERKSTOFF nonsymmetric, nonsymmetrical
**untauglich** *adj* ARCH & TRAGW improper, unsuitable
**unten**: **untere Bewehrung** *f* ARCH & TRAGW bottom reinforcement; **nach ~ biegend** *adj* ARCH & TRAGW sagging, BETON, WERKSTOFF sagging bending; **untere Heizleistung** *f* HEIZ & BELÜFT lower calorific value; **~ lackiert** *adj* ARCH & TRAGW, OBERFLÄCHE bottom-glazed; **untere Lage** *f* ARCH & TRAGW bottom reinforcement; **unteres Sammelbecken** *nt* UMWELT lower storage basin; **unteres Speicherbecken** *nt* UMWELT lower storage basin; **untere Treppenwange** *f* ARCH & TRAGW rough string
**Unterbau** *m* ARCH & TRAGW foundation, substructure, BETON base course, INFR & ENTW foundation, subbase, substratum
**unterbauen** *vt* ARCH & TRAGW underpin
**Unterbeton** *m* BETON backing concrete, subconcrete
**Unterboden** *m* ARCH & TRAGW subfloor, HEIZ & BELÜFT underfloor, INFR & ENTW subsoil; **Unterbodenbelüftung** *f* ARCH & TRAGW subfloor ventilation
**Unterbringung** *f* ARCH & TRAGW, INFR & ENTW accommodation
**unterbrochen**: **~er Träger** *m* ARCH & TRAGW tailpiece
**Unterdecke** *f* ARCH & TRAGW suspended ceiling
**unterdimensionieren** *vt* ARCH & TRAGW underdesign
**unterdimensioniert** *adj* ARCH & TRAGW underdesigned
**Unterdruck** *m* HEIZ & BELÜFT negative pressure
**Unterfahren** *nt* ARCH & TRAGW dead shoring
**unterfangen** *vt* ARCH & TRAGW underpin
**Unterfangen** *nt* ARCH & TRAGW dead shoring
**Unterfangung** *f* ARCH & TRAGW underpinning, vertical shoring
**Unterflur-** *in cpds* INFR & ENTW buried; **Unterflurhydrant** *m* ABWASSER, INFR & ENTW underground hydrant; **Unterflurkanal** *m* ABWASSER, ARCH & TRAGW *im Boden* buried duct, underfloor duct
**unterführen** *vt* INFR & ENTW pass under
**Unterführung** *f* INFR & ENTW subway (*BrE*), underbridge, underpass
**Unterfüllung** *f* BETON subsealing, *Beton* underfilling
**Unterfußboden** *m* ARCH & TRAGW subfloor
**unterfüttern** *vt* HOLZ line
**Unterfütterung** *f* HOLZ firring, furring
**untergehängt**: **~e Decke** *f* ARCH & TRAGW suspended ceiling
**Untergeschoß** *nt* ARCH & TRAGW basement storey (*BrE*), basement story (*AmE*)
**Untergestell** *nt* BAUMASCHIN trailer
**untergetaucht** *adj* ABWASSER submerged
**untergraben** *vt* INFR & ENTW undermine
**Untergrund** *m* ARCH & TRAGW *Fundament* base, soil, INFR & ENTW subgrade, subsoil; **Untergrundbahn** *f* INFR & ENTW subway (*AmE*), underground (*BrE*), underground railroad (*AmE*), underground railway (*BrE*)
**Untergruppe** *f* ELEKTR subgroup
**Untergurt** *m* ARCH & TRAGW *Fachwerk* bottom chord, bottom flange, girt, BAUMASCHIN girt, BETON, INFR & ENTW, STAHL, WERKSTOFF bottom chord, bottom flange, girt; **Untergurtstab** *m* STAHL bottom chord member
**unterhalten** *vt* ARCH & TRAGW, BAURECHT, INFR & ENTW maintain
**unterhöhlen** *vt* INFR & ENTW undercut, undermine
**unterhöhlt** *adj* INFR & ENTW undercut
**unterirdisch** *adj* ABWASSER, ARCH & TRAGW, ELEKTR, HEIZ & BELÜFT, INFR & ENTW below-ground, buried, underground; **~e Abfallbeseitigung** *f* INFR & ENTW, UMWELT underground waste disposal; **~e Konstruktion** *f* INFR & ENTW underground structure; **~e Lagerung** *f* INFR & ENTW underground storage; **~es Wasser** *nt* INFR & ENTW, UMWELT underground water; **~er Wasserfluß** *m* INFR & ENTW underground water flow
**Unterkante** *f* ARCH & TRAGW bottom edge
**unterkellert**: **nicht ~es Gebäude** *nt* ARCH & TRAGW nonbasement building
**Unterkonstruktion** *f* ARCH & TRAGW substructure, WERKSTOFF *für Putzauftrag* furring
**Unterkunft** *f* ARCH & TRAGW, INFR & ENTW accommodation
**Unterlage** *f* ARCH & TRAGW base, bed, support, INFR & ENTW underlay, ballast, bed, WERKSTOFF ballast
**Unterlagsbahn** *f* WERKSTOFF base deck
**Unterlagsblech** *nt* BESCHLÄGE backplate
**Unterlagscheibe** *f* BESCHLÄGE washer
**unterlegen** *vt* HOLZ line
**Unterlegstreifen** *m* WERKSTOFF backing strip
**untermauern** *vt* ARCH & TRAGW underpin
**Unterpulverschweißen** *nt* INFR & ENTW, STAHL submerged arc welding
**Unterputz** *m* NATURSTEIN first coat, backing plaster, pricking-up coat, undercoat, WERKSTOFF coarse stuff **Unterputzinstallation** *f* ABWASSER *Kabel, Rohre*, ELEKTR *Schalter, Steckdosen*, HEIZ & BELÜFT *Kabel, Rohre* concealed installation, flush mounting **Unterputzkabel** *nt* ARCH & TRAGW, ELEKTR concealed cable; **Unterputzleerrohr** *nt* ELEKTR concealed conduit; **Unterputzrohr** *nt* ELEKTR concealed pipe
**Untersatz** *m* ARCH & TRAGW, BAUMASCHIN, HEIZ & BELÜFT, HOLZ, INFR & ENTW, STAHL mounting
**Unterschicht** *f* ARCH & TRAGW backup, INFR & ENTW back-up, substratum
**unterschneiden** *vt* ARCH & TRAGW underream
**Unterschneidung** *f* ARCH & TRAGW undercut, weather drip, weather groove
**Unterseite** *f* ARCH & TRAGW underside, *eines Hobels* sole
**Untersicht** *f* ARCH & TRAGW bottom view, underview, soffit, underside, *bei Wölbungen* intrados
**unterspannt** *adj* ARCH & TRAGW trussed with sag rods; **~er Balken** *m* ARCH & TRAGW, HOLZ trussed beam
**unterspülen** *vt* INFR & ENTW underwash
**Unterspülung** *f* INFR & ENTW undermining
**Unterstrom** *m* UMWELT undercurrent
**Unterströmung** *f* UMWELT undercurrent
**unterstützen** *vt* ARCH & TRAGW support
**Unterstützung** *f* ARCH & TRAGW support, BAURECHT, INFR & ENTW, WERKSTOFF aid
**untersuchen** *vt* ARCH & TRAGW analyse (*BrE*), analyze (*AmE*), examine, investigate, INFR & ENTW, WERKSTOFF analyse (*BrE*), analyze (*AmE*)
**Untersuchung** *f* ARCH & TRAGW, BAURECHT, INFR & ENTW examination, investigation; **Untersuchungsprobe** *f* INFR & ENTW test sample
**Untertagedeponie** *f* (*UTD*) UMWELT subsurface repository, underground depot
**Unterteil** *nt* ARCH & TRAGW, INFR & ENTW foot
**unterteilen** *vt* ARCH & TRAGW subdivide, split into
**Untertunnelung** *f* INFR & ENTW tunneling (*AmE*), tunnelling (*BrE*)

**Unterwaschung** *f* INFR & ENTW undermining
**Unterwasser** *nt* INFR & ENTW *Schleuse* tail water; **Unterwasseratomexplosion** *f* UMWELT underwater atomic explosion; **Unterwasserausgrabung** *f* INFR & ENTW underwater excavation; **Unterwasserbeton** *m* BETON underwater concrete, submerged concrete; **Unterwasserbohranlage** *f* BAUMASCHIN, INFR & ENTW marine-drilling rig; **Unterwassergründung** *f* INFR & ENTW underwater foundation; **Unterwasserschneidbrenner** *m* BAUMASCHIN underwater cutting blowpipe; **Unterwasserschweißen** *nt* STAHL underwater welding; **Unterwassertunnel** *m* INFR & ENTW underwater tunnel
**Unterzeichnung** *f*: **~ der Verträge** BAURECHT exchange of contracts
**Unterzug** *m* ARCH & TRAGW binding beam, joist, sleeper, summer, bearer, sill, BETON binding beam, HOLZ binding beam, joist, trussing, INFR & ENTW transom, STAHL binding beam; **Unterzugbalken** *m* ARCH & TRAGW bridging piece
**ununterbrochen** *adj* ARCH & TRAGW continuous
**unverbindlich** *adj* BAURECHT without obligation
**unverbrennbar** *adj* BAURECHT, INFR & ENTW, WERKSTOFF incombustible
**unverkleidet** *adj* ARCH & TRAGW exposed
**unverputzt** *adj* NATURSTEIN unplastered
**unverrottbar** *adj* HOLZ, OBERFLÄCHE, UMWELT, WERKSTOFF antirot, imputrescible, rotproof
**unverrückbar** *adj* ARCH & TRAGW immovable
**unverschieblich** *adj* ARCH & TRAGW, BETON, INFR & ENTW, STAHL nonsway; **~es Kehlbalkendach** *nt* ARCH & TRAGW, HOLZ collar roof; **~er Rahmen** *m* ARCH & TRAGW, BETON, INFR & ENTW, STAHL nonsway frame
**unverwechselbar**: **~e Bezeichnung** *f* ELEKTR unmistakable labeling (*AmE*), unmistakable labelling (*BrE*)
**unverwüstlich** *adj* WERKSTOFF resilient
**unvollständig** *adj* ARCH & TRAGW, BAURECHT *Unterlagen* incomplete
**unwesentlich**: **~e Außenbauten** *m pl* ARCH & TRAGW minor external buildings
**UpM** *abbr* (*Umdrehungen pro Minute*) ABWASSER, BAUMASCHIN, HEIZ & BELÜFT rpm (*revolutions per minute*)
**Urea-Formaldehyd** *nt* (*UF*) HOLZ, UMWELT urea formaldehyde (*UF*); **Urea-Formaldehydschaum** *m* DÄMMUNG urea formaldehyde foam
**Urheberrecht** *nt*: **~ an Plänen** ARCH & TRAGW, BAURECHT copyright of the plans
**Urinalbecken** *nt* ABWASSER urinal
**ursprünglich**: **~er Zustand** *m* INFR & ENTW *Baugelände* original state; **den ~en Zustand wiederherstellen** *phr* INFR & ENTW restore the original state
**UTD** *abbr* (*Untertagedeponie*) UMWELT subsurface repository, underground depot
**U-Verschluß** *m* ARCH & TRAGW running trap
**UV-Strahlung** *f* UMWELT UV radiation
**UW-Schweißen** *nt* STAHL underwater welding

# V

**Vakuum** *nt* ARCH & TRAGW, WERKSTOFF vacuum; **Vakuumbeton** *m* BETON vacuum concrete; **Vakuumentwässerung** *f* BETON, INFR & ENTW vacuum dewatering; **Vakuumfilter** *m* HEIZ & BELÜFT vacuum filter; **Vakuumhartlöten** *nt* ARCH & TRAGW vacuum brazing; **Vakuumlanze** *f* INFR & ENTW vacuum lance; **Vakuumpumpe** *f* HEIZ & BELÜFT vacuum pump

**variabel** *adj* ARCH & TRAGW, BAURECHT variable

**Variable** *f* ARCH & TRAGW, BAURECHT variable

**Variante** *f* ARCH & TRAGW, BAURECHT variant, *Änderung* modification

**Variation** *f* ARCH & TRAGW, BAURECHT variation

**VDE** *abbr* (*Verband Deutscher Elektrotechniker*) ELEKTR association of German electrotechnicians

**VDE-Verbandszeichen** *nt* ELEKTR *Prüfzeichen* VDE test mark

**VDI** *abbr* (*Verband Deutscher Ingenieure*) ARCH & TRAGW, HEIZ & BELÜFT association of German engineers

**Vegetation** *f* INFR & ENTW vegetation

**Ventil** *nt* ABWASSER valve, cock, HEIZ & BELÜFT valve

**Ventilation** *f* HEIZ & BELÜFT ventilation; **Ventilationsleistung** *f* HEIZ & BELÜFT ventilation rate

**Ventilator** *m* ARCH & TRAGW fan, BESCHLÄGE, HEIZ & BELÜFT fan, ventilating fan, ventilator; **Ventilatorantrieb** *m* HEIZ & BELÜFT fan drive; **Ventilatorheizung** *f* BESCHLÄGE fan heater, HEIZ & BELÜFT downflow heater, fan heater; **Ventilatorschalldämpfer** *m* DÄMMUNG, HEIZ & BELÜFT fan silencer, fan sound damper; **Ventilatorschutzkorb** *m* HEIZ & BELÜFT fan guard

**ventilieren** *vt* HEIZ & BELÜFT ventilate

***Ventil***: **Ventilschlüssel** *m* ABWASSER cock key

**Venturiwäscher** *m* UMWELT venturi scrubber

**Veranda** *f* ARCH & TRAGW porch, veranda

**veränderlich** *adj* ARCH & TRAGW, BAURECHT variable; **~es Druckpotential** *nt* INFR & ENTW variable head, WERKSTOFF falling head; **~e Größe** *f* ARCH & TRAGW, BAURECHT variable size

**verändern** *vt* ARCH & TRAGW change, convert, modify, BAURECHT alter

**verändert** *adj* BAURECHT altered

**Veränderung** *f* ARCH & TRAGW modification

**verankern** *vt* ARCH & TRAGW, STAHL anchor, grapple, tie

**Verankerung** *f* ARCH & TRAGW anchorage, staying, BETON anchorage, INFR & ENTW anchorage, staying, NATURSTEIN clamping, pinning, STAHL anchorage; **Verankerungseisen** *nt* INFR & ENTW, STAHL, WERKSTOFF anchor bar; **Verankerungskraft** *f* ARCH & TRAGW anchorage force; **Verankerungsseil** *nt* ARCH & TRAGW, BESCHLÄGE guy rope, stay rope

**veranschlagen** *vt* ARCH & TRAGW, BAURECHT, INFR & ENTW calculate, estimate, value

**verantwortlich** *adj* ARCH & TRAGW accountable, responsible, BAURECHT accountable, liable

**Verantwortlichkeit** *f* ARCH & TRAGW accountability, responsibility, BAURECHT liability

**Verantwortung** *f* ARCH & TRAGW accountability, responsibility, BAURECHT liability

**verarbeitbar** *adj* ARCH & TRAGW, BETON, WERKSTOFF workable

**Verarbeitbarkeit** *f* BETON, WERKSTOFF workability

**verarbeiten** *vt* BESCHLÄGE handle, WERKSTOFF manufacture, process, treat

**Verarbeitung** *f* ARCH & TRAGW workmanship, WERKSTOFF handling

**verarbeitungsfähig** *adj* BETON, OBERFLÄCHE, WERKSTOFF workable

***Verarbeitung***: **Verarbeitungsrichtlinien** *f pl* OBERFLÄCHE, WERKSTOFF application specification, processing guidelines, processing specifications

**Verband** *m* ARCH & TRAGW bracing, BAURECHT association, NATURSTEIN bond; **~ Deutscher Elektrotechniker** (*VDE*) ELEKTR association of German electrotechnicians; **~ Deutscher Ingenieure** (*VDI*) ARCH & TRAGW, HEIZ & BELÜFT association of German engineers

**Verbau** *m* INFR & ENTW pit boards

**verbessern** *vt* ARCH & TRAGW improve

**Verbesserung** *f* ARCH & TRAGW improvement

**verbesserungsfähig** *adj* ARCH & TRAGW improvable

**verbinden** *vt* ARCH & TRAGW connect, link, join, BETON *zu fester Masse* concrete, NATURSTEIN bond

**Verbinden** *nt* ARCH & TRAGW connection, link, INFR & ENTW, NATURSTEIN, OBERFLÄCHE, WERKSTOFF bonding

**Verbindung** *f* ABWASSER connection, ARCH & TRAGW joint, junction, ELEKTR connection, HOLZ joint, junction, INFR & ENTW *Gemisch* compound; **Verbindungsbau** *m* ARCH & TRAGW connecting building, connecting structure; **Verbindungsbolzen** *m* BESCHLÄGE connection bolt; **Verbindungseinheit** *f* HOLZ, INFR & ENTW, STAHL connector unit; **Verbindungsgang** *m* ARCH & TRAGW connecting passage; **Verbindungslasche** *f* BESCHLÄGE backplate; **Verbindungsleitung** *f* ABWASSER connecting line, connection line; **Verbindungsmauern** *f pl* NATURSTEIN junction walls; **Verbindungsmittel** *nt* BETON, HOLZ, STAHL, WERKSTOFF combing agent, connecting device; **Verbindungsschlupf** *m* HOLZ joint slip; **Verbindungsschraube** *f* BESCHLÄGE, STAHL, WERKSTOFF connecting bolt, connecting screw; **Verbindungsstange** *f* STAHL, WERKSTOFF connecting bar; **Verbindungsstück** *nt* ABWASSER connection piece, ARCH & TRAGW tie, BETON waler, waling, HEIZ & BELÜFT connection piece, INFR & ENTW ranger, waler, waling, WERKSTOFF ranger; **Verbindungsstutzen** *m* HEIZ & BELÜFT connecting piece

**verblassen** *vi* OBERFLÄCHE, WERKSTOFF fade

**verblatten** *vt* HOLZ splice

**Verblattung** *f* HOLZ scarf joint

**verbleichen** *vi* OBERFLÄCHE, WERKSTOFF fade away

**Verblend-** *in cpds* NATURSTEIN cladding, facing

**verblenden** *vt* ARCH & TRAGW face, mask, veneer, line, BAUMASCHIN *Fenster* screen, NATURSTEIN veneer, *mit Ziegeln* brick

**Verblender** *m* ARCH & TRAGW, INFR & ENTW, NATURSTEIN facing block
**verblendet** *adj* ARCH & TRAGW, INFR & ENTW lined, NATURSTEIN faced
***Verblend-***: **Verblendmauer** *f* NATURSTEIN face wall; **Verblendmauerwerk** *nt* BETON, NATURSTEIN faced brickwork, facework, facing work; **Verblendplatte** *f* WERKSTOFF cladding panel; **Verblendstein** *m* INFR & ENTW, NATURSTEIN, WERKSTOFF facing block, fair-faced brick
**Verblendung** *f* ARCH & TRAGW, INFR & ENTW, NATURSTEIN, STAHL facework, facing, lining
**verbolzen** *vt* ARCH & TRAGW, BESCHLÄGE, HOLZ bolt
**verbolzt** *adj* ARCH & TRAGW, BESCHLÄGE, HOLZ bolted
**Verbolzung** *f* ARCH & TRAGW, BESCHLÄGE, HOLZ bolting
**Verbrauch** *m* HEIZ & BELÜFT, INFR & ENTW, WERKSTOFF consumption; **Verbrauchsberechnung** *f* ARCH & TRAGW, HEIZ & BELÜFT calculation of consumption; **Verbrauchskurve** *f* HEIZ & BELÜFT, INFR & ENTW, WERKSTOFF consumption curve; **Verbrauchssteuer** *f* BAURECHT excise tax
**Verbreitung** *f* ARCH & TRAGW, OBERFLÄCHE distribution, *Dispersion* dispersion, spreading
**Verbrennung** *f* UMWELT incineration; **Verbrennungsanlage** *f* INFR & ENTW incinerator plant; **Verbrennungsgas** *nt* UMWELT exhaust gas; **Verbrennungsrückstand** *m* UMWELT incineration residue, incineration ash; **Verbrennungsschlacke** *f* UMWELT incineration slag
**Verbretterung** *f* HOLZ furring
**Verbund** *m* INFR & ENTW, NATURSTEIN, OBERFLÄCHE, WERKSTOFF bond, bonding; **im ~ gemauertes Mauerwerk** *phr* NATURSTEIN bonded masonry; **Verbundanker** *m* STAHL shear connector; **Verbundbalken** *m* WERKSTOFF composite beam; **Verbundbauweise** *f* ARCH & TRAGW composite method of construction; **Verbundbetrieb** *m* BAURECHT, ELEKTR combined grid operation; **Verbunddecke** *f* ARCH & TRAGW, BETON, INFR & ENTW, STAHL, WERKSTOFF composite floor, composite flooring
**verbunden** *adj* ARCH & TRAGW jointed, HOLZ, NATURSTEIN bonded; **~es Mauerwerk** *nt* NATURSTEIN bonded masonry
***Verbund***: **Verbundfenster** *nt* ARCH & TRAGW composite window, BESCHLÄGE sash window, WERKSTOFF composite window; **Verbundfläche** *f* WERKSTOFF bonding area; **Verbundfundament** *nt* ARCH & TRAGW combined footing; **Verbundglas** *nt* WERKSTOFF laminated glass, multilayer glass; **Verbundkonstruktion** *f* ARCH & TRAGW, BETON, INFR & ENTW, STAHL composite structure; **Verbundlänge** *f* BETON *Spannbeton* transmission length; **Verbundplatte** *f* WERKSTOFF composite board, composite slab, sandwich board, sandwich panel; **Verbundprofil** *nt* WERKSTOFF composite profile; **Verbundquerschnitt** *m* ARCH & TRAGW, BETON, INFR & ENTW, WERKSTOFF composite cross section; **Verbundsäule** *f* ARCH & TRAGW combination column, *Formstahl, Beton* composite column; **Verbundschicht** *f* BETON knitting layer; **Verbundschornstein** *m* ARCH & TRAGW compound chimney; **Verbundsicherheitsglas** *nt* (*VS-Glas*) BAURECHT, WERKSTOFF laminated safety glass; **Verbundstein** *m* NATURSTEIN, WERKSTOFF interlocking paver; **Verbundstütze** *f* ARCH & TRAGW, BETON, INFR & ENTW, STAHL, WERKSTOFF composite column; **Verbundträger** *m* ARCH & TRAGW, BETON, INFR & ENTW, STAHL, WERKSTOFF composite beam, composite girder, compound girder; **Verbundtragwerk** *nt* ARCH & TRAGW composite supporting structure; **Verbundwerkstoff** *m* WERKSTOFF composite material; **Verbundwerkstoff** *m* **mit Wabenkern** WERKSTOFF honeycomb sandwich material; **Verbundwirkung** *f* ARCH & TRAGW composite action
**verchromt** *adj* OBERFLÄCHE, STAHL, WERKSTOFF chrome-plated, chromium-plated
**Verchromung** *f* OBERFLÄCHE, STAHL, WERKSTOFF chrome plating, chromium plating
**Verdämmen** *nt*: **~ von Bohrlöchern** BAUMASCHIN, INFR & ENTW tamping
**verdampfen** *vti* ARCH & TRAGW, HEIZ & BELÜFT, WERKSTOFF evaporate
**Verdampfer** *m* HEIZ & BELÜFT evaporator
**Verdampfung** *f* ARCH & TRAGW, HEIZ & BELÜFT, WERKSTOFF evaporation (*E*)
**verdeckt** *adj* ARCH & TRAGW concealed, covered, hidden; **~es Lichtbogenschweißen** *nt* STAHL submerged arc welding
**verderben** *vt* ARCH & TRAGW, INFR & ENTW spoil
**verdichten** *vt* INFR & ENTW compact, *Straße* pack, WERKSTOFF concentrate
**Verdichter** *m* BAUMASCHIN compactor, vibrator, HEIZ & BELÜFT condenser
**Verdichterventil** *nt* BAUMASCHIN compressor valve
**verdichtet** *adj* WERKSTOFF compacted; **~e Erde** *f* ARCH & TRAGW, INFR & ENTW rammed earth
**Verdichtung** *f* HEIZ & BELÜFT, INFR & ENTW, UMWELT compaction, compression; **Verdichtungsanlage** *f* UMWELT compacting plant, landfill compactor, packer unit; **Verdichtungsdeponie** *f* UMWELT tipping with compaction (*BrE*)
**verdichtungsfähig** *adj* INFR & ENTW compactible, WERKSTOFF compressible, *Boden, Sand* compactible
***Verdichtung***: **Verdichtungsfaktor** *m* INFR & ENTW, WERKSTOFF compaction index; **Verdichtungsgerät** *nt* BAUMASCHIN compaction machine; **Verdichtungspfahl** *m* INFR & ENTW compaction pile; **Verdichtungsversuch** *m* INFR & ENTW compaction test; **Verdichtungswalze** *f* BAUMASCHIN compaction roller
**Verdingungsordnung** *f* BAURECHT contracting regulations; **~ für Bauleistungen** BAURECHT contracting regulations for award of public work contracts; **~ für die Vergabe von Bauleistungen** (*VOB*) BAURECHT German regulations for contracts and execution of construction works
**Verdöbelung** *f* HOLZ peg
**verdrahten** *vt* ELEKTR wire
**Verdrahtung** *f* ELEKTR wiring
**verdrehen** *vt* ARCH & TRAGW twist, BESCHLÄGE *Lampe* adjust
**verdrillen** *vt* ARCH & TRAGW twist
**verdübeln** *vt* BESCHLÄGE dowel, HOLZ dowel, key, INFR & ENTW dowel
**verdübelt** *adj* HOLZ, NATURSTEIN dowelled
**Verdübelung** *f* ABWASSER plugging
**Verdunkelung** *f* ARCH & TRAGW blackout
**verdünnen** *vt* OBERFLÄCHE dilute
**Verdünnen** *nt* ARCH & TRAGW reducing
**Verdünner** *m* OBERFLÄCHE thinner, WERKSTOFF solvent, thinner
**Verdünnung** *f* OBERFLÄCHE dilution

**verdunsten** *vi* ARCH & TRAGW, HEIZ & BELÜFT, INFR & ENTW, UMWELT, WERKSTOFF evaporate
**Verdunstung** *f* ARCH & TRAGW, HEIZ & BELÜFT, INFR & ENTW, UMWELT, WERKSTOFF evaporation (*E*); **Verdunstungsverlust** *m* ARCH & TRAGW, HEIZ & BELÜFT, INFR & ENTW, WERKSTOFF evaporation loss, loss by evaporation
**Veredelung** *f* STAHL *Leichtmetall* ageing
**Vereinbarung** *f* BAURECHT agreement; **Vereinbarungsform** *f* BAURECHT form of agreement
**Vereinfachung** *f* ARCH & TRAGW, BETON, INFR & ENTW, STAHL, WERKSTOFF idealization, simplification
**Vereinigung** *f* BAURECHT association
**verfahrbar** *adj* BAUMASCHIN moveable
**Verfahren** *nt* ARCH & TRAGW method, procedure, process, technique, BAURECHT process, INFR & ENTW process, technique
**verfallen 1.** *adj* ARCH & TRAGW dilapidated; **2.** *vi* BAURECHT *finanzielle Mittel* expire
**verfalzt** *adj* STAHL interlocked
**Verfalzung** *f* STAHL interlocking joint
**verfaulen** *vi* BETON, HOLZ, UMWELT rot
**verfestigen** *vt* ARCH & TRAGW, HOLZ nail, INFR & ENTW compact, *Boden* consolidate, UMWELT solidify
**verfestigt**: **~es Material** *nt* WERKSTOFF solidified material
**Verfestigung** *f* UMWELT solidification; **~ von Abfällen** UMWELT solidifying of waste; **Verfestigungsmittel** *nt* UMWELT solidifying agent; **Verfestigungsprodukt** *nt* UMWELT solidified waste, solidified product, solidified material; **Verfestigungsverfahren** *nt* UMWELT solidification technique
**Verformung** *f* ARCH & TRAGW, INFR & ENTW, WERKSTOFF deformation; **Verformungsebene** *f* ARCH & TRAGW, WERKSTOFF plane deformation; **Verformungsgeschwindigkeit** *f* WERKSTOFF rate of deformation; **Verformungsmessung** *f* INFR & ENTW, WERKSTOFF strain measurement; **Verformungsmodul** *m* ARCH & TRAGW, INFR & ENTW, WERKSTOFF modulus of deformation
**verfügbar** *adj* ARCH & TRAGW, INFR & ENTW, WERKSTOFF available
**Verfügbarkeit** *f* ARCH & TRAGW, INFR & ENTW, WERKSTOFF availability
**verfugen** *vt* NATURSTEIN join, point
**Verfugen** *nt* NATURSTEIN jointing, pointing
**Verfügung** *f* BAURECHT regulation
**Verfüllbeton** *m* BETON packing
**Verfüllen** *nt* ARCH & TRAGW, INFR & ENTW refilling
**verfüllt**: **~e Deponie** *f* UMWELT complete fill
**Verfüllung** *f* UMWELT filling, backfilling
**Vergabe** *f* BAURECHT *Vertrag* award; **~ von Aufträgen** BAURECHT award of contracts
**Vergasermotor** *m* BAUMASCHIN, UMWELT gasoline engine (*AmE*), petrol engine (*BrE*)
**vergeben** *vt* BAURECHT *Auftrag* award
**vergießen** *vt* BETON pour, NATURSTEIN *mit Mörtel* grout, OBERFLÄCHE seal, WERKSTOFF mold (*AmE*), mould (*BrE*)
**vergipsen** *vt* NATURSTEIN plaster
**vergittern** *vt* ARCH & TRAGW *Fenster* screen
**Vergitterung** *f* STAHL lacing
**verglasen** *vt* ARCH & TRAGW glaze
**Verglasen** *nt* ARCH & TRAGW glazing
**verglast** *adj* ARCH & TRAGW glazed; **~e Laterne** *f* ARCH & TRAGW glazed lantern
**Verglasung** *f* ARCH & TRAGW glazing
**Vergleichskostenanalyse** *f* BAURECHT comparative cost analysis
**Vergraben** *nt* UMWELT *von Müll* land burial
**vergraben**: **~er Schatz** *m* BAURECHT buried treasure
**Vergrößerungsstück** *nt* ABWASSER, HEIZ & BELÜFT increaser
**Vergußmasse** *f* NATURSTEIN grouting compound, WERKSTOFF compound, joint cement
**Vergußmaterial** *nt* NATURSTEIN grouting compound
**vergüten** *vt* STAHL temper
**vergütet**: **~es Glas** *nt* INFR & ENTW, WERKSTOFF annealed glass
**Vergütung** *f* OBERFLÄCHE, STAHL, WERKSTOFF ageing
**verhalten**: **sich ~** *v refl* ARCH & TRAGW, INFR & ENTW, WERKSTOFF behave
**Verhalten** *nt* ARCH & TRAGW, INFR & ENTW, WERKSTOFF behavior (*AmE*), behaviour (*BrE*)
**Verhältnis** *nt* ARCH & TRAGW, INFR & ENTW, WERKSTOFF ratio; **~ zwischen Förderhöhe und Widerstand** UMWELT lift-to-drag ratio
**verhärten 1.** *vt* WERKSTOFF harden, set; **2. sich verhärten** *v refl* WERKSTOFF set
**Verhinderung** *f* BAURECHT, UMWELT prevention
**Verhütung** *f*: **~ der Wasserverschmutzung** UMWELT prevention of water pollution
**verjüngen** *vt* ARCH & TRAGW reduce
**verjüngt** *adj* ARCH & TRAGW *Holzbau* splayed, HOLZ beveled (*AmE*), bevelled (*BrE*)
**Verkabelung** *f* ELEKTR cabling, wiring, HEIZ & BELÜFT cabling
**verkalkt** *adj* ABWASSER, HEIZ & BELÜFT scaled
**Verkalkung** *f* WERKSTOFF furring
**verkämmen** *vt* HOLZ cog
**verkanten** *vt* WERKSTOFF bend out of line
**Verkehr** *m* ARCH & TRAGW, BAURECHT, INFR & ENTW traffic; **Verkehrsanlagen** *f pl* INFR & ENTW traffic facilities; **Verkehrsknotenpunkt** *m* INFR & ENTW traffic junction; **Verkehrslast** *f* ARCH & TRAGW, INFR & ENTW live load, rolling load, superimposed load, traffic load, traveling load (*AmE*), travelling load (*BrE*); **Verkehrsleuchtnagel** *m* INFR & ENTW *auf der Straße* reflecting stud; **Verkehrsschild** *nt* BAURECHT, INFR & ENTW traffic sign; **Verkehrssicherheit** *f* BAURECHT traffic safety; **Verkehrssicherheitseinrichtungen** *f pl* BAURECHT, INFR & ENTW traffic safety facilities; **Verkehrsweg** *m* INFR & ENTW traffic route
**verkeilen** *vt* NATURSTEIN key
**Verkeilung** *f* BAUMASCHIN, NATURSTEIN keying
**verkitten** *vt* BETON cement, OBERFLÄCHE seal
**verklammern** *vt* ARCH & TRAGW grapple, joggle, *Holz, Stein* cramp
**verkleben** *vt* BAUMASCHIN, BESCHLÄGE tape, BETON cement, OBERFLÄCHE glue
**Verkleben** *nt* BETON, INFR & ENTW, NATURSTEIN, OBERFLÄCHE, WERKSTOFF bonding, cementing
**verklebt**: **~e Stöße** *m pl* HOLZ glued joints
**verkleiden** *vt* ARCH & TRAGW face, jacket, line, case, BESCHLÄGE case, HEIZ & BELÜFT jacket, HOLZ plank, INFR & ENTW line, revet, OBERFLÄCHE *Material* surface
**verkleidet** *adj* ARCH & TRAGW cased, lined, BESCHLÄGE cased, BETON casing, HOLZ paneled (*AmE*), panelled (*BrE*), INFR & ENTW lined, WERKSTOFF faced
**Verkleidung** *f* ARCH & TRAGW lining, BESCHLÄGE,

BETON casing, HOLZ paneling (*AmE*), panelling (*BrE*), INFR & ENTW *Mauerwerk* revetment, lining, NATURSTEIN facework, facing, *Mauerwerk* revetment, STAHL casing; **Verkleidungsmaterial** *nt* WERKSTOFF sheeting; **Verkleidungsplatte** *f* WERKSTOFF cladding panel, facing slab; **Verkleidungstafel** *f* WERKSTOFF cladding panel, facing slab
**verkleinert**: **~er Maßstab** *m* ARCH & TRAGW, INFR & ENTW tapering scale
**Verkleinerungsmaßstab** *m* ARCH & TRAGW reduction scale
**verkupfern** *vt* OBERFLÄCHE, STAHL copper
**verkupfert** *adj* OBERFLÄCHE, STAHL coppered; **~er Draht** *m* STAHL coppered wire
**Verkupferung** *f* OBERFLÄCHE, STAHL copper plating
**verkürzt**: **~er Bogen** *m* ARCH & TRAGW skeen arch, skene arch
**Verladerampe** *f* ARCH & TRAGW, INFR & ENTW loading bay
**Verladesilo** *m*: **~ für bituminöses Mischgut** BAUMASCHIN charging silo for mix
**Verlagerung** *f* INFR & ENTW displacement
**Verlandung** *f* INFR & ENTW aggradation
**verlängern** *vt* ARCH & TRAGW extend, prolongate, stretch, WERKSTOFF stretch
**verlängert**: **~es Gestänge** *nt* BESCHLÄGE *Fensteröffner* extended stays
**Verlängerungsstück** *nt* WERKSTOFF extension piece
**verlaschen** *vt* HOLZ fishplate, fish
**Verlaschen** *nt* HOLZ fishing
**Verlaschung** *f* HOLZ fished joint, fishing
**Verlattung** *f* HOLZ lathing
**verlegen** *vt* ELEKTR *Kabel* install, lay, INFR & ENTW pave
**Verlegung** *f* ARCH & TRAGW, ELEKTR, INFR & ENTW installation, laying
**verleimen** *vt* HOLZ bond, glue, OBERFLÄCHE bond
**Verleimen** *nt* INFR & ENTW, NATURSTEIN, OBERFLÄCHE, WERKSTOFF bonding
**verleimt** *adj* HOLZ, NATURSTEIN bonded
**verloren**: **~e Betonschalung** *f* BETON permanent concrete shuttering; **~e Schalung** *f* BETON, INFR & ENTW permanent formwork, permanent shuttering
**Verlust** *m* INFR & ENTW, WERKSTOFF loss; **~ oder Kosten** *phr* BAURECHT loss or expense
**Vermarkungspunkt** *m*: **~ der Landesvermessung** ARCH & TRAGW, INFR & ENTW ordnance bench mark
**vermascht** *adj* WERKSTOFF meshed
**vermauern** *vt* NATURSTEIN block up, wall up, brick up, wall in
**Vermeidung** *f*: **~ von Lärmbelästigung** UMWELT prevention of noise pollution
**vermessen** *vt* ARCH & TRAGW, INFR & ENTW survey
**Vermessen** *nt* ARCH & TRAGW, INFR & ENTW surveying
**Vermesser** *m* ARCH & TRAGW, INFR & ENTW surveyor
**Vermessung** *f* ARCH & TRAGW measurement, survey, surveying, topographical survey, INFR & ENTW mensuration, survey; **Vermessungsgrundlinie** *f* INFR & ENTW transit line; **Vermessungsinstrument** *nt* BAUMASCHIN Y-level; **Vermessungsstab** *m* ARCH & TRAGW station pole; **Vermessungsstange** *f* INFR & ENTW stadia; **Vermessungswesen** *nt* INFR & ENTW surveying
**Vermiculite** *m* WERKSTOFF vermiculite
**vermischen** *vt* BETON, NATURSTEIN, WERKSTOFF blend
**Vermischung** *f*: **~ von Konservierungsmitteln** UMWELT, WERKSTOFF mixing preservative products
**Vermittler** *m* BAURECHT *von Berufs wegen* agent
**Vermittlungsstelle** *f* BAURECHT agency
**Vermodern** *nt* HOLZ rotting
**vermörteln** *vt* NATURSTEIN fix in mortar
**Vermörteln** *nt* NATURSTEIN grouting
**Vermörtelung** *f* NATURSTEIN *Dachziegel* mortar jointing
**vernageln** *vt* ARCH & TRAGW, HOLZ nail
**vernetzt** *adj* INFR & ENTW, WERKSTOFF crosslinked
**vernickelt** *adj* OBERFLÄCHE nickel-plated
**vernünftig**: **~e Sorgfalt** *f* BAURECHT reasonable care
**Verordnung** *f*: **~ über elektrische Anlagen in explosionsgefährdeten Standorten** BAURECHT regulations for electrical plants installed in hazardous locations
**Verpackung** *f* UMWELT packaging; **Verpackungsabfall** *m* UMWELT packaging waste; **Verpackungsmaterial** *nt* UMWELT packaging material; **Verpackungsmüll** *m* UMWELT packaging waste
**Verpflichtungsbedingungen** *f pl* BAURECHT conditions of engagement
**verpressen** *vt* BETON, INFR & ENTW inject, NATURSTEIN inject, *mit Mörtel* pressure-grout
**Verputz** *m* NATURSTEIN, WERKSTOFF plaster
**verputzen** *vt* NATURSTEIN parget, plaster, render
**Verputzen** *nt* NATURSTEIN plaster work, WERKSTOFF plastering, pargetting, rendering
**verriegeln** *vt* BESCHLÄGE bolt
**Verriegelung** *f* ARCH & TRAGW keeper; **Verriegelungsbolzen** *m* BESCHLÄGE locking bolt; **Verriegelungseinrichtung** *f* BESCHLÄGE lock staple
**verrippt** *adj* ARCH & TRAGW, BETON, HEIZ & BELÜFT, STAHL, WERKSTOFF ribbed
**verrödeln** *vt* BETON tiewire
**Verrohrung** *f* ABWASSER piping, BESCHLÄGE piping, tubing; **Verrohrungsseil** *nt* BAUMASCHIN casing line
**verrottbar** *adj* UMWELT, WERKSTOFF putrescible; **nicht ~** *adj* HOLZ, OBERFLÄCHE, UMWELT, WERKSTOFF antirot, imputrescible, rotproof; **~er Stoff** *m* UMWELT putrescible matter
**verrotten** *vi* BETON, HOLZ, UMWELT rot
**verrottet**: **~es Holz** *nt* HOLZ, WERKSTOFF decayed timber
**verrücken** *vt* ARCH & TRAGW *Gebäude* dislodge
**Versagenswinkel** *m* INFR & ENTW, WERKSTOFF angle of failure
**Versammlungsraum** *m* ARCH & TRAGW, INFR & ENTW assembly room
**Versatznietung** *f* STAHL zigzag riveting
**Versäuerung** *f* UMWELT *des Bodens* acidification
**Versäumnisse**: **~ des benannten Subunternehmers** *phr* BAURECHT nominated sub-contractor defaults
**verschalen** *vt* ARCH & TRAGW face, line, BETON board, clad, HOLZ board, clad, plank, *mit Leisten* batten, INFR & ENTW line
**verschalt** *adj* ARCH & TRAGW lined, BETON, HOLZ cladded, INFR & ENTW lined
**Verschalung** *f* ARCH & TRAGW sheathing, lining, BESCHLÄGE casing, BETON casing, formwork, HOLZ cladding, timbering, planking, INFR & ENTW lining, *mit Bohlen* planking
**verschieblich**: **~es Kehlbalkendach** *nt* ARCH & TRAGW, HOLZ collar roof

**Verschiebung** *f* ARCH & TRAGW *räumlich* offset, INFR & ENTW displacement, translation
**verschiefern** *vt* ARCH & TRAGW split into thin sheets
**Verschlag** *m* HOLZ boarding, crate
**Verschlammung** *f* UMWELT sludge accumulation
**verschleifen** *vt* BAUMASCHIN grind
**Verschleiß** *m* INFR & ENTW wear
**verschleißfest** *adj* INFR & ENTW hardwearing, WERKSTOFF wear-resistant
**Verschleißfestigkeit** *f* BETON, WERKSTOFF abrasion resistance
***Verschleiß***: **Verschleißschicht** *f* ARCH & TRAGW wearing course, INFR & ENTW wearing surface, surface dressing, topping, *Straße* finishing layer; **Verschleißwert** *m* ARCH & TRAGW wear rate
**verschließen** *vt* BESCHLÄGE lock, OBERFLÄCHE seal
**Verschließen** *nt* OBERFLÄCHE sealing
**Verschluß** *m* BESCHLÄGE lock; **Verschlußdecke** *f* OBERFLÄCHE *Straßenbau* seal coat; **Verschlußkappe** *f* ARCH & TRAGW cap; **Verschlußloch** *nt* ABWASSER plughole; **Verschlußschraube** *f* BESCHLÄGE screw plug
**verschmutzt** *adj* ABWASSER, BAURECHT, UMWELT polluted; **~e Luft** *f* BAURECHT, UMWELT polluted air; **~es Regenwasser** *nt* UMWELT polluted rainwater; **~es Wasser** *nt* ABWASSER, BAURECHT, UMWELT polluted water
**Verschmutzung** *f* BAURECHT pollution; **~ durch Feststoffe** UMWELT material pollution; **Verschmutzungsgefahr** *f* BAURECHT, INFR & ENTW danger of contamination; **Verschmutzungsgrad** *m* UMWELT degree of pollution; **Verschmutzungsgrad** *m* **des Wassers** UMWELT pollutional index; **Verschmutzungsüberwachung** *f* BAURECHT, INFR & ENTW pollution control
**verschneiden** *vt* BETON, WERKSTOFF mix
**Verschnitt** *m* HOLZ, STAHL offcut
**verschönern** *vt* ARCH & TRAGW improve
**Verschönerung** *f* ARCH & TRAGW improvement
**verschrauben** *vt* BAUMASCHIN screw
**verschraubt** *adj* ARCH & TRAGW, BESCHLÄGE, HOLZ bolted
**Verschraubung** *f* ARCH & TRAGW bolting, union, BESCHLÄGE, HOLZ bolting, NATURSTEIN, STAHL, WERKSTOFF screwed connection
**Verschwertung** *f* ARCH & TRAGW cross stays, cross stud, X-bracing, HOLZ, STAHL cross stud
**versenken** *vt* STAHL *Nieten* dimple
**Versenken** *nt* ABWASSER plunging
**versenkt** *adj* ARCH & TRAGW recessed; **~es Schloß** *nt* BESCHLÄGE dormant lock
**versetzbar**: **~e Wand** *f* BESCHLÄGE demountable partition
**versetzen** *vt* ARCH & TRAGW relocate
**versetzt** *adj* ARCH & TRAGW, BETON staggered; **~e Fuge** *f* ARCH & TRAGW staggered installation, broken joint; **~e Geschosse** *nt pl* ARCH & TRAGW split levels, staggered floors, staggered storeys (*BrE*), staggered stories (*AmE*); **~e Stockwerke** *nt pl* ARCH & TRAGW split levels, staggered floors, staggered storeys (*BrE*), staggered stories (*AmE*)
**Versetzung** *f* INFR & ENTW displacement
**verseucht**: **~er Boden** *m* BAURECHT, INFR & ENTW, UMWELT contaminated soil; **~es Land** *nt* BAURECHT, INFR & ENTW, UMWELT contaminated land
**Verseuchung** *f* UMWELT contamination; **Verseuchungsgefahr** *f* BAURECHT, INFR & ENTW, UMWELT contamination hazard
**Versicherung** *f* BAURECHT insurance
**Versickern** *nt* ABWASSER seepage, percolation, INFR & ENTW seepage
**Versickerung** *f* ABWASSER, INFR & ENTW, UMWELT *Abwasserreinigung* infiltration; **Versickerungsgraben** *m* ABWASSER, INFR & ENTW seepage trench; **Versickerungsschacht** *m* INFR & ENTW absorbing well
**versiegeln** *vt* OBERFLÄCHE seal
**versiegelt** *adj* BAURECHT *Sicherheit*, OBERFLÄCHE sealed
**Versiegelung** *f* OBERFLÄCHE sealing; **Versiegelungsmittel** *nt* ABWASSER, OBERFLÄCHE sealant
**Versiegler** *m* OBERFLÄCHE sealer
**Versorgung** *f* ABWASSER, ELEKTR, HEIZ & BELÜFT supply; **Versorgungseinrichtung** *f* ABWASSER, ARCH & TRAGW, ELEKTR, HEIZ & BELÜFT, INFR & ENTW utility; **Versorgungskanal** *m* ARCH & TRAGW, INFR & ENTW service tunnel; **Versorgungsleitung** *f* ABWASSER supply line, ARCH & TRAGW utility line, ELEKTR feeder, supply line, HEIZ & BELÜFT, INFR & ENTW supply line; **Versorgungsnetz** *nt* INFR & ENTW supply system
**verspachteln** *vt* OBERFLÄCHE fill, seal
**Verspachteln** *nt* ARCH & TRAGW, OBERFLÄCHE filling, sealing, trowel application
**Verspannung** *f* ARCH & TRAGW bracing, stay
**versperren** *vt* INFR & ENTW barricade
**verstärken** *vt* ARCH & TRAGW strengthen, HOLZ fish, WERKSTOFF reinforce
**Verstärken** *nt* ARCH & TRAGW reinforcing, strengthening
**verstärkt**: **~er Zug** *m* ARCH & TRAGW forced draft (*AmE*), forced draught (*BrE*)
**Verstärkung** *f* ARCH & TRAGW reinforcement, bracing, stiffening, strengthening, strutting, WERKSTOFF reinforcement; **Verstärkungsband** *nt* NATURSTEIN *Putz* reinforcing tape; **Verstärkungsblech** *nt* ARCH & TRAGW, HOLZ, WERKSTOFF reinforcing plate, reinforcing sheet; **Verstärkungsbogen** *m* BAURECHT safety arch; **Verstärkungspfeiler** *m* ARCH & TRAGW, NATURSTEIN counterfort; **Verstärkungsrippe** *f* ARCH & TRAGW stiffening rib
**versteifen** *vt* ARCH & TRAGW brace, stiffen, strut, *mittels Lehrbogen* cradle
**Versteifen** *nt* ARCH & TRAGW strutting
**versteift**: **~er Rahmen** *m* ARCH & TRAGW trussed frame
**Versteifung** *f* ARCH & TRAGW bracing, cradling, staying, stiffening, strutting; **Versteifungsrippe** *f* ARCH & TRAGW stiffening rib
**Versteinerung** *f*: **~ von Schlämmen** UMWELT sludge petrification
**Versteinungsmittel** *nt* INFR & ENTW, NATURSTEIN *Bodenverbesserung* grouting compound
**verstellbar** *adj* BAUMASCHIN, BESCHLÄGE, HEIZ & BELÜFT adjustable; **~e Regale** *nt pl* BESCHLÄGE adjustable shelving
**verstellen** *vt* ARCH & TRAGW, BAUMASCHIN adjust
**verstemmen** *vt* NATURSTEIN caulk
**verstiften** *vt* BESCHLÄGE, HOLZ dowel
**verstopfen** *vt* ABWASSER, ARCH & TRAGW, INFR & ENTW plug
**verstopft** *adj* ABWASSER clogged, clogged up
**Verstopfung** *f* ABWASSER occlusion, *Rohr* choking, *von Rohren* obstruction

**Verstreben** *nt* ARCH & TRAGW strutting
**Verstrebungsbalken** *m* ARCH & TRAGW, BAUMASCHIN straining beam, straining piece
**verstreichen** *vt* NATURSTEIN trowel
**Versuch** *m* INFR & ENTW test; **Versuchsanlage** *f* WERKSTOFF test rig; **Versuchsbaggern** *nt* INFR & ENTW trial dredging; **Versuchsbau** *m* ARCH & TRAGW, INFR & ENTW experimental building; **Versuchsbohrung** *f* INFR & ENTW, WERKSTOFF exploratory boring, trial boring; **Versuchsdurchführung** *f* ARCH & TRAGW, INFR & ENTW, WERKSTOFF experimentation; **Versuchsergebnis** *nt* INFR & ENTW test result; **Versuchspfahl** *m* INFR & ENTW test pile; **Versuchsstück** *nt* INFR & ENTW test piece; **Versuchsverdichtung** *f* INFR & ENTW trial compaction
**verteilen** *vt* ARCH & TRAGW spread, INFR & ENTW distribute
**Verteiler** *m* ABWASSER, HEIZ & BELÜFT manifold; **Verteilerbewehrung** *f* BETON temperature reinforcement; **Verteilerkasten** *m* ELEKTR junction box, distribution box; **Verteilerstab** *m* BETON distance between rafters, distribution rod, INFR & ENTW distribution bar; **Verteilerstäbe** *m pl* BETON secondary reinforcement
**Verteilung** *f* WERKSTOFF distribution; **Verteilungsbewehrung** *f* BETON temperature reinforcement; **Verteilungszahl** *f* INFR & ENTW distribution factor
**vertiefen** *vt* ARCH & TRAGW recess, hollow
**Vertiefen** *nt* ARCH & TRAGW recessing
**vertieft**: **~ anbringen** *vt* ARCH & TRAGW recess
**Vertiefung** *f* ARCH & TRAGW hollow, NATURSTEIN *in einem Ziegel* frog
**vertikal**: **~es Abscheren** *nt* ARCH & TRAGW, BETON, STAHL, WERKSTOFF vertical shear; **~e Drainage** *f* INFR & ENTW vertical drainage; **~e Durchlässigkeit** *f* WERKSTOFF vertical permeability
**Vertikal**: **Vertikalanordnung** *f* ARCH & TRAGW vertical alignment; **Vertikalebene** *f* INFR & ENTW vertical plane; **Vertikallast** *f* ARCH & TRAGW, HOLZ vertical load; **Vertikalschub** *m* INFR & ENTW shear; **Vertikalstab** *m* ARCH & TRAGW upright, vertical bar, vertical member; **Vertikalverband** *m* ARCH & TRAGW vertical bracing
**Vertrag** *m* BAURECHT agreement, contract; **einen ~ abschließen** *phr* BAURECHT conclude a contract
**verträglich** *adj* WERKSTOFF compatible
**Verträglichkeit** *f* WERKSTOFF compatibility
***Vertrag***: **Vertragsabschluß** *m* BAURECHT conclusion of contract; **Vertragsabsichtserklärung** *f* BAURECHT letter of intent; **Vertragsauflösung** *f* BAURECHT contract termination; **Vertragsbedingungen** *f pl* BAURECHT conditions of contract; **Vertragsbeendigung** *f* BAURECHT termination of the contract; **Vertragsfristen** *f pl* BAURECHT contract deadline; **Vertragsgesetz** *nt* BAURECHT law of the contract; **Vertragsgrundbedingungen** *f pl* BAURECHT standard terms of contract
**vertragsmäßig** *adj* BAURECHT according to contract
***Vertrag***: **Vertragsstrafe** *f* BAURECHT contractual penalty, penalty contract, stipulated penalty; **Vertragsstrafe** *f* **bei Terminüberschreitung** BAURECHT contractual time penalty; **Vertragsunterzeichnung** *f* BAURECHT signing of a contract; **Vertragsverwaltung** *f* BAURECHT contract administration; **Vertragswert** *m* BAURECHT contract value; **Vertragszeichnung** *f* BAURECHT contract drawing
**Vertrauensgrenzen** *f pl* ARCH & TRAGW tidual limits
**Vertreter** *m* BAURECHT *Repräsentant* agent
**Vertretungbefugnis** *f* BAURECHT agency
**verunreinigen** *vt* INFR & ENTW contaminate, UMWELT pollute
**Verunreinigung** *f* BAURECHT, INFR & ENTW pollution, UMWELT contamination; **~ des Grundwassers** BAURECHT, INFR & ENTW, UMWELT groundwater contamination, groundwater pollution; **Verunreinigungsquelle** *f* UMWELT pollution source; **Verunreinigungssubstanz** *f* INFR & ENTW, UMWELT, WERKSTOFF contaminant
**Verursacherprinzip** *nt* UMWELT polluter-pays principle
**Verwahrung** *f* STAHL flashing; **Verwahrungsblech** *nt* STAHL sheet metal flashing
**Verwaltung** *f* ARCH & TRAGW management, BAURECHT administration; **Verwaltungsgebäude** *nt* BAURECHT administration building; **Verwaltungsgebühr** *f* BAURECHT administrative fee; **Verwaltungsgericht** *nt* BAURECHT administrative tribunal; **Verwaltungoorgan** *nt* BAURECHT administrative body; **Verwaltungsstelle** *f* BAURECHT administrative agency
**Verweildauer** *f* UMWELT retention time
**Verweilzeit** *f* UMWELT residence time
**verwendbar** *adj* BAURECHT applicable, WERKSTOFF usable; **~e Nebenprodukte** *nt pl* UMWELT usable by-products
**Verwendung** *f* ARCH & TRAGW use, utilization, INFR & ENTW use
**Verwerfen** *nt* WERKSTOFF warping
**Verwerfung** *f* INFR & ENTW fault; **Verwerfungsfläche** *f* INFR & ENTW fault plane; **Verwerfungszone** *f* INFR & ENTW fault zone
**verwertbar** *adj* UMWELT recyclable; **nicht ~er Rückstand** *m* UMWELT waste product
**Verwertung** *f* ARCH & TRAGW, BAURECHT, INFR & ENTW utilization; **Verwertungsquote** *f* UMWELT recycling rate
**verwindungsfrei** *adj* ARCH & TRAGW torsion-proof
**verwindungssteif** *adj* ARCH & TRAGW *verwindungsfrei* torsion-proof
**Verwindungstest** *m* ARCH & TRAGW torsion test
**verwittern** *vi* OBERFLÄCHE, WERKSTOFF weather
**Verwitterung** *f* OBERFLÄCHE, WERKSTOFF weathering; **Verwitterungserde** *f* INFR & ENTW residual soil; **Verwitterungsfläche** *f* OBERFLÄCHE area of deep weathering; **Verwitterungstest** *m* WERKSTOFF weathering test
**verworfen**: **~es Holz** *nt* HOLZ warped timber
**verzahnen** *vt* ARCH & TRAGW joggle, HOLZ indent, NATURSTEIN key
**Verzahnung** *f* ARCH & TRAGW joggle, indentation, BAUMASCHIN keying, *Mauerwerk* toothing, HOLZ indented joint, indentation, indenting, NATURSTEIN toothing, keying
**Verzapfung** *f* HOLZ morticing, mortising, tenon joint
**Verzeichnis** *nt* ARCH & TRAGW, BAURECHT list
**Verziehen** *nt* WERKSTOFF warping
**verziert** *adj* ARCH & TRAGW adorned, decorated, INFR & ENTW adorned; **~er Sturz** *m* BESCHLÄGE platband
**Verzierung** *f* ARCH & TRAGW decoration, ornament, adornment, INFR & ENTW adornment

**verzinken** *vt* OBERFLÄCHE galvanize
**verzinkt** *adj* OBERFLÄCHE galvanized
**Verzinkung** *f* OBERFLÄCHE galvanizing, zinc coating
**verzinnen** *vt* OBERFLÄCHE tin
**Verzinnen** *nt* STAHL tinning
**verzogen**: **~e Stufe** *f* ARCH & TRAGW dancing step
**verzögern** *vt* BETON, UMWELT, WERKSTOFF retard
**verzögert**: **~es Schließen** *nt* BESCHLÄGE *Tür* delayed closing
**Verzögerung** *f* BAURECHT delay; **Verzögerungsänderungen** *f pl* BAURECHT delay changes; **Verzögerungsdauer** *f* WERKSTOFF retardation time; **Verzögerungsmittel** *nt* BETON, WERKSTOFF retarding agent
**Verzweigung** *f* ARCH & TRAGW junction, INFR & ENTW bifurcation; **Verzweigungsspannung** *f* INFR & ENTW, WERKSTOFF bifurcation stress
**Vestibül** *nt* ARCH & TRAGW anteroom, vestibule
**Viadukt** *m* INFR & ENTW viaduct
**Vibration** *f* ARCH & TRAGW, DÄMMUNG vibration
**vibrationsfrei**: **~e Montage** *f* ARCH & TRAGW, DÄMMUNG antivibration mounting
***Vibration***: **Vibrationsramme** *f* BAUMASCHIN vibrating tamper, vibration driver, vibration ram; **Vibrationssieb** *nt* BAUMASCHIN vibrating screen; **Vibrationsverdichter** *m* BAUMASCHIN vibrating compactor; **Vibrationswalze** *f* BAUMASCHIN vibrating roller
**Vibrator** *m* BAUMASCHIN vibrator
**vibrierend**: **~e Maschinen** *f pl* ARCH & TRAGW, BAUMASCHIN vibrating machinery
**Vibriertisch** *m* BAUMASCHIN vibrating table, vibrator table
**vieleckig** *adj* ARCH & TRAGW polygonal
**Viel-**: **Vieleckmauerwerk** *nt* NATURSTEIN random rubble masonry; **Vieleckverband** *m* NATURSTEIN polygonal bond
**vielfach**: **~e Entwicklung** *f* UMWELT multiple development
***Viel-***: **Vielfachbogensperre** *f* INFR & ENTW *Staumauer* multiple-arch dam
**vielgeschossig** *adj* ARCH & TRAGW, INFR & ENTW high-rise
***Viel-***: **Vielkeilverzahnung** *f* HOLZ spline bushing; **Vielzellenglas** *nt* DÄMMUNG, WERKSTOFF foam glass
**viereckig** *adj* ARCH & TRAGW four-cornered, four-sided, square; **~er Hof** *m* ARCH & TRAGW quadrangle
**Viereckrahmen** *m* ARCH & TRAGW square frame
**Vierendeelstütze** *f* ARCH & TRAGW Vierendeel column
**Vierendeelträger** *m* ARCH & TRAGW, WERKSTOFF open-web girder, Vierendeel girder
**vierflügelig** *adj* ARCH & TRAGW *Fenster* four-leaved
**Vierkant-** *in cpds* BESCHLÄGE square; **Vierkantkopfschraube** *f* BESCHLÄGE coach screw; **Vierkantlochung** *f* STAHL square hole; **Vierkantrohr** *nt* WERKSTOFF square tube; **Vierkantschraube** *f* ARCH & TRAGW square bolt; **Vierkantstab** *m* WERKSTOFF square bar; **Vierkantventil** *nt* ABWASSER cock with square head
**Vierpunktlagerung** *f* ARCH & TRAGW four-point bearing
**vierseitig** *adj* ARCH & TRAGW four-cornered
**Viertel** *nt* HOLZ, INFR & ENTW, NATURSTEIN, WERKSTOFF quarter; **Viertelabsatz** *m* ARCH & TRAGW *Treppe* quarter space; **Viertelkreissims** *m* NATURSTEIN ovolo; **Viertelstab** *m* ARCH & TRAGW quarter round; **Viertelstein** *m* NATURSTEIN queen closer, WERKSTOFF one-quarter brick; **Viertelziegel** *m* NATURSTEIN queen closer
**Vinyllack** *m* OBERFLÄCHE vinyl lacquer
**Visier-** *in cpds* ARCH & TRAGW, BAUMASCHIN, INFR & ENTW sighting; **Visiereinrichtung** *f* BAUMASCHIN, INFR & ENTW sight
**Visieren** *nt* INFR & ENTW sighting
***Visier-***: **Visierfernrohr** *nt* ARCH & TRAGW, BAUMASCHIN sighting telescope; **Visiergerüst** *nt* ARCH & TRAGW, BAUMASCHIN, INFR & ENTW sight rail; **Visiertafel** *f* BAUMASCHIN, BESCHLÄGE *Vermessung* boning rod
**Viskoelastizität** *f* WERKSTOFF visco-elasticity
**Viskoplastizität** *f* WERKSTOFF visco-plasticity
**viskos**: **~e Deformation** *f* WERKSTOFF viscous deformation
**Viskosimeter** *nt* BAUMASCHIN viscosimeter
**Viskosität** *f* WERKSTOFF viscosity; **Viskositätsbeiwert** *m* WERKSTOFF coefficient of viscosity
**Vlies** *nt* WERKSTOFF nonwoven fabric
**VOB** *abbr* (*Verdingungsordnung für die Vergabe von Bauleistungen*) BAURECHT German regulations for contracts and execution of construction works
**Vogelperspektive** *f* INFR & ENTW bird's-eye view
**voll**: **~ eingespannt** *adj* ARCH & TRAGW *Balken, Träger* fully restrained; **~er Schubanschluß** *m* ARCH & TRAGW, BETON, INFR & ENTW, WERKSTOFF *Verbundträger* full shear connection
**Voll-**: **Vollast** *f* ELEKTR full load; **Vollbinder** *m* NATURSTEIN perpend stone
**vollflächig** *adj* ARCH & TRAGW throughout **~e Mauer** *f* ARCH & TRAGW dead wall
***Voll-***: **Vollgeschoß** *nt* ARCH & TRAGW full storey (*BrE*), full story (*AmE*); **Vollgeschoßwohnung** *f* ARCH & TRAGW floor-through dwelling (*AmE*); **Vollgestängebohren** *nt* BAUMASCHIN boring by percussion with rods; **Vollholz** *nt* HOLZ solid wood
**völlig**: **~e Einspannung** *f* ARCH & TRAGW, INFR & ENTW fixed support
**vollimprägnieren** *vt* HOLZ saturate
***Voll-***: **Vollinie** *f* ARCH & TRAGW firm line; **Vollkunststoffplatte** *f* BESCHLÄGE, WERKSTOFF all-plastic board; **Vollkunststofftür** *f* ARCH & TRAGW, BESCHLÄGE all-plastic door; **Vollmacht** *f* BAURECHT agency; **Vollplastikplatte** *f* BESCHLÄGE, WERKSTOFF all-plastic board; **Vollplatte** *f* ARCH & TRAGW solid slab
**vollständig**: **~ einbetoniert** *adj* ARCH & TRAGW, BETON, INFR & ENTW, STAHL, WERKSTOFF fully encased; **~e Verdübelung** *f* ARCH & TRAGW, BETON, INFR & ENTW, WERKSTOFF full shear connection
***Voll-***: **Vollstein** *m* WERKSTOFF *Mauerwerk* solid brick
**vollverglast** *adj* ARCH & TRAGW fully glazed
***Voll-***: **Vollwandträger** *m* ARCH & TRAGW, STAHL plate girder; **Vollzapfen** *m* ARCH & TRAGW through tenon; **Vollziegel** *m* NATURSTEIN, WERKSTOFF solid brick
**Volumen** *nt* ARCH & TRAGW volume, WERKSTOFF bulk; **Volumenabnahme** *f* WERKSTOFF volume decrease; **Volumenänderung** *f* WERKSTOFF volume change; **Volumenerhöhung** *f* WERKSTOFF volume increase; **Volumenstrom** *m* HEIZ & BELÜFT *Luftleistung des Ventilators* air-handling capacity; **Volumenverlust** *m* INFR & ENTW, WERKSTOFF loss of volume
**volumetrisch**: **~er Wirkungsgrad** *m* UMWELT volumetric efficiency
**Volute** *f* ARCH & TRAGW volute

**Vor-** *pref* ARCH & TRAGW, INFR & ENTW, OBERFLÄCHE preliminary; **Voranalyse** *f* ARCH & TRAGW, INFR & ENTW, OBERFLÄCHE, UMWELT preliminary analysis; **Voranstrich** *m* OBERFLÄCHE precoat, priming, prime coat, undercoat; **Vorarbeiten** *f pl* ARCH & TRAGW, INFR & ENTW preliminary works; **Vorarbeiter** *m* ARCH & TRAGW foreman

**vorausgeplant** *adj* ARCH & TRAGW, BAURECHT, INFR & ENTW projected

**Voraussetzung** *f* ARCH & TRAGW, BAURECHT, INFR & ENTW premise

**Vorauszahlung** *f* BAURECHT advance payment

**vorbehandeln** *vt* BETON, HOLZ, INFR & ENTW pretreat, OBERFLÄCHE prefinish, pretreat

**vorbehandelt** *adj* BETON, HOLZ, INFR & ENTW, OBERFLÄCHE pretreated

***Vor-***: **Vorbehandlung** *f* BETON, HOLZ pretreatment, INFR & ENTW, OBERFLÄCHE preliminary treatment, preparatory treatment, pretreatment

**vorbelasten** *vt* INFR & ENTW preload

***Vor-***: **Vorbereitung** *f* ARCH & TRAGW, INFR & ENTW preparation

**vorbeugend**: **~er Brandschutz** *m* BAURECHT preventive fire protection; **~e Wartung** *f* ARCH & TRAGW preventive maintenance

***Vor-***: **Vorbohrloch** *nt* UMWELT mousehole; **Vorchor** *m* ARCH & TRAGW, INFR & ENTW antechoir; **Vordach** *nt* ARCH & TRAGW canopy, porch roof; **Vordachlattung** *f* HOLZ verge batten

**vordehnen** *vt* BETON prestrain

**Vorder-** *in cpds* ARCH & TRAGW face, front, BETON, INFR & ENTW, NATURSTEIN, WERKSTOFF face; **Vorderansicht** *f* ARCH & TRAGW front view, *eines Gebäudes* facade; **Vorderfront** *f* ARCH & TRAGW facade; **Vorderkante** *f* ARCH & TRAGW front edge; **Vorderseite** *f* ARCH & TRAGW face, front

***Vor-***: **Vordruck** *m* ABWASSER, HEIZ & BELÜFT, INFR & ENTW, WERKSTOFF admission pressure

**voreilend**: **~er Kämpferdruck** *m* ARCH & TRAGW leading abutment pressure

***Vor-***: **Voreinstellung** *f* ELEKTR, HEIZ & BELÜFT presetting; **Vorentwurf** *m* ARCH & TRAGW preliminary design

**vorfertigen** *vt* ARCH & TRAGW, BETON, STAHL, WERKSTOFF precast, prefabricate

**Vorfertigung** *f* ARCH & TRAGW, BETON, STAHL, WERKSTOFF *Betonbau* precasting; **~ im Betonwerk** BETON factory precasting

***Vor-***: **Vorfilter** *nt* ABWASSER precleaner; **Vorflutdrän** *m* ABWASSER main drain; **Vorfluter** *m* UMWELT receiving water

**vorgefertigt** *adj* ARCH & TRAGW, WERKSTOFF prefabricated

**vorgehängt**: **~e Dachrinne** *f* ARCH & TRAGW bracket-mounted roof gutter, eaves gutter

**vorgenäßt** *adj* NATURSTEIN prewetted

**vorgesehen**: **~er Raum** *m* **für Rollstuhl** ARCH & TRAGW, BAURECHT wheelchair space

**vorgespannt** *adj* BETON prestressed, tensioned, INFR & ENTW, WERKSTOFF prestressed; **~es Glas** *nt* ARCH & TRAGW tempered glass, toughened safety glass

**Vorhang** *m* BESCHLÄGE curtain

**Vorhängeschloß** *nt* BESCHLÄGE padlock

***Vorhang***: **Vorhangfassade** *f* ARCH & TRAGW, NATURSTEIN, STAHL, WERKSTOFF curtain wall; **Vorhangstange** *f* BESCHLÄGE curtain rod; **Vorhangwand** *f* ARCH & TRAGW curtain wall; **Vorhangwandrahmen** *m* ARCH & TRAGW curtain wall frame

***Vor-***: **Vorhof** *m* ARCH & TRAGW, INFR & ENTW atrium, forecourt; **Vorklärbecken** *nt* INFR & ENTW, UMWELT pretreatment tank, primary settlement basin; **Vorkonsolidierung** *f* INFR & ENTW preconsolidation; **Vorkopfkipper** *m* BAUMASCHIN front dumper

**vorkragen 1.** *vt* ARCH & TRAGW corbel; **2.** *vi* ARCH & TRAGW project

**vorkragend**: **~es Dach** *nt* ARCH & TRAGW, HOLZ cantilever roof; **~e Mauerschicht** *f* ARCH & TRAGW cordon; **~es Mauerwerk** *nt* NATURSTEIN corbel masonry; **~e Platte** *f* ARCH & TRAGW cantilever slab; **~es Teil** *nt* ARCH & TRAGW jutting piece

***Vor-***: **Vorkragung** *f* ARCH & TRAGW overhang, projection, NATURSTEIN corbeling (*AmE*), corbelling (*BrE*)

**vorkühlen** *vt* HEIZ & BELÜFT precool

***Vor-***: **Vorlauf** *m* HEIZ & BELÜFT supply

**vorläufig**: **~er Kostenvoranschlag** *m* ARCH & TRAGW preliminary cost estimate; **~e Lagerung** *f* UMWELT *von Müll* temporary storage; **~e Öffnung** *f* NATURSTEIN temporary opening

***Vor-***: **Vorlauftemperatur** *f* HEIZ & BELÜFT supply temperature; **Vormauerziegel** *m* INFR & ENTW hard-burnt brick, NATURSTEIN face brick, facing brick, well-burnt brick, WERKSTOFF hard-burnt brick; **Vormischsilo** *m* BAUMASCHIN prebatching bin; **Vornorm** *f* ARCH & TRAGW, BAURECHT, INFR & ENTW, WERKSTOFF initial standard, tentative standard; **Vorort** *m* INFR & ENTW suburb

**Vorortmischen** *nt* BETON mix in place

**vorragend**: **~e Treppe** *f* ARCH & TRAGW cantilever staircase

**Vorrat** *m* ARCH & TRAGW, WERKSTOFF stock; **Vorratsbehälter** *m* HEIZ & BELÜFT storage tank

***Vor-***: **Vorraum** *m* ARCH & TRAGW anteroom, hall, vestibule; **Vorreiber** *m* BESCHLÄGE casement fastener, sash lock; **Vorreiniger** *m* ABWASSER precleaner

**Vorrichtung** *f* BAUMASCHIN appliance

**Vorsatz**: **Vorsatzbeton** *m* BETON, INFR & ENTW, WERKSTOFF face concrete, facing concrete; **Vorsatzziegel** *m* NATURSTEIN facing brick

***Vor-***: **Vorschaltgerät** *nt* ELEKTR *Leuchtstoffröhre* ballast

**vorschlagen** *vt* ARCH & TRAGW suggest

**Vorschlaghammer** *m* BAUMASCHIN sledgehammer

**vorschreiben** *vt* ARCH & TRAGW specify, BAURECHT prescribe, specify

**Vorschrift** *f* BAURECHT code, instruction, regulation, specification

**vorschriftsgemäß** *adj* BAURECHT according to regulations, according to specification

**vorschriftsmäßig** *adj* BAURECHT as prescribed

***Vor-***: **Vorschweißflansch** *m* ABWASSER welding neck flange

**vorsehen** *vt* ARCH & TRAGW provide

**vorsetzen** *vt* ARCH & TRAGW, BETON, HOLZ, STAHL attach

**vorspannen** *vt* BETON prestress, precompress, tension, INFR & ENTW, WERKSTOFF prestress

***Vor-***: **Vorspannen** *nt* **ohne Verbund** BETON no-bond prestressing, nonbond tensioning; **Vorspannpfahl** *m* BETON, INFR & ENTW, WERKSTOFF prestressed pile; **Vorspannung** *f* BETON prestress, precompression, INFR & ENTW, WERKSTOFF prestress; **Vorspannungsverlust** *m* BETON loss of prestress;

**Vorsprung** *m* ARCH & TRAGW jut, projection, prominence, nose
**vorstehen** *vi* ARCH & TRAGW project
**vorstehend**: **~e Falle** *f* BESCHLÄGE *Schloß* projecting latch; **~er Rand** *m* STAHL flange
***Vor-***: **Vorstreichfarbe** *f* OBERFLÄCHE undercoat; **Vorstudie** *f* ARCH & TRAGW, BAURECHT, INFR & ENTW preliminary study; **Vortreiben** *nt* INFR & ENTW *Tunnelbau* forcing; **Vortriebschild** *m* ARCH & TRAGW *Tunnelbau* shield; **Vorversuch** *m* INFR & ENTW preliminary test
**vorwalzen** *vt* ARCH & TRAGW rough down
**vorwärmen** *vt* HEIZ & BELÜFT preheat
***Vor-***: **Vorwärmer** *m* HEIZ & BELÜFT preheater
**vorzeitig**: **~es Abbinden** *nt* BETON, INFR & ENTW, WERKSTOFF false setting
**Voute** *f* ARCH & TRAGW haunch; **Voutendecke** *f* ARCH & TRAGW, BETON, INFR & ENTW arched floor
**VS-Glas** *nt* (*Verbundsicherheitsglas*) BAURECHT, WERKSTOFF laminated safety glass
**V-Stab** *m* ARCH & TRAGW upright, *Fachwerk* vertical
**Vulkan** *m* INFR & ENTW volcano; **Vulkanfiber** *f* WERKSTOFF vulcanized fiber (*AmE*), vulcanized fibre (*BrE*)
**vulkanisch** *adj* INFR & ENTW volcanic; **~e Asche** *f* INFR & ENTW volcanic ash; **~er Ausstoß** *m* INFR & ENTW volcanic eruption; **~er Boden** *m* INFR & ENTW volcanic soil; **~es Erdbeben** *nt* INFR & ENTW volcanic earthquake; **~es Gebiet** *nt* INFR & ENTW volcanic zone; **~es Gestein** *nt* INFR & ENTW volcanic rock; **~e Schlacke** *f* INFR & ENTW volcanic slag; **~er Ton** *m* INFR & ENTW volcanic clay
**vulkanisiert** *adj* NATURSTEIN vulcanized
**Vulkanit** *m* INFR & ENTW, UMWELT igneous rocks
***Vulkan***: **Vulkanschlot** *m* INFR & ENTW, UMWELT *geologisch* conduit

**Waage** *f* ARCH & TRAGW balance

**waagerecht** *adj* ARCH & TRAGW, INFR & ENTW horizontal; **~e Stiefe** *f* ARCH & TRAGW *temporär* flying shore

**Wabe-** *in cpds* ARCH & TRAGW honeycomb; **Wabendämmplatte** *f* DÄMMUNG honeycomb insulating board; **Wabenkern** *m* ARCH & TRAGW *Verbundkonstruktion* honeycomb core; **Wabenstruktur** *f* ARCH & TRAGW honeycomb, honeycomb structure, BETON honeycomb

**Wagenkipper** *m* BAUMASCHIN tipper truck

**Wahrscheinlichkeit** *f* ARCH & TRAGW, INFR & ENTW probability

**Walm** *m* ARCH & TRAGW hip; **Walmdach** *nt* ARCH & TRAGW, HOLZ hip roof, hipped roof; **Walmgewölbe** *nt* ARCH & TRAGW *Kirchenbau* cloister vault; **Walmsparren** *m* HOLZ, INFR & ENTW hip rafter

**Walz-** *in cpds* ARCH & TRAGW, BAUMASCHIN, INFR & ENTW, STAHL, WERKSTOFF rolled, rolling

**Walze** *f* BAUMASCHIN roller, WERKSTOFF cylinder

**Walzen** *nt* INFR & ENTW rolling; **Walzenwehr** *nt* INFR & ENTW roller weir

***Walz-***: **Walzstahl** *m* STAHL rolled steel; **Walzstahlprofil** *nt* ARCH & TRAGW, STAHL, WERKSTOFF rolled structural steel; **Walzstahlträger** *m* ARCH & TRAGW rolled steel joist; **Walzträger** *m* ARCH & TRAGW rolled steel joist; **Walzwerk** *nt* BAUMASCHIN rolling mill

**Wand** *f* ARCH & TRAGW wall; **Wandabsteifung** *f* ARCH & TRAGW wall shoring; **Wandanker** *m* ARCH & TRAGW, BESCHLÄGE, NATURSTEIN wall tie; **Wandanschluß** *m* NATURSTEIN wall junction; **Wandanschlußblech** *nt* STAHL soaker; **Wandanschlußleiste** *f* BESCHLÄGE fillet; **Wandanschlußprofil** *nt* NATURSTEIN wall junction profile; **Wandanstrich** *m* OBERFLÄCHE wall coat, wall paint

**wandartig**: **~er Träger** *m* ARCH & TRAGW deep beam

***Wand***: **Wandausgußbecken** *nt* ABWASSER wall-mounted sink; **Wandbatterie** *f* ABWASSER vertical-mounted mixer faucet (*AmE*), vertical-mounted mixer tap (*BrE*); **Wandbaustoffe** *m pl* ARCH & TRAGW, NATURSTEIN walling; **Wandbehang** *m* ARCH & TRAGW wall hanging, BESCHLÄGE hanging; **Wandbelag** *m* NATURSTEIN vertical tilework, OBERFLÄCHE wall covering; **Wandbogen** *m* ARCH & TRAGW wall arch; **Wanddicke** *f* ARCH & TRAGW wall thickness; **Wanddruck** *m* INFR & ENTW wall pressure; **Wanddurchbruch** *m* ARCH & TRAGW wall breakthrough, wall opening; **Wanddurchführung** *f* ARCH & TRAGW wall bushing, wall duct; **Wandeinbauaxialventilator** *m* HEIZ & BELÜFT flush-mounted axial fan; **Wandeinfluß** *m* INFR & ENTW wall effect

**Wanderlast** *f* INFR & ENTW moving load

***Wand***: **Wandfarbe** *f* OBERFLÄCHE wall paint; **Wandfläche** *f* ARCH & TRAGW wall area; **Wandfuge** *f* ARCH & TRAGW wall joint; **Wandhalterung** *f* ARCH & TRAGW wall bracket, wall holdfast

**wandhängend**: **~es WC** *nt* ABWASSER wall-hung toilet, wall-mounted toilet

***Wand***: **Wandinnenseite** *f* ARCH & TRAGW interior wall surface; **Wandisolierung** *f* DÄMMUNG wall insulation; **Wandkonsole** *f* ARCH & TRAGW wall bracket; **Wandkran** *m* BAUMASCHIN wall crane

**Wandler** *m* ELEKTR transducer

***Wand***: **Wandleuchte** *f* BESCHLÄGE, ELEKTR wall-mounted light fixture; **Wandlufterhitzer** *m* HEIZ & BELÜFT wall-mounted fan heater; **Wandnische** *f* ARCH & TRAGW wall recess; **Wandpfeiler** *m* ARCH & TRAGW pilaster, wall pillar; **Wandplatte** *f* ARCH & TRAGW wall panel, wall slab, HOLZ, WERKSTOFF furring tile; **Wandputz** *m* NATURSTEIN wall plaster; **Wandreibung** *f* INFR & ENTW wall friction; **Wandsafe** *m* BESCHLÄGE wall safe; **Wandsäule** *f* ARCH & TRAGW wall post; **Wandschrank** *m* HOLZ built-in cupboard (*BrE*), closet (*AmE*); **Wandsockelleiste** *f* WERKSTOFF dado molding (*AmE*), dado moulding (*BrE*); **Wandstärke** *f* ARCH & TRAGW wall thickness; **Wandsystem** *nt* ARCH & TRAGW walling; **Wandtafel** *f* ARCH & TRAGW *Bauteil* wall panel; **Wandtürpuffer** *m* BESCHLÄGE wall-mounted door stop; **Wandurinal** *nt* ABWASSER urinal, wall-hung urinal, wall-mounted urinal; **Wandverglasung** *f* ARCH & TRAGW wall glazing; **Wandverkleidung** *f* ARCH & TRAGW, OBERFLÄCHE wall cladding, wall covering; **Wandwange** *f* ARCH & TRAGW wall string

**Wange** *f* ARCH & TRAGW, WERKSTOFF cheek, side plate, string board, stringer; **~ mit eingestemmten Stufen** ARCH & TRAGW housed string; **Wangenmauer** *f* ARCH & TRAGW *Treppe* string wall; **Wangentreppe** *f* ARCH & TRAGW stringer staircase

**Wanne** *f* ABWASSER tub, INFR & ENTW sump pan; **Wannengriff** *m* BESCHLÄGE grab bar; **Wannengründung** *f* ARCH & TRAGW tanking

**Warenaufzug** *m* ARCH & TRAGW, BAUMASCHIN, INFR & ENTW goods elevator (*AmE*), goods lift (*BrE*)

**Warm-** *in cpds* ARCH & TRAGW, INFR & ENTW, WERKSTOFF warm; **Warmdach** *nt* ARCH & TRAGW, INFR & ENTW externally insulated roof, nonventilated flat roof

**Wärme** *f* ABWASSER, DÄMMUNG, HEIZ & BELÜFT, UMWELT heat; **Wärmeabgabe** *f* HEIZ & BELÜFT thermal output; **Wärmeausdehnung** *f* ABWASSER, ARCH & TRAGW, BETON, INFR & ENTW, STAHL, WERKSTOFF heat expansion, thermal expansion; **Wärmeausnutzung** *f* HEIZ & BELÜFT heat efficiency, heat utilization; **Wärmeausstrahlung** *f* HEIZ & BELÜFT heat radiation; **Wärmeaustausch** *m* HEIZ & BELÜFT heat exchange, interchange of heat; **Wärmeaustauscher** *m* HEIZ & BELÜFT, UMWELT heat exchanger; **Wärmebedarf** *m* ARCH & TRAGW, BAURECHT, HEIZ & BELÜFT calorific requirement, heat demand, heat requirement; **Wärmebeiwert** *m* DÄMMUNG, HEIZ & BELÜFT temperature coefficient; **Wärmebelastung** *f* UMWELT thermal load, thermal pollution; **Wärmebelastungsberechnungen** *f pl* HEIZ & BELÜFT heat load calculations

**wärmebeständig** *adj* WERKSTOFF heatproof

***Wärme***: **Wärmebeständigkeit** *f* WERKSTOFF heat resistance, thermal stability; **Wärmebilanz** *f* HEIZ &

BELÜFT thermal balance; **Wärmebrücke** *f* ARCH & TRAGW, DÄMMUNG thermal bridge

**wärmedämmend** *adj* DÄMMUNG, UMWELT heat-insulating

***Wärme***: **Wärmedämmforderung** *f* BAURECHT, DÄMMUNG, UMWELT heat insulation requirement; **Wärmedämmschicht** *f* DÄMMUNG, UMWELT heat-insulating layer; **Wärmedämmstoff** *m* DÄMMUNG, INFR & ENTW, UMWELT, WERKSTOFF heat insulation material, thermal insulation material; **Wärmedämmung** *f* DÄMMUNG, HEIZ & BELÜFT INFR & ENTW, UMWELT, WERKSTOFF heat insulation, lagging, thermal insulation; **Wärmedämmungsarbeiten** *f pl* DÄMMUNG, UMWELT heat insulation work; **Wärmedehnung** *f* ABWASSER heat expansion, thermal expansion, ARCH & TRAGW, BETON thermal expansion, INFR & ENTW heat expansion, thermal expansion, STAHL thermal expansion, WERKSTOFF heat expansion, thermal expansion; **Wärmedehnungsfuge** *f* ABWASSER, INFR & ENTW *Rohr* thermal expansion joint; **Wärmedehnzahl** *f* WERKSTOFF coefficient of thermal expansion (*CTE*); **Wärmedurchgangskoeffizient** *m* INFR & ENTW, WERKSTOFF heat transition coefficient; **Wärmedurchgangswiderstand** *m* DÄMMUNG, WERKSTOFF heat resistance, thermal resistance; **Wärmedurchgangszahl** *f* HEIZ & BELÜFT overall coefficient of heat transfer, INFR & ENTW heat transition coefficient, WERKSTOFF heat transfer coefficient, heat transition coefficient; **Wärmedurchlässigkeitszahl** *f* WERKSTOFF thermal transmission factor; **Wärmedurchlaßkoeffizient** *m* DÄMMUNG u-value; **Wärmedurchlaßwiderstand** *m* DÄMMUNG, WERKSTOFF thermal resistance

**wärmeempfindlich** *adj* WERKSTOFF heat-sensitive, thermosensitive

***Wärme***: **Wärmeenergie** *f* HEIZ & BELÜFT heat energy, thermal energy; **Wärmeentwicklung** *f* HEIZ & BELÜFT heat build-up; **Wärmeermüdung** *f* WERKSTOFF thermal fatigue; **Wärmefluß** *m* DÄMMUNG, HEIZ & BELÜFT heat flux; **Wärmeflußrichtung** *f* ARCH & TRAGW, DÄMMUNG, HEIZ & BELÜFT heat flow; **Wärmefühler** *m* BAURECHT *Brandschutz*, BESCHLÄGE heat detector

**wärmegedämmt** *adj* DÄMMUNG thermally insulated; **~es Haus** *nt* ARCH & TRAGW thermally insulated house

***Wärme***: **Wärmehaushalt** *m* HEIZ & BELÜFT heat balance; **Wärme- und Kältetechnik** *f* HEIZ & BELÜFT heating and refrigerating engineering; **Wärmekapazität** *f* DÄMMUNG, HEIZ & BELÜFT heat capacity; **Wärmeklasse** *f* WERKSTOFF *Isolierstoff* thermal class; **Wärmeleistung** *f* DÄMMUNG thermal performance, HEIZ & BELÜFT thermal output; **Wärmeleitfähigkeit** *f* DÄMMUNG, WERKSTOFF heat conductivity, thermal conductivity; **Wärmeleitung** *f* DÄMMUNG, WERKSTOFF thermal conduction; **Wärmeleitzahl** *f* DÄMMUNG thermal transmittance coefficient; **Wärmemenge** *f* DÄMMUNG, HEIZ & BELÜFT heat content; **Wärmepumpe** *f* HEIZ & BELÜFT heat pump

**wärmereflektierend** *adj* DÄMMUNG heat-reflecting

***Wärme***: **Wärmeregler** *m* HEIZ & BELÜFT thermostat; **Wärmerückgewinnung** *f* HEIZ & BELÜFT heat recovery

**wärmerückstrahlend** *adj* DÄMMUNG heat-reflecting

***Wärme***: **Wärmeschutz** *m* DÄMMUNG, UMWELT heat insulation; **Wärmeschutzglas** *nt* ARCH & TRAGW, DÄMMUNG, WERKSTOFF heat protection glass

**wärmeschutztechnisch**: **~e Prüfung** *f* INFR & ENTW, WERKSTOFF thermal insulation test

***Wärme***: **Wärmeschutzwert** *m* WERKSTOFF thermal insulation value; **Wärmespeicher** *m* HEIZ & BELÜFT heat accumulator; **Wärmespeicherung** *f* DÄMMUNG, HEIZ & BELÜFT heat storage; **Wärmesperre** *f* DÄMMUNG heat barrier; **Wärmestoßspannung** *f* DÄMMUNG thermal shock; **Wärmestrahlung** *f* HEIZ & BELÜFT thermal radiation; **Wärmeströmung** *f* HEIZ & BELÜFT convection current; **Wärmeträger** *m* HEIZ & BELÜFT heat carrier, heat transfer medium; **Wärmeträgermedium** *m* HEIZ & BELÜFT heat transfer medium, heat carrier; **Wärmeübergangswiderstand** *m* WERKSTOFF heat transmission resistance; **Wärmeüberschuß** *m* HEIZ & BELÜFT excess heat; **Wärmeüberträger** *m* HEIZ & BELÜFT heat exchanger; **Wärmeübertragung** *f* HEIZ & BELÜFT heat transfer; **Wärmeverformung** *f* WERKSTOFF deformation under heat; **Wärmeverlust** *m* DÄMMUNG, UMWELT heat loss, loss of heat; **Wärmeverlustberechnung** *f* ARCH & TRAGW, HEIZ & BELÜFT calculation of heat loss, computation of heat loss; **Wärmeverteilung** *f* HEIZ & BELÜFT, INFR & ENTW distribution of heat, distribution of temperature; **Wärmewiedergewinnung** *f* HEIZ & BELÜFT heat recovery; **Wärmewirkungsgrad** *m* HEIZ & BELÜFT heat efficiency, heat utilization; **Wärmezirkulation** *f* HEIZ & BELÜFT thermal circulation

**warmgenietet** *adj* STAHL hot-riveted

**warmgewalzt** *adj* WERKSTOFF hot-rolled

***Warm-***: **Warmluftfrontkondensation** *f* DÄMMUNG warm front condensation; **Warmluftheizung** *f* HEIZ & BELÜFT hot-air heating, warm-air heating

**Warmwasser** *nt* ABWASSER, ELEKTR, HEIZ & BELÜFT hot water; **Warmwasserbereiter** *m* HEIZ & BELÜFT instantaneous water heater; **Warmwasserbereitungsanlage** *f* HEIZ & BELÜFT water-heating plant; **Warmwasserleitung** *f* ABWASSER, HEIZ & BELÜFT hot water pipe; **Warmwasserspeicher** *m* ABWASSER hot water tank, ELEKTR, HEIZ & BELÜFT boiler, hot water tank, storage water heater; **Warmwasserversorgung** *f* ABWASSER, HEIZ & BELÜFT hot water supply

**Warn-** *in cpds* BAURECHT, BESCHLÄGE, ELEKTR, INFR & ENTW warning; **Warnanlage** *f* BAURECHT, BESCHLÄGE, ELEKTR alarm system, warning system; **Warnmeldung** *f* BAURECHT, ELEKTR warning; **Warnpflicht** *f* BAURECHT duty to warn; **Warnschild** *nt* BAURECHT, INFR & ENTW danger sign

**Warnung** *f* BAURECHT alarm, warning, BESCHLÄGE alarm, ELEKTR warning

**Wartehalle** *f* ARCH & TRAGW waiting hall

**Wartung** *f* ARCH & TRAGW attendance, maintenance, BAURECHT maintenance, HEIZ & BELÜFT service, INFR & ENTW maintenance; **Wartungsanleitung** *f* BAUMASCHIN, HEIZ & BELÜFT service manual; **Wartungsschacht** *m* ARCH & TRAGW maintenance manhole; **Wartungsweg** *m* ARCH & TRAGW, INFR & ENTW runway

**Wasch-** *in cpds* ABWASSER washing; **Waschbecken** *nt* ABWASSER wash basin

**waschbeständig** *adj* OBERFLÄCHE washable, *Farbe* washproof

***Wasch-***: **Waschbeton** *m* BETON concrete exposed aggregate finish, exposed aggregate concrete, washed concrete, INFR & ENTW, WERKSTOFF exposed aggregate concrete

**waschfest** *adj* OBERFLÄCHE washable, *Farbe* washproof

***Wasch-***: **Waschmaschine** *f* ABWASSER washing machine; **Waschputz** *m* NATURSTEIN scrubbed plaster, acidwashed plaster, OBERFLÄCHE, WERKSTOFF acidwashed plaster; **Waschsieb** *nt* BAUMASCHIN washing screen; **Waschtisch** *m* ABWASSER wash basin; **Waschtrommel** *f* BAUMASCHIN washing drum

**Wasser** *nt* ABWASSER water; **Wasserabdichtung** *f* ABWASSER, DÄMMUNG waterproofing; **Wasserabgabe** *f* BAURECHT water charge; **Wasserablaufrinne** *f* ABWASSER stormwater gutter; **Wasserabscheidebauwerk** *nt* ABWASSER water extraction structure; **Wasserabsenkung** *f* INFR & ENTW groundwater lowering

**wasserabsorbierend** *adj* WERKSTOFF water-absorbing

***Wasser***: **Wasserabsorption** *f* WERKSTOFF water absorption; **Wasserader** *f* INFR & ENTW water vein; **Wasserarmatur** *f* BESCHLÄGE water fittings; **Wasseraufbereitung** *f* ABWASSER *bei Abwässern* water treatment, BAURECHT water conditioning, INFR & ENTW water conditioning, water treatment, UMWELT *bei Abwässern* water purification; **Wasseraufbereitungsanlage** *f* ABWASSER water treatment plant; **Wasseraufnahme** *f* WERKSTOFF water absorption; **Wasseraufnahmeversuch** *m* WERKSTOFF absorption test; **Wasserbauprojekt** *nt* BAURECHT, INFR & ENTW hydraulic engineering project; **Wasserbauwerk** *nt* INFR & ENTW hydraulic structure; **Wasserbecken** *nt* ABWASSER, INFR & ENTW water basin; **Wasserbedarf** *m* INFR & ENTW water demand; **Wasserbehälter** *m* ABWASSER cistern, INFR & ENTW water reservoir; **Wasserbeschaffenheit** *f* BAURECHT, INFR & ENTW water condition; **Wasserbilanz** *f* BAURECHT water balance; **Wasserbindungsvermögen** *nt* WERKSTOFF water-holding capacity; **Wasserdampfdiffusion** *f* ARCH & TRAGW, DÄMMUNG damp; **Wasserdampfdurchlässigkeit** *f* DÄMMUNG, WERKSTOFF water vapor permeability (*AmE*), water vapour permeability (*BrE*)

**wasserdicht 1.** *adj* DÄMMUNG watertight, WERKSTOFF waterproof; **2. ~ machen** *vt* DÄMMUNG waterproof

***wasserdicht***: **~e Bitumenisolierung** *f* **eines Kellergeschosses** INFR & ENTW asphalt tanking

***Wasser***: **Wasserdichtigkeit** *f* DÄMMUNG watertightness; **Wasserdruck** *m* ABWASSER, INFR & ENTW water pressure, UMWELT hydraulic thrust; **Wasserdurchlaß** *m* ABWASSER, INFR & ENTW *unter Straße* culvert

**wasserdurchlässig** *adj* WERKSTOFF permeable to vapor (*AmE*), permeable to vapour (*BrE*), pervious to water

***Wasser***: **Wassereindringung** *f* WERKSTOFF water penetration; **Wassereintritt** *m* UMWELT water intake; **Wassereinzugsgebiet** *nt* ABWASSER, INFR & ENTW *Wasserversorgung, Brunnen* intake area; **Wasserenthärter** *m* ABWASSER water softener; **Wasserenthärtungsanlage** *f* ABWASSER, INFR & ENTW water softening unit; **Wasserentnahme** *f* ABWASSER water intake; **Wasserentsalzung** *f* ABWASSER, INFR & ENTW desalination; **Wasserfarbe** *f* ABWASSER water-based paint; **Wasserfaß** *nt* ABWASSER water butt

**wasserführend**: **~e Schicht** *f* INFR & ENTW water-bearing stratum

**wassergebunden** *adj* INFR & ENTW *Straßenbau* waterbound

***Wasser***: **Wassergehalt** *m* ABWASSER, INFR & ENTW, WERKSTOFF water content

**wassergesättigt** *adj* WERKSTOFF water-logged

***Wasser***: **Wasserglasfarbe** *f* OBERFLÄCHE silicate paint; **Wasserglaskitt** *m* BETON, WERKSTOFF water glass cement; **Wassergüte** *f* ABWASSER, UMWELT water quality; **Wasserhahn** *m* ABWASSER faucet (*AmE*), tap (*BrE*), bibcock, water cock, ARCH & TRAGW spigot; **Wasserhaltung** *f* ARCH & TRAGW, INFR & ENTW groundwater lowering, unwatering; **Wasserhebung** *f* INFR & ENTW water lifting; **Wasserkraft** *f* ELEKTR, HEIZ & BELÜFT, INFR & ENTW, UMWELT hydroelectric power, hydroelectricity, water power; **Wasserkraftwerk** *nt* ELEKTR, HEIZ & BELÜFT, INFR & ENTW, UMWELT hydroelectric generating station, hydroelectric power plant, hydroelectric power station, water power station; **Wasserkreislauf** *m* INFR & ENTW water cycle; **Wasserlauf** *m* ABWASSER water course; **Wasserleitung** *f* ABWASSER water line

**wasserlöslich** *adj* ABWASSER water-soluble; **~es Flußmittel** *nt* ABWASSER water-soluble flux

***Wasser***: **Wassermengenwirtschaft** *f* BAURECHT water balance; **Wassermeßgefäß** *nt* BETON water measuring tank

**wässern** *vt* ABWASSER water

***Wasser***: **Wassernase** *f* ARCH & TRAGW water drip, weather drip, weather groove, gorge; **Wasseroberfläche** *f* INFR & ENTW water surface; **Wasserqualität** *f* ABWASSER, UMWELT water quality; **Wassersäule** *f* INFR & ENTW water column, UMWELT head of water; **Wasserschenkel** *m* ARCH & TRAGW, BESCHLÄGE weather groove, weather drip, throat, *Fenster, Tür* water drip; **Wasserschicht** *f* INFR & ENTW water layer; **Wasserschräge** *f* ARCH & TRAGW weathering; **Wasserschutzgebiet** *nt* BAURECHT, INFR & ENTW water protection area; **Wasserspannung** *f* WERKSTOFF water tension; **Wasserspeicher** *m* ABWASSER cistern; **Wasserspeier** *m* ARCH & TRAGW rainspout, waterspout, BESCHLÄGE gargoyle; **Wassersperre** *f* DÄMMUNG, INFR & ENTW water barrier; **Wasserspiegel** *m* INFR & ENTW water level, water table; **Wasserspülung** *f* ABWASSER, BESCHLÄGE flushing, water cistern

**Wasserstand** *m* INFR & ENTW water level; **Wasserstandsanzeiger** *m* BAUMASCHIN, INFR & ENTW water level indicator; **Wasserstandsdifferenz** *f* INFR & ENTW water level difference

***Wasser***: **Wasserstrahl** *m* INFR & ENTW water jet; **Wassertanker** *m* ABWASSER water tanker; **Wassertiefe** *f* INFR & ENTW, UMWELT water depth; **Wasserturbine** *f* UMWELT hydroturbine, water turbine; **Wasserturm** *m* ABWASSER water tower; **Wasserüberlauf** *m* ABWASSER water overflow pipe; **Wasseruhr** *f* ABWASSER water meter; **Wasserumwälzung** *f* ABWASSER water circulation

**wasserundurchlässig** *adj* ABWASSER, DÄMMUNG watertight, WERKSTOFF impermeable to water

***Wasser***: **Wasserundurchlässigkeit** *f* ABWASSER watertightness; **Wasserventil** *nt* ABWASSER water valve; **Wasserverlust** *m* INFR & ENTW loss of water; **Wasserverschmutzung** *f* BAURECHT, INFR & ENTW water pollution; **Wasserversorgung** *f* ABWASSER, INFR & ENTW, UMWELT water supply; **Wasserversorgung** *f* **für Feuerbekämpfung** BAURECHT fire mains; **Wasserverteilung** *f* ABWASSER water distribution

**wasserverunreinigend**: **~er Stoff** *m* UMWELT water pollutant

***Wasser***: **Wasserverunreinigung** *f* UMWELT water pollution; **Wasservorkommen** *nt* INFR & ENTW water resource; **Wasserwagen** *m* BAUMASCHIN water truck; **Wasserwirtschaft** *f* BAURECHT, INFR & ENTW water economy; **Wasserwirtschaftswesen** *nt* INFR & ENTW hydraulic engineering; **Wasserzähler** *m* ABWASSER water meter; **Wasserzementwert** *m* (*W/Z Wert*) ABWASSER, BETON, NATURSTEIN, WERKSTOFF water-cement ratio (*w/c ratio*); **Wasserzirkulation** *f* ABWASSER water circulation; **Wasserzulauf** *m* ABWASSER, UMWELT water intake

**wäßrig**: **~er Ausfluß** *m* UMWELT aqueous effluent; **~es Holzschutzmittel** *nt* HOLZ, OBERFLÄCHE, WERKSTOFF aqueous wood preservative

**Wattierung** *f* DÄMMUNG, HEIZ & BELÜFT wadding

**WC**: **WC-Anlage** *f* ABWASSER, INFR & ENTW toilet installations; **WC-Brille** *f* BESCHLÄGE toilet seat; **WC-Kabine** *f* INFR & ENTW toilet cubicle

**Webkante** *f* WERKSTOFF selvage, selvedge

**Wechsel** *m* BESCHLÄGE *Schloß* latch lever, HOLZ trimmer; **Wechselbalken** *m* ARCH & TRAGW header, trimmer, trimmer beam; **Wechselbiegefestigkeit** *f* STAHL, WERKSTOFF alternate bending strength; **Wechselklima** *nt* INFR & ENTW, WERKSTOFF alternating climate; **Wechsellast** *f* ARCH & TRAGW, INFR & ENTW, WERKSTOFF alternating load; **Wechselschaltung** *f* ELEKTR two-way wiring; **Wechselspannung** *f* ARCH & TRAGW, INFR & ENTW, WERKSTOFF alternating stress; **Wechselsprung** *m* INFR & ENTW hydraulic jump; **Wechselstab** *m* ARCH & TRAGW *Fachwerk* counter brace; **Wechselstrom** *m* ELEKTR, HEIZ & BELÜFT alternating current; **Wechselstrommaschine** *f* ELEKTR, UMWELT alternator; **Wechselwirkung** *f* INFR & ENTW, WERKSTOFF interaction

**Weg** *m* INFR & ENTW distance; **Wegekoffer** *m* INFR & ENTW pavement bed (*BrE*), sidewalk bed (*AmE*); **Wegerecht** *nt* BAURECHT, INFR & ENTW right-of-way; **Wegplatte** *f* NATURSTEIN pad

**wegräumen** *vt* ARCH & TRAGW clear away

***Weg***: **Wegunterführung** *f* INFR & ENTW undergrade crossing

**Wegwerffilter** *m* ABWASSER, BESCHLÄGE disposable filter

**Wehr** *nt* INFR & ENTW, UMWELT dam, weir; **Wehrabschöpfer** *m* UMWELT weir skimmer

**weich** *adj* WERKSTOFF flexible, soft

**Weich-** *in cpds* HOLZ, WERKSTOFF soft

***weich***: **~er Boden** *m* INFR & ENTW weak soil, bad ground; **~er Ton** *m* INFR & ENTW soft clay

**Weiche** *f* INFR & ENTW *Gleise* points (*BrE*), switch (*AmE*)

***Weich-***: **Weichfaserplatte** *f* HOLZ softboard; **Weichholz** *nt* HOLZ softwood; **Weichlot** *nt* WERKSTOFF soft solder

**weichlöten** *vt* WERKSTOFF solder

***Weich-***: **Weichlöten** *nt* WERKSTOFF soft soldering, soldering; **Weichmacher** *m* WERKSTOFF plasticizer; **Weichschaum** *m* WERKSTOFF flexible foam plastic; **Weichstahl** *m* STAHL mild steel

**weiß**: **~er Betonbelag** *m* BETON concrete whitetop overlay

**Weiß-**: **Weißasbest** *m* WERKSTOFF chrysotile; **Weißblech** *nt* STAHL tinplate; **Weißblechabfall** *m* UMWELT tinplate waste

**weißen** *vt* OBERFLÄCHE whiten, whitewash

**weißgeschält** *adj* HOLZ completely peeled

***Weiß-***: **Weißkalk** *m* WERKSTOFF fat lime; **Weißleim** *m* HOLZ casein glue

**weit** *adj* INFR & ENTW, WERKSTOFF ample

**Weite** *f* ARCH & TRAGW width

**weiter** *adj* ARCH & TRAGW, BAURECHT additional

**Weiterbehandlung** *f* BAURECHT, BETON, HOLZ, OBERFLÄCHE, STAHL aftertreatment

**Weiterverwertung** *f* BAURECHT, INFR & ENTW, UMWELT reclamation

**weitgespannt** *adj* ARCH & TRAGW long-span

**weitsäulig** *adj* ARCH & TRAGW diastyle

***weit***: **~ überhängendes Dach** *nt* ARCH & TRAGW umbrella roof

**Weitwinkelspion** *m* BESCHLÄGE *Türguckloch* viewer, wide-angle door viewer, wide-angle judas

**Well-** *in cpds* ARCH & TRAGW, STAHL, WERKSTOFF corrugated; **Wellasbesttafel** *f* WERKSTOFF corrugated asbestos panel; **Wellasbestzement** *m* WERKSTOFF corrugated asbestos cement; **Wellasbestzementdachbelag** *m* BETON, WERKSTOFF asbestos cement corrugated roof covering; **Wellbitumenpapier** *nt* WERKSTOFF corrugated asphalt paper; **Wellblech** *nt* STAHL corrugated metal, corrugated iron, corrugated sheet iron; **Wellblechbauten** *f pl* STAHL corrugated metal structures; **Welldach** *nt* ARCH & TRAGW corrugated roof; **Welldraht** *m* WERKSTOFF corrugated wire; **Welldrahtglas** *nt* BAURECHT wired glass, WERKSTOFF corrugated wire glass

**Welle** *f* ARCH & TRAGW roller, BAUMASCHIN axle, *Maschine* shaft, BESCHLÄGE roller, INFR & ENTW wave, *Rolladen* roller, OBERFLÄCHE roller; **Wellenausbreitung** *f* UMWELT wave propagation; **Wellenberg** *m* UMWELT wave crest; **Wellenbewegungsenergie** *f* **pro Meter Woge** UMWELT wave momentum per meter of crest (*AmE*), wave momentum per metre of crest (*BrE*); **Wellenfirsthaube** *f* WERKSTOFF corrugated ridge capping, corrugated ridge covering; **Wellenfortpflanzung** *f* UMWELT wave propagation; **Wellenfront** *f* INFR & ENTW wave front; **Wellengeschwindigkeit** *f* WERKSTOFF wave velocity; **Wellenkamm** *m* UMWELT wave crest

***Well-***: **Welleternit** *nt* WERKSTOFF corrugated eternit; **Wellfaserzement** *m* WERKSTOFF corrugated asbestos cement; **Wellfaserzementdachbelag** *m* BETON, WERKSTOFF asbestos cement corrugated roof covering; **Wellpappe** *f* WERKSTOFF corrugated paper; **Wellrohr** *nt* WERKSTOFF corrugated tube

**welsch**: **~e Haube** *f* ARCH & TRAGW imperial roof

**Wende-** *in cpds* INFR & ENTW turning; **Wendefläche** *f* INFR & ENTW turning bay; **Wendeflügel** *m* ARCH & TRAGW pivoted sash

**Wendel-** *in cpds* ARCH & TRAGW spiral; **Wendelstufe** *f* ARCH & TRAGW spiral winder, tapered tread, winder; **Wendeltreppe** *f* ARCH & TRAGW caracole, corkscrew staircase, corkscrew stairs, helical staircase, spiral stairs

**Wendelung** *f* ARCH & TRAGW *Treppe* winding
**Wende**: **Wendeplatz** *m* INFR & ENTW turning bay; **Wendepunkt** *m* INFR & ENTW turning point; **Wenderadius** *m* INFR & ENTW turning radius
**Wendigkeit** *f* INFR & ENTW versatility
**Werk** *nt* INFR & ENTW factory, work; **Werkbank** *f* ARCH & TRAGW, BAUMASCHIN bench, workbench; **Werkkanal** *m* UMWELT headrace canal; **Werkleimung** *f* HOLZ factory bonding; **Werkplan** *m* ARCH & TRAGW, INFR & ENTW working drawing
**werkseitig**: ~ **eingebaut** *adj* ELEKTR *Verteiler* factory-mounted
***Werk***: **Werkstattverbindung** *f* HOLZ, STAHL workshop connection; **Werkstein** *m* BESCHLÄGE, INFR & ENTW, NATURSTEIN, WERKSTOFF artificial stone; **Werksteinmauerwerk** *nt* ARCH & TRAGW, NATURSTEIN ashlar masonry, ashlaring, hewn stone masonry, natural stone masonry; **Werkstoff** *m* UMWELT materials, WERKSTOFF material; **Werkstoffnutzungszyklus** *m* UMWELT utilization cycle of materials; **Werksvorfertigung** *f* BETON off-site casting
**Werkzeug** *nt* BAUMASCHIN tool, utensil
**Wert** *m* ARCH & TRAGW, BAURECHT, INFR & ENTW value; **Wertsteigerung** *f* **von Grund und Boden** BAURECHT betterment; **Wertstoffrückgewinnung** *f* UMWELT resource recovery
**wesentlich** *adj* BAURECHT elemental; ~**e Kostenanalyse** *f* BAURECHT elemental cost analysis
**Wetter** *nt* UMWELT weather
**wetterbeständig** *adj* OBERFLÄCHE, WERKSTOFF weather-resistant, weatherproof
***Wetter***: **Wetterbrett** *nt* HOLZ barge board; **Wetterdach** *nt* ARCH & TRAGW penthouse, *Laderampe* station roof; **Wetterdaten** *nt pl* BAURECHT weather data; **Wetterfahne** *f* INFR & ENTW vane; **Wettermanschette** *f* ARCH & TRAGW weathering collar; **Wetterschenkel** *m* ARCH & TRAGW water drip, weather drip, weather groove; **Wetterschürze** *f* ARCH & TRAGW weather boarding; **Wetterschutzabdeckung** *f* BESCHLÄGE weathering; **Wetterschutzdach** *nt* ARCH & TRAGW, BESCHLÄGE hood; **Wetterseite** *f* ARCH & TRAGW side opposed to the weather
**Wichte** *f* WERKSTOFF specific weight
**Wickelfalzrohr** *nt* ABWASSER, ARCH & TRAGW folded spiral-seam tube
**wickeln** *vt* INFR & ENTW wind
**Widder** *m* BAUMASCHIN hydraulic ram
**Widerhakenbolzen** *m* BESCHLÄGE barbed bolt
**Widerlager** *nt* ARCH & TRAGW bearing pad, BETON abutment, INFR & ENTW abutment, bearing pad, NATURSTEIN, STAHL abutment; **Widerlagerabdeckung** *f* ARCH & TRAGW abutment flashing; **Widerlagerdruck** *m* BETON, INFR & ENTW, STAHL, WERKSTOFF abutment pressure; **Widerlagerstein** *m* INFR & ENTW, NATURSTEIN, WERKSTOFF abutment stone
**Widerstand** *m* ELEKTR resistance; **Widerstandsbeiwert** *m* (*DD*) UMWELT coefficient of drag, WERKSTOFF resistance coefficient
**widerstandsfähig** *adj* ELEKTR resistant, WERKSTOFF fast, resistant, proof
***Widerstand***: **Widerstandsfähigkeit** *f* WERKSTOFF resistance; **Widerstandsmoment** *nt* ARCH & TRAGW, HOLZ moment of resistance; **Widerstandsnahtschweißen** *nt* STAHL resistance seam welding; **Widerstandsschweißen** *nt* STAHL resistance welding; **Widerstandsschweißung** *f* STAHL resistance welding; **Widerstandsstumpfschweißen** *nt* STAHL resistance butt welding
**Wieder-** *in cpds* ARCH & TRAGW re-; **Wiederaufbau** *m* ARCH & TRAGW rebuilding, reconstruction, redevelopment
**wiederaufbauen** *vt* ARCH & TRAGW rebuild, reconstruct
***Wieder-***: **Wiederaufbauen** *nt* ARCH & TRAGW reconstruction
**wiederbelasten** *vt* ARCH & TRAGW, INFR & ENTW reload
**wiedererwärmen** *vt* HEIZ & BELÜFT reheat
**wiedergewinnen** *vt* UMWELT *Rohstoffe* recover
***Wieder-***: **Wiedergewinnung** *f* BAURECHT, INFR & ENTW recycling, UMWELT recycling, *von Rohstoffen* recovery; **Wiedergewinnungsanlage** *f* BAURECHT, INFR & ENTW, UMWELT reclamation plant, recycling plant, resource recovery plant
**wiedergewonnen**: ~**es Öl** *nt* UMWELT recovered oil; ~**es Polypropylen** *nt* WERKSTOFF recycled polypropylene
**wiederherstellen** *vt* ARCH & TRAGW renew, restore
***Wieder-***: **Wiederherstellung** *f* ARCH & TRAGW redevelopment; **Wiederherstellung** *f* **von Stromversorgung** ELEKTR reconnecting electrical supplies; **Wiederinbesitznahme** *f* BAURECHT repossession; **Wiederkonsolidierung** *f* INFR & ENTW reconsolidation; **Wiedernutzbarmachung** *f* BAURECHT, UMWELT rehabilitation
**wiederurbargemacht**: ~**es Gebiet** *nt* UMWELT reclaimed area
***Wieder-***: **Wiederurbarmachung** *f* UMWELT recultivation
**wiederverwendbar** *adj* WERKSTOFF reusable
**wiederverwenden** *vt* UMWELT reuse
***Wieder-***: **Wiederverwendung** *f* UMWELT reuse
**wiederverwendungsfähig**: ~**e Schalung** *f* BETON fit-up
**wiederverwertbar** *adj* UMWELT recyclable, reusable; ~**es Abfallprodukt** *nt* UMWELT reusable waste product
**wiederverwerten** *vt* UMWELT recycle, reuse
***Wieder-***: **Wiederverwertung** *f* UMWELT recovery, *von Abfallstoffen* recycling
**wiegen** *vt* WERKSTOFF balance
**wild**: ~**e Müllablagerung** *f* UMWELT open dump, uncontrolled tipping (*BrE*)
**windabwärts** *adv* INFR & ENTW downwind
**Wind**: **durch** ~ **erzeugt** *adj* ARCH & TRAGW wind-induced; **Windaussteifung** *f* ARCH & TRAGW sprocket piece, wind bracing, HOLZ cross lath, wind bracing, STAHL cross lath; **Windbelastung** *f* ARCH & TRAGW wind loading; **Windbrett** *nt* ARCH & TRAGW side board; **Winddruck** *m* ARCH & TRAGW, INFR & ENTW wind pressure
**Winde** *f* BAUMASCHIN gin, jack, winch
***Wind***: **Windenergie** *f* UMWELT wind energy; **Windfang** *m* ARCH & TRAGW porch, draft lobby (*AmE*), draught lobby (*BrE*), INFR & ENTW draft lobby (*AmE*), draught lobby (*BrE*); **Windfangtür** *f* ARCH & TRAGW vestibule door; **Windflügel** *m* INFR & ENTW vane; **Windgeschwindigkeit** *f* UMWELT wind velocity; **Windgeschwindigkeitsdaten** *pl* UMWELT wind speed data; **Windgeschwindigkeitsmesser** *m* UMWELT wind gage (*AmE*), wind gauge (*BrE*); **Windkanal** *m* INFR & ENTW wind tunnel; **Windkraft** *f* UMWELT wind

force, wind power; **Windlast** *f* ARCH & TRAGW *Leeseite* wind loading, INFR & ENTW wind load; **Windlatte** *f* ARCH & TRAGW sprocket piece, HOLZ cross lath; **Windmotorpumpe** *f* UMWELT windmill pump; **Windmühle** *f* UMWELT windmill; **Windmühlenflügel** *m* UMWELT windmill vane; **Windmühlenrad** *nt* UMWELT windmill; **Windrad** *nt* UMWELT windmill; **Windrispe** *f* ARCH & TRAGW, HOLZ cross lath; **Windrute** *f* ARCH & TRAGW sprocket piece, HOLZ cross lath; **Windscheibe** *f* ARCH & TRAGW *zur Lastverteilung* shear wall; **Windschutz** *m* ARCH & TRAGW, UMWELT wind protection; **Windsichter** *m* UMWELT air separator, air classifier, air separation plant; **Windsichtung** *f* UMWELT airstream sorting; **Windsog** *m* ARCH & TRAGW, INFR & ENTW wind suction, wind uplift; **Windsogkraft** *f* ARCH & TRAGW, INFR & ENTW wind uplift force; **Windstille** *f* UMWELT still air; **Windstrebe** *f* ARCH & TRAGW, HOLZ, INFR & ENTW wind brace; **Windströmung** *f* ARCH & TRAGW, INFR & ENTW wind flow; **Windturbinenpumpe** *f* UMWELT windmill pump

**Windung** *f* ARCH & TRAGW turn, *Treppe* winding, ELEKTR turn, winding

***Wind***: **Windverband** *m* ARCH & TRAGW lateral bracing, transverse bracing, wind sway bracing, HOLZ, INFR & ENTW, STAHL cross bracing, wind brace, wind bracing

**Winkel** *m* ABWASSER, ARCH & TRAGW, HEIZ & BELÜFT, INFR & ENTW, STAHL angle; **Winkelband** *nt* BESCHLÄGE angle hinge; **Winkelbeschleunigung** *f* INFR & ENTW angular acceleration; **Winkeleckleiste** *f* BESCHLÄGE nosing; **Winkelfenster** *nt* ARCH & TRAGW splayed window; **Winkelgeschwindigkeit** *f* UMWELT angular velocity; **Winkelgetriebe** *nt* BAUMASCHIN miter gear (*AmE*), mitre gear (*BrE*); **Winkelgrad** *m* ARCH & TRAGW angle degree; **Winkelkopf** *m* ARCH & TRAGW cross-staff head; **Winkellasche** *f* BESCHLÄGE knee brace; **Winkelprofil** *nt* BESCHLÄGE, BETON, INFR & ENTW, STAHL, WERKSTOFF angle bar; **Winkelstahlpfette** *f* HOLZ, INFR & ENTW, STAHL, WERKSTOFF angled purlin; **Winkelstück** *nt* ABWASSER *Rohr*, BESCHLÄGE bend; **Winkelstufe** *f* ARCH & TRAGW angle step, solid rectangular step, tread and riser; **Winkelstütze** *f* STAHL angular bracket

**Winterdienstmaschinen und Winterdienstgeräte** *f pl* BAUMASCHIN machinery and equipment for winter

**Wintergarten** *m* ARCH & TRAGW, HEIZ & BELÜFT *als thermischer Puffer* conservatory

**Wippkran** *m* BAUMASCHIN luffing crane

**Wippsäge** *f* BAUMASCHIN jigsaw

**Wirbel** *m* ABWASSER whirlpool; **Wirbelbett** *nt* HEIZ & BELÜFT fluidized bed; **Wirbelkammer** *f* INFR & ENTW swirling chamber; **Wirbelkammerdrossel** *f* INFR & ENTW swirling chamber control outlet; **Wirbelströmung** *f* ABWASSER, INFR & ENTW, UMWELT turbulence, turbulent flow

**Wirkungsgrad** *m* ARCH & TRAGW efficiency factor

**Wirtschaftsprüfer** *m* BAURECHT auditor

**wischbeständig** *adj* OBERFLÄCHE *Farbe* wipe-resistant

**Witterung** *f* UMWELT weather

**witterungsbeständig** *adj* OBERFLÄCHE weather-resistant, WERKSTOFF weatherproof, weather-resistant

**witterungsgeführt**: **~e Regelung** *f* HEIZ & BELÜFT *Heizung* weather-dependent control

***Witterung***: **Witterungsverhältnisse** *nt pl* UMWELT atmospheric conditions

**Wohn-** *in cpds* ARCH & TRAGW, INFR & ENTW residential; **Wohndichte** *f* INFR & ENTW housing density; **Wohnetage** *f* ARCH & TRAGW apartment floor; **Wohnhaus** *nt* ARCH & TRAGW residential building

**Wohnhausbau**: **nicht zum ~ gehörend** *phr* ARCH & TRAGW, INFR & ENTW nondomestic

***Wohn-***: **Wohn- und Nutzfläche** *f* ARCH & TRAGW, INFR & ENTW floor space; **Wohnsiedlung** *f* INFR & ENTW estate; **Wohnsitz** *m* ARCH & TRAGW home; **Wohnturm** *m* INFR & ENTW tower block

**Wohnung** *f* ARCH & TRAGW apartment, flat (*BrE*), home; **Wohnungsbau** *m* ARCH & TRAGW, INFR & ENTW domestic architecture, house building, housing, housing construction; **Wohnungseinheit** *f* ARCH & TRAGW, INFR & ENTW unit of accommodation; **Wohnungstrennwand** *f* ARCH & TRAGW party wall

**wölben** *vt* ARCH & TRAGW curve, *Dach* vault, WERKSTOFF bend

**Wölbung** *f* ARCH & TRAGW arch, coving, vault, crowning, INFR & ENTW, WERKSTOFF coving; **Wölbungsrücken** *m* ARCH & TRAGW extrados

**Wolkenkratzer** *m* ARCH & TRAGW, INFR & ENTW skyscraper

**wolkig** *adj* OBERFLÄCHE, WERKSTOFF cloudy

**Wolle** *f* DÄMMUNG wool

**Wrasen** *m* HEIZ & BELÜFT vapor (*AmE*), vapour (*BrE*); **Wrasenabzug** *m* HEIZ & BELÜFT air chimney, air flue; **Wrasenrohr** *nt* HEIZ & BELÜFT *Lüftungsrohr in Küche* vapor pipe (*AmE*), vapour pipe (*BrE*)

**Wuchtbaum** *m* BAUMASCHIN lifter

**Wulst** *m* ARCH & TRAGW bead, BAUMASCHIN flange, WERKSTOFF bulge

**Wurfweite** *f*: **~ eines Regners** ABWASSER, BAURECHT sprinkler reach

**Wurzel** *f* INFR & ENTW root; **Wurzelbeseitigung** *f* INFR & ENTW uprooting

**W/Z**: **~ Wert** *m* (*Wasserzementwert*) ABWASSER, BETON, NATURSTEIN, WERKSTOFF w/c ratio (*water-cement ratio*)

# X Y Z

**X-Achse** *f* ARCH & TRAGW x-axis

**Y-Achse** *f* ARCH & TRAGW y-axis
**Yard** *nt* INFR & ENTW yard

**Zähflüssigkeit** *f* WERKSTOFF viscosity
**Zähigkeit** *f* WERKSTOFF consistence, consistency
**Zahl** *f* ARCH & TRAGW figure; **Zahlenkombinationsschloß** *nt* BESCHLÄGE combination lock
**Zähler** *m* ELEKTR counter, meter; **Zählertafel** *f* ELEKTR meter board
**Zahn** *m* BAUMASCHIN, BESCHLÄGE, HOLZ, NATURSTEIN tooth; **Zahnankerplatte** *f* BESCHLÄGE, HOLZ toothed-plate connector; **Zahnhobel** *m* BAUMASCHIN tooth plane, toothing plane; **Zahnrad** *nt* BAUMASCHIN gear wheel; **Zahnspachtel** *m* BAUMASCHIN *Fliesenarbeiten* serrated trowel; **Zahnstein** *m* NATURSTEIN *Mauerwerk* toothing stone; **Zahntriebtürschließer** *m* BESCHLÄGE rack-and-pinion door closer
**Zahnung** *f* NATURSTEIN toothing
**Zange** *f* BAUMASCHIN tongs
**Zapfen** *m* ARCH & TRAGW tenon, *eines Hahns* spigot, BESCHLÄGE pintle, gudgeon, *Tür* pivot, HOLZ tenon, INFR & ENTW, STAHL, WERKSTOFF lug; **um einen ~ drehen** *phr* INFR & ENTW pivot; **Zapfenloch** *nt* HOLZ mortise; **Zapfenlochverbindung** *f* INFR & ENTW slot mortise joint; **Zapfenschlitz** *m* ARCH & TRAGW, through mortise; **Zapfenstreichmaß** *nt* HOLZ mortise gage (*AmE*), mortise gauge (*BrE*); **Zapfenverbindung** *f* HOLZ mortise and tenon joint
**Zapfstelle** *f* ABWASSER tapping point
**Zarge** *f* ARCH & TRAGW, HOLZ *Fenster & Türen*, INFR & ENTW, STAHL frame
**Zaun** *m* INFR & ENTW fence; **Zaunpaneel** *nt* INFR & ENTW, WERKSTOFF fence panel; **Zaunpfosten** *m* INFR & ENTW, WERKSTOFF fence post
**Zedernschindel** *f* HOLZ, WERKSTOFF cedar shingle
**Zehneck** *nt* ARCH & TRAGW decagon
**zehnsäulig** *adj* ARCH & TRAGW decastyle
**Zeichenbrett** *nt* ARCH & TRAGW drawing board, drawing table
**Zeichenpapier** *nt* ARCH & TRAGW plotting paper
**Zeichnung** *f* ARCH & TRAGW drawing
**Zeit** *f* ARCH & TRAGW, BAURECHT, WERKSTOFF time; **Zeitkonsolidationskurve** *f* WERKSTOFF time consolidation curve
**zeitlich**: **~e Schwankung** *f* UMWELT temporal fluctuation, temporal variation; **~ unbegrenzt** *adj* BAURECHT open-ended
***Zeit***: **Zeitschalter** *m* BAURECHT, ELEKTR time switch; **Zeitschaltuhr** *f* BESCHLÄGE, ELEKTR automatic timer; **Zeitschwingfestigkeit** *f* WERKSTOFF fatigue life; **Zeitverlängerung** *f* BAURECHT extension of time, *Vertrag* extension
**zeitweilig** *adj* ARCH & TRAGW temporary; **~e Belastung** *f* ARCH & TRAGW, BETON, INFR & ENTW, STAHL temporary load
***Zeit***: **Zeitzugabe** *f* ARCH & TRAGW, BAURECHT, INFR & ENTW extra time allowance

**Zelle** *f* ARCH & TRAGW cell, *Raumzelle* cubicle, WERKSTOFF *Hohlkastenträger* compartment; **Zellenbauweise** *f* ARCH & TRAGW cellular construction, cellular design, cubicle construction; **Zellenmodul** *nt* ARCH & TRAGW modular unit; **Zellenmörtel** *m* NATURSTEIN cellular mortar; **Zellenziegel** *m* NATURSTEIN, WERKSTOFF cellular brick
**Zell-Polyurethan** *nt*: **~ hoher Dichte** DÄMMUNG high-density polyurethane foam
**Zellstoff** *m* UMWELT paper pulp; **Zellstoffwatte** *f* DÄMMUNG, HEIZ & BELÜFT wadding
**Zeltdach** *nt* ARCH & TRAGW tent roof
**Zeltstoff** *m* ARCH & TRAGW, WERKSTOFF canvas
**Zement** *m* BETON cement; **~ für Bauzwecke** BETON, INFR & ENTW, WERKSTOFF masonry cement; **Zementanteil** *m* BETON, NATURSTEIN cement content; **Zementbeton** *m* BETON cement concrete; **Zementbrei** *m* NATURSTEIN grouting compound; **Zementdachstein** *m* BETON, NATURSTEIN cement roofing tile; **Zementersatz** *m* BETON cement substitute; **Zementestrich** *m* BETON cement floor; **Zementfabrik** *f* BAUMASCHIN, INFR & ENTW cement mill, cement plant; **Zementfestigkeit** *f* BETON cement strength
**zementgebunden** *adj* BETON, WERKSTOFF cement-bound
***Zement***: **Zementgehalt** *m* BETON, NATURSTEIN cement content; **Zementglattstrich** *m* ARCH & TRAGW cement trowel finish; **Zementgüte** *f* BETON cement grade
**zementhaltig** *adj* BETON cement-based
**Zementierdruck** *m* NATURSTEIN injection pressure
**zementieren** *vt* BETON cement
**Zementierpumpe** *f* BAUMASCHIN injection pump
**Zementierung** *f* BETON, NATURSTEIN cementation, cementing
**Zementkalkputz** *m* NATURSTEIN lime-cement plaster
**Zementkanone** *m*: **mit der ~ angeworfener Mörtel** NATURSTEIN pneumatically-applied mortar
***Zement***: **Zementkorn** *nt* BETON cement grain; **Zementleim** *m* BETON cement paste, neat cement; **Zementmilch** *f* WERKSTOFF slurry; **Zementmörtel** *m* BETON, NATURSTEIN, WERKSTOFF cement mortar; **Zementmühle** *f* BAUMASCHIN, INFR & ENTW cement mill; **Zementnorm** *f* BAURECHT, BETON cement standard; **Zementpaste** *f* BETON cement paste, neat cement; **Zementplatte** *f* BETON cement tile; **Zementprobe** *f* BETON, WERKSTOFF cement test; **Zementprüfung** *f* BETON, WERKSTOFF cement test; **Zementputz** *m* (*CPL*) NATURSTEIN cement plaster (*CEM PLAS*)
**zementreich**: **~er Beton** *m* BETON cement-rich concrete
***Zement***: **Zementschlamm** *m* BETON cement grout, neat cement; **Zementschleier** *m* BETON laitance, cement film, NATURSTEIN cement skin, OBERFLÄCHE cement film, laitance; **Zementschwinden** *nt* BETON shrinkage in cement; **Zementsperrschicht** *f* DÄMMUNG cement-based waterproof coating; **Zementstaub** *m* BETON cement dust; **Zement-**

**verpressung** *f* BETON, INFR & ENTW cement injection; **Zement-Wasser-Verhältnis** *nt* BETON, NATURSTEIN cement-water ratio; **Zementwerk** *nt* BAUMASCHIN cement plant, cement mill; **Zement-Zuschlagstoff-Verhältnis** *nt* BETON cement-aggregate ratio
**Zenitwinkel** *m* UMWELT zenith angle
**zentral** *adj* ARCH & TRAGW central
**Zentral-** *in cpds* ARCH & TRAGW, BESCHLÄGE, HEIZ & BELÜFT, INFR & ENTW central
***zentral***: **~er Obelisk** *m* INFR & ENTW central spine
**zentralbeheizt** *adj* HEIZ & BELÜFT centrally heated
**Zentrale** *f* HEIZ & BELÜFT central station
***Zentral-***: **Zentralheizung** *f* HEIZ & BELÜFT central heating; **Zentralhof** *m* ARCH & TRAGW central court; **Zentralschließanlage** *f* BESCHLÄGE central master-keyed system; **Zentralverschluß** *m* BESCHLÄGE central locking
**zentriert** *adj* ARCH & TRAGW centered (*AmE*), centred (*BrE*)
**Zentrifugal-** *in cpds* ABWASSER, BAUMASCHIN, UMWELT centrifugal; **Zentrifugalfettabscheider** *m* ABWASSER centrifugal grease interceptor; **Zentrifugalkraft** *f* UMWELT centrifugal force; **Zentrifugalpumpe** *f* BAUMASCHIN centrifugal pump
**Zentrifugieren** *nt* UMWELT centrifuging, centrifugation
**Zentrum** *nt* ARCH & TRAGW, INFR & ENTW center (*AmE*), centre (*BrE*)
**Zeolith** *m* BETON, WERKSTOFF zeolite; **Zeolithzementverbundstoff** *m* BETON zeolite cement composite
**zerkeilen** *vt* ARCH & TRAGW split with wedges
**Zerkleinerer** *m* BAUMASCHIN, UMWELT crusher, pulverizer
**zerkleinern** *vt* BAUMASCHIN *Abfall, Beton* comminute, WERKSTOFF break, crush, pulverize
**zerkleinert** *adj* WERKSTOFF crushed
**Zerkleinerung** *f* UMWELT crushing; **Zerkleinerungsanlage** *f* BAUMASCHIN crusher unit; **Zerkleinerungsmaschine** *f* BAUMASCHIN crusher, crushing machine; **Zerkleinerungsmaschinen** *f pl* BAUMASCHIN *für Baustoffe* crushing machinery
**zerlegt**: **~er Bügel** *m* BETON split hoop, split loop
**zermahlen** *vt* NATURSTEIN, WERKSTOFF mill
**Zerrbalken** *m* ARCH & TRAGW flexible foundation beam
**zerreißen** *vt* ARCH & TRAGW, INFR & ENTW tear
**Zerreißen** *nt* UMWELT *von Abfällen* crushing
**zerspringen 1.** *vt* WERKSTOFF break; **2.** *vi* BETON, NATURSTEIN, WERKSTOFF crack
**zerstäuben** *vt* WERKSTOFF pulverize
**Zerstäubung** *f* UMWELT *von Abfällen* grinding, WERKSTOFF atomization
**zerstörungsfrei**: **~es Testen** *nt* WERKSTOFF nondestructive testing; **~e Untersuchungstechniken** *f pl* BAURECHT nondestructive inspection techniques
**Zertifikat** *nt* BAURECHT certificate
**Zeugnis** *nt* BAURECHT certificate, evidence
**Zickzackpunktschweißung** *f* STAHL staggered spot welding
**Ziegel** *m* NATURSTEIN brick; **Ziegelbau** *m* ARCH & TRAGW, NATURSTEIN clay brick building; **Ziegelbrennmuster** *nt* NATURSTEIN, WERKSTOFF kiss marks on bricks; **Ziegelerde** *f* WERKSTOFF brick earth; **Ziegelformat** *nt* NATURSTEIN brick facing, brick size; **Ziegelgewölbe** *nt* ARCH & TRAGW, NATURSTEIN brick vault; **Ziegelgröße** *f* NATURSTEIN brick size; **Ziegelkamin** *m* NATURSTEIN brick fireplace; **Ziegellager** *nt* NATURSTEIN brickfield; **Ziegellieferant** *m* NATURSTEIN brick supplier; **Ziegelmaß** *nt* NATURSTEIN brick dimension; **Ziegelmauer** *f* NATURSTEIN brick wall; **Ziegelmuster** *nt* NATURSTEIN brick pattern
**Ziegeln**: **mit ~ mauern** *phr* NATURSTEIN brick
***Ziegel***: **Ziegelofen** *m* HEIZ & BELÜFT, NATURSTEIN brick kiln; **Ziegelpflaster** *nt* INFR & ENTW, NATURSTEIN brick pavement; **Ziegelpflasterung** *f* INFR & ENTW, NATURSTEIN brick paving; **Ziegelrabitz** *m* WERKSTOFF clay lathing, clayed wire mesh; **Ziegelsplitt** *m* ARCH & TRAGW, NATURSTEIN, UMWELT brick chippings, WERKSTOFF crushed brick aggregate; **Ziegelsplittbeton** *m* WERKSTOFF broken brick concrete, crushed brick aggregate concrete; **Ziegelsteinverband** *m* ARCH & TRAGW, NATURSTEIN brickwork; **Ziegelton** *m* NATURSTEIN brick clay, WERKSTOFF loam; **Ziegelverkleidung** *f* NATURSTEIN brick facing; **Ziegelwand** *f* NATURSTEIN brick wall
**Zieh-** *in cpds* ARCH & TRAGW, BAUMASCHIN, BESCHLÄGE, UMWELT drawing, pulling
**ziehen** *vt* ARCH & TRAGW draw
**Zieher** *m* BESCHLÄGE *Schloß* pull
***Zieh-***: **Ziehgriff** *m* BESCHLÄGE pull; **Ziehklinge** *f* BAUMASCHIN scraper, spokeshave; **Ziehschütze** *f* UMWELT sliding sluice
**Ziel** *nt* ARCH & TRAGW objective; **Zielachse** *f* INFR & ENTW collimation line
**Zielen** *nt* ARCH & TRAGW sighting
***Ziel***: **Zielfernrohr** *nt* ARCH & TRAGW, BAUMASCHIN sighting telescope; **Zielvertrag** *m* BAURECHT target contract
**Zier-** *in cpds* ARCH & TRAGW, BESCHLÄGE, BETON, HOLZ, STAHL decorative, ornamental; **Zierbeschläge** *m pl* BESCHLÄGE decorative fittings; **Zierbeton** *m* BETON ornamental concrete; **Zierbogen** *m* ARCH & TRAGW decorative arch; **Zierdecke** *f* ARCH & TRAGW decorated ceiling; **Zierfliese** *f* ARCH & TRAGW, BESCHLÄGE, NATURSTEIN decorative tile; **Ziergiebel** *m* ARCH & TRAGW decorative gable, *Gothik* gablet, HOLZ pediment; **Ziergitter** *nt* STAHL decorative gable; **Zierleiste** *f* ARCH & TRAGW, BESCHLÄGE fillet, molding (*AmE*), moulding (*BrE*), HOLZ molding (*AmE*), moulding (*BrE*), decorative batten; **Ziertür** *f* ARCH & TRAGW, BESCHLÄGE decorated door
**Ziffer** *f* ARCH & TRAGW figure
**Zimmer** *nt* ARCH & TRAGW chamber, room; **Zimmerarbeiten** *f pl* HOLZ carpenter's work
**Zimmerei** *f* HOLZ carpentry
**Zimmerer** *m* HOLZ carpenter
***Zimmer***: **Zimmerhandwerk** *nt* HOLZ carpentry
**Zimmermann** *m* HOLZ carpenter; **Zimmermannsbleistift** *m* BAUMASCHIN carpenter's pencil; **Zimmermannshammer** *m* BAUMASCHIN, BESCHLÄGE, HOLZ claw hammer
**zimmermannsmäßig**: **~e Dachkonstruktion** *f* HOLZ timber roof structure
***Zimmer***: **Zimmerwerkstatt** *f* HOLZ carpenter's shop; **Zimmerwerkzeug** *nt* BAUMASCHIN carpenter's tool
**Zink** *nt* WERKSTOFF zinc; **Zinkdruckguß** *m* WERKSTOFF zinc diecasting
**Zinkenverbindung** *f* HOLZ finger joint
***Zink***: **Zinkfarbe** *f* OBERFLÄCHE zinc paint; **Zinkschutzfarbe** *f* OBERFLÄCHE zinc paint; **Zinkstaubfarbe** *f* OBERFLÄCHE cold-galvanizing paint

**Zinn** *nt* WERKSTOFF tin
**Zinne** *f* ARCH & TRAGW, NATURSTEIN battlement; **Zinnen** *f pl* ARCH & TRAGW, NATURSTEIN battlement; **Zinnenkranz** *m* ARCH & TRAGW crenelation (*AmE*), crenellation (*BrE*); **Zinnenturm** *m* ARCH & TRAGW, INFR & ENTW battlement tower
**Zinnlöten** *nt* WERKSTOFF soldering
**Zinnober** *m* OBERFLÄCHE *Farbe* cinnabar
**Zirkel** *m* ARCH & TRAGW, BAUMASCHIN, INFR & ENTW compass
**Zirkulation** *f* HEIZ & BELÜFT circulation; **Zirkulationsleitung** *f* HEIZ & BELÜFT circulation line
**Zisterne** *f* ABWASSER, INFR & ENTW cistern
**Zitadelle** *f* ARCH & TRAGW citadel
**Zivilisationsmüll** *m* UMWELT waste products of civilization
**Zoll** *m* ABWASSER, ARCH & TRAGW, INFR & ENTW, WERKSTOFF inch; **Zollstock** *m* BAUMASCHIN carpenter's gage (*AmE*), carpenter's gauge (*BrE*)
**Zone** *f* ARCH & TRAGW zone
**Zubehör** *nt* BESCHLÄGE accessories
**zubilligen** *vt* BAURECHT allow
**Zubringer** *m* INFR & ENTW *Förderband* feeder line, *Straße* feeder road; **Zubringerstraße** *f* INFR & ENTW feeder road
**Zuckerrohrfaserdämmung** *f* DÄMMUNG cane fiber insulation (*AmE*), cane fibre insulation (*BrE*)
**Zufahrt** *f* ARCH & TRAGW, INFR & ENTW access, approach; **~ für Rettungsdienste** INFR & ENTW access for emergency services; **Zufahrtsrampe** *f* ARCH & TRAGW, INFR & ENTW access ramp; **Zufahrtsstraße** *f* ARCH & TRAGW, INFR & ENTW access road, approach road, slip road; **Zufahrtstor** *nt* ARCH & TRAGW, INFR & ENTW access gate
**zufällig**: **~er Ausfluß** *m* UMWELT accidental discharge
**zufriedenstellend** *adj* ARCH & TRAGW *Qualität* satisfactory
**Zufuhr** *f* HEIZ & BELÜFT intake
**Zuführung** *f* ABWASSER inlet, ELEKTR supply, INFR & ENTW inlet; **Zuführungskabel** *nt* ELEKTR power supply cable; **Zuführungsleitung** *f* ELEKTR power supply line; **Zuführungsschnecke** *f* BAUMASCHIN charging screw
**Zug** *m* ARCH & TRAGW tension; **Zugabfangung** *f* ELEKTR *Kabel* traction relief
**Zugang** *m* ARCH & TRAGW, INFR & ENTW access, approach
**zugänglich** *adj* ARCH & TRAGW accessible, open, BAUMASCHIN, BAURECHT, INFR & ENTW accessible; **~e Reservenquelle** *f* UMWELT accessible resource base
***Zugang***: **Zugangsstollen** *m* INFR & ENTW *Tunnelbau* adit; **Zugangstunnel** *m* ARCH & TRAGW, INFR & ENTW access tunnel
***Zug***: **Zuganker** *m* ARCH & TRAGW tension bar, tie bar, tie rod, stay; **Zugarmierung** *f* BETON tensile reinforcement; **Zugband** *nt* ARCH & TRAGW tension bar, tie bar
**zugbeansprucht** *adj* ARCH & TRAGW, BETON, STAHL, WERKSTOFF subject to tension; **~e Konstruktion** *f* ARCH & TRAGW tension structure
**zugbelastbar** *adj* WERKSTOFF tensile
***Zug***: **Zugbewehrung** *f* BETON tensile reinforcement; **Zugbinder** *m* NATURSTEIN through binder; **Zugdeichsel** *f* ARCH & TRAGW, BAUMASCHIN, STAHL tongue; **Zugdiagonale** *f* HOLZ diagonal tie
**zugeben** *vt* BETON, INFR & ENTW, WERKSTOFF *Zusatzstoffe, Zuschlagstoffe* add
**zugelassen**: **~es Produkt** *nt* INFR & ENTW, WERKSTOFF approved product
***Zug***: **Zugentlastung** *f* ELEKTR *Kabel* traction relief; **Zugfestigkeit** *f* WERKSTOFF tensile strength; **Zugglied** *nt* ARCH & TRAGW tension bar, tie bar; **Zuggurt** *m* ARCH & TRAGW tension flange, tension boom, HOLZ, INFR & ENTW, STAHL tension flange; **Zugkraft** *f* ARCH & TRAGW, BAUMASCHIN, ELEKTR traction, INFR & ENTW tensile force; **Zuglasche** *f* HOLZ fishplate; **Zuglast** *f* INFR & ENTW tensile load; **Zugluft** *f* ARCH & TRAGW draft air (*AmE*), draught air (*BrE*); **Zugluftschutz** *m* DÄMMUNG, HEIZ & BELÜFT draft excluder (*AmE*), draught excluder (*BrE*); **Zugluftumleiter** *m* DÄMMUNG, HEIZ & BELÜFT draft diverter (*AmE*), draught diverter (*BrE*); **Zugmaschine** *f* BAUMASCHIN tractor; **Zugpfahl** *m* INFR & ENTW tension pile; **Zugseil** *nt* ARCH & TRAGW stay, BAUMASCHIN haulage rope; **Zugspannung** *f* ARCH & TRAGW, WERKSTOFF tensile stress; **Zugstab** *m* ARCH & TRAGW *Zugprüfung*, INFR & ENTW tension bar, tension rod, tie bar; **Zugstange** *f* ARCH & TRAGW, STAHL tie, traction rod; **Zugstrebe** *f* HOLZ diagonal tie; **Zugverhältnisse** *nt pl* ARCH & TRAGW *Schornstein* flue conditions; **Zugversagen** *nt* WERKSTOFF tension failure; **Zugversuch** *m* INFR & ENTW, WERKSTOFF tension test
**Zuhaltung** *f* BESCHLÄGE *Schloß* tumbler; **Zuhaltungsschloß** *nt* BESCHLÄGE tumbler lock
**zukleben** *vt* BAUMASCHIN, BESCHLÄGE tape
**Zulage** *f* ARCH & TRAGW, BAURECHT, INFR & ENTW extra charge, surcharge
**zulassen** *vt* BAURECHT admit, allow, permit
**zulässig** *adj* ARCH & TRAGW, BAUMASCHIN, BAURECHT, BESCHLÄGE, HOLZ, INFR & ENTW, STAHL, UMWELT, WERKSTOFF admissible, allowable, permissible; **~e Abweichung** *f* ARCH & TRAGW, BAURECHT, BETON, INFR & ENTW, STAHL, WERKSTOFF allowable tolerance, allowance, permissible variation, tolerance; **~e Biegetragfähigkeit** *f* ARCH & TRAGW, BETON, HOLZ, INFR & ENTW, STAHL, WERKSTOFF design bending resistance; **~er Druck** *m* BETON, HOLZ, INFR & ENTW, STAHL allowable pressure; **~e Last** *f* ARCH & TRAGW, BAUMASCHIN allowable load; **~e Leimart** *f* BESCHLÄGE, HOLZ permissible adhesive type; **~e Pfahllast** *f* INFR & ENTW allowable pile load; **~e plastische Schubtragfähigkeit** *f* ARCH & TRAGW design plastic shear resistance; **~e Spannung** *f* BETON, HOLZ, INFR & ENTW, STAHL allowable stress; **~e Traglast** *f* BETON, HOLZ, INFR & ENTW, STAHL, WERKSTOFF allowable bearing capacity
**Zulassungsbescheid** *m* ARCH & TRAGW, BAURECHT certificate of approval
**Zulauf** *m* ABWASSER, HEIZ & BELÜFT *Flüssigkeit in einer Pumpe* admission, INFR & ENTW inlet
**Zuleitung** *f* ABWASSER, HEIZ & BELÜFT, INFR & ENTW inlet pipe, supply pipe; **Zuleitungsrohr** *nt* ABWASSER, HEIZ & BELÜFT, INFR & ENTW inlet pipe, supply pipe
**Zuluft** *f* HEIZ & BELÜFT supply air
**zumauern** *vt* NATURSTEIN block up, wall up, brick up, wall in
**Zunahme** *f* ARCH & TRAGW, BAURECHT *Kosten* increase
**Zündschnur** *f* INFR & ENTW *Sprengen* firing tape
**zunehmend**: **~e Erwärmung** *f* UMWELT incremental heating

**Zunge** *f* ARCH & TRAGW tongue
**Zuputzen** *nt* NATURSTEIN plastering-in
**Zurichten** *nt* HOLZ dressing
**Zurichthammer** *m* BAUMASCHIN maul
**zurückgesetzt** *adj* ARCH & TRAGW recessed
**zurückgewinnen** *vt* UMWELT *Rohstoffe* recover
**zurückgezogen**: **~er Zustand** *m* BESCHLÄGE *Falle des Schlosses* retracted position
**zurückhalten** *vt* ABWASSER, ARCH & TRAGW, BETON, INFR & ENT, NATURSTEIN, UMWELT, WERKSTOFF retain
**zurücksetzen** *vt* ARCH & TRAGW recess, set back
**Zusammenbau** *m* ARCH & TRAGW *Balkenträger* framing
**zusammenbauen** *vt* HEIZ & BELÜFT, HOLZ, STAHL *aus Einzelteilen* assemble
**Zusammenbruch** *m* ARCH & TRAGW collapse
**zusammendrückbar**: **nicht ~** *adj* INFR & ENTW, WERKSTOFF incompressible
**zusammenfallen** *vi* ARCH & TRAGW *Untergrund* subside
**zusammenfügen** *vt* ARCH & TRAGW join, join on to, HOLZ splice
**Zusammenfügen** *nt* ARCH & TRAGW joining, HOLZ splice
**zusammengesetzt**: **~er Flansch** *m* ARCH & TRAGW, INFR & ENTW, STAHL compound flange; **~e Spannungen** *f pl* ARCH & TRAGW, HOLZ combined stresses; **nicht ~er Stahlflansch** *m* ARCH & TRAGW, INFR & ENTW, STAHL noncomposite steel flange
**zusammenklappbar** *adj* ARCH & TRAGW, WERKSTOFF collapsible
**zusammenkneifen** *vt* STAHL punch
**zusammenpassen** *vi* ARCH & TRAGW match
**zusammensetzen** *vt* ARCH & TRAGW build up
**Zusammensetzung** *f* WERKSTOFF composition
**Zusammenziehen** *nt* WERKSTOFF contraction
**zusammenziehen**: **sich ~** *v refl* WERKSTOFF contract
**Zusammenziehungskoeffizient** *m* WERKSTOFF contraction coefficient
**Zusatz** *m* BAURECHT addition, BETON admixture, INFR & ENTW addition, WERKSTOFF addition, *in Beton, Farben- und Beschichtungsmitteln* admixture; **Zusatzabkommen** *nt* BAURECHT additional agreement; **Zusatzanmeldung** *f* BAURECHT additional application; **Zusatzbelastung** *f* ARCH & TRAGW additional loading, additional load; **Zusatzheizung** *f* ARCH & TRAGW, HEIZ & BELÜFT additional heating; **Zusatzklausel** *f* BAURECHT additional clause; **Zusatzlast** *f* ARCH & TRAGW, INFR & ENTW complementary load, superimposed load
**zusätzlich** *adj* ARCH & TRAGW, INFR & ENTW extra, supplementary; **~er Ausgang** *m* ARCH & TRAGW, BAURECHT alternative exit; **~e Bauleistungen** *f pl* ARCH & TRAGW, BAURECHT extras; **~e Heizungsanlage** *f* ARCH & TRAGW, HEIZ & BELÜFT auxiliary heating system; **~e Leistungen** *f pl* BAURECHT auxiliary work; **~er Ventilator** *m* ELEKTR, HEIZ & BELÜFT extension fan
***Zusatz***: **Zusatzmittel** *nt* BETON additive, INFR & ENTW, WERKSTOFF addition, additive; **Zusatzpumpe** *f* ABWASSER, ARCH & TRAGW, BAUMASCHIN booster pump; **Zusatzspannglied** *nt* ARCH & TRAGW, BETON auxiliary reinforcement; **Zusatzstoff** *m* BETON, INFR & ENTW, WERKSTOFF additive; **Zusatzstraße** *f* ARCH & TRAGW, BAURECHT, INFR & ENTW auxiliary lane
**Zuschauer** *m* INFR & ENTW spectator; **Zuschauerplatz** *m* ARCH & TRAGW, INFR & ENTW spectator seat; **Zuschauerraum** *m* ARCH & TRAGW, INFR & ENTW auditorium; **Zuschauerstehfläche** *f* BAURECHT, INFR & ENTW standing spectator area
**Zuschlag** *m* BAURECHT acceptance of tender, INFR & ENTW, WERKSTOFF addition; **Zuschlagablöseversuch** *m* WERKSTOFF *Bitumen* aggregate stripping test; **Zuschlagablösung** *f* BETON, WERKSTOFF aggregate stripping; **Zuschlagstoff** *m* BETON aggregate, filler, UMWELT additive, WERKSTOFF aggregate
**Zuschneiden** *nt* WERKSTOFF cutting to size
**Zusicherung** *f* BAURECHT assurance
**zuspitzen** *vt* ARCH & TRAGW, INFR & ENTW taper, tip up
**Zustöpseln** *nt* ABWASSER plugging
**zuteilen** *vt* BAURECHT allocate
**Zuteilen** *nt* BAURECHT allocation
**Zuteilung** *f* BAURECHT allotment
**Zutritt** *m* ARCH & TRAGW, INFR & ENTW approach
**zuverlässig** *adj* BAURECHT reliable, safe, secure
**zuweisen** *vt* BAURECHT allocate
**Zuweisen** *nt* BAURECHT allocation
**Zwangsentlüftung** *f* HEIZ & BELÜFT forced ventilation
**Zwangsmischer** *m* BAUMASCHIN, BETON compulsory mixer
**Zweck** *m* ARCH & TRAGW objective; **Zweckfähigkeit** *f* WERKSTOFF fitness for purpose; **Zweckforschung** *f* UMWELT applied research
**zweckgebaut** *adj* ARCH & TRAGW, INFR & ENTW purpose-built
**zweiachsig** *adj* ARCH & TRAGW, INFR & ENTW, WERKSTOFF biaxial
**zweibeinig** *adj* HOLZ two-legged
**zweidimensional** *adj* ARCH & TRAGW, INFR & ENTW two-dimensional
**Zweifeldrahmen** *m* ARCH & TRAGW two-bay frame
**zweifeldrig** *adj* ARCH & TRAGW two-bay, two-span
**zweiflügelig**: **~es Fenster** *nt* ARCH & TRAGW double-sash window; **~es Gebäude** *nt* ARCH & TRAGW double-winged building; **~e Tür** *f* ARCH & TRAGW double-leaf door
**Zweigelenkbogen** *m* ARCH & TRAGW two-hinged arch, two-pinned arch
**zweihängig** *adj* ARCH & TRAGW *Dach* ridged
**Zweikammerstein** *m* NATURSTEIN two-cell hollow block
**zweilagig** *adj* ARCH & TRAGW, NATURSTEIN double-layered; **~er Anstrich** *m* OBERFLÄCHE two-coat work; **~er Putz** *m* NATURSTEIN two-coat work
**zweiläufig** *adj* ARCH & TRAGW two-flight; **~e Treppe** *f* ARCH & TRAGW double-flight staircase
**Zweimetallverbinder** *m* STAHL two-metal connector
**zweischalig**: **~e Wand** *f* ARCH & TRAGW, DÄMMUNG, NATURSTEIN WERKSTOFF cavity wall, cavity wall insulation
**zweischiffig** *adj* ARCH & TRAGW two-bay, two-span
**zweischnittig** *adj* ARCH & TRAGW, HOLZ double shear
**zweiseitig** *adj* ARCH & TRAGW double-sided; **~ aufschlagende Tür** *f* ARCH & TRAGW double-action door; **~ eingespannter Balken** *m* ARCH & TRAGW fixed end beam; **~er Pflasterhammer** *m* BAUMASCHIN double-ended sledgehammer; **~er Spundhobel** *m* BAUMASCHIN double-ended match plane
**zweistielig** *adj* HOLZ two-legged; **~ abgestrebtes Pfettendach** *nt* HOLZ *mit Gebindeverstärkungen* queen-post truss roof; **~ strebenloses Pfettendach**

*nt* ARCH & TRAGW *mit Gebindeverstärkungen* queen-post truss roof
**zweistöckig** *adj* ARCH & TRAGW two-storey (*BrE*), two-story (*AmE*), two-tier
**zweistufig**: **~e Ausschreibung** *f* BAURECHT two-stage tendering
**zweite**: **zweiter Brennraum** *m* (*SCC*) HEIZ & BELÜFT, UMWELT afterburner chamber, secondary combustion chamber (*SCC*); **zweites Kellergeschoß** *nt* ARCH & TRAGW subbasement
**zweitourig**: **~es Schloß** *nt* BESCHLÄGE double-turn lock
**Zweiwegeventil** *nt* BESCHLÄGE, UMWELT two-way valve
**Zweiweghahn** *m* ABWASSER two-way cock
**zweizügig**: **~es Kabelformstück** *nt* ELEKTR two-way cable conduit, two-way cable duct
**Zwerggiebel** *m* ARCH & TRAGW *Gothik* gablet
**Zwickel** *m* ARCH & TRAGW pendentive, spandrel
**Zwillingspumpe** *f* ABWASSER, BAUMASCHIN twin pump
**Zwillingsträger** *m* ARCH & TRAGW double girder
**Zwinge** *f* ARCH & TRAGW, BAUMASCHIN, HOLZ clamp
**Zwingnuß** *f* BESCHLÄGE *Nußband an Schloß* counter-flap hinge
**Zwischen-** *in cpds* UMWELT intermediate; **Zwischenanstrich** *m* OBERFLÄCHE intermediate coat; **Zwischenaussteifung** *f* ARCH & TRAGW, BETON, INFR & ENTW, STAHL, WERKSTOFF *Verbundträger* intermediate stiffener; **Zwischenbehälter** *m* ABWASSER, BESCHLÄGE tundish; **Zwischenbewertung** *f* BAURECHT interim valuation; **Zwischenboden** *m* ARCH & TRAGW raised floor; **Zwischendecke** *f* ARCH & TRAGW intermediate ceiling, suspended ceiling; **Zwischenfeld** *nt* ARCH & TRAGW interior span
**zwischengelegt** *adj* ARCH & TRAGW inserted
***Zwischen-***: **Zwischengeschoß** *nt* ARCH & TRAGW mezzanine; **Zwischenlage** *f* ARCH & TRAGW intermediate layer; **Zwischenlager** *nt* ARCH & TRAGW, INFR & ENTW intermediate bearing; **Zwischenlagerplatz** *m* UMWELT refuse transfer station; **Zwischenlagerung** *f* ARCH & TRAGW, INFR & ENTW intermediate storage
**zwischenlegen** *vt* ARCH & TRAGW insert
***Zwischen-***: **Zwischenmauer** *f* ARCH & TRAGW party wall; **Zwischenpodest** *nt* ARCH & TRAGW intermediate landing, half pace; **Zwischenpumpwerk** *nt* INFR & ENTW intermediate pumping station
**Zwischenraum** *m* ARCH & TRAGW interspace, interstice, space, spacing; **mit ~ anordnen** *phr* ARCH & TRAGW space
***Zwischen-***: **Zwischenschicht** *f* ARCH & TRAGW intermediate layer; **Zwischensparren** *m* ARCH & TRAGW, HOLZ common rafter; **Zwischenstück** *nt* ARCH & TRAGW spacer block; **Zwischenstütze** *f* ARCH & TRAGW intermediate support; **Zwischenträger** *m* ARCH & TRAGW intermediate beam, joist, HOLZ joist; **Zwischenwand** *f* ARCH & TRAGW partition, BESCHLÄGE baffle plate; **Zwischenzahlung** *f* ARCH & TRAGW, BAURECHT interim payment
**zwölfsäulig** *adj* ARCH & TRAGW dodecastyle
**Zyklopen-** *in cpds* NATURSTEIN cyclopean; **Zyklopenbeton** *m* BETON cyclopean block; **Zyklopenmauer** *f* NATURSTEIN cyclopean wall; **Zyklopenmauerwerk** *nt* NATURSTEIN cyclopean masonry, polygonal masonry, random rubble masonry; **Zyklopenstein** *m* BETON, NATURSTEIN cyclopean block
**Zylinder** *m* ARCH & TRAGW, BESCHLÄGE, WERKSTOFF cylinder; **~ für abweichende Türstärken** BESCHLÄGE cylinder for different door thicknesses; **Zylinderdruckprüfung** *f* BETON, INFR & ENTW, WERKSTOFF cylinder test; **Zylindereinsteckmöbelschloß** *nt* BESCHLÄGE cylinder mortise cabinet lock; **Zylinderkastenfallenschloß** *nt* BESCHLÄGE cylinder rim latch lock; **Zylinderkastenriegelschloß** *nt* BESCHLÄGE cylinder rim deadlock; **Zylinderkopfniete** *f* BESCHLÄGE cheese-head rivet; **Zylinderkreissäge** *f* BAUMASCHIN, HOLZ annular saw; **Zylindermöbelschloß** *nt* BESCHLÄGE cylinder cabinet lock; **Zylinderpfahl** *m* ARCH & TRAGW, INFR & ENTW foundation cylinder; **Zylinderrohr** *nt* HEIZ & BELÜFT, WERKSTOFF cylindrical pipe; **Zylinderschale** *f* ARCH & TRAGW, INFR & ENTW cylindrical shell; **Zylinderschloß** *nt* BAUMASCHIN pin tumbler, BESCHLÄGE cylinder lock; **Zylinderverlängerung** *f* BESCHLÄGE *Schloß* cylinder extension; **Zylindervorhängeschloß** *nt* BESCHLÄGE cylinder padlock

# English–German
# Englisch–Deutsch

# A

**abandon** *vt* CONST LAW aufgeben
**abandonment** *n* CONST LAW Aufgabe *f*
**abbey**: ~ **court** *n* ARCH & BUILD, INFR & DES Abteihof *m*
**able-bodied** *adj* CONST LAW gesund, nicht behindert
**abolish** *vt* CONST LAW aufheben
**abolition** *n* CONST LAW Aufhebung *f*
**above**: ~**-grade** *adj* ARCH & BUILD, CONCR oberirdisch; ~**-ground** *adj* ARCH & BUILD, CONCR oberirdisch; ~**-ground hydrant** *n* INFR & DES, WASTE WATER Überflurhydrant *m*
**abrade** *vt* ARCH & BUILD, CONCR, MAT PROP abreiben, STEEL abschleifen, schleifen, STONE abreiben, TIMBER abreiben, abschleifen, schleifen
**abrasion** *n* ARCH & BUILD, CONCR, MAT PROP Abrieb *m*, Schleifwirkung *f*, STEEL Schleifwirkung *f*, SURFACE Abrieb *m*, TIMBER Schleifwirkung *f*; ~**-proof** *adj* SURFACE abriebfest, abriebbeständig; ~ **resistance** *n* CONCR, MAT PROP Verschleißfestigkeit *f*, SURFACE Abriebfestigkeit *f*; ~ **test** *n* MAT PROP Abriebprobe *f*
**abrasive** *adj* MAT PROP, STEEL, TIMBER abschleifend; ~ **blasting** *n* SURFACE Abstrahlen *nt*; ~ **paper** *n* INFR & DES, MAT PROP Sandpapier *nt*, Schmirgelpapier *nt*; ~ **sand** *n* MAT PROP Schleifsand *m*; ~ **surface** *n* MAT PROP Schleiffläche *f*; ~ **wear** *n* BUILD MACHIN Abriebverschleiß *m*
**abreuvoir** *n* CONCR, STONE Mörtelfuge *f*
**abridge** *vt* INFR & DES abkürzen
**absolute**: ~ **density** *n* INFR & DES, MAT PROP absolute Dichte *f*; ~ **humidity** *n* HEAT & VENT, INFR & DES, MAT PROP absolute Feuchte *f*, absolute Feuchtigkeit *f*
**absorb** *vt* INFR & DES, MAT PROP, SURFACE absorbieren
**absorbed**: ~ **water** *n* INFR & DES, MAT PROP absorbiertes Wasser *nt*
**absorbency** *n* ENVIRON, INFR & DES, MAT PROP, SURFACE Absorption *f*
**absorbent 1.** *adj* ENVIRON, INFR & DES, MAT PROP absorbierend, SURFACE aufsaugend, saugfähig; **2.** *n* ENVIRON, INFR & DES, MAT PROP, SURFACE Absorptionsmittel *nt*
***absorbent***: ~ **base** *n* SURFACE saugfähiger Untergrund *m*; ~ **belt skimmer** *n* ENVIRON absorbierendes Förderband *nt*
**absorber** *n* SURFACE Absorber *m*; ~ **plate** *n* HEAT & VENT, INFR & DES *solar collector* Absorberplatte *f*
**absorbing**: ~ **material** *n* MAT PROP Absorptionsmaterial *nt*; ~ **well** *n* ARCH & BUILD Sickerschacht *m*, INFR & DES Sickerschacht *m*, Versickerungsschacht *m*
**absorption** *n* ENVIRON, INFR & DES, MAT PROP Absorption *f*, SOUND & THERMAL Schlucken *nt*, *of impacts* Dämpfung *f*, SURFACE Absorption *f*; ~ **capacity** *n* INFR & DES, MAT PROP, SURFACE Absorptionsvermögen *nt*; ~ **test** *n* MAT PROP Wasseraufnahmeversuch *m*
**abstract** *adj* ARCH & BUILD abstrakt
**abut** *vt* ARCH & BUILD, CONCR, INFR & DES, STEEL, STONE stoßen
**abutment** *n* ARCH & BUILD, CONCR, INFR & DES, STEEL, STONE Widerlager *nt*; ~ **flashing** *n* ARCH & BUILD Widerlagerabdeckung *f*; ~ **pier** *n* CONCR, STEEL Strebepfeiler *m*; ~ **pressure** *n* CONCR, INFR & DES, MAT PROP, STEEL Widerlagerdruck *m*; ~ **stone** *n* INFR & DES, MAT PROP, STONE Widerlagerstein *m*
**Abyssinian**: ~ **well** *n* INFR & DES Rammbrunnen *m*
**academy** *n* ARCH & BUILD, INFR & DES Akademie *f*
**acanthus** *n* ARCH & BUILD Akanthus *m*; ~ **frieze** *n* ARCH & BUILD Akanthusfries *m*; ~ **leaf** *n* ARCH & BUILD Akanthusblatt *nt*
**accelerate** *vt* BUILD MACHIN, CONCR, ENVIRON, MAT PROP, STEEL beschleunigen
**accelerated** *adj* BUILD MACHIN, ENVIRON, MAT PROP, STEEL beschleunigt; ~ **composting** *n* ENVIRON beschleunigte Kompostierung *f*; ~ **weathering test** *n* MAT PROP abgekürzter Wetterbeständigkeitsversuch *m*
**accelerating**: ~ **additive** *n* MAT PROP Beschleuniger *m*; ~ **admixture** *n* CONCR Beschleuniger *m*
**acceleration** *n* BUILD MACHIN, CONCR, ENVIRON, MAT PROP, STEEL Beschleunigung *f*; ~ **lane** *n* INFR & DES *roads* Überholspur *f*
**accelerator** *n* CONCR, MAT PROP, STONE Abbindebeschleuniger *m*, Beschleuniger *m*
**accept** *vt* ARCH & BUILD *load*, CONST LAW *of job, responsibility, risk* aufnehmen, übernehmen, abnehmen
**acceptable** *adj* CONST LAW akzeptabel, annehmbar
**acceptance** *n* ARCH & BUILD, CONST LAW *of building* Abnahme *f*, Aufnahme *f*, *of job, responsibility, risk* Übernahme *f*; ~ **certificate** *n* ARCH & BUILD, CONST LAW Abnahmebescheinigung *f*; ~ **documents** *n pl* ARCH & BUILD, CONST LAW Abnahmeunterlagen *f pl*; ~ **of tender** *n* CONST LAW Zuschlag *m*; ~ **of work** *n* CONST LAW Bauabnahme *f*
**accepted**: ~ **load** *n* ARCH & BUILD aufgenommene Last *f*
**access** *n* ARCH & BUILD, CONST LAW, INFR & DES Einstieg *m*, *for vehicles* Zufahrt *f*, *for pedestrians* Zugang *m*; ~ **for emergency services** *n* INFR & DES Zufahrt *f* für Rettungsdienste; ~ **for fire services** *n* ARCH & BUILD, CONST LAW Feuerwehrzufahrt *f*; ~ **for wheelchairs** *n* ARCH & BUILD, CONST LAW Rollstuhlzugang *m*; ~ **gallery** *n* INFR & DES Laubengang *m*; ~ **gate** *n* ARCH & BUILD, INFR & DES Zufahrtstor *nt*; ~ **hatch** *n* ARCH & BUILD Einstiegluke *f*; ~ **hook** *n* CONST LAW, INFR & DES Klettereisen *nt*, Steigeisen *nt*
**accessible** *adj* ARCH & BUILD, BUILD MACHIN, CONST LAW, INFR & DES zugänglich; ~ **resource base** *n* ENVIRON zugängliche Reservenquelle *f*
***access***: ~ **ladder** *n* INFR & DES, MAT PROP, STEEL Einstiegsleiter *f*; ~ **opening** *n* ARCH & BUILD, BUILD MACHIN Einstiegsöffnung *f*
**accessories** *n pl* BUILD HARDW Zubehör *nt*
**accessory**: ~ **plates** *n pl* BUILD HARDW Hilfsplatten *f pl*
***access***: ~ **ramp** *n* ARCH & BUILD, INFR & DES Zufahrtsrampe *f*; ~ **road** *n* ARCH & BUILD, INFR & DES Zufahrtsstraße *f*; ~ **to site** *n* ARCH & BUILD, INFR & DES Baustellenzufahrt *f*; ~ **tunnel** *n* ARCH & BUILD, INFR & DES Zugangstunnel *m*; ~ **way** *n* ARCH & BUILD, INFR & DES Einfahrt *f*

**accident** *n* CONST LAW Unfall *m*
**accidental**: **~ discharge** *n* ENVIRON zufälliger Ausfluß *m*, *of oils, chemicals* havariebedingter Ausfluß *m*
***accident***: **~ prevention** *n* CONST LAW Unfallverhütung *f*
**accommodation** *n* ARCH & BUILD, INFR & DES *action* Unterbringung *f*, *lodgings* Unterkunft *f*; **~ stair** *n* ARCH & BUILD, INFR & DES Einwohnertreppenhaus *nt*
**according**: **~ to contract** *adj* CONST LAW vertragsmäßig; **~ to regulations** *adj* CONST LAW vorschriftsgemäß; **~ to scale** *adj* ARCH & BUILD, CONST LAW, INFR & DES maßstäblich, maßstabsgerecht; **~ to specification** *adj* CONST LAW vorschriftsgemäß
**accordion**: **~ door** *n* ARCH & BUILD, BUILD HARDW Harmonikatür *f*; **~ partition** *n* ARCH & BUILD, BUILD HARDW Faltwand *f*
**account** *n* ARCH & BUILD, CONST LAW Rechnung *f*
**accountability** *n* ARCH & BUILD, CONST LAW Verantwortlichkeit *f*, Verantwortung *f*
**accountable** *adj* ARCH & BUILD, CONST LAW rechenschaftspflichtig, verantwortlich
**accountancy** *n* CONST LAW Buchhaltung *f*
**accountant** *n* CONST LAW *book-keeper* Buchhalter *m*
***account***: **~ book** *n* ARCH & BUILD, CONST LAW Geschäftsbuch *nt*
**accounting** *n* CONST LAW Buchführung *f*, Rechnungswesen *nt*
**ACC**: **~ test** *n* INFR & DES, MAT PROP Quellvermögen *nt*, Bentonitzahl *f*
**accumulate** *vt* CONST LAW, INFR & DES, MAT PROP akkumulieren, ansammeln
**accumulation** *n* CONST LAW, INFR & DES, MAT PROP Akkumulation *f*, Ansammlung *f*
**accumulator** *n* BUILD MACHIN, ELECTR Akkumulator *m*, HEAT & VENT, INFR & DES Sammler *m*
**accuracy** *n* ARCH & BUILD, INFR & DES, MAT PROP Genauigkeit *f*; **~ of level** *n* ARCH & BUILD, INFR & DES, MAT PROP Planebenheit *f*; **~ of measurement** *n* ARCH & BUILD, INFR & DES, MAT PROP Meßgenauigkeit *f*; **~ of shape** *n* ARCH & BUILD, INFR & DES, MAT PROP Formgenauigkeit *f*
**accurate** *adj* ARCH & BUILD, BUILD MACH, CONCR, CONST LAW, INFR & DES, MAT PROP, STEEL exakt, genau, präzis
**acetylene** *n* MAT PROP, STEEL Azetylen *nt*; **~ blowpipe** *n* BUILD MACHIN, STEEL Azetylenschweißbrenner *m*; **~ generator** *n* BUILD MACHIN, STEEL Azetylenentwickler *m*
**acid 1.** *adj* ENVIRON, INFR & DES, MAT PROP, STEEL, SURFACE sauer; **2.** *n* ENVIRON, INFR & DES, MAT PROP, STEEL, SURFACE Säure *f*
***acid***: **~ aerosol** *n* ENVIRON saures Aerosol *nt*; **~ attack** *n* INFR & DES, STEEL, SURFACE Säureangriff *m*; **~ bath** *n* STEEL, SURFACE Säurebad *nt*; **~ concentration** *n* ENVIRON Säurekonzentration *f*, *activity* Eindampfen *nt* einer Säure; **~ content** *n* ENVIRON, INFR & DES, MAT PROP Säuregehalt *m*; **~ deposit** *n* ENVIRON saure Ablagerung *f*; **~ earth** *n* ENVIRON saure Erde *f*; **~ embossing** *n* STEEL, SURFACE Mattätzen *nt*; **~ etching** *n* STEEL, SURFACE Ätzung *f*; **~ fallout** *n* ARCH & BUILD, ENVIRON, INFR & DES saurer Niederschlag *m*, saurer Regen *m*; **~ fog** *n* ENVIRON Säurenebel *m*
**acidic**: **~ area** *n* ENVIRON angesäuerte Bodenfläche *f*; **~ particle** *n* ENVIRON saures Teilchen *nt*; **~ precursor** *n* ENVIRON Säurenvorstufe *f*; **~ rain** *n* ARCH & BUILD, ENVIRON, INFR & DES saurer Niederschlag *m*, saurer Regen *m*
**acidification** *n* ENVIRON Ansäuerung *f*, *agriculture* Versäuerung *f*
**acidified**: **~ lake** *n* ENVIRON saurer See *m*
**acidify** *vt* ENVIRON ansäuern, säuern
**acidifying** *adj* ENVIRON ansäuernd
**acidity** *n* ENVIRON, INFR & DES, MAT PROP, STEEL, SURFACE Säuregehalt *m*; **~ level** *n* ENVIRON Säuregrad *m*
***acid***: **~ lake** *n* ENVIRON saurer See *m*; **~ loading** *n* ENVIRON Säurebelastung *f*; **~ neutralizing** *n* ENVIRON Säureneutralisation *f*; **~-neutralizing capacity** *n* ENVIRON Säureneutralisierungsvermögen *nt*; **~ particle** *n* ENVIRON saures Teilchen *nt*; **~ pollution** *n* ENVIRON saure Umweltverschmutzung *f*; **~ precipitation** *n* ENVIRON Säurefällung *f*
**acidproof** *adj* ENVIRON, MAT PROP, STEEL, SURFACE säurebeständig; **~ paint** *n* BUILD HARDW, ENVIRON, STEEL, SURFACE Säureschutzfarbe *f*, säurefeste Farbe *f*
***acid***: **~ rain** *n* ARCH & BUILD, ENVIRON, INFR & DES saurer Niederschlag *m*, saurer Regen *m*; **~ resistance** *n* ENVIRON, MAT PROP, STEEL, SURFACE Säurebeständigkeit *f*; **~-resistant** *adj* ENVIRON, MAT PROP, STEEL, SURFACE säurebeständig, säurefest; **~-resistant brick** *n* INFR & DES, MAT PROP, STONE säurefester Klinker *m*; **~-resisting paint** *n* MAT PROP säurefeste Farbe *f*; **~ runoff** *n* ENVIRON saurer Abfluß *m*; **~ shock** *n* ENVIRON Säureschock *m*; **~ snow** *n* ENVIRON saurer Schnee *m*; **~ soil** *n* ENVIRON, INFR & DES, MAT PROP saurer Boden *m*; **~ stress** *n* ENVIRON Säurebelastung *f*; **~-stressed** *adj* ENVIRON durch Säure belastet; **~ tolerance** *n* ENVIRON Säureverträglichkeit *f*, Säurewiderstandsfähigkeit *f*; **~ treatment** *n* STEEL, SURFACE Säuern *nt*
**acidwash** *vt* MAT PROP, STONE, SURFACE absäuern
**acidwashed**: **~ plaster** *n* MAT PROP, STONE, SURFACE Waschputz *m*
***acid***: **~ waste** *n* INFR & DES, MAT PROP saurer Abfall *m*; **~ waste gas** *n* ENVIRON saures Abgas *nt*; **~ water** *n* ENVIRON saures Wasser *nt*
**acoustical**: **~ behavior** *AmE*, **~ behaviour** *BrE* *n* MAT PROP, SOUND & THERMAL akustisches Verhalten *nt*
**acoustic**: **~ blanket** *n* MAT PROP, SOUND & THERMAL Akustikmatte *f*; **~ board** *n* MAT PROP, SOUND & THERMAL Akustikplatte *f*; **~ building material** *n* INFR & DES, MAT PROP, SOUND & THERMAL Akustikbaustoff *m*; **~ ceiling** *n* INFR & DES, SOUND & THERMAL Akustikdecke *f*; **~ fencing** *n* INFR & DES, SOUND & THERMAL Lärmschutzzaun *m*; **~ impedance** *n* INFR & DES, MAT PROP akustische Impedanz *f*; **~ insulating materials** *n* ENVIRON, SOUND & THERMAL Schallisolationsmaterial *nt*; **~ lining** *n* INFR & DES, MAT PROP, SOUND & THERMAL Akustikverkleidung *f*; **~ material** *n* MAT PROP, SOUND & THERMAL Schallschluckmaterial *nt*
**acoustics** *n pl* SOUND & THERMAL Akustik *f*
**acoustic**: **~ testing of buildings** *n* MAT PROP, SOUND & THERMAL bauakustische Prüfung *f*; **~ velocity log** *n* MAT PROP, SOUND & THERMAL Schallgeschwindigkeitsprofil *nt*
**acquisition** *n* ARCH & BUILD, CONST LAW Erwerb *m*; **~ of property** *n* ARCH & BUILD, CONST LAW Eigentumserwerb *m*

**acquit** *vt* ARCH & BUILD, CONST LAW, INFR & DES entbinden, entlasten
**acquittal** *n* ARCH & BUILD, CONST LAW, INFR & DES Entlasten *nt*
**acquitted** *adj* ARCH & BUILD, CONST LAW, INFR & DES entlastet
**acrid**: ~ **odor** *AmE*, ~ **odour** *BrE n* ENVIRON beißender Geruch *m*
**acroterion** *n* ARCH & BUILD Akroterion *nt*
**acrylic** *n* INFR & DES, MAT PROP, SURFACE Acryl *nt*; ~ **concrete** *n* CONCR, MAT PROP Acrylharzbeton *m*; ~ **paint** *n* SURFACE Acrylfarbe *f*; ~ **resin** *n* MAT PROP Acrylharz *nt*; ~ **resin dispersion** *n* INFR & DES, MAT PROP, SURFACE Acrylharzdispersion *f*; ~ **resin emulsion** *n* INFR & DES, MAT PROP, SURFACE Acrylharzemulsion *f*
**action**: ~ **of building** *n* ARCH & BUILD Bau *m*; ~ **of frost** *n* ARCH & BUILD, CONCR, HEAT & VENT, INFR & DES, SURFACE Frosteinwirkung *f*
**activated**: ~ **carbon** *n* ENVIRON, INFR & DES Aktivkohle *f*; ~ **carbon treatment** *n* ENVIRON, INFR & DES Aktivkohlebehandlung *f*; ~ **sludge** *n* ENVIRON, INFR & DES Belebtschlamm *m*; ~ **sludge plant** *n* ENVIRON, INFR & DES Belebtschlammanlage *f*; ~ **sludge process** *n* ENVIRON, INFR & DES Belebtschlammverfahren *nt*; ~ **sludge tank** *n* CONCR Belüftungsbecken *nt*, ENVIRON, INFR & DES Belebtschlammbecken *nt*, Belüftungsbecken *nt*
**activation**: ~ **of sludge** *n* ENVIRON Schlammbelebung *f*
**activator** *n* CONCR Beschleuniger *m*
**active**: ~ **carbon** *n* ENVIRON, INFR & DES Aktivkohle *f*; ~ **carbon absorption** *n* ENVIRON Aktivkohleabsorption *f*; ~ **carbon cartridge** *n* HEAT & VENT Aktivkohlepatrone *f*; ~ **earth pressure** *n* INFR & DES aktiver Erddruck *m*; ~ **ingredients of preservative products** *n pl* MAT PROP aktive Bestandteile *m pl* der Konservierungsmittel; ~ **leaf** *n* BUILD HARDW Gehflügel *m*, Türflügel *m*; ~ **pressure** *n* INFR & DES, MAT PROP aktiver Druck *m*; ~ **Rankine pressure** *n* INFR & DES, MAT PROP aktiver Druck *m* nach Rankin; ~ **solar heating** *n* HEAT & VENT aktive Solarheizung *f*; ~ **solar system** *n* ENVIRON aktives Sonnensystem *nt*
**actual**: ~ **dimension** *n* ARCH & BUILD, INFR & DES Istmaß *nt*; ~ **load** *n* INFR & DES tatsächliche Belastung *f*; ~ **value** *n* HEAT & VENT Istwert *m*
**actuator** *n* HEAT & VENT Stellelement *nt*
**acute**: ~ **arch** *n* ARCH & BUILD überhöhter Spitzbogen *m*; ~ **effect** *n* ENVIRON akute Wirkung *f*
**adapt** *vt* ARCH & BUILD, BUILD MACHIN, CONCR, CONST LAW, ELECTR, HEAT & VENT, INFR & DES, STEEL angleichen, anpassen
**adaptation** *n* ARCH & BUILD, BUILD MACHIN, CONCR, CONST LAW, ELECTR, INFR & DES, STEEL Angleichung *f*
**adapter** *n* BUILD HARDW, BUILD MACHIN, CONCR, ELECTR, HEAT & VENT *pipe*, MAT PROP, STEEL Adapter *m*, Paßstück *nt*
**add** *vt* ARCH & BUILD hinzufügen, CONCR *additives, aggregates* hinzufügen, zugeben, CONST LAW hinzufügen, INFR & DES *additives, aggregates* hinzufügen, zugeben, *attachment* einfügen, anfügen, beifügen, MAT PROP *additives, aggregates* hinzufügen, zugeben, *attachment* anfügen, beifügen, einfügen; ~ **to** *vt* ARCH & BUILD, CONST LAW, INFR & DES, MAT PROP ergänzen
**added** *adj* INFR & DES, MAT PROP eingefügt
**addition** *n* ARCH & BUILD *to building* Anbau *m*, Erweiterung *f*, Gebäudeerweiterung *f*, *document* Nachtrag *m*, CONST LAW *supplement* Zusatz *m*, Nachtrag *m*, *adding* Hinzufügung *f*, *enlargement* Ergänzung *f*, Erweiterung *f*, INFR & DES *to building* Anbau *m*, Zuschlag *m*, Nachtrag *m*, Zusatzmittel *nt*, Beimengung *f*, Zusatz *m*, Erweiterung *f*, MAT PROP *admixture* Beimengung *f*, Zusatz *m*, Zusatzmittel *nt*, Zuschlag *m*
**additional** *adj* ARCH & BUILD, CONST LAW, INFR & DES ergänzend, hinzukommend, nachträglich, weiter; ~ **agreement** *n* CONST LAW Zusatzabkommen *nt*; ~ **application** *n* CONST LAW Zusatzanmeldung *f*; ~ **change key** *n* BUILD HARDW Mehrschlüssel *m*; ~ **charge** *n* CONST LAW Aufschlag *m*; ~ **charges** *n pl* CONST LAW Mehrkosten *pl*; ~ **clause** *n* CONST LAW Zusatzklausel *f*; ~ **heating** *n* ARCH & BUILD, HEAT & VENT Zusatzheizung *f*; ~ **load** *n* ARCH & BUILD Zusatzbelastung *f*; ~ **loading** *n* ARCH & BUILD Zusatzbelastung *f*; ~ **plasterwork** *n* STONE Beiputzarbeiten *f pl*, Beiputzen *nt*
**additive** *n* CONCR Betonzusatz *m*, Zusatzmittel *nt*, Zusatzstoff *m*, ENVIRON Additiv *nt*, Zuschlagstoff *m*, INFR & DES, MAT PROP Zusatzmittel *nt*, Zusatzstoff *m*
**add**: ~ **a storey** *phr BrE* ARCH & BUILD aufstocken; ~ **a story** *AmE see add a storey BrE*
**adhere 1.** *vi* INFR & DES, MAT PROP, SURFACE anhaften, ankleben, haften; ~ **firmly** STEEL gut haften; **2.** ~ **to** *vt* CONST LAW befolgen, einhalten
**adhesion** *n* INFR & DES, MAT PROP, SURFACE Adhäsion *f*, Haft-, Haftfähigkeit *f*, Haftvermögen *nt*; ~ **power** *n* INFR & DES, MAT PROP, SURFACE Haftfähigkeit *f*, Haftvermögen *nt*; ~ **promoter** *n* MAT PROP, SURFACE Haftvermittler *m*; ~**-promoting agent** *n* MAT PROP Haftmittel *nt*
**adhesive 1.** *adj* INFR & DES, MAT PROP, SURFACE haftend, klebrig; **2.** *n* INFR & DES, MAT PROP, SURFACE, TIMBER Kleber, Klebstoff *m m*
***adhesive***: ~ **anchor** *n* MAT PROP Klebeanker *m*; ~ **label** *n* MAT PROP *on technical equipment* Aufkleber *m*; ~ **masking tape** *n* INFR & DES, MAT PROP, SURFACE Abdeckband *nt*; ~ **tape** *n* MAT PROP, SOUND & THERMAL Klebeband *nt*
**adiabatic**: ~ **curing** *n* STONE adiabatische Nachbehandlung *f*
**adit** *n* INFR & DES *tunnel* Zugangsstollen *m*
**adjacent**: ~ **property** *n* ARCH & BUILD angrenzender Grundbesitz *m*; ~ **span** *n* ARCH & BUILD, CONCR, INFR & DES, STEEL angrenzende Spannweite *f*
**adjoining** *adj* CONST LAW, INFR & DES, MAT PROP angrenzend, benachbart
**adjust** *vt* ARCH & BUILD ausrichten, einpassen, einstellen, verstellen, BUILD HARDW *set correctly* regulieren, verdrehen, justieren, einpassen, einstellen, BUILD MACHIN *repair, retouch* ausbessern, *set correctly* verstellen, CONCR *reinforcement* ausrichten, CONST LAW *put in order* schlichten, in Ordnung bringen, einstellen, berichtigen, regeln, ELECTR *set point* einstellen, HEAT & VENT *set correctly* regulieren, einpassen, einstellen, justieren, INFR & DES, MAT PROP einstellen, STEEL angleichen, STONE, SURFACE ausbessern
**adjustable** *adj* BUILD HARDW, BUILD MACHIN, HEAT & VENT einstellbar, verstellbar; ~ **shelving** *n* BUILD HARDW verstellbare Regale *nt pl*; ~ **submersion weir** *n* ENVIRON regulierbares Grundwehr *nt*

**adjusting** *n* BUILD MACHIN, STONE, SURFACE Ausbessern *nt*
**adjustment** *n* ARCH & BUILD Änderung *f*, BUILD HARDW Einstellung *f*, Justierung *f*, Regelung *f*, BUILD MACHIN *correct setting* Justierung *f*, Regulierung *f*, *calibration* Eichung *f*, *adjusting device* Einstellvorrichtung *f*, Angleichung *f*, CONCR Angleichung *f*, CONST LAW Änderung *f*, Anpassung *f*, Ausgleich *m*, Beilegung *f*, Kontenabstimmung *f*, Regelung *f*, Angleichung *f*, Berichtigung *f*, ELECTR Angleichung *f*, HEAT & VENT Regulierung *f*, INFR & DES Angleichung *f*, Ausgleich *m*, MAT PROP Einstellung *f*, STEEL Angleichung *f*
**administration** *n* CONST LAW Verwaltung *f*; ~ **building** *n* CONST LAW Verwaltungsgebäude *nt*
**administrative**: ~ **agency** *n* CONST LAW Verwaltungsstelle *f*; ~ **body** *n* CONST LAW Verwaltungsorgan *nt*; ~ **fee** *n* CONST LAW Verwaltungsgebühr *f*; ~ **tribunal** *n* CONST LAW Verwaltungsgericht *nt*
**admissible** *adj* ARCH & BUILD, BUILD HARDW, BUILD MACHIN, CONST LAW, ENVIRON, INFR & DES, MAT PROP, STEEL, TIMBER zulässig
**admission** *n* CONST LAW Aufnahme *f*, HEAT & VENT, WASTE WATER *of liquid in a pump* Zulauf *m*; ~ **pressure** *n* HEAT & VENT, INFR & DES, MAT PROP, WASTE WATER *pump* Vordruck *m*
**admit** *vt* CONST LAW *allow to enter, permit* zulassen, *recognize* anerkennen
**admitted**: ~ **for use by the construction supervising authority** *phr* CONST LAW bauaufsichtlich zugelassen
**admixture** *n* CONCR, MAT PROP Beimischung *f*, Zusatz *m*
**adobe** *n* ARCH & BUILD Lehmziegel *m*
**adorn** *vt* ARCH & BUILD, INFR & DES verzieren
**adorned** *adj* ARCH & BUILD, INFR & DES verziert
**adornment** *n* ARCH & BUILD, INFR & DES Verzierung *f*
**adsorb** *vt* ENVIRON, INFR & DES, MAT PROP, STONE, SURFACE adsorbieren
**adsorbent** *n* ENVIRON, INFR & DES, MAT PROP, STONE, SURFACE Adsorptionsmittel *nt*
**adsorption** *n* ENVIRON, INFR & DES, MAT PROP, STONE, SURFACE Adsorption *f*, Anlagerung *f*; ~ **capacity** *n* ENVIRON, INFR & DES, MAT PROP, STONE, SURFACE Adsorptionsvermögen *nt*; ~ **efficiency** *n* ENVIRON, INFR & DES, MAT PROP, STONE, SURFACE Adsorptionswirkung *f*; ~ **phenomenon** *n* ENVIRON, INFR & DES, MAT PROP Adsorptionsphänomen *nt*; ~ **rate** *n* ENVIRON, INFR & DES, MAT PROP, STONE, SURFACE Adsorptionsgeschwindigkeit *f*; ~ **test** *n* ENVIRON, INFR & DES, MAT PROP Adsorptionstest *m*
**adsorptivity** *n* ENVIRON, INFR & DES, MAT PROP, STONE, SURFACE Adsorptionsvermögen *nt*
**advance**: ~ **payment** *n* CONST LAW Vorauszahlung *f*
**adverse**: ~ **slope** *n* ARCH & BUILD, INFR & DES Gegenneigung *f*
**aedicula** *n* ARCH & BUILD Ädikula *f*
**aerate** *vt* ARCH & BUILD, BUILD MACHIN, CONCR, HEAT & VENT, INFR & DES, STEEL, WASTE WATER belüften
**aerated**: ~ **cement floor** *n* CONCR, SURFACE Porenestrich *m*; ~ **concrete** *n* CONCR *partitions* Gasbeton *m*, Porenbeton *m*, Schaumbeton *m*, MAT PROP *partitions* Gasbeton *m*; ~ **lagoon** *n* ENVIRON Oxidationsteich *m*
**aeration** *n* ARCH & BUILD, BUILD MACHIN, CONCR, HEAT & VENT, INFR & DES, STEEL Belüftung *f*, Luftzufuhr *f*, WASTE WATER Belüftung *f*; ~ **method** *n* HEAT & VENT, WASTE WATER Belüftungsverfahren *nt*; ~ **plant** *n* INFR & DES Belüftungsanlage *f*; ~ **tank** *n* CONCR Belüftungsbecken *nt*, ENVIRON, INFR & DES Belebtschlammbecken *nt*, Belüftungsbecken *nt*
**aerator** *n* BUILD MACHIN, ENVIRON, INFR & DES, MAT PROP Belüfter *m*
**aerial** *n* BUILD MACHIN, ELECTR Antenne *f*; ~ **cableway** *n* INFR & DES, STEEL Drahtseilbahn *f*; ~ **mast** *n* ELECTR, STEEL Antennenmast *m*; ~ **railway** *n* BUILD MACHIN Seilwinde *f*, INFR & DES, STEEL Drahtseilbahn *f*; ~ **socket** *n* ELECTR Antennensteckdose *f*
**aerobe** *n* ENVIRON, INFR & DES, MAT PROP Aerobier *m*
**aerobic** *adj* ENVIRON, INFR & DES, MAT PROP aerob
**aerobically**: ~ **digested sludge** *n* ENVIRON aerob stabilisierter Schlamm *m*
**aerobic**: ~ **bacteria** *n pl* ENVIRON aerobe Bakterien *f pl*; ~ **decomposition** *n* ENVIRON aerobe Zersetzung *f*; ~ **degradation** *n* ENVIRON aerober Abbau *m*; ~ **fermentation** *n* ENVIRON aerobe Gärung *f*; ~ **sewage treatment** *n* ENVIRON Abwasserbehandlung *f* mittels aerober Reinigung; ~ **sludge stabilization** *n* ENVIRON aerobe Schlammstabilisierung *f*; ~ **treatment process** *n* ENVIRON aerobes Behandlungsverfahren *nt*
**aerobiology** *n* ENVIRON Aerobiologie *f*
**aerobiosis** *n* ENVIRON Aerobiose *f*
**aerodynamic**: ~ **force** *n* ARCH & BUILD, ENVIRON, INFR & DES aerodynamische Kraft *f*; ~ **power** *n* ARCH & BUILD, ENVIRON, INFR & DES aerodynamische Kraft *f*
**aeroelasticity** *n* INFR & DES Aeroelastizität *f*
**aerosol** *n* ENVIRON Aerosol *nt*, schwebestoffhaltige Luft *f*; ~ **dispenser** *n* ENVIRON Aerosolpackung *f*
**aesthetic** *adj* ARCH & BUILD, SURFACE ästhetisch
**aesthetically**: ~ **correct** *adj* ARCH & BUILD, SURFACE ästhetisch korrekt; ~ **pleasing** *adj* ARCH & BUILD, SURFACE formschön
**affidavit** *n* CONST LAW eidliche Erklärung *f*
**affinity** *n* ARCH & BUILD, INFR & DES, MAT PROP Affinität *f*
**afterburner**: ~ **chamber** *n* ENVIRON Nachbrennkammer *f* (*SCC*), zweiter Brennraum *m* (*SCC*), HEAT & VENT zweiter Brennraum *m* (*SCC*)
**aftertreatment** *n* CONCR, CONST LAW, STEEL, SURFACE, TIMBER Nachbehandlung *f*, Weiterbehandlung *f*
**after**: ~ **use** *n* ENVIRON Folgenutzung *f*
**age 1.** *n* CONCR, MAT PROP, SURFACE *of concrete* Betonalter *nt*; **2.** *vi* CONCR, MAT PROP, STEEL, SURFACE altern
**ageing** *n* CONCR Alterung *f*, MAT PROP Vergütung *f*, STEEL Alterung *f*, Veredelung *f*, Vergütung *f*, SURFACE Vergütung *f*
**agency** *n* CONST LAW *establishment* Vermittlungsstelle *f*, *proxy* Vertretungbefugnis *f*, Vollmacht *f*, Stellvertretung *f*
**agenda** *n* ARCH & BUILD, CONST LAW Tagesordnung *f*
**agent** *n* CONST LAW *representative* Vertreter *m*, *intermediary* Vermittler *m*, *real estate* Makler *m*, MAT PROP Mittel *nt*
**age**: ~**-proof** *adj* CONCR, MAT PROP, STEEL alterungsbeständig
**agglomerate 1.** *n* INFR & DES, MAT PROP Agglomerat *nt*; **2.** *vt* INFR & DES, MAT PROP anhäufen
**aggradation** *n* INFR & DES Verlandung *f*
**aggregate** *n* CONCR, MAT PROP Zuschlagstoff *m*, STEEL *amount* Anhäufung *f*, Menge *f*; ~ **exposure** *n* CONCR Freiliegen *nt* der Zuschlagstoffe; ~ **stripping** *n*

CONCR, MAT PROP Zuschlagablösung *f*; ~ **stripping test** *n* MAT PROP *bitumen* Zuschlagablöseversuch *m*
**aggressive** *adj* MAT PROP aggressiv; ~ **substance** *n* ENVIRON, MAT PROP, SURFACE Schadstoff *m*; ~ **to building material** *adj* CONST LAW, MAT PROP, SURFACE baustoffschädlich; ~ **water** *n* INFR & DES, MAT PROP aggressives Wasser *nt*
**agitate** *vt* BUILD MACHIN rühren, rütteln, CONCR, INFR & DES, MAT PROP, SURFACE rühren
**agitating**: ~ **equipment** *n* BUILD MACHIN, CONCR, INFR & DES, MAT PROP, SURFACE Rühreinrichtung *f*
**agitation** *n* BUILD MACHIN, CONCR, INFR & DES, MAT PROP, SURFACE Rühren *nt*
**agitator** *n* BUILD MACHIN, CONCR, SURFACE Rührer *m*, Rührwerk *nt*; **~-type compulsory mixer** *n* BUILD MACHIN, CONCR *for manufacture of concrete products* Rührwerkszwangsmischer *m*
**agora** *n* ARCH & BUILD Agora *f*
**agreement** *n* CONST LAW *treaty* Abkommen *nt*, Vereinbarung *f*, Vertrag *m*; ~ **certificate** *n* CONST LAW Abkommensbescheinigung *f*
**agricultural** *adj* ARCH & BUILD, INFR & DES landwirtschaftlich; ~ **building** *n* ARCH & BUILD landwirtschaftliches Gebäude *nt*; ~ **drainage** *n* INFR & DES landwirtschaftliche Drainage *f*
**aid** *n* CONST LAW, INFR & DES, MAT PROP Hilfe *f*, Unterstützung *f*
**aiming**: ~ **position** *n* ARCH & BUILD, SURFACE Richtstellung *f*
**air 1.** *n* HEAT & VENT Luft *f*; **2. ~-condition** *vt* ARCH & BUILD, HEAT & VENT klimatisieren
**air**: **~-blown mortar** *n* STONE Spritzmörtel *m*
**airborne**: ~ **dust** *n* ENVIRON, MAT PROP Flugstaub *m*; ~ **microorganisms** *n pl* ENVIRON, HEAT & VENT schwebende Mikroorganismen *m pl*; ~ **sound** *n* CONST LAW, INFR & DES, MAT PROP, SOUND & THERMAL Luftschall *m*; ~ **sound insulation** *n* INFR & DES, MAT PROP, SOUND & THERMAL Luftschalldämmung *f*; ~ **sound level** *n* CONST LAW, INFR & DES, MAT PROP, SOUND & THERMAL Luftschallpegel *m*
**air**: **~-break contactor** *n* ELECTR Luftschütz *m*; ~ **change** *n* HEAT & VENT Luftwechsel *m*; ~ **chimney** *n* HEAT & VENT Entlüftungsrohr *nt*, Wrasenabzug *m*; ~ **circulation** *n* HEAT & VENT Luftumwälzung *f*; ~ **classifier** *n* ENVIRON Luftklassierer *m*, Luftsortierer *m*, Windsichter *m*; ~ **compressor** *n* BUILD MACHIN, HEAT & VENT Druckluftkompressor *m*, Luftkompressor *m*; **~-conditioned** *adj* ARCH & BUILD, HEAT & VENT klimatisiert; ~ **conditioner** *n* ARCH & BUILD, HEAT & VENT Klimaanlage *f*, Klimagerät *nt*; ~ **conditioning** *n* ARCH & BUILD, BUILD MACHIN, HEAT & VENT *action* Klimatisierung *f*, *apparatus* Klimaanlage *f*, Klimagerät *nt*; **~-conditioning duct** *n* ARCH & BUILD Klimakanal *m*, BUILD MACHIN, HEAT & VENT Klimakanal *m*; **~-conditioning equipment** *n* ARCH & BUILD, BUILD MACHIN, HEAT & VENT Klimaanlage *f*, Klimagerät *nt*; **~-conditioning unit** *n* ARCH & BUILD, BUILD MACHIN, HEAT & VENT Klimagerät *nt*; ~ **contamination** *n* CONST LAW, ENVIRON, INFR & DES Luftverschmutzung *f*; ~ **content** *n* INFR & DES, MAT PROP Luftgehalt *m*; ~ **cooler** *n* BUILD MACHIN, HEAT & VENT Luftkühler *m*; ~ **cooling** *n* HEAT & VENT Luftkühlung *f*
**aircraft**: ~ **hangar** *n* ARCH & BUILD, INFR & DES Flugzeughalle *f*; ~ **waste gas** *n* ENVIRON Flugzeugabgas *nt*
**air**: **~-cured** *adj* MAT PROP luftgehärtet; ~ **curtain door** *n* HEAT & VENT Luftschleiertür *f*; ~ **deflector** *n* HEAT & VENT Luftlenkblech *nt*; ~ **dehumidifier** *n* HEAT & VENT Entfeuchter *m*; ~ **distribution** *n* HEAT & VENT Luftverteilung *f*; **~-dried** *adj* MAT PROP luftgetrocknet; ~ **drying** *n* MAT PROP Lufttrocknung *f*; ~ **duct** *n* HEAT & VENT Luftkanal *m*, Lüftungskanal *m*; **~-entrained concrete** *n* CONCR Luftporenbeton *m*; **~-entrained mortar** *n* INFR & DES, MAT PROP, STONE Luftporenmörtel *m*; **~-entraining admixture** *n* CONCR, INFR & DES, MAT PROP, STONE Luftporenbildner *m*; **~-entraining agent** *n* CONCR, INFR & DES, MAT PROP, STONE Luftporenbildner *m*; ~ **entrainment** *n* CONCR, INFR & DES, MAT PROP, STONE Porenbildung *f*; ~ **flotation** *n* ENVIRON Luftblasenflotation *f*; ~ **flow** *n* HEAT & VENT Luftstrom *m*; ~ **flow measurement** *n* HEAT & VENT Luftstrommessung *f*; ~ **flue** *n* HEAT & VENT Entlüftungsrohr *nt*, Wrasenabzug *m*
**airfoil** *n* ENVIRON Tragfläche *f*
**air**: **~-free** *adj* ARCH & BUILD, HEAT & VENT entlüftet; ~ **gap** *n* BUILD HARDW, HEAT & VENT Luftzwischenraum *m* (*LZR*); ~ **hammer** *n* BUILD MACHIN Druckluftramme *f*; **~-handling capacity** *n* CONCR, HEAT & VENT *fan* Luftleistung *f*, Volumenstrom *m*; **~-hardening** *adj* CONCR, MAT PROP, STONE *mortar, concrete* lufthärtend; **~-hardening lime** *n* MAT PROP, STONE Luftkalk *m*; ~ **heater** *n* HEAT & VENT Lufterhitzer *m*; ~ **heating** *n* HEAT & VENT Luftheizung *f*; ~ **hose** *n* HEAT & VENT *compressed air* Luftschlauch *m*; ~ **humidity** *n* ARCH & BUILD, CONCR, HEAT & VENT, SURFACE Luftfeuchtigkeit *f*
**airing** *n* HEAT & VENT Belüftung *f*, Entlüftung *f*
**air**: ~ **inlet** *n* ARCH & BUILD, HEAT & VENT Lufteinlaß *m*
**airless** *adj* MAT PROP, SURFACE luftlos; ~ **paint spraying** *n* SURFACE luftloses Farbspritzen *nt*
**air**: ~ **lock** *n* ARCH & BUILD, HEAT & VENT Luftschleuse *f*; ~ **mass** *n* ENVIRON atmosphärische Masse *f*, HEAT & VENT Masse *f* der Luft; ~ **mortar** *n* MAT PROP, STONE Luftporenmörtel *m*; ~ **opening** *n* ARCH & BUILD, HEAT & VENT Lüftungsöffnung *f*; ~ **outlet** *n* ARCH & BUILD Luftauslaß *m*; ~ **permeability** *n* ARCH & BUILD Luftdurchgängigkeit *f*; ~ **pipes** *n pl* HEAT & VENT Luttenrohre *f pl*; ~ **pocket** *n* CONCR Luftnest *nt*; ~ **poisoning** *n* ENVIRON Luftvergiftung *f*; ~ **pollutant** *n* ENVIRON Luftschadstoff *m*; **~-polluting** *adj* ENVIRON luftverunreinigend; **~-polluting substance** *n* ENVIRON luftverunreinigender Stoff *m*; ~ **pollution** *n* CONST LAW, ENVIRON, INFR & DES Luftverschmutzung *f*, Luftverunreinigung *f*; ~ **pollution control** *n* CONST LAW, ENVIRON Lufthygiene *f*; ~ **pollution emission** *n* ENVIRON Luftverunreinigungsemission *f*; ~ **pollution episode** *n* ENVIRON Luftverunreinigungsgefahrensituation *f*; ~ **pollution forecast** *n* ENVIRON Luftverschmutzungsvorhersage *f*; ~ **pollution incident** *n* ENVIRON Luftverunreinigungsereignis *nt*; ~ **preheater** *n* HEAT & VENT Luftvorwärmer *m*; ~ **pressure** *n* HEAT & VENT Luftdruck *m*; ~ **purity** *n* ENVIRON, HEAT & VENT Luftreinheit *f*; ~ **quality data** *n pl* ENVIRON Luftqualitätsdaten *nt pl*; ~ **quality measurement** *n* ENVIRON Messung *f* der Luftqualität; ~ **quality standard** *n* ENVIRON Luftqualitätsnorm *f*; **~-raid shelter** *n* ARCH & BUILD Luftschutzbunker *m*;

~ **rammer** *n* BUILD MACHIN Druckluftstampfer *m*; ~-**reducing agent** *n* CONCR, INFR & DES, MAT PROP Luftporenminderer *m*; ~ **separation plant** *n* ENVIRON Luftklassierer *m*, Luftsortierer *m*, Windsichter *m*; ~ **separator** *n* ENVIRON Luftabscheider *m*, Luftsortierer *m*, Windsichter *m*; ~-**setting** *adj* MAT PROP, STONE luftabbindend; ~ **shed** *n* ENVIRON Luftausfällung *f*; ~ **shutter** *n* BUILD HARDW, HEAT & VENT Luftklappe *f*; ~ **space** *n* HEAT & VENT, SOUND & THERMAL Lufthohlraum *m*
**airstream**: ~ **sorting** *n* ENVIRON Windsichtung *f*
***air***: ~ **stripping** *n* ENVIRON Ausstrippen *nt* mit Luft; ~ **temperature** *n* HEAT & VENT, INFR & DES Lufttemperatur *f*; ~ **terminal** *n* ARCH & BUILD Flughafenabfertigungsgebäude *nt*
**airtight** *adj* ARCH & BUILD, ENVIRON, HEAT & VENT, MAT PROP, SOUND & THERMAL luftdicht
**airtightness** *n* ARCH & BUILD, ENVIRON, HEAT & VENT, MAT PROP, SOUND & THERMAL Luftdichtigkeit *f*
***air***: ~ **to air heat pump** *n* HEAT & VENT Luft-Luft-Wärmepumpe *f*; ~ **valve** *n* ENVIRON Luftventil *nt*; ~ **velocity** *n* ARCH & BUILD, HEAT & VENT Luftgeschwindigkeit *f*; ~ **voids** *n pl* CONCR, INFR & DES, MAT PROP, SOUND & THERMAL, STONE Luftporen *f pl*; ~ **voids content** *n* CONCR, INFR & DES, MAT PROP, SOUND & THERMAL, STONE Luftporengehalt *m*; ~ **voids ratio** *n* CONCR, INFR & DES, MAT PROP, SOUND & THERMAL, STONE Luftporenanteil *m*
**aisle** *n* ARCH & BUILD, INFR & DES *church* Schiff *nt*, Seitenschiff *nt*, *theatre, cinema* Gang *m*
**alabaster** *n* MAT PROP Alabaster *m*
**alarm** *n* BUILD HARDW, CONST LAW, ELECTR *alert* Warnung *f*, *fire* Alarm *m*, Melder *m*, ; ~ **system** *n* BUILD HARDW, CONST LAW, ELECTR Alarmanlage *f*, Warnanlage *f*
**alcohol**: ~ **resistance** *n* MAT PROP, SURFACE Alkoholbeständigkeit *f*
**alcove** *n* ARCH & BUILD Alkoven *m*
**algae** *n pl*, ENVIRON, HEAT & VENT, INFR & DES, MAT PROP, SURFACE Algen *f pl*; ~ **growth** *n* ENVIRON, HEAT & VENT, INFR & DES, MAT PROP, SURFACE Algenwuchs *m*
**algicide** *n* ENVIRON, HEAT & VENT, INFR & DES, MAT PROP, SURFACE Algenvernichtungsmittel *nt*
**align** *vt* ARCH & BUILD *arrange* ausrichten, einrichten, *buildings* ausfluchten, *put in line* in eine Linie bringen, CONCR *reinforcement* ausrichten
**alignment** *n* ARCH & BUILD, INFR & DES Baufluchtlinie *f*, Fluchtlinie *f*
**aliphatic**: ~ **hydrocarbon** *n* MAT PROP aliphatischer Kohlenwasserstoff *m*
**alkali** *n* INFR & DES, MAT PROP, SURFACE Alkali *nt*
**alkaline** *adj* INFR & DES, MAT PROP, SURFACE Alkali-, alkalisch, basisch, laugenhaft; ~ **fermentation** *n* ENVIRON Methangärung *f*; ~ **soil** *n* INFR & DES, MAT PROP alkalischer Boden *m*
**alkalinity** *n* ENVIRON, INFR & DES, MAT PROP Alkalität *f*
***alkali***: ~ **paint stripper** *n* MAT PROP, SURFACE alkalisches Abbeizmittel *nt*; ~ **reactivity** *n* MAT PROP, SURFACE Alkaliempfindlichkeit *f*; ~ **resistance** *n* MAT PROP, SURFACE Alkalibeständigkeit *f*; ~ **silicon reaction** *n* (*ASR*) CONCR Alkali-Silikon-Reaktion *f*
**alkyd**: ~ **paint** *n* MAT PROP Alkydharzfarbe *f*, Alkydharzgrundierung *f*, SURFACE Alkydharzgrundierung *f*; ~ **primer** *n* MAT PROP, SURFACE Alkydharzgrundierung *f*

**all-** *pref* ARCH & BUILD, BUILD HARDW, STEEL Ganz-
**all-glass**: ~ **wall** *n* ARCH & BUILD, STEEL Ganzglaswand *f*
**all-in**: ~ **ballast** *n* MAT PROP unklassifizierte Zuschlagstoffe *m pl*
**allocate** *vt* CONST LAW zuteilen, zuweisen
**allocation** *n* CONST LAW Zuteilung *f*, Zuweisung *f*; ~ **of expenses** *n* CONST LAW Umlage *f* von Gemeinkosten
**allochthonous**: ~ **matter** *n* ENVIRON bodenfremde Substanz *f*
**allot** *vt* CONST, LAW zuteilen
**allotment** *n* CONST LAW *allocation* Zuteilung *f*, *plot* Parzelle *f*, INFR & DES *for gardening* Schrebergarten *m*, *plot* Parzelle *f*
**allow** *vt* CONST LAW *authorize, permit* zubilligen, zulassen, *grant* gewähren
**allowable** *adj* ARCH & BUILD, BUILD HARDW, BUILD MACHIN, CONST LAW, ENVIRON, INFR & DES, MAT PROP, STEEL, TIMBER zulässig; ~ **bearing capacity** *n* CONCR, INFR & DES, MAT PROP, STEEL, TIMBER zulässige Traglast *f*; ~ **load** *n* ARCH & BUILD, BUILD MACHIN zulässige Last *f*; ~ **pile load** *n* INFR & DES zulässige Pfahllast *f*; ~ **pressure** *n* CONCR, INFR & DES, STEEL, TIMBER zulässiger Druck *m*; ~ **stress** *n* CONCR, INFR & DES, STEEL, TIMBER zulässige Spannung *f*; ~ **tolerance** *n* ARCH & BUILD, CONCR, INFR & DES, MAT PROP, STEEL zulässige Abweichung *f*
**allowance** *n* CONST LAW Toleranz *f*, zulässige Abweichung *f*; ~ **for extra-heavy work** *n* ARCH & BUILD, BUILD HARDW, CONST LAW Erschwerniszulage *f*
**alloy** *n* MAT PROP, STEEL Legierung *f*
**alloyed** *adj* MAT PROP, STEEL legiert
**alloying**: ~ **material** *n* MAT PROP, STEEL Legierungsstoff *m*
***alloy***: ~ **steel** *n* MAT PROP, STEEL legierter Stahl *m*
**all-plastic**: ~ **board** *n* BUILD HARDW, MAT PROP Vollkunststoffplatte *f*, Vollplastikplatte *f*; ~ **door** *n* ARCH & BUILD, BUILD HARDW Vollkunststofftür *f*
**all-purpose**: ~ **mobile excavator** *n* BUILD MACHIN Universalmobilbagger *m*; ~ **primer** *n* SURFACE Universalgrundierung *f*
**all-round**: ~ **visibility** *n* ARCH & BUILD Rundumsicht *f*
**all-steel**: ~ **design** *n* STEEL Ganzstahlkonstruktion *f*
**alluvial** *adj* INFR & DES, MAT PROP Anschwemm-, alluvial, angeschwemmt; ~ **deposits** *n pl* ENVIRON Schlammablagerungen *f pl*; ~ **sand** *n* INFR & DES, MAT PROP Anschwemmsand *m*; ~ **soil** *n* INFR & DES, MAT PROP Anschwemmboden *m*
**alluvium** *n* INFR & DES Schwemmland *nt*
**all-wood**: ~ **door** *n* ARCH & BUILD, BUILD HARDW, TIMBER Ganzholztür *f*
**alter** *vt* ARCH & BUILD *structure* umbauen, ändern, CONST LAW *contract, terms* verändern
**alteration** *n* ARCH & BUILD Änderung *f*, Umbau *m*, CONST LAW Änderung *f*, Umänderung *f*
**altered** *adj* ARCH & BUILD umgebaut, CONST LAW verändert
**alternate**: ~ **bending strength** *n* MAT PROP, STEEL Wechselbiegefestigkeit *f*
**alternating**: ~ **climate** *n* INFR & DES, MAT PROP Wechselklima *nt*; ~ **current** *n* ELECTR, HEAT & VENT Wechselstrom *m*; ~ **load** *n* ARCH & BUILD, INFR & DES, MAT PROP Wechsellast *f*; ~ **stress** *n* ARCH & BUILD, INFR & DES, MAT PROP Wechselspannung *f*
**alternative**: ~ **access** *n* ARCH & BUILD, CONST LAW Alternativzugang *m*; ~ **entrance** *n* ARCH & BUILD,

CONST LAW Alternativeingang *m*; ~ **escape route** *n* ARCH & BUILD, CONST LAW Alternativfluchtweg *m*; ~ **exit** *n* ARCH & BUILD, CONST LAW zusätzlicher Ausgang *m*; ~ **means of escape** *n* ARCH & BUILD, CONST LAW Alternativfluchtweg *m*

**alternator** *n* ELECTR, ENVIRON Drehstromlichtmaschine *f*, Wechselstrommaschine *f*

**altimeter** *n* BUILD MACHIN Höhenmesser *m*

**altitude** *n* ARCH & BUILD, INFR & DES Höhenlage *f*

**alumina**: ~ **mortar** *n* STONE Tonerdemörtel *m*

**aluminium** *n BrE* ARCH & BUILD, INFR & DES, MAT PROP, STEEL Aluminium *nt*; ~ **blind** *n BrE* BUILD HARDW Aluminiumblende *f*, Aluminiumsonnenblende *f*; ~ **coping** *n BrE* ARCH & BUILD, MAT PROP, STEEL Aluminiumabdeckung *f*; ~ **facing** *n BrE* ARCH & BUILD, INFR & DES, MAT PROP Aluminiumverkleidung *f*; ~ **finish** *n BrE* STEEL Aluminiumoberfläche *f*; ~ **flashing** *n BrE* STEEL Aluminiumverwahrung *f*; ~ **foil** *n BrE* MAT PROP, SOUND & THERMAL Aluminiumfolie *f*; ~ **front** *n BrE* ARCH & BUILD, STEEL Aluminiumfassade *f*; ~ **glazing bar** *n BrE* ARCH & BUILD Aluminiumdach *nt*, STEEL Aluminiumdach *nt*, Aluminiumsprosse *f*; ~ **hinge** *n BrE* BUILD HARDW, STEEL Aluminiumband *nt*; ~ **nail** *n* BUILD HARDW, STEEL Aluminiumnagel *m*; ~ **roof cladding** *n BrE* ARCH & BUILD, STEEL Aluminiumdach *nt*; ~ **sash** *n BrE* STEEL Aluminiumfensterflügel *m*; ~ **scrap** *n BrE* ENVIRON Aluminiumschrott *m*; ~ **street lighting mast** *n BrE* ELECTR, STEEL Aluminiumstraßenleuchtenmast *m*; ~ **trim** *n BrE* BUILD HARDW, STEEL Aluminiumblende *f*

**aluminum** *AmE see aluminium BrE*

**ambient** *adj* ENVIRON, HEAT & VENT, INFR & DES, SOUND & THERMAL umgebend; ~ **air** *n* ENVIRON umgebende Luft *f*; ~ **air emission standard** *n* ENVIRON Immissionsgrenzwert *m* der Luft; ~ **air quality standard** *n* ENVIRON Immissionsgrenzwert *m* der Luft; ~ **noise** *n* INFR & DES, SOUND & THERMAL Umgebungslärm *m*; ~ **pollutant concentration** *n* ENVIRON Konzentration *f* der Umweltschadstoffe; ~ **temperature** *n* HEAT & VENT, INFR & DES Umgebungstemperatur *f*

**amelioration** *n* ARCH & BUILD, CONST LAW, INFR & DES Bodenverbesserung *f*, Melioration *f*

**American**: ~ **Society of Heating, Refrigeration and Air Conditioning Engineers** *n AmE* (*ASHRAE AmE*) HEAT & VENT Amerikanische Gesellschaft *f* der Heizungs-, Kühlungs- und Klimatechniker

**amine**: ~ **formaldehyde resin** *n* MAT PROP Amin-Formaldehyd-Harz *nt*

**ammonia** *n* ENVIRON, INFR & DES Ammoniak *nt*

**amount 1.** *n* ARCH & BUILD *quantity* Menge *f*, *sum total* Betrag *m*, *result* Ergebnis *nt*, BUILD HARDW *result* Ergebnis *nt*, *quantity* Menge *f*, *sum total* Betrag *m*; **2.** ~ **to** *vi* ARCH & BUILD, BUILD HARDW *total* sich ergeben

**amperage** *n* ELECTR Amperezahl *f*

**ampere** *n* ELECTR Ampere *nt*

**amphibole**: ~ **asbestos** *n* MAT PROP Amphibolasbest *m*

**ample** *adj* INFR & DES, MAT PROP geräumig, weit

**amplification** *n* INFR & DES, MAT PROP Amplifikation *f*

**amplitude** *n* ELECTR, STEEL Amplitude *f*; ~ **of vibration** *n* ENVIRON Schwingstärke *f*, Schwingungsweite *f*

**anaerobic** *adj* ENVIRON, INFR & DES, MAT PROP anaerob

**anaerobic**: ~ **digestion** *n* ENVIRON anaerobe Faulung *f*, anaerobe Gärung *f*; ~ **fermentation** *n* ENVIRON anaerobe Gärung *f*; ~ **lagoon** *n* ENVIRON Faulteich *m*, anaerober Teich *m*

**analogous** *adj* ARCH & BUILD, INFR & DES analog

**analogy** *n* ARCH & BUILD, INFR & DES Analogie *f*

**analyse** *vt BrE* ARCH & BUILD, CONST LAW, INFR & DES, MAT PROP analysieren, *evaluate* auswerten, *results* statisch berechnen, *investigate* untersuchen

**analysis** *n* ARCH & BUILD, CONST LAW, INFR & DES, MAT PROP Analyse *f*, Auswertung *f*

**analyze** *AmE see analyse BrE*

**anatomical**: ~ **waste** *n* ENVIRON infektiöser Abfall *m*, pathogener Abfall *m*

**anchor 1.** *n* ARCH & BUILD, CONCR, INFR & DES, STEEL Anker *m*; **2.** *vt* ARCH & BUILD verankern, CONCR, INFR & DES ankern, STEEL verankern

**anchorage** *n* ARCH & BUILD, CONCR, INFR & DES, STEEL Verankerung *f*; ~ **force** *n* ARCH & BUILD Verankerungskraft *f*

***anchor***: ~ **bar** *n* INFR & DES, MAT PROP, STEEL Verankerungseisen *nt*; ~ **bolt** *n* CONCR, INFR & DES, MAT PROP, STEEL Ankerbolzen *m*; ~ **channel** *n* INFR & DES, MAT PROP, STEEL Ankerschiene *f*; ~ **head** *n* INFR & DES Ankerkopf *m*

**anchoring**: ~ **bolt** *n* CONCR, INFR & DES, MAT PROP, STEEL Ankerbolzen *m*; ~ **rail** *n* INFR & DES, MAT PROP, STEEL Ankerschiene *f*; ~ **wire** *n* ARCH & BUILD, CONCR, MAT PROP, STEEL Abspanndraht *m*

***anchor***: ~ **nail** *n* INFR & DES, MAT PROP, STEEL Ankernagel *m*; ~ **plate** *n* INFR & DES, MAT PROP, STEEL Ankerplatte *f*; ~ **point** *n* ARCH & BUILD Abspannpunkt *m*; ~ **pulling test** *n* INFR & DES, MAT PROP Ankerausziehversuch *m*; ~ **rod** *n* ARCH & BUILD, INFR & DES, STEEL Ankerstab *m*; ~ **wall** *n* ARCH & BUILD, INFR & DES Ankerwand *f*

**anemometer** *n* BUILD MACHIN Luftgeschwindigkeitsmesser *m*

**angle** *n* ARCH & BUILD *geometry* Winkel *m*, HEAT & VENT, INFR & DES, STEEL, WASTE WATER Winkel *m*, *in pipe* Kniestück *nt*; ~ **bar** *n* BUILD HARDW, CONCR, INFR & DES, MAT PROP, STEEL Winkelprofil *nt*; ~ **bead** *n* CONCR, INFR & DES, MAT PROP, STEEL Eckschiene *f*; ~-**body valve** *n* HEAT & VENT Eckabsperrventil *nt*; ~ **brace** *n* INFR & DES, MAT PROP, TIMBER Bandholz *nt*, Bug *m*; ~ **degree** *n* ARCH & BUILD Winkelgrad *m*

**angled**: ~ **purlin** *n* ARCH & BUILD, INFR & DES, MAT PROP, STEEL, TIMBER Winkelstahlpfette *f*, geneigte Pfette *f*; ~ **rafter** *n* ARCH & BUILD, INFR & DES, TIMBER Gratbalken *m*, Gratsparren *m*, geneigter Sparren *m*

***angle***: ~ **of failure** *n* INFR & DES, MAT PROP Versagenswinkel *m*; ~ **of friction** *n* INFR & DES, MAT PROP Reibungswinkel *m*; ~ **hinge** *n* BUILD HARDW Klavierband *nt*, Winkelband *nt*; ~ **of incidence** *n* ARCH & BUILD, ENVIRON, INFR & DES Einfallwinkel *m*; ~ **of inclination** *n* ARCH & BUILD, INFR & DES Neigungswinkel *m*; ~ **of internal friction** *n* INFR & DES innerer Reibungswinkel *m*; ~ **quoin** *n* CONCR, INFR & DES, MAT PROP, STONE Eckstein *m*; ~ **rafter** *n* ARCH & BUILD, MAT PROP, TIMBER Gratbalken *m*, Gratsparren *m*; ~ **of repose** *n* ARCH & BUILD, INFR & DES Ruhewinkel *m*, natürlicher Böschungswinkel *m*; ~ **of rotation** *n* INFR & DES Drehwinkel *m*; ~ **of shear** *n* INFR & DES, MAT PROP Scherwinkel *m*; ~ **shut-off valve** *n* HEAT & VENT Eckabsperrventil *nt*, Eckventil *nt*; ~ **of slope** *n* ARCH & BUILD, INFR & DES Neigungswinkel *m*; ~ **step** *n* ARCH & BUILD

Winkelstufe *f*; ~ **of view** *n* ARCH & BUILD, INFR & DES Blickwinkel *m*
**angular**: ~ **acceleration** *n* INFR & DES Winkelbeschleunigung *f*; ~ **bracket** *n* STEEL Winkelstütze *f*; ~ **soffit bracket** *n* BUILD HARDW Sturzfutterwinkel *m*; ~ **velocity** *n* ENVIRON Kreisfrequenz *f*, Winkelgeschwindigkeit *f*
**anhydride**: ~ **screed** *n* MAT PROP Anhydritstrich *m*
**anhydrite** *n* MAT PROP Anhydrit *nt*; ~ **binder** *n* MAT PROP, STONE Anhydritbinder *m*; ~ **mortar** *n* MAT PROP, STONE Anhydritmörtel *m*
**animal**: ~ **glue** *n* MAT PROP Tierleim *m*
**anneal** *vt* INFR & DES, MAT PROP, STEEL glühen
**annealed**: ~ **glass** *n* INFR & DES, MAT PROP vergütetes Glas *nt*; ~ **soft wire** *n* INFR & DES, MAT PROP ausgeglühter Weichdraht *m*
**annex** *AmE see annexe BrE*
**annexe** *n BrE* ARCH & BUILD, CONST LAW, INFR & DES *to document* Anlage *f*, Nachtrag *m*, Anhang *m*, Anbau *m*, *to building* Erweiterungsbau *m*, Erweiterung *f*, Nebengebäude *nt*
**annual**: ~ **consumption** *n* ELECTR, HEAT & VENT, INFR & DES Jahresverbrauch *m*
**annular**: ~ **arch** *n* ARCH & BUILD, STONE Kranzbogen *m*; ~ **saw** *n* BUILD MACHIN, TIMBER Zylinderkreissäge *f*; ~ **shake** *n* TIMBER Ringkluft *f*; ~ **vault** *n* ARCH & BUILD, STONE Ringgewölbe *nt*
**anodic**: ~ **oxidation** *n* SURFACE Eloxierung *f*
**anodization** *n* SURFACE Eloxierung *f*, anodische Oxidierung *f*
**anodize** *vt* SURFACE anodisieren
**anodized** *adj* SURFACE eloxiert
**antechoir** *n* ARCH & BUILD, INFR & DES Vorchor *m*
**antenna** *n* BUILD MACHIN, ELECTR Antenne *f*; ~ **mast** *n* ELECTR, STEEL Antennenmast *m*; ~ **pipe mount** *n* BUILD HARDW, ELECTR Antennenhalterung *f*; ~ **socket** *n* ELECTR Antennensteckdose *f*
**anteroom** *n* ARCH & BUILD Vestibül *nt*, Vorraum *m*
**anthropogenic**: ~ **acidification** *n* ENVIRON anthropogen bedingte Übersäuerung *f*, menschlich bedingte Übersäuerung *f*
**anticapillary**: ~ **course** *n* ARCH & BUILD, INFR & DES *road* kapillarbrechende Schicht *f*
**anticorrosive**: ~ **agent** *n* INFR & DES, MAT PROP Korrosionsschutzmittel *nt*
**anticrack**: ~ **reinforcement** *n* CONCR Rißbewehrung *f*
**antidazzle** *adj* SURFACE blendfrei
**antidrill**: ~ **feature** *n* BUILD HARDW *lock*, BUILD MACHIN Anbohrschutz *m*, Aufbohrschutz *m*
**antidrumming**: ~ **agent** *n* MAT PROP, SOUND & THERMAL Entdröhnungsmittel *nt*
**antiflooding**: ~ **valve** *n* WASTE WATER Rückstauverschluß *m*
**antifreeze** *n* BUILD MACHIN, HEAT & VENT, INFR & DES, MAT PROP Frostschutz *m*; ~ **agent** *n* BUILD MACHIN, HEAT & VENT, INFR & DES, MAT PROP Frostschutzmittel *nt*; ~ **layer** *n* HEAT & VENT, INFR & DES, MAT PROP Frostschutzschicht *f*
**antiknock**: ~ **additive** *n* BUILD MACHIN, ENVIRON Antiklopfmittel *nt*
**antioxidant** *n* MAT PROP Antioxidans *nt*
**antipanic**: ~ **bolt lock** *n* BUILD HARDW, CONST LAW Panikriegelfallenschloß *nt*
**antique**: ~ **glass** *n* MAT PROP Antikglas *nt*
**antirot** *adj* ENVIRON, MAT PROP, SURFACE, TIMBER *decay-resistant* fäulnisbeständig, fäulnisfest, nicht verrottbar, unverrottbar
**antirust**: ~ **primer** *n* MAT PROP, STEEL, SURFACE Rostschutzgrundierung *f*
**antishrinkage**: ~ **admixture** *n* MAT PROP Schwindschutzzusatz *m*
**antisiphoning**: ~ **device** *n* WASTE WATER Rohrbelüfter *m*, Rohrentlüfter *m*
**antiskid** *adj* SURFACE rutschfest, trittsicher
**antiskinning**: ~ **agent** *n* MAT PROP Hautverhinderungsmittel *nt*
**antistress**: ~ **agent** *n* SURFACE Entspanner *m*, Entspannungsmittel *nt*
**antisun**: ~ **glass** *n* MAT PROP Sonnenschutzglas *nt*
**antivacuum**: ~ **device** *n* HEAT & VENT, WASTE WATER Rohrbelüfter *m*, Rohrentlüfter *m*
**antivibration**: ~ **mounting** *n* ARCH & BUILD, SOUND & THERMAL vibrationsfreie Montage *f*
**anvil** *n* STEEL Amboß *m*
**apartment** *n* ARCH & BUILD Appartement *nt*, Wohnung *f*; ~ **floor** *n* ARCH & BUILD Wohnetage *f*
**aperture** *n* ARCH & BUILD Öffnung *f*
**apex** *n* ARCH & BUILD, STONE Scheitel *m*; ~ **joint** *n* ARCH & BUILD, STONE Scheitelfuge *f*; ~ **stone** *n* ARCH & BUILD, STONE Giebelschlußstein *m*
**apparent**: ~ **cohesion** *n* INFR & DES, MAT PROP Scheinkohäsion *f*; ~ **density** *n* INFR & DES, MAT PROP Rohdichte *f*, Schüttdichte *f*, scheinbare Dichte *f*
**appearance** *n* ARCH & BUILD Aussehen *nt*
**appendix** *n* ARCH & BUILD, CONST LAW, INFR & DES Anhang *m*
**appliance** *n* ARCH & BUILD Einrichtung *f*, BUILD MACHIN Einrichtung *f*, Gerät *nt*, Vorrichtung *f*, INFR & DES Einrichtung *f*
**applicable** *adj* CONST LAW *regulations* einschlägig, verwendbar
**application** *n* CONST LAW Anmeldung *f*, Antrag *m*, SURFACE Anwendung *f*, Aufbringen *nt*; ~ **by brushing** *n* SURFACE Aufstreichen *nt*; ~ **method** *n* SURFACE Auftragsverfahren *nt*; ~ **rate** *n* SURFACE Auftragsmenge *f*; ~ **specification** *n* SURFACE Verarbeitungsrichtlinien *f pl*
**applied** *adj* CONST LAW, INFR & DES angewandt; ~ **method** *n* ARCH & BUILD, INFR & DES Anwendungsmethode *f*; ~ **research** *n* ENVIRON Zweckforschung *f*
**apply 1.** *vt* SURFACE *paint, varnish* auftragen; ~ **for** CONST LAW *seek permission* anmelden; **2.** ~ **to** *vi* CONST LAW *refer to* gelten für, sich beziehen auf
**apprentice** *n* CONST LAW Lehrling *m*
**apprenticeship** *n* CONST LAW Lehrjahre *nt pl*
**approach** *n* ARCH & BUILD, INFR & DES *to building* Zufahrt *f*, Zugang *m*, Zutritt *m*, *to problem* Lösungsweg *m*; ~ **road** *n* ARCH & BUILD, INFR & DES Zufahrtsstraße *f*
**appropriate** *adj* ARCH & BUILD, CONST LAW angemessen, geeignet, passend
**approval** *n* ARCH & BUILD, CONST LAW Erlaubnis *f*, Genehmigung *f*; ~ **note** *n* ARCH & BUILD, CONST LAW Freigabevermerk *m*
**approved** *adj* ARCH & BUILD, CONST LAW *design or type* bauartgeprüft; ~ **list** *n* ARCH & BUILD, CONST LAW genehmigte Liste *f*; ~ **product** *n* INFR & DES, MAT PROP zugelassenes Produkt *nt*
**approximate** *adj* ARCH & BUILD, INFR & DES angenähert, näherungsweise, ungefähr; ~ **assumption** *n* ARCH & BUILD, INFR & DES Näherungsannahme *f*; ~ **cost** *n*

ARCH & BUILD, CONST LAW, INFR & DES Kostenschätzung *f*; ~ **value** *n* ARCH & BUILD, INFR & DES, MAT PROP Richtwert *m*
**approximation**: ~ **method** *n* ARCH & BUILD, INFR & DES Näherungsverfahren *nt*
**apron** *n* ARCH & BUILD Frontschürze *f*; ~ **flashing** *n* ARCH & BUILD, MAT PROP Fugendeckblech *nt*; ~ **wall** *n* ARCH & BUILD Fensterbrüstung *f*
**apse**: ~ **arch** *n* ARCH & BUILD Absidenbogen *m*; ~ **arch impost** *n* ARCH & BUILD Absidenbogenkämpfer *m*
**aptitude**: ~ **test** *n* CONST LAW, INFR & DES, MAT PROP Eignungsprüfung *f*
**aquatic**: ~ **acidification** *n* ENVIRON, WASTE WATER Übersäuerung des Wassers *f*
**aqueduct** *n* ENVIRON, INFR & DES Aquädukt *nt*
**aqueous**: ~ **corrosion** *n* SURFACE Feuchtigkeitskorrosion *f*; ~ **effluent** *n* ENVIRON wäßriger Ausfluß *m*; ~ **wood preservative** *n* MAT PROP, SURFACE, TIMBER wäßriges Holzschutzmittel *nt*
**aquifer** *n* ENVIRON Aquifer *m*
**arabesque** *n* ARCH & BUILD Arabeske *f*
**aragonite** *n* MAT PROP Aragonit *m*
**arbitration** *n* CONST LAW Schiedsverfahren *nt*
**arbitrator** *n* CONST LAW Schlichter *m*
**arbor** *n* ARCH & BUILD, BUILD HARDW, BUILD MACHIN *grinding wheel* Dorn *m*
**arcade** *n* ARCH & BUILD Arkade *f*, Arkatur *f*, Bogengang *m*
**arcaded**: ~ **court** *n* ARCH & BUILD Arkadenhof *m*
**arcature** *n* ARCH & BUILD Arkatur *f*
**arc**: ~ **of a circle** *n* ARCH & BUILD *geometry* Kreisbogen *m*; ~ **cutting** *n* STEEL Lichtbogenschneiden *nt*
**arch 1.** *n* ARCH & BUILD Bogen *m*, Wölbung *f*, INFR & DES Bogen *m*; **2.** *vt* ARCH & BUILD krümmen, überwölben; **3.** *vi* ARCH & BUILD, INFR & DES, MAT PROP sich krümmen
***arch***: ~ **bay** *n* ARCH & BUILD Bogenfeld *nt*; ~ **bearing** *n* ARCH & BUILD, STONE Bogenauflager *nt*; ~ **center line** *AmE*, ~ **centre line** *BrE n* ARCH & BUILD Bogenachse *f*
**arched** *adj* ARCH & BUILD, CONCR, INFR & DES, STONE, TIMBER bogenförmig, gewölbt; ~ **beam** *n* ARCH & BUILD, CONCR, INFR & DES, STONE, TIMBER Bogenträger *m*; ~ **boom** *n* ARCH & BUILD, INFR & DES gekrümmte Gurtung *f*; ~ **dam** *n* ARCH & BUILD, INFR & DES Bogenstaumauer *f*; ~ **floor** *n* ARCH & BUILD, CONCR, INFR & DES Voutendecke *f*; ~ **girder** *n* ARCH & BUILD, CONCR, INFR & DES, STONE, TIMBER Bogenträger *m*; ~ **head** *n* ARCH & BUILD, INFR & DES Bogenabschluß *m*; ~ **lintel** *n* ARCH & BUILD, INFR & DES Segmentbogen *m*; ~ **recess** *n* ARCH & BUILD, INFR & DES Bogennische *f*; ~ **roof** *n* ARCH & BUILD, INFR & DES Bogendach *nt*, Bogengewölbe *nt*; ~ **vault** *n* ARCH & BUILD, INFR & DES Bogengewölbe *nt*
***arch***: ~**-gravity dam** *n* ARCH & BUILD, INFR & DES Bogengewichtsmauer *f*
**architect** *n* ARCH & BUILD Architekt *m*; ~ **in charge** *n* ARCH & BUILD, INFR & DES bauleitender Architekt *m*; ~ **on site** *n* ARCH & BUILD, INFR & DES bauleitender Architekt *m*
**architectural** *adj* ARCH & BUILD, CONST LAW architektonisch; ~ **acoustics** *n* SOUND & THERMAL Bauakustik *f*; ~ **award** *n* ARCH & BUILD Architekturpreis *m*; ~ **drawing** *n* ARCH & BUILD Architektenzeichnung *f*; ~ **education** *n* ARCH & BUILD Architekturausbildung *f*; ~ **feature** *n* ARCH & BUILD Baumerkmal *f*; ~ **style** *n* ARCH & BUILD Baustil *m*; ~ **work** *n* ARCH & BUILD Architektenleistung *f*
**architecture** *n* ARCH & BUILD Architektur *f*, Baukunst *f*
**architrave** *n* ARCH & BUILD Architrav *m*, STONE Natursteinsturz *m*, TIMBER Architrav *m*
***arch***: ~ **radius** *n* ARCH & BUILD, INFR & DES Bogenradius *m*; ~ **rise** *n* ARCH & BUILD, INFR & DES Bogenstichhöhe *f*; ~ **springer** *n* ARCH & BUILD, STONE Bogenkämpferstein *m*; ~ **stone** *n* ARCH & BUILD, STONE Schlußstein *m*; ~ **stress** *n* ARCH & BUILD, INFR & DES Bogenspannung *f*; ~ **thrust** *n* ARCH & BUILD, INFR & DES Gewölbeschub *m*
**archway** *n* ARCH & BUILD Bogengang *m*
**arc**: ~ **weld** *n* MAT PROP, STEEL Lichtbogenschweißen *nt*; ~ **welding** *n* MAT PROP, STEEL Lichtbogenschweißen *nt*
**area** *n* ARCH & BUILD *surface area* Grundfläche *f*, Fläche *f*, Oberfläche *f*, *tract of land* Geländeabschnitt *m*, CONST LAW Gebiet *nt*, INFR & DES *district* Gebiet *nt*, Oberfläche *f*, Fläche *f*, Gegend *f*, SURFACE Oberfläche *f*; ~ **of application** *n* ARCH & BUILD, INFR & DES Anwendungsbereich *m*; ~ **of deep weathering** *n* SURFACE Verwitterungsfläche *f*; ~ **emission source** *n* ENVIRON Emissionsquelle *f*
**areal**: ~ **excavation** *n* INFR & DES Flächenaushub *m*
***area***: ~ **load** *n* ARCH & BUILD, INFR & DES Flächenlast *f*
**areal**: ~ **scaffolding** *n* INFR & DES Flächengerüst *nt*
***area***: ~ **of pure air** *n* ENVIRON Reinluftgebiet *nt*
**areas**: ~ **particularly exposed to lightning** *n pl* ENVIRON blitzgefährdete Gebiete *nt pl*
***area***: ~ **of support** *n* ARCH & BUILD, CONCR, STEEL, STONE Auflagerfläche *f*
**arenaceous** *adj* ENVIRON, MAT PROP sandhaltig, sandig
**argillaceous** *adj* ENVIRON, INFR & DES, MAT PROP tonartig, tonhaltig, tonig; ~ **rock** *n* INFR & DES, MAT PROP, STONE Tongestein *nt*
**argillite** *n* INFR & DES, MAT PROP, STONE Argillit *m*, Tongestein *nt*
**arm** *n* ARCH & BUILD Arm *m*, Träger *m*, ELECTR Abzweigung *f*, INFR & DES Träger *m*; ~ **of eccentricity** *n* ARCH & BUILD, TIMBER Außermittigkeit *f*
**armored** *AmE see armoured BrE*
**armoring** *AmE see armouring BrE*
**armor**: ~ **plate** *n AmE* INFR & DES, MAT PROP, STEEL Panzerblech *nt*
**armoured** *adj BrE* ELECTR, INFR & DES, MAT PROP, STEEL, bewehrt, gepanzert; ~ **cable** *n BrE* ELECTR, INFR & DES, MAT PROP Panzerkabel *nt*, bewehrtes Kabel *nt*, gepanzertes Kabel *nt*; ~ **conduit** *n BrE* ELECTR *for protection of power cable* Panzerrohr *nt*; ~ **glass** *n BrE* BUILD HARDW, MAT PROP Drahtglas *nt*
**armouring** *n BrE* ELECTR *of cable* Bewehrung *f*
**aromatic**: ~ **compound** *n* MAT PROP aromatische Verbindung *f*
**arrange** *vt* ARCH & BUILD, INFR & DES *align* ausrichten, *order* anordnen, CONCR ausrichten
**arrangement** *n* ARCH & BUILD Anordnung *f*, Aufbau *m*, CONCR Ausrichtung *f*, INFR & DES Anordnung *f*, Aufbau *m*; ~ **of girders** *n* ARCH & BUILD, INFR & DES Trägeranordnung *f*; ~ **of windows** *n* ARCH & BUILD, INFR & DES Befensterung *f*, Fensteranordnung *f*
**array**: ~ **blanket** *n* ENVIRON Solarzellenlaken *nt*
**arrester**: ~ **device** *n* CONST LAW, ELECTR *lightning protection* Fangeinrichtung *f*; ~ **line** *n* CONST LAW,

ELECTR *lightning protection* Fangleitung *f*; ~ **rod** *n* CONST LAW, ELECTR *lightning protection* Fangstange *f*

**arris** *n* ARCH & BUILD Gratlinie *f*; ~ **rail** *n* ARCH & BUILD, TIMBER Firstpfette *f*

**arrival** *n* ARCH & BUILD Ankunft *f*; ~ **lounge** *n* ARCH & BUILD Ankunftshalle *f*; ~ **platform** *n* ARCH & BUILD Ankunftsbahnsteig *m*

**artesian** *adj* ENVIRON, INFR & DES artesisch; ~ **pressure** *n* ENVIRON, INFR & DES artesischer Druck *m*; ~ **well** *n* ENVIRON, INFR & DES artesischer Brunnen *m*

**articles**: ~ **of apprenticeship** *n pl* CONST LAW Lehrvertrag *m*

**articulated**: ~ **beam** *n* ARCH & BUILD Gelenkträger *m*, *half-timbering* Gerber-Fachwerkbinder *m*; ~ **elevation** *n* ARCH & BUILD aufgegliederte Fassade *f*; ~ **lorry** *n BrE* (*cf semitrailer truck AmE*) BUILD MACHIN Sattelschlepper *m*

**articulation** *n* ARCH & BUILD, BUILD HARDW, STEEL Gelenk *nt*

**artificial** *adj* MAT PROP *turf* Kunst-, künstlich; ~ **acidification** *n* ENVIRON künstliche Übersäuerung *f*; ~ **ageing** *n* ENVIRON künstliche Alterung *f*, MAT PROP künstliches Altern *nt*; ~ **lighting** *n* ELECTR künstliche Beleuchtung *f*; ~ **marble** *n* MAT PROP, STONE Kunstmarmor *m*; ~ **soil stabilization** *n* ARCH & BUILD, INFR & DES, MAT PROP Baugrundverbesserung *f*; ~ **stone** *n* BUILD HARDW, INFR & DES Betonwerkstein *m*, Werkstein *m*, MAT PROP, STONE Werkstein *m*; ~ **ventilation** *n* HEAT & VENT künstliche Lüftung *f*; ~ **weathering** *n* MAT PROP künstliche Bewitterung *f*

**artistic**: ~ **design** *n* ARCH & BUILD, INFR & DES künstlerische Gestaltung *f*; ~ **expression** *n* ARCH & BUILD, INFR & DES künstlerischer Ausdruck *m*; ~ **monument** *n* INFR & DES Kunstdenkmal *nt*

**Art Nouveau** *n* ARCH & BUILD Jugendstil *m*

**arts**: ~ **center** *AmE*, ~ **centre** *BrE n* ARCH & BUILD Kulturzentrum *nt*

**asbestos** *n* CONST LAW, MAT PROP Asbest *m*; ~ **board** *n* MAT PROP Asbestplatte *f*; ~ **cement** *n* CONCR Asbestzement *m*, MAT PROP Asbestzement *m*, Eternit® *nt*; ~ **cement board** *n* CONCR, INFR & DES, MAT PROP Berliner Verbau *m*, Berliner Welle *f*, KW-Platte *f*; ~ **cement corrugated roof covering** *n* CONCR, MAT PROP Wellasbestzementdachbelag *m*, Wellfaserzementdachbelag *m*; ~ **cement goods** *n pl* CONCR, MAT PROP Asbestzementwaren *f pl*; ~ **cement lining** *n* CONCR, MAT PROP *fire protection* Asbestzementmörtel *m*, Asbestzementverkleidung *f*; ~ **cement panel** *n* CONCR, MAT PROP Asbestzementtafel *f*; ~ **cement partition** *n* CONCR, MAT PROP Asbestzementtrennwand *f*; ~ **cement product** *n* CONCR, MAT PROP Asbestzementerzeugnis *nt*; ~ **cement shingle** *n* CONCR, MAT PROP Asbestzementschindel *f*; ~**-containing** *adj* CONST LAW, MAT PROP asbesthaltig; ~ **cord** *n* MAT PROP Asbestschnur *f*; ~ **dust** *n* CONST LAW, MAT PROP Asbeststaub *m*; ~ **fiber** *AmE*, ~ **fibre** *BrE n* CONST LAW, MAT PROP Asbestfaser *f*; ~ **gloves** *n pl* MAT PROP Asbesthandschuhe *m pl*; ~ **insulating board** *n* MAT PROP Asbestdämmplatte *f*; ~ **panel** *n* MAT PROP Asbesttafel *f*; ~ **pipe** *n* MAT PROP Asbestrohr *nt*; ~ **product** *n* MAT PROP Asbesterzeugnis *nt*; ~ **PVC floor tile** *n* MAT PROP Asbest-PVC-Bodenplatte *f*; ~ **rock** *n* MAT PROP Asbestgestein *nt*; ~ **roofing sheet** *n* MAT PROP Asbestdachplatte *f*; ~ **sheet** *n* MAT PROP Asbestplatte *f*; ~ **wool** *n* MAT PROP Asbestwolle *f*

**as-built**: ~ **drawing** *n* ARCH & BUILD, CONST LAW Bestandszeichnung *f*, Revisionsplan *m*, Revisionszeichnung *f*

**ash** *n* ENVIRON, HEAT & VENT *burnt material* Asche *f*, TIMBER *tree* Esche *f*; ~ **and combustion residue** *n* ENVIRON Asche- und Verbrennungsrückstand *m*; ~ **content** *n* MAT PROP Aschegehalt *m*; ~**-free** *adj* ENVIRON aschefrei

**ashlar** *n* ARCH & BUILD, INFR & DES, MAT PROP, STONE Naturstein *m*, Naturwerkstein *m*, Quader *m*

**ashlaring** *n* ARCH & BUILD, INFR & DES, STONE Natursteinmauerwerk *nt*, Quadermauerwerk *nt*, Werksteinmauerwerk *nt*

***ashlar***: ~ **masonry** *n* ARCH & BUILD, INFR & DES, STONE Natursteinmauerwerk *nt*, Quadermauerwerk *nt*, Werksteinmauerwerk *nt*

**ASHRAE** *abbr AmE* (*American Society of Heating, Refrigeration and Air Conditioning Engineers AmE*) HEAT & VENT ASHRAE (*Amerikanische Gesellschaft der Heizungs-, Kühlungs- und Klimatechniker*)

***ash***: ~ **veneer** *n* TIMBER Eschefurnier *nt*

**aspect** *n* ARCH & BUILD *appearance* Aspekt *m*, Aussehen *nt*, Erscheinung *f*, CONST LAW Aspekt *m*, INFR & DES *appearance* Aussehen *nt*, Erscheinung *f*, *element* Aspekt *m*, *point of view* Gesichtspunkt *m*

**asphalt 1.** *n* ARCH & BUILD Asphalt *m*, Bitumen *nt*, CONCR Asphalt *m*, INFR & DES, MAT PROP, SURFACE Asphalt *m*, Bitumen *nt*; **2.** *vt* INFR & DES, MAT PROP, SURFACE asphaltieren

***asphalt***: ~**-based** *adj* INFR & DES, MAT PROP, SURFACE bitumenhaltig; ~**-coated gravel** *n* INFR & DES, MAT PROP Bitukies *m*, Bitumenkies *m*; ~**-compatible** *adj* MAT PROP bitumenverträglich; ~ **concrete** *n* CONCR Asphaltbeton *m*, INFR & DES Asphaltbeton *m*, Bitumenbeton *m*, MAT PROP Asphaltbeton *m*; ~ **crusher** *n* INFR & DES Asphaltbrecher *m*; ~ **floor** *n* ARCH & BUILD, MAT PROP Asphaltfußboden *m*, Asphaltfußbodenbelag *m*, Asphaltstrich *m*; ~ **flooring** *n* ARCH & BUILD, MAT PROP Asphaltfußboden *m*, Asphaltfußbodenbelag *m*, Asphaltstrich *m*

**asphaltic** *adj* ARCH & BUILD, INFR & DES, MAT PROP, SURFACE asphaltisch, bituminös; ~ **binder** *n* INFR & DES, MAT PROP Asphaltbinder *m*; ~ **concrete finisher** *n* BUILD MACHIN, CONCR, INFR & DES Asphaltbetonfertiger *m*; ~ **felt** *n* INFR & DES, MAT PROP Teerpappe *f*; ~ **floor covering** *n* ARCH & BUILD, INFR & DES, MAT PROP Asphaltfußboden *m*, Asphaltfußbodenbelag *m*, Asphaltstrich *m*; ~ **hot adhesive compound** *n* INFR & DES, MAT PROP Bitumenheißklebemasse *f*; ~ **mastic** *n* INFR & DES, MAT PROP Gußasphalt *m*; ~ **mixture** *n* INFR & DES, MAT PROP Schwarzmaterial *nt*, bituminöses Mischgut *nt*; ~ **pavement** *n* INFR & DES, MAT PROP Asphaltdecke *f*; ~ **road binder** *n* INFR & DES, MAT PROP bituminöses Bindemittel *nt*; ~ **street pavement** *n* INFR & DES, MAT PROP bituminöse Fahrbahndecke *f*; ~ **tile** *n* INFR & DES, MAT PROP Asphaltplatte *f*

***asphalt***: ~**-impregnated** *adj* MAT PROP bitumengetränkt

**asphalting** *n* INFR & DES, MAT PROP, SURFACE Asphaltieren *nt*

***asphalt***: ~ **mastic** *n* INFR & DES, MAT PROP Asphaltmastix *f*, Asphaltmühle *f*; ~ **pavement** *n* INFR & DES Asphaltdecke *f*, Bitumenbelag *m*, Schwarzdecke *f*,

MAT PROP Asphaltdecke *f*; **~-resistant** *adj* INFR & DES, MAT PROP bitumenbeständig; **~ roofing felt** *n* INFR & DES, MAT PROP Bitumendachbahn *f*; **~-saturated** *adj* MAT PROP bitumengetränkt; **~-saturated paper** *n* INFR & DES, MAT PROP Bitumenpapier *nt*; **~ sheet** *n* INFR & DES, MAT PROP Bitumenschweißbahn *f*; **~ sheeting** *n* INFR & DES, MAT PROP Bitumenschweißbahn *f*; **~ surfacing** *n* INFR & DES Asphaltdecke *f*, Deckasphaltschicht *f*, MAT PROP Asphaltdecke *f*; **~ tanking** *n* INFR & DES wasserdichte Bitumenisolierung *f* eines Kellergeschosses; **~ tape** *n* MAT PROP Bitumenband *nt*, Bitumenbinde *f*; **~-treated chipboard** *n* TIMBER Bitumenholzfaserplatte *f*

**ASR** *abbr* (*alkali silicon reaction*) CONCR Alkali-Silikon-Reaktion *f*

**assay** *n* MAT PROP Metallprobe *f*; **~ balance** *n* BUILD MACHIN Prüfwaage *f*

**assemblage**: **~ point** *n* HEAT & VENT, STEEL, TIMBER Knoten *m*

**assemble** *vt* ARCH & BUILD aneinanderfügen, BUILD MACHIN einbauen, HEAT & VENT *fit together* aneinanderfügen, *from parts* montieren, INFR & DES einbauen, STEEL *fit together* aneinanderfügen, *from parts* montieren, zusammenbauen, *install* einbauen, TIMBER *from parts* montieren, einbauen, aneinanderfügen, zusammenbauen

**assembling** *n* HEAT & VENT, STEEL, TIMBER Einbauen *nt*

**assembly** *n* ARCH & BUILD, BUILD MACHIN Einbau *m*, HEAT & VENT *installation* Montage *f*, Einbau *m*, *unit* Baueinheit *f*, INFR & DES Einbau *m*, STEEL *installation* Einbau *m*, *unit* Baueinheit *f*, TIMBER *unit* Baueinheit *f*, Einbau *m*; **~ room** *n* ARCH & BUILD, INFR & DES Versammlungsraum *m*; **~ schedule** *n* CONCR, CONST LAW, STEEL Montageplan *m*

**assimilative**: **~ capacity** *n* ENVIRON Selbstreinigungskraft *f*

**association** *n* CONST LAW Gesellschaft *f*, Verband *m*, Vereinigung *f*; **~ of German engineers** *n* ARCH & BUILD, HEAT & VENT Verband *m* Deutscher Ingenieure (*VDI*)

**assort** *vt* MAT PROP *recycling* sortieren

**assume** *vt* ARCH & BUILD *load*, CONST LAW, INFR & DES annehmen

**assumed**: **~ load** *n* ARCH & BUILD Bemessungslast *f*, Lastannahme *f*

**assumption** *n* ARCH & BUILD *of load*, CONST LAW, INFR & DES Annahme *f*

**assurance** *n* CONST LAW Garantie *f*, *certainty* Sicherheit *f*, *guarantee* Bürgschaft *f*, *insurance* Versicherung *f*, *promise* Zusicherung *f*

**atmosphere** *n* ENVIRON, INFR & DES Atmosphäre *f*

**atmospheric**: **~ acidity** *n* ENVIRON atmosphärische Säurekapazität *f*; **~ chemical process** *n* ENVIRON chemischer Prozeß *m* in der Atmosphäre; **~ chemistry** *n* ENVIRON Luftchemie *f*; **~ concentration** *n* ENVIRON Konzentration *f* in der Atmosphäre; **~ conditions** *n pl* ENVIRON Witterungsverhältnisse *nt pl*; **~ fallout** *n* ENVIRON atmosphärischer Niederschlag *m*; **~ inversion** *n* ENVIRON Inversionsschicht *f*; **~ lifetime** *n* ENVIRON Lebensdauer *f* der Atmosphäre; **~ loading** *n* ENVIRON atmosphärische Luftbelastung *f*; **~ moisture** *n* ARCH & BUILD, CONCR, HEAT & VENT, SURFACE Luftfeuchtigkeit *f*; **~ obscurity** *n* ENVIRON atmosphärische Verdunklung *f*; **~ phenomenon** *n* ENVIRON atmosphärische Erscheinung *f*; **~ pollution** *n* CONST LAW, ENVIRON, INFR & DES Luftverschmutzung *f*; **~ precipitation** *n* ENVIRON atmosphärischer Niederschlag *m*; **~ pressure** *n* INFR & DES Luftdruck *m*; **~ scrubbing** *n* ENVIRON, HEAT & VENT atmosphärischer Auswaschvorgang *m*; **~ sulfur** *AmE*, **~ sulphur** *BrE* *n* ENVIRON atmosphärischer Schwefel *m*

**atom** *n* MAT PROP Atom *nt*

**atomic**: **~ weight** *n* MAT PROP Atomgewicht *nt*

**atomization** *n* MAT PROP Zerstäubung *f*

**atomize** *vt* MAT PROP zerstäuben

**atrium** *n* ARCH & BUILD, INFR & DES Atrium *nt*, Vorhof *m*

**attach** *vt* ARCH & BUILD, CONCR, STEEL, TIMBER anbringen, aufsetzen, *secure* befestigen, vorsetzen

**attached**: **~ garage** *n* ARCH & BUILD Anbaugarage *f*

**attaching**: **~ plate** *n* CONCR, STEEL, TIMBER Anschraubplatte *f*

**attachment** *n* BUILD HARDW Anbaugerät *nt*

**attachments**: **~ for wheeled and crawler tractors** *n pl* BUILD MACHIN Anbaugerät *nt* für Rad- und Raupenschlepper

**attack** *n* CONCR, MAT PROP, SURFACE, TIMBER Befall *m*

**attendance** *n* ARCH & BUILD Bedienung *f*, Wartung *f*

**attenuate** *vt* ARCH & BUILD, ELECTR, SOUND & THERMAL *noise* dämpfen

**attenuation** *n* ARCH & BUILD, ELECTR, SOUND & THERMAL *noise* Dämpfung *f*

**Atterberg**: **~ limits** *n pl* INFR & DES Atterbergsche Konsistenzgrenzen *f pl*

**attic** *n* ARCH & BUILD *garret* Bühne *f*, Dachboden *m*, Dachgeschoß *nt*, Speicher *m*; **~ access hatch** *n* ARCH & BUILD Dachbodeneinstiegsluke *f*

**attrition** *n* ARCH & BUILD, MAT PROP, SURFACE Abrieb *m*, Verschleiß *m*

**audibility**: **~ range** *n* ARCH & BUILD Audiometrie *f*, CONST LAW, INFR & DES Hörbereich *m*, SOUND & THERMAL Audiometrie *f*, Hörbereich *m*

**audiometry** *n* ARCH & BUILD, SOUND & THERMAL Audiometrie *f*

**auditor** *n* CONST LAW Rechnungsprüfer *m*, Wirtschaftsprüfer *m*

**auditorium** *n* ARCH & BUILD, INFR & DES Zuschauerraum *m*

**auger** *n* BUILD MACHIN Erdbohrer *m*, Holzbohrer *m*, Schnecke *f*, INFR & DES Erdbohrer *m*, TIMBER Holzbohrer *m*; **~ gimlet** *n* BUILD MACHIN Frettbohrer *m*

**augite**: **~ porphyry** *n* MAT PROP Augitporphyr *m*

**aula** *n* ARCH & BUILD, INFR & DES Aula *f*

**austenitic**: **~ stainless steel** *n* STEEL austenitischer Edelstahl *m*

**authority** *n* CONST LAW, INFR & DES Behörde *f*

**authorization** *n* ARCH & BUILD, CONST LAW Genehmigung *f*

**authorize** *vt* ARCH & BUILD, CONST LAW genehmigen

**autoclaved**: **~ aerated concrete product** *n* CONCR Autoklav-Schaumbeton-Erzeugnis *nt*

**autogenous**: **~ welding** *n* STEEL autogenes Schweißen *nt*

**automate** *vt* ARCH & BUILD, BUILD MACHIN automatisieren

**automated** *adj* ARCH & BUILD, BUILD MACHIN automatisiert

**automatic** *adj* ARCH & BUILD, BUILD MACHIN automatisch; **~ closing** *n* BUILD MACHIN automatisches Schließen *nt*; ~ **control** *n* BUILD MACHIN automatische Steuerung *f*; ~ **door closer** *n* BUILD HARDW, CONST LAW automatischer Türschließer *m*; ~ **fire alarm** *n* BUILD HARDW, ELECTR automatische Brandmeldeanlage *f*; ~ **fire door** *n* BUILD HARDW, CONST LAW automatische Brandschutztür *f*; ~ **fire door closer** *n* BUILD HARDW, CONST LAW automatischer Brandschutztürschließer *m*; ~ **operation** *n* BUILD MACHIN automatischer Betrieb *m*; ~ **release mechanism** *n* CONST LAW automatischer Auslösemechanismus *m*; ~ **scales** *n pl* ARCH & BUILD automatische Waage *f*; ~ **self-closing device** *n* BUILD HARDW, CONST LAW automatischer Türschließer *m*; ~ **slab grinder** *n* ARCH & BUILD Plattenschleifautomat *m*; ~ **sorting** *n* ENVIRON mechanische Trennung *f*; ~ **tile grinder** *n* BUILD MACHIN Plattenschleifautomat *m*; ~ **timer** *n* BUILD HARDW, ELECTR Zeitschaltuhr *f*

**automation** *n* ARCH & BUILD, BUILD MACHIN Automation *f*, Automatisierung *f*, ELECTR, HEAT & VENT Automation *f*

**automatize** *vt* ARCH & BUILD, BUILD MACHIN automatisieren

**auxiliary**: ~ **bearing** *n* ARCH & BUILD, INFR & DES Hilfsauflager *nt*; ~ **building** *n* ARCH & BUILD, INFR & DES Nebengebäude *nt*; ~ **construction** *n* ARCH & BUILD, CONST LAW, INFR & DES Hilfskonstruktion *f*; ~ **force** *n* ARCH & BUILD *statical calculation* Hilfskraft *f*; ~ **ground connection** *n* ELECTR *lightning protection* Hilfserder *m*; ~ **heating system** *n* ARCH & BUILD, HEAT & VENT zusätzliche Heizungsanlage *f*; ~ **lane** *n* ARCH & BUILD, CONST LAW, INFR & DES Entlastungsstraße *f*, Nebenstraße *f*, Zusatzstraße *f*; ~ **plastering work** *n* STONE Beiputzarbeiten *f pl*, Beiputzen *nt*; ~ **reinforcement** *n* ARCH & BUILD, CONCR Hilfsbewehrung *f*, Zusatzspannglied *nt*; ~ **work** *n* CONST LAW zusätzliche Leistungen *f pl*

**availability** *n* ARCH & BUILD, CONST LAW, INFR & DES, MAT PROP *disposability* Verfügbarkeit *f*, *supply* Angebot *nt*

**available** *adj* ARCH & BUILD, INFR & DES, MAT PROP erhältlich, verfügbar; ~ **from stock** *adj* ARCH & BUILD, MAT PROP lieferbar ab Lager

**avalanche** *n* ARCH & BUILD, INFR & DES Lawine *f*; ~ **gallery** *n* ARCH & BUILD, INFR & DES Lawinengalerie *f*; ~**-proof** *adj* ARCH & BUILD, INFR & DES lawinengeschützt; ~ **protector** *n* ARCH & BUILD Lawinenschutz *m*; ~ **screen** *n* ARCH & BUILD, INFR & DES Lawinenwehr *f*

**average 1.** *adj* ARCH & BUILD, INFR & DES, MAT PROP durchschnittlich; **2.** *n* ARCH & BUILD, INFR & DES, MAT PROP Durchschnitt *m*, Mittel *nt*, Mittelwert *m*; **3.** ~ **out** *vt* ARCH & BUILD, INFR & DES, MAT PROP ausgleichen

***average***: ~ **consumption** *n* ARCH & BUILD, HEAT & VENT, INFR & DES Durchschnittsverbrauch *m*; ~ **cost** *n* ARCH & BUILD, INFR & DES Durchschnittskosten *pl*; ~ **daily output** *n* ENVIRON durchschnittliche Tagesleistung *f*; ~ **grain diameter** *n* CONCR durchschnittlicher Korndurchmesser *m*; ~ **monthly dose** *n* ENVIRON mittlere Monatsdosis *f*; ~ **strength** *n* CONCR, MAT PROP durchschnittliche Festigkeit *f*; ~ **wind speed** *n* ENVIRON durchschnittliche Windgeschwindigkeit *f*, INFR & DES Durchschnittswindgeschwindigkeit *f*

**avoidance**: ~ **of emissions** *n* ENVIRON Emissionsvermeidung *f*

**award 1.** *n* CONST LAW *of contract* Vergabe *f*; **2.** *vt* CONST LAW *contract* vergeben

***award***: ~ **of the contract** *n* CONST LAW Auftragserteilung *f*; ~ **of contracts** *n* CONST LAW Vergabe *f* von Aufträgen; ~ **of damages** *n* CONST LAW Schadenersatz *m*

**awning** *n* ARCH & BUILD Markise *f*, Sonnenblende *f*

**ax** *AmE see axe BrE*

**axe 1.** *n BrE* BUILD MACHIN Axt *f*, Beil *nt*, TIMBER Axt *f*; **2.** *vt BrE* BUILD MACHIN behauen, STONE *quarry stone* aufstocken, bossieren

**axhammer** *n* BUILD MACHIN Mauerhammer *m*

**axial** *adj* ARCH & BUILD, CONCR, INFR & DES, MAT PROP, STEEL achsrecht, axial, mittig; ~ **compressive force** *n* ARCH & BUILD, CONCR, INFR & DES, MAT PROP, STEEL axiale Druckkraft *f*; ~ **fan** *n* BUILD HARDW, HEAT & VENT Axiallüfter *m*, Axialventilator *m*; ~ **force** *n* ARCH & BUILD, CONCR, INFR & DES, MAT PROP, STEEL Axialkraft *f*; ~ **load** *n* ARCH & BUILD, CONCR, INFR & DES, MAT PROP, STEEL Axiallast *f*

**axially**: ~ **loaded** *adj* ARCH & BUILD, CONCR, INFR & DES, MAT PROP, STEEL mittig belastet

**axial**: ~ **pressure** *n* ARCH & BUILD, CONCR, INFR & DES, MAT PROP, STEEL mittiger Druck *m*; ~ **rib** *n* ARCH & BUILD Scheitelrippe *f*; ~ **shear** *n* ARCH & BUILD, CONCR, INFR & DES, MAT PROP, STEEL Axialschub *m*; ~ **thrust bearing** *n* ENVIRON Axiallager *nt*; ~ **velocity** *n* ENVIRON Axialgeschwindigkeit *f*

**axis** *n* ARCH & BUILD, INFR & DES Achse *f*, Mittellinie *f*; ~ **of rotation** *n* ARCH & BUILD Rotationsachse *f*, INFR & DES Drehachse *f*, Rotationsachse *f*

**axle** *n* BUILD MACHIN Achse *f*, Welle *f*; ~ **load** *n* BUILD MACHIN Achsdruck *m*, Achslast *f*; ~ **pulley** *n* BUILD MACHIN Achsscheibe *f*

**axonometric**: ~ **cut-away section** *n* ARCH & BUILD, HEAT & VENT, TIMBER axonometrischer Schnitt *m*; ~ **drawing** *n* ARCH & BUILD axonometrische Zeichnung *f*

**azimuth** *n* ARCH & BUILD, INFR & DES Azimut *m*, Richtungswinkel *m*, MAT PROP Richtungswinkel *m*

**azure**: ~**-colored** *AmE*, ~**-coloured** *BrE adj* MAT PROP azurblau gefärbt

# B

**back 1.** *adj* ARCH & BUILD hinter; **2.** *n* ARCH & BUILD Hinterseite *f*, Rückseite *f*

**backed**: **~-up masonry** *n* STONE hintermauertes Mauerwerk *nt*

**backfill 1.** *n* ARCH & BUILD, CONCR, INFR & DES, STONE Auffüllung *f*, Hinterfüllung *f*; **2.** *vt* ARCH & BUILD, CONCR, INFR & DES, STONE auffüllen, hinterfüllen

***backfill***: **~ concrete** *n* ARCH & BUILD, CONCR, INFR & DES Auffüllbeton *m*, Hinterfüllbeton *m*

**backfilling** *n* ARCH & BUILD, CONCR Auffüllung *f*, Hinterfüllung *f*, ENVIRON Verfüllung *f*, INFR & DES, STONE Auffüllung *f*, Hinterfüllung *f*

**backflow** *n* WASTE WATER Rückfluß *m*; **~ preventer** *n* WASTE WATER Rückflußverhinderer *m*

**background**: **~ concentration** *n* ENVIRON Grundkonzentration *f*; **~ pollution** *n* ENVIRON grundlegende Verschmutzung *f*

***back***: **~ gutter** *n* WASTE WATER Schornsteindachrinne *f*

**backing** *n* INFR & DES, STONE Grundgefüge *nt*, Trägerschicht *f*; **~ concrete** *n* CONCR Unterbeton *m*; **~ masonry** *n* ARCH & BUILD, STONE Hintermauerung *f*; **~ plaster** *n* STONE Unterputz *m*; **~ strip** *n* MAT PROP Unterlegstreifen *m*

**backplate** *n* BUILD HARDW Unterlagsblech *nt*, Verbindungslasche *f*

**backpressure** *n* WASTE WATER Gegendruck *m*, Rückstau *m*, Rückstaudruck *m*; **~ level** *n* WASTE WATER Rückstauebene *f*; **~ preventer** *n* WASTE WATER Rückstauverhinderer *m*

**backsaw** *n* BUILD MACHIN Bogensäge *f*

***back***: **~ stairs** *n pl* ARCH & BUILD, INFR & DES Hintertreppe *f*; **~-to-back** *adj* ARCH & BUILD aufeinanderfolgend, Rücken an Rücken

**backup** *n* ARCH & BUILD, INFR & DES Füllschicht *f*, Unterschicht *f*; **~ heating system** *n* HEAT & VENT Ersatzheizungsanlage *f*; **~ masonry** *n* ARCH & BUILD, STONE Hintermauerung *f*

***back***: **~ view** *n* ARCH & BUILD, INFR & DES Rückansicht *f*

**backwash**: **~ shut-off valve** *n* WASTE WATER Rückstauverschluß *m*

**backwater** *n* INFR & DES, WASTE WATER Stauwasser *nt*; **~ effect** *n* ENVIRON, INFR & DES, WASTE WATER Rückstauwirkung *f*; **~ level** *n* INFR & DES, WASTE WATER Stauwasserhöhe *f*; **~ zone** *n* INFR & DES, WASTE WATER Stauwasserzone *f*

**backyard** *n* ARCH & BUILD Hinterhof *m*

**bacterial**: **~ contamination** *n* ENVIRON, INFR & DES Bakterienverseuchung *f*, bakterielle Verseuchung *f*; **~ corrosion** *n* INFR & DES, MAT PROP, SURFACE biologische Korrosion *f*

**bacteriological**: **~ purification** *n* ENVIRON bakteriologische Reinigung *f*; **~ treatment** *n* ENVIRON bakteriologische Behandlung *f*

**bad**: **~ ground** *n* INFR & DES weicher Boden *m*

**baffle**: **~ board** *n* MAT PROP Schutzbrett *nt*; **~ brick** *n* STONE Abweisstein *m*; **~ collector** *n* ENVIRON Schallschirm *m*; **~ pier** *n* CONCR, INFR & DES, STEEL Energievernichtungspfeiler *m*; **~ plate** *n* ARCH & BUILD Schlagleiste *f*, BUILD HARDW Leitblech *nt*, Prellplatte *f*, Zwischenwand *f*, Ablenkblech *nt*, Umlenkblech *nt*, Umlenkplatte *f*, HEAT & VENT Prallblech *nt*, Ablenkplatte *f*, INFR & DES, MAT PROP, STEEL, TIMBER Schlagleiste *f*

**bag 1.** *n* MAT PROP Sack *m*; **2.** *vt* MAT PROP absacken, in Säcke verpacken

**baggage** *n* ARCH & BUILD, INFR & DES Gepäck *nt*; **~ elevator** *n AmE* (*cf baggage lift BrE*) ARCH & BUILD, INFR & DES Gepäckaufzug *m*; **~-handling counter** *n* ARCH & BUILD, INFR & DES Gepäckschalter *m*; **~ lift** *n BrE* (*cf baggage elevator AmE*) ARCH & BUILD, INFR & DES Gepäckaufzug *m*; **~ reclaim area** *n* ARCH & BUILD, INFR & DES Gepäckausgabebereich *m*

**bagged**: **~ cement** *n* CONCR Sackzement *m*

**bagging**: **~ scale** *n* ARCH & BUILD Absackwaage *f*, MAT PROP Einsackwaage *f*

**bail** *vt* INFR & DES ausschöpfen

**bake** *vt* SURFACE einbrennen, TIMBER trocknen

**baked** *adj* SURFACE eingebrannt, TIMBER getrocknet

**bakelite**: **~ paint** *n* SURFACE Bakelitfarbe *f*

**baking** *n* SURFACE *paint* Einbrennen *nt*; **~ varnish** *n* SURFACE Einbrennlack *m*, Emaillelack *m*

**balance 1.** *n* ARCH & BUILD *scales* Waage *f*, ENVIRON *ecological* Gleichgewicht *nt*, HEAT & VENT Abgleich *m*; **2.** *vt* HEAT & VENT abgleichen, MAT PROP wiegen

***balance***: **~ bridge** *n* INFR & DES Klappbrücke *f*

**balanced** *adj* HEAT & VENT *fan* ausgewuchtet

***balance***: **~ method** *n* ARCH & BUILD Ausgleichverfahren *nt*; **~ point** *n* HEAT & VENT Abstimmung *f*; **~ step** *n* INFR & DES ausgeglichene Trittstufenfläche *f*

**balancing**: **~ of masses** *n* INFR & DES Massenausgleichung *m*; **~ weight** *n* ARCH & BUILD, BUILD MACHIN Gegengewicht *nt*

**balcony** *n* ARCH & BUILD Balkon *m*; **~ door** *n* ARCH & BUILD, BUILD HARDW Balkontür *f*; **~ girder** *n* ARCH & BUILD, CONCR, STEEL, TIMBER Balkonträger *m*; **~ support** *n* ARCH & BUILD, CONCR, STEEL, TIMBER Balkonträger *m*

**balk** *AmE see baulk BrE*

**ball** *n* WASTE WATER Kugel *f*

**ballast** *n* CONCR Schotter *m*, ELECTR *fluorescent lamp* Drossel *f*, Vorschaltgerät *nt*, INFR & DES, MAT PROP Bettung *f*, Unterlage *f*, Kiesschotter *m*

**ballasting**: **~ material** *n* MAT PROP Grobzuschlagstoffe *m pl*

***ball***: **~ cock** *n* WASTE WATER Kugelhahn *m*, Kugelventil *nt*; **~ hinge** *n* BUILD HARDW Kugelgelenk *nt*; **~ impact test** *n* MAT PROP Kugelschlagprobe *f*

**ballistic**: **~ separation** *n* ENVIRON Schleudertrennung *f*, ballistische Sichtung *f*; **~ sorter** *n* ENVIRON ballistische Auslesevorrichtung *f*; **~ sorting** *n* ENVIRON *of refuse* ballistische Sortierung *f*

***ball***: **~ mill** *n* BUILD MACHIN Kugelmühle *f*

**ballroom** *n* ARCH & BUILD, INFR & DES Festsaal *m*

***ball***: **~ and socket joint** *n* MAT PROP, STEEL Kugelgelenk *nt*; **~ valve** *n* WASTE WATER Kugelhahn *m*, Kugelventil *nt*; **~ valve float** *n* WASTE WATER Kugelventilschwimmer *m*

**baluster** *n* BUILD HARDW Baluster *m*, Geländerpfosten *m*
**balustrade** *n* ARCH & BUILD, BUILD HARDW, CONCR, STEEL, TIMBER Balustrade *f*, Brüstung *f*, Geländer *nt*
**band** *n* ARCH & BUILD *decoration* Band *nt*, BUILD MACHIN Band *nt*, *chimney* Fallrohr *nt*, MAT PROP Band *nt*; ~ **chain** *n* BUILD MACHIN *surveying* Meßkette *f*; ~ **conveyor** *n* BUILD MACHIN Förderband *nt*; ~ **saw** *n* ARCH & BUILD, TIMBER Bandsäge *f*
**banister** *n* TIMBER Geländerpfosten *m*
**bank 1.** *n* CONST LAW Bank *f*, INFR & DES Bank *f*, Damm *m*; **2.** *vt* CONCR, INFR & DES überhöhen; ~ **up** ARCH & BUILD andämmen, aufdämmen, aufschütten, INFR & DES andämmen
***bank***: ~ **approval** *n* CONST LAW Bankakzept *nt*, Bankgenehmigung *f*; ~ **credit** *n* CONST LAW Bankkredit *m*; ~ **draft** *n* CONST LAW Bankwechsel *m*
**banked** *adj* CONCR, INFR & DES überhöht
**bank**: ~**-funded** *adj* CONST LAW bankfinanziert; ~ **guarantee** *n* CONST LAW Bankbürgschaft *f*
**banking** *n* CONCR, INFR & DES Überhöhung *f*
**bankruptcy** *n* CONST LAW Bankrott *m*, Konkurs *m*
**banquette** *n* ARCH & BUILD Bankett *nt*, INFR & DES Bankett *nt*, Gehweg *m*
**bar** *n* ARCH & BUILD Stange *f*, CONCR, INFR & DES Bewehrungsstab *m*, MAT PROP Stab *m*, Bewehrungsstab *m*, STEEL Bewehrungsstab *m*, Stab *m*, Stange *f*
**barb**: ~ **bolt** *n* BUILD HARDW, CONCR, MAT PROP Steinschraube *f*
**barbed**: ~ **bolt** *n* BUILD HARDW Widerhakenbolzen *m*; ~ **wire** *n* MAT PROP, STEEL Stacheldraht *m*; ~ **wire fence** *n* INFR & DES Stacheldrahtzaun *m*
***bar***: ~ **bender** *n* CONCR Eisenbieger *m*
**bare** *adj* ELECTR, MAT PROP blank; ~ **copper down conductor** *n* ELECTR *lightning protection* blanke Ableitung *f* aus Kupfer
**barge**: ~ **board** *n* TIMBER Ortgangbrett *nt*, Stirnbrett *nt*, Wetterbrett *nt*
***bar***: ~ **grate** *n* ARCH & BUILD Gitterrost *m*, Stabgitter *nt*; ~ **intersection** *n* STEEL Stahlüberkreuzung *f*; ~ **iron** *n* CONCR, INFR & DES, MAT PROP, STEEL Bewehrungsstab *m*; ~ **iron list** *n* ARCH & BUILD, CONCR, STEEL Stahlliste *f*
**barite** *n* MAT PROP Schwerspat *m*; ~ **concrete** *n* ARCH & BUILD, CONCR Schwerbeton *m*
**baritic**: ~ **cement** *n* CONCR, MAT PROP Barytzement *m*
**barked**: ~ **timber** *n* TIMBER Schälholz *nt*
***bar***: ~ **mat** *n* ARCH & BUILD, CONCR, INFR & DES, MAT PROP, STEEL Betonstahlmatte *f*, Bewehrungsmatte *f*, TIMBER Bewehrungsmatte *f*
**barometrical**: ~ **height measurement** *n* ARCH & BUILD barometrische Höhenmessung *f*
**baroque** *adj* ARCH & BUILD, INFR & DES barock; ~ **castle** *n* ARCH & BUILD, INFR & DES Barockschloß *nt*; ~ **church** *n* ARCH & BUILD, INFR & DES Barockkirche *f*
**barracks** *n pl* ARCH & BUILD, INFR & DES Kaserne *f*
**barrage** *n* INFR & DES Sperrmauer *f*, Stauwehr *nt*
**barrel** *n* BUILD HARDW Faß *nt*, *of lock* Hülse *f*, *of roller* Rolle *f*; ~ **bolt** *n* BUILD HARDW Torschließbolzen *m*; ~ **vault** *n* ARCH & BUILD, STONE Tonnengewölbe *nt*
**barricade 1.** *n* INFR & DES Barrikade *f*, Hindernis *nt*, Sperre *f*; **2.** *vt* INFR & DES versperren
**barrier** *n* CONST LAW *safety*, SOUND & THERMAL Absperrung *f*; ~ **fence** *n* STEEL, TIMBER Absperrung *f*; ~ **layer** *n* ENVIRON, MAT PROP, SOUND & THERMAL Sperrschicht *f*; ~ **material** *n* ARCH & BUILD, MAT PROP Sperrstoff *m*; ~ **membrane** *n* ARCH & BUILD, MAT PROP, SOUND & THERMAL *sealing layer* Sperrfolie *f*
**barrow** *n* BUILD MACHIN Schubkarre *f*, Schubkarren *m*; ~ **way** *n* ARCH & BUILD Karrbohle *f*, Laufbohle *f*, Laufbrett *nt*, TIMBER Karrbohle *f*
***bar***: ~ **schedule** *n* ARCH & BUILD, CONCR, STEEL Stahlliste *f*, Bewehrungsliste *f*; ~ **spacer** *n* CONCR Abstandhalter *m* für Bewehrungsstahl; ~ **spacing** *n* CONCR Abstand *m* zwischen Bewehrungsstäben
**basal** *adj* INFR & DES an der Basis; ~ **drainage blanket** *n* INFR & DES Sohlendrainageschicht *f*
**basalt** *n* ENVIRON, MAT PROP Basalt *m*; ~ **chippings** *n pl* MAT PROP Basaltsplitt *m*; ~ **paver** *n* MAT PROP Basaltpflasterstein *m*
**bascule**: ~**-bolt** *n* BUILD HARDW *lock* Baskülverschluß *m*
**base** *n* ARCH & BUILD *foundation* Grundlage *f*, Untergrund *m*, *of wall, pillar* Basis *f*, *of solid* Grundfläche *f*, Sockel *m*, Fuß *m*, *underlay* Unterlage *f*, CONCR *of wall, pillar* Sockel *m*, INFR & DES *of solid* Grundfläche *f*, *of wall, pillar* Fuß *m*, *foundation* Grundlage *f*, Basis *f*
**baseboard** *n AmE* (*cf skirting board BrE*) ARCH & BUILD, BUILD HARDW, TIMBER Fußleiste *f*, Scheuerleiste *f*, Sockelleiste *f*
***base***: ~ **cation** *n* ENVIRON Basenkation *nt*; ~ **coat** *n* SURFACE Grundierschicht *f*; ~ **course** *n* CONCR Tragschicht *f*, Unterbau *m*, INFR & DES *street* Tragschicht *f*, STONE Binderschicht *f*; ~ **deck** *n* MAT PROP Unterlagsbahn *f*; ~ **fabric** *n* MAT PROP *plaster* Trägergewebe *nt*; ~ **level** *n* ARCH & BUILD, INFR & DES Gründungssohle *f*; ~ **line** *n* ARCH & BUILD Grundlinie *f*
**basement** *n* ARCH & BUILD Keller *m*, Kellergeschoß *nt*; ~ **boiler room** *n* ARCH & BUILD, HEAT & VENT Heizungskeller *m*; ~ **door** *n* ARCH & BUILD, BUILD HARDW Kellertür *f*; ~ **entrance** *n* ARCH & BUILD Kellereingang *m*; ~ **grating** *n* ARCH & BUILD, BUILD HARDW, STEEL Kellergitter *nt*; ~ **stairs** *n* ARCH & BUILD Kellertreppe *f*; ~ **storey** *n BrE* ARCH & BUILD Untergeschoß *nt*; ~ **story** *AmE see basement storey BrE*
***base***: ~ **plate** *n* ARCH & BUILD Fußplatte *f*, Grundplatte *f*; ~ **point** *n* ARCH & BUILD Fußpunkt *m*; ~ **sealing** *n* ENVIRON *depot* Basisabdichtung *f*; ~ **wall** *n* ARCH & BUILD Grundmauer *f*
**basic**: ~ **assumption** *n* ARCH & BUILD Grundabmessungen *f pl*; ~ **design** *n* ARCH & BUILD Grundausführung *f*; ~ **dimension** *n* ARCH & BUILD Baurichtmaß *nt*, Nennmaß *nt*; ~ **ingredient** *n* ARCH & BUILD Grundbestandteil *m*; ~ **insulation** *n* SOUND & THERMAL Grundisolierung *f*; ~ **law** *n* CONST LAW Grundgesetz *nt*; ~ **load** *n* ARCH & BUILD Grundlast *f*; ~ **module** *n* ARCH & BUILD Grundbaustein *m*; ~ **shear strength** *n* ARCH & BUILD, CONCR, MAT PROP Grundschubspannung *f*; ~ **SI unit** *n* ARCH & BUILD, INFR & DES, MAT PROP SI-Grundeinheit *f*; ~ **size** *n* ARCH & BUILD Grundmaß *nt*; ~ **training** *n* CONST LAW Grundausbildung *f*; ~ **wage** *n* CONST LAW Grundlohn *m*
**basin** *n* INFR & DES, WASTE WATER Becken *nt*
**basis** *n* ARCH & BUILD, INFR & DES Basis *f*, Grundlage *f*
**basket**: ~ **arch** *n* ARCH & BUILD Korbbogen *m*; ~ **guard** *n* ELECTR *light fixture* Schutzkorb *m*; ~ **handle arch** *n* ARCH & BUILD Korbbogen *m*

**bas-relief** *n* ARCH & BUILD, INFR & DES Flachrelief *nt*
**bastion** *n* ARCH & BUILD, INFR & DES, STONE Bastion *f*
**batardeau** *n* ARCH & BUILD, INFR & DES Fangdamm *m*
**batch 1.** *n* CONCR, MAT PROP, WASTE WATER Charge *f*; **2.** *vt* CONCR, MAT PROP, WASTE WATER dosieren
**batcher** *n* ARCH & BUILD Dosieranlage *f*
***batch***: **~ number** *n* MAT PROP *tiles, paint* Losnummer *f*
**bath**: **~ overflow** *n* WASTE WATER Badüberlauf *m*
**bathroom** *n* ARCH & BUILD Bad *nt*, Badezimmer *nt*
**batten 1.** *n* BUILD HARDW Leiste *f*, MAT PROP Dachlatte *f*, Leiste *f*, TIMBER Dachlatte *f*, Lagerholz *nt*, Leiste *f*; **2.** *vt* TIMBER *with slats* befestigen, verschalen
**battening** *n* TIMBER Holzverlattung *f*
***batten***: **~ plate** *n* ARCH & BUILD, INFR & DES, STEEL, TIMBER Bindeblech *nt*; **~ wall** *n* TIMBER Bretterwand *f*
**batter**: **~ board** *n* ARCH & BUILD, INFR & DES Schnurbrett *nt*, Schnurgerüst *nt*; **~ level** *n* ARCH & BUILD Neigungsmesser *m*; **~ wall** *n* ARCH & BUILD Schrägwand *f*
**battery** *n* WASTE WATER *mixing valve* Batterie *f*
**battlement** *n* ARCH & BUILD, STONE Mauerzacke *f*, Zinne *f*, *crenellations* Zinnen *f pl*; **~ tower** *n* ARCH & BUILD, INFR & DES Zinnenturm *m*
**baulk** *n BrE* ARCH & BUILD, CONCR, STEEL, TIMBER Balken *m*
**bauxite** *n* MAT PROP Bauxit *m*
**bauxitic**: **~ cement** *n* CONCR, MAT PROP Bauxitzement *m*
**bay** *n* ARCH & BUILD *compartment between pillars* Fach *nt*, Feld *nt*, Schiff *nt*, *machine shop* Abteilung *f*, INFR & DES *cove* Bucht *f*; **~ system** *n* ARCH & BUILD Rastersystem *nt*; **~ window** *n* ARCH & BUILD Erkerfenster *nt*
**BCS** *abbr* (*British Calibration Service*) CONST LAW BCS (*Britische Anstalt für Kalibrierung*)
**beach**: **~ gravel** *n* MAT PROP Strandkies *m*
**beacon 1.** *n* ARCH & BUILD Signal *nt*; **2.** *vt* CONST LAW markieren
**bead 1.** *n* ARCH & BUILD Wulst *m*; **2. ~ over** *vt* TIMBER umbördeln
**beam** *n* ARCH & BUILD, CONCR, STEEL, TIMBER Balken *m*, Querriegel *m*, Riegel *m*, Tragbalken *m*, Träger *m*; **~ anchor** *n* CONCR, INFR & DES, MAT PROP, STEEL, TIMBER Balkenanker *m*; **~ bending** *n* ARCH & BUILD Balkenbiegung *f*; **~ bottom** *n* ARCH & BUILD, CONCR, STEEL Balkenuntersicht *f*; **~ crossing** *n* CONCR, INFR & DES, STEEL, TIMBER Balkenkreuzung *f*; **~ cross section** *n* CONCR, INFR & DES, MAT PROP, STEEL Balkenquerschnitt *f*; **~ design** *n* ARCH & BUILD, CONCR, STEEL, TIMBER Balkenbemessung *f*; **~ distance** *n* TIMBER Balkenabstand *m*; **~ end** *n* ARCH & BUILD, TIMBER Balkenkopf *m*; **~ floor** *n* ARCH & BUILD, TIMBER Balkendecke *f*; **~ head** *n* ARCH & BUILD, TIMBER Balkenkopf *m*; **~ joint** *n* TIMBER Balkenstoß *m*
**beamless** *adj* ARCH & BUILD, CONCR, STEEL, TIMBER balkenfrei
***beam***: **~ mid-line** *n* TIMBER Balkenachse *f*, halbe Balkenhöhe *f*; **~ profile** *n* TIMBER Balkenprofil *nt*; **~ roof** *n* TIMBER Balkendach *nt*; **~ test** *n* ARCH & BUILD, INFR & DES, MAT PROP Balkenbruchprüfung *f*, Balkenprüfung *f*; **~ tie** *n* CONCR, INFR & DES, MAT PROP, STEEL, TIMBER Balkenanker *m*
**bear** *vt* ARCH & BUILD, INFR & DES, MAT PROP, STEEL, TIMBER lagern, stützen, tragen
**bearer** *n* ARCH & BUILD *for landing, staircase* Podestträger *m*, *main beam* Haupttragbalken *m*, *support* Stütze *f*, Unterzug *m*, Treppenpodestträger *m*, Träger *m*
**bearing** *n* ARCH & BUILD Lager *nt*, Lagerung *f*, *supporting* Stützen *nt*, Tragen *nt*, *direction* Richtungswinkel *m*, INFR & DES, MAT PROP Richtungswinkel *m*; **~ area** *n* ARCH & BUILD Auflagerfläche *f*, Gründungsfläche *f*, CONCR, STEEL, STONE Auflagerfläche *f*; **~ bar** *n* ARCH & BUILD Auflageflanschstab *m*; **~ bed** *n* ARCH & BUILD *geology* Tragschicht *f*; **~ bush** *n* ARCH & BUILD Lagerbuchse *f*; **~ capacity** *n* ARCH & BUILD Tragfähigkeit *f*; **~ course** *n* ARCH & BUILD, INFR & DES, MAT PROP Tragschicht *f*; **~ distance** *n* ARCH & BUILD, INFR & DES Spannweite *f*, Stützweite *f*, lichte Weite *f*; **~ frame** *n* ARCH & BUILD Tragrahmen *m*, INFR & DES Tragrahmen *m*, *manhole* Aufsetzkranz *m*; **~ load** *n* ARCH & BUILD Bodendruck *m*, Bodenpressung *f*; **~ pad** *n* ARCH & BUILD, INFR & DES Widerlager *nt*; **~ plate** *n* ARCH & BUILD, INFR & DES Stützplatte *f*, Tragplatte *f*; **~ pressure** *n* ARCH & BUILD Bodenpressung *f*, Lagerdruck *m*; **~ rail** *n* ARCH & BUILD Sattelschiene *f*, Tragschiene *f*; **~ reaction** *n* ARCH & BUILD, INFR & DES Anschlußkraft *f*, Auflagerkraft *f*; **~ structure** *n* ARCH & BUILD, CONCR, STEEL Tragekonstruktion *f*; **~ surface** *n* ARCH & BUILD *pipe, cable*, INFR & DES Auflage *f*; **~ wall** *n* ARCH & BUILD tragende Wand *f*
**bed 1.** *n* ARCH & BUILD, INFR & DES *foundation* Bett *nt*, Lagerbett *nt*, Unterlage *f*; **2. ~ in** *vt* ARCH & BUILD, INFR & DES, STONE einbetten; **~ out** ARCH & BUILD auspflanzen
**bedded** *adj* ARCH & BUILD, INFR & DES, MAT PROP, STONE eingebettet
**bedding** *n* ARCH & BUILD, CONCR, INFR & DES, MAT PROP Bettung *f*; **~ concrete** *n* CONCR Bettungsbeton *m*; **~ course** *n* CONCR Bettungsmörtelschicht *f*; **~ module** *n* ARCH & BUILD, INFR & DES, MAT PROP Bettungsmodul *m*
***bed***: **~ joint** *n* ARCH & BUILD, STONE *masonry* Lagerfuge *f*; **~ load** *n* INFR & DES Geschiebe *nt*; **~ plate** *n* ARCH & BUILD Bettungsplatte *f*, INFR & DES Grundplatte *f*; **~ rock** *n* INFR & DES gewachsener Fels *m*
**beech** *n* TIMBER Buche *f*; **~ parquetry** *n* ARCH & BUILD, TIMBER Buchenparkett *nt*; **~ shingle** *n* ARCH & BUILD, TIMBER Buchenholzschindel *f*
**beech-wood** *n* ARCH & BUILD, MAT PROP, TIMBER Buchenholz *nt*; **~ shingle** *n* ARCH & BUILD, TIMBER Buchenholzschindel *f*
**beetle** *n* BUILD MACHIN Steinramme *f*, *plaster work* Holzhammer *m*
**behave** *vi* ARCH & BUILD, INFR & DES, MAT PROP verhalten
**behavior** *AmE see behaviour BrE*
**behaviour** *n BrE* ARCH & BUILD, INFR & DES, MAT PROP Verhalten *nt*
**bell** *n* ARCH & BUILD Glocke *f*, BUILD HARDW Klingel *f*, ELECTR Schelle *f*, MAT PROP, WASTE WATER Muffe *f*; **~ cote** *n* ARCH & BUILD Giebeltürmchen *nt*; **~ frame** *n* ARCH & BUILD Glockenstuhl *m*
**bellow**: **~ expansion joint** *n* WASTE WATER *pipe* Kompensator *m*
**bellows** *n pl* BUILD HARDW Blasebalg *m*
***bell***: **~ and spigot joint** *n* MAT PROP Glockenmuffenverbindung *f*, Muffenverbindung *f*, WASTE WATER Glockenmuffenverbindung *f*; **~ and spigot**

**pipe** *n* MAT PROP, WASTE WATER Glockenmuffenrohr *nt*

**belly 1.** *n* MAT PROP Ausbauchung *f*; **2. ~ out** *vi* INFR & DES, MAT PROP anschwellen

**below**: **~-cloud scavenging** *n* ENVIRON Reinigungsfällung *f*; **~-ground** *adj* ARCH & BUILD, ELECTR, HEAT & VENT, INFR & DES, WASTE WATER unterirdisch

**belt** *n* BUILD MACHIN Band *nt*, Riemen *m*; **~ conveyor** *n* BUILD MACHIN Bandförderer *m*, Förderband *nt*; **~ course** *n* ARCH & BUILD Gurtgesims *nt*; **~ drive** *n* BUILD MACHIN Bandantrieb *m*, Riementrieb *m*; **~ stretcher** *n* BUILD MACHIN Riemenspanner *m*; **~-type proportioner** *n* BUILD MACHIN Banddosiereinrichtung *f*

**bench** *n* ARCH & BUILD Berme *f*, Absatz *m*, Bankett *nt*, Bank *f*, Werkbank *f*, BUILD MACHIN Werkbank *f*, Bank *f*, INFR & DES *embankment* Berme *f*, *seat* Bank *f*, Bankett *nt*

**benched** *adj* ARCH & BUILD abgetreppt; **~ foundation** *n* ARCH & BUILD, CONCR, INFR & DES abgetrepptes Fundament *nt*

**benching** *n* ARCH & BUILD, INFR & DES, STONE Abtreppung *f*

***bench***: **~ lathe** *n* BUILD MACHIN Tischdrehbank *f*; **~ mark** *n* ARCH & BUILD Festpunkt *m*, INFR & DES Festpunkt *m*, Höhenfestpunkt *m*; **~ plane** *n* BUILD MACHIN Tischhobel *m*; **~ screw** *n* BUILD MACHIN Bankschraube *f*; **~ shears** *n pl* BUILD MACHIN Bockschere *f*; **~ stop** *n* BUILD MACHIN *carpentry* Haltestock *m*; **~ test** *n* MAT PROP Laborversuch *m*; **~ vice** *n BrE* BUILD MACHIN Bankschraubstock *m*; **~ vise** *AmE see bench vice BrE*

**bend 1.** *n* ARCH & BUILD Biegung *f*, *sharp* Knickbereich *m*, Krümmung *f*, Knick *m*, BUILD HARDW *pipe* Winkelstück *nt*, Krümmer *m*, Rohrkrümmer *m*, Rohrbogen *m*, INFR & DES Biegung *f*, WASTE WATER *pipe* Krümmer *m*, Rohrbogen *m*, Rohrkrümmer *m*, Winkelstück *nt*; **2.** *vt* ARCH & BUILD, CONCR, INFR & DES biegen, MAT PROP wölben; **~ at right angles** HEAT & VENT kröpfen; **~ on edge** MAT PROP hochkantbiegen; **~ out of line** MAT PROP verkanten; **~ up** MAT PROP aufbiegen

**bender** *n* CONCR *concrete reinforcement* Biegetisch *m*

**bending** *n* ARCH & BUILD, CONCR Biegen *nt*; **~ allowance** *n* MAT PROP Biegetoleranz *f*; **~ angle** *n* CONCR, MAT PROP Biegewinkel *m*; **~ crack** *n* CONCR, MAT PROP, STEEL Biegeriß *m*; **~ cycles** *n pl* MAT PROP Biegewechselzahl *f*; **~ dimension** *n* ARCH & BUILD, MAT PROP Biegemaß *nt*; **~ failure** *n* CONCR, MAT PROP, STEEL Biegebruch *m*; **~ list** *n* MAT PROP Biegeliste *f*; **~ moment** *n* ARCH & BUILD, INFR & DES, MAT PROP, STEEL Biegemoment *nt*; **~ moment diagram** *n* ARCH & BUILD, MAT PROP, TIMBER Biegemomentdiagramm *nt*; **~ press** *n* ARCH & BUILD Biegepresse *f*; **~ radius** *n* ARCH & BUILD, CONCR, INFR & DES, MAT PROP, STEEL Biegeradius *m*; **~ rigidity** *n* ARCH & BUILD, CONCR, INFR & DES, MAT PROP, STEEL Biegesteifigkeit *f*; **~ specifications** *n pl* ARCH & BUILD, MAT PROP Biegeangaben *f pl*; **~ stress** *n* ARCH & BUILD, CONCR, INFR & DES, MAT PROP, STEEL, TIMBER Biegespannung *f*; **~ test** *n* MAT PROP Biegeversuch *m*; **~ yard** *n* CONCR Biegeplatz *m*

***bend***: **~ point** *n* MAT PROP Biegepunkt *m*, STEEL Biegestelle *f*

**bendproof** *adj* MAT PROP, STEEL biegesteif

**bend**: **~-resistant** *adj* MAT PROP, STEEL biegesteif

**bent** *adj* ARCH & BUILD, INFR & DES, MAT PROP, STEEL gebogen, gekrümmt

**bentonite** *n* CONCR Bentonit *nt*

**Berlin**: **~ black** *n* SURFACE Berliner Schwarz *nt*

**berm** *n* ARCH & BUILD Bankett *nt*, INFR & DES Bankett *nt*, Berme *f*

**Bernoulli**: **~ equation** *n* ARCH & BUILD Bernoulli-Gleichung *f*

**Bessemer**: **~ steel** *n* MAT PROP Bessemerstahl *m*

**betterment** *n* CONST LAW Wertsteigerung *f* von Grund und Boden

**bevel 1.** *n* ARCH & BUILD, TIMBER Abfasung *f*, Abschrägung *f*, Fase *f*, Gehrung *f*; **2.** *vt* ARCH & BUILD, INFR & DES abschrägen, TIMBER abfasen, abkanten, abschrägen

**beveled** *AmE see bevelled BrE*

***bevel***: **~-headed bolt** *n* BUILD HARDW Schrägkopfriegel *m*

**bevelled** *adj BrE* ARCH & BUILD abgeschrägt, TIMBER abgeschrägt, schräg, verjüngt; **~ cogging** *n* INFR & DES schräger Kamm *m*; **~ halving** *n* TIMBER Ecküberblattung *f* mit schrägem Schnitt; **~ joint** *n* TIMBER schräger Stoß *m*

**BFB** *abbr* (*broad-flanged beam*) ARCH & BUILD, CONCR, STEEL Breitflanschträger *m*

**BFW** *abbr* (*boiler feed water*) HEAT & VENT Kesselspeisewasser *nt*

**biaxial** *adj* ARCH & BUILD, INFR & DES, MAT PROP zweiachsig

**bibcock** *n* WASTE WATER *sink* Ablaß *m*, Wasserhahn *m*

**bicycle**: **~ track** *n* INFR & DES Radweg *m*

**bid** *n* ARCH & BUILD Angebot *nt*, Offerte *f*, CONST LAW Angebot *nt*, Bauangebot *nt*, Offerte *f*, INFR & DES Angebot *nt*

**bidding**: **~ period** *n* CONST LAW Ausschreibungsdauer *f*

**bifurcation** *n* INFR & DES Gabelung *f*, Verzweigung *f*; **~ stress** *n* INFR & DES, MAT PROP Verzweigungsspannung *f*

**bilinear**: **~ stress-strain relationship** *n* CONCR, INFR & DES, MAT PROP bilineares Druck-Dehnungs-Verhältnis *nt*

**bill**: **~ of quantity** *n* ARCH & BUILD, CONST LAW Massenverzeichnis *nt*

**bimetal**: **~ strip** *n* MAT PROP, STEEL Bimetallstreifen *m*

**bind** *vt* CONCR, MAT PROP binden

**binder** *n* ARCH & BUILD Binder *m*, Bügel *m*, ENVIRON Bindemittel *nt*, MAT PROP Bindemittel *nt*, Binder *m*, STONE Strecker *m*; **~ content** *n* MAT PROP Bindemittelanteil *m*; **~ course** *n* STONE Asphaltbinderschicht *f*

**binders**: **~ and joists** *n pl* ARCH & BUILD Gebälk *nt*, TIMBER Balkenlage *f*, Gebälk *nt*, Holzgebälk *nt*

***binder***: **~ soil** *n* INFR & DES, MAT PROP bindiges Erdmaterial *nt*

**binding**: **~ agent** *n* MAT PROP Bindemittel *nt*; **~ beam** *n* ARCH & BUILD Bundbalken *m*, Unterzug *m*, CONCR, STEEL Unterzug *m*, TIMBER Bundbalken *m*, Unterzug *m*; **~ material** *n* MAT PROP Bindemittel *nt*; **~ rafter** *n* ARCH & BUILD Bindersparren *m*, Pfette *f*, BUILD HARDW Pfette *f*, TIMBER Bindersparren *m*, Pfette *f*; **~ stone** *n* STONE *masonry* Binderstein *m*; **~ wire** *n* CONCR Bindedraht *m*, Rädeldraht *m*, Rödeldraht *m*

**bioaeration** *n* HEAT & VENT Belüftung *f*

**biocenosis** *AmE see biocoenosis*

**biochemical**: **~ tracer** *n* ENVIRON biochemischer Indikator *m*

**biocide** *n* CONST LAW, HEAT & VENT, INFR & DES, WASTE WATER Biozid *nt*, Schädlingsbekämpfungsmittel *nt*
**biocoenosis** *n BrE* ENVIRON Biozönose *f*
**biodegradability** *n* ENVIRON, INFR & DES, MAT PROP biologische Abbaubarkeit *f*
**biodegradable** *adj* ENVIRON, INFR & DES, MAT PROP biologisch abbaubar; **~ substance** *n* ENVIRON, INFR & DES, MAT PROP biologisch abbaubare Substanz *f*; **~ waste** *n* ENVIRON, INFR & DES, MAT PROP biologisch abbaubarer Abfall *m*
**biodegradation** *n* ENVIRON, MAT PROP biologischer Abbau *m*
**biodeterioration** *n* ENVIRON, MAT PROP biologischer Verfall *m*
**biofilter** *n* ENVIRON Biofilter *nt*
**biogas** *n* ENVIRON Biogas *nt*, Faulgas *nt*, INFR & DES, MAT PROP Faulgas *nt*
**bioindicator** *n* ENVIRON Bioindikator *m*
**biological** *adj* ENVIRON biologisch; **~ agent** *n* ENVIRON biologisches Agens *nt*; **~ clarification plant** *n* ENVIRON, INFR & DES biologische Kläranlage *f*; **~ degradation** *n* ENVIRON biologischer Abbau *m*; **~ energy conversion** *n* ENVIRON biologische Umwandlung *f*; **~ equilibrium** *n* ENVIRON biologisches Gleichgewicht *nt*; **~ filter** *n* ENVIRON biologisches Filter *nt*, INFR & DES biologische Tropfkörperanlage *f*; **~ indicator** *n* ENVIRON Bioindikator *m*; **~ oxygen demand** *n* (*BOD*) ENVIRON, INFR & DES, MAT PROP biologischer Sauerstoffbedarf *m* (*BSB*); **~ treatment** *n* ENVIRON biologische Behandlung *f*; **~ waste** *n* ENVIRON Bioabfall *m*; **~ waste composting** *n* ENVIRON Biomüllkompostierung *f*
**biomass** *n* INFR & DES Biomasse *f*
**biostabilizer** *n* ENVIRON Biostabilisator *m*
**biota** *n* ENVIRON Biotop *nt*
**biowaste**: **~ compost** *n* ENVIRON Biomüllkompost *m*
**birch**: **~ wood** *n* ARCH & BUILD, MAT PROP, TIMBER Birkenholz *nt*
**bird's-eye**: **~ view** *n* INFR & DES Vogelperspektive *f*
**bit** *n* BUILD HARDW Schlüsselbart *m*
**bitumen** *n* ARCH & BUILD, INFR & DES, MAT PROP Asphalt *m*, Bitumen *nt*, SURFACE Asphalt *m*; **~-coated gravel** *n* INFR & DES, MAT PROP Bitukies *m*, Bitumenkies *m*; **~ concrete** *n* CONCR, INFR & DES, MAT PROP Asphaltbeton *m*; **~ emulsion** *n* CONCR, MAT PROP Bitumenemulsion *f*; **~-impregnated** *adj* MAT PROP bitumengetränkt; **~ mastic** *n* CONCR, MAT PROP Bitumenmastix *m*; **~ mortar** *n* CONCR, MAT PROP Asphaltmörtel *m*, Bitumenmörtel *m*; **~ pavement** *n* INFR & DES Bitumenbelag *m*; **~-saturated** *adj* MAT PROP bitumengetränkt; **~ sheet roofing** *n* MAT PROP, SOUND & THERMAL Bitumendachpappe *f*; **~ spraying machine** *n* INFR & DES *for road maintenance* Bitumenspritzgerät *nt*; **~ without fillers** *n* MAT PROP ungefülltes Bitumen *nt*
**bituminization** *n* INFR & DES, MAT PROP, SURFACE Asphaltieren *nt*, Bituminieren *nt*
**bituminize** *vt* INFR & DES, MAT PROP, SURFACE asphaltieren, bituminieren
**bituminized** *adj* INFR & DES, MAT PROP, SURFACE asphaltiert, bituminisiert; **~ felt** *n* INFR & DES, MAT PROP, SURFACE Bitumenpappe *f*; **~ paper** *n* INFR & DES, MAT PROP, SURFACE Bitumenpappe *f*
**bituminous** *adj* ARCH & BUILD bituminös, INFR & DES, MAT PROP, SURFACE bitumenhaltig, bituminös; **~ base course** *n* INFR & DES, MAT PROP bituminöser Unterbau *m*; **~ binder** *n* INFR & DES, MAT PROP bituminöses Bindemittel *nt*; **~ buildings preservative** *n* ARCH & BUILD, INFR & DES, MAT PROP bituminöses Bautenschutzmittel *nt*; **~ carriageway pavement** *n BrE* (*cf bituminous street pavement AmE*) INFR & DES, MAT PROP bituminöse Fahrbahndecke *f*; **~ coat** *n* INFR & DES, MAT PROP, SURFACE Bitumenanstrich *m*, Bitumenaufstrich *m*, bituminöser Schutzanstrich *m*; **~ coating compound** *n* INFR & DES, MAT PROP, SURFACE bituminöse Beschichtungsmasse *f*; **~ compound** *n* MAT PROP Bitumenmasse *f*; **~ concrete** *n* CONCR, INFR & DES, MAT PROP Asphaltbeton *m*, Schwarzbeton *m*, bituminöser Beton *m*; **~ damp-proofing agent** *n* SURFACE bituminöses Sperrmittel *nt*; **~ paint** *n* MAT PROP, SURFACE Bitumenfarbe *f*; **~ pavement** *n* INFR & DES Asphaltdecke *f*, Bitumenbelag *m*, Schwarzdecke *f*, MAT PROP Asphaltdecke *f*; **~ rock** *n* STONE Asphaltgestein *nt*; **~ roofing felt** *n* MAT PROP, SOUND & THERMAL Bitumendachpappe *f*; **~ sheeting** *n* SURFACE *waterproofing* Bitumenbahn *f*, Schweißbahn *f*; **~ street pavement** *n AmE* (*cf bituminous carriageway pavement BrE*) INFR & DES, MAT PROP bituminöse Fahrbahndecke *f*; **~ surface finisher** *n* BUILD MACHIN Schwarzdeckenfertiger *m*; **~ waterproofing coat** *n* SURFACE bituminöser Sperranstrich *m*
**blackboard** *n* ARCH & BUILD, BUILD HARDW schwarzes Brett *nt*
***black***: **~ bolt** *n* STEEL rohe Schraube *f*; **~ chrome** *n* MAT PROP Schwarzchrom *nt*
**blackout** *n* ARCH & BUILD Verdunkelung *f*
***black***: **~ top** *n* INFR & DES, MAT PROP Schwarzdecke *f*; **~ water** *n* ENVIRON öliges Wasser *nt*
**blank** *n* MAT PROP Rohling *m*; **~ cap** *n* BUILD HARDW, WASTE WATER Endverschluß *m*; **~ cover** *n* BUILD HARDW, HEAT & VENT, WASTE WATER Blinddeckel *m*; **~ flange** *n* HEAT & VENT Blindflansch *m*
**blast** *vi* INFR & DES *in quarry* sprengen
**blast**: **~-cleaned** *adj* SURFACE abgestrahlt
**blast furnace** *n* CONCR, STEEL Hochofen *m*; **~ cement** *n* CONCR, MAT PROP Hochofenzement *m*, Hüttenzement *m*; **~ slag** *n* STEEL Hochofenschlacke *f*
**blasting** *n* INFR & DES *rock* Sprengen *nt*, Sprengung *f*
**bleach 1.** *n* MAT PROP Bleichstoff *m*, SURFACE *process* Ausbluten *nt*, *substance* Bleichstoff *m*; **2.** *vt* MAT PROP, SURFACE bleichen
**bleaching** *n* MAT PROP, SURFACE Bleichen *nt*; **~ agent** *n* MAT PROP, SURFACE Bleichstoff *m*
**bleeding** *n* CONCR, MAT PROP *concrete* Bluten *nt*
**blend** *vt* CONCR, MAT PROP, STONE vermischen
**blended** *adj* MAT PROP *carpet, fabric*, SURFACE meliert; **~ hydraulic cement** *n* CONCR gemischter hydraulischer Zement *m*
**blending**: **~ valve** *n* WASTE WATER Mischbatterie *f*
**blind** *n* BUILD HARDW Jalousie *f*, Sonnenblende *f*; **~ arch** *n* ARCH & BUILD Blendbogen *m*; **~ door** *n* ARCH & BUILD Blindtür *f*; **~ flange** *n* HEAT & VENT Blindflansch *m*; **~ hole** *n* ARCH & BUILD Blindbohrung *f*, *drilling* Sackloch *nt*
**blinding** *n* MAT PROP Oberflächenabschluß *m*; **~ concrete** *n* CONCR Ausgleichbeton *m*
***blind***: **~ pit** *n* INFR & DES, MAT PROP *mining* Blindschacht *m*; **~ rivet** *n* MAT PROP, STEEL Blindniete *f*; **~ shaft** *n* INFR & DES, MAT PROP Blindschacht *m*; **~ vault** *n* ARCH & BUILD Scheingewölbe *nt*; **~ wall** *n*

ARCH & BUILD durchgehendes Mauerwerk *nt*; ~ **window** *n* ARCH & BUILD Blindfenster *nt*
**blister** *n* CONCR, SURFACE *paint* Blase *f*
**blistering** *n* CONCR, SURFACE *paint* Blasenbildung *f*
**bloated**: ~ **pearlite** *n* MAT PROP Blähperlit *m*
**block 1.** *n* ARCH & BUILD Block *m*, INFR & DES, STONE Block *m*, Mauerblock *m*, Mauerstein *m*, TIMBER Block *m*; **2.** ~ **off** *vt* WASTE WATER absperren; ~ **up** ARCH & BUILD, INFR & DES, STONE vermauern, zumauern
**blockboard** *n* TIMBER Stäbchenplatte *f*, Tischlerplatte *f*
***block***: ~ **bond** *n* STONE Blockverband *m*; ~ **of buildings** *n* ARCH & BUILD, INFR & DES Gebäudekomplex *m*
**blocking**: ~ **off** *n* WASTE WATER Absperren *nt*
***block***: ~ **load** *n* INFR & DES, MAT PROP Blocklast *f*; ~ **pavement** *n* INFR & DES Steinpflaster *nt*; ~ **plan** *n* ARCH & BUILD, INFR & DES Übersichtsplan *m*
**blooming** *n* SURFACE Anlaufen *nt*
**blow**: ~ **out** *vt* HEAT & VENT, INFR & DES ausblasen; ~ **up** *vt* INFR & DES *rock* sprengen
**blowlamp** *n* STEEL Lötlampe *f*
**blowtorch** *n* STEEL Lötlampe *f*
**blueing** *n* SURFACE Bläuung *f*, TIMBER Blaufäule *f*
**blueprint** *n* ARCH & BUILD Blaupause *f*, Lichtpause *f*, Mutterpause *f*
**blueprinting** *n* ARCH & BUILD Blaupausen *nt*
**blue**: ~ **stain** *n* TIMBER Blaufäule *f*; ~ **vitriol** *n* MAT PROP Kupfervitriol *nt*
**board 1.** *n* MAT PROP, TIMBER Bohle *f*, Brett *nt*; **2.** *vt* ARCH & BUILD schalen, CONCR verschalen, INFR & DES schalen, TIMBER schalen, verschalen; ~ **up** CONCR mit Holzverschalung versehen
***board***: ~ **core** *n* MAT PROP Plattenkern *m*
**boarding** *n* ARCH & BUILD Bretterverkleidung *f*, Schalung *f*, CONCR, INFR & DES Schalung *f*, TIMBER Bretterverkleidung *f*, Verschlag *m*, *concrete* Schalung *f*
***board***: ~ **insulation** *n* SOUND & THERMAL Plattendämmung *f*
**boardmarked** *adj* ARCH & BUILD, CONCR *concrete* schalungsrauh
***board***: ~ **roof** *n* TIMBER Bretterdach *nt*
**boat**: ~ **scaffold** *n* BUILD MACHIN Hängebaugerüst *nt*
**bob** *n* ARCH & BUILD Senkblei *nt*, BUILD MACHIN Lot *nt*, Senkblei *nt*
**BOD** *abbr* (*biological oxygen demand*) ENVIRON, INFR & DES, MAT PROP BSB (*biologischer Sauerstoffbedarf*)
**boiler** *n* ELECTR, HEAT & VENT *hot water tank* Warmwasserspeicher *m*, Heißwasserspeicher *m*, *steam boiler* Dampfkessel *m*, Boiler *m*, Kessel *m*, WASTE WATER Boiler *m*, Heißwasserspeicher *m*; ~ **control** *n* ELECTR, HEAT & VENT Kesselregelung *f*; ~ **feed water** *n* (*BFW*) HEAT & VENT Kesselspeisewasser *nt*; ~ **house** *n* ARCH & BUILD Kesselhaus *nt*; ~ **manufacturer** *n* HEAT & VENT Kesselbauer *m*; ~ **pipe** *n* HEAT & VENT Siederohr *nt*; ~ **plate** *n* HEAT & VENT Kesselblech *nt*; ~ **thermostat** *n* BUILD HARDW, HEAT & VENT Kesselthermostat *nt*
**boiling**: ~ **range** *n* MAT PROP Siedebereich *m*
**bolster** *n* TIMBER Aufsattelung *f*, Sattelholz *nt*
**bolt 1.** *n* ARCH & BUILD Bolzen *m*, Schraube *f*, Dorn *m*, BUILD HARDW *door latch, door bolt* Türriegel *m*, Dorn *m*, Riegelbolzen *m*, MAT PROP Bolzen *m*, Riegelbolzen *m*, Schraube *f*; **2.** *vt* ARCH & BUILD verbolzen, BUILD HARDW festschrauben, verriegeln, verbolzen, TIMBER verbolzen
***bolt***: ~ **connection** *n* BUILD HARDW Bolzenverbindung *f*; ~ **driving tool** *n* BUILD MACHIN Bolzenschlaggerät *nt*
**bolted** *adj* ARCH & BUILD, BUILD HARDW, TIMBER verbolzt, verschraubt
***bolt***: ~ **with handle** *n* MAT PROP Stangenverschluß *m*
**bolting** *n* ARCH & BUILD, BUILD HARDW, TIMBER Verbolzung *f*, Verschraubung *f*
***bolt***: ~ **and nut** *n* BUILD HARDW Durchsteckschraube *f*, Mutter *f* und Bolzen *m*, Mutterschraube *f*
**bombproof** *adj* ARCH & BUILD bombensicher
**Bommer**: ~ **hinge** *n* BUILD HARDW *hardware* Bommerband *nt*
**bond 1.** *n* STONE Verbund *m*, Verband *m*; **2.** *vt* STONE verbinden, SURFACE, TIMBER verleimen
**bonded** *adj* STONE, TIMBER verbunden, verleimt; ~ **masonry** *n* STONE im Verbund gemauertes Mauerwerk, verbundenes Mauerwerk *nt*; ~ **wood construction** *n* ARCH & BUILD, TIMBER Holzleimbau *m*
**bonder** *n* STONE, TIMBER Strecker *m*
***bond***: ~**-improving** *adj* INFR & DES, MAT PROP, SURFACE haftverbessernd
**bonding** *n* INFR & DES, MAT PROP, STONE, SURFACE *composite* Verbund *m*, *process using adhesives* Verbinden *nt*, Verkleben *nt*, *adhesion* Haftung *f*, *gluing* Verleimen *nt*; ~ **area** *n* MAT PROP Haftbrücke *f*, Verbundfläche *f*; ~ **course** *n* STONE Binderschicht *f*, Streckerschicht *f*, *flooring* Haftbrücke *f*; ~ **finish** *n* STONE Haftputz *m*; ~ **plaster** *n* STONE Haftgipsputz *m*; ~ **strength** *n* ARCH & BUILD, INFR & DES, MAT PROP, SURFACE Haftfestigkeit *f*
***bond***: ~ **length** *n* ARCH & BUILD Einbindelänge *f*, Haftlänge *f*, CONCR *prestressed concrete* Einbindelänge *f*, INFR & DES Haftlänge *f*; ~ **strength** *n* ARCH & BUILD, INFR & DES, MAT PROP, SURFACE Haftfestigkeit *f*
**boning**: ~ **rod** *n* BUILD HARDW *surveying*, BUILD MACHIN Visiertafel *f*
**bonnet**: ~ **hinge** *n* BUILD HARDW Haubenscharnier *nt*
**bonus**: ~ **for early completion** *n* CONST LAW Prämie *f* für vorzeitige Fertigstellung
**boom** *n* BUILD MACHIN Flansch *m*, Gurt *m*, *of crane* Ausleger *m*; ~ **crane** *n* BUILD MACHIN Auslegerkran *m*; ~ **plate** *n* BUILD MACHIN Gurtblech *nt*; ~ **position** *n* ARCH & BUILD Auslegerstellung *f*
**booster**: ~ **pump** *n* ARCH & BUILD, BUILD MACHIN, WASTE WATER Druckerhöhungspumpe *f*, Zusatzpumpe *f*; ~ **pump system** *n* ARCH & BUILD, INFR & DES, WASTE WATER Druckerhöhungsanlage *f*
**boot** *n* HEAT & VENT *bottom of drainpipe* Schutzkappe *f*
**borax** *n* MAT PROP Borax *m*; ~ **bead** *n* MAT PROP Boraxperle *f*
**border 1.** *n* ARCH & BUILD Umrandung *f*, Rand *m*, INFR & DES Rabatte *f*; **2.** *vt* ARCH & BUILD, TIMBER einfassen, säumen
***border***: ~ **tile** *n* ARCH & BUILD, INFR & DES, MAT PROP, WASTE WATER Randfliese *f*
**bore 1.** *n* ENVIRON Bohrung *f*, HEAT & VENT *of tube*, INFR & DES, WASTE WATER Nennweite *f*; **2.** *vt* BUILD MACHIN, INFR & DES abteufen, bohren
***bore***: ~ **bit** *n* BUILD MACHIN Bohrmeißel *m*
**borehole** *n* BUILD MACHIN, ENVIRON, HEAT & VENT,

INFR & DES Bohrloch *nt*; ~ **pump** *n* BUILD MACHIN Bohrlochpumpe *f*
**boreholing** *n* INFR & DES Lochbohrung *f*
**borer** *n* BUILD MACHIN Bohrer *m*
***bore***: ~ **rod** *n* BUILD MACHIN Bohrstange *f*
**boring** *n* BUILD MACHIN Bohren *nt*, Lochen *nt*, INFR & DES Bohren *nt*; ~ **by percussion** *n* BUILD MACHIN Schlagbohren *nt*; ~ **by percussion with rods** *n* BUILD MACHIN Vollgestängebohren *nt*; ~ **by shot drills** *n* BUILD MACHIN Schrotbohren *nt*; ~ **contractor** *n* ARCH & BUILD Bohrunternehmer *m*; ~ **equipment for pile foundations** *n* ARCH & BUILD Bohrausrüstung *f* für Pfahlgründungen; ~ **rod** *n* BUILD MACHIN Bohrstange *f*, WASTE WATER *measurement technique* Lochlehre *f*; ~ **site** *n* INFR & DES Bohransatzpunkt *m*, Bohrstelle *f*
**boss** *n* ARCH & BUILD, MAT PROP Bosse *f*
**bossage** *n* STONE, TIMBER Bossenwerk *nt*
**bossing**: ~ **mallet** *n* BUILD MACHIN Holzschlaghammer *m*
**bottle** *n* BUILD MACHIN, WASTE WATER Flasche *f*; ~ **bank** *n* ENVIRON Altglascontainer *m*, Glascontainer *m*, Glasbehälter *m*; ~ **trap** *n* WASTE WATER Flaschensiphon *m*, Flaschenverschluß *m*
**bottom** *n* ARCH & BUILD, INFR & DES *of canyon, well* Sohle *f*, *of receptacle* Boden *m*, WASTE WATER *of pipe, trench* Sohle *f*; ~ **chord** *n* ARCH & BUILD *framework*, CONCR, INFR & DES, MAT PROP, STEEL Untergurt *m*; ~ **chord member** *n* STEEL Untergurtstab *m*; ~ **edge** *n* ARCH & BUILD Unterkante *f*; ~ **flange** *n* ARCH & BUILD, CONCR, INFR & DES, MAT PROP, STEEL Untergurt *m*; ~ **of foundation** *n* INFR & DES Fundamentsohle *f*; ~**-glazed** *adj* ARCH & BUILD, SURFACE unten lackiert; ~**-hinged fanlight** *n* ARCH & BUILD, TIMBER *semicircular window over a door* Kippoberlicht *nt*; ~**-hinged sash** *n* BUILD HARDW *window*, TIMBER Kippflügel *m*; ~**-hinged skylight** *n* ARCH & BUILD *exterior window*, TIMBER Kippoberlicht *nt*; ~**-hinged transom light** *n* ARCH & BUILD *window over a door*, TIMBER Kippoberlicht *nt*; ~**-hinged window** *n* ARCH & BUILD, TIMBER Kippfenster *nt*; ~ **plate** *n* ARCH & BUILD, BUILD HARDW, INFR & DES Fundamentplatte *f*; ~ **reinforcement** *n* ARCH & BUILD untere Bewehrung *f*, untere Lage *f*; ~ **sealing** *n* ENVIRON Basisabdichtung *f*, INFR & DES Sohlenabdichtung *f*; ~ **step** *n* ARCH & BUILD Antrittsstufe *f*; ~ **view** *n* ARCH & BUILD Untersicht *f*; ~ **width** *n* INFR & DES *trench* Sohlbreite *f*
**boulder** *n* INFR & DES Geröllblock *m*
**boundary** *n* ARCH & BUILD, CONST LAW, INFR & DES Begrenzung *f*, Grenze *f*; ~ **arch** *n* ARCH & BUILD, INFR & DES Randbogen *m*; ~ **beam** *n* ARCH & BUILD, INFR & DES, TIMBER Randbalken *m*; ~ **condition** *n* CONST LAW, ENVIRON Begrenzungsbedingung *f*; ~ **determination** *n* INFR & DES Grenzenbestimmung *f*; ~ **fence** *n* ARCH & BUILD, CONST LAW, INFR & DES Einfriedung *f*, Umzäunung *f*; ~ **layer** *n* ENVIRON Grenzschicht *f*; ~ **line** *n* ARCH & BUILD Begrenzungslinie *f*; ~ **mark** *n* ARCH & BUILD Grenzmarkierung *f*; ~ **post** *n* ARCH & BUILD Grenzpfosten *m*; ~ **of the site** *n* ARCH & BUILD, CONST LAW Baustellenbegrenzung *f*; ~ **stone** *n* CONST LAW, INFR & DES Grenzstein *m*; ~ **wall** *n* CONST LAW, INFR & DES Grundstücksmauer *f*
**bow**: ~ **calipers** *AmE*, ~ **callipers** *BrE* *n pl* BUILD MACHIN Bogendickenmesser *m*; ~ **compass** *n* BUILD MACHIN Bogenzirkel *m*; ~ **spring compasses** *n pl* BUILD MACHIN Bogenfederzirkel *m*
**bowstring**: ~ **bridge** *n* INFR & DES Bogenbrücke *f* mit Zugband; ~ **girder** *n* ARCH & BUILD Bogenträger *m* mit Zugband
**box** *n* ARCH & BUILD, INFR & DES, TIMBER Kasten *m*; ~ **culvert** *n* ARCH & BUILD Kastendurchlaß *m*, Kastendüker *m*; ~ **foundation** *n* CONCR, INFR & DES Kastenfundament *nt*; ~ **girder** *n* ARCH & BUILD, CONCR, INFR & DES Hohlkastenträger *m*, Hohlträger *m*, Kastenträger *m*, STEEL Hohlkastenträger *m*; ~ **gutter** *n* INFR & DES, MAT PROP Kastenrinne *f*; ~ **lock** *n* BUILD HARDW Kastenschloß *nt*; ~ **plate girder** *n* ARCH & BUILD, STEEL Kastenblechträger *m*; ~ **profile** *n* ARCH & BUILD, STEEL Kastenprofil *nt*, Kastenquerschnitt *m*; ~ **shutter** *n* BUILD HARDW, TIMBER Klappladen *m*; ~ **staple** *n* ARCH & BUILD Schließklappe *f*, Schloßkasten *m*
**brace 1.** *n* ARCH & BUILD *stay plate* Bindeblech *nt*, *reinforcement* Steife *f*, Fachwerkstrebe *f*, INFR & DES, STEEL Bindeblech *nt*, TIMBER Fachwerkstrebe *f*, *stay plate* Bindeblech *nt*, *strut* Steife *f*; **2.** *vt* ARCH & BUILD absteifen, aussteifen, versteifen
**bracing** *n* ARCH & BUILD Absteifung *f*, Aussteifung *f*, Verband *m*, *reinforcement* Verstärkung *f*, *steadying* Verspannung *f*, Versteifung *f*; ~ **wire** *n* ARCH & BUILD Spanndraht *m*
**bracket** *n* ARCH & BUILD, CONCR Konsole *f*, STONE, TIMBER Kragstein *m*; ~**-mounted roof gutter** *n* ARCH & BUILD vorgehängte Dachrinne *f*
**branch** *n* HEAT & VENT, WASTE WATER Abzweigung *f*; ~ **cable** *n* HEAT & VENT Stichkabel *nt*; ~ **line** *n* INFR & DES, WASTE WATER Hausanschlußleitung *f*, Stichleitung *f*
**brand**: ~ **name** *n* CONST LAW, MAT PROP Markenname *m*
**brass** *n* MAT PROP Messing *nt*; ~ **bar** *n* BUILD HARDW Messingstab *m*; ~ **hinge** *n* BUILD HARDW Messingband *nt*; ~ **round-head wood screw** *n* BUILD HARDW Messingrundkopfschraube *f*; ~ **solder** *n* MAT PROP Messinglot *nt*
**braze** *vt* MAT PROP, STEEL, SURFACE hartlöten
**brazed** *adj* MAT PROP, STEEL, SURFACE hartgelötet; ~ **joint** *n* MAT PROP, STEEL, SURFACE Lötnaht *f*
**brazing** *n* MAT PROP, STEEL, SURFACE Hartlöten *nt*; ~ **solder** *n* MAT PROP, STEEL Hartlot *nt*
**BRE** *abbr* (*Building Research Establishment*) CONST LAW BRE (*Institution für Forschung im Bauwesen*)
**breadth** *n* ARCH & BUILD, INFR & DES, MAT PROP Breite *f*
**break 1.** *vt* MAT PROP zerkleinern; **2.** *vi* MAT PROP zerspringen; ~ **open** INFR & DES *pavement* aufbrechen
**breaking**: ~ **elongation** *n* ARCH & BUILD, INFR & DES, MAT PROP Bruchdehnung *f*; ~ **load** *n* ARCH & BUILD, INFR & DES, MAT PROP Bruchbelastung *f*, Bruchlast *f*, Grenzbelastung *f*, Grenzlast *f*; ~ **point** *n* ARCH & BUILD, INFR & DES, MAT PROP Starrpunkt *m*; ~ **stress** *n* ARCH & BUILD, INFR & DES, MAT PROP Bruchspannung *f*; ~ **test** *n* MAT PROP Bruchversuch *m*
**breakproof** *adj* MAT PROP bruchsicher, bruchfest
**breakthrough** *n* ARCH & BUILD, BUILD MACHIN, INFR & DES, MAT PROP, SURFACE Durchbruch *m*
**breast** *n* ARCH & BUILD Brust *f*, Brüstung *f*, BUILD HARDW, CONCR Brüstung *f*, INFR & DES Brust *f*, STEEL, TIMBER Brüstung *f*
**breastsummer** *n* ARCH & BUILD Rähmstück *nt*, langer Sturzbalken *m*

***breast***: ~ **wall** *n* ARCH & BUILD Futtermauer *f*, Brüstungsmauer *f*, INFR & DES Brüstungsmauer *f*
**breastwork** *n* ARCH & BUILD *usually breast high* Umwehrung *f*
**breath** *n* ENVIRON Atem *m*
**breathing** *n* ENVIRON Atmung *f*; ~ **apparatus** *n* CONST LAW, ENVIRON Atemschutzgerät *nt*; ~ **capability** *n* ENVIRON, MAT PROP Atmungsvermögen *nt*; ~ **capacity** *n* ENVIRON Atemgrenzwert *m*; ~ **protection system** *n* ENVIRON Atemschutzsystem *nt*
**breccia** *n* INFR & DES Brekzie *f*
**bressumer** *n* ARCH & BUILD, INFR & DES Sturzbalken *m*
**brick 1.** *n* STONE Mauerstein *m*, Mauerziegel *m*, Ziegel *m*; **2.** *vt* ARCH & BUILD, STONE mit Ziegeln mauern, verblenden; ~ **in** ARCH & BUILD, STONE einmauern; ~ **up** ARCH & BUILD mauern, STONE aufmauern, mauern, *hole, entrance* vermauern, zumauern
***brick***: ~ **architecture** *n* ARCH & BUILD Backsteinarchitektur *f*; ~ **building** *n* ARCH & BUILD, STONE Backsteinbau *m*; ~**-built** *adj* ARCH & BUILD, STONE gemauert; ~ **chippings** *n pl* ARCH & BUILD, ENVIRON, STONE Ziegelsplitt *m*; ~ **clay** *n* MAT PROP Ziegelton *m*; ~ **dimension** *n* STONE Ziegelmaß *nt*; ~ **earth** *n* MAT PROP Ziegelerde *f*
***bricked***: ~**-over vault** *n* ARCH & BUILD, STONE übermauertes Gewölbe *nt*; ~ **up** *adj* ARCH & BUILD, STONE gemauert
***brick***: ~ **facing** *n* STONE Ziegelformat *nt*, Ziegelverkleidung *f*
**brickfield** *n* STONE Ziegellager *nt*
***brick***: ~ **fireplace** *n* INFR & DES, STONE Ziegelkamin *m*; ~ **kiln** *n* HEAT & VENT, STONE Ziegelofen *m*
**bricklayer**: ~**'s trowel** *n* BUILD HARDW, STONE Maurerkelle *f*
***brick***: ~**-on-edge course** *n* STONE Rollschicht *f*; ~ **packaging machine** *n* ARCH & BUILD, BUILD MACHIN Steinpaketierungsanlage *f*; ~ **pattern** *n* STONE Ziegelmuster *nt*; ~ **pavement** *n* INFR & DES, STONE Ziegelpflaster *nt*; ~ **paving** *n* INFR & DES, STONE Ziegelpflasterung *f*; ~ **size** *n* STONE Ziegelformat *nt*, Ziegelgröße *f*; ~ **structure** *n* ARCH & BUILD, STONE gemauertes Bauteil *nt*; ~ **supplier** *n* STONE Ziegellieferant *m*; ~ **trowel** *n* BUILD MACHIN Maurerkelle *f*; ~ **vault** *n* ARCH & BUILD, STONE Ziegelgewölbe *nt*; ~ **wall** *n* ARCH & BUILD, STONE Ziegelmauer *f*, Ziegelwand *f*
**brickwork** *n* ARCH & BUILD, STONE Backsteinverband *m*, Ziegelsteinverband *m*
**bridge 1.** *n* ELECTR, INFR & DES Brücke *f*; **2.** *vt* ELECTR, INFR & DES überbrücken
***bridge***: ~ **arch** *n* ARCH & BUILD, INFR & DES Brückenbogen *m*; ~ **crane** *n* BUILD MACHIN, INFR & DES Brückenkran *m*; ~ **paver** *n* BUILD MACHIN, STONE Brückensteinpflaster *nt*; ~ **pier** *n* CONCR, STEEL, STONE Brückenpfeiler *m*; ~ **plate** *n* STEEL Brückenblech *nt*; ~ **railing** *n* BUILD HARDW Brückengeländer *nt*; ~ **record book** *n* CONST LAW, INFR & DES Brückenbuch *nt*; ~ **steel** *n* STEEL Brückenbaustahl *m*; ~ **truss** *n* ARCH & BUILD, INFR & DES Brückenfachwerkträger *m*
**bridging** *n* ARCH & BUILD *of cracks, gaps* Spreizen *nt* zwischen Querbalken, *of river* Überbrückung *f*, ELECTR Überbrückung *f*; ~ **piece** *n* ARCH & BUILD Unterzugbalken *m*
**bright**: ~ **annealed wire** *n* STEEL blankgeglühter Draht *m*; ~ **galvanized** *adj* SURFACE glanzverzinkt

**brine** *n* INFR & DES, MAT PROP Salzsole *f*
**Brinell**: ~ **hardness** *n* MAT PROP, STEEL Brinellhärte *f*; ~ **hardness test** *n* STEEL Kugelprobe *f* nach Brinell
**British**: ~ **Calibration Service** *n* (*BCS*) CONST LAW Britische Anstalt *f* für Kalibrierung (*BCS*); ~ **Standard** *n* (*BS*) ARCH & BUILD, CONCR, INFR & DES, STEEL, TIMBER britische Norm *f* (*BS*); ~ **Standards Institution** *n* (*BSI*) ARCH & BUILD, CONST LAW, INFR & DES, MAT PROP, STEEL, TIMBER Britisches Institut *nt* für Normung (*BSI*)
**brittle**: ~ **fracture** *n* MAT PROP, STEEL Sprödbruch *m*
**brittleness** *n* MAT PROP, STEEL Sprödigkeit *f*
**brittle**: ~ **point** *n* MAT PROP Starrpunkt *m*
**broach** *n* ARCH & BUILD Spritzturm *m*
**broad** *adj* ARCH & BUILD, BUILD MACHIN, MAT PROP breit; ~**-flanged** *adj* MAT PROP, STEEL breitflanschig; ~**-flanged beam** *n* (*BFB*) ARCH & BUILD, CONCR, STEEL Breitflanschträger *m*
**broken**: ~ **brick concrete** *n* MAT PROP Ziegelsplittbeton *m*; ~ **joint** *n* ARCH & BUILD versetzte Fuge *f*; ~ **material** *n* MAT PROP Brechgut *nt*, gebrochenes Material *nt*; ~ **rock** *n* MAT PROP Schotter *m*; ~ **stone** *n* ARCH & BUILD Bruchstein *m*, Schotter *m*, MAT PROP Schotter *m*, STONE Bruchstein *m*
**broker** *n* CONST LAW *real estate agent, property agent* Grundstücksmakler *m*, Immobilienhändler *m*, Immobilienmakler *m*, Makler *m*
**bronze** *n* MAT PROP Bronze *f*; ~ **hinge** *n* BUILD HARDW Bronzeband *nt*
**broom** *n* CONCR, STONE Besen *m*; ~ **finish** *n* CONCR *concrete surface treatment*, STONE *plaster* Besenstrich *m*; ~ **finishing** *n* CONCR *concrete surface treatment* Besenabzug *m*
**brooming** *n* CONCR *concrete surface treatment*, STONE Besenstrich *m*
**brow** *n* ARCH & BUILD Rand *m*
**browpiece** *n* ARCH & BUILD, MAT PROP Türsturz *m*
**brush 1.** *n* SURFACE Bürste *f*; **2.** *vt* SURFACE bürsten
**brushability** *n* SURFACE Streichfähigkeit *f*
**brushable** *adj* SURFACE streichfähig
***brush***: ~ **application** *n* SURFACE Streichen *nt*
**brushing** *n* SURFACE Bürsten *nt*, Anstreichen *nt*
***brush***: ~ **painting** *n* SURFACE Pinselauftrag *m*
**BS** *abbr* (*British Standard*) ARCH & BUILD, CONCR, INFR & DES, STEEL, TIMBER BS (*britische Norm*)
**BSI** *abbr* (*British Standards Institution*) ARCH & BUILD, CONST LAW, INFR & DES, MAT PROP, STEEL, TIMBER BSI (*Britisches Institut für Normung*)
**bubble** *n* CONCR, SURFACE Blase *f*
**buck** *n* BUILD MACHIN Sägebock *m*, TIMBER Sägebock *m*, Türzarge *f*
**bucket** *n* BUILD HARDW Eimer *m*, BUILD MACHIN Baggerschaufel *f*, Kübel *m*, INFR & DES *mud trap in manhole* Eimer *m*; ~**-bearing ring** *n* INFR & DES *manhole* Eimertragring *m*; ~ **chain** *n* BUILD MACHIN Baggerkette *f*; ~ **conveyor** *n* ARCH & BUILD Becherwerk *nt*, BUILD MACHIN Becherwerk *nt*, Kübelfördergerät *nt*; ~ **dredger** *n* BUILD MACHIN Schaufelbagger *m*; ~ **elevator** *n* ARCH & BUILD Becherwerk *nt*, BUILD MACHIN Becherwerk *nt*, Eimerkettenaufzug *m*; ~ **excavator** *n* BUILD MACHIN Eimerkettenbagger *m*; ~ **foundation** *n* ARCH & BUILD, CONCR Köcherfundament *nt*; ~ **handle joint** *n* ARCH & BUILD, STONE Gerbergelenk *nt*; ~ **ladder** *n* BUILD MACHIN Becherleiter *f*; ~ **wheel** *n* BUILD

MACHIN Schaufelrad *nt*; **~-wheel excavator** *n* BUILD MACHIN Schaufelradbagger *m*

**buckle 1.** *vt* ARCH & BUILD, INFR & DES, MAT PROP einknicken, krümmen; **2.** *vi* ARCH & BUILD, INFR & DES, MAT PROP sich krümmen; **~ out** ARCH & BUILD, INFR & DES, MAT PROP ausknicken

**buckling** *n* CONCR, INFR & DES, MAT PROP, STEEL, TIMBER Knickung *f*; **~ load** *n* ARCH & BUILD, CONCR, INFR & DES, MAT PROP, STEEL Knicklast *f*; **~ resistance** *n* ARCH & BUILD, CONCR, INFR & DES, MAT PROP, STEEL Knickfestigkeit *f*; **~ risk** *n* ARCH & BUILD, CONCR, INFR & DES, MAT PROP, STEEL, TIMBER Knickgefahr *f*

**budget** *n* CONST LAW Budget *nt*

**buffability** *n* SURFACE *floor polish* Polierbarkeit *f*

**buff**: **~-colored brick** *AmE*, **~-coloured brick** *BrE* *n* STONE lederfarbener Ziegelstein *m*

**buffer** *n* ARCH & BUILD *crane*, CONST LAW, STEEL Puffer *m*

**build 1.** *n* SURFACE *coatings* Dicke *f*; **2.** *vt* ARCH & BUILD, INFR & DES bauen, errichten; **~ up** ARCH & BUILD aufbauen, zusammensetzen; **~ in** ARCH & BUILD, INFR & DES einbauen; **~ an extension** ARCH & BUILD, INFR & DES anbauen

**buildability** *n* ARCH & BUILD, HEAT & VENT, INFR & DES, TIMBER Baubarkeit *f*

**builder** *n* ARCH & BUILD Erbauer *m*, Baufachmann *m*, Bauunternehmer *m*, CONST LAW Bauunternehmer *m*

**building** *n* ARCH & BUILD *edifice* Bau *m*, Bauwerk *nt*, Gebäude *nt*, *business* Bauwesen *nt*, CONST LAW Erbau *m*, INFR & DES Bau *m*, Bauwerk *nt*, Gebäude *nt*; **~ activity** *n* ARCH & BUILD Bautätigkeit *f*

**Building**: **~ Acts** *n pl* CONST LAW Baugesetze *nt pl*; **~ Research Establishment** *n* (*BRE*) CONST LAW Institution *f* für Forschung im Bauwesen (*BRE*)

***building***: **~ administration** *n* CONST LAW Bauverwaltung *f*; **~ alteration** *n* ARCH & BUILD Gebäudeumbau *m*; **~ area** *n* INFR & DES bebaute Fläche *f*; **~ authority** *n* CONST LAW Baubehörde *f*; **~ block** *n* STONE Baustein *m*; **~ block module** *n* ARCH & BUILD Bauplatte *f*, Installationszelle *f*, INFR & DES Installationszelle *f*; **~ board** *n* MAT PROP Bauplatte *f*; **~ boom** *n* ARCH & BUILD, INFR & DES Bauhochkonjunktur *f*; **~ column** *n* CONCR, STEEL, TIMBER Gebäudestütze *f*; **~ complex** *n* ARCH & BUILD, INFR & DES Gebäudekomplex *m*; **~ construction** *n* ARCH & BUILD Bauausführung *f*, Baukonstruktion *f*, Hochbau *m*, INFR & DES Hochbau *m*; **~ contract** *n* ARCH & BUILD, CONST LAW, INFR & DES Bauvertrag *m*; **~ costs** *n pl* ARCH & BUILD, CONST LAW, INFR & DES Baukosten *f pl*, Erstellungskosten *pl*; **~ crane** *n* BUILD MACHIN Baukran *m*; **~ deficiency** *n* ARCH & BUILD Bauschaden *m*; **~ dimension** *n* ARCH & BUILD, INFR & DES Bauabmessung *f*; **~ drainage** *n* ARCH & BUILD, WASTE WATER Hausentwässerung *f*; **~ drawing** *n* ARCH & BUILD Bauzeichnung *f*; **~ engineer** *n* ARCH & BUILD, INFR & DES Bauingenieur *m*; **~ engineering** *n* ARCH & BUILD, INFR & DES Bautechnik *f*; **~ estate** *n* ARCH & BUILD, CONST LAW, INFR & DES Baugrundstück *nt*; **~ experience** *n* ARCH & BUILD, INFR & DES Bauerfahrung *f*; **~ expert** *n* ARCH & BUILD, CONST LAW, INFR & DES Bausachverständiger *m*; **~ firm** *n* ARCH & BUILD Bauunternehmen *nt*; **~ form** *n* ARCH & BUILD, INFR & DES, TIMBER Gebäudeform *f*; **~ foundation** *n* ARCH & BUILD Gebäudefundament *nt*; **~ ground** *n* ARCH & BUILD, CONST LAW, INFR & DES Baugelände *nt*, Bauplatz *m*; **~ hygiene** *n* ARCH & BUILD, CONST LAW Bauhygiene *f*; **~ index** *n* ARCH & BUILD, CONST LAW, INFR & DES Preisindex *m* im Bauwesen; **~ industry** *n* ARCH & BUILD, CONST LAW, INFR & DES Bauindustrie *f*; **~ inspection** *n* ARCH & BUILD Bauprüfung *f*, CONST LAW Bauaufsichtsbehörde *f*, Bauprüfung *f*, INFR & DES Bauprüfung *f*; **~ inspector** *n* CONST LAW Bauaufsichtsbeamte *f*, Bauaufsichtsbeamter *m*, Bauaufsichtsbeauftragte *f*, Bauaufsichtsbeauftragter *m*; **~ insulating foil** *n* INFR & DES, MAT PROP, SOUND & THERMAL Baudämmfolie *f*; **~ lime** *n* MAT PROP, STONE Baukalk *m*; **~ line** *n* ARCH & BUILD, CONST LAW, INFR & DES Baulinie *f*, Flucht *f*, Gebäudeflucht *f*; **~ lot** *n* *AmE* ARCH & BUILD, CONST LAW, INFR & DES Bauplatz *m*; **~ machinery** *n* BUILD MACHIN Baumaschinen *f pl*; **~ maintenance** *n* ARCH & BUILD, CONST LAW, INFR & DES Gebäudeunterhaltung *f*, Gebäudewartung *f*; **~ material** *n* ARCH & BUILD, MAT PROP Baumaterial *nt*, Baustoff *m*; **~ materials industry** *n* MAT PROP Baustoffindustrie *f*; **~ materials quality** *n* CONST LAW Baugüte *f*; **~ materials quality control** *n* CONST LAW Baugüteüberwachung *f*; **~ method** *n* ARCH & BUILD, CONST LAW, INFR & DES Bauweise *f*; **~ moisture** *n* CONCR, MAT PROP, STONE Baufeuchte *f*; **~ number** *n* ARCH & BUILD, INFR & DES Gebäudenummer *f*, Hausnummer *f*; **~ occupants** *n* ARCH & BUILD, INFR & DES Gebäudebelegung *f*; **~ office** *n* ARCH & BUILD, INFR & DES Baubüro *nt*; **~ paper** *n* MAT PROP, SOUND & THERMAL Baupappe *f*; **~ part** *n* ARCH & BUILD Gebäudeteil *m*; **~ permission** *n* CONST LAW Baugenehmigung *f*; **~ permit** *n* CONST LAW Baugenehmigung *f*; **~ phase** *n* ARCH & BUILD Baustadium *nt*, Bauphase *f*, CONST LAW Bauphase *f*, Baustadium *nt*, INFR & DES Bauphase *f*; **~ pit** *n* INFR & DES Baugrube *f*; **~ pit blasting** *n* INFR & DES Baugrubensprengung *f*; **~ pit lining** *n* ARCH & BUILD, INFR & DES Baugrubenverkleidung *f*; **~ pit sheeting work** *n* ARCH & BUILD, INFR & DES Baugrubenverbauarbeiten *f pl*, Baugrubenverkleidungsarbeiten *f pl*; **~ planning** *n* CONST LAW, INFR & DES Bauplanung *f*; **~ plaster** *n* MAT PROP Baugips *m*; **~ plot** *n* ARCH & BUILD, CONST LAW, INFR & DES Bauplatz *m*; **~ plumber** *n* HEAT & VENT, WASTE WATER Bauklempner *m*; **~ price index** *n* ARCH & BUILD, CONST LAW, INFR & DES Baupreisindex *m*; **~ principles** *n pl* ARCH & BUILD, CONST LAW, INFR & DES Baugrundsätze *m pl*; **~ project** *n* ARCH & BUILD, CONST LAW, INFR & DES Bauvorhaben *nt*; **~ proposal** *n* ARCH & BUILD, CONST LAW, INFR & DES Bauantrag *m*, Baugesuch *nt*; **~ protecting agents** *n pl* SURFACE Bautenschutzmittel *nt pl*; **~ regulation** *n* ARCH & BUILD, CONST LAW, INFR & DES Bauvorschrift *f*; **~ regulations** *n pl* ARCH & BUILD Bauordnung *f*, CONST LAW, INFR & DES Bauordnung *f*, Bauvorschriften *f pl*; **~ regulations approved** *adj* CONST LAW bauaufsichtlich zugelassen; **~ research** *n* ARCH & BUILD, CONST LAW, INFR & DES Bauforschung *f*; **~ restriction** *n* CONST LAW Baubeschränkung *f*; **~ sand** *n* MAT PROP Bausand *m*; **~ services** *n pl* ELECTR, HEAT & VENT, WASTE WATER technische Gebäudeausrüstung *f*; **~ sewer** *n* WASTE WATER Gebäudeanschlußleitung *f*; **~ site** *n* *BrE* (*cf job site AmE*) ARCH & BUILD, CONST LAW, INFR & DES Baugrundstück *nt*, Bauplatz *m*, Baustelle *f*; **~ skeleton** *n* ARCH & BUILD Gebäudegerippe *nt*;

~ **slab** *n* CONCR Bauplatte *f*; ~ **society** *n* CONST LAW Bausparkasse *f*

**buildings**: ~ **and structures** *n pl* ARCH & BUILD, INFR & DES bauliche Anlagen *f pl*

***building***: ~ **stone** *n* STONE Naturbaustein *m*; ~ **supervision** *n* ARCH & BUILD, CONST LAW Bauüberwachung *f*; ~ **trade** *n* INFR & DES Baugewerbe *nt*; ~ **wall** *n* ARCH & BUILD Hauswand *f*; ~ **ware** *n* ARCH & BUILD, MAT PROP Baumaterial *nt*; ~ **work** *n* ARCH & BUILD Bauleistung *f*; ~ **worker** *n* ARCH & BUILD Bauarbeiter *m*; ~ **works** *n pl* ARCH & BUILD, CONST LAW, INFR & DES Bauarbeiten *f pl*

**built**: ~**-in** *adj* ARCH & BUILD, INFR & DES eingebaut; ~**-in cupboard** *n BrE* (*cf closet AmE*) INFR & DES, TIMBER Einbauschrank *m*, Schrank *m*, Wandschrank *m*; ~**-in kitchen** *n* INFR & DES, TIMBER Einbauküche *f*; ~**-in kitchen fittings** *n pl* ARCH & BUILD, INFR & DES, WASTE WATER Einbauküchenfittings *nt pl*; ~**-in units** *n pl* BUILD HARDW, TIMBER, INFR & DES feste Einbauten *m pl*

**bulb**: ~ **of pressure** *n* INFR & DES Druckzwiebel *f*

**bulge 1.** *n* MAT PROP Ausbauchung *f*, Wulst *m*, Ausbeulen *nt*, SURFACE Ausbeulen *nt*; **2.** *vi* INFR & DES anschwellen, MAT PROP aufweiten, ausbeulen, anschwellen, SURFACE aufweiten, ausbeulen

**bulk** *n* MAT PROP Masse *f*, Volumen *nt*

***bulk***: ~ **collection** *n* ENVIRON Sperrmüllabfuhr *f*; ~ **density** *n* INFR & DES, MAT PROP Rohdichte *f*; ~ **deposition** *n* ENVIRON lose Ablagerung *f*

**bulkiness** *n* MAT PROP Sperrigkeit *f*

**bulking** *n* MAT PROP Quellvolumen *nt*

***bulk***: ~ **material** *n* MAT PROP Massengut *nt*, Schüttgut *nt*

**bulky** *adj* MAT PROP sperrig

**bulldozer** *n* BUILD MACHIN Planierraupe *f*

**bulletin**: ~ **board** *n* ARCH & BUILD schwarzes Brett *nt*, BUILD HARDW Anschlagtafel *f*, schwarzes Brett *nt*

**bulletproof** *adj* MAT PROP kugelsicher, schußsicher; ~ **glass** *n* ARCH & BUILD, MAT PROP Panzerglas *nt*

**bullhead**: ~ **rivet** *n* BUILD HARDW Doppelkopfniete *f*

**bullion** *n* MAT PROP Butzenscheibe *f*

**bull's-eye**: ~ **glass** *n* MAT PROP Ochsenauge *nt*, Rundfenster *nt*

**bungalow** *n* ARCH & BUILD Bungalow *m*

**bungled**: ~ **work** *n* CONST LAW Pfusch *m*, Pfuscharbeit *f*

**bunker** *n* ARCH & BUILD Bunker *m*, INFR & DES Bunker *m*, Luftschutzraum *m*

**buoyancy** *n* ARCH & BUILD, ENVIRON, INFR & DES, MAT PROP Auftrieb *m*; ~ **protection** *n* ARCH & BUILD, INFR & DES Auftriebssicherung *f*

**burglar**: ~ **alarm system** *n* CONST LAW, ELECTR Einbruchmeldeanlage *f*

**burglarproof** *adj* ARCH & BUILD, CONST LAW einbruchsicher, diebstahlsicher

**buried** *adj* ARCH & BUILD unterirdisch, ELECTR, HEAT & VENT, INFR & DES, WASTE WATER Unterflur-, erdverlegt, unterirdisch; ~ **duct** *n* ARCH & BUILD *in the ground* Unterflurkanal *m*, Bodenkanal *m*, HEAT & VENT Bodenkanal *m*, WASTE WATER Unterflurkanal *m*; ~ **tank** *n* HEAT & VENT, WASTE WATER Erdtank *m*; ~ **treasure** *n* CONST LAW vergrabener Schatz *m*

**burlap** *n* MAT PROP Sackleinen *nt*

**burn** *vt* ARCH & BUILD, ENVIRON, HEAT & VENT, MAT PROP brennen

**burner**: ~ **capacity** *n* ENVIRON, HEAT & VENT Brennerleistung *f*

**burning**: ~ **kiln** *n* MAT PROP Brennofen *m*

**burnt**: ~ **brick** *n* STONE gebrannter Ziegel *m*

**burst** *vi* INFR & DES, MAT PROP platzen

**bursting**: ~ **strength** *n* INFR & DES, MAT PROP Berstfestigkeit *f*

**bush** *n* MAT PROP *for pipes and cables* Buchse *f*, Durchführung *f*; ~ **hammering** *n* BUILD MACHIN Stockhammer *m*, CONCR, STONE Stocken *nt*

**bushing** *n* MAT PROP *for pipes and cables* Buchse *f*, Durchführung *f*

**business** *n* ARCH & BUILD, CONST LAW, ENVIRON, INFR & DES Gewerbe *nt*; ~ **tax** *n* CONST LAW, INFR & DES Gewerbesteuer *f*

**butt 1.** *n* BUILD HARDW Stoß *m*, stumpfes Ende *nt*; **2.** ~**-joint** *vt* ARCH & BUILD, STEEL, STONE, TIMBER stumpfstoßen

***butt***: ~ **end** *n* BUILD HARDW Stirnfläche *f*

**butterfly**: ~ **cock** *n* ENVIRON Flügelhahn *m*; ~ **nut** *n* BUILD HARDW Flügelmutter *f*; ~ **screw** *n* BUILD HARDW Flügelschraube *f*; ~ **valve** *n* ENVIRON Flügelhahn *m*, HEAT & VENT, WASTE WATER Absperrklappe *f*, Drosselklappenventil *nt*

***butt***: ~ **hinge** *n* BUILD HARDW Einstemmband *nt*, *part of lock* Fischband *nt*, Fitschband *nt*, Fitsche *f*; ~ **joint** *n* ARCH & BUILD Stoßfuge *f*, Stumpfstoß *m*, STEEL, STONE, TIMBER Stumpfstoß *m*; ~**-jointed** *adj* ARCH & BUILD, STEEL, STONE, TIMBER stumpf gestoßen

**button** *n* BUILD HARDW, ELECTR Knopf *m*; ~**-head bolt** *n* BUILD HARDW Rundkopfbolzen *m*

**buttress 1.** *n* ARCH & BUILD, CONCR, INFR & DES, STONE Bogenpfeiler *m*, Strebepfeiler *m*, Stützpfeiler *m*; ~ **clearance** *n* ARCH & BUILD, CONCR, INFR & DES, STONE lichter Strebepfeilerabstand *m*; ~ **pinnacle** *n* ARCH & BUILD *church* Fiale *f*; ~ **span** *n* ARCH & BUILD, INFR & DES Spannweite *f* des Bogens; **2.** *vt* ARCH & BUILD, CONCR, INFR & DES, STONE stützen

***butt***: ~ **strap** *n* ARCH & BUILD Lasche *f*; ~ **weld** *n* ARCH & BUILD, HEAT & VENT, MAT PROP, STEEL Stumpfschweißung *f*

**buyer** *n* MAT PROP Käufer *m*, Käuferin *f*

**by-law** *n* CONST LAW Gemeindeordnung *f*, Ortsstatut *nt*

**bypass** *n* HEAT & VENT Umgehung *f*, INFR & DES Umgehungsstraße *f*, Umleitung *f*; ~ **channel** *n* HEAT & VENT Umgehungskanal *m*, Seitenkanal *m*; ~ **pipe** *n* HEAT & VENT, WASTE WATER Umgehungsrohr *nt*; ~ **valve** *n* HEAT & VENT, WASTE WATER Bypassventil *nt*

**by-product** *n* ENVIRON Nebenprodukt *nt*

**byroad** *n* ARCH & BUILD, CONST LAW, INFR & DES Nebenstraße *f*, Seitenstraße *f*

**byway** *n* ARCH & BUILD, CONST LAW, INFR & DES Nebenstraße *f*, Seitenstraße *f*

# C

**c** *abbr* (*concentration*) ENVIRON, INFR & DES, MAT PROP c (*Konzentration*)

**cabin** *n* TIMBER Blockhaus *nt*

**cabinet** *n* TIMBER Schrank *m*; **~ lock** *n* BUILD HARDW Möbelschloß *nt*; **~-maker** *n* TIMBER Möbelschreiner *m*; **~-making** *n* TIMBER Kunsttischlerei *f*, Möbeltischlerei *f*

**cable** *n* BUILD MACHIN Seil *nt*, ELECTR Kabel *nt*; **~ assembly** *n* BUILD HARDW, ELECTR Kabelmontage *f*; **~ channel** *n* ELECTR Kabelrinne *f*; **~ clamp** *n* ELECTR Kabelschelle *f*; **~ clip** *n* ELECTR Kabelschelle *f*; **~ code** *n* ELECTR Kabelcode *m*; **~ conduit** *n* ELECTR Elektroinstallationsrohr *nt*, Kabelschutzrohr *nt*; **~ connection** *n* ELECTR Kabelverbindung *f*; **~ cover** *n* ELECTR Kabelabdeckhaube *f*, Kabelabdeckstein *m*; **~ crane** *n* BUILD MACHIN Kabelkran *m*; **~ cutting machine** *n* BUILD MACHIN Drahtseilsägemaschine *f*; **~ detector** *n* BUILD MACHIN Kabelsuchgerät *nt*; **~ drum** *n* ELECTR Kabeltrommel *f*; **~ duct** *n* ELECTR Kabelkanal *m*; **~ entry** *n* ELECTR Kabeleinführung *f*; **~ housing** *n* ELECTR Kabelgehäuse *nt*; **~ ladder** *n* BUILD HARDW Kabelleiter *f*, ELECTR Kabelleiter *f*, Kabelsteigtrasse *f*; **~ laying** *n* ELECTR Kabelverlegung *f*; **~-laying truck with drum** *n* BUILD MACHIN Kabeltrommeltransport- und -verlegewagen *m*; **~ line** *n* ELECTR Kabelführung *f*; **~ marker** *n* BUILD HARDW, ELECTR Kabelmerkstein *m*; **~ marking tape** *n* ELECTR Trassenband *nt*; **~ roof** *n* ARCH & BUILD Hängedach *nt*; **~ routing** *n* ELECTR Kabelführung *f*; **~ run** *n* ELECTR Kabelführung *f*; **~ shaft** *n* ARCH & BUILD Kabelschacht *m*; **~ sleeve** *n* ELECTR Kabelmuffe *f*; **~ store** *n* ARCH & BUILD *under the ground floor* Kabelkeller *m*, BUILD HARDW, ELECTR Kabellager *nt*; **~ support** *n* ELECTR, HEAT & VENT Kabelunterstützung *f*; **~ tray** *n* ELECTR Kabelträger *m*, Kabelpritsche *f*; **~ trench** *n* ELECTR, INFR & DES Kabelgraben *m*; **~ trunking** *n* ELECTR, HEAT & VENT Kabelhauptleitung *f*; **~ warning tape** *n* ELECTR Kabelwarnband *nt*

**cabling** *n* ELECTR, HEAT & VENT Verkabelung *f*; **~ system** *n* ELECTR, HEAT & VENT Kabelsystem *nt*

**CAD** *abbr* (*computer-aided design*) ARCH & BUILD, INFR & DES computerunterstütztes Entwerfen *nt*

**cadmium**: **~ coating** *n* SURFACE Kadmiumbeschichtung *f*; **~ finish** *n* SURFACE Kadmiumbeschichtung *f*

**cafeteria** *n* ARCH & BUILD, INFR & DES Cafeteria *f*

**caisson** *n* INFR & DES Caisson *m*, Druckluftkammer *f*, Senkkasten *m*; **~ disease** *n* CONST LAW, INFR & DES Senkkastenkrankheit *f*; **~ foundation** *n* INFR & DES Druckluftgründung *f*, Senkkastengründung *f*

**calcareous** *adj* MAT PROP, WASTE WATER kalkhaltig, kalkig; **~ tuff** *n* MAT PROP Kalktuff *m*

**calcium**: **~ oxide** *n* MAT PROP gebrannter Kalk *m*

**calculate** *vt* ARCH & BUILD, CONST LAW, INFR & DES *estimate* veranschlagen, *work out* berechnen, errechnen

**calculated** *adj* ARCH & BUILD, CONST LAW, INFR & DES berechnet

**calculating**: **~ formula** *n* ARCH & BUILD, INFR & DES Rechenformel *f*

**calculation** *n* ARCH & BUILD, CONST LAW, INFR & DES Berechnung *f*, Kalkulation *f*; **~ of area** *n* ARCH & BUILD, INFR & DES Flächenberechnung *f*; **~ of consumption** *n* ARCH & BUILD, HEAT & VENT Verbrauchsberechnung *f*; **~ of heat loss** *n* ARCH & BUILD, HEAT & VENT Wärmeverlustberechnung *f*; **~ of storage capacity** *n* ARCH & BUILD, INFR & DES Stauraumbemessung *f*

**calculator** *n* ARCH & BUILD Tischrechner *m*

**calibrate** *vt* ARCH & BUILD, BUILD HARDW, CONST LAW eichen, einstellen, ELECTR *setpoint*, HEAT & VENT einstellen, INFR & DES, MAT PROP eichen, einstellen

**calibrated**: **~ watershed** *n* ENVIRON geeichtes Wasserrückhaltebecken *nt*

**calibration** *n* ARCH & BUILD, BUILD HARDW, BUILD MACHIN, CONST LAW Eichung *f*, ELECTR Kalibrierung *f*, HEAT & VENT Einstellung *f*, INFR & DES, MAT PROP Eichung *f*, Einstellung *f*

**calidarium** *n* ARCH & BUILD *Roman baths* Kalidarium *nt*

**caliper** *AmE see calliper BrE*

**calliper** *n BrE* BUILD MACHIN Lehre *f*

**calorific**: **~ requirement** *n* ARCH & BUILD, CONST LAW, HEAT & VENT Wärmebedarf *m*; **~ value** *n* ENVIRON, HEAT & VENT Brennwert *m*, Heizwert *m*, Heizleistung *f*

**calotte** *n* ARCH & BUILD, INFR & DES Kalotte *f*, Kugelsegment *nt*

**camber** *n* ARCH & BUILD, CONCR, INFR & DES *arching* Aufwölbung *f*, *road surface* Überhöhung *f*

**cambered** *adj* ARCH & BUILD, CONCR, INFR & DES gekrümmt, gewölbt, überhöht, STONE, TIMBER gewölbt

**campanile** *n* INFR & DES Glockenturm *m*

**canal** *n* HEAT & VENT, INFR & DES, WASTE WATER Kanal *m*

**canalization** *n* INFR & DES Kanalisierung *f*

**canalize** *vt* INFR & DES kanalisieren

***canal***: **~ lock** *n* INFR & DES Kanalschleuse *f*

**cane**: **~ fiber insulation** *AmE*, **~ fibre insulation** *BrE* *n* SOUND & THERMAL Zuckerrohrfaserdämmung *f*

**canopy** *n* ARCH & BUILD Vordach *nt*

**cant** *n* ARCH & BUILD, TIMBER Abschrägung *f*

**canted** *adj* ARCH & BUILD, TIMBER abgeschrägt

**cantilever** *n* ARCH & BUILD *boom gibbet* Kragarm *m*, Auskragung *f*, Ausleger *m*, *jib* Ausladung *f*, CONCR Auskragung *f*, STEEL Auskragung *f*, Ausleger *m*; **~ beam** *n* ARCH & BUILD, CONCR, TIMBER Freiträger *m*, Kragträger *m*; **~ bridge** *n* INFR & DES Auslegerbrücke *f*

**cantilevered** *adj* ARCH & BUILD freitragend

**cantilevering** *adj* ARCH & BUILD, BUILD HARDW, BUILD MACHIN, MAT PROP, STONE auskragend; **~ length** *n* ARCH & BUILD Kraglänge *f*

***cantilever***: **~ landing** *n* ARCH & BUILD, TIMBER Kragtreppenpodest *nt*; **~ load** *n* ARCH & BUILD Kragarmlast *f*, Kraglast *f*, INFR & DES Kraglast *f*;

~ **plate** *n* ARCH & BUILD, CONCR Kragplatte *f*; ~ **roof** *n* ARCH & BUILD, TIMBER Kragdach *nt*, vorkragendes Dach *nt*; ~ **slab** *n* ARCH & BUILD vorkragende Platte *f*, Kragplatte *f*, CONCR Kragplatte *f*; ~ **staircase** *n* ARCH & BUILD, TIMBER vorragende Treppe *f*

***cant***: ~ **strip** *n* TIMBER *eaves board* Aufschiebling *m*

**canvas** *n* ARCH & BUILD Segeltuch *nt*, Zeltstoff *m*, MAT PROP Zeltstoff *m*

**cap 1.** *n* ARCH & BUILD Verschlußkappe *f*, *of roof ridge* Glocke *f*; **2.** *vt* ARCH & BUILD, STONE abdecken

**capacity** *n* ARCH & BUILD Kapazität *f*, *volumetric capacity or content* Fassungsvermögen *nt*, *water capacity* Rauminhalt *m*

***cap***: ~ **flashing** *n* ARCH & BUILD Kappstreifen *m*, STEEL oberer Anschluß *m*

**capillary**: ~ **action** *n* INFR & DES *soil* Kapillarwirkung *f*; ~ **attraction** *n* ARCH & BUILD, INFR & DES Porensaugwirkung *f*; ~ **fitting** *n* HEAT & VENT, INFR & DES Kapillarfitting *f*

**capital** *n* ARCH & BUILD *of column* Kapitell *nt*, CONST LAW *funds* Kapital *nt*; ~ **costs** *n pl* CONST LAW Kapitalkosten *pl*

**capped** *adj* ARCH & BUILD, STONE abgedeckt

***cap***: ~ **piece** *n* ARCH & BUILD, STONE Kappstreifen *m*

**capping** *n* ARCH & BUILD, STONE Abdeckung *f*

**capstan**: ~**-headed screw** *n* BUILD HARDW Knebelschraube *f*; ~ **screw** *n* BUILD HARDW Kreuzlochschraube *f*

**capstone** *n* STONE Abdeckstein *m*, *vault* Schlußstein *m*

**car** *n AmE* (*cf wagon BrE*) BUILD MACHIN *railway* Lore *f*

**caracole** *n* ARCH & BUILD Wendeltreppe *f*

**carbon** *n* ENVIRON, MAT PROP Kohlenstoff *m*

**carbonate**: ~ **pockets** *n pl* CONCR Carbonatnester *nt pl*

***carbon***: ~ **content** *n* MAT PROP Kohlenstoffanteil *m*, Kohlenstoffgehalt *m*; ~ **dioxide** *n* ($CO_2$) ENVIRON Kohlendioxid *nt* ($CO_2$); ~ **dioxide greenhouse effect** *n* ENVIRON Kohlendioxidtreibhauseffekt *m*

**carbonitriding** *n* MAT PROP Karbonitrierung *f*

***carbon***: ~ **monoxide** *n* (*CO*) ENVIRON Kohlenmonoxid *nt* (*CO*)

**carbonyl**: ~ **sulfide** *AmE*, ~ **sulphide** *BrE n* ENVIRON Carbonylsulfid *nt*

**carcass** *n* ARCH & BUILD *of a building* Gerippe *nt*, Rohbau *m*, CONCR Gerippe *nt*

**carcassing** *n* ARCH & BUILD, TIMBER Rohbauarbeiten *f pl*

***car***: ~ **fragmentation plant** *n* ENVIRON Autoverschrottungsanlage *f*

**cargo**: ~ **tank** *n* ENVIRON Ladetank *m*

***car***: ~ **park** *n BrE* (*cf parking lot AmE*) INFR & DES Parkplatz *m*

**carpenter** *n* TIMBER Zimmermann *m*, Zimmerer *m*; ~**'s gage** *AmE*, ~**'s gauge** *BrE n* BUILD MACHIN Zollstock *m*; ~**'s pencil** *n* BUILD MACHIN Zimmermannsbleistift *m*; ~**'s shop** *n* TIMBER Zimmerwerkstatt *f*; ~**'s tool** *n* BUILD MACHIN Zimmerwerkzeug *nt*; ~**'s work** *n* TIMBER Zimmerarbeiten *f pl*

**carpentry** *n* TIMBER Holzrohbau *m*, Zimmerhandwerk *nt*, Zimmerei *f*

**carport** *n* INFR & DES Carport *m*, überdeckter Autoabstellplatz *m*

**Carrara**: ~ **marble** *n* MAT PROP Carraramarmor *m*

**carriageway** *n BrE* (*cf highway AmE*) INFR & DES Fahrbahn *f*; ~ **slab** *n BrE* INFR & DES Fahrbahnplatte *f*

**carry** *vt* ARCH & BUILD *support* lagern, stützen, tragen, *transport* befördern, BUILD MACHIN, MAT PROP befördern; ~ **off** *vt* HEAT & VENT ableiten, WASTE WATER *heat* ableiten, *water* abführen

**carrying** *adj* ARCH & BUILD stützend; ~ **capacity** *n* ENVIRON Tragfähigkeit *f*, Tragvermögen *nt*; ~ **off** *n* WASTE WATER Abfuhr *f*

***car***: ~ **spaces** *n* INFR & DES Autoplätze *m pl*

**cartridge**: ~**-firing studder** *n* BUILD MACHIN, CONCR Bolzensetzwerkzeug *nt*

**carve** *vt* STONE behauen, TIMBER schnitzen

**carved**: ~ **ornament** *n* ARCH & BUILD, TIMBER Schnitzornament *nt*, *church* plastische Verzierung *f*

**carving** *n* STONE Steinmetzarbeit *f*, TIMBER Schnitzerei *f*

**caryatid** *n* ARCH & BUILD Karyatide *f*

**cascade** *n* INFR & DES Absturz *m*, Kaskade *f*, WASTE WATER Absturz *m*, Fallstufe *f*, Kaskade *f*

**case 1.** *n* BUILD HARDW Schloßkasten *m*; **2.** *vt* ARCH & BUILD, BUILD HARDW, CONCR, STEEL verkleiden

**cased** *adj* ARCH & BUILD, BUILD HARDW ummantelt, verkleidet; ~ **beam** *n* ARCH & BUILD, CONCR, INFR & DES, STEEL *composite structure* betonummantelter Träger *m*; ~ **column** *n* ARCH & BUILD, INFR & DES ummantelte Stütze *f*; ~ **pile** *n* ARCH & BUILD, INFR & DES Mantelpfahl *m*

**casein** *n* MAT PROP, SURFACE, TIMBER Kasein *nt*; ~ **binder** *n* SURFACE Kaseinbindemittel *nt*, Kaseinbinder *m*; ~ **distemper** *n* SURFACE Kaseinfarbe *f*; ~ **glue** *n* MAT PROP Kaseinleim *m*, TIMBER Weißleim *m*

***case***: ~ **law** *n* CONST LAW Fallrecht *nt*

**casemate** *n* ARCH & BUILD Kasematte *f*

**casement** *n* ARCH & BUILD Fensterflügel *m*, BUILD HARDW Drehflügel *m*, Sturm-, Fensterflügel *m*; ~ **door** *n* ARCH & BUILD, BUILD HARDW, TIMBER Fenstertür *f*; ~ **fastener** *n* BUILD HARDW Einreiber *m*, Fensterreiber *m*, Vorreiber *m*; ~ **frame** *n* TIMBER Flügelrahmen *m*; ~ **stay** *n* BUILD HARDW *window* Sturmhaken *m*; ~ **window** *n* ARCH & BUILD Drehflügelfenster *nt*, BUILD HARDW Drehflügelfenster *nt*, Flügelfenster *nt*

**caserne** *n* ARCH & BUILD, INFR & DES Kaserne *f*

**casing 1.** *adj* CONCR verkleidet; **2.** *n* ARCH & BUILD, BUILD HARDW, CONCR Ummantelung *f*, Umhüllung *f*, Verkleidung *f*, *formwork* Verschalung *f*, STEEL Umhüllung *f*, Verkleidung *f*, Ummantelung *f*

***casing***: ~ **cutter** *n* BUILD MACHIN Rohrschneider *m*; ~ **elevator** *n* BUILD MACHIN Rohrelevator *m*; ~ **expander** *n* BUILD MACHIN Rohrdichter *m*; ~ **head** *n* BUILD MACHIN Rohrkopf *m*; ~ **line** *n* BUILD MACHIN Verrohrungsseil *nt*; ~ **pipe** *n* ARCH & BUILD, BUILD HARDW Schalrohr *nt*; ~ **spears** *n pl* BUILD MACHIN Backenfänger *m*

**casino** *n* ARCH & BUILD, INFR & DES Kasino *nt*

**cassette**: ~ **ceiling** *n* ARCH & BUILD, CONCR, TIMBER Kassettendecke *f*

**cast** *vt* CONCR betonieren, gießen, MAT PROP, STEEL gießen; ~ **in** *vt* CONCR einbetonieren

**cast**: ~ **aluminium alloy** *n BrE* MAT PROP, STEEL Aluminiumgußlegierung *f*; ~ **aluminum alloy** *AmE see cast aluminium alloy BrE*

**castellated**: ~ **nut** *n* BUILD HARDW Kronenmutter *f*

**caster**: ~**-equipped container** *n* ENVIRON Abfallcontainer *m*

**cast**: ~ **glass** *n* MAT PROP Gußglas *nt*, gebogenes Glas *nt*; **~-in casings** *n pl* ARCH & BUILD, CONCR einbetoniertes Gehäuse *nt*; **~-in fixing rails** *n* CONCR, STEEL einbetonierte Befestigungsschienen *f pl*
**casting** *n* BUILD MACHIN, CONCR, MAT PROP, STEEL Guß *m*, Gußstück *nt*, Gußteil *nt*; ~ **concrete** *n* CONCR Betonieren *nt*; ~ **machine** *n* BUILD MACHIN Gießmaschine *f*; ~ **machine for artificial marble** *n* BUILD MACHIN Kunstmarmorgießmaschine *f*; ~ **machine for polymer concrete** *n* BUILD MACHIN Polymerbetongießmaschine *f*; ~ **marks** *n pl* ARCH & BUILD, CONCR Betoniereindrücke *m pl*; ~ **mortar** *n* MAT PROP Gießmörtel *m*; ~ **resin** *n* MAT PROP Gießharz *nt*
**cast**: **~-in gutter** *n* CONCR einbetonierte Rinne *f*; **~-in-place concrete** *n* ARCH & BUILD, CONCR Ortbeton *m*; **~-in-place concrete pile** *n* INFR & DES Ortpfahl *m*; ~ **iron** *n* (*CI*) MAT PROP Gußeisen *nt*; **~-iron boiler** *n* HEAT & VENT, STEEL Gußeisenkessel *m*; **~-iron cover** *n* INFR & DES *manhole*, STEEL Gußeisendeckel *m*; **~-iron drainpipe** *n* INFR & DES, STEEL, WASTE WATER Gußeisenabflußrohr *nt*; **~-iron elbow** *n* BUILD HARDW gußeiserner Krümmer *m*; **~-iron flange** *n* STEEL Gußeisenflansch *m*; **~-iron girder** *n* ARCH & BUILD, STEEL Gußeisenträger *m*; **~-iron joint** *n* BUILD HARDW Gußeisengelenk *nt*; **~-iron outlet** *n* INFR & DES, STEEL, WASTE WATER Gußauslauf *m*; **~-iron pipe** *n* (*CIP*) MAT PROP, STEEL Gußeisenrohr *nt*, gußeisernes Rohr *nt*; **~-iron radiator** *n* HEAT & VENT, STEEL gußeiserner Heizkörper *m*; ~ **stone** *n* MAT PROP, STONE Betonwerkstein *m*, Kunststein *m*
**catch 1.** *n* BUILD HARDW Arretierung *f*, MAT PROP *lock* Arretierung *f*, Sperre *f*; **2.** ~ **up** *vt* CONST LAW einholen
***catch***: **~-basin** *n* INFR & DES Auffangbecken *nt*, Sammelgrube *f*, *grit chamber* Sandfangbecken *nt*; ~ **drain** *n* HEAT & VENT, INFR & DES, WASTE WATER Sammelkanal *m*, Sammler *m*
**catchment**: ~ **area** *n* ENVIRON, INFR & DES Einzugsgebiet *nt*
**caterpillar**: ~ **chain** *n* BUILD MACHIN Gleiskette *f*, Raupenkette *f*; ~ **tractor** *n* BUILD MACHIN Raupe *f*
**cathodic**: ~ **corrosion protection system** *n* CONCR, ELECTR, STEEL, SURFACE kathodische Korrosionsschutzanlage *f*
**cationic**: ~ **wetting agent** *n* SURFACE kationisches Netzmittel *nt*
**Catnic**: ~ **steel lintel**® *n* STEEL Catnic-Stahlsturz® *m*
**caulk** *vt* ARCH & BUILD durchzeichnen, pausen, HEAT & VENT abdichten, STONE verstemmen, SURFACE, WASTE WATER abdichten
**caulking** *n* HEAT & VENT Abdichten *nt*, MAT PROP Fugendichtung *f*, Bleidichtung *f*, WASTE WATER Abdichten *nt*; ~ **strip** *n* HEAT & VENT, WASTE WATER Dichtungsband *nt*, Abdichtungsband *nt*
**causeway** *n* INFR & DES Dammstraße *f*
**cavitation** *n* CONCR, INFR & DES, MAT PROP Hohlraumbildung *f*, Kavitation *f*; ~ **damage** *n* CONCR, INFR & DES, MAT PROP Kavitationsschaden *m*
**cavity** *n* ARCH & BUILD, INFR & DES Hohlraum *m*, Aushöhlung *f*; ~ **barrier** *n* ARCH & BUILD, CONST LAW Hohlraumverschluß *m*; ~ **block** *n* MAT PROP, STONE Hohlblockstein *m*; ~ **fill** *n* ARCH & BUILD, MAT PROP, SOUND & THERMAL Hohlraumfüllung *f*; ~ **in concrete** *n* ARCH & BUILD, CONCR Betonnest *nt*; ~ **insulation** *n* ARCH & BUILD, STONE Hohlraumisolierung *f*; ~ **wall** *n* ARCH & BUILD, STONE zweischalige Wand *f*; ~ **wall dampproof course** *n* SOUND & THERMAL, STONE Hohlraumfeuchtigkeitssperre *f*; ~ **wall DPC** *n* SOUND & THERMAL, STONE Hohlraumfeuchtigkeitssperre *f*; ~ **wall insulation** *n* MAT PROP, SOUND & THERMAL Hohlraumisolierung *f*, zweischalige Wand *f*
**CB** *abbr* (*cast brass*) MAT PROP Gußmessing *nt*
**cedar**: ~ **shingle** *n* MAT PROP, TIMBER Zedernschindel *f*
**ceiling** *n* ARCH & BUILD Decke *f*, Raumdecke *f*; ~ **boarding** *n* CONCR, INFR & DES, TIMBER Deckenschalung *f*; ~ **diffuser** *n* HEAT & VENT Deckenluftverteiler *m*; ~ **fan** *n* HEAT & VENT Deckenventilator *m*; ~ **finish** *n* MAT PROP Deckenoberflächenbehandlung *f*; ~ **grille** *n* BUILD HARDW, HEAT & VENT Deckengitter *nt*; ~ **heating** *n* HEAT & VENT Deckenheizung *f*; ~ **height** *n* ARCH & BUILD Deckenhöhe *f*; ~ **hook** *n* BUILD HARDW Deckenhaken *m*; ~ **illumination** *n* BUILD HARDW, ELECTR Deckenbeleuchtung *f*; ~ **joist** *n* ARCH & BUILD, TIMBER Deckenbalken *m*; ~ **level** *n* ARCH & BUILD Deckenniveau *nt*; ~ **lighting** *n* BUILD HARDW, ELECTR Deckenbeleuchtung *f*; ~ **lining** *n* TIMBER Deckenverkleidung *f*; **~-mounted air heater** *n* HEAT & VENT Deckenlufterhitzer *m*; **~-mounted fan** *n* HEAT & VENT Deckenventilator *m*; ~ **panel** *n* BUILD HARDW Deckenfeld *nt*, Deckentafel *f*, TIMBER Deckentafel *f*; ~ **paneling** *AmE*, ~ **panelling** *BrE* *n* ARCH & BUILD, TIMBER Paneeldecke *f*; ~ **plaster** *n* STONE Deckenputz *m*; ~ **stone** *n* MAT PROP Deckenstein *m*; ~ **suspension** *n* ARCH & BUILD, STEEL Deckenabhängung *f*; ~ **tile** *n* MAT PROP Deckenplatte *f*; ~ **ventilator** *n* HEAT & VENT Deckenlüfter *m*
**cell** *n* ARCH & BUILD Kammer *f*, Zelle *f*, ENVIRON Kammer *f*
**cella** *n* ARCH & BUILD Cella *f*
**cellar** *n* ARCH & BUILD Keller *m*, Kellergeschoß *nt*; ~ **door** *n* ARCH & BUILD, BUILD HARDW Kellertür *f*; ~ **entrance** *n* ARCH & BUILD Kellereingang *m*; ~ **grating** *n* ARCH & BUILD, BUILD HARDW, STEEL Kellergitter *nt*; ~ **shelter** *n* ARCH & BUILD, INFR & DES Schutzkeller *m*; ~ **stairs** *n* ARCH & BUILD Kellertreppe *f*; ~ **vault** *n* ARCH & BUILD, STONE Kellergewölbe *nt*; ~ **wall** *n* ARCH & BUILD Kellergeschoßwand *f*
***cell***: ~ **method** *n* ENVIRON Fächermethode *f*
**cellular**: ~ **brick** *n* MAT PROP, STONE Zellenziegel *m*; ~ **concrete** *n* CONCR Gasbeton *m*, Porenbeton *m*, MAT PROP Gasbeton *m*; ~ **construction** *n* ARCH & BUILD Zellenbauweise *f*; ~ **design** *n* ARCH & BUILD Zellenbauweise *f*; ~ **mortar** *n* STONE Zellenmörtel *m*; ~ **rubber sealing** *n* MAT PROP, SOUND & THERMAL Moosgummiabdichtung *f*
**cement 1.** *n* CONCR Zement *m*, *binder* Bindemittel *nt*; **2.** *vt* CONCR, STONE zementieren, *bond* verkleben, *with filler, putty* verkitten
***cement***: ~ **adhesive** *n* CONCR Dünnbettkleber *m*; **~-aggregate ratio** *n* CONCR Zement-Zuschlagstoff-Verhältnis *nt*
**cementation** *n* CONCR, STONE Zementierung *f*
**cement**: **~-based** *adj* CONCR zementhaltig; **~-based waterproof coating** *n* SOUND & THERMAL Zementsperrschicht *f*; **~-bound** *adj* CONCR, MAT PROP zementgebunden; ~ **concrete** *n* CONCR Zementbeton *m*; ~ **content** *n* CONCR, STONE Zementanteil *m*, Zementgehalt *m*; ~ **dust** *n* CONCR Zementstaub *m*; ~ **film** *n* CONCR, SURFACE Zementschleier *m*; ~ **floor** *n* CONCR Zementestrich *m*; ~ **grade** *n* CONCR Zement-

güte *f*; ~ **grain** *n* CONCR Zementkorn *nt*; ~ **grout** *n* CONCR Zementschlamm *m*

**cementing** *n* CONCR, STONE Zementierung *f*, *bonding* Verkleben *nt*; ~ **material** *n* CONCR Bindemittel *nt*, Kleber *m*

***cement***: ~ **injection** *n* CONCR, INFR & DES Zementverpressung *f*; ~-**lime mortar** *n* CONCR, STONE Kalkzementmörtel *m*; ~ **mill** *n* BUILD MACHIN, INFR & DES Zementfabrik *f*, Zementmühle *f*, Zementwerk *nt*; ~ **mortar** *n* CONCR, MAT PROP, STONE Zementmörtel *m*; ~ **paste** *n* CONCR Zementleim *m*, Zementpaste *f*; ~ **plant** *n* BUILD MACHIN, INFR & DES Zementfabrik *f*, Zementwerk *nt*; ~ **plaster** *n* (*CEM PLAS*) STONE Zementputz *m* (*CPL*); ~-**rich concrete** *n* CONCR zementreicher Beton *m*; ~ **roofing tile** *n* CONCR, STONE Zementdachstein *m*; ~ **skin** *n* STONE Zementschleier *m*; ~ **standard** *n* CONCR, CONST LAW Zementnorm *f*; ~ **strength** *n* CONCR Zementfestigkeit *f*; ~ **substitute** *n* CONCR Zementersatz *m*; ~ **test** *n* CONCR, MAT PROP Zementprobe *f*, Zementprüfung *f*; ~ **tile** *n* CONCR Zementplatte *f*; ~ **trowel finish** *n* ARCH & BUILD Zementglattstrich *m*; ~-**water ratio** *n* CONCR, STONE Zement-Wasser-Verhältnis *nt*

**CEM PLAS** *abbr* (*cement plaster*) STONE CPL (*Zementputz*)

**CEN** *abbr* (*European Committee for Standardisation*) ARCH & BUILD, CONST LAW, INFR & DES CEN (*Europäisches Komitee für Normung*)

**CENELEC** *abbr* (*European Body of Electrical Standards*) CONST LAW, ELECTR CENELEC (*Europäischer Ausschuß für Elektrizitätsnormung*)

**center** *AmE see centre BrE*

**centered** *AmE see centred BrE*

**central** *adj* ARCH & BUILD, BUILD HARDW, HEAT & VENT, INFR & DES zentral, mittig; ~ **arch** *n* ARCH & BUILD Mittelbogen *m*; ~ **axis** *n* ARCH & BUILD Mittelachse *f*, Symmetrieachse *f*; ~ **bay** *n* ARCH & BUILD Mitteljoch *nt*; ~ **column** *n* ARCH & BUILD, CONCR, STEEL, STONE Mittelsäule *f*; ~ **corridor** *n* ARCH & BUILD Mittelgang *m*; ~ **court** *n* ARCH & BUILD Zentralhof *m*; ~ **courtyard** *n* ARCH & BUILD Mittelhof *m*, Atrium *nt*; ~ **fire alarm station** *n* ELECTR, INFR & DES Brandmeldezentrale *f*; ~ **gable** *n* INFR & DES, STONE Mittelgiebel *m*; ~ **heating** *n* HEAT & VENT Zentralheizung *f*; ~ **loading** *n* ARCH & BUILD mittige Belastung *f*; ~ **locking** *n* BUILD HARDW Zentralverschluß *m*

**centrally**: ~ **heated** *adj* HEAT & VENT zentralbeheizt

***central***: ~ **masterkeyed system** *n* BUILD HARDW Zentralschließanlage *f*; ~ **position** *n* ARCH & BUILD Mittellage *f*; ~ **purlin** *n* ARCH & BUILD, TIMBER Mittelpfette *f*; ~ **spine** *n* INFR & DES *planning* zentraler Obelisk *m*; ~ **station** *n* HEAT & VENT Zentrale *f*; ~ **structure** *n* ARCH & BUILD Mittelbau *m*; ~ **vault** *n* ARCH & BUILD, STONE Mittelgewölbe *nt*

**centre** *n BrE* ARCH & BUILD, INFR & DES Mittelpunkt *m*, Zentrum *nt*; ~ **bowl sink** *n BrE* BUILD HARDW Spüle *nt* mit Mittelbecken

**centred** *adj BrE* ARCH & BUILD zentriert

***centre***: ~ **of gravity** *n BrE* ARCH & BUILD, INFR & DES, MAT PROP Schwerpunkt *m*; ~-**hung window** *n BrE* ARCH & BUILD *horizontally or vertically* Drehkippfenster *nt*; ~ **line** *n BrE* ARCH & BUILD, INFR & DES Achslinie *f*, Achse *f*; ~ **purlin** *n BrE* ARCH & BUILD, TIMBER Mittelpfette *f*; ~ **span** *n BrE* ARCH & BUILD Mittelfeld *nt*, *of vaulted building, church* Mittelschiff *nt*, INFR & DES Mittelfeld *nt*; ~-**to-centre distance** *n BrE* ARCH & BUILD Achsmaß *nt*, Achsenabstand *m*

**centrifugal**: ~ **casting** *n* MAT PROP Schleuderguß *m*; ~ **fan** *n* BUILD HARDW, HEAT & VENT Radialventilator *m*; ~ **force** *n* ENVIRON Fliehkraft *f*, Zentrifugalkraft *f*; ~ **governor** *n* BUILD MACHIN, HEAT & VENT Fliehkraftregler *m*; ~ **grease interceptor** *n* WASTE WATER Zentrifugalfettabscheider *m*

**centrifugally**: ~ **cast** *adj* MAT PROP *pipe* geschleudert; ~ **cast concrete pipe** *n* CONCR, INFR & DES Schleuderbetonrohr *nt*

***centrifugal***: ~ **machine for concrete pipes** *n* BUILD MACHIN Schleudermaschine *f* für Betonrohre; ~ **pump** *n* BUILD MACHIN Zentrifugalpumpe *f*, Kreiselpumpe *f*, HEAT & VENT Kreiselpumpe *f*

**centrifugation** *n* BUILD MACHIN Schleudern *nt*, ENVIRON Zentrifugieren *nt*

**centrifuge 1.** *n* BUILD MACHIN Schleuder *f*, ENVIRON Zentrifuge *f*; **2.** *vt* BUILD MACHIN schleudern, ENVIRON zentrifugieren

**centrifuging** *n* BUILD MACHIN Schleudern *nt*, ENVIRON Zentrifugieren *nt*

**centring** *n* STONE Lehrgerüst *nt*

**centroid** *n* ARCH & BUILD, INFR & DES, MAT PROP Schwerpunkt *m*

**ceramic**: ~ **aggregate** *n* MAT PROP keramischer Zuschlagstoff *m*; ~ **facing** *n* MAT PROP Keramikverkleidung *f*; ~ **floor tile** *n* MAT PROP Keramikbodenfliese *f*; ~ **material** *n* MAT PROP Keramikwerkstoff *m*; ~ **pipe** *n* ARCH & BUILD, MAT PROP, WASTE WATER Steinzeugrohr *nt*

**ceramics** *n pl* MAT PROP Keramik *f*

**ceramic**: ~ **shower tray** *n* BUILD HARDW, MAT PROP, WASTE WATER Keramikduschwanne *f*; ~ **tile** *n* MAT PROP Keramikfliese *f*; ~ **wall tile** *n* MAT PROP Keramikwandfliese *f*; ~ **washbasin** *n* BUILD HARDW, MAT PROP, WASTE WATER Keramikwaschbecken *nt*

**ceresit** *n* MAT PROP Ceresit *nt*

**certificate** *n* CONST LAW Zertifikat *nt*, Zeugnis *nt*, Nachweis *m*; ~ **of acceptance** *n* CONST LAW Abnahmeschein *m*; ~ **of apprenticeship** *n* CONST LAW Lehrbrief *m*; ~ **of approval** *n* CONST LAW Zulassungsbescheid *m*

**certification** *n* CONST LAW Bescheinigung *f*

**cessation** *n* ENVIRON Anhalten *nt*

**cesspit** *n* INFR & DES, WASTE WATER Senkgrube *f*

**cesspool** *n* INFR & DES Grube *f*, WASTE WATER Senkgrube *f*; ~ **and sewer cleaning vehicle** *n* BUILD MACHIN, INFR & DES Gruben- und Kanalreinigungsfahrzeug *nt*

**CFC** *abbr* (*chlorofluorocarbon*) ENVIRON FCKW (*Fluorchlorkohlenwasserstoff*)

**chain** *n* ARCH & BUILD *surveying* Meßkette *f*, Kette *f*, BUILD HARDW, BUILD MACHIN Kette *f*, INFR & DES Kette *f*, Meßkette *f*, MAT PROP Kette *f*; ~ **barrier** *n* INFR & DES Kettenabsperrung *f*; ~ **bridge** *n* INFR & DES Kettenbrücke *f*; ~ **dredger** *n* BUILD MACHIN Kratzbagger *m*

**chaining** *n* INFR & DES Entfernungsmessen *nt* mit einer Kette

***chain***: ~ **shackle** *n* BUILD HARDW *lock* Kettenbügel *m*

**chalk** *n* MAT PROP Kreide *f*

**chalking** *n* SURFACE Abfärben *nt*, Auskreiden *nt*

**chalkproof** *adj* MAT PROP auskreidungsbeständig

**chalk**: **~ resistance** *n* SURFACE Abführbeständigkeit *f*
**chamber** *n* ARCH & BUILD Kammer *f*, Raum *m*, Zimmer *nt*, ENVIRON Kammer *f*
**chamfer 1.** *n* ARCH & BUILD Abfasung *f*, Abschrägung *f*, INFR & DES Abschrägung *f*, STONE, TIMBER Abfasung *f*, Abschrägung *f*, Fase *f*; **2.** *vt* ARCH & BUILD abfasen, abschrägen, schrägen, INFR & DES abschrägen, STONE, TIMBER abfasen, abschrägen
**chamfered**: **~ edge** *n* STONE, TIMBER Fase *f*; **~ joint** *n* ARCH & BUILD, TIMBER Schrägfuge *f*
**chamfer**: **~ stop** *n* ARCH & BUILD, STONE, TIMBER Stechfase *f*
**chamotte** *n* MAT PROP Schamotte *f*; **~ lining** *n* HEAT & VENT, MAT PROP Schamotteauskleidung *f*; **~ pipe** *n* MAT PROP Schamotterohr *nt*
**chancel** *n* ARCH & BUILD Altarraum *m*; **~ aisle** *n* ARCH & BUILD, INFR & DES Altarschiff *nt*; **~ neck** *n* ARCH & BUILD Chorhals *m*
**change** *vt* ARCH & BUILD verändern
**change**: **~ in color** *AmE*, **~ in colour** *BrE n* SURFACE Farbtonänderung *f*; **~ in length** *n* ARCH & BUILD Längenänderung *f*; **~ in pressure** *n* ARCH & BUILD, BUILD MACHIN, HEAT & VENT, INFR & DES Druckveränderung *f*; **~ of material** *n* MAT PROP Materialveränderung *f*
**changeover**: **~ switch** *n* ELECTR Umschalter *m*; **~ valve** *n* HEAT & VENT Umschaltventil *nt*
**changes**: **~ in the law** *n* CONST LAW Gesetzesänderungen *f pl*
**change**: **~ of use** *n* CONST LAW Nutzungsänderung *f*
**changing**: **~ facility** *n* ARCH & BUILD, INFR & DES Umkleideeinrichtung *f*; **~ room** *n* ARCH & BUILD, INFR & DES Umkleideraum *m*
**channel 1.** *n* HEAT & VENT Kanal *m*, INFR & DES *large, wide* Kanal *m*, *smaller, narrower* Gerinne *nt*, WASTE WATER Kanal *m*; **2.** *vt* ENVIRON lenken
**channeling** *AmE see channelling BrE*
**channel**: **~ iron** *n* STEEL U-Eisen *nt*
**channelling** *n BrE* ARCH & BUILD Kanalbildung *f*, *grooving* Auskehlung *f*, Kannelieren *nt*, Riffeln *nt*, ENVIRON Kanalbildung *f*, INFR & DES Kanalbildung *f*, Kannelieren *nt*, TIMBER Auskehlung *f*
**Channel**: **~ Tunnel** *n* INFR & DES Kanaltunnel *m*
**chapel** *n* ARCH & BUILD, INFR & DES Kapelle *f*
**chapter** *n* CONST LAW Kapitel *nt*
**characteristic**: **~ data** *n* ARCH & BUILD, ELECTR, HEAT & VENT, INFR & DES, MAT PROP Kenndaten *nt pl*; **~ impedance** *n* SOUND & THERMAL Schallimpedanz *f*
**charge 1.** *n* ARCH & BUILD Last *f*, *loading* Ladung *f*, Belastung *f*, *costs* Kosten *pl*, CONST LAW Kosten *pl*, ELECTR Aufladung *f*, *loading* Ladung *f*, INFR & DES Belastung *f*, Kosten *pl*; **2.** *vt* ARCH & BUILD belasten, ELECTR *battery, tool* aufladen, INFR & DES belasten
**chargeable** *adj* CONST LAW belastbar
**charging**: **~ period** *n* BUILD MACHIN, HEAT & VENT Aufladezeit *f*; **~ screw** *n* BUILD MACHIN Förderschnecke *f*, Zuführungsschnecke *f*, INFR & DES Förderschnecke *f*; **~ silo for mix** *n* BUILD MACHIN Verladesilo *m* für bituminöses Mischgut
**charring** *n* ARCH & BUILD, CONCR, MAT PROP Scharrieren *nt*
**chart** *n* ARCH & BUILD, CONST LAW Diagramm *nt*, Karte *f*, Tafel *f*, INFR & DES Karte *f*
**chase** *n* STONE Schlitz *m*
**check 1.** *n* CONST LAW Prüfung *f*, Überprüfung *f*, Überwachung *f*; **2.** *vt* CONST LAW prüfen, überprüfen, überwachen
**checkerboard**: **~ bond** *AmE see chequerboard bond BrE*
**checkered**: **~ plate** *AmE see chequered plate BrE*
**checkerwork** *AmE see chequerwork BrE*
**checking** *n* CONST LAW Prüfung *f*, Überprüfung *f*, Überwachung *f*
**check**: **~ valve** *n* BUILD HARDW Absperrventil *nt*, BUILD MACHIN Rückschlagventil *nt*, ENVIRON Absperrventil *nt*, Rückschlagventil *nt*, WASTE WATER Rückschlagventil *nt*, Rückstauklappe *f*
**cheek** *n* ARCH & BUILD Seitenfläche *f*, Wange *f*, MAT PROP Wange *f*
**cheese**: **~-head rivet** *n* BUILD HARDW Zylinderkopfniete *f*; **~-head screw** *n* BUILD HARDW Linsenkopfschraube *f*
**chemical**: **~ agent** *n* ENVIRON chemischer Wirkstoff *m*; **~ attack** *n* SURFACE Chemikalienangriff *m*; **~ closet** *n* BUILD HARDW chemische Toilette *f*; **~ composition** *n* MAT PROP chemische Zusammensetzung *f*; **~ damp-proof course** *n* (*chemical dpc*) MAT PROP, STONE chemische Feuchtigkeitssperre *f*; **~ dpc** *n* (*chemical dampproof course*) MAT PROP, STONE chemische Feuchtigkeitssperre *f*; **~ formula** *n* MAT PROP chemische Formel *f*; **~ hardening** *n* SURFACE chemisches Polieren *nt*; **~ oxygen demand** *n* (*COD*) ENVIRON, INFR & DES chemischer Sauerstoffbedarf *m* (*CSB*); **~ precipitation** *n* INFR & DES, MAT PROP chemische Ausfällung *f*; **~ property** *n* MAT PROP chemische Eigenschaft *f*
**chemicals** *n pl* MAT PROP Chemikalien *f pl*
**chemical**: **~ stabilization** *n* INFR & DES chemische Bodenverfestigung *f*; **~ surface treatment** *n* SURFACE chemische Oberflächenbehandlung *f*; **~ toilet** *n* BUILD HARDW chemische Toilette *f*; **~ waste** *n* CONST LAW, INFR & DES chemische Abfälle *m pl*; **~ water analysis** *n* INFR & DES, MAT PROP, WASTE WATER chemische Wasseruntersuchung *f*
**chequerboard**: **~ bond** *n BrE* STONE Schachbrettverband *m*
**chequered**: **~ plate** *n BrE* MAT PROP Grobblech *nt*, Riffelblech *nt*, STEEL Riffelblech *nt*
**chequerwork** *n BrE* STONE Schachbrettmuster *nt*
**chest**: **~ -high** *adj* ARCH & BUILD brusthoch
**Chicago**: **~ window** *n* ARCH & BUILD Chicagoer Fenster *nt*
**chill** *vt* MAT PROP *cast iron* härten
**chilling** *n* MAT PROP *cast iron* Härten *nt*
**chimney** *n* ARCH & BUILD *house, factory* Esse *f*, Kamin *m*, Schlot *m*, Schornstein *m*, INFR & DES *factory* Schlot *m*, Schornstein *m*, STONE Schornstein *m*; **~ base** *n* STONE Schornsteinfuß *m*; **~ block** *n* MAT PROP, STONE Kaminformstein *m*; **~ breast** *n* ARCH & BUILD Kaminvorsprung *m*; **~ brick** *n* MAT PROP Schornsteinziegel *m*; **~ builder** *n* ARCH & BUILD Schornsteinbauer *m*; **~ component** *n* MAT PROP Schornsteinelement *nt*; **~ cross section** *n* ARCH & BUILD Schornsteinquerschnitt *m*; **~ design** *n* ARCH & BUILD Schornsteinbemessung *f*; **~ fan** *n* HEAT & VENT Schornsteinventilator *m*; **~ flashing** *n* ARCH & BUILD Kaminverwahrung *f*; **~ flue** *n* ARCH & BUILD, STONE Schornsteinzug *m*; **~ head** *n* ARCH & BUILD Essenkopf *m*, Kaminhut *m*, Kaminkopf *m*; **~ junction** *n* ARCH & BUILD Schornsteinanschluß *m*; **~ ladder** *n* ARCH & BUILD Schornsteinleiter *f*; **~ lining** *n* STEEL, STONE

Schornsteinauskleidung *f*; ~ **mantle** *n* ARCH & BUILD Schornsteinmantel *m*; ~ **outlet** *n* ARCH & BUILD, STEEL Schornsteinöffnung *f*; ~ **pipe** *n* ARCH & BUILD, STEEL Kaminrohr *nt*, Schornsteinrohr *nt*; ~ **top** *n* ARCH & BUILD, STEEL Schornsteinaufsatz *m*, Schornsteinhaube *f*, Schornsteinkappe *f*

**chipboard** *n* MAT PROP, TIMBER Holzspanplatte *f*, Spanplatte *f*

**chipper** *n* BUILD MACHIN Meißelhammer *m*

**chipping** *n* SURFACE Abspringen *nt*; ~ **concrete** *n* CONCR Splittbeton *m*

**chippings** *n pl* MAT PROP Splitt *m*

**chisel** *vt* STONE, TIMBER ausstemmen

**chlorinate** *vt* ENVIRON chlorieren, INFR & DES, WASTE WATER chloren

**chlorinated**: ~ **lime** *n* MAT PROP Chlorkalk *m*; ~ **rubber** *n* MAT PROP Chlorkautschuk *m*; ~ **rubber coat** *n* SURFACE Chlorkautschukanstrich *m*; ~ **rubber lacquer** *n* SURFACE Chlorkautschuklack *m*; ~ **rubber paint** *n* SURFACE Chlorkautschukfarbe *f*

**chlorination** *n* ENVIRON Chlorierung *f*, INFR & DES, WASTE WATER Chlorung *f*

**chlorine** *n* ENVIRON, MAT PROP, WASTE WATER Chlor *nt*

**chlorofluorocarbon** *n* (*CFC*) ENVIRON Fluorchlorkohlenwasserstoff *m* (*FCKW*)

**chlorosulfonyl**: ~ **isocyanate** *n AmE see chlorosulphonyl isocyanate BrE*

**chlorosulphonyl**: ~ **isocyanate** *n BrE* (*CSI*) ENVIRON Chlorschwefelisocyanat *nt* (*CSI*)

**choir** *n* ARCH & BUILD *church* Chor *m*; ~ **aisle** *n* ARCH & BUILD Chorseitenschiff *nt*; ~ **arch** *n* ARCH & BUILD Chorbogen *m*; ~ **termination** *n* ARCH & BUILD Chorgewölbe *nt*

**choking** *n* WASTE WATER *pipe* Verstopfung *f*

**chord** *n* ARCH & BUILD Gurt *m*

**CHP** *abbr* (*combined heat and power*) HEAT & VENT kombinierte Wärme *f* und Energie *f*

**chromate** *n* SURFACE Chromat *nt*

**chromating**: ~ **process** *n* MAT PROP, STEEL, SURFACE Chromatierverfahren *nt*

**chrome** *n* MAT PROP Chrom *nt*; ~ **coat** *n* MAT PROP, STEEL, SURFACE Chromüberzug *m*; ~ **finish** *n* MAT PROP, STEEL, SURFACE Chromüberzug *m*; ~-**nickel** *n* MAT PROP Chromnickel *nt*; ~-**nickel steel** *n* MAT PROP, STEEL Chrom-Nickel-Stahl *m* (*Cr-Ni-Stahl*); ~ **pigment** *n* MAT PROP, STEEL, SURFACE Chrompigment *nt*; ~-**plated** *adj* MAT PROP, STEEL, SURFACE verchromt; ~ **plating** *n* MAT PROP, STEEL, SURFACE Verchromung *f*; ~ **red** *n* SURFACE Chromrot *nt*; ~ **steel** *n* MAT PROP, STEEL Chromstahl *m*

**chromium** *n* MAT PROP Chrom *nt* ; ~ **coat** *n* MAT PROP, STEEL, SURFACE Chromüberzug *m*; ~ **finish** *n* MAT PROP, STEEL, SURFACE Chromüberzug *m*; ~-**nickel** *n* MAT PROP Chromnickel *nt*; ~-**nickel steel** *n* MAT PROP, STEEL Chrom-Nickel-Stahl *m* (*Cr-Ni-Stahl*); ~-**plated** *adj* MAT PROP, STEEL, SURFACE verchromt; ~ **plating** *n* MAT PROP, STEEL, SURFACE Verchromung *f*; ~ **red** *n* SURFACE Chromrot *nt*; ~ **steel** *n* MAT PROP, STEEL Chromstahl *m*

**chronic**: ~ **effect** *n* ENVIRON chronische Wirkung *f*

**chrysotile** *n* MAT PROP *asbestos* Chrysotil *m*, Weißasbest *m*

**chute** *n* WASTE WATER Gerinne *nt*, Rinne *f*; ~ **spillway** *n* ENVIRON Stromschnelleüberlauf *m*

**CI** *abbr* (*cast iron*) MAT PROP Gußeisen *nt*

**CIB** *abbr* (*International Council for Building Research Studies and Documentation*) CONST LAW CIB (*Internationaler Rat für Forschung, Studium und Dokumentation des Bauwesens*)

**CIE**: ~ **diagram** *n* (*International Commission on Illumination diagram*) ARCH & BUILD, INFR & DES Normfarbtafel *f*

**cill** *n BrE* ARCH & BUILD Fensterbank *f*

**cinder** *n* MAT PROP Schlacke *f*; ~ **aggregate** *n* MAT PROP Schlackenzuschlagstoff *m*; ~ **block** *n* MAT PROP Schlackenbetonstein *m*; ~ **concrete** *n* CONCR, MAT PROP Schlackenbeton *m*

**cinema**: ~ **building** *n* ARCH & BUILD Kinogebäude *nt*

**cinnabar** *n* SURFACE *paint* Zinnober *m*

**CIP** *abbr* (*cast-iron pipe*) MAT PROP, STEEL Gußeisenrohr *nt*, gußeisernes Rohr *nt*

**cipher**: ~ **lock** *n* BUILD HARDW Drucktastenschloß *nt*

**circle** *n* ARCH & BUILD, INFR & DES Kreis *m*

**circuit** *n* ELECTR Stromkreis *m*, HEAT & VENT, WASTE WATER Kreislauf *m*; ~ **breaker** *n* ELECTR Ausschalter *m*; ~ **connection** *n* ELECTR Schaltverbindung *f*; ~ **diagram** *n* ELECTR Stromlaufplan *m*, Schaltplan *m*

**circular**: ~ **arc** *n* ARCH & BUILD Kreisbogen *m*; ~ **beam** *n* ARCH & BUILD, CONCR Ringbalken *m*; ~ **chamber** *n* ARCH & BUILD kreisförmige Kammer *f*; ~ **chimney** *n* ARCH & BUILD, CONCR, STEEL Rundschornstein *m*; ~ **column** *n* ARCH & BUILD Rundsäule *f*; ~ **corridor** *n* ARCH & BUILD Rundgang *m*; ~ **girder** *n* ARCH & BUILD Ringträger *m*; ~ **ground plan** *n* ARCH & BUILD kreisförmiger Grundriß *m*; ~ **hollow steel column** *n* ARCH & BUILD, CONCR, INFR & DES, MAT PROP, STEEL hohle Rundstahlstütze *f*; ~ **load** *n* ARCH & BUILD, INFR & DES Kreislast *f*; ~ **manhole** *n* INFR & DES, WASTE WATER Rundschacht *m*; ~ **manway** *n* INFR & DES, WASTE WATER Rundschacht *m*; ~ **saw** *n* BUILD MACHIN Kreissäge *f*; ~ **vault** *n* ARCH & BUILD Kreisgewölbe *nt*; ~ **walkway** *n* ARCH & BUILD Rundgehweg *m*

**circulate** *vt* HEAT & VENT umwälzen

**circulation** *n* HEAT & VENT Zirkulation *f*, Umwälzung *f*, INFR & DES Kreislauf *m*; ~ **line** *n* HEAT & VENT Umwälzleitung *f*, Zirkulationsleitung *f*; ~ **loss** *n* HEAT & VENT Umwälzverlust *m*; ~ **pump** *n* HEAT & VENT Umwälzpumpe *f*

**circumference** *n* ARCH & BUILD Umfang *m*

**cistern** *n* INFR & DES Zisterne *f*, WASTE WATER Wasserbehälter *m*, Wasserspeicher *m*, Zisterne *f*; ~ **float** *n* WASTE WATER Spülkastenschwimmer *m*

**citadel** *n* ARCH & BUILD Zitadelle *f*

**city**: ~ **hall** *n* ARCH & BUILD, INFR & DES Rathaus *nt*; ~ **supply** *n* ELECTR, HEAT & VENT, WASTE WATER Stadtversorgung *f*

**civil**: ~ **engineer** *n* ARCH & BUILD, INFR & DES Bauingenieur *m*; ~ **engineering** *n* INFR & DES Tiefbautechnik *f*

**CL** *abbr* (*lift coefficient*) ENVIRON CL (*Auftriebsbeiwert, Auftriebszahl*)

**clad** *vt* CONCR verschalen, SURFACE plattieren, TIMBER verschalen

**cladded** *adj* CONCR verschalt, SURFACE plattiert, TIMBER verschalt; ~ **steel plate** *n* MAT PROP plattiertes Stahlblech *nt*

**cladding** *n* CONCR Verschalung *f*, SURFACE *metal* Plattierung *f*, TIMBER Verschalung *f*

***cladding***: ~ **panel** *n* MAT PROP Verblendplatte *f*, Verkleidungsplatte *f*, Verkleidungstafel *f*

**claim** *n* CONST LAW Anspruch *m*; **~ for compensation** *n* CONST LAW Schadenersatzanspruch *m*
**clamp 1.** *n* ARCH & BUILD Zwinge *f*, BUILD MACHIN Bauklammer *f*, Zwinge *f*, ELECTR Schelle *f*, TIMBER Zwinge *f*; **2.** *vt* STONE verankern
**clamping** *n* STONE Verankerung *f*; **~ iron** *n* BUILD MACHIN Bauklammer *f*
***clamp***: **~-ring** *n* ELECTR Bügelschelle *f*
**clapboard** *n* TIMBER Schindel *f*
**clarification** *n* ARCH & BUILD Klarstellung *f*, INFR & DES Reinigung *f*; **~ basin** *n* ENVIRON Klärbecken *nt*, Kreisbecken *nt*, INFR & DES, WASTE WATER Klärbecken *nt*; **~ plant** *n* ENVIRON, INFR & DES Abwasserbehandlungsanlage *f*, Abwasserkläranlage *f*, Kläranlage *f*, Klärwerk *nt*, WASTE WATER Kläranlage *f*; **~ tank** *n* ENVIRON, INFR & DES, WASTE WATER Klärbecken *nt*
**clarify** *vt* ARCH & BUILD klären, INFR & DES reinigen
**classification** *n* MAT PROP Einstufung *f*
**classify** *vt* MAT PROP einstufen
**clause** *n* CONST LAW Klausel *f*
**claw**: **~ bar** *n* BUILD HARDW Brechstange *f*, BUILD MACHIN Kuhfuß *m*; **~ hammer** *n* BUILD HARDW, BUILD MACHIN, TIMBER Zimmermannshammer *m*
**clay** *n* INFR & DES Klei *m*, MAT PROP Lehm *m*, *argillaceous earth* Tonerde *f*, *for pottery* Ton *m*; **~ barrier** *n* ENVIRON Lehmwall *m*; **~ block** *n* STONE Lehmquader *m*; **~ brick building** *n* ARCH & BUILD, STONE Backsteinbau *m*, Ziegelbau *m*
**clayed**: **~ wire mesh** *n* MAT PROP Drahtziegelgewebe *nt*, Ziegelrabitz *m*
**clayey** *adj* ENVIRON, MAT PROP tonartig, tonhaltig, tonig; **~ sandstone** *n* INFR & DES Tonsandstein *m*
***clay***: **~ floor** *n* ARCH & BUILD Lehmestrich *m*, Lehmfußboden *m*; **~ lathing** *n* MAT PROP Drahtziegelgewebe *nt*, Ziegelrabitz *m*; **~ mortar** *n* MAT PROP Tonmörtel *m*; **~ pantile** *n* MAT PROP, STONE Pfannenziegel *m*; **~ pipe** *n* WASTE WATER Tonrohr *nt*; **~ pit** *n* MAT PROP Tongrube *f*; **~ pot** *n* MAT PROP Hourdi *m*, Tonhohlplatte *f*, STONE Tonhohlplatte *f*
**clean 1.** *adj* ENVIRON, TIMBER rein, astfrei; **2.** *vt* MAT PROP, SURFACE reinigen
***clean***: **~ air** *n* HEAT & VENT, INFR & DES Reinluft *f*
**Clean**: **~ Air Act** *n* ENVIRON Gesetz *nt* zur Reinhaltung der Luft
**cleaned** *adj* ENVIRON rein; **~ gas** *n* ENVIRON Abluft *f*
**cleaner** *n* MAT PROP, SURFACE Reiniger *m*, Reinigungsmittel *nt*
**cleaning** *n* MAT PROP, SURFACE Reinigung *f*; **~ agent** *n* MAT PROP, SURFACE Reiniger *m*, Reinigungsmittel *nt*
**cleanout**: **~ cover** *n* WASTE WATER Ausputzdeckel *m*, Putzdeckel *m*, Reinigungsdeckel *m*; **~ fitting** *n* WASTE WATER Reinigungsformstück *nt*; **~ opening** *n* WASTE WATER Reinigungsöffnung *f*, Reinigungsverschluß *m*; **~ piece** *n* WASTE WATER Putzstück *nt*; **~ plug** *n* WASTE WATER Reinigungsstopfen *m*
**clean**: **~ rain** *n* ENVIRON Reinregen *m*; **~ room** *n* ARCH & BUILD, HEAT & VENT Reinraum *m*; **~ technology** *n* ENVIRON abfallarme Technologie *f*, saubere Technologie *f*, umweltfreundliche Technologie *f*
**clear 1.** *adj* ARCH & BUILD klar, licht, MAT PROP, SURFACE klar, farblos; **2.** *vt* INFR & DES *empty* räumen, *expose, uncover, lay bare* freilegen; **~ away** ARCH & BUILD wegräumen; **~ out** ARCH & BUILD *soil* abtragen
**clearance** *n* ARCH & BUILD *headroom* Abbruch *m*, lichte Höhe *f*, lichtes Maß *nt*, INFR & DES lichte Höhe *f*, lichtes Maß *nt*; **~ height and width** *n* ARCH & BUILD Durchfahrtshöhe und -breite *f*, Durchgangshöhe und -breite *f*; **~ order** *n* CONST LAW Abrißerlaß *m*; **~ space** *n* TIMBER *for hardware* Ausnehmung *f*
**clear**: **~ dimension** *n* ARCH & BUILD, INFR & DES Lichtmaß *nt*, lichtes Maß *nt*; **~ glass** *n* ARCH & BUILD, MAT PROP Klarglas *nt*; **~ height** *n* ARCH & BUILD, INFR & DES lichte Höhe *f*
**clearing** *n* ARCH & BUILD *emptying* Räumung *f*; **~ operations** *n pl* ARCH & BUILD Schaffung *f* der Baufreiheit
**clear**: **~ lacquer** *n* SURFACE Klarlack *m*; **~ opening** *n* ARCH & BUILD, INFR & DES lichte Öffnung *f*; **~ opening dimensions** *n pl* ARCH & BUILD, INFR & DES Rohbaulichtmaß *nt*; **~ opening width** *n* ARCH & BUILD, INFR & DES lichte Öffnungsbreite *f*; **~ span** *n* ARCH & BUILD, INFR & DES lichte Weite *f*; **~ varnish** *n* SURFACE Klarlack *m*; **~ width** *n* ARCH & BUILD, INFR & DES lichte Breite *f*
**cleat** *n* ARCH & BUILD, TIMBER Knagge *f*
**cleated**: **~ joint** *n* ARCH & BUILD, TIMBER Knagganschluß *m*, Laschenanschluß *m*
**cleave** *vt* MAT PROP spalten
**cleaver** *n* BUILD MACHIN Spalthammer *m*
**cleaving** *n* MAT PROP Spalten *nt*; **~ saw** *n* BUILD MACHIN Spaltsäge *f*
**clevis** *n* STEEL Gabelkopf *m*
**client** *n* ARCH & BUILD, CONST LAW Auftraggeber *m* (*AG*); **~'s interest** *n* CONST LAW Klienteninteresse *nt*
**climate** *n* ENVIRON, HEAT & VENT Klima *nt*
**clinker** *n* ENVIRON Müllschlacke *f*, Schlacke *f*, MAT PROP Klinker *m*, Klinkerstein *m*, Klinkerziegel *m*, Müllschlacke *f*, STONE Klinker *m*, Klinkerstein *m*, Klinkerziegel *m*; **~ concrete** *n* CONCR Schlackenbeton *m*; **~ slab** *n* MAT PROP, STONE Klinkerplatte *f*
**clinometer** *n* INFR & DES Neigungsmesser *m*
**clip** *n* BUILD HARDW, ELECTR Halter *m*, Schelle *f*
**cloaca** *n* WASTE WATER Senkgrube *f*
**clock**: **~ system** *n* ELECTR Uhrenanlage *f*
**clod** *n* MAT PROP Erdklumpen *m*
**clog** *vti* WASTE WATER verstopfen; **~ up** *vti* WASTE WATER verstopfen
**clogged** *adj* WASTE WATER verstopft; **~ up** *adj* WASTE WATER verstopft
**cloister**: **~ vault** *n* ARCH & BUILD *in church* Walmgewölbe *nt*
**closed**: **with ~ cells** *adj* MAT PROP *plastic* geschlossenzellig; **~ circular pipeline** *n* HEAT & VENT, INFR & DES, WASTE WATER *pipe* Ringleitung *f*; **~ traverse** *n* INFR & DES geschlossener Polygonzug *m*
**closely**: **~ spaced** *adj* ARCH & BUILD engstehend
**closer** *n* STONE Riemchen *nt*
**closet** *n AmE* ARCH & BUILD (*cf cupboard BrE*) Abstellkammer *f*, TIMBER (*cf built-in cupboard BrE*) Einbauschrank *m*, Schrank *m*, Wandschrank *m*
**close**: **~-up** *n* ARCH & BUILD Nahansicht *f*
**closing**: **~ power** *n* ARCH & BUILD *door closer* Schließkraft *f*
**cloth**: **~ backing** *n* MAT PROP Gewebeträgermaterial *nt*; **~ base** *n* MAT PROP Gewebeträgermaterial *nt*; **~ insert** *n* MAT PROP Gewebeeinlage *f*; **~ lathing** *n* MAT PROP *plaster* Gittergewebe *nt*, Rabitzgewebe *nt*, STONE Gittergewebe *nt*
**clothoid** *n* ARCH & BUILD, INFR & DES Klothoide *f*
**cloth**: **~-reinforced** *adj* MAT PROP gewebeverstärkt
**cloudy** *adj* MAT PROP, SURFACE wolkig

**clout** *n* MAT PROP Breitkopfstift *m*; ~ **nail** *n* MAT PROP Breitkopfstift *m*
**clustered**: ~ **column** *n* ARCH & BUILD Säulenbündel *nt*
**clutch** *n* BUILD MACHIN Kupplung *f*
**CO** *abbr* (*carbon monoxide*) ENVIRON CO (*Kohlenmonoxid*)
**coach**: ~ **screw** *n* BUILD HARDW Vierkantkopfschraube *f*
**coagulant** *n* ENVIRON Fällmittel *nt*, INFR & DES, MAT PROP Fällungsmittel *nt*
**coagulation** *n* ENVIRON Ausfällung *f*, INFR & DES, MAT PROP Fällung *f*, Koagulation *f*; ~ **tank** *n* INFR & DES, WASTE WATER Fällungsbecken *nt*
**coal** *n* BUILD HARDW Kohle *f*; ~ **chute** *n* BUILD HARDW Kohlenrutsche *f*; ~-**fired boiler** *n* ENVIRON, HEAT & VENT Kohlenkessel *m*; ~ **manhole cover** *n* BUILD HARDW Kohlenschachtabdeckung *f*
**coarse** *adj* INFR & DES, MAT PROP grob; ~ **aggregate** *n* MAT PROP grobe Zuschlagstoffe *m pl*, grober Zuschlag *m*; ~ **fraction** *n* MAT PROP Grobfraktion *f*, Grobkorn *nt*; ~-**grained** *adj* MAT PROP *soil, sand* grobkörnig; ~ **gravel** *n* MAT PROP grober Kies *m*, Grobkies *m*; ~ **gravel filling** *n* INFR & DES Grobkiesschüttung *f*; ~ **mortar** *n* STONE Grobmörtel *m*; ~-**pitch screw** *n* BUILD HARDW grobgängige Schraube *f*; ~ **plaster** *n* STONE Rohputz *m*; ~ **sand** *n* MAT PROP Grobsand *m*, Schottersand *m*; ~ **screen** *n* ENVIRON Grobrechen *m*; ~ **stuff** *n* CONCR Kalksandmörtel *m*, MAT PROP Kalksandmörtel *m*, Unterputz *m*, STONE Kalksandmörtel *m*
**coastal**: ~ **defence** *BrE*, ~ **defense** *AmE n* INFR & DES Küstenschutz *m*
**coat 1.** *n* ARCH & BUILD, SURFACE Anstrich *m*, Schicht *f*; **2.** *vt* ARCH & BUILD beschichten, MAT PROP, SURFACE beschichten, kaschieren, überziehen
**coated** *adj* ARCH & BUILD, SURFACE beschichtet; ~ **on both sides** *adj* SURFACE beidseitig beschichtet
**coating** *n* MAT PROP *process* Beschichtung *f*, SURFACE *paint* Beschichtung *f*, Anstreichen *nt*, Beschichten *nt*, *protective layer* Umhüllen *nt*, Umhüllung *f*, *with laminates* Kaschieren *nt*; ~ **clay** *n* CONCR Engobeton *m*; ~ **compound** *n* SURFACE Beschichtungsmasse *f*; ~ **defect** *n* MAT PROP Beschichtungsdefekt *m*, SURFACE Anstrichfehler *m*; ~ **resin** *n* SURFACE Beschichtungsharz *nt*; ~ **thickness** *n* SURFACE Schichtdicke *f*
***coat***: ~ **of paint** *n* SURFACE Anstrich *m*, Farbanstrich *m*
**coaxial**: ~ **cable** *n* ELECTR Koaxialkabel *nt*
**cobalt** *n* MAT PROP Kobalt *nt*; ~ **blue** *n* SURFACE *pigment* Kobaltblau *nt*; ~ **green** *n* SURFACE *pigment* Kobaltgrün *nt*; ~ **violet** *n* SURFACE *pigment* Kobaltviolett *nt*
**cobble** *n* MAT PROP *pavement* Kopfstein *m*
**cobblestone** *n* MAT PROP *pavement* Kopfstein *m*
**cobweb**: ~ **pattern** *n* ARCH & BUILD Spinnwebenmuster *nt*
**cock** *n* WASTE WATER Hahn *m*, Ventil *nt*
**cocking**: ~ **piece** *n* ARCH & BUILD Rispe *f*
***cock***: ~ **key** *n* WASTE WATER Ventilschlüssel *m*; ~ **loft** *n* ARCH & BUILD Spitzboden *m*; ~ **with square head** *n* WASTE WATER Vierkantventil *nt*
**COD** *abbr* (*chemical oxygen demand*) ENVIRON, INFR & DES CSB (*chemischer Sauerstoffbedarf*)
**code** *n* CONST LAW Vorschrift *f*
**codeposition** *n* ENVIRON gemischte Ablagerung *f*
***code***: ~ **of practice** *n* CONST LAW Ausführungsbestimmungen *f pl*, Richtlinie *f*
**codification** *n* ARCH & BUILD, CONST LAW, INFR & DES, MAT PROP Kodifizierung *f*
**codify** *vt* ARCH & BUILD, CONST LAW, INFR & DES, MAT PROP kodifizieren
**coding**: ~ **legend** *n* ARCH & BUILD Legende *f*
**codisposal** *n* ENVIRON gemischte Ablagerung *f*; ~ **landfill** *n* ENVIRON Mischdeponie *f*
**coefficient** *n* CONCR Koeffizient *m*, MAT PROP Beiwert *m*, Koeffizient *m*; ~ **of drag** *n* ENVIRON Luftwiderstandsbeiwert *m* (*DD*), Widerstandsbeiwert *m* (*DD*); ~ **of elasticity** *n* CONCR, MAT PROP, STEEL E-Modul *m* (*Elastizitätsmodul*), Elastizitätsmodul *m* (*E-Modul*); ~ **of friction** *n* MAT PROP Reibungsbeiwert *m*; ~ **of linear expansion** *n* MAT PROP linearer Ausdehnungsbeiwert *m*; ~ **of permeability** *n* ENVIRON Durchlässigkeitsbeiwert *m*, Durchlässigkeitskoeffizient *m*, Kf-Wert *m*, MAT PROP Durchlässigkeitskoeffizient *m*; ~ **of plasticity** *n* MAT PROP Bildsamkeit *f*; ~ **of thermal conductivity** *n* MAT PROP, SOUND & THERMAL spezifische Wärmeleitfähigkeit *f*; ~ **of thermal expansion** *n* (*CTE*) MAT PROP Wärmedehnzahl *f*; ~ **of torque** *n* ARCH & BUILD Drehmomentskoeffizient *m*; ~ **of volume expansion** *n* MAT PROP Raumausdehnungskoeffizient *m*; ~ **of viscosity** *n* MAT PROP Viskositätsbeiwert *m*
**coffer 1.** *n* ARCH & BUILD Fach *nt*, Kassette *f*, Kassettenfeld *nt*; **2.** *vt* ARCH & BUILD *ceiling* in Kassetten teilen, kassettierren
***coffer***: ~ **dam** *n* INFR & DES Kastenfangdamm *m*
**coffered** *adj* ARCH & BUILD in Kassetten geteilt, kassettiert; ~ **ceiling** *n* ARCH & BUILD, CONCR, TIMBER Kassettendecke *f*
**cog** *vt* TIMBER verkämmen
**coherence** *n* INFR & DES, MAT PROP *soil* Bindigkeit *f*
**cohesion** *n* MAT PROP Kohäsion *f*
**coiled** *adj* STEEL gerollt
**coir** *n* MAT PROP Kokosfaser *f*; ~ **insulating material** *n* MAT PROP, SOUND & THERMAL Kokosfaserdämmstoff *m*; ~ **mat** *n* MAT PROP, SOUND & THERMAL Kokosfasermatte *f*; ~-**reinforced gypsum plank** *n* MAT PROP, SOUND & THERMAL Kokosfaserdiele *f*
**coke**: ~ **boiler** *n* HEAT & VENT Kokskessel *m*; ~-**fired** *adj* HEAT & VENT koksgefeuert
**cold 1.** *adj* HEAT & VENT kalt; **2.** *n* HEAT & VENT Kälte *f*; **3.** ~-**galvanize** *vt* SURFACE kaltverzinken
***cold***: ~ **air** *n* HEAT & VENT Kaltluft *f*; ~ **air duct** *n* HEAT & VENT Kaltluftkanal *m*; ~ **asphaltic coating** *n* SURFACE Bitumenkaltaufstrich *m*; ~ **bending** *n* STEEL Kaltbiegen *nt*; ~ **coating** *n* SURFACE Kaltanstrich *m*; ~-**curing** *n* SURFACE Kalthärtung *f*; ~-**curing resin** *n* MAT PROP Reaktionsharz *nt*; ~-**galvanizing paint** *n* SURFACE Zinkstaubfarbe *f*; ~ **glue** *n* MAT PROP Kaltleim *m*; ~-**hardening** *adj* MAT PROP kalthärtend; ~-**pourable** *adj* MAT PROP kaltvergießbar; ~-**poured** *adj* MAT PROP kaltvergossen; ~ **requirement** *n* HEAT & VENT Kältebedarf *m*; ~-**rolled** *adj* MAT PROP, STEEL kaltgewalzt; ~-**rolled beam** *n* MAT PROP kaltgewalzter Balken *m*; ~-**rolled steel** *n* MAT PROP kaltgewalzter Stahl *m*; ~ **rolling** *n* MAT PROP Kaltwalzen *nt*; ~ **room** *n* ARCH & BUILD, HEAT & VENT Kühlraum *m*; ~-**setting** *adj* MAT PROP kalthärtend

**cold-setting**: ~ **adhesive** *n* MAT PROP Reaktionsklebstoff *m*
***cold***: ~ **spot** *n* HEAT & VENT, SOUND & THERMAL kalter Punkt *m*; ~**-storage door** *n* ARCH & BUILD Kühlraumtür *f*; ~ **store** *n* ARCH & BUILD, HEAT & VENT Kühlraum *m*; ~ **water** *n* WASTE WATER Kaltwasser *nt*; ~ **water conduit** *n* WASTE WATER Kaltwasserleitung *f*; ~ **water pipe** *n* WASTE WATER Kaltwasserleitung *f*; ~ **water supply** *n* WASTE WATER Kaltwasserversorgung *f*; ~ **water tap** *n* BUILD HARDW, WASTE WATER Kaltwasserhahn *m*; ~ **weather concreting** *n* CONCR Kaltwetterbetonieren *nt*; ~**-workability** *n* MAT PROP Kaltverarbeitbarkeit *f*
**collapse 1.** *n* ARCH & BUILD Einsturz *m*, Zusammenbruch *m*, INFR & DES, MAT PROP Einsturz *m*; **2.** *vi* ARCH & BUILD, INFR & DES, MAT PROP einstürzen
***collapse***: ~ **load** *n* ARCH & BUILD, INFR & DES, MAT PROP Einsturzlast *f*
**collapsible** *adj* ARCH & BUILD zusammenklappbar, MAT PROP demontierbar, zusammenklappbar
**collar** *n* HEAT & VENT, WASTE WATER *pipe* Manschette *f*, Überschiebmuffe *f*; ~ **beam** *n* ARCH & BUILD, TIMBER Kehlbalken *m*; ~ **beam roof** *n* ARCH & BUILD, TIMBER Kehlbalkendach *nt*; ~ **beam truss** *n* ARCH & BUILD, TIMBER Kehlbalkenbinder *m*; ~ **roof** *n* ARCH & BUILD, TIMBER *stiffened against side sway* unverschiebliches Kehlbalkendach *nt*, Kehlbalkendach *nt*, *with side sway* verschiebliches Kehlbalkendach *nt*
**collateral**: ~ **warranty** *n* CONST LAW kollaterale Gewährleistung *f*
**collecting**: ~ **electrode** *n* ENVIRON Niederschlagselektrode *f*; ~ **funnel** *n* HEAT & VENT Auffangtrichter *m*; ~ **hopper** *n* HEAT & VENT Auffangtrichter *m*; ~ **pipeline** *n* HEAT & VENT Sammelrohrleitung *f*; ~ **pit** *n* INFR & DES Sammelgrube *f*; ~ **reservoir** *n* HEAT & VENT, WASTE WATER Auffangschale *f*, Auffangwanne *f*
**collection** *n* ENVIRON *gathering* Einsammeln *nt*, *removal* Abfuhr *f*, Abholen *nt*; ~ **tank** *n* ENVIRON Sammeltank *m*; ~ **of waste oil** *n* ENVIRON Einsammeln *nt* von Altöl; ~ **of waste paper** *n* ENVIRON Altpapiersammlung *f*
**collector** *n* ENVIRON Kollektor *m*, INFR & DES, WASTE WATER Sammelkanal *m*, Sammler *m*; ~ **efficiency** *n* ENVIRON Kollektorleistungsvermögen *nt*
**college** *n* ARCH & BUILD, INFR & DES College *nt*
**collimate** *vt* INFR & DES kollimieren
**collimation**: ~ **line** *n* INFR & DES Peillinie *f*, Sehlinie *f*, Zielachse *f*
**colloidal** *adj* CONCR, MAT PROP, SURFACE unlöslich; ~ **concrete** *n* CONCR, MAT PROP, SURFACE Kolloidbeton *m*; ~ **mortar** *n* CONCR, MAT PROP, SURFACE Kolloidmörtel *m*
**colloid**: ~ **mortar** *n* CONCR, MAT PROP, SURFACE Kolloidmörtel *m*
**colonnade** *n* ARCH & BUILD Arkade *f*, Kolonnade *f*
**color** *AmE see colour BrE*
**colored**: ~ **glass** *AmE see coloured glass BrE*
**colorfast** *AmE see colourfast BrE*
**coloring**: ~ **matter** *AmE see colouring matter BrE*
**colorless** *AmE see colourless BrE*
**colour** *n BrE* SURFACE Farbe *f*; ~ **chart** *n BrE* SURFACE Farbenkarte *f*; ~**-coded** *adj BrE* ELECTR, HEAT & VENT farbig gekennzeichnet; ~ **coding** *n BrE* ELECTR, HEAT & VENT Farbmarkierung *f*
**coloured**: ~ **glass** *n BrE* MAT PROP Buntglas *nt*, farbiges Glas *nt*; ~ **marble** *n BrE* MAT PROP Buntmarmor *m*
**colourfast** *adj BrE* MAT PROP farbecht
**colouring**: ~ **matter** *n BrE* MAT PROP Farbmittel *nt*
**colourless** *adj BrE* MAT PROP, SURFACE farblos
***colour***: ~ **retention** *n BrE* MAT PROP, SURFACE Farbhaltung *f*; ~ **selection** *n BrE* SURFACE Farbwahl *f*; ~ **temperature** *n BrE* SURFACE Farbentemperatur *f*
**column** *n* ARCH & BUILD, CONCR, STEEL, STONE Pfeiler *m*, Stütze *f*, *architectural feature* Säule *f*
**columnar**: ~ **portal** *n* ARCH & BUILD Säulenportal *nt*
***column***: ~ **base** *n* ARCH & BUILD, CONCR, STEEL Stützenfuß *m*; ~ **bearing** *n* ARCH & BUILD Stützenauflage *f*; ~ **capital** *n* ARCH & BUILD Säulenkapitell *nt*; ~ **cross section** *n* ARCH & BUILD, CONCR, STEEL, TIMBER Säulenquerschnitt *m*; ~ **diameter** *n* ARCH & BUILD Säulendurchmesser *m*
**columned**: ~ **hall** *n* ARCH & BUILD Säulenhalle *f*
***column***: ~ **footing** *n* ARCH & BUILD, CONCR Stützenfundament *nt*; ~**-free** *adj* ARCH & BUILD stützenfrei; ~ **grid** *n* ARCH & BUILD Stützenraster *m*; ~ **head** *n* ARCH & BUILD Stützenkopf *m*, Säulenkopf *m*; ~ **lining** *n* ARCH & BUILD, SOUND & THERMAL, STONE Stützenverkleidung *f*; ~ **pedestal** *n* ARCH & BUILD Säulensockel *m*; ~ **radiator** *n* HEAT & VENT Röhrenheizkörper *m*; ~ **shape** *n* ARCH & BUILD Säulenform *f*; ~ **spacing** *n* ARCH & BUILD Stützenabstand *m*
**comb** *vt* STONE kämmen
**combination**: ~ **column** *n* ARCH & BUILD Verbundsäule *f*; ~ **lock** *n* ARCH & BUILD Kombinationsschloß *nt*, BUILD HARDW Kombinationsschloß *nt*, Zahlenkombinationsschloß *nt*; ~ **safety lock** *n* BUILD HARDW Kombinationssicherheitsschloß *nt*
**combined**: ~ **drainage** *n* INFR & DES, WASTE WATER Mischentwässerung *f*; ~ **drainage system** *n* ENVIRON, INFR & DES, WASTE WATER Mischkanalisation *f*, Mischsystem *nt*; ~ **footing** *n* ARCH & BUILD *several columns* Verbundfundament *nt*, gemeinsames Fundament *nt*; ~ **grid operation** *n* CONST LAW, ELECTR Verbundbetrieb *m*; ~ **heating and hot water supply system** *n* HEAT & VENT kombinierte Raumheizung *f* und Warmwasserversorgung *f*; ~ **heat and power** *n* (*CHP*) HEAT & VENT kombinierte Wärme *f* und Energie *f*; ~ **master key system** *n* ARCH & BUILD *lock* kombinierte Hauptschlüsselanlage *f*; ~ **sewage** *n* INFR & DES, WASTE WATER Mischwasser *nt*; ~ **sewer** *n* INFR & DES, WASTE WATER Mischwassersammler *m*; ~ **sewerage system** *n* ENVIRON, INFR & DES, WASTE WATER Mischkanalisation *f*, Mischsystem *nt*; ~ **stresses** *n pl* ARCH & BUILD, TIMBER zusammengesetzte Spannungen *f pl*
**combing** *n* STONE Kämmen *nt*; ~ **agent** *n* CONCR, MAT PROP, STEEL Verbindungsmittel *nt*
**comb**: ~ **rendering** *n* STONE Kammputz *m*
**combustibility** *n* CONST LAW, MAT PROP, SOUND & THERMAL Brennbarkeit *f*
**combustible** *adj* CONST LAW, MAT PROP, SOUND & THERMAL brennbar; ~ **material** *n* ENVIRON brennbares Material *nt*; ~ **waste** *n* ENVIRON brennbarer Abfall *m*
**combustion**: ~ **chamber** *n* HEAT & VENT Brennkammer *f*; ~ **rammer** *n* BUILD MACHIN Explosionsramme *f*
**commencement**: ~ **of setting** *n* CONCR Bindebeginn *m*

**commercial** *adj* MAT PROP handelsüblich; **~ area** *n* CONST LAW, INFR & DES gewerbliche Nutzfläche *f*; **~ building** *n* ARCH & BUILD, CONST LAW, INFR & DES gewerbliches Gebäude *nt*; **~ length** *n* MAT PROP handelsübliche Länge *f*; **~ waste** *n* CONST LAW, ENVIRON gewerblicher Abfall *m*, Industrieabfall *m*
**comminute** *vt* BUILD MACHIN *waste, concrete* zerkleinern
**commode**: **~ step** *n* ARCH & BUILD gebogene Antrittsstufe *f*
**common**: **to be in ~ practice** *phr* ARCH & BUILD allgemein üblich sein; **~ flue** *n* ARCH & BUILD *chimney*, HEAT & VENT Sammelzug *m*; **~ lighting** *n* ELECTR Allgemeinbeleuchtung *f*; **~ parts** *n pl* CONST LAW, INFR & DES gemeinschaftliche Teile *m pl*; **~ rafter** *n* ARCH & BUILD Zwischensparren *m*, TIMBER Bindersparren *m*, Zwischensparren *m*; **~ stairs** *n pl* ARCH & BUILD, CONST LAW allgemeines Fluchttreppenhaus *nt*
**communal** *adj* CONST LAW kommunal; **~ laundry** *n* ARCH & BUILD Gemeinschaftswaschküche *f*
**community** *n* CONST LAW *municipality* Gemeinde *f*, *organized body* Gemeinwesen *nt*; **~ center** *AmE*, **~ centre** *BrE n* INFR & DES Gemeindezentrum *nt*; **~ planning** *n* ENVIRON, INFR & DES Siedlungsplanung *f*
**commuter** *n* INFR & DES Pendler *m*
**compact 1.** *adj* INFR & DES kompakt, **2.** *vt* ENVIRON kompaktieren, verdichten, verfestigen, HEAT & VENT verdichten, INFR & DES kompaktieren, verdichten, verfestigen, MAT PROP verdichten
**compact**: **~ air-conditioning unit** *n* HEAT & VENT Kompaktklimagerät *nt*
**compacted** *adj* ENVIRON kompaktiert, verdichtet, verfestigt, HEAT & VENT verdichtet, INFR & DES kompaktiert, verdichtet, verfestigt, MAT PROP verdichtet
**compactible** *adj* ENVIRON, INFR & DES, MAT PROP *soil, sand* verdichtungsfähig
**compacting**: **~ plant** *n* ENVIRON Verdichtungsanlage *f*
**compaction** *n* ENVIRON, HEAT & VENT Verdichtung *f*, INFR & DES Kompression *f*, Kompaktierung *f*, Verdichtung *f*, *of waste* Müllverdichtung *f*; **~ index** *n* INFR & DES, MAT PROP Verdichtungsfaktor *m*; **~ machine** *n* BUILD MACHIN Verdichtungsgerät *nt*; **~ pile** *n* INFR & DES Verdichtungspfahl *m*; **~ roller** *n* BUILD MACHIN Verdichtungswalze *f*; **~ test** *n* INFR & DES Verdichtungsversuch *m*, Proctorversuch *m*
**compactness**: **~ of the ground** *n* INFR & DES Lagerungsdichte *f*
**compactor** *n* BUILD MACHIN Bodenverfestigungsgerät *nt*, Stampfer *m*, Verdichter *m*; **~ and vibrating compactor for the manufacture of concrete** *n* BUILD MACHIN Stampfmaschine *f* und Rüttelstampfmaschine *f* zur Betonsteinherstellung
**companion**: **~ flange** *n* HEAT & VENT Gegenflansch *m*
**comparative**: **~ cost analysis** *n* CONST LAW Vergleichskostenanalyse *f*
**compartment** *n* ARCH & BUILD *section* Abschnitt *m*, *small room* Kammer *f*, *panel* Fach *nt*, ENVIRON Kammer *f*, MAT PROP *hollow beam* Zelle *f*; **~ wall** *n* ARCH & BUILD, CONST LAW Brandmauer *f*, Brandwand *f*
**compass** *n* ARCH & BUILD, BUILD MACHIN Zirkel *m*, INFR & DES *for direction-finding* Kompaß *m*, *for drawing* Zirkel *m*; **~ brick** *n* STONE Keilziegel *m*, Ringziegel *m*, Radialziegel *m*; **~ header** *n* STONE Radialstrecker *m*; **~ plane** *n* BUILD MACHIN Rundhobel *m*; **~ rafter** *n* TIMBER gekrümmter Sparren *m*; **~ roof** *n* ARCH & BUILD Runddach *nt*, Tonnendach *nt*; **~ saw** *n* BUILD MACHIN Stichsäge *f*
**compatibility** *n* MAT PROP Verträglichkeit *f*
**compatible** *adj* MAT PROP verträglich
**compensation**: **~ joint** *n* ARCH & BUILD Ausgleichsfuge *f*
**complementary**: **~ angle** *n* ARCH & BUILD Ergänzungswinkel *m*, Komplementärwinkel *m*; **~ load** *n* ARCH & BUILD, INFR & DES Zusatzlast *f*
**complete** *vt* ARCH & BUILD fertigstellen
**completed** *adj* ARCH & BUILD fertiggestellt
**complete**: **~ fill** *n* ENVIRON abgeschlossene Deponie *f*, verfüllte Deponie *f*
**completely**: **~ peeled** *adj* TIMBER weißgeschält
**completion** *n* ARCH & BUILD Fertigstellung *f*, CONST LAW *of building* Fertigstellung *f*, *of property sale* Besitzübergabe *f*; **~ date** *n* ARCH & BUILD, CONST LAW, INFR & DES Fertigstellungstermin *m*; **~ work** *n* ARCH & BUILD *interior work* Ausbauarbeiten *f pl*
**complex** *n* ARCH & BUILD, CONST LAW, INFR & DES Komplex *m*
**component** *n* ARCH & BUILD Bauelement *nt*, INFR & DES, MAT PROP Bestandteil *m*
**composite**: **~ action** *n* ARCH & BUILD Verbundwirkung *f*; **~ beam** *n* ARCH & BUILD, CONCR, INFR & DES Verbundträger *m*, MAT PROP Verbundbalken *m*, Verbundträger *m*, STEEL Verbundträger *m*; **~ board** *n* MAT PROP Verbundplatte *f*, SOUND & THERMAL Sandwichplatte *f*; **~ column** *n* ARCH & BUILD Verbundstütze *f*, *sectional steel and concrete* Verbundsäule *f*, CONCR, INFR & DES, MAT PROP, STEEL Verbundstütze *f*; **~ cross section** *n* ARCH & BUILD, CONCR, INFR & DES, MAT PROP Verbundquerschnitt *m*; **~ floor** *n* ARCH & BUILD, CONCR, INFR & DES, MAT PROP, STEEL Verbunddecke *f*; **~ flooring** *n* ARCH & BUILD, CONCR, INFR & DES, MAT PROP, STEEL Verbunddecke *f*; **~ flooring finish** *n* ARCH & BUILD, MAT PROP Bodenverbundfliese *f*; **~ flooring tile** *n* ARCH & BUILD, MAT PROP Bodenverbundfliese *f*; **~ girder** *n* ARCH & BUILD, CONCR, INFR & DES, MAT PROP, STEEL Verbundträger *m*; **~ material** *n* MAT PROP Verbundwerkstoff *m*; **~ method of construction** *n* ARCH & BUILD Verbundbauweise *f*; **~ profile** *n* MAT PROP Verbundprofil *nt*; **~ slab** *n* ARCH & BUILD, CONCR, INFR & DES *composite beam* Stahlbetongurt *m*, MAT PROP Verbundplatte *f*, STEEL *composite beam* Stahlbetongurt *m*; **~ structure** *n* ARCH & BUILD, CONCR, INFR & DES, STEEL Verbundkonstruktion *f*; **~ supporting structure** *n* ARCH & BUILD Verbundtragwerk *nt*; **~ window** *n* ARCH & BUILD, MAT PROP Verbundfenster *nt*
**composition** *n* MAT PROP Aufbau *m*, Zusammensetzung *f*; **~ flooring** *n* ARCH & BUILD fugenloser Fußboden *m*
**compound** *n* INFR & DES Verbindung *f*, MAT PROP Vergußmasse *f*; **~ chimney** *n* ARCH & BUILD Verbundschornstein *m*; **~ flange** *n* ARCH & BUILD, INFR & DES, STEEL zusammengesetzter Flansch *m*; **~ girder** *n* ARCH & BUILD, CONCR, INFR & DES, MAT PROP, STEEL Verbundträger *m*
**compregnated**: **~ wood** *n* TIMBER Schichtpreßholz *nt*
**compress** *vt* ENVIRON, HEAT & VENT verdichten, MAT PROP komprimieren
**compressed**: **~ air** *n* MAT PROP Druckluft *f*; **~ asphalt**

*n* INFR & DES Stampfasphalt *m*; ~ **concrete** *n* CONCR Druckbeton *m*, Stampfbeton *m*; ~ **cork** *n* MAT PROP Preßkork *m*; ~ **gas cylinder** *n* ARCH & BUILD *for welding* Druckgasflasche *f*; ~ **peat** *n* MAT PROP Preßtorf *m*

**compressible** *adj* MAT PROP verdichtungsfähig

**compression** *n* ENVIRON, HEAT & VENT Verdichtung *f*, MAT PROP Kompression *f*; **~-bending stress** *n* ARCH & BUILD, CONCR, INFR & DES, MAT PROP, STEEL, TIMBER Druckbiegespannung *f*; ~ **boom** *n* ARCH & BUILD, STEEL, TIMBER Druckgurt *m*; ~ **cube** *n* MAT PROP Probewürfel *m*; ~ **fitting** *n* HEAT & VENT, WASTE WATER Rohrverschraubung *f*; ~ **flange** *n* ARCH & BUILD, MAT PROP, STEEL, TIMBER Druckgurt *m*; ~ **force** *n* ARCH & BUILD, CONCR, STEEL, STONE, TIMBER Druckkraft *f*; ~ **joint** *n* ARCH & BUILD Preßfuge *f*; ~ **pile** *n* INFR & DES Druckpfahl *m*; ~ **resistance** *n* ARCH & BUILD, CONCR, MAT PROP, STEEL, TIMBER Drucktragfähigkeit *f*, Druckwiderstand *m*; **~-resistant** *adj* CONCR, MAT PROP druckfest; ~ **ring** *n* ARCH & BUILD Druckring *m*; ~ **strength** *n* ARCH & BUILD, CONCR, MAT PROP, STEEL, TIMBER Druckfestigkeit *f*; ~ **test** *n* MAT PROP Kompressionsversuch *m*

**compressive**: ~ **force** *n* ARCH & BUILD, CONCR, STEEL, STONE, TIMBER Druckkraft *f*, Drucklast *f*; ~ **reinforcement** *n* ARCH & BUILD Druckbewehrung *f*

**compressor** *n* BUILD MACHIN, HEAT & VENT Kompressor *m*; ~ **valve** *n* BUILD MACHIN Verdichterventil *nt*

**compulsory**: ~ **mixer** *n* BUILD MACHIN, CONCR Zwangsmischer *m*

**computation** *n* ARCH & BUILD Rechentechnik *f*, Berechnung *f*, INFR & DES Berechnung *f*; ~ **of heat loss** *n* ARCH & BUILD, HEAT & VENT Wärmeverlustberechnung *f*

**computer** *n* ARCH & BUILD, INFR & DES Computer *m*; **~-aided design** *n* (*CAD*) ARCH & BUILD, INFR & DES computerunterstütztes Entwerfen *nt*; ~ **calculation** *n* ARCH & BUILD, INFR & DES Computerberechnung *f*; ~ **model validation** *n* ARCH & BUILD, INFR & DES Computermodellvalidation *f*; ~ **prediction** *n* ARCH & BUILD, INFR & DES Computervoraussage *f*; ~ **simulation** *n* ARCH & BUILD, INFR & DES Computersimulation *f*

**concave**: ~ **molding** *AmE*, ~ **moulding** *BrE n* ARCH & BUILD Hohlkehle *f*

**conceal** *vt* ARCH & BUILD verdecken

**concealed** *adj* ARCH & BUILD verdeckt, ELECTR, HEAT & VENT, WASTE WATER *cables, pipes* unter Putz; ~ **cable** *n* ARCH & BUILD, ELECTR Unterputzkabel *nt*; ~ **conduit** *n* ELECTR Unterputzleerrohr *nt*; ~ **edge band** *n* TIMBER Einleimer *m*; ~ **installation** *n* ELECTR, HEAT & VENT, WASTE WATER *cables, pipes* Unterputzinstallation *f*; ~ **lighting** *n* ELECTR indirekte Beleuchtung *f*; ~ **pipe** *n* ELECTR Unterputzrohr *nt*

**concentrate** *vt* MAT PROP verdichten

**concentrated**: ~ **load** *n* ARCH & BUILD, CONCR, MAT PROP konzentrierte Last *f*

**concentration** *n* (*c*) ENVIRON, INFR & DES, MAT PROP Konzentration *f* (*c*)

**concertina**: ~ **door** *n* ARCH & BUILD, BUILD HARDW Harmonikatür *f*

**conclude**: ~ **a contract** *phr* CONST LAW einen Vertrag abschließen

**conclusion**: ~ **of contract** *n* CONST LAW Vertragsabschluß *m*

**concourse** *n* ARCH & BUILD, INFR & DES Freifläche *f*

**concrete 1.** *n* CONCR Beton *m*; **2.** *vt* CONCR *cover* betonieren, *bond* verbinden

***concrete***: ~ **accelerated curing** *n* CONCR beschleunigtes Betonabbinden *nt*; ~ **admixture** *n* CONCR Betonzusatz *m*; ~ **aggregate** *n* CONCR Betonzuschlagstoff *m*; ~ **arch** *n* CONCR Betonbogen *m*; ~ **area** *n* CONCR Betonfläche *f*; ~ **backing** *n* CONCR, INFR & DES Betonhinterfüllung *f*; ~ **base** *n* CONCR Betonsockel *m*; **~-batching plant** *n* CONCR Betondosieranlage *f*; ~ **binder** *n* CONCR Betonbindemittel *nt*; ~ **block** *n* CONCR Betonblockstein *m*; ~ **breaker** *n* BUILD MACHIN Betonbrecher *m*, CONCR Betonaufbruchhammer *m*; ~ **building** *n* ARCH & BUILD, CONCR Betongebäude *nt*; ~ **cable cover** *n* CONCR Betonkabelabdeckstein *m*; ~ **cantilever** *n* CONCR Betonkragarm *m*; ~ **capping** *n* CONCR Betonmauerdeckung *f*; ~ **carriageway** *n* CONCR, INFR & DES Betonfahrbahn *f*; ~ **centrifugal casting** *n* CONCR Betonschleuderverfahren *nt*; ~ **chimney** *n* CONCR Betonkamin *m*, Betonschornstein *m*; ~ **chord** *n* CONCR Betongurt *m*; ~ **collar** *n* CONCR *surround* Betonkranz *m*; ~ **column** *n* CONCR, INFR & DES Betonstütze *f*, Betonsäule *f*; ~ **component** *n* ARCH & BUILD, CONCR, INFR & DES, MAT PROP Betonfertigteil *nt*; ~ **composition** *n* CONCR Betonzusammensetzung *f*; ~ **consistency** *n* CONCR Betonkonsistenz *f*, MAT PROP Betonsteife *f*; ~ **cover** *n* ARCH & BUILD, CONCR Betondeckung *f*, Betonüberdeckung *f*; ~ **crown** *n* CONCR Betonkranz *m*; ~ **curb** *AmE see concrete kerb BrE*; ~ **disintegration** *n* CONCR Betonkranz *m*; ~ **dowel** *n* BUILD HARDW, CONCR Betondübel *m*; ~ **drain pipe** CONCR, INFR & DES, WASTE WATER Betonentwässerungsrohr *nt*; **durability** *n* CONCR Betonhaltbarkeit *f*; ~ **efflorescence** *n* CONCR Betonausblühung *f*; **~-encased** *adj* ARCH & BUILD, CONCR, INFR & DES, MAT PROP, STEEL *composite column* betonumhüllt, betonummantelt, einbetoniert; ~ **encasement** *n* CONCR Betonmantel *m*, Betonummantelung *f*; ~ **engineering** *n* CONCR Betontechnik *f*; ~ **exposed aggregate finish** *n* CONCR Waschbeton *m*; **factory** *n* CONCR, INFR & DES Betonwerk *nt*; ~ **fast track paving** *n* CONCR Betondeckenherstellung *f* im Schnellverfahren; **~-filled** *adj* CONCR betongefüllt; ~ **finish** *n* CONCR, STEEL Betonoberfläche *f*; ~ **fitting** *n* MAT PROP Betonformstück *m*; ~ **flange** *n* ARCH & BUILD, CONCR Betonflansch *m*, Betongurt *m*; ~ **floor** *n* CONCR Betonfußboden *m*; ~ **floor slab** *n* CONCR Betonfußbodenplatte *f*; ~ **flue** *n* CONCR Betonzug *m*; ~ **foam** *n* CONCR Betonschaum *m*; ~ **formwork** *n* STEEL, TIMBER Betonschalung *f* Schalung *f*; ~ **foundation** *n* ARCH & BUILD Betonfundament *nt*; ~ **girder** *n* ARCH & BUILD, CONCR Betonträger *m*; ~ **goods** *n pl* CONCR Betonware *f*; ~ **hardener** *n* CONCR Betonhärter *m*; ~ **hip tile** *n* CONCR Betongratstein *m*; ~ **hydration** *n* CONCR Betonhydratation *f*; **~-impregnating agent** *n* CONCR Betonimprägniermittel *nt*; ~ **insert** *n* CONCR Betondübel *m*; ~ **joint** *n* ARCH & BUILD, CONCR, INFR & DES Betongelenk *nt*; ~ **kerb** *n BrE* CONCR, INFR & DES Betonbordstein *m*; **~lane finisher** *n* BUILD MACHIN Betonstraßenfertiger *m*; ~ **lintel** *n* CONCR Betonsturz *m*; ~ **manhole** *n* CONCR, INFR & DES Betonschacht *m*; ~ **manufacture** *n* CONCR Betonherstellung *f*; ~ **masonry** *n* CONCR Betonmauerwerk *nt*; ~ **mixer** *n* BUILD MACHIN, CONCR Betonmischer *m*; ~ **mixing plant** *n* CONCR Betonmischanlage *f*; ~ **mold** *AmE*, ~ **mould** *BrE n*

CONCR, TIMBER Betonform *f*; ~ **pantile** *n* CONCR Betondachpfanne *f*; ~ **patching** *n* CONCR, INFR & DES Betonausbesserung *f*, Betonreparatur *f*, Betonsanierung *f*; ~ **pavement** *n* CONCR, INFR & DES *civil engineering* Betondecke *f*; ~ **pile** *n* INFR & DES Betonpfahl *m*; ~ **pipe** *n* CONCR Betonrohr *nt*; ~ **plastic settlement crack** *n* CONCR plastischer Betonsetzungsriß *m*; ~ **pocket** *n* ARCH & BUILD, CONCR Betonnest *nt*; ~ **post** *n* CONCR Betonpfosten *m*; ~ **preservative** CONCR, SURFACE Betonschutzmittel *nt*; ~ **production** *n* CONCR Betonherstellung *f*; ~ **protection** *n* SURFACE Betonschutz *m*; ~ **pump** *n* BUILD MACHIN Betonpumpe *f*; ~ **quality** *n* CONCR Betongüte *f*; ~ **railroad tie** *n AmE* (*cf concrete railway sleeper BrE*) CONCR, INFR & DES Betonschienenschwelle *f*; ~ **railway sleeper** *n BrE* (*cf concrete railroad tie AmE*) CONCR, INFR & DES Betonschienenschwelle *f*; ~ **and reinforced concrete work** *n* CONCR Beton- und Stahlbetonarbeiten *f pl*; ~ **roofing tile** *n* CONCR Betondachstein *m*; ~ **sample** CONCR Betonprobewürfel *m*; ~ **sewer** *n* INFR & DES, WASTE WATER Betonabwasserleitung *f*; ~ **shell** *n* ARCH & BUILD, CONCR, INFR & DES Betonschale *f*; ~ **site sampling** *n* CONCR örtliche Betonprobenentnahme *f*; ~ **skeleton construction** *n* ARCH & BUILD, CONCR, INFR & DES Betonskelettkonstruktion *f*; ~ **slab** *n* ARCH & BUILD, CONCR *composite beam* Betonflansch *m*, Betongurt *m*, Betonplatte *f*, INFR & DES Betondecke *f*; ~ **specification** *n* CONCR Betonvorschrift *f*; ~ **strength** *n* CONCR, MAT PROP Betonfestigkeit *f*; ~ **stress** *n* CONCR, MAT PROP Betonspannung *f*; ~ **supporting framework** *n* ARCH & BUILD, CONCR, INFR & DES Betontragwerk *nt*; ~ **table** *n* CONCR Betontabelle *f*; ~ **temperature** *n* CONCR Betontemperatur *f*; ~ **test** *n* CONCR, MAT PROP Betonprüfung *f*; ~ **thickness** *n* CONCR Betondicke *f*; ~ **tilt-up fast factory construction** *n* CONCR schnelle Beton-Tilt-up-Fabrikherstellung *f*; ~ **topping** *n* CONCR Aufbeton *m*, Betondruckplatte *f*; ~ **wall** *n* CONCR Betonwand *f*; ~ **wearing surface** *n* SURFACE Betonverschleißoberfläche *f*; ~ **whitetop overlay** *n* CONCR weißer Betonbelag *m*; ~ **work** *n* CONCR Betonarbeiten *f pl*; ~ **workability** *n* CONCR Betonverarbeitbarkeit *f*

**concreting** *n* CONCR Betoneinbringung *f*, Betonierung *f*, Betonieren *nt*; ~ **plant** *n* INFR & DES Betonieranlage *f*

**condemn** *vt* ARCH & BUILD abrißreif erklären, CONST LAW abbruchreif erklären

**condemned** *adj* ARCH & BUILD abrißreif, CONST LAW abbruchreif

**condensate** *n* ENVIRON, HEAT & VENT, SOUND & THERMAL Kondensat *nt*, Kondenswasser *nt*, Schwitzwasser *nt*, Tauwasser *nt*; ~ **piping** *n* HEAT & VENT Kondensatleitung *f*

**condensation** *n* ENVIRON, HEAT & VENT, SOUND & THERMAL Kondensation *f*; ~ **control** *n* HEAT & VENT Kondenswasserregelung *f*; ~ **gutter** *n* HEAT & VENT Kondenswasserrinne *f*; ~ **nucleus** *n* ENVIRON Kondensationskern *m*; ~ **nucleus counter** *n* ENVIRON Kondensationskernzähler *m*; ~ **water** *n* HEAT & VENT, SOUND & THERMAL Kondensat *nt*, Kondenswasser *nt*, Schwitzwasser *nt*, Tauwasser *nt*

**condense** *vti* ENVIRON, HEAT & VENT, SOUND & THERMAL kondensieren

**condenser** *n* ELECTR Kondensator *m*, HEAT & VENT Verdichter *m*, Kondensator *m*

**condition 1.** *n* CONST LAW Auflage *f*; **2.** *vt* ENVIRON *waste* aufbereiten, HEAT & VENT klimatisieren

**conditioned** *adj* ENVIRON *waste* aufbereitet, HEAT & VENT klimatisiert; ~ **air** *n* HEAT & VENT klimatisierte Luft *f*

**conditioner** *n* HEAT & VENT Klimatgerät *nt*

**conditioning**: ~ **duct** *n* ARCH & BUILD Klimakanal *m*, BUILD MACHIN, HEAT & VENT Klimakanal *m*, Klimatechnik *f*; ~ **unit** *n* HEAT & VENT Klimagerät *nt*; ~ **of waste** *n* CONST LAW, ENVIRON Müllaufbereitung *f*

**conditions**: ~ **of contract** *n pl* CONST LAW Vertragsbedingungen *f pl*; ~ **of engagement** *n pl* CONST LAW Verpflichtungsbedingungen *f pl*

**conduct** *vt* ELECTR, MAT PROP leiten

**conductive** *adj* ELECTR, MAT PROP leitfähig; ~ **flooring** *n* ELECTR, MAT PROP ableitfähiger Fußbodenbelag *m*; ~ **material** *n* ELECTR, MAT PROP leitfähiges Material *nt*

**conductivity** *n* ELECTR, MAT PROP Leitfähigkeit *f*

**conductor** *n* ELECTR Leiter *m*; ~ **clearance** *n* ELECTR *lightning protection* Abstand *m* der Ableitung ; ~ **loop** *n* ELECTR *lightning protection* Auffangring *m*

**conduit** *n* ELECTR Installationsrohr *nt*, Leerrohr *nt*, *channel* Kanal *m*, ENVIRON *geology* Vulkanschlot *m*, *pipe* Rohrleitung *f*, *for insulation* Isolierrohr *nt*, HEAT & VENT Leitung *f*, Rohr *nt*, INFR & DES Vulkanschlot *m*, MAT PROP Rohr *nt*, Leitung *f*, WASTE WATER Rohrleitung *f*, *pipe* Rohr *nt*, Leitung *f*, *protective tube* Schutzrohr *nt*; ~ **pipe** *n* ENVIRON, WASTE WATER Leitungsrohr *nt*; ~ **tile** *n* ELECTR Kabelformstein *m*, Kabelkanalstein *m*

**cone** *n* ARCH & BUILD Kegel *m*; ~ **head rivet** *n* BUILD HARDW Kegelkopfniet *m*; ~ **penetration test** *n* MAT PROP Kegeleindringungsversuch *m*; ~ **vault** *n* ARCH & BUILD Kegelgewölbe *nt*

**conference** *n* ARCH & BUILD *site conferences* Besprechung *f*; ~ **room** *n* ARCH & BUILD Konferenzzimmer *nt*, Sitzungsraum *m*, INFR & DES Besprechungszimmer *nt*

**confined**: ~ **ground water** *n* INFR & DES gespanntes Grundwasser *nt*

**conglomerate** *n* INFR & DES Konglomerat *nt*

**conical**: ~ **roof** *n* ARCH & BUILD Kegeldach *nt*

**conic**: ~ **section** *n* ARCH & BUILD Kegelschnitt *m*

**coniferous**: ~ **wood** *n* MAT PROP, TIMBER Nadelholz *nt*

**connect** *vt* ARCH & BUILD, ELECTR, WASTE WATER verbinden

**connected**: ~ **load** *n* ELECTR Anschlußwert *m*

**connecting**: ~ **bar** *n* MAT PROP, STEEL Verbindungsstange *f*; ~ **bolt** *n* BUILD HARDW, MAT PROP, STEEL Verbindungsschraube *f*; ~ **building** *n* ARCH & BUILD Verbindungsbau *m*; ~ **corridor** *n* ELECTR Anschlußleitung *f*; ~ **device** *n* CONCR, MAT PROP, STEEL, TIMBER Verbindungsmittel *nt*; ~ **line** *n* WASTE WATER Anschlußleitung *f*, Verbindungsleitung *f*; ~ **passage** *n* ARCH & BUILD Durchgang *m*, Verbindungsgang *m*; ~ **piece** *n* HEAT & VENT Verbindungsstutzen *m*; ~ **screw** *n* BUILD HARDW, MAT PROP, STEEL Verbindungsschraube *f*; ~ **structure** *n* ARCH & BUILD Verbindungsbau *m*

**connection** *n* ARCH & BUILD Verbinden *nt*, ELECTR Anschluß *m*, Verbindung *f*, WASTE WATER Verbindung *f*; ~ **bolt** *n* BUILD HARDW Verbindungsbolzen *m*; ~ **box** *n* ELECTR Klemmdose *f*; ~ **dimension** *n*

ARCH & BUILD Anschlußmaß *nt*; ~ **joint** *n* ARCH & BUILD Anschlußfuge *f*; ~ **line** *n* HEAT & VENT, WASTE WATER Anschlußleitung *f*, Verbindungsleitung *f*; ~ **lug** *n* ELECTR Anschlußfahne *f*; ~ **piece** *n* HEAT & VENT, WASTE WATER Stutzen *m*, Verbindungsstück *nt*
**connector** *n* ELECTR, HEAT & VENT Anschlußklemme *f pl*, MAT PROP Dübel *m*; ~ **axis** *n* ARCH & BUILD, TIMBER Achse *f* der Verbindungsmittel; ~ **unit** *n* INFR & DES, STEEL, TIMBER Verbindungseinheit *f*
**conservation** *n* ARCH & BUILD, CONST LAW, ENVIRON, INFR & DES Umweltschutz *m*; ~ **order** *n* CONST LAW, ENVIRON Umwelterhaltungsbescheid *m*
**conservatory** *n* ARCH & BUILD, HEAT & VENT *as thermal buffer* Wintergarten *nt*; ~ **as thermal buffer** *n* ENVIRON, SOUND & THERMAL Konservatorium *nt* als thermischer Puffer
**consistency** *n* CONCR, INFR & DES Konsistenz *f*, MAT PROP Konsistenz *f*, Zähigkeit *f*
**console** *n* CONCR, MAT PROP, STEEL Konsole *f*, STONE Kragstein *m*, TIMBER Konsole *f*
**consolidate** *vt* INFR & DES *soil* verfestigen
**consolidation** *n* INFR & DES Bodenverdichtung *f*, Bodenverfestigung *f*, Konsolidation *f*
**constant**: ~**-volume sampling** *n* ENVIRON Probeentnahmen *f pl* gleicher Menge
**constituent** *n* INFR & DES, MAT PROP Bestandteil *m*
**construct** *vt* ARCH & BUILD errichten, konstruieren, INFR & DES errichten
**constructing** *n* CONST LAW Erbau *m*
**construction** *n* ARCH & BUILD *constructing* Bauausführung *f*, *edifice* Bauwerk *nt*, Konstruktion *f*, Bau *m*, INFR & DES Bau *m*, Bauwerk *nt*
**constructional**: ~ **defect** *n* ARCH & BUILD Baumangel *m*; ~ **drawing** *n* ARCH & BUILD Bauzeichnung *f*; ~ **element** *n* MAT PROP Bauteil *nt*; ~ **standard** *n* ARCH & BUILD, CONST LAW, INFR & DES Baunorm *f*; ~ **steel** *n* ARCH & BUILD, INFR & DES, MAT PROP, STEEL Baustahl *m*; ~ **timber** *n BrE* (*cf construction lumber AmE*) TIMBER Bauholz *nt*; ~ **tolerance** *n* ARCH & BUILD *permissible deviation of constructional elements* Bautoleranz *f*; ~ **unit** *n* MAT PROP Bauelement *nt*, Montageelement *nt*
***construction***: ~ **authority** *n* CONST LAW Baubehörde *f*; ~ **chemistry** *n* MAT PROP Bauchemie *f*; ~ **costs** *n pl* ARCH & BUILD, CONST LAW, INFR & DES Baukosten *f pl*; ~ **drawing** *n* ARCH & BUILD Bauzeichnung *f*; ~ **engineer** *n* ARCH & BUILD, INFR & DES Bauingenieur *m*; ~ **engineering** *n* ARCH & BUILD, INFR & DES Bautechnik *f*; ~ **equipment** *n* BUILD MACHIN Baumaschinen *f pl*; ~ **expert** *n* ARCH & BUILD Baufachmann *m*; ~ **firm** *n* CONST LAW Baufirma *f*, Bauunternehmung *f*; ~ **glass** *n* MAT PROP Bauglas *nt*; ~ **gravel** *n* MAT PROP Baukies *m*; ~ **ground** *n* ARCH & BUILD Baugelände *nt*, CONST LAW Baugelände *nt*, Bauland *nt*, INFR & DES Baugelände *nt*; ~ **industry** *n* ARCH & BUILD, CONST LAW, INFR & DES Bauindustrie *f*, Bausektor *m*, Bauwirtschaft *f*; ~ **joint** *n* ARCH & BUILD, CONCR Arbeitsfuge *f*, Betonierfuge *f*, Dehnungsfuge *f*, INFR & DES Dehnungsfuge *f*; ~ **lumber** *n AmE* (*cf constructional timber BrE*) TIMBER Bauholz *nt*; ~ **material** *n* ARCH & BUILD, MAT PROP Baumaterial *nt*, Baustoff *m*; ~ **method** *n* ARCH & BUILD, CONST LAW, INFR & DES Bauweise *f*; ~ **noise** *n* CONST LAW Baulärm *m*; ~ **office** *n* ARCH & BUILD, INFR & DES Baubüro *nt*; ~ **of parks and cemeteries** *n* INFR & DES Anlagen- und Friedhofsbau *m*; ~ **plan** *n* ARCH & BUILD Bauentwurf *m*, Bauplan *m*, INFR & DES Bauentwurf *m*; ~ **price** *n* ARCH & BUILD Baupreis *m*; ~ **price index** *n* ARCH & BUILD, CONST LAW, INFR & DES Baupreisindex *m*; ~ **regulations** *n pl* CONST LAW, INFR & DES Baubestimmungen *f pl*, Bauvorschriften *f pl*; ~ **sand** *n* MAT PROP Bausand *m*; ~ **schedule** *n* ARCH & BUILD, CONST LAW, INFR & DES Terminplan *m*; ~ **sheeting** *n* MAT PROP Baufolie *f*; ~ **sign** *n* CONST LAW Bauschild *nt*; ~ **standard** *n* ARCH & BUILD, CONST LAW, INFR & DES Baunorm *f*; ~ **steel** *n* ARCH & BUILD, INFR & DES, MAT PROP, STEEL Baustahl *m*; ~ **supervision** *n* ARCH & BUILD, CONST LAW Bauaufsicht *f*, Bauüberwachung *f*, Bauleitung *f*; ~ **system** *n* ARCH & BUILD Bausystem *nt*; ~ **time** *n* ARCH & BUILD Bauzeit *f*; ~ **waste** *n* ARCH & BUILD Bauschutt *m*, ENVIRON Bauschutt *m*, Baustellenabfall *m*, INFR & DES, MAT PROP Bauschutt *m*; ~ **of windows** *n* ARCH & BUILD Fensterbau *m*; ~ **work** *n* ARCH & BUILD Bauarbeit *f*, Bauarbeiten *f pl*, Baumaßnahme *f*, CONST LAW, INFR & DES Bauarbeiten *f pl*
**consult** *vt* CONST LAW, INFR & DES *documents* einsehen, *person* sich besprechen mit, konsultieren
**consultation** *n* CONST LAW, INFR & DES Beratung *f*, Konsultation *f*
**consulting**: ~ **engineer** *n* ARCH & BUILD, INFR & DES beratender Ingenieur *m*; ~ **engineers** *n pl* ARCH & BUILD, INFR & DES Ingenieurbüro *nt*
**consumer**: ~ **waste** *n* ENVIRON, INFR & DES Hausmüll *m*
**consumption** *n* HEAT & VENT, INFR & DES, MAT PROP Verbrauch *m*; ~ **curve** *n* HEAT & VENT, INFR & DES, MAT PROP Verbrauchskurve *f*
**contact** *n* ARCH & BUILD Berührung *f*, Kontakt *m*, BUILD HARDW, INFR & DES, MAT PROP Berührung *f*; ~ **adhesive** *n* ARCH & BUILD, MAT PROP Kontaktkleber *m*; ~ **area** *n* ARCH & BUILD Berührungsfläche *f*, Kontaktfläche *f*; ~ **breaker** *n* ELECTR Kontaktunterbrecher *m*; ~ **date** *n* CONST LAW Kontakttermin *m*
**contactor** *n* ELECTR Schütz *nt*
**container** *n* ARCH & BUILD Behälter *m*, INFR & DES Reservoir *nt*, Sammler *m*
**containing**: ~ **asbestos fibers** *AmE*, ~ **asbestos fibres** *BrE adj* MAT PROP asbestfaserhaltig
**contaminant** *n* ENVIRON, INFR & DES *impurity* Verunreinigungssubstanz *f*, *of water* Schmutzstoff *m*, *noxious substance* Schadstoff *m*, MAT PROP *noxious substance* Schadstoff *m*, *impurity* Verunreinigungssubstanz *f*, *of water* Schmutzstoff *m*
**contaminate** *vt* ENVIRON, INFR & DES, MAT PROP verunreinigen
**contaminated**: ~ **land** *n* CONST LAW, ENVIRON, INFR & DES verseuchtes Land *nt*; ~ **site** *n* ENVIRON kontaminierter Standort *m*; ~ **soil** *n* CONST LAW, ENVIRON, INFR & DES verseuchter Boden *m*; ~ **water** *n* ENVIRON radioaktiv kontaminiertes Wasser *nt*
**contamination** *n* ENVIRON Verunreinigung *f*, Verseuchung *f*; ~ **control** *n* CONST LAW, ENVIRON, INFR & DES Kontaminationsüberwachung *f*; ~ **fallout** *n* ENVIRON Fremdstoffniederschlag *m*; ~ **hazard** *n* CONST LAW, ENVIRON, INFR & DES Verseuchungsgefahr *f*
**content** *n* ARCH & BUILD Rauminhalt *m*
**contingency**: ~ **cost** *n* ARCH & BUILD, CONST LAW, INFR & DES Kosten *pl* für unvorhergesehene Arbeiten; ~ **item** *n* ARCH & BUILD Eventualposition *f*
**continuous** *adj* ARCH & BUILD durchgehend,

durchlaufend, ununterbrochen; ~ **beam** *n* ARCH & BUILD Durchlaufträger *m*, Mehrfeldträger *m*; ~ **floor** *n* ARCH & BUILD durchgehende Decke *f*; ~ **footing** *n* ARCH & BUILD Bankett *nt*, Streifenfundament *nt*, CONCR Streifenfundament *nt*, INFR & DES Bankett *nt*; ~ **foundation** *n* ARCH & BUILD Bankett *nt*, Streifenfundament *nt*, CONCR Streifenfundament *nt*, INFR & DES Bankett *nt*; ~ **frame** *n* ARCH & BUILD Mehrfeldrahmen *m*, Durchlaufrahmen *m*; ~ **galvanizing** *n* SURFACE Bandverzinkung *f*; ~ **girder** *n* ARCH & BUILD durchgehender Träger *m*, durchlaufender Unterzug *m*; ~ **glass surface** *n* MAT PROP durchgehend verglaste Fläche *f*; ~ **handrail** *n* ARCH & BUILD, BUILD HARDW durchgehender Handlauf *m*; ~ **hinge** *n* BUILD HARDW Klavierband *nt*; ~ **load** *n* ARCH & BUILD, INFR & DES, MAT PROP Dauerlast *f*

**continuously**: ~ **galvanized** *adj* SURFACE bandverzinkt

**continuous**: ~ **mixer** *n* BUILD MACHIN Durchlaufmischer *m*, Stetigmischer *m*; ~ **prestressed concrete girder** *n* ARCH & BUILD, CONCR, INFR & DES durchlaufender Spannbetonträger *m*; ~ **prestressed girder** *n* ARCH & BUILD, CONCR, INFR & DES durchlaufender Spannbetonträger *m*; ~ **purlin** *n* TIMBER Durchlaufpfette *f*; ~ **reinforcement** *n* CONCR durchlaufende Bewehrung *f*; ~ **rod reinforcement** *n* CONCR Durchlaufstabarmierung *f*; ~ **slab** *n* ARCH & BUILD, CONCR Durchlaufplatte *f*, Mehrfeldplatte *f*; ~ **truss** *n* ARCH & BUILD durchgehendes Fachwerk *nt*

**contour** *n* ARCH & BUILD *outline* Umriß *m*, *surveying* Höhenlinie *f*; ~ **line** *n* ARCH & BUILD, INFR & DES Höhenschichtlinie *f*; ~ **map** *n* ARCH & BUILD Höhenplan *m*, Schichtenkarte *f*, INFR & DES Höhenplan *m*

**contract 1.** *n* CONST LAW Vertrag *m*; **2.** *vt* CONST LAW *enter into an agreement* schließen; **3.** *vi* MAT PROP *constrict* sich zusammenziehen, *shrink* schrumpfen

***contract***: ~ **administration** *n* CONST LAW Vertragsverwaltung *f*; ~ **deadline** *n* CONST LAW Vertragsfristen *f pl*; ~ **drawing** *n* CONST LAW Vertragszeichnung *f*

**contracting**: ~ **regulations** *n pl* CONST LAW Verdingungsordnung *f*; ~ **regulations for award of public work contracts** *n* CONST LAW Verdingungsordnung *f* für Bauleistungen

**contraction** *n* ARCH & BUILD, CONCR Schwinden *nt*, MAT PROP Kontraktion *f*, Schwinden *nt*, *constriction* Zusammenziehen *nt*, *shrinkage* Schrumpfung *f*; ~ **coefficient** *n* MAT PROP Zusammenziehungskoeffizient *m*; ~ **crack** *n* ARCH & BUILD, CONCR, ELECTR, INFR & DES Schrumpfriß *m*; ~ **joint** *n* ARCH & BUILD Kontraktionsfuge *f*, Schwindfuge *f*, CONCR, INFR & DES Schwindfuge *f*

**contractor** *n* ARCH & BUILD Bauunternehmer *m*, CONST LAW Auftragnehmer *m*, Bauunternehmer *m*; ~**'s estimate** *n* ARCH & BUILD Baukostenvoranschlag *m*; ~**'s site office** *n* ARCH & BUILD Baubüro *nt*, Bauleitungsbüro *nt* des Auftragnehmers, INFR & DES Baubüro *nt*

***contract***: ~ **termination** *n* CONST LAW Vertragsauflösung *f*

**contractual**: ~ **penalty** *n* CONST LAW Vertragsstrafe *f*; ~ **time penalty** *n* CONST LAW Vertragsstrafe *f* bei Terminüberschreitung

***contract***: ~ **value** *n* CONST LAW Vertragswert *m*

**control** *n* ARCH & BUILD, CONST LAW Kontrolle *f*, Überwachung *f*, ELECTR Steuerung *f*, HEAT & VENT Steuer-, Steuerung *f*; ~ **instrument** *n* ELECTR, HEAT & VENT Regelgerät *nt*

**controlled**: ~ **dumping** *n* ENVIRON geordnete Deponie *f*, kontrollierte Müllablagerung *f*, kontrolliertes Abladen *nt* von Schutt, überwachtes Abladen *nt* von Schutt

**controller** *n* ELECTR, HEAT & VENT Regeleinrichtung *f*, Regler *m*

**controlling**: ~ **dimension** *n* ARCH & BUILD Baurichtmaß *nt*; ~ **equipment** *n* ELECTR, HEAT & VENT Regeleinrichtung *f*, Regelung *f*

***control***: ~ **room** *n* ELECTR Leitzentrale *f*, Schaltwarte *f*; ~ **slide valve** *n* HEAT & VENT Steuerschieber *m*; ~ **system** *n* ELECTR, HEAT & VENT Regeleinrichtung *f*; ~ **valve** *n* HEAT & VENT Regulierventil *nt*, Steuerventil *nt*

**convection**: ~ **current** *n* HEAT & VENT Wärmeströmung *f*; ~ **heating** *n* HEAT & VENT Konvektionsheizung *f*

**convector** *n* HEAT & VENT Konvektor *m*; ~ **heating element** *n* HEAT & VENT Konvektionsheizungsgerät *nt*, Konvektorheizkörper *m*

**convenient** *adj* ARCH & BUILD, CONST LAW *easy* bequem, *suitable* passend, geeignet, angemessen

**conventional** *adj* ARCH & BUILD herkömmlich; ~ **construction method** *n* ARCH & BUILD, INFR & DES herkömmliche Bauweise *f*

**conversion** *n* ARCH & BUILD Umbau *m*, bauliche Veränderung *f*; ~ **factor** *n* ARCH & BUILD Umrechnungsfaktor *m*

**convert** *vt* ARCH & BUILD *building* umbauen, verändern, *measurement* umrechnen

**convex**: ~ **tile** *n* MAT PROP, STONE Mönch *m*, Preiße *f*

**convey** *vt* ARCH & BUILD, BUILD MACHIN befördern, MAT PROP befördern, transportieren

**conveyance** *n* CONST LAW Übertragung *f*

**conveyancing** *n* CONST LAW Eigentumsübertragung *f*, Übertragung *f*

**conveying** *n* ARCH & BUILD, BUILD MACHIN, MAT PROP Beförderung *f*, Transport *m*; ~ **belt** *n* BUILD MACHIN Förderband *nt*; ~ **equipment for heavy clay industry** *n* BUILD MACHIN Förderanlage *f* für die grobkeramische Industrie

**conveyor** *n* BUILD MACHIN Förderer *m*; ~ **belt skimmer** *n* ENVIRON Förderbandskimmer *m*

**cool**: ~ **down** *vti* HEAT & VENT abkühlen

**coolant** *n* ENVIRON, HEAT & VENT, MAT PROP Kühlmittel *nt*

**cooler** *n* HEAT & VENT Kühler *m*

**cooling 1.** *adj* HEAT & VENT kühlend; **2.** *n* HEAT & VENT Kühlung *f*

***cooling***: ~ **air** *n* HEAT & VENT Kühlluft *f*; ~ **load** *n* HEAT & VENT Kühllast *f*, Kühlwasserkreislauf *m*; ~ **water** *n* HEAT & VENT Kühlwasser *nt*

**coordination** *n* ARCH & BUILD Abstimmung *f*

**copal** *n* SURFACE Kopal *m*; ~ **resin** *n* SURFACE Kopalharz *m*

**cope 1.** *n* ARCH & BUILD, STONE Abwässerung *f*, Krone *f*; **2.** *vt* ARCH & BUILD abdecken, STONE abdecken, aufmauern, bedecken

**coping** *n* ARCH & BUILD, STONE *on top of masonry walls* Abdecken *nt*, Aufmauern *nt*; ~ **stone** *n* ARCH & BUILD, STONE Abdeckstein *m*, Deckstein *m*

**copolymer** *n* MAT PROP Copolymer *nt*, Mischpolymerisat *nt*

**copper 1.** *n* MAT PROP, STEEL Kupfer *nt*; **2.** *vt* STEEL, SURFACE verkupfern

***copper***: ~ **alloy** *n* MAT PROP Kupferlegierung *f*; ~ **bar** *n* ELECTR, MAT PROP Kupferschiene *f*; ~ **bend** *n* HEAT & VENT Kupferkrümmer *m*; ~ **bit** *n* BUILD MACHIN, ELECTR Lötkolben *m*; ~ **bus bar** *n* ELECTR Kupfersammelschiene *f*; ~ **capillary fitting** *n* ELECTR, HEAT & VENT, MAT PROP Kupferkapillarfitting *nt*; ~ **cladding** *n* ELECTR Kupfermantel *m*; ~**-clad earth rod** *n BrE* (*cf copper-clad ground rod AmE*) ELECTR *lightning protection* kupferummantelter Staberder *m*, kupferüberzogener Staberder *m*; ~**-clad ground rod** *n AmE* (*cf copper-clad earth rod BrE*) ELECTR *lightning protection* kupferummantelter Staberder *m*, kupferüberzogener Staberder *m*; ~**-clad roof** *n* ARCH & BUILD, STEEL Kupferdach *nt*; ~ **cross** *n* HEAT & VENT Kupferkreuzstück *nt*

**coppered** *adj* STEEL, SURFACE verkupfert; ~ **wire** *n* STEEL verkupferter Draht *m*

***copper***: ~ **elbow** *n* STEEL Kupferbogen *m*; ~ **fin pipe** *n* HEAT & VENT Kupferlamellenrohr *nt*; ~ **fitting** *n* MAT PROP, WASTE WATER Kupferfitting *nt*, Kupferverbindungsstück *nt*; ~ **nail** *n* MAT PROP Kupfernagel *m*; ~ **pipe** *n* HEAT & VENT, WASTE WATER Kupferrohr *nt*; ~ **plating** *n* SURFACE Kupferplattierung *f*, Verkupferung *f*; ~ **rivet** *n* MAT PROP Kupferniet *m*; ~ **roof** *n* ARCH & BUILD, STEEL Kupferdach *nt*; ~ **roof gutter** *n* STEEL, WASTE WATER Kupferdachrinne *f*; ~ **roofing** *n* ARCH & BUILD, STEEL Kupferbedachung *f*, Kupferdach *nt*; ~**-sheathed cable** *n* ELECTR Kupfermantelkabel *nt*; ~ **sheet** *n* MAT PROP Kupferblech *nt*, Kupferdraht *m*; ~ **wire** *n* STEEL Kupferdraht *m*; ~ **wire mesh** *n* MAT PROP Kupfermatte *f*

**copyright**: ~ **of the plans** *n* ARCH & BUILD, CONST LAW Urheberrecht *nt* an Plänen

**corbel 1.** *n* ARCH & BUILD, STONE Konsole *f*, Kragstein *m*; **2.** *vt* ARCH & BUILD vorkragen, ausladen, auskragen; ~ **out** ARCH & BUILD auskragen

**corbeling** *n AmE*, **corbelling** *BrE* **1.** *adj* ARCH & BUILD, BUILD HARDW, BUILD MACHIN, MAT PROP, STONE auskragend; **2.** *n* ARCH & BUILD, STONE auskragende Schicht *f*, Vorkragung *f*

***corbel***: ~ **masonry** *n* ARCH & BUILD, STONE vorkragendes Mauerwerk *nt*; ~ **piece** *n* ARCH & BUILD, TIMBER Aufsattelung *f*, Sattelholz *nt*; ~ **step** *n* ARCH & BUILD Giebelstufe *f*; ~ **vault** *n* ARCH & BUILD Giebelstufe *f*, Kraggewölbe *nt*

**cord** *n* BUILD HARDW Schnur *f*

**cordon** *n* ARCH & BUILD Mauerband *nt*, vorkragende Mauerschicht *f*, STONE Mauerband *nt*

**core 1.** *n* ARCH & BUILD Gebäudekern *m*, Kern *m*, INFR & DES Dichtungskern *f*; **2.** *vt* ARCH & BUILD, MAT PROP lochen

**coreboard** *n* TIMBER Tischlerplatte *f*

***core***: ~ **cross section** *n* ARCH & BUILD Kernquerschnitt *m*

**cored** *adj* MAT PROP gelocht; ~ **block masonry** *n* STONE Lochsteinmauerwerk *nt*; ~ **brick** *n* MAT PROP, STONE Lochziegel *m*; ~ **panel** *n* MAT PROP Hohlplatte *f*, Hohltafel *f*, Lochplatte *f*; ~ **paper** *n* MAT PROP Lochpappe *f*

***core***: ~ **driller** *n* BUILD MACHIN Kernbohrmaschine *f*; ~ **drilling** *n* INFR & DES Kernbohrung *f*; ~ **insulation** *n* SOUND & THERMAL Kerndämmung *f*; ~ **nail** *n* BUILD HARDW Kernnagel *m*; ~ **slab** *n* MAT PROP Hohlplatte *f*, Lochplatte *f*

**coring** *n* MAT PROP *bricks* Lochreihe *f*, Lochung *f*

**cork** *n* MAT PROP Kork *m*; ~ **backing** *n* MAT PROP Korkunterlage *f*; ~**-based** *adj* MAT PROP korkhaltig

**corkboard** *n* MAT PROP Korkplatte *f*

***cork***: ~ **covering** *n* SOUND & THERMAL Korkbelag *m*; ~ **flooring** *n* MAT PROP Korkbodenbelag *m*; ~ **insulation** *n* SOUND & THERMAL Korkdämmung *f*

**corkscrew**: ~ **staircase** *n* ARCH & BUILD Wendeltreppe *f*; ~ **stairs** *n pl* ARCH & BUILD Wendeltreppe *f*

***cork***: ~ **tile** *n* MAT PROP Korkfliese *f*

**corner** *n* ARCH & BUILD, HEAT & VENT, INFR & DES Ecke *f*; **at the** ~ *phr* ARCH & BUILD an der Ecke; ~ **angle** *n* ARCH & BUILD Eckwinkel *m*; ~ **balcony** *n* ARCH & BUILD Eckbalkon *m*; ~ **bead** *n* BUILD HARDW Eckprofil *nt*, STONE Eckprofil *nt*, Eckschutzleiste *f*, Eckenschutzschiene *f*, Kantenschutz *m*, Kantenschutzleiste *f*, Kantenschutzprofil *nt*, Putzeckleiste *f*; ~ **bench** *n* TIMBER Eckverband *m*; ~ **bracing** *n* ARCH & BUILD Eckversteifung *f*; ~ **cleat** *n* ARCH & BUILD, BUILD HARDW Eckklemme *f*; ~ **column** *n* ARCH & BUILD Eckstütze *f*; ~ **fireplace** *n* ARCH & BUILD Eckkamin *m*; ~ **house** *n* ARCH & BUILD Eckhaus *nt*; ~ **and nodal points** *n pl* ARCH & BUILD Eck- und Knotenpunkte *m pl*; ~ **pillar** *n* ARCH & BUILD Eckpfeiler *m*; ~ **post** *n* ARCH & BUILD Eckpfosten *m*; ~ **region** *n* INFR & DES Eckbereich *m*

**corners**: ~ **of the room** *n pl* ARCH & BUILD Eckpunkte *m pl* des Raumes

**cornerstone** *n* CONCR, INFR & DES, MAT PROP, STONE Eckstein *m*; ~ **ceremony** *n* CONST LAW Grundsteinlegung *f*

**corner**: ~ **-welded** *adj* STEEL eckverschweißt

**cornice** *n* ARCH & BUILD *on a door* Gesims *nt*, Sims *m*, *principle moulding* Gurtgesims *nt*, Kranzgesims *nt*

**correct** *adj* CONST LAW richtig

**correction**: ~ **of deficiencies** *n* ARCH & BUILD, CONST LAW Mängelbehebung *f*, Mängelbeseitigung *f*

**corridor** *n* ARCH & BUILD, INFR & DES Flur *m*, Gang *m*, Korridor *m*

**corrode** *vti* ARCH & BUILD, ENVIRON, INFR & DES, MAT PROP, STEEL, SURFACE korrodieren

**corrosion** *n* ARCH & BUILD, ENVIRON, INFR & DES, MAT PROP, STEEL, SURFACE Korrosion *f*; ~ **control** *n* MAT PROP, STEEL, SURFACE Korrosionsbekämpfung *f*; ~ **damage** *n* MAT PROP, SURFACE Korrosionsschaden *m*; ~**-proof** *adj* MAT PROP, SURFACE korrosionsgeschützt; ~**-proofer** *n* MAT PROP, SURFACE Korrosionsschutz *m*; ~ **protection** *n* MAT PROP, SURFACE Korrosionsschutz *m*; ~ **protection coat** *n* MAT PROP, SURFACE Korrosionsschutzanstrich *m*; ~ **protection tape** *n* MAT PROP, SURFACE Korrosionsschutzbinde *f*; ~ **resistance** *n* MAT PROP, SURFACE Korrosionsbeständigkeit *f*; ~**-resistant** *adj* MAT PROP, SURFACE korrosionsbeständig

**corrosive** *adj* ARCH & BUILD, ENVIRON, INFR & DES korrosiv, MAT PROP, STEEL, SURFACE korrodierend, korrosiv, ätzend

**corrugated** *adj* ARCH & BUILD, MAT PROP, STEEL gewellt; ~ **asbestos cement** *n* MAT PROP Wellasbestzement *m*, Wellfaserzement *m*; ~ **asbestos cement sheet** *n* CONCR, MAT PROP Asbestzementwellplatte *f*; ~ **asbestos panel** *n* MAT PROP Wellasbesttafel *f*; ~ **asphalt paper** *n* MAT PROP Wellbitumenpapier *nt*; ~ **eternit** *n* MAT PROP Welleternit *nt*; ~ **iron** *n* STEEL Wellblech *nt*; ~ **metal** *n* STEEL Wellblech *nt*; ~ **metal structures** *n pl* STEEL Wellblechbauten *f pl*; ~ **paper** *n* MAT PROP Wellpappe *f*; ~ **polyester board** *n* MAT

PROP Polyesterwellplatte *f*; ~ **ridge capping** *n* MAT PROP Wellenfirsthaube *f*; ~ **ridge covering** *n* MAT PROP Wellenfirsthaube *f*; ~ **roof** *n* ARCH & BUILD Welldach *nt*; ~ **sheet iron** *n* STEEL Wellblech *nt*; ~ **tube** *n* MAT PROP Wellrohr *nt*; ~ **wire** *n* MAT PROP Welldraht *m*; ~ **wire glass** *n* MAT PROP Welldrahtglas *nt*

**corundum**: ~ **stone** *n* MAT PROP Korundstein *m*

**cost** *n* ARCH & BUILD, CONST LAW, INFR & DES Kosten *pl*, Preis *m*; ~ **benefit comparison** *n* ARCH & BUILD, CONST LAW, INFR & DES Kostenvergleich *m*; ~ **benefit program** *AmE*, ~ **benefit programme** *BrE n* ARCH & BUILD, INFR & DES Kostennutzungsprogramm *nt*; ~ **calculation** *n* ARCH & BUILD, CONST LAW Kostenberechnung *f*; ~ **check** *n* ARCH & BUILD, INFR & DES Kostenüberprüfung *f*; ~ **comparison** *n* ARCH & BUILD, CONST LAW, INFR & DES Kostenvergleich *m*; ~ **control** *n* ARCH & BUILD, INFR & DES Kostenkontrolle *f*; ~-**effective** *adj* ARCH & BUILD, INFR & DES kosteneffektiv; ~ **estimate** *n* ARCH & BUILD, CONST LAW, INFR & DES Kostenschätzung *f*

**costly** *adj* ARCH & BUILD aufwendig

***cost***: ~ **reimbursement contract** *n* CONST LAW Rückvergütungsvertrag *m*

**cottered**: ~ **joint** *n* ARCH & BUILD Splintverbindung *f*

**cotton**: ~ **cloth** *n* MAT PROP Baumwollgewebe *nt*; ~ **insulation** *n* SOUND & THERMAL Baumwollisolierung *f*

**$CO_2$** *abbr* (*carbon dioxide*) ENVIRON $CO_2$ (*Kohlendioxid*)

***$CO_2$***: ~ **emission index** *n* ENVIRON, HEAT & VENT $CO_2$ Emissionsindex *m*

**council** *n* CONST LAW, INFR & DES Behörde *f*; ~ **housing** *n BrE* (*cf public housing units AmE*) INFR & DES sozialer Wohnungsbau *m*; ~ **tax** *n BrE* CONST LAW, INFR & DES häusliche Steuer *f*

**counter** *n* ELECTR *meter* Zähler *nt*, TIMBER *in shop* Kundenschalter *m*, Ladentisch *m*; ~ **batten** *n* MAT PROP, TIMBER Dachlatte *f*; ~ **brace** *n* ARCH & BUILD *truss* Wechselstab *m*; ~ **ceiling** *n* ARCH & BUILD, BUILD HARDW abgehängte Decke *f*; ~ **current method** *n* INFR & DES Gegenstromprinzip *nt*

**countercurrent**: ~ **mixer** *n* BUILD MACHIN *manufacture of concrete products* Gegenstrommischer *m*

**counterflange** *n* HEAT & VENT Gegenflansch *m*

**counterflap**: ~ **hinge** *n* BUILD HARDW *lock* Zwingnuß *f*

**counterfloor** *n* ARCH & BUILD Blindboden *m*

**counterfort** *n* ARCH & BUILD, STONE Mauerpfeiler *m*, Strebepfeiler *m*, Verstärkungspfeiler *m*

**counterlath** *n* ARCH & BUILD, BUILD HARDW, STONE Putzleiste *f*

**counterlathing** *n* ARCH & BUILD, STONE Berliner Lattendecke *f*, Putzträger *m* aus Leisten, TIMBER Konterlattung *f*

**counterpoise**: ~ **bridge** *n* INFR & DES Klappbrücke *f*

**counterslope** *n* INFR & DES Gegenböschung *f*

**countersunk**: ~ **button-head rivet** *n* BUILD HARDW Senkkopfniet *m*; ~ **head-bolt** *n* ARCH & BUILD Bolzen *m* mit versenktem Kopf; ~ **riveting** *n* STEEL Senkkopfvernietung *f*

**counterweight** *n* BUILD MACHIN *crane, door* Ausgleichsgewicht *nt*

**couple** *n* TIMBER Gebinde *nt*, Sparrengebinde *nt*, Sparrenpaar *nt*; ~ **close roof** *n* ARCH & BUILD Satteldach *nt*

**coupler** *n* HEAT & VENT *pipe* Muffe *f*, Überschiebmuffe *f*, WASTE WATER Überschiebmuffe *f*

***couple***: ~ **roof** *n* ARCH & BUILD, TIMBER einfaches Sparrendach *nt*

**coupling** *n* HEAT & VENT Kupplung *f*

**course** *n* INFR & DES Lage *f*, Schicht *f*, *progression* Lauf *m*, STONE Lage *f*, Schicht *f*

**coursed**: ~ **masonry** *n* MAT PROP, STONE Bruchsteinmauerwerk *nt*, Schichtenmauerwerk *nt*, geschichtetes Bruchsteinmauerwerk *nt*

**coursing** *n* STONE *masonry* Schichtanordnung *f*

**court** *n* CONST LAW Gericht *nt*, INFR & DES Hof *m*; ~ **entrance** *n* INFR & DES Hofeingang *m*

**courthouse** *n* INFR & DES Gerichtsgebäude *nt*

**courtyard** *n* INFR & DES Hof *m*

**cove** *n* ARCH & BUILD, INFR & DES, MAT PROP Hohlkehle *f*

**coved**: ~ **skirting** *n* ARCH & BUILD, INFR & DES, MAT PROP Sockel *m* mit Hohlkehle, Sockelleiste *f* mit Hohlkehle

**cover 1.** *n* INFR & DES Überdeckung *f*, SURFACE Abdekkung *f*, Belag *m*; **2.** *vt* ARCH & BUILD abdecken, verdecken, CONST LAW abdecken, INFR & DES überdekken, SURFACE abdecken, belegen

**coverage** *n* ARCH & BUILD, CONCR Betondeckung *f*, INFR & DES Ergiebigkeit *f*, überdachte Fläche *f*, MAT PROP Ergiebigkeit *f*

**covered** *adj* ARCH & BUILD abgedeckt, verdeckt, CONST LAW abgedeckt; ~ **drain** *n* ENVIRON überdeckter Abzugsgraben *m*; ~ **passage** *n* ARCH & BUILD überdachter Gang *m*; ~ **vault** *n* ARCH & BUILD Domikalgewölbe *nt*, geschlossene Rundzelle *f*; ~ **walk** *n* ARCH & BUILD Laubengang *m*; ~ **walkway** *n* ARCH & BUILD gedeckter Fußgängerweg *m*; ~ **way** *n* ARCH & BUILD gedeckter Weg *m*

***cover***: ~ **fillet** *n* BUILD HARDW Deckleiste *f*, Paßleiste *f*, STEEL Abdeckleiste *f*, TIMBER Abdeckleiste *f*, Deckleiste *f*, Paßleiste *f*

**covering** *n* ARCH & BUILD Abdeckung *f*, Belag *m*, SURFACE Abdeckung *f*, Deckschicht *f*

***covering***: ~ **frame** *n* BUILD HARDW *roof window* Eindeckrahmen *m*; ~ **grid** *n* MAT PROP, STEEL Abdeckgitter *nt*; ~ **joint** *n* ARCH & BUILD Deckfuge *f*; ~ **layer** *n* ARCH & BUILD, INFR & DES Deckschicht *f*; ~ **material** *n* ENVIRON, MAT PROP Abdeckmaterial *nt*; ~ **plate** *n* MAT PROP, STEEL Abdeckblech *nt*

***cover***: ~ **plate** *n* MAT PROP Abdeckblech *nt*, STEEL Abdeckblech *nt*, Abdeckrost *m*, Deckplatte *f*; ~ **strip** *n* BUILD HARDW Deckleiste *f*, TIMBER Fugenleiste *f*, Deckleiste *f*

**coving** *n* ARCH & BUILD, INFR & DES, MAT PROP Kehlung *f*, Wölbung *f*

***CO***: ~-**warning system** *n* CONST LAW, ELECTR, HEAT & VENT CO-Warnanlage *f*

**cowl** *n* BUILD HARDW Schachtabdeckung *f*, Schornsteinaufsatz *m*

**crack 1.** *n* CONCR, MAT PROP, STONE Riß *m*, Spalt *m*, Sprung *m*; **2.** *vi* CONCR, MAT PROP, STONE reißen, zerspringen

***crack***: ~ **bridging** *n* STONE Rißüberbrückung *f*; ~-**covering reinforcement tape** *n* STONE risseüberbrückendes Armierungsgewebe *nt*

**cracked**: ~ **flexural stiffness** *n* ARCH & BUILD, CONCR, INFR & DES, MAT PROP Biegesteifigkeit *f* nach Rißbildung

***crack***: ~ **filler** *n* TIMBER Holzkitt *m*; ~ **formation** *n* CONCR, MAT PROP, STONE Rißbildung *f*

**cracking** *n* CONCR, MAT PROP, STONE Rißbildung *f*; ~ **risk** *n* CONCR, STONE Rißgefahr *f*; ~ **stress** *n* CONCR, INFR & DES, MAT PROP, STONE Rißspannung *f*

**crack**: ~-**proof** *adj* CONCR, MAT PROP rißsicher; ~ **width** *n* CONCR, STONE Rißbreite *f*

**cradle 1.** *n* ARCH & BUILD, BUILD MACHIN, INFR & DES Hängebühne *f*, Hängerüstung *f*, hängendes Gerüst *nt*; **2.** *vt* ARCH & BUILD versteifen

***cradle***: ~ **machine** *n* BUILD MACHIN Reinigungswagen *m*; ~ **vault** *n* ARCH & BUILD, STONE Tonnengewölbe *nt*

**cradling** *n* ARCH & BUILD Lehrbogen *m*, Versteifung *f*, INFR & DES, STEEL, TIMBER Lehrbogen *m*

**cramp 1.** *n* BUILD HARDW Klammer *f*, MAT PROP Krampe *f*, *to couple wood, stone* Klammer *f*; **2.** *vt* ARCH & BUILD *wood, stone* verklammern

***cramp***: ~ **iron** *n* MAT PROP *for connecting stones, wood* Bankeisen *nt*

**crane** *n* BUILD MACHIN Kran *m*; ~ **bridge** *n* BUILD MACHIN Kranbrücke *f*; ~-**displaceable formwork** *n* CONCR kranversetzbare Verschalung *f*; ~ **driving** *n* BUILD MACHIN Kranfahren *nt*; ~ **gantry** *n* BUILD MACHIN Krangerüst *nt*; ~ **girder** *n* BUILD MACHIN Kranbahnträger *m*; ~ **hook** *n* BUILD MACHIN Kranlasthaken *m*; ~ **operator** *n* BUILD MACHIN Kranführer *m*; ~ **rail** *n* BUILD MACHIN Kranschiene *f*; ~ **runway** *n* BUILD MACHIN Kranbahn *f*

**craneway** *n* BUILD MACHIN Kranbahn *f*

**crank** *vt* HEAT & VENT kröpfen

**crate** *n* ARCH & BUILD, INFR & DES Kasten *m*, TIMBER Kasten *m*, Verschlag *m*

**crawler** *n* BUILD MACHIN Planierraupe *f*; ~ **dozer** *n* BUILD MACHIN Planierraupe *f*; ~ **excavator** *n* BUILD MACHIN Raupenbagger *m*; ~ **tractor** *n* BUILD MACHIN Planierraupe *f*

**crawlway** *n* ARCH & BUILD Kriechkeller *m*

**craze** *n* MAT PROP Haarriß *m*

**crazing** *n* MAT PROP Haarrisse *m pl*

**creation** *n* ARCH & BUILD Erschaffung *f*

**creative** *adj* ARCH & BUILD schöpferisch

**creep 1.** *n* ARCH & BUILD, CONCR, INFR & DES, MAT PROP, STEEL, Kriechen *nt*; ~ **of concrete** *n* ARCH & BUILD, CONCR Betonkriechen *nt*; ~ **effect** *n* ARCH & BUILD, CONCR, INFR & DES, MAT PROP, STEEL Kriecheinfluß *m*; **2.** *vi* ARCH & BUILD, CONCR, INFR & DES, MAT PROP, STEEL kriechen

**creeping**: ~ **rafter** *n* TIMBER Schiftsparren *m*

***creep***: ~ **modulus** *n* CONCR Kriechmodul *m*; ~ **rate** *n* MAT PROP Kriechgeschwindigkeit *f*; ~ **resistance** *n* ARCH & BUILD, CONCR, MAT PROP, STEEL Kriechfestigkeit *f*

**Cremona**: ~ **diagram** *n* ARCH & BUILD *determination of forces* Cremona-Plan *m*

**crenelated** *AmE see crenellated BrE*

**crenelation** *AmE see crenellation BrE*

**crenellated** *adj BrE* ARCH & BUILD kreneliert

**crenellation** *n BrE* ARCH & BUILD Zinnenkranz *m*

**creosote 1.** *n* TIMBER Kreosot *nt*; **2.** *vt* TIMBER kreosotieren

**crest** *n* ARCH & BUILD Krone *f*, Scheitel *m*, Mauerkrone *f*; ~ **tile** *n* ARCH & BUILD, INFR & DES, MAT PROP, STONE Firstziegel *m*

**criterion** *n* ARCH & BUILD, CONST LAW, INFR & DES Gesichtspunkt *m*, Kriterium *nt*

**critical** *adj* ARCH & BUILD *load* kritisch; ~ **cross-section** *n* ARCH & BUILD, CONCR, INFR & DES, MAT PROP kritischer Querschnitt *m*; ~ **length** *n* ARCH & BUILD, CONCR, INFR & DES, MAT PROP kritische Länge *f*; ~ **load** *n* ARCH & BUILD, CONCR, STEEL, TIMBER kritische Last *f*; ~ **location** *n* ARCH & BUILD kritischer Standort *m*; ~ **path analysis** *n* CONST LAW Kritische-Pfad-Analyse *f*; ~ **path method** *n* CONST LAW Kritische-Pfad-Methode *f*; ~ **path network** *n* CONST LAW Netzplantechnik *f* nach CPM; ~ **value** *n* ARCH & BUILD Grenzwert *m*; ~ **wind velocity** *n* ARCH & BUILD, INFR & DES kritische Windgeschwindigkeit *f*

**crocket** *n* ARCH & BUILD *medieval ornament* Steinblume *f*, Krabbe *f*, Knolle *f*, Kriechblume *f*; ~ **capital** *n* ARCH & BUILD Knospenkapitell *nt*, Kriechblumenkapitell *nt*, Steinblumenkapitell *nt*

**cross** *n* HEAT & VENT Kreuz *nt*, Kreuzstück *nt*

**crossband** *n* TIMBER Absperrfurnier *nt*, Blindfurnier *nt*, Sperrfurnier *nt*

**crossbar** *n* ARCH & BUILD, INFR & DES Querbalken *m*, Querriegel *m*, Querstab *m*, Querstange *f*, Querstück *nt*, Traverse *f*, *impost* Kämpfer *m*

***cross***: ~ **beam** *n* ARCH & BUILD Querbalken *m*; ~ **bond** *n* STONE *masonry* Kreuzverband *m*; ~ **bracing** *n* ARCH & BUILD, CONCR, MAT PROP Queraussteifung *f*, STEEL, TIMBER Windverband *m*, Kreuzverband *m*, Queraussteifung *f*, *shipbuilding* Querverstrebung *f*; ~ **connection** *n* STEEL, TIMBER Kreuzverbindung *f*; ~ **connector** *n* BUILD HARDW, STEEL, TIMBER Kreuzverbinder *m*; ~-**frame member** *n* ARCH & BUILD, STEEL, TIMBER Riegel *m*; ~ **garnet** *n* BUILD HARDW T-Band *nt*; ~ **girder** *n* ARCH & BUILD, STEEL, TIMBER Querbalken *m*, Querträger *m*; ~-**grained wood** *n* TIMBER Hirnholz *nt*

**crosshead** *n* ARCH & BUILD, CONCR *prestressed concrete* Spanntraverse *f*

**crossing** *n* INFR & DES Kreuzung *f*, *road junction* Straßenkreuzung *f*; ~ **area** *n* ARCH & BUILD Kreuzungsbereich *m*, *steel girder* Kreuzung *f*, INFR & DES Kreuzungsbereich *m*; ~ **point** *n* ARCH & BUILD, INFR & DES Kreuzungspunkt *m*

***cross***: ~ **joint** *n* HEAT & VENT *pipe* Kreuzstück *nt*, STONE Kreuzfuge *f*, Kreuzstoß *m*, TIMBER Kreuzverzapfung *f*; ~ **lath** *n* ARCH & BUILD Windrispe *f*, STEEL Windaussteifung *f*, TIMBER Sturmlatte *f*, Windrispe *f*, Windlatte *f*, Windaussteifung *f*, Windrute *f*

**crosslinked** *adj* INFR & DES, MAT PROP vernetzt

**Cross**: ~ **method** *n* ARCH & BUILD, INFR & DES *structural analysis* Cross-Verfahren *nt*, Momentenausgleichsverfahren *nt*

***cross***: ~ **mounting** *n* ELECTR Kreuzschaltung *f*; ~ **plate** *n* ARCH & BUILD Joch *nt*; ~ **profile** *n* ARCH & BUILD Querprofil *nt*; ~ **rafter** *n* TIMBER Quersparren *m*; ~ **reinforcement** *n* ARCH & BUILD, CONCR Querbewehrung *f*; ~ **rib** *n* ARCH & BUILD Kreuzrippe *f*

**crossroad** *n* INFR & DES Querstraße *f*

***cross***: ~ **section** *n* ARCH & BUILD *Straßenbau* Querschnittsfläche *f*, MAT PROP, STEEL, TIMBER Querschnitt *m*; ~-**sill** *n* ARCH & BUILD *framework* Schwelle *f*; ~-**staff head** *n* ARCH & BUILD Kreuzscheibe *f*, Winkelkopf *m*; ~ **stay** *n* ARCH & BUILD Diagonalstrebe *f*, Querversteifung *f*, CONCR, MAT PROP, STEEL Querversteifung *f*, TIMBER Diagonalstrebe *f*; ~ **stays** *n* ARCH & BUILD Verschwertung *f*; ~ **stud** *n* ARCH & BUILD, STEEL, TIMBER Kreuzstrebe *f*, Querstrebe *f*, Verschwertung *f*; ~ **traversing** *n* BUILD MACHIN *crane* Katzfahren *nt*; ~-**vault** *n* ARCH & BUILD,

INFR & DES, TIMBER Kreuzgewölbe *nt*; **~-ventilated** *adj* HEAT & VENT quergelüftet
**crosswalk** *n* INFR & DES Fußgängerübergang *m*
**crosswall** *n* ARCH & BUILD Querwand *f*; **~ construction** *n* ARCH & BUILD Schottenbauweise *f*
**crossway** *n* INFR & DES Querstraße *f*
**crowbar** *n* BUILD MACHIN Brecheisen *nt*
**crowd**: **~-induced** *adj* ARCH & BUILD durch Menschenmenge verursacht
**crown** *n* ARCH & BUILD *arch, vault* Scheitel *m*, *coping stone* Deckplatte *f*, Krone *f*, Haupt *nt*, *frame* Aufsatz *m*
**crowning** *n* ARCH & BUILD Wölbung *f*
***crown***: **~ post** *n* ARCH & BUILD *roof timbering* Firstpfosten *m*; **~ tile** *n* ARCH & BUILD, MAT PROP Firststein *m*; **~-tile roof** *n* ARCH & BUILD Kronendach *nt*, Ritterdach *nt*
**crowstep** *n* ARCH & BUILD Giebelstufe *f*
**cruciform**: **~ ground plan** *n* ARCH & BUILD kreuzförmiger Grundriß *m*
**crude** *adj* ENVIRON, MAT PROP roh
**crumble**: **~ away** *vi* MAT PROP abbröckeln
**crush** *vt* MAT PROP brechen, zerkleinern
**crushed** *adj* MAT PROP gebrochen, zerkleinert; **~ brick aggregate** *n* MAT PROP Ziegelsplitt *m*; **~ brick aggregate concrete** *n* MAT PROP Ziegelsplittbeton *m*; **~ material** *n* MAT PROP gebrochenes Material *nt*, Brechgut *nt*; **~ rock** *n* MAT PROP gebrochenes Gestein *nt*, Schotter *m*; **~ rock fine aggregate** *n* MAT PROP Brechsand *m*; **~ stone** *n* MAT PROP Schotter *m*; **~ stone base course** *n* ARCH & BUILD Schottertragschicht *f*; **~ stone sand** *n* MAT PROP Brechsand *m*; **~ stone subbase** *n* ARCH & BUILD Schottertragschicht *f*
**crusher** *n* BUILD MACHIN Brecher *m*, Zerkleinerer *m*, Zerkleinerungsmaschine *f*, ENVIRON Zerkleinerer *m*; **~ dust** *n* ARCH & BUILD, CONST LAW Bruchstaub *m*; **~ unit** *n* BUILD MACHIN Zerkleinerungsanlage *f*
**crushing** *n* MAT PROP *compaction* Zerkleinerung *f*, Zerreißen *nt*; **~ equipment** *n* BUILD MACHIN, INFR & DES, MAT PROP Brechwerk *nt*; **~ load** *n* ARCH & BUILD, INFR & DES, MAT PROP Scheitelbruchlast *f*; **~ machine** *n* BUILD MACHIN Brecher *m*, Zerkleinerungsmaschine *f*; **~ machinery** *n* BUILD MACHIN *for building materials* Zerkleinerungsmaschinen *f pl*; **~ strength** *n* CONCR, MAT PROP, STEEL, TIMBER Druckfestigkeit *f*
**crystal** *n* MAT PROP Kristall *m*
**CSI** *abbr* (*chlorosulphonyl isocyanate BrE*) ENVIRON CSI (*Chlorschwefelisocyanat*)
**CTE** *abbr* (*coefficient of thermal expansion*) MAT PROP Wärmedehnzahl *f*
**cubage** *n* ARCH & BUILD Rauminhalt *m*
**cubical**: **~ content** *n* ARCH & BUILD umbauter Raum *m*; **~ contents** *n* ARCH & BUILD Rauminhalt *m*
**cubicle** *n* ARCH & BUILD Zelle *f*, *small room or closet* Kabine *f*; **~ construction** *n* ARCH & BUILD Zellenbauweise *f*
**cubic**: **~ yardage** *n* ARCH & BUILD umbauter Raum *m*
**cul-de-sac** *n* INFR & DES Sackgasse *f*
**cullet** *n* ENVIRON Glasbruch *m*
**Cullmann**: **~'s method** *n* INFR & DES Cullmannsche E-Linie *f*
**cultivate** *vt* INFR & DES *outdoor facilities* bearbeiten
**cultivation** *n* INFR & DES *horticultural* Anpflanzung *f*, Bearbeitung *f*
**culvert** *n* INFR & DES, WASTE WATER *crossing under road, sidewalk* Durchlaßkanal *m*, Düker *m*, Wasserdurchlaß *m*
**cumulative**: **~ discharge** *n* ENVIRON kumulativer Ausfluß *m*
**cupboard** *n* *BrE* (*cf closet AmE*) ARCH & BUILD Abstellkammer *f*
**cup**: **~ head rivet** *n* BUILD HARDW Flachrundkopfniet *m*
**cupola** *n* ARCH & BUILD Kuppel *f*; **~ surface** *n* ARCH & BUILD Kuppeloberfläche *f*
**curb** *n* *AmE* (*cf kerb BrE*) INFR & DES *along street* Bordeinfassung *f*, Bordkante *f*, Bordstein *m*, MAT PROP Bordstein *m*; **~ roof** *n* ARCH & BUILD gebrochenes Mansardendach *nt*
**curbstone** *n* *AmE* (*cf kerbstone BrE*) INFR & DES Bordstein *m*, Randstein *m*, Rinnstein *m*, Schramme *f*, MAT PROP Bordstein *m*, Randstein *m*, Rinnstein *m*
**cure** *vt* CONCR nachbehandeln, *concrete, mortar, by keeping damp* die Härtung fördern
**curing** *n* CONCR Nachbehandeln *nt*, Nachbehandlung *f*; **~ agent** *n* CONCR Abdeckmatte *f*; **~ compound** *n* MAT PROP *concrete surface* Dichtungsmittel *nt*
**current** *n* ELECTR elektrischer Strom *m*, Strom *m*, INFR & DES, WASTE WATER Strömung *f*; **~ meter** *n* ENVIRON Strommesser *m*
**curtail** *n* ARCH & BUILD geschwungene Antrittsstufe *f*; **~ step** *n* ARCH & BUILD geschwungene Antrittsstufe *f*
**curtain** *n* BUILD HARDW Gardine *f*, Vorhang *m*; **~ rod** *n* BUILD HARDW Vorhangstange *f*; **~ wall** *n* ARCH & BUILD Vorhangfassade *f*, Vorhangwand *f*, INFR & DES Dichtungsschleier *m*, MAT PROP Vorhangfassade *f*, SOUND & THERMAL Dichtungsschleier *m*, STEEL, STONE Vorhangfassade *f*; **~ wall frame** *n* ARCH & BUILD Vorhangwandrahmen *m*
**curve 1.** *n* ARCH & BUILD, INFR & DES *roads, paths, railways* Biegung *f*, Krümmung *f*; **2.** *vt* ARCH & BUILD krümmen, wölben; **3.** *vi* ARCH & BUILD, INFR & DES, MAT PROP sich krümmen
**curved** *adj* ARCH & BUILD, INFR & DES gekrümmt; **~ beam** *n* ARCH & BUILD, CONCR gekrümmter Balken *m*; **~ brickwork** *n* STONE Bogenmauerwerk *nt*; **~ girder** *n* ARCH & BUILD, CONCR Bogenträger *m*, gekrümmter Träger *m*, INFR & DES, STONE, TIMBER Bogenträger *m*; **~ roof** *n* ARCH & BUILD gebogener Sparren *m*, geschweiftes Dach *nt*, TIMBER gebogener Sparren *m*; **~ sided triangle** *n* ARCH & BUILD *church* Bogendreieck *nt*; **~ slab** *n* ARCH & BUILD Nase *f*, gekrümmte Platte *f*
**cushion**: **~ head** *n* BUILD MACHIN *pile foundations* Rammhaube *f*
**cusped**: **~ arch** *n* ARCH & BUILD Nasenbogen *m*
**cuspidation** *n* ARCH & BUILD *Gothic architecture* Nasenverzierung *f*
**cusping** *n* ARCH & BUILD *Gothic architecture* Nasenverzierung *f*
**customer** *n* ARCH & BUILD Auftraggeber *m* (*AG*), Kunde *m*, CONST LAW Auftraggeber *m* (*AG*); **~ service area** *n* ARCH & BUILD, INFR & DES *counters* Schalterhalle *f*, *tills* Kassenraum *m*
**cut 1.** *n* ARCH & BUILD, INFR & DES Schnitt *m*; **2.** *vt* ARCH & BUILD, INFR & DES *earth* abtragen, MAT PROP schneiden; **~ down** INFR & DES *tree* fällen; **~ to length** MAT PROP ablängen
***cut***: **~ glass** *n* ARCH & BUILD geschliffenes Glas *nt*; **~ nail** *n* MAT PROP Schnittnagel *m*

**cutoff**: ~ **cock** *n* HEAT & VENT, WASTE WATER Absperrhahn *m*

**cutout** *n* ELECTR Sicherung *f*, TIMBER Ausklinkung *f*

***cut***: ~ **roof** *n* ARCH & BUILD, TIMBER abgestumpftes Dach *nt*, Flachdach *nt*; ~ **stone** *n* STONE Haustein *m*, Quader *m*

**cutter** *n* BUILD MACHIN Schneidewerkzeug *nt*; ~ **disc** *n* *BrE* BUILD MACHIN Schneidescheibe *f*; ~ **disk** *AmE see cutter disc BrE*

**cutting 1.** *adj* BUILD MACHIN schneidend; **2.** *n* ARCH & BUILD, INFR & DES *earth* Abtrag *m*, Abtragen *nt*, Abtragung *f*, MAT PROP Schneiden *nt*

***cutting***: ~ **diamond** *n* BUILD MACHIN Diamantschneider *m*; ~ **edge** *n* BUILD MACHIN Schneidkante *f*, Schneide *f*, MAT PROP Schnittkante *f*; ~ **foot** *n* BUILD MACHIN *pile foundation* Schneidfuß *m*; ~ **gage** *AmE*, ~ **gauge** *BrE* *n* BUILD MACHIN einstellbares Schneidmesser *nt*; ~ **list** *n* TIMBER Schnittliste *f*; ~ **pliers** *n pl* BUILD MACHIN Schneidzange *f*; ~ **to size** *n* MAT PROP Zuschneiden *nt*

**cut**: ~ **to fit** *adj* STEEL angepaßt

**cutwater** *n* INFR & DES *bridge* Pfeilerkopf *m*

**cyclopean**: ~ **block** *n* CONCR Zyklopenbeton *m*, Zyklopenstein *m*, STONE Zyklopenstein *m*; ~ **masonry** *n* STONE Zyklopenmauerwerk *nt*; ~ **wall** *n* CONCR, STONE Zyklopenmauer *f*

**cylinder** *n* ARCH & BUILD, BUILD HARDW Zylinder *m*, MAT PROP Rolle *f*, *barrel drum* Trommel *f*, *roller* Walze *f*, Zylinder *m*; ~ **cabinet lock** *n* BUILD HARDW Zylindermöbelschloß *nt*; ~ **extension** *n* BUILD HARDW *lock* Zylinderverlängerung *f*; ~ **for different door thicknesses** *n* BUILD HARDW Zylinder *m* für abweichende Türstärken; ~ **lock** *n* BUILD HARDW Zylinderschloß *nt*; ~ **mortise cabinet lock** *n* BUILD HARDW Zylindereinsteckmöbelschloß *nt*; ~ **padlock** *n* BUILD HARDW Zylindervorhängeschloß *nt*; ~ **rim deadlock** *n* BUILD HARDW Zylinderkastenriegelschloß *nt*; ~ **rim latch lock** *n* BUILD HARDW Zylinderkastenfallenschloß *nt*; ~ **test** *n* CONCR, INFR & DES, MAT PROP Zylinderdruckprüfung *f*

**cylindrical**: ~ **chimney** *n* ARCH & BUILD, CONCR, STEEL Rundschornstein *m*; ~ **column** *n* ARCH & BUILD Rundstütze *f*; ~ **manhole** *n* INFR & DES, WASTE WATER Rundschacht *m*; ~ **pipe** *n* HEAT & VENT, MAT PROP Zylinderrohr *nt*; ~ **shell** *n* ARCH & BUILD, INFR & DES Zylinderschale *f*; ~ **steel column** *n* STEEL Rundstahlstütze *f*

# D

**dado** *n* ARCH & BUILD *of room wall* Sockel *m*; **~ base** *n* ARCH & BUILD Sockelfuß *m*; **~ molding** *AmE*, **~ moulding** *BrE n* MAT PROP Wandsockelleiste *f*; **~ tile** *n* STONE, WASTE WATER Sockelfliese *f*

**daily**: **~ consumption** *n* ARCH & BUILD, INFR & DES Tagesverbrauchsmenge *f*; **~ cover** *n* ENVIRON arbeitstägliche Abdeckung *f*; **~ sun path** *n* ARCH & BUILD, ENVIRON, HEAT & VENT, INFR & DES täglicher Sonnenweg *m*

**dam** *n* ENVIRON, INFR & DES Damm *m*, Wehr *nt*, Staumauer *f*, *reservoir embankment* Staudamm *m*, *storage basin of reservoir* Talsperre *f*

**damage 1.** *n* CONST LAW Defekt *m*, Schädigung *f*, Schaden *m*; **2.** *vt* CONST LAW beschädigen

***damage***: **~ by efflorescence** *n* CONCR Salpeterfraß *m*

**damaged** *adj* CONST LAW schadhaft, beschädigt

***damage***: **~ due to humidity** *n* INFR & DES, SOUND & THERMAL Feuchtigkeitsschaden *m*; **~ due to penetration of moisture** *n* CONST LAW, INFR & DES Durchfeuchtungsschaden *m*

**damp 1.** *adj* ARCH & BUILD, HEAT & VENT, INFR & DES, MAT PROP, SOUND & THERMAL, STEEL feucht; **2.** *n* ARCH & BUILD, MAT PROP, SOUND & THERMAL Wasserdampfdiffusion *f*; **3.** *vt* SOUND & THERMAL dämpfen

**dampen** *vt* ARCH & BUILD, MAT PROP, SOUND & THERMAL, STEEL anfeuchten

**damper** *n* HEAT & VENT Schornsteinschieber *m*, *gas regulator* Drosselklappe *f*, Drosselventil *nt*

**damping** *n* ARCH & BUILD, MAT PROP, STEEL Anfeuchten *nt*, SOUND & THERMAL *of sound* Dämpfung *f*; **~ capacity** *n* SOUND & THERMAL Dämpfungsvermögen *nt*

**dampness** *n* ARCH & BUILD, HEAT & VENT, INFR & DES, MAT PROP, SOUND & THERMAL, STEEL Feuchte *f*, Feuchtigkeit *f*

**dampproof** *adj* ARCH & BUILD, HEAT & VENT, INFR & DES, MAT PROP, SOUND & THERMAL feuchtigkeitsbeständig; **~ course** *n* (*DPC*) MAT PROP, SOUND & THERMAL, STONE Feuchtigkeitssperrschicht *f*; **~ equipment** *n* WASTE WATER Feuchtraumausstattung *f*

**dampproofing** *n* ARCH & BUILD, HEAT & VENT, INFR & DES, MAT PROP, SOUND & THERMAL Feuchtigkeitsisolierung *f*

***dampproofing***: **~ additive** *n* MAT PROP, SOUND & THERMAL Sperrzusatz *m*; **~ course** *n* MAT PROP, SOUND & THERMAL, STONE Feuchtigkeitssperrschicht *f*; **~ material** *n* MAT PROP, SOUND & THERMAL Sperrstoff *m*; **~ paint** *n* MAT PROP, SOUND & THERMAL, SURFACE Sperranstrichmittel *nt*

**dampproof**: **~ lighting fixture** *n* ELECTR, MAT PROP FR-Leuchte *f*, Feuchtraumleuchte *f*; **~ membrane** *n* (*DPM*) MAT PROP, SOUND & THERMAL Dichtungshaut *f*

**damp**: **~ room** *n* ARCH & BUILD, SOUND & THERMAL Feuchtraum *m* (*FR*)

**dancing**: **~ step** *n* ARCH & BUILD verzogene Stufe *f*

**danger** *n* ARCH & BUILD, CONST LAW, INFR & DES Gefahr *f*; **~ class** *n* CONST LAW Gefahrenklasse *f*; **~ of contamination** *n* CONST LAW, INFR & DES Verschmutzungsgefahr *f*

**dangerous** *adj* ARCH & BUILD, CONST LAW, INFR & DES gefährlich; **~ structure** *n* ARCH & BUILD, CONST LAW, INFR & DES gefährliche Konstruktion *f*

***danger***: **~ sign** *n* CONST LAW, INFR & DES Warnschild *nt*

**dappled** *adj* MAT PROP gefleckt

**darby** *n* MAT PROP Abziehlatte *f*

**dark**: **~ room** *n* ARCH & BUILD Dunkelkammer *f*

**data**: **~ sheet** *n* ARCH & BUILD, CONST LAW, INFR & DES Datenblatt *nt*

**date** *n* ARCH & BUILD Datum *nt*, Termin *m*, CONST LAW Termin *m*

**datum** *n* ARCH & BUILD Bezugshöhe *f*, *reference quantity* Bezugsgröße *f*, *reference position* Bezugspunkt *m*, INFR & DES Bezugshöhe *f*, *reference quantity* Bezugsgröße *f*, *bench mark* Bezugspunkt *m*; **~ level** *n* ARCH & BUILD Bezugshöhe *f*, Marke *f*, INFR & DES Bezugshöhe *f*; **~ line** *n* ARCH & BUILD, INFR & DES Bezugslinie *f*; **~ plane** *n* ARCH & BUILD, INFR & DES Bezugsebene *f*; **~ point** *n* ARCH & BUILD, INFR & DES Bezugspunkt *m*

**daub 1.** *n* STONE *plaster* Bewurf *m*; **2.** *vt* STONE *plaster* mit Putz bewerfen, bewerfen

**daylight** *n* ARCH & BUILD Tageslicht *nt*; **~ illumination** *n* ARCH & BUILD Tageslichtbeleuchtung *f*

**dayroom** *n* ARCH & BUILD, INFR & DES Aufenthaltsraum *m*

**daywork**: **~ accounts** *n pl* CONST LAW Tagesarbeitsbücher *nt pl*; **~ sheets** *n pl* CONST LAW Tagesarbeitsblätter *nt pl*

**dB** *abbr* (*decibel*) ENVIRON, INFR & DES, MAT PROP, SOUND & THERMAL dB (*Dezibel*) *nt*

**DCF** *abbr* (*discounted cash flow*) CONST LAW abgezinster Cash-flow *m*, diskontierter Cash-flow *m*

**DD** *abbr* (*deep-drawn*) MAT PROP tiefgezogen

**de-ionize** *vt* MAT PROP, WASTE WATER entionisieren

**deadbolt** *n* BUILD HARDW Schloßriegel *m*, *lock* Absteller *m*

**dead**: **~ dike** *n* ENVIRON ungenutzte Stauung *f*

**deaden** *vt* SOUND & THERMAL auffüllen, dämmen, dämpfen, isolieren

**dead**: **~-end** *n* INFR & DES Sackgasse *f*; **~-end railroad station** *n AmE*, **~-end railway station** *BrE n* INFR & DES Sackbahnhof *m*

**deadening 1.** *adj* SOUND & THERMAL dämpfend; **2.** *n* SOUND & THERMAL Dämpfung *f*, Dämpfen *nt*

**dead**: **~ floor** *n* ARCH & BUILD schallschluckender Boden *m*, Blendboden *m*, Fehlboden *m*, SOUND & THERMAL schallschluckender Boden *m*

**deadline** *n* CONST LAW Frist *f*

**dead**: **~ load** *n* ARCH & BUILD, INFR & DES, MAT PROP Eigengewicht *nt*; **~-load deflection** *n* ARCH & BUILD Eigenlastdurchbiegung *f*

**deadlock** *n* BUILD HARDW klinkenloses Schloß *nt*

**deadman** *n* ARCH & BUILD, INFR & DES Ankerblock *m*, Bodenverankerung *f*, Erdanker *m*, STEEL Erdanker *m*

**dead**: **~ shore** *n* ARCH & BUILD *bracing* Stempel *m*; **~ shoring** *n* ARCH & BUILD Unterfahren *nt*, Unterfangen *nt*; **~ wall** *n* ARCH & BUILD vollflächige Mauer *f*, blinde Mauer *f*; ~ **weight** *n* ARCH & BUILD, INFR & DES, MAT PROP Eigengewicht *nt*, Eigenmasse *f*; ~ **wire** *n* ELECTR tote Leitung *f*

**deal** *n* TIMBER Bohle *f*, Planke *f*, Diele *f*

**deambulatory** *n* ARCH & BUILD Chorumgang *m*, Umgangsgewölbe *nt*

**debris** *n* ARCH & BUILD Trümmer *pl*, HEAT & VENT Absatz *m*, INFR & DES Schutt *m*, WASTE WATER Absatz *m*; ~ **cone** *n* INFR & DES Schuttkegel *m*; ~ **utilization** *n* INFR & DES Trümmerverwertung *f*

**debur** *vt* STEEL entgraten

**decagon** *n* ARCH & BUILD Zehneck *nt*

**decantation** *n* ENVIRON Absetzklärung *f*; ~ **rate** *n* ENVIRON, INFR & DES, MAT PROP Sinkgeschwindigkeit *f*

**decanter** *n* ENVIRON Abklärgefäß *nt*, Absetzgefäß *nt*, INFR & DES Absetzgefäß *nt*

**decastyle** *adj* ARCH & BUILD zehnsäulig

**decayed**: ~ **timber** *n* MAT PROP, TIMBER verrottetes Holz *nt*

**decibel** *n* (*dB*) ENVIRON, INFR & DES, MAT PROP, SOUND & THERMAL Dezibel *nt* (*dB*)

**decimetric**: ~ **system** *n* ARCH & BUILD *measurement* Dezimetersystem *nt*

**deck** *n* INFR & DES Deck *nt*, *bridge* Fahrbahn *f*, Fahrbahnplatte *f*; ~ **bridge** *n* INFR & DES Brücke *f* mit obenliegender Fahrbahn

**decking** *n* ARCH & BUILD Bohlenbelag *m*, Abdeckung *f*, INFR & DES, TIMBER Bohlenbelag *m*

**declination** *n* INFR & DES Deklination *f*

**decoiled** *adj* STEEL ungerollt

**decolorization** *n AmE*, **decolourization** *n BrE* SURFACE Entfärbung *f*

**decomposition**: ~ **of forces** *n* ARCH & BUILD, INFR & DES Kräftezerlegung *f*

**decompression** *n* BUILD MACHIN, HEAT & VENT, WASTE WATER Druckabnahme *f*; ~ **sickness** *n* CONST LAW, INFR & DES Senkkastenkrankheit *f*

**decontaminate** *vt* CONST LAW, ENVIRON, INFR & DES, MAT PROP *soil* entseuchen

**decontaminated** *adj* CONST LAW, ENVIRON, INFR & DES, MAT PROP *soil* entseucht

**decontamination** *n* CONST LAW, ENVIRON, INFR & DES, MAT PROP *of soil* Dekontaminierung *f*

**décor** *n* ARCH & BUILD Dekor *nt*

**decorate** *vt* ARCH & BUILD verzieren

**decorated** *adj* ARCH & BUILD verziert; ~ **ceiling** *n* ARCH & BUILD Zierdecke *f*; ~ **door** *n* ARCH & BUILD, BUILD HARDW Dekortür *f*, Ziertür *f*; ~ **style** *n* ARCH & BUILD, INFR & DES dekorierter Stil *m*

**decoration** *n* ARCH & BUILD *ornament* Verzierung *f*, *style* Dekor *nt*

**decorative** *adj* ARCH & BUILD , BUILD HARDW, CONCR, STEEL, TIMBER dekorativ; ~ **arch** *n* ARCH & BUILD Zierbogen *m*; ~ **batten** *n* TIMBER Zierleiste *f*; ~ **coat** *n* SURFACE Dekorbeschichtung *f*; ~ **finish** *n* SURFACE Dekorbeschichtung *f*; ~ **fittings** *n pl* BUILD HARDW Zierbeschläge *m pl*; ~ **gable** *n* ARCH & BUILD Ziergiebel *m*, STEEL Ziergitter *nt*; ~ **tile** *n* ARCH & BUILD, BUILD HARDW, STONE Zierfliese *f*

**decorator** *n* ARCH & BUILD, SURFACE Maler *m*

**decrease 1.** *n* BUILD MACHIN Abfall *m*; ~ **in pressure** *n* BUILD MACHIN, HEAT & VENT, WASTE WATER Druckabfall *m*; **2.** *vi* HEAT & VENT, WASTE WATER abfallen

**decreaser** *n* HEAT & VENT, WASTE WATER Reduktionsstück *nt*

**deep**: ~ **basement construction** *n* ARCH & BUILD, INFR & DES Tiefgeschoßkonstruktion *f*; ~ **beam** *n* ARCH & BUILD wandartiger Träger *m*, hoher Balkenträger *m*; ~**-drawn** *adj* (*DD*) MAT PROP tiefgezogen; ~ **well** *n* INFR & DES Tiefbrunnen *m*; ~ **workings** *n pl* INFR & DES Tiefbau *m*

**defect** *n* ARCH & BUILD, CONST LAW Defekt *m*, Mangel *m*, HEAT & VENT, INFR & DES, WASTE WATER Störung *f*

**defective** *adj* ARCH & BUILD, CONST LAW, HEAT & VENT, INFR & DES, WASTE WATER fehlerhaft, mangelhaft, schadhaft

**defectively**: ~ **designed** *adj* ARCH & BUILD, CONST LAW, HEAT & VENT, INFR & DES, WASTE WATER fehlerhaft geplant

**defects**: ~ **date** *n* CONST LAW Defektdatum *nt*

**deferrization** *n* ENVIRON Enteisenung *f*

**deficiencies** *n pl* ARCH & BUILD, CONST LAW Mängel *m pl*

**definable** *adj* ARCH & BUILD, CONST LAW bestimmbar

**define** *vt* ARCH & BUILD, CONST LAW, INFR & DES abgrenzen, bestimmen, definieren

**defined** *adj* ARCH & BUILD bestimmt, festgelegt, CONST LAW festgelegt, INFR & DES bestimmt

**definition** *n* ARCH & BUILD, CONST LAW Abgrenzung *f*, Begriffsbestimmung *f*, Definition *f*, INFR & DES Bestimmung *f*

**deflect** *vt* ARCH & BUILD durchbiegen, sich durchbiegen, MAT PROP durchbiegen

**deflection** *n* ARCH & BUILD, MAT PROP Durchbiegung *f*, Abweichung *f*; ~ **curve** *n* MAT PROP *beam, girder* Biegelinie *f*; ~ **transducer** *n* ARCH & BUILD, INFR & DES, MAT PROP, STEEL Durchbiegungsmeßwertaufnehmer *m*

**deflectometer** *n* BUILD MACHIN Durchbiegungsmesser *m*

**deflector** *n* ENVIRON Deflektor *m*; ~ **cap** *n* HEAT & VENT Deflektorhaube *f*

**deformation** *n* ARCH & BUILD, INFR & DES, MAT PROP Verformung *f*; ~ **during burning** *n* INFR & DES, MAT PROP Brandverzug *m*; ~ **under heat** *n* MAT PROP Wärmeverformung *f*

**degasification** *n* ENVIRON Entgasung *f*

**degassing** *n* ENVIRON Entgasung *f*

**degradable** *adj* ENVIRON abbaubar

**degreasing** *n* WASTE WATER Entfettung *f*

**degree**: ~ **of accuracy** *n* INFR & DES, MAT PROP Genauigkeitsgrad *m*; ~ **of clarification** *n* INFR & DES, MAT PROP Reinigungsgrad *m*; ~ **day** *n* CONST LAW *energy*, HEAT & VENT Gradtag *m*; ~ **of exposure** *n* HEAT & VENT, MAT PROP Aussetzungsgrad *m*; ~ **of freedom** *n* ARCH & BUILD, INFR & DES Freiheitsgrad *m*; ~ **of hardness** *n* MAT PROP Härtegrad *m*; ~ **of indeterminacy** *n* ARCH & BUILD, INFR & DES Unbestimmtheitsgrad *m*; ~ **of pollution** *n* ENVIRON Verschmutzungsgrad *m*; ~ **of statical indeterminacy** *n* INFR & DES Grad *m* der statischen Unbestimmtheit

**dehumidifier** *n* HEAT & VENT Entfeuchter *m*

**dehumidify** *vt* HEAT & VENT, MAT PROP entfeuchten

**dehydrate** *vt* HEAT & VENT, INFR & DES dehydrieren, entwässern

**dehydration** *n* HEAT & VENT, INFR & DES Dehydration *f*, Entwässerung *f*; ~ **of sludge** *n* ENVIRON Schlammentwässerung *f*, Schlammverdickung *f*, INFR & DES Schlammentwässerung *f*

**delay 1.** *n* CONST LAW Verzögerung *f*; **2.** *vt* CONST LAW verzögern
**delayed**: **~ closing** *n* BUILD HARDW *door* verzögertes Schließen *nt*
**deleterious**: **~ substance** *n* CONST LAW, MAT PROP Schadstoff *m*
**deliver** *vt* ARCH & BUILD, INFR & DES, MAT PROP liefern
**delivery** *n* ARCH & BUILD, INFR & DES, MAT PROP Anlieferung *f*, Lieferung *f*; **~ door** *n* ARCH & BUILD, INFR & DES Lieferantentür *f*; **~ pipe** *n* ENVIRON, HEAT & VENT, WASTE WATER Druckleitung *f*; **~ ramp** *n* ARCH & BUILD, INFR & DES Anlieferungsrampe *f*; **~ room** *n* ARCH & BUILD, INFR & DES Lieferantenraum *m*
**demi**: **~-column** *n* BUILD HARDW Halbsäule *f*
**demineralization** *n* ENVIRON, WASTE WATER Entsalzung *f*; **~ technique** *n* ENVIRON *waste* Entsalzungsverfahren *nt*
**demineralize** *vt* ENVIRON, WASTE WATER entsalzen
**demineralizer** *n* ENVIRON, WASTE WATER Entsalzungsgerät *nt*
**demolish** *vt* CONST LAW, ENVIRON, INFR & DES abreißen, niederreißen
**demolition** *n* CONST LAW, ENVIRON, INFR & DES Abbruch *m*, Abriß *m*; **~ permit** *n* ARCH & BUILD, CONST LAW, ENVIRON Abbrucherlaubnis *f*; **~ rubbish** *n* ARCH & BUILD, ENVIRON, INFR & DES Abbruchmaterial *nt*, Schutt *m*; **~ waste** *n* ARCH & BUILD Bauschutt *m*, Abbruchabfall *m*, ENVIRON Abbruchabfall *m*, Abbruchmaterial *nt*, Bauschutt *m*, INFR & DES Bauschutt *m*, Schutt *m*; **~ work** *n* ARCH & BUILD, ENVIRON, INFR & DES Abbrucharbeit *f*
**demonstration**: **~ kitchen** *n* ARCH & BUILD Musterküche *f*
**demountable**: **~ partition** *n* BUILD HARDW versetzbare Wand *f*; **~ structure** *n* ARCH & BUILD, BUILD HARDW demontierbare Konstruktion *f*
**demulsifier** *n* ENVIRON Demulgator *m*, Emulsionsspalter *m*
**denitrification** *n* ENVIRON, INFR & DES, WASTE WATER Denitrifikation *f*
**densified**: **~ plywood** *n* TIMBER Preßsperrholz *nt*
**density** *n* MAT PROP Raumgewicht *nt*
**denticulated** *adj* ARCH & BUILD feingezahnt
**deodorization** *n* WASTE WATER Geruchsbeseitigung *f*
**Department**: **~ of the Environment** *n BrE* CONST LAW Umweltschutzministerium *nt*
**departure**: **~ lounge** *n* ARCH & BUILD, INFR & DES Abflughalle *f*
**depolluted**: **~ water** *n* ENVIRON, WASTE WATER gereinigtes Wasser *nt*
**deposit 1.** *n* ENVIRON Ablagerung *f*, INFR & DES Absetzen *nt*, Bodensatz *m*, MAT PROP Bodensatz *m*; **2.** *vt* ENVIRON ablagern, INFR & DES absetzen
***deposit***: **~ build-up** *n* ENVIRON, MAT PROP Ablagerung *f*
**deposited**: **~ matter** *n* ENVIRON Sinkgut *nt*, Sinkstoff *m*
**deposition** *n* ENVIRON Sedimentablagerung *f*, *mining* Ablagerung *f*; **~ rate** *n* ENVIRON Ablagerungsrate *f*; **~ of silt** *n* ENVIRON Schlickablagerung *f*; **~ value** *n* ENVIRON Ablagerungswert *m*; **~ velocity** *n* ENVIRON *of radioactive particles* Ablagerungsgeschwindigkeit *f*
**depreciation** *n* CONST LAW Abschreibung *f*
**depth** *n* ARCH & BUILD, BUILD MACHIN, INFR & DES, MAT PROP Höhe *f*, Tiefe *f*; **~ factor** *n* ARCH & BUILD, TIMBER Tiefenfaktor *m*; **~ of foundation** *n* ARCH & BUILD Fundamenttiefe *f*, Gründungstiefe *f*, INFR & DES Fundamenttiefe *f*; **~ of frost penetration** *n* ARCH & BUILD, INFR & DES Frostgrenze *f*, Frosttiefe *f*; **~ of penetration** *n* INFR & DES, MAT PROP, SURFACE Eindringungstiefe *f*
**derrick** *n* BUILD MACHIN Derrick *m*, Mastenkran *m*, Hebebock *m*, *drilling rig* Bohrturm *m*, ENVIRON Bohrturm *m*; **~ crane** *n* BUILD MACHIN Derrickkran *m*; **~ girt** *n* BUILD MACHIN Turmverbindung *f*; **~ post** *n* BUILD MACHIN Derrickmast *m*
**derusting**: **~ agent** *n* MAT PROP, SURFACE Rostentferner *m*
**desalinate** *vt* INFR & DES, WASTE WATER entsalzen
**desalination** *n* INFR & DES, WASTE WATER Wasserentsalzung *f*
**descale** *vt* HEAT & VENT entkalken, SURFACE entzundern, WASTE WATER entkalken
**descaling** *n* HEAT & VENT Entkalken *nt*, SURFACE Entzundern *nt*, WASTE WATER Entkalken *nt*
**description**: **~ of work** *n* ARCH & BUILD, CONST LAW Baubeschreibung *f*
**design** *n* ARCH & BUILD, INFR & DES, MAT PROP *plan* Plan *m*, *development* Entwurf *m*, Design *nt*, Dessin *nt*, *of constructions* Konstruktion *f*, SURFACE Dessin *nt*; **~ assumption** *n* ARCH & BUILD, CONCR, INFR & DES, STEEL, TIMBER Bemessungsannahme *f*
**designation** *n* ARCH & BUILD, CONST LAW Bezeichnung *f*
***design***: **~ bending resistance** *n* ARCH & BUILD, CONCR, INFR & DES, MAT PROP, STEEL, TIMBER zulässige Biegetragfähigkeit *f*; **~ capacity** *n* ARCH & BUILD Ausbaugröße *f*; **with ~ certification** *phr* CONST LAW bauartzugelassen; **~ criterion** *n* ARCH & BUILD Bemessungkriterium *nt*; **~ data** *n pl* ARCH & BUILD Entwurfsdaten *nt pl*; **~ drawing** *n* ARCH & BUILD Entwurfszeichnung *f*
**designed** *adj* ARCH & BUILD, CONST LAW, INFR & DES, MAT PROP rechnerisch; **~ stress-strain diagram** *n* ARCH & BUILD, CONCR, MAT PROP rechnerische Spannungsdehnungslinie *f*
**designer** *n* ARCH & BUILD, INFR & DES Planer *m*
***design***: **~ error** *n* ARCH & BUILD Bemessungsfehler *m*, Planungsfehler *m*; **~ load** *n* ARCH & BUILD, CONCR, INFR & DES, STEEL Lastannahme *f*, Lastansatz *m*; **~ objective** *n* ARCH & BUILD, INFR & DES Designziel *nt*; **~ plastic shear resistance** *n* ARCH & BUILD zulässige plastische Schubtragfähigkeit *f*; **~ pressure** *n* HEAT & VENT Bemessungsdruck *m*; **~ resistance** *n* ARCH & BUILD, CONCR, INFR & DES, MAT PROP Entwurfstragfähigkeit *f*, Entwurfswiderstand *m*; **~ table** *n* INFR & DES Bemessungstabelle *f*, Bemessungstafel *f*; **~ temperature** *n* HEAT & VENT Bemessungstemperatur *f*; **~ temperature map** *n* HEAT & VENT, INFR & DES Klimakarte *f*; **~ wind load** *n* ARCH & BUILD, INFR & DES Bemessungswindlast *f*
**destroyed**: **~ by fire** *adj* ARCH & BUILD, CONST LAW abgebrannt, niedergebrannt
**desulfurization** *AmE see desulphurization BrE*
**desulfurize** *AmE see desulphurize BrE*
**desulfurized** *AmE see desulphurized BrE*
**desulphurization** *n BrE* ENVIRON Entschwefelung *f*
**desulphurize** *vt BrE* ENVIRON entschwefeln
**desulphurized** *adj BrE* ENVIRON entschwefelt
**detachable** *adj* STEEL, TIMBER abschraubbar
**detached** *adj* ARCH & BUILD, INFR & DES freistehend; **~ building** *n* ARCH & BUILD, INFR & DES freistehendes Gebäude *nt*

**detail** *n* ARCH & BUILD, INFR & DES Detail *nt*; **~ drawing** *n* ARCH & BUILD, INFR & DES Detailzeichnung *f*
**detailed** *adj* ARCH & BUILD, CONST LAW, INFR & DES detailliert; **~ estimate** *n* ARCH & BUILD, CONST LAW, INFR & DES detaillierter Kostenvoranschlag *m*; **~ plans** *n pl* ARCH & BUILD, CONST LAW Bestimmung *f*, detaillierte Pläne *m pl*, INFR & DES detaillierte Pläne *m pl*
**detergents** *n pl* WASTE WATER Detergentien *f pl*
**determination** *n* ARCH & BUILD, CONST LAW, INFR & DES Bestimmung *f*
**determine** *vt* ARCH & BUILD, CONST LAW, INFR & DES bestimmen; **~ statically** *vt* INFR & DES statisch berechnen
**detin** *vt* ENVIRON entzinnen
**detinning** *n* ENVIRON Entzinnung *f*
**developed**: **~ building land** *n* ARCH & BUILD, INFR & DES erschlossenes Bauland *nt*
**development** *n* ARCH & BUILD Abwicklung *f*, CONST LAW, INFR & DES Bebauung *f*
**deviation** *n* ARCH & BUILD Abweichung *f*
**dew** *n* ENVIRON, HEAT & VENT Tau *m*
**dewater** *vt* ENVIRON, INFR & DES entwässern
**dewatered**: **~ sludge** *n* ENVIRON entwässerter Schlamm *m*; **~ waste** *n* ENVIRON entwässerter Abfall *m*
**dewatering** *n* ENVIRON, INFR & DES Entwässerung *f*
**dewdrop**: **~ glass** *n* MAT PROP Tautropfenglas *nt*
***dew***: **~ point** *n* HEAT & VENT, SOUND & THERMAL Taupunkt *m*
**DF** *abbr* (*drinking fountain*) INFR & DES Trinkbrunnen *m*
**diagenesis** *n* ENVIRON Diagenese *f*
**diagenetic** *adj* ENVIRON diagenetisch
**diagonal** *adj* ARCH & BUILD, CONCR, INFR & DES, MAT PROP, STEEL, TIMBER diagonal; **~ bar** *n* MAT PROP Diagonalstab *m*; **~ bracing** *n* ARCH & BUILD Diagonalverband *m*, STEEL Schrägverband *m*, Diagonalverband *m*, *of joists* Diagonalaussteifung *f*, *truss* Dreiecksverband *m*; **~ compression** *n* ARCH & BUILD Diagonaldruck *m*
**diagonal**: **~ masonry bond** *n* STONE Schränkverband *m*; **~ member** *n* ARCH & BUILD, STEEL Diagonale *f*; **~ rod** *n* ARCH & BUILD, TIMBER Diagonalstrebe *f*, Strebe *f*; **~ strut** *n* ARCH & BUILD, STEEL Kreuzstrebe *f*, TIMBER Druckstab *m*, Druckstrebe *f*, Kreuzstrebe *f*; **~ tie** *n* TIMBER Zugdiagonale *f*, Zugstrebe *f*
**diamond** *n* ARCH & BUILD Raute *nt*, BUILD HARDW, BUILD MACHIN Diamant *m*; **~ drill** *n* BUILD MACHIN Diamantbohrer *m*; **~ pattern** *n* ARCH & BUILD Rautenmuster *nt*; **~ pencil** *n* ARCH & BUILD, BUILD HARDW Glaserdiamant *m*; **~ point** *n* BUILD HARDW Diamantspitze *f*; **~-shaped** *adj* ARCH & BUILD rautenförmig; **~ wire lattice** *n* BUILD HARDW Rautendrahtgitter *nt*
**diaphragm** *n* SOUND & THERMAL Abdichtung *f*; **~ wall** *n* ENVIRON, INFR & DES, SOUND & THERMAL, TIMBER Dichtungswand *f*
**diastyle** *adj* ARCH & BUILD weitsäulig
**die** *n* ARCH & BUILD Sockelschaft *m*; **~-casting metal** *n* STEEL Druckgußwerkstoff *m*
**difference** *n* ARCH & BUILD, MAT PROP Differenz *f*
**differential**: **~ equation** *n* ARCH & BUILD Differentialgleichung *f*; **~ force** *n* HEAT & VENT Differentialdruck *m*, Differentialkraft *f*; **~ head** *n* ENVIRON, INFR & DES, WASTE WATER Differenzdruckhöhe *f*; **~ pressure** *n* ENVIRON Differenzdruck *m*; **~ thermal analysis** *n* (*DTA*) ENVIRON Differentialthermoanalyse *f* (*DTA*); **~ thermostat** *n* ELECTR, HEAT & VENT, SOUND & THERMAL Differentialthermostat *m*
**diffuse**: **~ sound field** *n* SOUND & THERMAL diffuses Schallfeld *nt*
**diffusion** *n* ENVIRON Ausbreitung *f*, SOUND & THERMAL Diffusion *f*; **~ coefficient** *n* SOUND & THERMAL Diffusionsbeiwert *m*; **~-tight** *adj* SOUND & THERMAL diffusionsdicht
**digested**: **~ sludge** *n* ENVIRON Faulschlamm *m*
**digester** *n* ENVIRON Faulbehälter *m*; **~ gas** *n* ENVIRON Biogas *nt*, Faulgas *nt*, INFR & DES, MAT PROP Faulgas *nt*
**digestion**: **~ chamber** *n* INFR & DES, WASTE WATER Schlammfaulraum *m*; **~ deposit** *n* ENVIRON Rottedeponie *f*; **~ plant** *n* ENVIRON Faulanlagen *f pl*; **~ plants** *n pl* WASTE WATER Faulanlagen *f pl*; **~ sump** *n* ENVIRON Faulbehälter *m*, Faulraum *m*, Schlammfaulbehälter *m*; **~ tank** *n* ENVIRON Faulbehälter *m*, Faulraum *m*; **~ tower** *n* INFR & DES, WASTE WATER Faulturm *m*
**dike** *n* ARCH & BUILD Graben *m*, ENVIRON Entwässerungskanal *m*, Graben *m*, Stauung *f*, INFR & DES Deich *m*, Graben *m*
**dilapidated** *adj* ARCH & BUILD baufällig, verfallen
**dilute** *vt* SURFACE verdünnen
**dilution** *n* SURFACE Verdünnung *f*
**dimension** *n* ARCH & BUILD Abmessung *f*
**dimensional**: **~ analysis** *n* ARCH & BUILD Dimensionierung *f*; **~ tolerance** *n* ARCH & BUILD Maßtoleranz *f*; **~ variation** *n* ARCH & BUILD Maßabweichung *f*
**dimmer** *n* ELECTR Dimmer *m*
**dimple** *vt* STEEL *rivets* versenken
**DIN** *abbr* (*German Industrial Standard*) ARCH & BUILD, CONST LAW, INFR & DES DIN (*Deutsche Industrienorm*)
**dip** *n* ENVIRON *geology* Einfallen *nt*, HEAT & VENT Durchhang *m*, INFR & DES Einfallen *nt*, SURFACE Tauchen *nt*; **~ coating** *n* SURFACE Tauchlackierung *f*
**dipping**: **~ varnish** *n* SURFACE Tauchlack *m*
**dip**: **~-primed** *adj* SURFACE tauchgrundiert
**direct** *adj* ENVIRON, HEAT & VENT direkt; **~ conversion** *n* ENVIRON Direktumwandlung *f*; **~ coupling** *n* ENVIRON direkte Kupplung *f*; **~ heating** *n* HEAT & VENT Direktheizung *f*
**direction**: **~ of circulation** *n* HEAT & VENT Strömungsrichtung *f*; **~ of rotation** *n* BUILD MACHIN Umlaufrichtung *f*; **~ of thrust** *n* ARCH & BUILD, CONCR, INFR & DES, MAT PROP Schubrichtung *f*
**direct**: **~ light** *n* ELECTR direktes Licht *nt*; **~ lighting** *n* ELECTR direkte Beleuchtung *f*; **~ radiation** *n* ENVIRON direkte Strahlung *f*; **~ solar gain** *n* ENVIRON, HEAT & VENT direkter Solargewinn *m*; **~ support** *n* ARCH & BUILD, CONCR direkte Lagerung *f*; **~ transmission** *n* SOUND & THERMAL direkte Übertragung *f*
**dirt** *n* ENVIRON Schmutz *m*; **~ filter** *n* ENVIRON, WASTE WATER Schmutzfilter *m*; **~-repellent** *adj* SURFACE schmutzabweisend
**disabled** *adj* ARCH & BUILD, CONST LAW, INFR & DES behindert; **~ facilities** *n pl* ARCH & BUILD, CONST LAW, INFR & DES Einrichtungen *f pl* für Behinderte
**disassemble** *vt* ARCH & BUILD, ELECTR, INFR & DES, WASTE WATER demontieren
**disassembly** *n* ARCH & BUILD, ELECTR, INFR & DES, WASTE WATER Demontage *f*
**discharge 1.** *n* ARCH & BUILD, CONST LAW Entlastung *f*, ELECTR Entladung *f*, ENVIRON *sewage* Einleitung *f*,

INFR & DES, MAT PROP Entlastung *f*; **2.** *vt* ARCH & BUILD, CONST LAW entlasten, ELECTR entladen, ENVIRON *water* ablassen, INFR & DES entlasten; **3.** *vi* ENVIRON, WASTE WATER ausfließen

***discharge***: ~ **coefficient** *n* ENVIRON Ausflußzahl *f*; ~ **lamp** *n* ELECTR Entladungslampe *f*; ~ **system** *n* ENVIRON Einleitungsanlage *f*; ~ **valve** *n* BUILD HARDW *halon*, HEAT & VENT Ausströmöffnung *f*

**discharging**: ~ **arch** *n* ARCH & BUILD *masonry* Entlastungsbogen *m*

**disconnect** *vt* ELECTR abschalten, HEAT & VENT abkuppeln

**discounted**: ~ **cash flow** *n* (*DCF*) CONST LAW abgezinster Cash-flow *m*, diskontierter Cash-flow *m*

**disc**: ~ **skimmer** *n BrE* ENVIRON Scheibenabschöpfer *m*

**disinfectant** *n* HEAT & VENT, WASTE WATER Entkeimer *m*

**disk**: ~ **skimmer** *AmE see disc skimmer BrE*

**dislodge** *vt* ARCH & BUILD verrücken

**dismantle** *vt* INFR & DES demontieren

**dismantling** *n* INFR & DES Demontage *f*

**dismount** *vt* ARCH & BUILD, ELECTR, INFR & DES, WASTE WATER demontieren

**dispersant** *n* ENVIRON Dispergens *nt*, Dispersant *nt*, SURFACE Dispergens *nt*

**disperse** *vt* ENVIRON dispergieren, SURFACE dispergieren, fein verteilen

**dispersed**: ~ **resin** *n* SURFACE dispergiertes Harz *nt*

**dispersion** *n* ARCH & BUILD, SURFACE Verbreitung *f*; ~ **paint** *n* SURFACE, WASTE WATER Dispersionsfarbe *f*

**displacement** *n* INFR & DES Verlagerung *f*, Verschiebung *f*, Versetzung *f*; ~ **of support** *n* INFR & DES Auflagerverschiebung *f*

**display**: ~ **window** *n* ARCH & BUILD Ladenfenster *nt*, Schaufenster *nt*

**disposable**: ~ **filter** *n* BUILD HARDW, WASTE WATER Wegwerffilter *m*

**disposal** *n* CONST LAW Entsorgung *f*, ENVIRON *of organic matter* Beseitigung *f*, Entsorgung *f*, INFR & DES Entsorgung *f*; ~ **line** *n* INFR & DES Entsorgungsleitung *f*; ~ **route** *n* ENVIRON Entsorgungsweg *m*; ~ **zone** *n* ENVIRON Endlager *nt*, Endlagerungsstätte *f*

**dispute** *n* CONST LAW Disput *m*

**dissolved**: ~ **matter** *n* MAT PROP gelöste Stoffe *m pl*; ~ **oxygen** *n* MAT PROP gelöster Sauerstoff *m*

**distance** *n* ARCH & BUILD Abstand *m*, INFR & DES Abstand *m*, Distanz *f*, Entfernung *f*, Strecke *f*, Weg *m*; ~ **between girders** *n* ARCH & BUILD, INFR & DES Trägerabstand *m*; ~ **between rafters** *n* ARCH & BUILD Sparrenabstand *m*, CONCR Verteilerstab *m*, TIMBER Sparrenabstand *m*; ~ **block** *n* ARCH & BUILD, ELECTR, HEAT & VENT, WASTE WATER Abstandhalter *m*; ~ **piece** *n* CONCR Abstandhalter *m*

**distemper** *n* MAT PROP, SURFACE Leimfarbe *f*, Temperafarbe *f*

**distil** *vt BrE* ENVIRON destillieren

**distill** *vt AmE* ENVIRON destillieren

**distillation** *n* ENVIRON Destillation *f*

**distribute** *vt* INFR & DES, MAT PROP verteilen

**distributed**: ~ **load** *n* ARCH & BUILD Streckenlast *f*, INFR & DES Flächenlast *f*, Streckenlast *f*

**distribution** *n* ARCH & BUILD Verbreitung *f*, INFR & DES, MAT PROP Verteilung *f*, SURFACE Verbreitung *f*; ~ **bar** *n* INFR & DES Verteilerstab *m*; ~ **box** *n* ELECTR Verteilerkasten *m*; ~ **factor** *n* INFR & DES Verteilungszahl *f*; ~ **of heat** *n* HEAT & VENT, INFR & DES Wärmeverteilung *f*; ~ **rod** *n* CONCR Verteilerstab *m*; ~ **of temperature** *n* HEAT & VENT, INFR & DES Wärmeverteilung *f*

**district** *n* CONST LAW, INFR & DES *region* Gebiet *nt*; ~ **heat** *n* INFR & DES Fernwärme *f*; ~ **heating** *n* HEAT & VENT, INFR & DES Fernheizung *f*; ~ **heating line** *n* HEAT & VENT, INFR & DES Fernheizleitung *f*; ~ **heating plant** *n* HEAT & VENT, INFR & DES Fernheizwerk *nt*; ~ **heat supply** *n* INFR & DES Fernwärmeversorgung *f*; ~ **steam heating** *n* HEAT & VENT, INFR & DES Dampffernheizung *f*

**disturb** *vt* INFR & DES, MAT PROP stören

**ditch** *n* ARCH & BUILD, ENVIRON, INFR & DES *drainage* Graben *m*; ~ **refilling** *n* INFR & DES Baugrubenverfüllung *f*

**division** *n* ARCH & BUILD Aufteilung *f*, Gliederung *f*

**dodecastyle** *adj* ARCH & BUILD zwölfsäulig

**dogleg**: ~ **stairs** *n pl* ARCH & BUILD gegenläufige Treppe *f*

**dogtooth** *n* ARCH & BUILD Hundezahnornament *nt*

**dolomite**: ~ **brick** *n* STONE Dolomitstein *m*; ~ **marble** *n* STONE Dolomitmarmor *m*

**dolomitic**: ~ **lime** *n* MAT PROP Dolomitkalk *m*

**dome** *n* ARCH & BUILD Kappe *f*, Kuppel *f*

**domed** *adj* ARCH & BUILD kuppelförmig; ~ **roof** *n* ARCH & BUILD Kuppeldach *nt*; ~ **structure** *n* ARCH & BUILD Kuppelbauwerk *nt*

***dome***: ~ **ring** *n* ARCH & BUILD Kuppelring *m*; ~ **roof** *n* ARCH & BUILD Kuppeldach *nt*

**domestic**: ~ **appliance** *n* ELECTR Haushaltsgerät *nt*; ~ **architecture** *n* ARCH & BUILD, INFR & DES Wohnungsbau *m*; ~ **consumption** *n* INFR & DES Hausverbrauch *m*; ~ **drinking water** *n* INFR & DES, WASTE WATER häusliches Trinkwasser *nt*; ~ **heating** *n* HEAT & VENT Heizung *f* für Wohnhäuser; ~ **service facilities** *n pl* ELECTR, HEAT & VENT, WASTE WATER haustechnische Anlagen *f pl*; ~ **timber** *n* TIMBER einheimisches Holz *nt*; ~ **waste** *n* ENVIRON, INFR & DES Hausmüll *m*; ~ **waste water** *n* INFR & DES, WASTE WATER häusliches Abwasser *nt*; ~ **water heating** *n* HEAT & VENT häusliche Wasserheizung *f*; ~ **wood** *n* TIMBER einheimisches Holz *nt*

**domical**: ~ **vault** *n* ARCH & BUILD Domikalgewölbe *nt*

**door** *n* BUILD HARDW Tür *f*; ~ **accessories** *n pl* BUILD HARDW Türzubehör *nt*; ~ **bar** *n* BUILD HARDW Türstange *f*; ~ **bolt** *n* BUILD HARDW Türriegel *m*; ~ **buck** *n* BUILD HARDW, TIMBER Türzarge *f*; ~ **bumper** *n* BUILD HARDW Türpuffer *m*; ~ **case** *n* BUILD HARDW Türfutter *nt*, Türzarge *f*; ~ **casing** *n* BUILD HARDW, TIMBER Türzarge *f*; ~ **chain** *n* BUILD HARDW Türkette *f*; ~ **closer** *n* BUILD HARDW *fire door* Türschließer *m*; ~ **fittings** *n pl* BUILD HARDW Türbeschläge *m pl*; ~ **frame** *n* BUILD HARDW, TIMBER Türrahmen *m*, Türzarge *f*; ~ **front** *n* BUILD HARDW Stirnseite *f* der Tür; ~ **furniture** *n* BUILD HARDW Türbeschläge *m pl*; ~ **handle** *n* BUILD HARDW Drücker *m*, Griff *m*, Klinke *f*; ~ **hinge** *n* BUILD HARDW Türband *nt*; ~ **knob** *n* BUILD HARDW Türknauf *m*; ~ **knocker** *n* BUILD HARDW Türklopfer *m*; ~ **leaf** *n* BUILD HARDW Türblatt *nt*; ~ **lining** *n* INFR & DES, MAT PROP, SOUND & THERMAL Türverkleidung *f*; ~ **lintel** *n* ARCH & BUILD, MAT PROP Türsturz *m*; ~ **opener** *n* BUILD HARDW Türöffner *m*, Türdrücker *m*; ~ **panel** *n* BUILD HARDW Türfüllung *f*, Türverkleidung *f*; ~ **post** *n* BUILD HARDW Türpfosten

*m*; ~ **reveal** *n* ARCH & BUILD, BUILD HARDW Türleibung *f*
**doorsill** *n* ARCH & BUILD Türschwelle *f*
***door***: ~ **size** *n* ARCH & BUILD Türgröße *f*
**doorway** *n* ARCH & BUILD Türöffnung *f*
**Doric** *adj* ARCH & BUILD dorisch; ~ **capital** *n* ARCH & BUILD dorisches Kapitell *nt*; ~ **column** *n* ARCH & BUILD dorische Säule *f*; ~ **style** *n* ARCH & BUILD dorischer Stil *m*
**dormant**: ~ **lock** *n* BUILD HARDW versenktes Schloß *nt*
**dormitory**: ~ **town** *n* INFR & DES Schlafstadt *f*
**dosage**: ~ **pump** *n* WASTE WATER Dosierpumpe *f*
**dose**: ~ **response** *n* ENVIRON Dosiswirkung *f*
**double**: ~**-action door** *n* ARCH & BUILD zweiseitig aufschlagende Tür *f*; ~ **archway** *n* INFR & DES Doppeldurchfahrt *f*; ~ **bay** *n* WASTE WATER Doppelabzweig *m*; ~ **bucket collector** *n* ENVIRON Doppelschaufelabscheider *m*; ~ **calipers** *AmE*, ~ **callipers** *BrE n pl* BUILD MACHIN Doppelzirkel *m*; ~ **channel section** *n* MAT PROP, STEEL Doppel-U-Profil *nt*; ~ **cross timber** *n* TIMBER Doppelsturzriegel *m*; ~**-ended match plane** *n* BUILD MACHIN zweiseitiger Spundhobel *m*; ~**-ended sledgehammer** *n* BUILD MACHIN zweiseitiger Pflasterhammer *m*; ~ **false ceiling** *n* ARCH & BUILD doppelte Zwischendecke *f*; ~**-flight staircase** *n* ARCH & BUILD zweiläufige Treppe *f*; ~ **floor** *n* ARCH & BUILD Doppelboden *m*; ~ **girder** *n* ARCH & BUILD Zwillingsträger *m*; ~ **glazing** *n* ARCH & BUILD Doppelverglasung *f*; ~ **lap** *n* ARCH & BUILD, INFR & DES Doppelüberlappung *f*; ~ **lap plain tile** *n* ARCH & BUILD, INFR & DES doppelüberlappender, glatter Ziegel *m*; ~**-layered** *adj* ARCH & BUILD, STONE zweilagig; ~**-leaf door** *n* ARCH & BUILD zweiflügelige Tür *f*; ~**-pitched roof** *n* ARCH & BUILD gebrochenes Mansardendach *nt*; ~ **pitch roof** *n* ARCH & BUILD abgewalmtes Mansardendach *nt*; ~ **reed lathing** *n* INFR & DES, MAT PROP doppeltes Rohrgewebe *nt*; ~ **rib** *n* ARCH & BUILD, MAT PROP Doppelrippe *f*; ~**-sash window** *n* ARCH & BUILD zweiflügeliges Fenster *nt*; ~ **shear** *adj* ARCH & BUILD, TIMBER zweischnittig; ~**-sided** *adj* ARCH & BUILD zweiseitig; ~**-skin partition** *n* ARCH & BUILD Doppelstehfalz *m*; ~ **stretcher** *n* STONE Doppelläufer *m*; ~ **symmetric** *adj* ARCH & BUILD, CONCR, INFR & DES, MAT PROP, STEEL doppelsymmetrisch; ~ **symmetry** *n* ARCH & BUILD, CONCR, INFR & DES, MAT PROP, STEEL Doppelsymmetrie *f*; ~**-threaded screw** *n* MAT PROP Doppelgewindeschraube *f*; ~**-turn lock** *n* BUILD HARDW zweitouriges Schloß *nt*; ~ **wall** *n* STONE Doppelwand *f*; ~**-walled** *adj* ARCH & BUILD doppelwandig; ~ **window** *n* ARCH & BUILD Doppelfenster *nt*; ~**-winged building** *n* ARCH & BUILD zweiflügeliges Gebäude *nt*
**dovetail** **1.** *n* TIMBER Schwalbenschwanz *m*; **2.** *vt* TIMBER einschwalben
***dovetail***: ~ **joint** *n* TIMBER Schwalbenschwanzverbindung *f*; ~ **tie** *n* STONE Schwalbenschwanzmaueranker *m*
**dowel** **1.** *n* BUILD HARDW, INFR & DES, STONE, TIMBER Dübel *m*, Stift *m*; **2.** *vt* BUILD HARDW, INFR & DES, STONE, TIMBER verdübeln, verstiften
***dowel***: ~ **hole** *n* BUILD HARDW Dübelloch *nt*
**dowelled** *adj* INFR & DES, STONE, TIMBER verdübelt; ~ **connection** *n* STONE Dübelverbindung *f*
***dowel***: ~ **pin** *n* BUILD HARDW Dübel *m*, Paßstift *m*
**downdraught** *n* HEAT & VENT Falluft *f*
**downflow**: ~ **heater** *n* HEAT & VENT Ventilatorheizung *f*
**downhill**: ~ **slope** *n* ARCH & BUILD, INFR & DES Abhang *m*
**downlighter** *n* ARCH & BUILD, ELECTR Spotlicht *nt*
**downpipe** *n* HEAT & VENT, WASTE WATER Fallrohr *nt*
**downspout** *n* WASTE WATER Fallrohr *nt*
**downstream** *adj* ENVIRON, INFR & DES stromabwärts
**downward**: ~ **gradient** *n* INFR & DES *incline* Gefälle *nt*
**downwind** **1.** *adj* ENVIRON, INFR & DES dem Wind abgekehrt; **2.** *adv* ENVIRON mit dem Wind, INFR & DES windabwärts
**DPC** *abbr* (*dampproof course*) SOUND & THERMAL, STONE Feuchtigkeitssperrschicht *f*
**DPM** *abbr* (*dampproof membrane*) MAT PROP, SOUND & THERMAL Dichtungshaut *f*
**draft** *AmE see draught BrE*
**drain** **1.** *n* HEAT & VENT, INFR & DES, WASTE WATER Dole *f*, Entwässerungsgraben *m*, Entwässerungsrohr *nt*; **2.** *vt* HEAT & VENT entleeren, INFR & DES entwässern, WASTE WATER entleeren, entwässern; **3.** *vi* HEAT & VENT, INFR & DES, WASTE WATER ablaufen
**drainage** *n* HEAT & VENT, INFR & DES, WASTE WATER Drainage *f*, Entwässerung *f*; ~ **channel** *n* INFR & DES Entwässerungsrinne *f*; ~ **duct** *n* INFR & DES Entwässerungskanal *m*; ~ **gutter** *n* INFR & DES Abflußrinne *f*, Entwässerungsrinne *f*, WASTE WATER Abflußrinne *f*; ~ **layer** *n* ENVIRON *of depot*, INFR & DES Entwässerungsschicht *f*; ~ **line** *n* INFR & DES Entwässerungsleitung *f*; ~ **pipe** *n* WASTE WATER Abwasserrohr *nt*; ~ **trench** *n* INFR & DES Drainagegraben *m*, Drängraben *m*, Entwässerungsgraben *m*; ~ **ways** *n pl* INFR & DES Entwässerungskanäle *m pl*, Entwässerungsleitungen *f pl*
***drain***: ~ **cock** *n* WASTE WATER Ablaßhahn *m*
**drained** *adj* HEAT & VENT, INFR & DES, WASTE WATER entleert
**drainpipe** *n* INFR & DES, WASTE WATER Abflußrohr *nt*, Abwasserrohr *nt*
***drain***: ~ **system** *n* ENVIRON, INFR & DES, WASTE WATER Kanalisation *f*; ~ **trap** *n* WASTE WATER Geruchsverschluß *m*; ~ **water** *n* ENVIRON, INFR & DES Abwasser *nt*, Schmutzwasser *nt*, WASTE WATER Schmutzwasser *nt*
**draught** *n BrE* ARCH & BUILD Kaminzug *m*, HEAT & VENT *chimney* Kaminzug *m*, Luftzug *m*; ~ **air** *n BrE* ARCH & BUILD Zugluft *f*; ~ **diverter** *n BrE* HEAT & VENT, SOUND & THERMAL Zugluftumleiter *m*; ~ **excluder** *n BrE* HEAT & VENT, SOUND & THERMAL Zugluftschutz *m*; ~ **lobby** *n BrE* ARCH & BUILD, INFR & DES Windfang *m*
**draw** *vt* ARCH & BUILD *a plan* zeichnen; ~ **to scale** *vt* ARCH & BUILD maßstäblich zeichnen
**drawing** *n* ARCH & BUILD Plan *m*, Zeichnung *f*; ~ **board** *n* ARCH & BUILD Zeichenbrett *nt*; ~ **table** *n* ARCH & BUILD Zeichenbrett *nt*; ~ **to scale** *n* ARCH & BUILD maßstabsgerechte Zeichnung *f*
**drawn**: ~ **to scale** *adj* ARCH & BUILD maßstäblich gezeichnet
**dredging**: ~ **pump** *n* BUILD MACHIN Baggerpumpe *f*
**dress** *vt* STONE behauen, TIMBER zurichten
**dressed**: ~ **stone** *n* STONE behauener Naturstein *m*
**dressing** *n* STONE Behauen *nt*, TIMBER Zurichten *nt*
**drill** *vt* BUILD MACHIN, INFR & DES *borehole* bohren
**drilling** *n* BUILD MACHIN, INFR & DES Bohren *nt*; ~ **chart** *n* ARCH & BUILD Bohrdiagramm *nt*; ~ **rig** *n* BUILD

MACHIN, ENVIRON Bohrgestell *nt*, Bohrturm *m*; **~ spoon** *n* INFR & DES Bohrlöffel *m*; **~ tool** *n* BUILD MACHIN Bohrwerkzeug *nt*
**drill**: **~ pin** *n* BUILD MACHIN Bohrstift *m*; **~ stem test** *n* ENVIRON Drillstem-Test *m*
**drinking**: **~ fountain** *n* (*DF*) INFR & DES Trinkbrunnen *m*; **~ water** *n* INFR & DES Trinkwasser *nt*; **~ water supply** *n* INFR & DES Trinkwasserversorgung *f*
**drip** *n* ENVIRON Tropf *m*; **~ cock** *n* WASTE WATER Entwässerungshahn *m*; **~ cup** *n* BUILD HARDW Tropfbecher *m*, Tropfzylinder *m*
**drive** *vt* BUILD HARDW *screw* einschrauben
**driven**: **~ well** *n* INFR & DES Rammbrunnen *m*
**driving**: **~ depth** *n* ARCH & BUILD, INFR & DES Einschlagtiefe *f*
**drop** *n* BUILD MACHIN Abfall *m*; **~ in pressure** *n* BUILD MACHIN, HEAT & VENT, WASTE WATER Druckabfall *m*; **~ in temperature** *n* ENVIRON, HEAT & VENT Temperaturabfall *m*; **~ latches** *n pl* BUILD HARDW *lock* Fallenriegel *m*; **~ pile hammer** *n* BUILD MACHIN Freifallramme *f*
**drum**: **~ washer** *n* BUILD MACHIN Trommelwascher *m*
**dry**: **~ area** *n* ARCH & BUILD, INFR & DES kapillarbrechende Schicht *f*; **~ desulfurization process** *AmE*, **~ desulphurization process** *BrE n* ENVIRON Trockenentschwefelungsprozeß *m*
**drying**: **~ bed** *n* INFR & DES Trockenbett *nt*; **~ out** *n* ARCH & BUILD, CONCR, MAT PROP, TIMBER Austrocknen *nt*; **~ times** *n pl* MAT PROP Trockenzeiten *f pl*
**dry**: **~ joint** *n* HEAT & VENT, STONE mörtellose Fuge *f*; **~ lining system** *n* ARCH & BUILD, INFR & DES, MAT PROP, STONE Trockenputzsystem *nt*; **~ rot** *n* ARCH & BUILD, MAT PROP, TIMBER Trockenfäule *f*; **~-sludge disposal site** *n* ENVIRON Trockenschlammdeponie *f*; **~ sprinkler system** *n* ARCH & BUILD Trockensprinklersystem *nt*; **~ steam** *n* ENVIRON Trockendampf *m*
**DTA** *abbr* (*differential thermal analysis*) ENVIRON DTA (*Differentialthermoanalyse*) *f*
**dual**: **~ carriageway** *n* INFR & DES Schnellstraße *f*
**ductile** *adj* MAT PROP geschmeidig
**ductwork** *n* INFR & DES Kanalsystem *nt*
**dummy**: **~ rafter** *n* TIMBER Blindsparren *m*
**dump 1.** *n* ENVIRON, INFR & DES, Bodenkippe *f*, Deponie *f*, Müllabladeplatz *m*, Mülldeponie *f*, Müllhalde *f*, Standort der Deponie *m*; **2.** *vt* ENVIRON, INFR & DES abkippen
**dumper** *n* BUILD MACHIN Muldenkipper *m*
**dump**: **~ ground** *n* ENVIRON, INFR & DES Müllabladeplatz *m*, Mülldeponie *f*, Müllhalde *f*
**dumping** *n* ENVIRON, INFR & DES Auskippen *nt*, Ausschütten *nt*, Müllabladen *nt*; **~ bucket** *n* BUILD MACHIN Kippkübel *m*; **~ fees** *n pl* CONST LAW, INFR & DES Schuttabladegebühren *f pl*; **~ ground** *n* ENVIRON, INFR & DES Deponie *f*, Müllabladeplatz *m*, Mülldeponie *f*, Müllhalde *f*; **~ site** *n* ENVIRON, INFR & DES Deponiegelände *nt*, Mülldeponie *f*, Schuttabladeplatz *m*; **~ station** *n* BUILD MACHIN Abladestation *f*
**dump**: **~ site** *n* ENVIRON, INFR & DES Müllabladeplatz *m*, Mülldeponie *f*, Müllhalde *f*; **~ skip** *n* BUILD MACHIN Abkippförderkorb *m*
**dune** *n* INFR & DES Düne *f*
**duplicate** *n* CONST LAW Duplikat *nt*
**durability** *n* MAT PROP Dauerfestigkeit *f*
**dust** *n* ENVIRON Staub *m*
**dustbin** *n BrE* (*cf garbage can AmE*) ENVIRON *outside house* Mülltonne *f*
**dust**: **~ content** *n* ENVIRON Staubanteil *m*; **~ and noise control** *n* CONST LAW, INFR & DES, SOUND & THERMAL Staub- und Lärmschutz *m*; **~ particle** *n* ENVIRON Staubteilchen *nt*; **~ preventer** *n* MAT PROP, SURFACE Staubschutzmittel *nt*; **~ protection** *n* CONST LAW, INFR & DES Staubschutz *m*
**duty** *n* CONST LAW Pflicht *f*; **~ of care** *n* CONST LAW Sorgepflicht *f*; **~ to warn** *n* CONST LAW Warnpflicht *f*
**dynamic** *adj* INFR & DES dynamisch
**dynamically**: **~ loaded beam** *n* ARCH & BUILD, INFR & DES dynamisch belasteter Balken *m*
**dynamic**: **~ buckling** *n* ARCH & BUILD, INFR & DES dynamisches Knicken *nt*; **~ force** *n* INFR & DES dynamische Kraft *f*; **~ load** *n* INFR & DES dynamische Last *f*; **~ loading** *n* ARCH & BUILD, INFR & DES dynamische Belastung *f*; **~ strength** *n* ARCH & BUILD, INFR & DES dynamische Festigkeit *f*; **~ test** *n* MAT PROP dynamische Untersuchung *f*; **~ viscosity** *n* MAT PROP dynamische Zähigkeit *f*

# E

**E** *abbr* (*evaporation*) ARCH & BUILD, ENVIRON, HEAT & VENT, INFR & DES, MAT PROP E (*Evaporation*)
**early**: ~ **strength** *n* INFR & DES, MAT PROP Frühfestigkeit *f*
**earth 1.** *n* ARCH & BUILD, ELECTR (*cf ground AmE*) Erde *f*, Erdung *f*, INFR & DES Boden *m*, Grund *m*, Erdreich *nt*, Erdboden *m*, Erde *f*; **2.** *vt BrE* (*cf ground AmE*) ELECTR erden
***earth***: ~ **auger** *n* BUILD MACHIN, INFR & DES Erdbohrer *m*; ~ **body** *n* INFR & DES Erdkörper *m*; ~ **box** *n BrE* (*cf ground box AmE*) ELECTR Erdungsbox *f*; ~ **busbar** *n BrE* (*cf ground bus bar AmE*) ELECTR Erdungsschiene *f*; ~ **compacting equipment** *n* INFR & DES Bodenverdichtungsgerät *nt*; ~ **conditions** *n pl* (*cf ground conditions AmE*) INFR & DES Bodenverhältnisse *nt pl*; ~ **conductor** *n BrE* (*cf ground conductor AmE*) ELECTR Erdungsleitung *f*; ~ **construction** *n* INFR & DES Erdkonstruktion *f*; ~ **dam** *n* INFR & DES Erddamm *m*; ~ **displacement** *n* INFR & DES Bodenbewegung *f*; ~ **edge** *n BrE* (*cf ground edge AmE*) ARCH & BUILD geschliffene Kante *f*
**earthenware**: ~ **pipe** *n* WASTE WATER Steingutrohr *nt*, Tonrohr *nt*
***earth***: ~ **fault circuit interrupter** *n BrE* (*cf ground fault circuit interrupter AmE*) ELECTR Fehlerstromschutzschalter *m*; ~ **fill** *n* INFR & DES Erdaufschüttung *f*; ~ **glass** *n BrE* (*cf ground glass AmE*) MAT PROP Mattglas *nt*; ~ **improvement** *n BrE* (*cf ground improvement AmE*) ARCH & BUILD Baugrundverbesserung *f*
**earthing** *n BrE* (*cf grounding AmE*) ELECTR Erdung *f*, Schutzerdung *f*; ~ **field** *n BrE* (*cf grounding field AmE*) ELECTR Erdungsfeld *nt*; ~ **pipe clamp** *n BrE* (*cf grounding pipe clamp AmE*) ELECTR Erdungsrohrschelle *f*; ~ **plate** *n BrE* (*cf grounding plate AmE*) ELECTR Erdungsplatte *f*; ~ **resistance** *n BrE* (*cf grounding resistance AmE*) ELECTR Ableitungswiderstand *m*, Erdableitwiderstand *m*, Erdungswiderstand *m*; ~ **strip** *n BrE* (*cf grounding strip AmE*) ELECTR Banderder *m*; ~ **system** *n BrE* (*cf grounding system AmE*) ELECTR Erdungsanlage *f*; ~ **well** *n BrE* (*cf grounding well AmE*) ELECTR Erdungsmeßschacht *m*
***earth***: ~ **lead** *n BrE* ELECTR Erddraht *m*; ~ **line** *n BrE* ELECTR Grundleitung *f*; ~ **movement** *n* ARCH & BUILD, INFR & DES Erdbewegung *f*; ~ **moving plant** *n* BUILD MACHIN Erdbewegungsanlage *f*; ~ **plate** *n BrE* ELECTR Erdungsplatte *f*; ~ **pressure** *n* INFR & DES Bodendruck *m*; ~ **pressure cell** *n* INFR & DES Erddruckmeßdose *f*; ~ **pressure coefficient** *n* INFR & DES Erddruckkoeffizient *m*; ~ **pressure diagram** *n* INFR & DES Erddruckdiagramm *nt*; ~ **pressure line** *n* INFR & DES Stellungslinie *f*; ~ **pressure at rest** *n* INFR & DES Erdruhedruck *m*
**earthquake** *n* INFR & DES Erdbeben *nt*; ~ **measurement** *n* INFR & DES Erdbebenmessung *f*; **~-proof** *adj* INFR & DES erdbebensicher
***earth***: ~ **rammer** *n* BUILD MACHIN Erdstampfer *m*; ~ **reservoir** *n* INFR & DES Erdbecken *nt*; **~-retaining wall** *n* INFR & DES Erdstützmauer *f*; ~ **rod** *n BrE* ELECTR Staberder *m*, Tiefenerder *m*; **~'s core** *n* ENVIRON Erdkern *m*; **~'s crust** *n* ENVIRON Erdkruste *f*, Lithosphäre *f*; ~ **slope** *n* INFR & DES Erdböschung *f*; ~ **tester** *n BrE* ELECTR (*cf ground tester AmE*) Erdungsmeßgerät *nt*
**earthwork** *n* ARCH & BUILD, INFR & DES Bodenbewegung *f*, Erdbau *m*, Erdbauarbeiten *f pl*
**earthworking**: ~ **machinery** *n* BUILD MACHIN Erdbaumaschinen *f pl*
**easily**: ~ **accessible** *adj* BUILD MACHIN leicht zugänglich
**easy**: ~ **fit** *n* BUILD HARDW Gleitpassung *f*
**eaves** *n pl* ARCH & BUILD, TIMBER Dachtraufe *f*, Traufkante *f*, Traufe *f*; ~ **board** *n* ARCH & BUILD, INFR & DES, MAT PROP Traufbohle *f*, TIMBER Aufschiebling *m*, Stirnbrett *nt*, Traufbohle *f*, Traufbrett *nt*; ~ **gutter** *n* ARCH & BUILD Dachrinne *f*, Traufrinne *f*, vorgehängte Dachrinne *f*, WASTE WATER Dachrinne *f*; ~ **height** *n* ARCH & BUILD Traufhöhe *f*; ~ **trough** *n* ARCH & BUILD, WASTE WATER Dachrinne *f*
**ebb**: ~ **generation** *n* ENVIRON Ebbekrafterzeugung *f*; ~ **tide** *n* ENVIRON Ebbeströmung *f*, Tideablauf *m*
**EC** *abbr* (*Eurocode*) ARCH & BUILD EC (*Eurocode*), CONST LAW (*European Commission, Eurocode*) KEG (*Kommission der Europäischen Gemeinschaft*), EC (*Eurocode*), INFR & DES, MAT PROP, STEEL EC (*Eurocode*)
**eccentric**: ~ **connection** *n* ARCH & BUILD, CONCR, INFR & DES, MAT PROP, STEEL, TIMBER außermittige Verbindung *f*
**eccentricity** *n* ARCH & BUILD, TIMBER Außermittigkeit *f*
**eccentric**: ~ **load** *n* ARCH & BUILD, TIMBER exzentrische Last *f*
***EC***: ~ **conformity mark** *n* CONST LAW EC-Übereinstimmungszeichen *nt*
**ecological** *adj* ENVIRON ökologisch; ~ **awareness** *n* ENVIRON Umweltbewußtsein *nt*; ~ **balance** *n* ENVIRON ökologisches Gleichgewicht *nt*; ~ **disaster** *n* ENVIRON Umweltkatastrophe *f*, ökologischer Zusammenbruch *m*; ~ **factor** *n* ENVIRON Umweltfaktor *m*; ~ **menace** *n* ENVIRON Umweltgefahr *f*
**ecology** *n* ENVIRON Ökologie *f*
**ecosystem** *n* ENVIRON Ökosystem *nt*
**edge**: ~ **distance** *n* TIMBER Randabstand *m*; ~ **plate** *n* BUILD HARDW Türkantenschoner *m*; ~ **pressure** *n* STONE Kantenpressung *f*; ~ **region** *n* ARCH & BUILD, INFR & DES Randbereich *m*
**edging**: ~ **board** *n* INFR & DES, TIMBER Blende *f*, Dachabschlußblende *f*, Dachrandprofil *nt*
**edifice** *n* ARCH & BUILD, INFR & DES Bau *m*, Bauwerk *nt*, Gebäude *nt*
**educational**: ~ **building** *n* ARCH & BUILD Ausbildungsgebäude *nt*
**eel**: ~ **ladder** *n* INFR & DES Aalleiter *f*
**effect**: ~ **of creep** *n* CONCR, INFR & DES Kriecheinwirkung *f*; ~ **of frost** *n* ARCH & BUILD, CONCR, HEAT & VENT, INFR & DES, SURFACE Frosteinwirkung *f*
**effective**: ~ **depth** *n* ARCH & BUILD, INFR & DES Nutzhöhe *f*; ~ **head** *n* ENVIRON Druckhöhe *f*,

nutzbares Gefälle *nt*; **~ lateral stiffness** *n* ARCH & BUILD, CONCR, INFR & DES, MAT PROP effektive Biegesteifigkeit *f*, effektive Knicksteifigkeit *f*; **~ length** *n* ARCH & BUILD, CONCR, STEEL, TIMBER Knicklänge *f*; **~ modulus** *n* CONCR, INFR & DES, MAT PROP effektiver Modul *m*; **~ sound pressure** *n* ENVIRON, SOUND & THERMAL effektiver Schalldruck *m*; **~ width** *n* ARCH & BUILD, CONCR, INFR & DES, MAT PROP mitwirkende Gurtweite *f*

**effect**: **~ of shrinkage** *n* ARCH & BUILD, CONCR, MAT PROP, STEEL Schwindeinfluß *m*

**efficiency** *n* ARCH & BUILD, BUILD MACHIN Leistungsfähigkeit *f*; **~ factor** *n* ARCH & BUILD Wirkungsgrad *m*

**efflorescence** *n* MAT PROP Ausblühung *f*

**efflorescent** *adj* MAT PROP ausblühend

**effluent** *n* ENVIRON, WASTE WATER Abfluß *m*, Abwasser *nt*; **~ controller** *n* ENVIRON, WASTE WATER Abflußregler *m*; **~ sludge** *n* ENVIRON, INFR & DES, WASTE WATER Abwasserschlamm *m*; **~ slurry** *n* ENVIRON, INFR & DES, WASTE WATER Abwasserschlamm *m*

**elastic**: **~ analysis** *n* ARCH & BUILD, CONCR, INFR & DES, MAT PROP, STEEL elastische Berechnung *f*; **~ compression** *n* INFR & DES, MAT PROP elastische Zusammendrückung *f*; **~ constant** *n* INFR & DES elastische Konstante *f*; **~ equilibrium** *n* INFR & DES, MAT PROP elastisches Equilibrium *nt*

**elasticity** *n* MAT PROP Elastizität *f*

**elastic**: **~ law** *n* ARCH & BUILD Elastizitätsgesetz *nt*; **~ limit** *n* ARCH & BUILD, CONCR, MAT PROP, STEEL Elastizitätsgrenze *f*; **~ modulus** *n* CONCR, MAT PROP, STEEL E-Modul *m*, Elastizitätsmodul *m*; **~-perfectly-plastic** *adj* ARCH & BUILD, INFR & DES elastisch-perfekt-plastisch; **~-plastic method** *n* ARCH & BUILD, CONCR, STEEL elastisch-plastische Methode *f*; **~ properties** *n pl* INFR & DES, MAT PROP elastische Eigenschaften *f pl*; **~ rebound** *n* INFR & DES elastische Rückbildung *f*; **~ resistance** *n* ARCH & BUILD, CONCR, INFR & DES, MAT PROP, STEEL elastische Tragfähigkeit *f*; **~ resistance to bending** *n* ARCH & BUILD, CONCR, INFR & DES, MAT PROP, STEEL elastische Momententragfähigkeit *f*; **~ section properties** *n pl* ARCH & BUILD, CONCR, INFR & DES, MAT PROP elastische Querschnitteigenschaften *f pl*; **~ support** *n* ARCH & BUILD, CONCR, INFR & DES, MAT PROP, STEEL elastische Lagerung *f*, elastische Unterstützung *f*; **~ theory** *n* ARCH & BUILD, CONCR, INFR & DES, MAT PROP, STEEL Elastizitätstheorie *f*, elastische Theorie *f*

**elastomer** *n* ARCH & BUILD, INFR & DES, MAT PROP Elastomer *nt*; **~ support** *n* ARCH & BUILD, INFR & DES, MAT PROP Elastomerlager *nt*

**elbow**: **~ joint** *n* BUILD HARDW, WASTE WATER *pipe* Kniestück *nt*; **~ rail** *n* BUILD HARDW Armauflage *f*; **~ union** *n* WASTE WATER Krümmerüberwurf *m*

**electric** *adj* ELECTR elektrisch

**electrical** *adj* ELECTR elektrisch; **~ sounding** *n* ELECTR, INFR & DES elektrische Sondierung *f*

**electric**: **~ arc welding** *n* ELECTR, STEEL Lichtbogenschweißen *nt*

**electron** *n* ELECTR, ENVIRON Elektron *nt*; **~ capture detector** *n* ENVIRON Elektroneneinfangdetektor *m*

**electronic**: **~ control system** *n* ELECTR elektronisches Überwachungssystem *nt*

**electro**: **~-osmotic-method** *n* ENVIRON, WASTE WATER Elektro-Osmose-Verfahren *nt*

**electrostatic**: **~ precipitator** *n* (*ESP*) ENVIRON elektrostatischer Staubabscheider *m* (*ESA*)

**element** *n* INFR & DES Element *nt*

**elemental** *adj* CONST LAW wesentlich; **~ cost analysis** *n* CONST LAW wesentliche Kostenanalyse *f*

**elevate** *vt* ARCH & BUILD anheben

**elevated**: **~ tank** *n* INFR & DES Hochreservoir *nt*

**elevating**: **~ equipment** *n* BUILD MACHIN Hebezeug *nt*

**elevation** *n* ARCH & BUILD, ENVIRON *lifting process* Anhebung *f*, *height* Höhe *f*, *upright projection* Aufriß *m*, *of water* Anstieg *m*; **~ angle** *n* ARCH & BUILD, ENVIRON Höhenwinkel *m*

**elevator** *n AmE* (*cf lift BrE*) ARCH & BUILD, BUILD HARDW, ELECTR Aufzug *m*; **~ door** *n AmE* (*cf lift door BrE*) ARCH & BUILD, BUILD HARDW Aufzugstür *f*

**ell** *n* INFR & DES Rohrbogen *m*

**ellipse** *n* ARCH & BUILD, STEEL Ellipse *f*

**elliptical**: **~ arch** *n* ARCH & BUILD, STEEL Ellipsenbogen *m*

**elm** *n* TIMBER Ulme *f*

**elutriation** *n* ENVIRON Auslaugung *f*; **~ test** *n* ENVIRON, STEEL Schlämmversuch *m*

**embankment** *n* ENVIRON *waste disposal site* Grubenwand *f*, INFR & DES *roadside* Straßendamm *m*

**embed** *vt* ARCH & BUILD, INFR & DES, MAT PROP, STONE einbauen, einbetten

**embedded** *adj* ARCH & BUILD, INFR & DES, MAT PROP, STONE eingebettet

**embedding** *n* ARCH & BUILD, INFR & DES, MAT PROP, STONE Einbettung *f*

**embedment** *n* ARCH & BUILD, INFR & DES, MAT PROP Einbindung *f*

**emergency** *n* CONST LAW Notfall *m*; **~ call lamp** *n* CONST LAW, ELECTR Ruflampe *f*; **~ exit** *n* CONST LAW, INFR & DES Notausgang *m*; **~ function** *n* CONST LAW Gefahrenfunktion *f*; **~ generator** *n* ELECTR Notgenerator *m*; **~ illumination** *n* CONST LAW, ELECTR Sicherheitsbeleuchtung *f*; **~ key** *n* CONST LAW Notschlüssel *m*; **~ key function** *n* CONST LAW Notschlüsselfunktion *f*; **~ lighting** *n* CONST LAW, ELECTR Notbeleuchtung *f*, Sicherheitsbeleuchtung *f*; **~ stairs** *n pl* CONST LAW, INFR & DES Fluchttreppe *f*; **~ water supply** *n* INFR & DES Notwasserversorgung *f*

**emission** *n* ENVIRON Emission *f*; **~ data** *n* ENVIRON Emissionsdaten *nt pl*; **~ inventory** *n* ENVIRON Emissionsverzeichnis *nt*; **~ point** *n* ENVIRON Emissionsort *m*; **~ source** *n* ENVIRON Emissionsquelle *f*; **~ standard** *n* ENVIRON Emissionsstandard *m*

**emissivity** *n* ENVIRON Emissionsvermögen *nt*

**emittance** *n* ENVIRON Emittanz *f*

**empirical** *adj* ARCH & BUILD, INFR & DES empirisch; **~ value** *n* INFR & DES Erfahrungswert *m*

**employ** *vt* ARCH & BUILD *make use of* anwenden, gebrauchen, CONST LAW *personnel* beschäftigen

**empty**: **~ weight** *n* MAT PROP Leergewicht *nt*

**emulsion** *n* SURFACE Emulsion *f*; **~ paint** *n* SURFACE Emulsionsfarbe *f*

**enamel** **1.** *n* MAT PROP *varnish* Emaillelack *m*, *glaze* Glasur *f*, *paint varnish* Lackfarbe *f*, *vitreous* Emaille *nt*, SURFACE Lackfarbe *f*; **2.** MAT PROP, SURFACE emaillieren

**encastré** *adj* ARCH & BUILD, INFR & DES eingebaut; **~ beam** *n* ARCH & BUILD, INFR & DES, TIMBER eingespannter Träger *m*

**enclosed**: **~ space** *n* ARCH & BUILD umbauter Raum *m*

**enclosing**: **~ wall** *n* ARCH & BUILD *outer wall of building* Außenwand *f*, Umfassungsmauer *f*, STONE Außenwand *f*

**enclosure** *n* ARCH & BUILD, CONST LAW, INFR & DES Einfriedung *f*, Umfassung *f*, Umzäunung *f*; ~ **wall** *n* ARCH & BUILD, CONST LAW, INFR & DES Einfriedungsmauer *f*, Umfassungswand *f*

**end**: ~-**anchored reinforcement** *n* ARCH & BUILD, INFR & DES Endverankerung *f* der Bewehrung; ~ **anchoring** *n* ARCH & BUILD, INFR & DES Endverankerung *f*; ~ **bearing** *n* ARCH & BUILD, INFR & DES Endauflager *nt*, Endauflagerung *f*; ~-**bearing capacity** *n* ARCH & BUILD, INFR & DES Spitzentragfähigkeit *f*; ~-**bearing pile** *n* ARCH & BUILD, INFR & DES Spitzendruckpfahl *m*; ~ **bearing support** *n* ARCH & BUILD, INFR & DES Endlager *nt*; ~ **distance** *n* TIMBER *parallel to fibre* Endabstand *m*; ~ **moment** *n* ARCH & BUILD, INFR & DES Einspannmoment *nt*, Endmoment *nt*; ~-**of-life vehicle** *n* ENVIRON Altfahrzeug *nt*; ~ **osmosis** *n* INFR & DES Endosmose *f*

**endothermic**: ~ **effect** *n* ENVIRON, INFR & DES endothermischer Effekt *m*

**end**: ~ **rafter** *n* ARCH & BUILD, INFR & DES Endpfette *f*; ~ **of sleeper** *n* ARCH & BUILD, INFR & DES Schwellenkopf *m*; ~ **span** *n* ARCH & BUILD, CONCR, INFR & DES, MAT PROP, STEEL, TIMBER Endfeld *nt*, Endträger *m*; ~ **zone** *n* ARCH & BUILD, INFR & DES *reinforcement* Endverankerungsbereich *m*

**energy** *n* ARCH & BUILD, ENVIRON, INFR & DES Energie *f*; ~ **absorption** *n* ARCH & BUILD, INFR & DES Energieaufnahme *f*; ~ **balance** *n* ARCH & BUILD, INFR & DES Energiebilanz *f*; ~ **budget** *n* ARCH & BUILD, ENVIRON, INFR & DES Energiehaushalt *m*; ~ **conservation** *n* ARCH & BUILD, ENVIRON, HEAT & VENT, INFR & DES Energieerhaltung *f*; ~ **consumption** *n* ARCH & BUILD, ELECTR, ENVIRON, HEAT & VENT, INFR & DES Energieverbrauch *m*; ~ **efficiency** *n* ENVIRON, HEAT & VENT, INFR & DES Energiewirtschaftlichkeit *f*; ~ **exchange** *n* ENVIRON, HEAT & VENT, INFR & DES Energieaustausch *m*; ~ **extraction** *n* ENVIRON Energiegewinnung *f*; ~ **loss** *n* HEAT & VENT Energieverlust *m*; ~ **pattern factor** *n* ENVIRON Energiemusterfaktor *m*; ~ **recovery** *n* ENVIRON, HEAT & VENT *from waste* Energierückgewinnung *f*, energetische Verwertung *f*; ~ **recovery factor** *n* ENVIRON Energierückgewinnungsfaktor *m*; ~ **resources** *n pl* ENVIRON Energieressourcen *f pl*; ~-**saving technology** *n* ENVIRON energiesparende Technologie *f*; ~ **supply** *n* ENVIRON Energieversorgung *f*; ~ **technology** *n* ENVIRON Energietechnik *f*

**engineer** *n* ARCH & BUILD, INFR & DES Ingenieur *m*, Techniker *m*

**engineering** *n* ARCH & BUILD Technik *f*, INFR & DES Ingenieurwesen *nt*, Technik *f*, technische Planung *f*; ~ **brick** *n* MAT PROP, STONE Hartbrandstein *m*, Klinker *m*, Klinkerstein *m*; ~ **room** *n* ARCH & BUILD, INFR & DES Technikraum *m*

**engine**: ~ **room** *n* ARCH & BUILD, BUILD MACHIN Maschinenraum *m*

**enrockment** *n* INFR & DES, MAT PROP Packwerk *nt*, Steinauflage *f*, Steinschüttung *f*

**entablature** *n* ENVIRON Gebälk *nt*, TIMBER Gebälk *nt*, Säulengebälk *nt*

**enthalpy** *n* ENVIRON Enthalpie *f*

**entrance** *n* ARCH & BUILD, INFR & DES Eingang *m*

**entropy** *n* ENVIRON Entropie *f*

**ENV** *abbr* (*European initial standard, European tentative standard*) ARCH & BUILD, CONST LAW, INFR & DES, MAT PROP ENV (*Europäische Vornorm*)

**environment** *n* ARCH & BUILD, CONST LAW Milieu *nt*, Umgebung *f*, Umwelt *f*, ENVIRON Umwelt *f*, INFR & DES Milieu *nt*, Umgebung *f*, Umwelt *f*

**environmental**: ~ **assessment** *n* ENVIRON Umwelteinschätzung *f*; ~ **compatibility** *n* ENVIRON Umweltverträglichkeit *f*; ~ **conditions** *n pl* ENVIRON Umweltbedingungen *f pl*; ~ **disaster** *n* ENVIRON Umweltkatastrophe *f*; ~ **geology** *n* ENVIRON, INFR & DES Umweltgeologie *f*; ~ **geotechnology** *n* ENVIRON, INFR & DES Umweltgeotechnik *f*; ~ **guide** *n* CONST LAW, ENVIRON Umweltführer *m*; ~ **impact** *n* ENVIRON Umweltbelastung *f*

**environmentalism** *n* ARCH & BUILD, CONST LAW, ENVIRON, INFR & DES Umweltschutz *m*

**environmental**: ~ **law** *n* CONST LAW, ENVIRON Umweltgesetz *nt*

**environmentally**: ~ **friendly** *adj* ENVIRON umweltfreundlich

**environmental**: ~ **management scheme** *n* CONST LAW, ENVIRON Umweltverwaltungsprojekt *nt*; ~ **planning** *n* ENVIRON Umweltplanung *f*; ~ **protection** *n* ARCH & BUILD, CONST LAW, ENVIRON, INFR & DES Umweltschutz *m*; ~ **protection agency** *n* CONST LAW, ENVIRON Umweltschutzbehörde *f*; ~ **target** *n* ENVIRON Umweltziel *nt*

**EOTA** *abbr* (*European Organisation for Technical Approvals*) CONST LAW EOTA (*Europäische Organisation für Technische Zulassung*)

**EOTC** *abbr* (*European Organisation for Testing and Certification*) CONST LAW EOTC (*Europäische Organisation für Prüfung und Normung*)

**epicenter** *AmE see epicentre BrE*

**epicentre** *n BrE* INFR & DES Epizentrum *nt*

**episode** *n* ENVIRON Gefahrensituation *f*

**EPM** *abbr* (*equivalent per million*) ENVIRON EPM (*Äquivalent je Million*)

**equalizing**: ~ **ring** *n* CONCR Ausgleichsring *m*

**equal**: ~-**sided angle** *n* BUILD HARDW gleichschenkliges Winkeleisen *nt*

**equilibration** *n* BUILD MACHIN Gewichtsausgleich *m*

**equilibrium** *n* ARCH & BUILD Ausgleich *m*, Gleichgewicht *nt*, Ruhelage *f*, ENVIRON Gleichgewicht *nt*, INFR & DES Ausgleich *m*, Gleichgewicht *nt*, Ruhelage *f*; ~ **conditions** *n pl* ARCH & BUILD, INFR & DES Gleichgewichtsbedingungen *f pl*; ~ **equation** *n* ARCH & BUILD, INFR & DES Gleichgewichtsgleichung *f*; ~ **of forces** *n* ARCH & BUILD, INFR & DES Kräftegleichgewicht *nt*; ~ **limit** *n* ARCH & BUILD, INFR & DES Gleichgewichtsgrenze *f*; ~ **model** *n* ARCH & BUILD, INFR & DES Gleichgewichtsmodell *nt*; ~ **position** *n* ARCH & BUILD, INFR & DES Gleichgewichtszustand *m*; ~ **of stresses** *n* ARCH & BUILD, INFR & DES Spannungsgleichgewicht *nt*; ~ **tide** *n* ENVIRON Gleichgewichtszustand *m* des Gezeitenwechsels

**equinoctial**: ~ **tide** *n* ENVIRON Äquinoktialgezeiten *pl*

**equipment** *n* BUILD MACHIN Ausrüstung *f*, Ausstattung *f*

**equipotential** *n* ARCH & BUILD, ELECTR, INFR & DES, MAT PROP Äquipotential *nt*; ~ **bonding** *n* ELECTR Potentialausgleich *m*; ~ **busbar** *n* ELECTR Potentialausgleichsschiene *f*; ~ **line** *n* ARCH & BUILD, ELECTR, INFR & DES, MAT PROP Äquipotentiallinie *f*

**equivalent** *adj* ARCH & BUILD, INFR & DES äquivalent, gleichwertig; ~-**beam method** *n* ARCH & BUILD, INFR

& DES Ersatzbalkenmethode *f*; ~ **force** *n* ARCH & BUILD, INFR & DES Ersatzkraft *f*; ~ **length** *n* ARCH & BUILD, INFR & DES, TIMBER Äquivalentlänge *f*; ~ **model** *n* ARCH & BUILD, INFR & DES Äquivalentmodell *nt*; ~ **per million** *n* (*EPM*) ENVIRON Äquivalent je Million *nt* (*EPM*); ~ **standard smoke** *n* ENVIRON Äquivalenznormalruß *m*

**erect** *vt* ARCH & BUILD aufstellen, INFR & DES aufstellen, errichten

**erection** *n* ARCH & BUILD Aufstellung *f*, INFR & DES *installation* Aufstellung *f*, Errichtung *f*, Montage *f*

**ergonomic**: ~ **design** *n* ARCH & BUILD, HEAT & VENT, INFR & DES ergonomisches Design *nt*

**erode** *vt* ARCH & BUILD, ENVIRON, INFR & DES, MAT PROP erodieren

**erodibility** *n* ARCH & BUILD, ENVIRON, INFR & DES, MAT PROP Erodierbarkeit *f*

**erosion** *n* ARCH & BUILD, ENVIRON, INFR & DES, MAT PROP Erosion *f*; ~ **protection** *n* INFR & DES Erosionsschutz *m*

**erosive**: ~ **capacity** *n* ARCH & BUILD, ENVIRON, INFR & DES, MAT PROP Erosionsvermögen *nt*

**erratic**: ~ **structure** *n* INFR & DES erratische Struktur *f*; ~ **subsoil** *n* INFR & DES erratische Bodenverhältnisse *nt pl*

**erroneous** *adj* CONST LAW irrtümlich

**error**: ~ **theory** *n* ARCH & BUILD Fehlertheorie *f*

**escalator** *n* INFR & DES Rolltreppe *f*

**escape**: ~ **lighting** *n* CONST LAW, ELECTR Notbeleuchtung *f*; ~ **route** *n* CONST LAW Fluchtweg *m*; ~ **staircase** *n* CONST LAW Fluchttreppe *f*; ~ **way** *n* CONST LAW Fluchtweg *m*

**escutcheon** *n* BUILD HARDW Schlüsselschild *nt*

**ESP** *abbr* (*electrostatic precipitator*) ENVIRON ESA (*elektrostatischer Staubabscheider*)

**espagnolette** *n* BUILD HARDW *window ironmongery* Stangenverschluß *m*

**estate** *n* CONST LAW Grundbesitz *m*, INFR & DES *housing estate* Wohnsiedlung *f*, *large site* großes Grundstück *nt*; ~ **agent** *n* *BrE* (*cf real estate agent AmE*) CONST LAW Grundstücksmakler *m*, Immobilienhändler *m*, Immobilienmakler *m*; ~ **drainage** *n* WASTE WATER Grundstücksentwässerung *f*

**estimate 1.** *n* ARCH & BUILD, CONST LAW, INFR & DES Schätzung *f*, Voranschlag *m*; **2.** *vt* ARCH & BUILD, CONST LAW, INFR & DES *costs* abschätzen, veranschlagen

***estimate***: ~ **of costs** *n* ARCH & BUILD, CONST LAW, INFR & DES Kostenschätzung *f*, Kostenvoranschlag *m*

**etching** *n* SURFACE Ätzen *nt*

**eternit**® *n* MAT PROP Eternit® *nt*

**Euler**: ~**'s cases** *n pl* ARCH & BUILD, CONCR, INFR & DES, MAT PROP, STEEL, TIMBER Euler Fälle *m pl*; ~**'s critical tension** *n* ARCH & BUILD, CONCR, INFR & DES, MAT PROP, STEEL, TIMBER Eulersche Knickspannung *f*; ~**'s formula** *n* ARCH & BUILD, CONCR, INFR & DES, MAT PROP, STEEL, TIMBER Eulersche Knickformel *f*

**Eurocode** *n* (*EC*) ARCH & BUILD, CONST LAW, INFR & DES, MAT PROP, STEEL Eurocode *m* (*EC*)

**Euronorm** *n* ARCH & BUILD, CONST LAW, INFR & DES Euronorm *f*

**European**: ~ **Body of Electrical Standards** *n* (*CENELEC*) CONST LAW, ELECTR Europäischer Ausschuß *m* für Elektrizitätsnormung (*CENELEC*); ~ **Commission** *n* (*EC*) CONST LAW Kommission *f* der Europäischen Gemeinschaft (*KEG*); ~ **Committee for Standardisation** *n* (*CEN*) ARCH & BUILD, CONST LAW, INFR & DES Europäisches Komitee *nt* für Normung (*CEN*); ~ **initial standard** *n* (*ENV*) ARCH & BUILD, CONST LAW, INFR & DES, MAT PROP Europäische Vornorm *f* (*ENV*); ~ **Organisation for Technical Approvals** *n* (*EOTA*) CONST LAW Europäische Organisation *f* für Technische Zulassung (*EOTA*); ~ **Organisation for Testing and Certification** *n* (*EOTC*) CONST LAW Europäische Organisation *f* für Prüfung und Normung (*EOTC*); ~ **standard** *n* ARCH & BUILD, CONST LAW, INFR & DES Europäische Vorschrift *f*; ~ **Technical Specifications** *n pl* ARCH & BUILD, CONST LAW, INFR & DES Europäische Technische Vorschriften *f pl*; ~ **tentative standard** *n* (*ENV*) ARCH & BUILD, CONST LAW, INFR & DES, MAT PROP Europäische Vornorm *f* (*ENV*)

**Eurostar**® *n* INFR & DES Eurostar® *m*

**eustatic**: ~ **movement** *n* INFR & DES eustatische Bewegung *f*

**eutrophication** *n* ENVIRON Eutrophierung *f*

**evacuate** *vt* CONST LAW evakuieren

**evacuated**: ~ **tube collector** *n* HEAT & VENT evakuierter Rohrkollektor *m*

**evacuation** *n* CONST LAW Evakuierung *f*; ~ **elevator** *n* *AmE* (*cf evacuation lift BrE*) CONST LAW, INFR & DES Evakuierungsaufzug *m*; ~ **lift** *BrE* *n* (*cf evacuation elevator AmE*) CONST LAW, INFR & DES Evakuierungsaufzug *m*

**evaluate** *vt* ARCH & BUILD, CONST LAW, INFR & DES, MAT PROP auswerten, bewerten

**evaluation** *n* ARCH & BUILD, CONST LAW, INFR & DES, MAT PROP Auswertung *f*, Bewertung *f*

**evaporate** *vi* ARCH & BUILD, ENVIRON, HEAT & VENT, INFR & DES, MAT PROP evaporieren, verdampfen, verdunsten

**evaporation** *n* (*E*) ARCH & BUILD, ENVIRON, HEAT & VENT, INFR & DES, MAT PROP Evaporation *f* (*E*), Verdampfung *f*, Verdunstung *f*; ~ **loss** *n* ARCH & BUILD, HEAT & VENT, INFR & DES, MAT PROP Verdunstungsverlust *m*

**evaporator** *n* HEAT & VENT Verdampfer *m*

**even** *adj* ARCH & BUILD, INFR & DES Flach-, eben, flach, MAT PROP eben; ~ **with the ground** *adj* ARCH & BUILD, CONCR ebenerdig

**evenly**: ~ **graded** *adj* MAT PROP gleichmäßig gekörnt

**evenness** *n* ARCH & BUILD, INFR & DES, MAT PROP Planebenheit *f*

**evidence 1.** *n* CONST LAW Beweis *m*, Nachweis *m*, Zeugnis *nt*; **2.** *vt* CONST LAW nachweisen

**exact** *adj* ARCH & BUILD, BUILD MACHIN, CONCR, CONST LAW, INFR & DES, MAT PROP, STEEL *dimensions* exakt, genau

**examination** *n* ARCH & BUILD, CONST LAW, INFR & DES *investigation* Prüfung *f*, Untersuchung *f*, Überprüfung *f*

**examine** *vt* ARCH & BUILD, CONST LAW, INFR & DES prüfen, untersuchen, überprüfen

**excavate** *vt* ENVIRON, INFR & DES ausheben, *dig* abtragen, ausgraben, *deepen* ausbaggern, *road construction* auskoffern

**excavated**: ~ **area** *n* ENVIRON Abtrag *m*, Aushubfläche *f*, INFR & DES Aushubfläche *f*; ~ **material** *n* ENVIRON Aushub *m*, Aushubmaterial *nt*, INFR & DES Abtrag *m*, Aushub *m*, Aushubmaterial *nt*; ~ **soil** *n* ENVIRON, INFR & DES Bodenaushub *m*

**excavate**: ~ **soil true to profile** *phr* INFR & DES Boden profilgerecht lösen
**excavation** *n* ENVIRON, INFR & DES Ausschachtung *f*, *civil engineering* Aushub *m*, *digging* Bodenaushub *m*, Abtrag *m*, Abtragen *nt*, *tunnel* Ausbruch *m*; ~ **area** *n* INFR & DES Aushubgebiet *nt*; ~ **blasting** *n* INFR & DES Aushubsprengung *f*; ~ **depth** *n* INFR & DES Aushubtiefe *f*; ~ **for foundation** *n* INFR & DES Fundamentaushub *m*; ~ **support** *n* ARCH & BUILD Aushubsverstrebung *f*
**excavator** *n* BUILD MACHIN Bagger *m*, Trockenbagger *m*; ~ **bucket** *n* BUILD MACHIN Baggereimer *m*; ~**-loader** *n* BUILD MACHIN Ladebagger *m*
**excelsior**: ~ **building slab** *n* MAT PROP Holzwollebauplatte *f*; ~ **concrete** *n* MAT PROP Holzwollebeton *m*
**exception** *n* ARCH & BUILD Ausnahme *f*
**excess**: ~ **heat** *n* HEAT & VENT Wärmeüberschuß *m*
**excessive**: ~ **increase** *n* HEAT & VENT übermäßiger Zuwachs *m*
**excess**: ~ **load** *n* INFR & DES Überlast *f*; ~ **pressure** *n* INFR & DES Überdruck *m*
**exchange**: ~ **contracts** *phr* CONST LAW die Verträge unterzeichnen
**exchangeable**: ~ **cation** *n* ENVIRON auswechselbares Kation *nt*
**exchange**: ~ **of contracts** *n* CONST LAW Unterzeichnung *f* der Verträge; ~ **of members** *n* ARCH & BUILD Stabvertauschung *f*
**exchanger** *n* HEAT & VENT Austauscher *m*
**excise**: ~ **tax** *n* CONST LAW Verbrauchssteuer *f*
**exciter** *n* ENVIRON Erreger *m*
**execute** *vt* ARCH & BUILD, BUILD MACHIN, INFR & DES, MAT PROP ausführen, durchführen
**execution** *n* ARCH & BUILD, BUILD MACHIN, INFR & DES, MAT PROP Ausführung, *f* Durchführung *f*; ~ **of construction work** *n* ARCH & BUILD, INFR & DES Bauabwicklung *f*
**exhaust 1.** *n* BUILD MACHIN Auspuff *m*, Auslaß *m*, HEAT & VENT, WASTE WATER Auslaß *m*; **2.** *vt* HEAT & VENT, WASTE WATER entleeren
***exhaust***: ~ **air** *n* HEAT & VENT, INFR & DES Abluft *f*, Fortluft *f*; ~ **air damper** *n* BUILD HARDW, HEAT & VENT Abluftklappe *f*, Fortluftklappe *f*
**exhausted** *adj* HEAT & VENT, WASTE WATER entleert
**exhauster** *n* HEAT & VENT, WASTE WATER Entlüfter *m*
***exhaust***: ~ **fan** *n* BUILD HARDW Abluftventilator *m*, HEAT & VENT Abluftventilator *m*, Entlüfter *m*, WASTE WATER Entlüfter *m*; ~ **gas** *n* ENVIRON Abgas *nt*, Verbrennungsgas *nt*, HEAT & VENT Abgas *nt*; ~ **gas duct** *n* HEAT & VENT Abgaskanal *m*; ~ **hood** *n* BUILD HARDW, HEAT & VENT *over range* Dunstabzugshaube *f*
**exhaustion** *n* ARCH & BUILD *joinery dust* Absaugung *f*
***exhaust***: ~ **pipe** *n* BUILD HARDW Abzugsrohr *nt*, ENVIRON, HEAT & VENT Abzugsrohr *nt*, Entlüftungsrohr *nt*; ~ **vapor** *AmE*, ~ **vapour** *BrE* Brüdendampf *m*
**exhibition** *n* ARCH & BUILD, INFR & DES Ausstellung *f*
**existing**: ~ **plant** *n* ENVIRON Betrieb *m*
**exit** *n* ARCH & BUILD, HEAT & VENT, INFR & DES Ausgang *m*, Austritt *m*; ~ **air** *n* HEAT & VENT, INFR & DES Abluft *f*, Fortluft *f*; ~ **holes** *n pl* ELECTR, HEAT & VENT Ausgangsöffnungen *f pl*
**expand 1.** *vt* ARCH & BUILD, BUILD MACHIN dehnen, MAT PROP dehnen, *plastics* aufschäumen; **2.** *vi* MAT PROP expandieren, sich ausdehnen
**expandable** *adj* MAT PROP ausdehnbar
**expanded** *adj* MAT PROP expandiert, ausgeweitet, *plastics* aufgeschäumt, *rubber, plastic* porig-zellig, SOUND & THERMAL aufgeschäumt; ~ **cinder concrete** *n* CONCR, MAT PROP Bimsbeton *m*; ~ **concrete** *n* CONCR Blähbeton *m*, MAT PROP Blähbeton *m*, Leichtbeton *m*; ~ **cork** *n* MAT PROP expandierter Kork *m*; ~ **mesh** *n* MAT PROP, STEEL Streckmetall *nt*; ~ **metal** *n* MAT PROP, STEEL Streckmetall *nt*; ~ **plastic** *n* MAT PROP Schaumstoff *m*; ~ **rubber band** *n* MAT PROP, SOUND & THERMAL Schaumgummiband *nt*
**expanding**: ~ **foam** *n* INFR & DES, MAT PROP, SOUND & THERMAL Bauschaum *m*
**expansion** *n* ARCH & BUILD, BUILD MACHIN, HEAT & VENT, INFR & DES, MAT PROP Aufblähen *nt*, Ausdehnung *f*, Dehnung *f*; ~ **bearing** *n* ARCH & BUILD, INFR & DES bewegliches Auflager *nt*; ~ **bend** *n* HEAT & VENT *pipe* Dehnungsbogen *m*, Dehnungsschlaufe *f*, Ausgleichsbogen *m*, Dehnungsschleife *f*, INFR & DES Ausgleichsbogen *m*; ~ **bolt** *n* MAT PROP Spreizdübel *m*; ~ **crack** *n* INFR & DES Dehnungsriß *m*, Treibriß *m*, MAT PROP Dehnungsriß *m*; ~ **force** *n* ARCH & BUILD, INFR & DES, MAT PROP Ausdehnungskraft *f*; ~ **joint** *n* ARCH & BUILD Ausdehnungsfuge *f*, Bewegungsfuge *f*, Dehnungsfuge *f*, Trennfuge *f*, CONCR Dehnungsfuge *f*, INFR & DES Ausdehnungsfuge *f*, Bewegungsfuge *f*, Dehnungsfuge *f*, Trennfuge *f*; ~ **joint filler** *n* ARCH & BUILD, INFR & DES, MAT PROP Fugeneinlage *f*; ~ **joint sealing** *n* ARCH & BUILD, INFR & DES Dehnfugenabdichtung *f*; ~ **tank** *n* HEAT & VENT Ausdehnungsgefäß *nt*; ~ **vessel** *n* HEAT & VENT Ausdehnungsgefäß *nt*; ~ **work** *n* ARCH & BUILD Erweiterungsarbeiten *f pl*; ~ **zone** *n* HEAT & VENT Dehnungszone *f*
**expected**: ~ **construction costs** *n pl* ARCH & BUILD erwartete Baukosten *pl*
**experiment 1.** *n* ARCH & BUILD, INFR & DES, MAT PROP Experiment *nt*; **2.** *vi* ARCH & BUILD, INFR & DES, MAT PROP experimentieren
**experimental** *adj* ARCH & BUILD, INFR & DES, MAT PROP experimentell; ~ **building** *n* ARCH & BUILD, INFR & DES Experimentalbau *m*, Versuchsbau *m*; ~ **investigation** *n* ARCH & BUILD, INFR & DES, MAT PROP experimentelle Untersuchung *f*; ~ **method** *n* ARCH & BUILD, INFR & DES, MAT PROP experimentelle Methode *f*
**experimentation** *n* ARCH & BUILD, INFR & DES, MAT PROP Versuchsdurchführung *f*
**expertise** *n* ARCH & BUILD, CONST LAW Sachkenntnis *f*
**expire** *vi* CONST LAW *contract, offer* ungültig werden, *fund* verfallen, *guarantee* ablaufen
**exploded**: ~ **isometric** *adj* ARCH & BUILD, INFR & DES auseinandergezogen-isometrisch
**exploit** *vt* ENVIRON, INFR & DES ausbeuten
**exploitation** *n* ENVIRON, INFR & DES Ausbeutung *f*, Exploitation *f*; ~ **of geothermal energy** *n* ENVIRON Ausbeutung *f* der geothermalen Energie
**exploratory**: ~ **boring** *n* INFR & DES, MAT PROP Versuchsbohrung *f*; ~ **dig** *n* INFR & DES Bodenschürfung *f*; ~ **sounding** *n* INFR & DES, MAT PROP Probesondierung *f*
**explosion** *n* INFR & DES, MAT PROP Explosion *f*; ~ **charge** *n* INFR & DES Sprengladung *f*; ~ **pressure** *n* ARCH & BUILD, HEAT & VENT, INFR & DES, MAT PROP Explosionsdruck *m*; ~**-proof** *adj* ARCH & BUILD, ELECTR, INFR & DES, MAT PROP explosionsgeschützt (*ex-geschützt*), explosionssicher; ~**-proof design** *n* ARCH & BUILD, ELECTR, INFR & DES, MAT PROP

explosionssichere Ausführung *f*; **~ wave** *n* INFR & DES, MAT PROP Explosionswelle *f*

**explosive**: **~ effect** *n* INFR & DES Sprengwirkung *f*

**expose** *vt* ARCH & BUILD freilegen, MAT PROP *fire* aussetzen

**exposed** *adj* ARCH & BUILD ungeschützt, freiliegend, unverkleidet; **~ aggregate concrete** *n* CONCR, INFR & DES, MAT PROP Waschbeton *m*; **~ aggregate finish** *n* CONCR, INFR & DES, MAT PROP freigelegte Betonoberfläche *f* mit Zuschlagstoffen; **~ concrete** *n* CONCR, INFR & DES, MAT PROP Sichtbeton *m*; **~ concrete finish** *n* CONCR, INFR & DES, MAT PROP Sichtbeton *m*; **~ concrete formwork** *n* CONCR Sichtschalung *f*, Sichtbetonschalung *f*, TIMBER Sichtbetonschalung *f*, Sichtschalung *f*; **~ drainpipe** *n* HEAT & VENT, WASTE WATER freigelegtes Entwässerungsrohr *nt*; **~ masonry** *n* INFR & DES, STONE Sichtmauerwerk *nt*; **~ position** *n* ARCH & BUILD, INFR & DES ungeschützte Lage *f*; **~ wiring** *n* ELECTR, HEAT & VENT Aufputzinstallation *f*

**exposition** *n* ARCH & BUILD, INFR & DES Ausstellung *f*

**exposure** *n* INFR & DES Aufschluß *m*, MAT PROP *fire, weathering, radiation* Aussetzung *f*

**expropriate** *vt* CONST LAW enteignen

**expropriating**: **~ order** *n* CONST LAW Enteignungsbeschluß *m*; **~ proceedings** *n pl* CONST LAW Enteignungsverfahren *nt*

**expropriation** *n* CONST LAW Enteignung *f*

**exsiccate** *vti* MAT PROP austrocknen

**exsiccation** *n* MAT PROP Austrocknung *f*

**extend 1.** *vt* ARCH & BUILD erweitern, verlängern, ausbauen; **2.** *vi* ARCH & BUILD sich erstrecken

**extendable** *adj* ARCH & BUILD *building* erweiterbar, ausbaufähig, *table flap* einschiebbar, ausziehbar

**extended**: **~ stays** *n pl* BUILD HARDW *window opener* verlängertes Gestänge *nt*; **~ time test** *n* MAT PROP Dauerversuch *m*, Langzeitversuch *m*

**extensible** *adj* ARCH & BUILD ausbaufähig

**extension** *n* ARCH & BUILD *enlargement* Erweiterungsbau *m*, Erweiterung *f*, Anbau *m*, Ausbau *m*, BUILD MACHIN Dehnung *f*, CONST LAW *of contract* Zeitverlängerung *f*, HEAT & VENT Dehnung *f*, INFR & DES *enlargement* Erweiterung *f*, *annexe* Anbau *m*, Erweiterungsbau *m*, Ausdehnung *f*, Dehnung *f*, MAT PROP Ausdehnung *f*, Dehnung *f*; **~ bellows** *n* HEAT & VENT, WASTE WATER *pipe* Dehnungsausgleicher *m*; **~-compensating member** *n* HEAT & VENT, WASTE WATER *pipe* Dehnungsausgleicher *m*; **~ fan** *n* ELECTR, HEAT & VENT zusätzlicher Ventilator *m*; **~ joint filler** *n* SOUND & THERMAL Fugeneinlage *f*; **~ piece** *n* MAT PROP Verlängerungsstück *nt*; **~ of time** *n* CONST LAW Zeitverlängerung *f*; **~ of time claim** *n* CONST LAW Anspruch *m* bei Zeitverlängerung ; **~ tube** *n* HEAT & VENT Aufsteckhülse *f*

**extensometer** *n* BUILD MACHIN, INFR & DES, MAT PROP Dehnungsmesser *m*

**extent** *n* ARCH & BUILD *of work*, INFR & DES Ausmaß *nt*, Umfang *m*

**exterior**: **~ corridor** *n* ARCH & BUILD Laubengang *m*; **~ corrosion** *n* INFR & DES, MAT PROP Außenkorrosion *f*; **~ door** *n* ARCH & BUILD Außentür *f*; **~ lighting** *n* ELECTR Außenbeleuchtung *f*; **~ plaster** *n* ARCH & BUILD, MAT PROP, STONE Außenputz *m*, Außenverputz *m*; **~ render** *n* ARCH & BUILD, MAT PROP, STONE Außenputz *m*, Außenverputz *m*; **~ ring earth connection** *n BrE* (*cf exterior ring ground connection AmE*) ELECTR *lightning protection* äußerer Ringerder *m*; **~ ring ground connection** *n AmE* (*cf exterior ring earth connection BrE*) ELECTR *lightning protection* äußerer Ringerder *m*

**external**: **~ cavity wall** *n* STONE äußere Schale *f*; **~ dimension** *n* ARCH & BUILD, INFR & DES äußere Abmessung *f*; **~ features** *n pl* ARCH & BUILD, INFR & DES Außenanlagen *f pl*; **~ force** *n* ARCH & BUILD, INFR & DES äußere Kraft *f*; **~ friction** *n* ARCH & BUILD, INFR & DES äußere Reibung *f*; **~ hydrant** *n* CONST LAW *fire fighting* Außenhydrant *m*; **~ input** *n* ENVIRON Fremdzufuhr *f*; **~ load** *n* ARCH & BUILD, INFR & DES äußere Last *f*

**externally**: **~ insulated roof** *n* ARCH & BUILD, INFR & DES Warmdach *nt*

**external**: **~ moment** *n* ARCH & BUILD, INFR & DES äußeres Moment *nt*; **~ pressure** *n* ARCH & BUILD, INFR & DES Außendruck *m*; **~ rendering** *n* ARCH & BUILD Außenputz *m*, Außenverputz *m*, MAT PROP, STONE Außenputz *m*; **~ stress** *n* ARCH & BUILD, INFR & DES Außenspannung *f*; **~ string** *n* ARCH & BUILD *stairs* Außenwange *f*, Treppenlochwange *f*, Öffnungswange *f*; **~ surface** *n* SOUND & THERMAL äußere Oberfläche *f*; **~ thread** *n* MAT PROP Außengewinde *nt*; **~ torsion** *n* ARCH & BUILD, INFR & DES äußere Torsion *f*; **~ transverse force** *n* ARCH & BUILD, INFR & DES äußere Querkraft *f*; **~ vibrator** *n* BUILD MACHIN Außenrüttler *m*, Schalungsrüttler *m*; **~ wall** *n* ARCH & BUILD, STONE Außenmauer *f*, Außenwand *f*; **~ wall insulation** *n* MAT PROP, SOUND & THERMAL Außenmauerisolierung *f*, Außenwandisolierung *f*; **~ works** *n pl* ARCH & BUILD, INFR & DES Außenarbeiten *f pl*

**extinguish** *vt* CONST LAW löschen

**extinguisher** *n* CONST LAW Löscher *m*

**extinguishing**: **~ agent** *n* CONST LAW Löschmittel *nt*

**extra** *adj* ARCH & BUILD, INFR & DES *additional* zusätzlich; **~-bright** *adj* SURFACE hochglanzpoliert; **~ charge** *n* ARCH & BUILD, CONST LAW, INFR & DES Zulage *f*

**extract** *vt* INFR & DES entziehen, entnehmen, MAT PROP entziehen, WASTE WATER entnehmen

**extracted** *adj* INFR & DES, MAT PROP, WASTE WATER entnommen

**extraction** *n* INFR & DES, MAT PROP, WASTE WATER Entziehung *f*; **~ resistance** *n* INFR & DES, MAT PROP Ausziehwiderstand *m*

**extractor** *n* BUILD MACHIN, ENVIRON, HEAT & VENT Auszieher *m*, Entnahmegerät *nt*, Herausheber *m*; **~ hood** *n* ARCH & BUILD *roof* Dachhaube *f*, BUILD HARDW Ablufthaube *f*, HEAT & VENT Dachhaube *f*, Ablufthaube *f*

**extract**: **~ ventilation shaft** *n* ENVIRON, HEAT & VENT Entlüftungsschacht *m*; **~ ventilation unit** *n* ENVIRON, HEAT & VENT, WASTE WATER Entlüfter *m*

**extrados** *n* ARCH & BUILD Gewölberücken *m*, Wölbungsrücken *m*

**extras** *n pl* ARCH & BUILD, CONST LAW zusätzliche Bauleistungen *f pl*

**extra**: **~ time allowance** *n* ARCH & BUILD, CONST LAW, INFR & DES Zeitzugabe *f*; **~ work** *n* ARCH & BUILD, CONST LAW, INFR & DES Mehraufwand *m*

**extrude** *vti* INFR & DES, MAT PROP, STEEL *metal, plastic* extrudieren

**extruded** *adj* INFR & DES, MAT PROP, STEEL *metal, plastic* Strang-, extrudiert, stranggepreßt; **~ clay roof tile** *n* INFR & DES, MAT PROP, STONE Strangdachziegel *m*;

~ **interlocking tile** *n* INFR & DES, MAT PROP, STONE Strangfalzziegel *m*; ~ **profile** *n* MAT PROP Strangpreßprofil *nt*; ~ **rigid foam** *n* MAT PROP extrudierter Hartschaum *m*; ~ **section** *n* MAT PROP Strangpreßprofil *nt*
**extrusion** *n* INFR & DES, MAT PROP, STEEL Extrusion *f*
**exudation** *n* CONCR Ausschwitzung *f*, STONE Ausschwitzung *f*, Salpeterfraß *m*
**exude** *vti* CONCR, STONE ausschwitzen
**eye** *n* ARCH & BUILD Öse *f*
**eyebar** *n* STEEL *bridge, roof* Augenstab *m*
**eye**: ~ **level** *n* ARCH & BUILD Augenhöhe *f*; ~ **protection equipment** *n* BUILD HARDW, CONST LAW Augenschutzausrüstung *f*; ~ **shower** *n* ENVIRON Augendusche *f*

# F

**fabric** *n* MAT PROP Gewebe *nt*, Stoff *m*
**fabricate** *vt* MAT PROP fabrizieren
**fabricated** *adj* MAT PROP fabriziert
***fabric***: **~ filter** *n* ENVIRON *cloth* Tuchfilter *m*, Staubfilter *m*, Stoffilter *nt*, *fibrous* Gewebefilter *m*, INFR & DES *fibrous* Gewebefilter *m*; **~ roof** *n* ARCH & BUILD, INFR & DES, MAT PROP Stoffdach *nt*
**facade** *n* ARCH & BUILD *of building* Fassade *f*, Vorderfront *f*, Vorderansicht *f*; **~ scaffolding** *n* ARCH & BUILD Fassadengerüst *nt*
**face 1.** *n* ARCH & BUILD Vorderseite *f*; **2.** *vt* ARCH & BUILD, INFR & DES verkleiden, verschalen, verblenden
***face***: **~ arch** *n* ARCH & BUILD Stirnbogen *m*; **~ of arch** *n* ARCH & BUILD Stirnseite *f* eines Bogens; **~ brick** *n* STONE Blendziegel *m*, Vormauerziegel *m*; **~ concrete** *n* CONCR, INFR & DES, MAT PROP Vorsatzbeton *m*
**faced** *adj* ARCH & BUILD verkleidet, MAT PROP verkleidet, STONE verblendet; **~ brickwork** *n* CONCR, STONE Verblendmauerwerk *nt*
***face***: **~ grain direction** *n* MAT PROP oberflächige Faserrichtung *f*; **~ joint** *n* STONE sichtbare Fuge *f*
**facelift** *vt* ARCH & BUILD *room, building* renovieren
***face***: **~ piece** *n* INFR & DES *pit boards* Endspreize *f*, Stirnspreize *f*; **~ string** *n* ARCH & BUILD *stairs* Lichtwange *f*, Treppenlochwange *f*, Öffnungswange *f*; **~ veneer** *n* TIMBER Deckfurnier *nt*, Außenfurnier *nt*; **~ wall** *n* STONE Frontwand *f*, Verblendmauer *f*
**facework** *n* ARCH & BUILD Verblendung *f*, CONCR Verblendmauerwerk *nt*, INFR & DES, STEEL Verblendung *f*, STONE Verblendmauerwerk *nt*, Verblendung *f*, Verkleidung *f*
**facility** *n* ARCH & BUILD Anlage *f*, Einrichtung *f*, BUILD MACHIN Einrichtung *f*, INFR & DES Anlage *f*, Einrichtung *f*
**facing** *n* ARCH & BUILD, INFR & DES, STEEL Verblendung *f*, STONE Verblendung *f*, Verkleidung *f*, Sichtfläche *f*; **~ arch** *n* ARCH & BUILD Frontbogen *m*, Fassadenbogen *m*; **~ block** *n* ARCH & BUILD Verblender *m*, INFR & DES Verblendstein *m*, Verblender *m*, MAT PROP Verblendstein *m*, STONE Verblendstein *m*, Verblender *m*; **~ brick** *n* STONE Blendstein *m*, Vormauerziegel *m*, Vorsatzziegel *m*; **~ concrete** *n* CONCR, INFR & DES, MAT PROP Sichtbeton *m*, Vorsatzbeton *m*; **~ joint** *n* STONE Sichtfuge *f*, Fassadenfuge *f*; **~ masonry** *n* STONE Blendmauerwerk *nt*; **~ rail** *n* ARCH & BUILD Anlaufschiene *f*; **~ slab** *n* MAT PROP Verkleidungsplatte *f*, Verkleidungstafel *f*; **~ stone** *n* STONE Fassadenstein *m*, Naturstein *m* zur Wandverkleidung; **~ wall** *n* STONE Blendmauer *f*; **~ work** *n* CONCR, STONE Putzlage *f*, Verblendmauerwerk *nt*
**factor** *n* ARCH & BUILD, INFR & DES, MAT PROP Faktor *m*
**factory** *n* INFR & DES Fabrik *f*, Werk *nt*; **~ bonding** *n* TIMBER Werkleimung *f*; **~-built** *adj* MAT PROP fabrikfertig; **~-mounted** *adj* ELECTR *distribution panel* werkseitig eingebaut; **~ precasting** *n* CONCR Vorfertigung *f* im Betonwerk
**fade** *vi* MAT PROP *colour*, SURFACE verblassen; **~ away** *vi* MAT PROP, SURFACE verbleichen
**fagot** *n* INFR & DES, MAT PROP Faschine *f*
**fagotting** *n* INFR & DES *dam construction* Faschinenpackwerk *nt*, Faschinenverbauung *f*, Faschinenbau *m*
**Fahrenheit** *n* HEAT & VENT Fahrenheit *nt*; **~ scale** *n* HEAT & VENT Fahrenheitskala *f*
**faience** *n* STONE Steingut *nt*
**fail** *vi* ARCH & BUILD zu Bruch gehen
**failure** *n* BUILD MACHIN, INFR & DES, MAT PROP Ausfall *m*, Bruch *m*, Defekt *m*; **~ circle** *n* INFR & DES, MAT PROP Bruchkreis *m*; **~ condition** *n* INFR & DES, MAT PROP Bruchbedingung *f*; **~ criterium** *n* INFR & DES, MAT PROP Bruchkriterium *nt*; **~ curve** *n* INFR & DES, MAT PROP Bruchkurve *f*; **~ hypothesis** *n* INFR & DES, MAT PROP Bruchhypothese *f*; **~ limit** *n* INFR & DES, MAT PROP Bruchgrenze *f*; **~ load** *n* ARCH & BUILD, INFR & DES, MAT PROP *limit of load stress* Grenzlast *f*, Grenzbelastung *f*, *rupture* Bruchkreis *m*, *fully factored load* Bruchlast *f*, *collapse load* Einsturzlast *f*, Bruchbelastung *f*; **~ moment** *n* INFR & DES, MAT PROP Bruchmoment *nt*; **~ plane** *n* INFR & DES, MAT PROP Bruchfläche *f*; **~ stress** *n* ARCH & BUILD, INFR & DES, MAT PROP Bruchspannung *f*; **~ test** *n* INFR & DES, MAT PROP Bruchtest *m*; **~ warning light** *n* CONST LAW, ELECTR, INFR & DES, MAT PROP Ausfallwarnleuchte *f*
**fair**: **~-faced** *adj* ARCH & BUILD bündig, glatt, MAT PROP glatt; **~-faced brick** *n* INFR & DES, MAT PROP, STONE Sichtmauerstein *m*, Verblendstein *m*; **~-faced mansory** *n* INFR & DES Sichtmauerwerk *nt*; **~-faced masonry** *n* STONE Sichtmauerwerk *nt*
**fall 1.** *n* INFR & DES Gefälle *nt*; **2. ~ in** *vi* ARCH & BUILD einstürzen
***fall***: **~ of earth** *n* INFR & DES Erdeinsturz *m*; **~ of ground** *n* INFR & DES Geländeneigung *f*
**falling**: **~ gradient** *n* ARCH & BUILD, INFR & DES Gefälle *nt*, Neigung *f*; **~ head** *n* MAT PROP veränderliches Druckpotential *nt*; **~ tide** *n* ENVIRON Ebbe *f*
**fallout** *n* ENVIRON *radioactive* Fallout *m*, Niederschlag *m*; **~ shelter** *n* ARCH & BUILD, CONCR Nuklearschutzbunker *m*
***fall***: **~ pipe** *n* WASTE WATER Falleitung *f*, Fallrohr *nt*
**falls** *n pl* HEAT & VENT Fallrohre *nt pl*
**false**: **~-bearing** *adj* ARCH & BUILD nicht satt aufliegend; **~ ceiling** *n* ARCH & BUILD, BUILD HARDW abgehängte Decke *f*; **~ cohesion** *n* INFR & DES, MAT PROP falsche Kohäsion *f*; **~ edge** *n* TIMBER Aufdoppelung *f*; **~ floor** *n* ARCH & BUILD Blindboden *m*, Doppelboden *m*; **~ gable** *n* ARCH & BUILD *churches* Blendgiebel *m*; **~ hip** *n* ARCH & BUILD, TIMBER Krüppelwalm *m*; **~ hip roof** *n* ARCH & BUILD, TIMBER Krüppelwalmdach *nt*; **~ setting** *n* CONCR, INFR & DES, MAT PROP vorzeitiges Abbinden *nt*
**falsework** *n* ARCH & BUILD, INFR & DES Lehrgerüst *nt*, Schalgerüst *nt*
**fan** *n* ARCH & BUILD, BUILD HARDW, HEAT & VENT Fächer *m*, Gebläse *nt*, Lüfter *m*, Ventilator *m*
**fancy**: **~ sheet metal** *n* STEEL Dessinblech *nt*
***fan***: **~ drive** *n* HEAT & VENT Ventilatorantrieb *m*
**fang**: **~ bolt** *n* BUILD HARDW Bolzen *m*
***fan***: **~ guard** *n* HEAT & VENT Lüfterhaube *f*, Ventilator-

schutzkorb *m*; ~ **heater** *n* BUILD HARDW, HEAT & VENT Ventilatorheizung *f*
**fanlight** *n* ARCH & BUILD *over door* Oberlicht *nt*, Türoberlicht *nt*
***fan***: ~ **silencer** *n* HEAT & VENT, SOUND & THERMAL Ventilatorschalldämpfer *m*; ~ **sound damper** *n* HEAT & VENT, SOUND & THERMAL Ventilatorschalldämpfer *m*; ~ **vault** *n* ARCH & BUILD *church* Fächergewölbe *nt*
**fascia** *n* ARCH & BUILD Faszie *f*, Gurtsims *m*, BUILD HARDW, MAT PROP, TIMBER Leiste *f*; ~ **board** *n* ARCH & BUILD Simsbrett *nt*, Traufbrett *nt*, TIMBER Stirnbrett *nt*
**fascine** *n* INFR & DES, MAT PROP Faschine *f*; ~ **work** *n* INFR & DES Faschinenpackwerk *nt*, Faschinenbau *m*, Faschinenverbauung *f*
**fast** *adj* MAT PROP widerstandsfähig
**fastener** *n* ARCH & BUILD Halterung *f*; ~ **hole** *n* ARCH & BUILD Befestigungsloch *nt*
**fastening** *n* ARCH & BUILD Befestigung *f*, Bügel *m*; ~ **clamp** *n* MAT PROP Befestigungsklemme *f*; ~ **rail** *n* MAT PROP Befestigungsschiene *f*; ~ **structure** *n* ARCH & BUILD Befestigungskonstruktion *f*
**fat**: ~ **clay** *n* INFR & DES, MAT PROP fetter Ton *m*
**fatigue** *n* MAT PROP Ermüdung *f*, Ermüdungserscheinungen *f pl*; ~ **behavior** *AmE*, ~ **behaviour** *BrE* *n* INFR & DES, MAT PROP Ermüdungsverhalten *nt*; ~ **failure** *n* INFR & DES, MAT PROP Ermüdungsversagen *nt*; ~ **life** *n* MAT PROP Zeitschwingfestigkeit *f*; ~ **limit** *n* INFR & DES, MAT PROP Dauerfestigkeit *f*, Ermüdungsgrenze *f*; ~ **strength** *n* INFR & DES, MAT PROP Dauerschwingfestigkeit *f*; ~ **yield limit** *n* ARCH & BUILD, CONST LAW, MAT PROP Dauerdehngrenze *f*
**fat**: ~ **lime** *n* MAT PROP Fettkalk *m*, Weißkalk *m*
**fatness**: ~ **of concrete** *n* CONCR Fettheit *f* des Betons
**faucet** *n* *AmE* (*cf tap BrE*) WASTE WATER Hahn *m*, Wasserhahn *m*; ~ **joint** *n* WASTE WATER kurzes Verbindungsstück *nt*
**fault** *n* CONST LAW Fehler *m*, Mangel *m*, INFR & DES Verwerfung *f*; ~ **plane** *n* INFR & DES Verwerfungsfläche *f*
**faulty** *adj* ARCH & BUILD, CONST LAW schadhaft, mangelhaft
***fault***: ~ **zone** *n* INFR & DES Verwerfungszone *f*
**faying** *adj* ARCH & BUILD satt anliegend
**feasibility** *n* ARCH & BUILD, INFR & DES Durchführbarkeit *f*; ~ **study** *n* ARCH & BUILD, INFR & DES Durchführbarkeitsstudie *f*
**feasible** *adj* ARCH & BUILD, INFR & DES durchführbar
**feather** *n* TIMBER Feder *f*, Federkeil *m*; **~-edged brick** *n* ARCH & BUILD Keilziegel *m*; ~ **joint** *n* TIMBER gefederte Verbindung *f*; ~ **tongue** *n* TIMBER Federkeil *m*
**feature** *n* ARCH & BUILD Besonderheit *f*, Merkmal *nt*
**Federal**: ~ **States Working Group on Waste** *n* ENVIRON *in Germany* Länderarbeitsgemeinschaft *f* für Abfall (*LAGA*)
**fee** *n* ARCH & BUILD, CONST LAW Gebühr *f*
**feed** *vt* ELECTR, WASTE WATER einspeisen
**feeder** *n* ELECTR Speise-, Speiseleitung *f*, Versorgungsleitung *f*, WASTE WATER Einspeisung *f*, Speise-; ~ **basin** *n* WASTE WATER Speisebecken *nt*; ~ **line** *n* ELECTR *road, railway* Speiseleitung *f*, INFR & DES *conveyor* Zubringer *m*; ~ **road** *n* INFR & DES Zubringer *m*, Zubringerstraße *f*
**feed**: **~-in** *n* ELECTR, WASTE WATER Einspeisung *f*
**feldspar** *n* INFR & DES, MAT PROP Feldspat *m*
**feldspathic**: ~ **sandstone** *n* MAT PROP feldspathaltiger Sandstein *m*
**felt 1.** *n* MAT PROP Filz *m*; **2.** *vt* STONE *plaster* filzen
***felt***: **~-and-gravel roof** *n* ARCH & BUILD, INFR & DES Kiespreßdach *nt*
**felted** *adj* STONE *plaster* gefilzt
**female**: ~ **thread** *n* STEEL Innengewinde *nt*
**fen** *n* INFR & DES *flat land often artificially drained* Fenn *nt*, *marshy region* Marsch *f*, Moor *nt*, Morast *m*, *swamp* Sumpf *m*
**fence 1.** *n* INFR & DES Zaun *m*; ~ **panel** *n* INFR & DES, MAT PROP Zaunpaneel *nt*; ~ **post** *n* INFR & DES, MAT PROP Umzäunungspfahl *m*, Zaunpfosten *m*; ~ **wall** *n* ARCH & BUILD Einfriedungsmauer *f*, Umfassungsmauer *f*, CONST LAW, INFR & DES Einfriedungsmauer *f*; **2.** *vt* INFR & DES einzäunen
**fencing** *n* INFR & DES Einzäunung *f*
**fender**: ~ **beam** *n* ARCH & BUILD Prellbalken *m*; ~ **pile** *n* ARCH & BUILD Prellpfahl *m*, Streichpfahl *m*
**fenestration** *n* ARCH & BUILD Fensteranordnung *f*, *fitting of windows* Befensterung *f*, *layout and distribtion of windows* Fensteraufteilung *f*
**fermentation**: ~ **gas** *n* ENVIRON Biogas *nt*, Faulgas *nt*, INFR & DES, MAT PROP Faulgas *nt*; ~ **of refuse** *n* ENVIRON Abfallgärung *f*
**ferrous**: ~ **scrap** *n* ENVIRON Eisenschrott *m*
**festoon** *n* ARCH & BUILD *ornament* Gehänge *nt*
**fiber** *AmE see fibre BrE*
**fiberboard** *AmE see fibreboard BrE*
**fiberglass** *AmE see fibreglass BrE*
***fiber***: ~ **insulating material** *AmE see fibre insulating material BrE*
**fibre** *n BrE* MAT PROP Faser *f*, Fiber *f*
**fibreboard** *n BrE* MAT PROP, SOUND & THERMAL Faserplatte *f*
**fibreglass** *n BrE* MAT PROP Fiberglas *nt*, Glasfaser *f*; **~-reinforced plastics** *n pl BrE* MAT PROP glasfaserverstärkte Kunststoffe *m pl*
***fibre***: ~ **insulating material** *n BrE* INFR & DES, MAT PROP, SOUND & THERMAL Faserdämmstoff *m*; ~ **insulation material** *n BrE* INFR & DES, MAT PROP, SOUND & THERMAL Faserdämmstoff *m*
**fibrous**: ~ **peat** *n* INFR & DES faserhaltiger Torf *m*
**field** *n* INFR & DES Feld *nt*; ~ **data** *n* INFR & DES Felddaten *pl*; ~ **moment** *n* ARCH & BUILD, INFR & DES Feldmoment *nt*; ~ **survey** *n* ARCH & BUILD, INFR & DES Standortvermessung *f*
**figure** *n* ARCH & BUILD *digit* Zahl *f*, *numeral* Ziffer *f*
**figured**: ~ **glass** *n* MAT PROP Ornamentglas *nt*, Profilglas *nt*
**fill 1.** *n* INFR & DES *earth* angeschüttetes Gelände *nt*, Anschüttung *f*, Aufschüttung *f*, Ausschüttung *f*; **2.** *vt* ARCH & BUILD ausfüllen, füllen, CONCR ausspachteln, ENVIRON verfüllen, STONE ausspachteln, SURFACE auftragen, ausspachteln, füllen, verspachteln; ~ **in** STONE *masonry, tiling* ausfugen; ~ **up** ARCH & BUILD *trench* auffüllen; ~ **up with masonry** STONE beimauern
**filled** *adj* ARCH & BUILD ausgefüllt, gefüllt, CONCR, STONE ausgespachtelt, SURFACE aufgetragt, ausgespachtelt, gefüllt, verspachtelt; ~ **ground** *n* INFR & DES *earth* Ausschüttung *f*, Aufschüttung *f*, Schüttung *f*, angeschüttetes Gelände *nt*; **~-up ground** *n* INFR & DES *earth* Ausschüttung *f*, Aufschüttung *f*, Schüttung *f*, angeschüttetes Gelände *nt*
**filler** *n* CONCR Zuschlagstoff *m*, MAT PROP Füllmaterial

*nt*, Füllstoff *m*, STEEL Futterblech *nt*, TIMBER Futterstück *nt*, Füllholz *nt*; **~ block floor** *n* ARCH & BUILD Füllkörperdecke *f*; **~ content** *n* MAT PROP Füllergehalt *m*, Fülleranteil *m*; **~ joist** *n* ARCH & BUILD Deckenträger *m*, Querträger *m*; **~ tile** *n* ARCH & BUILD Deckenfüllkörper *m*; **~ wall** *n* ARCH & BUILD Tafelwand *f*, Ausfachung *f*

**fillet** *n* ARCH & BUILD Zierleiste *f*, Kehlleiste *f*, Saumleiste *f*, *windows* Fensterdichtleiste *f*, BUILD HARDW Hohlkehle *f*, Wandanschlußleiste *f*, Zierleiste *f*, Deckleiste *f*, MAT PROP *tiling work* Riemchen *nt*, TIMBER Deckleiste *f*; **~ gutter** *n* BUILD HARDW Schornsteinblechrinne *f*; **~ joint** *n* TIMBER Kehlnaht *f*

**filling** *n* ARCH & BUILD Verspachteln *nt*, ENVIRON Verfüllung *f*, INFR & DES Bodenaufschüttung *f*, *earth* Aufschüttung *f*, Auftrag *m*, MAT PROP Füllstoff *m*, Füllmaterial *nt*, Bodenaufschüttung *f*, *earth* Auftrag *m*, SURFACE Verspachteln *nt*; **~ compound** *n* MAT PROP Spachtelmasse *f*, Füllmasse *f*; **~ in** *n* STONE Ausfugen *nt*; **~ masonry** *n* ARCH & BUILD, STONE Ausmauerung *f*; **~ material** *n* MAT PROP Füllmasse *f*, Spachtelmasse *f*; **~ of openings with humus soil** *n* INFR & DES *gardening* Humusverfüllung *f* der Hohlräume; **~ rod** *n* TIMBER Futterleiste *f*; **~ trowel** *n* BUILD MACHIN Fugenkelle *f*

***fill***: **~ insulation** *n* SOUND & THERMAL lose Dämmung *f*

**fillister** *n* BUILD HARDW *window* Falzhobel *m*, Kittfalz *m*

***fill***: **~ mass** *n* ENVIRON Deponiegut *nt*, Lagergut *nt*; **~ material** *n* INFR & DES, MAT PROP Schüttmaterial *nt*

**film** *n* MAT PROP *plastic* Folie *f*

**filter** *n* HEAT & VENT, INFR & DES, MAT PROP, WASTE WATER Filter *m*; **~ bed** *n* INFR & DES Filterbett *nt*; **~ drain** *n* INFR & DES Filterrohr *nt*, WASTE WATER Filterrohr *nt*, Sickerrohr *nt*, *drainpipe* Sickerleitung *f*

**filtered** *adj* HEAT & VENT, INFR & DES, MAT PROP, WASTE WATER filtriert, gefiltert

***filter***: **~ fabric** *n* MAT PROP Filtergewebe *nt*; **~-fan unit** *n* HEAT & VENT Filterventilatoreinheit *f*; **~ flow** *n* INFR & DES Filterströmung *f*; **~ gravel** *n* INFR & DES, MAT PROP Filterkies *m*

**filtering** *n* HEAT & VENT, INFR & DES, MAT PROP, WASTE WATER Filtern *nt*, Filtrieren *nt*; **~ unit** *n* ENVIRON Filteranlage *f*

***filter***: **~ layer** *n* INFR & DES Dränschicht *f*, Filterschicht *f*; **~ material** *n* INFR & DES Filtermaterial *nt*; **~ pipe** *n* INFR & DES, WASTE WATER Filterrohr *nt*; **~ plant** *n* INFR & DES Filteranlage *f*; **~ run** *n* HEAT & VENT, SOUND & THERMAL Filterlaufzeit *f*, Filterstandzeit *f*; **~ sand** *n* INFR & DES, MAT PROP Filtersand *m*; **~ sediment** *n* INFR & DES, WASTE WATER Filtersediment *nt*; **~ velocity** *n* INFR & DES, MAT PROP, WASTE WATER Filtergeschwindigkeit *f*

**filtrate** *n* HEAT & VENT Filtrat *nt*

**filtration** *n* HEAT & VENT, INFR & DES, MAT PROP, WASTE WATER Filtration *f*, Filtrierung *f*; **~ plant** *n* INFR & DES Filtrationsanlage *f*; **~ rate** *n* WASTE WATER Filtergeschwindigkeit *f*

**fin** *n* HEAT & VENT Lamelle *f*

**final**: **~ coat** *n* STONE Oberputz *m*; **~ cover** *n* ENVIRON *of landfill* Endabdeckung *f*, Oberflächenversiegelung *f*; **~ design** *n* ARCH & BUILD Ausführungsunterlagen *f pl*, INFR & DES Ausführungsunterlagen *f pl*, endgültiger Entwurf *m*; **~ drawing** *n* ARCH & BUILD, INFR & DES Ausführungszeichnung *f*; **~ exit** *n* CONST LAW, INFR & DES Endausgang *m*; **~ float for road surfacing** *n* CONCR Glättbohle *f* für Oberbeton oder Deckenschluß; **~ grade** *n* INFR & DES Feinplanum *nt*, Fundamentsohle *f*; **~ grading** *n* INFR & DES Feinplanierung *f*; **~ inspection** *n* ARCH & BUILD, CONST LAW, INFR & DES Endabnahme *f*; **~ payment** *n* CONST LAW, INFR & DES Endzahlung *f*, Schlußzahlung *f*; **~ plan** *n* ARCH & BUILD Ausführungsplan *m*; **~ planning documents** *n* ARCH & BUILD, INFR & DES Ausführungsunterlagen *f pl*; **~ pressure** *n* WASTE WATER *pump* Enddruck *m*; **~ pre-stress** *n* CONCR *pre-stressed concrete* Endvorspannung *f*, Endvorspannkraft *f*; **~ rendering** *n* STONE Deckputz *m*, Oberputz *m*, Feinputz *m*; **~ settling tank** *n* INFR & DES Nachklärbecken *nt*; **~ stability** *n* INFR & DES Endstandsicherheit *f*; **~ storage** *n* ENVIRON Endlagerung *f*

**fine** *adj* ARCH & BUILD *coat* dünn, MAT PROP dünn, *sand, gravel* Fein-, fein, SURFACE *coat* dünn; **~ adjustment** *n* BUILD MACHIN Feinregelung *f*; **~ aggregate** *n* CONCR Feinzuschlag *m*, feinkörniger Zuschlag *m*; **~ chippings** *n pl* MAT PROP Feinsplitt *m*

**finegrading** *n* INFR & DES Feinplanierung *f*

***fine***: **~-grained** *adj* MAT PROP feinkörnig; **~-grained fracture** *n* MAT PROP feinkörniger Bruch *m*; **~-grained powder** *n* MAT PROP feinkörniges Pulver *nt*; **~ gravel** *n* MAT PROP Feinkies *m*; **~ gravel bed** *n* INFR & DES Feinkiesbett *nt*

**finely**: **~ crushed basalt chippings** *n pl* MAT PROP Basaltedelsplitt *m*

***fine***: **~-mesh** *adj* MAT PROP feinmaschig; **~ plane** *n* BUILD MACHIN *plaster* Putzhobel *m*; **~ sand** *n* MAT PROP feinkörniger Sand *m*; **~ screen** *n* ENVIRON Feinrechen *m*; **~ stuff** *n* STONE *plaster* Tünchsandputz *m*

**finger**: **~ joint** *n* TIMBER Zinkenverbindung *f*; **~ plate** *n* BUILD HARDW Schloßschutzblech *nt*

**finish 1.** *n* ARCH & BUILD, INFR & DES Oberfläche *f*, SURFACE Oberflächenbelag *m*, Oberflächenbeschaffenheit *f*, Oberfläche *f*; **2.** *vt* STONE *plaster* abziehen, SURFACE die Oberfläche bearbeiten

***finish***: **~ coat** *n* STONE Deckputz *m*, Feinputz *m*, SURFACE Deckanstrich *m*, Fertiganstrich *m*, Schlußanstrich *m*, Schutzanstrich *m*, letzter Anstrich *m*

**finished**: **~ floor** *n* ARCH & BUILD Fertigfußboden *m*; **~ grade** *n* INFR & DES Erdplanum *nt*

**finisher** *n* BUILD MACHIN Deckenfertiger *m*, *roads* Straßenfertiger *m*

**finishing** *n* SURFACE Nachbearbeitung *f*; **~ coat** *n* STONE Feinputz *m*, Deckputz *m*, SURFACE *of paintwork* Schutzanstrich *m*, letzter Anstrich *m*; **~ coat plaster** *n* STONE Deckputz *m*; **~ layer** *n* ARCH & BUILD Deckschicht *f*, INFR & DES *road* Deckschicht *f*, Verschleißschicht *f*; **~ machine** *n* BUILD MACHIN Deckenfertiger *m*, *roads* Straßenfertiger *m*; **~ trades** *n pl* ARCH & BUILD Ausbaugewerke *nt pl*; **~ trowel** *n* BUILD MACHIN Glätter *m*; **~ work** *n* ARCH & BUILD Ausbauarbeiten *pl*, SURFACE Oberflächenbearbeitung *f*, Oberflächengestaltung *f*

***finish***: **~ plaster** *n* STONE Feinputz *m*

**finite**: **~-difference method** *n* ARCH & BUILD, INFR & DES finite Differenzmethode *f*; **~ element** *n* ARCH & BUILD, INFR & DES finites Element *nt*; **~-element method** *n* ARCH & BUILD, INFR & DES finite Elementmethode *f*

**finned**: **~ absorber plate** *n* HEAT & VENT gerippte Absorberplatte *f*; **~ radiator** *n* HEAT & VENT

Lamellenheizkörper *m*, Konvektor *m*, Rippenheizkörper *m*
**fir** *n* TIMBER Tanne *f*
**fire** *n* ARCH & BUILD, CONST LAW Brand *m*, Feuer *nt*; **~ alarm** *n* CONST LAW, ELECTR Brandmelder *m*, Feuermelder *m*; **~ alarm system** *n* CONST LAW, ELECTR Brandmeldeanlage *f*, Feuermeldeanlage *f*; **~ blanket** *n* BUILD HARDW, CONST LAW Feuerdecke *f*; **~ brick** *n* STONE Schamottestein *m*, feuerfester Ziegel *m*; **~ bucket** *n* BUILD HARDW, CONST LAW Feuereimer *m*; **~ cabinet** *n* ARCH & BUILD, BUILD HARDW Feuerlöschkasten *m*; **~ classification** *n* CONST LAW, MAT PROP Brandklasse *f*, Feuerwiderstandsklasse *f*; **~ clay** *n* MAT PROP Feuerton *m*, feuerfester Ton *m*; **~ compartment** *n* ARCH & BUILD, CONST LAW Brandabschnitt *m*
**fired** *adj* MAT PROP gebrannt, gefeuert
***fire***: **~ damage** *n* CONST LAW, MAT PROP Feuerschaden *m*; **~-damaged building** *n* ARCH & BUILD, CONST LAW feuerbeschädigtes Gebäude *nt*; **~ damper** *n* HEAT & VENT Feuerschutzklappe *f*; **~ detector** *n* CONST LAW, ELECTR Feueranzeiger *f*; **~ door** *n* ARCH & BUILD Brandschutztür *f*, Feuerschutztür *f*, BUILD HARDW, CONST LAW Brandschutztür *f*; **~ endurance** *n* MAT PROP Feuerwiderstandsfähigkeit *f*; **~ escape** *n* CONST LAW Fluchtweg *m*; **~ exit** *n* CONST LAW, INFR & DES Notausgang *m*; **~ extinguisher** *n* BUILD HARDW Feuerlöscher *m*; **~ extinguisher box** *n* ARCH & BUILD, BUILD HARDW Feuerlöschkasten *m*; **~ extinguishing equipment** *n* BUILD MACHIN, CONST LAW Löscheinrichtung *f*
**firefighter**: **~'s ladder** *n* CONST LAW, INFR & DES Feuerleiter *f*
**firefighting**: **~ lift** *n* CONST LAW, INFR & DES Aufzug *m* zur Feuerbekämpfung, feuergeschützter Aufzug *m*; **~ lobby** *n* CONST LAW, INFR & DES feuergeschütztes Vestibül *nt*; **~ shaft** *n* CONST LAW, INFR & DES feuergeschützter Schacht *m*; **~ stair** *n* CONST LAW, INFR & DES feuergeschütztes Treppenhaus *nt*
***fire***: **~ hazard** *n* CONST LAW Brandgefahr *f*; **~ hydrant** *n* BUILD HARDW Feuerhahn *m*; **~ ladder** *n* CONST LAW Feuerleiter *f*; **~ load** *n* CONST LAW Brandlast *f*; **~ lobby** *n* ARCH & BUILD, CONST LAW Brandabschnitt *m*; **~ mains** *n pl* CONST LAW Wasserversorgung *f* für Feuerbekämpfung; **~ notice** *n* ARCH & BUILD, CONST LAW Feueranweisung *f*; **~ partition** *n* ARCH & BUILD, CONST LAW Brandmauer *f*, Brandschott *nt*
**fireplace** *n* ARCH & BUILD Kamin *m*, offene Feuerstelle *f*
**fireplug** *n* *AmE* (*cf hydrant BrE*) CONST LAW, WASTE WATER *firefighting* Hydrant *m*
***fire***: **~ policy** *n* CONST LAW Feuerrichtlinie *f*; **~ precautions** *n pl* CONST LAW Brandschutzmaßnahmen *f pl*, Brandsicherheitsmaßnahmen *f pl*
**fireproof 1.** *adj* CONST LAW feuerbeständig, MAT PROP feuerbeständig, feuerfest; **2.** *vt* MAT PROP feuerbeständig machen, feuerfest machen
***fireproof***: **~ building material** *n* ARCH & BUILD, CONST LAW, INFR & DES, MAT PROP, SOUND & THERMAL feuerfester Baustoff *m*; **~ door** *n* ARCH & BUILD, BUILD HARDW, CONST LAW Brandschutztür *f*
**fireproofing**: **~ coat** *n* SURFACE Brandschutzbeschichtung *f*, Brandschutzüberzug *m*
**fireproof**: **~ wall** *n* ARCH & BUILD, CONST LAW, INFR & DES, SOUND & THERMAL Brandmauer *f*, Brandschott *nt*
**fire**: **~-protected** *adj* CONST LAW, ELECTR, INFR & DES feuergeschützt; **~-protected circuit** *n* CONST LAW feuergeschützter Stromkreis *m*; **~ protection** *n* CONST LAW Brandschutz *m*; **~ rating** *n* CONST LAW, MAT PROP Brandklasse *f*, Feuerwiderstandsklasse *f*; **~ resistance** *n* CONST LAW, MAT PROP, SOUND & THERMAL Feuerbeständigkeit *f*, Feuerwiderstandsfähigkeit *f*; **~-resistant** *adj* CONST LAW, MAT PROP feuerbeständig; **~-resistant ceiling** *n* ARCH & BUILD, BUILD HARDW, CONST LAW Brandschutzdecke *f*; **~-resistant door** *n* ARCH & BUILD, BUILD HARDW, CONST LAW Brandschutztür *f*; **~-resistant shutter** *n* CONST LAW Brandabschluß *m*, Brandabschnittswand *f*; **~-resisting floor** *n* ARCH & BUILD, CONST LAW feuerbeständiger Boden *m*; **~-retardant** *adj* MAT PROP feuerhemmend; **~ risk** *n* CONST LAW Brandgefahr *f*, Brandrisiko *nt*, Feuergefahr *f*, Feuerrisiko *nt*; **~ stop** *n* CONST LAW Feuerversiegelung *f*; **~ test** *n* BUILD HARDW, CONST LAW Brandversuch *m*; **~ wall** *n* ARCH & BUILD, CONST LAW Brandmauer *f*, Brandschott *nt*
**firing** *n* MAT PROP, STONE Brennen *nt*; **~ tape** *n* INFR & DES *blasting* Zündschnur *f*
**firm** *adj* MAT PROP fest; **~ clay** *n* INFR & DES, MAT PROP fester Ton *m*; **~ ground** *n* ARCH & BUILD, INFR & DES fester Grund *m*; **~ line** *n* ARCH & BUILD Vollinie *f*
***fir***: **~ plank** *n* TIMBER Tannenholzbohle *f*
**firring** *n* TIMBER Futterholz *nt*, Unterfütterung *f*
**first**: **~ adjustment** *n* HEAT & VENT Ersteinregulierung *f*; **~ coat** *n* STONE Unterputz *m*, SURFACE Grundanstrich *m*; **~ floor** *n* *AmE* (*cf ground floor BrE*) ARCH & BUILD Erdgeschoß *nt*; **~-floor plan** *n* *AmE* (*cf ground-floor plan BrE*) ARCH & BUILD Erdgeschoßplan *m*, Grundriß *m*; **~ order theory** *n* ARCH & BUILD, CONCR, INFR & DES, MAT PROP, STEEL Theorie *f* I. Ordnung; **~ and second finishes** *n* MAT PROP, SURFACE erste und zweite Oberflächenbehandlungen *f pl*
**fish 1.** *n* TIMBER Lasche *f*; **2.** *vt* TIMBER verstärken, verlaschen
***fish***: **~ beam** *n* ARCH & BUILD Fischbauchträger *m*; **~-bellied** *adj* ARCH & BUILD fischbauchig; **~-bellied flap** *n* ARCH & BUILD Fischbauchklappe *f*; **~-bellied girder** *n* ARCH & BUILD Fischbauchträger *m*; **~ bolt** *n* MAT PROP Laschenschraube *f*, Laschenbolzen *m*
**fished**: **~ joint** *n* TIMBER Laschenverbindung *f*, Verlaschung *f*, Laschenstoß *m*
**fishing** *n* TIMBER Verlaschung *f*, Verlaschen *nt*
***fish***: **~ pass** *n* ENVIRON, INFR & DES Fischgerinne *nt*, Fischgraben *m*, Fischleiter *f*
**fishplate 1.** *n* TIMBER Zuglasche *f*, Lasche *f*; **2.** *vt* TIMBER verlaschen
**fishtail**: **~ ducting** *n* HEAT & VENT Fischschwanzrohrverlegung *f*
**fissure 1.** *n* ARCH & BUILD *in wall*, INFR & DES, MAT PROP Riß *m*, Spalt *m*; **2.** *vi* MAT PROP rissig werden
**fit 1.** *n* ARCH & BUILD Passung *f*; **2.** *vt* ARCH & BUILD anbringen, montieren, CONCR, STEEL, TIMBER anbringen
**fit**: **~ for traffic** *adj* ARCH & BUILD, INFR & DES befahrbar; **~ in** *vt* ARCH & BUILD einbauen, einpassen, BUILD HARDW, HEAT & VENT einpassen
**fitness**: **~ for purpose** *n* MAT PROP Zweckfähigkeit *f*
**fitted** *adj* ARCH & BUILD eingebaut, eingepaßt, INFR & DES eingebaut; **~ carpet** *n* BUILD HARDW Teppichboden *m*
**fitter** *n* ARCH & BUILD Installateur *m*, Monteur *m*, HEAT & VENT Monteur *m*

**fitting** *n* BUILD HARDW Beschlag *m*, HEAT & VENT *for pipe* Fitting *nt*, Rohrverbindungsstück *nt*, *mounting* Armatur *f*; **~ dimensions** *n pl* ARCH & BUILD Anschlagmaße *nt pl*; **~ instructions** *n pl* BUILD HARDW Anschlaganleitung *f*

**fittings** *n pl* BUILD HARDW Garnitur *f*

***fit***: **~-up** *n* CONCR Schalungstafel *f*, wiederverwendungsfähige Schalung *f*

**fix** *vt* ARCH & BUILD *secure* anbringen, befestigen, einspannen, BUILD MACHIN einspannen, CONCR, INFR & DES, STEEL anbringen, einspannen, STONE *in a wall* einmauern, TIMBER anbringen

**fixed** *adj* ARCH & BUILD fest, feststehend; **~ arch** *n* ARCH & BUILD eingespannter Bogen *m*; **~ datum** *n* ARCH & BUILD *surveying* Festpunkt *m*, Bezugshöhe *f*, Fixpunkt *m*, INFR & DES Bezugswert *m*, Bezugshöhe *f*, Festpunkt *m*, Fixpunkt *m*; **~ end** *n* ARCH & BUILD eingespanntes Trägerende *nt*, Fixpunkt *m*, INFR & DES Fixpunkt *m*; **~ end beam** *n* ARCH & BUILD beidseitig eingespannter Balken *m*, zweiseitig eingespannter Balken *m*; **~ glazing** *n* ARCH & BUILD Festverglasung *f*; **~ point** *n* ARCH & BUILD, INFR & DES Festpunkt *m*; **~ post** *n* ARCH & BUILD Standmast *m*; **~ roller sluice gate** *n* ENVIRON fixiertes Walzenwehr *nt*; **~ support** *n* ARCH & BUILD, INFR & DES völlige Einspannung *f*

**fixing** *n* BUILD HARDW Befestigung *f*; **~ block** *n* ARCH & BUILD Dübelstein *m*; **~ bolt** *n* MAT PROP Befestigungsschraube *f*; **~ clamp** *n* MAT PROP Befestigungspratze *f*; **~ device** *n* MAT PROP Befestigungsvorrichtung *f*; **~ frame** *n* ARCH & BUILD Befestigungsrahmen *m*; **~ point** *n* ARCH & BUILD Einspannstelle *f*, Befestigungsstelle *f*; **~ screw** *n* MAT PROP Befestigungsschraube *f*

**fix**: **~ in mortar** *phr* STONE vermörteln

**fixtures**: **~ and fittings** *n pl* BUILD HARDW Installationsobjekte *nt pl*, festeingebaute Wohnungsgegenstände *m pl*

**flag** *n* INFR & DES, STONE Gehwegplatte *f*, Natursteinplatte *f*

**flagging** *n* INFR & DES *flagstones* Plattenbelag *m*, *action* Plattenverlegen *nt*, STONE *flagstones* Plattenbelag *m*

**flagstone** *n* INFR & DES, STONE Gehwegplatte *f*, Natursteinplatte *f*; **~ pavement** *n BrE* (*cf flagstone sidewalk AmE*) INFR & DES Gehwegplattenbelag *m*; **~ sidewalk** *n AmE* (*cf flagstone pavement BrE*) INFR & DES Gehwegplattenbelag *m*

**flake**: **1. ~ graphite** *n* MAT PROP Lamellengraphit *m*; **2. ~ off** *vi* CONCR, INFR & DES, MAT PROP, STONE absplittern

**flame**: **~ cutter** *n* BUILD MACHIN Schneidbrenner *m*

**flameproof** *adj* INFR & DES, MAT PROP flammsicher

**flame**: **~-resistant** *adj* ARCH & BUILD, CONST LAW, INFR & DES, MAT PROP flammwidrig; **~-retardant** *adj* ARCH & BUILD, CONST LAW, INFR & DES, MAT PROP, SOUND & THERMAL, TIMBER flammenhemmend; **~-retardant treatment** *n* ARCH & BUILD, CONST LAW, MAT PROP, SOUND & THERMAL, TIMBER flammenhemmende Behandlung *f*

**flamesprayed**: **~ zinc coating** *n* SURFACE Flammspritzverzinkung *f*

**flame**: **~ tube** *n* HEAT & VENT Flammrohr *nt*

**flange** *n* BUILD MACHIN Ausleger *m*, Trägerflansch *m*, Wulst *m*, HEAT & VENT *of tube or pipe end* Flansch *m*, STEEL Gurt *m*, Flansch *m*, vorstehender Rand *m*; **~ bend** *n* HEAT & VENT Flanschkrümmer *m*

**flanged**: **~ adapter** *n* HEAT & VENT Flanschpaßstück *nt*; **~ cast-iron pipe** *n* BUILD HARDW Flanschgußrohr *nt*; **~ connection** *n* HEAT & VENT Flanschstutzen *m*, Flanschverbindung *f*, STEEL Flanschverbindung *f*; **~ nut** *n* MAT PROP Bundmutter *f*; **~ plate** *n* BUILD HARDW Bördelblech *nt*; **~ socket** *n* HEAT & VENT Flanschmuffe *f*, E-Stück *nt*

***flange***: **~ pipe** *n* BUILD HARDW, HEAT & VENT Flanschrohr *nt*; **~ plate** *n* BUILD HARDW *girder* Gurtplatte *f*; **~ tile** *n* BUILD HARDW Deckplatte *f*; **~ tube** *n* BUILD HARDW, HEAT & VENT Flanschrohr *nt*

**flank** *n* ARCH & BUILD Bankett *nt*, Flanke *f*, INFR & DES *of road* Bankett *nt*; **~ of an arch** *n* ARCH & BUILD Bogenachse *f*, Bogenhälfte *f*, Bogenschenkel *m*

**flanking**: **~ transmission** *n* SOUND & THERMAL Flankenübertragung *f*

***flank***: **~ wall** *n* ARCH & BUILD, STONE Giebelmauer *f*, Giebelwand *f*

**flap** *n* ARCH & BUILD, HEAT & VENT Klappe *f*; **~ door** *n* ARCH & BUILD Falltür *f*; **~ gate** *n* ENVIRON Klapptor *nt*; **~ hinge** *n* BUILD HARDW Scharnierband *nt*, Klappenscharnier *nt*; **~ tile** *n* ARCH & BUILD, MAT PROP *roofing* Kremper *m*, Krempziegel *m*; **~ trap** *n* WASTE WATER Rückschlagklappe *f*

**flared**: **~ head** *n* ARCH & BUILD Pilzkopf *m*

**flashed**: **~ glass** *n* MAT PROP *coated with metallic oxide* Überfangglas *nt*

**flashing** *n* MAT PROP Abdeckblech *nt*, STEEL Abdeckblech *nt*, Verwahrung *f*, *on a chimney* Schürze *f*

**flat 1.** *adj* ARCH & BUILD Flach-, eben, flach, INFR & DES eben, flach, MAT PROP eben, SURFACE matt; **2.** *n BrE* ARCH & BUILD Wohnung *f*

***flat***: **~ arch** *n* ARCH & BUILD gerader Bogen *m*, Flachbogen *m*; **~-beaded profile** *n* MAT PROP, STEEL Flachwulstprofil *nt*; **~ clay roofing tile** *n* MAT PROP Flachdachpfanne *f*, Flachziegel *m*, TIMBER Flachziegel *m*; **~ joint** *n* STONE bündige Mauerwerksfuge *f*, Schnittfuge *f*; **~ plate solar collector** *n* HEAT & VENT Flachsolarkollektor *m*; **~ roof** *n* ARCH & BUILD Flachdach *nt*; **~ roofing tile** *n* MAT PROP Flachdachpfanne *f*, Flachziegel *m*, TIMBER Flachziegel *m*; **~ roof outlet** *n* WASTE WATER Flachdachablauf *m*; **~ roof system** *n* ARCH & BUILD Flachdachaufbau *m*; **~ safety glass** *n* INFR & DES, MAT PROP flaches Sicherheitsglas *nt*; **~ steel** *n* MAT PROP, STEEL Flacheisen *nt*, Flachstahl *m*, Bandstahl *m*

**flatten** *vt* INFR & DES *ground* abflachen, STEEL *sheet metal* richten

**flattened** *adj* INFR & DES *ground* abgeflacht, STEEL *sheet metal* gerichtet

**flattening** *n* INFR & DES *ground* Abflachung *f*, STEEL *sheet metal* Richten *nt*

***flat***: **~ top** *n* ARCH & BUILD Flachboden *m*; **~ trowel** *n* BUILD MACHIN Spachtel *m*; **~ truss** *n* ARCH & BUILD Parallelfachwerk *nt*, parallelgurtiges Fachwerk *nt*

**flaw** *n* MAT PROP Mangel *m*

**flawed** *adj* MAT PROP mangelhaft

**flawless** *adj* MAT PROP einwandfrei

**flèche** *n* ARCH & BUILD *gothic style* Spitzturm *m*

**fletton** *n* STONE Fletton *nt*

**flexibility** *n* MAT PROP Flexibilität *f*, Biegsamkeit *f*

**flexible** *adj* MAT PROP flexibel, weich, elastisch, biegsam; **~ flue liner** *n* HEAT & VENT, INFR & DES, MAT PROP flexible Abzugsverkleidung *f*; **~ foam plastic** *n* MAT PROP Weichschaum *m*; **~ foundation** *n* ARCH & BUILD, INFR & DES flexibles Fundament *nt*; **~ foundation beam** *n* ARCH & BUILD Zerrbalken *m*; **~ smoke seal** *n*

CONST LAW, INFR & DES, MAT PROP flexible Rauchdichtung *f*
**flexion** *n* ARCH & BUILD, INFR & DES Biegung *f*, Krümmung *f*
**flexural** *adj* STONE biegefähig
**flexurally**: ~ **rigid** *adj* MAT PROP, STEEL biegesteif
**flexural**: ~ **stiffness** *n* ARCH & BUILD, CONCR, INFR & DES, MAT PROP, STEEL Biegesteifigkeit *f*; ~ **strength** *n* ARCH & BUILD, INFR & DES, MAT PROP, STEEL Biegefestigkeit *f*
**flexure** *n* ARCH & BUILD, INFR & DES Durchbiegung *f*
**flight**: ~ **of stairs** *n* ARCH & BUILD Treppenlauf *m*; ~ **width** *n* ARCH & BUILD *of stairs* Laufbreite *f*
**flitched**: ~ **beam** *n* TIMBER Dübelbalken *m*
**flitch**: ~ **plate** *n* BUILD HARDW Sandwichplatte *f*
**float 1.** *n* BUILD MACHIN Glättkelle *f*, Abziehbrett *nt*, STONE Glättkelle *f*; **2.** *vt* ARCH & BUILD abreiben, CONCR glätten, abreiben, MAT PROP abreiben, STONE aufspachteln, glätten, abreiben, TIMBER abreiben; ~ **off** STONE abschwemmen; **3.** *vi* ARCH & BUILD, BUILD MACHIN, INFR & DES schwimmen
***float***: ~ **coat** *n* STONE Glattstrich *m*
**floated**: ~ **screed** *n* CONCR abgezogener Estrich *m*
***float***: ~ **glass** *n* MAT PROP Floatglas *nt*
**floating** *adj* ARCH & BUILD, BUILD MACHIN, INFR & DES schwimmend; ~ **boom** *n* ENVIRON Schwimmschirm *m*; ~ **caisson foundation** *n* CONCR Schwimmkastengründung *f*; ~ **crane** *n* BUILD MACHIN schwimmender Kran *m*; ~ **derrick** *n* BUILD MACHIN schwimmender Derrick *m*; ~ **flooring** *n* ARCH & BUILD, SOUND & THERMAL, STONE schwimmender Estrich *m*; ~ **foundation** *n* CONCR, INFR & DES Schwimmgründung *f*; ~ **layer** *n* ARCH & BUILD, SOUND & THERMAL, STONE schwimmender Estrich *m*; ~ **pile foundation** *n* INFR & DES, STEEL schwimmende Pfahlgründung *f*; ~ **pipeline** *n* INFR & DES, STEEL schwimmende Rohrleitung *f*; ~ **structure** *n* INFR & DES schwimmendes Bauwerk *nt*; ~ **support** *n* BUILD MACHIN, INFR & DES Schwimmbühne *f*
***float***: ~ **and set** *n* STONE dreilagiger Putz *m*, Reibeputz *m*
**flocculating**: ~ **agent** *n* INFR & DES, MAT PROP Ausflokkungsmittel *nt*
**flocculation** *n* ENVIRON Ausflockung *f*, Flockung *f*, MAT PROP Ausflockung *f*
**flood 1.** *n* ENVIRON, HEAT & VENT Flut *f*; **2.** *vt* ENVIRON, HEAT & VENT überschwemmen
***flood***: ~ **control** *n* ARCH & BUILD, ENVIRON, INFR & DES Hochwasserschutz *m*; ~ **damage** *n* ARCH & BUILD, INFR & DES Hochwasserschaden *m*
**flooding** *n* ENVIRON, HEAT & VENT Überschwemmung *f*
**floodlight**: ~ **mast** *n* INFR & DES Flutlichtmast *m*
***flood***: ~ **loss** *n* ENVIRON Flutverlust *m*; ~ **plane** *n* ENVIRON, INFR & DES Überschwemmungsgebiet *nt*; ~ **protection** *n* ARCH & BUILD, ENVIRON, INFR & DES Hochwasserschutz *m*
**floor 1.** *n* ARCH & BUILD Etage *f*, Geschoßdecke *f*, Stockwerk *nt*, MAT PROP Fußboden *m*, *of basin* Sohle *f*; **2.** *vt* ARCH & BUILD, STONE pflastern, TIMBER dielen
***floor***: ~ **arch** *n* ARCH & BUILD Sohlengewölbe *nt*, scheitrechter Bogen *m*; ~ **area** *n* ARCH & BUILD Grundfläche *f*; ~ **beam** *n* ARCH & BUILD Deckenbalken *m*, Deckenträger *m*, TIMBER Deckenbalken *m*; ~ **boards** *n pl* TIMBER Fußbodenbretter *nt pl*; ~ **channel** *n* ARCH & BUILD, HEAT & VENT Bodenkanal *m*; ~ **closer** *n* BUILD HARDW *door* Bodentürschließer *m*; ~ **covering** *n* ARCH & BUILD, BUILD HARDW, MAT PROP, STONE Bodenbelag *m*, Fußbodenbelag *m*; ~ **decking** *n* ARCH & BUILD, BUILD HARDW, MAT PROP, STONE Bodenbelag *m*, Fußbodenbelag *m*; ~ **drain** *n* ENVIRON, HEAT & VENT Ablauf *m*, WASTE WATER Ablauf *m*, Bodenablauf *m*; ~ **grating** *n* STEEL Bodenrost *m*; ~ **guide** *n* BUILD HARDW *sliding door* Schiebetürführung *f*, Fußbodenführung *f*; ~ **gully** *n* ENVIRON, HEAT & VENT Ablauf *m*, WASTE WATER Bodenablauf *m*, Ablauf *m*; ~ **heating** *n* ARCH & BUILD, HEAT & VENT Fußbodenheizung *f*
**flooring** *n* ARCH & BUILD Fußbodenbelag *m*, BUILD HARDW Dielung *f*, Fußbodenbelag *m*, STONE, TIMBER Fußbodenbelag *m*; ~ **joint** *n* ARCH & BUILD Bodenfuge *f*; ~ **layout** *n* ARCH & BUILD Deckenauslegung *f*; ~ **nail** *n* BUILD HARDW Fußbodennagel *m*; ~ **tile** *n* ARCH & BUILD, BUILD HARDW Bodenfliese *f*, Fußbodenfliese *f*, Fußbodenplatte *f*, CONCR Fußbodenplatte *f*, INFR & DES, MAT PROP Bodenfliese *f*, Fußbodenfliese *f*, Fußbodenplatte *f*
***floor***: ~ **inlet** *n* ENVIRON, HEAT & VENT Ablauf *m*, WASTE WATER Bodenablauf *m*, Ablauf *m*; ~ **insulation** *n* SOUND & THERMAL Bodendämmung *f*, Fußbodenisolierung *f*; ~ **joist** *n* ARCH & BUILD Deckenbalken *m*, Deckenträger *m*, Deckenunterzug *m*, BUILD HARDW Deckenunterzug *m*, TIMBER Lagerholz *nt*, Deckenbalken *m*; ~ **lamp** *n* ELECTR Stehlampe *f*; ~ **master key** *n* BUILD HARDW Gruppenschlüssel *m*; ~ **pavement** *n* ARCH & BUILD, BUILD HARDW, MAT PROP, STONE Fußbodenbelag *m*; ~ **recess** *n* ARCH & BUILD Bodenaussparung *f*, Bodenvertiefung *f*; ~ **screed conveyor** *n* BUILD MACHIN Estrichförderer *m*; ~ **slab** *n* ARCH & BUILD, BUILD HARDW, CONCR, INFR & DES, MAT PROP Bodenplatte *f*, Fußbodenplatte *f*; ~ **space** *n* ARCH & BUILD, INFR & DES Grundfläche *f*, Nutzfläche *f*, Wohn- und Nutzfläche *f*; ~ **stop** *n* ARCH & BUILD, BUILD HARDW Türanschlag *m*, Türstopper *m*; ~ **system** *n* INFR & DES *of bridge* Brückenrost *m*; ~**-through dwelling** *n* *AmE* ARCH & BUILD Vollgeschoßwohnung *f*; ~ **tile** *n* ARCH & BUILD, BUILD HARDW Bodenfliese *f*, Fußbodenfliese *f*, Fußbodenplatte *f*, CONCR Fußbodenplatte *f*, INFR & DES, MAT PROP Bodenfliese *f*, Fußbodenfliese *f*, Fußbodenplatte *f*; ~**-to-floor height** *n* ARCH & BUILD Geschoßhöhe *f*; ~ **track** *n* BUILD HARDW Bodenschiene *f*
**flow 1.** *n* ENVIRON, HEAT & VENT, INFR & DES, MAT PROP, WASTE WATER Durchfluß *m*, Durchlauf *m*, Fließ-, Fluß *m*, Strom *m*, Strömung *f*; **2.** *vi* ENVIRON, HEAT & VENT, INFR & DES, MAT PROP, WASTE WATER fließen, strömen; ~ **out** ENVIRON, WASTE WATER ausfließen; ~ **through** ARCH & BUILD, CONCR, HEAT & VENT durchlaufen
***flow***: ~ **chart** *n* INFR & DES, WASTE WATER Fließdiagramm *nt*; ~ **coefficient** *n* INFR & DES Durchflußkoeffizient *m*; ~ **formula** *n* INFR & DES, MAT PROP Fließformel *f*; ~ **heater** *n* HEAT & VENT, WASTE WATER Durchlauferhitzer *m*
**flowing**: ~ **tide** *n* ENVIRON Flutströmung *f*
***flow***: ~**-line construction** *n* ARCH & BUILD Fließfertigung *f*; ~**-type calorifier** *n* HEAT & VENT, WASTE WATER Durchlauferhitzer *m*; ~ **velocity** *n* INFR & DES, WASTE WATER Fließgeschwindigkeit *f*; ~ **of waste** *n* ENVIRON Abfallfluß *m*, Abfallstrom *m*
**fluctuation**: ~ **of load** *n* ARCH & BUILD Belastungsschwankung *f*

**flue** *n* ARCH & BUILD Fuchs *m*; **~ conditions** *n* ARCH & BUILD Zugverhältnisse *nt pl*; **~ gas desulfurization** *AmE*, **~ gas desulphurization** *BrE* *n* ENVIRON Rauchgasentschwefelung *f*; **~-gas-tight** *adj* ARCH & BUILD rauchgasdicht; **~ lining** *n* SOUND & THERMAL, STONE Fuchsauskleidung *f*

**fluid 1.** *adj* ENVIRON, INFR & DES, MAT PROP flüssig; **2.** *n* INFR & DES, MAT PROP Flüssigkeit *f*

***fluid***: **~ chromatography** *n* ENVIRON Flüssigchromatographie *f*

**fluidized**: **~ bed** *n* ENVIRON Flüssigbett *nt*, HEAT & VENT Wirbelbett *nt*

***fluid***: **~ mechanics** *n* INFR & DES Strömungslehre *f*

**flume** *n* ENVIRON Ablaufkanal *m*, Gerinne *nt*, INFR & DES Gerinne *nt*

**fluorescent 1.** *adj* ARCH & BUILD, ELECTR fluoreszierend; **2.** *n* ELECTR Leuchtstoff *m*

***fluorescent***: **~ lamp** *n* ELECTR Leuchtstofflampe *f*, Leuchtstoffröhre *f*; **~ lighting** *n* ARCH & BUILD, ELECTR Leuchtstoffbeleuchtung *f*

**fluoridation** *n* ENVIRON, WASTE WATER Fluoridierung *f*

**fluorspar** *n* INFR & DES Flußspat *m*

**fluosilicate**: **~ treatment** *n* SURFACE Fluatierung *f*

**flush 1.** *adj* ARCH & BUILD *door* bündig, *facade* fluchteben, ELECTR *switch, socket* unter Putz, INFR & DES *kerb* eingelassen; **2.** *vt* ENVIRON, HEAT & VENT, WASTE WATER spülen, durchspülen; **~ out** ENVIRON ausspülen

**flush**: **~ with the adjacent areas** *phr* ARCH & BUILD flächenbündig; **~ arrangement of joints** *n* ARCH & BUILD Bündigkeit *f* von Stößen; **~ bracket** *n* MAT PROP Flachstütze *f*; **~ curb** *AmE see flush kerb BrE*

**flushing** *n* BUILD HARDW Wasserspülung *f*, ENVIRON Spülung *f*, HEAT & VENT, WASTE WATER Spülen *nt*, Spülung *f*, Wasserspülung *f*; **~ aid** *n* HEAT & VENT, WASTE WATER Spülhilfe *f*; **~ cistern** *n* WASTE WATER Spülkasten *m*; **~ pipe** *n* WASTE WATER Spülrohr *nt*; **~ tank** *n* WASTE WATER Spülkasten *m*; **~ valve** *n* WASTE WATER Druckspüler *m*

**flush**: **~ kerb** *n BrE* INFR & DES Tiefbordstein *m*; **~ lock** *n* BUILD HARDW Kantenschloß *nt*; **~-mounted** *adj* ARCH & BUILD *door* bündig, ELECTR *switch, socket* unter Putz montiert; **~-mounted axial fan** *n* HEAT & VENT Wandeinbauaxialventilator *m*; **~ mounting** *n* ELECTR *of switch, socket* Unterputzinstallation *f*; **~-mounting frame** *n* ARCH & BUILD Einputzrahmen *m*; **~ valve** *n* WASTE WATER Druckspüler *m*; **~ weld** *n* STEEL Flachschweißnaht *f*

**flute** *n* MAT PROP Nut *f*, Riffelung *f*

**fluted**: **~ column** *n* ARCH & BUILD kannelierte Säule *f*

**fluting** *n* MAT PROP Nuten *nt*, Riffeln *nt*

**fluvial** *adj* ENVIRON, INFR & DES fluvial; **~ clay** *n* ENVIRON, INFR & DES, MAT PROP Flußton *m*; **~ deposit** *n* ENVIRON, INFR & DES Flußablagerung *f*; **~ sand** *n* ENVIRON, INFR & DES Flußsand *m*; **~ sediment** *n* ENVIRON, INFR & DES Flußablagerung *f*

**flux** *n* MAT PROP *welding* Schmelzmittel *nt*

**fly**: **~ ash** *n* ENVIRON, MAT PROP Flugasche *f*

**flyer** *n* ARCH & BUILD Blockstufe *f*

**flyers** *n pl* ARCH & BUILD Außentreppe *f*, freitragende Treppe *f*

**flying**: **~ bond** *n* STONE märkischer Verband *m*; **~ bridge** *n* INFR & DES fliegende Brücke *f*; **~ buttress** *n* ARCH & BUILD Strebepfeiler *m*; **~ scaffold** *n* ARCH & BUILD, BUILD MACHIN, INFR & DES Hängegerüst *nt*, hängendes Gerüst *nt*; **~ shore** *n* ARCH & BUILD Hilfsstütze *f*, waagerechte Stiefe *f*

**flyover** *n* INFR & DES Hochstraße *f*

**fly**: **~ screen** *n* BUILD HARDW Fliegengitter *nt*

**foam 1.** *n* MAT PROP Schaum *m*; **2.** *vt* MAT PROP aufschäumen; **3.** *vi* CONCR, HEAT & VENT, MAT PROP, WASTE WATER schäumen

***foam***: **~ in** *vt* ARCH & BUILD *window* einschäumen

***foam***: **~ concrete** *n* CONCR Gasbeton *m*, MAT PROP Gasbeton *m*, Porenbeton *m*

**foamed** *adj* MAT PROP porig, *plastic, rubber* aufgeschäumt, SOUND & THERMAL aufgeschäumt; **~ in place** *adj* ARCH & BUILD *window* eingeschäumt

***foam***: **~ glass** *n* MAT PROP Foamglas *nt*, Schaumglas *nt*, Vielzellenglas *nt*, SOUND & THERMAL Schaumglas *nt*, Vielzellenglas *nt*; **~ rubber tape** *n* MAT PROP, SOUND & THERMAL Schaumgummiband *nt*

**foil** *n* MAT PROP *metal* Folie *f*

**fold 1.** *n* STEEL Abkantung *f*, Falz *m*; **2.** *vt* STEEL *sheet metal* abkanten

**folded**: **~-plate structure** *n* ARCH & BUILD Faltwerk *nt*; **~ spiral-seam tube** *n* ARCH & BUILD, WASTE WATER Wickelfalzrohr *nt*

**folding** *n* STEEL Abkanten *nt*, Falzen *nt*; **~ door** *n* ARCH & BUILD, BUILD HARDW Falttor *nt*, Falttür *f*, Flügeltür *f*, Harmonikatür *f*; **~ ladder** *n* ARCH & BUILD, BUILD HARDW Einschubtreppe *f*, Scherentreppe *f*; **~ partition** *n* ARCH & BUILD, BUILD HARDW Faltwand *f*; **~ shutter** *n* ARCH & BUILD Fensterladen *m*, BUILD HARDW Fensterladen *m*, Klappladen *m*, TIMBER Klappladen *m*; **~ staircase** *n* BUILD HARDW Einschubtreppe *f*, Scherentreppe *f*

**foot** *n* ARCH & BUILD, INFR & DES *lower section* Unterteil *nt*, Fuß *m*, *of wall, pillar* Sockel *m*, *bottom* Sohle *f*

**footbridge** *n* INFR & DES Fußgängerbrücke *f*

**footfall**: **~ sound** *n* SOUND & THERMAL Trittschall *m*; **~ sound insulation** *n* SOUND & THERMAL Trittschalldämmung *f*

**foothold** *n* ARCH & BUILD Auftrittsbreite *f*

**footing** *n* ARCH & BUILD *building* Gründung *f*, *foundations* Fundament *nt*, *of wall* Sockel *m*, INFR & DES Auflager *nt*, Fundament *nt*, Gründung *f*, *of wall* Sockel *m*; **~ block** *n* ARCH & BUILD Fundamentblock *m*; **~ trench** *n* ARCH & BUILD, INFR & DES Fundamentgraben *m*

***foot***: **~ scraper mat** *n* BUILD HARDW Fußabstreifmatte *f*

**footstep** *n* SOUND & THERMAL Laufschritt *m*; **~ sound** *n* SOUND & THERMAL Trittschall *m*; **~ sound insulation** *n* SOUND & THERMAL Trittschalldämmung *f*

***foot***: **~ valve** *n* WASTE WATER Bodenventil *nt*, Fußventil *nt*; **~ wall** *n* INFR & DES *tunnelling* Sohle *f*

**force** *n* ARCH & BUILD, INFR & DES Kraft *f*

**forced**: **~ draft** *AmE*, **~ draught** *BrE* *n* ARCH & BUILD Saugzug *m*, verstärkter Zug *m*

***force***: **~ diagram** *n* ARCH & BUILD, INFR & DES Kräfteplan *m*

***forced***: **~ ventilation** *n* HEAT & VENT Zwangsentlüftung *f*

**forcing** *n* BUILD HARDW Führung *f*, INFR & DES *tunnelling* Vortreiben *nt*; **~ pump** *n* WASTE WATER Druckpumpe *f*

**forecourt** *n* ARCH & BUILD, INFR & DES Vorhof *m*

**foreign** *adj* ENVIRON fremd; **~ emissions** *n pl* ENVIRON

Fremdemissionen *f pl*; ~ **water** *n* INFR & DES Schichtwasser *nt*

**foreman** *n* ARCH & BUILD Vorarbeiter *m*, Polier *m*

**forestry**: ~ **research** *n* ENVIRON Forstwirtschaftsforschung *f*

**forked**: ~ **pipe** *n* WASTE WATER Gabelrohr *nt*; ~ **tie** *n* ARCH & BUILD Gabelanker *m*; ~ **wood** *n* TIMBER Gabelholz *nt*

**fork**: ~**-lift truck** *n* BUILD MACHIN Gabelstapler *m*, Hubstapler *m*

**form 1.** *n* ARCH & BUILD Gestalt *f*, Schalung *f*, Form *f*, CONCR Schalung *f*, CONST LAW Formular *nt*, INFR & DES Form *f*; **2.** *vt* ARCH & BUILD formen, bilden, CONCR einschalen, INFR & DES bilden, formen

***form***: ~ **of agreement** *n* CONST LAW Vereinbarungsform *f*

**format** *n* ARCH & BUILD Format *nt*

**formation**: ~ **of condensate** *n* HEAT & VENT, WASTE WATER Kondenswasserentstehung *f*; ~ **level** *n* INFR & DES Feinplanumshöhe *f*, Planum *nt*

**formboard** *n* ARCH & BUILD, CONCR Schalbrett *nt*

***form***: ~ **and function** *n* ARCH & BUILD, INFR & DES Form *f* und Funktion *f*

**forming** *n* ARCH & BUILD, CONCR, INFR & DES Formgebung *f*

***form***: ~ **oil** *n* CONCR Schalungsöl *nt*; ~ **paste** *n* CONCR Schalungspaste *f*; ~ **removal** *n* CONCR Ausschalen *nt*, Entschalen *nt*

**forms**: ~ **of stress** *n pl* ARCH & BUILD, CONCR, TIMBER Spannungsarten *f pl*

***form***: ~ **stripping** *n* CONCR Ausschalen *nt*, Entschalen *nt*

**formwork** *n* ARCH & BUILD Schalung *f*, CONCR Betonschalung *f*, Schalung *f*, Verschalung *f*, Einschalung *f*, Deckenschalung *f*, INFR & DES, TIMBER Deckenschalung *f*; ~ **board** *n* ARCH & BUILD, CONCR Schalbrett *nt*; ~ **lube** *n* *AmE* (*cf formwork oil BrE*) CONCR Schalöl *nt*; ~ **lube stains** *n* *AmE* (*cf formwork oil stains BrE*) CONCR Schalölflecken *m pl*; ~ **lubricant** *n* CONCR Schalöl *nt*; ~ **lubricant stains** *n pl* CONCR Schalölflecken *m pl*; ~ **oil** *n* *BrE* (*cf formwork lube AmE*) CONCR Schalöl *nt*; ~ **oil stains** *n pl* *BrE* (*cf formwork lube stains AmE*) CONCR Schalölflecken *m pl*; ~ **panel** *n* CONCR Schalungstafel *f*; ~ **removal** *n* CONCR Ausschalen *nt*

**fouling** *n* ENVIRON Faulung *f*

**foul**: ~ **water** *n* ENVIRON, INFR & DES Schmutzwasser *nt*, WASTE WATER Abwasser *nt*, Schmutzwasser *nt*; ~ **water sewer** *n* INFR & DES, WASTE WATER Schmutzwasserkanal *m*

**foundation** *n* ARCH & BUILD, INFR & DES *base, footing* Sockel *m*, *footing* Gründung *f*, Unterbau *m*, *of building* Fundament *nt*; ~ **base** *n* ARCH & BUILD, INFR & DES Fundamentsohle *f*; ~ **block** *n* ARCH & BUILD Blockfundament *nt*, Einzelfundament *nt*, Fundamentblock *m*, Fundamentklotz *m*, CONCR Einzelfundament *nt*, INFR & DES Blockfundament *nt*, Einzelfundament *nt*, Fundamentblock *m*, Fundamentklotz *m*; ~ **bolt** *n* BUILD HARDW Fundamentschraube *f*; ~ **cylinder** *n* ARCH & BUILD, INFR & DES Zylinderpfahl *m*; ~ **depth** *n* ARCH & BUILD, INFR & DES Fundamenttiefe *f*, Gründungstiefe *f*; ~ **drawing** *n* ARCH & BUILD, INFR & DES Fundamentplan *m*; ~ **earthing** *n* *BrE* (*cf foundation grounding AmE*) ELECTR Fundamenterder *m*; ~ **grounding** *n* *AmE* (*cf foundation earthing BrE*) ELECTR Fundamenterder *m*; ~ **level** *n* ARCH & BUILD, INFR & DES Gründungssohle *f*; ~ **pad** *n* ARCH & BUILD, CONCR, INFR & DES Einzelfundament *nt*; ~ **pile** *n* ARCH & BUILD, INFR & DES Gründungspfahl *m*; ~ **plate** *n* ARCH & BUILD, BUILD HARDW, INFR & DES Fundamentplatte *f*; ~ **pressure** *n* ARCH & BUILD, INFR & DES Erddruck *m*, Gründungsdruck *m*; ~ **raft** *n* ARCH & BUILD, BUILD HARDW, CONCR, INFR & DES Fundamentplatte *f*, Plattenfundament *nt*; ~ **slab** *n* ARCH & BUILD, INFR & DES Fundamentplatte *f*, Gründungsplatte *f*; ~ **soil** *n* ARCH & BUILD, INFR & DES Baugrund *m*; ~ **stone** *n* ARCH & BUILD Grundstein *m*; ~ **strip** *n* ARCH & BUILD Fundamentstreifen *m*; ~ **trench** *n* ARCH & BUILD, INFR & DES Fundamentgraben *m*; ~ **width** *n* ARCH & BUILD, INFR & DES Fundamentbreite *f*; ~ **work** *n* ARCH & BUILD, INFR & DES Fundamentarbeit *f*

**foundry** *n* MAT PROP, STEEL Hütte *f*

**fountain** *n* BUILD HARDW, INFR & DES, WASTE WATER Brunnen *m*

**four**: ~**-centered arch** *AmE*, ~**-centred arch** *BrE* *n* ARCH & BUILD *Tudor arch* gedrückter Spitzbogen *m*; ~**-cornered** *adj* ARCH & BUILD viereckig, vierseitig; ~**-leaved** *adj* ARCH & BUILD *window* vierflügelig; ~**-point bearing** *n* ARCH & BUILD Vierpunktlagerung *f*; ~**-sided** *adj* ARCH & BUILD viereckig

**fraction** *n* INFR & DES, MAT PROP Anteil *m*; ~ **weight** *n* INFR & DES, MAT PROP Gewichtsanteil *m*

**fracture** *n* MAT PROP Bruch *m*; ~**-proof** *adj* MAT PROP bruchsicher

**frame 1.** *n* ARCH & BUILD, INFR & DES, STEEL *of door, window* Zarge *f*, Gerüst *nt*, *of machine* Gestell *nt*, *carcass of building* Skelett *nt*, Rahmen *m*, TIMBER Zarge *f*, WASTE WATER *street inlet* Aufsatz *m*; **2.** *vt* ARCH & BUILD umrahmen

***frame***: ~ **analysis** *n* ARCH & BUILD, INFR & DES Rahmenberechnung *f*, Rahmenstatik *f*; ~ **bridge** *n* INFR & DES Fachwerkbrücke *f*; ~ **construction** *n* ARCH & BUILD, INFR & DES Skelettbauweise *f*

**framed**: ~ **door** *n* ARCH & BUILD Rahmentür *f*; ~ **floor** *n* ARCH & BUILD Doppelfußboden *m*; ~ **partition** *n* ARCH & BUILD Gerippetrennwand *f*, *inner wall* Gerippewand *f*; ~ **roof** *n* ARCH & BUILD Fachwerkdach *nt*; ~ **structure** *n* ARCH & BUILD, INFR & DES Skelettbau *m*; ~ **wall** *n* ARCH & BUILD Riegelwand *f*, Fachwerkwand *f*

***frame***: ~ **house** *n* ARCH & BUILD Fachwerkhaus *nt*; ~ **of joists** *n* ARCH & BUILD Gebälk *nt*, TIMBER Balkenlage *f*, Gebälk *nt*; ~ **leg** *n* TIMBER Rahmenschenkel *m*

**frameless** *adj* ARCH & BUILD rahmenlos, rahmenfrei; ~ **construction** *n* ARCH & BUILD, CONCR Scheibenbauweise *f*

***frame***: ~ **member** *n* ARCH & BUILD Fachwerkstab *m*; ~ **and panel construction** *n* ARCH & BUILD, INFR & DES Fertigteilbauweise *f* mit Skelett und Ausfachungsplatten; ~ **piece** *n* TIMBER Rahmenschenkel *m*

**framework** *n* ARCH & BUILD Fachwerk *nt*, Stabwerk *nt*, Skelett *nt*, INFR & DES Fachwerk *nt*, Skelett *nt*, MAT PROP, TIMBER Fachwerk *nt*; ~ **wall** *n* STONE Riegelwand *f*

**framing** *n* ARCH & BUILD Zusammenbau *m*, BUILD HARDW *of partition wall* Umrahmung *f*

**Francis**: ~ **turbine** *n* BUILD MACHIN Francis-Turbine *f*

**Franki**: ~**-pile** *n* INFR & DES Franki-Bohrpfahl *m*

**free**: ~ **beam** *n* ARCH & BUILD freier Träger *m*; ~ **bearing** *n* ARCH & BUILD freies Auflager *nt*; ~ **of**

**cavities** *phr* CONCR hohlraumfrei; **~ delivery** *n* CONST LAW Lieferung *f* frei Verwendungsstelle
**freedom**: **~ from load** *n* ARCH & BUILD Lastenfreiheit *f*
**free**: **~ end** *n* ARCH & BUILD, CONCR, INFR & DES, MAT PROP, STEEL freies Ende *nt*; **~-fall boring** *n* BUILD MACHIN Freifallbohrung *f*; **~-fall drilling** *n* BUILD MACHIN Freifallbohrung *f*; **~-falling stamp** *n* BUILD MACHIN Freifallstanze *f*; **~ ground water** *n* INFR & DES ungespanntes Grundwasser *nt*
**freely**: **~ supported** *adj* BUILD HARDW frei aufliegend
**free**: **~ pore-water** *n* INFR & DES, MAT PROP freies Porenwasser *nt*; **~ roller sluice gate** *n* ENVIRON freies Walzenwehr *nt*; **~ space** *n* ARCH & BUILD, INFR & DES freier Raum *m*; **~-span system** *n* ARCH & BUILD, INFR & DES Freispannsystem *nt*; **~-standing** *adj* BUILD HARDW freistehend; **~-standing column** *n* ARCH & BUILD Freisäule *f*
**freestone**: **~ masonry** *n* ARCH & BUILD, STONE Quadermauerwerk *nt*
**free**: **~ support** *n* ARCH & BUILD freies Auflager *nt*; **~ surface** *n* ARCH & BUILD, INFR & DES freie Oberfläche *f*; **~ vibration** *n* SOUND & THERMAL freie Schwingung *f*
**freeway** *n AmE* (*cf motorway BrE*) INFR & DES Autobahn *f*
**freeze** *vti* HEAT & VENT, INFR & DES, MAT PROP, WASTE WATER gefrieren
**freezer**: **~ plant** *n* HEAT & VENT Tiefkühlanlage *f*; **~ room** *n* ARCH & BUILD Tiefkühlraum *m*, Gefrierraum *m*
**freezing** *n* HEAT & VENT Tiefkühlen *nt*, Gefrieren *nt*; **~ point** *n* INFR & DES, MAT PROP Gefrierpunkt *m*; **~ process** *n* INFR & DES, MAT PROP Gefrierverfahren *nt*; **~ and thawing test** *n* MAT PROP Frost-Tau-Versuch *m*
**French**: **~ door** *n* ARCH & BUILD, TIMBER Fenstertür *f*, Glastür *f*, französisches Fenster *nt*; **~ drain** *n* WASTE WATER Sammeldrän *m*; **~ truss** *n* ARCH & BUILD Polonceauträger *m*; **~ walnut** *n* TIMBER *colour* französischer Nußbaum *m*; **~ window** *n* ARCH & BUILD, TIMBER Fenstertür *f*
**frequency** *n* ARCH & BUILD Häufigkeit *f*, ELECTR Frequenz *f*, INFR & DES Häufigkeit *f*, SOUND & THERMAL Frequenz *f*; **~ curve** *n* SOUND & THERMAL Frequenzkurve *f*, Häufigkeitskurve *f*; **~ meter** *n* BUILD MACHIN Frequenzmesser *m*, SOUND & THERMAL Frequenzmesser *m*, Häufigkeitsmesser *m*
**freshly**: **~ mixed concrete** *n* CONCR, MAT PROP Frischbeton *m*
**fresh**: **~ sludge** *n* ENVIRON, INFR & DES Frischschlamm *m*; **~ water** *n* ENVIRON, INFR & DES Frischwasser *nt*, Süßwasser *nt*
**freshwater**: **~ lens** *n* INFR & DES Süßwasserlinse *f*; **~ supply** *n* INFR & DES, WASTE WATER Frischwasserversorgung *f*
**friable** *adj* INFR & DES *soil* rollig, nichtbindig, kohäsionslos, nichtkohäsiv, MAT PROP kohäsionslos; **~ rock** *n* INFR & DES, MAT PROP bröckeliges Gestein *nt*
**friction** *n* INFR & DES, MAT PROP Reibung *f*
**frictional**: **~ drag** *n* ARCH & BUILD, INFR & DES *windload* Reibungszugkraft *f*; **~ loss** *n* ENVIRON, INFR & DES, MAT PROP Reibungsverlust *m*; **~ resistance** *n* ARCH & BUILD, INFR & DES, MAT PROP Reibungswiderstand *m*; **~ stress** *n* ARCH & BUILD, INFR & DES, MAT PROP Reibungsspannung *f*
***friction***: **~ circle** *n* INFR & DES, MAT PROP Reibungskreis *m*; **~ coefficient** *n* ARCH & BUILD, INFR & DES, MAT PROP Reibungskoeffizient *m*; **~ connection** *n* ARCH & BUILD, CONCR, INFR & DES, MAT PROP, STEEL Reibungsverbund *m*; **~ force** *n* ARCH & BUILD, INFR & DES, MAT PROP Reibungskraft *f*; **~-grip bolt** *n* STEEL gleitfeste Schraube *f*, hochfest vorgespannte Schraube *f* (*HV-Schraube*); **~-grip bolt connection** *n* ARCH & BUILD, STEEL HV-Schraubenverbindung *f*, gleitfeste Schraubenverbindung *f*; **~-grip bolting** *n* ARCH & BUILD, STEEL gleitfeste Schraubenverbindung *f*, HV-Schraubenverbindung *f*
**frictionless** *adj* INFR & DES, MAT PROP reibungslos
***friction***: **~ loss** *n* ENVIRON, INFR & DES, MAT PROP Reibungsverlust *m*; **~ pile** *n* INFR & DES, MAT PROP Reibungspfahl *m*; **~ test** *n* INFR & DES, MAT PROP Reibungstest *m*
**frieze** *n* ARCH & BUILD Fries *m*
**frog** *n* STONE *in brick* Aushöhlung *f*, Mulde *f*, Vertiefung *f*; **~ rammer** *n* BUILD MACHIN Explosionsstampfer *m*
**front** *n* ARCH & BUILD Brust *f*, Front *f*, Stirnseite *f*, Vorderseite *f*, INFR & DES Brust *f*
**frontage** *n* ARCH & BUILD Gebäudeflucht *f*, Straßenfront *f*, Fassade *f*, CONST LAW, INFR & DES Gebäudeflucht *f*
***front***: **~ dumper** *n* BUILD MACHIN Vorkopfkipper *m*, Kopfkipper *m*; **~ edge** *n* ARCH & BUILD Vorderkante *f*
**frontispiece** *n* ARCH & BUILD Frontispiz *m*
***front***: **~ loader** *n* BUILD MACHIN Frontlader *m*; **~ portion** *n* ARCH & BUILD, INFR & DES Brust *f*; **~ view** *n* ARCH & BUILD Vorderansicht *f*
**frost** *n* ARCH & BUILD, CONCR, HEAT & VENT Frost *m*, INFR & DES Frost *m*, Reif *m*, MAT PROP Frost *m*; **~ blanket** *n* INFR & DES *road construction* Frostschutzschicht *f*; **~ blanket gravel** *n* MAT PROP Frostschutzkies *m*; **~ damage** *n* ARCH & BUILD Frostschaden *m*; **~ heave** *n* INFR & DES Frostaufbruch *m*; **~ layer** *n* HEAT & VENT, INFR & DES, MAT PROP Frostschutzschicht *f*; **~ line** *n* ARCH & BUILD, INFR & DES Frostgrenze *f*, Frosttiefe *f*; **~ penetration depth** *n* ARCH & BUILD, INFR & DES Frosteindringtiefe *f*; **~-proof** *adj* INFR & DES, MAT PROP frostunempfindlich; **~-protected** *adj* ARCH & BUILD, INFR & DES, MAT PROP, SOUND & THERMAL *foundation, pipe* frostsicher, frostfrei; **~ protection** *n* ARCH & BUILD, INFR & DES, MAT PROP, SOUND & THERMAL Frostschutz *m*; **~ protection agent** *n* BUILD MACHIN, HEAT & VENT, INFR & DES, MAT PROP Frostschutzmittel *nt*; **~ resistance** *n* CONCR, INFR & DES, MAT PROP Frostbeständigkeit *f*; **~-resistant** *adj* CONCR, INFR & DES, MAT PROP frostbeständig; **~-susceptible** *adj* INFR & DES, MAT PROP frostempfindlich
**frozen** *adj* HEAT & VENT, INFR & DES, MAT PROP, WASTE WATER gefroren; **~ soil** *n* ARCH & BUILD, INFR & DES, MAT PROP gefrorener Boden *m*
**fuel** *n* MAT PROP Brennstoff *m*, Heizmaterial *nt*, *oil* Heizöl *nt*; **~ consumption** *n* HEAT & VENT Brennstoffverbrauch *m*; **~-heated** *adj* HEAT & VENT ölbeheizt; **~ heating** *n* HEAT & VENT Ölheizung *f*; **~-resistant** *adj* MAT PROP ölbeständig; **~ separator** *n* WASTE WATER Heizölabscheider *m*; **~ tank** *n* HEAT & VENT Heizöltank *m*
**fulcrum**: **~ bracket** *n* ARCH & BUILD Hebelträger *m*
**Fuller's**: **~ chalk** *n* MAT PROP Fullerkreide *f*; **~ earth** *n* MAT PROP Fullererde *f*

**full**: **~-height atrium** *n* ARCH & BUILD Atrium *nt* in ganzer Höhe; **~ load** *n* ELECTR Vollast *f*; **~ shear connection** *n* ARCH & BUILD, CONCR, INFR & DES, MAT PROP *composite structure* voller Schubanschluß *m*, vollständige Verdübelung *f*; **~ storey** *n BrE* ARCH & BUILD Vollgeschoß *nt*; **~ story** *AmE see full storey BrE*

**fully** *adv* ARCH & BUILD, BUILD HARDW, STEEL Ganz-

**fully**: **~ encased** *adj* ARCH & BUILD, CONCR, INFR & DES, MAT PROP, STEEL *composite column* vollständig einbetoniert; **~ glazed** *adj* ARCH & BUILD vollverglast; **~ glazed door** *n* BUILD HARDW Ganzglastür *f*; **~ glazed facade** *n* STEEL Ganzglasfassade *f*; **~ insulated wiring** *n* ELECTR berührungssichere Verdrahtung *f*; **~ restrained** *adj* ARCH & BUILD *beam, girder* voll eingespannt; **~ restrained beam** *n* ARCH & BUILD beidseitig eingespannter Balken *m*

**fumes** *n pl* ENVIRON Rauch *m*, Rauchgas *nt*, Rauchgase *nt pl*

**function** *n* ARCH & BUILD Funktion *f*, Tätigkeit *f*

**fundamental**: **~ properties** *n pl* MAT PROP Grundeigenschaften *f pl*

**funding**: **~ of projects** *n* CONST LAW Projektfinanzierung *f*

**fungal**: **~ attack** *n* MAT PROP Pilzbefall *m*

**fungicide** *n* CONST LAW, MAT PROP Fungizid *nt*, pilztötendes Mittel *nt*

**funicular**: **~ railway** *n* BUILD MACHIN, INFR & DES Seilbahn *f*

**funnel** *n* BUILD MACHIN *chimney* Schornstein *m*, Schlot *m*, *for pouring* Trichter *m*, *for ventilation* Lichtschacht *m*, Lüftungsschacht *m*

**furnace** *n* HEAT & VENT Ofen *m*, Feuerungsanlage *f*, INFR & DES Brennanlage *f*

**furnish**: **~ and install** *phr* ARCH & BUILD liefern und montieren

**furniture** *n* TIMBER Möbel *nt pl*

**furring** *n* CONST LAW Putzabstandshalter *m*, MAT PROP Unterkonstruktion *f*, *of boiler* Verkalkung *f*, STEEL *expanded metal* Gipsputzunterlage *f*, STONE Putzabstandshalter *m*, TIMBER Futterholz *nt*, Verbretterung *f*, Unterfütterung *f*; **~ piece** *n* TIMBER Futterholz *nt*; **~ tile** *n* MAT PROP, TIMBER Wandplatte *f*

**furrow** *n* ARCH & BUILD Rille *f*

**fur**: **~ up** *vi* MAT PROP *boiler* verkalken

**fuse**: **~ cartridge** *n* ELECTR Sicherungspatrone *f*; **~ link** *n* ELECTR Sicherungseinsatz *m*

**fusibility** *n* MAT PROP Schmelzbarkeit *f*

**fusible**: **~ link** *n* BUILD HARDW *fire door protection* Schmelzpatrone *f*

# G

**g** *abbr* (*gram, gramme*) ARCH & BUILD, INFR & DES , MAT PROP g (*Gramm*)

**gabbro** *n* MAT PROP Gabbro *m*

**gable** *n* ARCH & BUILD Giebel *m*; **~ column** *n* ARCH & BUILD Giebelstütze *f*

**gabled**: **~ dormer** *n* ARCH & BUILD Dacherker *m*, Dachfenster *nt*, Giebelgaube *f*, BUILD HARDW Dachfenster *nt*; **~ dormer window** *n* ARCH & BUILD Dacherker *m*, Dachfenster *nt*, Giebelgaube *f*, BUILD HARDW Dachfenster *nt*

***gable***: **~ and eaves edges** *n pl* ARCH & BUILD Giebel- und Traufkanten *f pl*; **~ end** *n* ARCH & BUILD, STONE Giebelwand *f*; **~ peak** *n* ARCH & BUILD Giebelspitze *f*; **~ roof** *n* ARCH & BUILD Satteldach *nt*, Giebeldach *nt*

**gablet** *n* ARCH & BUILD *gothic* Ziergiebel *m*, Zwerggiebel *m*

***gable***: **~ wall** *n* ARCH & BUILD Giebelseite *f*; **~ window** *n* ARCH & BUILD Giebelfenster *m*

**gage** *AmE see gauge BrE*

**gaged**: **~ brickwork** *AmE see gauged brickwork BrE*

**gaging** *AmE see gauging BrE*

**gain** *n* INFR & DES *excavation work* Schrägung *f*

**gallery** *n* ARCH & BUILD Galerie *f*, Empore *f*, Laufgang *m*

**galvanize** *vt* MAT PROP, SURFACE galvanisieren, verzinken

**galvanized** *adj* MAT PROP, SURFACE galvanisiert, verzinkt

**galvanizing** *n* MAT PROP, SURFACE Galvanisierung *f*, Verzinkung *f*

**gambrel**: **~ roof** *n* ARCH & BUILD, TIMBER Mansardendach *nt*

**gamma**: **~ probe** *n* INFR & DES, MAT PROP Gammasonde *f*

**Gang-Nail**(R) *n* INFR & DES, MAT PROP, STEEL Gang-Nail(R) *nt*; **~ galvanized connector plate**(R) *n* INFR & DES, MAT PROP, STEEL Gang-Nail verzinkte Nagelplatte(R) *f*

**gangway** *n* ARCH & BUILD Gang *m*

**gantry** *n* ARCH & BUILD schweres Baugerüst *nt*, BUILD MACHIN Gerüst *nt*, Kranportal *nt*, schweres Baugerüst *nt*; **~ crane** *n* BUILD MACHIN Portalkran *m*

**gap** *n* ARCH & BUILD Hohlraum *m*, Lücke *f*, *between buildings* Baulücke *f*; **~ grading** *n* MAT PROP Ausfallkörnung *f*

**garage** *n* ARCH & BUILD Garage *f*

**garbage** *n AmE* (*cf rubbish BrE*) ARCH & BUILD Abfall *m*, Bauschutt *m*, CONST LAW Abfall *m*, Müll *m*, ENVIRON, INFR & DES Abfall *m*, Bauschutt *m*, Müll *m*; **~ can** *n AmE* ENVIRON (*cf rubbish bin BrE*), ENVIRON (*cf dustbin BrE*) *in public place or inside house* Abfallbehälter *m*, Abfalleimer *m*, Mülleimer *m*, *outside house* Mülltonne *f*; **~ chute** *n AmE* ARCH & BUILD, ENVIRON Müllabwurfschacht *m*, Müllschlukker *m*; **~ collection** *n AmE* CONST LAW, ENVIRON Müllabfuhr *f*, Müllsammlung *f*, Sammlung *f* von Hausmüll; **~ container** *n AmE* ENVIRON Müllbehälter *m*, Mülltonne *f*; **~ disposal** *n AmE* CONST LAW, ENVIRON Müllabfuhr *f*, Müllsammlung *f*, Sammlung *f* von Hausmüll; **~ incineration plant** *n AmE* ENVIRON MVA (*Müllverbrennungsanlage*); **~ press** *n AmE* BUILD MACHIN, ENVIRON Müllpresse *f*; **~ truck** *n AmE* BUILD MACHIN, ENVIRON Müllabfuhrwagen *m*, Müllsammelfahrzeug *nt*

**gargoyle** *n* BUILD HARDW *feature of Gothic churches* Wasserspeier *m*

**garnet**: **~ hinge** *n* BUILD HARDW Türangel *f*

**garret** *n* ARCH & BUILD Dachboden *m*

**gas** *n* ENVIRON, HEAT & VENT, MAT PROP Gas *nt*; **~ bottle** *n* INFR & DES, MAT PROP, STEEL Gasflasche *f*; **~ burner** *n* HEAT & VENT Gasbrenner *m*; **~ calorific value** *n* HEAT & VENT, INFR & DES, MAT PROP Gasheizwert *m*; **~ concrete** *n* CONCR, MAT PROP Blähbeton *m*, Gasbeton *m*; **~ conduit** *n* HEAT & VENT Gasleitung *f*; **~ connection** *n* HEAT & VENT *to mains supply* Gasversorgungsleitung *f*; **~ consumption** *n* HEAT & VENT Gasverbrauch *m*; **~ cylinder** *n* INFR & DES, MAT PROP, STEEL Gasflasche *f*; **~ desulfurization** *AmE*, **~ desulphurization** *BrE n* ENVIRON Gasentschwefelung *f*

**gaseous**: **~ medium** *n* ENVIRON gasförmiges Medium *nt*

**gas**: **~-fired** *adj* HEAT & VENT gasgefeuert; **~-fired central heating** *n* HEAT & VENT Gaszentralheizung *f*; **~ firing** *n* HEAT & VENT Gasfeuerung *f*; **~ fittings** *n* HEAT & VENT Gasarmatur *f*; **~ heating** *n* HEAT & VENT Gasheizung *f*; **~ installation** *n* HEAT & VENT Gasinstallation *f*

**gasket** *n* HEAT & VENT, INFR & DES, MAT PROP, SOUND & THERMAL, WASTE WATER Dichtung *f*

**gasketted**: **~ cover strip** *n* ARCH & BUILD, MAT PROP abgedichtete Abdeckleiste *f*

***gas***: **~ mains** *n pl* HEAT & VENT, INFR & DES Hauptgasleitung *f*; **~ meter** *n* HEAT & VENT Gasmesser *m*

**gasoline**: **~ engine** *n AmE* (*cf petrol engine BrE*) BUILD MACHIN, ENVIRON Benzinmotor *m*, Vergasermotor *m*; **~ separator** *n AmE* (*cf petrol separator BrE*) WASTE WATER Benzinabscheider *m*

***gas***: **~ pipe** *n* HEAT & VENT Gasrohr *nt*, Gasleitung *f*; **~ pipeline** *n* INFR & DES Gasrohrleitung *f*; **~ pliers** *n pl* BUILD MACHIN Gasrohrzange *f*; **~ pressure** *n* HEAT & VENT, INFR & DES Gasdruck *m*

**gasproof** *adj* HEAT & VENT, INFR & DES, MAT PROP gasdicht

***gas***: **~ recovery** *n* ENVIRON Gasrückgewinnung *f*; **~ supply company** *n* CONST LAW, HEAT & VENT Gasversorgungsgesellschaft *f*; **~ supply system** *n* HEAT & VENT, INFR & DES Gasversorgungsanlage *f*; **~ tap** *n* BUILD HARDW Gashahn *m*

**gastight** *adj* HEAT & VENT, INFR & DES, MAT PROP gasdicht

***gas***: **~ vapor recovery plant** *n AmE* (*cf petrol vapour recovery plant BrE*) ENVIRON Benzindampfrückgewinnungsanlage *f*; **~ vent** *n* HEAT & VENT Gasabzug *m*

**gate** *n* ARCH & BUILD Tor *nt*, BUILD HARDW Schleusentor *nt*, Staukörper *m*; **~ hook** *n* BUILD HARDW Metallbolzen *m*; **~ latch** *n* BUILD HARDW Torriegel

*m*; ~ **valve** *n* HEAT & VENT Absperrklappe *f*, WASTE WATER Absperrklappe *f*, Absperrschieber *m*
**gateway** *n* ARCH & BUILD Tordurchfahrt *f*, Torweg *m*
**gauge 1.** *n BrE* ARCH & BUILD Pegel *m*, Richtprofil *nt*, Meßfühler *m*, Meßlatte *f*, BUILD MACHIN Lehre *f*, Meßlehre *f*, HEAT & VENT Meßfühler *m*, INFR & DES Richtprofil *nt*, Meßlatte *f*, Pegel *m*; **2.** *vt BrE* CONCR, MAT PROP, STONE *mortar* anmachen
***gauge***: ~ **box** *n BrE* BUILD MACHIN Maßkasten *m*; ~ **brick** *n BrE* STONE Bogenziegel *m*, Keilziegel *m*
**gauged**: ~ **brickwork** *n BrE* STONE Maßmauerwerk *nt*
***gauge***: ~ **piece** *n BrE* ELECTR *for circuit breakers* Paßeinsatz *m*
**gauging** *n BrE* CONCR, MAT PROP, STONE *of mortar* Anmachen *nt*; ~ **board** *n BrE* STONE Mischbrett *nt*, Mischpodest *nt*; ~ **water** *n BrE* CONCR, STONE *mortar* Anmachwasser *nt*
**gazebo** *n* INFR & DES Laube *f*
**gazump** *vt BrE infrml* CONST LAW übervorteilen beim Grundbesitzankauf
**gear**: ~ **wheel** *n* BUILD MACHIN Zahnrad *nt*
**Geiger**: ~ **counter** *n* ENVIRON Geigerzähler *m*
**general**: ~ **diffused light** *n* ELECTR freistrahlende Leuchte *f*; ~ **method of design** *n* ARCH & BUILD, CONCR, INFR & DES, STEEL allgemeine Bemessungsmethode *f*; ~ **plan** *n* ARCH & BUILD, INFR & DES Übersichtsplan *m*; ~ **public** *n* CONST LAW Öffentlichkeit *f*; ~ **view** *n* ARCH & BUILD, INFR & DES Überblick *m*
**generate** *vt* ELECTR, HEAT & VENT erzeugen
**generation** *n* ELECTR, HEAT & VENT Erzeugung *f*
**generator** *n* ELECTR Generator *m*
**geodesics** *n pl* INFR & DES geodätische Tragwerke *nt pl*
**geodesy** *n* INFR & DES Geodäsie *f*
**geodetic**: ~ **surveying** *n* INFR & DES Landvermessung *f*
**geofabric** *n* INFR & DES, MAT PROP Geovlies *nt*
**geographic**: ~ **variation** *n* ENVIRON geographische Abweichung *f*
**geological**: ~ **map** *n* INFR & DES geologischer Plan *m*; ~ **profile** *n* INFR & DES geologisches Profil *nt*; ~ **survey** *n* CONST LAW Bodengutachten *nt*, INFR & DES Bodengutachten *nt*, geologische Aufnahme *f*
**geology** *n* INFR & DES Geologie *f*
**geomembrane** *n* INFR & DES, MAT PROP Geomembrane *f*; ~ **liner** *n* INFR & DES, MAT PROP Geomembranverkleidung *f*
**geomorphology** *n* INFR & DES Morphologie *f*
**geosynthetic**: ~ **liner** *n* INFR & DES, MAT PROP geosynthetische Verkleidung *f*
**geotextile**: ~ **laboratory** *n* INFR & DES, MAT PROP Labor *nt* für Geotextilien
**geotextiles** *n pl* INFR & DES, MAT PROP Geotextilien *f pl*
**geothermal** *adj* ENVIRON geothermisch; ~ **cycle** *n* ENVIRON geothermischer Kreislauf *m*; ~ **drilling equipment** *n* ENVIRON geothermische Bohrausrüstung *f*; ~ **energy** *n* ENVIRON, HEAT & VENT geothermische Energie *f*; ~ **field** *n* ENVIRON geothermisches Feld *nt*; ~ **gradient** *n* ENVIRON geothermische Tiefenstufe *f*; ~ **plant** *n* ENVIRON geothermische Anlage *f*; ~ **power** *n* ENVIRON, HEAT & VENT geothermische Energie *f*; ~ **resources** *n pl* ENVIRON geothermische Quellen *f pl*; ~ **step** *n* ENVIRON geothermische Tiefenstufe *f*
**German**: ~ **Industrial Standard** *n* (*DIN*) ARCH & BUILD, CONST LAW, INFR & DES Deutsche Industrienorm *f* (*DIN*); ~ **Industrial Standards Institute** *n* ARCH & BUILD, CONST LAW, INFR & DES Deutsches Institut *nt* für Normung (*DIN-Institut*); ~ **regulations for contracts and execution of construction works** *phr* CONST LAW Verdingungsordnung *f* für die Vergabe von Bauleistungen (*VOB*)
**germ**: ~-**free** *adj* ENVIRON, MAT PROP keimfrei
**gimlet** *n* BUILD MACHIN *auger bit* Schneckenbohrer *m*, *hand brace, drill* Handbohrer *m*, Nagelbohrer *m*
**gin** *n* BUILD MACHIN *drilling* Dreibein *nt*, Göpel *m*, Hebezeug *nt*, Winde *f*
**girder** *n* ARCH & BUILD Balken *m*, Tragbalken *m*, Träger *m*, CONCR, STEEL, TIMBER Balken *m*; ~ **bridge** *n* ARCH & BUILD, INFR & DES Trägerbrücke *f*; ~ **forms** *n* CONCR Trägerschalung *f*; ~ **spacing** *n* ARCH & BUILD, INFR & DES Trägerabstand *m*
**girt** *n* ARCH & BUILD, BUILD MACHIN, CONCR, INFR & DES, MAT PROP, STEEL Untergurt *m*
**give**: ~ **way** *vi* ARCH & BUILD, MAT PROP nachgeben
**glacier** *n* INFR & DES Gletscher *m*
**gland**: ~ **packing** *n* WASTE WATER Dichtungspackung *f*, Packungsstopfbuchse *f*, Stopfbuchsenpackung *f*
**glare** *n* ARCH & BUILD, INFR & DES Blendung *f*
**glass** *n* MAT PROP Glas *nt*; ~ **block** *n* MAT PROP Glasbaustein *m*; ~ **cutter** *n* BUILD MACHIN Glasschneider *m*; ~ **fabric** *n* MAT PROP Glasfasergewebe *nt*; ~ **fiber** *n AmE*, ~ **fibre** *BrE n* MAT PROP Glasfaser *f*; ~ **fibre formwork** *n BrE* ARCH & BUILD Glasfaserschalung *f*; ~ **fibre gutter** *n BrE* WASTE WATER Glasfaserdachrinne *f*; ~ **fibre mat** *n BrE* MAT PROP Glasvlies *nt*; ~ **fibre quilt** *n BrE* MAT PROP, SOUND & THERMAL Glasfasermatte *f*; ~ **fibre reinforced** *adj BrE* MAT PROP *plastic* glasfaserverstärkt; ~ **fibre reinforced cement** *n BrE* CONCR, MAT PROP glasfaserbewehrter Zement *m*; ~ **fibre reinforced cement cladding** *n BrE* ENVIRON glasfaserbewehrte Zementverkleidung *f*; ~ **fibre reinforced plastics** *n pl BrE* MAT PROP glasfaserverstärkte Kunststoffe *m pl*; ~ **front** *n* ARCH & BUILD Glasfassade *f*; ~ **pane** *n* ARCH & BUILD Glasscheibe *f*; ~ **paper** *n* MAT PROP Glaspapier *nt*, feines Glaspapier *nt*; ~ **recycling** *n* ENVIRON Altglasverwertung *f*; ~ **roundel** *n* ARCH & BUILD Butzenscheibe *f*; ~ **tile** *n* ARCH & BUILD, MAT PROP *roofing* Glasziegel *m*; ~ **type** *n* MAT PROP Glasart *f*, Glassorte *f*; ~ **wadding** *n* MAT PROP Glaswatte *f*; ~ **wool** *n* MAT PROP, SOUND & THERMAL Glaswolle *f*
**Glauber**: ~**'s salt** *n* HEAT & VENT, MAT PROP Glaubersalz *nt*
**glauconite** *n* MAT PROP Glaukonit *nt*
**glaze 1.** *n* SURFACE *of paint* Lasur *f*, *on pottery* Glasur *f*; **2.** *vt* ARCH & BUILD *with glass* verglasen, SURFACE *pottery* glasieren
**glazed** *adj* ARCH & BUILD *with glass* verglast, SURFACE *pottery* glasiert; ~ **door** *n* ARCH & BUILD Fenstertür *f*, Glastür *f*, französisches Fenster *nt*; ~ **lantern** *n* ARCH & BUILD verglaste Laterne *f*; ~ **sash** *n* ARCH & BUILD, BUILD HARDW Fensterglasflügel *m*
**glazier** *n* ARCH & BUILD Glaser *m*; ~**'s diamond** *n* ARCH & BUILD, BUILD HARDW Glaserdiamant *m*
**glazing** *n* ARCH & BUILD Verglasen *nt*, Verglasung *f*; ~ **bar** *n* ARCH & BUILD Fenstersprosse *f*
**gloss** *n* SURFACE Glanz *m*
**glue 1.** *n* MAT PROP, SURFACE, TIMBER Kleber *m*, Leim *m*; **2.** *vt* MAT PROP, SURFACE kleben, verkleben, TIMBER leimen, verleimen; ~ **on** SURFACE, TIMBER aufkleben;

~ **on throughout** SURFACE, TIMBER ganzflächig aufkleben
***glue***: ~ **brush** *n* BUILD MACHIN Leimbürste *f*
**glued**: ~ **joints** *n pl* TIMBER verklebte Stöße *m pl*
**glue**: ~**-laminated** *adj* (*glulam*) TIMBER brettverleimt, schichtverleimt; ~**-nail joint** *n* TIMBER Leim-Nagel-Verbindung *f*; ~ **priming** *n* SURFACE, TIMBER Leimgrundierung *f*
**glulam** *abbr* (*glue-laminated*) TIMBER brettverleimt, schichtverleimt
***glulam***: ~ **timber** *n* TIMBER Leimholz *nt*, Schichtpreßholz *nt*
**gneiss** *n* INFR & DES, MAT PROP Gneis *m*
**going** *n* ARCH & BUILD projektierter Treppenauftritt *m*
**goliath**: ~ **crane** *n* BUILD MACHIN Schwerlastkran *m*
**good**: ~ **bearing soil** *n* INFR & DES tragfähiger Boden *m*, tragfähiger Naturgrund *m*; ~ **practice** *n* CONST LAW berufliche Sorgfalt *f*; **in** ~ **repair** *adj* ARCH & BUILD in gutem Zustand
**goods**: ~ **elevator** *n AmE* (*cf goods lift BrE*) ARCH & BUILD, BUILD MACHIN, INFR & DES Lastenaufzug *m*, Warenaufzug *m*; ~ **lift** *BrE n* (*cf goods elevator AmE*) ARCH & BUILD, BUILD MACHIN, INFR & DES Lastenaufzug *m*, Warenaufzug *m*
**gorge** *n* ARCH & BUILD Hohlkehle *f*, Wassernase *f*
**gouge** *n* BUILD MACHIN *woodworking* Kehlbeitel *m*, STEEL Hohleisen *nt*, Hohlmeißel *m*
**grab** *n* BUILD MACHIN Baggerkorb *m*, Greifer *m*; ~ **bar** *n* BUILD HARDW *near bathtub or shower* Griffstange *f*, Stangengriff *m*, Wannengriff *m*; ~ **bucket** *n* BUILD MACHIN Greiferkorb *m*; ~ **crane** *n* BUILD MACHIN Greifbagger *m*; ~ **dredger** *n* BUILD MACHIN Greiferbagger *m*, Schwimmbagger *m*
**grade 1.** *n* ARCH & BUILD Geländehöhe *f*, INFR & DES Geländehöhe *f*, Geländeoberfläche *f*, Gefälle *nt*, MAT PROP Güteklasse *f*, Qualität *f*; **2.** *vt* ARCH & BUILD, INFR & DES einebnen, MAT PROP sortieren, klassieren; ~ **up** ARCH & BUILD, MAT PROP *building* aufwerten
**grade**: **on** ~ *adj* ARCH & BUILD auf dem Erdboden, SURFACE bodengleich; **at** ~ *adj* ARCH & BUILD, CONCR ebenerdig; ~ **beam** *n* ARCH & BUILD Gründungsschwelle *f*, Fundamentbalken *m*; ~ **of concrete** *n* CONCR Betongüte *f*
**graded** *adj* MAT PROP klassiert, sortiert
**grader** *n* BUILD MACHIN Planiergerät *nt*, Sortiergerät *nt*
**gradient** *n* ARCH & BUILD, INFR & DES Gefälle *nt*
**grading** *n* ARCH & BUILD, INFR & DES, MAT PROP *according to particle size* Einstufung *f*, Klassierung *f*, Körnung *f*, Planierarbeiten *f pl*; ~ **curve** *n* MAT PROP *gravel, sand* Kornverteilungskurve *f*, Siebkurve *f*
**grain** *n* INFR & DES, MAT PROP Korn *nt*; ~ **arrangement** *n* INFR & DES, MAT PROP Kornanordnung *f*; ~ **character** *n* INFR & DES, MAT PROP Kornbeschaffenheit *f*; ~**-cut timber** *n* TIMBER Hirnholz *nt*; ~ **diameter** *n* INFR & DES, MAT PROP Korndurchmesser *m*
**grained** *adj* MAT PROP gemasert, SURFACE genarbt
***grain***: ~ **encapsulation** *n* ENVIRON Einbindung *f*
**graining** *n* TIMBER Masern *nt*, Maserung *f*
***grain***: ~ **mixture** *n* INFR & DES, MAT PROP Körnungsgemisch *nt*; ~ **shape** *n* INFR & DES, MAT PROP Kornform *f*; ~ **size** *n* INFR & DES, MAT PROP Korngröße *f*; ~**-size classification** *n* INFR & DES, MAT PROP Korngrößeneinteilung *f*; ~**-size distribution** *n* MAT PROP Korngrößenverteilung *f*; ~**-size fraction** *n* MAT PROP Kornfraktion *f*; ~**-size range** *n* MAT PROP Kornfraktion *f*; ~ **skeleton** *n* MAT PROP Korngerüst *nt*; ~ **structure** *n* MAT PROP Kornstruktur *f*; ~ **surface** *n* MAT PROP Kornoberfläche *f*
**grainy** *adj* INFR & DES, MAT PROP körnig
**gram**, **gramme** *n* (*g*) ARCH & BUILD, INFR & DES, MAT PROP Gramm *nt* (*g*)
**grandmaster**: ~ **keyed system** *n* BUILD HARDW, CONST LAW *locking system* Hauptschlüsselanlage *f*
**granite** *n* INFR & DES, MAT PROP Granit *m*
**granolithic**: ~ **concrete** *n* MAT PROP Hartbeton *m*
**granular** *adj* INFR & DES, MAT PROP körnig; ~ **fraction** *n* INFR & DES, MAT PROP körniger Anteil *m*
**granularity** *n* INFR & DES, MAT PROP Körnigkeit *f*
**granular**: ~ **soil** *n* INFR & DES, MAT PROP körniger Boden *m*
**granulate** *vt* BUILD MACHIN *concrete or stone surfaces*, CONCR aufspitzen, stocken, STONE aufspitzen
**granulated**: ~ **cinder** *n* MAT PROP Hüttensand *m*, granulierte Hochofenschlacke *f*
**graphical**: ~ **analysis** *n* CONST LAW graphische Analyse *f*
**graphite** *n* MAT PROP Graphit *m*
**grapple** *vt* ARCH & BUILD verklammern, verankern
**grappling**: ~ **of an arch** *n* ARCH & BUILD Bogenverankerung *f*
**grass** *n* INFR & DES Gras *nt*
**grassed**: ~ **area** *n* INFR & DES Grasboden *m*, Grünfläche *f*
***grass***: ~ **paver** *n* INFR & DES, MAT PROP Rasengitterstein *m*, Rasenstein *m*; ~ **slope** *n* INFR & DES Grasböschung *f*
**grate** *n* BUILD HARDW, STEEL Gitter *nt*, Gitterrost *nt*; ~ **area** *n* ARCH & BUILD Rostfläche *f*; ~ **bar** *n* STEEL Gitterstab *m*
**grating** *n* BUILD HARDW, STEEL Gitterrost *m*, Gitter *nt*
**gravel** *n* INFR & DES, MAT PROP Kies *m*; ~ **bank** *n* INFR & DES, MAT PROP Kiesbank *f*; ~ **bed** *n* INFR & DES, MAT PROP Kiesschicht *f*; ~ **bottom** *n* INFR & DES, MAT PROP Kiessohle *f*; ~ **chippings** *n pl* MAT PROP Kiessplitt *m*; ~ **concrete** *n* CONCR Kiesbeton *m*; ~ **deposit** *n* INFR & DES, MAT PROP Kiesablagerung *f*; ~ **drain** *n* INFR & DES, MAT PROP Kiesdrain *m*; ~ **fill** *n* INFR & DES Kiesverfüllung *f*; ~**-filled drain trench** *n* WASTE WATER Rigole *f*; ~ **filter** *n* INFR & DES, MAT PROP Kiesfilter *m*; ~ **filter layer** *n* ARCH & BUILD, ENVIRON, INFR & DES Kiesfilterschicht *f*; ~ **fraction** *n* INFR & DES, MAT PROP Kiesfraktion *f*
**graveling** *AmE*, **gravelling** *n BrE* ARCH & BUILD Bekiesung *f*
**gravelly**: ~ **loam** *n* INFR & DES, MAT PROP kieshaltiger Lehm *m*; ~ **soil** *n* INFR & DES, MAT PROP Kiesboden *m*, kieshaltiger Boden *m*
***gravel***: ~ **packing** *n* ARCH & BUILD, INFR & DES Kiesfüllung *f*, Kiespacklage *f*; ~ **pit** *n* INFR & DES Kiesgrube *f*; ~ **pocket** *n* ARCH & BUILD, CONCR Kiesnest *nt*; ~ **sample** *n* INFR & DES, MAT PROP Kiesprobe *f*; ~ **stop** *n* INFR & DES *gib on flat roof* Kiesfang *m*, Dachabschlußblende *f*, Blende *f*, Dachrandprofil *nt*, TIMBER Dachabschlußblende *f*, Dachrandprofil *nt*; ~ **substructure** *n* INFR & DES Kiesunterbau *m*
**gravitation** *n* ENVIRON, INFR & DES, MAT PROP Erdanziehungskraft *f*
**gravitational**: ~ **force** *n* ENVIRON, INFR & DES, MAT PROP Schwerkraft *f*
**gravity** *n* ENVIRON, INFR & DES Gravitation *f*, Schwer-

gewicht *nt*; ~ **abutment** *n* CONCR, INFR & DES Gewichtswiderlager *nt*; ~ **dam** *n* INFR & DES Schwergewichtsmauer *f*, Schwergewichtsdamm *m*; ~ **drainage** *n* INFR & DES, WASTE WATER Schwerkraftabfluß *m*; ~ **flow** *n* INFR & DES, WASTE WATER Schwerkraftströmung *f*; ~ **force** *n* INFR & DES, MAT PROP Schwerkraft *f*; ~ **heating** *n* HEAT & VENT Schwerkraftheizung *f*; ~ **incline** *n* INFR & DES *railways* Ablauframpe *f*; ~ **retaining wall** *n* CONCR, INFR & DES, STONE Schwergewichtsstützmauer *f*; ~ **separation** *n* INFR & DES Trennung *f* durch Schwerkraft; ~ **system** *n* HEAT & VENT, INFR & DES, WASTE WATER Schwerkraftsystem *nt*

**gray**: ~ **cast iron** *AmE see grey cast iron BrE*

**graystone**: ~ **lime plaster** *AmE see greystone lime plaster BrE*

**GRC** *abbr* (*glass fiber reinforced cement AmE*, *glass fibre reinforced cement BrE*), CONCR, MAT PROP glasfaserbewehrter Zement *m*

***GRC***: ~ **cladding** *n* ENVIRON glasfaserbewehrte Zementverkleidung *f*

**grease 1.** *n* MAT PROP Fett *nt*; ~ **filter** *n* BUILD HARDW *of exhaust hood* Fettfilter *nt*; ~**-resistant** *adj* MAT PROP fettbeständig; ~ **separator** *n* ENVIRON, WASTE WATER Fettabscheider *m*; **2.** *vt* ARCH & BUILD fetten

**green** *adj* ENVIRON grün, MAT PROP *mortar, concrete* feucht, frisch, *bricks* ungebrannt, STONE ungebrannt, SURFACE grün; ~ **building** *n* ENVIRON grünes Gebäude *nt*; ~ **concrete** *n* CONCR, MAT PROP Frischbeton *m*

**greenhouse** *n* ARCH & BUILD Gewächshaus *nt*, Treibhaus *nt*; ~ **effect** *n* ENVIRON Treibhauseffekt *m*

***green***: ~ **stone** *n* MAT PROP Grünstein *m*

**grey**: ~ **cast iron** *n BrE* MAT PROP Grauguß *m*

**greystone**: ~ **lime plaster** *n BrE* STONE Schwarzkalkputz *m*

**grid** *n* BUILD HARW Gitter *nt*, Gitterrost *m*, INFR & DES Raster *m*, STEEL Gitter *nt*, Gitterrost *m*; ~ **crusher** *n* ENVIRON Gitterzerkleinerer *m*; ~ **iron** *n* STEEL Gitterrost *m*, *grate* Rost *m*; ~ **line** *n* ARCH & BUILD, INFR & DES Rasterlinie *f*

**grillage** *n* BUILD HARD, STEEL Gitter *nt*, Gitterrost *m*; ~ **foundation** *n* ARCH & BUILD, INFR & DES Gründungsrost *m*

**grille** *n* BUILD HARDW, STEEL Gitter *nt*, Gitterrost *m*,

**grind** *vt* BUILD MACHIN schleifen, verschleifen, ENVIRON *refuse* zerstäuben

**grinder** *n* BUILD MACHIN Schleifmaschine *f*

**grinding** *n* BUILD MACHIN Schleifen *nt*, Mahlen *nt*, ENVIRON *of refuse* Zerstäubung *f*; ~ **and polishing line** *n* BUILD MACHIN *marble* Schleif- und Polierstraße *f*; ~ **and polishing machine with flexible shaft** *n* BUILD MACHIN Schleif- und Poliermaschine *f* mit biegsamer Welle; ~ **wheel** *n* BUILD MACHIN Schleifscheibe *f*

**grindstone** *n* BUILD MACHIN Schleifstein *m*

**grip** *n* SURFACE Haftfähigkeit *f*, Haftvermögen *nt*, Griffigkeit *f*; ~ **handle** *n* BUILD HARDW Stoßgriff *m*; ~ **length of nail** *n* ARCH & BUILD Nagelhaftlänge *f*

**grit** *n* MAT PROP abstumpfende Stoffe *m pl*, scharfkantiger Sand *m*, *blast cleaning* Strahlsand *m*, *stone* Splitt *m*; ~ **box** *n* INFR & DES, WASTE WATER Schlammeimer *m*; ~ **chamber** *n* ENVIRON Sandfanganlage *f*, Sandfänger *m*, INFR & DES Schlammfang *m*; ~ **spreader** *n* BUILD MACHIN *for road maintenance* Splittstreuer *m*; ~ **trap** *n* INFR & DES Straßenablauf *m*

**groin** *n* INFR & DES Grat *m*

**groove** *n* ARCH & BUILD Kerbe *f*, Nut *f*, Ausnutung *f*, Kehle *f*, *architecture* Rille *f*

**grooved** *adj* MAT PROP genutet; ~ **and tongued joint** *n* TIMBER Nut- und Federverspundung *f*

**grooving**: ~ **iron** *n* BUILD HARDW, BUILD MACHIN Kröseleisen *nt*; ~ **plane** *n* BUILD MACHIN Nuthobel *m*, Spundhobel *m*

**gross**: ~ **area** *n* ENVIRON *of collector* Gesamtfläche *f*; ~ **storey area** *n BrE* ARCH & BUILD Bruttogeschoßfläche *f*; ~ **story area** *AmE see gross storey area BrE*

**ground 1.** *adj* MAT PROP geschliffen; **2.** *n* ARCH & BUILD Erde *f*, Grund *m*, ELECTR (*cf earth BrE*) Erdung *f*, Erde *f*, INFR & DES *soil, earth* Boden *m*, Erdreich *nt*, Grund *m*, Erde *f*, Gelände *nt*, Erdboden *m*, STEEL Erde *f*, Grund *m*; **3.** *vt AmE* (*cf earth BrE*) ELECTR erden

**ground**: **at ~ level** *adj* ARCH & BUILD, CONCR ebenerdig; ~ **anchor** *n* ARCH & BUILD, INFR & DES, STEEL Erdanker *m*; ~ **auger** *n* BUILD MACHIN, INFR & DES Erdbohrer *m*; ~ **beam** *n* ARCH & BUILD Grundbalken *m*; ~ **box** *n AmE* (*cf earth box BrE*) ELECTR Erdungsbox *f*

**groundbreaking** *n* ARCH & BUILD erster Spatenstich *m*

***ground***: ~ **bus bar** *n AmE* (*cf earth busbar BrE*) ELECTR Erdungsschiene *f*; ~ **conditions** *n pl* (*cf earth conditions*) INFR & DES Bodenverhältnisse *nt pl*; ~ **conductor** *n AmE* (*cf earth conductor BrE*) ELECTR Erdungsleitung *f*; ~ **edge** *n AmE* (*cf earth edge BrE*) ARCH & BUILD geschliffene Kante *f*; ~ **fault circuit interrupter** *n AmE* (*cf earth fault circuit interrupter BrE*) ELECTR FI-Schutzschalter *m*, Fehlerstromschutzschalter *m*; ~ **floor** *n BrE* (*cf first floor AmE*) ARCH & BUILD Erdgeschoß *nt*; ~**-floor plan** *n BrE* (*cf first-floor plan AmE*) ARCH & BUILD Erdgeschoßplan *m*, Grundriß *m*; ~ **glass** *n AmE* (*cf earth glass BrE*) MAT PROP Mattglas *nt*; ~ **improvement** *n* (*cf earth improvement BrE*) ARCH & BUILD, INFR & DES, MAT PROP Baugrundverbesserung *f*

**grounding** *n AmE* (*cf earthing BrE*) ELECTR Erdung *f*, Schutzerdung *f*; ~ **field** *n AmE* (*cf earthing field BrE*) ELECTR Erdungsfeld *nt*; ~ **pipe clamp** *n AmE* (*cf earthing pipe clamp BrE*) ELECTR Erdungsrohrschelle *f*; ~ **plate** *n AmE* (*cf earthing plate BrE*) ELECTR Erdungsplatte *f*; ~ **resistance** *n AmE* (*cf earthing resistance BrE*) ELECTR Ableitungswiderstand *m*, Erdableitwiderstand *m*, Erdungswiderstand *m*; ~ **strip** *n AmE* (*cf earthing strip BrE*) ELECTR Banderder *m*; ~ **system** *n AmE* (*cf earthing system BrE*) ELECTR Erdungsanlage *f*; ~ **well** *n AmE* (*cf earthing well BrE*) ELECTR Erdungsmeßschacht *m*

***ground***: ~ **lead** *n AmE* ELECTR Erddraht *m*; ~ **level** *n* ARCH & BUILD, INFR & DES Geländehöhe *f*, Geländeoberfläche *f*; ~ **line** *n* ELECTR, WASTE WATER Grundleitung *f*; ~ **plate** *n* ARCH & BUILD Bodenschwelle *f*, Schwelle *f*, ELECTR Erdungsplatte *f*; ~ **pressure** *n* ARCH & BUILD, INFR & DES Bodenpressung *f*; ~ **rod** *n AmE* ELECTR Staberder *m*, Tiefenerder *m*; ~ **settlement** *n* INFR & DES Geländesetzung *f*

**groundsill** *n* ARCH & BUILD Fußholz *nt*, Grundschwelle *f*

***ground***: ~**-supported concrete floor slabs** *n pl*

CONCR erdgelagerte Betonbodenplatten *f pl*; **~-supported in-situ concrete floors** *n pl* CONCR erdgelagerte Ortbetonböden *m pl*; **~ tester** *n AmE* (*cf earth tester BrE*) ELECTR Erdungsmeßgerät *nt*

**groundwater** *n* ENVIRON, INFR & DES Grundwasser *nt*; **~ contamination** *n* CONST LAW, ENVIRON, INFR & DES Grundwasserverschmutzung *f*, Verunreinigung *f* des Grundwassers; **~ current** *n* INFR & DES Grundwasserströmung *f*, Grundwasserstrom *m*; **~ drainage** *n* INFR & DES Grundwasserdrainage *f*; **~ hydraulics** *n pl* INFR & DES Grundwasserhydraulik *f*; **~ layer** *n* INFR & DES Grundwasserschicht *f*; **~ lens** *n* INFR & DES Grundwasserlinse *f*; **~ level** *n* ARCH & BUILD, INFR & DES Grundwasserhöhe *f*; **~ lowering** *n* ARCH & BUILD Wasserhaltung *f*, INFR & DES Grundwasserhaltung *f*, Wasserabsenkung *f*, Wasserhaltung *f*; **~ pollution** *n* CONST LAW, ENVIRON, INFR & DES Grundwasserverschmutzung *f*, Verunreinigung *f* des Grundwassers; **~ protection** *n* ARCH & BUILD, CONST LAW, ENVIRON, INFR & DES Grundwasserschutz *m*; **~ pumping** *n* CONST LAW, INFR & DES Grundwasserabpumpen *nt*; **~ quality** *n* CONST LAW, INFR & DES, MAT PROP Grundwassergüte *f*

**groundwork** *n* ARCH & BUILD, INFR & DES Fundament *nt*, Gründung *f*

**grout** *vt* STONE *with mortar* vergießen, abdichten

**grout**: **~ anchor** *n* INFR & DES Injektionsanker *m*

**grouting** *n* ARCH & BUILD, BUILD MACHIN Einpressen *nt*, STONE Vermörteln *nt*, *of mortar* Einpressen *nt*; **~ compound** *n* ARCH & BUILD Versteinungsmittel *nt*, STONE Dichtungsschlämme *f*, Zementbrei *m*, Vergußmaterial *nt*, Vergußmasse *f*, Versteinungsmittel *nt*

**grout**: **~ injection** *n* BUILD MACHIN, INFR & DES, STONE Einpressen *nt*

**growth** *n* INFR & DES Aufwuchs *m*

**groyne** *n* INFR & DES Buhne *f*

**guarantee 1.** *n* CONST LAW Garantie *f*, Gewährleistung *f*; **2.** *vt* CONST LAW garantieren, gewährleisten

**guarantor** *n* CONST LAW Bürge *m*

**guard 1.** *n* ARCH & BUILD Schutzvorrichtung *f*, BUILD MACHIN Schutzvorrichtung *f*, Abweiser *m*, CONST LAW, INFR & DES Schutzvorrichtung *f*; **2.** *vt* ARCH & BUILD, BUILD MACHIN, CONST LAW, INFR & DES schützen

**guarding** *n* ARCH & BUILD, BUILD MACHIN, CONST LAW, INFR & DES Schutzvorrichtung *f*

***guard***: **~ rail** *n* ARCH & BUILD, BUILD HARDW, STEEL, TIMBER Geländer *nt*

**gudgeon** *n* BUILD HARDW Zapfen *m*

**guide**: **~ board** *n* ARCH & BUILD, INFR & DES Hinweisschild *nt*; **~ fillet** *n* ARCH & BUILD Führungsleiste *f*; **~ groove** *n* TIMBER *of sliding cupboard door* Führungsnut *f*

**guideline** *n* CONST LAW Richtlinie *f*

**guide**: **~ pole** *n* BUILD HARDW Führungsstange *f*; **~ post** *n* INFR & DES Leitpfosten *m*; **~ pulley** *n* BUILD MACHIN Führungsrolle *f*; **~ rail** *n* BUILD HARDW *door, window* Leitschiene *f*, *of door, window* Führungsschiene *f*

**gully** *n* INFR & DES, WASTE WATER Gully *m*, Regenwasserablauf *m*

***gully***: **~ cover** *n* INFR & DES, WASTE WATER Gullydeckel *m*, Rinnendeckel *m*; **~ trap** *n* INFR & DES Schmutzfang *m*

**gun 1.** *n* BUILD MACHIN Spritze *f*; **2.** *vt* CONCR torkretieren

**gunite** *n* MAT PROP Spritzbeton *m*, Torkretbeton *m*

**gunning** *n* CONCR Torkretieren *nt*

**gunny**: **~ cloth** *n* MAT PROP Jutetuch *nt*

**gusset**: **~ plate** *n* STEEL, TIMBER Stützblech *nt*, Knotenblech *nt*, Lasche *f*

**gutter** *n* ARCH & BUILD *on roof* Dachrinne *f*, INFR & DES Rinne *f*, *in street* Straßenrinne *f*, WASTE WATER *on roof* Dachrinne *f*; **~ board** *n* ARCH & BUILD, INFR & DES, MAT PROP Traufbohle *f*, Traufbrett *nt*, TIMBER Traufbohle *f*; **~ bracket** *n* BUILD HARDW Dachrinnenhalter *m*, Rinnhaken *m*, Rinnenhaken *m*, Rinnenhalter *m*; **~ clamp** *n* BUILD HARDW Dachrinnenhalter *m*, Rinnenhalter *m*; **~ hanger** *n* BUILD HARDW Dachrinnenhalter *m*, Rinnenhalter *m*; **~ outlet** *n* ARCH & BUILD, BUILD HARDW Rinnenstutzen *m*; **~ pipe** *n* ARCH & BUILD, BUILD HARDW Fallrohr *nt*; **~ ridge** *n* WASTE WATER Rinnenfirst *m*; **~ stop end** *n* WASTE WATER Dachrinnenendstück *nt*

**guy** *n* INFR & DES Abspannung *f*, Seilverspannung *f*

**guying** *n* ARCH & BUILD Seilverspannung *f*, INFR & DES Abspannung *f*, Seilverspannung *f*

***guy***: **~ rope** *n* ARCH & BUILD, BUILD HARDW Abspannseil *nt*, Verankerungsseil *nt*

**gymnasium** *n* ARCH & BUILD, INFR & DES Turnhalle *f*

**gypsum** *n* MAT PROP Gips *m*; **~ plaster** *n* MAT PROP Gipskartonplatte *f*, Gipsunterputz *m*; **~ slag cement** *n* CONCR, MAT PROP Gipsschlackenzement *m*

# H

**habitable** *adj* ARCH & BUILD, CONST LAW bewohnbar
**hacking**: **~ knife** *n* BUILD HARDW, BUILD MACHIN Kabelmesser *nt*, Kittentfernungsmesser *nt*; **~ off** *n* BUILD HARDW, INFR & DES, STONE, SURFACE Aufrauhen *nt*
**hack**: **~ off** *vt* BUILD HARDW, INFR & DES, STONE, SURFACE aufrauhen
**hair**: **~ crack** *n* INFR & DES, MAT PROP, STEEL Haarriß *m*, Mikroriß *m*
**half**: **~-beam** *n* ARCH & BUILD Halbbalken *m*, Halbholzbalken *m*; **~-brick** *adj* STONE halbsteindick; **~-mortice cabinet lock** *n* (*see half-mortise cabinet lock*) BUILD HARDW Einlaßmöbelschloß *nt*; **~-mortise cabinet lock** *n* BUILD HARDW Einlaßmöbelschloß *nt*; **~ pace** *n* ARCH & BUILD Zwischenpodest *nt*; **~-round bar** *n* STEEL Halbrundstahl *m*, Segmentstahl *m*; **~-round gutter** *n* WASTE WATER halbrunde Dachrinne *f*; **~-round rod** *n* STEEL Halbrundstange *f*; **~-round tile** *n* ARCH & BUILD, MAT PROP, STONE halbrunder Ziegel *m*; **~-timber** *n* TIMBER Halbholzbalken *m*, Halbbalken *m*; **~-timbered house** *n* ARCH & BUILD Fachwerkhaus *nt*; **~ truss** *n* ARCH & BUILD Halbbinder *m*; **~-turn stairs** *n pl* ARCH & BUILD halbgedrehte Treppe *f*
**hall** *n* ARCH & BUILD *of house* Flur *m*, *auditorium* Saal *m*, *lobby of hotel* Vorraum *m*, Korridor *m*, *sportshall, gymnasium* Halle *f*, INFR & DES Korridor *m*; **~-type building** *n* ARCH & BUILD, INFR & DES Gebäude *nt* in Hallenbauweise
**hallway** *n* ARCH & BUILD *in house* Flur *m*, Korridor *m*, INFR & DES Korridor *m*
**halved**: **~ joint** *n* ARCH & BUILD Blattung *f*, TIMBER Überblattung *f*, Anblattung *f*
**halving** *n* TIMBER Überblattung *f*, Anblattung *f*
**hammer 1.** *n* BUILD MACHIN Bär *m*, Hammer *m*, Ramme *f*; **2.** *vt* BUILD MACHIN, SURFACE hämmern
***hammer***: **~ beam** *n* ARCH & BUILD Stichbalken *m*; **~ crusher** *n* BUILD MACHIN, ENVIRON Hammerzerkleinerer *m*; **~ drill** *n* BUILD MACHIN Schlagbohrer *m*; **~-finish enamel** *n* SURFACE Hammerschlaglack *m*
**hammering** *n* BUILD MACHIN, SURFACE Hämmern *nt*
***hammer***: **~ mill** *n* ENVIRON Hammermühle *f*
**hammertone**: **~ enamel** *n* SURFACE Hammerschlaglack *m*
**hand 1.** *n* BUILD MACHIN Hand *f*; **2.** *vt* **~ over** ARCH & BUILD, CONST LAW übergeben
**handbarrow** *n* BUILD MACHIN Trage *f*
**handchain** *n* BUILD MACHIN Handkette *f*
***hand***: **~-formed brick** *n* STONE Handstrichziegel *m*
**handing**: **~ over** *n* ARCH & BUILD, CONST LAW Bauübergabe *f*, *key* Übergabe *f*
**handle 1.** *n* BUILD HARDW Griff *m*, Handgriff *m*, *of vice* Knebel *m*; **2.** *vt* MAT PROP verarbeiten
***handle***: **~ collar** *n* BUILD MACHIN Griffzwinge *f*
**handling** *n* MAT PROP Verarbeitung *f*; **~ load** *n* TIMBER Transportbelastung *f*; **~ materials** *n pl* BUILD MACHIN Fördergüter *nt pl*; **~ platform** *n* ARCH & BUILD Ladebühne *f*, Umschlagbühne *f*

***hand***: **~-made brick** *n* STONE Handstrichziegel *m*; **~-molded brick** *AmE*, **~-moulded brick** *BrE* *n* STONE Handstrichziegel *m*; **~-operated fire extinguisher** *n* CONST LAW Handfeuerlöscher *m*; **~-operated shut-off valve** *n* HEAT & VENT Handabsperrklappe *f*; **~ operation** *n* BUILD MACHIN Handbetrieb *m*
**handover** *n* ARCH & BUILD Bauübergabe *f*
**handrail** *n* ARCH & BUILD, BUILD HARDW Handlauf *m*
**handsaw** *n* BUILD MACHIN Fuchsschwanz *m*, Handsäge *f*
***hand***: **~ scraper** *n* BUILD MACHIN Handschrapper *m*; **~ tamper** *n* BUILD MACHIN Handstampfer *m*; **~ wheel** *n* HEAT & VENT Handrad *nt*
**hangar** *n* ARCH & BUILD, INFR & DES Flugzeughalle *f*, Hangar *m*
**hanger** *n* MAT PROP Aufhängevorrichtung *f*
**hanging** *n* BUILD HARDW Wandbehang *m*
***hanging***: **~ bridge** *n* INFR & DES Hängebrücke *f*; **~ gutter** *n* ARCH & BUILD, WASTE WATER Hängerinne *f*; **~ post** *n* ARCH & BUILD, BUILD HARDW hängender Torpfosten *m*; **~ scaffold** *n* ARCH & BUILD, BUILD MACHIN, INFR & DES Hängebühne *f*, Hängegerüst *nt*; **~ shelf** *n* BUILD HARDW Hängeregal *nt*; **~ stage** *n* ARCH & BUILD, BUILD MACHIN, INFR & DES Hängegerüst *nt*; **~ stairs** *n pl* ARCH & BUILD Freiträgertreppe *f*, Kragtreppe *f*; **~ steps** *n pl* ARCH & BUILD Kragtreppe *f*, eingespannte Stufen *f pl*; **~ stormwater gutter** *n* ARCH & BUILD, WASTE WATER Hängerinne *f*; **~ truss** *n* ARCH & BUILD Hängewerk *nt*
**harbor** *AmE see harbour BrE*
**harbour** *n BrE* INFR & DES Hafen *m*; **~ railway** *n* INFR & DES Hafenbahn *f*, Hafengleis *nt*
**hard** *adj* INFR & DES, MAT PROP hart
**hardboard** *n* MAT PROP, TIMBER Hartfaserplatte *f*
***hard***: **~-burnt** *adj* MAT PROP hartgebrannt; **~-burnt brick** *n* INFR & DES, MAT PROP, STONE Hartbrandziegel *m*, Vormauerziegel *m*; **~ clay** *n* INFR & DES, MAT PROP harter Ton *m*
**hardcore** *n* CONCR Packlage *f*
**harden 1.** *vt* CONCR erhärten, MAT PROP erhärten, verhärten, STONE erhärten; **2.** *vi* CONCR hart werden, härten, MAT PROP hart werden, härten, *concrete, mortar* abbinden
**hardening** *n* CONCR, MAT PROP, STONE Erhärten *nt*, Härten *nt*; **~ accelerator** *n* CONCR, MAT PROP Erhärtungsbeschleuniger *m*; **~ agent** *n* SURFACE *paint* Härtemittel *nt*, Härter *m*; **~ graph** *n* INFR & DES, MAT PROP Härtungskurve *f*
**hard**: **~ fiber board** *AmE*, **~ fibre board** *BrE* *n* TIMBER Holzfaserplatte *f*
**hardly**: **~ inflammable** *adj* MAT PROP schwer entflammbar, schwer entzündlich
**hardness** *n* INFR & DES, MAT PROP Härte *f*; **~ scale** *n* MAT PROP Härteskala *f*; **~ test** *n* MAT PROP Härtetest *m*
**hard**: **~ rock** *n* INFR & DES, MAT PROP Hartgestein *nt*; **~ rubber bar** *n* MAT PROP Hartgummistab *m*; **~ rub-**

**ber pipe** *n* MAT PROP Hartgummileitung *f*; **~ soil** *n* INFR & DES, MAT PROP harter Boden *m*
**hardstand** *n* CONCR, INFR & DES Betonabstellplatz *m*
**hardware** *n* BUILD HARDW Baubeschlag *m*, Beschlag *m*, Kleineisenzeug *nt*, MAT PROP Kleineisenzeug *nt*; **~ cutouts** *n pl* BUILD HARDW Band- und Schloßfräsungen *f pl*; **~ kit** *n* BUILD HARDW Beschlagsatz *m*; **~ work** *n* ARCH & BUILD Beschlagsarbeit *f*
**hard**: **~ water** *n* ENVIRON, INFR & DES, WASTE WATER hartes Wasser *nt*
**hardwearing** *adj* INFR & DES verschleißfest, MAT PROP strapazierfähig
**hardwood** *n* MAT PROP, TIMBER Hartholz *nt*
**harmfulness** *n* ENVIRON Schädlichkeit *f*
**harmful**: **~ substance** *n* ENVIRON, INFR & DES, MAT PROP Schadstoff *m*; **~ to the environment** *phr* ENVIRON umweltschädlich
**harmonic**: **~ vibration** *n* SOUND & THERMAL harmonische Schwingung *f*
**harmonize** *vt* ARCH & BUILD angleichen, anpassen, harmonisch anordnen, BUILD MACHIN, CONCR, CONST LAW, ELECTR, INFR & DES, STEEL angleichen
**hasp** *n* BUILD HARDW *locksmithing* Haspe *f*
**hatch 1.** *n* ARCH & BUILD Durchreiche *f*, Luke *f*; **2.** *vt* ARCH & BUILD schraffieren
**hatching** *n* ARCH & BUILD Schraffur *f*
**haul** *vt* BUILD MACHIN transportieren; **~ away** *vt* ARCH & BUILD, BUILD MACHIN, ENVIRON *soil, rubbish, waste*, INFR & DES abfahren
**haulage** *n* BUILD MACHIN Förderung *f*, Transport *m*; **~ rope** *n* BUILD MACHIN Zugseil *nt*, Förderseil *nt*
**hauling**: **~ machine** *n* BUILD MACHIN Fördermaschine *f*
**haunch** *n* ARCH & BUILD *tapered* Schräge *f*, Bogenschenkel *m*, Gewölbeschenkel *m*, Voute *f*, INFR & DES Schräge *f*
**haunched** *adj* ARCH & BUILD, CONCR, INFR & DES, MAT PROP, STEEL betonummantelt; **~ mortice and tenon joint** *n* ARCH & BUILD, TIMBER Holzbalkenzugankerverbindung *f*; **~ mortise and tenon joint** *n* (*cf haunched mortice and tenon joint*) ARCH & BUILD, TIMBER Holzbalkenzugankerverbindung *f*
**Häusler**: **~ roofing** *n* ARCH & BUILD Holzzementdach *nt*
**hawk** *n* STONE *plaster* Tünchscheibe *f*
**hazard** *n* CONST LAW, INFR & DES Gefahr *f*
**hazardous** *adj* CONST LAW gefährlich; **~ location** *n* CONST LAW explosionsgefährdeter Betriebsraum *m*; **~ material** *n* ENVIRON gefährliche Stoffe *m pl*; **~ sewage** *n* CONST LAW, ENVIRON, INFR & DES gefährliches Schmutzwasser *nt*; **~ waste** *n* CONST LAW, ENVIRON, INFR & DES Sondermüll *m*, gefährliche Abfälle *m pl*; **~ waste collection** *n* ENVIRON Sondermülleinsammlung *f*; **~ waste landfill** *n* ENVIRON Sondermülldeponie *f*
**H-beam** *n* ARCH & BUILD, CONCR, STEEL Breitflanschträger *m*, Doppel-T-Träger *m*, H-Träger *m*
**HCFC** *abbr* (*hydrochlorofluorocarbon*) ENVIRON HFCK (*Hydrofluorchlorkohlenwasserstoff*)
**HDPE** *abbr* (*high-density polyethylene*) MAT PROP PE hart (*Polyethylen hoher Dichte*)
**head** *n* ARCH & BUILD Kopf *m*, Streich-, oberes Ende *nt*, *on door, window* Sturz *m*, *pillar* Säule *f*, *crown tile* Firstziegel *m*, *of pressure* Fallhöhe *f*, *of a pump* Förderhöhe *f*, INFR & DES *hydraulic* Druckhöhe *f*, Firstziegel *m*, MAT PROP, STONE Firstziegel *m*; **~ beam** *n* ARCH & BUILD Oberholm *m*; **~ end loader** *n* BUILD MACHIN Frontlader *m*
**header** *n* ARCH & BUILD Binderstein *m*, Wechselbalken *m*, *perpend* Kopfstein *m*, STONE Binder *m*, Strecker *m*, *perpend* Kopfstein *m*; **~ bond** *n* ARCH & BUILD Kopfverband *m*, STONE Binderverband *m*, Streckerverband *m*, Kopfverband *m*; **~ course** *n* ARCH & BUILD Binderlage *f*, Kopfsteinschicht *f*; **~ joint** *n* TIMBER Stumpffuge *f*
***head***: **~ gate** *n* ENVIRON Obertor *nt*, oberes Schleusentor *nt*; **~ mason** *n* ARCH & BUILD Polier *m*, Kapo *m*; **~ packing** *n* BUILD MACHIN Rammhaube *f*; **~ piece** *n* ARCH & BUILD Oberschwelle *f*, Streichbalken *m*; **~ plate** *n* ARCH & BUILD Oberschwelle *f*, Streichbalken *m*
**headrace**: **~ canal** *n* ENVIRON Hangkanal *m*, Werkkanal *m*
**headroom** *n* ARCH & BUILD, INFR & DES Durchfahrtshöhe *f*, Durchgangshöhe *f*, Kopfhöhe *f*, lichte Höhe *f*
***head***: **~ runner** *n* ARCH & BUILD Bundbalken *m*, TIMBER Bundbalken *m*, Rahmholz *nt*, Rähm *m*; **~ tree** *n* TIMBER Sattelholz *nt*, Aufsattelung *f*; **~ of water** *n* ENVIRON Wassersäule *f*; **~ of water over weir** *n* INFR & DES Überströmungshöhe *f*
**headway** *n* ARCH & BUILD, INFR & DES Durchfahrtshöhe *f*
**health**: **~ and comfort** *n* HEAT & VENT Gesundheit *f* und Komfort
**Health**: **~ and Safety Executive** *n* (*HSE BrE*) CONST LAW Unfallverhütungsausschuß *m*; **~ and Safety at Work Act** *n* CONST LAW Unfallverhütungsvorschrift *f* an der Arbeitsstelle
**heap** *n* ARCH & BUILD Haufen *m*
**hearth** *n* ARCH & BUILD, HEAT & VENT Feuerstellenboden *m*
**hearting** *n* STONE Ausfüllung *f*, Kernfüllung *f*
**heart**: **~ trowel** *n* ARCH & BUILD, BUILD MACHIN Herzblattpolierschaufel *f*
**heartwood** *n* MAT PROP, TIMBER Kernholz *nt*
**heat 1.** *n* ENVIRON, HEAT & VENT, SOUND & THERMAL, WASTE WATER Hitze *f*, Wärme *f*; **2.** *vt* ENVIRON, HEAT & VENT, SOUND & THERMAL heizen, WASTE WATER erhitzen, *water* erwärmen
***heat***: **~ accumulator** *n* HEAT & VENT Wärmespeicher *m*; **~ balance** *n* HEAT & VENT Wärmehaushalt *m*; **~ barrier** *n* SOUND & THERMAL Wärmesperre *f*; **~ build-up** *n* HEAT & VENT Wärmeentwicklung *f*; **~ capacity** *n* HEAT & VENT, SOUND & THERMAL Wärmekapazität *f*; **~ carrier** *n* HEAT & VENT Wärmeträgermedium *nt*, Wärmeträger *m*; **~ conductivity** *n* MAT PROP, SOUND & THERMAL Wärmeleitfähigkeit *f*; **~ content** *n* HEAT & VENT, SOUND & THERMAL Wärmemenge *f*; **~ demand** *n* ARCH & BUILD, CONST LAW, HEAT & VENT Wärmebedarf *m*; **~ detector** *n* BUILD HARDW, CONST LAW *fire protection* Wärmefühler *m*; **~ efficiency** *n* HEAT & VENT Wärmeausnutzung *f*, Wärmewirkungsgrad *m*; **~ energy** *n* HEAT & VENT Wärmeenergie *f*
**heater** *n* HEAT & VENT Heizgerät *nt*
***heat***: **~ exchange** *n* HEAT & VENT Wärmeaustausch *m*; **~ exchanger** *n* ENVIRON Wärmeaustauscher *m*, HEAT & VENT Wärmeüberträger *m*, Wärmeaustauscher *m*; **~ expansion** *n* ARCH & BUILD, CONCR, INFR & DES, MAT PROP, STEEL, WASTE WATER Wärmeausdehnung *f*, Wärmedehnung *f*; **~ flow** *n* ARCH & BUILD, HEAT & VENT, SOUND & THERMAL Wärmeflußrichtung *f*; **~ flux** *n* HEAT & VENT, SOUND & THERMAL Wärme-

fluß *m*; ~ **of hydration** *n* CONCR, STONE Hydratationswärme *f*

**heating** *n* ENVIRON, HEAT & VENT, SOUND & THERMAL, WASTE WATER Heiz-, Heizung *f*; ~ **circuit** *n* HEAT & VENT Heizkreis *m*; ~ **convector** *n* HEAT & VENT Konvektor *m*; ~ **element** *n* HEAT & VENT Heizkörper *m*; ~ **load** *n* HEAT & VENT Heizlast *f*; ~ **melter** *n* BUILD MACHIN Glühschmelzer *m*; ~ **period between seasons** *n* HEAT & VENT Übergangsheizperiode *f*; ~ **pipe** *n* HEAT & VENT Heizungsrohr *nt*; ~ **and refrigerating engineering** *n* HEAT & VENT Wärme- und Kältetechnik *f*; ~ **surface** *n* HEAT & VENT Heizfläche *f*; ~ **system** *n* HEAT & VENT Heizungsanlage *f*; ~ **tube** *n* HEAT & VENT Heizungsrohr *nt*

**heat**: ~**-insulating** *adj* ENVIRON, SOUND & THERMAL wärmedämmend; ~**-insulating layer** *n* ENVIRON, SOUND & THERMAL Wärmedämmschicht *f*; ~**-insulating plaster** *n* ENVIRON, STONE Isolierputz *m*; ~ **insulation** *n* ENVIRON, SOUND & THERMAL Wärmedämmung *f*, Wärmeschutz *m*; ~ **insulation material** *n* ENVIRON, INFR & DES, MAT PROP, SOUND & THERMAL Wärmedämmstoff *m*; ~ **insulation requirement** *n* CONST LAW, ENVIRON, SOUND & THERMAL Wärmedämmforderung *f*; ~ **insulation work** *n* ENVIRON, SOUND & THERMAL Wärmedämmungsarbeiten *f pl*; ~ **load calculations** *n pl* HEAT & VENT Wärmebelastungsberechnungen *f pl*; ~ **loss** *n* ENVIRON, SOUND & THERMAL Wärmeverlust *m*; ~ **pollution** *n* ENVIRON Umweltverschmutzung *f* durch Wärme

**heatproof** *adj* MAT PROP hitzefest, wärmebeständig, hitzebeständig

***heat***: ~ **protection glass** *n* ARCH & BUILD, MAT PROP, SOUND & THERMAL Wärmeschutzglas *nt*; ~ **pump** *n* HEAT & VENT Wärmepumpe *f*; ~ **radiation** *n* HEAT & VENT Wärmeausstrahlung *f*; ~ **recovery** *n* HEAT & VENT Wärmerückgewinnung *f*, Wärmewiedergewinnung *f*; ~**-reflecting** *adj* SOUND & THERMAL wärmereflektierend, wärmerückstrahlend; ~ **requirement** *n* ARCH & BUILD, CONST LAW, HEAT & VENT Wärmebedarf *m*; ~ **resistance** *n* MAT PROP Wärmebeständigkeit *f*, Wärmedurchgangswiderstand *m*, SOUND & THERMAL Wärmedurchgangswiderstand *m*; ~**-sensitive** *adj* MAT PROP wärmeempfindlich; ~ **sink** *n* HEAT & VENT Abstrahlblock *m*; ~ **storage** *n* HEAT & VENT, SOUND & THERMAL Wärmespeicherung *f*; ~ **transfer** *n* HEAT & VENT Wärmeübertragung *f*; ~ **transfer coefficient** *n* MAT PROP Wärmedurchgangszahl *f*; ~ **transfer medium** *n* HEAT & VENT Wärmeträger *m*, Wärmeträgermedium *m*; ~ **transition coefficient** *n* INFR & DES, MAT PROP Wärmedurchgangskoeffizient *m*, Wärmedurchgangszahl *f*, k-Wert *m*; ~ **transmission resistance** *n* MAT PROP Wärmeübergangswiderstand *m*; ~**-treated glass** *n* MAT PROP Einscheibensicherheitsglas *nt*; ~ **utilization** *n* HEAT & VENT Wärmeausnutzung *f*, Wärmewirkungsgrad *m*

**heavy** *adj* INFR & DES schwer; ~ **clay** *n* INFR & DES, MAT PROP schwerer Ton *m*; ~**-duty** *adj* INFR & DES, MAT PROP hochbeanspruchbar; ~**-duty carpet** *n* INFR & DES, MAT PROP industrieller Teppich *m*; ~**-duty lintel** *n* INFR & DES, MAT PROP hochbeanspruchbarer Sturz *m*; ~**-duty mortise lock** *n* BUILD HARDW Hochleistungseinsteckschloß *nt*; ~ **metal** *n* CONST LAW, ENVIRON, MAT PROP Schwermetall *nt*; ~ **soil** *n* INFR & DES schwerer Boden *m*

**height** *n* ARCH & BUILD Höhe *f*; ~ **above sea level** *n* ARCH & BUILD Höhe *f* über dem Meeresspiegel; ~ **above zero level** *n* INFR & DES Kote *f* über Normalnull (*Kote über NN*); ~ **adjustable** *adj* BUILD HARDW *hinge* höhenverstellbar

**heighten** *vt* ARCH & BUILD erhöhen, *building* aufstocken

**heightening** *n* ARCH & BUILD *building* Aufstockung *f*, Erhöhen *nt*

***height***: ~ **indication** *n* ARCH & BUILD, INFR & DES Höhenkote *f*; ~ **of instrument** *n* ARCH & BUILD, INFR & DES *surveying* Instrumentenhöhe *f*; ~ **notation** *n* ARCH & BUILD, INFR & DES Kote *f*; ~ **of suspension** *n* ARCH & BUILD, INFR & DES Abhängehöhe *f*

**helical**: ~ **hinge** *n* BUILD HARDW Federband *nt*; ~ **staircase** *n* ARCH & BUILD Spindeltreppe *f*, Wendeltreppe *f*

**helipad** *n* ARCH & BUILD, INFR & DES *on roof of building* Heliport *m*, Hubschrauberlandeplatz *m*

**heliport** *n* ARCH & BUILD, INFR & DES Heliport *m*, Hubschrauberlandeplatz *m*

**helistop** *n* ARCH & BUILD, INFR & DES Heliport *m*, Hubschrauberlandeplatz *m*

**helmet** *n* BUILD MACHIN Rammhaube *f*, Schlaghaube *f*

**helm**: ~ **roof** *n* ARCH & BUILD Helmdach *nt*, Rautendach *nt*, Rhombendach *nt*

**helper** *n* ARCH & BUILD Hilfsarbeiter *m*

**hem** *vt* ARCH & BUILD, TIMBER einfassen, säumen

**hemicycle** *n* ARCH & BUILD Halbkreis *m*

**Heraklith**: ~ **insulating board** *n* INFR & DES, MAT PROP Heraklithplatte *f*

**herringbone**: ~ **pattern** *n* TIMBER *parquetry* Fischgrätenmuster *nt*, Fischgrätenform *f*

**hew** *vt* INFR & DES *tree* fällen, STONE hauen

**hewer** *n* BUILD MACHIN *of stone, wood* Hauer *m*

**hewn**: ~ **stone** *n* ARCH & BUILD, STONE Naturwerkstein *m*; ~ **stone facing** *n* ARCH & BUILD, STONE Steinverblendung *f*; ~ **stone masonry** *n* ARCH & BUILD, STONE Werksteinmauerwerk *nt*

**H-girder** *n* ARCH & BUILD, CONCR, STEEL H-Träger *m*

**hidden** *adj* ARCH & BUILD verdeckt

**high** *adj* ARCH & BUILD, INFR & DES hoch; ~**-bond action** *n* CONCR, INFR & DES, MAT PROP starke Verbindung *f*; ~**-density housing** *n* ARCH & BUILD hochintensiver Wohnungsbau *m*; ~**-density polyethylene** *n* (*HDPE*) MAT PROP Polyethylen *nt* hoher Dichte (*PE hart*); ~**-density polyurethane foam** *n* SOUND & THERMAL Zell-Polyurethan *nt* hoher Dichte; ~**-early-strength** *adj* STONE *cement* frühhochfest

**highest**: ~ **high water level** *n* INFR & DES höchster Hochwasserstand *m*

***high***: ~ **flood** *n* INFR & DES Hochwasserstand *m*; ~**-frequency finisher** *n* BUILD MACHIN Hochfrequenzfertiger *m*; ~**-grade energy** *n* HEAT & VENT hochwertige Energie *f*; ~**-grade fireclay** *n* MAT PROP Edelfeuerton *m*; ~ **head** *n* ENVIRON hohe Wassersäule *f*; ~**-level railroad** *n AmE* (*cf high-level railway BrE*) INFR & DES Hochbahn *f*; ~**-level railway** *BrE n* (*cf high-level railroad AmE*) INFR & DES Hochbahn *f*

**highly**: ~ **absorbent plaster base** *n* STONE stark saugender Putzgrund *m*; ~ **compressed** *adj* MAT PROP hochverdichtet; ~ **corrosion-proof** *adj* MAT PROP hochkorrosionsfest; ~ **impact-resistant** *adj* MAT PROP *plastic* hochschlagfest; ~ **inflammable** *adj* CONST LAW, MAT PROP feuergefährlich, leicht entflammbar, leicht entzündbar, leicht entzündlich; **~ resinous** *adj* TIMBER harzreich; **~ resinous wood** *n* TIMBER harzreiches Holz *nt*; **~ shock-resistant** *adj* MAT PROP hochschlagfest

***high*: ~-pitched roof** *n* ARCH & BUILD Steildach *nt*; **~-pressure** *adj* HEAT & VENT *impeller* druckstark; **~-pressure steam** *n* HEAT & VENT, INFR & DES Hochdruckdampf *m*; **~-pressure tank** *n* BUILD MACHIN Hochdruckbehälter *m*; **~-quality** *adj* MAT PROP hochwertig; **~-rate filtration** *n* ENVIRON Hochleistungsfilterung *f*; **~-rise** *adj* ARCH & BUILD, INFR & DES vielgeschossig; **~-rise building** *n BrE* ARCH & BUILD, INFR & DES Hochhaus *nt*; **~-speed track** *n* INFR & DES *railways* Hochgeschwindigkeitsspur *f*; **~-speed train** *n* INFR & DES Hochgeschwindigkeitszug *m*; **~-strength** *adj* MAT PROP *cement* hochfest; **~-strength bolt** *n* ARCH & BUILD, MAT PROP hochfester Bolzen *m*; **~-strength bronze** *n* MAT PROP Sonderbronze *f*; **~-strength friction grip** *n* ARCH & BUILD, STEEL HV-Schraubenverbindung *f*, hochfest verschraubte Verbindung *f* (*HV-Verbindung*); **~-tensile bolted structural joint** *n* ARCH & BUILD, STEEL HV-Schraubenverbindung *f*, gleitfeste Schraubenverbindung *f*, hochfest verschraubte Verbindung *f* (*HV-Verbindung*); **~ tide** *n* ENVIRON Flut *f*, INFR & DES Hochwasser *nt*; **~ tide mark** *n* INFR & DES Hochwassermarke *f*; **~ water level** *n* INFR & DES Hochwasserspiegel *m*, Hochwasserstand *m*; **~-water ordinary spring tide** *n* (*HWOST*) ENVIRON normale Springzeitflut *f*

**highway** *n* (*cf carriageway BrE*) INFR & DES Fernverkehrsstraße *f*, Fahrbahn *f*; **~ slab** *n AmE* INFR & DES Fahrbahnplatte *f*

**high: ~-yield stress steel** *n* ARCH & BUILD, MAT PROP hochbelastbarer Stahl *m*

**hinge** *n* ARCH & BUILD Gelenk *nt*, BUILD HARDW *for windows, doors* Band *nt*, *flap hinge* Scharnierband *nt*, *for doors only* Türangel *f*, *hinged joint* Gelenk *nt*, Scharnier *nt*, STEEL Gelenk *nt*; **~ bolt** *n* BUILD HARDW Gelenkbolzen *m*

**hinged: ~ column** *n* ARCH & BUILD Pendelstütze *f*; **~ grating** *n* STEEL Klapprost *m*; **~ ridge purlin** *n* TIMBER Gerberpfette *f*

***hinge*: ~ joint** *n* BUILD HARDW Gelenkverbindung *f*, Scharnier *nt*; **~ pin** *n* BUILD HARDW Drehstift *m*, Scharnierstift *m*; **~ pocket** *n* BUILD HARDW, TIMBER Bandtasche *f*; **~ post** *n* ARCH & BUILD, BUILD HARDW hängender Torpfosten *m*

**hip** *n* ARCH & BUILD Walm *m*

**hipped: ~ roof** *n* ARCH & BUILD, TIMBER Walmdach *nt*

***hip*: ~ rafter** *n* INFR & DES Walmsparren *m*, TIMBER Gratbalken *m*, Gratsparren *m*, Walmsparren *m*; **~ and ridge roof** *n* ARCH & BUILD, TIMBER Gratsparrendach *nt*; **~ roof** *n* ARCH & BUILD, TIMBER Walmdach *nt*; **~ tile** *n* ARCH & BUILD, INFR & DES, MAT PROP, STONE Firstziegel *m*, Gratziegel *m*; **~ and valley roof** *n* ARCH & BUILD eingeschnittenes Walmdach *nt*

**hoarding** *n* INFR & DES Bauzaun *m*

**hogbacked: ~ bridge** *n* INFR & DES Halbparabelbrücke *f*

**hogging: ~ bend** *n* ARCH & BUILD, CONCR, INFR & DES, MAT PROP Hochbiegen *nt*; **~ bending** *adj* ARCH & BUILD, CONCR, INFR & DES, MAT PROP nach oben biegend; **~ moment** *n* ARCH & BUILD, CONCR, INFR & DES, MAT PROP, STEEL Aufwölbungsmoment *nt*; **~ transverse bending** *n* ARCH & BUILD, CONCR, INFR & DES, STEEL *composite beam* Aufwölbungsquerbiegung *f*

**hoist 1.** *n* BUILD MACHIN Hebezeug *nt*; **2.** *vt* ARCH & BUILD, BUILD MACHIN, WASTE WATER heben

**hoisting** *n* ARCH & BUILD, BUILD MACHIN, WASTE WATER Heben *nt*; **~ height** *n* BUILD MACHIN, INFR & DES Förderhöhe *f*, Hubhöhe *f*; **~ speed** *n* BUILD MACHIN Hubgeschwindigkeit *f*

**holding: ~ angle** *n* MAT PROP Haltewinkel *m*; **~ bolt** *n* BUILD HARDW, BUILD MACHIN Haltebolzen *m*

**hole** *n* ARCH & BUILD, BUILD MACHIN, INFR & DES, MAT PROP, SURFACE Loch *nt*, *aperture, opening* Durchbruch *m*; **~ in the ozone layer** *n* ENVIRON Ozonloch *nt*

**hollow 1.** *adj* ARCH & BUILD, MAT PROP hohl; **2.** *n* ARCH & BUILD, MAT PROP Aushöhlung *f*, Hohlkörper *m*, Vertiefung *f*, INFR & DES Aushöhlung *f*; **3.** *vt* ARCH & BUILD, TIMBER *wood* auskehlen, vertiefen; **~ out** ARCH & BUILD, TIMBER aushöhlen

***hollow*: ~ block** *n* MAT PROP Hohlkörper *m*; **~ brick** *n* MAT PROP Hohlstein *m*, STONE Hohlziegel *m*; **~ brick ceiling** *n* ARCH & BUILD Hohlkörperdecke *f*; **~ clay block** *n* MAT PROP Tonhohlplatte *f*, Hourdi *m*; **~ clay pot** *n* STONE Tonhohlplatte *f*; **~ concrete block** *n* MAT PROP Betonhohlblockstein *m*; **~-core block** *n* MAT PROP, STONE Hohlblockstein *m*; **~-core plank** *n* MAT PROP Stegzementdiele *f*, Stegdiele *f*, Hohldiele *f*; **~ filler block** *n* MAT PROP Deckenhohlkörper *m*; **~ filler block floor** *n* ARCH & BUILD Hohlkörperdecke *f*, Füllkörperdecke *f*; **~-gaged brick** *AmE*, **~-gauged brick** *BrE n* MAT PROP, STONE Hourdi *m*, Tonhohlplatte *f*

**hollowing** *n* ARCH & BUILD, TIMBER Auskehlung *f*

**hollowness** *n* MAT PROP Hohlheit *f*

***hollow*: ~ section** *n* STEEL Hohlprofil *nt*; **~ shape** *n* STEEL Hohlprofil *nt*; **~ wall box** *n* ELECTR Hohlwanddose *f*

**home** *n* ARCH & BUILD Heim *nt*, Wohnung *f*, Wohnsitz *m*, CONST LAW Heim *nt*

**homogenous** *adj* INFR & DES, MAT PROP homogen

**honeycomb** *n* ARCH & BUILD, CONCR *cavity* Betonnest *nt*, Wabenstruktur *f*; **~ brick** *n* MAT PROP, STONE Gitterziegel *m*, Hochlochziegel *m* (*HLZ*); **~ core** *n* ARCH & BUILD *composite structure* Wabenkern *m*; **~ insulating board** *n* SOUND & THERMAL Wabendämmplatte *f*; **~ masonry** *n* STONE, TIMBER durchbrochenes Mauerwerk *nt*; **~ rib** *n* ARCH & BUILD durchbrochene Rippe *f*; **~ sandwich material** *n* MAT PROP Verbundwerkstoff *m* mit Wabenkern; **~ structure** *n* ARCH & BUILD Wabenstruktur *f*

**hood** *n* ARCH & BUILD Wetterschutzdach *nt*, BUILD HARDW *on flue, vent* Abzugshaube *f*, *wall coping* Mauerkappe *f*, *weathering protection* Wetterschutzdach *nt*

**hook** *n* BUILD HARDW, MAT PROP Haken *m*; **~ bolt** *n* BUILD HARDW *lock* Hakenfalle *f*

**Hooke: ~'s law** *n* ARCH & BUILD, INFR & DES, MAT PROP Hookesches Gesetz *nt*

***hook*: ~ track** *n* BUILD MACHIN *of crane* Hakenweg *m*

**hoop** *n* MAT PROP, STEEL Bügel *m*, Bandstahl *m*

**hooped: ~ column** *n* ARCH & BUILD, CONCR, INFR & DES *reinforced column* bügelbewehrte Säule *f*, ringarmierte Säule *f*, ringbewehrte Säule *f*; **~ concrete** *n* CONCR Sinterbeton *m*

**hopper** *n* BUILD MACHIN Einfülltrichter *m*, Schütt-

gutbehälter *m*; ~ **head** *n* ARCH & BUILD Einlauf *m*; ~ **window** *n* ARCH & BUILD, TIMBER Kippfenster *nt*
**horizontal** *adj* ARCH & BUILD, INFR & DES horizontal, waagerecht; ~ **bracing** *n* ARCH & BUILD, STEEL, TIMBER Längsverband *m*; ~ **core slab** *n* MAT PROP Langlochplatte *f*; ~ **coring block** *n* MAT PROP, STONE Langlochstein *m*; ~ **coring brick** *n* MAT PROP, STONE Langlochziegel *m*; ~ **joint** *n* ARCH & BUILD, STONE Lagerfuge *f*
**horizontally**: ~ **pivoted sach window** *n* BUILD HARDW Schwingflügel *m*; ~ **pivoted sash** *n* ARCH & BUILD Schwingflügel *m*; ~ **pivoted sash window** *n* ARCH & BUILD Schwingfenster *nt*
**horizontal**: ~ **member** *n* ARCH & BUILD, CONCR, INFR & DES, STEEL Rahmenriegel *m*; ~ **shear** *n* ARCH & BUILD, BUILD MACHIN Schubkraft *f*, Horizontalschub *m*
**horseshoe**: ~ **arch** *n* ARCH & BUILD Hufeisengewölbe *nt*
**hose** *n* BUILD HARDW, WASTE WATER Schlauch *m*; ~ **clip** *n* BUILD HARDW Schlauchhalter *m*, WASTE WATER Schlauchhalter *m*, Schlauchschelle *f*, Schlauchklemme *f*; ~ **connection** *n* BUILD HARDW Schlauchanschluß *m*; ~ **nozzle** *n* BUILD HARDW, WASTE WATER Schlauchtülle *f*; ~ **reel** *n* BUILD HARDW, WASTE WATER Schlauchtrommel *f*
**hospice** *n* ARCH & BUILD, INFR & DES Fremdenheim *nt*, Hospiz *nt*
**hot**: ~ **adhesive compound** *n* MAT PROP Heißklebemasse *f*; ~-**air heating** *n* HEAT & VENT Warmluftheizung *f*; ~ **asphalt** *n* MAT PROP Heißbitumen *nt*; ~-**dip galvanized** *adj* SURFACE feuerverzinkt; ~-**dip galvanizing** *n* STEEL, SURFACE Feuerverzinken *nt*; ~-**dipped galvanized steel** *n* STEEL feuerverzinkter Stahl *m*; ~-**dipping process** *n* SURFACE Schmelztauchverfahren *nt*
**hotel** *n* ARCH & BUILD, INFR & DES Hotel *nt*
**hothouse** *n* ARCH & BUILD Treibhaus *nt*, Gewächshaus *nt*
**hot**: ~-**riveted** *adj* STEEL warmgenietet; ~-**rolled** *adj* MAT PROP warmgewalzt; ~-**tarred** *adj* MAT PROP heiß geteert; ~ **water** *n* ELECTR, HEAT & VENT, WASTE WATER Warmwasser *nt*; ~ **water pipe** *n* HEAT & VENT, WASTE WATER Warmwasserleitung *f*; ~ **water reservoir** *n* ELECTR, HEAT & VENT, WASTE WATER Heißwasserspeicher *m*; ~ **water supply** *n* HEAT & VENT, WASTE WATER Warmwasserversorgung *f*; ~ **water tank** *n* ELECTR, HEAT & VENT, WASTE WATER Boiler *m*, Heißwasserspeicher *m*, Warmwasserspeicher *m*
**hourly**: ~ **solar radiation** *n* HEAT & VENT stündliche Solarstrahlung *f*
**hours**: ~ **of work** *n pl* CONST LAW Arbeitsstunden *f pl*
**house** *n* ARCH & BUILD Haus *nt*; ~-**builders' scheme** *n* CONST LAW Bauherrenmodell *nt*; ~ **building** *n* ARCH & BUILD, INFR & DES Wohnungsbau *m*; ~ **connection** *n* ELECTR, HEAT & VENT, WASTE WATER Hausanschluß *m*; ~ **connection box** *n* ELECTR Hausanschlußkasten *m*
**housed**: ~ **joint** *n* ARCH & BUILD Anschluß *m* mit Versatz, TIMBER Anschluß *m* mit Versatz, Einarmzapfverbindung *f*
***house***: ~ **drainage** *n* ARCH & BUILD, WASTE WATER Hausentwässerung *f*
***housed***: ~ **string** *n* ARCH & BUILD Wange *f* mit eingestemmten Stufen
**household** *n* ENVIRON, INFR & DES Haushalt *m*; ~ **refuse** *n* ENVIRON, INFR & DES Hausmüll *m*; ~ **waste water** *n* ENVIRON Haushaltsabwässer *nt pl*
***house***: ~ **paint** *n* MAT PROP, SURFACE Bautenanstrichfarbe *f*, Fassadenfarbe *f*; ~ **service connection** *n* ELECTR Hausanschlußleitung *f*
**housing** *n* ARCH & BUILD Einhausung *f*, *for statues* Nische *f*, *casing* Einschließung *f*, Gehäuse *nt*, *groove* Nut *f*, *building of houses* Wohnungsbau *m*, ENVIRON Gehäuse *nt*, INFR & DES *building of houses* Wohnungsbau *m*; ~ **construction** *n* ARCH & BUILD, INFR & DES Wohnungsbau *m*; ~ **density** *n* INFR & DES Wohndichte *f*
**HSE** *abbr BrE* (*Health and Safety Executive*) CONST LAW Unfallverhütungsausschuß *m*
**H-section** *n* MAT PROP H-Profil *nt*, Breitflanschprofil *nt*
**hub**: ~ **with bronze bearing** *n* BUILD HARDW *lock* bronzeringumbuchste Nuß *f*
**humid** *adj* ARCH & BUILD, HEAT & VENT, INFR & DES, MAT PROP, SOUND & THERMAL feucht
**humidification** *n* HEAT & VENT Befeuchtung *f*
**humidifier** *n* HEAT & VENT Befeuchter *m*
**humidify** *vt* HEAT & VENT befeuchten
**humidity** *n* ARCH & BUILD, HEAT & VENT, INFR & DES, MAT PROP, SOUND & THERMAL Feuchte *f*, Feuchtigkeit *f*; ~ **barrier** *n* SOUND & THERMAL Feuchtigkeitssperrschicht *f*; ~ **content** *n* INFR & DES, MAT PROP, TIMBER Feuchtigkeitsgehalt *m*
**humid**: ~ **room** *n* ARCH & BUILD Feuchtraum *m*
**humus** *n* ARCH & BUILD Mutterboden *m*, INFR & DES Humus *m*, Mutterboden *m*, SOUND & THERMAL Humus *m*; ~ **content** *n* INFR & DES, MAT PROP Humusgehalt *m*
**HWOST** *abbr* (*high-water ordinary spring tide*) ENVIRON normale Springzeitflut *f*
**hybrid**: ~ **concrete construction** *n* ARCH & BUILD, MAT PROP Hybrid-Betonkonstruktion *f*
**hydrant** *n BrE* (*cf fireplug AmE*) CONST LAW, WASTE WATER *firefighting* Hydrant *m*
**hydrate** *vt* INFR & DES, MAT PROP hydratisieren
**hydration** *n* INFR & DES, MAT PROP Hydratation *f*; ~ **heat** *n* CONCR, STONE Hydratationswärme *f*; ~ **water** *n* CONCR, INFR & DES, MAT PROP Hydratationswasser *nt*
**hydraulic**: ~ **binder** *n* MAT PROP hydraulisches Bindemittel *nt*; ~ **cement** *n* CONCR, MAT PROP hydraulischer Zement *m*; ~ **conductivity** *n* ENVIRON Durchlässigkeitsbeiwert *m*; ~ **efficiency** *n* ENVIRON hydraulischer Wirkungsgrad *m*; ~ **engineering** *n* INFR & DES Wasserwirtschaftswesen *nt*; ~ **engineering project** *n* CONST LAW, INFR & DES Wasserbauprojekt *nt*; ~ **fill** *n* INFR & DES eingeschlämmte Erdmassen *f pl*, eingespülte Erdmassen *f pl*; ~ **fracturing** *n* ENVIRON hydraulische Rißbildung *f*; ~ **head** *n* INFR & DES hydraulische Druckhöhe *f*; ~ **jump** *n* INFR & DES Wechselsprung *m*; ~ **lift** *n* ARCH & BUILD, INFR & DES hydraulischer Lift *m*; ~ **model** *n* ENVIRON, INFR & DES, MAT PROP hydraulisches Modell *nt*; ~ **model test** *n* INFR & DES, MAT PROP hydraulischer Modellversuch *m*; ~ **press** *n* BUILD MACHIN Hydraulikpresse *f*; ~ **pressure** *n* ARCH & BUILD hydraulischer Druck *m*; ~ **ram** *n* BUILD MACHIN Widder *m*, ENVIRON Hydropulsor *m*, hydraulischer Widder *m*
**hydraulics** *n pl* INFR & DES, WASTE WATER Hydraulik *f*
**hydraulic**: ~ **structure** *n* INFR & DES Wasserbauwerk

*nt*; ~ **thrust** *n* ENVIRON Wasserdruck *m*, hydrostatischer Druck *m*

**hydrocarbon** *n* ENVIRON Kohlenwasserstoff *m*

**hydrochlorofluorocarbon** *n* (*HCFC*) ENVIRON Hydrofluorchlorkohlenwasserstoff *m* (*HFCK*)

**hydrodynamic** *adj* INFR & DES, WASTE WATER hydrodynamisch; ~ **model** *n* INFR & DES, WASTE WATER hydrodynamisches Modell *nt*

**hydroelectric**: ~ **generating station** *n* ELECTR, ENVIRON, HEAT & VENT, INFR & DES Wasserkraftwerk *nt*

**hydroelectricity** *n* ELECTR, ENVIRON, HEAT & VENT, INFR & DES Wasserkraft *f*

**hydroelectric**: ~ **power** *n* ELECTR, ENVIRON, HEAT & VENT, INFR & DES Wasserkraft *f*; ~ **power plant** *n* ELECTR, ENVIRON, HEAT & VENT, INFR & DES Wasserkraftwerk *nt*; ~ **power station** *n* ELECTR, ENVIRON, HEAT & VENT, INFR & DES Wasserkraftwerk *nt*

**hydrography** *n* ENVIRON, INFR & DES Gewässerkunde *f*, Hydrographie *f*

**hydrolizing**: ~ **tank** *n* ENVIRON Abwasserfaulraum *m*

**hydrostatic**: ~ **tube balance** *n* BUILD MACHIN Schlauchwaage *f*

**hydrothermal**: ~ **process** *n* ENVIRON hydrothermaler Prozeß *m*, hydrothermales Verfahren *nt*

**hydroturbine** *n* ENVIRON Wasserturbine *f*

**hygrometer** *n* HEAT & VENT, INFR & DES, MAT PROP Feuchtigkeitsmesser *m*

**hyperbolic**: ~ **paraboloid shell** *n* ARCH & BUILD, INFR & DES hyperbolische Paraboloidschale *f* (*HP-Schale*)

**hyperstatic** *adj* ARCH & BUILD, INFR & DES statisch unbestimmt

# I

**ice** *n* INFR & DES Eis *nt*
**ICE** *abbr* (*Institution of Civil Engineers*) CONST LAW, INFR & DES Institut *nt* für Bauingenieure
***ice***: **~ apron** *n* INFR & DES *upstream from bridge* Eisbrecher *m*; **~ crystal** *n* INFR & DES, MAT PROP Eiskristall *nt*; **~ layer** *n* INFR & DES, MAT PROP Eisschicht *f*; **~ protection** *n* INFR & DES, MAT PROP Eisschutz *m*; **~ wall** *n* ARCH & BUILD Frostmauer *f*, Frostschürze *f*
**idealization** *n* ARCH & BUILD, CONCR, INFR & DES, MAT PROP, STEEL Vereinfachung *f*
**ideal**: **~ velocity** *n* ENVIRON Idealgeschwindigkeit *f*
**identified**: **~ resources** *n pl* ENVIRON identifizierte Quellen *f pl*
**idle**: **~ running** *n* CONST LAW Leerlauf *m*; **~ time** *n* ARCH & BUILD CONST LAW , INFR & DES Totzeit *f*, unproduktive Zeit *f*
**igneous**: **~ rock** *n* INFR & DES Erstarrungsgestein *nt*; **~ rocks** *n* ENVIRON, INFR & DES Ergußgestein *nt*, Vulkanit *m*
**ignition**: **~ loss** *n* ENVIRON Glühverlust *m*
**illuminated**: **~ exit sign** *n* ELECTR Ausgangstransparent *nt*
**illumination** *n* ELECTR Beleuchtung *f*
**Imhoff**: **~ cone** *n* INFR & DES, MAT PROP, WASTE WATER Absetzglas *nt*, Imhoff-Trichter *m*, Sedimentierglas *nt* nach Imhoff; **~ tank** *n* ENVIRON, INFR & DES, WASTE WATER Imhoffbrunnen *m*
**immovable** *adj* ARCH & BUILD unverrückbar
**impact** *n* ARCH & BUILD, INFR & DES Anprall *m*; **~ force** *n* ARCH & BUILD, INFR & DES Anprallkraft *f*
**impaction** *n* ENVIRON Aufschlag *m*, Impaktion *f*
***impact***: **~ load** *n* INFR & DES Anprallast *f*; **~ on groundwater** *n* ENVIRON Grundwasserbelastung *f*; **~ resistance** *n* ARCH & BUILD, CONCR, MAT PROP, STEEL Schlagfestigkeit *f*; **~-resistant** *adj* MAT PROP schlagfest; **~ of soil** *n* ENVIRON Bodenbelastung *f*; **~ sound** *n* SOUND & THERMAL Körperschall *m*; **~ sound insulation** *n* SOUND & THERMAL Körperschalldämmung *f*; **~ test** *n* MAT PROP Schlagversuch *m*, Schlagprobe *f*
**impaired** *adj* ARCH & BUILD, CONST LAW beeinträchtigt
**impediment** *n* ARCH & BUILD Behinderung *f*, Erschwernis *f*
**impeller** *n* HEAT & VENT, INFR & DES *fan* Laufrad *nt*
**impenetrable** *adj* ARCH & BUILD undurchdringbar; **~ paint coat** *n* SOUND & THERMAL, SURFACE Isolieranstrich *m*
**imperfection** *n* ARCH & BUILD, INFR & DES, MAT PROP Imperfektion *f*
**imperial**: **~ roof** *n* ARCH & BUILD welsche Haube *f*, Kaiserdach *nt*
**impermeability** *n* INFR & DES, MAT PROP Dichtheit *f*, Impermeabilität *f*, Undurchlässigkeit *f*
**impermeable** *adj* INFR & DES undurchlässig, MAT PROP undurchlässig, *container, structure* dicht; **~ soil** *n* INFR & DES, MAT PROP undurchlässiger Boden *m*; **~ to water** *adj* MAT PROP wasserundurchlässig
**impervious** *adj* INFR & DES, MAT PROP undurchlässig; **~ course** *n* SOUND & THERMAL Sperrschicht *f*
**implied**: **~ agreement** *n* CONST LAW implizierte Übereinkunft *f*
**imposed**: **~ load** *n* ARCH & BUILD, INFR & DES Nutzlast *f*
**impost** *n* ARCH & BUILD, INFR & DES Auflage *f*, Kämpfer *m*
**impregnate** *vt* SOUND & THERMAL, TIMBER imprägnieren, tränken
**impregnated** *adj* SOUND & THERMAL, TIMBER getränkt, imprägniert
**impregnating**: **~ primer** *n* SURFACE Einlaßgrund *m*
**impregnation** *n* SOUND & THERMAL, SURFACE, TIMBER Imprägnierung *f*, Tränkung *f*; **~ of wood** *n* TIMBER Holzimprägnierung *f*
**improper** *adj* ARCH & BUILD ungeeignet, untauglich, unangemessen, falsch
**improvable** *adj* ARCH & BUILD verbesserungsfähig
**improve** *vt* ARCH & BUILD verbessern, verschönern
**improvement** *n* ARCH & BUILD Verbesserung *f*, Verschönerung *f*
**impulse**: **~ turbine** *n* ENVIRON Aktionsturbine *f*, Gleichdruckturbine *f*
**imputrescible** *adj* MAT PROP, SURFACE, TIMBER fäulnisbeständig, fäulnisfest, nicht verrottbar, unverrottbar
**inch** *n* ARCH & BUILD, INFR & DES, MAT PROP, WASTE WATER Zoll *m*
**incidentals** *n pl* MAT PROP Kleinmaterial *nt*
**incinerate** *vt* ENVIRON verbrennen
**incineration** *n* ENVIRON Verbrennung *f*; **~ ash** *n* ENVIRON Verbrennungsrückstand *m*; **~ residue** *n* ENVIRON Verbrennungsrückstand *m*; **~ slag** *n* ENVIRON Verbrennungsschlacke *f*; **~ train** *n* ENVIRON thermisches Verbrennungsverfahren *nt*
**incinerator** *n* ENVIRON Abfallverbrennungsofen *m*; **~ charge** *n* ENVIRON Brenngut *nt*; **~ plant** *n* INFR & DES Verbrennungsanlage *f*
**inclination** *n* ARCH & BUILD, INFR & DES Neigung *f*, Schräge *f*; **~ of slope** *n* INFR & DES Böschungsneigung *f*
**incline** *n* ARCH & BUILD, INFR & DES Neigung *f*, Schräge *f*
**inclined** *adj* ARCH & BUILD, INFR & DES geneigt, schrägliegend; **~ bar** *n* CONCR Schrägstab *m*, aufgebogener Stab *m*; **~ plane** *n* ARCH & BUILD schiefe Ebene *f*
**incoherent** *adj* INFR & DES, MAT PROP inkohärent
**incombustible** *adj* CONST LAW, ENVIRON, INFR & DES, MAT PROP nicht brennbar, unverbrennbar
**incomplete** *adj* ARCH & BUILD *document*, CONST LAW unvollständig
**incompressible** *adj* INFR & DES, MAT PROP inkompressibel, nicht zusammendrückbar
**incorporate** *vt* ARCH & BUILD, ENVIRON, STONE einbinden, einbauen
**incorporated** *adj* ARCH & BUILD, ENVIRON, STONE eingebunden

**increase** *n* ARCH & BUILD, CONST LAW *in cost* Erhöhung *f*, Steigerung *f*, Zunahme *f*
**increaser** *n* HEAT & VENT, WASTE WATER Vergrößerungsstück *nt*
***increase***: ~ **of tension** *n* ARCH & BUILD Spannungserhöhung *f*
**increment** *n* ARCH & BUILD, INFR & DES Maßsprung *m*, Schrittweite *f*
**incremental**: ~ **heating** *n* ENVIRON zunehmende Erwärmung *f*
**indefinite** *adj* ARCH & BUILD unbestimmt
**indemnity** *n* CONST LAW Sicherstellung *f*
**indent** *vt* ARCH & BUILD, TIMBER verzahnen
**indentation** *n* ARCH & BUILD, TIMBER Einkerbung *f*, Einschnitt *m*, Verzahnung *f*
**indented**: ~ **joint** *n* TIMBER Verzahnung *f*
**indenting** *n* ARCH & BUILD, TIMBER Verzahnung *f*
**independent**: ~ **crane** *n* BUILD MACHIN freistehender Kran *m*; ~ **footing** *n* ARCH & BUILD, CONCR, INFR & DES Einzelfundament *nt*
**indeterminable** *adj* ARCH & BUILD unbestimmbar
**indeterminacy** *n* ARCH & BUILD Unbestimmtheit *f*
**indeterminate** *adj* ARCH & BUILD unbestimmt
**indetermined** *adj* ARCH & BUILD unbestimmt
**index**: ~ **bolt** *n* BUILD HARDW Rastbolzen *m*, MAT PROP Schaltbolzen *m*
**indicating**: ~ **label** *n* BUILD HARDW Hinweisschild *nt*
**indication** *n* BUILD MACHIN Anzeige *f*; ~ **error** *n* BUILD MACHIN Anzeigefehler *m*; ~ **range** *n* BUILD MACHIN Anzeigebereich *m*; ~ **of suitability** *n* INFR & DES, MAT PROP Eignungsnachweis *m*
**indirect**: ~ **artificial lighting** *n* ELECTR indirekte künstliche Beleuchtung *f*; ~ **gain** *n* HEAT & VENT indirekter Gewinn *m*; ~ **support** *n* CONCR indirekte Lagerung *f*
**individual**: ~ **footing** *n* ARCH & BUILD, CONCR, INFR & DES Einzelfundament *nt*
**indoor**: ~ **air quality** *n* HEAT & VENT Raumluftqualität *f*; ~ **installations** *n pl* ELECTR, HEAT & VENT, WASTE WATER Gebäudeinstallation *f*; ~ **swimming pool** *n* ARCH & BUILD, INFR & DES Hallenbad *nt*
**inducing**: ~ **flow** *n* ELECTR induzierender Durchfluß *m*
**induction**: ~ **pipe** *n* HEAT & VENT Ansaugrohr *nt*
**industrial**: ~ **building** *n* INFR & DES Industriebau *m*, Industriegebäude *nt*; ~ **building method** *n* ARCH & BUILD, INFR & DES industrielle Bauweise *f*; ~ **construction method** *n* ARCH & BUILD, INFR & DES industrielle Bauweise *f*; ~ **effluent** *n* ENVIRON, INFR & DES, WASTE WATER gewerbliches Abwasser *nt*; ~ **estate** *n* INFR & DES Industriegebiet *nt*; ~ **heating** *n* INFR & DES, WASTE WATER industrielle Heizungsanlage *f*
**industrialized**: ~ **construction method** *n* ARCH & BUILD, INFR & DES industrielle Bauweise *f*
**industrial**: ~ **landfill** *n* ENVIRON Industriemülldeponie *f*; ~ **plant** *n* INFR & DES Industrieanlage *f*; ~ **process** *n* ENVIRON Betriebsablauf *m*; ~ **railroad** *n AmE* (*cf industrial railway BrE*) INFR & DES Industriegleis *nt*; ~ **railway** *n BrE* (*cf industrial railroad AmE*) INFR & DES Industriegleis *nt*; ~ **sewage** *n* ENVIRON, INFR & DES, WASTE WATER gewerbliches Abwasser *nt*; ~ **sewage sludge** *n* ENVIRON, INFR & DES industrieller Klärschlamm *m*; ~ **waste** *n* CONST LAW gewerblicher Abfall *m*, ENVIRON Gewerbeabfall *m*, Industriemüll *m*, gewerblicher Abfall *m*; ~ **waste water** *n* ENVIRON, INFR & DES Industrieabwasser *nt*; ~ **water** *n* CONST LAW, ENVIRON, INFR & DES, WASTE WATER Brauchwasser *nt*
**industry** *n* ENVIRON, INFR & DES Industrie *f*
**inelastic** *adj* ARCH & BUILD, CONCR, INFR & DES, MAT PROP unelastisch; ~ **deformation** *n* ARCH & BUILD, CONCR, INFR & DES, MAT PROP plastische Verformung *f*
**inert** *adj* ENVIRON, INFR & DES inert
**inertia**: ~ **force** *n* ARCH & BUILD, ENVIRON, INFR & DES Beharrungsvermögen *nt*, Trägheitskraft *f*
**inert**: ~ **material** *n* ENVIRON inertes Material *nt*; ~ **waste** *n* ENVIRON inerter Abfall *m*
**infectious**: ~ **waste** *n* ENVIRON infektiöser Abfall *m*, pathogener Abfall *m*
**inferior**: ~ **purlin** *n* TIMBER Fußpfette *f*, Sparrenschwelle *f*, Fußschwelle *f*
**infill** *n* MAT PROP Füllmaterial *nt*, Füllstoff *m*; ~ **block** *n* MAT PROP *in ceiling* Füllkörper *m*; ~ **brickwork** *n* ARCH & BUILD, STONE Ausmauerung *f*
**infiller**: ~ **masonry** *n* ARCH & BUILD, STONE *part of framework* Ausmauerung *f*
**infilling** *n* STONE Ausfachung *f*; ~ **wall** *n* STONE Ausfachung *f*, Ausfüllung *f*
***infill***: ~ **panel** *n* MAT PROP Füllelement *nt*
**infiltration** *n* ENVIRON, INFR & DES, WASTE WATER *purification* Infiltration *f*, Versickerung *f*; ~ **basin** *n* ENVIRON Anreicherungsbecken *nt*, Sickerbecken *nt*; ~ **coefficient** *n* INFR & DES, MAT PROP Infiltrationsbeiwert *m*; ~ **water** *n* INFR & DES, MAT PROP Infiltrationswasser *nt*
**inflammable** *adj* CONST LAW, MAT PROP entflammbar, entzündbar, entzündlich
**inflected**: ~ **arch** *n* ARCH & BUILD Gegenbogen *m*
**infrared**: ~ **thermography** *n* MAT PROP infrarote Thermographie *f*
**ingot**: ~ **steel** *n* MAT PROP Flußstahl *m*
**ingrain**: ~ **wallpaper** *n* MAT PROP Rauhfasertapete *f*
**inhabitant** *n* CONST LAW Bewohner *m*, Einwohner *m*
**inherently**: ~ **safe** *adj* ELECTR eigensicher
**inherent**: ~ **strength** *n* MAT PROP Eigenfestigkeit *f*
**initial**: ~ **cost** *n* ARCH & BUILD, CONST LAW, INFR & DES Anlagekosten *pl*; ~ **pressure** *n* WASTE WATER *of pump* Anfangsdruck *m*; ~ **prestress force** *n* CONCR anfängliche Vorspannungskraft *f*; ~ **sediment** *n* INFR & DES, MAT PROP Anfangssetzung *f*; ~ **set** *n* CONCR Erstarrungsbeginn *m*; ~ **settlement** *n* MAT PROP Anfangssetzung *f*; ~ **shrinkage** *n* CONCR, INFR & DES, MAT PROP Trockenschwinden *nt*; ~ **standard** *n* ARCH & BUILD, CONST LAW, INFR & DES, MAT PROP Vornorm *f*; ~ **strength** *n* INFR & DES, MAT PROP Anfangsfestigkeit *f*; ~ **tearing resistance** *n* MAT PROP Einreißfestigkeit *f*; ~ **void ratio** *n* INFR & DES, MAT PROP Anfangsporenziffer *f*
**inject** *vt* BUILD MACHIN, CONCR, INFR & DES, STONE einpressen, verpressen
**injection** *n* BUILD MACHIN, CONCR, INFR & DES, STONE Einpressen *nt*, Injektion *f*; ~ **material** *n* INFR & DES, MAT PROP Einpreßgut *nt*; ~ **method** *n* CONCR, INFR & DES, STONE Einpreßverfahren *nt*; ~ **pressure** *n* STONE Zementierdruck *m*; ~ **pump** *n* BUILD MACHIN Einspritzpumpe *f*, Einpreßpumpe *f*, Zementierpumpe *f*
**inlet** *n* ARCH & BUILD Einlauf *m*, Eintritt *m*, HEAT & VENT Einlaß *m*, INFR & DES, WASTE WATER Einfüll-, Einlaß *m*, Zuführung *f*, Zulauf *m*; ~ **connector** *n* INFR & DES, WASTE WATER Einfüllstutzen *m*; ~ **pipe** *n* HEAT & VENT, INFR & DES, WASTE WATER Einlaßrohr *nt*,

Zuleitung *f*, Zuleitungsrohr *nt*; **~ strainer** *n* INFR & DES, WASTE WATER Einlaufsieb *nt*; **~ temperature** *n* ENVIRON Eintrittstemperatur *f*
**inner** *adj* ARCH & BUILD, BUILD HARDW, BUILD MACHIN, CONCR, ELECTR, INFR & DES, STONE, SURFACE, WASTE WATER innere; **~ corner** *n* ARCH & BUILD Inneneck *nt*; **~ leaf** *n* STONE *of cavity wall* innere Schale *f*; **~ span** *n* ARCH & BUILD Innenweite *f*
**inorganic** *adj* INFR & DES, MAT PROP anorganisch; **~ silt** *n* INFR & DES, MAT PROP anorganischer Schluff *m*; **~ soil** *n* INFR & DES, MAT PROP anorganischer Boden *m*
**inquiry** *n* CONST LAW Erkundigung *f*, Nachfrage *f*
**insect**: **~ attack** *n* TIMBER Befall *m* von Insekten; **~ screen** *n* BUILD HARDW Insektenschutzgitter *nt*
**insert 1.** *n* ARCH & BUILD Einlage *f*; **2.** *vt* ARCH & BUILD einlegen, zwischenlegen
**inserted** *adj* ARCH & BUILD eingelegt, zwischengelegt; **~ ceiling** *n* ARCH & BUILD, SOUND & THERMAL Einschubdecke *f*
**inside 1.** *adv* ARCH & BUILD innen; **2.** *n* ARCH & BUILD Innenseite *f*
***inside***: **~ cylinder** *n* BUILD HARDW Innenzylinder *m*
**in-situ**: **~ concrete** *n* ARCH & BUILD, CONCR Ortbeton *m*; **~ foam** *n* SOUND & THERMAL Ortschaum *m*
**inspect** *vt* ARCH & BUILD *document* einsehen, CONST LAW inspizieren
**inspection** *n* ARCH & BUILD *of materials* Überprüfung *f*, *examination* Prüfung *f*, *viewing* Besichtigung *f*, Kontrolle *f*, Durchsicht *f*, BUILD MACHIN, CONST LAW Beaufsichtigung *f*, Überprüfung *f*, Kontrolle *f*, Durchsicht *f*, INFR & DES Überprüfung *f*; **~ fitting** *n* INFR & DES, WASTE WATER Revisionsstück *nt*; **~ manhole** *n* INFR & DES, WASTE WATER Kontrollschacht *m*, Revisionsschacht *m*; **~ opening** *n* ARCH & BUILD, INFR & DES, WASTE WATER Kontrollöffnung *f*; **~ piece** *n* INFR & DES, WASTE WATER Revisionsstück *nt*; **~ pit** *n* ARCH & BUILD, INFR & DES *for maintenance work* Arbeitsgrube *f*; **~ sheet** *n* ARCH & BUILD, CONST LAW, INFR & DES Prüfprotokoll *nt*
**inspector** *n* ARCH & BUILD, CONST LAW, INFR & DES Bauaufseher *m*
**install** *vt* ARCH & BUILD anbringen, montieren, einbauen, CONCR anbringen, ELECTR installieren, *cable* verlegen, HEAT & VENT installieren, STEEL, TIMBER anbringen, WASTE WATER installieren
**installation** *n* ARCH & BUILD *of cable* Verlegung *f*, *fitting* Installation *f*, *assembly* Aufstellung *f*, Einbau *m*, *mounting* Montage *f*, BUILD MACHIN Einbau *m*, ELECTR *of cable* Verlegung *f*, HEAT & VENT Einbau *m*, INFR & DES Aufstellung *f*, Verlegung *f*, Installation *f*, Montage *f*, Einbau *m*, STEEL, TIMBER Einbau *m*; **~ chase** *n* STONE Installationsschlitz *m*; **~ plan** *n* ARCH & BUILD, INFR & DES Aufstellungsplan *m*; **~ work** *n* ARCH & BUILD, HEAT & VENT, INFR & DES, WASTE WATER Installationsarbeiten *f pl*, Montagearbeiten *f pl*
**installed**: **~ capacity** *n* ELECTR installierte Leistung *f*
**instantaneous**: **~ concentration** *n* ENVIRON Grundkonzentration *f*; **~ heater** *n* HEAT & VENT, WASTE WATER Durchlauferhitzer *m*; **~ water heater** *n* HEAT & VENT Warmwasserbereiter *m*
**Institute**: **~ for Building Technology** *n* CONST LAW Institut *nt* für Bautechnik (*IfBt*)
**institution** *n* INFR & DES Institution *f*
**Institution**: **~ of Civil Engineers** *n* (*ICE*) CONST LAW, INFR & DES Institut *nt* für Bauingenieure
**instruction** *n* CONST LAW Instruktion *f*, *explanation, assistance* Anleitung *f*, *order* Vorschrift *f*; **~ of staff** *n* CONST LAW Einweisung *f* von Personal
**insulance** *n* ELECTR, MAT PROP, SOUND & THERMAL, SURFACE Isolierwert *m*
**insulant** *n* ELECTR, MAT PROP Dämmstoff *m*, Isoliermaterial *nt*, *waterproofing* Sperrstoff *m*, SOUND & THERMAL, SURFACE Dämmstoff *m*
**insulate** *vt* ELECTR, MAT PROP, SOUND & THERMAL, SURFACE dämmen, isolieren
**insulated** *adj* ELECTR, MAT PROP, SOUND & THERMAL, SURFACE gedämmt, isoliert; **~ blinds** *n pl* SOUND & THERMAL isolierte Jalousien *f pl*
**insulating** *adj* ELECTR, MAT PROP, SOUND & THERMAL, SURFACE dämmend, isolierend, nichtleitend; **~ asphalt felt** *n* MAT PROP, SOUND & THERMAL Bitumensperrpappe *f*; **~ compound** *n* MAT PROP, SOUND & THERMAL Isoliermasse *f*; **~ concrete** *n* CONCR, MAT PROP Isolierbeton *m*; **~ course** *n* SOUND & THERMAL Dämmschicht *f*; **~ felt** *n* MAT PROP Sperrpappe *f*, Dämmfilz *m*; **~ glass** *n* MAT PROP, SOUND & THERMAL Isolierglas *nt*; **~ jacket** *n* MAT PROP, SOUND & THERMAL Isoliermantel *m*; **~ layer** *n* MAT PROP, SOUND & THERMAL Dämmschicht *f*, Isolierschicht *f*; **~ material** *n* ELECTR, MAT PROP, SOUND & THERMAL Dämmstoff *m*; **~ paint coat** *n* SOUND & THERMAL, SURFACE Isolieranstrich *m*; **~ slab** *n* MAT PROP, SOUND & THERMAL Dämmatte *f*, Dämmplatte *f*, Isolationsplatte *f*
**insulation** *n* ELECTR, ENVIRON, MAT PROP Isolierung *f*, SOUND & THERMAL, SURFACE Isolierung *f*, Sperrung *f*, Dämmung *f*, *waterproofing* Isolation *f*; **~ resistance** *n* ELECTR Isolierwert *m*; **~ thickness** *n* SOUND & THERMAL Isolierdicke *f*; **~ value** *n* SOUND & THERMAL Isolierwert *m*, Dämmwert *m*
**insurance**: **~ against loss or damage to works** *n pl* CONST LAW Bauschäden- und Verlustversicherung *f*
**intake** *n* HEAT & VENT, INFR & DES, WASTE WATER Einlaß *m*, Zufuhr *f*; **~ area** *n* INFR & DES *water supply, well*, WASTE WATER *sewage* Wassereinzugsgebiet *nt*; **~ grille** *n* HEAT & VENT Ansauggitter *nt*; **~ louver** *AmE*, **~ louvre** *BrE* *n* HEAT & VENT Ansauggitter *nt*; **~ screen** *n* BUILD HARDW, INFR & DES Einlaufrost *m*
**integrally**: **~ colored** *AmE*, **~ coloured** *BrE* *adj* MAT PROP durchgefärbt
**intelligent**: **~ building** *n* ARCH & BUILD, INFR & DES intelligentes Gebäude *nt*; **~ site** *n* ARCH & BUILD, INFR & DES intelligente Baustelle *f*
**intensity**: **~ of illumination** *n* ARCH & BUILD Beleuchtungsstärke *f*; **~ of noise** *n* CONST LAW, SOUND & THERMAL Lärmstärke *f*; **~ of rainfall** *n* INFR & DES Regendichte *f*
**interaction** *n* INFR & DES, MAT PROP Interaktion *f*, Wechselwirkung *f*; **~ diagram** *n* ARCH & BUILD, CONCR, INFR & DES, MAT PROP, STEEL Interaktionsdiagramm *nt*
**intercepting**: **~ drain** *n* WASTE WATER Sammelkanal *m*, Sammler *m*; **~ sewer** *n* INFR & DES, WASTE WATER Sammelkanal *m*
**interceptor**: **~ sewer** *n* ENVIRON Abwassersammler *m*, Hauptsammelkanal *m*, Hauptsammler *m*, INFR & DES, WASTE WATER Hauptsammler *m*
**interchange**: **~ of heat** *n* HEAT & VENT Wärmeaustausch *m*

**intercolumniation** *n* ARCH & BUILD Säulenabstand *m*
**intercom**: **~ system** *n* ELECTR Gegensprechanlage *f*
**intercostal**: **~ girder** *n* ARCH & BUILD Interkostalträger *m*, Sekundärträger *m*
**interference** *n* INFR & DES Interferenz *f*
**interim**: **~ payment** *n* ARCH & BUILD, CONST LAW Abschlagzahlung *f*, Zwischenzahlung *f*; **~ valuation** *n* CONST LAW Zwischenbewertung *f*
**interior**: **~ decoration** *n* ARCH & BUILD Innenarchitektur *f*, Raumgestaltung *f*; **~ design** *n* ARCH & BUILD Raumgestaltung *f*, Innenarchitektur *f*; **~ door** *n* ARCH & BUILD Innentür *f*; **~ finishing coat** *n* SURFACE Innendeckanstrich *m*; **~ furnishings** *n* BUILD HARDW Innenausstattung *f*; **~ lightning protection** *n* ELECTR innerer Blitzschutz *m*; **~ perspective** *n* ARCH & BUILD, INFR & DES Innenperspektive *f*; **~ plaster** *n* STONE Innenputz *m*; **~ ring earth** *n BrE* (*cf interior ring ground AmE*) ELECTR innerer Ringerder *m*; **~ ring ground** *n AmE* (*cf interior ring earth BrE*) ELECTR innerer Ringerder *m*; **~ span** *n* ARCH & BUILD Innenfeld *nt*, Mittelfeld *nt*, Zwischenfeld *nt*, INFR & DES Mittelfeld *nt*; **~ wall** *n* ARCH & BUILD Innenwand *f*; **~ wall plaster** *n* STONE Innenwandputz *m*; **~ wall surface** *n* ARCH & BUILD Wandinnenseite *f*; **~ work** *n* ARCH & BUILD Innenausbau *m*, Ausbauarbeiten *f pl*; **~ work ratio** *n* ARCH & BUILD Ausbauverhältnis *nt*
**interlocked** *adj* STEEL verfalzt
**interlocking**: **~ joint** *n* STEEL Verfalzung *f*; **~ paver** *n* MAT PROP, STONE Verbundstein *m*; **~ tile** *n* MAT PROP Falzziegel *m*
**intermediate**: **~ beam** *n* ARCH & BUILD Zwischenträger *m*; **~ bearing** *n* ARCH & BUILD, INFR & DES Zwischenlager *nt*; **~ ceiling** *n* ARCH & BUILD Zwischendecke *f*; **~ coat** *n* SURFACE Zwischenanstrich *m*; **~ landing** *n* ARCH & BUILD Zwischenpodest *nt*; **~ layer** *n* ARCH & BUILD Zwischenlage *f*, Zwischenschicht *f*; **~ pumping station** *n* INFR & DES Zwischenpumpwerk *nt*; **~ stiffener** *n* ARCH & BUILD, CONCR, INFR & DES, MAT PROP, STEEL *composite beam* Zwischenaussteifung *f*; **~ storage** *n* ARCH & BUILD, INFR & DES Zwischenlagerung *f*; **~ support** *n* ARCH & BUILD Zwischenstütze *f*
**internal**: **~ carcass** *n* ARCH & BUILD Innengerippe *nt*; **~ cavity wall** *n* STONE innere Schale *f*; **~ combustion rammer** *n* BUILD MACHIN Explosionsramme *f*; **~ diameter** *n* INFR & DES, WASTE WATER Innendurchmesser *m*; **~ dimension** *n* ARCH & BUILD, INFR & DES innere Abmessung *f*; **~ finishing work** *n* ARCH & BUILD Innenausbau *m*; **~ force** *n* ARCH & BUILD, INFR & DES innere Kraft *f*; **~ forces** *n pl* ARCH & BUILD, CONCR, INFR & DES, MAT PROP, STEEL, TIMBER Schnittkräfte *f pl*; **~ friction** *n* ARCH & BUILD, INFR & DES innere Reibung *f*; **~ lining** *n* ARCH & BUILD, SOUND & THERMAL innere Verkleidung *f*; **~ load** *n* ARCH & BUILD, INFR & DES innere Last *f*; **~ moment** *n* ARCH & BUILD, INFR & DES inneres Moment *nt*; **~ pressure** *n* ARCH & BUILD, INFR & DES Innendruck *m*; **~ source** *n* ENVIRON Inlandsquelle *f*; **~ stress** *n* ARCH & BUILD, INFR & DES Eigenspannung *f*, Innenspannung *f*; **~ surface** *n* SOUND & THERMAL innere Oberfläche *f*; **~ thread** *n* MAT PROP Innengewinde *nt*; **~ torsion** *n* ARCH & BUILD, INFR & DES innere Torsion *f*; **~ transverse force** *n* ARCH & BUILD, INFR & DES innere Querkraft *f*; **~ vibration** *n* CONCR Innenrüttlung *f*; **~ vibrator** *n* BUILD MACHIN Innenrüttler *m*, Tauchrüttler *m*, Flaschenrüttler *m*; **~ wall plaster** *n* STONE Innenwandputz *m*
**International**: **~ Commission on Illumination diagram** *n* (*CIE diagram*) ARCH & BUILD, INFR & DES Normfarbtafel *f*; **~ Council for Building Research Studies and Documentation** *n* (*CIB*) CONST LAW Internationaler Rat *m* für Forschung, Studium und Dokumentation des Bauwesens (*CIB*); **~ Standards Organization** *n* (*ISO*) CONST LAW Internationale Organisation *f* für Standardisierung (*ISO*)
**intersection** *n* ARCH & BUILD, ELECTR, INFR & DES *crossroads* Straßenkreuzung *f*, Knotenpunkt *m*, Schnittpunkt *m*
**interspace** *n* ARCH & BUILD, INFR & DES Zwischenraum *m*, Abstand *m*
**interstice** *n* ARCH & BUILD Lücke *f*, Zwischenraum *m*, Hohlraum *m*
**interstitial**: **~ condensation** *n* HEAT & VENT, MAT PROP interstitielle Kondensation *f*
**intertie**: **~ beam** *n* ARCH & BUILD, INFR & DES Querbalken *m*, Sturzriegel *m*
**interval** *n* ARCH & BUILD Intervall *nt*
**intrados** *n pl* ARCH & BUILD Bogenleibung *f*, Gewölbeleibung *f*, *of vault* Untersicht *f*
**intruder**: **~-proof** *adj* ARCH & BUILD, CONST LAW einbruchsicher; **~ protection** *n* BUILD HARDW, CONST LAW *lock* Aufsperrsicherung *f*
**intrusion**: **~ detection system** *n* CONST LAW, ELECTR Einbruchmeldeanlage *f*; **~ mortar** *n* CONCR, STONE Einpreßmörtel *m*
**intumescence** *n* INFR & DES, MAT PROP Anschwellen *nt*
**intumescent**: **~ strip** *n* INFR & DES, MAT PROP anschwellende Dichtung *f*
**inventory** *n* ARCH & BUILD Bestandsaufnahme *f*
**invert** *n* WASTE WATER Sohle *f*
**inverted**: **~ arch** *n* ARCH & BUILD Fundamentgewölbe *nt*, Gegengewölbe *nt*; **~ cavetto** *n* ARCH & BUILD, INFR & DES Anlauf *m*, Rampe *f*
**investigate** *vt* ARCH & BUILD, CONST LAW, INFR & DES untersuchen
**investigation** *n* ARCH & BUILD, CONST LAW, INFR & DES Untersuchung *f*; **~ of deficiencies** *n* ARCH & BUILD Bestandsaufnahme *f* der Mängel
**investment** *n* CONST LAW Anlage *f*
**invitation**: **~ to tender** *n* ARCH & BUILD, CONST LAW, INFR & DES Ausschreibung *f*
**invoice** *n* ARCH & BUILD, CONST LAW Rechnung *f*
**Ionian**: **~ order** *n* ARCH & BUILD ionische Säulenanordung *f*
**iron** *n* MAT PROP Eisen *nt*; **~ bar cutter** *n* BUILD MACHIN Betonstahlschere *f*; **~ bridge** *n* STEEL Eisenbrücke *f*; **~ cross section** *n* MAT PROP Eisenquerschnitt *m*; **~ girder** *n* ARCH & BUILD, STEEL Eisenträger *m*; **~ pipe** *n* STEEL Eisenrohr *nt*; **~ Portland cement** *n* MAT PROP Eisenportlandzement *m* (*EPZ*); **~ tie** *n* ARCH & BUILD Kopfanker *m*, Schlauder *m*
**ironware** *n* BUILD HARDW, MAT PROP Eisenbeschläge *m pl*, Kleineisenzeug *nt*
**ironwork** *n* BUILD HARDW, MAT PROP *craft, activity* Schmiedearbeit *f*, *objects* Eisenbeschläge *m pl*
**irradiance** *n* ENVIRON Bestrahlungsdichte *f*, Strahlungsintensität *f*
**irradiation** *n* ENVIRON Bestrahlung *f*
**irregular** *adj* ARCH & BUILD, INFR & DES unregelmäßig
**irreversible** *adj* ARCH & BUILD, INFR & DES irreversibel

**irrigated**: ~ **area** *n* ENVIRON, INFR & DES Bewässerungsfläche *f*

**irrigation** *n* ENVIRON, INFR & DES Bewässerung *f*, Riesel-; ~ **ditch** *n* ENVIRON, INFR & DES Bewässerungsgraben *m*; ~ **engineering** *n* ENVIRON, INFR & DES Bewässerungstechnik *f*; ~ **field** *n* ENVIRON, INFR & DES Rieselfeld *nt*; ~ **furrow** *n* ENVIRON, INFR & DES Rieselfurche *f*; ~ **line** *n* ENVIRON, INFR & DES Bewässerungsleitung *f*

**ISO** *abbr* (*International Standards Organization*) CONST LAW ISO (*Internationale Organisation für Standardisierung*)

**isolate** *vt* ELECTR isolieren, SOUND & THERMAL absperren

**isolated**: ~ **building** *n* ARCH & BUILD, INFR & DES freistehendes Gebäude *nt*; ~ **footing** *n* ARCH & BUILD, CONCR, INFR & DES Einzelfundament *nt*

**isolation** *n* ELECTR, SOUND & THERMAL Absperrung *f*, Isolierung *f*

**isometric**: ~ **projection** *n* ARCH & BUILD, INFR & DES maßgleiche Darstellung *f*; ~ **representation** *n* ARCH & BUILD, INFR & DES maßgleiche Darstellung *f*; ~ **view** *n* INFR & DES isometrische Ansicht *f*

**isosceles** *adj* INFR & DES gleichschenklig

**isostasy** *n* ARCH & BUILD, ENVIRON Isostasie *f*

**isostatic** *adj* ARCH & BUILD, ENVIRON statisch bestimmt

**isotherm** *n* HEAT & VENT, INFR & DES, MAT PROP Isotherme *f*

**isotropic** *adj* INFR & DES, MAT PROP isotrop

**item** *n* ARCH & BUILD, CONST LAW Gegenstand *m*, Einzelheit *f*

**iteration** *n* ARCH & BUILD Iteration *f*

# J

**jack** *n* BUILD MACHIN Hebevorrichtung *f*, Winde *f*, ELECTR Dose *f*; **~ arch** *n* ARCH & BUILD gerader Bogen *m*, Flachbogen *m*

**jacket 1.** *n* ARCH & BUILD, HEAT & VENT Mantel *m*, Umhüllung *f*, Ummantelung *f*; **2.** *vt* ARCH & BUILD, HEAT & VENT ummanteln, verkleiden

***jacket***: **~ pipe** *n* HEAT & VENT Mantelrohr *nt*

**jacking**: **~ anchorage** *n* CONCR, INFR & DES Spannverankerung *f*; **~ block** *n* CONCR, INFR & DES Spannblock *m*

**jackknife**: **~ drilling mast** *n* BUILD MACHIN Klappbohrmast *m*, Faltbohrmast *m*

***jack***: **~ plane** *n* BUILD MACHIN Rauhhobel *m*; **~ rafter** *n* BUILD MACHIN, TIMBER Schifter *m*, Schiftsparren *m*; **~ rod** *n* CONCR Kletterstange *f*, Gleitstange *f*

**jamb** *n* ARCH & BUILD, TIMBER *of door* Gewändepfosten *m*, Türpfosten *m*, *of window* Fenstergewände *nt*; **~ lining** *n* BUILD HARDW Türfutter *nt*, **~ stone** *n* ARCH & BUILD Eckpfeiler *m*; **~ wall** *n* ARCH & BUILD, CONCR, STONE Drempelwand *f*, Kniestock *m*

**jaw**: **~ crusher** *n* BUILD MACHIN Backenbrecher *m*, Backenquetsche *f*

**JCB®** *n BrE* BUILD MACHIN *excavator* JCB® *f*

**jerry**: **~ builder** *n* CONST LAW Pfuscher *m*; **~-built** *adj* CONST LAW gepfuscht

**jetcrete®** *n* CONCR Spritzbeton *m*

**jet**: **~ pipe nozzle** *n* BUILD MACHIN Strahlrohrmundstück *nt*

**jetty** *n* ARCH & BUILD, INFR & DES *in harbour* Hafendamm *m*, Mole *f*

**jet**: **~ valve** *n* HEAT & VENT Düsenventil *nt*

**jib** *n* BUILD MACHIN *of crane* Ausleger *m*; **~ crane** *n* ARCH & BUILD, BUILD MACHIN Auslegerkran *m*, Brückendrehkran *m*; **~ door** *n* ARCH & BUILD Geheimtür *f*; **~ position** *n* BUILD MACHIN Auslegerstellung *f*

**jigging**: **~ conveyor** *n* BUILD MACHIN Schaukelförderer *m*, Schüttelförderer *m*; **~ and vibrating screen** *n* BUILD MACHIN Schwing- und Vibrationssieb *nt*; **~ washer** *n* BUILD MACHIN Schwingwascher *m*

**jigsaw** *n* BUILD MACHIN Wippsäge *f*

**job**: **~ site** *n AmE* ARCH & BUILD (*cf building site*), CONST LAW (*cf building site*), INFR & DES (*cf site*) Bauplatz *m*, Baustelle *f*; **~ site installations** *n pl AmE* (*cf site installations BrE*) ARCH & BUILD Baustelleneinrichtung *f*; **~ site mobilization** *n AmE* (*cf site mobilization BrE*) ARCH & BUILD Baustelleneinrichtung *f*; **~ site mobilization plan** *n AmE* (*cf site mobilization plan BrE*) ARCH & BUILD Baustelleneinrichtungsplan *m*; **~ site office** *n AmE* (*cf site office BrE*) ARCH & BUILD, INFR & DES Baubüro *nt*, Baustellenbüro *nt*

**jog** *n* ARCH & BUILD Ausklinkung *f*

**joggle 1.** *n* ARCH & BUILD *beam linkage* Dübel *m*, *joint* Falz *m*, Verzahnung *f*; **2.** *vt* ARCH & BUILD kröpfen, verklammern, verzahnen

***joggle***: **~ joint** *n* STONE Steinverklammerung *f*

**join** *vt* ARCH & BUILD *add* aneinanderfügen, stoßen, zusammenfügen, *link* verbinden, HEAT & VENT, STEEL aneinanderfügen, STONE ausfugen, verfugen, TIMBER aneinanderfügen; **~ on to** *vt* ARCH & BUILD zusammenfügen

**joiner** *n* TIMBER Tischler *m*; **~'s bench** *n* BUILD MACHIN Hobelbank *f*; **~'s clamp** *n* BUILD MACHIN Schraubzwinge *f*; **~'s glue** *n* MAT PROP, TIMBER Tischlerleim *m*; **~'s work** *n* TIMBER Tischlerarbeit *f*

**joinery** *n* TIMBER Bauschreinerei *f*, Bautischlerei *f*, Schreinerei *f*, Tischlerei *f*

**joining** *n* ARCH & BUILD Zusammenfügen *nt*, STONE Ausfugen *nt*

**joint** *n* ARCH & BUILD Fuge *f*, Stoß *m*, Verbindung *f*, ELECTR Abzweigdose *f*, INFR & DES Fuge *f*, Stoß *m*, STONE Fuge *f*, TIMBER Verbindung *f*; **~ bolt** *n* BUILD HARDW Gelenkbolzen *m*; **~ box** *n* BUILD HARDW Abzweigdose *f*; **~ cement** *n* MAT PROP Fugenkitt *m*, Vergußmasse *f*; **~ concrete** *n* CONCR Fugenbeton *m*; **~ cutter** *n* BUILD MACHIN Fugenschneider *m*, Fugenmesser *nt*; **~ cutting** *n* CONCR Fugenschnitt *m*

**jointed** *adj* ARCH & BUILD gegliedert, gelenkig, verbunden

**jointer** *n* ARCH & BUILD, STONE Fugeisen *nt*

***joint***: **~ filler** *n* ARCH & BUILD, INFR & DES, MAT PROP, SOUND & THERMAL Fugeneinlage *f*

**jointing** *n* STONE Ausstreichen *nt*, Verfugen *nt*; **~ compound** *n* MAT PROP Ausfugmasse *f*, Fugenfüller *m*; **~ plane** *n* BUILD MACHIN Teilungsfläche *f*; **~ and pointing tools** *n* STONE Fugenwerkzeug *nt*

**jointless** *adj* ARCH & BUILD fugenlos

***joint***: **~ lining** *n* ARCH & BUILD, INFR & DES, MAT PROP, SOUND & THERMAL Fugeneinlage *f*; **~ mortar** *n* MAT PROP, STONE Fugenmörtel *m*; **~ packing** *n* STONE Fugendichtung *f*; **~ pattern** *n* STONE Fugeneinteilung *f*; **~ permeability coefficient** *n* INFR & DES, MAT PROP, STONE Fugendurchlaßkoeffizient *m*; **~ plaster** *n* STONE Fugengips *m*; **~ plate** *n* STEEL Anschlußblech *nt*; **~ raking** *n* STONE Fugenreißen *nt*; **~ sealer** *n* MAT PROP Fugenmasse *f*; **~ sealing** *n* STONE Fugendichtung *f*; **~ sealing compound** *n* MAT PROP, SURFACE Fugenvergußmasse *f*; **~ sealing strip** *n* ARCH & BUILD, INFR & DES, MAT PROP, SOUND & THERMAL Fugeneinlage *f*; **~ sewage treatment plant** *n* ENVIRON, WASTE WATER Sammelkläranlage *f*; **~ slip** *n* TIMBER Verbindungsschlupf *m*; **~ tape** *n* MAT PROP Fugenband *nt*; **~ venture** *n* CONST LAW Arbeitsgemeinschaft *f*

**joist** *n* ARCH & BUILD, CONCR, STEEL Balken *m*, TIMBER *binding joint* Unterzug *m*, *for ceiling* Balken *m*, Deckenbalken *m*, *beam, girder* Träger *m*, *intermediate girder* Zwischenträger *m*; **~ ceiling** *n* ARCH & BUILD, TIMBER Balkendecke *f*, Balkenträgerdecke *f*; **~ end** *n* ARCH & BUILD, TIMBER Balkenkopf *m*; **~ hanger** *n* STONE, TIMBER Balkenschuh *m*

**jolted**: **~ concrete** *n* CONCR Rüttelbeton *m*

**judas** *n* BUILD HARDW Türgucker *m*; **~ hole** *n* BUILD HARDW Türgucker *m*

**jumbo** *n* BUILD MACHIN fahrbares Gerüst *nt*; **~ brick** *n* MAT PROP großformatiger Ziegelstein *m*

**jumper** *n* BUILD MACHIN Bohrmeißel *m*; **~ cable** *n* ELECTR Starthilfekabel *nt*

**junction** *n* ARCH & BUILD Verbindung *f*, Knotenpunkt *m*, *in road and rail communications* Verzweigung *f*, *in supply lines* Anschluß *m*, INFR & DES *crossroads* Straßenkreuzung *f*, Knotenpunkt *m*, TIMBER Verbindung *f*; ~ **box** *n* ELECTR Verteilerkasten *m*; ~ **line** *n* ELECTR Anschlußleitung *f*; ~ **plate** *n* TIMBER Knotenblech *nt*; ~ **walls** *n pl* STONE Verbindungsmauern *f pl*

**junk** *n* ARCH & BUILD, CONST LAW, ENVIRON, INFR & DES Abfall *m*, Abfallstoff *m*; ~ **iron** *n* STEEL Eisenschrott *m*

**jut** *n* ARCH & BUILD Vorsprung *m*

**jutting**: ~ **piece** *n* ARCH & BUILD vorkragendes Teil *nt*, überstehendes Geschoß *nt*

# K

**kcal** *abbr* (*kilocalorie*) HEAT & VENT kcal (*Kilokalorie*)
**keel**: **~ arch** *n* ARCH & BUILD Kielbogen *m*, persischer Bogen *m*
**keen** *adj* BUILD MACHIN *edge* scharf
**Keene**: **~'s cement**(R) *n* MAT PROP Marmorgips *m*
**keenness** *n* BUILD MACHIN *of cutting edge of tool* Schärfe *f*
**keep** *vt* ARCH & BUILD aufbewahren, lagern
**keeper** *n* ARCH & BUILD *in lock* Schließblech *nt*, Verriegelung *f*
**kerb** *n BrE* (*cf curb AmE*) INFR & DES *along street* Bordeinfassung *f*, Bordkante *f*, Bordstein *m*, MAT PROP Bordstein *m*
**kerbstone** *n BrE* (*cf curbstone AmE*) INFR & DES, MAT PROP *along street, drive, road* Bordstein *m*, Randstein *m*, Rinnstein *m*
**key 1.** *n* BUILD HARDW Schlüssel *m*; **2.** *vt* BUILD MACHIN, STONE verzahnen, verkeilen, TIMBER verdübeln
**keybit** *n* BUILD HARDW Schlüsselbart *m*
***key***: **~ drop** *n* BUILD HARDW Schlüssellochabdeckung *f*
**keyed**: **~ pointing** *n* STONE Hohlkehlenverfugung *f*
**keyhole** *n* BUILD HARDW Schlüsselloch *nt*; **~ plate** *n* BUILD HARDW Schlüsselschild *nt*; **~ saw** *n* BUILD MACHIN Stichsäge *f*
**keying** *n* BUILD MACHIN, STONE Verkeilung *f*, Verzahnung *f*
**keyline** *n* ARCH & BUILD Schlüssellinie *f*
***key***: **~-operated switch** *n* ELECTR Schlüsselschalter *m*; **~ plan** *n* ARCH & BUILD, INFR & DES Übersichtsplan *m*; **~ plate** *n* BUILD HARDW Schlüsselschild *nt*
**keystone** *n* STONE *of arch, vault* Schlußstein *m*
**keyway** *n* ARCH & BUILD schlitzartige Aussparung *f*, BUILD HARDW *in lock* Schlüsselkanal *m*, BUILD MACHIN Keilnut *f*, STONE Mauerschlitz *m*
**kg** *abbr* (*kilogram, kilogramme*) ARCH & BUILD, INFR & DES, MAT PROP kg (*Kilogramm*)
**K**: **~ glass** *n* MAT PROP K-Glas *nt*
**kick**: **~ plate** *n* BUILD HARDW *of door* Stoßblech *nt*
**kilo** *n* (*kilogram, kilogramme*) ARCH & BUILD , INFR & DES, MAT PROP Kilo *nt* (*Kilogramm*)
**kilocalorie** *n* (*kcal*) HEAT & VENT Kilokalorie *f* (*kcal*)
**kilogram** *n* (*kg, kilo*)ARCH & BUILD, INFR & DES, MAT PROP Kilogramm *nt* (*Kilo, kg*)
**kilogramme** (*see kilogram*)
***kilo***: **~-newton** *n* (*kN*) ARCH & BUILD, INFR & DES, MAT PROP Kilonewton *nt* (*kN*)
**kilowatt** *n* (*kW*) ELECTR, HEAT & VENT Kilowatt *nt* (*kW*); **~-hour** *n* (*kWh*) ELECTR, HEAT & VENT Kilowattstunde *f* (*kWh*)
**kinetic**: **~ energy** *n* INFR & DES kinetische Energie *f*; **~ friction** *n* INFR & DES kinetische Reibung *f*
**kingbolt** *n* STEEL Hängesäule *f*
**king**: **~ closer** *n* ARCH & BUILD, STONE Dreiviertelquartier *nt*, Dreiviertelziegel *m*; **~ post** *n* ARCH & BUILD Hängesäule *f*; **~ post truss** *n* ARCH & BUILD, TIMBER einfaches Hängewerk *nt*, *unbraced* einstieliges strebenloses Pfettendach *nt*
**kiss**: **~ marks on bricks** *n pl* MAT PROP, STONE Ziegelbrennmuster *nt*
**kN** *abbr* (*kilo-newton*) ARCH & BUILD, INFR & DES, MAT PROP kN (*Kilonewton*)
**knead** *vt* INFR & DES, MAT PROP kneten
**kneadable** *adj* INFR & DES, MAT PROP knetbar
**kneading**: **~ test** *n* INFR & DES, MAT PROP Knettest *m*
**knee** *n* WASTE WATER Knie *nt*, Kniestück *nt*; **~ brace** *n* BUILD HARDW Winkellasche *f*, TIMBER Kopfband *nt*; **~ bracket plate** *n* BUILD HARDW Knotenpunktverbindung *f*
**knife**: **~-cut veneer** *n* TIMBER Messerfurnier *nt*
**knifing**: **~ filler** *n* MAT PROP Spachtelmasse *f*
**knitting**: **~ layer** *n* CONCR Verbundschicht *f*, Bindeschicht *f*
**knoll** *n* INFR & DES Kuppe *f*
**knot** *n* TIMBER Ast *m*; **~ borer** *n* BUILD MACHIN, TIMBER Astbohrer *m*; **~ hole** *n* TIMBER Astloch *nt*
**knotless** *adj* TIMBER astfrei
**knuckle** *n* BUILD HARDW Scharniergelenk *nt*
**k-value** *n* INFR & DES, MAT PROP k-Wert *m*
**kW** *abbr* (*kilowatt*) ELECTR, HEAT & VENT kW (*Kilowatt*)
**kWh** *abbr* (*kilowatt-hour*) ELECTR, HEAT & VENT kWh (*Kilowattstunde*)

# L

**label** *n* BUILD HARDW Bezeichnungsschild *nt*, Schild *nt*, Markierung *f*
**laboratory** *n* ARCH & BUILD, ELECTR, HEAT & VENT, INFR & DES, WASTE WATER Labor *nt*; ~ **investigation** *n* INFR & DES, MAT PROP Laboruntersuchung *f*; ~ **result** *n* INFR & DES, MAT PROP Laborresultat *nt*; ~ **test** *n* INFR & DES, MAT PROP Labortest *m*
**labor**: *AmE see labour BrE*
**labour**: ~ **charges** *n pl BrE* ARCH & BUILD, CONST LAW, INFR & DES Arbeitskosten *pl*; ~ **costs** *n pl BrE* ARCH & BUILD, CONST LAW, INFR & DES Arbeitskosten *pl*
**lacing** *n* CONCR Bewehrungsverteilung *f*, STEEL Vergitterung *f*
**lacquer** *n* MAT PROP, SURFACE Lack *m*, *varnish, stain* Lackfarbe *f*
**ladder** *n* BUILD MACHIN Leiter *f*; ~ **scaffolding** *n* BUILD MACHIN Leitergerüst *nt*
**ladle** *n* BUILD MACHIN Gießlöffel *m*
**lagging** *n* HEAT & VENT Wärmedämmung *f*
**lag**: ~ **screw** *n* BUILD HARDW Ankerbolzen *m*, Ankerschraube *f*, Fundamentschraube *f*
**laid**: ~**-dry** *adj* STONE *masonry* trockengemauert; ~ **to falls** *adj* SOUND & THERMAL im Gefälle verlegt
**laitance** *n* CONCR, SURFACE Zementschleier *m*
**lamb**: ~**'s wool roll** *n* SURFACE Lammfellrolle *f*
**laminar** *adj* INFR & DES, MAT PROP laminar; ~ **flow** *n* HEAT & VENT, INFR & DES, MAT PROP Laminarströmung *f*
**laminate** *vt* ARCH & BUILD beschichten, INFR & DES, MAT PROP kaschieren, SURFACE *material* beschichten
**laminated** *adj* ARCH & BUILD, INFR & DES, MAT PROP beschichtet, mehrschichtig; ~ **board** *n* MAT PROP Mehrschichtenplatte *f*; ~ **glass** *n* MAT PROP Verbundglas *nt*; ~ **insulating glass** *n* MAT PROP Mehrscheibenisolierglas *nt*; ~ **plastic board** *n* MAT PROP Schichtpreßstoffplatte *f*; ~ **safety glass** *n* CONST LAW, MAT PROP Sicherheitsverbundglas *nt*, Verbundsicherheitsglas *nt* (*VS-Glas*); ~ **soil** *n* INFR & DES, MAT PROP laminierter Boden *m*; ~ **timber** *n* TIMBER Schichtpreßholz *nt*; ~ **wired glass** *n* MAT PROP Drahtverbundsicherheitsglas *nt*, Stahlfadenverbundglas *nt*; ~ **wood** *n* TIMBER Schichtpreßholz *nt*, lamelliertes Holz *nt*
**laminating** *n* ARCH & BUILD, INFR & DES, MAT PROP Kaschieren *nt*
**lamp** *n* ELECTR Lampe *f*, Leuchtmittel *nt*
**lancet**: ~ **arch** *n* ARCH & BUILD Spitzbogen *m*, gotischer Bogen *m*
**land** *n* CONST LAW, ENVIRON, INFR & DES Land *nt*; ~ **boundary** *n* ARCH & BUILD, CONST LAW, INFR & DES Grundstücksgrenze *f*; ~ **burial** *n* ENVIRON *of waste, refuse* Vergraben *nt*; ~ **clearing** *n* CONST LAW, INFR & DES Grundstücksräumung *f*; ~ **consolidation** *n* CONST LAW, INFR & DES *of farmland* Flurbereinigung *f*; ~ **degradation** *n* ENVIRON, INFR & DES Devastierung *f*; ~ **development** *n* CONST LAW, INFR & DES Erschließung *f* von Baugelände; ~ **development fees** *n pl* CONST LAW Erschließungskosten *pl*; ~ **disturbance** *n* ENVIRON, INFR & DES Devastierung *f*; ~ **drainage** *n* CONST LAW, INFR & DES Landentwässerung *f*
**landfill** *n* ENVIRON, INFR & DES *process* Landauffüllung *f*, *site* Deponie *f*, Mülldeponie *f*; ~ **cell** *n* ENVIRON Kassette *f*, Polder *m*, INFR & DES Polder *m*; ~ **compactor** *n* ENVIRON Kompaktor *m*, Müllverdichter *m*, Verdichtungsanlage *f*, INFR & DES Kompaktor *m*; ~ **design** *n* ENVIRON, INFR & DES Deponietyp *m*
**landfilling** *n* ENVIRON geordnetes Ablagern *nt*
***landfill***: ~ **site** *n* ENVIRON, INFR & DES Deponiestandort *m*; ~ **type** *n* ENVIRON, INFR & DES Deponietyp *m*
**landholder** *n* CONST LAW Grundeigentümer *m*
**landing** *n* ARCH & BUILD Treppenabsatz *m*, Treppenpodest *nt*; ~ **area** *n* INFR & DES Landeplatz *m*; ~ **step** *n* ARCH & BUILD Podeststufe *f*; ~ **trimmer** *n* BUILD MACHIN Podestwechselbalken *m*
***land***: ~ **leveling** *AmE*, ~ **levelling** *BrE n* CONST LAW, INFR & DES Geländeplanierung *f*; ~ **line** *n* ELECTR, INFR & DES Überlandleitung *f*
**landlord** *n* CONST LAW *house owner* Hausbesitzer *m*, Hauseigentümer *m*
**landmark** *n* CONST LAW, INFR & DES Grenzstein *m*
***land***: ~ **measuring** *n* INFR & DES Feldmessung *f*; ~ **measuring chain** *n* INFR & DES Feldmeßkette *f*; ~ **owner** *n* CONST LAW Grundbesitzer *m*, Grundeigentümer *m*; ~ **ownership** *n* CONST LAW Grundbesitz *m*, Grundeigentum *nt*; ~ **pollutant** *n* ENVIRON bodenverunreinigender Stoff *m*; ~ **pollution** *n* ENVIRON Bodenverunreinigung *f*; ~ **purchase** *n* CONST LAW Grunderwerb *m*; ~ **reclamation** *n* CONST LAW, INFR & DES Landwiedergewinnung *f*; ~ **reform** *n* CONST LAW, INFR & DES Bodenreform *f*; ~ **register** *n BrE* (*cf register of real estates AmE*) CONST LAW Grundbuch *nt*, Kataster *m*; ~ **registry** *n BrE* CONST LAW Grundbuchamt *nt*; ~ **registry entry** *n BrE* CONST LAW Grundbucheintragung *f*; ~ **registry search** *n BrE* CONST LAW Grundbuchüberprüfung *f*; ~ **restoration** *n* CONST LAW, INFR & DES Rekultivieren *nt*
**landscape** *n* INFR & DES Landschaft *f*; ~ **architecture** *n* INFR & DES Landschaftsgärtnerei *f*; ~ **gardener** *n* INFR & DES Landschaftsgärtner *m*, Landschaftsgärtnerin *f*
**landscaper** *n* INFR & DES Landschaftsgärtner *m*, Landschaftsgärtnerin *f*
**landscaping** *n* INFR & DES landschaftsgärtnerische Arbeiten *f pl*; ~ **work** *n* INFR & DES landschaftsgärtnerische Arbeiten *f pl*
***land***: ~ **survey** *n* CONST LAW, INFR & DES topographische Landaufnahme *f*; ~ **surveying** *n* INFR & DES Landvermessung *f*, Terrainaufnahme *f*; ~ **uplift** *n* INFR & DES Bodenhebung *f*
**lane** *n* INFR & DES *of road* Fahrspur *f*
**lantern** *n* ARCH & BUILD Laterne *f*; ~ **light** *n* ARCH & BUILD Laterne *f*, *on roof* Dachlaterne *f*; ~ **tower** *n* ARCH & BUILD Dachaufsatz *m*
**lap** *n* ARCH & BUILD Falz *m*, Überlappung *f*; ~ **joint** *n*

INFR & DES, STEEL, TIMBER Überlappungsfuge *f*, Überlappungsstoß *m*

**lapped**: ~ **scarf** *n* ARCH & BUILD überlappte Verbindung *f*

**laptop**: ~ **computer** *n* ARCH & BUILD Tischrechner *m*

**larder** *n* ARCH & BUILD Speisekammer *f*

**large** *adj* MAT PROP, STONE groß; ~ **block** *n* STONE Großblock *m*; ~ **format** *n* STONE *of brick* Großformat *nt*, Dickformat *nt*; ~ **panel** *n* MAT PROP Großplatte *f*; ~-**panel construction method** *n* ARCH & BUILD Großplattenbauweise *f*; ~ **sett** *n* STONE Großpflaster *nt*

**Larssen**: ~ **sheet pile** *n* INFR & DES Larssenbohle *f*

**laser** *n* BUILD MACHIN Lasergerät *nt*

**lasting** *adj* ARCH & BUILD, INFR & DES dauerhaft

**latch**: ~ **bolt** *n* BUILD HARDW *of lock* Falle *f*; ~ **bolt stop** *n* BUILD HARDW *of lock* Fallenfeststeller *m*; ~ **lever** *n* BUILD HARDW *of lock* Wechsel *m*; ~ **lock** *n* BUILD HARDW Schnappschloß *nt*; ~ **pin** *n* BUILD HARDW Raststift *m*

**latent** *adj* INFR & DES, MAT PROP latent; ~ **hydraulic binders** *n pl* HEAT & VENT latente hydraulische Bindemittel *nt pl*

**lateral** *adj* ARCH & BUILD, CONCR, INFR & DES, MAT PROP, STEEL quer, seitlich; ~ **bracing** *n* ARCH & BUILD Windverband *m*; ~ **compression** *n* INFR & DES, MAT PROP Querzusammendrückung *f*; ~ **expansion** *n* MAT PROP Querdehnung *f*; ~ **extension** *n* MAT PROP Querdehnung *f*; ~ **force** *n* ARCH & BUILD, CONCR, INFR & DES, MAT PROP, STEEL Querkraft *f*; ~ **force resistance** *n* ARCH & BUILD, CONCR, INFR & DES, MAT PROP, STEEL Querkrafttragfähigkeit *f*; ~ **friction** *n* INFR & DES, MAT PROP Querreibung *f*; ~ **inclination** *n* ARCH & BUILD Querneigung *f*; ~ **load** *n* ARCH & BUILD, INFR & DES Seitenlast *f*

**laterally**: ~ **restrained** *adj* ARCH & BUILD, CONCR, INFR & DES, STEEL seitlich festgehalten; ~ **stable** *adj* ARCH & BUILD, CONCR, INFR & DES, MAT PROP, STEEL seitlich stabil

**lateral**: ~ **movement** *n* ARCH & BUILD, INFR & DES, TIMBER Seitenbewegung *f*; ~ **pressure** *n* ARCH & BUILD, INFR & DES, MAT PROP Seitendruck *m*; ~ **reinforcing structure** *n* ARCH & BUILD, CONCR, INFR & DES, MAT PROP, STEEL Seitenversteifung *f*; ~ **restraint** *n* ARCH & BUILD, CONCR, INFR & DES, MAT PROP seitliche Halterung *f*; ~ **stability** *n* ARCH & BUILD, CONCR, INFR & DES, MAT PROP, STEEL seitliche Stabilität *f*; ~ **strain** *n* CONCR, INFR & DES, MAT PROP, STEEL Querverformung *f*; ~ **support** *n* ARCH & BUILD, INFR & DES seitliche Abstützung *f*; ~-**torsional buckling** *n* ARCH & BUILD, INFR & DES, MAT PROP, STEEL seitliches Torsionsausknicken *nt*

**latex**: ~ **paint** *n* SURFACE Latexfarbe *f*

**lath** *n* TIMBER Latte *f*

**lathe** *n* BUILD MACHIN Drehbank *f*

**lathing** *n* TIMBER Verlattung *f*, Holzverlattung *f*

***lath***: ~ **nail** *n* BUILD HARDW Lattenstift *m*; ~ **partition** *n* TIMBER Lattenverschlag *m*

**lathwood** *n* TIMBER Lattenholz *nt*

**lathwork** *n* STONE Putzträger *m*

**latitude** *n* HEAT & VENT Breite *f*

**lattice** *n* ARCH & BUILD, INFR & DES, MAT PROP, TIMBER Fachwerk *nt*, Gitter *nt*; ~ **beam** *n* TIMBER Gitterträger *m*; ~ **bridge** *n* TIMBER Gitterbrücke *f*

**latticed**: ~ **column** *n* TIMBER Gitterstütze *f*

***lattice***: ~ **fence** *n* TIMBER Lattenzaun *m*; ~ **girder** *n* ARCH & BUILD, TIMBER Fachwerkträger *m*, Gitterträger *m*; ~ **shell** *n* TIMBER Gitterschale *f*; ~ **tower** *n* TIMBER Stahlgittermast *m*; ~ **truss** *n* ARCH & BUILD, INFR & DES, TIMBER Gitterbalken *m*, Kreuzfachwerkbinder *m*

**latticework** *n* TIMBER Fachwerk *nt*, Gitterwerk *nt*

**launder** *n* ARCH & BUILD Gerinne *nt*

**lava** *n* INFR & DES Lava *f*

**law**: ~ **of the contract** *n* CONST LAW Vertragsgesetz *nt*

**lay** *vt* ARCH & BUILD legen, ELECTR, INFR & DES *cable* verlegen, TIMBER *parquet* legen; ~ **bare** *vt* ARCH & BUILD *cable, pipe* freilegen; ~ **bricks** *vt* ARCH & BUILD, STONE mauern; ~ **in** *vt* ARCH & BUILD *sealing tape* einlegen

**lay**: ~-**by** *n* INFR & DES Haltespur *f*, Parkstreifen *m*

**layer** *n* INFR & DES Lage *f*, SURFACE Schicht *f*; ~-**construction** *n* INFR & DES Lagenschüttung *f*; ~ **depth** *n* INFR & DES Schichttiefe *f*; ~ **thickness** *n* INFR & DES Schichtdicke *f*

**laying** *n* ARCH & BUILD, ELECTR, INFR & DES Verlegung *f*; ~ **of foundation stone** *n* CONST LAW Grundsteinlegung *f*

**lay**: ~-**on grating** *n* STEEL aufgelegter Gitterrost *m*

**layout** *n* ARCH & BUILD Anordnung *f*, Auslegung *f*, Plan *m*

**LDPE** *abbr* (*low-density polyethylene*) MAT PROP PE weich (*Polyethylen niedriger Dichte*)

**leach** *vt* ENVIRON, INFR & DES, MAT PROP auslaugen

**leachate**: ~ **detection layer** *n* ENVIRON, INFR & DES, MAT PROP Auslaugkontrollschicht *f*

**leached** *adj* ENVIRON, INFR & DES, MAT PROP ausgelaugt

**leaching** *n* ENVIRON, INFR & DES, MAT PROP Auslaugen *nt*; ~ **property** *n* ENVIRON Auslaugverfahren *nt*; ~ **test** *n* ENVIRON Auslaugtest *m*, Elutionsversuch *m*

**lead 1.** *n* MAT PROP, STEEL Blei *nt*; **2.** *vt* STEEL ausloten

***lead***: ~ **additives** *n pl* ENVIRON Bleizusatzstoffe *m pl*; ~-**containing pigments** *n pl* MAT PROP, SURFACE bleihaltige Pigmente *nt pl*; ~ **filler** *n* MAT PROP Bleieinlage *f*; ~ **glass** *n* MAT PROP Bleiglas *nt*; ~ **glazing** *n* MAT PROP Bleiverglasung *f*; ~-**in** *n* ELECTR Einführung *f*

**leading**: ~ **abutment pressure** *n* ARCH & BUILD voreilender Kämpferdruck *m*

***lead***: ~ **joint** *n* MAT PROP Bleidichtung *f*; ~ **packing** *n* MAT PROP Bleidichtung *f*; ~ **saddle** *n* STEEL Bleiauflager *nt*; ~ **sheet** *n* STEEL Bleiblech *nt*; ~ **wedge** *n* MAT PROP, Bleikeil *m*; ~ **welding** *n* MAT PROP Bleischweißen *nt*

**leaf** *n* ARCH & BUILD Blatt *nt*, Flügel *m*

**leafage** *n* ARCH & BUILD Blattwerk *nt*

***leaf***: ~ **door** *n* BUILD HARDW Flügeltür *f*; ~ **trap** *n* ARCH & BUILD Laubfang *m*

**leak 1.** *n* ENVIRON, HEAT & VENT, INFR & DES, WASTE WATER Leck *nt*; **2.** *vi* ENVIRON, HEAT & VENT, INFR & DES, WASTE WATER auslaufen, durchsickern; ~ **out** ENVIRON, HEAT & VENT, INFR & DES, WASTE WATER ausfließen

**leakage** *n* ELECTR *of current* Leckstrom *m*, ENVIRON Leck *nt*, Leckage *f*, HEAT & VENT, INFR & DES, WASTE WATER Undichtigkeit *f*; ~ **current** *n* ELECTR Leckstrom *m*; ~ **detector** *n* WASTE WATER Abhorchgerät *nt*; ~ **loss** *n* INFR & DES Leckverlust *m*; ~ **water** *n* ENVIRON, INFR & DES, WASTE WATER Leckwasser *nt*, Sickerwasser *nt*

***leak***: ~ **detection** *n* ENVIRON Leckbestimmung *f*;

~ **detector** *n* HEAT & VENT, WASTE WATER Leckanzeigegerät *nt*, Lecksucher *m*
**leakproof** *adj* ARCH & BUILD, ENVIRON, HEAT & VENT, INFR & DES, WASTE WATER auslaufsicher, lecksicher
**lean 1.** *adj* CONCR, MAT PROP mager; **2.** *vi* INFR & DES, MAT PROP sich stützen, WASTE WATER sich neigen
***lean***: ~ **clay** *n* INFR & DES, MAT PROP magerer Ton *m*; ~ **concrete** *n* CONCR, MAT PROP Magerbeton *m*
**leaning** *adj* ARCH & BUILD schief
***lean***: ~ **mix** *n* CONCR, MAT PROP Magermischung *f*; **~-to** *n* ARCH & BUILD Pultanbau *m*; **~-to roof** *n* ARCH & BUILD, TIMBER Pultdach *nt*
**lease** *n* CONST LAW Erbbaurecht *nt*
**leather** *n* MAT PROP Leder *nt*
**leathering** *n* MAT PROP Ledermanschette *f*
**leave**: ~ **open** *vt* ARCH & BUILD, TIMBER aussparen
**ledge** *n* ARCH & BUILD Sims *m*, Absatz *m*
**ledger** *n* TIMBER Riegel *m*, Querholz *nt*, Sperrleiste *f*, *scaffold* Längsstange *f*
**left**: **~-hand door closer** *n* BUILD HARDW linkswirkender Türschließer *m*
**legally**: ~ **valid** *adj* CONST LAW rechtskräftig
**legionnaire**: **~'s disease** *n* ENVIRON, HEAT & VENT Legionärskrankheit *f*
**legislation** *n* CONST LAW Gesetzgebung *f*
**leisure**: ~ **center** *AmE*, ~ **centre** *BrE* *n* INFR & DES Freizeitzentrum *nt*; ~ **facilities** *n pl* INFR & DES Freizeitanlagen *f pl*; ~ **time** *n* INFR & DES Freizeit *f*
**length** *n* ARCH & BUILD Länge *f*; ~ **of channel** *n* ENVIRON, INFR & DES, WASTE WATER Kanallänge *f*; ~ **of piston stroke** *n* BUILD MACHIN Kolbenhublänge *f*; ~ **of step** *n* ARCH & BUILD Stufenbreite *f*
**lessee** *n* CONST LAW Erbbauberechtigter *m*
**lessor** *n* CONST LAW Grundbesitzer *m*, Grundeigentümer *m*
**lethal**: ~ **concentration** *n* ENVIRON tödliche Konzentration *f*; ~ **effect** *n* ENVIRON tödliche Wirkung *f*
**letter** *vt* ARCH & BUILD beschriften
**lettering** *n* ARCH & BUILD Beschriftung *f*
**letter**: ~ **of intent** *n* CONST LAW Vertragsabsichtserklärung *f*
**level 1.** *adj* ARCH & BUILD Flach-, eben, flach, bündig, INFR & DES flach, MAT PROP eben; **2.** *n* ARCH & BUILD Höhe *f*, Kote *f*, *instrument* Richtwaage *f*, Niveau *nt*, Ebene *f*, Richtscheit *m*, Richtlatte *f*, BUILD MACHIN Richtwaage *f*, INFR & DES Kote *f*, Ebene *f*; **3.** *vt* ARCH & BUILD ausgleichen, INFR & DES planieren, einebnen; ~ **out** INFR & DES *soil* einebnen
**level**: ~ **with the floor** *phr* ARCH & BUILD fußbodenbündig
**leveling** *n AmE*, **levelling** *n BrE* ARCH & BUILD, INFR & DES Einebnen *nt*, Höhenaufnahme *f*, Nivellierung *f*, Nivellieren *nt*; ~ **alidade** *n BrE* BUILD MACHIN, INFR & DES *surveying* Kippregel *f*; ~ **compass** *n BrE* BUILD MACHIN Kompaß *m*; ~ **concrete** *n BrE* CONCR Ausgleichbeton *m*; ~ **course** *n BrE* INFR & DES Ausgleichsschicht *f*; ~ **instrument** *n BrE* ARCH & BUILD, BUILD MACHIN Nivellierinstrument *nt*; ~ **layer** *n BrE* INFR & DES Ausgleichsschicht *f*; ~ **machine** *n BrE* BUILD MACHIN Planiermaschine *f*; ~ **point** *n BrE* ARCH & BUILD, BUILD MACHIN Nivellierpunkt *m*; ~ **pole** *n BrE* BUILD HARDW, BUILD MACHIN, INFR & DES Nivellierlatte *f*; ~ **rod** *n BrE* INFR & DES *surveying* Nivellierkreuz *nt*, Planierstange *f*; ~ **staff** *n BrE* BUILD HARDW, BUILD MACHIN, INFR & DES *surveying* Nivellierlatte *f*
**level**: ~ **with the pavement** *phr* ARCH & BUILD belagsbündig
**lever** *n* ARCH & BUILD Hebel *m*; ~ **arm** *n* ARCH & BUILD, BUILD MACHIN, INFR & DES Hebelarm *m*; ~ **handle** *n* BUILD HARDW Drücker *m*, Klinke *f*; ~ **spring** *n* BUILD HARDW *of lock* Stiftfeder *f*
**lewis** *n* BUILD MACHIN Keilklaue *f*, Steinwolf *m*
**liability** *n* CONST LAW Haftpflicht *f*, Verantwortlichkeit *f*, Verantwortung *f*
**liable** *adj* CONST LAW verantwortlich
**lie** *n* INFR & DES *of land* Lage *f*
**life**: ~ **cycle costing** *n* CONST LAW Lebenszykluskostenberechnung *f*
**lift 1.** *n* (*cf elevator AmE*) ARCH & BUILD Aufzug *m*, Auftrieb *m*, BUILD HARDW Aufzug *m*, BUILD MACHIN Hub *m*, Hubhöhe *f*, ELECTR Aufzug *m*, ENVIRON Auftrieb *m*, INFR & DES Hubhöhe *f*, Auftrieb *m*, Hub *m*, MAT PROP Auftrieb *m*, WASTE WATER *of pump* Hubhöhe *f*, Hub *m*; **2.** *vt* ARCH & BUILD, BUILD MACHIN, INFR & DES, WASTE WATER ausheben, heben
***lift***: ~ **bridge** *n* INFR & DES Hebebrücke *f*; ~ **coefficient** *n* (*CL*) ENVIRON Auftriebsbeiwert *m* (*CL*), Auftriebszahl *f* (*CL*); ~ **door** *n BrE* (*cf elevator door AmE*) ARCH & BUILD, BUILD HARDW Aufzugstür *f*
**lifter** *n* BUILD MACHIN Heber *m*, Wuchtbaum *m*
***lift***: ~ **gate** *n* ARCH & BUILD Hubtor *nt*
**lifting** *n* ARCH & BUILD , BUILD MACHIN Anheben, *nt* Heben *nt*, Hochheben *nt*, INFR & DES, WASTE WATER Heben *nt*; ~ **bridge** *n* INFR & DES Hubbrücke *f*; ~ **chain** *n* BUILD MACHIN Hubkette *f*; ~ **device** *n* BUILD MACHIN Hebevorrichtung *f*; ~ **door** *n* ARCH & BUILD Hebetür *f*; ~ **equipment** *n* BUILD MACHIN Hebezeug *nt*; ~ **force** *n* BUILD MACHIN, INFR & DES Hebekraft *f*; ~ **height** *n* BUILD MACHIN, INFR & DES Hubhöhe *f*; ~ **key** *n* ARCH & BUILD, WASTE WATER *of manhole cover* Aushebeschlüssel *m*; **~-off** *n* ARCH & BUILD, INFR & DES Abheben *nt*; ~ **press** *n* BUILD MACHIN Hubpresse *f*
***lift***: **~-off** *n* ARCH & BUILD Ausheben *nt*; **~-off guard** *n* ARCH & BUILD, WASTE WATER *of manhole cover* Aushebesicherung *f*; ~ **pipe** *n* HEAT & VENT Hubrohr *nt*; ~ **slab** *n* ARCH & BUILD Hubplatte *f*; **~-slab method** *n* ARCH & BUILD Hubplattenverfahren *nt*; **~-to-drag ratio** *n* ENVIRON Verhältnis *nt* zwischen Förderhöhe und Widerstand
**light 1.** *adj* ARCH & BUILD, MAT PROP *in colour* hell, *in weight* leicht; **2.** *n* ELECTR Licht *nt*
***light***: ~ **aggregate** *n* MAT PROP Leichtzuschlagstoff *m*; ~ **building board** *n* MAT PROP Leichtbauplatte *f*; ~ **construction method** *n* ARCH & BUILD Leichtbauweise *f*; ~ **cupola** *n* ARCH & BUILD Lichtkuppel *f*, Dachkuppel *f*, Oberkuppel *f*; **~-fastness** *n* SURFACE Lichtechtheit *f*; ~ **fixture** *n* ELECTR Beleuchtungskörper *m*, Leuchte *f*
**lighting** *n* ELECTR Beleuchtung *f*; ~ **installation** *n* ELECTR Beleuchtungsanlage *f*; ~ **row** *n* ELECTR Lichtband *nt*; ~ **system** *n* ELECTR Beleuchtungsanlage *f*
**lightning** *n* ELECTR Blitz *m*; ~ **arrester** *n* ELECTR Fangeinrichtung *f*; ~ **arrester rod** *n* ELECTR Blitzauffangstange *f*; ~ **conductor holder** *n* ELECTR Dachleitungshalter *m*; ~ **down conductor** *n* ELECTR Blitzschutzableitung *f*; ~ **globe** *n* ELECTR *lightning arrester* Blitzkugel *f*; ~ **strike** *n* ELECTR Blitzeinschlag *m*
***light***: ~ **shaft** *n* ARCH & BUILD Lichtschacht *m*; ~ **steel**

**construction** *n* ARCH & BUILD, STEEL Stahlleichtbau *m*

**lightweight** *adj* MAT PROP Leicht-, leicht; **~ aggregate** *n* MAT PROP Leichtzuschlagstoff *m*; **~ building board** *n* MAT PROP Leichtbauplatte *f*; **~ clay brick** *n* MAT PROP, STONE Leichtziegel *m*; **~ concrete** *n* ARCH & BUILD, CONCR Leichtbeton *m*, Leichtgewichtsbeton *m*; **~ construction** *n* ARCH & BUILD Leichtbau *m*; **~ metal construction** *n* ARCH & BUILD, STEEL Metallleichtbau *m*; **~ plaster** *n* STONE Leichtverputz *m*; **~-profiled PVC roof** *n* ARCH & BUILD, MAT PROP profiliertes Leichtgewicht-PVC-Dach *nt*; **~ steel construction** *n* ARCH & BUILD, STEEL Stahlleichtbau *m*; **~ timber truss** *n* ARCH & BUILD, TIMBER leichter Holzbinder *m*

**ligneous**: **~ asbestos** *n* STEEL Holzasbest *m*

**lignite** *n* MAT PROP Lignit *nt*

**lignivorous** *adj* ENVIRON, TIMBER *pest* holzfressend

**like**: **~-grained concrete** *n* CONCR Einkornbeton *m*

**limba** *n* TIMBER *African wood* Limba *nt*

**lime** *n* STONE Kalk *m*; **~-cement plaster** *n* STONE Kalkzementputz *m*, Zementkalkputz *m*; **~ content** *n* MAT PROP, STONE Kalkgehalt *m*; **~ floor** *n* MAT PROP, STONE Kalkestrich *m*; **~ leaching** *n* MAT PROP, STONE Auskalkung *f*; **~ mortar** *n* CONCR, STONE Kalkmörtel *m*; **~ plaster** *n* STONE Kalkputz *m*; **~ putty** *n* STONE fetter Kalkputzmörtel *m*

**limestone** *n* STONE Kalkstein *m*; **~-filled concrete** *n* CONCR Beton *m* mit Kalksteinzuschlag

**limewash 1.** *n* SURFACE Kalktünche *f*; **2.** *vt* SURFACE schlämmen, tünchen, kalken

**limewashing** *n* SURFACE Kalken *nt*, Schlämmen *nt*, Tünchen *nt*

**limewhite** *vt* SURFACE schlämmen, tünchen, kalken

**liming** *n* SURFACE Kalken *nt*, Schlämmen *nt*, Tünchen *nt*

**limit** *n* ARCH & BUILD, CONST LAW, INFR & DES Grenz-, Grenze *f*

**limitation** *n* ARCH & BUILD Begrenzung *f*, CONST LAW Begrenzung *f*, Einschränkung *f*, Beschränkung *f*, INFR & DES Begrenzung *f*; **~ of contractor's liabilities** *n* CONST LAW Haftungsbegrenzung *f* des Bauunternehmers

***limit***: **~ design** *n* ARCH & BUILD, CONCR, INFR & DES, MAT PROP Traglastverfahren *nt*; **~ of elasticity** *n* ARCH & BUILD, CONCR, MAT PROP, STEEL Elastizitätsgrenze *f*; **~ indicator** *n* HEAT & VENT Grenzwertgeber *m*

**limiting**: **~ concentration** *n* ENVIRON Grenzkonzentration *f*

***limit***: **~ of liability** *n* CONST LAW Haftpflichtgrenze *f*; **~ load** *n* ARCH & BUILD, INFR & DES, MAT PROP Grenzbelastung *f*, Grenzlast *f*, Traglast *f*; **~ of load stress** *n* ARCH & BUILD, INFR & DES, MAT PROP Grenzbelastung *f*; **~ stop** *n* BUILD MACHIN *of crane* Anschlag *m*; **~ value indicator** *n* HEAT & VENT Grenzwertgeber *m*

**line 1.** *n* ARCH & BUILD, ELECTR, HEAT & VENT, INFR & DES, WASTE WATER Leitung *f*, Linie *f*; **2.** *vt* ARCH & BUILD verschalen, schalen, auskleiden, verblenden, verkleiden, ENVIRON *landfill* abdichten, INFR & DES *face* auskleiden, *lag* verschalen, *with sheeting* schalen, *clad* verkleiden, TIMBER schalen, unterfüttern, unterlegen; **3. ~ out** *vi* ARCH & BUILD, BUILD MACHIN, CONCR, MAT PROP, STONE anreißen, WASTE WATER ausfluchten

**linear** *adj* ARCH & BUILD, INFR & DES geradlinig, linear; **~ load** *n* ARCH & BUILD, INFR & DES Streckenlast *f*; **~ thermal expansion** *n* CONCR, INFR & DES, MAT PROP, STEEL lineare Wärmeausdehnung *f*

***line***: **~ of buckets** *n* BUILD MACHIN *conveyor* Schaufelreihe *f*

**lined** *adj* ARCH & BUILD, INFR & DES verblendet, *faced* ausgekleidet, *cladding* verkleidet, *lagged* verschalt

***line***: **~ entry** *n* ELECTR Einspeisung *f*; **~ of influence** *n* ARCH & BUILD Einflußlinie *f*; **~ of levels** *n* INFR & DES Gradientenzug *m*; **~ load** *n* ARCH & BUILD, INFR & DES Linienlast *f*, Streckenlast *f*; **~ pin** *n* BUILD HARDW Schnurnagel *m*

**liner** *n* ARCH & BUILD, INFR & DES, MAT PROP Auskleidung *f*; **~ sheet** *n* ENVIRON Dichtungsbahn *f*

***line***: **~ of sight** *n* INFR & DES Sehlinie *f*

**lining** *n* ARCH & BUILD *facing* Auskleidung *f*, *cladding* Verkleidung *f*, Futter *nt*, Füllung *f*, *veneer* Verschalung *f*, Verblendung *f*, BUILD HARDW Füllung *f*, ENVIRON *of a landfill* Abdichtung *f*, INFR & DES *veneer* Verschalung *f*, Verkleidung *f*, Auskleidung *f*, Futter *nt*, Füllung *f*, Verblendung *f*, MAT PROP Auskleidung *f*, STONE Mauermantel *m*; **~ wall** *n* ARCH & BUILD, STONE Futtermauer *f*

**link 1.** *n* ARCH & BUILD Verbinden *nt*; **2.** *vt* ARCH & BUILD verbinden

**linoleum** *n* MAT PROP Linoleum *nt*

**lintel** *n* ARCH & BUILD, INFR & DES Sturz *m*, Sturzbalken *m*, Sturzriegel *m*, Sturzträger *m*, STONE Sturz *m*; **~ beam** *n* INFR & DES durchlaufender Sturzbalken *m*

**lip** *n* ARCH & BUILD Ausguß *m*, Überlaufkante *f*; **~ sealing** *n* BUILD HARDW Lippendichtung *f*

**liquid 1.** *adj* INFR & DES, MAT PROP Flüssig-, flüssig; **2.** *n* INFR & DES, MAT PROP Flüssigkeit *f*

**liquidated**: **~ damages** *n pl* CONST LAW bezifferter Schadenersatz *m*

***liquid***: **~ limit test** *n* INFR & DES, MAT PROP Fließgrenzversuch *m*; **~ manure** *n* INFR & DES Jauche *f*; **~ sludge** *n* ENVIRON Flüssigschlamm *m*, Naßschlamm *m*, INFR & DES Flüssigschlamm *m*; **~ waste** *n* ENVIRON flüssiger Abfall *m*; **~ waterproofing agent** *n* SOUND & THERMAL Sperrflüssigkeit *f*

**list** *n* ARCH & BUILD, CONST LAW Liste *f*, Verzeichnis *nt*; **~ of defects** *n* ARCH & BUILD, CONST LAW Mängelliste *f*; **~ of deficiencies** *n* ARCH & BUILD, CONST LAW Mängelliste *f*

**listed**: **~ building** *n BrE* CONST LAW, INFR & DES denkmalgeschütztes Bauwerk *nt*

**listel** *n* ARCH & BUILD, BUILD HARDW Riemchen *nt*, Leistchen *nt*

***list***: **~ price** *n* INFR & DES, MAT PROP Listenpreis *m*

**lithosphere** *n* ENVIRON Lithosphäre *f*

**litter** *n* ENVIRON Gerümpel *nt*, *on street* Straßenkehricht *m*

**live**: **~ load** *n* ARCH & BUILD, INFR & DES *civil engineering* Nutzlast *f*, Verkehrslast *f*; **~ wire** *n* ELECTR stromführender Leiter *m*

**living**: **~ allowance** *n* CONST LAW Auslösung *f*

**lixiviate** *vt* ENVIRON, INFR & DES auslaugen

**lixiviation** *n* ENVIRON, INFR & DES Auslaugung *f*

**load 1.** *n* ARCH & BUILD, CONCR, INFR & DES, MAT PROP Beanspruchung *f*, Belastung *f*, Last *f*, *of composite beam* Belastungsschlupfverhalten *nt*; **2.** *vt* ARCH & BUILD, BUILD MACHIN, CONCR, INFR & DES, MAT PROP beanspruchen, beladen, belasten, laden

**load**: **~-bearing** *adj* ARCH & BUILD, INFR & DES tragend; **~-bearing brick** *n* ARCH & BUILD statisch mitwirkender Ziegel *m*; **~-bearing capacity** *n* ARCH & BUILD, INFR & DES Tragfähigkeit *f*, Tragkraft *f*, Tragvermögen *nt*; **~-bearing structure** *n* ARCH & BUILD Tragwerk *nt*, Tragkonstruktion *f*, tragende Konstruktion *f*, Stützkonstruktion *f*, INFR & DES Tragkonstruktion *f*, Tragwerk *nt*; **~-bearing studs** *n pl* ARCH & BUILD, MAT PROP tragende Bundsäulen *f pl*; **~-bearing wall** *n* ARCH & BUILD tragende Wand *f*, INFR & DES Tragmauer *f*; **~ category** *n* ARCH & BUILD, INFR & DES Laststufe *f*; **~ combination** *n* ARCH & BUILD, INFR & DES Lastkombination *f*; **~ compensation** *n* ARCH & BUILD, INFR & DES Belastungsausgleich *m*; **~-deflection curve** *n* INFR & DES Lastverformungsdiagramm *nt*; **~ distribution** *n* INFR & DES Lastverteilung *f*; **~ distribution plate** *n* INFR & DES Lastverteilungsplatte *f*; **~ duration** *n* INFR & DES Lastdauer *f*

**loaded** *adj* ARCH & BUILD, INFR & DES beansprucht, beladen, belastet; **~ concrete** *n* MAT PROP Strahlenschutzbeton *m*, Schwerstbeton *m*

**loader** *n* BUILD MACHIN Ladegerät *nt*

***load***: **~ group** *n* ARCH & BUILD, INFR & DES Beanspruchungsgruppe *f*; **~ hook** *n* BUILD MACHIN Lasthaken *m*

**loading** *n* ARCH & BUILD, BUILD MACHIN, INFR & DES Laden *nt*; **~ assumption** *n* ARCH & BUILD, INFR & DES Lastannahme *f*; **~ bay** *n* ARCH & BUILD, INFR & DES Ladeplatz *m*, Verladerampe *f*; **~ bridge** *n* INFR & DES Ladebrücke *f*; **~ case** *n* ARCH & BUILD, INFR & DES Belastungsfall *m*; **~ condition** *n* ARCH & BUILD Lastfall *m*; **~ control** *n* INFR & DES Belastungskontrolle *f*; **~ history** *n* ARCH & BUILD, CONCR, INFR & DES, MAT PROP Belastungsgeschichte *f*; **~ hopper** *n* ENVIRON Mülleinfülltrichter *m*; **~ mechanism** *n* ENVIRON Beschickungseinrichtung *f*; **~ platform** *n* ARCH & BUILD, INFR & DES Laderampe *f*; **~ points** *n pl* ARCH & BUILD, TIMBER Lastangriffspunkte *m pl*; **~ ramp** *n* ARCH & BUILD, INFR & DES Belade- und Entladerampe *f*, Laderampe *f*; **~ shovel** *n* BUILD MACHIN Hublader *m*, Ladeschaufel *f*; **~ weight** *n* ARCH & BUILD, INFR & DES *of lift, elevator* Belastungsgewicht *nt*

***load***: **~ at rupture** *n* ARCH & BUILD, INFR & DES, MAT PROP Bruchlast *f*; **~ scheme** *n* ARCH & BUILD Belastungsfall *m*, Lastfall *m*, INFR & DES Belastungsfall *m*; **~ settlement diagram** *n* INFR & DES, MAT PROP Lastsetzungsdiagramm *nt*; **~-sharing concept** *n* ARCH & BUILD Lastverteilungsprinzip *nt*; **~-swelling diagram** *n* INFR & DES, MAT PROP Lastschwellungsdiagramm *nt*; **~ test** *n* ARCH & BUILD, INFR & DES Belastungsprobe *f*, Belastungsversuch *m*; **~ transmission** *n* ARCH & BUILD, INFR & DES Lastübertragung *f*

**loam** *n* INFR & DES, MAT PROP Lehm *m*, Ziegelton *m*; **~ content** *n* INFR & DES, MAT PROP Lehmgehalt *m*; **~ core** *n* INFR & DES Lehmkern *m*; **~ layer** *n* INFR & DES Lehmschicht *f*; **~ pit** *n* INFR & DES Lehmgrube *f*

**loamy** *adj* INFR & DES, MAT PROP lehmig; **~ sand** *n* INFR & DES, MAT PROP lehmiger Sand *m*; **~ soil** *n* INFR & DES, MAT PROP lehmiger Boden *m*

**lobby** *n* ARCH & BUILD Lobby *f*, Diele *f*

**local**: **~ authority** *n* CONST LAW Kommunalbehörde *f*; **~ conditions** *n pl* ARCH & BUILD örtliche Gegebenheiten *f pl*; **~ district heating** *n* HEAT & VENT örtliche Fernheizung *f*; **~ search** *n* CONST LAW örtliche behördliche Überprüfung *f*

**locate** *vt* ARCH & BUILD Standort festlegen, einpeilen, trassieren, ELECTR trassieren

**location** *n* ARCH & BUILD *place* Lage *f*, Ort *m*, Standort *m*, *tracing* Trassierung *f*, ELECTR Trassierung *f*; **~ of the line** *n* ELECTR Trassierung *f*; **~ line of the cable** *n* ELECTR Kabeltrasse *f*; **~ plan** *n* ARCH & BUILD, CONST LAW, INFR & DES Lageplan *m*

**lock 1.** *n* BUILD HARDW Schloß *nt*, Verschluß *m*, Arretierung *f*, MAT PROP Arretierung *f*; **2.** *vt* BUILD HARDW verschließen, absperren

***lock***: **~ bolt** *n* BUILD HARDW Riegel *m*; **~ bush** *n* BUILD HARDW Steckbuchse *f*, Überwurf *m*; **~ case** *n* BUILD HARDW Schließkasten *m*, Schloßkasten *m*; **~ casing** *n* BUILD HARDW Schloßkasten *m*, Schließkasten *m*; **~ cylinder** *n* BUILD HARDW Schließzylinder *m*; **~ fitting** *n* BUILD HARDW Schließbeschlag *m*; **~ fore-end** *n* BUILD HARDW Schloßstulp *m*; **~ for flush door** *n* BUILD HARDW Schloß *nt* mit Stulp auf Mitte; **~ for rebated door** *n* BUILD HARDW Schloß *nt* mit einseitigem Stulp

**locking** *n* BUILD HARDW Absperren *nt*; **~ bolt** *n* BUILD HARDW Verriegelungsbolzen *m*; **~ handle** *n* BUILD HARDW Griffverschluß *m*, Knebelgriff *m*; **~ hook** *n* BUILD HARDW Schließhaken *m*; **~ system** *n* BUILD HARDW Schließanlage *f*

***lock***: **~ knob** *n* BUILD HARDW Einrastknopf *m*; **~ plate** *n* BUILD HARDW Schließblech *nt*

**lockplate**: **~ width** *n* BUILD HARDW Stulpbreite *f*

***lock***: **~ rail** *n* BUILD HARDW Türquerriegel *m*; **~ saw** *n* BUILD HARDW, BUILD MACHIN Lochsäge *f*

**lockseaming** *n* STEEL Falzen *nt*

**lockset** *n* BUILD HARDW Schloßgarnitur *f*

***lock***: **~ staple** *n* BUILD HARDW Verriegelungseinrichtung *f*

**loess** *n* INFR & DES, MAT PROP Löß *m*; **~ clay** *n* MAT PROP, STONE Lößlehm *m*

**loft** *n* ARCH & BUILD *attic* Dachgeschoß *nt*, *garret* Speicher *m*

**loggia** *n* ARCH & BUILD Hauslaube *f*, *recessed balcony* Loggia *f*

**logistics**: **~ of disposal** *n* ENVIRON Entsorgungslogistik *f*

**log**: **~ road** *n* INFR & DES Knüppeldamm *m*

**long**: **~ columns** *n pl* ARCH & BUILD überschlanke Säulen *f pl*; **~-distance** *adj* INFR & DES fern; **~-distance conveyor** *n* BUILD MACHIN, INFR & DES Langstreckenfördergerät *nt*; **~-distance conveyor belt** *n* BUILD MACHIN Bandanlage *f* für Langstrekkenförderung; **~-distance heating line** *n* HEAT & VENT, INFR & DES Fernheizleitung *f*; **~-distance steam heating** *n* HEAT & VENT, INFR & DES Dampffernheizung *f*

**longitudinal**: **~ axis** *n* INFR & DES Längsachse *f*; **~ bar** *n* ARCH & BUILD, CONCR, INFR & DES, MAT PROP Längseisen *nt*; **~ beam** *n* ARCH & BUILD Längsträger *m*; **~ bracing** *n* ARCH & BUILD, STEEL, TIMBER Längsverband *m*; **~ compression joint** *n* ARCH & BUILD Längspreßfuge *f*; **~ construction joint** *n* ARCH & BUILD Längsbetonierfuge *f*; **~ deformation** *n* CONCR, INFR & DES, MAT PROP, STEEL Längsverformung *f*; **~ fold** *n* STEEL Längsfalz *m*; **~ girder** *n* ARCH & BUILD Längsträger *m*; **~ joint** *n* ARCH & BUILD Längsfuge *f*; **~ seam** *n* STEEL Längsnaht *f*; **~ section** *n* ARCH & BUILD Längsschnitt *m*; **~ shear** *n*

ARCH & BUILD, CONCR, INFR & DES, MAT PROP Längsabscheren *nt*; ~ **slip** *n* ARCH & BUILD, CONCR, INFR & DES, MAT PROP Längsschlupf *m*; ~ **spacing** *n* ARCH & BUILD, CONCR Abstand *m* der Längseisen

***long***: ~**-span** *adj* ARCH & BUILD mit großer Spannweite, weitgespannt; ~**-term effect** *n* CONST LAW Langzeitwirkung *f*

**loop** *n* BUILD HARDW Schlinge *f*

**looped**: ~ **fabric** *n* MAT PROP Schlingengewebe *nt*, Schlingenware *f*

**loose** *adj* INFR & DES, MAT PROP lose, locker; ~ **chippings** *n pl* INFR & DES, MAT PROP Rollsplitt *m*; ~ **material** *n* INFR & DES, MAT PROP loses Material *nt*

**loosen** *vt* INFR & DES, MAT PROP auflockern

**loosened** *adj* INFR & DES, MAT PROP locker; ~ **soil** *n* INFR & DES, MAT PROP lockere Erde *f*

***loose***: ~**-packed** *adj* INFR & DES, MAT PROP locker gelagert; ~**-pin hinge** *n* BUILD HARDW Scharnier *nt* mit lösbaren Bolzen; ~ **sand** *n* INFR & DES, MAT PROP locker gelagerter Sand *m*; ~ **tongue** *n* TIMBER Einsteckfeder *f*

**lorry** *n BrE* (*cf truck AmE*) BUILD MACHIN Lastkraftwagen *m* (*LKW*)

**loss** *n* INFR & DES, MAT PROP Verlust *m*; ~ **by evaporation** *n* ARCH & BUILD, HEAT & VENT, INFR & DES, MAT PROP Verdunstungsverlust *m*; ~ **of head** *n* BUILD MACHIN Druckverlust *m*, HEAT & VENT, INFR & DES, WASTE WATER Druckhöhenverlust *m*, Druckverlust *m*, Gefälleverlust *m*; ~ **of heat** *n* ENVIRON, SOUND & THERMAL Wärmeverlust *m*; ~ **or expense** *phr* CONST LAW Verlust oder Kosten; ~ **of prestress** *n* CONCR Vorspannungsverlust *m*, Spannungsabbau *m*, Spannungsabfall *m*; ~ **of voltage** *n* ELECTR Spannungsabfall *m*; ~ **of volume** *n* INFR & DES, MAT PROP Volumenverlust *m*; ~ **of water** *n* INFR & DES Wasserverlust *m*; ~ **of weight** *n* INFR & DES, MAT PROP Gewichtsverlust *m*

**lot** *n* CONST LAW Los *nt*, Parzelle *f*, INFR & DES Parzelle *f*, Landstück *nt*

**loudness** *n* SOUND & THERMAL Lautstärke *f*

**louver** *AmE see louvre BrE*

**louvered** *AmE see louvred BrE*

**louvre** *n BrE* HEAT & VENT Lüftungsgitter *nt*

**louvred** *adj BrE* ARCH & BUILD geteilt, mit Luftschlitzen

***louvre***: ~ **door** *n BrE* ARCH & BUILD Lamellentür *f*, Schindeltür *f*, Tür *f* mit Luftschlitzfüllung; ~ **vent** *n BrE* HEAT & VENT Ausblasjalousie *f*; ~ **window** *n BrE* ARCH & BUILD Jalousiefenster *nt*

**low**: ~**-cost assembly** *n* ARCH & BUILD kostengünstige Montage *f*; ~**-density polyethylene** *n* (*LDPE*) MAT PROP Polyethylen *nt* niedriger Dichte (*PE weich*); ~**-energy consumption house** *n* HEAT & VENT Haus *nt* mit niedrigem Energieverbrauch

**lower**: ~ **calorific value** *n* HEAT & VENT untere Heizleistung *f*

**lowering**: ~ **of the groundwater level** *n* ARCH & BUILD, INFR & DES Grundwasserabsenkung *f*

**lower**: ~ **storage basin** *n* ENVIRON unteres Sammelbecken *nt*, unteres Speicherbecken *nt*

***low***: ~**-grade energy** *n* HEAT & VENT minderwertige Energie *f*; ~**-gradient** *adj* INFR & DES flach geneigt; ~ **head** *n* ENVIRON tiefe Wassersäule *f*; ~**-level noise** *n* SOUND & THERMAL niedriger Lärmpegel *m*; ~ **luster** *AmE*, ~ **lustre** *BrE n* SURFACE Mattglanz *m*; ~**-noise** *adj* BUILD MACHIN geräuscharm; ~ **pressure** *n* HEAT & VENT Niederdruck *m*; ~**-rise** *adj* ARCH & BUILD niedrig; ~**-rise housing** *n* ARCH & BUILD, INFR & DES erdgeschossiger Wohnungsbau *m*; ~**-shrink** *adj* MAT PROP schwindarm; ~**-slump concrete** *n* CONCR steifer Beton *m*; ~**-temperature stability** *n* MAT PROP Kältebeständigkeit *f*; ~ **tide** *n* ENVIRON Ebbe *f*; ~**-waste technology** *n* ENVIRON abfallarme Technologie *f*; ~ **water/cement mix** *n* (*low w/c mix*) MAT PROP Betonmischung *f* mit niedrigem Wasser-Zement-Faktor; ~ **water level** *n* INFR & DES Niedrigwasserstand *m*; ~ **w/c mix** *n* (*low water/cement mix*) MAT PROP Betonmischung *f* mit niedrigem Wasser-Zement-Faktor

**lubricant** *n* MAT PROP Schmiermittel *nt*

**lubricate** *vt* MAT PROP schmieren

**lubricating**: ~ **device** *n* BUILD MACHIN Schmiervorrichtung *f*

**luffing**: ~ **crane** *n* BUILD MACHIN Wippkran *m*

**lug** *n* ELECTR Öse *f*, INFR & DES, MAT PROP, STEEL Zapfen *m*; ~ **bolt** *n* INFR & DES, MAT PROP, STEEL Flacheisenzapfen *m*

**luggage** *n* ARCH & BUILD, INFR & DES Gepäck *nt*; ~ **elevator** *n AmE* (*cf luggage lift AmE*) ARCH & BUILD Gepäckaufzug *m*; ~ **lift** *BrE n* (*cf luggage elevator AmE*) ARCH & BUILD Gepäckaufzug *m*

**lumber** *n AmE* (*cf timber BrE*) TIMBER Bauholz *nt*, Nutzholz *nt*, Schnittholz *nt*

**luminaire** *n* ELECTR Leuchte *f*

**luminous** *adj* ARCH & BUILD, ELECTR leuchtend; ~ **ceiling** *n* ARCH & BUILD, ELECTR Leuchtdecke, *f* Lichtdecke *f*

**lump**: ~**-free** *adj* CONCR, INFR & DES, MAT PROP klumpenfrei; ~ **sum** *n* ARCH & BUILD, CONST LAW Pauschale *f*; ~ **sum contract** *n* ARCH & BUILD, CONST LAW Pauschalvertrag *m*; ~ **sum settlement** *n* CONST LAW Pauschalabfindung *f*

**lumpy** *adj* INFR & DES, MAT PROP klumpig; ~ **soil** *n* INFR & DES klumpige Erde *f*

**luster** *n AmE*, **lustre** *n BrE* SURFACE Glanz *m*

**lux** *n* ELECTR, INFR & DES, MAT PROP *SI unit of illumination* Lux *nt*

# M

**m** *abbr* (*metre BrE*) ARCH & BUILD m (*Meter*)
**macadam** *n* INFR & DES *road surface* Schotter *m*, Schotterdecke *f*, MAT PROP *substance* Makadam *m*
**macadamization** *n* INFR & DES Makadamisierung *f*
**macadamize** *vt* INFR & DES makadamisieren, schottern
**macerate** *vt* ENVIRON, INFR & DES, MAT PROP einweichen
**maceration** *n* ENVIRON, INFR & DES, MAT PROP Einweichen *nt*
**macerator** *n* BUILD HARDW, ENVIRON, MAT PROP Einweichanlage *f*
**machine**: **~-applied plaster** *n* MAT PROP Spritzputz *m*, STONE Spritzbewurf *m*, Spritzputz *m*, Maschinenputz *m*
**machined** *adj* BUILD MACHIN, MAT PROP bearbeitet
**machine**: **~ foundation** *n* CONCR, INFR & DES Maschinenfundament *nt*; **~ room** *n* ARCH & BUILD, BUILD MACHIN Maschinenraum *m*
**machinery** *n* BUILD MACHIN, INFR & DES Maschinen *f pl*, Maschinenanlage *f*; **~ and equipment for road maintenance** *n* BUILD MACHIN Maschinen *f pl* und Geräte *nt pl* für die Straßeninstandhaltung; **~ and equipment for winter** *n* BUILD MACHIN Winterdienstmaschinen *f pl* und Winterdienstgeräte; **~ for recycling of building materials** *n* BUILD MACHIN Aufbereitungsmaschinen *f pl* für Baustoffe
**machine**: **~-straightened sheet** *n* MAT PROP, STEEL maschinell gerichtetes Blech *nt*
**made**: **~ ground** *n* INFR & DES Auffüllung *f*; **~-up** *adj* INFR & DES aufgeschüttet; **~-up ground** *n* INFR & DES Aufschüttung *f*
**magnesite** *n* INFR & DES, MAT PROP Magnesit *m*; **~ brick** *n* INFR & DES, STONE Magnesitstein *m*, Magnesitziegel *m*; **~ composition** *n* MAT PROP Steinholz *nt*; **~ flooring** *n* ARCH & BUILD Steinholzfußboden *m*, fugenloser Fußboden *m*
**magnesium**: **~ chloride** *n* INFR & DES, MAT PROP Magnesiumchlorid *nt*
**magnet** *n* ENVIRON Magnet *m*
**magnetic**: **~ separation** *n* ENVIRON Magnetabscheidung *f*, *of refuse* Magnetsortierung *f*; **~ separator** *n* ENVIRON Magnetabscheider *m*
**magnitude** *n* INFR & DES Größenordnung *f*
**main** *n* ELECTR, HEAT & VENT, INFR & DES, WASTE WATER Hauptleitung *f*; **~ access** *n* INFR & DES Hauptzugang *m*; **~ beam** *n* ARCH & BUILD, TIMBER Hauptbalken *m*; **~ bearing member** *n* INFR & DES Haupttragglied *nt*; **~ collector** *n* ENVIRON, INFR & DES, WASTE WATER Hauptsammler *m*, Sammelkanal *m*; **~ column** *n* ARCH & BUILD, INFR & DES Hauptstütze *f*; **~ contractor** *n* ARCH & BUILD Generalauftragnehmer *m*, CONST LAW, INFR & DES Hauptunternehmer *m*; **~ distribution panel** *n* ELECTR Hauptverteiler *m*; **~ drain** *n* ENVIRON, INFR & DES Hauptsammler *m*, WASTE WATER Hauptsammler *m*, Vorflutdrän *m*; **~ earth box** *n BrE* (*cf main ground box AmE*) ELECTR Haupterdungskasten *m*; **~ entrance** *n* ARCH & BUILD, INFR & DES Haupteingang *m*; **~ gable** *n* ARCH & BUILD, STONE Mittelgiebel *m*; **~ girder** *n* ARCH & BUILD, INFR & DES Hauptträger *m*; **~ ground box** *n AmE* (*cf main earth box BrE*) ELECTR Haupterdungskasten *m*; **~ lift** *n* BUILD MACHIN *of crane* Haupthub *m*; **~ line** *n* ELECTR, HEAT & VENT, INFR & DES, WASTE WATER Hauptleitung *f*; **~ pier** *n* ARCH & BUILD, INFR & DES Hauptpfeiler *m*; **~ pipe** *n* INFR & DES, WASTE WATER Hauptleitungsrohr *nt*; **~ reinforcement** *n* CONCR, INFR & DES Hauptbewehrung *f*
**mains** *n pl* ELECTR, INFR & DES Netz *nt*, Stromversorgungsnetz *nt*
***main***: **~ sewer** *n* ENVIRON, INFR & DES, WASTE WATER Hauptsammler *m*, Sammelkanal *m*; **~ shut-off valve** *n* WASTE WATER Hauptabsperrung *f*
***mains***: **~ power supply** *n* ELECTR Netzbetrieb *nt*
***main***: **~ storage yard** *n* ARCH & BUILD Hauptlagerplatz *m*; **~ structure** *n* ARCH & BUILD Hauptbauwerk *nt*; **~ supply line** *n* ELECTR Hauptverteilleitung *f*; **~ switch** *n* ELECTR Hauptschalter *m*
**maintain** *vt* ARCH & BUILD, CONST LAW, INFR & DES instandhalten, unterhalten
**maintenance** *n* ARCH & BUILD, CONST LAW, INFR & DES Instandhaltung *f*, Pflege *f*, Wartung *f*; **~ cradle** *n* BUILD MACHIN Instandhaltungshängebühne *f*; **~ manhole** *n* ARCH & BUILD Wartungsschacht *m*; **~ pit** *n* ARCH & BUILD Montagegrube *f*
***main***: **~ wall** *n* ARCH & BUILD tragende Wand *f*; **~ works** *n pl* ARCH & BUILD Rohbau *m*
**maisonette** *n* ARCH & BUILD Maisonette *f*
**major**: **~ source** *n* ENVIRON Hauptquelle *f*
**make**: **~ good** *vt* BUILD MACHIN, STONE, SURFACE ausbessern
**make**: **~-up tank** *n* BUILD MACHIN Ausgleichbehälter *m*
**making**: **~ good** *n* BUILD MACHIN, STONE, SURFACE Ausbessern *nt*
**malleability** *n* MAT PROP, STEEL Schmiedbarkeit *f*
**malleable**: **~ cast iron** *n* MAT PROP, STEEL Temperguß *m*
**mammoth**: **~ pump** *n* BUILD MACHIN Mammutpumpe *f*
**management** *n* ARCH & BUILD Verwaltung *f*, Leitung *f*; **~ contract** *n* CONST LAW Managementvertrag *m*
**manager** *n* ARCH & BUILD Leiter *m*, Manager *m*
**maneuver** *AmE see manoeuvre BrE*
**manhole** *n* ARCH & BUILD Schacht *m*, INFR & DES, WASTE WATER Einstiegsschacht *m*, Mannloch *nt*, Schacht *m*; **~ bottom** *n* INFR & DES, WASTE WATER Schachtsohle *f*; **~ cover** *n* INFR & DES, WASTE WATER Schachtdeckel *m*; **~ covering** *n* INFR & DES, WASTE WATER Schachtabdeckung *f*; **~ door** *n* ARCH & BUILD Schachttür *f*; **~ ring** *n* ENVIRON, INFR & DES, WASTE WATER Schachtring *m*; **~ top ring** *n* INFR & DES, WASTE WATER Auflagering *m*
**manifold** *n* HEAT & VENT, WASTE WATER Verteiler *m*
**manoeuvre** *vt BrE* ARCH & BUILD manövrieren
**manometer** *n* BUILD MACHIN, HEAT & VENT Manometer *nt*
**manpower** *n* ARCH & BUILD Arbeitskraft *f*; **~ planning** *n* CONST LAW Arbeitskraftplanung *f*

**mansard**: ~ **roof** *n* ARCH & BUILD, TIMBER Mansardendach *nt*
**man**: ~**-sized** *adj* WASTE WATER *sewer* begehbar
**manual** *n* ARCH & BUILD Handbuch *nt*; ~ **labor** *AmE*, ~ **labour** *BrE* *n* ARCH & BUILD Handarbeit *f*; ~ **penetrometer** *n* BUILD MACHIN Handsondiergerät *nt*; ~ **separation** *n* ENVIRON *of waste* Handsortierung *f*; ~ **sorting** *n* ENVIRON *of waste* Handsortierung *f*
**manufacture 1.** *n* MAT PROP *fabrication* Fertigung *f*, *generation* Erzeugung *f*, *production* Produktion *f*, Herstellung *f*; **2.** *vt* MAT PROP erzeugen, fertigen, herstellen, produzieren, verarbeiten
**manufacturer** *n* BUILD MACHIN Hersteller *m*, Erzeuger *m*
**map 1.** *n* ARCH & BUILD, CONST LAW, INFR & DES Karte *f*; **2.** *vt* ARCH & BUILD, CONST LAW, INFR & DES trassieren
**mapper** *n* ARCH & BUILD, CONST LAW, INFR & DES *surveying* Kartierer *m*
**mapping** *n* ARCH & BUILD, CONST LAW, INFR & DES Kartierung *f*; ~ **out** *n* ARCH & BUILD Aufnahme *f*
**marble** *n* STONE Marmor *m*; ~ **dust** *n* MAT PROP, STONE Marmorstaub *m*
**margin** *n* ARCH & BUILD Grenze *f*, Rand *m*
**marginal**: ~ **strip** *n* INFR & DES *at roadside* Randstreifen *m*
***margin***: ~ **tile** *n* ARCH & BUILD Ortstein *m*, Randziegel *m*, Traufstein *m*, STONE Randziegel *m*
**marigraph** *n* BUILD MACHIN Flutschreiber *m*
**marina** *n* INFR & DES Marina *f*
**marine**: ~ **clay** *n* INFR & DES, MAT PROP, STONE Meerton *m*; ~ **deposit** *n* INFR & DES marine Ablagerung *f*; ~**-drilling rig** *n* BUILD MACHIN, INFR & DES Unterwasserbohranlage *f*; ~ **pollution** *n* ENVIRON Meeresverschmutzung *f*; ~ **salt** *n* INFR & DES, MAT PROP Meersalz *nt*; ~ **sand** *n* INFR & DES Meeressand *m*; ~ **sediment** *n* INFR & DES marine Ablagerung *f*; ~ **sewage disposal** *n* ENVIRON Abwassereinleitung *f* ins Meer
**marking** *n* ARCH & BUILD Markierung *f*, Kennzeichnung *f*; ~ **coat** *n* INFR & DES Markierungsanstrich *m*; ~ **material** *n* MAT PROP Markierungsstoff *m*; ~ **off** *n* ARCH & BUILD Anreißen *nt*; ~**-paint** *n* SURFACE Markierungsfarbe *f*; ~ **table** *n* ARCH & BUILD Anreißtisch *m*
**marl** *n* INFR & DES, MAT PROP Mergel *m*
**marly**: ~ **sandstone** *n* INFR & DES, MAT PROP Mergelsandstein *m*; ~ **soil** *n* INFR & DES, MAT PROP Mergelboden *m*
**marsh** *n* INFR & DES Marsch *f*, Moor *nt*, Morast *m*, Sumpf *m*
**marshland** *n* INFR & DES Marschland *nt*, Moorboden *m*, Sumpfboden *m*
**mask** *vt* ARCH & BUILD verblenden
**mason** *n* STONE Maurer *m*
**masonry** *n* ARCH & BUILD, STONE Gemäuer *nt*, Mauerwerk *nt*, *working with stone* Mauerwerksarbeiten *f pl*; ~ **backup** *n* ARCH & BUILD, STONE Hintermauerung *f*; ~ **cement** *n* CONCR, INFR & DES, MAT PROP Mauermörtelzement *m*, Zement *m* für Bauzwecke; ~ **dam** *n* INFR & DES, STONE gemauerter Damm *m*; ~ **drill** *n* BUILD MACHIN Mauerbohrer *m*; ~ **nail** *n* BUILD HARDW Stahlnagel *m*; ~ **paint** *n* INFR & DES, MAT PROP, SURFACE Maueranstrichfarbe *f*; ~ **wall** *n* STONE Mauer *f*; ~ **wall slot** *n* STONE Mauerschlitz *m*; ~ **work** *n* ARCH & BUILD, STONE Mauerarbeiten *f pl*
**mass** *n* ARCH & BUILD Masse *f*, Menge *f*; ~ **concrete** *n* CONCR, INFR & DES Kernbeton *m*, Massenbeton *m*, Massivbeton *m*, unbewehrter Beton *m*; ~ **transportation** *n* BUILD MACHIN Massentransport *m*
**mast** *n* INFR & DES Mast *m*
**master** *n* ARCH & BUILD *person* Meister *m*, *original* Original *nt*, Muster *nt*, CONST LAW Original *nt*; ~ **builder** *n* ARCH & BUILD Baumeister *m*, Generalunternehmer *m*; ~ **key** *n* BUILD HARDW, CONST LAW Generalschlüssel *m*, Hauptschlüssel *m*; ~**-keyed system** *n* BUILD HARDW Schließanlage *f*; ~ **key system** *n* BUILD HARDW, CONST LAW Hauptschlüsselanlage *f*; ~ **mason** *n* ARCH & BUILD, STONE Maurermeister *m*; ~ **plan** *n* CONST LAW Bebauungsplan *m*, Flächennutzungsplan *m*, Generalbebauungsplan *m*, Hauptbebauungsplan *m*; ~ **switch** *n* ELECTR Hauptschalter *m*
**mastic** *n* MAT PROP Mastixharz *nt*; ~ **asphalt** *n* INFR & DES, MAT PROP Asphaltmastix *f*; ~ **flooring** *n* ARCH & BUILD, MAT PROP Asphaltfußboden *m*, Asphaltfußbodenbelag *m*, Asphaltstrich *m*
**mat 1.** *n* MAT PROP Matte *f*; **2.** *adj* MAT PROP, SURFACE matt
**match 1.** *vt* ARCH & BUILD, BUILD MACHIN, CONCR, ELECTR, STEEL angleichen; **2.** *vi* ARCH & BUILD zusammenpassen
**matchboard** *n* TIMBER Nut- und Federbrett *nt*, Profilbrett *nt*, Spundbrett *nt*
**matchboarding** *n* TIMBER Riemen *m*; ~ **floor** *n* TIMBER Holzriemenboden *m*
**matched**: ~ **joint** *n* TIMBER Nutverbindung *f*
**matching** *n* TIMBER Spundholzlage *f*; ~ **machine** *n* BUILD MACHIN Spund- und Nutmaschine *f*; ~ **plane** *n* BUILD MACHIN Spundhobel *m*
**material** *n* MAT PROP Material *nt*, Werkstoff *m*, Stoff *m*; ~ **behavior** *AmE*, ~ **behaviour** *BrE* *n* INFR & DES, MAT PROP Materialverhalten *nt*; ~ **characteristics** *n pl* INFR & DES, MAT PROP Materialkennwerte *m pl*; ~ **of limited combustibility** *n* CONST LAW, INFR & DES, MAT PROP Materialien *nt pl* mit begrenzter Brennbarkeit; ~ **pollution** *n* ENVIRON Verschmutzung *f* durch Feststoffe; ~ **properties** *n pl* INFR & DES, MAT PROP Materialeigenschaften *f pl*; ~ **recovery** *n* ENVIRON Materialrückgewinnung *f*
**materials** *n pl* ENVIRON Werkstoff *m*
***material***: ~ **separation operation** *n* ENVIRON Stofftrennprozeß *m*
**mathematical** *adj* ARCH & BUILD, INFR & DES mathematisch; ~ **equation** *n* ARCH & BUILD, INFR & DES mathematische Gleichung *f*; ~ **model** *n* ARCH & BUILD, INFR & DES Rechenmodell *nt*
***mat***: ~ **reinforcement** *n* CONCR, INFR & DES Bewehrungsnetz *nt*, Netzbewehrung *f*
**matrix** *n* ARCH & BUILD Matrix *f*, MAT PROP Einbettungsmasse *f*, Grundmasse *f*, Grundgefüge *nt*
**matter** *n* ARCH & BUILD, INFR & DES, MAT PROP Materie *f*, Stoff *m*
**mattock** *n* BUILD MACHIN Breithacke *f*, spitzer Maurerhammer *m*, STEEL Haueisen *nt*
**mattress** *n* ARCH & BUILD, CONCR Betonplatte *f*
**mature 1.** *vt* CONCR, SURFACE *concrete, mortar* aushärten; **2.** *vi* CONCR, SURFACE *concrete, paint, bitumen* altern
**maul** *n* BUILD MACHIN Zurichthammer *m*, schwerer Holzhammer *m*

**maximum** *adj* ARCH & BUILD, ELECTR, INFR & DES größte, höchste, maximal

***maximum***: **~ axial thrust** *n* ARCH & BUILD, ENVIRON maximaler Axialdruck *m*; **~ bending moment** *n* (*max. M*) ARCH & BUILD, CONCR maximales Biegemoment *nt* (*max. M*); **~ depth** *n* INFR & DES größte Tiefe *f*; **~ hogging moment** *n* ARCH & BUILD, CONCR, INFR & DES, MAT PROP, STEEL maximales Aufwölbungsmoment *nt*; **~ permissible level** *n* (*MPL*) CONST LAW, ENVIRON, INFR & DES höchstzulässiges Niveau *nt*; **~ power at rated wind speed** *n* ENVIRON Höchstleistung *f* bei Nennwindgeschwindigkeit; **~ pressure** *n* INFR & DES maximaler Druck *m*; **~ shaft speed** *n* ENVIRON maximale Wellengeschwindigkeit *f*; **~ total load** *n* ARCH & BUILD, INFR & DES maximale Gesamtbelastung *f*; **~ welding current** *n* ELECTR Höchstschweißstrom *m*

**max.**: **~ M** *abbr* (*maximum bending moment*) ARCH & BUILD, CONCR max. M (*maximales Biegemoment*)

**Maxwell**: **~ diagram** *n* ARCH & BUILD Maxwellscher Kräfteplan *m*

**meandering** *adj* INFR & DES gewunden, schlangenförmig

**mean**: **~ radiant temperature** *n* CONST LAW Durchschnittsstrahlungstemperatur *f*; **~ sea level** *n* INFR & DES Normalnull *f*

**means**: **~ of escape** *n pl* ARCH & BUILD, CONST LAW, INFR & DES Fluchtmöglichkeiten *f pl*; **~ of escape in case of fire** *n* CONST LAW Fluchtwege *m pl* bei Feuer; **~ of ventilation** *n pl* ARCH & BUILD, HEAT & VENT Belüftungsmöglichkeiten *f pl*

**mean**: **~ tidal range** *n* ENVIRON mittlerer Tidehub *m*; **~ value** *n* ARCH & BUILD, INFR & DES, MAT PROP Mittelwert *m*, Durchschnitt *m*, Durchschnittswert *m*; **~ water** *n* INFR & DES Mittelwasser *nt*; **~ water level** *n* INFR & DES mittlerer Wasserstand *m*; **~ wind speed** *n* ENVIRON mittlere Windgeschwindigkeit *f*

**measure** *n* ARCH & BUILD, CONST LAW *action* Maßnahme *f*

**measured**: **~ drawings** *n pl* ARCH & BUILD, INFR & DES maßstabsgerechte Bauaufnahmezeichnungen *f pl*; **~ work** *n* ARCH & BUILD, INFR & DES Aufmaß *nt*

***measure***: **~ of length** *n* ARCH & BUILD Längenmaß *nt*

**measurement** *n* ARCH & BUILD Messung *f*, Vermessung *f*, ELECTR, INFR & DES, MAT PROP Messung *f*; **~ error** *n* ARCH & BUILD, INFR & DES, MAT PROP Meßfehler *m*; **~ period** *n* ARCH & BUILD, INFR & DES, MAT PROP Meßzeitraum *m*

***measure***: **~ of redevelopment** *n* ENVIRON Sanierungsmaßnahme *f*

**measuring** *n* ARCH & BUILD, INFR & DES, MAT PROP Messung *f*; **~ accuracy** *n* ARCH & BUILD, BUILD MACHIN, INFR & DES, MAT PROP Meßgenauigkeit *f*; **~ apparatus** *n* BUILD MACHIN Meßapparat *m*; **~ chain** *n* BUILD MACHIN Meßkette *f*; **~ and control equipment** *n* ELECTR, HEAT & VENT Meß-, Steuerungs- und Regeleinrichtungen *f pl* (*MSR-Einrichtungen*); **~ electrode** *n* ARCH & BUILD, INFR & DES, MAT PROP Meßelektrode *f*; **~ range** *n* BUILD MACHIN Meßbereich *m*; **~ tank** *n* BUILD MACHIN *concrete* Meßgefäß *nt*; **~ tape** *n* ARCH & BUILD Bandmaß *nt*

**mechanical**: **~ behavior** *AmE*, **~ behaviour** *BrE* *n* MAT PROP mechanisches Verhalten *nt*; **~ cipher lock** *n* BUILD HARDW mechanisches Drucktastenschloß *nt*; **~ collector** *n* ENVIRON mechanischer Abscheider *m*; **~ composting** *n* ENVIRON Schnellkompostierung *f*, beschleunigte Kompostierung *f*, geschlossene Kompostierung *f*; **~ connection** *n* ARCH & BUILD, CONCR, INFR & DES, MAT PROP, STEEL mechanischer Verbund *m*; **~ efficiency** *n* ENVIRON mechanischer Wirkungsgrad *m*; **~ failure** *n* INFR & DES, MAT PROP mechanisches Versagen *nt*; **~ floor** *n* ARCH & BUILD Installationsgeschoß *nt*; **~ grab** *n* BUILD MACHIN mechanisch betriebener Greifer *m*

**mechanically**: **~ ventilated** *adj* HEAT & VENT mechanisch belüftet

**mechanical**: **~ properties** *n pl* INFR & DES, MAT PROP mechanische Eigenschaften *f pl*; **~ separation** *n* ENVIRON automatische Müllsortierung *f*, mechanische Sortierung *f*, mechanische Trennung *f*; **~ services** *n pl* ELECTR, HEAT & VENT, WASTE WATER technische Gebäudeausrüstung *f*, *domestic appliances* Haustechnik *f*; **~ transmission** *n* BUILD MACHIN, ENVIRON mechanische Kraftübertragung *f*; **~ wear** *n* BUILD MACHIN mechanische Abnutzung *f*

**mechanics**: **~ of materials** *n* ARCH & BUILD, INFR & DES, MAT PROP Festigkeitslehre *f*

**mechanism** *n* BUILD HARDW, BUILD MACHIN Mechanismus *m*

**medium**: **~-grained** *adj* INFR & DES, MAT PROP mittelkörnig; **~ gravel** *n* INFR & DES, MAT PROP Mittelkies *m*; **~-heavy** *adj* MAT PROP mittelschwer; **~ pressure** *n* HEAT & VENT Mitteldruck *m*; **~-pressure boiler** *n* HEAT & VENT Mitteldruckdampfkessel *m*; **~-rise** *adj* ARCH & BUILD mittelhoch; **~ silt** *n* INFR & DES, MAT PROP Mittelschluff *m*; **~-sized mosaic** *n* STONE Mittelmosaik *nt*

**meet** *vt* ARCH & BUILD *requirements* erfüllen

**meeting** *n* ARCH & BUILD Fuge *f*, Stoß *m*; **~ rail** *n* ARCH & BUILD *double-hung window* Querfries *m*

**mega**: **~-newton** *n* (*MN*) ARCH & BUILD, INFR & DES, MAT PROP Meganewton *nt* (*MN*)

**megapond** *n* (*Mp*) ARCH & BUILD, INFR & DES, MAT PROP Megapond *nt* (*Mp*)

**melamine** *n* INFR & DES, MAT PROP, SURFACE, TIMBER Melamin *nt*; **~ adhesive** *n* INFR & DES, MAT PROP, TIMBER Melaminleim *m*; **~ formaldehyde resin** *n* INFR & DES, MAT PROP, SURFACE, TIMBER Melamin-Formaldehydharz *nt*; **~ resin coated** *adj* SURFACE melaminharzbeschichtet

**melamine/urea**: **~ formaldehyde** *n* (*MF/UF*) ENVIRON, MAT PROP, TIMBER Melamin/Urea-Formaldehyd *nt* (*MF/UF*)

**member** *n* ARCH & BUILD Teil *nt*, Bauteil *nt*, Konstruktionsteil *nt*, Element *nt*

**membrane**: **~ curing compound** *n* CONCR *concrete surface* Dichtungsmittel *nt*; **~ protection** *n* INFR & DES Membranschutz *m*

**memorandum** *n* ARCH & BUILD Baubesprechungsprotokoll *nt*

**mensuration** *n* INFR & DES Messung *f*, Vermessung *f*, *of earthwork* Massenberechnung *f*

**mercury** *n* CONST LAW, ENVIRON, INFR & DES, MAT PROP Quecksilber *nt*; **~ vapor** *AmE*, **~ vapour** *BrE* *n* MAT PROP Quecksilberdampf *m*

**mesh** *n* BUILD MACHIN Siebmasche *f*, MAT PROP Masche *f*, Maschengitter *nt*

**meshed** *adj* MAT PROP vermascht

***mesh***: **~ lath** *n* MAT PROP, STONE *plaster* Gittergewebe *nt*, Gittergewebematte *f*; **~ size** *n* MAT PROP Maschenweite *f*; **~ wire** *n* BUILD HARDW Drahtgeflecht *nt*, MAT

PROP Maschendrahtgewebe *nt*, Drahtgitter *nt*, Drahtgeflecht *nt*; **~ wire fence** *n* INFR & DES Maschendrahtzaun *m*

**metabolic**: **~ waste** *n* ENVIRON Stoffwechselschlacken *f pl*

**metal** *n* INFR & DES *road surface* Schotter *m*, MAT PROP Metall *nt*; **~ anchor** *n* MAT PROP Metallanker *m*; **~ angle bead** *n* BUILD HARDW, STEEL Kantenschutzwinkel *m*; **~ building material** *n* MAT PROP Metallbaumaterial *nt*; **~ cladding** *n* STEEL Metallauskleidung *f*; **~-clad door** *n* ARCH & BUILD, CONST LAW metallausgekleidete Tür *f*; **~ cloth** *n* STEEL Drahtgewebe *nt*; **~ coat** *n* MAT PROP, STEEL, SURFACE Metallbelag *m*, Metallüberzug *m*; **~ coating** *n* STEEL, SURFACE Metallbeschichtung *f*; **~ column** *n* ARCH & BUILD, STEEL Metallstütze *f*; **~ construction** *n* STEEL Metallbau *m*; **~ construction work** *n* ARCH & BUILD, STEEL Metallbauarbeiten *f pl*; **~ core** *n* MAT PROP Metallkern *m*; **~ cramp** *n* MAT PROP, STEEL Metallklammer *f*; **~ decking** *n* MAT PROP, STEEL Metallabdeckung *f*

**metaled**: **~ road** *AmE see metalled road BrE*

**metal**: **~-enclosed** *adj* STEEL metallgekapselt; **~ façade** *n* ARCH & BUILD, STEEL Metallfassade *f*; **~ fatigue** *n* MAT PROP Metallermüdung *f*; **~ foil** *n* MAT PROP Metallfolie *f*; **~ foil insertion** *n* MAT PROP Metallfolieneinlage *f*; **~ hanger** *n* MAT PROP, STEEL Metallschuh *m*; **~ inert gas welding** *n* STEEL Metallinertgasschweißen *nt*; **~ insertion** *n* MAT PROP Metalleinlage *f*

**metalization** *AmE see metallization BrE*

***metal***: **~ lathing** *n* MAT PROP, STEEL Metallputzträger *m*

**metalled**: **~ road** *n BrE* INFR & DES Schotterstraße *f*

**metallic**: **~ structure** *n* ARCH & BUILD, STEEL Metallbauwerk *nt*

***metal***: **~ lining** *n* STEEL Metallauskleidung *f*

**metallization** *n BrE* MAT PROP, STEEL, SURFACE Metallbelag *m*

***metal***: **~ primer** *n* SURFACE Metallgrundierung *f*, Metallgrundierungsmittel *nt*; **~ rail** *n* MAT PROP Metallschiene *f*; **~ rung** *n* BUILD MACHIN *ladder* Metallsprosse *f*; **~ separator** *n* ENVIRON *magnetic separation* Metallsortieranlage *f*; **~ structure** *n* STEEL Metallbau *m*; **~ stud partition** *n* ARCH & BUILD, STEEL Metallskelettwand *f*; **~ stud wall** *n* ARCH & BUILD, STEEL Metallskelettwand *f*; **~ tie** *n* MAT PROP Metallanker *m*; **~ timber connector** *n* STEEL, TIMBER Holzmetallverbindung *f*; **~ valley** *n* STEEL Metalldachkehle *f*; **~ wall anchor** *n* MAT PROP, STEEL Metallmaueranker *m*; **~ waste** *n* ENVIRON Metallabfall *m*, Metallschrott *m*; **~ window** *n* STEEL Metallfenster *nt*

**meter** *n* ARCH & BUILD *measuring device* Zähler *m*, Meßgerät *nt*, *AmE* (*see metre BrE*) ; **~ board** *n* ELECTR Zählertafel *f*

**methane**: **~ digestion** *n* ENVIRON Methangärung *f*; **~ fermentation** *n* ENVIRON Methangärung *f*; **~ gas** *n* ENVIRON Methan *nt*

**method** *n* ARCH & BUILD Methode *f*, Verfahren *nt*; **~ of joints** *n* ARCH & BUILD Knotenpunktverfahren *nt*, Rundschnittverfahren *nt*; **~ of measurement** *n* ARCH & BUILD, INFR & DES Meßverfahren *nt*; **~ of moment distribution** *n* ARCH & BUILD, INFR & DES Cross-Verfahren *nt*, Methode *f* wiederholter Momentenverteilung, Momentenausgleichsverfahren *nt*; **~ of sections** *n* ARCH & BUILD Schnittverfahren *nt*

**methylene**: **~ blue test** *n* ENVIRON *method to determine putrefaction of water* Methylenblautest *m*

**meticulous**: **~ inspection** *n* ARCH & BUILD genaue Überprüfung *f*

**metre** *n BrE* ARCH & BUILD Meter *m* (*m*)

**metric**: **~ system** *n* ARCH & BUILD, CONST LAW, INFR & DES metrisches System *nt*

**mezzanine** *n* ARCH & BUILD Zwischengeschoß *nt*, Halbgeschoß *nt*, Mezzanin *nt*

**MF/UF** *abbr* (*melamine/urea formaldehyde*) ENVIRON, MAT PROP, TIMBER MF/UF (*Melamin/Urea-Formaldehyd*)

**mica** *n* MAT PROP Glimmer *m*

**micaceous**: **~ clay** *n* MAT PROP Glimmerton *m*

***mica***: **~ slate** *n* MAT PROP Glimmerschiefer *m*

**microclimate** *n* ENVIRON Mikroklima *nt*

**micron** *n* MAT PROP Mikron *nt*

**microorganism** *n* ENVIRON, INFR & DES, MAT PROP Mikroorganismus *m*

**micropollutant** *n* ENVIRON, INFR & DES, MAT PROP Mikroverschmutzer *m*

**micropollution** *n* ENVIRON, INFR & DES, MAT PROP Mikroverschmutzung *f*

**microscope** *n* INFR & DES Mikroskop *nt*

**microstructure** *n* MAT PROP Mikrostruktur *f*

**mid**: **~-depth** *n* ARCH & BUILD, CONCR, INFR & DES, MAT PROP Mitte *f*

**middle** *n* ARCH & BUILD Mitte *f*; **~ muntin** *n* ARCH & BUILD *window* Mittelsprosse *f*; **~ purlin** *n* ARCH & BUILD, TIMBER Mittelpfette *f*; **~ rail** *n* ARCH & BUILD Mittelschiene *f*; **~ span** *n* ARCH & BUILD, INFR & DES Mittelfeld *nt*; **~ third** *n* ARCH & BUILD mittleres Drittel *nt*

**midfeather** *n* ARCH & BUILD Schornsteinzunge *f*

**midspan** *n* ARCH & BUILD, CONCR Feldmitte *f*, Trägermitte *f*

**mid-third** *n* ARCH & BUILD Kern *m*

**mild**: **~ steel** *n* STEEL Flußstahl *m*, Weichstahl *m*; **~ steel pipe** *n* STEEL Flußstahlrohr *nt*

**mileometer** *n* BUILD MACHIN, INFR & DES, MAT PROP Entfernungsmesser *m*, Odometer *nt*

**milk**: **~ glass** *n* MAT PROP Milchglas *nt*

**mill** *vt* MAT PROP zermahlen, STONE behauen, fräsen, zermahlen

**millimeter** *AmE see millimetre BrE*

**millimetre** *n BrE* ARCH & BUILD Millimeter *m* (*mm*)

**millimetre**: **~ paper** *n BrE* ARCH & BUILD Millimeterpapier *nt*

**mined**: **~ space** *n* ENVIRON bergmännisch hergestellter Hohlraum *m*

**mineral** **1.** *adj* MAT PROP mineralisch; **2.** *n* MAT PROP Mineral *nt*

***mineral***: **~ fiber** *AmE*; **~ fibre** *BrE n* MAT PROP Mineralfaser *f*; **~ fibre filling** *n BrE* MAT PROP, SOUND & THERMAL Mineralfasereinlage *f*; **~ fibre mat** *n BrE* MAT PROP, SOUND & THERMAL Mineralfasermatte *f*; **~ fibre slab** *n BrE* MAT PROP, SOUND & THERMAL Mineralfaserplatte *f*

**mineralogy** *n* INFR & DES Mineralogie *f*, MAT PROP Mineralogie *f*, Gesteinskunde *f*

***mineral***: **~ orange** *n* MAT PROP, SURFACE Bleimennige *f*; **~ pitch** *n* INFR & DES, STONE Erdpech *nt*, Naturasphalt *m*; **~ red** *n* MAT PROP, SURFACE Bleimennige *f*; **~ tar** *n* INFR & DES Erdteer *m*; **~ wool** *n* MAT PROP Mineralwolle *f*

**minimize** *vt* ARCH & BUILD auf ein Minimum reduzieren

**minimum**: ~ **cross section** *n* ARCH & BUILD, ELECTR, HEAT & VENT Mindestquerschnitt *m*; ~ **depth** *n* ARCH & BUILD, INFR & DES Mindesttiefe *f*; ~ **quality** *n* MAT PROP Mindestgüte *f*; ~ **welding current** *n* ELECTR Mindestschweißstrom *m*

**minor**: ~ **external buildings** *n pl* ARCH & BUILD unwesentliche Außenbauten *m pl*

**mirror** *n* MAT PROP Spiegel *m*; ~ **glass** *n* MAT PROP Spiegelglas *nt*; ~ **method** *n* ARCH & BUILD spiegeloptisches Verfahren *nt*; ~ **reflector** *n* ELECTR *light fixture* Spiegelreflektor *m*

**misaligned** *adj* ARCH & BUILD schlecht fluchtend

**mission**: ~ **tiling** *n* ARCH & BUILD, STONE Mönch-Nonne-Ziegeldeckung *f*

**mist**: ~ **eliminator** *n* ENVIRON Tropfenabscheider *m*

**miter** *AmE see mitre BrE*

**mitered** *AmE see mitred BrE*

**mitre** *n BrE* BUILD MACHIN Gehrung *f*; ~ **board** *n BrE* BUILD MACHIN Gehrungsschnittlehre *f*; ~ **box** *n BrE* BUILD MACHIN Gehrungsschnittlehre *f*; ~**-cutting machine** *n BrE* BUILD MACHIN Gehrungsstanzmaschine *f*

**mitred** *adj BrE* TIMBER, STEEL, STONE auf Gehrung geschnitten; ~ **steel lintel** *n BrE* STEEL, STONE auf Gehrung geschnittener Stahlsturz *m*

***mitre***: ~ **gear** *n BrE* BUILD MACHIN Winkelgetriebe *nt*; ~ **joint** *n BrE* BUILD MACHIN Gehrstoß *m*, Gehrungsfuge *f*; ~ **sill** *n BrE* ARCH & BUILD, BUILD MACHIN Drempel *m*; ~ **square** *n BrE* BUILD MACHIN Gehrungswinkel *m*, festes Gehrungsdreieck *nt*

**mix** *vt* CONCR, MAT PROP, STONE anmachen, mischen, verschneiden

**mixed**: ~**-grained** *adj* MAT PROP gemischtkörnig; ~ **soil** *n* INFR & DES Mischboden *m*

**mixer** *n* BUILD MACHIN Mischer *m*, CONCR Mischmaschine *f*; ~ **conveyor** *n* BUILD MACHIN, CONCR Betontransportfahrzeug *nt*, Transportmischer *m*; ~ **platform** *n* BUILD MACHIN Mischerbühne *f*; ~ **tap** *n BrE* (*cf mixing faucet AmE*) WASTE WATER Einlochbatterie *f*, *bathroom* Mischbatterie *f*; ~ **truck** *n* BUILD MACHIN, CONCR Betontransportfahrzeug *nt*, Transportmischer *m*

**mixing** *n* CONCR, MAT PROP, STONE Anmachen *nt*, Mischen *nt*; ~ **drum** *n* BUILD MACHIN Mischtrommel *f*; ~ **faucet** *n AmE* (*cf mixer tap BrE*) WASTE WATER Einlochbatterie *f*, Mischbatterie *f*; ~ **liquid** *n* CONCR, STONE Anmachflüssigkeit *f*; ~ **preservative products** *n* ENVIRON, MAT PROP Vermischung *f* von Konservierungsmitteln; ~ **time** *n* CONCR Mischdauer *f*, Mischzeit *f*; ~ **water** *n* CONCR, STONE Anmachwasser *nt*

**mix**: ~ **in place** *n* CONCR Vorortmischen *nt*; ~ **proportions** *n pl* CONCR, MAT PROP Mischungsverhältnis *nt*

**mixture** *n* MAT PROP Gemisch *nt*, Mischung *f*; ~ **ratio** *n* CONCR, MAT PROP Mischungsverhältnis *nt*

**mm** *abbr* (*millimetre BrE*) ARCH & BUILD mm (*Millimeter*)

**MN** *abbr* (*mega-newton*) ARCH & BUILD, INFR & DES, MAT PROP MN (*Meganewton*)

**mobile** *adj* BUILD MACHIN beweglich, ortsveränderlich, fahrbar; ~ **concrete factory** *n* BUILD MACHIN fliegendes Betonwerk *nt*; ~ **crane** *n* BUILD MACHIN Autokran *m*

**mock**: ~**-up** *n* INFR & DES Lehrmodell *nt*

**model** *n* ARCH & BUILD Modell *nt*, Entwurf *m*; ~ **analysis** *n* ARCH & BUILD Modellstatik *f*; ~ **scale** *n* INFR & DES, MAT PROP Modellmaßstab *m*; ~ **study** *n* INFR & DES, MAT PROP Modellstudie *f*; ~ **test** *n* INFR & DES, MAT PROP Modelltest *m*

**modern** *adj* ARCH & BUILD modern

**modernization** *n* ARCH & BUILD Modernisierung *f*

**modernize** *vt* ARCH & BUILD modernisieren

**modes**: ~ **of control** *n pl* BUILD MACHIN, ELECTR, HEAT & VENT Kontrollmodi *m pl*

**modification** *n* ARCH & BUILD, CONST LAW Modifikation *f*, Variante *f*, Änderung *f*, Veränderung *f*

**modify** *vt* ARCH & BUILD ändern, verändern, CONST LAW ändern

**modular** *adj* ARCH & BUILD nach dem Baukastenprinzip gebaut, in Modulbauweise, modular; ~ **building system** *n* ARCH & BUILD Baukastenbauweise *f*, Modulbauweise *f*, Baukastenprinzip *nt*; ~ **building unit** *n* WASTE WATER Installationsblock *m*, Installationszelle *f*; ~ **coordination** *n* ARCH & BUILD Modulordnung *f*; ~ **design** *n* ARCH & BUILD Baukastensystem *nt*; ~ **dimensions** *n pl* ARCH & BUILD Rastermaß *nt*; ~ **gap** *n* ARCH & BUILD Modulspiel *nt*; ~ **grid** *n* ARCH & BUILD Bandraster *m*, Flächenmuster *nt*, Systemnetz *nt*, Systemliniengitter *nt*, Systemliniennetz *nt*; ~ **grid light fixture** *n* ELECTR Bandrasterleuchte *f*; ~ **grid system** *n* ARCH & BUILD Bandrastersystem *nt*; ~ **line** *n* ARCH & BUILD Systemlinie *f*; ~ **plane** *n* ARCH & BUILD Modulrasterebene *f*; ~ **point** *n* ARCH & BUILD Rasterpunkt *m*; ~ **space grid** *n* ARCH & BUILD dreidimensionaler Raster *m*, Raumraster *m*; ~ **unit** *n* ARCH & BUILD Zellenmodul *nt*

**module** *n* ARCH & BUILD Baustein *m*, Rastergrundmaß *nt*

**modulus** *n* ARCH & BUILD Modul *m*; ~ **of deformation** *n* ARCH & BUILD, INFR & DES, MAT PROP Verformungsmodul *m*; ~ **of elasticity** *n* CONCR, MAT PROP, STEEL E-Modul *m*, Elastizitätsmodul *m*; ~ **of longitudinal deformation** *n* ARCH & BUILD, CONCR, INFR & DES, MAT PROP, STEEL Ausdehnungsmodul *m*; ~ **of plasticity** *n* ARCH & BUILD, INFR & DES, MAT PROP Plastizitätsmodul *m*

**Mohr**: ~**'s circle** *n* INFR & DES Mohrscher Kreis *m*

**moist** *adj* CONCR, HEAT & VENT, INFR & DES, MAT PROP, STONE, SURFACE, SOUND & THERMAL feucht

**moisten** *vt* CONCR, STONE, SURFACE befeuchten, benetzen

**moist**: ~ **subsoil** *n* INFR & DES feuchter Untergrund *m*; ~ **unit weight** *n* MAT PROP Feuchtraumgewicht *nt*

**moisture** *n* ARCH & BUILD, HEAT & VENT, INFR & DES, MAT PROP, SOUND & THERMAL nichtdrückendes Wasser *nt*, Feuchte *f*, Feuchtigkeit *f*; ~ **content** *n* HEAT & VENT, INFR & DES, MAT PROP, TIMBER Feuchtigkeitsgehalt *m*; ~ **content control** *n* INFR & DES, MAT PROP Feuchtigkeitsgehaltkontrolle *f*; ~ **index** *n* INFR & DES, MAT PROP Feuchtigkeitsindex *m*; ~ **loss** *n* INFR & DES, MAT PROP Feuchtigkeitsverlust *m*; ~**-proof** *adj* MAT PROP, SOUND & THERMAL feuchtigkeitsbeständig; ~**-proof roofing sheet** *n* MAT PROP, SOUND & THERMAL Dachdichtungsbahn *f*

**mold** *AmE see mould BrE*

**molded** *AmE see moulded BrE*

**molding** *AmE see moulding BrE*

**moldy**: ~ **odor** *AmE see mouldy odour BrE*

**molecular** *adj* INFR & DES, MAT PROP molekular; ~ **analysis** *n* MAT PROP Molekularanalyse *f*; ~ **forces** *n pl* INFR & DES, MAT PROP Molekular-

kräfte *f pl*; **~ structure** *n* INFR & DES, MAT PROP Molekularstruktur *f*
**mole**: **~ drainage** *n* INFR & DES Schlitzdränung *f*
**moment** *n* ARCH & BUILD Moment *nt*; **~ area** *n* ARCH & BUILD, INFR & DES Momentenfläche *f*; **~ distribution method** *n* ARCH & BUILD, INFR & DES Momentenausgleichsverfahren *nt*; **~ equilibrium** *n* ARCH & BUILD, INFR & DES Momentengleichgewicht *nt*; **~ of flexion** *n* ARCH & BUILD, INFR & DES, MAT PROP, STEEL Biegemoment *nt*; **~ of inertia** *n* ARCH & BUILD Trägheitsmoment *nt*; **~ pole** *n* ARCH & BUILD Momentenpunkt *m*, Drehpunkt *m*; **~ of resistance** *n* ARCH & BUILD, TIMBER Widerstandsmoment *nt*; **~ at support** *n* ARCH & BUILD, INFR & DES Stützmoment *nt*
**monial** *n* ARCH & BUILD Fensterpfosten *m*
**monitor 1.** *n* BUILD MACHIN, CONST LAW Überwachungsgerät *nt*; **2.** *vt* CONST LAW überwachen, ENVIRON kontrollieren, INFR & DES überwachen
**monitoring** *n* CONST LAW, INFR & DES Überwachung *f*; **~ after site closure** *n* ENVIRON Nachsorge *f*; **~ equipment** *n* CONST LAW, INFR & DES Überwachungsanlage *f*; **~ network** *n* INFR & DES Überwachungsnetz *nt*; **~ well** *n* ENVIRON, INFR & DES Beobachtungsbrunnen *m*, Kontrollbrunnen *m*
**monolith** *n* ARCH & BUILD Monolit *m*, Menhir *m*
**monolithic**: **~ connection** *n* ARCH & BUILD, CONCR, INFR & DES, MAT PROP monolithische Verbindung *f*; **~ construction method** *n* ARCH & BUILD monolithische Bauweise *f*
**monopitch**: **~ roof** *n* ARCH & BUILD Pultdach *nt*, Schleppdach *nt*, TIMBER Pultdach *nt*
**monorail** *n* INFR & DES Einschienenbahn *f*, Einschienenhängebahn *f*
**mono**: **~-symmetric** *adj* ARCH & BUILD, CONCR, INFR & DES, MAT PROP, STEEL einfach symmetrisch
**monthly**: **~ sun paths** *n pl* ENVIRON, HEAT & VENT Sonnenwege *m pl* im jeweiligen Monat
**moor** *n* INFR & DES Moorboden *m*
**moraine** *n* INFR & DES Moräne *f*
**morainic**: **~ filter layer** *n* ENVIRON *disposal site* Moränenfilterschicht *f*
**mordant** *n* SURFACE Beize *f*, Beizmittel *nt*
**mortar** *n* MAT PROP, STONE Mörtel *m*; **~ additive** *n* MAT PROP, STONE Mörtelzusatz *m*; **~ analysis** *n* STONE Mörtelanalyse *f*; **~ bed** *n* STONE Mörtelbett *nt*; **~ class** *n* MAT PROP Mörtelgruppe *f*; **~ droppings** *n pl* STONE Mörtelabfälle *m pl*; **~ intrusion** *n* STONE Mörteleinpressung *f*; **~ jointing** *n* STONE *of roof tiles* Vermörtelung *f*; **~ mixer** *n* BUILD MACHIN Mörtelmischmaschine *f*; **~ mixing** *n* STONE Mörtelmischen *nt*; **~ pigment** *n* STONE Mörtelpigment *nt*; **~ ratios** *n pl* STONE Mörtelverhältnisse *nt pl*
**mortgage** *n* CONST LAW Hypothek *f*
**mortgagee** *n* CONST LAW Hypothekengläubiger *m*
**mortgagor** *n* CONST LAW Hypothekenschuldner *m*
**mortise 1.** *n* STONE, TIMBER Falz *m*, Fuge *f*, Stemmloch *nt*, Zapfenloch *nt*; **2.** *vt* STONE ausstemmen, TIMBER ausstemmen, stemmen, verzapfen
***mortise***: **~ chisel** *n* BUILD MACHIN, TIMBER Stemmeisen *nt*, Stemmeißel *m*; **~ deadlock** *n* BUILD HARDW, TIMBER Einsteckschloß *nt*; **~ gage** *AmE*, **~ gauge** *BrE n* TIMBER Zapfenstreichmaß *nt*; **~ latch** *n* BUILD HARDW Einsteckfallenschloß *nt*; **~ lock** *n* BUILD HARDW, TIMBER Einsteckschloß *nt*; **~ and tenon joint** *n* TIMBER Zapfenverbindung *f*
**mortising** *n* STONE, TIMBER Einstemmen *nt*, Stemm-, Verzapfung *f*; **~ machine** *n* BUILD MACHIN, TIMBER Stemmaschine *f*
**mosaic** *n* ARCH & BUILD, STONE Mosaik *nt*; **~ flooring** *n* ARCH & BUILD, STONE Mosaikbelag *m*, Mosaikfußboden *m*; **~ paving** *n* ARCH & BUILD, INFR & DES, STONE Mosaikpflaster *nt*
**motion**: **~ detector** *n* ELECTR Bewegungsmelder *m*; **~ sensor** *n* ELECTR Bewegungsmelder *m*
**motor** *n* BUILD MACHIN Motor *m*; **~ grader** *n* BUILD MACHIN Motorstraßenhobel *m*; **~ roller** *n* BUILD MACHIN Motorwalze *f*
**motorway** *n* (*cf freeway AmE*) INFR & DES Schnellstraße *f*, Autobahn *f*
**mottled** *adj* MAT PROP, SURFACE meliert
**mould 1.** *n BrE* BUILD MACHIN *concrete test cubes* Form *f*, Gußform *f*, HEAT & VENT, SURFACE Schimmel *m*; **2.** *vt BrE* MAT PROP formen, vergießen
**moulded** *adj BrE* MAT PROP geformt; **~ fibre board** *n BrE* MAT PROP, TIMBER Hartfaserplatte *f*; **~ glass** *n BrE* MAT PROP Preßglas *nt*; **~ laminated plastic** *n BrE* MAT PROP Schichtpreßstoff *m*
***mould***: **~ growth** *n BrE* HEAT & VENT Schimmelwuchs *m*
**moulding** *n BrE* ARCH & BUILD Zierleiste *f*, Fries *m*, Gesims *nt*, *groove* Kehlung *f*, BUILD HARDW, MAT PROP, TIMBER Leiste *f*, Zierleiste *f*; **~ machine** *n BrE* BUILD MACHIN, TIMBER Leistenhobelmaschine *f*; **~ machine for prefabricated concrete elements** *n BrE* BUILD MACHIN Formmaschine *f* für Betonfertigbauteile
***mould***: **~ oil** *n BrE* CONCR Schalungsöl *nt*; **~-resistant** *adj BrE* MAT PROP schimmelbeständig
**mouldy**: **~ odour** *n BrE* HEAT & VENT Schimmelgeruch *m*
**mount** *vt* ARCH & BUILD, HEAT & VENT, INFR & DES anbringen, montieren, einbauen, CONCR, STEEL, TIMBER anbringen
**mountain** *n* INFR & DES Berg *m*
**mounting** *n* ARCH & BUILD *process* Einbau *m*, Einspannung *f*, Montage *f*, *support, base* Untersatz *m*, Sockel *m*; **~ accessories** *n pl* BUILD HARDW Einbaugarnitur *f*; **~ anchor** *n* MAT PROP Montageanker *m*; **~ angle** *n* MAT PROP Haltewinkel *m*; **~ base** *n* MAT PROP Tragplatte *f*; **~ bracket** *n* BUILD HARDW Haltelasche *f*; **~ device** *n* BUILD MACHIN Einspannvorrichtung *f*, MAT PROP Befestigung *f*, Befestigungsvorrichtung *f*; **~ frame** *n* MAT PROP Einbaurahmen *m*; **~ height** *n* ARCH & BUILD Ausführungshöhe *f*; **~ material** *n* MAT PROP Befestigungsmaterial *nt*; **~ plate** *n* MAT PROP Montageplatte *f*; **~ strap** *n* MAT PROP Befestigungslasche *f*; **~ template** *n* BUILD HARDW Montageschablone *f*; **~ work** *n* ARCH & BUILD, INFR & DES Montagearbeiten *f pl*
**mousehole** *n* ENVIRON Vorbohrloch *nt*
**mouth** *n* INFR & DES Mündung *f*
**moveable** *adj* BUILD MACHIN beweglich, verfahrbar
**move**: **~ the center line** *AmE*, **~ the centre line** *BrE phr* INFR & DES Achse verschieben
**movement** *n* ARCH & BUILD, MAT PROP Bewegung *f*; **~ joint** *n* ARCH & BUILD, INFR & DES Bewegungsfuge *f*
**moving**: **~ load** *n* INFR & DES Wanderlast *f*, bewegliche Last *f*; **~ platform** *n* INFR & DES bewegliche Arbeitsbühne *f*; **~ staircase** *n* INFR & DES Rolltreppe *f*; **~ stairway** *n* INFR & DES Rolltreppe *f*; **~ traffic** *n* INFR & DES fließender Verkehr *m*

**Mp** *abbr* (*megapond*) ARCH & BUILD, INFR & DES, MAT PROP Mp (*Megapond*)
**MPL** *abbr* (*maximum permissible level*) CONST LAW, ENVIRON, INFR & DES höchstzulässiges Niveau *nt*
**MSW** *abbr* (*municipal solid waste*) ENVIRON fester Siedlungsabfall *m*
**mud** *n* HEAT & VENT, INFR & DES, MAT PROP Schlamm *m*, Schlick *m*; ~ **bank** *n* INFR & DES Schlammbank *f*; ~ **deposit** *n* INFR & DES Schlammablagerung *f*
**muddy**: ~ **sand** *n* INFR & DES, MAT PROP schlammiger Sand *m*; ~ **soil** *n* INFR & DES, MAT PROP schlammiger Boden *m*
***mud***: ~ **layer** *n* INFR & DES, MAT PROP Schlammschicht *f*; ~ **sample** *n* INFR & DES, MAT PROP Schlammprobe *f*
**mudsill** *n* ARCH & BUILD Schlammschwelle *f*
**muffle** *vt* SOUND & THERMAL dämpfen
**muffler** *n* *AmE* (*cf silencer BrE*) SOUND & THERMAL Schalldämpfer *m*
**multibarrier**: ~ **principle** *n* ENVIRON Mehrschichtprinzip *nt*
**multibeam** *adj* ARCH & BUILD, TIMBER mehrbalkig
**multicellular**: ~ **glass** *n* MAT PROP Schaumglas *nt*
**multicomponent**: ~ **epoxy resin** *n* MAT PROP Mehrkomponentenepoxidharz *nt*
**multidisciplinary**: ~ **team** *n* ARCH & BUILD, INFR & DES multidiziplinarisches Team *nt*
**multilayer**: ~ **glass** *n* MAT PROP Verbundglas *nt*
**multiple**: ~**-arch dam** *n* ARCH & BUILD Bogenpfeilermauer *f*, INFR & DES Gewölbpfeilermauer *f*, Bogenpfeilermauer *f*, Staumauer *f*, Vielfachbogensperre *f*; ~ **currencies** *n pl* CONST LAW mehrere Währungen *f pl*; ~ **development** *n* ENVIRON vielfache Entwicklung *f*; ~**-hearth incinerator** *n* ENVIRON Etagenofen *m*; ~**-point earthing** *n* *BrE* (*cf multiple-point grounding AmE*) ELECTR Mehrpunkterdung *f*; ~**-point grounding** *n* *AmE* (*cf multiple-point earthing BrE*) ELECTR Mehrpunkterdung *f*; ~ **rib pillar** *n* ARCH & BUILD Bündelpfeiler *m*; ~ **series** *n* ELECTR *connection* Gruppenschaltung *f*; ~ **shear connection** *n* ARCH & BUILD, INFR & DES mehrschnittige Verbindung *f*; ~ **shear joint** *n* ARCH & BUILD, INFR & DES mehrschnittiges Gelenk *nt*
**multispan** *adj* ARCH & BUILD mehrfeldrig
**multistorey** *adj* *BrE* ARCH & BUILD mehrgeschossig, mehrstöckig; ~ **building** *n* *BrE* ARCH & BUILD, INFR & DES mehrstöckiges Gebäude *nt*; ~ **car park** *n* *BrE* INFR & DES Parkhaus *nt*
**multistory** *AmE see multistorey BrE*
**municipal**: ~ **sewage** *n* ENVIRON kommunales Abwasser *nt*; ~ **sewage works** *n* ENVIRON kommunale Kläranlage *f*; ~ **sewer system** *n* WASTE WATER städtische Kanalnetz *nt*; ~ **solid waste** *n* (*MSW*) ENVIRON fester Siedlungsabfall *m*; ~ **waste** *n* ENVIRON Hausmüll *m*, Siedlungsabfall *m*, kommunaler Abfall *m*, INFR & DES Hausmüll *m*; ~ **waste landfill** *n* ENVIRON Hausmülldeponie *f*
**munnion** *n* ARCH & BUILD Fenstersprosse *f*
**muntin** *n* ARCH & BUILD Sprosse *f*
**mushroom**: ~ **valve** *n* HEAT & VENT Pilzventil *nt*, Tellerventil *nt*

# N

**N** *abbr* (*newton*) ARCH & BUILD, INFR & DES, MAT PROP N (*Newton*)

**NACCB** *abbr BrE* (*National Accreditation Council for Certification Bodies*) CONST LAW NACCS (*Nationaler Akkreditationsrat für Normungsausschüsse*)

**nail 1.** *n* ARCH & BUILD, BUILD MACHIN, MAT PROP, STEEL, TIMBER Nagel *m*; **2.** *vt* ARCH & BUILD, BUILD MACHIN, MAT PROP, STEEL, TIMBER annageln, mitnageln, nageln, verfestigen, vernageln

***nail***: **~ claw** *n* BUILD HARDW, TIMBER Nagelklaue *f*, Nagelzieheisen *nt*; **~ diameter** *n* MAT PROP, TIMBER Nageldurchmesser *m*; **~ distance** *n* MAT PROP, TIMBER Nagelabstand *m*

**nailed** *adj* ARCH & BUILD, STEEL, TIMBER genagelt; **~ beam** *n* ARCH & BUILD, TIMBER genagelter Träger *m*; **~ connection** *n* TIMBER Nagelverbindung *f*; **~ framework** *n* ARCH & BUILD, TIMBER Nagelbinder *m*; **~ plywood gusset** *n* STEEL, TIMBER genageltes Sperrholzknotenblech *nt*; **~ truss** *n* ARCH & BUILD, TIMBER Nagelbinder *m*

***nail***: **~ extractor** *n* BUILD MACHIN Nagelzieher *m*; **~ head** *n* MAT PROP Nagelkopf *m*; **~ length** *n* MAT PROP Nagellänge *f*; **~ point** *n* MAT PROP Nagelspitze *f*; **~ puller** *n* BUILD MACHIN Nagelzieher *m*; **~ punch** *n* BUILD MACHIN Nageltreiber *m*; **~ roof truss** *n* ARCH & BUILD Nageldachbinder *m*; **~ set** *n* BUILD MACHIN Handdurchschläger *m*

**NAMAS** *abbr BrE* (*National Measurement Accreditation Service*) CONST LAW NAMAS (*Nationaler Messungs-Akkreditationsservice*)

**nap** *n* MAT PROP Noppe *f*

**narrow** *adj* ARCH & BUILD eng, schmal

**National**: **~ Accreditation Council for Certification Bodies** *n BrE* (*NACCB*) CONST LAW Nationaler Akkreditationsrat *m* für Normungsausschüsse (*NACCS*); **~ Laboratory Testing Accreditation Service** *n BrE* (*NATLAS*) CONST LAW Staatliche Anstalt *f* für die Beglaubigung von Laboruntersuchungen (*NATLAS*); **~ Measurement Accreditation Service** *n BrE* (*NAMAS*) CONST LAW Nationaler Messungs-Akkreditationsservice *m* (*NAMAS*); **~ Physical Laboratory** *n BrE* (*NPL*) CONST LAW Staatliches Physikalisches Labor *nt* (*NPL*)

**NATLAS** *abbr BrE* (*National Laboratory Testing Accreditation Service*) CONST LAW NATLAS (*Staatliche Anstalt für die Beglaubigung von Laboruntersuchungen*)

**NATM**® *abbr* (*new Austrian tunnelling method*) INFR & DES NATM® (*neue österreichische Tunnelbauweise*)

**natural** *adj* ARCH & BUILD, CONCR, CONST LAW, ENVIRON, HEAT & VENT, INFR & DES, MAT PROP, SOUND & THERMAL natürlich, TIMBER schalungsrauh; **~ cement** *n* INFR & DES, MAT PROP Naturzement *m*; **~ cleft** *adj* STONE bruchrauh; **~ color** *AmE*, **~ colour** *BrE adj* MAT PROP naturfarben, natürliche Farbe *f*; **~ convection** *n* HEAT & VENT natürliche Konvektion *f*; **~ convective loop** *n* HEAT & VENT natürliche Konvektionsschleife *f*; **~ curb** *n AmE* (*cf natural kerb BrE*) MAT PROP Naturbordstein *m*; **~ curbstone** *n AmE* (*cf natural kerbstone AmE*) MAT PROP Naturbordstein *m*; **~ environment** *n* ENVIRON natürliche Umwelt *f*; **~ fall** *n* INFR & DES natürliches Gefälle *nt*; **~ filtration** *n* INFR & DES natürliche Filtrierung *f*; **~ finish** *adj* TIMBER *veneer* naturbelassen; **~ foundation** *n* ARCH & BUILD tragfähiger Boden *m*, tragfähiger Naturgrund *m*; **~ frequency** *n* HEAT & VENT, INFR & DES, SOUND & THERMAL Eigenfrequenz *f*; **~ gas** *n* ENVIRON, HEAT & VENT, INFR & DES Erdgas *nt*, Reingas *nt*; **~ kerb** *BrE* (*cf natural curb AmE*) MAT PROP Naturbordstein *m*; **~ kerbstone** *BrE* (*cf natural curbstone AmE*) MAT PROP Naturbordstein *m*

**naturally**: **~ acid lake** *n* ENVIRON natursaurer See *m*; **~ split** *adj* MAT PROP *stone* bruchrauh

***natural***: **~ pavement** *n* STONE Naturpflaster *nt*; **~ pozzolana** *n* CONCR, MAT PROP natürliche Puzzolanerde *f*; **~ purification** *n* ENVIRON Selbstreinigung *f*, INFR & DES natürliche Reinigung *f*; **~ resources** *n pl* INFR & DES Bodenschätze *m pl*; **~ sand** *n* MAT PROP Natursand *m*, Betonsand *m*; **~ slate** *n* MAT PROP Naturschiefer *m*; **~ slope** *n* ARCH & BUILD natürliche Böschung *f*, INFR & DES natürlicher Böschungswinkel *m*; **~ soil** *n* INFR & DES Naturboden *m*; **~ stone** *n* INFR & DES, MAT PROP, STONE Naturstein *m*; **~ stone masonry** *n* ARCH & BUILD, STONE Natursteinmauerwerk *nt*, *ashlar work* Quadermauerwerk *nt*, Werksteinmauerwerk *nt*; **~ ventilation** *n* INFR & DES natürliche Belüftung *f*; **~ water content** *n* INFR & DES, MAT PROP natürlicher Wasserinhalt *m*

**nature** *n* INFR & DES Natur *f*, MAT PROP Eigenschaft *f*; **~ of the ground** *n* INFR & DES Bodenbeschaffenheit *f*; **~ of soil** *n* INFR & DES Bodenart *f*

**neap**: **~ tide** *n* ENVIRON Nipptide *f*

**neat**: **~ cement** *n* CONCR Zementschlamm *m*, *paste* Zementpaste *f*, Zementleim *m*; **~ lime** *n* CONCR, MAT PROP, STONE Kalkschlämme *m pl*, reiner Kalkmörtel *m*

**neck** *n* ARCH & BUILD, BUILD HARDW, HEAT & VENT Hals *m*; **~ molding** *AmE*, **~ moulding** *BrE n* ARCH & BUILD Säulenhals *m*

**needle 1.** *n* ARCH & BUILD Abstützbalken *m*, BUILD MACHIN Nadel *f*; **2.** *vt* ARCH & BUILD absteifen

***needle***: **~ beam** *n* ARCH & BUILD, INFR & DES Spundpfahl *m*, Abstützbalken *m*; **~ felt** *n* ARCH & BUILD, INFR & DES, MAT PROP Nadelfilz *m*

**negative**: **~ bending moment** *n* MAT PROP, TIMBER negatives Biegemoment *nt*; **~ equity** *n* CONST LAW negativer Immobilienwert *m*; **~ pressure** *n* ARCH & BUILD negativer Druck *m*, HEAT & VENT Unterdruck *m*; **~ reinforcement** *n* CONCR Bewehrung *f* im Bereich negativer Momente

**neighborhood** *AmE see neighbourhood BrE*

**neighboring** *AmE see neighbouring BrE*

**neighbourhood** *n BrE* ARCH & BUILD Nachbarschaft *f*

**neighbouring** *adj BrE* ARCH & BUILD benachbart

**neoprene** *n* ARCH & BUILD, MAT PROP Neopren *nt*; **~ bearing pad** *n* ARCH & BUILD, MAT PROP Neoprenlagerkissen *nt*; **~ gasket** *n* ARCH & BUILD, MAT PROP

Neoprendichtungsband *nt*, Neoprendichtungsprofil *nt*; ~ **sealing** *n* MAT PROP Neoprendichtungsband *nt*, Neoprendichtungsprofil *nt*
**nervure** *n* ARCH & BUILD *vault* Seitenrippe *f*
**net** *adj* ARCH & BUILD, INFR & DES, MAT PROP netto; ~ **area** *n* ARCH & BUILD, INFR & DES Nettofläche *f*; ~ **capacity** *n* ELECTR Nutzleistung *f*
**netting** *n* MAT PROP Geflecht *nt*, STEEL Geflecht *nt*, *wire* Netzflechtwerk *nt*
**network** *n* CONST LAW, ELECTR, HEAT & VENT, INFR & DES, MAT PROP, WASTE WATER Netz *nt*, Netzwerk *nt*; ~ **diagram** *n* CONST LAW Netzwerkdiagramm *nt*; ~ **planning** *n* INFR & DES Netzwerkplanung *f*
**neutral** *adj* ARCH & BUILD, MAT PROP neutral; ~ **axis** *n* ARCH & BUILD Nullachse *f*, neutrale Achse *f*
**neutralization** *n* ENVIRON Neutralisierung *f*
**neutralize** *vti* ENVIRON neutralisieren
**neutralizer** *n* ENVIRON Neutralisationsmittel *nt*
**neutralizing**: ~ **agent** *n* ENVIRON Neutralisationsmittel *nt*
**neutral**: ~ **surface** *n* MAT PROP neutrale Oberfläche *f*; ~ **wire** *n* ELECTR Nulleiter *m*
**new** *adj* ARCH & BUILD neu; ~ **Austrian tunnelling method** *n* (*NATM*®) INFR & DES neue österreichische Tunnelbauweise *f* (*NATM*®); ~ **building** *n* ARCH & BUILD Neubau *m*
**newel** *n* ARCH & BUILD, TIMBER Austrittspfosten *m*, Spindel *f*, Treppenpfosten *m*; ~ **cap** *n* ARCH & BUILD *stairs* Spindelkappe *f*; ~ **post** *n* ARCH & BUILD Antrittspfosten *m*, TIMBER massive Treppenspindel *f*
**new**: ~ **plant** *n* ENVIRON neue Betriebsanlage *f*
**newton** *n* (*N*) ARCH & BUILD, INFR & DES, MAT PROP Newton *nt* (*N*)
**niche** *n* ARCH & BUILD Nische *f*
**nickel** *n* MAT PROP, SURFACE Nickel *nt*; ~ **oxide coating** *n* MAT PROP, SURFACE Nickeloxydbeschichtung *f*; ~**-plated** *adj* MAT PROP, SURFACE vernickelt; ~ **silver** *n* MAT PROP Neusilber *nt*
**night**: ~ **current** *n* ELECTR Nachtstrom *m*; ~ **economy** *n* HEAT & VENT *heating* Nachtabsenkung *f*; ~ **economy feature** *n* HEAT & VENT *heating* Nachtabsenkung *f*; ~ **insulation** *n* SOUND & THERMAL Nachtisolierung *f*; ~ **storage heating** *n* HEAT & VENT Nachtspeicherheizung *f*, Nachtstromspeicherheizung *f*, Speicherheizung *f*
**nipper**: ~ **pliers** *n* BUILD MACHIN Kneifzange *f*
**nippers** *n pl* BUILD MACHIN Greifer *m*, *pincers* Kneifzange *f*, *stone tongs, lifting appliance* Steingreifer *m*
**nipple** *n* BUILD MACHIN Anschlußstück *nt*, Nippel *m*
**niter** *AmE see nitre BrE*
**nitrate** *n* ENVIRON, MAT PROP Nitrat *nt*
**nitre** *n BrE* ENVIRON, MAT PROP Salpeter *m*
**nitric**: ~ **acid** *n* ENVIRON, MAT PROP Salpetersäure *f*; ~ **oxide** *n* ENVIRON, MAT PROP Stickoxid *nt*
**nitrocellulose**: ~ **lacquer** *n* SURFACE Nitrolack *m*
**nitrogen** *n* ENVIRON, MAT PROP Stickstoff *m*; ~ **dioxide** *n* ENVIRON, MAT PROP Stickstoffdioxid *nt*; ~ **oxide** *n* ENVIRON , MAT PROP Stickoxid *nt*; ~ **pentoxide** *n* ENVIRON, MAT PROP Stickstoffpentoxid *nt*; ~ **peroxide** *n* ENVIRON, MAT PROP Stickstoffdioxid *nt*
**noble**: ~ **metal** *n* ENVIRON, MAT PROP Edelmetall *nt*
**no-bond**: ~ **prestressing** *n* CONCR Vorspannen *nt* ohne Verbund; ~ **tensioning** *n* CONCR Vorspannen *nt* ohne Verbund
**nodal**: ~ **joint** *n* TIMBER *truss* Knotenpunktverbindung *f*; ~ **point** *n* ARCH & BUILD, INFR & DES Knoten *m*, Knotenpunkt *m*
**nodular**: ~ **cast iron** *n* MAT PROP Sphäroguß *m*, Kugelgraphitguß *m*
**no-fines**: ~ **concrete** *n* CONCR entfeinter Beton *m*, haufwerksporiger Beton *m*
**nog** *n* TIMBER Holzpflock *m*, *wooden dowel* Holzdübel *m*, *wooden peg* Holzstift *m*
**nogging** *n* ARCH & BUILD Ausfachung *f*, Ausmauerung *f*, STONE Ausmauerung *f*, TIMBER Holzdübel *m*; ~ **piece** *n* ARCH & BUILD Holzriegel *m*
**noise** *n* CONST LAW, ENVIRON, INFR & DES, SOUND & THERMAL Geräusch *nt*, Lärm *m*; ~ **abatement** *n*, CONST LAW, ENVIRON, SOUND & THERMAL Lärmschutz *m*; ~**-absorbent** *adj* SOUND & THERMAL geräuschdämpfend; ~ **barrier** *n* SOUND & THERMAL Lärmschutzsperre *f*; ~ **control** *n* CONST LAW, ENVIRON, SOUND & THERMAL Lärmbekämpfung *f*; ~ **insulation** *n* SOUND & THERMAL Lärmschutz *m*; ~ **intensity** *n* SOUND & THERMAL Geräuschstärke *f*; ~ **level** *n* SOUND & THERMAL Lärmpegel *m*; ~ **measurement** *n* CONST LAW, ENVIRON Geräuschmessung *f*; ~ **nuisance** *n* CONST LAW, ENVIRON Lärmbelästigung *f*; ~ **pollution** *n* CONST LAW, ENVIRON Lärmbelästigung *f*; ~ **protection** *n* SOUND & THERMAL Lärmschutz *m*; ~ **protection embankment** *n* INFR & DES Lärmschutzwall *m*; ~ **protection window** *n* ARCH & BUILD, SOUND & THERMAL Lärmschutzfenster *nt*, Schallschutzfenster *nt*; ~ **reduction** *n* CONST LAW, ENVIRON, SOUND & THERMAL Geräuschminderung *f*; ~ **source** *n* ENVIRON Lärmquelle *f*; ~ **suppression** *n* ENVIRON, SOUND & THERMAL Lärmbekämpfung *f*
**noisy** *adj* SOUND & THERMAL laut
**nominal**: ~ **cross section** *n* ARCH & BUILD Nennquerschnitt *m*; ~ **current** *n* ELECTR Nennstrom *m*; ~ **dimension** *n* ARCH & BUILD Nennmaß *nt*; ~ **measure** *n* TIMBER Rauhmaß *nt*; ~ **pressure** *n* HEAT & VENT, WASTE WATER Nenndruck *m*; ~ **size** *n* ARCH & BUILD Nennmaß *nt*; ~ **stress** *n* ARCH & BUILD, TIMBER Nennspannung *f*; ~ **width** *n* ARCH & BUILD Nennweite *f*
**nominated**: ~ **sub-contractor** *n* CONST LAW benannter Subunternehmer *m*; ~ **sub-contractor defaults** *n pl* CONST LAW Versäumnisse *nt pl* des benannten Subunternehmers
**nonacidic**: ~ **lake** *n* ENVIRON nicht saurer See *m*
**nonaggressive** *adj* INFR & DES nicht aggressiv
**nonbasement**: ~ **building** *n* ARCH & BUILD nicht unterkellertes Gebäude *nt*
**nonbiodegradable**: ~ **packaging** *n* ENVIRON persistente Packstoffe *m pl*; ~ **waste** *n* ENVIRON biologisch nicht abbaubarer Abfall *m*
**noncohesive** *adj* INFR & DES, MAT PROP inkohärent, kohäsionslos, nicht kohäsiv
**noncombustible** *adj* CONST LAW, ENVIRON, INFR & DES, MAT PROP nicht brennbar; ~ **residue** *n* ENVIRON inertisierter Rückstand *m*
**noncomplanar**: ~ **forces** *n pl* ARCH & BUILD nicht regelmäßig verteilte Kräfte *f pl*
**noncomposite**: ~ **steel flange** *n* ARCH & BUILD, INFR & DES, STEEL nicht zusammengesetzter Stahlflansch *m*
**noncompostable**: ~ **waste** *n* ENVIRON nicht kompostierbarer Abfall *m*
**nonconductive** *adj* MAT PROP nichtleitend

**nondomestic** *adj* ARCH & BUILD, INFR & DES nicht zum Wohnhausbau gehörend
**nonferrous**: ~ **metal** *n* MAT PROP Nichteisenmetall *nt*, NE-Metall *nt*
**nonhardening** *adj* MAT PROP dauerplastisch
**nonplastic** *adj* MAT PROP nicht plastisch, rollig
**nonpoisonous** *adj* MAT PROP ungiftig
**nonporous** *adj* MAT PROP porenfrei
**nonpotable**: ~ **water** *n* CONST LAW, ENVIRON, INFR & DES, WASTE WATER Brauchwasser *nt*
**nonprestressed** *adj* CONCR schlaff bewehrt
**nonrecoverable**: ~ **waste** *n* ENVIRON nicht rückgewinnbarer Abfall *m*
**nonreinforced** *adj* CONCR, MAT PROP nicht bewehrt, unarmiert, unbewehrt; ~ **concrete** *n* CONCR unbewehrter Beton *m*
**nonreturn**: ~ **valve** *n* BUILD MACHIN, ENVIRON, WASTE WATER Rückschlagventil *nt*
**nonskid** *adj* MAT PROP, SURFACE rutschfest, gleitsicher
**nonstandard** *adj* INFR & DES, MAT PROP nicht auf Lager
**nonstructural** *adj* ARCH & BUILD nichttragend
**nonsway** *adj* ARCH & BUILD, CONCR, INFR & DES, STEEL unverschieblich; ~ **frame** *n* ARCH & BUILD, CONCR, INFR & DES, STEEL unverschieblicher Rahmen *m*
**nonsymmetric** *adj* ARCH & BUILD, CONCR, INFR & DES, MAT PROP, STEEL unsymmetrisch
**nonsymmetrical** *adj* ARCH & BUILD, CONCR, INFR & DES, MAT PROP, STEEL unsymmetrisch
**nontoxic** *adj* ENVIRON, MAT PROP ungiftig
**nontreated** *adj* MAT PROP unbehandelt
**nonuniform** *adj* ARCH & BUILD, CONCR, INFR & DES, MAT PROP, STEEL ungleichförmig
**nonventilated** *adj* MAT PROP *roof* einschalig; ~ **flat roof** *n* ARCH & BUILD, INFR & DES Warmdach *nt*
**nonvolatile**: ~ **contents** *n pl* SURFACE *paint* nicht flüchtiger Inhalt *m*
**nonwashable**: ~ **distemper** *n* MAT PROP, SURFACE Leimfarbe *f*
**nonwaste**: ~ **technology** *n* (*NWT*) ENVIRON saubere Technologie *f*, umweltfreundliche Technologie *f*
**nonwoven**: ~ **fabric** *n* MAT PROP Faservlies *nt*, Vlies *nt*
**normal** *adj* ARCH & BUILD, BUILD HARDW, CONCR, ENVIRON, HEAT & VENT, INFR & DES, MAT PROP Normal-, normal; ~ **concrete** *n* CONCR Normalbeton *m*; ~ **format** *n* ARCH & BUILD, MAT PROP Normalformat *nt* (*NF*); ~**-weight concrete** *n* CONCR, INFR & DES, MAT PROP Normalgewichtsbeton *m*
**northlight**: ~ **roof** *n* ARCH & BUILD Sägezahndach *nt*, Sheddach *nt*
**nose** *n* ARCH & BUILD Vorsprung *m*, HEAT & VENT Bolzenverbindung *f*
**nosing** *n* BUILD HARDW *door bolt* Mitnehmer *m*, Winkeleckleiste *f*, abgerundete Kantenschutzschiene *f*, *of steps* Kantenschutzschiene *f*; ~ **line** *n* ARCH & BUILD Stufenkantenlinie *f*
**notation** *n* MAT PROP Kennzeichnung *f*
**notch 1.** *n* ARCH & BUILD Aussparung *f*, Kerbe *f*, Nut *f*; **2.** *vt* ARCH & BUILD, TIMBER ausklinken, aussparen, einschneiden, einzapfen
**notched** *adj* MAT PROP gekerbt
**notching** *n* TIMBER Aussparen *nt*
***notch***: ~ **joint** *n* ARCH & BUILD Kerbverbindung *f*; ~ **sensitivity** *n* MAT PROP Kerbempfindlichkeit *f*
**notice**: ~ **of award** *n* ARCH & BUILD Auftragsschreiben *nt*; ~ **of determination** *n* CONST LAW Bestimmungsbescheid *m*
**notices** *n pl* CONST LAW Anzeigen *f pl*
**notional** *adj* ARCH & BUILD fiktiv; ~ **building** *n* ARCH & BUILD, INFR & DES theoretisches Gebäude *nt*
**nozzle** *n* ENVIRON, HEAT & VENT Düse *f*; ~ **flange** *n* HEAT & VENT Düsenflansch *m*
**NPL** *abbr BrE* (*National Physical Laboratory*) CONST LAW NPL (*Staatliches Physikalisches Labor*)
**nuclear**: ~ **power plant** *n* ENVIRON, INFR & DES Atomkraftwerk *nt* (*AKW*); ~ **power station** *n* ENVIRON, INFR & DES Atomkraftwerk *nt* (*AKW*); ~ **waste** *n* CONST LAW, ENVIRON radioaktiver Abfall *m*
**number** *n* ARCH & BUILD Stückzahl *f*, *quantity* Anzahl *f*; ~ **of air changes** *n* HEAT & VENT Luftwechselzahl *f*
**numerical**: ~ **analysis** *n* ARCH & BUILD, INFR & DES numerische Analyse *f*; ~ **method** *n* ARCH & BUILD, INFR & DES numerische Methode *f*; ~ **model** *n* ARCH & BUILD, INFR & DES numerisches Modell *nt*
**nut** *n* BUILD HARDW Mutter *f*
**NWT** *abbr* (*nonwaste technology*) ENVIRON saubere Technologie *f*, umweltfreundliche Technologie *f*

# O

**oak** *n* TIMBER *wood* Eiche *f*, Eichenholz *nt*; **~ timber** *n* TIMBER Eichenholz *nt*
**objective** *n* ARCH & BUILD Ziel *nt*, Zweck *m*
**oblique**: **~-angled slab** *n* ARCH & BUILD, CONCR, INFR & DES schiefwinklige Platte *f*; **~ arch** *n* ARCH & BUILD Schiefbogen *m*; **~ bridge** *n* INFR & DES schiefwinklige Brücke *f*; **~ coordinates** *n pl* ARCH & BUILD, INFR & DES schiefwinklige Koordinaten *f pl*
**oblong**: **~ hole** *n* STEEL Langloch *nt*
**obscured**: **~ glass** *n* MAT PROP Trübglas *nt*
**observation** *n* ARCH & BUILD, ENVIRON, INFR & DES, WASTE WATER Beobachtung *f*, Überwachung *f*; **~ grid** *n* INFR & DES Festpunktnetz *nt*; **~ point** *n* INFR & DES Beobachtungspunkt *m*; **~ tube** *n* WASTE WATER Beobachtungsrohr *nt*; **~ well** *n* ENVIRON, INFR & DES Beobachtungsbrunnen *m*, Kontrollbrunnen *m*
**obstacle** *n* INFR & DES Hemmnis *nt*, Hindernis *nt*
**obstruction** *n* ARCH & BUILD, CONST LAW Hemmnis *nt*, Hindernis *nt*, WASTE WATER *in pipes* Verstopfung *f*
**obtuse**: **~ angle** *n* ARCH & BUILD, INFR & DES Stumpfwinkel *m*
**occlusion** *n* WASTE WATER Verstopfung *f*
**occupancy** *n* CONST LAW Besitzergreifung *f*
**occupant** *n* CONST LAW Bewohner *m*
**occupational**: **~ diseases** *n pl* CONST LAW, ENVIRON Betriebskrankheiten *f pl*
**occupy** *vt* CONST LAW *building* belegen, nutzen
**ocean**: **~ dumping** *n* ENVIRON Seeverklappung *f*; **~ floor** *n* INFR & DES Meeresboden *m*
**octagonal**: **~ space frame** *n* TIMBER Achtkantraumfachwerk *nt*
**odometer** *n* BUILD MACHIN, INFR & DES, MAT PROP Entfernungsmesser *m*, Odometer *nt*
**odor** *AmE*, **odour** *n BrE* ENVIRON, WASTE WATER Geruch *m*; **~ control** *n BrE* ENVIRON Geruchsbekämpfung *f*; **~ emission** *n BrE* ENVIRON Geruchsemission *f*; **~ nuisance** *n BrE* ENVIRON Geruchsbelästigung *f*; **~ trap** *n BrE* WASTE WATER Geruchsverschluß *m*
**off-center** *AmE*, **off-centre** *adj BrE* ARCH & BUILD seitlich versetzt
**offcut** *n* STEEL, TIMBER Verschnitt *m*
**offer 1.** *n* ARCH & BUILD, CONST LAW, INFR & DES Angebot *nt*; Offerte *f*; **2.** *vt* ARCH & BUILD, CONST LAW anbieten
**official**: **~ approval** *n* CONST LAW amtliche Genehmigung *f*; **~ responsible for hazardous goods** *n* ENVIRON Gefahrgutbeauftragter *m*
**off-peak**: **~ electricity heaters** *n pl* ELECTR, HEAT & VENT Nachtspeicherheizkörper *m*; **~ electricity heating** *n* ELECTR, HEAT & VENT Nachtspeicherheizung *f*, Nachtstromspeicherheizung *f*; **~ electricity supply** *n* ELECTR, HEAT & VENT Nachtstromversorgung *f*; **~ electricity supply meter** *n* ELECTR, HEAT & VENT Nachtstromzähler *m*
**off-road**: **~ truck for earthwork** *n* BUILD MACHIN geländegängiges Fahrzeug *nt* für den Erdbau
**offset** *n* ARCH & BUILD, STONE, WASTE WATER Absatz *m*, Schwanenhals *m*, Sprungrohr *nt*, *movement, displacement* Verschiebung *f*, *recess* Rücksprung *m*
**offshore**: **~ drilling** *n* INFR & DES Off-shore-Bohren *nt*; **~ pipeline** *n* INFR & DES, STEEL Off-shore-Rohrleitung *f*; **~ structure** *n* CONCR, INFR & DES, STEEL Off-shore-Konstruktion *f*
**off-site** *adj* ARCH & BUILD außerhalb der Baustelle
**off-site**: **~ casting** *n* CONCR Werksvorfertigung *f*
**ogee** *n* ARCH & BUILD, WASTE WATER Karnies *nt*; **~ gutter** *n* ARCH & BUILD, WASTE WATER Karniesrinne *f*; **~ plane** *n* BUILD MACHIN Karnieshobel *m*
**ogive** *n* ARCH & BUILD Spitzbogen *m*
**ohm** *n* ELECTR Ohm *nt*
**oil 1.** *n* BUILD HARDW, BUILD MACHIN, HEAT & VENT, MAT PROP, SURFACE, WASTE WATER Öl *nt*; **2.** *vt* BUILD HARDW, BUILD MACHIN, MAT PROP, SURFACE ölen, *furniture* einölen, mit Öl tränken
***oil***: **~-based paint** *n* MAT PROP, SURFACE Ölfarbe *f*; **~ collector** *n* BUILD HARDW, INFR & DES, MAT PROP Ölabscheider *m*; **~-containing waste water** *n* ENVIRON ölhaltiges Abwasser *nt*; **~-contaminated waters** *n pl* ENVIRON ölverseuchte Gewässer *nt pl*; **~-cooled** *adj* BUILD MACHIN ölgekühlt; **~ film** *n* MAT PROP Ölfilm *m*; **~ firing** *n* HEAT & VENT Ölfeuerung *f*; **~ furnace** *n* HEAT & VENT Ölfeuerung *f*, Ölheizung *f*; **~ heating** *n* HEAT & VENT Ölheizung *f*; **~ paint** *n* MAT PROP, SURFACE Ölfarbe *f*; **~-polluted waste water** *n* ENVIRON, INFR & DES ölverschmutztes Abwasser *nt*; **~ pollution emergency** *n* ENVIRON Ölverschmutzungsnotfall *m*; **~ regeneration plant** *n* ENVIRON Ölaufbereitungsanlage *f*; **~ removal** *n* ENVIRON Entölen *nt*, Ölabscheidung *f*; **~-resistant** *adj* MAT PROP ölbeständig; **~ rig** *n* INFR & DES Bohrinsel *f*, Bohrplattform *f*; **~ separation** *n* ENVIRON Entölen *nt*, Ölabscheidung *f*; **~ separator** *n* ENVIRON, WASTE WATER Ölabscheider *m*; **~ spill** *n* ENVIRON Ölverschmutzung des Meeres *f*; **~ waste** *n* ENVIRON Ölrückstände *m pl*, Ölabfall *m*
**old** *adj* ENVIRON Alt-, alt; **~ deposit** *n* ENVIRON Altablagerung *f*
**oleiferous**: **~ waste water** *n* ENVIRON ölhaltiges Abwasser *nt*
**oleoresinous**: **~ paint** *n* SURFACE Naturholzfarbe *f*
**Omega**: **~ method** *n* ARCH & BUILD, CONCR, INFR & DES, STEEL, TIMBER Omega-Verfahren *nt*
**omitted**: **~-size fraction** *n* MAT PROP Ausfallkörnung *f*
**one**: **~-coat** *adj* STONE *plaster* einlagig; **~-hole mixer** *n* WASTE WATER Einlochbatterie *f*; **~-hole shower head** *n* WASTE WATER Einlochdüsenbrausekopf *m*; **~-man operation** *n* BUILD MACHIN Einmannbedienung *f*; **~-pipe system** *n* WASTE WATER Einrohrsystem *nt*; **~-quarter brick** *n* MAT PROP Viertelstein *m*; **~-wheel roller** *n* BUILD MACHIN Einradwalze *f*
**oozing**: **~ basin** *n* WASTE WATER Sickerbecken *nt*
**opal**: **~ glass** *n* MAT PROP Opalglas *nt*, *obscured, frosted* Trübglas *nt*, *opaque* Mattglas *nt*; **~ louver** *AmE*, **~ louvre** *BrE* *n* ELECTR *light fixture* opale Abdeckhaube *f*

**opaque**: ~ **coat** *n* SURFACE Deckanstrich *m*; ~ **glass** *n* MAT PROP Opakglas *nt*
**open** *adj* ARCH & BUILD, HEAT & VENT, INFR & DES offen, *uncovered* ungeschützt, nicht überdacht, *accessible* zugänglich; ~ **air** *n* ARCH & BUILD Freiluft *f*; **~-air plant** *n* INFR & DES Freiluftanlage *f*; **~-air stadium** *n* INFR & DES Freistadium *nt*; **~-air structure** *n* ARCH & BUILD Freiluftbauwerk *nt*; **~-air swimming pool** *n* INFR & DES Freibad *nt*; ~ **area** *n* ARCH & BUILD, CONST LAW Freifläche *f*; ~ **caisson** *n* CONCR, INFR & DES Mantelkasten *m*, offener Senkkasten *m*; ~ **channel** *n* INFR & DES, WASTE WATER Halbschale *f*, offenes Gerinne *nt*; ~ **drain** *n* ENVIRON offener Abzugsgraben *m*, INFR & DES Entwässerungsgraben *m*; ~ **dump** *n* ENVIRON ungeordnete Deponie *f*, wilde Müllablagerung *f*; **~-ended** *adj* CONST LAW zeitlich unbegrenzt; ~ **formwork** *n* CONCR Sparschalung *f*; ~ **foundation method** *n* INFR & DES offene Gründung *f*
**opening** *n* ARCH & BUILD Öffnung *f*, Durchbruch *m*, BUILD MACHIN, INFR & DES, MAT PROP, SURFACE Durchbruch *m*; ~ **angle** *n* ARCH & BUILD *door* Öffnungswinkel *m*, BUILD HARDW *door* Öffnungsmaß *nt*; ~ **width** *n* ARCH & BUILD Öffnungsbreite *f*
***open***: ~ **newel stairs** *n* ARCH & BUILD offene Spindeltreppe *f*; ~ **sewer** *n* ENVIRON offener Abzugsgraben *m*; **~-sided building** *n* ARCH & BUILD Gebäude *nt* ohne Umfassungswände; ~ **space** *n* CONST LAW Freifläche *f*; ~ **spatial planning** *n* ARCH & BUILD, INFR & DES Großraumplanung *f*; ~ **string stairs** *n* ARCH & BUILD aufgesattelte Treppe *f*; ~ **tender** *n* ARCH & BUILD, CONST LAW, INFR & DES öffentliches Ausschreiben *nt*; ~ **wall string** *n* ARCH & BUILD *stairs* Sattelwange *f*; **~-web girder** *n* ARCH & BUILD, MAT PROP, TIMBER Fachwerkbinder *m*, Vierendeelträger *m*; ~ **well** *n* ARCH & BUILD Treppenauge *nt*
**openwork**: ~ **chimney top** *n* ARCH & BUILD durchbrochener Schornsteinaufsatz *m*; ~ **spire** *n* ARCH & BUILD durchbrochene Turmspitze *f*; ~ **tracery** *n* ARCH & BUILD durchbrochenes Maßwerk *nt*
**operating**: ~ **chain** *n* BUILD HARDW *blind* Bedienungskette *f*; ~ **characteristics** *n pl* BUILD MACHIN Betriebskennwerte *m pl*; ~ **costs** *n pl* ARCH & BUILD, BUILD MACHIN, CONST LAW Betriebskosten *pl*; ~ **face** *n* ENVIRON Deponieoberfläche *f*; ~ **handle** *n* BUILD HARDW Bedienungsgriff *m*; ~ **hour** *n* BUILD MACHIN Betriebsstunde *f*; ~ **and maintenance manuals** *n pl* ARCH & BUILD Betriebs- und Wartungshandbücher *nt pl*; ~ **manual** *n* BUILD MACHIN Bedienungsanleitung *f*; ~ **pressure** *n* HEAT & VENT, WASTE WATER Betriebsdruck *m*; ~ **rod** *n* BUILD HARDW Bedienungsstange *f*; ~ **test** *n* CONST LAW Betriebsprüfung *f*; ~ **weight** *n* BUILD MACHIN Betriebsgewicht *nt*
**operation** *n* ARCH & BUILD, BUILD MACHIN, INFR & DES Betrieb *m*
**operational**: ~ **dependability** *n* BUILD MACHIN, CONST LAW Betriebssicherheit *f*; ~ **error** *n* ENVIRON betriebsbedingter Fehler *m*; ~ **reliability** *n* BUILD MACHIN, CONST LAW Betriebssicherheit *f*
**operator** *n* BUILD MACHIN Anlagenfahrer *m*, CONST LAW Maschinist *m*
**opposing**: ~ **traffic** *n* INFR & DES Gegenverkehr *m*
**optical**: ~ **probe** *n* MAT PROP optischer Fühler *m*; ~ **sorter** *n* ENVIRON optischer Abscheider *m*
**optimization** *n* ARCH & BUILD, CONST LAW, INFR & DES Optimierung *f*
**optimize** *vt* ARCH & BUILD, CONST LAW, INFR & DES optimieren
**optimum**: ~ **compaction** *n* INFR & DES, MAT PROP optimale Verdichtung *f*; ~ **compression** *n* INFR & DES, MAT PROP optimale Verdichtung *f*; ~ **moisture content** *n* CONCR, MAT PROP optimaler Wassergehalt *m*
**opus**: ~ **rusticum** *n* STONE Rustikaverband *m*, Bukkelquaderverband *m*, *rustic work* Bossenwerk *nt*
**order 1.** *n* ARCH & BUILD *type of architecture* Säulenordnung *f*, CONST LAW *commission, request* Auftrag *m*, *sequence* Reihenfolge *f*; **2.** *vt* ARCH & BUILD, CONST LAW bestellen
***order***: ~ **variations** *n pl* CONST LAW Bestellungsabweichungen *f pl*
**ordinary**: ~ **cement** *n* CONCR Normalzement *m*; ~ **Portland cement** *n* CONCR, MAT PROP gewöhnlicher Portlandzement *m*
**ordnance**: ~ **bench mark** *n* ARCH & BUILD, INFR & DES Vermarkungspunkt *m* der Landesvermessung
**Ordnance**: ~ **Survey map** *n* INFR & DES amtliche Vermessungskarte *f*
**organic** *adj* ENVIRON, INFR & DES, MAT PROP organisch; ~ **chemistry** *n* MAT PROP organische Chemie *f*; ~ **clay** *n* INFR & DES organischer Ton *m*; ~ **compound** *n* ENVIRON, INFR & DES, MAT PROP organische Verbindung *f*; ~ **matter** *n* ENVIRON organischer Stoff *m*, INFR & DES, MAT PROP organisches Material *nt*; ~ **salts** *n pl* MAT PROP, organische Salze *nt pl*; ~ **soil** *n* INFR & DES, MAT PROP organischer Boden *m*; ~ **solvent** *n* MAT PROP, SURFACE organisches Lösungsmittel *nt*; ~ **waste** *n* ENVIRON organischer Abfall *m*
**oriel** *n* ARCH & BUILD Erker *m*
**orientation** *n* ARCH & BUILD, INFR & DES Orientierung *f*
**original**: ~ **document** *n* CONST LAW Original *nt*; ~ **state** *n* INFR & DES *site* ursprünglicher Zustand *m*
**ornament** *n* ARCH & BUILD Ornament *nt*, Verzierung *f*
**ornamental** *adj* ARCH & BUILD, BUILD HARDW, CONCR, STEEL, TIMBER schmückend, zierend; ~ **concrete** *n* CONCR Schmuckbeton *m*, Zierbeton *m*
**orthotropic** *adj* ARCH & BUILD orthotrop
**oscillation** *n* ARCH & BUILD, MAT PROP, SOUND & THERMAL Schwingung *f*; ~ **amplitude** *n* SOUND & THERMAL Schwingungsamplitude *f*
**osmosis** *n* ENVIRON, INFR & DES, MAT PROP Osmose *f*
**osmotic**: ~ **pressure** *n* ENVIRON, INFR & DES, MAT PROP osmotischer Druck *m*
**outbond**: ~ **brick** *n* STONE Läufer *m*, Strecker *m*
**outcrop** *n* INFR & DES Ausstrich *m*
**outdoor**: ~ **facilities** *n pl* ARCH & BUILD Außenanlagen *f pl*, INFR & DES Freiluftanlage *f*, Außenanlagen *f pl*; ~ **sensor** *n* HEAT & VENT Außenfühler *m*; ~ **staircase** *n* ARCH & BUILD Außentreppe *f*; ~ **stairs** *n* ARCH & BUILD Außentreppe *f*; ~ **storage area** *n* INFR & DES Freilagerfläche *f*; ~ **thermostat** *n* HEAT & VENT Außenthermostat *m*
**outer**: ~ **corner** *n* ARCH & BUILD Außeneck *nt*; ~ **edge** *n* ARCH & BUILD Außenkante *f*; ~ **leaf** *n* ARCH & BUILD Außenschale *f*, STONE *of cavity wall* äußere Schale *f*; ~ **string** *n* ARCH & BUILD Freiwange *f*, Treppenlochwange *f*; ~ **veneer** *n* TIMBER Außenfurnier *nt*, Deckfurnier *nt*; ~ **window** *n* ARCH & BUILD, TIMBER Außenfenster *nt*
**outlet** *n* BUILD MACHIN Auslaß *m*, ELECTR Steckdose *f*, ENVIRON, HEAT & VENT, WASTE WATER Abfluß *m*, Ablauf *m*, Auslaß *m*; ~ **elbow** *n* WASTE WATER

Auslaufbogen *m*; ~ **opening** *n* BUILD HARDW, HEAT & VENT Ausströmöffnung *f*; ~ **temperature** *n* ENVIRON, HEAT & VENT Austrittstemperatur *f*; ~ **vent** *n* HEAT & VENT, WASTE WATER Dunstrohr *nt*; ~ **vent tile** *n* STONE Dunstrohrziegel *m*, Entlüftungsziegel *m*, WASTE WATER Dunstrohrziegel *m*

**outline** *n* ARCH & BUILD Umriß *m*; ~ **map** *n* ARCH & BUILD, INFR & DES Übersichtskarte *f*; ~ **planning** *n* INFR & DES Umrißplanung *f*

**output** *n* ELECTR, HEAT & VENT Leistung *f*

**outside 1.** *adv* ARCH & BUILD draußen; **2.** *n* ARCH & BUILD Außenseite *f*

***outside***: ~ **air** *n* HEAT & VENT Außenluft *f*; ~ **air temperature** *n* HEAT & VENT Außenlufttemperatur *f*; ~ **cylinder** *n* BUILD HARDW Außenzylinder *m*; ~ **diameter** *n* ARCH & BUILD, INFR & DES, WASTE WATER Außendurchmesser *m*; ~ **pressure** *n* WASTE WATER Außendruck *m*; ~ **window** *n* ARCH & BUILD, TIMBER Außenfenster *nt*

**oval**: ~ **knob** *n* BUILD HARDW Ovaltürknopf *m*; **~-shaped sewer pipe** *n* HEAT & VENT, INFR & DES, MAT PROP Eiprofil *nt*

**overall**: ~ **coefficient of heat transfer** *n* HEAT & VENT Wärmedurchgangszahl *f*; ~ **efficiency** *n* ENVIRON, HEAT & VENT Gesamtwirkungsgrad *m*; ~ **height** *n* ARCH & BUILD Bauhöhe *f*; ~ **length** *n* ARCH & BUILD Baulänge *f*

**overbridge** *n* INFR & DES Überführung *f*

**overburned**: ~ **brick** *n* STONE überbrannter Ziegel *m*

**overcharge** *n* HEAT & VENT, WASTE WATER Überlastung *f*

**overcoat 1.** *n* ARCH & BUILD Oberbekleidung *f*; **2.** *vt* ARCH & BUILD beschichten, überdecken, SURFACE beschichten

**overdefined** *adj* ARCH & BUILD überbestimmt

**overdimensioning** *n* ARCH & BUILD, INFR & DES Überdimensionierung *f*

**overdue** *adj* ARCH & BUILD überfällig

**overestimate 1.** *n* ARCH & BUILD, CONST LAW, INFR & DES Überbewertung *f*; **2.** *vt* ARCH & BUILD, CONST LAW, INFR & DES überbewerten, überschätzen

**overfall**: **~-type fish pass** *n* ENVIRON Überfallfischgerinne *nt*

**overflow 1.** *n* HEAT & VENT, INFR & DES, WASTE WATER Überlauf *m*; **2.** *vi* HEAT & VENT, INFR & DES, WASTE WATER überlaufen

***overflow***: ~ **edge** *n* INFR & DES, WASTE WATER Überlaufkante *f*; ~ **gutter** *n* INFR & DES, WASTE WATER Überlaufrinne *f*; ~ **pipe** *n* HEAT & VENT, INFR & DES, WASTE WATER Überlaufrohr *nt*

**overhand**: ~ **bricklaying** *n* STONE Mauern *nt* von der Außenseite

**overhang** *n* ARCH & BUILD Auskragung *f*, Ausladung *f*, Vorkragung *f*, Überhang *m*, CONCR, MAT PROP, STEEL, STONE Auskragung *f*

**overhanging** *adj* ARCH & BUILD auskragend, ausladend, freitragend, BUILD HARDW, BUILD MACHIN, CONCR, MAT PROP, STEEL, STONE auskragend; ~ **beam** *n* ARCH & BUILD, CONCR, TIMBER Kragträger *m*; ~ **wall** *n* ARCH & BUILD Überhangwand *f*

**overhead** *adv* ARCH & BUILD, INFR & DES oben

**overhead**: ~ **clearance** *n* ARCH & BUILD, INFR & DES Durchfahrtshöhe *f*; ~ **costs** *n pl* ARCH & BUILD, CONST LAW Gemeinkosten *pl*; ~ **crane** *n* BUILD MACHIN, INFR & DES Brückenkran *m*, Laufkran *m*; ~ **door closer** *n* BUILD HARDW obenliegender Türschließer *m*; ~ **light** *n* ARCH & BUILD Oberlicht *nt*; ~ **line** *n* ELECTR, INFR & DES Freileitung *f*, Oberleitung *f*

**overheads** *n pl* ARCH & BUILD, CONST LAW Gemeinkosten *pl*

**overheated** *adj* HEAT & VENT überheizt

**overheating** *n* HEAT & VENT Überheizung *f*

**overlap 1.** *n* ARCH & BUILD, INFR & DES Überdeckung *f*, Überlappung *f*; **2.** *vi* ARCH & BUILD, INFR & DES einander überdecken, überlappen

**overlapping** *adj* ARCH & BUILD, INFR & DES überlappend; ~ **joint** *n* ARCH & BUILD überlappender Stoß *m*, INFR & DES Überlappungsstoß *m*

**overlay** *n* INFR & DES Belag *m*, Überzug *m*

**overlaying** *n* INFR & DES Belag *m*, Überzug *m*

**overleap**: ~ **joint** *n* INFR & DES Überblattung *f*

**overload 1.** *n* ARCH & BUILD Überlast *f*, Überladung *f*, ELECTR Überlast *f*; **2.** *vt* ARCH & BUILD überladen, überlasten, ELECTR überlasten

***overload***: ~ **device** *n* ELECTR Überstromvorrichtung *f*

**overlying** *adj* ARCH & BUILD, INFR & DES überlagernd

**overpackaging** *n* ENVIRON Umverpackung *f*

**overshadowing** *n* ARCH & BUILD, HEAT & VENT Überschattung *f*

**oversize** *vt* ARCH & BUILD, INFR & DES überdimensionieren

**oversized** *adj* ARCH & BUILD, INFR & DES überdimensioniert, übergroß

**oversizing** *n* ARCH & BUILD, INFR & DES Überdimensionierung *f*

**overspeed**: ~ **control** *n* ENVIRON Überdrehzahlkontrolle *f*

**overstress 1.** *n* ARCH & BUILD Überbeanspruchung *f*; **2.** *vt* ARCH & BUILD überbeanspruchen, überlasten

**overtile** *n* MAT PROP, STONE Mönch *m*

**overtime** *n* CONST LAW Überstunden *f pl*

**overturning**: ~ **moment** *n* ARCH & BUILD, ENVIRON, INFR & DES, TIMBER Kippmoment *nt*

**ovolo** *n* BUILD MACHIN Echinus *m*, konvexer Stab *m*, STONE *quarter-circle shape* Viertelkreissims *m*

**owner** *n* CONST LAW Eigentümer *m*

**ownership** *n* CONST LAW Besitz *m*

**oxidation** *n* ENVIRON, MAT PROP, SURFACE Oxidation *f*; ~ **ditch** *n* ENVIRON Oxidationsgraben *m*; ~ **pond** *n* ENVIRON Oxidationsteich *m*

**oxidizing**: ~ **agent** *n* ENVIRON, MAT PROP Oxidationsmittel *nt*; ~ **flame** *n* STEEL oxidierende Flamme *f*

**oxyacetylene**: ~ **blowpipe** *n* BUILD MACHIN Autogenbrenner *m*, STEEL Autogenbrenner *m*, Azetylensauerstoffbrenner *m*; ~ **welding** *n* STEEL Autogenschweißen *nt*

**oxyarc**: ~ **cutting** *n* STEEL Sauerstofflichtbogenschneiden *nt*

**oxygen** *n* BUILD MACHIN, INFR & DES, MAT PROP, STEEL Sauerstoff *m*; ~ **arc cutting** *n* STEEL Sauerstofflichtbogenschneiden *nt*; ~ **arc welding** *n* STEEL Lichtbogensauerstoffschweißen *nt*; ~ **content** *n* INFR & DES, MAT PROP Sauerstoffgehalt *m*; ~ **generator** *n* BUILD MACHIN Sauerstofferzeuger *m*; ~ **lance** *n* BUILD MACHIN Sauerstofflanze *f*; ~ **lancing** *n* BUILD MACHIN Brennbohren *nt*, Sauerstoffbohren *nt*; ~ **requirement** *n* INFR & DES Sauerstoffbedarf *m*

**ozone** *n* ENVIRON Ozon *nt*; ~ **layer** *n* ENVIRON Ozonschicht *f*

# P

**pack** *vt* INFR & DES *road* stopfen, verdichten, STONE *crushed rock* stopfen
**packaging** *n* ENVIRON Verpackung *f*; **~ material** *n* ENVIRON Packstoff *m*, Verpackungsmaterial *nt*; **~ waste** *n* ENVIRON Verpackungsabfall *m*, Verpakkungsmüll *m*
**packer**: **~ lorry** *n* BUILD MACHIN, ENVIRON Preßmüllwagen *m*, fahrbarer Verdichter *m*; **~ unit** *n* ENVIRON Kompaktor *m*, Müllverdichter *m*, Verdichtungsanlage *f*, INFR & DES Kompaktor *m*
**packing** *n* CONCR Verfüllbeton *m*, HEAT & VENT Abdichtung *f*, Dichtung *f*, MAT PROP Kornpackung *f*; **~ piece** *n* TIMBER Futterholz *nt*, Füllholz *nt*, Futter *nt*; **~ rubber** *n* MAT PROP Auflage *f*
**pad** *n* MAT PROP Polster *nt*, Puffer *m*, STONE Straßenplatte *f*, Wegplatte *f*; **~ foundation** *n* ARCH & BUILD Flächengründung *f*
**padlock** *n* BUILD HARDW Vorhängeschloß *nt*; **~ hasp** *n* BUILD HARDW *lock* Überfalle *f*
**padstone** *n* ARCH & BUILD Auflagerstein *m*
**pail** *n* BUILD MACHIN Kübel *m*
**paint 1.** *n* MAT PROP, SURFACE Anstrichfarbe *f*, Farbe *f*; **2.** *vt* MAT PROP, SURFACE lackieren, streichen
***paint***: **~-burning lamp** *n* SURFACE Farbabbrennlampe *f*
**painted** *adj* MAT PROP, SURFACE gestrichen, lackiert
**painter** *n* ARCH & BUILD, SURFACE Maler *m*
**painting** *n* SURFACE Malerarbeiten *f pl*, Anstrich *m*, *process* Lackieren *nt*, Anstreichen *nt*; **~ specification** *n* MAT PROP Farbanwendungsspezifikation *f*; **~ work** *n* SURFACE Anstreichen *nt*, Anstrich *m*, Malerarbeiten *f pl*, *coating* Lackieren *nt*
***paint***: **~ sludge** *n* ENVIRON Lackschlamm *m*; **~ sprayer** *n* BUILD MACHIN, SURFACE Farbspritzpistole *f*; **~ stripper** *n* BUILD MACHIN, SURFACE Abbeizer *m*, Ablauger *m*
**palace** *n* ARCH & BUILD Palast *m*, Schloß *nt*
**paling** *n* ARCH & BUILD, INFR & DES Geländerausfachung *f*, Rundlingszaun *m*, TIMBER *latticed fencing* Lattenzaun *m*
**palisade** *n* INFR & DES, TIMBER Palisadenzaun *m*
**pan**: **~ ceiling** *n* ARCH & BUILD, CONCR, TIMBER Kassettendecke *f*
**panel 1.** *n* ARCH & BUILD *ceiling* Fach *nt*, BUILD HARDW *wainscot* Türfüllung *f*, Deckentafel *f*, MAT PROP Tafel *f*, Platte *f*, TIMBER *framework* Feld *nt*, *wainscot* Paneel *nt*, Täfelung *f*, Deckentafel *f*; **2.** *vt* ARCH & BUILD, TIMBER täfeln
***panel***: **~ construction** *n* ARCH & BUILD Tafelbauweise *f*
**paneled** *AmE see panelled BrE*
***panel***: **~-frame construction** *n* ARCH & BUILD Skelettplattenbauweise *f*
**paneling** *AmE see panelling BrE*
**panelled** *adj BrE* ARCH & BUILD, TIMBER verkleidet; **~ ceiling** *n BrE* ARCH & BUILD, CONCR, TIMBER Kassettendecke *f*, getäfelte Decke *f*
**panelling** *n BrE* ARCH & BUILD, TIMBER *sheathing* Holztäfelung *f*, Täfelung *f*, *veneer wall* Verkleidung *f*, *veneering* Beplankung *f*
***panel***: **~ wall** *n* ARCH & BUILD Tafelwand *f*

**panic**: **~ bolt** *n* BUILD HARDW, CONST LAW Panikverschluß *m*; **~ handle set** *n* BUILD HARDW, CONST LAW Panikdrückergarnitur *f*
**pantile** *n* ARCH & BUILD, MAT PROP, STONE, TIMBER Dachpfanne *f*, Dachziegel *m*, Pfanne *f*
**pantiling** *n* ARCH & BUILD, MAT PROP, STONE, TIMBER Dachdeckung *f*
**pantry** *n* ARCH & BUILD Speisekammer *f*
**paper 1.** *n* ENVIRON, MAT PROP Papier *nt*; **2.** *vt* ARCH & BUILD, SURFACE *room, wall* tapezieren
**paperboard** *n* MAT PROP Pappe *f*
***paper***: **~ collection** *n* ENVIRON Altpapiersammlung *f*
**papering** *n* ARCH & BUILD, SURFACE Tapezierarbeiten *f pl*
***paper*** **~ pulp** *n* ENVIRON Zellstoff *m*; **~ towel dispenser** *n* WASTE WATER Papierhandtuchspender *m*, **~ web** *n* MAT PROP Papierbahn *f*
**parabolic** *adj* ARCH & BUILD, INFR & DES parabolisch
**paraboloid** *adj* ARCH & BUILD, INFR & DES paraboloid
**paragraph** *n* CONST LAW Paragraph *m*
**parallel** *adj* ARCH & BUILD, TIMBER parallel; **~ chord truss** *n* TIMBER Parallelfachwerk *nt*; **~ girder** *n* ARCH & BUILD Parallelträger *m*; **~ gutter** *n* ARCH & BUILD, WASTE WATER Kastenrinne *f*
**parameter** *n* ARCH & BUILD, INFR & DES, MAT PROP Parameter *m*
**parapet** *n* ARCH & BUILD, BUILD HARDW, CONCR, INFR & DES, STEEL, TIMBER Brüstung *f*, Brüstungsmauer *f*; **~ cross beam** *n* ARCH & BUILD Brüstungsholm *m*; **~ gutter** *n* ARCH & BUILD, WASTE WATER Dachrinne *f*; **~ height** *n* ARCH & BUILD Brüstungshöhe *f*; **~ slab** *n* ARCH & BUILD Brüstungsplatte *f*; **~ wall** *n* ARCH & BUILD, INFR & DES Brüstungsmauer *f*
**parent**: **~ rock** *n* INFR & DES Muttergestein *nt*
**parget 1.** *n* STONE *ornamental plaster* Putz *m*, Stuck *m*; **2.** *vt* STONE Stuckarbeit ausführen, verputzen
**paring**: **~ chisel** *n* BUILD MACHIN Hobelmeißel *m*, Schäleisen *nt*; **~ machine** *n* BUILD MACHIN Schälmaschine *f*
**park 1.** *n* INFR & DES Park *m*, Parkanlage *f*; **2.** *vti* INFR & DES parken
***park***: **~-and-ride lots** *n pl AmE* (*cf park-and-ride sites BrE*) INFR & DES Park-and-Ride-Parkplätze *m pl* (*PR-Parkplätze*); **~-and-ride sites** *BrE n pl* (*cf park-and-ride lots AmE*) INFR & DES Park-and-Ride-Parkplätze *m pl* (*PR-Parkplätze*)
**parker**: **~ screw** *n* MAT PROP Blechtreibschraube *f*
**parking**: **~ area** *n* INFR & DES Parkfläche *f*, Parkplatz *m*; **~ bay** *n* INFR & DES Parkbucht *f*; **~ lot** *n AmE* (*cf car park BrE*) INFR & DES Parkplatz *m*
**parquet**: **~ flooring** *n* TIMBER Parkettbodenbelag *m*
**parquetry** *n* TIMBER Parkettfußboden *m*, Parkett *nt*
**part**: **~ by volume** *n* INFR & DES, MAT PROP Raumteil *nt*; **~ by weight** *n* INFR & DES, MAT PROP Gewichtsteil *nt*
**partial**: **~ acceptance** *n* CONST LAW, INFR & DES Teilannahme *f*; **~ hip** *n* ARCH & BUILD, TIMBER Krüppelwalm *m*, Schopfwalm *m*
**partially**: **~ cased section** *n* ARCH & BUILD, CONCR, INFR & DES, STEEL teilweise einbetoniertes Profil *nt*

**partial**: ~ **safety factor** *n* ARCH & BUILD, CONST LAW, INFR & DES Teilsicherheitsfaktor *m*; ~ **shear connection** *n* ARCH & BUILD, CONCR, INFR & DES, MAT PROP *of composite beam* teilweise Verdübelung *f*, teilweiser Schubanschluß *m*
**particle** *n* MAT PROP Teilchen *nt*; ~ **board** *n* MAT PROP, TIMBER Holzspanplatte *f*, Spanplatte *f*; ~ **density** *n* MAT PROP Teilchendichte *f*; ~ **size** *n* MAT PROP Teilchengröße *f*; ~**-size distribution** *n* MAT PROP Korngrößenverteilung *f*; ~**-size distribution curve** *n* MAT PROP Sieblinie *f*
**particulate**: ~ **collection** *n* ENVIRON Rauchgasentstaubung *f*, Staubabscheidung *f*; ~ **material** *n* ENVIRON, HEAT & VENT Partikel *f*, Teilchen *nt pl*; ~ **matter** *n* ENVIRON Schwebstoffteilchen *nt*
**parting** *n* INFR & DES Trennschicht *f*
**partition 1.** *n* ARCH & BUILD, INFR & DES Raumteiler *m*, Trennwand *f*, Zwischenwand *f*; **2.** *vt* ARCH & BUILD, INFR & DES aufteilen, teilen, trennen
**partitioning** *n* ARCH & BUILD, INFR & DES Abtrennung *f*, Raumteilung *f*
**partner** *n* ARCH & BUILD, CONST LAW, INFR & DES Partner *m*
**part**: ~**-payment** *n* CONST LAW Teilzahlung *f*
**party**: ~ **wall** *n* ARCH & BUILD Grundstücksbegrenzungsmauer *f*, *partition* Wohnungstrennwand *f*, Zwischenmauer *f*, Brandmauer *f*, CONST LAW Brandmauer *f*
**pass 1.** *vt* HEAT & VENT, INFR & DES durchströmen; ~ **over** INFR & DES *road* überführen; ~ **under** INFR & DES unterführen; **2.** ~ **through** *vi* INFR & DES durchsickern
**passage** *n* ARCH & BUILD Durchgang *m*
**passenger**: ~ **elevator** *n AmE* (*cf passenger lift BrE*) ARCH & BUILD, BUILD MACHIN, INFR & DES Personenaufzug *m*; ~ **lift** *n BrE* (*cf passenger elevator AmE*) ARCH & BUILD, BUILD MACHIN, INFR & DES Personenaufzug *m*
**passing**: ~ **column** *n* ARCH & BUILD *through several storeys* durchgehende Säule *f*; ~ **lane** *n* INFR & DES Überholspur *f*; ~ **pillar** *n* ARCH & BUILD, BUILD MACHIN durchgehender Ständer *m*; ~ **post** *n* ARCH & BUILD, BUILD MACHIN durchgehender Ständer *m*
**passive**: ~ **earth pressure** *n BrE* (*cf passive ground pressure AmE*) INFR & DES passiver Erddruck *m*; ~ **ground pressure** *n AmE* (*cf passive earth pressure BrE*) INFR & DES passiver Erddruck *m*; ~ **pressure** *n* INFR & DES passiver Druck *m*; ~ **Rankin pressure** *n* INFR & DES passiver Rankindruck *m*; ~ **system** *n* ENVIRON passives System *nt*
**paste** *n* MAT PROP Paste *f*
**pasteboard** *n* MAT PROP Schichtenpappe *f*
**pasty**: ~ **waste** *n* ENVIRON pastöser Abfall *m*
**patent 1.** *n* CONST LAW Patent *nt*; **2.** *vt* CONST LAW patentieren lassen
**patented** *adj* CONST LAW patentiert
***patent***: ~ **glazing** *n* ARCH & BUILD Trockenverglasung *f*, kittlose Verglasung *f*
**pathological**: ~ **waste** *n* ENVIRON infektiöser Abfall *m*, pathogener Abfall *m*
**path**: ~ **to earth** *n BrE* (*cf path to ground AmE*) ELECTR *lightning protection* Ableitung *f*; ~ **to ground** *AmE n* (*cf path to earth BrE*) ELECTR *lightning protection* Ableitung *f*
**patio** *n* ARCH & BUILD Patio *m*, *feature of Spanish-American buildings* Innenhof *m*
**patten** *n* ARCH & BUILD Säulenfuß *m*
**pattern** *n* ARCH & BUILD, INFR & DES Dessin *nt*, MAT PROP *model, guide* Muster *nt*, *design* Dessin *nt*, SURFACE Dessin *nt*
**pave** *vt* INFR & DES befestigen, pflastern, verlegen
**paved**: ~ **area** *n* INFR & DES Flächenbefestigung *f*
**pavement** *n* (*cf sidewalk AmE*) INFR & DES *pedestrian walkway* Fußgängerweg *m*, Bürgersteig *m*, *covering* Decke *f*, *paving material* Straßenbefestigung *f*, Gehweg *m*, *surfacing* obere Tragschicht *f*, Belag *m*, Flächenbefestigung *f*, STONE Pflasterbelag *m*, Straßenpflaster *nt*; ~ **bed** *n BrE* INFR & DES Wegekoffer *m*; ~ **design** *n BrE* INFR & DES Deckenbemessung *f*; ~ **expansion joint** *n BrE* INFR & DES, STONE Gehwegplatte *f*; ~ **paving flag** *n BrE* (*cf sidewalk paving flag AmE*) INFR & DES Gehwegplatte *f*, MAT PROP Gehwegplatte *f*, Pflasterbelag *m*, Straßenpflaster *nt*; ~**-quality concrete** *n BrE* CONCR Straßendeckenbeton *m*; ~ **surface evenness** *n BrE* INFR & DES Ebenheit *f* der Straßendecke
**pavilion** *n* ARCH & BUILD Pavillon *m*
**paving** *n* INFR & DES, MAT PROP *road* Belag *m*, Decke *f*, STONE *road* Befestigung *f*, Pflaster *nt*; ~ **stone** *n* INFR & DES, STONE Gehwegplatte *f*
**pavior** *AmE*, **paviour** *n BrE* ARCH & BUILD Pflasterer *m*, Steinsetzer *m*
**pay**: ~ **stipulated fine** *phr* CONST LAW Konventionalstrafe zahlen
**PE** *abbr* (*polyethylene*) MAT PROP PE (*Polyethylen*)
**peak** *n* ARCH & BUILD, ENVIRON, HEAT & VENT, INFR & DES, MAT PROP Spitze *f*; ~ **capacity** *n* HEAT & VENT Spitzenleistung *f*; ~ **concentration** *n* ENVIRON Spitzenkonzentration *f*; ~ **demand** *n* ELECTR Spitzenbedarf *m*; ~ **strength** *n* INFR & DES, MAT PROP Spitzenstärke *f*
**peat** *n* INFR & DES, MAT PROP Torf *m*; ~ **bog** *n* INFR & DES Torfmoor *nt*; ~ **deposit** *n* INFR & DES Torfablagerung *f*; ~ **layer** *n* INFR & DES Torfschicht *f*
**peaty** *adj* INFR & DES, MAT PROP torfhaltig; ~ **soil** *n* INFR & DES, MAT PROP torfhaltige Erde *f*
**pebble** *n* MAT PROP, STONE Kiesel *m*, Kieselstein *m*; ~ **dash** *n* STONE Steinputz *m*; ~ **pavement** *n* STONE Kleinpflaster *nt*
**pebbles** *n pl* MAT PROP, STONE Kiesel *m pl*
***pebble***: ~ **stone** *n* MAT PROP, STONE Geröllstein *m*, Geröll *nt*, grober Kies *m*, Flußkies *m*
**pedestal** *n* ARCH & BUILD *of monument, statue* Sockel *m*
**pedestrian** *n* INFR & DES Fußgänger *m*; ~ **access** *n* INFR & DES Fußgängerzugang *m*; ~ **bridge** *n* INFR & DES Fußgängerbrücke *f*; ~ **concourse** *n* INFR & DES Gehfläche *f*
**pediment** *n* STONE, TIMBER Ziergiebel *m*
**peel** *vt* TIMBER schälen
**peg 1.** *n* TIMBER Dollen *m*, Verdöbelung *f*; **2.** ~ **out** *vt* ARCH & BUILD, CONST LAW *area* abstecken, festsetzen
**pegging** *adj* ARCH & BUILD Absteck-
**pelletization** *n* ENVIRON Pelletisierung *f*
**pellicular**: ~ **water** *n* INFR & DES Häutchenwasser *nt*
**penalty**: ~ **contract** *n* CONST LAW Vertragsstrafe *f*
**pendant**: ~**-framed bridge** *n* INFR & DES Hängewerksbrücke *f*
**pendentive** *n* ARCH & BUILD Zwickel *m*, Hängezwickel *m*, Pendentif *nt*
**pendulum**: ~ **suspension** *n* ARCH & BUILD Pendelaufhängung *f*
**penetrate 1.** *vt* ARCH & BUILD, INFR & DES, MAT PROP,

SURFACE durchdringen; **2.** *vi* ARCH & BUILD, INFR & DES eindringen, einsickern, WASTE WATER einsickern
**penetrating**: ~ **stopper** *n* SURFACE Einlaßgrund *m*
**penetration** *n* ARCH & BUILD, BUILD MACHIN, INFR & DES, MAT PROP, SURFACE *breach, opening* Durchbruch *m*, Eindringen *nt*, Penetration *f*; ~ **depth** *n* INFR & DES, MAT PROP Eindringtiefe *f*; ~**-grade asphalt** *n* MAT PROP Heißbitumen *nt*; ~ **load curve** *n* INFR & DES, MAT PROP Eindringlastkurve *f*; ~ **record** *n* INFR & DES, MAT PROP Rammprotokoll *nt*; ~ **resistance** *n* INFR & DES, MAT PROP Eindringwiderstand *m*
**penetrometer** *n* BUILD MACHIN Eindringtiefenmesser *m*, Penetrometer *nt*
**penthouse** *n* ARCH & BUILD *apartment* Penthouse *nt*, *shed* Wetterdach *nt*; ~ **roof** *n* ARCH & BUILD Schleppdach *nt*
**pent**: ~ **roof** *n* ARCH & BUILD Halbdach *nt*
**pen**: ~ **trough** *n* ENVIRON Druckrohrleitung *f*
**perch** *n* ARCH & BUILD Kragstein *m*
**percolate** *vti* ARCH & BUILD, ENVIRON, INFR & DES, WASTE WATER durchsickern, einsickern
**percolated** *adj* ARCH & BUILD, ENVIRON, INFR & DES, WASTE WATER durchgesickert, eingesickert
**percolating**: ~ **filter** *n* ENVIRON Tropfkörper *m*, INFR & DES Tropfkörperanlage *f*; ~ **water** *n* ENVIRON, INFR & DES Sickerwasser *nt*
**percolation** *n* ARCH & BUILD, ENVIRON, INFR & DES, WASTE WATER *seepage, infiltration* Einsickern *nt*, Sikkerung *f*, Versickern *nt*
**perfect** *adj* ARCH & BUILD einwandfrei
**perforate** *vt* BUILD HARDW, BUILD MACHIN, MAT PROP *with single hole* durchstecken, lochen, *with row of holes* perforieren, STEEL lochen, TIMBER *with row of roles* perforieren, *with single hole* durchstecken, lochen
**perforated** *adj* MAT PROP gelocht; ~ **brick** *n* MAT PROP, STONE Gitterziegel *m*, Lochziegel *m*, gelochter Backstein *m*; ~ **brick masonry** *n* STONE Lochsteinmauerwerk *nt*; ~ **paper** *n* MAT PROP perforiertes Papier *nt*; ~ **pipe** *n* INFR & DES geschlitztes Filterrohr *nt*; ~ **plate** *n* MAT PROP Lochblech *nt*; ~ **sandlime brick** *n* STONE Kalksandlochstein *m*; ~ **steel strip** *n* MAT PROP Lochband *nt*
**perform** *vti* ARCH & BUILD, BUILD MACHIN, INFR & DES, MAT PROP ausführen, leisten
**performance**: ~ **test** *n* ARCH & BUILD Eignungsprüfung *f*, Eignungstest *m*
**pergola** *n* ARCH & BUILD Pergola *f*
**perimeter** *n* ARCH & BUILD Umkreis *m*, *circumference* Umfang *m*, *span* Umfassung *f*, CONST LAW, INFR & DES *span* Umfassung *f*; ~ **column** *n* ARCH & BUILD Außenstütze *f*; ~ **fence** *n* ARCH & BUILD, INFR & DES Umfassungszaun *m*; ~ **insulation** *n* MAT PROP Außenwandisolierung *f*; ~ **system** *n* INFR & DES Ringkanalsystem *nt*; ~ **wall** *n* INFR & DES Umfassungswand *f*
**period** *n* ARCH & BUILD, CONST LAW, INFR & DES Periode *f*
**periodic** *adj* ARCH & BUILD, INFR & DES periodisch
**periphery** *n* INFR & DES Peripherie *f*, Umfang *m*, Umkreis *m*
**perlite** *n* MAT PROP, SOUND & THERMAL Perlit *m*; ~ **insulation** *n* SOUND & THERMAL Perlitisolierung *f*
**permanent** *adj* ARCH & BUILD, CONST LAW, ENVIRON, INFR & DES, MAT PROP, SOUND & THERMAL dauerhaft, fest verlegt; ~ **concrete shuttering** *n* CONCR verlorene Betonschalung *f*; ~ **deformation** *n* ARCH & BUILD, INFR & DES bleibende Verformung *f*; ~ **electrode** *n* ELECTR Dauerelektrode *f*; ~ **formwork** *n* CONCR, INFR & DES Dauerschalung *f*, verlorene Schalung *f*; ~ **load** *n* ARCH & BUILD, INFR & DES Eigenmasse *f*, Eigenlast *f*, *dead weight* ruhende Last *f*, MAT PROP Eigenmasse *f*, Eigenlast *f*
**permanently**: ~ **elastic** *adj* MAT PROP dauerelastisch; ~ **plastic** *adj* MAT PROP dauerplastisch
**permanent**: ~ **set** *n* ARCH & BUILD bleibende Senkung *f*; ~ **shuttering** *n* CONCR, HEAT & VENT, INFR & DES Dauerlüftung *f*, verlorene Schalung *f* ; ~ **waste storage** *n* ENVIRON Endlagerung *f* von Abfällen; ~**-way equipment** *n* INFR & DES Oberbaumaterial *nt*; ~ **weight** *n* ARCH & BUILD, INFR & DES, MAT PROP Eigenlast *f*, Eigenmasse *f*
**permeability** *n* ENVIRON, INFR & DES, MAT PROP, STONE, TIMBER Durchlässigkeit *f*; ~ **coefficient** *n* ENVIRON, MAT PROP Durchlässigkeitskoeffizient *m*; ~ **test** *n* MAT PROP Dichtigkeitsprüfung *f*
**permeable** *adj*, ENVIRON INFR & DES, MAT PROP, STONE, TIMBER durchlässig, permeabel; ~ **ground** *n* INFR & DES, MAT PROP durchlässiger Boden *m*; ~ **layer** *n* INFR & DES, MAT PROP durchlässige Schicht *f*; ~ **subsoil** *n* INFR & DES, MAT PROP durchlässiger Untergrund *m*; ~ **to vapor** *AmE*, ~ **to vapour** *BrE adj* MAT PROP wasserdurchlässig
**permissible** *adj* ARCH & BUILD, BUILD HARDW, BUILD MACHIN, CONST LAW, ENVIRON, INFR & DES, MAT PROP, STEEL, TIMBER zulässig; ~ **adhesive type** *n* BUILD HARDW, CONST LAW, TIMBER zulässige Leimart *f*; ~ **variation** *n* ARCH & BUILD, CONST LAW zulässige Abweichung *f*
**permit 1.** *n* CONST LAW, Genehmigung *f*, Bewilligung *f*; **2.** *vt* CONST LAW gestatten, zulassen
**perpend** *n* STONE Durchbinder *m*
**perpendicular** *adj* ARCH & BUILD, INFR & DES lotrecht, senkrecht
***perpend***: ~ **stone** *n* STONE Natureckstein *m*, Vollbinder *m*
**perron** *n* ARCH & BUILD Freitreppe *f*, Außentreppe *f*
**persistent**: ~ **oil** *n* ENVIRON persistentes Öl *nt*
**personnel** *n* ARCH & BUILD, CONST LAW Personal *nt*
**perspective**: ~ **drawing** *n* ARCH & BUILD, INFR & DES Perspektivzeichnung *f*
**perspiration** *n* HEAT & VENT, SOUND & THERMAL Schwitzwasser *nt*
**pertinent** *adj* CONST LAW *regulation* einschlägig
**pervious** *adj* CONCR, ENVIRON, INFR & DES, MAT PROP, STONE, TIMBER durchlässig; ~ **to steam** *adj* MAT PROP, SOUND & THERMAL dampfdurchlässig; ~ **to vapor** *AmE*, ~ **to vapour** *BrE adj* MAT PROP, SOUND & THERMAL dampfdurchlässig; ~ **to water** *adj* MAT PROP wasserdurchlässig
**PETRIFIX**: ~ **process** *n* ENVIRON PETRIFIX-Verfahren *nt* (*Verfestigungsverfahren für Sonderabfälle*)
**petrol**: ~ **engine** *n BrE* (*cf gasoline engine AmE*) BUILD MACHIN, ENVIRON Benzinmotor *m*, Vergasermotor *m*
**petroleum** *n* ENVIRON, INFR & DES Erdöl *nt*; ~ **product** *n* ENVIRON, INFR & DES Erdölerzeugnis *nt*
**petrol**: ~ **separator** *n BrE* (*cf gasoline separator AmE*) WASTE WATER Benzinabscheider *m*; ~ **vapour recovery plant** *n BrE* (*cf gas vapor recovery plant AmE*) ENVIRON Benzindampfrückgewinnungsanlage *f*
**PF** *abbr* (*phenol-formaldehyde*) MAT PROP, TIMBER PF (*Phenol-Formaldehyd*)

**PF/RF** *abbr* (*phenol/resorcinol-formaldehyde*) ENVIRON, MAT PROP, TIMBER PF/RF (*Phenol/Resorcin-Formaldehyd*)
**phantom**: ~ **dump** *n* ENVIRON ungenehmigte Deponie *f*
**phase**: ~ **separation** *n* ENVIRON Phasentrennung *f*
**pH**: ~ **drop** *n* ENVIRON pH-Abnahme *f*
**phenol** *n* MAT PROP, TIMBER Phenol *nt*; **~-formaldehyde** *n* (*PH*) ENVIRON, MAT PROP, TIMBER Phenol-Formaldehyd *nt* (*PF*)
**phenolic**: ~ **foam** *n* MAT PROP, SOUND & THERMAL Phenolschaumstoff *m*
***phenol***: ~ **resin** *n* TIMBER Phenolharz *nt*
**phenol/resorcinol**: **~-formaldehyde** *n* (*PF/RF*) ENVIRON, MAT PROP, TIMBER Phenol/Resorcin-Formaldehyd *nt* (*PF/RF*)
**photogrammetry** *n* INFR & DES Fotogrammetrie *f*, Meßbildverfahren *nt*
**pH**: **~-value** *n* ENVIRON, MAT PROP pH-Wert *m*
**physical** *adj* ENVIRON, INFR & DES, MAT PROP physikalisch; ~ **model** *n* INFR & DES, MAT PROP physikalisches Modell *nt*; ~ **properties** *n pl* INFR & DES, MAT PROP physikalische Eigenschaften *f pl*; ~ **stabilization** *n* ENVIRON mechanischer Einschluß *m*; ~ **weathering** *n* INFR & DES, MAT PROP physikalische Verwitterung *f*
**physico**: **~-chemical** *adj* INFR & DES, MAT PROP mechanisch-chemisch; **~-chemical stabilization** *n* INFR & DES mechanisch-chemische Bodenstabilisierung *f*
**piano**: ~ **hinge** *n* BUILD HARDW Klavierband *nt*, Scharnierband *nt*
**pick 1.** *n* BUILD MACHIN Pickel *m*; **2.** *vt* CONCR, STONE aufstocken
**picket** *n* ARCH & BUILD Absteckpfahl *m*, TIMBER Pfahl *m*, Pflock *m*
**picking** *n* STONE Abspitzen *nt*, SURFACE Abklopfen *nt*
**pickle** *vt* SURFACE ablaugen, beizen
**pickling** *n* SURFACE Abbeizen *nt*, Ätzen *nt*
**picture**: ~ **window** *n* ARCH & BUILD großflächiges Fenster *nt*, Panoramafenster *nt*
**pieced**: ~ **wood** *n* TIMBER angesetztes Holz *nt*
**piece**: ~ **list** *n* ARCH & BUILD Stückliste *f*
**piend** *n* ARCH & BUILD Gratlinie *f*
**pier** *n* ARCH & BUILD, INFR & DES *bridge* Pfeiler *m*, *foundation* Gründungspfahl *m*, Landebrücke *f*, Mole *f*; ~ **basilica** *n* ARCH & BUILD Pfeilerbasilika *f*; ~ **bond** *n* STONE Pfeilerverband *m*
**pierce** *vt* ARCH & BUILD durchbohren, durchbrechen
***pier***: ~ **foundation** *n* INFR & DES Pfeilergründung *f*; ~ **head** *n* ARCH & BUILD Pfeilerkopf *m*
**piezometer** *n* BUILD MACHIN, INFR & DES, MAT PROP Piezometer *nt*
**pigeon**: **~-hole masonry** *n* STONE durchbrochenes Mauerwerk *nt*
**pigment** *n* MAT PROP Pigment *nt*
**pigmented** *adj* MAT PROP durchgefärbt
***pigment***: ~ **sludge** *n* ENVIRON Pigmentschlamm *m*
**pilaster** *n* ARCH & BUILD Halbsäule *f*, Pilaster *m*, Wandpfeiler *m*; ~ **strip** *n* ARCH & BUILD Pfeilervorlage *f*
**pile 1.** *n* ARCH & BUILD, BUILD MACHIN, INFR & DES Pfahl *m*, Pfeiler *m*; **2.** *vt* ARCH & BUILD, BUILD MACHIN, INFR & DES *drive in* eintreiben, rammen, *stack* aufschichten, stapeln
***pile***: ~ **cap** *n* INFR & DES Pfahlkappe *f*, Pfahlkopfplatte *f*, Rammhaube *f*; ~ **cut-off level** *n* INFR & DES Pfahlabschnitthöhe *f*; ~ **deflection** *n* INFR & DES Pfahlabweichung *f*
**piled**: ~ **foundation** *n* ARCH & BUILD, INFR & DES Pfahlgründung *f*
***pile***: ~ **diameter** *n* INFR & DES Pfahldurchmesser *m*; ~ **drawer** *n* BUILD MACHIN, INFR & DES Pfahlzieher *m*; ~ **driver** *n* ARCH & BUILD, BUILD MACHIN Fallhammer *m*, Rammhammer *m*, Ramme *f*, Pfahlramme *f*, Rammbär *m*; ~ **driving** *n* ARCH & BUILD, INFR & DES Pfahlrammung *f*, Rammarbeiten *f pl*; **~-driving test** *n* INFR & DES Pfahlrammversuch *m*; ~ **extractor** *n* BUILD MACHIN, INFR & DES Pfahlzieher *m*; ~ **ferrule** *n* INFR & DES Pfahlring *m*, Pfahlzwinge *f*; ~ **foundation** *n* ARCH & BUILD, INFR & DES Pfahlgründung *f*; ~ **foundation grille** *n* ARCH & BUILD, INFR & DES Pfahlrost *m*; ~ **group** *n* ARCH & BUILD, INFR & DES Pfahlgründung *f*; ~ **head** *n* INFR & DES Pfahlkopf *m*; ~ **length** *n* INFR & DES Pfahllänge *f*; ~ **load** *n* INFR & DES Pfahllast *f*; ~ **load test** *n* INFR & DES Pfahllasttest *m*; ~ **plan** *n* INFR & DES Pfahlplan *m*; ~ **plank** *n* ARCH & BUILD, INFR & DES, STEEL Spundbohle *f*; ~ **reinforcement** *n* CONCR, INFR & DES Pfahlbewehrung *f*; ~ **resistance** *n* INFR & DES Pfahltragfähigkeit *f*; ~ **and sheet-pile driver** *n* BUILD MACHIN Pfahl- und Spundwandramme *f*; ~ **shoe** *n* ARCH & BUILD, INFR & DES, STEEL Pfahlschuh *m*; ~ **spacing** *n* INFR & DES Pfahlabstand *m*; **~-supported** *adj* INFR & DES mit Pfahlgründung; ~ **toe** *n* ENVIRON, STEEL Pfahlfuß *m*; ~ **wall** *n* ARCH & BUILD, INFR & DES Pfahlwand *f*
**pilework** *n* ARCH & BUILD Pfahlbau *m*
**piling** *n* ARCH & BUILD, BUILD MACHIN, INFR & DES *driving in* Pfahltreiben *nt*, Rammen *nt*, *stacking* Aufschichtung *f*, Stapeln *nt*; ~ **frame** *n* ARCH & BUILD, BUILD MACHIN Rammgerüst *nt*, Pfahlramme *f*; ~ **hammer** *n* ARCH & BUILD, BUILD MACHIN Rammbär *m*; ~ **up** *n* ARCH & BUILD Aufhäufen *nt*
**pillar** *n* ARCH & BUILD Pfeiler *m*, Ständer *m*, Säule *f*, *support* Stütze *f*, Tragsäule *f*, *freestanding* freistehender Pfeiler *m*, Freipfeiler *m*, Stützpfeiler *m*, CONCR Stützpfeiler *m*, INFR & DES Pfeiler *m*, Ständer *m*, freistehender Pfeiler *m*, Freipfeiler *m*, Stütze *f*, Tragsäule *f*, Stützpfeiler *m*, Säule *f*; ~ **hydrant** *n* INFR & DES, WASTE WATER Überflurhydrant *m*; ~ **stone** *n* CONCR, INFR & DES, MAT PROP, STONE Eckstein *m*
**pilot**: ~ **cutting** *n* INFR & DES Durchstich *m*; ~ **drift** *n* INFR & DES Richtstollen *m*
**pin 1.** *n* BUILD HARDW, MAT PROP, STEEL *of lock* Bolzen *m*, Dorn *m*, *pivot needle* Nadel *f*, *pegging device* Stift *m*; **2.** *vt* STONE verankern
**pincers** *n pl* BUILD MACHIN Kneifzange *f*
***pin***: ~ **chain** *n* BUILD HARDW Nietbolzenkette *f*
**pinch**: ~ **bar** *n* BUILD MACHIN Brecheisen *nt*, Brechstange *f*
**pinchcock** *n* BUILD MACHIN Quetschhahn *m*
***pin***: ~ **drift** *n* ARCH & BUILD, BUILD HARDW Dorn *m*
**pine** *n* TIMBER *wood* Kiefer *f*, Kiefernholz *nt*; ~ **wood** *n* TIMBER Kiefernholz *nt*
**pinhole**: ~ **corrosion** *n* MAT PROP Nadellochkorrosion *f*
**pinnacle** *n* ARCH & BUILD Fiale *f*, Spitztürmchen *nt*
**pinning** *n* STONE Verankerung *f*
***pin***: ~ **punch** *n* BUILD MACHIN Durchtreiber *m*;

~ **spanner** *n* BUILD MACHIN Stiftschlüssel *m*, Steckdorn *m*

**pintle** *n* BUILD HARDW Bolzen *m*, Zapfen *m*, *of gate lock* Spurzapfen *m*

***pin***: ~ **tumbler** *n* BUILD MACHIN Zylinderschloß *nt*; ~ **valve** *n* BUILD MACHIN Stiftventil *nt*; ~ **vice** *n BrE* BUILD MACHIN Feilkolben *m*, Stiftkolben *m*; ~ **vise** *AmE see pin vice BrE*

**pipe 1.** *n* BUILD HARDW, HEAT & VENT, MAT PROP, WASTE WATER Rohr *nt*, Röhre *f*; **2.** *vt* ENVIRON, HEAT & VENT, INFR & DES, WASTE WATER in Röhre leiten, *lay pipes* Rohre verlegen

***pipe***: ~ **bend** *n* WASTE WATER *conduit elbow* Knierohr *nt*, *quadrant pipe* Rohrbogen *m*, Rohrkrümmer *m*; ~ **blockage** *n* WASTE WATER Rohrverstopfung *f*; ~ **clamp** *n* MAT PROP Rohrschelle *f*; ~ **clay** *n* WASTE WATER Bindeton *m*; ~ **collar** *n* WASTE WATER Rohrmanschette *f*; ~ **connection** *n* WASTE WATER Rohranschluß *m*, Rohrverbindung *f*; ~ **coupling** *n* WASTE WATER Rohrverbindung *f*; ~ **cross section** *n* ARCH & BUILD, WASTE WATER Rohrquerschnitt *m*; ~ **culvert** *n* INFR & DES, WASTE WATER Rohrdurchlaß *m*; ~ **cutter** *n* WASTE WATER Rohrschneider *m*; ~ **diameter** *n* MAT PROP Rohrweite *f*; ~ **fitter** *n* WASTE WATER Rohrschlosser *m*; ~ **fitting** *n* MAT PROP Formstück *nt*; ~ **gasket** *n* WASTE WATER Rohrdichtung *f*; ~ **gradient** *n* ARCH & BUILD, INFR & DES, WASTE WATER Rohrgefälle *nt*; ~ **hanger** *n* MAT PROP Rohrschelle *f*; ~ **hook** *n* WASTE WATER Rohrhaken *m*; ~ **joint** *n* WASTE WATER Rohrmuffe *f*, Rohrverbindung *f*; ~ **knee** *n* WASTE WATER Rohrkrümmer *m*; ~ **laying** *n* WASTE WATER Rohrverlegung *f*, Rohrleitungsverlegung *f*

**pipeline** *n* ENVIRON, INFR & DES, WASTE WATER Hauptrohrleitung *f*, Pipeline *f*, Rohrstrang *m*, Rohrleitung *f*; ~ **construction** *n* INFR & DES Rohrleitungsbau *m*; ~ **heating** *n* ELECTR, WASTE WATER Rohrbegleitheizung *f*

***pipe***: ~ **reducer** *n* WASTE WATER Rohrreduzierstück *nt*; ~ **resistance** *n* INFR & DES Rohrwiderstand *m*; ~ **screwing** *n* WASTE WATER Rohrverschraubung *f*

**pipes**: ~ **and fittings** *n pl* BUILD HARDW Rohre *nt pl* und Armaturen

***pipe***: ~ **socket** *n* MAT PROP Rohrstutzen *m*; ~ **steel** *n* MAT PROP Rohrstahl *m*; ~ **strap** *n* WASTE WATER Rohrschelle *f*; ~ **system** *n* ARCH & BUILD, INFR & DES, WASTE WATER Rohrnetz *nt*; ~ **tap** *n* BUILD MACHIN Rohrgewindebohrer *m*; ~ **thread** *n* WASTE WATER Rohrgewinde *nt*; ~ **threader** *n* WASTE WATER Gewindedreher *m*; ~ **tongs** *n pl* BUILD MACHIN Rohrzange *f*; ~ **top** *n* WASTE WATER Rohrgraben *m*, Rohrscheitel *m*; ~ **twister** *n* WASTE WATER Rohrwickler *m*; ~ **union** *n* HEAT & VENT, WASTE WATER Anschlußstutzen *m*, Rohrverbindungsstück *nt*; ~ **vice** *n BrE* BUILD MACHIN Rohrschraubstock *m*; ~ **vise** *AmE see pipe vice BrE*

**pipework** *n* ARCH & BUILD, BUILD HARDW, INFR & DES, WASTE WATER Rohrbau *m*, Rohrleitungsnetz *nt*

***pipe***: ~ **wrench** *n* WASTE WATER Rohrschlüssel *m*, Rohrzange *f*

**piping** *n* ARCH & BUILD, BUILD HARDW, WASTE WATER Verrohrung *f*; ~ **plan** *n* ARCH & BUILD Rohrleitungsplan *m*

**piston** *n* BUILD MACHIN Kolben *m*; ~ **relief duct** *n* BUILD MACHIN Kolbenentlastungskanal *m*; ~ **stroke** *n* BUILD MACHIN Kolbenhub *m*

**pit** *n* INFR & DES Grube *f*; ~ **boards** *n pl* INFR & DES Verbau *m*

**pitch** *n* ARCH & BUILD *inclination, slant* Stichhöhe *f*, Neigungswinkel *m*, Pfeilhöhe *f*, Steigung *f*, Schräge *f*, ENVIRON Neigung *f*, INFR & DES *camber, rising height* Pfeilhöhe *f*, Steigung *f*, Stichhöhe *f*, Schräge *f*, Neigungswinkel *m*, MAT PROP *tar* Teer *m*; ~ **angle** *n* ARCH & BUILD, INFR & DES Steigungswinkel *m*

**pitched**: ~ **roof** *n* ARCH & BUILD geneigtes Dach *nt*, Pultdach *nt*, Schrägdach *nt*

**pitcher** *n* STONE Pflasterstein *m*

**pitching**: ~ **moment** *n* ARCH & BUILD, ENVIRON, INFR & DES, TIMBER Kippmoment *nt*

***pitch***: ~ **of links** *n* CONCR Bügelabstand *m*; ~ **of roof** *n* ARCH & BUILD, MAT PROP Dachschräge *f*; ~ **roof** *n* TIMBER Pultdach *nt*; ~ **of spiral** *n* CONCR *spiral reinforcement* Ganghöhe *f*

***pit***: ~ **gravel** *n* STONE Grubenkies *m*; ~ **work** *n* INFR & DES Grubenarbeit *f*

**pivot 1.** *n* BUILD HARDW *door* Angel *f*, Zapfen *m*; **2.** *vi* INFR & DES schwenken, um einen Zapfen drehen

***pivot***: ~ **bridge** *n* INFR & DES Drehbrücke *f*

**pivoted** *adj* BUILD HARDW drehzapfengelagert; ~ **sash** *n* ARCH & BUILD, BUILD HARDW Drehflügelfenster *nt*, Wendeflügel *m*

***pivot***: ~**-hung sash** *n* BUILD HARDW Drehfensterflügel *m*, Kippfensterflügel *m*; ~**-hung window** *n* ARCH & BUILD, BUILD HARDW, TIMBER Drehfenster *nt*, Kippfenster *nt*; ~ **window** *n* ARCH & BUILD Drehkippfenster *nt*

**placing**: ~ **of concrete** *n* CONCR Betoneinbringung *f*, Einbringung *f* des Betons, Betonierung *f*

**plain** *n* INFR & DES Ebene *f*; ~ **bars** *n* ARCH & BUILD, CONCR, STEEL glatte Bewehrungseisen *nt pl*; ~ **concrete** *n* CONCR unbewehrter Beton *m*; ~ **tile** *n* MAT PROP, TIMBER Biberschwanz *m*, Flachziegel *m*

**plaiting** *n* MAT PROP Flechtwerk *nt*, Geflecht *nt*

**plan 1.** *n* ARCH & BUILD Plan *m*; **2.** *vt* ARCH & BUILD planen, entwerfen

**plane 1.** *adj* ARCH & BUILD, INFR & DES, MAT PROP eben; **2.** *n* INFR & DES Ebene *f*, BUILD MACHIN *tool* Hobel *m*; **3.** *vt* BUILD MACHIN glätten, hobeln, INFR & DES planieren, einebnen, TIMBER hobeln

***plane***: ~ **deformation** *n* ARCH & BUILD, MAT PROP Verformungsebene *f*

**planed**: ~ **formwork** *n* TIMBER gehobelte Schalung *f*; ~ **shuttering** *n* CONCR, TIMBER gehobelte Schalung *f*

***plane***: ~ **iron** *n* BUILD MACHIN Hobeleisen *nt*, Hobelmesser *nt*

**planer** *n* BUILD MACHIN Hobelmaschine *f*

***plane***: ~ **stock** *n* BUILD MACHIN Hobelkasten *m*; ~ **of symmetry** *n* ARCH & BUILD, INFR & DES Symmetrieebene *f*; ~ **table** *n* INFR & DES *surveying* Meßtisch *m*

**planimeter** *n* ARCH & BUILD, BUILD MACHIN, INFR & DES Flächeninhaltsmesser *m*, *surveying* Planimeter *nt*

**planimetry** *n* ARCH & BUILD, BUILD MACHIN, INFR & DES Planimetrie *f*

**planing** *n* BUILD MACHIN, TIMBER *smoothing process* Polieren *nt*, Glätten *nt*, Planieren *nt*, Hobeln *nt*; ~ **machine** *n* BUILD MACHIN Hobelmaschine *f*; ~ **tool** *n* BUILD MACHIN Hobelwerkzeug *nt*

**planish** *vt* SURFACE aufweiten, ausbeulen

**planishing** *n* SURFACE Ausbeulen *nt*

**plank 1.** *n* ARCH & BUILD *running board* Laufbohle *f*, INFR & DES, MAT PROP, TIMBER Bohle *f*, Diele *f*, Planke *f*; **2.** *vt* ARCH & BUILD, INFR & DES schalen, TIMBER schalen, verkleiden, verschalen

***plank***: **~ covering** *n* ARCH & BUILD, INFR & DES, TIMBER Bohlenbelag *m*

**planked** *adj* ARCH & BUILD, INFR & DES, TIMBER beplankt

***plank***: **~ grating** *n* TIMBER Bohlenrost *m*

**planking** *n* ARCH & BUILD Bohlenbelag *m*, CONCR Holzschalung *f*, INFR & DES Bohlenbelag *m*, Dielung *f*, *shoring* Verschalung *f*, TIMBER Bohlenbelag *m*, Bretterschalung *f*, Holzschalung *f*, *shoring* Verschalung *f*

***plank***: **~ partition** *n* ARCH & BUILD Schalwand *f*; **~ revetment** *n* ARCH & BUILD, TIMBER Bretterverkleidung *f*

**plankton** *n* ENVIRON Plankton *nt*

***plank***: **~ truss** *n* ARCH & BUILD, TIMBER Brettbinder *m*, Nagelbinder *m*

**planned** *adj* ARCH & BUILD, CONST LAW, INFR & DES geplant; **~ environment** *n* ENVIRON Umweltplanung *f*

**planner** *n* ARCH & BUILD Planer *m*

**planning** *n* ARCH & BUILD, CONST LAW, INFR & DES Planung *f*; **~ grid** *n* ARCH & BUILD, INFR & DES Planungsraster *m*; **~ permission** *n* ARCH & BUILD, CONST LAW, INFR & DES Baugenehmigung *f*; **~ regulation** *n* ARCH & BUILD, CONST LAW, INFR & DES Bauvorschrift *f*

**plant** *n* ARCH & BUILD, INFR & DES *installation, works* Anlage *f*, Ausrüstung *f*; **~ layout** *n* ARCH & BUILD *of plant or construction* Auslegung *f*

**Plant**: **~ Protection Act** *n* ENVIRON Pflanzenschutzgesetz *nt*

**plasma** *n* BUILD MACHIN, STEEL Plasma *nt*; **~ arc cutting** *n* STEEL Plasmalichtbogenschneiden *nt*; **~ cutting** *n* BUILD MACHIN Plasmaschneiden *nt*, Plasmatrennen *nt*

**plaster 1.** *n* MAT PROP, STONE Mörtel *m*, Putz *m*, Verputz *m*, *mortar* Gipsmörtel *m*; **2.** *vt* STONE vergipsen, verputzen

***plaster***: **~ angle** *n* BUILD HARDW, MAT PROP, STONE Putzwinkel *m*

**plasterboard** *n* MAT PROP Gipskartonplatte *f*, Rigipsplatte[R] *f*

***plaster***: **~ floor** *n* STONE Gipsestrich *m*

**plasterer** *n* ARCH & BUILD, SURFACE Gipser *m*

**plastering** *n* MAT PROP, STONE Putzarbeiten *f pl*, Verputzen *nt*; **~-in** *n* STONE Beiputzen *nt*, Zuputzen *nt*; **~ machine** *n* BUILD MACHIN Putzmaschine *f*; **~ trowel** *n* BUILD MACHIN Putzkelle *f*

***plaster***: **~ of Paris** *n* STONE Stuckgips *m*; **~ reinforcement** *n* STONE Putzbewehrung *f*; **~ rock** *n* STONE Gipsgestein *nt*, Rohgips *m*; **~ stone** *n* STONE Gipsgestein *nt*, Rohgips *m*; **~ and stucco work** *n* STONE Putzarbeiten *f pl*; **~-throwing machine** *n* BUILD MACHIN Putzmaschine *f*; **~ work** *n* STONE Putzarbeiten *f pl*, Verputzen *nt*

**plastic 1.** *adj* ARCH & BUILD, ENVIRON, MAT PROP plastisch; **2.** *n* ARCH & BUILD, ENVIRON, MAT PROP Kunststoff *m*, Plastik *nt*

***plastic***: **~ adhesive tape** *n* MAT PROP Kunststoffklebeband *nt*; **~ analysis** *n* ARCH & BUILD, CONCR, INFR & DES, MAT PROP Traglastverfahren *nt*; **~ conduit** *n* MAT PROP Kunststoffrohr *nt*; **~ container** *n* ENVIRON Kunststoffverpackung *f*, Kunststoffgebinde *nt*; **~ deformation** *n* MAT PROP plastische Verformung *f*; **~ dispersion** *n* SURFACE Kunststoffdispersion *f*; **~ elongation** *n* MAT PROP plastische Dehnung *f*; **~ flow** *n* ARCH & BUILD, CONCR, INFR & DES, MAT PROP Kriechen *nt*, plastischer Fluß *m*; **~ foam** *n* MAT PROP Schaumstoff *m*; **~ foil** *n* BUILD HARDW Kunststoffolie *f*; **~ hinge** *n* ARCH & BUILD, CONCR, STEEL *composite beam* plastisches Gelenk *nt*; **~ hinge location** *n* ARCH & BUILD, CONCR, STEEL Stelle *f* des plastischen Gelenkes

**plasticity** *n* CONCR, MAT PROP Formbarkeit *f*, Plastizität *f*; **~ of concrete** *n* CONCR Betonplastizität *f*; **~ index** *n* MAT PROP Plastizitätsindex *m*

**plasticizer** *n* MAT PROP Weichmacher *m*

***plastic***: **~ limit** *n* INFR & DES, MAT PROP *rolling out of soil* Ausrollgrenze *f*, Plastizitätsgrenze *f*, Rollgrenze *f*; **~ limit test** *n* MAT PROP Ausrollversuch *m*; **~ neutral axis** *n* ARCH & BUILD, CONCR, MAT PROP plastische Nullinie *f*; **~ packing material** *n* ENVIRON Kunststoffverpackung *f*; **~ resistance moment** *n* ARCH & BUILD, CONCR, MAT PROP plastische Momententragfähigkeit *f*; **~ shear resistance** *n* ARCH & BUILD, CONCR, MAT PROP, STEEL plastische Schubtragfähigkeit *f*; **~-sheathed cable** *n* ELECTR Kunststoffmantelleitung *f*, Mantelleitung *f*

**plastics**: **~ recycling** *n* ENVIRON Kunststoffrecycling *nt*, Kunststoffverwertung *f*

***plastic***: **~ theory** *n* ARCH & BUILD, CONCR, INFR & DES, MAT PROP, STEEL Plastizitätstheorie *f*, Traglastverfahren *nt*; **~ waste** *n* ENVIRON Kunststoffabfall *m*

**plastification** *n* CONCR, MAT PROP Plastifizieren *nt*

**plastifying**: **~ admixture** *n* CONCR Betonverflüssiger *m*

**platband** *n* BUILD HARDW Kranzleiste *f*, verzierter Sturz *m*

**plate 1.** *n* ARCH & BUILD, BUILD HARDW, BUILD MACHIN, INFR & DES, MAT PROP, STEEL Blech *nt*, Tafel *f*, Platte *f*; **2.** *vt* MAT PROP, SURFACE beschichten

***plate***: **~ casting** *n* STEEL Plattenguß *m*

**plated** *adj* MAT PROP, SURFACE beschichtet

***plate***: **~ fixing** *n* BUILD HARDW, MAT PROP Plattenbefestigung *f*; **~ frame** *n* STEEL Blechrahmen *m*; **~ girder** *n* ARCH & BUILD Plattenträger *m*, Vollwandträger *m*, STEEL Blechträger *m*, Vollwandträger *m*; **~ glass** *n* MAT PROP Tafelglas *nt*, Flachglas *nt*, Spiegelglas *nt*; **~ load test** *n* INFR & DES Lastplattendruckversuch *m*, *soil* Lastplattenversuch *m*, Plattendruckversuch *m*; **~ shears** *n pl* BUILD MACHIN Blechschere *f*; **~ thickness** *n* STEEL Blechdicke *f*; **~ vibrator** *n* BUILD MACHIN Plattenrüttler *m*, Rüttelplatte *f*; **~ web girder** *n* ARCH & BUILD Blechstegträger *m*

**platform** *n* ARCH & BUILD Bühne *f*, Plattform *f*, *erecting desk* Arbeitsbühne *f*, *podium* Podest *nt*, BUILD MACHIN Bühne *f*; **~ weighing machine** *n* BUILD MACHIN Brückenwaage *f*

**plating** *n* MAT PROP, SURFACE Beschichten *nt*, Beschichtung *f*

**playground** *n* INFR & DES Spielplatz *m*; **~ construction** *n* INFR & DES Spielplatzbau *m*

**plexiglass** *n* MAT PROP Plexiglas *nt*

**plinth** *n* ARCH & BUILD Säulenplatte *f*, *of monument, statue* Sockel *m*; **~ wall** *n* ARCH & BUILD Sockelmauer *f*

**plot 1.** *n* CONST LAW Grundstück *nt*; **2.** *vt* ARCH & BUILD *on map, graph, plan* auftragen, aufzeichnen, darstellen

**plotting** *n* ARCH & BUILD Anreißen *nt*; **~ paper** *n* ARCH & BUILD Millimeterpapier *nt*, Zeichenpapier *nt*
**plough** *n BrE* BUILD MACHIN Kehlhobel *m*, Nuthobel *m*
**ploughed**: **~-and-feathered joint** *n BrE* ARCH & BUILD Federverbindung *f*; **~-and-tongued-joint** *n BrE* ARCH & BUILD Spundverbindung *f*
***plough***: **~ plane** *n BrE* BUILD MACHIN Nuthobel *m*, Spundhobel *m*
**plow** *AmE see plough BrE*
**plowed**: **~-and-feathered joint** *AmE see ploughed-and-feathered joint BrE*
**plug 1.** *n* ELECTR Stecker *m*, MAT PROP *dowel, pin* Dübel *m*, TIMBER Dollen *m*, WASTE WATER Dübel *m*, *bung* Spund *m*, *pin, peg* Stift *m*, *stopper* Stopfen *m*; **2.** *vt* ARCH & BUILD, INFR & DES verstopfen, TIMBER dübeln, WASTE WATER stopfen, verstopfen
***plug***: **~ cock** *n* WASTE WATER Drehregelventil *nt*; **~ gage** *AmE*, **~ gauge** *BrE n* WASTE WATER *measuring instrument* Lochlehre *f*
**plugging** *n* TIMBER *dowelling, pinning* Dübeln *nt*, *dowelled joint* Verdübelung *f*, WASTE WATER *insulating layer* schalldichte Zwischenschicht *f*, *stopping* Zustöpseln *nt*
**plughole** *n* WASTE WATER Verschlußloch *nt*
***plug***: **~ tap** *n* BUILD MACHIN, WASTE WATER Drehregelventil *nt*, Gewindenachbohrer *m*
**plumb 1.** *adj* ARCH & BUILD lotrecht, senkrecht; **off ~** *adj* ARCH & BUILD nicht lotgerecht, nicht senkrecht; **2.** *adv* ARCH & BUILD senkrecht; **3.** *n* ARCH & BUILD, BUILD MACHIN Senkblei *nt*, Senklot *nt*
***plumb***: **~ bob** *n* ARCH & BUILD, BUILD MACHIN Schnurlot *nt*, Senkblei *nt*
**plumber** *n* ARCH & BUILD, HEAT & VENT, STEEL, WASTE WATER Installateur *m*, Klempner *m*, Blechner *m*, Flaschner *m*, Spengler *m*; **~'s solder** *n* MAT PROP Lötzinn *nt*; **~'s work** *n* ARCH & BUILD, STEEL, WASTE WATER Blechnerarbeiten *f pl*, Flaschnerarbeiten *f pl*, Spenglerarbeiten *f pl*
**plumbing** *n* ARCH & BUILD, HEAT & VENT, STEEL, WASTE WATER *pipefitting* Installation *f*, Klempnerarbeiten *f pl*, *soldering* Löten *nt*; **~ core** *n* HEAT & VENT Installationskern *m*; **~ unit** *n* ARCH & BUILD Naßzelle *f*, HEAT & VENT Installationsblock *m*, Installationszelle *f*, WASTE WATER Naßzelle *f*; **~ wall** *n* ARCH & BUILD *for piping* Installationswand *f*
***plumb***: **~ joint** *n* ARCH & BUILD gelötete Blechverbindung *f*; **~ line** *n* ARCH & BUILD, BUILD MACHIN Lotschnur *f*, Schnurlot *nt*
**plummet** *n* ARCH & BUILD, BUILD MACHIN Lot *nt*, Richtblei *nt*
**plunger** *n* BUILD MACHIN Tauchkolben *m*; **~ elevator** *n* BUILD MACHIN Saugheber *m*
**plunging** *n* WASTE WATER Eintauchen *nt*, Versenken *nt*
**plywood** *n* TIMBER Sperrholz *nt*
**pneumatically**: **~-applied mortar** *n* STONE mit der Zementkanone angeworfener Mörtel *m*
**pneumatic**: **~ classification** *n* ENVIRON pneumatisches Sortieren *nt*; **~ drill** *n* BUILD MACHIN Druckluftbohrer *m*; **~ drive** *n* BUILD MACHIN Druckluftantrieb *m*; **~ hammer** *n* BUILD MACHIN Preßlufthammer *m*; **~ hammer drill** *n* BUILD MACHIN Druckluftschlagbohrer *m*; **~ riveting** *n* STEEL Druckluftnietung *f*; **~ sorter** *n* ENVIRON pneumatische Sortieranlage *f*; **~-tired roller** *AmE*, **~-tyred roller** *BrE n* BUILD MACHIN *road construction* Gummiradwalze *f*

**pocket** *n* CONCR *hollow* Hohlraum *m*, MAT PROP *for hardware* Tasche *f*; **~ of loose gravel** *n* ARCH & BUILD, CONCR Kiesnest *nt*
**point 1.** *n* ARCH & BUILD, ELECTR Punkt *m*; **2.** *vt* STONE *joints* verfugen, ausfugen
**pointed**: **~ arch** *n* ARCH & BUILD Spitzbogen *m*
***point***: **~ of ignition** *n* MAT PROP Flammpunkt *m*
**pointing** *n* STONE Verfugen *nt*, Ausfugen *nt*, *joints* Ausstreichen *nt*
***point***: **~ of junction** *n* ARCH & BUILD Knoten *m*
**points** *n pl BrE* (*cf switch AmE*) INFR & DES *railtrack* Weiche *f*; **~ of contraflexure** *n pl* ARCH & BUILD Punkte *m pl* der Gegenbiegung
**poison** *n* CONST LAW, ENVIRON, INFR & DES Gift *nt*
**poisonous**: **~ waste** *n* ENVIRON Giftmüll *m*
**Poisson**: **~'s ratio** *n* CONCR, INFR & DES, MAT PROP Poissonsche Querdehnzahl *f*, Querdehnzahl *f*
**poker**: **~ vibrator** *n* BUILD MACHIN Innenrüttler *m*
**pole** *n* INFR & DES, TIMBER Mast *m*, Pfahl *m*; **~ plate** *n* TIMBER Auflageholz *nt*, Fußholz *nt*; **~ spacing** *n* INFR & DES Mastabstand *m*
**polish** *vt* BUILD MACHIN, MAT PROP, SURFACE polieren, bohnern
**polished** *adj* MAT PROP, SURFACE gebohnert, poliert; **~ brass** *n* MAT PROP, SURFACE poliertes Messing *nt*; **~ plate glass** *n* MAT PROP Kristallspiegelglas *nt*; **~ stone finish** *n* MAT PROP, SURFACE polierte Steinoberfläche *f*
**polishing** *n* MAT PROP, SURFACE Polieren *nt*; **~ machine with flexible shaft** *n* BUILD MACHIN Poliermaschine *f* mit biegsamer Welle
**poll** *n* BUILD MACHIN *of hammer* Breitende *nt*
**pollutant** *n* ENVIRON, INFR & DES, MAT PROP Schadstoff *m*, Schmutzstoff *m*; **~ deposition** *n* ENVIRON Schadstoffablagerung *f*; **~-impacted ground** *n* ENVIRON schadstoffbelastetes Erdreich *nt*
**pollute** *vt* ENVIRON verunreinigen
**polluted** *adj* CONST LAW, ENVIRON, WASTE WATER verschmutzt; **~ air** *n* CONST LAW, ENVIRON verschmutzte Luft *f*; **~ rainwater** *n* CONST LAW, ENVIRON verschmutztes Regenwasser *nt*; **~ water** *n* CONST LAW, ENVIRON, WASTE WATER verschmutztes Wasser *nt*
**polluter**: **~-pays principle** *n* ENVIRON Verursacherprinzip *nt*
**pollution** *n* CONST LAW Verunreinigung *f*, Verschmutzung *f*, ENVIRON Umweltverschmutzung *f*
**pollutional**: **~ index** *n* ENVIRON Verschmutzungsgrad *m* des Wassers
***pollution***: **~ burden** *n* ENVIRON Schadstoffbelastung *f*; **~ control** *n* CONST LAW Verschmutzungsüberwachung *f*, ENVIRON *air* Reinhaltung *f*, INFR & DES Verschmutzungsüberwachung *f*; **~ emitter** *n* ENVIRON Emissionsquelle *f*; **~ research** *n* ENVIRON Umweltverschmutzungsforschung *f*; **~ source** *n* ENVIRON Verunreinigungsquelle *f*
**polychloroprene**: **~ rubber** *n* MAT PROP Polychloroprenkautschuk *m*
**polycyclic**: **~ aromatic hydrocarbon** *n* ENVIRON polyzyklischer aromatischer Kohlenwasserstoff *m*
**polyester** *n* ENVIRON, MAT PROP, SURFACE Polyester *nt*; **~ board** *n* MAT PROP Polyesterplatte *f*; **~ foam** *n* ENVIRON Polyesterschaumstoff *m*; **~ paint** *n* SURFACE Polyesterfarbe *f*; **~ resin** *n* MAT PROP Polyesterharz *nt*
**polyethylene** *n* (*PE*) MAT PROP Polyethylen *nt* (*PE*); **~ foil** *n* MAT PROP Polyethylenfolie *f*

**polygonal** *adj* ARCH & BUILD polygonal, vieleckig; **~ bond** *n* STONE Vieleckverband *m*; **~ masonry** *n* STONE Zyklopenmauerwerk *nt*
**polygon**: **~ of forces** *n*, INFR & DES Kräftepolygon *m*, Kräftevieleck *nt*
**polymer** *n* MAT PROP Polymer *nt*
**polystyrene** *n* (*PS*) MAT PROP Polystyrol *nt* (*PS*)
**polyurethane** *n* (*PUR*) MAT PROP, SOUND & THERMAL Polyurethan *nt* (*PUR*); **~ rigid foam** *n* MAT PROP, SOUND & THERMAL PUR-Hartschaum *m*
**polyvinyl**: **~ chloride** *n* (*PVC*) MAT PROP, SURFACE Polyvinylchlorid *nt* (*PVC*)
**pond** *n* INFR & DES Teich *m*
**pondage** *n* ENVIRON Inhalt *m* der Kanalhaltung
**pontoon** *n* BUILD MACHIN, CONCR, STEEL Ponton *m*; **~ crane** *n* BUILD MACHIN Schwimmkran *m*
**pool** *n* ARCH & BUILD Pool *m*, INFR & DES Teich *m*
**poor**: **~ lime** *n* MAT PROP Magerkalk *m*; **~ soil** *n* ENVIRON, INFR & DES schlechter Boden *m*
**population** *n* CONST LAW, ENVIRON, WASTE WATER Bevölkerung *f*; **~ equivalence** *n* CONST LAW, ENVIRON, WASTE WATER *sewage* Einwohnergleichwert *m* (*EGW*)
**porch** *n* ARCH & BUILD Veranda *f*, *entrance hall* Eingangshalle *f*, *for door* Windfang *m*; **~ roof** *n* ARCH & BUILD Vordach *nt*
**pore** *n* CONCR, MAT PROP Pore *f*; **~ diameter** *n* CONCR Porendurchmesser *m*; **~ pressure** *n* CONCR Porendruck *m*; **~ size** *n* MAT PROP Porengröße *f*; **~-water content** *n* CONCR Porenwassergehalt *m*
**porosity** *n* CONCR, ENVIRON, INFR & DES, MAT PROP Porosität *f*, Hohlraumgehalt *m*, *permeability* Durchlässigkeit *f*; **~ test** *n* MAT PROP Porositätstest *m*
**porous** *adj* CONCR, ENVIRON, INFR & DES, MAT PROP durchlässig, porös; **~ concrete** *n* CONCR, MAT PROP Gasbeton *m*, Porenbeton *m*; **~ cover** *n* ENVIRON poröse Abdeckung *f*; **~ rock** *n* INFR & DES, MAT PROP poröses Gestein *nt*; **~ soil** *n* INFR & DES, MAT PROP poröser Boden *m*
**porphyry** *n* MAT PROP Porphyr *m*
**portable**: **~ crane** *n* BUILD MACHIN Fahrkran *m*; **~ hoisting platform** *n* BUILD MACHIN fahrbare Hubbühne *f*; **~ light** *n* BUILD HARDW Handleuchte *f*; **~ vice** *n* *BrE* BUILD MACHIN fahrbarer Schraubstock *m*; **~ vise** *AmE see portable vice BrE*
**portal** *n* ARCH & BUILD, CONCR *portico* Tor *nt*, Pforte *f*, Portal *nt*, Portikus *m*, INFR & DES Tunneleingang *m*, TIMBER *gate* Pforte *f*, Portal *nt*, *portico* Portikus *m*; **~ crane** *n* BUILD MACHIN Portalkran *m*; **~ frame** *n* CONCR, TIMBER Portalrahmen *m*; **~ frame bridge** *n* ARCH & BUILD, CONCR, INFR & DES Rahmenträgerbrücke *f*
**portico** *n* ARCH & BUILD Portikus *m*
**Portland**: **~ blast-furnace cement** *n* MAT PROP Eisenportlandzement *m* (*EPZ*)
**position 1.** *n* ARCH & BUILD, INFR & DES Position *f*, *location* Lage *f*, *placement* Stellung *f*; **2.** *vt* ARCH & BUILD, BUILD MACHIN, INFR & DES aufstellen
**possession** *n* CONST LAW Besitz *m*
**post** *n* ARCH & BUILD, BUILD MACHIN, CONCR *stanchion*, *stay post* Stiel *m*, Ständer *m*, Pfosten *m*; **~-and-beam structure** *n* ARCH & BUILD Ständerbau *m*, Ständerwerk *nt*; **~ bracket** *n* ARCH & BUILD Säulenkonsollager *nt*; **~ crane** *n* BUILD MACHIN Säulenkran *m*; **~ foundation** *n* ARCH & BUILD *fence* Pfostenfundament *nt*; **~ tensioning** *n* CONCR Nachspannen *nt*
**potable** *adj* WASTE WATER *water* trinkbar
**potassium**: **~ nitrate** *n* ENVIRON, MAT PROP Salpeter *m*
**potential** *n* ELECTR, INFR & DES, MAT PROP Potential *nt*; **~ energy** *n* MAT PROP potentielle Energie *f*; **~ flow** *n* ELECTR Potentialströmung *f*; **~ method** *n* INFR & DES Potentialmethode *f*
**pothole** *n* INFR & DES *in road* Schlagloch *nt*
**pounds**: **~ per square inch** *n pl* (*psi*) HEAT & VENT, WASTE WATER Pfund *nt* pro Quadratzoll
**pour** *vt* CONCR schütten, *cast* gießen, *cover* betonieren, *grout* vergießen; **~ in** *vt* CONCR eingießen; **~ out** *vt* CONCR ausgießen
**poured**: **~ in** *adj* CONCR eingegossen; **~-in-place** *adj* CONCR am Einbauort betoniert, ortbetoniert
**pouring** *n* CONCR Einbringen *nt*, Gießen *nt*, Schüttung *f*; **~ of concrete** *n* CONCR Betonieren *nt*, Betonierung *f*
**powder** *n* MAT PROP Puder *m*, Pulver *nt*; **~-coated** *adj* SURFACE pulverbeschichtet; **~ coating** *n* SURFACE Pulverlackbeschichtung *f*
**power** *n* BUILD MACHIN, ELECTR, ENVIRON, INFR & DES Kraft *f*, Leistung *f*; **~ basin** *n* ELECTR Staubecken *nt*; **~ circuit** *n* ELECTR Stromkreis *m*; **~ coefficient** *n* ENVIRON Leistungskoeffizient *m*; **~ consumption** *n* ARCH & BUILD, ELECTR, ENVIRON, HEAT & VENT, INFR & DES Energieverbrauch *m*; **~ curve** *n* ENVIRON Leistungskurve *f*; **~ density** *n* ENVIRON Leistungsdichte *f*; **~ drill** *n* BUILD MACHIN elektrische Bohrmaschine *f*
**powerhouse** *n* ENVIRON Kraftanlage *f*
***power***: **~ output** *n* ENVIRON Leistungsabgabe *f*; **~ requirement** *n* CONST LAW, ELECTR, INFR & DES Energiebedarf *m*; **~ shovel** *n* BUILD MACHIN Löffelbagger *m*; **~ station** *n* ARCH & BUILD, ELECTR, ENVIRON, INFR & DES Kraftwerk *nt*; **~ supply** *n* ELECTR, INFR & DES Stromführung *f*; **~ supply cable** *n* ELECTR Zuführungskabel *nt*; **~ supply line** *n* ELECTR Zuführungsleitung *f*; **~ supply unit** *n* ELECTR Netzgerät *nt*; **~ tower** *n* ENVIRON Kraftturm *m*
**pozzolanic**: **~ cement** *n* CONCR, MAT PROP Puzzolanzement *m*
**practical**: **~ application** *n* ARCH & BUILD, INFR & DES praktische Anwendung *f*; **~ experience** *n* ARCH & BUILD, INFR & DES Betriebserfahrung *f*
**preambles** *n pl* CONST LAW Einleitungen *f pl*
**prebatching**: **~ bin** *n* BUILD MACHIN Vormischsilo *m*
**precast 1.** *adj* CONCR, STEEL fertig; **2.** *vt* ARCH & BUILD, CONCR, MAT PROP, STEEL vorfertigen
***precast***: **~ concrete** *n* CONCR Fertigteilbeton *m*; **~ concrete beam** *n* ARCH & BUILD, CONCR Montagebetonbalken *m*; **~ concrete block** *n* CONCR Betonstein *m*
**precasting** *n* ARCH & BUILD, CONCR, MAT PROP, STEEL Vorfertigung *f*; **~ plant** *n* CONCR Betonfertigteilwerk *nt*
**precast**: **~ tile** *n* MAT PROP Formfliese *f*; **~ unit** *n* CONCR, MAT PROP Fertigsturz *m*, Fertigteil *nt*
**precipice** *n* INFR & DES Abgrund *m*
**precipitate** *vi* ENVIRON, INFR & DES ausfällen
**precipitation** *n* ENVIRON *chemical* Ausfällung *f*, *in atmosphere* atmosphärischer Niederschlag *m*, INFR & DES Niederschlag *m*; **~ collector** *n* ENVIRON Niederschlagssammler *m*; **~ event** *n* ENVIRON

Niederschlagsvorfall *m*; ~ **tank** *n* ENVIRON, INFR & DES Absetzbecken *nt*
**precipitator** *n* INFR & DES, WASTE WATER Abscheider *m*
**precise** *adj* ARCH & BUILD, BUILD MACHIN, CONCR, CONST LAW, INFR & DES, MAT PROP, STEEL genau
**precleaner** *n* WASTE WATER Vorreiniger *m*, Vorfilter *nt*
**precoat 1.** *n* SURFACE Voranstrich *m*; **2.** *vt* SURFACE grundieren
**precoated** *adj* SURFACE grundiert
**precompress** *vt* CONCR vorspannen
**precompression** *n* CONCR Vorspannung *f*
**preconsolidation** *n* INFR & DES Vorkonsolidierung *f*
**preconstruction**: ~ **conference** *n* ARCH & BUILD Bauvorbesprechung *f*; ~ **drawing** *n* ARCH & BUILD Baueingabeplan *m*
**precool** *vt* HEAT & VENT vorkühlen
**prefabricate** *vt* ARCH & BUILD, CONCR, MAT PROP, STEEL vorfertigen
**prefabricated** *adj* ARCH & BUILD, CONCR, MAT PROP, STEEL vorgefertigt; ~ **building member** *n* ARCH & BUILD, CONCR, MAT PROP Montageelement *nt*; ~ **building unit** *n* ARCH & BUILD, CONCR, MAT PROP, STEEL Fertigbauteil *nt*, Fertigteil *nt*; ~ **concrete unit** *n* ARCH & BUILD, CONCR, MAT PROP Betonfertigteil *nt*, Fertigteil *nt*; ~ **construction method** *n* ARCH & BUILD Elementbauweise *f*, Fertigbauweise *f*; ~ **element** *n* ARCH & BUILD, CONCR, MAT PROP, STEEL Fertigteil *nt*; ~ **house** *n* ARCH & BUILD Fertighaus *nt*; ~ **lintel** *n* ARCH & BUILD, CONCR, MAT PROP, STEEL Fertigsturz *m*
**prefinish** *vt* SURFACE vorbehandeln
**preformed**: ~ **gasket** *n* MAT PROP Fugenband *nt*; ~ **section** *n* ARCH & BUILD Schale *f*
**preheat** *vt* HEAT & VENT vorwärmen
**preheater** *n* HEAT & VENT Vorwärmer *m*
**preliminary**: ~ **analysis** *n* ARCH & BUILD, ENVIRON, INFR & DES, SURFACE Voranalyse *f*; ~ **building works** *n pl* ARCH & BUILD Rohbau *m*; ~ **cost estimate** *n* ARCH & BUILD vorläufiger Kostenvoranschlag *m*; ~ **design** *n* ARCH & BUILD Vorentwurf *m*; ~ **study** *n* ARCH & BUILD, CONST LAW, INFR & DES Vorstudie *f*; ~ **test** *n* INFR & DES Vorversuch *m*; ~ **treatment** *n* INFR & DES, SURFACE Vorbehandlung *f*; ~ **works** *n pl* ARCH & BUILD, INFR & DES Vorarbeiten *f pl*
**preload** *vt* INFR & DES vorbelasten
**premise** *n* ARCH & BUILD, CONST LAW, INFR & DES Voraussetzung *f*
**premises** *n pl* CONST LAW Grundstück *nt*
**premixed**: ~ **stuff** *n* MAT PROP, STONE Trockenputz *m*
**prepackaging** *n* ENVIRON Fertigpackung *f*
**preparation** *n* ARCH & BUILD Vorbereitung *f*, BUILD MACHIN, ENVIRON Aufbereitung *f*, INFR & DES Vorbereitung *f*, Aufbereitung *f*; ~ **plant** *n* ENVIRON, INFR & DES Aufbereitungsanlage *f*; ~ **process** *n* ENVIRON Aufbereitungsverfahren *nt*
**preparatory**: ~ **treatment** *n* INFR & DES, SURFACE Vorbehandlung *f*
**prepare** *vt* ARCH & BUILD, INFR & DES, MAT PROP aufbereiten, erzeugen, mischen; ~ **mortar** *vt* STONE Mörtel anmachen
**prescribe** *vt* CONST LAW vorschreiben
**prescribed**: **as** ~ *adv* CONST LAW vorschriftsmäßig
**preservation** *n* CONST LAW Konservierung *f*, SURFACE *by chemical means* Imprägnierung *f*, TIMBER Haltbarmachung *f*, Konservierung *f*, *maintenance* Erhaltung *f*; ~ **of structures** *n* SURFACE Bautenschutzmittel *nt*
**preservative** *n* SURFACE, TIMBER Konservierungsmittel *nt*; ~ **treatment** *n* SURFACE, TIMBER Schutzbehandlung *f*
**preserve** *vt* CONST LAW, SURFACE, TIMBER konservieren
**presetting** *n* ELECTR, HEAT & VENT Voreinstellung *f*
**press 1.** *n* BUILD MACHIN Presse *f*; **2.** *vt* BUILD MACHIN, INFR & DES drucken
**pressed**: ~ **glass** *n* MAT PROP Preßglas *nt*; ~ **metal edging** *n* BUILD HARDW Metallkante *f*; ~ **steel gutter** *n* MAT PROP, STEEL Preßblechdachrinne *f*
***press***: ~ **for the manufacture of concrete pipes** *n* BUILD MACHIN Presse *f* zur Betonrohrherstellung
**pressiometer** *n* INFR & DES Pressiometer *nt*
**pressure 1.** *n* ARCH & BUILD, BUILD MACHIN, ENVIRON, HEAT & VENT, INFR & DES, WASTE WATER Druck *m*; **2.** ~**-grout** *vt* STONE *with mortar* auspressen, verpressen
***pressure***: ~ **change** *n* ARCH & BUILD, BUILD MACHIN, HEAT & VENT, INFR & DES Druckveränderung *f*; ~ **coefficient** *n* ENVIRON Druckkoeffizient *m*; ~**-compensating tank** *n* ENVIRON, HEAT & VENT, INFR & DES Druckausgleichsbehälter *m*, Druckausdehnungsgefäß *nt*; ~ **compensation** *n* ARCH & BUILD, ENVIRON, HEAT & VENT Druckausgleich *m*; ~ **compensation layer** *n* INFR & DES Druckausgleichsschicht *f*; ~ **control** *n* HEAT & VENT, WASTE WATER Druckregelung *f*; ~ **controller** *n* HEAT & VENT, WASTE WATER Druckregler *m*; ~ **curve** *n* INFR & DES Druckkurve *f*; ~ **cylinder** *n* BUILD MACHIN Treibkessel *m*; ~ **difference** *n* HEAT & VENT, WASTE WATER Druckdifferenz *f*; ~ **distribution** *n* ARCH & BUILD, INFR & DES Druckverteilung *f*; ~ **drop** *n* BUILD MACHIN, ENVIRON, HEAT & VENT Druckabfall *m*, Druckabnahme *f*, Druckverlust *m*; ~ **equalisation** *n* HEAT & VENT Druckausgleich *m*; ~ **equalization** *n* ARCH & BUILD, ENVIRON Druckausgleich *m*; ~ **field** *n* INFR & DES Druckfeld *nt*; ~ **filter** *n* ENVIRON Druckfilter *nt*; ~ **fluctuation** *n* BUILD MACHIN Druckschwankung *f*; ~ **gage** *AmE*, ~ **gauge** *BrE* *n* BUILD MACHIN, HEAT & VENT Druckmesser *m*, Manometer *nt*; ~ **grouting** *n* STONE Auspreßverfahren *nt*, Injektionsverfahren *nt*; ~ **joint** *n* HEAT & VENT, WASTE WATER Druckstutzen *m*; ~ **line** *n* ARCH & BUILD *arch* Drucklinie *f*; ~ **load** *n* ARCH & BUILD Druckbelastung *f*; ~ **loss** *n* BUILD MACHIN, HEAT & VENT, WASTE WATER Druckverlust *m*; ~ **pickup** *n* HEAT & VENT, WASTE WATER Druckfühler *m*; ~ **pipe** *n* ENVIRON Druckleitung *f*, HEAT & VENT, WASTE WATER Druckrohr *nt*, Druckleitung *f*; ~**-proof design** *n* ARCH & BUILD druckfeste Ausführung *f*; ~ **ratio** *n* HEAT & VENT, WASTE WATER Druckverhältnis *nt*; ~ **reducer** *n* HEAT & VENT, WASTE WATER Druckminderer *m*; ~**-reducing valve** *n* HEAT & VENT, WASTE WATER Druckminderer *m*; ~ **regulator** *n* HEAT & VENT, WASTE WATER Druckregler *m*; ~**-resistant** *adj* ARCH & BUILD druckbeständig; ~ **rise** *n* HEAT & VENT, WASTE WATER Druckerhöhung *f*; ~ **stage** *n* HEAT & VENT, WASTE WATER Druckstufe *f*; ~ **tank** *n* HEAT & VENT, WASTE WATER Druckbehälter *m*; ~ **test** *n* HEAT & VENT, MAT PROP, WASTE WATER Druckprobe *f*; ~**-tight** *adj* HEAT & VENT, WASTE WATER druckdicht; ~ **transducer** *n* HEAT & VENT, WASTE WATER Druckfühler *m*; ~ **vessel** *n* HEAT & VENT, WASTE WATER Druckgefäß *nt*; ~ **welding** *n* STEEL Preßschweißen *nt*
**pressurized**: ~ **hot-water tank** *n* HEAT & VENT Druckspeicher *m*
**prestrain** *vt* CONCR vordehnen

**prestress 1.** *n* CONCR, INFR & DES, MAT PROP Vorspannung *f*; **2.** *vt* CONCR, INFR & DES, MAT PROP vorspannen

**prestressed** *adj* CONCR, INFR & DES, MAT PROP vorgespannt; **~ concrete** *n* CONCR, INFR & DES, MAT PROP Spannbeton *m*; **~ concrete wire** *n* CONCR, INFR & DES, MAT PROP Spanndraht *m*; **~ glass** *n* INFR & DES, MAT PROP Einscheibensicherheitsglas *nt*; **~ pile** *n* CONCR, INFR & DES, MAT PROP Vorspannpfahl *m*

**pretreat** *vt* CONCR, INFR & DES, SURFACE, TIMBER vorbehandeln

**pretreated** *adj* CONCR, INFR & DES, SURFACE, TIMBER vorbehandelt

**pretreatment** *n* CONCR, INFR & DES, SURFACE, TIMBER Vorbehandlung *f*; **~ tank** *n* ENVIRON, INFR & DES Vorklärbecken *nt*

**prevention** *n* CONST LAW, ENVIRON Verhinderung *f*; **~ of noise pollution** *n* ENVIRON Vermeidung *f* von Lärmbelästigung; **~ of water pollution** *n* ENVIRON Verhütung *f* der Wasserverschmutzung

**preventive**: **~ coating** *n* SURFACE Schutzschicht *f*; **~ fire protection** *n* CONST LAW vorbeugender Brandschutz *m*; **~ maintenance** *n* ARCH & BUILD vorbeugende Wartung *f*

**prewetted** *adj* STONE vorgenäßt

**price** *n* ARCH & BUILD, CONST LAW, INFR & DES Preis *m*

**pricking**: **~-up coat** *n* STONE Grobputzschicht *f*, Unterputz *m*

**primary** *adj* ARCH & BUILD, CONCR, ENVIRON, HEAT & VENT, INFR & DES, TIMBER primär; **~ air** *n* HEAT & VENT Primärluft *f*; **~ settlement basin** *n* ENVIRON, INFR & DES Vorklärbecken *nt*; **~ sewage treatment** *n* ENVIRON mechanische Abwasserreinigung *f*; **~ sludge** *n* ENVIRON, INFR & DES Primärschlamm *m*; **~ structure** *n* ARCH & BUILD Primärkonstruktion *f*; **~ truss** *n* CONCR, TIMBER Primärbinder *m*

**prime**: **~ coat** *vt* SURFACE grundieren

**prime**: **~ coat** *n* SURFACE Voranstrich *m*, Grundieranstrich *m*

**primer** *n* SURFACE Grundierung *f*

**priming** *n* SURFACE Voranstrich *m*, Grundanstrich *m*

**principal**: **~ beam** *n* ARCH & BUILD, TIMBER Hauptbalken *m*; **~ entrance** *n* ARCH & BUILD, INFR & DES Haupteingang *m*; **~ entrance storey** *n BrE* ARCH & BUILD, INFR & DES Geschoß *nt* mit dem Haupteingang; **~ entrance story** *AmE see principal entrance storey BrE*

**prismatic**: **~ diffuser** *n* ELECTR *light fixture* Prismenwanne *f*

**privy**: **~ tank** *n* ENVIRON Abwasserfaulraum *m*

**probability** *n* ARCH & BUILD, INFR & DES Wahrscheinlichkeit *f*; **~ of failure** *n* INFR & DES Bruchwahrscheinlichkeit *f*

**probe** *n* ARCH & BUILD, HEAT & VENT Meßfühler *m*

**problem**: **~ site** *n* ENVIRON Altlast *f*, kontaminierter Standort *m*

**procedure** *n* ARCH & BUILD, CONST LAW, INFR & DES Verfahren *nt*

**process 1.** *n* ARCH & BUILD, CONST LAW, INFR & DES Prozeß *m*, Verfahren *nt*; **2.** *vt* ENVIRON, MAT PROP bearbeiten, verarbeiten

**processing** *n* ENVIRON, MAT PROP Aufarbeitung *f*, Bearbeitung *f*, Verarbeitung *f*; **~ center for recyclable solid waste materials** *AmE*, **~ centre for recyclable solid waste materials** *BrE n* ENVIRON Aufarbeitungszentrum *nt* für wiederverwertbare feste Abfallmaterialien; **~ guidelines** *n pl* MAT PROP, SURFACE Verarbeitungsrichtlinien *f pl*; **~ plant** *n* ENVIRON Aufbereitungsanlage *f*; **~ specifications** *n pl* MAT PROP, SURFACE Verarbeitungsrichtlinien *f pl*

***process***: **~ waste** *n* CONST LAW, ENVIRON produktionsspezifischer Abfall *m*, *useless byproducts* Produktionsabfall *m*, gewerblicher Abfall *m*; **~ water** *n* CONST LAW, ENVIRON Brauchwasser *nt*, Nutzwasser *nt*, INFR & DES, WASTE WATER Brauchwasser *nt*

**Proctor**: **~ density** *n* INFR & DES, MAT PROP Proctordichte *f*; **~ test** *n* INFR & DES, MAT PROP Proctortest *m*

**produce** *vt* MAT PROP erzeugen

**product** *n* MAT PROP Produkt *nt*, Erzeugnis *nt*

**production** *n* MAT PROP Produktion *f*

**productivity** *n* ARCH & BUILD, MAT PROP Produktivität *f*

**professional**: **~ indemnity insurance** *n* CONST LAW berufliche Haftpflichtversicherung *f*

**profile 1.** *n* ARCH & BUILD, BUILD HARDW, INFR & DES, MAT PROP Profil *nt*; **2.** *vt* ARCH & BUILD, BUILD HARDW, INFR & DES, MAT PROP profilieren

***profile***: **~ cylinder** *n* BUILD HARDW Profilzylinder *m*

**profiled** *adj* ARCH & BUILD, BUILD HARDW, INFR & DES, MAT PROP profiliert; **~ rod** *n* CONCR, MAT PROP, STEEL Profilstange *f*; **~ sheet** *n* MAT PROP, STEEL Formblech *nt*; **~ steel deck** *n* CONCR, MAT PROP, STEEL Profilstahldeck *nt*; **~ wire** *n* MAT PROP, STEEL Dessindraht *m*, Formdraht *m*, Profildraht *m*

***profile***: **~ steel sheeting** *n* MAT PROP, STEEL Profilbleche *nt pl*

**profiling** *n* ARCH & BUILD, BUILD HARDW, INFR & DES, MAT PROP Profilmessung *f*

**profit** *n* CONST LAW, INFR & DES Gewinn *m*; **~ margin** *n* CONST LAW, INFR & DES Gewinnspanne *f*

**progress** *n* ARCH & BUILD Fortschritt *m*, Fortgang *m*; **~ chart** *n* ARCH & BUILD, INFR & DES Bauablaufplan *m*, Baufristenplan *m*; **~ of construction work** *n* ARCH & BUILD Baufortschritt *m*

**progressive**: **~ erosion** *n* INFR & DES fortschreitende Erosion *f*; **~ settlement** *n* ARCH & BUILD, INFR & DES fortschreitende Setzung *f*; **~ subsidence** *n* ARCH & BUILD, INFR & DES fortschreitende Setzung *f*

***progress***: **~ payment** *n* ARCH & BUILD Bauabschlagszahlung *f*; **~ report** *n* ARCH & BUILD Baufortschrittsbericht *m*

**project 1.** *n* ARCH & BUILD, CONST LAW, INFR & DES Bauvorhaben *nt*, Plan *m*, Projekt *nt*; **2.** *vt* ARCH & BUILD, CONST LAW, INFR & DES *plan in advance* planen, vorausplanen; **3.** ARCH & BUILD, CONST LAW, INFR & DES *protrude* vorkragen, vorstehen, auskragen

**projected** *adj* ARCH & BUILD, CONST LAW, INFR & DES geplant, vorausgeplant; **~ area** *n* ARCH & BUILD Projektionsfläche *f*; **~ window** *n* ARCH & BUILD, BUILD HARDW Lüftungsflügel *m*, Lüftungsflügelfenster *nt*

***project***: **~ engineer** *n* ARCH & BUILD, INFR & DES ausführender Ingenieur *m*

**projecting** *adj* ARCH & BUILD, BUILD HARDW, CONCR, STEEL, STONE auskragend, vorkragend, vorstehend; **~ end** *n* ARCH & BUILD Überstand *m*; **~ latch** *n* BUILD HARDW *lock* vorstehende Falle *f*; **~ platform** *n* ARCH & BUILD ausladende Plattform *f*; **~ scaffolding** *n* BUILD MACHIN Auslegergerüst *nt*

**projection** *n* ARCH & BUILD *overhang* Überstand *m*, Vorsprung *m*, *jut* Vorkragung *f*, Auskragung *f*, Überhang *m*, *plan* Projektion *f*, Abbildung *f*, CONCR, STEEL, STONE Auskragung *f*; **~ length** *n*

ARCH & BUILD Projektionslänge *f*; ~ **welding** *n* ARCH & BUILD Buckelschweißung *f*
***project***: ~ **management** *n* ARCH & BUILD, CONST LAW Projektleitung *f*; ~ **manager** *n* ARCH & BUILD, CONST LAW Projektleiter *m*; ~ **monitoring** *n* ARCH & BUILD, CONST LAW Projektüberwachung *f*
**prolongate** *vt* ARCH & BUILD verlängern
**prominence** *n* ARCH & BUILD Erhöhung *f*, Vorsprung *m*
**pronged**: ~ **shovel** *n* BUILD MACHIN gezackte Schaufel *f*
**proof 1.** *adj* ELECTR beständig, MAT PROP beständig, widerstandsfähig; **2.** *n* CONST LAW Nachweis *m*, Beweis *m*; **3.** *vt* SOUND & THERMAL abdichten, SURFACE imprägnieren
**proofed** *adj* SOUND & THERMAL abgedichtet, SURFACE imprägniert
**proofing** *n* SOUND & THERMAL Abdichtung *f*, SURFACE Imprägnierung *f*
**prop 1.** *n* ARCH & BUILD Ständer *m*; **2.** *vt* ARCH & BUILD abfangen, absteifen; ~**up** ARCH & BUILD abfangen, absteifen
**propagate** *vt* ENVIRON, MAT PROP ausbreiten
**propagation** *n* ENVIRON, MAT PROP Ausbreitung *f*; ~ **of pollutant** *n* ENVIRON Schadstoffausbreitung *f*
**propeller**: ~ **fan** *n* HEAT & VENT Propellergebläse *nt*, Lüftungsschraube *f*
**proper**: ~ **disposal** *n* CONST LAW, ENVIRON geordnete Ablagerung *f*
**property** *n* CONST LAW *possession* Eigentum *nt*, MAT PROP *characteristic* Eigenschaft *f*; ~ **line** *n* ARCH & BUILD, CONST LAW, INFR & DES Grundstücksgrenze *f*; ~ **tax** *n BrE* (*cf real estate tax AmE*) CONST LAW Grundsteuer *f*
**proportionality**: ~ **limit** *n* MAT PROP Proportionalitätsgrenze *f*
**proportioner** *n* BUILD MACHIN Dosierer *m*
**proportioning**: ~ **device** *n* BUILD MACHIN Dosiergerät *nt*; ~ **equipment** *n* BUILD MACHIN Dosieranlage *f*
**propping** *n* ARCH & BUILD Absteifung *f*, Abfangung *f*
**propwood** *n* TIMBER Stützholz *nt*
**prospection** *n* INFR & DES Prospektion *f*
**protected** *adj* CONST LAW, ELECTR, INFR & DES *against fire* feuergeschützt
**protection** *n* CONST LAW, ELECTR, INFR & DES Schutz *m*; ~ **against accidental contact** *n* ELECTR Berührungsschutz *m*; ~ **device** *n* ARCH & BUILD Umwehrung *f*; ~ **works** *n pl* CONST LAW, INFR & DES Schutzwerke *nt pl*
**protective** *adj* ARCH & BUILD, ELECTR, MAT PROP, STONE, SURFACE schützend; ~ **cage** *n* ARCH & BUILD *chimney ladder* Schutzkorb *m*; ~ **coating** *n* SURFACE Schutzanstrich *m*; ~ **conduit** *n* ELECTR Schutzrohr *nt*; ~ **conduit for underground cabling** *n* ELECTR Schutzrohr *nt* für Erdeinführungen von Leitungen; ~ **earth** *n BrE* (*cf protective ground AmE*) ELECTR Schutzerdung *f*; ~ **grating** *n* STEEL Schutzgitter *nt*; ~ **ground** *n AmE* (*cf protective earth BrE*) ELECTR Schutzerdung *f*; ~ **layer** *n* STONE Schutzschicht *f*; ~ **paint coat** *n* SURFACE Schutzanstrich *m*; ~ **system** *n* ELECTR Schutzart *f*; ~ **tape** *n* MAT PROP Schutzband *nt*, Schutzbinde *f*
**prototype**: ~ **building** *n* ARCH & BUILD Musterbau *m*
**protruding**: ~ **length** *n* ARCH & BUILD Kraglänge *f*; ~ **platform** *n* ARCH & BUILD ausladende Plattform *f*
**provide** *vt* ARCH & BUILD vorsehen, beistellen, bereitstellen
**provisional**: ~ **works** *n pl* CONST LAW, INFR & DES Behelfsarbeiten *f pl*
**PS** *abbr* (*polystyrene*) MAT PROP PS (*Polystyrol*)
**psi** *abbr* (*pounds per square inch*) HEAT & VENT, WASTE WATER Pfund *nt* pro Quadratzoll
**p-trap** *n* WASTE WATER P-Verschluß *m*
**public**: ~ **area** *n* CONST LAW, INFR & DES öffentliche Fläche *f*; ~ **building** *n* ARCH & BUILD, CONST LAW, INFR & DES öffentliches Gebäude *nt*; ~ **convenience** *n* ARCH & BUILD, INFR & DES öffentliche Bedürfnisanstalt *f*; ~ **hall** *n* ARCH & BUILD Bürgerhaus *nt*; ~ **health engineering** *n* CONST LAW, INFR & DES, WASTE WATER Gesundheitstechnik *f*, Sanitärtechnik *f*; ~ **housing units** *n AmE* (*cf council housing BrE*) INFR & DES sozialer Wohnungsbau *m*
**publicly**: ~ **financed housing** *n* CONST LAW sozialer Wohnungsbau *m*
**public**: ~ **road** *n* INFR & DES öffentliche Straße *f*; ~ **sewers** *n* ENVIRON, INFR & DES, WASTE WATER Kanalisation *f*; ~ **water supply** *n* CONST LAW, INFR & DES, WASTE WATER Gemeindewasserversorgung *f*, öffentliche Wasserversorgung *f*; ~ **way** *n* CONST LAW, INFR & DES öffentlicher Weg *m*; ~ **works** *n pl* ARCH & BUILD, CONST LAW, INFR & DES öffentliche Bauarbeiten *f pl*
**puddingstone** *n* MAT PROP Flintkonglomerat *nt*
**puddle** *vt* CONCR *compaction of concrete* stochern
**pugging** *n* SOUND & THERMAL Auffüllung *f*
**pull 1.** *n* BUILD HARDW Ziehgriff *m*, *lock* Zieher *m*; **2.** ~ **down** *vt* ARCH & BUILD *demolish* abbrechen; ~ **out** ARCH & BUILD *nail* herausziehen
***pull***: ~-**cord** *n* BUILD HARDW *blind* Schnurzug *m*
**pulley** *n* BUILD MACHIN Rolle *f*
**pulling**: ~ **down** *n* ARCH & BUILD Abbruch *m*
**pulp** *n* ENVIRON Brei *m*
**pulpwood** *n* ENVIRON Faserholz *nt*, Papierholz *nt*
**pulsate** *vi* ARCH & BUILD, INFR & DES *building* beben, pulsieren, ELECTR pulsieren
**pulsating** *adj* ELECTR pulsierend; ~ **load** *n* ARCH & BUILD, INFR & DES pulsierende Last *f*
**pulverize** *vt* MAT PROP pulverisieren, zerkleinern, zerstäuben
**pulverized**: ~ **fuel ash** *n* MAT PROP Flugasche *f*
**pulverizer** *n* BUILD MACHIN, ENVIRON Feinmühle *f*, Zerkleinerer *m*
**pumice** *n* CONCR, MAT PROP, SOUND & THERMAL, STONE Bims *m*, Bimsstein *m*; ~ **concrete** *n* CONCR, MAT PROP Bimsbeton *m*; ~ **concrete panel** *n* CONCR Bimsbetondiele *f*; ~ **masonry** *n* STONE Bimsmauerwerk *nt*; ~ **slab** *n* STONE Bimsplatte *f*
**pump 1.** *n* BUILD MACHIN, HEAT & VENT, WASTE WATER Pumpe *f*; **2.** *vt* BUILD MACHIN, HEAT & VENT, WASTE WATER pumpen
**pumpcrete** *n* CONCR Pumpbeton *m*
***pump***: ~ **dredger** *n* BUILD MACHIN Pumpenbagger *m*, Saugbagger *m*
**pumped**: ~ **concrete** *n* CONCR Pumpbeton *m*
**pumping** *n* BUILD & MACHIN, HEAT & VENT, WASTE WATER Pumpen *nt*; ~ **station** *n* INFR & DES, WASTE WATER Schöpfwerk *nt*; ~ **test** *n* INFR & DES Pumpversuch *m*
***pump***: ~ **station** *n* INFR & DES, WASTE WATER Hebeanlage *f*; ~ **strainer** *n* BUILD MACHIN Saugkorb *m*
**punch** *vt* BUILD MACHIN stanzen, STEEL stanzen, zusammenkneifen
**punched**: ~ **metal plate fastener** *n* STEEL, TIMBER

Nadelplatte *f*; ~ **plate** *n* STEEL Lochblech *nt*; ~**-plate screen** *n* BUILD MACHIN Siebblech *nt*
**puncturability** *n* ENVIRON Durchstoßfestigkeit *f*
**puncture** *n* ENVIRON Durchstoß *m*; ~ **resistance** *n* ENVIRON Durchstoßfestigkeit *f*
**PUR** *abbr* (*polyurethane*) MAT PROP, SOUND & THERMAL PUR (*Polyurethan*)
**purchase** *n* CONST LAW Ankauf *m*
**purification** *n* CONST LAW, ENVIRON, WASTE WATER Klärung *f*; ~ **capacity** *n* ENVIRON Reinigungsvermögen *nt*; ~ **plant** *n* ENVIRON, INFR & DES, WASTE WATER Kläranlage *f*, Klärwerk *nt*
**purify** *vt* CONST LAW, ENVIRON reinigen, WASTE WATER klären
**purlin** *n* ARCH & BUILD, BUILD HARDW *on roof* Pfette *f*, Dachrahmen *m*, Dachpfette *f*, STEEL, TIMBER Dachpfette *f*, Pfette *f*; ~ **joint** *n* ARCH & BUILD, STEEL, TIMBER Dachpfettenstoß *m*; ~ **nail** *n* BUILD HARDW Pfettennagel *m*, Sparrennagel *m*; ~ **post** *n* ARCH & BUILD, TIMBER Pfettenstützholz *nt*
**purpose**: ~**-built** *adj* ARCH & BUILD, INFR & DES zweckgebaut; ~**-made block** *n* MAT PROP Profilstein *m*; ~**-made concrete element** *n* MAT PROP Betonformstück *nt*; ~**-made tile** *n* MAT PROP Spezialfliese *f*
**push**: ~ **button** *n* ELECTR Taster *m*; ~ **handle** *n* BUILD HARDW Stoßgriff *m*
**put**: ~ **out of operation** *vt* ELECTR, INFR & DES außer Betrieb setzen; ~ **out of service** *vt* CONST LAW, WASTE WATER außer Betrieb setzen, stillegen; ~ **under cover** *vt* MAT PROP bedecken; ~ **up** *vt* ARCH & BUILD, INFR & DES errichten
**putlock** *n see putlog*
**putlog** *n* ARCH & BUILD Rüstholz *nt*, BUILD MACHIN, STONE Rüstholz *nt*, Rüststange *f*; ~ **hole** *n* STONE Gerüstloch *nt*, Rüstloch *nt*
**putrefaction** *n* INFR & DES Fäulnis *f*
**putrefactive**: ~ **fermentation** *n* INFR & DES Fäulnisgärung *f*
**putrescibility** *n* ENVIRON, MAT PROP Faulfähigkeit *f*
**putrescible** *adj* ENVIRON verrottbar; ~ **matter** *n* ENVIRON fäulnisfähiger Stoff *m*, verrottbarer Stoff *m*; ~ **sludge** *n* ENVIRON faulfähiger Schlamm *m*
**putty** *n* ARCH & BUILD, BUILD HARDW, MAT PROP Kitt *m*, Spachtelmasse *f*; ~ **joint** *n* ARCH & BUILD Kittfuge *f*, BUILD HARDW Spachtelverbindung *f*; ~ **knife** *n* BUILD HARDW, BUILD MACHIN Kittmesser *nt*; ~ **powder** *n* MAT PROP Polierasche *f*; ~ **rebate** *n* ARCH & BUILD Kittfalz *m*
**puzzle**: ~ **lock** *n* ARCH & BUILD, BUILD HARDW Kombinationsschloß *nt*
**PVC** *abbr* (*polyvinyl chloride*) MAT PROP, SURFACE PVC (*Polyvinylchlorid*)
**pylon** *n* ARCH & BUILD, BUILD HARDW, INFR & DES Brückenpfeiler *m*, Gittermast *m*, Pylon *m*
**pyrite** *n* INFR & DES Pyrit *m*
**pyrolysis** *n* ENVIRON Pyrolyse *f*, thermische Zersetzung *f*

# Q

**Q**: **~ factor** *n* (*quality factor*) ENVIRON Q-Faktor *m* (*Gütefaktor, Qualitätsfaktor*)
**quadrangle** *n* ARCH & BUILD viereckiger Hof *m*
**quadrant**: **~ pipe** *n* WASTE WATER Rohrkrümmer *m*, *bending tube* Bogenrohr *nt*
**quadrature** *n* ARCH & BUILD, INFR & DES Flächenberechnung *f*
**qualification**: **~ examination** *n* CONST LAW *staff* Eignungsprüfung *f*, Eignungstest *m*
**quality** *n* MAT PROP *characteristic, property* Güte *f*, *grade, sort* Qualität *f*, *specification* Gütebestimmungen *f pl*; **~ of aggregate** *n* MAT PROP Güte *f* des Zuschlagstoffs; **~ assessment** *n* CONST LAW Qualitätsbewertung *f*; **~ assurance** *n* CONST LAW, MAT PROP Qualitätssicherung *f*; **~ assurance system** *n* CONST LAW, MAT PROP Qualitätssicherungssystem *nt*; **~ class** *n* TIMBER Güteklasse *f*; **~ control** *n* CONST LAW, MAT PROP Gütesicherung *f*, Qualitätskontrolle *f*; **~ control association** *n* CONST LAW, MAT PROP Gütegemeinschaft *f*; **~ control mark** *n* CONST LAW, MAT PROP Gütezeichen *nt*; **~ description** *n* CONST LAW, MAT PROP Gütebeschreibung *f*; **~ factor** *n* (*Q factor*) ENVIRON Gütefaktor *m* (*Q-Faktor*), Qualitätsfaktor *m* (*Q-Faktor*); **~ requirements** *n pl* CONST LAW Gütebestimmungen *f pl*
**quantity** *n* ARCH & BUILD Quantität *f*, Menge *f*, Masse *f*; **~ survey** *n* ARCH & BUILD Aufmaß *nt*, Massenberechnung *f*, Massenermittlung *f*, INFR & DES Aufmaß *nt*; **~ surveying services** *n pl* ARCH & BUILD, CONST LAW, INFR & DES Massenermittlungsdienstleistungen *f pl*; **~ surveyor** *n* ARCH & BUILD Kostenplaner *m*, Baukostenkalkulator *m*, Baukostenkalkulatorin *f*, CONST LAW Massenberechner *m*, INFR & DES Kostenplaner *m*
**quarry**: **~-faced** *adj* STONE bruchrauh; **~-faced masonry** *n* STONE Bossenmauerwerk *nt*; **~ stone** *n* ARCH & BUILD, MAT PROP, STONE Bruchstein *m*; **~ stone work** *n* MAT PROP, STONE Bruchsteinmauerwerk *nt*; **~ tile** *n* INFR & DES, STONE Natursteinplatte *f*
**quarter** *n* ARCH & BUILD *area, district* Quartier *nt*, INFR & DES, MAT PROP, STONE, TIMBER *fraction* Viertel *nt*; **~ brick** *n* MAT PROP Riemchen *nt*, STONE Riemenstück *nt*; **~ round** *n* ARCH & BUILD, TIMBER Viertelstab *m*; **~ space** *n* ARCH & BUILD Viertelabsatz *m*; **~ timber** *n* TIMBER Kreuzholz *nt*
**quartz** *n* MAT PROP, STONE Quarz *m*; **~ clay** *n* MAT PROP Quarzton *m*; **~-containing** *adj* MAT PROP quarzhaltig; **~ grain** *n* MAT PROP Quarzkorn *nt*
**quartzite** *n* MAT PROP Quarzit *m*
***quartz***: **~ sand** *n* MAT PROP Quarzsand *m*
**quay** *n* CONCR, INFR & DES, STONE Kai *m*; **~ wall** *n* CONCR, INFR & DES, STONE Kaimauer *f*
**queen**: **~ closer** *n* STONE Viertelziegel *m*, Riemchen *nt*, Viertelstein *m*; **~-post truss roof** *n* ARCH & BUILD, TIMBER *braced rafters* zweistielig abgestrebtes Pfettendach *nt*, *unbraced, stiffened rafters* zweistielig strebenloses Pfettendach *nt*
**quench** *vt* MAT PROP abschrecken
**quicklime** *n* STONE gebrannter Kalk *m*
**quicksand** *n* MAT PROP Treibsand *m*
**quirk** *n* BUILD HARDW Hohlkehle *f*, Nut *f*
**quoin** *n* CONCR, INFR & DES, MAT PROP Eckstein *m*, STONE Eckstein *m*, Mauerecke *f*, TIMBER scharfe Kante *f*

# R

**rabbet** *n* ARCH & BUILD, BUILD HARDW, BUILD MACHIN Anschlag *m*, Falz *m*, Nut *f*
**rabbeted**: **~ joint** *n* ARCH & BUILD gefalzte Verbindung *f*
***rabbet***: **~ iron** *n* BUILD HARDW, BUILD MACHIN Kröseleisen *nt*; **~ ledge** *n* ARCH & BUILD *window* Schlagleiste *f*; **~ plane** *n* BUILD MACHIN Falzhobel *m*, Nuthobel *m*
**rabble**: **~ arm** *n* ENVIRON Krählarm *m*, Rührarm *m*
**raceway** *n* WASTE WATER Gerinne *nt*
**rack** *n* BUILD HARDW Gestell *nt*; **~-and-pinion door closer** *n* BUILD HARDW Zahntriebtürschließer *m*
**racking** *n* ARCH & BUILD, INFR & DES, STONE Abtreppung *f*; **~ force** *n* ARCH & BUILD, MAT PROP Dehnungskraft *f*
**radial** *adj* HEAT & VENT, MAT PROP radial; **~ fan** *n* BUILD HARDW, HEAT & VENT Radialventilator *m*; **~ pressure** *n* MAT PROP Radialdruck *m*; **~ stress** *n* MAT PROP Radialspannung *f*; **~ system** *n* WASTE WATER Radialnetz *nt*
**radiant**: **~ ceiling heating** *n* HEAT & VENT Deckenstrahlungsheizung *f*; **~ heat** *n* HEAT & VENT Strahlungswärme *f*; **~ panel heating** *n* HEAT & VENT Flächenheizung *f*
**radiate** *vti* HEAT & VENT, INFR & DES ausstrahlen
**radiation**: **~ protection** *n* ENVIRON Strahlenschutz *m*
**radiator** *n* ELECTR Radiator *m*, HEAT & VENT Heizkörper *m*, Radiator *m*; **~ element** *n* HEAT & VENT Heizkörperglied *nt*
**radioactive** *adj* CONST LAW, ENVIRON, MAT PROP radioaktiv; **~ fallout** *n* ENVIRON radioaktiver Niederschlag *m*; **~ pollution** *n* ENVIRON radioaktive Verschmutzung *f*; **~ substance** *n* ENVIRON radioaktiver Stoff *m*; **~ waste** *n* CONST LAW, ENVIRON Atommüll *m*, radioaktiver Abfall *m*
**radius** *n* ARCH & BUILD Radius *m*; **~ of gyration** *n* ARCH & BUILD Trägheitsradius *m*
**radon** *n* HEAT & VENT, MAT PROP Radon *nt*; **~ exclusion in buildings** *n* CONST LAW, HEAT & VENT, INFR & DES, MAT PROP Radonausschluß *m* in Gebäuden
**rafter** *n* ARCH & BUILD, TIMBER Sparren *m*; **~ head** *n* ARCH & BUILD, TIMBER Sparrenkopf *m*; **~ roof** *n* ARCH & BUILD, TIMBER Sparrendach *nt*; **~ system** *n* ARCH & BUILD, TIMBER Sparrenlage *f*; **~-to-purlin connector** *n* ARCH & BUILD, INFR & DES, STEEL Sparrenpfettenanker *m*; **~ trimmer** *n* ARCH & BUILD Schiftsparren *m*
**raft**: **~ foundation** *n* CONCR, INFR & DES Plattengründung *f*
**rag**: **~ bolt** *n* BUILD HARDW, CONCR, MAT PROP Steinschraube *f*
**rail** *n* BUILD HARDW *window* Sprosse *f*, INFR & DES Schiene *f*; **~ connection** *n* INFR & DES Schienenstoß *m*; **~ head** *n* INFR & DES Schienenkopf *m*
**railing** *n* ARCH & BUILD, BUILD HARDW, CONCR, STEEL, TIMBER Brüstung *f*, Geländer *nt*, Geländerpfosten *m*
***rail***: **~ jack** *n* BUILD MACHIN Schienenheber *m*; **~ joint** *n* INFR & DES Schienenstoß *m*
**railroad** *n* *AmE* (*cf railway BrE*) INFR & DES Eisenbahn *f*; **~ bridge** *n* *AmE* INFR & DES Eisenbahnbrücke *f*; **~ construction** *n* *AmE* INFR & DES Eisenbahnbau *m*; **~ tunnel** *n* *AmE* INFR & DES Eisenbahntunnel *m*; **~ underpass** *n* *AmE* INFR & DES Eisenbahnunterführung *f*
**railway** *n* *BrE* (*cf railroad AmE*) INFR & DES Eisenbahn *f*; **~ bridge** *n* *BrE* INFR & DES Eisenbahnbrücke *f*; **~ construction** *n* *BrE* INFR & DES Eisenbahnbau *m*; **~ tunnel** *n* *BrE* INFR & DES Eisenbahntunnel *m*; **~ underbridge** *n* *BrE* INFR & DES Eisenbahnunterführung *f*
**rain** *n* ENVIRON Regen *m*; **~ damage** *n* MAT PROP Regenschaden *m*; **~-drainage channel** *n* INFR & DES, WASTE WATER Regenwasserkanal *m*
**rainfall** *n* INFR & DES, WASTE WATER Niederschlag *m*; **~ per second per area** *n* ENVIRON, INFR & DES Regenspende *f*
***rain***: **~ gage** *AmE*, **~ gauge** *BrE* *n* BUILD MACHIN Niederschlagsmesser *m*, Pluviometer *nt*, Regenmeßgerät *nt*, ENVIRON Niederschlagsmesser *m*
**rainproof** *adj* MAT PROP, SURFACE regenundurchlässig
***rain***: **~ protection** *n* INFR & DES Regenschutz *m*; **~-repellent** *adj* MAT PROP, SURFACE regenabweisend; **~ repeller** *n* ARCH & BUILD Regenabweiser *m*; **~ resistance** *n* INFR & DES, MAT PROP Regenbeständigkeit *f*
**rainspout** *n* ARCH & BUILD Wasserspeier *m*
**rainwater** *n* INFR & DES, WASTE WATER Regenwasser *nt*; **~ collector** *n* BUILD MACHIN Niederschlagsmeßgerät *nt*; **~ downpipe** *n* WASTE WATER Regenwasserfallrohr *nt*; **~ head** *n* WASTE WATER Rinnenkasten *m*; **~ inlet** *n* WASTE WATER Regenwasserablauf *m*; **~ pipe** *n* WASTE WATER Fallrohr *nt*; **~ retention basin** *n* INFR & DES, WASTE WATER Regenrückhaltebecken *nt*
**raise** *vt* ARCH & BUILD, INFR & DES *building* aufstocken, *embankment, ground level* aufschütten, erhöhen; **~ on edge** *vt* ARCH & BUILD hochkanten
**raised** *adj* ARCH & BUILD, INFR & DES, SURFACE hoch, erhaben; **~ curb** *n* *AmE* (*cf raised kerb BrE*) INFR & DES Hochbordstein *m*; **~ floor** *n* ARCH & BUILD Zwischenboden *m*, Kabelboden *m*; **~ kerb** *n* *BrE* (*cf raised curb AmE*) INFR & DES Hochbordstein *m*
**raising** *n* ARCH & BUILD, INFR & DES Erhöhen *nt*, *of wall* Heben *nt*, Hochheben *nt*, Aufhöhung *f*
**raker** *n* BUILD MACHIN Kopfband *nt*, Kratzeisen *nt*, Räumlöffel *m*, Schrägbalken *m*; **~ pile** *n* ARCH & BUILD Diagonalverband *m*, Schrägpfahl *m*, INFR & DES Schrägpfahl *m*, STEEL Diagonalverband *m*
**raking**: **~ abutments** *n pl* INFR & DES, STONE schräge Widerlager *nt pl*; **~ bond** *n* STONE Schrägverband *m*, Fischgrätenverband *m*; **~ shore** *n* BUILD MACHIN Abstützbohle *f*
**ram** *n* BUILD MACHIN Ramme *f*
**rammed**: **~ earth** *n* ARCH & BUILD, INFR & DES verdichtete Erde *f*
**rammer** *n* BUILD MACHIN Stampfer *m*
**ramp** *n* ARCH & BUILD, INFR & DES Rampe *f*
**rampant**: **~ arch** *n* ARCH & BUILD steigender Bogen *m*
***ramp***: **~ incline** *n* ARCH & BUILD, INFR & DES Rampen-

aufgang *m*, Rampenschräge *f*; ~ **landfill** *n* ENVIRON Anböschung *f*
**random**: ~ **rubble** *n* ARCH & BUILD, INFR & DES, STONE unsortierter Bruchstein *m*; ~ **rubble fill** *n* ARCH & BUILD, INFR & DES, STONE Bruchsteinschüttung *f*; ~ **rubble masonry** *n* ARCH & BUILD, INFR & DES, STONE Vieleckmauerwerk *nt*, Polygonmauerwerk *nt*, Zyklopenmauerwerk *nt*; ~ **sample** *n* CONST LAW Stichprobe *f*; ~ **test** *n* CONST LAW Stichprobe *f*
**range 1.** *n* ARCH & BUILD, CONST LAW, INFR & DES Bereich *m*; **2.** ~ **in** *vt* ARCH & BUILD einfluchten
**ranged**: ~ **in** *adj* ARCH & BUILD eingefluchtet
***range***: ~ **pole** *n* INFR & DES Fluchtstab *m*
**ranger** *n* CONCR Brustholz *nt*, INFR & DES, MAT PROP Brustholz *nt*, Gurt *m*, Gurtholz *nt*, Riegel *m*, Verbindungsstück *nt*, TIMBER Brustholz *nt*
***range***: ~ **rod** *n* ARCH & BUILD Fluchtstab *m*
**ranging**: ~ **pole** *n* ARCH & BUILD Absteckstange *f*, Bake *f*, Fluchtstange *f*
**rapid**: ~ **fermentation** *n* ENVIRON Schnellkompostierung *f*, beschleunigte Kompostierung *f*, geschlossene Kompostierung *f*; ~ **filter** *n* WASTE WATER Schnellfilter *m*; ~ **frame system** *n* ARCH & BUILD, INFR & DES schnellerstellbares Rahmensystem *nt*; ~-**hardening cement** *n* CONCR, MAT PROP frühhochfester Zement *m*
**rasp** *n* BUILD MACHIN Raspe *f*, Reibeisen *nt*
**ratchet** *n* BUILD HARDW Sperrklinke *f*, BUILD MACHIN Ratsche *f*, Sperrhaken *m*; ~ **brace** *n* BUILD MACHIN Schraubstempel *m*; ~ **wrench** *n* BUILD MACHIN Spannschlüssel *m*
**rate** *vt* ARCH & BUILD abschätzen, bemessen, CONST LAW, INFR & DES abschätzen
**rate**: ~ **of absorption** *n* MAT PROP Absorptionsgeschwindigkeit *f*; ~ **of creep** *n* MAT PROP Kriechgeschwindigkeit *f*; ~ **of curing** *n* MAT PROP Erstarrungsgeschwindigkeit *f*; ~ **of deformation** *n* MAT PROP Verformungsgeschwindigkeit *f*
**rated**: ~ **welding current** *n* ELECTR Nennschweißstrom *m*
**rate**: ~ **of erosion** *n* INFR & DES, MAT PROP Erosionsgeschwindigkeit *f*; ~ **of progress** *n* ARCH & BUILD Bauablaufgeschwindigkeit *f*; ~ **of spread** *n* ARCH & BUILD Ausbreitungsgeschwindigkeit *f*
**ratio** *n* ARCH & BUILD, INFR & DES, MAT PROP Verhältnis *nt*
**raw** *adj* ENVIRON, MAT PROP roh; ~ **material** *n* MAT PROP Rohstoff *m*; ~ **refuse** *n* ENVIRON Rohmüll *m*; ~ **sewage** *n* CONST LAW, ENVIRON, INFR & DES, WASTE WATER Rohabwasser *nt*; ~ **sludge** *n* ENVIRON Rohschlamm *m*; ~ **water** *n* CONST LAW, ENVIRON, INFR & DES, WASTE WATER Brauchwasser *nt*
**RDF** *abbr* (*refuse-derived fuel*) ENVIRON Müllbrennstoff *m*
**re-** *pref* ARCH & BUILD Wieder-
**Re** *abbr* (*Reynolds number*) ENVIRON, INFR & DES, MAT PROP Re (*Reynoldszahl*)
**re-align** *vt* ARCH & BUILD *wall* wieder ausrichten
**re-equip** *vt* BUILD MACHIN neu ausrüsten, neu ausstatten
**react** *vi* MAT PROP reagieren
**reaction** *n* MAT PROP Reaktion *f*; ~ **turbine** *n* ENVIRON Reaktionsturbine *f*, Überdruckturbine *f*
**reading** *n* ELECTR, HEAT & VENT Ablesung *f*, INFR & DES Ablesung *f*, Ablesen *nt*
**ready**: ~-**mixed concrete** *n* CONCR Fertigbeton *m*, Transportbeton *m*; ~-**mixed stuff** *n* MAT PROP, STONE Trockenputz *m*; ~-**to-paint** *adj* SURFACE malerfertig
**real**: ~ **estate** *n* CONST LAW Grundeigentum *nt*, Landbesitz *m*; ~ **estate agent** *n* *AmE* (*cf estate agent BrE*) CONST LAW Grundstücksmakler *m*, Immobilienhändler *m*, Immobilienmakler *m*; ~ **estate tax** *n* *AmE* (*cf property tax BrE*) CONST LAW Grundsteuer *f*
**reamer** *n* BUILD MACHIN Reibahle *f*, Räumer *m*
**reaming**: ~ **iron** *n* BUILD MACHIN Aufreibdorn *m*
**rear** *adj* ARCH & BUILD Hinter-, hinter
**rearrangement** *n* ARCH & BUILD Neuordnung *f*
**rear**: ~ **side** *n* ARCH & BUILD Hinterseite *f*, Rückseite *f*; ~ **wall** *n* ARCH & BUILD Rückwand *f*
**reasonable**: ~ **care** *n* CONST LAW vernünftige Sorgfalt *f*
**reassemble** *vt* ARCH & BUILD wieder zusammenbauen
**rebate** *n* ARCH & BUILD Falz *m*, Anschlagfalz *m*, Kittfalz *m*
**rebated** *adj* ARCH & BUILD gefalzt; ~ **door** *n* BUILD HARDW Falztür *f*; ~ **joint** *n* ARCH & BUILD Falzfuge *f*, überfalzte Fuge *f*
***rebate***: ~ **for putty** *n* ARCH & BUILD Glasfalz *m*; ~ **ledge** *n* ARCH & BUILD *double-sash window* Schlagseite *f*; ~ **plane** *n* BUILD MACHIN Falzhobel *m*, Nuthobel *m*; ~ **pressure** *n* MAT PROP Anpreßdruck *m* im Glas
**rebuild** *vt* ARCH & BUILD umbauen, wiederaufbauen
**rebuilding** *n* ARCH & BUILD Wiederaufbau *m*
**receiving**: ~ **bin** *n* ENVIRON Lagerbunker *m*; ~ **bunker** *n* ENVIRON Aufnahmebunker *m*, Müllbunker *m*; ~ **water** *n* ENVIRON Vorfluter *m*
**receptacle** *n* ELECTR Steckdose *f*
**receptor** *n* ENVIRON Empfänger *m*; ~ **region** *n* ENVIRON Empfängerbereich *m*
**recess 1.** *n* ARCH & BUILD Nische *f*, Aussparung *f*, Rücksprung *m*; **2.** *vt* ARCH & BUILD aussparen, vertiefen, zurücksetzen, einlassen, vertieft anbringen, TIMBER aussparen
**recessed** *adj* ARCH & BUILD eingelassen, versenkt, ausgespart, zurückgesetzt, TIMBER ausgespart; ~ **light fixture** *n* ELECTR Einbauleuchte *f*
**recessing** *n* ARCH & BUILD Aussparen *nt*, Vertiefen *nt*, TIMBER Aussparen *nt*
**recharge** *vt* BUILD MACHIN *batteries* aufladen
**recirculated**: ~ **air** *n* HEAT & VENT Umluft *f*
**reclaimed**: ~ **area** *n* ENVIRON wiederurbargemachtes Gebiet *nt*
**reclamation** *n* CONST LAW, ENVIRON, INFR & DES Weiterverwertung *f*; ~ **of land** *n* CONST LAW, ENVIRON, INFR & DES Landerschließung *f*, Landgewinnung *f*; ~ **plant** *n* CONST LAW, ENVIRON, INFR & DES Rückgewinnungsanlage *f*, Wiedergewinnungsanlage *f*
**recommendation** *n* CONST LAW Empfehlung *f*
**recompact** *vt* INFR & DES *soil* nachverdichten
**reconditioning** *n* ARCH & BUILD, INFR & DES Aufarbeitung *f*
**reconnecting**: ~ **electrical supplies** *phr* ELECTR Wiederherstellung *f* von Stromversorgung
**reconsolidation** *n* INFR & DES Wiederkonsolidierung *f*
**reconstruct** *vt* ARCH & BUILD rekonstruieren, wiederaufbauen
**reconstruction** *n* ARCH & BUILD Umbau *m*, Wiederaufbau *m*, Rekonstruktion *f*, Wiederaufbauen *nt*
**recontamination** *n* CONST LAW, ENVIRON Rekontamination *f*

**record 1.** *n* ARCH & BUILD Aufzeichnung *f*; **2.** *vt* ARCH & BUILD aufzeichnen

***record***: **~ drawings** *n pl* CONST LAW Baubestandszeichnungen *f pl*

**recording**: **~ inspection findings** *phr* CONST LAW Aufzeichnung *f* der Untersuchungsergebnisse

**recover** *vt* ENVIRON *raw material* wiedergewinnen, zurückgewinnen, INFR & DES gewinnen

**recoverable**: **~ waste** *n* ENVIRON rückgewinnbarer Abfall *m*

**recovered**: **~ heat** *n* ENVIRON, HEAT & VENT rückgewonnene Wärme *f*; **~ oil** *n* ENVIRON wiedergewonnenes Öl *nt*, Ölregenerat *nt*; **~ pulp** *n* ENVIRON rückgewonnene Pulpe *f*

**recovery** *n* ENVIRON Wiederverwertung *f*, Wiedergewinnung *f*; **~ boiler** *n* ENVIRON Rückgewinnungskessel *m*; **~ device** *n* ENVIRON Aufarbeitungsvorrichtung *f*

**recreation**: **~ center** *AmE*, **~ centre** *BrE* *n* ARCH & BUILD, INFR & DES Erholungszentrum *nt*, Freizeiteinrichtung *f*; **~ room** *n* ARCH & BUILD, INFR & DES Aufenthaltsraum *m*

**rectangular** *adj* ARCH & BUILD rechtwinklig, rechteckig; **~ conduit** *n* ELECTR, HEAT & VENT Rechteckkanal *m*; **~ cross-section** *n* ARCH & BUILD rechteckiger Querschnitt *m*; **~ section gutter** *n* ARCH & BUILD, WASTE WATER Kastenrinne *f*; **~ triangle** *n* ARCH & BUILD rechtwinkliges Dreieck *nt*

**rectify** *vt* ARCH & BUILD begradigen

**recultivation** *n* ENVIRON *of land or water* Wiederurbarmachung *f*

**recyclable** *adj* ENVIRON recyclingfähig, verwertbar, wiederverwertbar

**recycle** *vt* ENVIRON recyceln, wiederverwerten

**recycled**: **~ paper** *n* ENVIRON Recyclingpapier *nt*; **~ polypropylene** *n* ENVIRON wiedergewonnenes Polypropylen *nt*

**recycle**: **~ sludge** *n* ENVIRON Rücklaufschlamm *m*

**recycling** *n* CONST LAW *salvage* Wiedergewinnung *f*, ENVIRON *of waste* Wiedergewinnung *f*, Recycling *nt*, Wiederverwertung *f*, Rückführung *f*, Rückgewinnung *f*, INFR & DES *salvage* Wiedergewinnung *f*; **~ economy** *n* ENVIRON Kreislaufwirtschaft *f*; **~ of inoculated compost** *n* ENVIRON Impfkompostrückführung *f*; **~ plant** *n* CONST LAW, ENVIRON, INFR & DES Recyclinganlage *f*, Wiedergewinnungsanlage *f*; **~ process** *n* ENVIRON Recyclingprozeß *m*; **~ rate** *n* ENVIRON Recyclingquote *f*, Verwertungsquote *f*; **~ of sludge** *n* ENVIRON Schlammverwertung *f*

**red**: **~ brass** *n* MAT PROP Rotguß *m*

**redesign 1.** *n* ARCH & BUILD, INFR & DES Umbemessung *f*; **2.** *vt* ARCH & BUILD, INFR & DES umarbeiten, umplanen

**redevelop** *vt* ARCH & BUILD neugestalten, sanieren

**redevelopment** *n* ARCH & BUILD Wiederaufbau *m*, Wiederherstellung *f*, Rekonstruktion *f*

***red***: **~ lead** *n* MAT PROP, SURFACE Bleimennige *f*; **~ mud** *n* ENVIRON Rotschlamm *m*; **~ ocher** *AmE*, **~ ochre** *BrE* *n* INFR & DES Rotocker *m*; **~ rod** *n* ENVIRON roter Stab *m*

**reduce** *vt* ARCH & BUILD verjüngen

**reducer** *n* ARCH & BUILD Reduzierstück *nt*, Übergangsstück *nt*, HEAT & VENT Übergangsstück *nt*, WASTE WATER Reduzierstück *nt*, Übergangsstück *nt*, Übergangsrohr *nt*

**reducing** *n* ARCH & BUILD *board* Verdünnen *nt*; **~ fitting** *n* HEAT & VENT, WASTE WATER Reduktionsstück *nt*; **~ pipe fitting** *n* HEAT & VENT, WASTE WATER Reduzierstück *nt*

**reduction**: **~ scale** *n* ARCH & BUILD Verkleinerungsmaßstab *m*

**redundancy** *n* ARCH & BUILD Grad *m* der statischen Unbestimmtheit

**redundant** *adj* ARCH & BUILD unbestimmt, überzählig; **~ bar** *n* ARCH & BUILD *statical calculation* überzähliger Stab *m*

**reed**: **~ lathing** *n* STONE Rohrgewebe *nt*, Rohrmatte *f*; **~ roof** *n* ARCH & BUILD Reetdach *nt*, Rieddach *nt*

**reface** *vt* ARCH & BUILD die Fassade erneuern, die Oberfläche erneuern

**reference**: **~ mark** *n* ARCH & BUILD *surveying*, INFR & DES Bezugspunkt *m*; **~ point** *n* ARCH & BUILD, INFR & DES Bezugspunkt *m*; **~ sound pressure** *n* ENVIRON Bezugsschalldruck *m*; **~ table** *n* ARCH & BUILD Nachschlagtabelle *f*; **~ value** *n* ARCH & BUILD, INFR & DES Bezugsgröße *f*

**refilling** *n* ARCH & BUILD, INFR & DES Auffüllung *f*, Verfüllen *nt*

**refinery**: **~ waste** *n* ENVIRON Raffinerierückstände *m pl*

**reflectance** *n* ENVIRON Rückstrahlungsvermögen *nt*

**reflecting**: **~ stud** *n* INFR & DES *in roadbed* Verkehrsleuchtnagel *m*

**reflector** *n* ELECTR *light fixture* Reflektor *m*

**refract** *vt* INFR & DES brechen

**refraction** *n* INFR & DES Refraktion *f*

**refrigerate** *vt* HEAT & VENT *chill* kühlen, *freeze* tiefkühlen

**refrigerating**: **~ machine** *n* HEAT & VENT Kältemaschine *f*

**refrigeration** *n* HEAT & VENT Kühlung *f*

**refurbish** *vt* ARCH & BUILD, INFR & DES auffrischen, aufpolieren, renovieren, aufarbeiten

**refurbishment** *n* ARCH & BUILD, INFR & DES Aufarbeitung *f*, Renovierung *f*, Auffrischung *f*

**refuse**: **~ bunker** *n* ENVIRON Lagerbunker *m*; **~ cell** *n* ENVIRON, INFR & DES Kassette *f*, Polder *m*; **~ collection service** *n* CONST LAW, ENVIRON Müllabfuhr *f*; **~ collection vehicle** *n* BUILD MACHIN, ENVIRON Müllfahrzeug *nt*, Müllwagen *m*, Müllabfuhrwagen *m*, Müllsammelfahrzeug *nt*; **~-derived fuel** *n* (*RDF*) ENVIRON Müllbrennstoff *m*; **~ disposal** *n* ENVIRON Abfallbeseitigung *f*; **~ disposal site** *n* ENVIRON Müllkippe *f*; **~ grinder** *n* ENVIRON Müllzerkleinerer *m*; **~ incineration** *n* ENVIRON Müllverbrennung *f*; **~ incineration plant** *n* ENVIRON Müllverbrennungsanlage *f* (*MVA*); **~ incinerator** *n* ENVIRON Abfallverbrennungsanlage *f*; **~ sack** *n* ENVIRON Müllsack *m*; **~ sack collection** *n* ENVIRON Einsammeln *nt* von Müllsäcken; **~ separation plant** *n* ENVIRON Abfallsortieranlage *f*, Müllsortierungsanlage *f*; **~ transfer station** *n* ENVIRON Zwischenlagerplatz *m*

**regeneration** *n* ENVIRON Regeneration *f*

**register** *n* ARCH & BUILD, HEAT & VENT Klappe *f*; **~ of hazardous substances** *n* ENVIRON Gefahrstoffkataster *nt*; **~ of real estates** *n* *AmE* (*cf land register BrE*) CONST LAW Grundbuch *nt*

**regrating**: **~ skin** *n* STONE *plaster* Besenwurf *m*

**regular** *adj* INFR & DES regelmäßig; **~ coursed ashlar stone work** *n* ARCH & BUILD, STONE Quadermauerwerk *nt*

**regulate** *vt* HEAT & VENT regulieren, einregulieren
**regulating**: **~ valve** *n* HEAT & VENT Regulierventil *nt*
**regulation** *n* CONST LAW Vorschrift *f*, Verfügung *f*
**regulations**: **~ for electrical plants installed in hazardous locations** *n pl* CONST LAW Verordnung *f* über elektrische Anlagen in explosionsgefährdeten Standorten
**regulator** *n* ELECTR, HEAT & VENT Regler *m*
**rehabilitation** *n* ARCH & BUILD Sanierung *f*, CONST LAW, ENVIRON *of land or water* Wiedernutzbarmachung *f*
**reheat** *vt* HEAT & VENT wiedererwärmen
**reinforce** *vt* ARCH & BUILD armieren, bewehren, verstärken, CONCR armieren, bewehren, MAT PROP verstärken
**reinforced** *adj* CONCR bewehrt, MAT PROP baustahlarmiert; **~ border** *n* MAT PROP Randverstärkung *f*; **~ brickwork** *n* STONE bewehrtes Mauerwerk *nt*; **~ concrete** *n* CONCR Stahlbeton *m*, bewehrter Beton *m*; **~ concrete floor** *n* ARCH & BUILD, CONCR Stahlbetondecke *f*; **~ soil** *n* INFR & DES bewehrte Erde *f*
**reinforcement** *n* ARCH & BUILD, CONCR, MAT PROP Armierung *f*, Bewehrung *f*, Verstärkung *f*; **~ of cover** *n* CONCR Abdeckungsverstärkung *f*; **~ mat** *n* ARCH & BUILD, CONCR, INFR & DES, MAT PROP, STEEL, TIMBER Baustahlmatte *f*, Bewehrungsmatte *f*, Armierungsmatte *f*; **~ steel** *n* CONCR Betonstahl *m*; **~ steel mesh** *n* ARCH & BUILD, CONCR, INFR & DES, MAT PROP, STEEL, TIMBER Armierungsmatte *f*, Bewehrungsmatte *f*, Baustahlmatte *f*
**reinforcing** *n* ARCH & BUILD Absteifung *f*, Aussteifung *f*, Bewehren *nt*, Verstärken *nt*; **~ bar** *n* ARCH & BUILD Bewehrungseisen *nt*, BUILD MACHIN Betoneisen *nt*, Bewehrungsstahl *m*, CONCR, MAT PROP Bewehrungseisen *nt*; **~ cage** *n* CONCR Bewehrungskorb *m*; **~ cross member** *n* CONCR Aussteifungsriegel *m*; **~ frame** *n* ARCH & BUILD Aussteifungsrahmen *m*; **~ plate** *n* ARCH & BUILD, MAT PROP, TIMBER Verstärkungsblech *nt*; **~ sheet** *n* ARCH & BUILD, MAT PROP, TIMBER Verstärkungsblech *nt*; **~ steel** *n* ARCH & BUILD Betonstahlmatte *f*, CONCR Betonstahl *m*, INFR & DES, MAT PROP, STEEL Betonstahlmatte *f*; **~ steel mesh** *n* ARCH & BUILD, CONCR, INFR & DES, MAT PROP, STEEL, TIMBER Baustahlmatte *f*, Bewehrungsmatte *f*, Betonstahlmatte *f*, Baustahlgewebe *nt*; **~ tape** *n* STONE *plaster* Verstärkungsband *nt*, Armierungsgewebe *nt*
**relative**: **~ humidity** *n* MAT PROP, SOUND & THERMAL relative Feuchtigkeit *f*; **~ water velocity** *n* ENVIRON relative Wassergeschwindigkeit *f*, HEAT & VENT relative Feuchtigkeit *f*
**relaxation** *n* CONCR Entspannung *f*, Relaxation *f*
**release** *n* BUILD HARDW, ELECTR *alarm* Auslösung *f*, ENVIRON Freisetzung *f*; **~ lube** *n* CONCR Schalungsöl *nt*; **~ paste** *n* CONCR Schalungspaste *f*; **~ pin** *n* BUILD HARDW *profile cylinder* Auslösenadel *f*
**releasing**: **~ hook** *n* BUILD HARDW Ausklinkhaken *m*
**releveling** *AmE*, **relevelling** *n BrE* INFR & DES *surveying* Neuaufnahme *f*
**reliable** *adj* BUILD MACHIN, CONST LAW zuverlässig, betriebssicher
**relief** *n* ARCH & BUILD, CONST LAW, INFR & DES, MAT PROP Entlastung *f*; **~ valve** *n* ENVIRON, HEAT & VENT Entlastungsventil *nt*, Sicherheitsventil *nt*
**relieving**: **~ layer** *n* ARCH & BUILD Druckausgleichsschicht *f*
**reload** *vt* ARCH & BUILD, INFR & DES wiederbelasten
**relocate** *vt* ARCH & BUILD versetzen, umstellen
**relocation** *n* ARCH & BUILD, INFR & DES Versetzung *f*, Umstellung *f*
**remainder** *n* ENVIRON Reststoff *m*
**remedial**: **~ building work** *n* ARCH & BUILD bauliche Abhilfemaßnahmen *f pl*; **~ measure** *n* ARCH & BUILD Abhilfemaßnahme *f*; **~ treatment** *n* ARCH & BUILD Abhilfemaßnahme *f*
**remetaling** *AmE*, **remetalling** *n BrE* INFR & DES *of roads* Aufbringen *nt* einer neuen Schotterschicht
**remote**: **~ control** *n* ELECTR Fernbedienung *f*, Fernsteuerung *f*; **~ sensor** *n* ELECTR Fernfühler *m*; **~ window controls** *n pl* BUILD HARDW Fensterfernbedienungsgeräte *nt pl*
**removal** *n* ARCH & BUILD *by vehicle* Abtransport *m*, *demolition* Beseitigung *f*, Entfernung *f*, ENVIRON *of suspended matter by sedimentation* Beseitigung *f*
**remover** *n* ARCH & BUILD *paint* Abbeizer *m*, SURFACE *solvent* Abbeizmittel *nt*, Beize *f*, Beizmittel *nt*
**removing** *n* ARCH & BUILD *dismantling* Entfernen *nt*, Ausbauen *nt*, Entlasten *nt*, CONST LAW *pressure*, INFR & DES Entlasten *nt*
**render 1.** *n* STONE Außenverputz *m*; **2.** *vt* STONE verputzen
**render**: **~, float and set** *n* (*RFS*) STONE dreilagiger Putz *m*
**rendering** *n* STONE Anwurf *m*, rauher Putz *m*
**renew** *vt* ARCH & BUILD erneuern, wiederherstellen
**renewable**: **~ energy** *n* ENVIRON regenerative Energie *f*
**renewal** *n* ARCH & BUILD, SURFACE Erneuerung *f*
**renovate** *vt* ARCH & BUILD, SURFACE erneuern, renovieren, umbauen, sanieren
**renovation** *n* ARCH & BUILD, SURFACE Erneuerung *f*, Renovierung *f*, Umbau *m*, Sanierung *f*; **~ coat** *n* SURFACE Renovierungsanstrich *m*
**rental**: **~ charge** *n* ARCH & BUILD, CONST LAW Mietkosten *pl*
**repaint** *vt* SURFACE neu streichen
**repainting** *n* SURFACE Erneuerungsanstrich *m*
**repair 1.** *n* ARCH & BUILD, BUILD MACHIN, SURFACE Reparatur *f*; **2.** *vt* ARCH & BUILD, BUILD MACHIN, SURFACE reparieren
**repairs** *n pl* ARCH & BUILD, INFR & DES Instandsetzungsarbeiten *f pl*
**repair**: **~ work** *n* ARCH & BUILD, INFR & DES Ausbesserungsarbeiten *f pl*
**repellent** *adj* MAT PROP abweisend
**replacement** *n* INFR & DES Ersatz *m*
**repoint** *vt* STONE wieder ausfugen, Fugen ausfüllen
**report 1.** *n* ARCH & BUILD Bericht *m*; **2.** *vti* ARCH & BUILD berichten
**repose** *n* ARCH & BUILD, INFR & DES Ruhe *f*
**repository** *n* ENVIRON *for radioactive waste* Endlagerstätte *f*, Deponie *f*, Lagerstätte *f*, INFR & DES Deponie *f*
**repossess** *vt* CONST LAW wieder in Besitz nehmen
**repossession** *n* CONST LAW Wiederinbesitznahme *f*
**represent** *vt* ARCH & BUILD darstellen
**representation** *n* ARCH & BUILD Darstellung *f*
**representing** *adj* ARCH & BUILD darstellend
**reprocess** *vt* ENVIRON aufarbeiten
**reprocessing** *n* ENVIRON Aufarbeitung *f*
**reprofiling** *n* ARCH & BUILD Neuprofilieren *nt*
**repulpable**: **~ adhesive** *n* ENVIRON einstampfbarer Klebstoff *m*
**repulsion** *n* INFR & DES Abstoß *m*

**request 1.** *n* ARCH & BUILD, CONST LAW Einholung *f*; **2.** *vt* ARCH & BUILD, CONST LAW einholen
**request**: **~ for bids** *n* ARCH & BUILD Einholung *f* von Angeboten
**required**: **~ hinge rotation** *n* ARCH & BUILD, CONCR, INFR & DES erforderliche Gelenkrotation *f*
**requirement** *n* CONST LAW Anforderung *f*, Forderung *f*
**research** *n* CONST LAW, INFR & DES Forschung *f*; **~ institute** *n* CONST LAW, INFR & DES Forschungsanstalt *f*; **~ program** *AmE*, **~ programme** *BrE* *n* CONST LAW, INFR & DES Forschungsprogramm *nt*
**reservoir** *n* ENVIRON, INFR & DES, WASTE WATER Sammelbecken *nt*; **~ sedimentation** *n* INFR & DES Stauraumsedimentierung *f*
**residence**: **~ time** *n* ENVIRON Verweilzeit *f*
**resident** *n* CONST LAW Anlieger *m*, Bewohner *m*
**residential**: **~ building** *n* ARCH & BUILD Wohnhaus *nt*
**residual**: **~ gas** *n* ENVIRON Gasrückstand *m*; **~ oil** *n* ENVIRON Altöl *nt*, Ölabfall *m*, Ölrückstände *m pl*; **~ soil** *n* INFR & DES Verwitterungserde *f*; **~ stress** *n* MAT PROP Restspannung *f*
**residue** *n* ENVIRON, MAT PROP Rückstand *m*; **~-derived energy** *n* ENVIRON Energie *f* aus Abfall; **~ landfill** *n* ENVIRON Reststoffdeponie *f*, Rückstandsdeponie *f*
**resilient** *adj* MAT PROP elastisch, unverwüstlich; **~ layer** *n* SOUND & THERMAL schallisolierende Schicht *f*
**resin** *n* MAT PROP, TIMBER Harz *nt*; **~-based mortar** *n* MAT PROP Kunstharzmörtel *m*
**resinoid**: **~-bonded** *adj* MAT PROP kunstharzgebunden
**resinous** *adj* MAT PROP harzhaltig
***resin***: **~ pocket** *n* TIMBER Harzeinschluß *m*
**resistance** *n* ELECTR Widerstand *m*, MAT PROP *hardiness* Beständigkeit *f*, *resistivity* Widerstandsfähigkeit *f*; **~ butt welding** *n* STEEL Widerstandsstumpfschweißen *nt*; **~ coefficient** *n* MAT PROP Widerstandsbeiwert *m*; **~ seam welding** *n* STEEL Widerstandsnahtschweißen *nt*; **~ to buckling** *n* ARCH & BUILD, CONCR, INFR & DES, MAT PROP, STEEL Knickfestigkeit *f*; **~ to creep** *n* ARCH & BUILD, CONCR, MAT PROP, STEEL Kriechfestigkeit *f*; **~ to heavy rain** *n* ARCH & BUILD, ELECTR Schlagregensicherheit *f*; **~ to impact** *n* ARCH & BUILD, CONCR, MAT PROP, STEEL Schlagfestigkeit *f*; **~ to shear** *n* ARCH & BUILD, CONCR, MAT PROP, STEEL, TIMBER Schubtragfähigkeit *f*; **~ welding** *n* STEEL Widerstandsschweißung *f*, Widerstandsschweißen *nt*
**resistant** *adj* ELECTR, MAT PROP beständig, widerstandsfähig; **~ to chemicals** *adj* MAT PROP chemikalienbeständig; **~ to heavy rain** *adj* ELECTR schlagregensicher; **~ to temperature changes** *adj* MAT PROP temperaturwechselbeständig
**resolution**: **~ of forces** *n* ARCH & BUILD, INFR & DES Kräftezerlegung *f*
**resonance** *n* ENVIRON, MAT PROP, SOUND & THERMAL Resonanz *f*; **~ screen** *n* ENVIRON Resonanzsieb *nt*
**resorcinol**: **~-formaldehyde** *n* (*RF*) ENVIRON, MAT PROP, TIMBER Resorcin-Formaldehyd *nt* (*RF*)
**resource** *n* ENVIRON, INFR & DES Hilfsquelle *f*; **~ recovery** *n* ENVIRON Rohstoffrückgewinnung *f*, Wertstoffrückgewinnung *f*; **~ recovery plant** *n* CONST LAW, ENVIRON, INFR & DES Wiedergewinnungsanlage *f*
**respiration** *n* ENVIRON Atmung *f*
**respirator** *n* INFR & DES Atmungsgerät *nt*
**respiratory**: **~ protection equipment** *n* ENVIRON Atemschutzgerät *nt*
**respire** *vt* ENVIRON atmen, respirieren
**response**: **~ temperature** *n* ELECTR Öffnungstemperatur *f*
**responsibility** *n* ARCH & BUILD Verantwortlichkeit *f*, Verantwortung *f*
**responsible** *adj* ARCH & BUILD verantwortlich
**rest** *n* ARCH & BUILD, INFR & DES Ruhe *f*
**resting**: **~ platform** *n* ARCH & BUILD *steel lattice tower*, BUILD MACHIN Ruheplattform *f*
**restoration** *n* ARCH & BUILD Restaurierung *f*, Wiederherstellung *f*
**restore** *vt* ARCH & BUILD restaurieren, wiederherstellen
**restore**: **~ the original state** *phr* INFR & DES den ursprünglichen Zustand wiederherstellen
**restrain** *vt* ARCH & BUILD, BUILD MACHIN, CONCR, INFR & DES, STEEL einspannen
**restrained** *adj* ARCH & BUILD, BUILD MACHIN, CONCR, INFR & DES, STEEL eingespannt
**restraining** *n* ARCH & BUILD, BUILD MACHIN, CONCR, HEAT & VENT, INFR & DES, STEEL, TIMBER Einspannung *f*; **~ straps** *n pl* BUILD MACHIN, STEEL Einspannungsbügel *m pl*
**restrict** *vt* CONST LAW *curtail* einschränken, *limit* begrenzen, *localize* beschränken
**restriction** *n* CONST LAW *curtailment* Einschränkung *f*, *limiting* Beschränkung *f*
**result** *n* INFR & DES Resultat *nt*
**resultant** *n* ARCH & BUILD, INFR & DES Resultierende *f*; **~ force** *n* ARCH & BUILD, INFR & DES resultierende Kraft *f*
**resurvey** *n* ARCH & BUILD Nachvermessung *f*
**retain** *vt* ARCH & BUILD, CONCR, ENVIRON, INFR & DES, MAT PROP, STONE, WASTE WATER stauen, zurückhalten
**retaining**: **~ basin** *n* INFR & DES Staubecken *nt*; **~ bolt** *n* MAT PROP Befestigungsschraube *f*; **~ dam** *n* INFR & DES Staudamm *m*; **~ ring** *n* MAT PROP Übersteckring *m*; **~ screw** *n* MAT PROP Befestigungsschraube *f*; **~ wall** *n* ARCH & BUILD Stützmauer *f*, CONCR, INFR & DES, STONE Stützwand *f*
**retard** *vt* CONCR, ENVIRON, MAT PROP verzögern
**retardation**: **~ time** *n* MAT PROP Verzögerungsdauer *f*
**retarder** *n* ENVIRON Reaktionsverzögerer *m*
**retarding**: **~ agent** *n* CONCR Verzögerungsmittel *nt*, ENVIRON Reaktionsverzögerer *m*, MAT PROP Verzögerungsmittel *nt*
**retention**: **~ basin** *n* INFR & DES Rückhaltebecken *nt*, Rücklagenwand *f*; **~ fee amount** *n* CONST LAW Einbehaltungsbetrag *m*; **~ money** *n* CONST LAW Einbehaltungssumme *f*, Geldeinbehaltung *f*; **~ of money** *n* CONST LAW Geldeinbehaltung *f*; **~ percentage** *n* CONST LAW Prozentsatz *m* der Einbehaltung; **~ time** *n* ENVIRON Verweildauer *f*; **~ of title** *n* CONST LAW Einbehaltung *f* des Rechtsanspruches
**reticulated**: **~ vault** *n* ARCH & BUILD Netzgewölbe *nt*; **~ window** *n* ARCH & BUILD Netzwerkfenster *nt*
**retractable** *adj* BUILD HARDW einschiebbar
**retracted**: **~ position** *n* BUILD HARDW *latch bolt of lock* zurückgezogener Zustand *m*
**retreat** *n* INFR & DES Rücksprung *m*
**retreatment** *n* INFR & DES Nachbehandlung *f*
**retrofit 1.** *n* ENVIRON Nachrüsten *nt*; **2.** *vt* ENVIRON nachrüsten
**retrogressive**: **~ erosion** *n* INFR & DES, MAT PROP, SURFACE rückwärtsschreitende Korrosion *f*
**return**: **~ flow** *n* HEAT & VENT Rücklauf *m*; **~ sludge** *n* ENVIRON Rücklaufschlamm *m*; **~ temperature** *n*

HEAT & VENT Rücklauftemperatur *f*; ~ **wall** *n* ARCH & BUILD Flügelwand *f*; ~ **water** *n* BUILD MACHIN, HEAT & VENT Rücklaufwasser *nt*

**reusable** *adj* ENVIRON, MAT PROP wiederverwertbar wiederverwendbar; ~ **waste product** *n* ENVIRON wiederverwertbares Abfallprodukt *nt*

**reuse 1.** *n* ENVIRON, MAT PROP Wiederverwendung *f*; **2.** *vt* ENVIRON, MAT PROP wiederverwenden, wiederverwerten

**reveal** *n* ARCH & BUILD *door, window* Leibung *f*

**revegetation** *n* ENVIRON, INFR & DES Rekultivierung *f*

**reversed**: ~ **arch** *n* ARCH & BUILD Gegenbogen *m*, Grundbogen *m*

**reverse**: ~ **drum** *n* BUILD MACHIN *mixer* Umkehrtrommel *f*; ~ **flow** *n* HEAT & VENT Rückfließen *nt*; ~ **flow filter** *n* ENVIRON Gegenstromfilter *nt*; ~ **gradient** *n* INFR & DES Gegengefälle *nt*; ~ **osmosis** *n* ENVIRON, MAT PROP Umkehrosmose *f*

**reversible**: ~ **cylinder** *n* BUILD HARDW *lock* umstellbarer Zylinder *m*; ~ **latch bolt** *n* BUILD HARDW *lock* umlegbare Falle *f*

**reversing**: ~**-drum mixer** *n* BUILD MACHIN Freifallmischer *m*

**revet** *vt* INFR & DES, STONE *masonry* verkleiden, *slope* befestigen

**revetment** *n* INFR & DES *masonry* Verkleidung *f*, *slope* Befestigung *f*, STONE *masonry* Verkleidung *f*; ~ **of slopes** *n* INFR & DES Böschungsverkleidung *f*, Uferbefestigung *f*, Uferschutz *m*; ~ **wall** *n* ARCH & BUILD Splitterschutzwand *f*

**revised**: ~ **drawing** *n* ARCH & BUILD, CONST LAW Revisionszeichnung *f*; ~ **tender** *n* CONST LAW Nachtragsangebot *nt*

**revolutions**: ~ **per minute** *n pl* (*rpm*) BUILD MACHIN, HEAT & VENT, WASTE WATER Drehzahl *f*, Umdrehungen *f pl* pro Minute (*UpM*)

**revolving**: ~ **door** *n* ARCH & BUILD Drehtür *f*; ~ **drum** *n* BUILD MACHIN, ENVIRON Drehtrommel *f*

**rework** *vt* CONST LAW nacharbeiten

**Reynolds**: ~ **number** *n* (*Re*) ENVIRON, INFR & DES, MAT PROP Reynoldszahl *f* (*Re*)

**RF** *abbr* (*resorcinol-formaldehyde*) ENVIRON, MAT PROP, TIMBER RF (*Resorcin-Formaldehyd*)

**RFS** *abbr* (*render, float and set*) STONE dreilagiger Putz *m*

**Rhine**: ~ **sand** *n* MAT PROP Rheinsand *m*

**rib** *n* ARCH & BUILD Rippe *f*, HEAT & VENT Lamelle *f*

**ribbed** *adj* ARCH & BUILD, CONCR, HEAT & VENT, MAT PROP, STEEL geriefelt, gerippt, verrippt; ~ **bar** *n* CONCR, STEEL geripptes Bewehrungseisen *nt*; ~ **ceiling** *n* ARCH & BUILD Rippendecke *f*; ~ **concrete floor** *n* ARCH & BUILD Betonrippendecke *f*, Rippenplatte *f*, Stahlbetonrippendecke *f*; ~ **floor** *n* ARCH & BUILD Rippendecke *f*; ~ **glass** *n* MAT PROP Riffelglas *nt*; ~ **radiator** *n* HEAT & VENT Lamellenheizkörper *m*, Konvektor *m*, Rippenheizkörper *m*

***rib***: ~ **factor** *n* CONCR, MAT PROP Ribbungsfaktor *m*; ~ **mesh** *n* STONE Rippenstreckmetall *nt*

**Richter**: ~ **scale** *n* INFR & DES Richterskala *f*

**ride** *vi* INFR & DES ausfahren, entlanggleiten

**ridge** *n* ARCH & BUILD First *m*, Grat *m*; ~ **beam** *n* ARCH & BUILD Firstbalken *m*; ~ **capping** *n* ARCH & BUILD Firstabdeckung *f*; ~ **capping tile** *n* ARCH & BUILD Firstkappe *f*; ~ **covering** *n* ARCH & BUILD Firstabdeckung *f*; ~ **covering tile** *n* ARCH & BUILD Firstkappe *f*

**ridged** *adj* ARCH & BUILD *roof* zweihängig

***ridge***: ~ **height** *n* ARCH & BUILD Firsthöhe *f*; ~ **joint** *n* ARCH & BUILD Firstpunkt *m*; ~ **line** *n* ARCH & BUILD Firstlinie *f*; ~ **piece** *n* ARCH & BUILD Firstbrett *nt*, Firststück *nt*; ~ **plate** *n* ARCH & BUILD Sattelblech *nt*; ~ **rib** *n* ARCH & BUILD Scheitelrippe *f*; ~ **roof** *n* ARCH & BUILD Giebeldach *nt*, Satteldach *nt*; ~ **starting tile** *n* ARCH & BUILD Firstanschlußziegel *m*; ~ **tile** *n* ARCH & BUILD, INFR & DES, MAT PROP, STONE Firstziegel *m*, Firststein *m*; ~ **ventilation cap** *n* HEAT & VENT Entlüftungsfirstkappe *f*

**rig 1.** *n* ARCH & BUILD, INFR & DES Ausrüstung *f*; **2.** *vt* ARCH & BUILD aufstellen, aufbauen, rüsten, BUILD MACHIN, INFR & DES aufstellen

**right 1.** *adj* ARCH & BUILD recht; **2.** *adv* ARCH & BUILD rechts

***right***: ~ **angle** *n* ARCH & BUILD rechter Winkel *m*; ~**-angled valve** *n* HEAT & VENT Eckventil *nt*, Eckabsperrventil *nt*; ~**-hand door closer** *n* BUILD HARDW rechtswirkender Türschließer *m*; ~**-hand lock** *n* BUILD HARDW Schloß *nt* DIN rechts; ~**-of-way** *n* CONST LAW, INFR & DES Wegerecht *nt*

**rigid** *adj* ARCH & BUILD, CONCR, INFR & DES, MAT PROP, SOUND & THERMAL fest, stark, steif; ~ **arch** *n* ARCH & BUILD eingespannter Bogen *m*; ~ **connection** *n* ARCH & BUILD, INFR & DES starre Verbindung *f*; ~ **construction** *n* ARCH & BUILD steife Konstruktion *f*; ~ **fiberglass** *AmE*, ~ **fibreglass** *BrE* *n* ARCH & BUILD, MAT PROP, SOUND & THERMAL steifes Fiberglas *nt*; ~ **foam** *n* MAT PROP Hartschaum *m*; ~ **foundation** *n* ARCH & BUILD, INFR & DES starres Fundament *nt*; ~ **frame** *n* MAT PROP Starrahmen *m*; ~ **frame bridge** *n* ARCH & BUILD, CONCR, INFR & DES Rahmenträgerbrücke *f*

**rigidity** *n* ARCH & BUILD, CONCR, INFR & DES, MAT PROP, SOUND & THERMAL Steifigkeit *f*

**rigid**: ~**-plastic theory** *n* CONCR, MAT PROP steifplastische Theorie *f*, steife Plastizitätstheorie *f*; ~ **structure** *n* ARCH & BUILD, INFR & DES starres Bauwerk *nt*

**rim**: ~ **beam** *n* ARCH & BUILD Randträger *m*; ~ **lock** *n* BUILD HARDW Aufschraubschloß *nt*, Aufsatzschloß *nt*

**ring** *n* ARCH & BUILD, BUILD MACHIN, ELECTR, INFR & DES Ring *m*; ~ **beam** *n* ARCH & BUILD Ringanker *m*, Ringbalken *m*, CONCR Ringbalken *m*; ~ **earth** *n* *BrE* (*cf ring ground AmE*) ELECTR Ringerder *m*; ~ **earth system** *n* *BrE* (*cf ring ground system AmE*) ELECTR Ringerdersystem *nt*; ~ **girder** *n* ARCH & BUILD Ringanker *m*, Ringbalken *m*, Ringträger *m*, CONCR Ringbalken *m*; ~ **ground** *n* *AmE* (*cf ring earth BrE*) ELECTR Ringerder *m*; ~ **ground system** *n* *AmE* (*cf ring earth system BrE*) ELECTR Ringerdersystem *nt*; ~**-shear test** *n* BUILD MACHIN, INFR & DES, MAT PROP Ringschertest *m*; ~**-shear tester** *n* BUILD MACHIN, INFR & DES, MAT PROP Ringschertester *m*, Ringschertestgerät *nt*

**rip** *vt* INFR & DES aufreißen

**ripper** *n* INFR & DES *roadworks* Aufreißer *m*

**ripping** *n* INFR & DES, MAT PROP Spalten *nt*

**rip**: ~**-rap** *n* INFR & DES Schüttsteine *m pl*, Steinpackung *f*, Steinschüttung *f*; ~ **saw** *n* BUILD MACHIN Längsschnittsäge *f*

**rise** *n* ARCH & BUILD *of arch, vault* Stichhöhe *f*, Pfeilhöhe *f*, *of ground* Anstieg *m*, Steigmaß *nt*, ENVIRON Anstieg *m*, INFR & DES *of arch, vault* Stichhöhe *f*

**riser** *n* ARCH & BUILD Setzstufe *f*, Futterbrett *nt*; **~ pipe** *n* WASTE WATER Steigrohr *nt*
***rise***: **~-run ratio** *n* ARCH & BUILD Steigungsverhältnis *nt*, Tritthöhe *f*
**rising**: **~ arch** *n* ARCH & BUILD abfallender Bogen *m*, steigender Bogen *m*; **~ damp** *n* ARCH & BUILD, SOUND & THERMAL, STONE aufsteigende Feuchtigkeit *f*; **~ gradient** *n* ARCH & BUILD Steigung *f*; **~ humidity** *n* ARCH & BUILD, SOUND & THERMAL, STONE aufsteigende Feuchtigkeit *f*; **~ main** *n* WASTE WATER Steigleitung *f*; **~ moisture** *n* ARCH & BUILD, SOUND & THERMAL, STONE aufsteigende Feuchtigkeit *f*; **~ wall** *n* ARCH & BUILD aufgehende Wand *f*
**risk**: **~ of erosion** *n* ENVIRON, INFR & DES Erosionsgefahr *f*; **~ of rot** *n* TIMBER Gefahr *f* von Holzfäule
**risks**: **~ from radon** *n pl* CONST LAW Radonrisiken *nt pl*
**river** *n* ENVIRON, INFR & DES, MAT PROP Fluß *m*; **~ bank** *n* INFR & DES Flußufer *nt*; **~ bed** *n* INFR & DES Flußbett *nt*; **~ gravel** *n* MAT PROP Flußkies *m*; **~ and lake protection** *n* CONST LAW, ENVIRON Gewässerschutz *m*
**rivet 1.** *n* BUILD HARDW, BUILD MACHIN, STEEL Niet *m*, Niete *f*; **2.** *vt* BUILD MACHIN, STEEL nieten; **~ on** BUILD MACHIN, STEEL aufnieten
***rivet***: **~ diameter** *n* STEEL Nietdurchmesser *m*; **~ dolly** *n* BUILD MACHIN Gegenhalter *m*
**riveted** *adj* BUILD MACHIN, STEEL aufgenietet, genietet; **~ joint** *n* BUILD MACHIN, STEEL Nietverbindung *f*; **~ plate** *n* BUILD HARDW, STEEL Nietplatte *f*
**riveter** *n* BUILD MACHIN, STEEL Nietmaschine *f*, Nieter *m*
***rivet***: **~ hammer** *n* BUILD MACHIN Niethammer *m*; **~ head** *n* BUILD HARDW, STEEL Nietkopf *m*
**riveting** *n* BUILD MACHIN, STEEL Nieten *nt*, Nietung *f*; **~ hammer** *n* BUILD MACHIN Niethammer *m*; **~ machine** *n* BUILD MACHIN Nietmaschine *f*; **~ pressure** *n* BUILD HARDW Niet *m*, STEEL Niet *m*, Nietdruck *m*; **~ set** *n* BUILD MACHIN Döpper *m*
***rivet***: **~ joint** *n* BUILD MACHIN, STEEL Nietverbindung *f*; **~ set** *n* BUILD MACHIN Döpper *m*, Nietkopfsetzer *m*; **~ snap** *n* BUILD MACHIN Döpper *m*
**road** *n* INFR & DES Straße *f*; **~ bed** *n* INFR & DES Fahrbahn *f*, Straßenunterbau *m*; **~ bed excavation** *n* INFR & DES Auskofferung *f*; **~ bridge** *n* INFR & DES Straßenbrücke *f*; **~ construction** *n* INFR & DES Straßenbau *m*; **~ construction machinery** *n* BUILD MACHIN Straßenbaumaschine *f*; **~ construction work** *n* INFR & DES Straßenbauarbeiten *f pl*; **~ crossing** *n* INFR & DES Straßenkreuzung *f*, Straßenkreuzungspunkt *m*; **~ drainage** *n* INFR & DES Straßenentwässerung *f*; **~ embankment** *n* INFR & DES Straßendamm *m*; **~ foundation** *n* INFR & DES Straßengründung *f*; **~ grader** *n* BUILD MACHIN Straßenrauhmaschine *f*; **~ groover** *n* BUILD MACHIN Straßenfräsmaschine *f*; **~ inlet** *n* INFR & DES Straßenablauf *m*; **~ marking** *n* INFR & DES Fahrbahnmarkierung *f*, Straßenmarkierung *f*; **~ marking machine** *n* BUILD MACHIN Fahrbahnmarkierungsgerät *nt*, Straßenmarkierungsgerät *nt*; **~ marking paint** *n* INFR & DES Markierungsfarbe *f*; **~ metal** *n* BUILD MACHIN Straßenschotter *m*; **~ painting** *n* BUILD MACHIN, INFR & DES Fahrbahnmarkierung *f*; **~ ripper** *n* BUILD MACHIN Straßenaufreißer *m*; **~ roller** *n* BUILD MACHIN Straßenwalze *f*; **~ setting out** *n* INFR & DES Straßenabsteckung *f*; **~ stone** *n* INFR & DES Pflasterstein *m*; **~ tunnel** *n* INFR & DES Straßentunnel *m*
**roadway** *n* INFR & DES Fahrbahn *f*
**roadworks** *n pl* INFR & DES Straßenbauarbeiten *f pl*
**rock** *n* ARCH & BUILD, INFR & DES, STONE Fels *m*, Gestein *nt*; **~ anchor** *n* INFR & DES Felsanker *m*
**rockbolt** *n* CONCR, INFR & DES, MAT PROP, STEEL Ankerbolzen *m*, Gesteinsanker *m*
***rock***: **~ borer** *n* BUILD MACHIN Gesteinsbohrer *m*; **~ breaker** *n* BUILD MACHIN Steinbrecher *m*; **~ dowel** *n* BUILD MACHIN Steindübel *m*; **~ drill** *n* BUILD MACHIN Bohrhammer *m*, Gesteinsbohrer *m*; **~ fill** *n* INFR & DES Steinschüttung *f*; **~ fill dam** *n* INFR & DES Steinschüttdamm *m*; **~ formation** *n* INFR & DES Felsformation *f*
**rocking**: **~ pier** *n* ARCH & BUILD Pendelstütze *f*
***rock***: **~ layer** *n* INFR & DES Felslage *f*; **~ material** *n* STONE Felsgestein *nt*; **~ mechanics** *n* INFR & DES Felsmechanik *f*; **~ quarry** *n* INFR & DES, STONE Steinbruch *m*; **~ salt** *n* INFR & DES Steinsalz *nt*; **~ surface** *n* INFR & DES Felsoberfläche *f*; **~ wool** *n* SOUND & THERMAL Stabeisen *nt*, Steinwolle *f*
**rocky**: **~ soil** *n* INFR & DES, MAT PROP mit Fels durchsetzter Boden *m*
**rod** *n* MAT PROP, STEEL Stab *m*, Stange *f*
**rolled**: **~ glass** *n* MAT PROP Flachglas *nt*; **~ steel** *n* MAT PROP Bandstahl *m*, Flacheisen *nt*, Flachstahl *m*, Walzstahl *m*; **~ steel joist** *n* ARCH & BUILD Walzstahlträger *m*, Walzträger *m*; **~ structural steel** *n* ARCH & BUILD, MAT PROP, STEEL Walzstahlprofil *nt*
**roller** *n* ARCH & BUILD, BUILD HARDW Welle *f*, BUILD MACHIN Walze *f*, INFR & DES, SURFACE Welle *f*; **~ bearing** *n* ARCH & BUILD Rollenlager *nt*; **~ bolt** *n* BUILD HARDW *lock* Rollfalle *f*; **~ bridge** *n* INFR & DES Rollbrücke *f*; **~ painting** *n* SURFACE Farbaufrollen *nt*; **~ shutter** *n* ARCH & BUILD, BUILD HARDW *for window, door* Rolladen *m*; **~ shutter housing** *n* ARCH & BUILD Rolladenkasten *m*; **~ weir** *n* INFR & DES Walzenwehr *nt*
**rolling** *n* INFR & DES Einwalzen *nt*, Walzen *nt*; **~ bearing** *n* BUILD MACHIN Rollenlager *nt*; **~ bridge** *n* INFR & DES Rollbrücke *f*; **~ load** *n* ARCH & BUILD, INFR & DES Betriebslast *f*, Verkehrslast *f*; **~ mill** *n* BUILD MACHIN Walzwerk *nt*; **~ shutter** *n* ARCH & BUILD Rolladentor *nt*; **~ shutter door** *n* ARCH & BUILD Rolltor *nt*; **~ stock** *n* BUILD MACHIN, INFR & DES Fuhrpark *m*
**roll**: **~-out container** *n* ENVIRON Müllcontainer *m*
**rolock** *n* STONE Rollschicht *f*
**Roman** *adj* ARCH & BUILD römisch
**Roman**: **~ cement** *n* MAT PROP Romanzement *m*
**roof** *n* ARCH & TRAGW, BUILD & HARDW, ELECTR, INFR & DES, HEAT & VENT, MAT PROP, SOUND & THERMAL, STEEL, TIMBER Dach *nt*
**roofage** *n* ARCH & BUILD *measurement* Dachfläche *f*
***roof***: **~ with air circulation** *n* ARCH & BUILD Kaltdach *nt*; **~ antenna** *n* ELECTR Dachantenne *f*; **~ area** *n* ARCH & BUILD Dachfläche *f*; **~ batten** *n* MAT PROP, TIMBER Dachlatte *f*; **~ battening** *n* ARCH & BUILD Lattung *f*; **~ boarding** *n* CONCR, MAT PROP, TIMBER Dachschalung *f*, Holzschalung *f*; **~ cladding** *n* ARCH & BUILD Dachdeckung *f*, Dacheindeckung *f*, Bedachung *f*, MAT PROP Dachdeckung *f*; **~ collector** *n* ELECTR, HEAT & VENT *solar heat* Dachkollektor *m*; **~ conductor holder** *n* ELECTR *lightning protection* Dachleitungshalter *m*; **~ covering** *n* ARCH & BUILD

*vapour barrier* Dachdeckung *f*, Dachhaut *f*, *with tiles* Dacheindeckung *f*, Bedachung *f*, INFR & DES *vapour barrier* Dachhaut *f*, MAT PROP Dachdeckung *f*, SOUND & THERMAL Dachhaut *f*; **~ edge** *n* ARCH & BUILD Dachkante *f*

**roofer** *n* ARCH & BUILD Dachdecker *m*

***roof***: **~ fan** *n* HEAT & VENT Dachlüfter *m*, Dachventilator *m*; **~ flashing** *n* STEEL Dachanschluß *m*, Dachanschlußstreifen *m*, Dachverwahrung *f*; **~ frame** *n* ARCH & BUILD, INFR & DES, MAT PROP, TIMBER Binder *m*, Dachbinder *m*, Fachwerkbinder *m*; **~ framework** *n* ARCH & BUILD, INFR & DES, MAT PROP, TIMBER Dachstuhl *m*; **~ framing** *n* ARCH & BUILD, INFR & DES, MAT PROP, TIMBER Binder *m*, Dachbinder *m*; **~ garden** *n* ARCH & BUILD Dachgarten *m*; **~ girder** *n* ARCH & BUILD, STEEL, TIMBER Dachträger *m*; **~ glazing** *n* ARCH & BUILD Dachverglasung *f*; **~ gully** *n* WASTE WATER Dachgully *m*; **~ heliport** *n* ARCH & BUILD, INFR & DES Dachhubschrauberlandeplatz *m*; **~ hook** *n* ARCH & BUILD Reparaturhaken *m*, BUILD HARDW Dachhaken *m*

**roofing** *n* ARCH & BUILD Dachdeckung *f*, Überdachung *f*, Bedachung *f*, Dacheindeckung *f*, MAT PROP Dachschalung *f*, Dachdeckung *f*, TIMBER Dachschalung *f*; **~ batten** *n* MAT PROP, TIMBER Dachlatte *f*; **~ felt** *n* MAT PROP Dachpappe *f*; **~ felt with granulated slate surface** *n* MAT PROP beschieferte Dachpappe *f*; **~ felt nail** *n* MAT PROP Dachpappennagel *m*; **~ gravel** *n* MAT PROP Dachkies *m*; **~ plank** *n* MAT PROP Dachdekkungsdiele *f*; **~ skin** *n* ARCH & BUILD, INFR & DES, SOUND & THERMAL Dachhaut *f*; **~ slate** *n* STONE Dachschiefer *m*; **~ tile** *n* INFR & DES, MAT PROP, STONE, TIMBER Dachziegel *m*, Dachpfanne *f*, Dachstein *m*; **~ work** *n* ARCH & BUILD Dacharbeiten *f pl*

***roof***: **~ inlet** *n* WASTE WATER Dachgully *m*; **~ insulation** *n* SOUND & THERMAL Dachdämmung *f*; **~ lath** *n* MAT PROP, TIMBER Dachlatte *f*

**rooflight** *n* ARCH & BUILD, BUILD HARDW Dachfenster *nt*, Dachlaterne *f*

**roofline** *n* ARCH & BUILD Dachsilhouette *f*

***roof***: **~ load** *n* ARCH & BUILD, INFR & DES Dachlast *f*; **~ loading** *n* INFR & DES Dachbelastung *f*; **~ membrane** *n* ARCH & BUILD, INFR & DES, SOUND & THERMAL Dachhaut *f*; **~-mounted bracket** *n* ARCH & BUILD Dachkonsole *f*; **~ opening** *n* ARCH & BUILD Dachdurchbruch *m*; **~ panel** *n* ARCH & BUILD Dachplatte *f*; **~ parapet** *n* ARCH & BUILD Attika *f*, Dachbrüstung *f*; **~ penetration** *n* ARCH & BUILD Dachdurchbruch *m*; **~ pitch** *n* ARCH & BUILD, MAT PROP Dachgefälle *nt*, Dachneigung *f*, Dachschräge *f*; **~ plate** *n* MAT PROP Dachstuhl-Auflageplatte *f*; **~ profile** *n* ARCH & BUILD Dachform *f*; **~ ridge** *n* ARCH & BUILD Dachfirst *m*; **~ shape** *n* ARCH & BUILD Dachform *f*; **~ sheathing** *n* ARCH & BUILD, MAT PROP Dachschale *f*, Dachdeckung *f*; **~ slope** *n* ARCH & BUILD, MAT PROP Dachschräge *f*; **~ structure** *n* ARCH & BUILD, INFR & DES, MAT PROP, TIMBER Dachkonstruktion *f*, Dachstuhl *m*; **~ structures** *n pl* ARCH & BUILD Dachaufbauten *m pl*; **~ surround** *n* ARCH & BUILD Dacheinfassung *f*; **~ terrace** *n* ARCH & BUILD Dachterrasse *f*; **~ tile** *n* TIMBER Dachziegel *m*; **~-top heliport** *n* ARCH & BUILD, INFR & DES Dachhubschrauberlandeplatz *m*

**rooftop**: **~ terrace garden** *n* ARCH & BUILD Dachgarten *m*

***roof***: **~ truss** *n* ARCH & BUILD, INFR & DES, MAT PROP, TIMBER *principal support* Dachbinder *m*, Dachstuhl *m*, Fachwerk *nt*; **~ type** *n* ARCH & BUILD Dachform *f*; **~ ventilator** *n* HEAT & VENT Dachventilator *m*, Dachlüfter *m*; **~ water** *n* WASTE WATER Dachwasser *nt*

**room** *n* ARCH & BUILD Raum *m*, Zimmer *nt*, BUILD MACHIN, HEAT & VENT, MAT PROP Raum *m*; **~ divider** *n* ARCH & BUILD Raumteiler *m*; **~-high** *adj* ARCH & BUILD *tiles* raumhoch; **~ temperature** *n* HEAT & VENT, MAT PROP Raumtemperatur *f*; **~ thermostat** *n* BUILD HARDW, HEAT & VENT Raumthermostat *m*

**root** *n* INFR & DES Wurzel *f*; **~ line mean square water level** *n* INFR & DES quadratischer Wasserstandmittelwert *m*

**rope** *n* MAT PROP Seil *nt*, Tau *nt*

**ropeway** *n* INFR & DES, STEEL Drahtseilbahn *f*

**rose** *n* BUILD HARDW *door* Rosette *f*

**rosette** *n* ARCH & BUILD Rosette *f*

**rot 1.** *n* MAT PROP, TIMBER Fäulnis *f*, Holzfäule *f*; **2.** *vti* CONCR, ENVIRON, MAT PROP, SURFACE, TIMBER verfaulen, verrotten, vermodern

**rotary 1.** *adj* BUILD MACHIN, ENVIRON, MAT PROP rotierend; **2.** *n AmE* (*cf roundabout BrE*) INFR & DES Kreisverkehr *m*

***rotary***: **~ crane** *n* BUILD MACHIN Drehkran *m*; **~ drill** *n* BUILD MACHIN, TIMBER Drehbohrer *m*; **~ drilling** *n* BUILD MACHIN Drehbohren *nt*, Drehspülbohren *nt*, INFR & DES Rotationsbohren *nt*; **~-drum mixer** *n* BUILD MACHIN Freifallmischer *m*; **~ furnace** *n* BUILD MACHIN, ENVIRON Drehrohrofen *m*, Rotationsofen *m*; **~ lever** *n* BUILD HARDW *lock* Drehriegel *m*; **~ lever lock** *n* BUILD HARDW Drehriegelverschluß *m*; **~ screen** *n* BUILD MACHIN Siebtrommel *f*; **~ switch** *n* ELECTR Drehschalter *m*; **~ traffic** *n* INFR & DES Kreisverkehr *m*; **~ valve** *n* ENVIRON Drehschieber *m*

**rotation** *n* ARCH & BUILD Rotation *f*, Drehung *f*, BUILD MACHIN *pump* Umdrehung *f*; **~ capacity** *n* ARCH & BUILD, CONCR, MAT PROP Rotationskapazität *f*

**rotor** *n* BUILD MACHIN *pump* Laufrad *nt*

**rotproof** *adj* MAT PROP, SURFACE, TIMBER fäulnisbeständig, fäulnisfest, nicht verrottbar, unverrottbar

***rot***: **~ protection** *n* MAT PROP Fäulnisschutz *m*

**rotting** *n* TIMBER Vermodern *nt*

**rough 1.** *adj* ARCH & BUILD *approximate* grob, ENVIRON roh, MAT PROP rauh, roh, TIMBER ungehobelt, grob; **2. ~-hew** *vt* STONE bossieren; **~ down** ARCH & BUILD rauhschleifen, vorwalzen

**rough**: **~-axed** *adj* STONE *natural stone* grob zugehauen; **~ calculation** *n* ARCH & BUILD Überschlagsrechnung *f*

**roughcast** *n* STONE *pebble dash* Steinputz *m*, *plaster* Kratzputz *m*, Rauhputz *m*, *rendering* rauher Putz *m*, Anwurf *m*; **~ glass** *n* MAT PROP Gußglas *nt*; **~ wired glass** *n* MAT PROP Gußdrahtglas *nt*

**rough**: **~ dressing** *n* ARCH & BUILD *of stone*, STONE Rohbehauen *nt*

**roughen** *vt* BUILD HARDW, INFR & DES, STONE, SURFACE aufrauhen

**roughening** *n* BUILD HARDW, INFR & DES, STONE, SURFACE Aufrauhen *nt*

**rough**: **~-grained** *adj* TIMBER grobkörnig; **~-hewn** *adj* TIMBER *wood* baumkantig

**roughness** *n* MAT PROP Rauheit *f*, Rauhigkeit *f*

**rough**: **~ opening dimensions** *n* ARCH & BUILD Rohbaulichtmaß *nt*; **~-shuttered** *adj* ARCH & BUILD,

CONCR schalungsrauh; ~ **string** *n* ARCH & BUILD untere Treppenwange *f*; ~ **walling** *n* STONE Rohmauerung *f*; ~ **wood** *n* TIMBER Grobholz *nt*; ~ **work** *n* ARCH & BUILD, TIMBER Rohbauarbeiten *f pl*

**round 1.** *adj* ARCH & BUILD Rund-, rund; **2.** ~ **off** *vt* ARCH & BUILD, INFR & DES abrunden; ~ **up** ARCH & BUILD, INFR & DES aufrunden

**roundabout** *n BrE* (*cf rotary AmE*) INFR & DES Kreisverkehr *m*

**round**: ~ **arch** *n* ARCH & BUILD Rundbogen *m*; ~**-arched** *adj* ARCH & BUILD rundbogig; ~ **bar** *n* MAT PROP Rundstab *m*; ~ **bolt** *n* BUILD HARDW *lock* Rundbolzenriegel *m*

**rounded** *adj* ARCH & BUILD abgerundet; ~**-off striking plate** *n* BUILD HARDW abgerundetes Schließblech *nt*

**roundel** *n* ARCH & BUILD rundes Fenster *nt*, Rundfenster *nt*, runde Nische *f*

**round**: ~ **hollow steel column** *n* ARCH & BUILD, CONCR, INFR & DES, MAT PROP, STEEL hohle Rundstahlstütze *f*; ~ **pipe** *n* MAT PROP Rundrohr *nt*; ~ **stair well** *n* ARCH & BUILD Treppenauge *nt*; ~ **steel** *n* MAT PROP Rundstahl *m*

**route** *n* ARCH & BUILD *cable, pipe* Trasse *f*, BUILD MACHIN *cable, pipe* Strecke *f*; ~ **mapping** *n* INFR & DES Linienführung *f*, Streckenführung *f*

**row** *n* ARCH & BUILD Reihe *f*

**rowlock** *n* STONE Rollschicht *f*

***row***: ~ **of windows** *n* ARCH & BUILD Lichtband *nt*

**rpm** *abbr* (*revolutions per minute*) BUILD MACHIN, HEAT & VENT, WASTE WATER UpM (*Umdrehungen pro Minute*)

**rub** *vt* ARCH & BUILD abreiben, abziehen, polieren, CONCR, MAT PROP, STONE, TIMBER abreiben; ~ **down** *vt* ARCH & BUILD, CONCR, MAT PROP, STONE, TIMBER abreiben

**rubber** *n* BUILD HARDW, BUILD MACHIN, MAT PROP, SURFACE Gummi *nt*, Kautschuk *m*; ~ **coating** *n* SURFACE Gummibelag *m*; ~ **collar** *n* BUILD HARDW Gummimanschette *f*; ~ **core** *n* MAT PROP Gummieinlage *f*; ~ **flooring** *n* MAT PROP Gummifußbodenbelag *m*; ~ **gasket** *n* BUILD HARDW Gummilippendichtung *f*; ~ **lip sealing** *n* BUILD HARDW Gummilippendichtung *f*; ~ **ply** *n* MAT PROP Gummieinlage *f*; ~ **profile** *n* MAT PROP Gummiprofil *nt*; ~**-tired roller** *AmE*, ~**-tyred roller** *BrE n* BUILD MACHIN Gummiradwalze *f*; ~**-wheel roller** *n* BUILD MACHIN Gummiradwalze *f*

**rubbish** *n BrE* (*cf garbage AmE*) ARCH & BUILD, CONST LAW, ENVIRON, INFR & DES Abfall *m*, Bauschutt *m*, Müll *m*; ~ **bin** *n BrE* (*cf garbage can AmE*) ENVIRON *in public place or inside house* Abfallbehälter *m*, Abfalleimer *m*, Mülleimer *m*; ~ **chute** *n BrE* ARCH & BUILD, ENVIRON Müllabwurfschacht *m*, Müllschlukker *m*; ~ **collection** *n BrE* CONST LAW Müllabfuhr *f*, ENVIRON Müllabfuhr *f*, Müllsammlung *f*; ~ **container** *n BrE* ENVIRON Müllbehälter *m*, Mülltonne *f*; ~ **disposal** *n BrE* CONST LAW Müllabfuhr *f*, ENVIRON Müllabfuhr *f*, Müllsammlung *f*, Sammlung *f* von Hausmüll; ~ **press** *n BrE* BUILD MACHIN, ENVIRON Müllpresse *f*

**rubble** *n* ARCH & BUILD *debris* Schutt *m*, Bruchstein *m*, Trümmer *pl*, Abbruchabfall *m*, ENVIRON *demolition detritus* Abbruchmaterial *nt*, Abbruchabfall *m*, MAT PROP *coarse gravel* Grobkies *m*, Schutt *m*, Steinschüttung *f*, STONE Bruchstein *m*; ~ **bedding** *n* INFR & DES Steinpackung *f*; ~ **drain** *n* INFR & DES, WASTE WATER Sickergraben *m*; ~ **layer** *n* INFR & DES Steinwurf *m*; ~ **masonry** *n* MAT PROP, STONE Bruchsteinmauerwerk *nt*

**rubblestone** *n* MAT PROP Bruchstein *m*

**ruin** *n* ARCH & BUILD Ruine *f*

**ruins** *n pl* ARCH & BUILD Ruinen *f pl*, Trümmer *pl*

**rule** *n* ARCH & BUILD *straight edge* Lineal *nt*, CONST LAW *regulation* Regel *f*

**ruler** *n* ARCH & BUILD Richtlatte *f*, Lineal *nt*

**run 1.** *n* ARCH & BUILD *foothold* Auftrittbreite *f*, *walking line* Lauflinie *f*; **2.** ~ **through** *vi* ARCH & BUILD, CONCR, HEAT & VENT, INFR & DES durchlaufen

**rung** *n* ARCH & BUILD, BUILD HARDW *of ladder* Sprosse *f*, Leitersprosse *f*

**runner** *n* ARCH & BUILD *beam* Gurtholz *nt*, *cross beam* Holm *m*, *header joist* Oberschwelle *f*, BUILD MACHIN *crane* Kranbahnträger *m*

**running** *n* ARCH & BUILD Einfluchten *nt*; ~ **bond** *n* ARCH & BUILD, STONE *masonry* Läuferverband *m*; ~ **joint** *n* ARCH & BUILD, CONCR, INFR & DES Dehnungsfuge *f*, Raumfuge *f*; ~ **trap** *n* ARCH & BUILD U-Verschluß *m*

**runoff** *n* ARCH & BUILD, ENVIRON, HEAT & VENT, WASTE WATER Oberflächenabfluß *m*, Ablauf *m*, *rainwater* Abfluß *m*; ~ **coefficient** *n* WASTE WATER Abflußbeiwert *m*

***run***: ~**-of-river scheme** *n* ENVIRON Flußprojektierung *f*; ~**-of-river station** *n* ENVIRON Laufkraftwerk *nt*

**runway** *n* ARCH & BUILD, INFR & DES *maintenance calls* Wartungsweg *m*; ~ **plank** *n* ARCH & BUILD Karrbohle *f*, MAT PROP Laufbohle *f*, TIMBER Karrbohle *f*

**rupture** *n* MAT PROP Bruch *m*; ~ **line** *n* MAT PROP Bruchlinie *f*

**rural** *adj* CONST LAW, INFR & DES ländlich; ~ **planning** *n* CONST LAW, INFR & DES Raumordnungsplanung *f*

**rust 1.** *n* STEEL, SURFACE Rost *m*; **2.** *vi* STEEL, SURFACE rosten; ~ **through** STEEL durchrosten

**rusted**: ~ **through** *adj* STEEL durchgerostet

***rust***: ~ **film** *n* STEEL Rostfilm *m*

**rustic** *adj* ARCH & BUILD rustikal

**rustication** *n* STONE Bossenwerk *nt*, roh behauenes Quaderwerk *nt*

**rustproofed** *adj* MAT PROP rostfest, STEEL, SURFACE rostgeschützt, rostfest, *stainless* nichtrostend, rostfrei

**rustproofing** *n* MAT PROP, STEEL, SURFACE Rostschutz *m*; ~ **agent** *n* MAT PROP, STEEL, SURFACE Rostschutzmittel *nt*; ~ **paint** *n* MAT PROP, STEEL, SURFACE Rostschutzfarbe *f*; ~ **primer** *n* MAT PROP, STEEL, SURFACE Rostschutzgrundierung *f*

# S

**saddle** *n* ARCH & BUILD Sattel *m*, TIMBER Sattelholz *nt*; **~ stone** *n* STONE Dachstein *m*
**safe** *adj* CONST LAW sicher, zuverlässig; **~ breakage** *n* MAT PROP sicheres Zerbrechen *nt*; **~ disposal** *n* ENVIRON geordnete Beseitigung *f*
**safety** *n* CONST LAW Sicherheit *f*; **~ adviser** *n* CONST LAW Sicherheitsberater *m*; **~ arch** *n* CONST LAW Entlastungsbogen *m*, Verstärkungsbogen *m*; **~ belt** *n* INFR & DES Schutzgurt *m*; **~ cage** *n* ARCH & BUILD *chimney ladder* Rückenschutz *m*; **~ catch** *n* BUILD HARDW *lift, elevator* Fangvorrichtung *f*; **~ device** *n* ARCH & BUILD, BUILD MACHIN, CONST LAW, INFR & DES Schutzvorrichtung *f*; **~ door** *n* CONST LAW Sicherheitstor *nt*, Sicherheitstür *f*; **~ factor** *n* ARCH & BUILD, CONST LAW, INFR & DES Sicherheitsbeiwert *m*, Sicherheitsfaktor *m*; **~ glass** *n* CONST LAW, MAT PROP Sicherheitsglas *nt*; **~ lighting** *n* CONST LAW, ELECTR Notbeleuchtung *f*, Panikbeleuchtung *f*, Sicherheitsbeleuchtung *f*; **~ limit** *n* CONST LAW Sicherheitsgrenze *f*; **~ lock** *n* BUILD HARDW, CONST LAW Sicherheitsschloß *nt*; **~ margin** *n* CONST LAW Sicherheitsspielraum *m*; **~ record** *n* CONST LAW Sicherheitsprotokoll *nt*; **~ solvent** *n* CONST LAW, SURFACE unbedenkliches Lösungsmittel *nt*; **~ strip** *n* INFR & DES Sicherheitsstreifen *m*; **~ valve** *n* ENVIRON, HEAT & VENT Sicherheitsventil *nt*
**safe**: **~ working arrangements** *n pl* CONST LAW sichere Arbeitsübereinkünfte *f pl*; **~ working load** *n* (*SWL*) ARCH & BUILD, CONST LAW, INFR & DES sichere Belastung *f*
**sag** *vi* ARCH & BUILD durchhängen
**sagged** *adj* ARCH & BUILD durchgebogen
**sagging 1.** *adj* ARCH & BUILD nach unten biegend; **2.** *n* ARCH & BUILD, INFR & DES Senkung *f*, Setzung *f*
***sagging***: **~ bend** *n* ARCH & BUILD, CONCR, MAT PROP Durchbiegen *nt*; **~ bending** *adj* CONCR, MAT PROP nach unten biegend
**saliferous** *adj* INFR & DES salzhaltig; **~ clay** *n* INFR & DES, MAT PROP Salzton *m*; **~ marl** *n* INFR & DES, MAT PROP Keupermergel *m*
**saline** *n* INFR & DES Saline *f*; **~ plant** *n* INFR & DES Salzwerk *nt*
**salinity** *n* ENVIRON Salzhaltigkeit *f*
**salt** *n* MAT PROP Salz *nt*; **~ bath brazing** *n* STEEL Salzbadlöten *nt*
**saltpeter** *AmE see saltpetre BrE*
**saltpetre** *n BrE* ENVIRON Salpeter *m*
**saltwater**: **~-proof** *adj* MAT PROP salzwasserbeständig
**sample 1.** *n* MAT PROP Muster *nt*; **2.** *vt* INFR & DES eine Probe entnehmen
***sample***: **~ bottle** *n* BUILD MACHIN Probeflasche *f*; **~ key** *n* BUILD HARDW Musterschlüssel *m*
**samples** *n pl* ARCH & BUILD Bemusterungsunterlagen *f pl*
**sampling** *n* ARCH & BUILD Bemusterung *f*, ENVIRON Probenahme *f*, INFR & DES Probeentnahme *f*; **~ point** *n* INFR & DES Probenentnahmestelle *f*
**sanatorium** *n BrE* INFR & DES Sanatorium *nt*

**sand 1.** *n* STONE Sand *m*; **2.** *vt* MAT PROP mit Sand abdecken
***sand***: **~ asphalt** *n* INFR & DES, STONE Sandasphalt *m*; **~ bank** *n* INFR & DES Sandbank *f*; **~ bedding** *n* INFR & DES Sandbett *nt*
**sandblast** *vt* SURFACE absanden
**sandblasted** *adj* SURFACE sandgestrahlt
**sandblasting** *n* STONE, SURFACE Sandstrahlen *nt*
***sand***: **~ blasting** *n* STONE, SURFACE Sandstrahlen *nt*; **~ content** *n* STONE Sandanteil *m*; **~ cushion foundation** *n* CONCR Sandpolstergründung *f*; **~ dam** *n* INFR & DES Sanddamm *m*; **~ deposit** *n* INFR & DES Sandablagerung *f*; **~ dune** *n* INFR & DES Sanddüne *f*; **~ equivalent** *n* MAT PROP Sandäquivalent *nt*; **~ filling** *n* ARCH & BUILD Sandschüttung *f*; **~ filter** *n* ENVIRON Sandfang *m*, Sandfilter *nt*; **~ formation** *n* INFR & DES Sandformation *f*; **~ traction** *n* STONE Sandfraktion *f*; **~ and gravel processing equipment** *n* BUILD MACHIN Sand- und Kiesaufbereitungsmaschine *f*; **~ inclusion** *n* INFR & DES, MAT PROP Sandeinschluß *m*; **~ layer** *n* INFR & DES Sandschicht *f*
**sandlime**: **~ brick** *n* STONE Kalksandstein *m*
***sand***: **~ sample** *n* STONE Sandprobe *f*; **~ slope** *n* INFR & DES Sandböschung *f*
**sandstone** *n* INFR & DES, STONE Sandstein *m*
**sand**: **~-surfaced** *adj* STONE besandet; **~ trap** *n* ENVIRON Sandfanganlage *f*, Sandfänger *m*, INFR & DES, WASTE WATER Sandabscheider *m*, Sandfang *m*
**sandwich**: **~ board** *n* ARCH & BUILD, MAT PROP Sandwichplatte *f*, Verbundplatte *f*
**sandwiched**: **~ truss** *n* ARCH & BUILD, TIMBER Brettbinder *m*
**sandwich**: **~ panel** *n* ARCH & BUILD, MAT PROP Sandwichplatte *f*, Verbundplatte *f*
**sandy**: **~ clay** *n* INFR & DES, MAT PROP Sandton *m*, magerer Ton *m*; **~ ground** *n* INFR & DES Sandboden *m*; **~ loam** *n* INFR & DES, MAT PROP sandiger Lehm *m*
**sanitarium** *AmE see sanatorium BrE*
**sanitary** *adj* CONST LAW hygienisch, sanitär, INFR & DES, WASTE WATER Sanitär-, sanitär; **~ china** *n* MAT PROP Sanitärkeramik *f*, Sanitärporzellan *nt*; **~ engineering** *n* CONST LAW, INFR & DES, WASTE WATER Sanitärtechnik *f*; **~ equipment** *n* ARCH & BUILD, INFR & DES, WASTE WATER sanitäre Einrichtungen *f pl*; **~ facilities** *n pl* ARCH & BUILD, INFR & DES Sanitäranlagen *f pl*; **~ installations** *n pl* WASTE WATER Sanitärinstallation *f*; **~ landfill** *n* ENVIRON Mülldeponie *f*, geordnete Deponie *f*, kontrollierte Müllablagerung *f*; **~ landfilling** *n* ENVIRON geordnete Ablagerung *f*; **~ module** *n* WASTE WATER Sanitärzelle *f*; **~ sewer** *n* WASTE WATER Schmutzwasserleitung *f*
**sanitation** *n* ARCH & BUILD, WASTE WATER sanitäre Einrichtungen *f pl*
**sapwood** *n* TIMBER Splintholz *nt*
**sarking** *n* MAT PROP Schieferunterlegschicht *f*; **~ felt** *n* MAT PROP Dachpappenunterlage *f*
**sash** *n* ARCH & BUILD *of window* Fensterflügelrahmen *m*, Flügelrahmen *m*, BUILD HARDW Fenster-

flügelrahmen *m*, Schiebefensterrahmen *m*; ~ **bar** *n* BUILD MACHIN Fenstersprosse *f*, Fensterstab *m*, Sprosseneisen *nt*; ~ **door** *n* BUILD MACHIN Glasfüllungstür *f*; ~ **fastener** *n* BUILD HARDW Schiebefensterfeststeller *m*; ~ **frame** *n* BUILD MACHIN Schieberahmen *m*; ~ **hardware** *n* BUILD HARDW Schiebefensterbeschläge *m pl*; ~ **lock** *n* BUILD HARDW Vorreiber *m*; ~ **putty** *n* ARCH & BUILD Glaserkitt *m*; ~ **rail** *n* ARCH & BUILD Fensterriegel *m*; ~ **window** *n* ARCH & BUILD Hubfenster *nt*, Schiebefenster *nt*, BUILD HARDW Verbundfenster *nt*

**satin**: ~-**finish glass** *n* ARCH & BUILD satiniertes Glas *nt*, MAT PROP satiniertes Glas *nt*, seidenmattes Glas *nt*

**satisfactory** *adj* ARCH & BUILD *quality* zufriedenstellend, ausreichend

**saturate** *vt* SURFACE *with water* tränken, TIMBER *wood preservation* vollimprägnieren

**saturated** *adj* MAT PROP getränkt; ~ **colors** *AmE*, ~ **colours** *BrE n pl* SURFACE satte Farben *f pl*; ~ **vapor pressure** *AmE*, ~ **vapour pressure** *BrE n* INFR & DES, MAT PROP, SOUND & THERMAL Sättigungsdruck *m*

**saturation** *n* MAT PROP, SOUND & THERMAL Sättigung *f*; ~ **point** *n* MAT PROP Sättigungspunkt *m*; ~ **ratio** *n* MAT PROP Sättigungsgrad *m*

**saucer**: ~ **dome** *n* ARCH & BUILD Lichtkuppel *f*

**saving** *n* ARCH & BUILD Einsparung *f*

**saw 1.** *n* BUILD MACHIN Säge *f*; **2.** ~ **out** *vt* TIMBER heraussägen

***saw***: ~ **bench** *n* BUILD MACHIN Sägebank *f*

**sawbuck** *n AmE* (*cf sawhorse BrE*) BUILD MACHIN, TIMBER Sägebock *m*

***saw***: ~ **cut** *n* BUILD MACHIN, TIMBER Sägeschnitt *m*

**sawdust** *n* BUILD MACHIN, TIMBER Sägemehl *nt*

**sawhorse** *n BrE* (*cf sawbuck AmE*) BUILD MACHIN, TIMBER Sägebock *m*

**sawing** *n* BUILD MACHIN Sägen *nt*; ~ **machine** *n* BUILD MACHIN Sägemaschine *f*

***saw***: ~ **log** *n* BUILD MACHIN Sägeblock *m*

**sawmill** *n* BUILD MACHIN, INFR & DES Sägewerk *nt*

***saw***: ~ **timber** *n* BUILD MACHIN Schneideholz *nt*

**sawtooth**: ~ **roof** *n* ARCH & BUILD Sheddach *nt*

**scabble** *vt* BUILD MACHIN, CONCR, STONE aufspitzen, aufstocken

**scabbled**: ~ **area** *n* CONCR, STONE gespitzte Fläche *f*

**scaffold 1.** *n* ARCH & BUILD, BUILD HARDW *building* Gerüst *nt*; **2.** *vt* ARCH & BUILD einrüsten; **3.** *vi* ARCH & BUILD ein Gerüst aufstellen, ein Gerüst bauen

***scaffold***: ~ **board** *n* ARCH & BUILD Gerüstbohle *f*

**scaffolding** *n* ARCH & BUILD Einrüsten *nt*, Gerüst *nt*

***scaffold***: ~ **pole** *n* ARCH & BUILD Gerüststange *f*

**scale** *n* ARCH & BUILD, HEAT & VENT, INFR & DES, MAT PROP, WASTE WATER Maßstab *m*, Kesselstein *m*; **to** ~ *adj* ARCH & BUILD, INFR & DES maßstabsgerecht; ~ **board** *n* TIMBER Furnierplatte *f*

**scaled** *adj* HEAT & VENT, WASTE WATER verkalkt

***scale***: ~ **factor** *n* ARCH & BUILD, INFR & DES Kräftemaßstab *m*, Maßstab *m*; ~ **model** *n* ARCH & BUILD, INFR & DES Maßstabsmodell *nt*

**scaling**: ~ **hammer** *n* BUILD MACHIN Entschuppungshammer *m*; ~ **off** *n* SURFACE Entzunderung *f*

**scalpings** *n pl* CONCR Mineralbeton *m*

**scantling** *n* CONCR, INFR & DES, MAT PROP Brustholz *nt*, TIMBER Kantholz *nt*, Brustholz *nt*, Kreuzholz *nt*

**scarf** *n* TIMBER Laschenverbindung *f*, Laschung *f*; ~ **joint** *n* TIMBER Verblattung *f*, Schäftverbindung *f*, Stumpfstoß *m*; ~ **jointing** *n* TIMBER Stumpfstoßen *nt*; ~ **tenon** *n* TIMBER Blattzapfen *m*

**scarification** *n* BUILD HARDW, INFR & DES, STONE, SURFACE *road* Aufrauhen *nt*, Aufreißen *nt*

**scarifier** *n* INFR & DES Abschäler *m*, Aufreißer *m*

**scarify** *vt* BUILD HARDW, INFR & DES, STONE, SURFACE aufrauhen, aufreißen

**scarp 1.** *n* INFR & DES Steilhang *m*; **2.** *vt* INFR & DES abböschen

**scatter** *vt* MAT PROP streuen

**SCC** *abbr* (*secondary combustion chamber*) ENVIRON, HEAT & VENT SCC (*Nachbrennkammer, zweiter Brennraum*)

**schedule** *n* ARCH & BUILD Liste *f*, Plan *m*, Programm *nt*, CONST LAW Liste *f*; **on** ~ *adj* ARCH & BUILD, CONST LAW termingemäß; ~ **of building occupancy** *n* INFR & DES *analysis of energy requirements* Belegungsplan *m*

**scheduled**: ~ **service** *n* ARCH & BUILD, BUILD MACHIN, CONST LAW regelmäßige Wartung *f*

**schematic**: ~ **drawing** *n* BUILD MACHIN, INFR & DES Schemabild *nt*

**schematization** *n* ARCH & BUILD, INFR & DES Schematisierung *f*

**scheme** *n* ARCH & BUILD, INFR & DES Ausmaß *nt*, Schema *nt*; ~ **arch** *n* ARCH & BUILD, MAT PROP Flachbogen *m*

**schist** *n* BUILD MACHIN, INFR & DES, MAT PROP Schiefer *m*

**schistose**: ~ **rock** *n* INFR & DES Schiefergestein *nt*

**scope** *n* ARCH & BUILD, CONST LAW, INFR & DES Ausmaß *nt*, Bereich *m*, Umfang *m*; ~ **of work** *n* ARCH & BUILD, INFR & DES Arbeitsumfang *m*, Leistungsumfang *m*

**score** *vt* ARCH & BUILD einkerben, BUILD HARDW, INFR & DES, STONE, SURFACE *plaster* aufrauhen

**scoria** *n* MAT PROP Schlacke *f*; ~ **brick** *n* STONE Schlackenstein *m*

**Scotch**: ~ **tape**® *n AmE* (*cf Sellotape® BrE*) ARCH & BUILD Tesafilm® *m*

**scotia** *n* ARCH & BUILD Hohlkehle *f*

**scour** *vt* INFR & DES auskolken, SURFACE ablaugen, beizen

**scour**: ~ **depth** *n* INFR & DES Auskolkungstiefe *f*

**scrap** *n* CONST LAW, ENVIRON, INFR & DES Schrott *m*; ~-**baling press** *n* BUILD MACHIN, ENVIRON Eisenschrottpresse *f*, Schrottpresse *f*

**scraped**: ~ **rendering** *n* STONE Kratzputz *m*; ~ **stucco** *n* STONE Kratzputz *m*, Schabputz *m*, Stockputz *m*

**scraper** *n* BUILD MACHIN Schaber *m*, Schrapper *m*, Ziehklinge *f*; ~ **conveyor** *n* BUILD MACHIN Kratzband *nt*; ~ **extractor** *n* ENVIRON Schaufelentnahmegerät *nt*

***scrap***: ~ **iron** *n* ENVIRON Eisenschrott *m*; ~ **lead** *n* ENVIRON Altblei *nt*; ~ **metal** *n* ENVIRON Altmetall *nt*; ~ **metal separation** *n* ENVIRON Entschrottung *f*; ~ **motor car** *n* ENVIRON Autowrack *nt*; ~ **processing** *n* ENVIRON, INFR & DES Schrottverwertung *f*; ~ **recovery** *n* ENVIRON, INFR & DES Schrottverwertung *f*; ~ **recycling** *n* ENVIRON, INFR & DES Schrottverwertung *f*; ~ **sorting** *n* ENVIRON Schrottsortierung *f*

**scrapyard** *n* ENVIRON Schrottplatz *m*

**scratch 1.** *n* BUILD MACHIN, MAT PROP Schramme *f*; **2.** *vt* SURFACE *plaster* aufkratzen

***scratch***: ~ **awl** *n* BUILD MACHIN Markiernadel *f*, Reißnadel *f*; ~ **brush** *n* BUILD MACHIN Drahtbürste

*f*; ~ **gage** *AmE*, ~ **gauge** *BrE* *n* BUILD MACHIN Streichmaß *nt*
**screed** *vt* ARCH & BUILD, CONCR abziehen
***screed***: ~ **board** *n* ARCH & BUILD, STONE Abziehbohle *f*, Glättbohle *f*; ~ **heating** *n* ARCH & BUILD, HEAT & VENT Fußbodenheizung *f*; ~ **height** *n* ARCH & BUILD Estrichstärke *f*; ~ **topping** *n* STONE Estrich *m*
**screen 1.** *n* BUILD MACHIN Sieb *nt*, ENVIRON *landfill* Sichtschutz *m*; **2.** *vt* ARCH & BUILD *with grating* vergittern, BUILD MACHIN *sieve* sieben, *windows* verblenden
***screen***: ~ **bar** *n* BUILD MACHIN Gitterstab *m*
**screened** *adj* ARCH & BUILD abgestuft, BUILD MACHIN *covered* geschützt, *gradated* abgestuft, *sieved* gesiebt
**screening** *n* BUILD MACHIN, ENVIRON Klassierung *f*, Sieben *nt*; ~ **equipment** *n* ENVIRON Sieb *nt*; ~ **fraction** *n* MAT PROP Kornfraktion *f*
**screenings** *n pl* ENVIRON Siebrest *m*, Siebrückstand *m*
***screening***: ~ **sheet** *n* BUILD MACHIN Siebblech *nt*
***screen***: ~ **residue** *n* MAT PROP Überlauf *m*, *filter* Rückstand *m*; ~ **wall** *n* ARCH & BUILD, BUILD MACHIN, INFR & DES Sichtschutzwand *f*, Blendmauer *f*, Gittermauer *f*
**screw 1.** *n* BUILD HARDW Schraube *f*, BUILD MACHIN Bolzen *m*, Schraub-, Schraube *f*; **2.** *vt* BUILD MACHIN verschrauben; ~ **off** STEEL, TIMBER abschrauben
***screw***: ~ **cap** *n* WASTE WATER Drehdeckel *m*, Schraubdeckel *m*; ~ **conveyor** *n* BUILD MACHIN Schneckenförderer *m*; **~-down cock** *n* BUILD MACHIN Niederschraubhahn *m*; **~-down stop valve** *n* BUILD MACHIN Niederschraubabsperrventil *nt*; **~-down valve** *n* BUILD MACHIN Niederschraubventil *nt*
**screwdriver** *n* BUILD MACHIN Schraubendreher *m*, Schraubenzieher *m*
**screwed**: ~ **connection** *n* MAT PROP, STEEL, STONE Schraubverbindung *f*, Verschraubung *f*; ~ **pipe joint** *n* HEAT & VENT, WASTE WATER Rohrverschraubung *f*
***screw***: ~ **elevator** *n* BUILD MACHIN Förderrohr *nt*, Senkrechtförderschnecke *f*; ~ **hole** *n* BUILD MACHIN Schraubloch *nt*; ~ **hook** *n* BUILD MACHIN Hakenschraube *f*; ~ **jack** *n* BUILD MACHIN Schraubenwinde *f*; ~ **nail** *n* BUILD HARDW Schraubennagel *m*; ~ **plug** *n* BUILD HARDW Verschlußschraube *f*, BUILD MACHIN Gewindestopfen *m*; ~ **pump** *n* BUILD MACHIN Schraubenpumpe *f*; ~ **socket** *n* HEAT & VENT, WASTE WATER Gewindemuffe *f*; ~ **tap** *n* BUILD MACHIN Gewindebohrer *m*; ~ **union** *n* BUILD MACHIN, HEAT & VENT Einschraubstutzen *m*
**scribe 1.** *n* TIMBER Anreißer *m*; **2.** *vt* ARCH & BUILD, BUILD MACHIN, CONCR, MAT PROP, STONE anreißen, anzeichnen
**scribing**: ~ **awl** *n* BUILD MACHIN Reißspitze *f*; ~ **gage** *AmE*, ~ **gauge** *BrE* *n* BUILD MACHIN Reißlehre *f*; ~ **iron** *n* BUILD MACHIN Markierungseisen *nt*
**scrim** *n* MAT PROP Gewebeverstärkung *f*
**scrubbable** *adj* SURFACE scheuerbeständig, scheuerfest
**scrubbed**: ~ **gas** *n* ENVIRON Abluft *f*; ~ **plaster** *n* STONE Waschputz *m*
**scum** *n* INFR & DES Schaum *m*
**scumble** *n* SURFACE Lasur *f*
**scutcheon** *n* ARCH & BUILD Schlüsselschild *nt*
**seabed** *n* ENVIRON, INFR & DES Meeresboden *m*, Meeresgrund *m*
**sea**: ~ **defence construction** *n* INFR & DES Küstenschutzkonstruktion *f*; ~ **depth** *n* INFR & DES Meerestiefe *f*; ~ **groin** *AmE*, ~ **groyne** *BrE* *n* INFR & DES Seebuhne *f*
**seal 1.** *n* BUILD HARDW Plombe *f*, Siegel *nt*; **2.** *vt* ENVIRON *landfill* abdichten, einkapseln, INFR & DES abdichten, SURFACE *floor* versiegeln, vergießen, verschließen, abdichten, *paint* verspachteln, *with putty* verkitten; ~ **with mortar** STONE mit Mörtel ausgießen
**sealant** *n* SURFACE, WASTE WATER Abdichtungsmittel *nt*, Versiegelungsmittel *nt*
***seal***: ~ **of approval** *n* CONST LAW amtliches Gütezeichen *nt*; ~ **coat** *n* SURFACE *roads* Verschlußdecke *f*
**sealed** *adj* CONST LAW, SURFACE plombiert, *security* versiegelt
**sealer** *n* SURFACE Absperrmittel *nt*, Dichtungsmittel *nt*, Versiegler *m*, Einlaßgrund *m*, Porenfüller *m*
**sea**: ~ **level** *n* ARCH & BUILD Meereshöhe *f*, ENVIRON, INFR & DES Meeresspiegel *m*; ~ **level datum plane** *n* INFR & DES Meeresspiegelbezugsebene *f*
**sealing** *n* ARCH & BUILD Verspachteln *nt*, ENVIRON Abdichten *nt*, *landfill* Einkapselung *f*, INFR & DES Abdichten *nt*, SURFACE Abdichten *nt*, Abdichtung *f*, Verschließen *nt*, Plombe *f*, Siegel *nt*, Verspachteln *nt*, Versiegelung *f*; ~ **agent** *n* INFR & DES, MAT PROP, SURFACE Abdichtungsmasse *f*; ~ **coat** *n* SOUND & THERMAL, SURFACE Isolieranstrich *m*; ~ **compound** *n* INFR & DES Abdichtungsmasse *f*, MAT PROP, SURFACE Abdichtungsmasse *f*, Dichtungsmasse *f*, Fugenvergußmasse *f*; ~ **core** *n* INFR & DES Dichtungskern *m*; ~ **end** *n* SURFACE Dichtungsmaterial *nt*, Endverschluß *m*; ~ **layer** *n* SOUND & THERMAL, SURFACE Dichtungsschicht *f*; ~ **material** *n* SURFACE Dichtungsmittel *nt*, WASTE WATER Dichtungsmaterial *nt*; ~ **ring** *n* WASTE WATER Dichtring *m*; ~ **sheet** *n* SURFACE Dichtungsmanschette *f*; ~ **strip** *n* ARCH & BUILD, INFR & DES, MAT PROP Fugeneinlage *f*, SOUND & THERMAL Dichtleiste *f*, Dichtungsstreifen *m*, Einlage *f*, Fugeneinlage *f*; ~ **works** *n pl* ARCH & BUILD, INFR & DES Abdichtungsarbeiten *f pl*
***seal***: ~ **sheeting** *n* SOUND & THERMAL Dichtungsbahn *f*
**seam 1.** *n* STEEL Naht *f*; **2.** *vt* STEEL falzen
**seamless**: ~ **pipe** *n* HEAT & VENT nahtloses Rohr *nt*
**seasoning** *n* MAT PROP, TIMBER Austrocknen *nt*
**seating** *n* ARCH & BUILD, CONCR, STEEL, STONE Auflagerfläche *f*
**sea**: ~ **wall** *n* INFR & DES Strandmauer *f*
**secondary**: ~ **access** *n* ARCH & BUILD, INFR & DES Nebenzugang *m*; ~ **air** *n* HEAT & VENT Nebenluft *f*; ~ **beam** *n* ARCH & BUILD Querträger *m*; ~ **combustion chamber** *n* (*SCC*) ENVIRON, HEAT & VENT Nachbrennkammer *f* (*SCC*), zweiter Brennraum *m* (*SCC*); ~ **reinforcement** *n* CONCR Verteilerstäbe *m pl*; ~ **sedimentation basin** *n* ENVIRON Nachklärbecken *nt*; ~ **settling tank** *n* ENVIRON Nachklärbecken *nt*; ~ **sewage treatment** *n* ENVIRON biologische Nachreinigung *f*, INFR & DES biologische Nachklärung *f*
**second**: ~ **floor** *n* *AmE* (*cf first floor BrE*) ARCH & BUILD *above ground floor* erste Etage *f*, erstes Stockwerk *nt*; ~ **order theory** *n* ARCH & BUILD, CONCR, INFR & DES, MAT PROP, STEEL Theorie *f* II. Ordnung
**secret**: ~ **gutter** *n* WASTE WATER eingebaute Dachrinne *f*
**section** *n* ARCH & BUILD Teil *m*, Abschnitt *m*, Schnitt *m*, STEEL Profil *nt*
**sectional**: ~ **axonometric drawing** *n* ARCH & BUILD, INFR & DES axonometrische Schnittzeichnung *f*;

~ **completion** *n* CONST LAW Abschnittsfertigstellung *f*; ~ **drawing** *n* ARCH & BUILD, INFR & DES Schnitt *m*, Schnittzeichnung *f*, *of building* Gebäudeschnittzeichnung *f*; ~ **sheet** *n* STEEL Profilblech *nt*; ~ **steel** *n* ARCH & BUILD, STEEL Profilstahl *m*; ~ **steel construction** *n* ARCH & BUILD, STEEL Profilstahlkonstruktion *f*; ~ **surround** *n* ARCH & BUILD Einfaßprofil *nt*

***section***: ~ **cutter** *n* BUILD MACHIN Profilschere *f*, Profilschneider *m*

**sector** *n* INFR & DES *of circle* Kreissektor *m*

**secure 1.** *adj* CONST LAW zuverlässig; **2.** *vt* CONST LAW sichern, sicherstellen

**security** *n* CONST LAW Sicherheit *f*; ~ **lock** *n* BUILD HARDW, CONST LAW Sicherheitsschloß *nt*; ~ **lock for lifting doors** *n* CONST LAW Hebetürsicherung *f*

**sedilia**: ~ **niche** *n* ARCH & BUILD *for group of three sedilia* Sediliennische *f*

**sediment** *n* INFR & DES, MAT PROP Ablagerung *f*, Bodensatz *m*

**sedimentation** *n* ENVIRON, INFR & DES Sedimentablagerung *f*; ~ **analysis** *n* INFR & DES Sedimentationsanalyse *f*; ~ **basin** *n* ENVIRON, INFR & DES Absetzbecken *nt*; ~ **machine** *n* BUILD MACHIN Schlämmgerät *nt*; ~ **test** *n* INFR & DES Absetzprobe *f*

***sediment***: ~ **tank** *n* ENVIRON, INFR & DES, WASTE WATER Klärbecken *nt*

**seed 1.** *vt* INFR & DES einsäen; **2.** *vi* INFR & DES sickern

**seep**: ~ **in** *vi* ARCH & BUILD, INFR & DES, WASTE WATER einsickern

**seepage** *n* ARCH & BUILD, INFR & DES, WASTE WATER Einsickern *nt*, Versickern *nt*; ~ **line** *n* INFR & DES Sickerlinie *f*; ~ **loss** *n* INFR & DES Sickerverlust *m*; ~ **path** *n* INFR & DES Sickerweg *m*; ~ **pressure** *n* INFR & DES Sickerströmungsdruck *m*; ~ **shaft** *n* INFR & DES, WASTE WATER Sickerschacht *m*; ~ **trench** *n* INFR & DES, WASTE WATER Sickergraben *m*, Versickerungsgraben *m*; ~ **water** *n* ENVIRON, INFR & DES Sickerwasser *nt*

**segment** *n* ARCH & BUILD Segment *nt*

**segmental**: ~ **arch** *n* ARCH & BUILD Segmentbogen *m*, Stichbogen *m*

**segregate** *vt* ENVIRON abscheiden

**segregation** *n* ENVIRON Abscheidung *f*; ~ **berm** *n* ENVIRON Berme *f*

**seism** *n* INFR & DES Erdbeben *nt*

**seismic** *adj* INFR & DES Spreng-, seismisch; ~ **activity** *n* INFR & DES seismische Aktivität *f*; ~ **area** *n* INFR & DES Erdbebengebiet *nt*; ~ **construction** *n* ARCH & BUILD, INFR & DES erdbebensicheres Bauen *nt*; ~ **design** *n* INFR & DES erdbebensichere Bemessung *f*; ~ **effects** *n pl* MAT PROP seismische Auswirkungen *f pl*; ~ **exploration method** *n* INFR & DES angewandte Seismik *f*, Sprengseismik *f*, seismische Bodenuntersuchung *f*; ~ **focus** *n* INFR & DES Erdbebenherd *m*; ~ **inspection** *n* INFR & DES seismische Bodenforschung *f*; ~ **reflections** *n pl* INFR & DES Reflexionsseismik *f*; ~ **survey** *n* INFR & DES seismischer Aufschluß *m*

**seismograph** *n* BUILD MACHIN, INFR & DES Erdbebenmesser *m*, Seismograph *m*

**seismographic**: ~ **observation** *n* INFR & DES seismographische Beobachtung *f*

**seismology** *n* INFR & DES Seismologie *f*

**seismometer** *n* BUILD MACHIN Erdbebenmesser *m*, Seismometer *nt*

**selection** *n* ARCH & BUILD Auswahl *f*

**selective**: ~ **collection** *n* ENVIRON getrennte Müllabfuhr *f*, getrennte Müllsammlung *f*; ~ **surface** *n* MAT PROP selektive Oberfläche *f*

**self**: ~-**adhesive** *adj* MAT PROP selbstklebend; ~ **build** *n* CONST LAW Eigenbau *m*; ~-**closing cock** *n* WASTE WATER automatisches Rohrventil *nt*; ~-**closing door** *n* BUILD HARDW Automatiktür *f*, selbstschließende Tür *f*; ~-**closing faucet** *n AmE* (*cf self-closing tap BrE*) BUILD HARDW Selbstschlußbatterie *f*; ~-**closing tap** *n BrE* (*cf self-closing faucet AmE*) BUILD HARDW Selbstschlußbatterie *f*; ~-**contained** *adj* MAT PROP freitragend; ~-**coring chisel** *n* BUILD MACHIN Automatikmeißel *m*; ~-**draining system** *n* ARCH & BUILD, INFR & DES Selbstentwässerungssystem *nt*; ~-**dumping bucket** *n* BUILD MACHIN Selbstentladeeimer *m*; ~-**extinguishing** *adj* MAT PROP selbstverlöschend; ~-**propelled roller** *n* BUILD MACHIN selbstfahrende Walze *f*; ~-**quenching** *adj* MAT PROP selbstlöschend; ~-**supporting** *adj* ARCH & BUILD freistehend, freitragend, selbsttragend; ~-**supporting partition** *n* ARCH & BUILD freistehende Zwischenwand *f*; ~-**tensioning** *adj* CONCR selbstspannend

**Sellotape**[R] *n BrE* (*cf Scotch tape*[R] *AmE*) ARCH & BUILD Tesafilm[R] *m*

**selvage** *n* MAT PROP Dachpappenrandstreifen *m*, Webkante *f*

**selvedge** *n* (*see selvage*) MAT PROP Webkante *f*

**semi-** *pref* ARCH & BUILD, INFR & DES Halb-

**semibasement** *n* ARCH & BUILD Souterrain *nt*

**semibeam** *n* ARCH & BUILD, CONCR, TIMBER Freiträger *m*, Kragbalken *m*, Kragträger *m*

**semicircle** *n* ARCH & BUILD Halbkreis *m*

**semicircular** *adj* ARCH & BUILD halbkreisförmig; ~ **arch** *n* ARCH & BUILD Halbkreisbogen *m*, Rundbogen *m*

**semidetached**: ~ **house** *n* ARCH & BUILD, INFR & DES Doppelhaushälfte *f*

**semigantry**: ~ **crane** *n* INFR & DES Halbportalkran *m*

**semigirder** *n* ARCH & BUILD Konsolbalken *m*

**semigloss** *n* ARCH & BUILD, SURFACE Halbglanz *m*

**semihydraulic**: ~ **lime** *n* MAT PROP Schwarzkalk *m*; ~ **lime plaster** *n* STONE Schwarzkalkputz *m*

**semirigid** *adj* ARCH & BUILD, INFR & DES *frame* halbsteif, MAT PROP *plastic* halbsteif, halbstarr

**semisolid** *adj* ENVIRON, MAT PROP halbfest, stichfest

**semitrailer**: ~ **truck** *n AmE* (*cf articulated lorry BrE*) BUILD MACHIN Sattelschlepper *m*

**semiwet**: ~ **sorting** *n* ENVIRON Halbfeuchttrennung *f*

**sensitivity** *n* BUILD MACHIN Empfindlichkeit *f*

**sensor** *n* ARCH & BUILD Meßfühler *m*, CONST LAW, ELECTR Sensor *m*, HEAT & VENT Meßfühler *m*

**separate** *vt* INFR & DES, WASTE WATER trennen

**separate**: ~ **sewage system** *n* WASTE WATER Trennverfahren *nt*

**separating**: ~ **agent** *n* MAT PROP Trennmittel *nt*; ~ **oil** *n* MAT PROP Schalöl *nt*

**separation** *n* ENVIRON Abtrennung *f*, Trennung *f*, INFR & DES Trenn-, Trennen *nt*, WASTE WATER Absonderung *f*, Trenn-, Trennen *nt*, Trennung *f*; ~ **of flow** *n* INFR & DES Ablösung *f*; ~ **joint** *n* ARCH & BUILD, INFR & DES Trennfuge *f*; ~ **layer** *n* ARCH & BUILD Trennschicht *f*; ~ **wall** *n* ARCH & BUILD Trennwand *f*

**separator** *n* INFR & DES, WASTE WATER Abscheider *m*, Trenner *m*

**septic**: ~ **tank** *n* ENVIRON Abwasserfaulraum *m*
**sequence**: ~ **of assembly** *n* CONST LAW Montageablauf *m*; ~ **of construction work** *n* ARCH & BUILD, INFR & DES Baufolge *f*; ~ **of operations** *n* ARCH & BUILD, INFR & DES Arbeitsablauf *m*; ~ **of strata** *n* INFR & DES, MAT PROP, STONE Schichtenfolge *f*
**series**: ~ **mounting** *n* ELECTR Reihenschaltung *f*, Serienschaltung *f*
**serpentinization** *n* ENVIRON Serpentinisierung *f*
**serrated**: ~ **trowel** *n* BUILD MACHIN *tiling* Zahnspachtel *m*
**service** *n* HEAT & VENT Betrieb *m*, *maintenance* Wartung *f*; ~ **elevator** *n AmE* (*cf service lift BrE*) ARCH & BUILD, BUILD MACHIN, INFR & DES Baustellenaufzug *m*, Lastenaufzug *m*; ~ **expectancy** *n* BUILD MACHIN, ELECTR, HEAT & VENT, MAT PROP, WASTE WATER Lebensdauer *f*; ~ **gangway** *n* ARCH & BUILD, INFR & DES Bedienungsgang *m*, Bedienungssteg *m*; ~ **life** *n* BUILD MACHIN, ELECTR, HEAT & VENT, MAT PROP, WASTE WATER Lebensdauer *f*; ~ **lift** *n BrE* (*cf service elevator AmE*) ARCH & BUILD, BUILD MACHIN, INFR & DES Baustellenaufzug *m*, Lastenaufzug *m*; ~ **line** *n* ELECTR, HEAT & VENT, WASTE WATER Hausanschlußleitung *f*; ~ **manual** *n* BUILD MACHIN, HEAT & VENT Wartungsanleitung *f*; ~ **staff** *n* ARCH & BUILD Bedienungstrupp *m*; ~ **switch cabinet** *n* ELECTR Hausanschlußkasten *m*; ~ **tunnel** *n* ARCH & BUILD, INFR & DES Versorgungskanal *m*; ~ **weight** *n* ELECTR Betriebsgewicht *nt*
**servicing** *n* ELECTR, HEAT & VENT Bedienung *f*
**servomotor** *n* BUILD MACHIN, ENVIRON Servomotor *m*
**set 1.** *n* BUILD HARDW Garnitur *f*, STONE dritte Putzlage *f*; **2.** *vt* ARCH & BUILD, BUILD HARDW, CONCR erhärten, CONST LAW, ELECTR, HEAT & VENT *setpoint* einstellen, INFR & DES setzen, einstellen, MAT PROP *setpoint* einstellen, erhärten, *toughen* härten, *leather* verhärten; **3.** *vi* CONCR hart werden, ENVIRON erstarren, MAT PROP *concrete, mortar* abbinden, *solidify* erstarren, *toughen* hart werden, sich verhärten
***set***: ~ **back** *vt* ARCH & BUILD nach hinten versetzen; ~ **in concrete** *vt* CONCR einbetonieren; ~ **out** *vt* ARCH & BUILD *surveying* abstecken; ~ **up** *vt* ARCH & BUILD aufbauen, montieren; ~ **up profiles** *vt* STONE Profile setzen, profilieren
**set**: ~ **back** *phr* ARCH & BUILD zurücksetzen; ~ **boards edgewise** *phr* ARCH & BUILD Bohlen hochkant verlegen; ~ **of door handles** *n* BUILD HARDW Drückergarnitur *f*; ~ **of handles** *n* BUILD HARDW Drückergarnitur *f*; ~ **of instruments** *n* ARCH & BUILD, BUILD MACHIN Besteck *nt*; ~**-off** *n* ARCH & BUILD Rücksprung *m*
**sett** *n* STONE *small paving stone* Pflasterstein *m*
**setting** *n* CONCR Erhärten *nt*, ENVIRON Erstarrung *f*, MAT PROP, STONE Erhärten *nt*; ~ **coat** *n* STONE dritte Putzlage *f*; ~ **jig** *n* BUILD MACHIN *drilling equipment* Lehre *f*; ~ **time** *n* CONCR, ENVIRON, MAT PROP Abbindzeit *f*
**settle 1.** *vt* CONST LAW schlichten, setzen; **2.** *vi* INFR & DES sacken
**settlement** *n* ARCH & BUILD, ENVIRON Bodensetzung *f*, Sackung *f*, INFR & DES Bodensenkung *f*, Setzung *f*, *place* Siedlung *f*, *subsidence*, Setzen *nt*, Bodensetzung *f*, Senkung *f*; ~ **crack** *n* ARCH & BUILD, INFR & DES Setzriß *m*; ~ **curve** *n* CONCR, MAT PROP Setzungskurve *f*; ~ **duration** *n* CONCR, INFR & DES, MAT PROP Setzungsdauer *f*; ~ **gage** *AmE*, ~ **gauge** *BrE n* BUILD MACHIN Setzungsmeßgerät *nt*; ~ **joint** *n* ARCH & BUILD, INFR & DES Setzfuge *f*; ~ **observation** *n* ARCH & BUILD, INFR & DES Setzungsbeobachtung *f*; ~ **stress** *n* INFR & DES Setzungsspannung *f*; ~ **of support** *n* ARCH & BUILD, INFR & DES Stützensenkung *f*; ~ **test** *n* INFR & DES Setzversuch *m*
**settling** *n* ENVIRON, INFR & DES *sedimentation* Absetzen *nt*, *subsidence* Bodensetzung *f*, Sackung *f*; ~ **basin** *n* ENVIRON, INFR & DES Absetzbecken *nt*; ~ **chamber** *n* ENVIRON Absetzkammer *f*, Beruhigungskammer *f*; ~ **tank** *n* ENVIRON, INFR & DES Absetzbecken *nt*; ~ **velocity** *n* ENVIRON, MAT PROP Absetzgeschwindigkeit *f*; ~ **vessel** *n* ENVIRON, INFR & DES Absetzgefäß *nt*
***set***: ~ **of tools** *n* BUILD MACHIN Handwerkszeug *nt*
**severe**: ~ **exposure conditions** *n pl* ARCH & BUILD, INFR & DES besonders korrosionsfördernde Einflüsse *m pl*
**sewage** *n* ENVIRON, INFR & DES, WASTE WATER Abwasser *nt*, Schmutzwasser *nt*; ~ **composition** *n* ENVIRON Abwasserzusammensetzung *f*; ~ **discharge** *n* ENVIRON Abwasserbeseitigung *f*, Abwassereinleitung *f*, INFR & DES, WASTE WATER Abwasserbeseitigung *f*; ~ **disposal** *n* ENVIRON, INFR & DES, WASTE WATER Abwasserbeseitigung *f*; ~ **flow** *n* ENVIRON Abwasseranfall *m*, Abwassermenge *f*, INFR & DES Abwasserfluß *m*; ~ **fungus** *n* ENVIRON Abwasserpilz *m*; ~ **installation** *n* INFR & DES Abwasseranlagen *f pl*; ~ **outfall** *n* ENVIRON Abwassereinleitungsstelle *f*; ~ **oxidation pond** *n* ENVIRON Oxidationsteich *m*; ~ **pipe** *n* ENVIRON Kanalisationsrohr *nt*, INFR & DES, WASTE WATER Abwasserleitung *f*, Abwasserrohr *nt*, Kanalisationsrohr *nt*; ~ **purification** *n* ENVIRON Abwasserklärung *f*, Abwasserreinigung *f*; ~ **purification plant** *n* ENVIRON, INFR & DES Abwasserkläranlage *f*; ~ **sludge** *n* ENVIRON, INFR & DES, WASTE WATER Klärschlamm *m*, *municipal* Abwasserschlamm *m*; ~ **treatment** *n* ENVIRON Abwasseraufbereitung *f*, Abwasserbehandlung *f*, Abwasserklärung *f*, Abwasserreinigung *f*, INFR & DES Abwasserbehandlung *f*; ~ **treatment plant** *n* ENVIRON Abwasserbehandlungsanlage *f*, Abwasserkläranlage *f*, Abwasserreinigungsanlage *f*, Kläranlage *f*, Klärwerk *nt*, INFR & DES Abwasserkläranlage *f*, Kläranlage *f*, Klärwerk *nt*, WASTE WATER Kläranlage *f*; ~ **treatment process** *n* ENVIRON Abwasserbehandlungsverfahren *nt*; ~ **treatment works** *n* ENVIRON Abwasserreinigungsanlage *f*; ~ **water disposal** *n* ENVIRON, INFR & DES, WASTE WATER Abwasserbeseitigung *f*; ~ **works** *n* ENVIRON, INFR & DES, WASTE WATER Kläranlage *f*, Klärwerk *nt*
**sewer** *n* INFR & DES, WASTE WATER Abflußkanal *m*, Abwasserkanal *m*, Abwasserleitung *f*, Siel *nt*
**sewerage**: ~ **charge** *n* CONST LAW Abwasserabgabe *f*; ~ **gas** *n* INFR & DES Kanalgas *nt*; ~ **system** *n* ENVIRON, INFR & DES, WASTE WATER Grundstücksentwässerung *f*, Kanalisation *f*, Stadtentwässerung *f*
***sewer***: ~ **brick** *n* STONE, WASTE WATER Kanalziegel *m*; ~ **cleaning** *n* INFR & DES, WASTE WATER Kanalreinigung *f*; ~ **culvert** *n* INFR & DES, WASTE WATER Kanaldüker *m*; ~ **jetting truck** *n* BUILD MACHIN, INFR & DES, WASTE WATER Hochdruckspülfahrzeug *nt* für die Kanalreinigung; ~ **maintenance** *n* INFR & DES, WASTE WATER Kanalwartung *f*; ~ **mud extractor** *n* BUILD MACHIN Kanalschlammabsauggerät *nt*; ~ **pipe** *n* ENVIRON, INFR & DES, WASTE WATER Abwasser-

leitung *f*, Abwasserrohr *nt*, Kanalisationsrohr *nt*; **~ system** *n* ENVIRON, INFR & DES, WASTE WATER Kanalisation *f*

**shackle** *n* BUILD HARDW *lock* Lastöse *f*, Bügel *m*

**shading**: **~ device** *n* ARCH & BUILD Sonnenschutzvorrichtung *f*

**shaft** *n* ARCH & BUILD Schacht *m*, Schaft *m*, Säulenrumpf *m*, BUILD MACHIN Welle *f*, INFR & DES, WASTE WATER Schacht *m*

**shaking**: **~ chute** *n* BUILD MACHIN Schüttelrinne *f*, Rütteltisch *m*

**shale** *n* BUILD MACHIN, INFR & DES, MAT PROP Schiefer *m*

**shallow** *adj* ARCH & BUILD niedrig, muldenförmig, INFR & DES muldenförmig, WASTE WATER flach; **~ bowl water closet** *n* BUILD HARDW, WASTE WATER Flachspülklosett *nt* (*Flachspül-WC*); **~ bowl WC** *n* BUILD HARDW, WASTE WATER Flachspül-WC *nt* (*Flachspülklosett*); **~ building pit** *n* INFR & DES Flachbaugrube *f*; **~ excavation** *n* INFR & DES *earthworks* Flachaushub *m*, Flachbaugrube *f*, *process* Flachbaggerung *f*

**shank** *n* ARCH & BUILD Schaft *m*, Säulenrumpf *m*

**shape 1.** *n* ARCH & BUILD Form *f*, INFR & DES Form *f*, Geometrie *f*, Profil *nt*, MAT PROP Profil *nt*; **2.** *vt* ARCH & BUILD, INFR & DES formen

**shaping** *n* ARCH & BUILD, CONCR, INFR & DES Formgebung *f*

**shave**: **~ hook** *n* BUILD HARDW Bleifeile *f*, Schab *nt*

**shaving** *n* ARCH & BUILD, TIMBER Hobelspan *m*

**shear 1.** *n* ARCH & BUILD Schub *m*, Schubspannung *f*, INFR & DES Scherspannung *f*, Scherung *f*, Schub *m*, Vertikalschub *m*; **2.** *vt* INFR & DES abscheren, auf Schub beanspruchen

***shear***: **~ area** *n* ARCH & BUILD, CONCR, MAT PROP, STEEL Schubfläche *f*; **~ buckling** *n* ARCH & BUILD, CONCR, MAT PROP, STEEL Schubknicken *nt*; **~ buckling resistance** *n* ARCH & BUILD, CONCR, MAT PROP, STEEL Schubknicktragfähigkeit *f*; **~ connector** *n* CONCR Schubverbindung *f*, *composite beam* Kopfbolzendübel *m*, STEEL Verbundanker *m*; **~ failure** *n* INFR & DES Grundbruch *m*; **~ force resistance** *n* ARCH & BUILD, CONCR, MAT PROP, STEEL Schubkrafttragfähigkeit *f*

**shearing**: **~ force** *n* ARCH & BUILD Scherkraft *f*; **~ stress** *n* ARCH & BUILD, INFR & DES Scherspannung *f*; **~ test** *n* ARCH & BUILD, INFR & DES, MAT PROP Scherprobe *f*, Scherversuch *m*

***shear***: **~ leg** *n* INFR & DES Dreibeinkran *m*; **~ moment** *n* ARCH & BUILD Schermoment *nt*; **~ panel** *n* ARCH & BUILD schubkraftübertragendes Panel *nt*, TIMBER schubkraftübertragende Täfelung *nt*; **~-resistant** *adj* ARCH & BUILD schubfest; **~ stress** *n* ARCH & BUILD Scherspannung *f*; **~ test** *n* INFR & DES Abscherversuch *m*; **~ wall** *n* ARCH & BUILD *load distribution* Windscheibe *f*, BUILD HARDW *load distribution* Scheibenwand *f*; **~ zone** *n* INFR & DES Scherzone *f*

**sheath** *n* CONCR Hüllrohr *nt*, ELECTR *cable* Mantel *m*

**sheathe** *vt* ARCH & BUILD umhüllen, ummanteln

**sheathed** *adj* ARCH & BUILD ummantelt

**sheathing** *n* ARCH & BUILD Verschalung *f*, Umhüllung *f*, Ummantelung *f*

**shed** *n* ARCH & BUILD Schuppen *m*; **~ dormer** *n* ARCH & BUILD Schleppgaube *f*; **~ roof** *n* ARCH & BUILD Halbdach *nt*, Pultdach *nt*, Flugdach *nt*, TIMBER Pultdach *nt*

**sheepsfoot**: **~ roller** *n* ARCH & BUILD, INFR & DES Schaffußwalze *f*

**sheet 1.** *n* MAT PROP Blech *nt*, Folie *f*, Platte *f*; **2. ~ out** *vt* ARCH & BUILD *civil engineering* auswalzen

***sheet***: **~ frame** *n* STEEL Tafelglas *nt*

**sheeting** *n* MAT PROP Bahnen *f pl*, Verkleidungsmaterial *nt*, *plastic* Folien *f pl*

***sheet***: **~ iron** *n* MAT PROP Eisenblech *nt*; **~ iron pipe** *n* STEEL Blechrohr *nt*; **~ lead** *n* STEEL Bleiblech *nt*, Tafelblei *nt*; **~ lining** *n* STEEL Blechauskleidung *f*, Blechverkleidung *f*; **~ metal flashing** *n* STEEL Blechanschluß *m*, Verwahrungsblech *nt*; **~ metal flashing piece** *n* STEEL Blechanschlußstreifen *m*; **~ metal jacket** *n* STEEL Blechummantelung *f*, Blechmantel *m*; **~ metal work** *n* STEEL Blechnerarbeiten *f pl*, Flaschnerarbeiten *f pl*, Schweißen *nt*; **~ pile** *n* ARCH & BUILD, INFR & DES Spundbohle *f*, Spundpfahl *m*, STEEL Spundbohle *f*; **~ pile anchorage** *n* INFR & DES Spundwandverankerung *f*; **~ piling** *n* ARCH & BUILD, INFR & DES Spundwand *f*, Spundwerk *nt*; **~-piling driver** *n* BUILD MACHIN Spundwandramme *f*; **~ shears** *n pl* BUILD MACHIN Blechschere *f*; **~ thickness** *n* MAT PROP Blechdicke *f*

**shelf** *n* TIMBER Fachboden *m*, Gestell *nt*

**shell** *n* ARCH & BUILD, BUILD MACHIN Hülle *f*, Mantel *m*, Schale *f*

**shellac** *n* MAT PROP, SURFACE Lackfarbe *f*, Schellack *m*

***shell***: **~ auger** *n* BUILD MACHIN, INFR & DES Löffelbohrer *m*; **~ concrete** *n* CONCR Schalenbeton *m*; **~ construction** *n* ARCH & BUILD Schalenbauweise *f*; **~ gimlet** *n* TIMBER Schneckenhandbohrer *m*; **~ roof** *n* ARCH & BUILD Schalendach *nt*; **~ structure** *n* ARCH & BUILD, INFR & DES Schalenkonstruktion *f*; **~ work** *n* ARCH & BUILD, TIMBER Rohbauarbeiten *f pl*

**shelter** *n* ARCH & BUILD Schutzdach *nt*, Schutzraum *m*, Überdeckung *f*

**shield** *n* ARCH & BUILD Schild *m*, Vortriebschild *m*; **~ driving method** *n* INFR & DES *tunnel construction* Schildvortrieb *m*

**shielding** *n* ELECTR, HEAT & VENT Abschirmung *f*

***shield***: **~ tunneling** *AmE*, **~ tunnelling** *BrE* *n* INFR & DES Schildbauweise *f*, Schildvortrieb *m*

**shingle** *n* MAT PROP *roof* Schindel *f*; **~ roof** *n* ARCH & BUILD Schindeldach *nt*

**shipping**: **~ container** *n* INFR & DES Schiffscontainer *m*

**shock**: **~ absorber** *n* BUILD MACHIN Stoßdämpfer *m*; **~-proof** *adj* MAT PROP stoßfest; **~ resistance** *n* ARCH & BUILD, CONCR, MAT PROP, STEEL Schlagfestigkeit *f*, Stoßfestigkeit *f*; **~-resistant** *adj* MAT PROP schlagfest, stoßfest

**shoe** *n* BUILD MACHIN *downpipe* Fallrohrauslauf *m*

**shoed**: **~ bar** *n* STEEL angeschuhte Stange *f*

**shoot** *vt* INFR & DES *detonate* sprengen, *plane* glatthobeln

**shop** *n* ARCH & BUILD Laden *m*

**shopping**: **~ center** *AmE*, **~ centre** *BrE* *n* INFR & DES Einkaufszentrum *nt*; **~ complex** *n* INFR & DES Einkaufszentrum *nt*

**shore** *n* ARCH & BUILD Strebe *f*, INFR & DES Ufer *nt*

**Shore**: **~ hardness** *n* MAT PROP Shorehärte *f*

**shoring** *n* ARCH & BUILD Abstützung *f*, Hilfsgerüst *nt*, Hilfsrüstung *f*

**short**: **~ ramp** *n* ARCH & BUILD, INFR & DES Anlauf *m*, Rampe *f*; **~-range aggregate concrete** *n* CONCR Einkornbeton *m*; **~-term loading** *n* CONCR, MAT PROP Kurzzeitbelastung *f*

**shot**: ~ **blasting** *n* INFR & DES Strahlreinigen *nt*
**shotcrete** *n* CONCR Spritzbeton *m*
**shoulder** *n* ARCH & BUILD Absatz *m*
**shouldered**: ~ **tenon** *n* TIMBER Brustzapfen *m*
**shovel** *n* BUILD MACHIN *excavator* Schaufel *f*, Löffel *m*; ~ **loader** *n* BUILD MACHIN Schaufellader *m*; ~ **work** *n* INFR & DES Handschachten *nt*, Schaufeln *nt*
**shower** *n* WASTE WATER Brause *f*, Dusche *f*, *rain* Regenschauer *m*; ~ **head** *n* WASTE WATER Duschkopf *m*; ~ **installation** *n* WASTE WATER Duschanlage *f*; ~ **outlet** *n* WASTE WATER Brauseauslauf *m*; ~ **receptor** *n* WASTE WATER Duschwanne *f*; ~ **of sparks** *n* INFR & DES Funkenregen *m*; ~ **tray** *n* WASTE WATER Duschwanne *f*; ~ **tub** *n* WASTE WATER Duschwanne *f*
**shredded**: ~ **refuse landfill** *n* ENVIRON Shredder-Abfälledeponie *f*
**shrink** *vi* MAT PROP schrumpfen; ~ **on** *vi* MAT PROP aufschrumpfen
**shrinkage** *n* ARCH & BUILD, CONCR, INFR & DES, MAT PROP Schrumpfen *nt*, Schwinden *nt*; ~ **crack** *n* ARCH & BUILD, CONCR Schwindriß *m*; ~ **deformation** *n* ARCH & BUILD, CONCR, MAT PROP, STEEL Schwindverformung *f*; ~ **effect** *n* ARCH & BUILD, CONCR, MAT PROP, STEEL Schwindeinfluß *m*; ~ **in cement** *n* CONCR Zementschwinden *nt*; ~ **joint** *n* ARCH & BUILD, CONCR, INFR & DES Schwindfuge *f*, Kontraktionsfuge *f*
**shrinking** *n* ARCH & BUILD, CONCR, MAT PROP Schrumpfen *nt*, Schwinden *nt*
**shrink**: ~**-on bushing** *n* WASTE WATER Schrumpfabschottung *f*; ~**-on collar** *n* WASTE WATER Schrumpfmanschette *f*; ~**-on sleeve** *n* WASTE WATER Schrumpfmuffe *f*
**shrinkproof** *adj* MAT PROP schwindbeständig
**shrunk**: ~**-on** *adj* MAT PROP aufgeschrumpft; ~**-on sleeve** *n* WASTE WATER Schrumpfmanschette *f*
**shunting**: ~ **engine** *n* *BrE* (*cf switch engine AmE*) BUILD MACHIN Rangierlok *f*; ~ **winch** *n* *BrE* (*cf switching winch AmE*) BUILD MACHIN Rangierwinde *f*
**shut** *vt* HEAT & VENT absperren, schließen; ~ **down** *vt* CONST LAW, INFR & DES stillegen; ~ **off** *vt* WASTE WATER absperren
**shutdown** *n* CONST LAW, INFR & DES Stillegung *f*
**shut**: ~**-off valve** *n* HEAT & VENT, WASTE WATER Absperrklappe *f*
**shutter 1.** *n* ARCH & BUILD *sealing device* beweglicher Abschluß *m*, *for window* Fensterladen *m*, *for window, door* Rolladen *m*, BUILD HARDW *doorbolt* Riegel *m*, *for window, door* Rolladen *m*; **2.** *vt* CONCR einschalen
***shutter***: ~ **cabinet** *n* TIMBER Rollschrank *m*
**shuttering** *n* CONCR Deckenschalung *f*, HEAT & VENT Schalung *f*, INFR & DES, TIMBER Deckenschalung *f*; ~ **panel** *n* STEEL Schalungstafel *f*; ~ **removal** *n* CONCR Ausschalen *nt*, Entschalen *nt*
**shutting**: ~ **off** *n* WASTE WATER Absperren *nt*; ~ **post** *n* INFR & DES Schließpfosten *m*
**SI** *abbr* (*Système International*) ARCH & BUILD, INFR & DES, MAT PROP SI (*Système International*)
**sick**: ~ **building syndrome** *n* CONST LAW, ENVIRON, INFR & DES krankes Gebäudesyndrom *nt*
**side** *n* ARCH & BUILD Schenkel *m*, Seite *f*, Seitenfläche *f*; ~ **board** *n* ARCH & BUILD Stirnbrett *nt*, Windbrett *nt*; ~ **face** *n* ARCH & BUILD Seitenansicht *f*
**sidehill**: ~ **cut** *n* INFR & DES Hangeinschnitt *m*
***side***: ~**-hung window** *n* ARCH & BUILD, BUILD HARDW Drehfenster *nt*, Drehflügelfenster *nt*; ~ **opposed to the weather** *n* ARCH & BUILD Wetterseite *f*; ~ **plate** *n* ARCH & BUILD Wange *f*, MAT PROP Seitenblech *nt*, Wange *f*; ~ **post** *n* ARCH & BUILD Kehlbalkenstütze *f*; ~ **pulley** *n* ARCH & BUILD, BUILD MACHIN Seitenwinde *f*; ~ **rabbet plane** *n* BUILD MACHIN Simshobel *m*; ~ **rail** *n* INFR & DES Leitplanke *f*; ~ **room** *n* ARCH & BUILD Nebenraum *m*; ~ **street** *n* INFR & DES Seitenstraße *f*; ~ **track** *n* BUILD HARDW *blind* Seitenführung *f*, Seitenführungsschiene *f*
**sidewalk** *n* *AmE* (*cf pavement BrE*) INFR & DES Bürgersteig *m*, Fußgängerweg *m*, Gehweg *m*; ~ **bed** *n* *AmE* INFR & DES Wegekoffer *m*; ~ **design** *n* *AmE* INFR & DES *roads* Deckenbemessung *f*; ~ **paving flag** *n* *AmE* (*cf pavement paving flag BrE*) INFR & DES Gehwegplatte *f*, MAT PROP Gehwegplatte *f*, Pflasterbelag *m*, Straßenpflaster *nt*; ~**-quality concrete** *n* *AmE* CONCR Straßendeckenbeton *m*; ~ **surface evenness** *n* *AmE* INFR & DES Ebenheit *f* der Straßendecke
**siding** *n* INFR & DES Anschlußgleis *nt*, Gleisanschluß *m*
**sieve** *n* BUILD HARDW Sieb *nt*; ~ **analysis** *n* MAT PROP Siebanalyse *f*; ~ **bottom** *n* BUILD MACHIN Siebboden *m*; ~ **fraction** *n* MAT PROP Kornfraktion *f*; ~ **refuse** *n* MAT PROP Siebschutt *m*
**sieving** *n* BUILD HARDW Sieben *nt*, ENVIRON Siebung *f*
**sight** *n* BUILD MACHIN, INFR & DES *surveying* Einstellung *f*, Visiereinrichtung *f*; ~ **check** *n* ARCH & BUILD, INFR & DES Sichtkontrolle *f*; ~ **distance** *n* ARCH & BUILD Sichtweite *f*
**sighted**: ~ **alidade** *n* ARCH & BUILD anvisierte Dioptrie *f*; ~ **level** *n* ARCH & BUILD anvisierte Höhe *f*
**sighting** *n* ARCH & BUILD Anvisieren *nt*, Zielen *nt*, INFR & DES Visieren *nt*; ~ **rod** *n* BUILD HARDW *surveying*, BUILD MACHIN, INFR & DES Nivellierlatte *f*; ~ **telescope** *n* ARCH & BUILD, BUILD MACHIN Visierfernrohr *nt*, Zielfernrohr *nt*
***sight***: ~ **rail** *n* ARCH & BUILD, BUILD MACHIN, INFR & DES Schnurbretter *nt pl*, Schnurgerüst *nt*, Visiergerüst *nt*; ~ **vane** *n* ARCH & BUILD, INFR & DES Kompaßdiopter *m*
**sign** *n* BUILD HARDW Schild *nt*
**signing**: ~ **of a contract** *n* CONST LAW Vertragsunterzeichnung *f*
**silencer** *n* *BrE* (*cf muffler AmE*) SOUND & THERMAL Schalldämpfer *m*
**silica** *n* INFR & DES Kieselerde *f*, Quarzsand *m*, MAT PROP Kieselerde *f*, Siliziumoxid *nt*, SOUND & THERMAL Quarzsand *m*
**silicate** *n* MAT PROP Silikat *nt*; ~ **glass** *n* MAT PROP Silikatglas *nt*; ~ **paint** *n* SURFACE Wasserglasfarbe *f*, S-Farbe *f*, Silikatfarbe *f*
**siliceous**: ~ **sand** *n* MAT PROP Quarzsand *m*
**silicon**: ~ **cell** *n* ENVIRON Silikonzelle *f*
**silicone** *n* MAT PROP Silikon *nt*
**sill** *n* ARCH & BUILD Schwelle *f*, Sohlbank *f*, Grundschwelle *f*, Türschwelle *f*, Unterzug *m*, *window* Fensterbank *f*; ~ **of framework** *n* ARCH & BUILD Fachwerkschwelle *f*; ~ **plate** *n* ARCH & BUILD Grundschwelle *f*, Schwellholz *nt*
**silo** *n* ARCH & BUILD Speicher *m*, BUILD MACHIN Silo *m*
**silt** *n* HEAT & VENT, INFR & DES, WASTE WATER Schlick *m*; ~ **box** *n* WASTE WATER Schlammeimer *m*, Schlammfang *m*; ~ **container** *n* ENVIRON Schlammsammelbehälter *m*; ~ **content** *n* MAT PROP Schluffgehalt *m*; ~ **layer** *n* INFR & DES Schluffschicht *f*
**silty**: ~ **soil** *n* INFR & DES schluffiger Boden *m*

**similar** *adj* INFR & DES ähnlich
**simple**: ~ **beam** *n* ARCH & BUILD, CONCR, STEEL Einfeldträger *m*, Träger *m* auf zwei Stützen; ~ **bridge** *n* INFR & DES Einfeldbrücke *f*; ~ **support** *n* ARCH & BUILD einfaches Auflager *nt*, einfache Auflagerung *f*
**simplification** *n* ARCH & BUILD, CONCR, INFR & DES, MAT PROP, STEEL Vereinfachung *f*
**simply**: ~ **supported** *adj* ARCH & BUILD frei aufgelagert, frei aufliegend; ~-**supported beam** *n* ARCH & BUILD, CONCR, STEEL Einfeldträger *m*, Träger *m* auf zwei Stützen
**simultaneous** *adj* BUILD MACHIN gleichzeitig
**single** *adj* ARCH & BUILD, CONCR einzeln; ~-**beam bridge crane** *n* BUILD MACHIN Einträgerlaufkran *m*; ~-**beam travel crane** *n* BUILD MACHIN Einträgerlaufkran *m*; ~ **bituminous surface treatment** *n* INFR & DES einfache bituminöse Oberflächenbehandlung *f*; ~-**branch pipe** *n* WASTE WATER Einfachabzweig *m*; ~-**coat** *adj* STONE, SURFACE einlagig; ~ **duct** *n* WASTE WATER einzelne Rohrleitung *f*; ~-**floor** *adj* ARCH & BUILD eingeschossig; ~ **floor** *n* ARCH & BUILD, TIMBER Balkendecke *f*, Balkenträgerdecke *f*; ~-**flow pump** *n* WASTE WATER einflutige Pumpe *f*; ~ **footing** *n* ARCH & BUILD, INFR & DES Blockfundament *nt*; ~-**grained structure** *n* MAT PROP Einkornstruktur *f*; ~-**latch bolt lock** *n* BUILD HARDW Einfallenschloß *nt*; ~-**layer** *adj* ARCH & BUILD, INFR & DES einlagig, einschichtig; ~-**leaf** *adj* ARCH & BUILD *wall* einschalig, einflügelig; ~-**leaf door** *n* ARCH & BUILD einflügelige Tür *f*; ~-**lever mixer** *n* WASTE WATER Einhebelbatterie *f*; ~-**lever mixing valve** *n* WASTE WATER Einhebelbatterie *f*; ~ **load** *n* ARCH & BUILD Einzellast *f*; ~-**pass stabilizer** *n* BUILD MACHIN Eingangmischer *m*; ~-**phase** *adj* ELECTR Einphasen-, einphasig; ~-**phase electric current** *n* ELECTR Einphasenstrom *m*; ~ **pitch roof** *n* ARCH & BUILD freitragendes Pultdach *nt*; ~-**point earthing** *n BrE* (*cf single-point grounding AmE*) ELECTR Punkterdung *f*; ~-**point grounding** *n AmE* (*cf single-point earthing BrE*) ELECTR Punkterdung *f*; ~-**post purlin roof** *n* ARCH & BUILD, TIMBER einstieliges strebenloses Pfettendach *nt*; ~ **rafter roof** *n* ARCH & BUILD, TIMBER einfaches Sparrendach *nt*; ~-**riveted joint** *n* STEEL einreihige Nietverbindung *f*; ~-**riveted lap joint** *n* STEEL einreihige Nietüberlappung *f*; ~ **roller** *n* BUILD MACHIN Einradwalze *f*; ~-**row** *adj* ARCH & BUILD einreihig; ~ **shear connection** *n* ARCH & BUILD, INFR & DES einschnittige Verbindung *f*; ~ **shear joint** *n* ARCH & BUILD, INFR & DES einschnittiges Gelenk *nt*; ~-**side formwork** *n* CONCR einhäuptige Schalung *f*; ~-**sized concrete** *n* CONCR Einkornbeton *m*; ~-**size gravel aggregate** *n* MAT PROP gleichkörniger Kieszuschlagstoff *m*; ~-**span beam** *n* ARCH & BUILD, CONCR, STEEL Einfeldträger *m*; ~-**storey** *adj BrE* ARCH & BUILD eingeschossig; ~-**story** *AmE see single-storey BrE*
**sink 1.** *n* INFR & DES Senke *f*, WASTE WATER Ausguß *m*, Spüle *f*; **2.** *vi* INFR & DES sinken
**sinking** *n* INFR & DES *soil* Senkung *f*
***sink***: ~ **unit** *n* WASTE WATER Spültisch *m*
**siphon 1.** *n* ENVIRON, WASTE WATER Flüssigkeitsheber *m*, Siphon *m*; **2.** *vt* ENVIRON, MAT PROP hebern
***siphon***: ~ **crest** *n* ENVIRON Siphonhöhe *f*; ~ **spillway** *n* ENVIRON Siphonüberlauf *m*
**site** *n* (*cf job site AmE*) ARCH & BUILD *location* Lage *f*, INFR & DES *building site* Bauplatz *m*, *location* Standort *m*, *piece of land* Gelände *nt*
**site**: **on** ~ *adv* ARCH & BUILD auf der Baustelle
***site***: ~ **accommodation** *n* ARCH & BUILD Bauunterkunft *f*; ~-**assembled** *adj* ARCH & BUILD, INFR & DES, STEEL baustellenmontiert; ~ **concrete** *n* ARCH & BUILD, CONCR Ortbeton *m*; ~ **criteria** *n pl* ENVIRON Standortkriterien *nt pl*; ~ **development** *n* INFR & DES Erschließung *f*; ~ **engineer** *n* ARCH & BUILD Bauleiter *m*; ~ **exploration** *n* ARCH & BUILD, INFR & DES Geländeerkundung *f*; ~ **facilities** *n pl* ARCH & BUILD Baustelleneinrichtung *f*; ~ **facilities program** *AmE*, ~ **facilities programme** *BrE nt* ARCH & BUILD Einrichtungsplan *m*; ~ **fence** *n* INFR & DES Bauzaun *m*; ~ **hut** *n* ARCH & BUILD Baustellenbaracke *f*, Bauhütte *f*; ~ **inspection** *n* ARCH & BUILD Begehung *f* einer Baustelle; ~ **installations** *n pl BrE* (*cf job site installations AmE*) ARCH & BUILD Baustelleneinrichtung *f*; ~ **measuring** *n* ARCH & BUILD, INFR & DES Aufmaß *nt*; ~ **meeting** *n* ARCH & BUILD Baustellenbesprechung *f*; ~ **mobilization** *n BrE* (*cf job site mobilization AmE*) ARCH & BUILD Baustelleneinrichtung *f*; ~ **mobilization plan** *n BrE* (*cf job site mobilization plan AmE*) ARCH & BUILD Baustelleneinrichtungsplan *m*; ~ **office** *n BrE* (*cf job site office AmE*) ARCH & BUILD, INFR & DES Baubüro *nt*, Baustellenbüro *nt*; ~ **plan** *n* ARCH & BUILD, CONST LAW, INFR & DES Lageplan *m*; ~ **progress photographs** *n pl* CONST LAW Baustellenfortschrittsfotografien *f pl*; ~ **road** *n* INFR & DES Baustellenstraße *f*; ~ **sampling** *n* ARCH & BUILD, INFR & DES Baugeländebodenprobenentnahme *f*; ~ **supervision** *n* INFR & DES örtliche Bauleitung *f*; ~ **survey** *n* ARCH & BUILD, CONST LAW, INFR & DES Ortsbesichtigung *f*; ~ **survey plan** *n* INFR & DES Geländeaufnahmeplan *m*; ~ **toilet** *n* WASTE WATER Bauabort *m*; ~-**welded** *adj* STEEL baustellengeschweißt
**siting** *n* ARCH & BUILD, INFR & DES Standortbestimmung *f*, Standortwahl *f*
***SI***: ~ **unit** *n* ARCH & BUILD, INFR & DES, MAT PROP SI-Einheit *f*
**size** *n* ARCH & BUILD Format *nt*, Größe *f*, Maß *nt*; ~ **of bore** *n* STEEL, TIMBER Bohrdurchmesser *m*; ~ **range** *n* INFR & DES Körnung *f*, MAT PROP Körnung *f*, Korngruppe *f*; ~ **when completed** *n* ARCH & BUILD Ausbaumaß *nt*
**sizing**: ~ **and washing machine** *n* BUILD MACHIN *for building materials* Klassier- und Waschmaschine *f*
**skeleton** *n* ARCH & BUILD Gebäudeskelett *nt*, Gerippe *nt*, Rohbauskelett *nt*, Skelett *nt*, CONCR Gerippe *nt*; ~ **construction** *n* ARCH & BUILD Skelettbauweise *f*; ~ **framing** *n* ARCH & BUILD, INFR & DES Skelett *nt*; ~ **girder** *n* ARCH & BUILD Skeletträger *m*
**skene**: ~ **arch** *n* ARCH & BUILD verkürzter Bogen *m*
**sketch** *n* ARCH & BUILD Skizze *f*
**skew** *n* ARCH & BUILD Giebelfußstein *m*; ~ **bridge** *n* INFR & DES schiefe Brücke *f*
**skewing** *n* INFR & DES Schrägstellung *f*
**skidproof** *adj* MAT PROP gleitsicher, rutschfest, trittsicher, SURFACE rutschfest
**skid**: ~ **track** *n* INFR & DES Hemmschiene *f*
**skill** *n* CONST LAW Fertigkeit *f*
**skilled**: ~ **worker** *n* ARCH & BUILD Facharbeiter *m*
**skim** *vt* SURFACE *remove* abtragen, *smooth* glätten; ~ **off** *vt* INFR & DES abschöpfen

**skimming**: ~ **tank** *n* ENVIRON, WASTE WATER Fettabscheider *m*
**skim**: ~ **off** *phr* ENVIRON abschöpfen
**skin** *n* SOUND & THERMAL Haut *f*
**skip** *n* BUILD MACHIN Kipper *m*, ENVIRON Abfuhrwagen *m*, Kipper *m*; ~ **lorry** *n BrE* (*cf skip truck AmE*) BUILD MACHIN Muldenkipper *m*, Muldentransporter *m*; ~ **truck** *n AmE* (*cf skip lorry BrE*) BUILD MACHIN Muldenkipper *m*, Muldentransporter *m*
**skirt** *n* TIMBER Kante *f*, Rand *m*
**skirting** *n BrE* ARCH & BUILD, BUILD HARDW, TIMBER Einfassung *f*, Fußleiste *f*; ~ **board** *n BrE* (*cf baseboard AmE*) ARCH & BUILD, BUILD HARDW, TIMBER Fußleiste *f*, Scheuerleiste *f*, Sockelleiste *f*
**skylight** *n* ARCH & BUILD Oberlicht *nt*, Laternenoberlicht *nt*, Dachaufsatz *m*, Dachfenster *nt*, Dachluke *f*, BUILD HARDW Dachfenster *nt*; ~ **turret** *n* ARCH & BUILD Dachlukenaufsatz *m*, Laterne *f*
**skyscraper** *n* ARCH & BUILD, INFR & DES Hochhaus *nt*, Wolkenkratzer *m*
**slab** *n* MAT PROP Platte *f*; ~**-and-beam ceiling** *n* ARCH & BUILD Plattenbalkendecke *f*; ~ **bridge** *n* INFR & DES Plattenbrücke *f*; ~ **footing** *n* ARCH & BUILD, CONCR, INFR & DES Plattenfundament *nt*; ~ **foundation** *n* ARCH & BUILD, CONCR, INFR & DES Plattenfundament *nt*; ~ **lining** *n* ARCH & BUILD Plattenverkleidung *f*; ~**-on-grade** *n* ARCH & BUILD nichtunterkellerte Fußbodenplatte *f*; ~ **partition** *n* ARCH & BUILD Plattentrennwand *f*, Plattenwand *f*; ~ **wall** *n* ARCH & BUILD Plattentrennwand *f*, Plattenwand *f*
**slag** *n* ENVIRON, MAT PROP Müllschlacke *f*; ~ **cement** *n* CONCR Hochofenzement *m*, Portlandzement *m*, MAT PROP Hochofenzement *m*, Schlackenzement *m*; ~ **concrete** *n* CONCR Schlackenbeton *m*; ~ **wool** *n* MAT PROP Schlackenwolle *f*, Hüttenwolle *f*
**slant** *n* ARCH & BUILD, INFR & DES Gefälle *nt*, Neigung *f*
**slanted** *adj* ARCH & BUILD schräg
**slanting** *adj* ARCH & BUILD schief, schräg
**slate** *n* BUILD MACHIN Schiefer *m*, Schieferplatte *f*, INFR & DES, MAT PROP Schiefer *m*; ~ **ax** *AmE*, ~ **axe** *BrE n* BUILD MACHIN Dachhammer *m*; ~ **clay** *n* MAT PROP Schieferton *m*; ~ **hammer** *n* BUILD MACHIN Dachhammer *m*; ~ **nail** *n* BUILD MACHIN Schiefernagel *m*
**slater** *n* ARCH & BUILD Dachdecker *m*
***slate***: ~ **roof cladding** *n* BUILD MACHIN Schieferbedachung *f*
**slatted**: ~ **curtain** *n* BUILD HARDW Lamellenvorhang *m*
**sledge** *vt* BUILD MACHIN hämmern
**sledgehammer** *n* BUILD MACHIN Pflasterhammer *m*, Vorschlaghammer *m*
**sledging** *n* BUILD MACHIN Hämmern *nt*
**sleeper** *n* ARCH & BUILD, CONCR, TIMBER Schwelle *f*, Unterzug *m*; ~**-carrying girder** *n* ARCH & BUILD Schwellenträger *m*
**sleeve** *n* HEAT & VENT, WASTE WATER Muffe *f*, Buchse *f*, Hülse *f*
**sleeveless** *adj* HEAT & VENT muffenlos
**slenderness** *n* ARCH & BUILD, INFR & DES Schlankheit *f*; ~ **ratio** *n* ARCH & BUILD, INFR & DES Schlankheitsgrad *m*
**slewing** *n BrE* BUILD MACHIN *crane* Drehen *nt*, Schwenken *nt*; ~ **crane** *n BrE* ARCH & BUILD Schwenkkran *m*, BUILD MACHIN Drehkran *m*, Schwenkkran *m*
**sliced**: ~ **veneer** *n* TIMBER Messerfurnier *nt*
**slice**: ~ **method** *n* INFR & DES Lamellenverfahren *nt*
**slidable**: ~ **lattice grate** *n* BUILD MACHIN Scherengitter *nt*
**slide**: ~ **gate** *n* ARCH & BUILD Schiebetor *nt*; ~**-in unit** *n* ELECTR Einschub *m*; ~ **ring packing** *n* WASTE WATER *pump* Gleitringdichtung *f*; ~ **ring sealing** *n* HEAT & VENT, WASTE WATER *pump* Gleitringdichtung *f*; ~ **valve** *n* WASTE WATER Schieber *m*
**slideway** *n* STEEL Gleitschiene *f*
**sliding 1.** *adj* ARCH & BUILD, INFR & DES, MAT PROP gleitend; **2.** *n* INFR & DES Geländebruch *m*
***sliding***: ~ **bearing** *n* BUILD MACHIN Gleitlager *nt*; ~ **door** *n* ARCH & BUILD Schiebetür *f*; ~ **folding door** *n* ARCH & BUILD, BUILD HARDW Harmonikatür *f*; ~ **formwork** *n* ARCH & BUILD, INFR & DES Gleitschalung *f*; ~ **friction** *n* MAT PROP Gleitreibung *f*; ~ **gate** *n* ARCH & BUILD Rolltor *nt*, Schiebetor *nt*; ~ **hatch** *n* ARCH & BUILD Schiebeluke *f*, Schiebefenster *nt*; ~ **joint** *n* WASTE WATER *pipe* Gleitfuge *f*; ~ **motion** *n* INFR & DES Gleitbewegung *f*; ~ **resistance** *n* INFR & DES Gleitwiderstand *m*, MAT PROP Gleitreibung *f*; ~ **sash** *n* ARCH & BUILD Schiebefenster *nt*; ~ **shuttering** *n* ARCH & BUILD Gleitschalung *f*; ~ **sluice** *n* ENVIRON Ziehschütze *f*; ~ **surface** *n* ARCH & BUILD Gleitfläche *f*; ~ **window** *n* ARCH & BUILD Schiebefenster *nt*
**slip** *vi* INFR & DES gleiten
**slip**: ~ **circle** *n* INFR & DES *soil mechanics* Bruchkreis *m*, Gleitkreis *m*; ~ **joint** *n* INFR & DES Gleitfuge *f*
**slippage** *n* ARCH & BUILD Rutschen *nt*, Schlupf *m*
***slip***: ~ **partition** *n* ARCH & BUILD Harmonikatrennwand *f*
**slippery**: ~ **flooring** *n* CONST LAW rutschiger Boden *m*
***slip***: ~ **plane** *n* INFR & DES Gleitfläche *f*; ~ **road** *n* ARCH & BUILD, INFR & DES Autobahnzubringer *m*, Zufahrtsstraße *f*; ~ **tongue joint** *n* ARCH & BUILD Federverbindung *f*
**slit**: ~ **box** *n* INFR & DES Schlammeimer *m*
**slope** *n* ARCH & BUILD *incline* Neigung *f*, Gefälle *nt*, INFR & DES Querneigung *f*, Steigung *f*, Böschung *f*, Abhang *m*, *pitch* Neigung *f*, Gefälle *nt*, Hang *m*; ~ **angle** *n* INFR & DES Böschungswinkel *m*; ~ **circle** *n* INFR & DES Böschungskreis *m*; ~ **compaction** *n* INFR & DES Böschungsverdichtung *f*
**sloped** *adj* ARCH & BUILD *building, area* geneigt, INFR & DES geböscht
***slope***: ~ **failure** *n* INFR & DES Böschungsbruch *m*; ~ **landfill** *n* ENVIRON Anböschung *f*; ~ **level** *n* INFR & DES Neigungsmesser *m*; ~ **line** *n* INFR & DES Fallinie *f*; ~ **method** *n* ENVIRON Anböschung *f*; ~ **mower** *n* BUILD MACHIN Böschungsmähgerät *nt*; ~ **protection** *n* CONST LAW, INFR & DES Böschungsschutz *m*, Böschungssicherung *f*; ~ **stability** *n* INFR & DES Böschungsstabilität *f*
**sloping** *adj* ARCH & BUILD *building, area* geneigt; ~ **arch** *n* INFR & DES abfallender Bogen *m*; ~ **concrete** *n* CONCR Gefällebeton *m*; ~ **screed** *n* STONE Gefälleestrich *m*
**slop**: ~ **tank** *n* ENVIRON Altöltank *m*
**slot** *n* ARCH & BUILD *groove* Kerbe *f*, Nut *f*, *slit* Schlitz *m*, Spalt *m*; ~ **mortise joint** *n* INFR & DES Zapfenlochverbindung *f*; ~ **pipe** *n* WASTE WATER Schlitzrohr *nt*
**slotted** *adj* ARCH & BUILD geschlitzt; ~ **fillister head screw** *n* BUILD HARDW Linsenschraube *f*; ~ **head**

**locking device** *n* BUILD HARDW Knebelverschluß *m*; **~ rivet** *n* BUILD HARDW Schlitzniet *m*

**slude**: **~ dewatering** *n* INFR & DES Schlammentwässerung *f*

**sludge** *n* ENVIRON, HEAT & VENT, INFR & DES, WASTE WATER Schlamm *m*, Schlick *m*; **~ accumulation** *n* ENVIRON Verschlammung *f*; **~ bed** *n* INFR & DES Schlammbett *nt*; **~ cake** *n* ENVIRON Schlammkuchen *m*; **~ composting** *n* ENVIRON, INFR & DES Schlammkompostierung *f*; **~ contact process** *n* ENVIRON Kontaktschlammverfahren *nt*; **~ dewatering** *n* ENVIRON Schlammentwässerung *f*, Schlammverdickung *f*; **~ digestion** *n* ENVIRON Schlammfaulung *f*; **~ digestion chamber** *n* INFR & DES, WASTE WATER Schlammfaulraum *m*; **~ digestion tank** *n* ENVIRON Schlammfaulbehälter *m*, INFR & DES Ausfaulgrube *f*; **~ drying** *n* ENVIRON Schlammtrocknung *f*; **~ drying bed** *n* ENVIRON Schlammtrockenbett *nt*; **~ petrification** *n* ENVIRON Versteinerung *f* von Schlämmen; **~ processing** *n* ENVIRON, INFR & DES Schlammaufbereitung *f*, Schlammbehandlung *f*; **~ rake** *n* ENVIRON Schlammräumer *m*; **~ removal** *n* ENVIRON Schlammbeseitigung *f*; **~ stabilization** *n* ENVIRON, INFR & DES Schlammstabilisierung *f*; **~ sump** *n* ENVIRON Schlammfang *m*, Schlammsammelbehälter *m*; **~ thickening** *n* ENVIRON, INFR & DES Schlammeindickung *f*

**sluice** *n* ENVIRON Schleuse *f*

**sluicegate** *n* ENVIRON oberes Schleusentor *nt*

**sluing** *AmE see slewing BrE*

**slump** *n* CONCR, INFR & DES *concrete testing* Ausbreitungsmaß *nt*, Ausbreitmaß *nt*, MAT PROP Ausbreitmaß *nt*, Setzmaß *nt*, Ausbreitungsmaß *nt*, Senkung *f*, Setz-, Absackung *f*; **~ cone** *n* MAT PROP Setzbecher *m*; **~ test** *n* CONCR, MAT PROP Ausbreitmaßprüfung *f*, Setzprobe *f*

**slurry** *n* ENVIRON Brei *m*, Flüssigschlamm *m*, Naßschlamm *m*, Schlamm *m*, *waste* Flüssigmist *m*, Gülle *f*, Jauche *f*, INFR & DES Flüssigschlamm *m*, MAT PROP *cement* Zementmilch *f*, Rohschlamm *m*; **~ seal** *n* INFR & DES Schlammversiegelung *f*; **~ trenching** *n* ENVIRON, INFR & DES Schlitzwandverfahren *nt*; **~ wall** *n* ENVIRON Dichtungswand *f*, Sperrwand *f*, Trennwand *f*, INFR & DES, SOUND & THERMAL, TIMBER Dichtungswand *f*

**small**: **~ cobble** *n* MAT PROP, STONE Kleinpflasterstein *m*; **~ cobbles** *n pl* MAT PROP Rollkies *m*

**smelter** *n* MAT PROP, STEEL Hütte *f*

**smeltery** *n* MAT PROP, STEEL Hütte *f*

**smithery** *n* STEEL Schmiedearbeit *f*, Schmiedehandwerk *nt*

**smith's**: **~ pliers** *n pl* BUILD MACHIN Schmiedezange *f*

**smithy** *n* STEEL Schmiede *f*

**smog** *n* ENVIRON Smog *m*

**smoke** *n* CONST LAW, HEAT & VENT Rauch *m*; **~ alarm** *n* BUILD HARDW, CONST LAW Rauchmelder *m*; **~ detector** *n* BUILD HARDW, CONST LAW Rauchmelder *m*; **~ flue** *n* ARCH & BUILD Fuchskanal *m*; **~ funnel** *n* ARCH & BUILD Rauchabzug *m*; **~ outlet** *n* ARCH & BUILD, CONST LAW Rauchabzug *m*, Rauchöffnung *f*; **~ pipe** *n* ARCH & BUILD, CONST LAW Rauchrohr *nt*; **~ test** *n* CONST LAW Rauchtest *m*; **~-tight** *adj* CONST LAW rauchdicht; **~-tight door** *n* ARCH & BUILD, BUILD HARDW, CONST LAW rauchdichte Tür *f*; **~ tube** *n* ARCH & BUILD Rauchrohr *nt*

**smooth** *adj* ARCH & BUILD, INFR & DES, MAT PROP glatt, eben; **~ formwork** *n* CONCR glatte Schalung *f*

**smoothing**: **~ board** *n* STONE Glättbohle *f*; **~ trowel** *n* BUILD MACHIN, STONE Glättkelle *f*

**smooth**: **~ roller** *n* BUILD MACHIN, INFR & DES Glattmantelwalze *f*, Stahlmantelwalze *f*; **~-surface roofing felt** *n* MAT PROP, SOUND & THERMAL unbesandete Pappe *f*; **~-surface roofing paper** *n* MAT PROP, SOUND & THERMAL unbesandete Pappe *f*

**snagging**: **~ list** *n* CONST LAW Restarbeitenliste *f*

**snap**: **~ header** *n* STONE Halbziegel *m*

**snow** *n* WASTE WATER Schnee *m*; **~ cover** *n* INFR & DES Schneedecke *f*; **~ guard** *n* ARCH & BUILD *roof* Schneefang *m*

**snowload** *n* ARCH & BUILD Schneelast *f*

***snow***: **~ plough** *n BrE* BUILD MACHIN Schneepflug *m*; **~ plow** *AmE see snow plough BrE*

**soakaway** *n* INFR & DES, WASTE WATER Sickergrube *f*, Sickerschacht *m*

**soaker** *n* STEEL Wandanschlußblech *nt*

**soap**: **~ dish** *n* BUILD HARDW Seifenschüssel *f*; **~ dispenser** *n* BUILD HARDW, WASTE WATER Seifenspender *m*

**society** *n* CONST LAW Gesellschaft *f*

**socket** *n* ELECTR, HEAT & VENT Buchse *f*, Muffe *f*, Steckdose *f*; **~ pipe** *n* HEAT & VENT, WASTE WATER Muffenrohr *nt*

**socle** *n* ARCH & BUILD *of monument, statue* Sockel *m*

**sod**: **~ square** *n* INFR & DES Rasenziegel *m*

**soffit** *n* ARCH & BUILD Untersicht *f*, *arch, vault* Leibung *f*

**soft** *adj* MAT PROP, TIMBER weich

**softboard** *n* TIMBER Weichfaserplatte *f*

**soft**: **~ clay** *n* INFR & DES weicher Ton *m*

**softened**: **~ water** *n* INFR & DES, WASTE WATER enthärtetes Wasser *nt*

**softening**: **~ plant** *n* INFR & DES, WASTE WATER Enthärteranlage *f*, Enthärtungsanlage *f*; **~ unit** *n* WASTE WATER Enthärtungsanlage *f*

**soft**: **~ solder** *n* MAT PROP Weichlot *nt*; **~ soldering** *n* MAT PROP Weichlöten *nt*

**softwood** *n* MAT PROP, TIMBER Nadelholz *nt*, Weichholz *nt*

**soil** *n* ARCH & BUILD, INFR & DES Boden *m*, Erdreich *nt*, Erdboden *m*, Erde *f*, Grund *m*, STEEL Erde *f*, Grund *m*; **~ aeration** *n* INFR & DES Bodenbelüftung *f*; **~ analysis** *n* INFR & DES Bodenanalyse *f*; **~ atmosphere concentration** *n* ENVIRON Bodenluftkonzentration *f*; **~ behavior** *AmE*, **~ behaviour** *BrE n* MAT PROP Bodenverhalten *nt*; **~ cementation** *n* INFR & DES, MAT PROP Bodenstabilisierung *f*; **~ character** *n* MAT PROP Bodeneigenschaften *f pl*; **~ characteristics** *n pl* MAT PROP Bodeneigenschaften *f pl*; **~ chemistry** *n* MAT PROP Bodenchemie *f*; **~ class** *n* INFR & DES Bodenklasse *f*; **~ classification** *n* MAT PROP Bodeneinstufung *f*; **~ compaction** *n* INFR & DES Bodenverdichtung *f*, Bodenverfestigung *f*; **~ composition** *n* ENVIRON Bodenbeschaffenheit *f*, Bodenzusammensetzung *f*; **~ conditions** *n pl* INFR & DES Bodenverhältnisse *nt pl*; **~ cover** *n* INFR & DES Erdandeckung *f*; **~ covering plants** *n pl* INFR & DES Bodendeckung *f*; **~ densification** *n* INFR & DES Bodenverdichtung *f*, Bodenverfestigung *f*; **~ displacement** *n* INFR & DES Bodenverdrängung *f*, Lösen *nt* und Fördern *nt* von Boden; **~ engineering**

*n* ARCH & BUILD Erdbau *m*, INFR & DES Erdbau *m*, Grundbau *m*; ~ **erosion** *n* INFR & DES Bodenabtrag *m*; ~ **examination** *n* ENVIRON, INFR & DES Baugrunduntersuchung *f*, Bodenuntersuchung *f*; ~ **filter** *n* ENVIRON Bodenfilter *nt*; ~ **humidity** *n* MAT PROP Bodenfeuchtigkeit *f*; ~ **irrigation** *n* INFR & DES Bodenbewässerung *f*; ~ **mechanics** *n pl* INFR & DES Bodenmechanik *f*; ~ **parameter** *n* INFR & DES Bodenparameter *m*; ~ **particle** *n* MAT PROP Bodenteilchen *nt*; ~ **placement** *n* INFR & DES Bodeneinbau *m*; ~ **pollutant** *n* ENVIRON bodenverschmutzender Stoff *m*; ~ **pollution** *n* CONST LAW, ENVIRON, INFR & DES Bodenverschmutzung *f*, Bodenverunreinigung *f*; ~ **pressure** *n* INFR & DES Bodendruck *m*, Bodenpressung *f*; ~ **sample** *n* ENVIRON, INFR & DES Bodenprobe *f*; ~ **stabilization** *n* INFR & DES, MAT PROP Bodenvermörtelung *f*; ~ **stratum** *n* INFR & DES Bodenschicht *f*; ~ **structure** *n* MAT PROP Bodenstruktur *f*; ~ **type** *n* INFR & DES Bodenart *f*; ~ **and vent pipe** *n* WASTE WATER Entwässerungs- und Lüftungsrohr *nt*

**solar** *adj* ENVIRON, INFR & DES solar; ~ **absorption coefficient** *n* ENVIRON Sonnenabsorptionskoeffizient *m*; ~ **absorptivity** *n* ENVIRON Sonnenabsorptionsvermögen *nt*; ~ **altitude** *n* ENVIRON Sonnenhöhe *f*, Sonnenstand *m*; ~ **altitude angle** *n* ENVIRON Erhebungswinkel *m* der Sonne; ~ **azimuth** *n* ENVIRON Sonnenazimut *m*; ~ **battery** *n* ENVIRON Sonnenbatterie *f*; ~ **cell** *n* ENVIRON Solarzelle *f*; ~ **collector** *n* ELECTR, ENVIRON, HEAT & VENT, INFR & DES Sonnenkollektor *m*, Solaranlage *f*, Sonnenwärmekollektor *m*; ~ **concentrator** *n* ENVIRON Sonnenwärmekonzentrator *m*, Strahlungsbündler *m*; ~ **constant** *n* ENVIRON Solarkonstante *f*; ~ **control** *n* ARCH & BUILD, HEAT & VENT Solarkontrolle *f*; ~ **distillation** *n* ENVIRON Destillation *f* mittels Sonnenenergie; ~ **dynamics** *n* ENVIRON Solardynamik *f*; ~ **energy** *n* ENVIRON Solarenergie *f*, Sonnenenergie *f*; ~ **energy plant** *n* ELECTR, ENVIRON, HEAT & VENT, INFR & DES Solaranlage *f*, Sonnenkollektor *m*; ~ **engineering** *n* ENVIRON Solartechnik *f*; ~ **farm** *n* ENVIRON Sonnenfarm *f*; ~ **furnace** *n* ENVIRON Sonnenofen *m*; ~ **gain** *n* HEAT & VENT Solargewinn *m*; ~ **heat** *n* ENVIRON Sonnenwärme *f*; ~ **heating system** *n* ENVIRON Solarheizungssystem *nt*

**solarimeter** *n* ENVIRON Solarimeter *nt*

**solar**: ~ **panel** *n* ENVIRON Solarzellenplatte *f*; ~ **plant** *n* ELECTR, ENVIRON, HEAT & VENT, INFR & DES Solaranlage *f*; ~ **pond** *n* ENVIRON Solarpond *nt*; ~**-powered** *adj* ENVIRON mit Sonnenenergie betrieben; ~ **radiation** *n* ENVIRON, INFR & DES Sonnenstrahlung *f*; ~ **technology** *n* ENVIRON Solartechnik *f*; ~ **thermoelectric conversion** *n* ENVIRON thermoelektrische Sonnenenergieumwandlung *f*; ~ **tower** *n* ENVIRON Sonnenturm *m*

**solder** *vt* MAT PROP löten, weichlöten

**soldered** *adj* ELECTR, HEAT & VENT, STEEL gelötet; ~ **joint** *n* ELECTR, HEAT & VENT, MAT PROP Lötverbindung *f*

**soldering** *n* MAT PROP Löt-, Weichlöten *nt*, Zinnlöten *nt*; ~ **blowpipe** *n* MAT PROP Lötrohr *nt*; ~ **iron** *n* STEEL Lötkolben *m*

**soldier** *n* STONE aufrechtstehender Ziegel *m*; ~ **string course** *n* STONE aufrechtstehende Gesimsziegelschicht *f*

**sole** *n* ARCH & BUILD *plane* Unterseite *f*; ~ **piece** *n* ARCH & BUILD, TIMBER Fußholz *nt*, Schwelle *f*; ~ **plate** *n* TIMBER Schwelle *f*

**solicitor** *n* CONST LAW Rechtsanwalt *m*; ~**'s fee** *n* CONST LAW Rechtsanwaltshonorar *nt*

**solid 1.** *adj* ARCH & BUILD massiv, *wood, stone* fest, MAT PROP fest; **2.** *n* ARCH & BUILD, INFR & DES Festkörper *m*, MAT PROP Feststoff *m*

***solid***: ~**-borne sound** *n* SOUND & THERMAL Körperschall *m*; ~ **brick** *n* MAT PROP, STONE Vollstein *m*, Vollziegel *m*; ~ **ceiling** *n* ARCH & BUILD Massivdecke *f*; ~ **construction** *n* ARCH & BUILD Massivbau *m*; ~ **floor** *n* ARCH & BUILD Massivboden *m*

**solidification** *n* ENVIRON Verfestigung *f*, INFR & DES Bodenverdichtung *f*, MAT PROP Erstarren *nt*, Bodenverfestigung *f*; ~ **technique** *n* ENVIRON Verfestigungsverfahren *nt*

**solidified**: ~ **material** *n* ENVIRON Umsetzungsprodukt *nt*, Verfestigungsprodukt *nt*, MAT PROP verfestigtes Material *nt*; ~ **product** *n* ENVIRON Umsetzungsprodukt *nt*, Verfestigungsprodukt *nt*; ~ **waste** *n* ENVIRON Umsetzungsprodukt *nt*, Verfestigungsprodukt *nt*

**solidify 1.** *vt* ENVIRON verfestigen; **2.** *vi* MAT PROP *liquids* erstarren

**solidifying**: ~ **agent** *n* ENVIRON Reaktionsmittel *nt*, Verfestigungsmittel *nt*; ~ **of waste** *n* ENVIRON Verfestigung *f* von Abfällen

**solidity** *n* MAT PROP Festigkeit *f*

***solid***: ~ **measure** *n* ARCH & BUILD Festmaß *nt*; ~ **newel stair** *n* ARCH & BUILD Spindeltreppe *f*; ~ **particle** *n* ENVIRON, INFR & DES, MAT PROP Feststoffteilchen *nt*; ~ **phase** *n* MAT PROP feste Phase *f*; ~ **rectangular step** *n* ARCH & BUILD Winkelstufe *f*, Blockstufe *f*; ~ **sandlime brick** *n* STONE Kalksandstein *m*

**solids**: ~ **content** *n* ENVIRON Feststoffgehalt *m*

***solid***: ~ **slab** *n* ARCH & BUILD Vollplatte *f*; ~ **state** *n* MAT PROP festes Stadium *nt*; ~ **step** *n* ARCH & BUILD Massivstufe *f*; ~ **waste** *n* CONST LAW feste Abfälle *m pl*, ENVIRON fester Abfall *m*, fester Abfallstoff *m*, Feststoffabfall *m*, INFR & DES feste Abfälle *m pl*; ~ **wood** *n* TIMBER Vollholz *nt*

**Solnhofer**: ~ **stone** *n* MAT PROP Solnhofer Platte *f*

**soluble** *adj* MAT PROP löslich, *in liquids* lösbar

**solvent** *n* ENVIRON, MAT PROP Lösungsmittel *nt*, Verdünner *m*; ~**-based wood preservative** *n* MAT PROP lösliches Holzschutzmittel *nt*; ~**-free** *adj* SURFACE *paint* lösungsmittelfrei

**soot** *n* CONST LAW, INFR & DES Ruß *m*

**sound** *n* SOUND & THERMAL Schall *m*, Ton *m*; ~**-absorbent material** *n* SOUND & THERMAL Schalldämmstoff *m*; ~ **absorption** *n* SOUND & THERMAL Schallabsorption *f*; ~**-boarded ceiling** *n* SOUND & THERMAL Fehlboden *m*; ~**-boarded floor** *n* ARCH & BUILD, SOUND & THERMAL Einschubdecke *f*; ~ **boarding** *n* ARCH & BUILD, SOUND & THERMAL Einschubdecke *f*; ~ **damping** *n* SOUND & THERMAL Schalldämpfung *f*; ~ **deadening** *n* SOUND & THERMAL Schalldämpfung *f*; ~**-deadening coating** *n* SOUND & THERMAL, SURFACE Antidröhnbeschichtung *f*

**sounding** *n* INFR & DES, MAT PROP Sondierung *f*; ~ **cone** *n* INFR & DES Sondenspitze *f*; ~ **graph** *n* INFR & DES Spitzendruckdiagramm *nt*; ~ **hammer** *n* SOUND & THERMAL Lärmtesthammer *m*; ~ **method** *n* INFR & DES Sondiermethode *f*; ~ **pipe** *n* INFR & DES

Peilrohr *nt*; ~ **rod** *n* INFR & DES Sondiergestänge *nt*; ~ **test** *n* INFR & DES, MAT PROP Probesondierung *f*

***sound***: ~ **insulation** *n* SOUND & THERMAL Schallschutz *m*; ~ **insulation material** *n* SOUND & THERMAL Schalldämmstoff *m*; ~ **intensity** *n* CONST LAW, SOUND & THERMAL Schallstärke *f*; ~ **level** *n* SOUND & THERMAL Schallpegel *m*; ~ **pollution** *n* CONST LAW, ENVIRON Lärmbelästigung *f*; ~ **pressure level** *n* ENVIRON, SOUND & THERMAL Schalldruckpegel *m*; ~ **pressure spectrum** *n* ENVIRON Schalldruckspektrum *nt*

**soundproof** *adj* SOUND & THERMAL schalldicht

**soundproofing** *n* SOUND & THERMAL Schallschutz *m*

***sound***: ~ **source** *n* ENVIRON Schallquelle *f*; ~ **transmittance** *n* ARCH & BUILD, CONST LAW, SOUND & THERMAL Schallübertragung *f*; ~ **velocity** *n* ARCH & BUILD, MAT PROP, SOUND & THERMAL Schallgeschwindigkeit *f*; ~ **wave** *n* SOUND & THERMAL Schallwelle *f*

**source** *n* ENVIRON, SOUND & THERMAL *sound* Quelle *f*; ~ **area** *n* ENVIRON, INFR & DES Quellbereich *m*; ~ **of errors** *n* ARCH & BUILD Fehlerquelle *f*; ~ **separation** *n* ENVIRON Abfallsortierung *f* am Anfallsort

**space 1.** *n* ARCH & BUILD *free area, void* Raum *m*, *between things* Zwischenraum *m*, Abstand *m*, INFR & DES Abstand *m*; **2.** *vt* ARCH & BUILD mit Zwischenraum anordnen

***space***: ~ **assignment plan** *n* CONST LAW, INFR & DES Aufstellungsplan *m*

**spaced**: ~ **column** *n* ARCH & BUILD mehrteilige Säule *f*

***space***: ~ **frame** *n* ARCH & BUILD, INFR & DES räumliches Tragwerk *nt*; ~ **framework** *n* ARCH & BUILD, INFR & DES räumliches Tragwerk *nt*, Raumtragwerk *nt*; ~ **lattice** *n* ARCH & BUILD, INFR & DES räumliches Gittertragwerk *nt*

**spacer** *n* ARCH & BUILD, CONCR, ELECTR, HEAT & VENT, TIMBER, WASTE WATER Abstandhalter *m*, Distanzhalter *m*, Abstandhalter *m* für Bewehrungsstahl ; ~ **block** *n* ARCH & BUILD, CONCR, ELECTR, TIMBER, WASTE WATER Abstandhalter *m*, Zwischenstück *nt*; ~ **clamp** *n* ELECTR Distanzklemme *f*; ~ **stay** *n* HEAT & VENT, TIMBER Abstandhalter *m*

**space**: **~-saving** *adj* ARCH & BUILD, INFR & DES raumsparend; ~ **structure** *n* ARCH & BUILD, INFR & DES räumliches Tragwerk *nt*, Raumtragwerk *nt*; ~ **truss** *n* ARCH & BUILD, INFR & DES Raumfachwerk *nt*, räumliches Fachwerk *nt*, räumliches Gittertragwerk *nt*; ~ **utilization** *n* ARCH & BUILD, CONST LAW, INFR & DES Raumnutzung *f*

**spacing** *n* ARCH & BUILD, INFR & DES Abstand *m*, Zwischenraum *m*; ~ **block** *n* ARCH & BUILD, CONCR, ELECTR, HEAT & VENT, TIMBER, WASTE WATER Abstandhalter *m*; ~ **of rivets** *n* STEEL Nietabstand *m*, Nietteilung *f*; ~ **stay** *n* ARCH & BUILD, CONCR, ELECTR, HEAT & VENT, TIMBER, WASTE WATER Abstandhalter *m*; ~ **of stirrups** *n* CONCR Bügelverlegung *f*

**spade** *n* BUILD MACHIN Spaten *m*

**spall** *vi* ARCH & BUILD abblättern, abplatzen

**spalling** *n* ARCH & BUILD Abplatzen *nt*, Abschlagen *nt*, CONCR, INFR & DES Abplatzen *nt*, STONE *stone* Absprengung *f*

**span 1.** *n* ARCH & BUILD *range* Feld *nt*, Spannweite *f*, Stützlänge *f*, Umfassung *f*, CONST LAW *range* Umfassung *f*, INFR & DES *range* Spannweite *f*, Umfassung *f*; **2.** *vt* ARCH & BUILD überspannen, überbrücken

***span***: ~ **ceiling** *n* ARCH & BUILD, TIMBER Balkendecke *f*

**spandrel** *n* ARCH & BUILD Zwickel *m*; ~ **panel** *n* ARCH & BUILD, TIMBER Außenwandplatte *f* zwischen Geschoßfenstern; ~ **step** *n* ARCH & BUILD Dreieckstufe *f*

**spanner** *n BrE* BUILD MACHIN Schraubenschlüssel *m*

**spanning** *n* ARCH & BUILD Überbrückung *f*

***span***: ~ **piece** *n* ARCH & BUILD, TIMBER Kehlbalken *m*; ~ **roof** *n* ARCH & BUILD gleichseitiges Giebeldach *nt*

**spare** *vt* ARCH & BUILD, TIMBER aussparen

**spark** *n* ELECTR Funken *m*; ~ **arrester** *n* ELECTR Funkenfang *m*

**spatial**: ~ **distribution** *n* ENVIRON räumliche Verteilung *f*; ~ **lattice girder** *n* ARCH & BUILD, INFR & DES Raumfachwerkträger *m*; ~ **pattern** *n* ENVIRON räumliche Struktur *f*; ~ **structure** *n* ARCH & BUILD, INFR & DES Raumtragwerk *nt*, räumliches Tragwerk *nt*; ~ **trend** *n* ENVIRON räumliche Tendenz *f*; ~ **variability** *n* ENVIRON räumliche Veränderlichkeit *f*

**spatter** *n* STEEL *welding* Spritzer *m*

**spatula** *n* BUILD MACHIN Spachtel *m*

**speaking**: ~ **rod** *n* ARCH & BUILD, INFR & DES *surveying* Meßlatte *f*

**spec** *n* (*specification*) ARCH & BUILD, CONST LAW Spezifikation *f*

**special**: ~ **design** *n* ARCH & BUILD, INFR & DES Sonderanfertigung *f*; ~ **experience** *n* ARCH & BUILD, CONST LAW Sachkenntnis *f*

**specialist** *n* ARCH & BUILD Fachmann *m*; ~ **journal** *n* ARCH & BUILD, INFR & DES Fachzeitschrift *f*; ~ **literature** *n* ARCH & BUILD, INFR & DES Fachliteratur *f*; ~ **service** *n* CONST LAW fachmännische Dienstleistung *f*; ~ **sub-contractor** *n* CONST LAW spezialisierter Subunternehmer *m*

**specials** *n pl* ARCH & BUILD, STONE Sondererzeugnisse *nt pl*

***special***: ~ **steel** *n* MAT PROP, STEEL Sonderstahl *m*; ~ **waste** *n* ENVIRON Sonderabfall *m*, Sondermüll *m*

**specification** *n* (*spec*) ARCH & BUILD Spezifikation *f*, CONST LAW Pflichtenheft *nt*, Vorschrift *f*, Spezifikation *f*

**specifications** *n pl* ARCH & BUILD, CONST LAW Baubeschreibung *f*, Leistungsbeschreibung *f*, Leistungsverzeichnis *nt* (*LV*)

***specification***: ~ **test** *n* CONST LAW, INFR & DES Materialprüfung *f* gemäß Spezifikationen

**specific**: ~ **capacity of a well** *n* ENVIRON spezifische Brunnenkapazität *f*; ~ **electrical resistivity** *n* ELECTR spezifischer elektrischer Widerstand *m*; ~ **gravity** *n* MAT PROP spezifisches Gewicht *nt*, Dichte *f*, Raumgewicht *nt*; ~ **heat** *n* HEAT & VENT, MAT PROP spezifische Wärme *f*; ~ **heat capacity** *n* HEAT & VENT, MAT PROP spezifische Wärmekapazität *f*; ~ **mass** *n* MAT PROP spezifische Masse *f*; ~ **speed** *n* ENVIRON spezifische Drehzahl *f*; ~ **thermal conductivity** *n* MAT PROP, SOUND & THERMAL spezifische Wärmeleitfähigkeit *f*; ~ **volume** *n* MAT PROP spezifisches Volumen *nt*; ~ **weight** *n* MAT PROP Wichte *f*; ~ **weight of building volume** *n* MAT PROP spezifisches Gewicht *nt* je cbm umbauten Raumes

**specify** *vt* ARCH & BUILD, CONST LAW detailliert beschreiben, genau beschreiben, spezifizieren, *prescribe* vorschreiben

**specimen** *n* MAT PROP Muster *nt*, Probe *f*

**spectator** *n* INFR & DES Zuschauer *m*; ~ **seat** *n* ARCH & BUILD, INFR & DES Zuschauerplatz *m*

**speed** *n* BUILD MACHIN, HEAT & VENT, INFR & DES,

WASTE WATER *pump, fan* Drehzahl *f*, Geschwindigkeit *f*; **~ control device** *n* ENVIRON Drehzahlregler *m*

**sphere** *n* ARCH & BUILD Kugel *f*

**spherical** *adj* ARCH & BUILD kugelförmig, INFR & DES sphärisch; **~ cap** *n* ARCH & BUILD, INFR & DES Kalotte *f*; **~ grain** *n* MAT PROP sphärisches Korn *nt*; **~ shell** *n* ARCH & BUILD, INFR & DES, MAT PROP Kugelschale *f*; **~ vault** *n* ARCH & BUILD, INFR & DES Deckenkuppel *f*, Kugelgewölbe *nt*

**spheroidal**: **~ cast iron** *n* MAT PROP Sphäroguß *m*

**spider**: **~ connection** *n* ARCH & BUILD Kreuzverbindung *f*

**spigot** *n* ARCH & BUILD *bung* Zapfen *m*, *socket joint* Spitzende *nt*, Einsteckende *nt*, *tap* Wasserhahn *m*; **~ joint** *n* WASTE WATER Muffenrohrverbindung *f*, Rohrsteckverbindung *f*; **~ and socket joint** *n* WASTE WATER Muffenverbindung *f*; **~ and socket joint pipes** *n pl* WASTE WATER Muffenverbindungsrohre *nt pl*

**spike** *n* ARCH & BUILD Hakennagel *m*, Mauerhaken *m*

**spile** *n* TIMBER Holzpfahl *m*, Holzpflock *m*, Spund *m*, kleiner Holzpfropfen *m*

**spiling** *n* ARCH & BUILD, CONCR, INFR & DES, MAT PROP, STEEL, TIMBER Bewehrungsmatte *f*, Rammen *nt*

**spillway** *n* ENVIRON, INFR & DES Überfallwehr *nt*, Überlauf *m*; **~ channel** *n* INFR & DES, WASTE WATER Überlaufkanal *m*; **~ pipe** *n* HEAT & VENT, INFR & DES, WASTE WATER Überlaufrohr *nt*

**spindle** *n* BUILD MACHIN Spindel *f*, Spindelstab *m*, MAT PROP, WASTE WATER *pump* Spindel *f*; **~ molding machine** *AmE*, **~ moulding machine** *BrE* *n* BUILD MACHIN *carpentry* Spindelfräsmaschine *f*

**spine**: **~ wall** *n* ARCH & BUILD Mittellängswand *f*

**spiral 1.** *adj* ARCH & BUILD spiralenförmig; **2.** *n* ARCH & BUILD Spirale *f*

***spiral***: **~ stairs** *n pl* ARCH & BUILD Schneckentreppe *f*, Wendeltreppe *f*; **~ winder** *n* ARCH & BUILD Wendelstufe *f*

**spire** *n* ARCH & BUILD Turmspitze *f*, Turmhelm *m*; **~ roof** *n* ARCH & BUILD Pyramidenturmdach *nt*

**spirit**: **~ lacquer** *n* SURFACE Spirituslack *m*; **~ level** *n* BUILD MACHIN Nivellierwaage *f*

**splashback** *n* BUILD HARDW Spritzwand *f*

**splashproof** *adj* BUILD MACHIN, MAT PROP spritzwassergeschützt

**splay** *vt* ARCH & BUILD, INFR & DES, TIMBER abschrägen

**splayed** *adj* ARCH & BUILD, TIMBER *carpentry* abgeschrägt, verjüngt; **~ joint** *n* ARCH & BUILD Schrägverblattung *f*; **~ miter joint** *AmE*, **~ mitre joint** *BrE n* ARCH & BUILD abgeschrägte Gehrungsfuge *f*; **~ scarf** *n* ARCH & BUILD schräges Blatt *nt*; **~ window** *n* ARCH & BUILD Winkelfenster *nt*

**splice 1.** *n* TIMBER gespleißte Stelle *f*, Längsverbindung *f*; **2.** *vt* TIMBER verblatten, zusammenfügen

***splice***: **~ joint** *n* ARCH & BUILD Blattung *f*, Laschenverbindung *f*; **~ plate** *n* TIMBER Stoßblech *nt*

**spline**: **~ bushing** *n* TIMBER Vielkeilverzahnung *f*; **~ cutting tool** *n* TIMBER Fräswerkzeug *nt* für Vielkeilverzahnung

**splinter** *n* ARCH & BUILD, TIMBER *wood* Span *m*, Splitter *m*

**split 1.** *adj* ARCH & BUILD geteilt, gespalten; **2.** *n* ARCH & BUILD Spalt *m*, *rafter* Spaltfuge *f*; **3. ~ into** *vt* ARCH & BUILD, INFR & DES aufteilen, unterteilen; **~ into thin sheets** ARCH & BUILD verschiefern; **~ with wedges** ARCH & BUILD spalten mit einem Keil, zerkeilen

***split***: **~ hoop** *n* CONCR zerlegter Bügel *m*; **~ levels** *n pl* ARCH & BUILD versetzte Geschosse *nt pl*, versetzte Stockwerke *nt pl*; **~ loop** *n* CONCR zerlegter Bügel *m*; **~ ring connector** *n* BUILD HARDW, TIMBER *one- or two-sided* Einlaßdübel *m*; **~-section extrusion** *n* MAT PROP Hilfsphasenextrusion *f*, Extrusion *f* im Sekundenbruchteil; **~-tiled roof** *n* ARCH & BUILD, TIMBER Spließdach *nt*

**splitting** *n* ARCH & BUILD Aufspaltung *f*, Spaltung *f*; **~ ax** *AmE*, **~ axe** *BrE n* BUILD MACHIN Spaltaxt *f*

**spoil 1.** *n* INFR & DES *civil engineering* überschüssiger Aushubboden *m*; **2.** *vt* ARCH & BUILD, INFR & DES verderben

**spoilage** *n* CONST LAW Ausschuß *m*, MAT PROP Streu- und Bruchverlust *m*

***spoil***: **~ area** *n* INFR & DES Abraumkippe *f*, Seitenablagerung *f*; **~ heap** *n* INFR & DES Halde *f*

**spokeshave** *n* BUILD MACHIN Lederhobel *m*, Ziehklinge *f*

**sponge**: **~ rubber sealing** *n* MAT PROP, SOUND & THERMAL Moosgummiabdichtung *f*; **~ sealing** *n* MAT PROP, SOUND & THERMAL Moosgummiabdichtung *f*

**spongy** *adj* MAT PROP schwammartig; **~ soil** *n* INFR & DES schwammartiger Boden *m*

**spoon**: **~ auger** *n* BUILD MACHIN, INFR & DES Löffelbohrer *m*, Probelöffel *m*

**sports**: **~ facilities** *n pl* ARCH & BUILD, INFR & DES Sportanlagen *f pl*

**sportsground** *n* INFR & DES Sportplatz *m*; **~ maintenance equipment** *n* BUILD MACHIN Sportplatzpflegegerät *nt*

**sports**: **~ stadium** *n* INFR & DES Sportstadion *nt*

**spot** *n* ARCH & BUILD, ELECTR Punkt *m*

**spot**: **on the ~** *adv* ARCH & BUILD vor Ort

***spot***: **~ welding** *n* STEEL Punktschweißung *f*, Punktschweißen *nt*

**spout** *n* WASTE WATER Speirohr *nt*, Abflußrohr *nt*

**spray 1.** *n* MAT PROP, SURFACE Spray *nt*, Sprühmittel *nt*; **2.** *vt* BUILD MACHIN, MAT PROP spritzen, SURFACE sprayen, spritzen, sprühen, *paint* aufspritzen

**sprayed**: **~ mortar** *n* MAT PROP, STONE Spritzputz *m*

**spraying** *n* MAT PROP Spritzen *nt*; **~ device** *n* BUILD MACHIN Spritze *f*; **~ gun** *n* BUILD MACHIN Spritzpistole *f*

**spray**: **~ on** *phr* STONE *mortar* anspritzen, SURFACE *paint* aufspritzen; **~ painting** *n* SURFACE Spritzlackierung *f*

**spread** *vt* ARCH & BUILD verteilen

**spreader** *n* BUILD MACHIN Streich-, Streichgerät *nt*

**spread**: **~ footing** *n* ARCH & BUILD Flächengründung *f*; **~ foundation** *n* ARCH & BUILD, INFR & DES Flachgründung *f*

**spreading** *n* ARCH & BUILD Verbreitung *f*, CONST LAW, INFR & DES Ausbreitung *f*, SURFACE Verbreitung *f*

**spring** *n* BUILD HARDW, BUILD MACHIN *part of mechanism* Feder *f*, ENVIRON, INFR & DES Brunnen *m*, *of water* Quelle *f*, WASTE WATER Brunnen *m*; **~ bolt** *n* BUILD HARDW Federbolzen *m*; **~ bolt lock** *n* BUILD HARDW Schnäpperschloß *nt*

**springer** *n* ARCH & BUILD *arch, vault* Kämpferstein *m*, INFR & DES *arch* Kämpfer *m*; **~ stone** *n* STONE Kämpferstein *m*

***spring***: **~ hanger pin** *n* BUILD HARDW Federbolzen *m*

**springing** *n* INFR & DES Bogenanfang *m*; **~ course** *n* INFR & DES Kämpferschicht *f*; **~ line** *n* ARCH & BUILD, INFR & DES Kämpferlinie *f*

***spring***: ~ **line** *n* ARCH & BUILD, INFR & DES Kämpferlinie *f*; ~ **lock** *n* BUILD HARDW Fallschloß *nt*, Federschloß *nt*; ~ **maximum of fallout** *n* ENVIRON Frühlingsmaximum *nt*; ~ **neap cycle** *n* ENVIRON Springnipptide-Zyklus *m*; ~ **tide** *n* ENVIRON Springtide *f*; ~ **tide mark** *n* ENVIRON, INFR & DES Springtide-Marke *f*; ~ **water** *n* INFR & DES Quellwasser *nt*

**sprinkler** *n* WASTE WATER Sprinkler *m*; ~ **area** *n* WASTE WATER Beregnungsfläche *f*; ~ **reach** *n* CONST LAW, WASTE WATER Wurfweite *f* eines Regners; ~ **system** *n* WASTE WATER Sprinkleranlage *f*

**sprinkling**: ~ **filter** *n* ENVIRON Tropfkörper *m*

**sprocket** *n* ARCH & BUILD Sturm *m*; ~ **piece** *n* ARCH & BUILD Sturmlatte *f*, Windaussteifung *f*, Windlatte *f*, Windrute *f*

**spun**: ~ **concrete** *n* CONCR Schleuderbeton *m*; ~ **glass** *n* MAT PROP, SOUND & THERMAL Glaswolle *f*

**spur** *n* ARCH & BUILD Dorn *m*, Mauervorsprung *m*, Steigeisen *nt*, Strebe *f*, Stütze *f*, BUILD HARDW, BUILD MACHIN Dorn *m*; ~ **post** *n* ARCH & BUILD Radabweiser *m*; ~ **stone** *n* INFR & DES Radabweiser *m*; ~ **tenon joint** *n* ARCH & BUILD Kurzzapfverbindung *f*; ~ **track** *n* INFR & DES Nebengleis *nt*

**square 1.** *adj* ARCH & BUILD quadratisch, viereckig; **2.** *n* ARCH & BUILD Blockstufe *f*, Quadrat *nt*; **3.** *vt* ARCH & BUILD *stones* behauen, TIMBER besäumen

***square***: ~ **bar** *n* MAT PROP Vierkantstab *m*; ~ **bolt** *n* ARCH & BUILD Vierkantschraube *f*; ~**-edged** *adj* ARCH & BUILD *wood* besäumt; ~ **edge preparation** *n* ARCH & BUILD *welding* Steilkantenvorbereitung *f*; ~ **frame** *n* ARCH & BUILD Viereckrahmen *m*; ~ **hole** *n* STEEL Vierkantlochung *f*; ~ **joint** *n* ARCH & BUILD rechtwinklige Verbindung *f*; ~ **rabbet plane** *n* BUILD MACHIN rechtwinkliger Falzhobel *m*; ~ **splice** *n* ARCH & BUILD rechteckige Verblattung *f*; ~ **staff** *n* ARCH & BUILD Rechteckdeckleiste *f*; ~ **timber** *n* TIMBER Kantholz *nt*; ~ **tube** *n* MAT PROP Vierkantrohr *nt*

**squaring** *n* ARCH & BUILD Rechtwinkligschneiden *nt*

**stability** *n* ARCH & BUILD Beständigkeit *f*, Stabilität *f*, ENVIRON Festigkeit *f*, INFR & DES Beständigkeit *f*, Tragfähigkeit *f*, Stabilität *f*, Standsicherheit *f*, SURFACE *substrate* Tragfähigkeit *f*; ~ **against gliding** *n* INFR & DES Gleitsicherheit *f*, Kippsicherheit *f*; ~ **against tilting** *n* ARCH & BUILD, CONST LAW Kippsicherheit *f*; ~ **analysis** *n* INFR & DES Stabilitätsanalyse *f*; ~ **conditions** *n* ARCH & BUILD, INFR & DES Stabilitätsbedingungen *f pl*; ~ **factor** *n* INFR & DES Stabilitätsfaktor *m*; ~ **study** *n* INFR & DES Stabilitätsstudie *f*; ~ **test** *n* INFR & DES Stabilitätsversuch *m*; ~ **under load** *n* ARCH & BUILD, INFR & DES Standfestigkeit *f*

**stabilization** *n* INFR & DES Stabilisierung *f*; ~ **pond** *n* ENVIRON Abwasserbecken *nt*

**stabilize** *vt* INFR & DES stabilisieren

**stabilizer** *n* MAT PROP Stabilisator *m*

**stable** *adj* ARCH & BUILD, INFR & DES stabil, standfest, standsicher; ~ **slope** *n* INFR & DES standfeste Böschung *f*; ~ **soil** *n* INFR & DES standfester Boden *m*

**stack** *n* ARCH & BUILD *pile* Stapel *m*, Kamin *m*, Schornstein *m*, Stoß *m*, WASTE WATER senkrechte Abwasserleitung *f*

**stacked**: ~ **wood** *n* TIMBER Schichtpreßholz *nt*

***stack***: ~ **effect** *n* HEAT & VENT Schornsteineffekt *m*

**stacking**: ~ **ground** *n* INFR & DES Stapelplatz *m*, Lagerfläche *f*; ~ **yard** *n* INFR & DES Lagerhof *m*

***stack***: ~ **pipe** *n* WASTE WATER Regenfallrohr *nt*; ~ **ventilation** *n* HEAT & VENT, WASTE WATER Abwasserrohrentlüftung *f*

**stadia** *n* INFR & DES Vermessungsstange *f*; ~ **surveying** *n* ARCH & BUILD tachometrische Vermessung *f*

**stadiometer** *n* INFR & DES *surveying* Stadiometer *nt*

**staff** *n* ARCH & BUILD *workforce* Personal *nt*, *surveying* Meßlatte *f*, CONCR Gipsmörtel *m*, CONST LAW Personal *nt*; ~ **holder** *n* ARCH & BUILD, INFR & DES Meßlattenträger *m*

**stage** *n* ARCH & BUILD, INFR & DES *phase* Stadium *nt*, Abschnitt *m*, *platform* Bühne *f*; ~ **of decomposition** *n* ENVIRON Abbaustufe *f*

**staggered** *adj* ARCH & BUILD, CONCR gestaffelt, versetzt; ~ **floors** *n pl* ARCH & BUILD versetzte Stockwerke *nt pl*, versetzte Geschosse *nt pl*; ~ **installation** *n* ARCH & BUILD versetzte Fuge *f*; ~ **spot welding** *n* STEEL Zickzackpunktschweißung *f*; ~ **storeys** *n pl* *BrE* ARCH & BUILD versetzte Geschosse *nt pl*, versetzte Stockwerke *nt pl*; ~ **stories** *AmE see staggered storeys BrE*

**staging** *n* ARCH & BUILD Arbeitsbrücke *f*, Gerüstbau *m*

**stagnant**: ~ **water** *n* INFR & DES stehendes Wasser *nt*

**stain 1.** *n* SURFACE *colourant* Beize *f*, *mark* Fleck *m*; **2.** *vt* SURFACE *wood* beizen

**stained** *adj* SURFACE gebeizt

**stainless** *adj* MAT PROP, STEEL, SURFACE rostfest, korrosionsbeständig; ~ **chromium steel** *n* MAT PROP rostfreier Chromstahl *m*; ~ **steel** *n* MAT PROP nichtrostender Stahl *m*, Edelstahl *m*; ~ **steel lintel** *n* ARCH & BUILD, STEEL rostfreier Stahlsturz *m*; ~ **steel sink** *n* WASTE WATER rostfreies Spülbecken *nt*

**stair** *n* ARCH & BUILD Treppenstufe *f*, Tritt *m*

**staircase** *n* ARCH & BUILD Treppenhaus *nt*, Treppe *f*

**stairlift** *n* ARCH & BUILD, BUILD MACHIN Treppenlift *m*

**stairs** *n pl* ARCH & BUILD Treppe *f*

***stair***: ~ **stringer** *n* ARCH & BUILD Treppenwange *f*

**stairway** *n* ARCH & BUILD Treppe *f*, Treppenhaus *nt*

**stake 1.** *n* ARCH & BUILD Pfosten *m*, Pfahl *m*; **2.** ~ **out** *vt* ARCH & BUILD abstecken

**stamp**: ~ **duty** *n* CONST LAW Stempelgebühr *f*

**stanchion** *n* ARCH & BUILD Pfosten *m*

**standard 1.** *adj* GEN normal; **2.** *n* ARCH & BUILD *column* Ständer *m*, Standard *m*, Stütze *f*, Pfahl *m*, Stiel *m*, BUILD HARDW, BUILD MACHIN, CONCR Standard *m*, CONST LAW Standard *m*, *average* Norm *f*, ENVIRON, HEAT & VENT, INFR & DES, MAT PROP Standard *m*

***standard***: ~ **air** *n* ENVIRON, HEAT & VENT Normalluft *f*

**Standard**: ~ **Construction Services Manual** *n* ARCH & BUILD Standardleistungsbuch *nt* für das Bauwesen

***standard***: ~ **deviation** *n* ARCH & BUILD Standardabweichung *f*

**standardization** *n* CONST LAW Normung *f*

***standard***: ~ **mounting** *n* BUILD HARDW Normalanschlag *m*

**Standard**: ~ **Penetration Test** *n* (*STP*) INFR & DES Rammsondierung *f*

***standard***: ~ **size** *n* ARCH & BUILD, MAT PROP Normalformat *nt* (*NF*); ~ **specification** *n* CONST LAW Normungsvorschrift *f*; ~ **terms of contract** *n pl* CONST LAW Vertragsgrundbedingungen *f pl*; ~ **test** *n* CONST LAW, INFR & DES Normversuch *m*

**standing** *adj* ARCH & BUILD stehend, aufrecht, INFR & DES stehend; ~ **seam** *n* STEEL Stehfalz *m*; ~ **spectator area** *n* CONST LAW, INFR & DES Zuschauerstehfläche *f*; ~ **timber** *n* TIMBER Stammholz *nt*

**standpipe** *n* INFR & DES Hydrant *m*, WASTE WATER Steigrohr *nt*

**staple** *n* BUILD HARDW *lock* Haspe *f*, Klammer *f*, Krampe *f*, MAT PROP Klammer *f*

**star**: **~ shake** *n* STEEL *wood* Sternriß *m*

**starter**: **~ bar** *n* ARCH & BUILD Anschlußbewehrungsstab *m*

**starting**: **~ step** *n* ARCH & BUILD Antritt *m*

**start**: **~ of work** *n* ARCH & BUILD Baubeginn *m*

**state** *n* ARCH & BUILD Beschaffenheit *f*; **~ of equilibrium** *n* ARCH & BUILD, INFR & DES Gleichgewichtszustand *m*; **~ of failure** *n* MAT PROP Bruchzustand *m*; **~-of-the-art** *adj* ARCH & BUILD hochmodern; **~-of-the-art technique** *n* ARCH & BUILD Spitzentechnik *f*, hochmoderne Technik *f*; **~-of-the-art technology** *n* ARCH & BUILD Spitzentechnologie *f*; **~ of rest** *n* ARCH & BUILD, BUILD MACHIN, INFR & DES Ruhezustand *m*; **~ of transition** *n* CONST LAW, INFR & DES Übergangszustand *m*

**static** *adj* ARCH & BUILD statisch, feststehend

**statical**: **~ indeterminacy** *n* ARCH & BUILD statische Unbestimmtheit *f*

**statically**: **~ admissible** *adj* ARCH & BUILD statisch zulässig; **~ definable** *adj* ARCH & BUILD statisch bestimmbar; **~ defined** *adj* ARCH & BUILD statisch bestimmt; **~ determinable** *adj* ARCH & BUILD statisch bestimmbar; **~ indeterminable** *adj* ARCH & BUILD statisch unbestimmbar; **~ indeterminate** *adj* ARCH & BUILD, INFR & DES statisch unbestimmt; **~ overdefined** *adj* ARCH & BUILD statisch überbestimmt; **~ overdetermined** *adj* ARCH & BUILD statisch überbestimmt

**static**: **~ calculation** *n* ARCH & BUILD, INFR & DES statische Berechnung *f*; **~ electricity** *n* ELECTR Reibungselektrizität *f*, statische Elektrizität *f*; **~ equation** *n* ARCH & BUILD, INFR & DES statische Gleichung *f*; **~ equilibrium** *n* ARCH & BUILD, INFR & DES statisches Gleichgewicht *nt*; **~ forces** *n pl* ARCH & BUILD, CONCR, INFR & DES, MAT PROP, STEEL, TIMBER Schnittkräfte *f pl*; **~ investigation** *n* ARCH & BUILD, INFR & DES statische Untersuchung *f*; **~ load** *n* INFR & DES statische Last *f*; **~ moment** *n* ARCH & BUILD, INFR & DES statisches Moment *nt*; **~ pressure** *n* ARCH & BUILD statischer Druck *m*; **~ proof-loading machine** *n* ARCH & BUILD statische Belastungstestanlage *f*; **~ requirements** *n pl* ARCH & BUILD statische Erfordernisse *nt pl*; **~ strength** *n* ARCH & BUILD statische Festigkeit *f*; **~ stress** *n* ARCH & BUILD statische Beanspruchung *f*; **~ test** *n* ARCH & BUILD statische Prüfung *f*

**stationary** *adj* ARCH & BUILD stationär, feststehend, ortsfest; **~ emission source** *n* ENVIRON ortsfeste Emissionsquelle *f*

**station**: **~ pole** *n* ARCH & BUILD Vermessungsstab *m*; **~ roof** *n* ARCH & BUILD *loading ramp* Wetterdach *nt*

**statistically** *adv* ARCH & BUILD, CONST LAW, ENVIRON, INFR & DES statistisch

**statistics** *n* ARCH & BUILD, CONST LAW, ENVIRON, INFR & DES Statistik *f*

**stay** *n* ARCH & BUILD *girder* Träger *m*, *pin* Bolzen *m*, *stanchion* Ständer *m*, Pfosten *m*, Strebe *f*, Stütze *f*, *steadying with guy* Verspannung *f*, *tension rod* Zuganker *m*, *tensioning rope* Spannseil *nt*, *traction rope* Zugseil *nt*; **~ bolt** *n* ARCH & BUILD Spange *f*, Stehbolzen *m*, BUILD HARDW Stehbolzen *m*

**staying** *n* ARCH & BUILD Absteifung *f*, Versteifung *f*, Verankerung *f*, INFR & DES Verankerung *f*

***stay***: **~ rod** *n* ARCH & BUILD, INFR & DES, STEEL Ankerstab *m*; **~ rope** *n* ARCH & BUILD, BUILD HARDW Verankerungsseil *nt*; **~ wire** *n* ARCH & BUILD, CONCR, MAT PROP, STEEL Abspanndraht *m*

**steam** *vi* ENVIRON, HEAT & VENT, SOUND & THERMAL dampfen

**steam**: **~ curing** *n* CONCR Dampfbehandlung *f*, Dampfhärten *nt*

**steamed**: **~ beech** *n* TIMBER gedämpfte Buche *f*

**steam**: **~-hardening process** *n* CONCR Dampfhärtungsprozeß *m*; **~ heating** *n* HEAT & VENT Dampfheizung *f*; **~ jet** *n* ENVIRON Dampfstrahl *m*; **~ pipe** *n* HEAT & VENT Dampfleitung *f*

**steel** *n* STEEL Stahl *m*; **~ band chain** *n* BUILD MACHIN, STEEL *surveying* Stahlbandkette *f*; **~ beam** *n* STEEL Stahlträger *m*; **~ cable** *n* STEEL Stahlkabel *nt*; **~ chimney** *n* STEEL Blechschornstein *m*; **~ construction** *n* ARCH & BUILD, STEEL Stahlbau *m*, Stahlkonstruktion *f*; **~ core** *n* STEEL Stahlkern *m*, Stahlseele *f*; **~ fabric** *n* ARCH & BUILD, CONCR, INFR & DES, MAT PROP, STEEL, TIMBER Bewehrungsmatte *f*

**steelfixer** *n* STEEL Stahlflechter *m*

***steel***: **~ fixing** *n* STEEL Bewehrungsarbeiten *f pl*; **~ flange** *n* ARCH & BUILD, STEEL Stahlgurt *m*; **~ forms** *n pl* CONCR, STEEL Stahlschalung *f*; **~ formwork** *n* CONCR, STEEL Stahlschalung *f*; **~ girder** *n* ARCH & BUILD Stahlträger *m*; **~ grade** *n* CONCR, MAT PROP, STEEL Stahlart *f*, Stahlsorte *f*; **~ insert** *n* STEEL Stahleinsatz *m*; **~ key** *n* BUILD HARDW Stahlschlüssel *m*; **~ lathing** *n* MAT PROP, STEEL Metallputzträger *m*; **~ lattice tower** *n* STEEL Stahlgitterturm *m*; **~ mesh** *n* STEEL Stahlgeflecht *nt*; **~ netting** *n* STEEL Stahlgeflecht *nt*; **~ pile** *n* STEEL Stahlpfahl *m*; **~ piling** *n* STEEL Spundwand *f*; **~ purlin** *n* ARCH & BUILD Stahlpfette *f*; **~ schedule** *n* ARCH & BUILD, CONCR, STEEL Stahlliste *f*; **~ scrap** *n* ENVIRON Stahlschrott *m*; **~ section** *n* ARCH & BUILD, STEEL Formstahl *m*, Profilstahl *m*; **~ shear studs** *n pl* ARCH & BUILD Stahlschubbolzen *m pl*; **~ sheet** *n* STEEL Stahlblech *nt*; **~ sheet piling** *n* INFR & DES, STEEL Stahlspundwand *f*; **~ skeleton construction** *n* ARCH & BUILD, STEEL Stahlskelettbau *m*; **~ skeleton structure** *n* ARCH & BUILD, STEEL Stahlskelettbau *m*; **~ trowel** *n* BUILD MACHIN Stahlkelle *f*; **~ web** *n* CONCR Betonstahlgewebe *nt*; **~ wire mesh** *n* CONCR Armierungsmatte *f*, Baustahlmatte *f*; **~ wire rope** *n* STEEL Stahldrahtseil *nt*

**steelwork** *n* MAT PROP, STEEL Stahlarbeiten *f pl*

**steelworks** *n* MAT PROP, STEEL Hütte *f*

**steep** *adj* ARCH & BUILD, INFR & DES steil; **~ gradient** *n* ARCH & BUILD steiler Abhang *m*

**steeple**: **~ head rivet** *n* BUILD HARDW, STEEL Spitzkopfniet *m*

**steepness** *n* ARCH & BUILD, INFR & DES Steilheit *f*

**steep**: **~ road** *n* ARCH & BUILD steile Straße *f*; **~ roof** *n* ARCH & BUILD Steildach *nt*; **~ slope** *n* ARCH & BUILD Steilböschung *f*, INFR & DES steile Neigung *f*

**stellar**: **~ pattern** *n* ARCH & BUILD Sternbild *nt*

**stem** *n* ARCH & BUILD Spindel *f*, Steg *m*

**stench**: **~ trap** *n* WASTE WATER Geruchsverschluß *m*

**step** *n* ARCH & BUILD Schritt *m*, *raised surface* Stufe *f*, Treppenstufe *f*

**stepback** *n* ARCH & BUILD, INFR & DES, STONE Abtreppung *f*

***step***: ~ **iron** *n* ARCH & BUILD, WASTE WATER *manhole* Steigeisen *nt*; ~ **joint** *n* ARCH & BUILD Überlappung *f*
**stepped** *adj* ARCH & BUILD abgestuft, abgetreppt, BUILD MACHIN abgestuft; ~ **DPC** *n* SOUND & THERMAL, STONE abgestufte Feuchtigkeitssperre *f*; ~ **gable** *n* ARCH & BUILD Staffelgiebel *m*
**stepping** *n* ARCH & BUILD *foundation*, INFR & DES, STONE Abtreppung *f*
**sterile** *adj* WASTE WATER steril
**sticky**: ~ **clay** *n* INFR & DES klebriger Ton *m*; ~ **limit** *n* MAT PROP Haftgrenze *f*
**stiff** *adj* ARCH & BUILD, CONCR, INFR & DES steif, MAT PROP biegesteif, steif, STEEL biegesteif; ~ **clay** *n* INFR & DES steifer Ton *m*
**stiffen** *vt* ARCH & BUILD aussteifen, versteifen
**stiffened**: ~ **suspension bridge** *n* INFR & DES Hängebrücke *f* mit steifer Fahrbahntafel
**stiffening** *n* ARCH & BUILD Absteifung *f*, Versteifung *f*, Aussteifung *f*, Verstärkung *f*; ~ **rib** *n* ARCH & BUILD Versteifungsrippe *f*, Verstärkungsrippe *f*
**stiff**: ~**-jointed** *adj* ARCH & BUILD mit biegesteifen Knoten, steifknotig
**stiffness** *n* ARCH & BUILD, INFR & DES, MAT PROP, SOUND & THERMAL Steifigkeit *f*
**stiff**: ~ **soil** *n* INFR & DES steifer Boden *m*
**stile** *n* ARCH & BUILD *door* Höhenfries *m*
**still**: ~ **air** *n* ENVIRON Windstille *f*, stillstehende Luft *f*
**stilling**: ~ **basin** *n* INFR & DES, WASTE WATER Tosbecken *nt*
**stilt** *n* ARCH & BUILD Stelze *f*
**stilted**: ~ **arch** *n* ARCH & BUILD Stelzbogen *m*, byzantinischer Bogen *m*
**stink**: ~ **trap** *n* WASTE WATER Siphon *m*, Röhrensiphon *m*
**stipulated**: ~ **penalty** *n* CONST LAW Vertragsstrafe *f*
**stirrup** *n* CONCR Bügel *m*
**stochastic** *adj* MAT PROP stochastisch; ~ **model** *n* INFR & DES stochastisches Modell *nt*
**stock** *n* ARCH & BUILD, MAT PROP Vorrat *m*
**stockade** *n* INFR & DES Palisade *f*
**stone** *n* ARCH & BUILD, STONE Gestein *nt*, Stein *m*; ~ **band saw** *n* BUILD MACHIN Steinbandsägemaschine *f*; ~ **bond** *n* STONE Steinverband *m*; ~ **breaker** *n* BUILD MACHIN Schotterbrecher *m*; ~ **carving** *n* STONE Steinbildhauerei *f*, Steinskulptur *f*; ~ **chippings** *n* STONE Splitt *m*; ~ **chips** *n pl* STONE Splitt *m*; ~ **construction** *n* STONE Steinbau *m*; ~ **crusher** *n* BUILD MACHIN Steinbrecher *m*; ~ **dam** *n* INFR & DES Steindamm *m*; ~ **dresser** *n* ARCH & BUILD, INFR & DES Steinmetz *m*; ~ **facing** *n* STONE Plattenverkleidung *f*; ~**-filled** *adj* STONE splittreich; ~**-filled asphaltic concrete pavement** *n* INFR & DES splittreiche Asphaltbetondeckschicht *f*; ~ **filling** *n* INFR & DES, MAT PROP Packwerk *nt*
**stonemason** *n* ARCH & BUILD, INFR & DES Steinmetz *m*
***stone***: ~ **mill** *n* INFR & DES Steinfräse *f*; ~ **packing** *n* INFR & DES Steinpackung *f*; ~ **pavement** *n* INFR & DES Steinpflaster *nt*; ~ **pit** *n* INFR & DES, STONE Steinbruch *m*; ~ **plaster** *n* STONE Steinputz *m*; ~ **quarry** *n* INFR & DES, STONE Steinbruch *m*; ~**-splitting machine** *n* BUILD MACHIN Steinspaltmaschine *f*; ~ **structure** *n* STONE Steinbauwerk *nt*, Steinbau *m*
**stoneware** *n* STONE Steinzeug *nt*, Steingut *nt*; ~ **bend** *n* WASTE WATER Steinzeugbogen *m*; ~ **branch** *n* WASTE WATER Steinzeugabzweig *m*; ~ **pipe** *n* STONE, WASTE WATER Steinzeugrohr *nt*; ~ **tile** *n* STONE Steinzeugplatte *f*, Steinzeugfliese *f*
***stone***: ~ **wood floor** *n* ARCH & BUILD Steinholzfußboden *m*
**stoneworking** *n* INFR & DES, STONE Steinbearbeitung *f*
**stony** *adj* INFR & DES steinig; ~ **ground** *n* INFR & DES steiniger Untergrund *m*
**stop 1.** *n* ARCH & BUILD *door* Anschlag *m*; **2.** *vt* ARCH & BUILD, SURFACE *moisture protection* dichten, sperren; ~ **with putty** ARCH & BUILD auskitten
***stop***: ~ **button** *n* ELECTR Abstellknopf *m*
**stopcock** *n* HEAT & VENT, WASTE WATER Absperrhahn *m*
***stop***: ~ **cushion** *n* ARCH & BUILD *door* Anschlagdämpfung *f*; ~ **log** *n* BUILD HARDW Staubalken *m*
**stopper** *n* SURFACE Spachtelmasse *f*
**stopping** *n* ARCH & BUILD, SURFACE Absperren *nt*, *with sealant* Dichten *nt*; ~ **knife** *n* BUILD MACHIN Spachtelmesser *nt*
***stop***: ~ **rail** *n* ARCH & BUILD Anschlagschiene *f*; ~ **screw** *n* BUILD HARDW Arretierschraube *f*; ~ **valve** *n* BUILD HARDW, ENVIRON Absperrventil *nt*
**storage**: ~ **area** *n* ENVIRON Ablagerungsplatz *m*; ~ **basin** *n* ENVIRON, INFR & DES, WASTE WATER Sammelbecken *nt*, Speicherbecken *nt*; ~ **chamber** *n* ENVIRON *for radioactive waste* Lagerkammer *f*; ~ **facility** *n* ENVIRON Lagerstätte *f*, Lagermöglichkeit *f*; ~ **heating** *n* HEAT & VENT Speicherheizung *f*; ~ **scheme** *n* ENVIRON Speicherschema *nt*; ~ **shed** *n* ARCH & BUILD Lagerschuppen *m*; ~ **site** *n* ENVIRON, INFR & DES Deponie *f*; ~ **space** *n* INFR & DES Speicher *m*; ~ **tank** *n* ENVIRON Sammelbehälter *m*, HEAT & VENT Vorratsbehälter *m*; ~ **water heater** *n* ELECTR, HEAT & VENT, WASTE WATER Boiler *m*, Heißwasserspeicher *m*, Warmwasserspeicher *m*
**store 1.** *n* ARCH & BUILD Lager *nt*; **2.** *vt* INFR & DES, MAT PROP lagern
**storey** *n BrE* ARCH & BUILD Etage *f*, Geschoß *nt*, Stockwerk *nt*
**storm**: ~ **drain** *n* ARCH & BUILD Straßenablauf *m*, INFR & DES Regenwasserleitung *f*; ~ **sewer** *n* INFR & DES, WASTE WATER Niederschlagswasserkanal *m*, Oberflächenwasserkanal *m*, Regenwasserkanal *m*
**stormwater** *n* ENVIRON, INFR & DES, SURFACE, WASTE WATER Dachwasser *nt*, Niederschlag *m*, Tagwasser *nt*, Oberflächenwasser *nt*, Regenwasser *nt*; ~ **deflector** *n* ARCH & BUILD Regenleiste *f*, Regenschutzschiene *f*; ~ **gutter** *n* WASTE WATER Regenrinne *f*, Wasserablaufrinne *f*; ~ **inlet** *n* WASTE WATER Regenwasserablauf *m*; ~ **pipe clamp** *n* BUILD HARDW Regenrohrschelle *f*; ~ **retention basin** *n* INFR & DES, WASTE WATER Regenrückhaltebecken *nt*; ~ **sewer** *n* INFR & DES, WASTE WATER Niederschlagswasserkanal *m*, Oberflächenwasserkanal *m*
**story** *AmE see storey BrE*
**stove** *n* HEAT & VENT Ofen *m*
**stoved**: ~ **enamel finish** *n* SURFACE Einbrennlackierung *f*
***stove***: ~**-enameled** *AmE*, ~**-enamelled** *BrE adj* SURFACE einbrennlackiert; ~ **pipe** *n* HEAT & VENT Ofenrohr *nt*
**stoving**: ~ **temperature** *n* SURFACE Einbrenntemperatur *f*
**STP** *abbr* (*Standard Penetration Test*) INFR & DES Rammsondierung *f*
**straight** *adj* ARCH & BUILD gerade, direkt; ~ **arch** *n*

ARCH & BUILD scheitrechter Bogen *m*, gerader Bogen *m*; **~ edge** *n* ARCH & BUILD gerade Kante *f*
**straighten** *vt* ARCH & BUILD begradigen, richten
**straightened**: **~ sheet** *n* STEEL gerichtetes Blech *nt*
**straightening** *n* ARCH & BUILD Richten *nt*
**straight**: **~ joint** *n* CONCR, STONE glatte Fuge *f*, TIMBER Stoßfuge *f*; **~ line** *n* ARCH & BUILD Gerade *f*; **~-pane hammer** *n* BUILD MACHIN Hammer *m* mit gerader Finne; **~-peen hammer** *n* BUILD MACHIN Hammer *m* mit gerader Finne
**strain 1.** *n* ARCH & BUILD, INFR & DES Beanspruchung *f*, Belastung *f*; **2.** *vt* ARCH & BUILD, BUILD MACHIN, INFR & DES, MAT PROP beanspruchen, dehnen, belasten
***strain***: **~ distribution** *n* MAT PROP Dehnungsverteilung *f*
**strainer** *n* INFR & DES, WASTE WATER Schmutzsieb *nt*, Sieb *nt*
***strain***: **~ gage** *AmE*, **~ gauge** *BrE* *n* BUILD MACHIN, INFR & DES, MAT PROP Dehnungsmesser *m*
**straining** *n* ARCH & BUILD, BUILD MACHIN Sprengen *nt*; **~ beam** *n* ARCH & BUILD, BUILD MACHIN, TIMBER Sprengstrebe *f*, Verstrebungsbalken *m*; **~ piece** *n* ARCH & BUILD, BUILD MACHIN Jochbalken *m*, Verstrebungsbalken *m*
***strain***: **~ measurement** *n* INFR & DES, MAT PROP Verformungsmessung *f*; **~ meter** *n* BUILD MACHIN, INFR & DES, MAT PROP Dehnungsmesser *m*; **~ modulus** *n* MAT PROP Dehnungskoeffizient *m*
**strand** *n* CONCR Spanndrahtbündel *nt*, MAT PROP Litze *f*, STEEL Spanndrahtbündel *nt*
**strap** *n* ARCH & BUILD Querriegel *m*, BUILD HARDW Schelle *f*
**S-trap** *n* WASTE WATER S-Verschluß *m*
***strap***: **~ bolt** *n* ARCH & BUILD Bügelschraube *f*; **~ hinge** *n* ARCH & BUILD Bandscharnier *nt*, BUILD HARDW Scharnierband *nt*; **~ steel** *n* STEEL Bandeisen *nt*
**strata**: **~ boundary** *n* INFR & DES Schichtengrenze *f*
**stratified** *adj* MAT PROP geschichtet; **~ soil** *n* INFR & DES geschichteter Boden *m*
**stratum** *n* INFR & DES Schicht *f*
**strawboard** *n* SOUND & THERMAL Preßstrohplatte *f*, Strohbauplatte *f*
**stream** *n* ENVIRON, INFR & DES, MAT PROP Fluß *m*, Strom *m*; **~ erosion** *n* INFR & DES Flußerosion *f*; **~ machine** *n* BUILD MACHIN Naßpresse *f*
**street** *n* BUILD MACHIN, INFR & DES Straße *f*
**streetcar** *n* *AmE* (*cf tramcar BrE*) INFR & DES Straßenbahn *f*
***street***: **~ cleaning** *n* ENVIRON, INFR & DES Straßenreinigung *f*; **~ furniture** *n* INFR & DES Straßenzubehör *nt*; **~ inlet** *n* INFR & DES Straßenablauf *m*; **~ lighting** *n* CONST LAW, ELECTR, INFR & DES Straßenbeleuchtung *f*; **~ line** *n* INFR & DES Straßenmarkierung *f*; **~ sweeping** *n* ENVIRON, INFR & DES Straßenreinigung *f*
**strength** *n* MAT PROP Festigkeit *f*
**strengthen** *vt* ARCH & BUILD verstärken
**strengthening** *n* ARCH & BUILD Verstärkung *f*, Verstärken *nt*
***strength***: **~ test** *n* INFR & DES, MAT PROP Festigkeitsprüfung *f*
**stress 1.** *n* ARCH & BUILD, INFR & DES Beanspruchung *f*, Belastung *f*, Spannung *f*; **2.** *vt* ARCH & BUILD beanspruchen, belasten, BUILD MACHIN beanspruchen, CONCR spannen, INFR & DES beanspruchen
***stress***: **~ analysis** *n* ARCH & BUILD Spannungsnachweis *m*, INFR & DES Spannungsermittlung *f*
**stressbed** *n* CONCR Spannbett *nt*
***stress***: **~-block** *n* ARCH & BUILD, CONCR, MAT PROP Druckblock *m*, Druckverteilungsblock *m*; **~ corrosion** *n* MAT PROP, STEEL Spannungskorrosion *f*; **~ crack** *n* MAT PROP, STEEL Spannungsriß *m*; **~ distribution** *n* ARCH & BUILD, CONCR, INFR & DES, MAT PROP, STEEL Spannungsverteilung *f*; **~-free** *adj* ARCH & BUILD, MAT PROP spannungsfrei
**stressing**: **~ bar** *n* ARCH & BUILD, CONCR, STEEL Spannstab *m*
***stress***: **~-strain diagram** *n* ARCH & BUILD, CONCR, INFR & DES, MAT PROP Druckdehnungsdiagramm *nt*; **~-strain relationship** *n* ARCH & BUILD, CONCR, INFR & DES, MAT PROP, STEEL Druckdehnungsbeziehung *f*; **~ theory** *n* ARCH & BUILD, INFR & DES Spannungslehre *f*
**stretch** *vt* ARCH & BUILD, BUILD MACHIN, MAT PROP dehnen, spannen, strecken, verlängern
**stretched** *adj* ARCH & BUILD, MAT PROP gedehnt
**stretcher** *n* STONE Läufer *m*, Läuferstein *m*; **~ bond** *n* ARCH & BUILD, STONE Läuferverband *m*; **~ course** *n* ARCH & BUILD, STONE Läuferlage *f*, Läuferschicht *f*
**stretching** *n* ARCH & BUILD Einspannen *nt*, Festspannen *nt*, Spann-; **~ bond** *n* ARCH & BUILD, STONE Läuferverband *m*; **~ course** *n* ARCH & BUILD, STONE Läuferschicht *f*; **~ wire** *n* ARCH & BUILD Spanndraht *m*
**strickle**: **~ board** *n* ARCH & BUILD Abstreichplatte *f*, Schablone *f*
**strike 1.** *n* BUILD HARDW *lock* Schließwinkel *m*, Schließbügel *m*, Schließkasten *m*; **2.** *vt* CONCR ausschalen, entschalen; **~ off** INFR & DES abstreichen, STONE *tiles, plaster* abschlagen
**striker**: **~ bar** *n* ARCH & BUILD Anschlagschiene *f*
**striking**: **~ framework** *n* ARCH & BUILD Tragwerksabbau *m*; **~ plate** *n* BUILD HARDW Schließblech *nt*
**string** *n* ARCH & BUILD Gesimsband *nt*, Mauerband *nt*, Treppenwange *f*, BUILD HARDW Schnur *f*, STONE Mauerband *nt*; **~ board** *n* ARCH & BUILD, MAT PROP Wange *f*; **~ course** *n* ARCH & BUILD Gurtgesims *nt*
**stringer** *n* ARCH & BUILD Wange *f*, Holzgurtgesims *nt*, Längsbalken *m*, Stützbalken *m*, Treppenwange *f*, MAT PROP Wange *f*; **~ staircase** *n* ARCH & BUILD Wangentreppe *f*
***string***: **~ piece** *n* ARCH & BUILD Streckbalken *m*; **~ wall** *n* ARCH & BUILD *stairs* Wangenmauer *f*; **~ wreath** *n* ARCH & BUILD *handrail* Kropfstück *nt*, Krümmling *m*
**strip 1.** *n* ARCH & BUILD Streifen *m*, Band *nt*; **2.** *vt* CONCR ablösen, ausschalen, entschalen, INFR & DES abtragen, abziehen, SURFACE abbeizen; **~ framework** ARCH & BUILD ausschalen
***strip***: **~ flooring** *n* TIMBER Stabparkettfußboden *m*, Stabfußboden *m*, Stäbchenparkett *nt*; **~ foundation** *n* ARCH & BUILD, CONCR Streifenfundament *nt*; **~ iron** *n* STEEL Bandeisen *nt*
**striplight** *n* ELECTR Lichtband *nt*
***strip***: **~ load** *n* ARCH & BUILD, INFR & DES Streifenbelastung *f*
**stripped** *adj* CONCR entschalt
**stripping** *n* INFR & DES Abtragen *nt*, *floor* Abstreifen *nt*; **~ time** *n* CONCR Schalungsfrist *f*, Ausschalfrist *f*
***strip***: **~ steel** *n* MAT PROP, STEEL Bandstahl *m*
**strong**: **~-back** *n* ARCH & BUILD Traverse *f*

**struck** *adj* ARCH & BUILD *glass* angelaufen, *scaffold* abgebaut

**structural** *adj* ARCH & BUILD, CONCR, INFR & DES, MAT PROP, STEEL baulich, konstruktiv; **~ alteration** *n* ARCH & BUILD, CONST LAW bauliche Änderung *f*; **~ analysis** *n* ARCH & BUILD, INFR & DES Baustatik *f*, Berechnung *f* und Bemessung *f* von Baugliedern, Statik *f*, statischer Nachweis *m*; **~ ceramics** *n pl* STONE Baukeramik *f*; **~ component** *n* ARCH & BUILD Baugruppe *f*; **~ concrete** *n* ARCH & BUILD, CONCR Bauwerksbeton *m*, konstruktiver Beton *m*; **~ design** *n* ARCH & BUILD Bauentwurf *m*, konstruktive Ausbildung *f*, CONST LAW Bauentwurf *m*, INFR & DES Bauentwurf *m*, Bemessung *f*; **~ designer** *n* ARCH & BUILD, INFR & DES Statiker *m*; **~ details** *n pl* ARCH & BUILD, INFR & DES konstruktive Details *nt pl*; **~ dimension** *n* ARCH & BUILD Baumaß *nt*; **~ element** *n* ARCH & BUILD Montageelement *nt*, Bauelement *nt*; **~ engineer** *n* ARCH & BUILD, INFR & DES Statiker *m*; **~ engineering** *n* ARCH & BUILD, INFR & DES Bauingenieurwesen *nt*, Baustatik *f*, Bautechnik *f*, Hochbau *m*, Ingenieurhochbau *m*; **~ fin** *n* ARCH & BUILD Konstruktionsrippe *f*; **~ form** *n* ARCH & BUILD konstruktive Form *f*; **~ grade steel** *n* STEEL Konstruktionsbaustahl *m*; **~ hierarchy** *n* ARCH & BUILD strukturelle Hierarchie *f*; **~ iron** *n* STEEL Baueisen *nt*, Flußeisen *nt*; **~ light concrete** *n* CONCR konstruktiver Leichtbeton *m*, Konstruktionsleichtbeton *m*, tragender Leichtbeton *m*; **~ lightweight concrete** *n* CONCR konstruktiver Leichtbeton *m*, Konstruktionsleichtbeton *m*, tragender Leichtbeton *m*; **~ module** *n* ARCH & BUILD Bauraster *m*, Raster *m*; **~ opening dimensions** *n pl* ARCH & BUILD *door frame* Bauöffnungsmaße *nt pl*; **~ reinforcement** *n* ARCH & BUILD, CONCR, INFR & DES konstruktive Bewehrung *f*; **~ steel** *n* ARCH & BUILD, INFR & DES, MAT PROP, STEEL Baustahl *m*, konstruktiv verlegter Baustahl *m*; **~ steel engineering** *n* ARCH & BUILD, STEEL Stahlbautechnik *f*; **~ steel section** *n* STEEL Baustahlprofil *nt*; **~ steel work** *n* ARCH & BUILD, STEEL Stahlbau *m*; **~ theory** *n* ARCH & BUILD, INFR & DES Baukonstruktionslehre *f*; **~ timber** *n* TIMBER Bauholz *nt*

**structure** *n* ARCH & BUILD Bau *m*, Baukörper *m*, Bauwerk *nt*, Konstruktion *f*, INFR & DES Bau *m*, Bauwerk *nt*, Struktur *f*, Gefüge *nt*; **~-borne sound** *n* SOUND & THERMAL Körperschall *m*; **~-borne sound absorber** *n* SOUND & THERMAL Körperschalldämpfer *m*

**structured** *adj* STEEL *sheet metal* strukturiert

**strut 1.** *n* ARCH & BUILD Kopfband *nt*, Sprengstrebe *f*, Bandholz *nt*, Steife *f*, Druckglied *nt*, BUILD MACHIN Sprengstrebe *f*, INFR & DES, MAT PROP Bug *m*, TIMBER Bug *m*, Kopfband *nt*, Sprengstrebe *f*; **2.** *vt* ARCH & BUILD versteifen, absteifen

**strutted**: **~ frame** *n* ARCH & BUILD, TIMBER Sprengwerk *nt*; **~ ridge purlin** *n* TIMBER Kopfbandpfette *f*; **~ roof** *n* TIMBER Sprengwerk *nt*

**strutting** *n* ARCH & BUILD Verstärkung *f*, Versteifung *f*, Versteifen *nt*, Verstreben *nt*; **~ board** *n* ARCH & BUILD Spannbohle *f*; **~ head** *n* ARCH & BUILD Strebenkopf *m*

**stub** *n* ARCH & BUILD Baumstumpf *m*, Stumpf *m*, Stutzen *m*; **~ cable** *n* ELECTR Abzweigkabel *nt*, Stichleitung *f*; **~ mortice** *n* (*see stub mortise*) ARCH & BUILD kurzes Zapfenloch *nt*; **~ mortise** *n* ARCH & BUILD kurzes Zapfenloch *nt*; **~ stack** *n* WASTE WATER Stichfallrohr *nt*; **~ tenon** *n* ARCH & BUILD kurzer Zapfen *m*

**stucco** *n* STONE Gipsputz *m*, Putz *m*, Stuck *m*; **~ ceiling** *n* STONE Gipsdecke *f*, Stuckdecke *f*; **~ mortar** *n* SURFACE Stuckmörtel *m*; **~ work** *n* STONE Stukkatur *f*, Stuckarbeit *f*

**stud** *n* ARCH & BUILD Bundständer *m*, *pillar stanchion* Stiel *m*, Bundstiel *m*, Bundpfosten *m*, *framework* Gerippe *nt*, *head post* Bundsäule *f*, Ständer *m*, CONCR Gerippe *nt*; **~ bolt** *n* ARCH & BUILD Gewindestift *m*

**studded**: **~ link cable chain** *n* STEEL Stegkette *f*

***stud***: **~ link** *n* ARCH & BUILD Steg *m*; **~ link cable chain** *n* ARCH & BUILD Stegkette *f*; **~ partition** *n* ARCH & BUILD Ständerwand *f*; **~ union** *n* ARCH & BUILD Ständerverbindung *f*; **~ wall** *n* ARCH & BUILD Gerippetrennwand *f*, Gerippewand *f*, Ständerwand *f*, Fachwerkwand *f*, STEEL, TIMBER Ständertrennwand *f*; **~ welding** *n* STEEL Bolzenschweißen *nt*; **~ welding gun** *n* STEEL Bolzenschweißpistole *f*

**study** *n* ARCH & BUILD, CONST LAW, INFR & DES Studie *f*

**stuff 1.** *adj* SOUND & THERMAL steif; **2.** *n* ARCH & BUILD, STONE Putzmasse *f*, Putzmörtel *m*

**stuffing** *n* ARCH & BUILD, BUILD HARDW, INFR & DES Packung *f*, Füllung *f*

**stump** *n* ARCH & BUILD Baumstumpf *m*, *of tower* Turmansatz *m*, INFR & DES Stubben *m*

**sturdy** *adj* ARCH & BUILD massiv; **~ design** *n* ARCH & BUILD robuste Ausführung *f*

**style** *n* ARCH & BUILD Stil *m*, Baustil *m*, Bauart *f*

**subbase** *n* ARCH & BUILD, CONCR Sauberkeitsschicht *f*, INFR & DES Unterbau *m*, Sauberkeitsschicht *f*, Frostschutzschicht *f*

**subbasement** *n* ARCH & BUILD zweites Kellergeschoß *nt*

**subconcrete** *n* CONCR Unterbeton *m*

**subcontractor** *n* CONST LAW Nachunternehmer *m*, Subunternehmer *m*

**subdivide** *vt* ARCH & BUILD unterteilen

**subfloor** *n* ARCH & BUILD Unterfußboden *m*, Unterboden *m*; **~ ventilation** *n* ARCH & BUILD Unterbodenbelüftung *f*

**subgrade** *n* INFR & DES Straßenbett *nt*, Untergrund *m*; **~ reaction** *n* INFR & DES Bodenpressung *f*

**subgroup** *n* ELECTR Untergruppe *f*

**subject**: **~ to** *adj* ARCH & BUILD, CONCR, MAT PROP, STEEL beansprucht; **~ to authorization** *adj* CONST LAW genehmigungspflichtig; **~ to bending** *adj* ARCH & BUILD, CONCR, MAT PROP, STEEL biegebeansprucht; **~ to buckling** *adj* ARCH & BUILD, CONCR, MAT PROP, STEEL knickbeansprucht; **~ to compression** *adj* ARCH & BUILD, CONCR, MAT PROP, STEEL druckbeansprucht; **~ to tension** *adj* ARCH & BUILD, CONCR, MAT PROP, STEEL zugbeansprucht

**submaster**: **~ key** *n* BUILD HARDW Gruppenschlüssel *m*

**submerged** *adj* WASTE WATER untergetaucht; **~ arc welding** *n* INFR & DES Unterpulverschweißen *nt*, STEEL Unterpulverschweißen *nt*, verdecktes Lichtbogenschweißen *nt*; **~ concrete** *n* CONCR Unterwasserbeton *m*

**submersible** *adj* WASTE WATER absenkbar

**subpopulation**: **~ collective dose** *n* ENVIRON Bevölkerungsteildosis *f*

**subsealing** *n* CONCR Unterfüllung *f*

**subsequent**: **~ installation** *n* ARCH & BUILD, HEAT & VENT Nachinstallation *f*

**subsequently**: **~ adjustable** *adj* ELECTR, HEAT & VENT nachträglich verstellbar
**subside** *vi* ARCH & BUILD *ground* sacken, sich senken, zusammenfallen
**subsidence** *n* ARCH & BUILD Setzung *f*, ENVIRON *ground* Einsinken *nt*, INFR & DES Setzung *f*; **~ of ground** *n* INFR & DES Bodensenkung *f*; **~ rubbish** *n* ARCH & BUILD Einsturzschutt *m*
**subsidiary**: **~ shaft** *n* ARCH & BUILD Nebensäule *f*, Hilfssäule *f*
**subsoil** *n* ARCH & BUILD Baugrund *m*, INFR & DES Untergrund *m*, Unterboden *m*, Boden *m*, Baugrund *m*; **~ expertise** *n* CONST LAW, INFR & DES Bodengutachten *nt*
**substance** *n* MAT PROP Substanz *f*, Stoff *m*, Material *nt*
**substandard** *adj* ARCH & BUILD unter der Norm, minderwertig
**substrate** *n* MAT PROP Trägerschicht *f*
**substratum** *n* INFR & DES Unterbau *m*, Unterschicht *f*, Tragschicht *f*
**substructure** *n* ARCH & BUILD Unterkonstruktion *f*, Unterbau *m*
**subsurface** *n* INFR & DES Schicht *f*; **~ investigation** *n* INFR & DES Baugrunduntersuchung *f*; **~ repository** *n* ENVIRON Untertagedeponie *f* (*UTD*)
**suburb** *n* INFR & DES Vorort *m*
**subway** *n AmE* (*cf underground BrE*) INFR & DES U-Bahn *f*, Untergrundbahn *f*, Unterführung *f*; **~ station** *n AmE* (*cf underground station BrE*) INFR & DES U-Bahnhof *m*
**suction** *n* ELECTR, HEAT & VENT Absaugen *nt*, Saugen *nt*; **~ filter** *n* ENVIRON, INFR & DES, WASTE WATER Saugfilter *nt*; **~ head** *n* INFR & DES Saughöhe *f*; **~ load** *n* ARCH & BUILD Sauglast *f*; **~ pipe** *n* HEAT & VENT, WASTE WATER Saugrohr *nt*; **~ port** *n* BUILD MACHIN, ENVIRON Saugmund *m*; **~ pump** *n* BUILD MACHIN, WASTE WATER Saugpumpe *f*; **~ side** *n* HEAT & VENT *pump, fan* Saugseite *f*; **~ water level** *n* WASTE WATER Saugwasserspiegel *m*
**suggest** *vt* ARCH & BUILD vorschlagen
**suitability** *n* ARCH & BUILD, INFR & DES Eignung *f*, Brauchbarkeit *f*; **~ of materials** *n* CONST LAW, MAT PROP Materialientauglichkeit *f*
**suitable** *adj* ARCH & BUILD geeignet, angemessen, brauchbar, passend, CONST LAW angemessen, INFR & DES brauchbar; **~ for wheelchairs** *adj* MAT PROP rollstuhlgeeignet
**sulfate** *AmE see sulphate BrE*
**sulfur** *AmE see sulphur BrE*
**sulfuric**: **~ acid** *AmE see sulphuric acid BrE*
**sulfurous** *AmE see sulphurous BrE*
**sulphate** *n BrE* ENVIRON, MAT PROP Sulfat *nt*; **~ attack** *n BrE* MAT PROP *concrete* Sulfatangriff *m*; **~-resistant cement** *n BrE* CONCR, MAT PROP sulfatbeständiger Zement *m*; **~-resisting Portland cement** *n BrE* CONCR, MAT PROP sulfatbeständiger Portlandzement *m*
**sulphur** *n BrE* ENVIRON, MAT PROP Schwefel *m*; **~ cement** *n BrE* MAT PROP Schwefelzement *m*; **~ dioxide** *n BrE* ENVIRON Schwefeldioxid *nt*; **~ dioxide reduction** *n BrE* ENVIRON Schwefeldioxidreduktion *f*
**sulphuric**: **~ acid** *n BrE* ENVIRON, MAT PROP Schwefelsäure *f*; **~ anhydride** *n BrE* ENVIRON Schwefelsäurenanhydrid *nt*
**sulphurous** *adj BrE* MAT PROP schwefelig; **~ acid** *n BrE* ENVIRON schweflige Säure *f*
***sulphur***: **~ oxide** *n BrE* ENVIRON Schwefeloxid *nt*; **~ recovery plant** *n BrE* ENVIRON Anlage *f* zur Schwefelrückgewinnung
**sum** *n* ARCH & BUILD Summe *f*, Betrag *m*
**summation** *n* INFR & DES Summierung *f*; **~ curve** *n* INFR & DES Summenganglinie *f*
**summer** *n* ARCH & BUILD Unterzug *m*; **~ beam** *n* ARCH & BUILD, INFR & DES Rähmstück *nt*, Sturzbalken *m*; **~ stone** *n* ARCH & BUILD Kragstein *m*; **~ tree** *n* TIMBER Geschoßquerbalken *m*
**summit** *n* ARCH & BUILD Gipfel *m*, Kuppe *f*, Spitze *f*
**sump**: **~ pan** *n* INFR & DES Sammelbehälter *m*, Wanne *f*
**sun**: **~-dried brick** *n* STONE Grünling *m*, Lehmbaustein *m*
**sunken**: **~ road** *n* INFR & DES Straße *f* in Tieflage
**sunk**: **~ freeway** *n AmE* (*cf sunk motorway*) INFR & DES Straße *f* in Tieflage; **~ motorway** *BrE n* (*cf sunk freeway AmE*) INFR & DES Straße *f* in Tieflage
**superelevated** *adj* INFR & DES *road, railtracks* überhöht
**superelevation** *n* INFR & DES *rails, roads* Überhöhung *f*
**superheating** *n* HEAT & VENT Überhitzen *nt*, Überhitzung *f*
**superhigh**: **~ frequency** *n* ELECTR superhohe Frequenz *f*
**superimposed**: **~ load** *n* ARCH & BUILD, INFR & DES Auflast *f*, Nutzlast *f*, Verkehrslast *f*, Zusatzlast *f*
**superplastisizer** *n* MAT PROP *concrete* starker Weichmacher *m*
**superposition** *n* INFR & DES Überlagerung *f*
**superpressure** *n* HEAT & VENT, WASTE WATER Überdruck *m*
**superstructure** *n* ARCH & BUILD, INFR & DES Überbau *m*, Aufbau *m*, Oberbau *m*
**supersulfated**: **~ cement** *AmE see supersulphated cement BrE*
**supersulphated**: **~ cement** *n BrE* CONCR, MAT PROP Gipsschlackenzement *m*, Sulfathüttenzement *m* (*SHZ*)
**supervise** *vt* ARCH & BUILD überwachen
**supervising**: **~ authority** *n* CONST LAW, INFR & DES Aufsichtsbehörde *f*
**supervision** *n* ARCH & BUILD, CONST LAW Aufsicht *f*, Bauleitung *f*, Überwachung *f*
**supplementary** *adj* ARCH & BUILD, INFR & DES zusätzlich; **~ boiler** *n* HEAT & VENT Hilfskessel *m*; **~ regulation** *n* CONST LAW, INFR & DES Ergänzungsbestimmung *f*; **~ structures** *n pl* ARCH & BUILD Ergänzungsbauten *m pl*; **~ work** *n* CONST LAW Nebenleistung *f*
**supplier** *n* CONST LAW Lieferant *m*
**supply 1.** *n* ARCH & BUILD Lieferung *f*, ELECTR Versorgung *f*, Zuführung *f*, HEAT & VENT Versorgung *f*, Vorlauf *m*, WASTE WATER Versorgung *f*; **2.** *vt* ARCH & BUILD liefern
***supply***: **~ air** *n* HEAT & VENT Zuluft *f*; **~ line** *n* ELECTR, HEAT & VENT, INFR & DES, WASTE WATER Versorgungsleitung *f*; **~ mains** *n pl* ELECTR, INFR & DES einspeisendes Netz *nt*; **~ pipe** *n* HEAT & VENT, INFR & DES, WASTE WATER Zuleitung *f*, Zuleitungsrohr *nt*; **~ system** *n* INFR & DES Versorgungsnetz *nt*; **~ temperature** *n* HEAT & VENT Vorlauftemperatur *f*
**support 1.** *n* ARCH & BUILD Auflage *f*, Stütze *f*, Abfangung *f*, Unterlage *f*, Unterstützung *f*, Ständer

*m*; **2.** *vt* ARCH & BUILD lagern, stützen, tragen, unterstützen

***support***: **~ conditions** *n pl* ARCH & BUILD Auflagerbedingungen *f pl*

**supported** *adj* ARCH & BUILD aufgelagert, aufliegend; **~ beam** *n* ARCH & BUILD Stützbalken *m*

**supporting** *adj* ARCH & BUILD stützend, tragend; **~ cable** *n* ARCH & BUILD Tragseil *nt*; **~ column** *n* ARCH & BUILD Tragsäule *f*; **~ frame** *n* ARCH & BUILD Auflagerrahmen *m*, Stützgerüst *nt*; **~ layer** *n* INFR & DES Stützschicht *f*; **~ structure** *n* ARCH & BUILD tragende Konstruktion *f*, Tragkonstruktion *f*, Stützkonstruktion *f*, INFR & DES Tragkonstruktion *f*; **~ wall** *n* ARCH & BUILD Stützmauer *f*

**surcharge** *n* ARCH & BUILD, CONST LAW Extrakosten *pl*, Zulage *f*, INFR & DES Zulage *f*, Überbelastung *f*

**surcharged**: **~ drain** *n* INFR & DES, WASTE WATER überbelasteter Kanal *m*

**surface 1.** *n* ARCH & BUILD, INFR & DES, SURFACE Oberfläche *f*, *surface area* Fläche *f*; **2.** *vt* ARCH & BUILD beschichten, SURFACE *material* beschichten, verkleiden, überziehen

***surface***: **~-acting agent** *n* MAT PROP, SURFACE oberflächenaktiver Stoff *m*; **~-active agent** *n* SURFACE Netzmittel *nt*; **~ area** *n* ENVIRON *of land, water* Oberflächenbereich *m*, *of reservoir* Staufläche *f*; **~ bearing** *n* ARCH & BUILD ebenes Lager *nt*; **~ characteristics** *n pl* ARCH & BUILD, CONCR, SURFACE Oberflächeneigenschaften *f pl*; **~ demarcation** *n* SURFACE *surveying* Oberflächenvermarkung *f*; **~ digging** *n* INFR & DES Flachbaggerung *f*; **~ drainage** *n* INFR & DES Oberflächenentwässerung *f*; **~ dressing** *n* INFR & DES Verschleißschicht *f*, *roads* Oberflächenbehandlung *f*; **~ earthing** *n BrE* (*cf surface grounding AmE*) ELECTR Oberflächenerdung *f*; **~ earthing electrode** *n BrE* (*cf surface grounding electrode AmE*) ELECTR Oberflächenerder *m*; **~ erosion** *n* INFR & DES Oberflächenerosion *f*; **~ finishing** *n* SURFACE Oberflächenbearbeitung *f*; **~ friction** *n* INFR & DES, MAT PROP Oberflächenreibung *f*; **~ grounding** *n AmE* (*cf surface earthing BrE*) ELECTR Oberflächenerdung *f*; **~ grounding electrode** *n AmE* (*cf surface earthing electrode BrE*) ELECTR Oberflächenerder *m*; **~ method** *n* ENVIRON *disposal technique* Oberflächenmethode *f*; **~-mounted** *adj* ELECTR, HEAT & VENT auf Putz, WASTE WATER auf Putz, auf Putz montiert; **~-mounted light fixture** *n* ELECTR Aufbauleuchte *f*; **~-mounted lock** *n* BUILD HARDW aufliegendes Schloß *nt*; **~ preparation** *n* SURFACE Oberflächenvorbehandlung *f*; **~ protection** *n* SURFACE Oberflächenschutz *m*; **~ resistance** *n* SOUND & THERMAL Oberflächenwiderstand *m*; **~ roughness** *n* SURFACE Oberflächenrauheit *f*; **~ runoff** *n* ENVIRON Abfluß *m* im Oberflächenbereich; **~ rust** *n* SURFACE Oberflächenrost *m*; **~ temperature** *n* HEAT & VENT, SURFACE Oberflächentemperatur *f*; **~ tension** *n* MAT PROP, SURFACE Oberflächenspannung *f*; **~ treatment** *n* SURFACE Oberflächenbehandlung *f*; **~ type** *n* ELECTR, HEAT & VENT, WASTE WATER Aufputzausführung *f*; **~-vibrating machine** *n* BUILD MACHIN Flächenrüttler *m*; **~ vibrator** *n* BUILD MACHIN Flächenrüttler *m*; **~ water** *n* ENVIRON, INFR & DES, SURFACE, WASTE WATER Oberflächenwasser *nt*, Straßenabwasser *nt*

**surge**: **~ tank** *n* ENVIRON, HEAT & VENT, INFR & DES Ausgleichbecken *nt*, Druckausgleichbehälter *m*

**surround** *n* ARCH & BUILD Einfassung *f*, Randeinfassung *f*, Abschluß *m*, Rand *m*, TIMBER Einfassung *f*

**surrounding** *adj* ENVIRON, HEAT & VENT, INFR & DES, SOUND & THERMAL umgebend

**surroundings** *n pl* ARCH & BUILD, CONST LAW, INFR & DES Umgebung *f*

**surveillance** *n* CONST LAW Überwachung *f*

**survey 1.** *n* ARCH & BUILD, INFR & DES Aufmaß *nt*, Überblick *m*, Aufnahme *f*, Baugutachten *nt*, Vermessung *f*, Übersicht *f*, Maßaufnahme *f*, *of site* Einmessen *nt*; **2.** *vt* ARCH & BUILD, INFR & DES *measure* vermessen, übersehen

***survey***: **~ of buildings and site** *n* ARCH & BUILD bauliche Aufnahme *f*

**surveying** *n* ARCH & BUILD Vermessung *f*, Meßkunde *f*, Vermessen *nt*, *inspection* Besichtigung *f*, INFR & DES Vermessen *nt*, Vermessungswesen *nt*, Meßkunde *f*; **~ aneroid barometer** *n* ARCH & BUILD Metallbarometer *nt*

**surveyor** *n* ARCH & BUILD Landmesser *m*, Vermesser *m*, amtlicher Inspektor *m*, INFR & DES Vermesser *m*; **~'s chain** *n* ARCH & BUILD Absteckkette *f*; **~'s compass** *n* ARCH & BUILD, BUILD MACHIN Stativkompaß *m*; **~'s level** *n* ARCH & BUILD, BUILD MACHIN Nivelliergerät *nt*; **~'s staff** *n* ARCH & BUILD Absteckpfahl *m*; **~'s tape** *n* ARCH & BUILD Meßband *nt*

***survey***: **~ of site** *n* INFR & DES Abmessen *nt* der Baustelle

**survival**: **~ wind speed** *n* ENVIRON funktionsfähige Windgeschwindigkeit *f*

**suspend** *vt* ARCH & BUILD herunterhängen lassen, *ceiling* abhängen

**suspended** *adj* ARCH & BUILD hängend, aufgehängt; **~ beam** *n* ARCH & BUILD Schwebeträger *m*, Einhängeträger *m*; **~ ceiling** *n* ARCH & BUILD Unterdecke *f*, Hängedecke *f*, untergehängte Decke *f*, Zwischendecke *f*, abgehängte Decke *f*, BUILD HARDW abgehängte Decke *f*; **~ joint** *n* ARCH & BUILD schwebender Schienenstoß *m*; **~ matter** *n* ENVIRON, MAT PROP Schutzbehandlung *f*; **~ particle** *n* MAT PROP Schwebteilchen *nt*; **~ pile** *n* ARCH & BUILD, INFR & DES Mantelreibungspfahl *m*, schwebender Pfahl *m*; **~ rail** *n* BUILD HARDW Hängeschiene *f*; **~ roof** *n* ARCH & BUILD Hängedach *nt*; **~ scaffold** *n* ARCH & BUILD, BUILD MACHIN, INFR & DES Hängegerüst *nt*, Hängerüstung *f*, hängendes Gerüst *nt*; **~ span** *n* ARCH & BUILD *bridge* Einhängefeld *nt*; **~ suspension** *n* ARCH & BUILD, STEEL Deckenabhängung *f*; **~ trolley** *n* BUILD MACHIN Hängelaufkatze *f*

**suspender** *n* TIMBER Hängestange *f*, Hängesäule *f*

**suspension** *n* INFR & DES Schweb-, MAT PROP Suspension *f*; **~ boom** *n* ARCH & BUILD Hängegurtung *f*; **~ bracket** *n* BUILD HARDW Aufhängeeisen *nt*, Traverse *f*; **~ bridge** *n* INFR & DES Hängebrücke *f*; **~ railroad** *n AmE* (*cf suspension railway*) INFR & DES Hängebahn *f*, Schwebebahn *f*; **~ railway** *n BrE* (*cf suspension railroad AmE*) INFR & DES Hängebahn *f*, Schwebebahn *f*; **~ truss** *n* ARCH & BUILD Hängewerk *nt*

**sustained**: **~ loading** *n* ARCH & BUILD, INFR & DES Dauerbelastung *f*, Langzeitbelastung *f*

**sustaining**: **~ fluid** *n* INFR & DES, MAT PROP Stützflüssigkeit *f*

**swamp** *n* INFR & DES Sumpf *m*, Sumpfgebiet *nt*
**swampy** *adj* INFR & DES sumpfig; ~ **ground** *n* INFR & DES sumpfiger Boden *m*; ~ **soil** *n* INFR & DES Sumpfboden *m*
**swan**: ~ **neck** *n* ARCH & BUILD Schwanenhals *m*, Sprungrohr *nt*, WASTE WATER Etagenbogen *m*, S-Rohr *nt*, Schwanenhals *m*, Sprungrohr *nt*
**sway**: ~ **bracing** *n* ARCH & BUILD Querverband *m*, Querversteifung *f*, CONCR, MAT PROP, STEEL Querversteifung *f*; ~ **stiffness** *n* ARCH & BUILD, INFR & DES, STEEL Schwingsteifigkeit *f*
**sweep** *vt* ARCH & BUILD *chimney* kehren
**swelling** *n* MAT PROP Schwellung *f*; ~ **agent** *n* MAT PROP Schwellmittel *nt*; ~ **capacity** *n* MAT PROP Schwellungsvermögen *nt*; ~ **clay** *n* INFR & DES Blähton *m*; ~ **curve** *n* MAT PROP Schwellkurve *f*; ~ **ground** *n* INFR & DES Schwellboden *m*; ~ **pressure** *n* MAT PROP Schwelldruck *m*; ~ **zone** *n* INFR & DES Schwellzone *f*
**swell**: ~ **test** *n* INFR & DES, MAT PROP Schwelltest *m*
**swimming**: ~ **bath** *n* INFR & DES Schwimmbad *nt*; ~ **pool** *n* INFR & DES Schwimmbad *nt*
**swing 1.** *n* ARCH & BUILD Pendel *nt*; **2.** *vi* ARCH & BUILD schwingen
***swing***: ~ **bridge** *n* INFR & DES Schwenkbrücke *f*, Drehbrücke *f*; ~ **door** *n* ARCH & BUILD Pendeltür *f*; ~ **gate** *n* ARCH & BUILD Drehtor *nt*, Pendeltor *nt*
**swinging**: ~ **chute** *n* BUILD MACHIN Schwenkrinne *f*; ~ **door** *n* ARCH & BUILD Pendeltür *f*; ~ **platform** *n* ARCH & BUILD, BUILD MACHIN Schwenkbühne *f*; ~ **post** *n* ARCH & BUILD Torpfosten *m*; ~ **round** *n* ARCH & BUILD *crane*, BUILD MACHIN Schwenkradius *m*
**swirling**: ~ **chamber** *n* INFR & DES Wirbelkammer *f*; ~ **chamber control outlet** *n* INFR & DES Wirbelkammerdrossel *f*
**switch 1.** *n* ELECTR Schalter *m*, INFR & DES (*cf points BrE*) *railtrack* Weiche *f*; **2.** *vt* ELECTR schalten
***switch***: ~ **cabinet** *n* ELECTR Schaltschrank *m*; ~ **cock** *n* WASTE WATER Schalthahn *m*; ~ **engine** *n AmE* (*cf shunting engine BrE*) BUILD MACHIN Rangierlok *f*
**switching**: ~ **winch** *n AmE* (*cf shunting winch BrE*) BUILD MACHIN Rangierwinde *f*
**swivel**: ~ **bridge** *n* INFR & DES Schwenkbrücke *f*
**swiveling** *AmE see swivelling BrE*
**swivelling** *adj BrE* INFR & DES, WASTE WATER schwenkbar; ~ **mixer tap** *n BrE* (*cf swiveling mixer faucet AmE*) WASTE WATER Schwenkbatterie *f*
**swivel**: ~ **mixer tap** *n* WASTE WATER Schwenkbatterie *f*
**SWL** *abbr* (*safe working load*) ARCH & BUILD, CONST LAW, INFR & DES sichere Belastung *f*
**symmetric** *adj* ARCH & BUILD symmetrisch
**symmetry** *n* ARCH & BUILD Symmetrie *f*
**synchronous**: ~ **speed** *n* ENVIRON Synchrongeschwindigkeit *f*
**synthetic** *adj* MAT PROP künstlich, synthetisch; ~ **enamel** *n* SURFACE Kunstharzlack *m*; ~ **lining** *n* ENVIRON künstliche Abdichtung *f*; ~ **membrane** *n* MAT PROP Kunststoffmembran *f*; ~ **resin** *n* SURFACE Kunstharz *nt*; ~**-resin-bound** *adj* SURFACE kunstharzgebunden; ~ **resin plaster** *n* SURFACE Kunstharzputz *m*; ~ **resin varnish** *n* SURFACE Kunstharzlack *m*
**syphon 1.** *n* (*see siphon*) ENVIRON, WASTE WATER Siphon *m*; **2.** *vt* (*see siphon*) ENVIRON, MAT PROP hebern
**system** *n* ARCH & BUILD System *nt*, Anordnung *f*, Anlage *f*, *pipes, wires* Netz *nt*, CONST LAW Anordnung *f*, System *nt*, ELECTR, HEAT & VENT Anlage *f*, INFR & DES Anlage *f*, Anordnung *f*, System *nt*
**systematic**: ~ **errors** *n pl* ARCH & BUILD systematische Fehler *m pl*
***system***: ~ **building construction** *n* ARCH & BUILD Fertigteilbau *m*
**Système**: ~ **International** *n* (*SI*) ARCH & BUILD, INFR & DES, MAT PROP Système *nt* International (*SI*)
***system***: ~ **of pipes** *n* ARCH & BUILD, INFR & DES, WASTE WATER Rohrnetz *nt*
**systems**: ~ **building** *n* ARCH & BUILD Systembau *m*

# T

**table** *n* ARCH & BUILD Tabelle *f*, TIMBER Tisch *m*; **~ of prices** *n* ARCH & BUILD, CONST LAW Preistabelle *f*
**tabletop** *n* ARCH & BUILD *kitchen* Arbeitsplatte *f*
**tacheometer** *n* ARCH & BUILD *surveying*, BUILD MACHIN Tachymeter *nt*
**tacheometry** *n* ARCH & BUILD *surveying* Tachymetrie *f*
**tack**: **~ coat** *n* SURFACE *roads* Bindeschicht *f*
**tackle** *n* BUILD MACHIN Flaschenzug *m*, Hebezeug *nt*
**tacky** *adj* SURFACE *paint* klebrig
**tail** *vt* STONE einbauen, einbinden
**tail**: **~ beam** *n* ARCH & BUILD Stichbalken *m*
**tailing** *n* ARCH & BUILD, STONE Einbauen *nt*, Einbinden *nt*
**tailings** *n pl* INFR & DES *screening residues* Überlaufgut *nt*; **~ dam** *n* INFR & DES Staumauer *f* aus übergroßen Schottersteinen; **~ pond** *n* ENVIRON Becken *nt* für Rückstände
**tailpiece** *n* ARCH & BUILD Muffenverbindung *f*, unterbrochener Träger *m*
**tail**: **~ water** *n* INFR & DES, WASTE WATER Stauwasser *nt*, Unterwasser *nt*
**take**: **~ down** *vt* ARCH & BUILD *scaffolding* abbauen, ausrüsten; **~ into account** *vt* ARCH & BUILD berücksichtigen; **~ out of service** *vt* BUILD MACHIN außer Betrieb nehmen
**talcumed** *adj* SURFACE talkumiert
**tallboy** *n* ARCH & BUILD hoher Schornsteinaufsatz *m*
**talus** *n* INFR & DES Hangschutt *m*
**tamp** *vt* BUILD MACHIN, INFR & DES stampfen
**tamped**: **~ concrete** *n* CONCR Druckbeton *m*, Stampfbeton *m*
**tamper** *n* BUILD MACHIN, INFR & DES Stampfbohle *f*, Stampfer *m*
**tamping** *n* BUILD MACHIN, INFR & DES Stampfen *nt*, *for blasting* Verdämmen *nt* von Bohrlöchern; **~ foot** *n* BUILD MACHIN Stempel *m*; **~ rod** *n* BUILD MACHIN, INFR & DES Stampfstange *f*
**tandem** *n* BUILD MACHIN, INFR & DES Tandem *nt*; **~ roller** *n* BUILD MACHIN Tandemwalze *f*; **~ vibrating roller** *n* BUILD MACHIN *road construction* Tandemvibrationswalze *f*
**tangent** *n* ARCH & BUILD, INFR & DES Tangente *f*; **~ point** *n* ARCH & BUILD, INFR & DES Tangentialpunkt *m*
**tank** *n* ARCH & BUILD Grundwasserwanne *f*, ENVIRON, HEAT & VENT, INFR & DES Behälter *m*, Tank *m*; **~-cleaning plant** *n* INFR & DES Gefäßreinigungsanlage *f*
**tanking** *n* ARCH & BUILD Wannengründung *f*, Grundwasserdichtungsschicht *f*
**tap** *n* (*cf faucet AmE*) ELECTR *conduit* Abzweig *m*, WASTE WATER Entnahmestelle *f*, *general domestic* Wasserhahn *m*, Hahn *m*, *out-flow* Auslaufventil *nt*
**tape 1.** *n* INFR & DES Maßband *nt*; **2.** *vt* BUILD HARDW, BUILD MACHIN verkleben, zukleben
**taper** *vt* ARCH & BUILD, INFR & DES, TIMBER abschrägen, zuspitzen
**tapered** *adj* ARCH & BUILD, INFR & DES spitz zulaufend; **~ haunch** *n* CONCR Auflagerschräge *f*; **~ pipe** *n* WASTE WATER Übergangsrohr *nt*; **~ tread** *n* ARCH & BUILD, STEEL, TIMBER Wendelstufe *f*, konisch zulaufender Treppentritt *m*
**tapering** *n* ARCH & BUILD, INFR & DES Einziehen *nt*, Schwächerwerden *nt*; **~ scale** *n* ARCH & BUILD, INFR & DES verkleinerter Maßstab *m*
**tapping**: **~ point** *n* WASTE WATER Entnahmestelle *f*, Zapfstelle *f*
***tap***: **~ water** *n* INFR & DES Leitungswasser *nt*
**tar** *n* INFR & DES, MAT PROP, SOUND & THERMAL, SURFACE Teer *m*; **~ boiler** *n* BUILD MACHIN, INFR & DES Teerkessel *m*
**target** *n* ARCH & BUILD *surveying* Fluchtstange *f*; **~ contract** *n* CONST LAW Zielvertrag *m*; **~ leveling rod** *AmE see target levelling rod BrE*; **~ leveling staff** *AmE see target levelling staff BrE*; **~ levelling rod** *n BrE* INFR & DES *surveying* Nivellierlatte *f* mit Anzeige; **~ levelling staff** *n BrE* INFR & DES *surveying* Nivellierlatte *f* mit Anzeige
**tariff** *n* ARCH & BUILD, CONST LAW, INFR & DES Tarif *m*
**tarmac 1.** *n* INFR & DES *civil engineering* Teermakadam *m*; **2.** *vt* INFR & DES, MAT PROP, SURFACE asphaltieren
**tarmacadam** *n* INFR & DES *civil engineering* Teermakadam *m*
**tarpaulin** *n* BUILD HARDW Segeltuchplane *f*, Teerleinwand *f*
**tarred**: **~ felt** *n* MAT PROP Teerpappe *f*
***tar***: **~-saturated paper** *n* MAT PROP, SOUND & THERMAL Teerbitumenpappe *f*; **~ sprayer** *n* BUILD MACHIN, INFR & DES Teerspritzmaschine *f*; **~ sprinkler** *n* BUILD MACHIN, INFR & DES Teerspritzgerät *nt*
**tax 1.** *n* CONST LAW Steuer *f*; **2.** *vt* CONST LAW besteuern
**taxation** *n* CONST LAW Steuer *f*
***tax***: **~ band** *n* CONST LAW steuerliche Einstufung *f*; **~ rate** *n* CONST LAW Steuersatz *m*
**T-beam** *n* ARCH & BUILD, STEEL T-Träger *m*
**tear** *vt* ARCH & BUILD, INFR & DES *material* zerreißen, *hole* reißen
***tear***: **~ resistance** *n* MAT PROP Einreißfestigkeit *f*; **~ strength** *n* MAT PROP Reißfestigkeit *f*, Einreißfestigkeit *f*
**technical** *adj* ARCH & BUILD, BUILD MACHIN, ELECTR, HEAT & VENT, INFR & DES, WASTE WATER technisch; **~ advisory service** *n* ARCH & BUILD, INFR & DES technische Beratung *f*; **~ college** *n* CONST LAW Berufsschule *f*; **~ equipment** *n* ELECTR, HEAT & VENT, WASTE WATER *domestic appliances* technische Einrichtung *f*, technische Gebäudeausstattung *f*
**Technical**: **~ Instruction on Waste Management** *n* ENVIRON Technische Anleitung *f* Abfall (*TAA*)
**technical**: **~ solution** *n* ARCH & BUILD technische Lösung *f*; **~ specifications** *n pl* ARCH & BUILD, BUILD MACHIN, CONST LAW, INFR & DES technische Angaben *f pl*; **~ staff** *n* ARCH & BUILD Fachpersonal *nt*
**technician** *n* ARCH & BUILD Techniker *m*, Technikerin *f*
**technique** *n* ARCH & BUILD, INFR & DES Technik *f*, Verfahren *nt*
**technology** *n* ARCH & BUILD, INFR & DES Technologie *f*

**telecommunication**: ~ **cable** *n* ELECTR, INFR & DES Fernmeldekabel *nt*

**telephone**: ~ **network** *n* ELECTR, INFR & DES Fernsprechnetz *nt*

**tellurometer** *n* BUILD MACHIN Tellurmesser *m*

**temper** *vt* ARCH & BUILD anfeuchten, CONCR anmachen, MAT PROP *mortar* anfeuchten, anmachen, STEEL tempern, anfeuchten, vergüten, STONE anmachen

**temperature** *n* ARCH & BUILD, CONCR, ENVIRON, HEAT & VENT, MAT PROP, SOUND & THERMAL Temperatur *f*; ~ **coefficient** *n* HEAT & VENT, SOUND & THERMAL Wärmebeiwert *m*; ~ **drop** *n* ENVIRON, HEAT & VENT Temperaturabfall *m*; ~ **effect** *n* ARCH & BUILD, CONCR, MAT PROP Temperatureinfluß *m*; ~ **gradient** *n* MAT PROP Temperaturgradient *m*; ~ **reinforcement** *n* CONCR Temperaturbewehrung *f*, Verteilungsbewehrung *f*, Verteilerbewehrung *f*; ~ **rise** *n* ENVIRON, HEAT & VENT Temperaturanstieg *m*

**tempered** *adj* STEEL gehärtet; ~ **glass** *n* ARCH & BUILD Einscheibensicherheitsglas *nt*, vorgespanntes Glas *nt*; ~ **safety glass** *n* CONST LAW, MAT PROP Sicherheitsglas *nt*

**tempering** *n* ARCH & BUILD, CONCR, MAT PROP *mortar*, STEEL, STONE Anfeuchten *nt*, Anmachen *nt*

**template** *n* ARCH & BUILD Schablone *f*

**temple** *n* ARCH & BUILD, INFR & DES Tempel *m*

**temporal**: ~ **fluctuation** *n* ENVIRON zeitliche Schwankung *f*; ~ **variation** *n* ENVIRON zeitliche Schwankung *f*

**temporary** *adj* ARCH & BUILD zeitweilig, provisorisch; ~ **bridge** *n* INFR & DES Behelfsbrücke *f*; ~ **deposit for hazardous waste** *n* ENVIRON Sonderabfallzwischenlager *nt*; ~ **installations** *n pl* ARCH & BUILD, INFR & DES Baubehelf *m*; ~ **load** *n* ARCH & BUILD, CONCR, INFR & DES, STEEL zeitweilige Belastung *f*; ~ **opening** *n* STONE vorläufige Öffnung *f*; ~ **storage** *n* ENVIRON *of refuse* vorläufige Lagerung *f*; ~ **stress** *n* ARCH & BUILD, CONCR, INFR & DES, STEEL Montagespannung *f*; ~ **structure** *n* ARCH & BUILD, CONST LAW, INFR & DES Behelfsbau *m*; ~ **water connection** *n* INFR & DES provisorischer Wasseranschluß *m*

**tender 1.** *n* ARCH & BUILD, CONST LAW, INFR & DES Angebot *nt*, Offerte *f*; **2.** *vt* CONST LAW anbieten

**tenderer** *n* CONST LAW Bieter *m*

**tendering** *n* ARCH & BUILD, CONST LAW, INFR & DES Ausschreibung *f*; ~ **date** *n* ARCH & BUILD, CONST LAW, INFR & DES Angebotsabgabetermin *m*

**tendon** *n* CONCR Spannglied *nt*

**tenon** *n* ARCH & BUILD, TIMBER Zapfen *m*; ~ **joint** *n* TIMBER Verzapfung *f*; ~ **saw** *n* ARCH & BUILD Ansatzsäge *f*, Fuchsschwanz *m*

**tensile** *adj* MAT PROP zugbelastbar, streckbar; ~ **force** *n* INFR & DES Zugkraft *f*; ~ **load** *n* INFR & DES Zuglast *f*; ~ **reinforcement** *n* CONCR Zugbewehrung *f*, Zugarmierung *f*; ~ **strength** *n* MAT PROP Zugfestigkeit *f*; ~ **strength of concrete** *n* ARCH & BUILD, CONCR Betonzugfestigkeit *f*; ~ **stress** *n* ARCH & BUILD, MAT PROP Zugspannung *f*

**tension 1.** *n* ARCH & BUILD Spannung *f*, Zug *m*; **2.** *vt* CONCR spannen, vorspannen

***tension***: ~ **bar** *n* ARCH & BUILD Zuganker *m*, *test piece* Zugstab *m*, *brace* Zugband *nt*, *rod* Zugglied *nt*, INFR & DES Zugstab *m*; ~ **boom** *n* ARCH & BUILD Zuggurt *m*; ~**-compensating member** *n* HEAT & VENT, WASTE WATER Dehnungsausgleicher *m*; ~ **curve** *n* MAT PROP Spannungskurve *f*

**tensioned** *adj* CONCR vorgespannt

***tension***: ~ **failure** *n* MAT PROP Zugversagen *nt*; ~ **flange** *n* ARCH & BUILD, INFR & DES, STEEL, TIMBER Zuggurt *m*

**tensioning** *n* ARCH & BUILD Spann-, Spannen *nt*; ~ **wire** *n* ARCH & BUILD Spanndraht *m*

***tension***: ~ **pile** *n* INFR & DES Zugpfahl *m*; ~ **rod** *n* ARCH & BUILD, INFR & DES Zugstab *m*; ~ **structure** *n* ARCH & BUILD zugbeanspruchte Konstruktion *f*; ~ **test** *n* INFR & DES, MAT PROP Zugversuch *m*

**tentative**: ~ **standard** *n* ARCH & BUILD, CONST LAW, INFR & DES, MAT PROP Vornorm *f*

**tent**: ~ **roof** *n* ARCH & BUILD Zeltdach *nt*

**term** *n* ARCH & BUILD Dauer *f*, CONST LAW Frist *f*, Dauer *f*, MAT PROP Dauer *f*

**terminal**: ~ **box** *n* ELECTR Anschlußkasten *m*, Klemmkasten *m*; ~ **building** *n* ARCH & BUILD, INFR & DES Empfangsgebäude *nt*

**termination**: ~ **of choir** *n* ARCH & BUILD Chorschluß *m*; ~ **of the contract** *n* CONST LAW Vertragsbeendigung *f*

**terms**: ~ **of delivery** *n pl* CONST LAW Lieferbedingungen *f pl*

**terrace 1.** *n* ARCH & BUILD Terrasse *f*; **2.** *vt* SURFACE treppenförmig anlegen

***terrace***: ~ **garden** *n* ARCH & BUILD Dachgarten *m*

**terracing** *n* INFR & DES Terrassieren *nt*, Anlegen *nt* von Terrassen

**terracotta** *n* STONE Terrakotta *f*

**terrazzo** *n* STONE Terrazzo *m*

**territory** *n* CONST LAW, INFR & DES Gebiet *nt*

**terry**: ~ **cloth** *n* MAT PROP *carpet* Schlingengewebe *nt*

**tertiary**: ~ **sewage treatment** *n* ENVIRON dritte Reinigungsstufe *f*

**tessera** *n* STONE Mosaikstein *m*

**test 1.** *n* ARCH & BUILD, BUILD MACHIN, CONST LAW, ELECTR, HEAT & VENT Prüfung *f*, INFR & DES, Prüfung *f*, Test *m*, Versuch *m*; **2.** *vt* CONST LAW, ELECTR, HEAT & VENT, INFR & DES prüfen

***test***: ~ **apparatus** *n* BUILD MACHIN Prüfvorrichtung *f*; ~ **certificate** *n* CONST LAW Prüfbescheinigung *f*, Prüfzeugnis *nt*; ~ **core** *n* INFR & DES *mining* Bohrkern *m*; ~ **cube** *n* CONCR Betonprobewürfel *m*, Probewürfel *m*; ~ **cylinder** *n* CONCR Probezylinder *m*

**testing**: ~ **kit** *n* INFR & DES Prüfausrüstung *f*

***test***: ~ **load** *n* INFR & DES Prüflast *f*; ~ **mark** *n* CONST LAW Prüfzeichen *nt*; ~ **method** *n* CONST LAW, INFR & DES Prüfverfahren *nt*; ~ **piece** *n* INFR & DES Versuchsstück *nt*; ~ **pile** *n* INFR & DES Versuchspfahl *m*; ~ **pit** *n* INFR & DES Schürfgrube *f*; ~ **pressure** *n* HEAT & VENT, WASTE WATER Probedruck *m*; ~ **probe** *n* INFR & DES Sonde *f*; ~ **pumping** *n* INFR & DES Pumpversuch *m*; ~ **report** *n* CONST LAW Prüfbericht *m*; ~ **result** *n* INFR & DES Versuchsergebnis *nt*; ~ **rig** *n* MAT PROP Versuchsanlage *f*; ~ **sample** *n* INFR & DES Untersuchungsprobe *f*; ~ **specifications** *n pl* CONST LAW Prüfvorschriften *f pl*, Prüfbestimmungen *f pl*; ~ **standard** *n* CONST LAW Prüfnorm *f*; ~ **symbol** *n* CONST LAW Prüfzeichen *nt*

**texture** *n* MAT PROP Struktur *f*

**textured**: ~ **and profiled finishes** *n pl* CONCR, SURFACE strukturierte und profilierte Oberflächen

**T-girder** *n* ARCH & BUILD, STEEL T-Träger *m*

**thatch 1.** *n* ARCH & BUILD *reed* Reet *nt*, *straw* Stroh *nt*; **2.** *vt* ARCH & BUILD mit Reet decken, mit Stroh decken

**thatchboard** *n* MAT PROP Strohplatte *f*
**thatched**: ~ **roof** *n* ARCH & BUILD Strohdach *nt*
**thawing**: ~ **point** *n* HEAT & VENT, SOUND & THERMAL Taupunkt *m*
**theme**: ~ **park** *n* INFR & DES Freizeitpark *m*
**theodolite** *n* INFR & DES, MAT PROP Theodolit *m*
**theoretical** *adj* ARCH & BUILD, INFR & DES theoretisch
**theory** *n* ARCH & BUILD, CONST LAW, INFR & DES Theorie *f*; ~ **of stability** *n* ARCH & BUILD, INFR & DES Stabilitätstheorie *f*; ~ **of structures** *n* ARCH & BUILD, INFR & DES Baustatik *f*, Statik *f*; ~ **of thin shells** *n* ARCH & BUILD, INFR & DES Schalentheorie *f*
**thermal**: ~ **balance** *n* HEAT & VENT Wärmebilanz *f*; ~ **bridge** *n* ARCH & BUILD, SOUND & THERMAL Kältebrücke *f*, Wärmebrücke *f*; ~ **buffer** *n* SOUND & THERMAL thermischer Puffer *m*; ~ **circulation** *n* HEAT & VENT Wärmezirkulation *f*; ~ **class** *n* MAT PROP *insulating material* Wärmeklasse *f*; ~ **conduction** *n* MAT PROP, SOUND & THERMAL Wärmeleitung *f*; ~ **conductivity** *n* MAT PROP, SOUND & THERMAL Wärmeleitfähigkeit *f*; ~ **convection** *n* HEAT & VENT thermische Konvektion *f*; ~ **diffusivity** *n* HEAT & VENT, SOUND & THERMAL Temperaturleitzahl *f*; ~ **discharge** *n* ENVIRON, SOUND & THERMAL Abwärme *f*; ~ **energy** *n* HEAT & VENT Wärmeenergie *f*; ~ **expansion** *n* ARCH & BUILD, CONCR, INFR & DES, MAT PROP, STEEL, WASTE WATER Wärmeausdehnung *f*, Wärmedehnung *f*; ~ **expansion joint** *n* INFR & DES, WASTE WATER *pipe* Wärmedehnungsfuge *f*; ~ **fatigue** *n* MAT PROP Wärmeermüdung *f*; ~ **insulation** *n* INFR & DES, MAT PROP, SOUND & THERMAL Wärmedämmung *f*; ~ **insulation material** *n* INFR & DES, MAT PROP, SOUND & THERMAL Wärmedämmstoff *m*; ~ **insulation test** *n* INFR & DES, MAT PROP wärmeschutztechnische Prüfung *f*; ~ **insulation value** *n* MAT PROP Wärmeschutzwert *m*; ~ **load** *n* ENVIRON Wärmebelastung *f*; ~ **loss** *n* ENVIRON, SOUND & THERMAL Abwärme *f*
**thermally**: ~ **induced buoyancy** *n* ENVIRON thermisch erzeugter Auftrieb *m*; ~ **insulated** *adj* SOUND & THERMAL wärmegedämmt; ~ **insulated house** *n* ARCH & BUILD wärmegedämmtes Haus *nt*
**thermal**: ~ **mass** *n* ENVIRON, SOUND & THERMAL, STONE thermisch wirksame Masse *f*; ~ **output** *n* HEAT & VENT Wärmeabgabe *f*, Wärmeleistung *f*; ~ **overload protection** *n* ELECTR, HEAT & VENT thermischer Überlastungsschutz *m*; ~ **performance** *n* SOUND & THERMAL Wärmeleistung *f*; ~ **pollution** *n* ENVIRON Wärmebelastung *f*; ~ **properties** *n pl* MAT PROP thermische Eigenschaften *f pl*; ~ **radiation** *n* HEAT & VENT Wärmestrahlung *f*; ~ **reaction** *n* MAT PROP thermische Reaktion *f*; ~ **resistance** *n* MAT PROP, SOUND & THERMAL Wärmedurchgangswiderstand *m*, Wärmedurchlaßwiderstand *m*; ~ **shock** *n* SOUND & THERMAL Wärmestoßspannung *f*; ~ **stability** *n* MAT PROP Wärmebeständigkeit *f*; ~ **stress** *n* MAT PROP Temperaturspannung *f*; ~ **transmission factor** *n* MAT PROP Wärmedurchlässigkeitszahl *f*; ~ **transmittance coefficient** *n* SOUND & THERMAL Wärmeleitzahl *f*
**thermit**: ~ **welding** *n* STEEL aluminothermisches Schweißen *nt*
**thermopane**: ~**-glazed** *adj* SOUND & THERMAL isolierverglast; ~ **glazing** *n* MAT PROP, SOUND & THERMAL Isolierglas *nt*, Isolierverglasung *f*
**thermoplastic**: ~ **materials** *n pl* MAT PROP thermoplastische Kunststoffe *m pl*
**thermoplastics** *n* MAT PROP Thermoplaste *m pl*
**thermosensitive** *adj* MAT PROP wärmeempfindlich
**thermosetting**: ~ **plastic** *n* SURFACE Duroplast *m*; ~ **polyester powder coating** *n* SURFACE thermofixierte Polyesterpulverbeschichtung *f*
**thermosiphon** *n* HEAT & VENT Thermosyphon *m*
**thermostat** *n* HEAT & VENT Thermostat *m*, Wärmeregler *m*
**thermostatic** *adj* HEAT & VENT thermostatisch; ~ **blending valve** *n* WASTE WATER Thermostatbatterie *f*; ~ **mixer** *n* WASTE WATER Thermostatbatterie *f*; ~ **radiator valve** *n* HEAT & VENT thermostatisches Heizkörperventil *nt*; ~ **valve** *n* HEAT & VENT, WASTE WATER Thermostatbatterie *f*, Thermostatventil *nt*
**thick** *adj* ARCH & BUILD dick, stark, CONCR, INFR & DES, MAT PROP, SOUND & THERMAL stark
**thickness** *n* ARCH & BUILD Dicke *f*, Stärke *f*
**thick**: ~**-walled** *adj* ARCH & BUILD dickwandig
**thin** *adj* ARCH & BUILD, MAT PROP, SURFACE dünn; ~**-bed method** *n* STONE *tiling* Dünnbettverfahren *nt*
**T-hinge** *n* BUILD HARDW T-Band *nt*
**thinner** *n* MAT PROP, SURFACE Verdünner *m*
***thin***: ~ **sheet** *n* MAT PROP Feinblech *nt*; ~ **shell** *n* ARCH & BUILD, INFR & DES dünnwandige Schale *f*; ~**-walled** *adj* ARCH & BUILD dünnwandig; ~ **window glass** *n* ARCH & BUILD Dünnglas *nt*
**Thiokol**[R] *n* INFR & DES, MAT PROP Thiokol[R] *nt*; ~**-based**[R] *adj* INFR & DES, MAT PROP auf Thiokolbasis[R]
**third**: ~ **party testing** *n* CONST LAW, MAT PROP Testen *nt* durch Dritte
**thoroughfare** *n BrE* (*cf thruway AmE*) ARCH & BUILD *passageway* Durchfahrt *f*, INFR & DES *through road* Hauptverkehrsader *f*, Schnellstraße *f*
**thread** *n* BUILD MACHIN, HEAT & VENT, MAT PROP, STEEL, WASTE WATER Gewinde *nt*; ~**-cutting oil** *n* STEEL Gewindeschneidöl *nt*; ~ **depth** *n* BUILD MACHIN Gewindetiefe *f*
**threaded** *adj* BUILD MACHIN aufschraubbar, einschraubbar; ~ **bolt** *n* BUILD HARDW Gewindebolzen *m*; ~ **bushing** *n* HEAT & VENT, WASTE WATER Einschraubstutzen *m*; ~ **connection** *n* HEAT & VENT, WASTE WATER Gewindeanschluß *m*; ~ **earthing sleeve** *n BrE* (*cf threaded grounding sleeve AmE*) ELECTR Erdungsbuchse *f*; ~ **grounding sleeve** *n AmE* (*cf threaded earthing sleeve BrE*) ELECTR Erdungsbuchse *f*; ~ **hose connection** *n* BUILD HARDW, WASTE WATER Schlauchverschraubung *f*; ~ **hose joint** *n* BUILD HARDW, WASTE WATER Schlauchverschraubung *f*; ~ **pipe** *n* WASTE WATER Gewinderohr *nt*; ~ **pipe connection** *n* WASTE WATER Rohrverschraubung *f*; ~ **rod** *n* BUILD HARDW Gewindestab *m*, Gewindestange *f*; ~ **sleeve joint** *n* HEAT & VENT, WASTE WATER Muffenverschraubung *f*; ~ **socket** *n* HEAT & VENT, WASTE WATER Einschraubstutzen *m*
**threebay** *adj* ARCH & BUILD dreifeldrig, dreischiffig
**three**: ~**-centered arch** *AmE*, ~**-centred arch** *BrE n* ARCH & BUILD Dreizentrenbogen *m*, Korbbogen *m*, elliptisches Gewölbe *nt*; ~**-coat plaster** *n* STONE dreilagiger Putz *m*; ~**-coat work** *n* STONE dreilagiger Putz *m*; ~**-core block** *n* MAT PROP Dreikammerstein *m*; ~**-dimensional** *adj* ARCH & BUILD, INFR & DES dreidimensional; ~**-hinged arch** *n* ARCH & BUILD Dreigelenkbogen *m*; ~**-layer** *adj* STONE dreilagig;

**~-ply wood** *n* TIMBER Furnierplatte *f*, Dreilagenholz *nt*; **~-pole** *adj* ELECTR dreipolig; **~-quarter brick** *n* ARCH & BUILD, STONE Dreiviertelziegel *m*; **~-way cock** *n* WASTE WATER Dreiwegehahn *m*; **~-way mixer** *n* WASTE WATER Dreiwegemischer *m*; **~-way tap** *n* WASTE WATER Dreiwegehahn *m*; **~-way valve** *n* WASTE WATER Dreiwegeventil *nt*; **~-wheeled roller** *n* BUILD MACHIN Dreiradwalze *f*

**threshold** *n* ARCH & BUILD Schwelle *f*; **~ limit value** *n* (*TLV*) ENVIRON höchstzulässige Konzentration *f* (*HZK*)

**throat** *n* ARCH & BUILD Eintrittsöffnung *f*, Halsstück *nt*, Wasserschenkel *m*, BUILD HARDW Wasserschenkel *m*

**throttle 1.** *n* BUILD MACHIN *on engine* Drossel *f*, *valve* Drosselklappe *f*; **2.** *vt* BUILD MACHIN, HEAT & VENT drosseln

**throttled** *adj* BUILD MACHIN, HEAT & VENT, WASTE WATER gedrosselt

***throttle***: **~ slide valve** *n* HEAT & VENT Drosselschieber *m*; **~ valve** *n* HEAT & VENT Drosselventil *nt*, Drosselklappe *f*

**through**: **~ binder** *n* STONE *masonry* Zugbinder *m*; **~ column** *n* ARCH & BUILD durchgehende Säule *f*; **~ crack** *n* STONE durchgehender Riß *m*; **~ mortise** *n* ARCH & BUILD Zapfenschlitz *m*

**throughout** *adj* ARCH & BUILD vollflächig

***through***: **~ pillar** *n* ARCH & BUILD, BUILD MACHIN durchgehender Ständer *m*; **~ stone** *n* ARCH & BUILD Binderstein *m*; **~ tenon** *n* ARCH & BUILD Vollzapfen *m*; **~-wall flashing** *n* STEEL durchgehender Anschluß *m*; **~-wall flashing piece** *n* STEEL durchgehender Anschlußstreifen *m*

**throw**: **~ back into alignment** *vt* ARCH & BUILD *wall* wieder ausrichten; **~ back to waste** *vt* INFR & DES als Abfall zurückhalten

**throwing**: **~ on** *n* STONE *plaster* Anwerfen *nt*, Anwurf *m*

**thrust** *n* ARCH & BUILD, INFR & DES Schub *m*; **~ bearing** *n* ARCH & BUILD Axialdrucklager *nt*

**thruway** *n AmE* (*cf thoroughfare BrE*) ARCH & BUILD *passageway* Durchfahrt *f*, INFR & DES *through road* Hauptverkehrsader *f*, Schnellstraße *f*

**thumb**: **~ screw** *n* BUILD HARDW Flügelschraube *f*

**tidal**: **~ basin** *n* ENVIRON Flutbecken *nt*; **~ current** *n* ENVIRON Gezeitenstrom *m*; **~ movement** *n* ENVIRON Gezeitenbewegung *f*; **~ power** *n* ENVIRON, INFR & DES Gezeitenenergie *f*, Gezeitenkraft *f*; **~ power plant** *n* ENVIRON, INFR & DES Gezeitenkraftwerk *nt*; **~ power station** *n* ENVIRON, INFR & DES Flutkraftwerk *nt*, Gezeitenkraftwerk *nt*; **~ prism** *n* ENVIRON Gezeitenprisma *nt*; **~ range** *n* ENVIRON Gezeitenbereich *m*

**tide** *n* ENVIRON Gezeit *f*, INFR & DES Tide *f*; **~ gage** *AmE*, **~ gauge** *BrE n* ENVIRON Gezeitenmesser *m*; **~ lock** *n* INFR & DES Flutschleuse *f*; **~ mill** *n* ENVIRON Gezeitenmühle *f*

**tidual**: **~ limits** *n pl* ARCH & BUILD Vertrauensgrenzen *f pl*

**tie 1.** *n* ARCH & BUILD Verbindungsstück *nt*, Zugstange *f*, STEEL Zugstange *f*; **2.** *vt* ARCH & BUILD, STEEL verankern

***tie***: **~ anchor** *n* ARCH & BUILD Gewölbeanker *m*; **~ bar** *n* ARCH & BUILD Ankereisen *nt*, Ankerstab *m*, Zuganker *m*, Zugband *nt*, Zugglied *nt*, Zugstab *m*, INFR & DES Ankerstab *m*, Zugstab *m*, STEEL Ankerstab *m*; **~ beam** *n* ARCH & BUILD Spannbalken *m*

**tied**: **~ arch** *n* ARCH & BUILD Bogen *m* mit Zugband

***tie***: **~ plate** *n* ARCH & BUILD, INFR & DES, STEEL, TIMBER Bindeblech *nt*; **~ rod** *n* ARCH & BUILD Zuganker *m*; **~ wall** *n* ARCH & BUILD aussteifende Trennwand *f*

**tiewire** *vt* CONCR verrödeln

***tie***: **~ wire** *n* CONCR Schalungsspanndraht *m*, Rödeldraht *m*

**tight** *adj* SOUND & THERMAL dicht; **~ corner** *n* ARCH & BUILD geschlossene Ecke *f*

**tighten** *vt* ARCH & BUILD spannen

***tight***: **~ fit** *adj* MAT PROP, TIMBER fest angepaßt; **~ gravel** *n* MAT PROP dichtgelagerter Kies *m*

**tightness** *n* SOUND & THERMAL Dichtheit *f*

***tight***: **~ sheathing** *n* ARCH & BUILD, INFR & DES Spundschalung *f*

**tile** *n* BUILD HARDW, INFR & DES, MAT PROP, STONE *on floor, wall* Fliese *f*, *on stove* Kachel *f*, *on roof* Dachziegel *m*

**tiled**: **~ stove** *n* HEAT & VENT Kachelofen *m*

***tile***: **~ flooring** *n* ARCH & BUILD, TIMBER Fliesenfußboden *m*; **~ hanging** *n* STONE Plattenwandverkleidung *f*; **~ pattern** *n* ARCH & BUILD Fliesenmuster *nt*, Fliesenraster *nt*, Fliesenbild *nt*

**tiler** *n* ARCH & BUILD, STONE Fliesenleger *m*, Plattenleger *m*

**tiling** *n* ARCH & BUILD Fliesenbelag *m*; **~ mortar** *n* MAT PROP Fliesenmörtel *m*; **~ work** *n* ARCH & BUILD Fliesenarbeiten *f pl*

**tilt 1.** *n* ARCH & BUILD, INFR & DES Neigung *f*; **2.** *vi* ARCH & BUILD umkippen

**tiltable**: **~ tower** *n* ENVIRON kippbarer Turm *m*

**tilted** *adj* ARCH & BUILD *building* geneigt

**tilting**: **~ drum mixer** *n* BUILD MACHIN Schrägtrommelmischer *m*, Kipptrommelmischer *m*; **~ fillet** *n* STONE Trauflatte *f*; **~ gate** *n* ENVIRON Klappschütz *nt*

***tilt***: **~ and turn window** *n* ARCH & BUILD, MAT PROP Drehkippflügel *m*; **~-up method** *n* ARCH & BUILD Aufkippbauweise *f*, Richtaufbauweise *f*, Tilt-up-Bauweise *f*

**timber 1.** *n BrE* (*cf lumber AmE*) TIMBER Bauholz *nt*, Nutzholz *nt*, Schnittholz *nt*; **2.** *vt* TIMBER *house* mit Fachwerk versehen

***timber***: **~ baseboard** *n AmE* (*cf timber skirting BrE*) MAT PROP, TIMBER Holzsockelleiste *f*; **~ bridge** *n* TIMBER Holzbrücke *f*; **~ decay** *n* TIMBER Holzverfall *m*; **~ dog** *n* TIMBER Holzklammer *f*

**timbered** *adj* TIMBER eingeschalt; **~ shaft** *n* TIMBER abgesteifter Schacht *m*

***timber***: **~ formwork** *n* CONCR, TIMBER Holzschalung *f*; **~ frame** *n* TIMBER Holzrahmen *m*; **~ frame panel** *n* TIMBER Holzrahmentafel *f*; **~ frame wall** *n* TIMBER Holzfachwerkwand *f*; **~ framework** *n* TIMBER Riegelwand *f*; **~ framing** *n* TIMBER Holzfachwerk *nt*; **~ girder** *n* ARCH & BUILD, TIMBER Holzträger *m*

**timbering** *n* TIMBER Holzwerk *nt*, *boarding* Schalung *f*, *formwork, scaffolding* Einschalung *f*, *lagging, lining* Holzverkleidung *f*, Verschalung *f*

***timber***: **~ jack** *n* TIMBER Holzbock *m*; **~ lintel** *n* TIMBER Holzsturz *m*; **~ moisture** *n* MAT PROP, TIMBER Holzfeuchte *f*, Holzfeuchtigkeit *f*; **~ pavement** *n* TIMBER Holzpflaster *nt*; **~ paving** *n* TIMBER Holzpflaster *nt*; **~ pile** *n* TIMBER Holzpfahl *m*; **~ pillar** *n* TIMBER *framework* Ständer *m*; **~ preservative** *n* MAT PROP, SURFACE, TIMBER Holzschutzmittel *nt*; **~ quality** *n* CONST LAW, TIMBER Holzgüte *f*; **~ raft** *n* TIMBER Holzfloß *nt*; **~ rafter** *n* ARCH & BUILD, TIMBER

Holzsparren *m*; **~ roof structure** *n* TIMBER zimmermannsmäßige Dachkonstruktion *f*; **~ sheet piling** *n* INFR & DES Holzspundwand *f*; **~ skirting** *n BrE* (*cf timber baseboard AmE*) MAT PROP, TIMBER Holzsockelleiste *f*; **~ splitting wedge** *n* TIMBER Holzspaltkeil *m*; **~ structure** *n* TIMBER Holzbauwerk *nt*; **~ truss** *n* TIMBER Holzfachwerkträger *m*

**timberwork** *n* ARCH & BUILD Gebälk *nt*, TIMBER *framing of joists* Balkenlage *f*, Gebälk *nt*, Holzgebälk *nt*

***timber***: **~ yard** *n* TIMBER Bauhof *m*, Holzplatz *m*

**time** *n* ARCH & BUILD, CONST LAW, MAT PROP Zeit *f*; **~ consolidation curve** *n* MAT PROP Zeitkonsolidationskurve *f*; **~ limit** *n* ARCH & BUILD, CONST LAW, INFR & DES Fertigstellungstermin *m*; **~ schedule** *n* ARCH & BUILD, CONST LAW, INFR & DES Bauzeitenplan *m*; **~ switch** *n* CONST LAW, ELECTR Zeitschalter *m*

**timetable** *n* ARCH & BUILD, CONST LAW, INFR & DES Ablaufplan *m*

**tin 1.** *n* MAT PROP Zinn *nt*; **2.** *vt* SURFACE verzinnen

**tinning** *n* STEEL Verzinnen *nt*

**tinplate** *n* STEEL Weißblech *nt*; **~ waste** *n* ENVIRON Weißblechabfall *m*

**tint 1.** *n* SURFACE Ton *m*; **2.** *vt* SURFACE *colour* abtönen, aufhellen, tönen

**tinted** *adj* SURFACE getönt; **~ plexiglass pane** *n* ARCH & BUILD getönte Plexiglasscheibe *f*

**tip 1.** *n* ARCH & BUILD *blow torch flame* Brennermundstück *nt*, Brennerspitze *f*, ENVIRON Müllhalde *f*, Standort *m* der Deponie, Bodenkippe *f*, INFR & DES Müllhalde *f*; **2.** *vt* BUILD MACHIN *tilt* kippen, ENVIRON *refuse* ablagern, INFR & DES *tilt* kanten, kippen; **~ up** ARCH & BUILD umkippen, zuspitzen, INFR & DES zuspitzen

***tip***: **~-and-turn hardware** *n* BUILD HARDW Drehkippverschluß *m*; **~-and-turn sash** *n* ARCH & BUILD, MAT PROP Drehkippflügel *m*; **~-filling method** *n* ENVIRON Füllart *f* der Deponie

**tipper** *n* BUILD MACHIN Kipper *m*, Lore *f*, ENVIRON Kipper *m*; **~ truck** *n* BUILD MACHIN Kipper *m*, Wagenkipper *m*, ENVIRON Abfuhrwagen *m*, Kipper *m*

**tipping** *n* BUILD MACHIN Kippen *nt*, Umkippen *nt*, INFR & DES Kippen *nt*; **~ bucket** *n* BUILD MACHIN Kippkübel *m*; **~ with compaction** *n BrE* ENVIRON Verdichtungsdeponie *f*; **~ device** *n* BUILD MACHIN Kippvorrichtung *f*; **~ site** *n BrE* ENVIRON Deponiegelände *nt*

**T-iron** *n* MAT PROP, STEEL T-Eisen *nt*

**tissue** *n* MAT PROP Gewebe *nt*

**TLV** *abbr* (*threshold limit value*) ENVIRON HZK (*höchstzulässige Konzentration*)

***TLV***: **~ in the free environment** *n* ENVIRON HZK *f* in der Umwelt; **~ in the workplace** *n* ENVIRON HZK *f* am Arbeitsplatz

**toe**: **~ wall** *n* ARCH & BUILD Böschungsmauer *f*

**toggle** *n* BUILD HARDW Knebel *m*

**toilet** *n* BUILD HARDW, INFR & DES, WASTE WATER Abort *m*, Klosett *nt*, Toilette *f*; **~ cubicle** *n* INFR & DES WC-Kabine *f*; **~ installations** *n pl* INFR & DES, WASTE WATER Toilettenanlage *f*, WC-Anlage *f*; **~ seat** *n* BUILD HARDW WC-Brille *f*

**TOK**: **~-joint ribbon** *n* ARCH & BUILD TOK-Band *nt*

**tolerance** *n* ARCH & BUILD Toleranz *f*, INFR & DES Toleranz *f*, zulässige Abweichung *f*

**toll**: **~ bridge** *n* INFR & DES Mautbrücke *f*

**tongs** *n pl* BUILD MACHIN Zange *f*

**tongue** *n* ARCH & BUILD Feder *f*, Lasche *f*, Zugdeichsel *f*, Zunge *f*, BUILD MACHIN, STEEL Zugdeichsel *f*; **~-and-groove joint** *n* TIMBER Nut- und Federverbindung *f*, *spline joint* Federverbindung *f*, Spundverbindung *f*

**tongued**: **~-and-grooved** *adj* TIMBER gespundet; **~ flooring** *n* TIMBER gefederter Dielenfußboden *m*

***tongue***: **~ plane** *n* ARCH & BUILD Spundhobel *m*

**tonguing**: **~-and-grooving** *n* TIMBER Spundung *f*; **~-and-grooving machine** *n* TIMBER Spundmaschine *f*; **~ iron** *n* STEEL Spundeisen *nt*; **~ plane** *n* STEEL Spundhobel *m*

**tool** *n* BUILD MACHIN Werkzeug *nt*

**tooled**: **~ concrete finish** *n* CONCR bearbeitete Betonoberfläche *f*

***tool***: **~ shed** *n* ARCH & BUILD Geräteschuppen *m*

**tooth** *n* BUILD HARDW, BUILD MACHIN, STONE, TIMBER Zahn *m*

**toothed**: **~-plate connector** *n* BUILD HARDW, TIMBER Zahnankerplatte *f*

**toothing** *n* BUILD MACHIN *masonry* Verzahnung *f*, STONE Zahnung *f*, Verzahnung *f*; **~ plane** *n* BUILD MACHIN Zahnhobel *m*; **~ stone** *n* STONE *masonry* Zahnstein *m*

***tooth***: **~ plane** *n* BUILD MACHIN Zahnhobel *m*

**top 1.** *n* ARCH & BUILD, CONCR, INFR & DES, MAT PROP, STEEL, SURFACE *upper part* Oberteil *nt*, Spitze *f*, oberes Ende *nt*; **2.** *vt* SURFACE bedecken

***top***: **~ beam** *n* ARCH & BUILD Kehlbalken *m*; **~ boom** *n* ARCH & BUILD, CONCR, INFR & DES, MAT PROP, STEEL Obergurt *m*; **~ chord** *n* ARCH & BUILD, CONCR, INFR & DES, MAT PROP, STEEL Obergurt *m*; **~ coat** *n* SURFACE Überzugslack *m*; **~ edge** *n* ARCH & BUILD Oberkante *f*; **~ flange** *n* ARCH & BUILD, CONCR, INFR & DES, MAT PROP, STEEL Obergurt *m*; **~-hat cover piece** *n* MAT PROP Hutabdeckung *f*; **~-hat section** *n* MAT PROP Hutquerschnitt *m*; **~-hinged sash window** *n* ARCH & BUILD Klappflügelfenster *nt*; **~-hung window** *n* ARCH & BUILD Klappflügelfenster *nt*; **~-lit** *adj* INFR & DES oben beleuchtet

**topographical**: **~ survey** *n* ARCH & BUILD Vermessung *f*

**topography** *n* ARCH & BUILD, INFR & DES Topographie *f*

***top***: **~ panel** *n* ARCH & BUILD Deckplatte *f*

**topped**: **~-out** *adj* ARCH & BUILD rohbaufertig

**topping** *n* ARCH & BUILD Deckschicht *f*, INFR & DES *road surface* Straßendecke *f*, Deckschicht *f*, Druckschicht *f*, Verschleißschicht *f*; **~-out ceremony** *n* ARCH & BUILD Richtfest *nt*; **~ slab** *n* ARCH & BUILD Druckplatte *f*

***top***: **~ rail** *n* ARCH & BUILD Rahmholz *nt*

**topsoil** *n* ARCH & BUILD Mutterboden *m*, INFR & DES Oberboden *m*, Mutterboden *m*; **~ filling** *n* INFR & DES Oberbodenauftrag *m*; **~ stripping** *n* ARCH & BUILD, INFR & DES Mutterbodenabtrag *m*

***top***: **~ surface** *n* TIMBER Kopffläche *f*, Hirnschnittfläche *f*

**torch** *n* STEEL *welding* Brenner *m*; **~ brazing** *n* BUILD MACHIN Brennerlöten *nt*, Gaslöten *nt*; **~ for MIG-MAG welding** *n* BUILD MACHIN MIG-MAG-Schweißbrenner *m*; **~ for plasma welding** *n* BUILD MACHIN Plasmaschweißbrenner *m*; **~ for TIG welding** *n* BUILD MACHIN TIG-Schweißbrenner *m*

**torsion** *n* ARCH & BUILD Torsion *f*

**torsional**: **~-flexural buckling** *n* ARCH & BUILD, CONCR, MAT PROP, STEEL Biegedrillknicken *nt*; **~-flexural**

**buckling analysis** *n* ARCH & BUILD, CONCR, MAT PROP, STEEL Biegedrillknicknachweis *m*; ~ **rigid** *n* ARCH & BUILD Torsionsspannung *f*; ~ **stress** *n* ARCH & BUILD Torsionsspannung *f*

**torsion**: ~**-proof** *adj* ARCH & BUILD verwindungsfrei, verwindungssteif; ~ **test** *n* ARCH & BUILD Verwindungstest *m*

**tor**: ~ **steel** *n* STEEL TOR-Stahl *m*

**torus**: ~ **roll flashing** *n* BUILD MACHIN Rundwulstabdeckung *f*

**total** *n* ARCH & BUILD Betrag *m*, Gesamtbetrag *m*, Summe *f*; ~ **area** *n* ARCH & BUILD Gesamtfläche *f*; ~ **building costs** *n pl* ARCH & BUILD, CONST LAW, INFR & DES Gesamtbaukosten *pl*; ~ **capacity** *n* ARCH & BUILD Gesamtinhalt *m*; ~ **cost** *n* CONST LAW Gesamtkosten *pl*; ~ **cross-section** *n* ARCH & BUILD, INFR & DES Gesamtquerschnitt *m*; ~ **deposition** *n* ENVIRON Gesamtablagerung *f*; ~ **design load** *n* ARCH & BUILD, INFR & DES Gesamtlast *f*; ~ **earth resistance** *n BrE* (*cf total ground resistance AmE*) ELECTR Gesamterdungswiderstand *m*; ~ **efficiency** *n* INFR & DES Gesamtwirkung *f*; ~ **evacuation** *n* CONST LAW, INFR & DES Gesamtevakuierung *f*; ~ **friction** *n* ARCH & BUILD, INFR & DES Gesamtreibung *f*; ~ **ground resistance** *n AmE* (*cf total earth resistance BrE*) ELECTR Gesamterdungswiderstand *m*; ~ **hardness** *n* MAT PROP Gesamthärte *f*; ~ **height** *n* ARCH & BUILD Bauhöhe *f*; ~ **loss** *n* CONST LAW, MAT PROP Gesamtverlust *m*; ~ **pressure** *n* ARCH & BUILD, INFR & DES Gesamtdruck *m*; ~ **weight** *n* ARCH & BUILD, INFR & DES Gesamtgewicht *nt*

**touching**: ~**-up** *n* CONCR Nacharbeiten *nt*; ~**-up of edges** *n* CONCR Nacharbeiten *nt* der Kanten

**toughened**: ~ **glass** *n* CONST LAW Hartglas *nt*; ~ **safety glass** *n* ARCH & BUILD Einscheibensicherheitsglas *nt*, vorgespanntes Glas *nt*

**towed**: ~ **grader** *n* BUILD MACHIN Anhängestraßenhobel *m*

**tower** *n* ARCH & BUILD Gittermast *m*, Turm *m*, BUILD HARDW, INFR & DES Gittermast *m*; ~ **base** *n* ARCH & BUILD Turmfuß *m*, Fuß *m* des Turmes; ~ **block** *n* INFR & DES Wohnturm *m*; ~ **bolt** *n* INFR & DES Schubriegel *m*; ~ **construction** *n* ARCH & BUILD Turmbau *m*; ~ **crane** *n* BUILD MACHIN, INFR & DES Turmkran *m*, Turmdrehkran *m*; ~ **pier** *n* INFR & DES Turmpfeiler *m*; ~ **ring earth** *n BrE* (*cf tower ring ground AmE*) ELECTR Turmringerder *m*; ~ **ring ground** *n AmE* (*cf tower ring earth BrE*) ELECTR Turmringerder *m*; ~ **shaft** *n* ARCH & BUILD Turmschaft *m*

**town** *n* INFR & DES Stadt *f*; ~ **planning** *n* INFR & DES Städteplanung *f*

**toxic** *adj* CONST LAW, ENVIRON giftig; ~ **agent** *n* CONST LAW, ENVIRON, INFR & DES Giftstoff *m*, chemischer Kampfstoff *m*

**toxicant** *n* CONST LAW, ENVIRON, INFR & DES Giftstoff *m*

**toxic**: ~ **by-product** *n* CONST LAW, INFR & DES giftiges Nebenprodukt *nt*; ~ **degradation product** *n* ENVIRON toxisches Abfallprodukt *nt*; ~ **effect** *n* ENVIRON toxische Wirkung *f*; ~ **matter** *n* CONST LAW, ENVIRON, INFR & DES Giftstoff *m*; ~ **substance** *n* ENVIRON Schadstoff *m*; ~ **waste** *n* CONST LAW giftiger Abfall *m*, ENVIRON toxische Abfälle *m pl*, Giftmüll *m*, INFR & DES giftiger Abfall *m*; ~ **waste disposal plant** *n* ENVIRON Giftmüllentsorgungsanlage *f*; ~ **waste product** *n* ENVIRON toxisches Abfallprodukt *nt*

**toxin** *n* CONST LAW, ENVIRON, INFR & DES Gift *nt*

**T-piece** *n* HEAT & VENT, WASTE WATER *pipe* T-Stück *nt*

**t-piece**: ~ **union** *n* ARCH & BUILD T-förmige Verbindung *f*

**T-pipe** *n* HEAT & VENT, WASTE WATER T-Rohr *nt*

**trabeated** *adj* ARCH & BUILD aus Balken gebaut

**trace**: ~ **element** *n* ENVIRON, MAT PROP Spurenelement *nt*

**traceried**: ~ **gable** *n* ARCH & BUILD Maßwerkgiebel *m*

**tracery** *n* ARCH & BUILD Maßwerk *nt*, Netzwerk *nt*

**track** *n* INFR & DES Gleis *nt*, Schiene *f*, *progression* Lauf *m*; ~ **ballast** *n* INFR & DES Gleisschotter *m*; ~ **base plates** *n pl* INFR & DES Gleistrageplatten *f pl*; ~ **bedding** *n* INFR & DES Gleisbett *nt*; ~ **connection** *n* INFR & DES Gleisverbindung *f*

**tracked**: ~ **tractor** *n* BUILD MACHIN Kettenzugmaschine *f*

***track***: ~ **laying** *n* INFR & DES Gleisverlegung *f*; ~ **line** *n* INFR & DES Gleistrasse *f*

**traction** *n* ARCH & BUILD, BUILD MACHIN, ELECTR Kraftschluß *m*, Zugkraft *f*; ~ **relief** *n* ELECTR *cable* Zugentlastung *f*, Zugabfangung *f*; ~ **rod** *n* ARCH & BUILD, STEEL Zugstange *f*; ~ **system** *n* ARCH & BUILD, INFR & DES Liniennetz *nt*

**tractor** *n* BUILD MACHIN Schlepper *m*, Traktor *m*, *traction engine* Zugmaschine *f*; ~**-drawn roller** *n* BUILD MACHIN Anhängewalze *f*; ~ **scraper** *n* BUILD MACHIN Motorschürfkübel *m*

**trade** *n* ARCH & BUILD, CONST LAW, ENVIRON, INFR & DES *commodities, consumer products* Handel *m*, *industry, business* Gewerbe *nt*, Gewerk *nt*; ~ **journal** *n* ARCH & BUILD, INFR & DES Fachzeitschrift *f*; ~ **waste** *n* CONST LAW, ENVIRON gewerblicher Abfall *m*

**traffic** *n* ARCH & BUILD, CONST LAW, INFR & DES Verkehr *m*; ~ **cone** *n* INFR & DES Leitkegel *m*; ~ **facilities** *n pl* INFR & DES Verkehrsanlagen *f pl*; ~ **junction** *n* INFR & DES Verkehrsknotenpunkt *m*; ~ **load** *n* ARCH & BUILD, INFR & DES Verkehrslast *f*; ~ **route** *n* INFR & DES Verkehrsweg *m*; ~ **safety** *n* CONST LAW Verkehrssicherheit *f*; ~ **safety facilities** *n pl* CONST LAW, INFR & DES Verkehrssicherheitseinrichtungen *f pl*; ~ **sign** *n* CONST LAW, INFR & DES Verkehrsschild *nt*

**trailer** *n* BUILD MACHIN Untergestell *nt*; ~ **truck** *n* BUILD MACHIN Schleppzug *m*

**trajectory**: ~ **of stress** *n* ARCH & BUILD, MAT PROP Spannungstrajektorie *f*, Spannungsweg *m*

**tramcar** *n BrE* (*cf streetcar AmE*) INFR & DES Straßenbahn *f*

**tram**: ~ **track** *n* INFR & DES Schienengleis *nt*

**tramway**: ~ **switch** *n* INFR & DES Straßenbahnweiche *f*, Pflasterweiche *f*

**transboundary**: ~ **movement of waste** *n* ENVIRON grenzüberschreitende Abfallverbringung *f*

**transducer** *n* ELECTR Wandler *m*

**transfer 1.** *n* CONST LAW Überweisung *f*; **2.** *vt* ARCH & BUILD abtragen, übertragen, CONST LAW überweisen

***transfer***: ~ **table** *n* ARCH & BUILD Schiebebühne *f*

**transformation** *n* ELECTR Umwandlung *f*; ~ **rate** *n* ENVIRON Umwandlungsrate *f*

**transformer** *n* ELECTR Transformator *m*; ~ **room** *n* ELECTR Transformatorraum *m*, Traforaum

**transit** *n* INFR & DES *surveying instrument*, MAT PROP Theodolit *m*

**transition** *n* CONST LAW, HEAT & VENT, INFR & DES, WASTE WATER Übergang *m*; ~ **curve** *n* INFR & DES Übergangsbogen *m*; ~ **curve length** *n* INFR & DES

Übergangsbogenlänge *f*; ~ **layer** *n* INFR & DES Übergangsschicht *f*; ~ **piece** *n* HEAT & VENT, WASTE WATER Reduktionsstück *nt*; ~ **time** *n* CONST LAW Übergangszeit *f*; ~ **zone** *n* INFR & DES Übergangszone *f*

***transit***: ~ **line** *n* INFR & DES Vermessungsgrundlinie *f*; **~-mixed concrete** *n* CONCR Fertigbeton *m*

**translation** *n* INFR & DES *shifting, transposition* Verschiebung *f*

**transmission** *n* ELECTR *alarm* Übertragung *f*; ~ **length** *n* CONCR *prestressed concrete* Verbundlänge *f*, Eintragungslänge *f*; ~ **line** *n* ELECTR, INFR & DES Hochspannungsleitung *f*, Stromleitung *f*; ~ **of load** *n* ARCH & BUILD, INFR & DES Lastübertragung *f*; ~ **of motion** *n* ARCH & BUILD Bewegungsübertragung *f*

**transom** *n* ARCH & BUILD Kämpferriegel *m*, *window* Querholz *nt*, INFR & DES Unterzug *m*; ~ **light** *n* ARCH & BUILD Türoberlicht *nt*, Oberlicht *nt*; ~ **mounting** *n* BUILD HARDW *door closer* Kopfmontage *f*

**transparent** *adj* ARCH & BUILD, MAT PROP, SURFACE durchsichtig, lichtdurchlässig; ~ **copy** *n* ARCH & BUILD Transparentpause *f*; ~ **glass** *n* ARCH & BUILD, MAT PROP Klarglas *nt*; ~ **roof** *n* ARCH & BUILD Lichtdach *nt*; ~ **varnish** *n* SURFACE *chemically drying* Klarlack *m*

**transport 1.** *n* BUILD MACHIN *conveyance* Beförderung *f*, Förderung *f*, Transport *m*; **2.** *vt* ARCH & BUILD, BUILD MACHIN, MAT PROP *convey* befördern, transportieren

**transportation**: ~ **system** *n* INFR & DES Transportsystem *nt*

**transported** *adj* BUILD MACHIN transportiert

**transshipment**: ~ **of hazardous goods** *n* ENVIRON Gefahrgutumschlag *m*

**transverse** *adj* ARCH & BUILD, CONCR, INFR & DES, MAT PROP, STEEL quer; ~ **bracing** *n* ARCH & BUILD Querverband *m*, Windverband *m*; ~ **girder** *n* ARCH & BUILD Querträger *m*; ~ **joint** *n* ARCH & BUILD Querfuge *f*; ~ **reinforcement** *n* CONCR Querbewehrung *f*; ~ **rib** *n* ARCH & BUILD Querrippe *f*; ~ **slope** *n* INFR & DES Querneigungsgefälle *nt*; ~ **stiffener** *n* ARCH & BUILD, CONCR, MAT PROP, STEEL Querversteifung *f*; ~ **stiffening** *n* ARCH & BUILD, CONCR, MAT PROP, STEEL, TIMBER Queraussteifung *f*; ~ **strain** *n* ARCH & BUILD, INFR & DES Scherspannung *f*, Schubspannung *f*

**trap** *n* ARCH & BUILD *backpressure valve* Rückstauklappe *f*, Schlußstein *m*, INFR & DES *receptacle* Abscheider *m*, WASTE WATER *collecting basin* Auffanggefäß *nt*, *collecting mechanism* Auffangvorrichtung *f*, *receptacle* Abscheider *m*; ~ **door** *n* ARCH & BUILD Bodenluke *f*, Falltür *f*; ~ **elbow** *n* WASTE WATER Siphonbogen *m*

**trapezoidal**: ~ **sheet metal** *n* STEEL Trapezblech *nt*

**trapped**: ~ **humidity** *n* ARCH & BUILD Baufeuchte *f*

**trash**: ~ **rack** *n* ENVIRON, WASTE WATER Rechen *m*

**trass**: ~ **cement** *n* CONCR Traßzement *m*

**travel 1.** *n* BUILD MACHIN *of piston* Lauf *m*; **2.** *vi* SOUND & THERMAL *waves* sich ausbreiten

**traveling**: *AmE see travelling BrE*

**travelling**: ~ **crab** *n BrE* BUILD MACHIN *crane* Laufkatze *f*; ~ **cradle** *n BrE* ARCH & BUILD, BUILD MACHIN, INFR & DES Hängegerüst *nt*, Umziehgerüst *nt*; ~ **crane** *n BrE* BUILD MACHIN Laufkran *m*; ~ **form** *n BrE* CONCR Gleitschalung *f*; ~ **ladder** *n BrE* INFR & DES Schiebeleiter *f*; ~ **load** *n BrE* ARCH & BUILD, INFR & DES Verkehrslast *f*; ~ **platform** *n BrE* INFR & DES Schiebebühne *f*; ~ **trolley** *n BrE* BUILD MACHIN *crane* Laufkatze *f*; ~ **winch** *n BrE* INFR & DES Laufkatze *f*

**traverse** *n* ARCH & BUILD Traverse *f*

**travertin** *n* STONE *marble* Travertin *m*

**tread** *n* ARCH & BUILD Trittfläche *f*, Stufe *f*, Stufenbreite *f*, *stairs* Auftritt *m*; ~ **and riser** *n* ARCH & BUILD Winkelstufe *f*

**treat** *vt* ENVIRON aufbereiten, MAT PROP behandeln, verarbeiten, WASTE WATER klären

**treatment** *n* BUILD MACHIN, ENVIRON, INFR & DES Aufbereitung *f*, MAT PROP Behandlung *f*; ~ **plant** *n* ENVIRON, INFR & DES Aufbereitungsanlage *f*; ~ **process** *n* ENVIRON Aufbereitungsverfahren *nt*; ~ **of sewage sludge** *n* ENVIRON Behandlung *f* von Klärschlamm

**tree**: ~ **nail** *n* TIMBER Holznagel *m*

**trellis** *n* TIMBER Flechtwerk *nt*, Gitter *nt*; ~ **post** *n* TIMBER Gitterpfosten *m*; ~ **work** *n* STONE, TIMBER Flechtwerk *nt*, durchbrochenes Mauerwerk *nt*

**tremie** *n* CONCR Schüttrohr *nt*, Elefantenrüssel *m*, *funnel* Betoniertrichter *m*

**trench** *n* ARCH & BUILD Baugrube *f*, Graben *m*, Rinne *f*, ENVIRON, INFR & DES Graben *m*; ~ **excavator** *n* ARCH & BUILD Grabenbagger *m*, BUILD MACHIN Grabenfräser *m*, Grabenbagger *m*; ~ **for pipes and cables** *n* INFR & DES Leitungsgraben *m*

**trenching** *n* ARCH & BUILD Grabenaushub *m*, Grabenherstellung *f*

***trench***: ~ **landfill** *n* ENVIRON Grube *f*; ~ **method** *n* ENVIRON Grabenmethode *f*; ~ **sheeting** *n* ARCH & BUILD Grabenverbau *m*, INFR & DES Kanaldiele *f*

**trenchwork** *n* ARCH & BUILD, INFR & DES Grabenarbeiten *f pl*

**trestle** *n* ARCH & BUILD, INFR & DES Gerüstbock *m*, Gestell *nt*, Pfahljoch *nt*; ~ **bridge** *n* INFR & DES Bockbrücke *f*; ~ **shore** *n* INFR & DES Bockstütze *f*

**trial**: ~ **boring** *n* INFR & DES Probebohrung *f*, Aufschlußbohrung *f*, Versuchsbohrung *f*, MAT PROP Versuchsbohrung *f*; ~ **compaction** *n* INFR & DES Versuchsverdichtung *f*; ~ **dredging** *n* INFR & DES Versuchsbaggern *nt*; ~ **and error method** *n* INFR & DES empirisches Ermittlungsverfahren *nt*; ~ **hole** *n* INFR & DES Schürfloch *nt*; ~ **pile driving** *n* INFR & DES Proberammung *f*; ~ **pit** *n* INFR & DES Schürfgrube *f*; ~ **run** *n* BUILD MACHIN, HEAT & VENT Probelauf *m*

**triangle** *n* ARCH & BUILD Dreieck *nt*; ~ **of forces** *n* ARCH & BUILD, INFR & DES Kräftedreieck *nt*

**triangular 1.** *adj* ARCH & BUILD, TIMBER dreieckig; **2.** *n* ARCH & BUILD Dreieck *nt*

***triangular***: ~ **arch** *n* ARCH & BUILD Dreieckbogen *m*, Giebelbogen *m*; ~ **cleat** *n* ARCH & BUILD, TIMBER Dreikantleiste *f*; ~ **fillet** *n* ARCH & BUILD, TIMBER Dreikantleiste *f*; ~ **truss** *n* ARCH & BUILD Dreiecksbinder *m*

**triangulated**: ~ **frame** *n* ARCH & BUILD, TIMBER Dreiecksrahmen *m*

**triangulation** *n* ARCH & BUILD Triangulation *f*, INFR & DES *surveying* Dreiecksvermessung *f*; ~ **network** *n* ARCH & BUILD Landesvermessungsnetz *nt*; ~ **point** *n* ARCH & BUILD Triangulationspunkt *m*

**triaxial**: ~ **compression cell** *n* BUILD MACHIN Dreiaxialdruckgerät *nt*; ~ **test** *n* INFR & DES *roads* Dreiaxialprüfung *f*

**tribunal**: ~ **proceedings** *n pl* CONST LAW Gerichtsverhandlungen *f pl*

**trickle** *vi* WASTE WATER sickern

**trickle**: ~ **pool** *n* WASTE WATER Sickergrube *f*; ~ **ventilation** *n* HEAT & VENT, SOUND & THERMAL gering-kontinuierliche Belüftung *f*
**trickling**: ~ **filter** *n* INFR & DES Tropfkörperanlage *f*
**trigonometrical**: ~ **function** *n* ARCH & BUILD trigonometrische Funktion *f*
**trim 1.** *n* ARCH & BUILD Abgleich *m*, BUILD HARDW, TIMBER Deckleiste *f*; **2.** *vt* ARCH & BUILD abgleichen
**trimmer** *n* ARCH & BUILD Formkachel *f*, Streichbalken *m*, Wechselbalken *m*, STEEL Auswechselung *f*, STONE Formkachel *f*, TIMBER Wechsel *m*, Auswechselung *f*; ~ **beam** *n* ARCH & BUILD Streichbalken *m*, Wechselbalken *m*
**trimming** *n* TIMBER Auswechselung *f*
**tringle** *n* BUILD HARDW, TIMBER Deckleiste *f*
**trip**: ~ **lever** *n* BUILD MACHIN Auslösehebel *m*
**triumphal**: ~ **arch** *n* ARCH & BUILD Triumphbogen *m*
**Trombe**: ~ **wall** *n* SOUND & THERMAL, STONE Trombe-Wand *f*
**tropical**: ~ **hardwood** *n* TIMBER tropisches Hartholz *nt*
**trouble** *n* HEAT & VENT, INFR & DES, WASTE WATER Störung *f*
**trough**: ~ **bridge** *n* INFR & DES Trogbrücke *f*, ~ **gutter** *n* INFR & DES *roof* Kastenrinne *f*
**trowel 1.** *n* BUILD MACHIN Kelle *f*, Glättkelle *f*, Spachtel *f*, STONE Glättkelle *f*; **2.** *vt* STONE verstreichen, glattstreichen; ~ **off** CONCR, STONE, SURFACE ausspachteln
***trowel***: ~ **application** *n* ARCH & BUILD, SURFACE Verspachteln *nt*; ~ **finish** *n* STONE Glattstrich *m*; ~**-finished layer** *n* STONE Glattstrich *m*; ~ **plaster** *n* STONE Kellenputz *m*, Kellenwurf *m*
**truck** *n AmE* BUILD MACHIN (*cf lorry BrE*), BUILD MACHIN (*cf car AmE*), Lastkraftwagen *m* (*LKW*), *railway* Lore *f*; ~ **agitator** *n* BUILD MACHIN Nachmischer *m*; ~**-mixed concrete** *n* CONCR Lieferbeton *m*; ~ **trailer** *n* BUILD MACHIN Schleppzug *m*
**true**: **in** ~ **alignment** *phr* ARCH & BUILD fluchtgerecht, genau fluchtend; ~ **cohesion** *n* MAT PROP, SURFACE echte Kohäsion *f*; ~ **to dimensions** *adj* ARCH & BUILD maßgerecht; ~ **to profile** *adj* INFR & DES *soil excavation* profilgemäß
**trunk** *n* INFR & DES *tree* Stamm *m*
**truss 1.** *n* ARCH & BUILD *roof* Binder *m*, Gebinde *nt*, *strut brace* Sprengwerk *nt*, *scaffolding* Fachwerk *nt*, INFR & DES, MAT PROP *scaffolding* Fachwerk *nt*; **2.** *vt* ARCH & BUILD, INFR & DES abstützen
***truss***: ~ **bridge** *n* TIMBER Fachwerkbrücke *f*
**trussed**: ~ **beam** *n* ARCH & BUILD, TIMBER Fachwerkträger *m*, unterspannter Balken *m*; ~ **frame** *n* ARCH & BUILD versteifter Rahmen *m*; ~ **girder** *n* ARCH & BUILD, TIMBER Fachwerkbinder *m*, Fachwerkträger *m*; ~ **rafter roof** *n* STONE Fachwerkbinderdachstuhl *m*; ~ **roof** *n* TIMBER Fachwerkbinderdach *nt*; ~ **with sag rods** *adj* ARCH & BUILD unterspannt; ~ **wooden beam** *n* ARCH & BUILD, TIMBER Fachwerkholzträger *m*
***truss***: ~ **frame** *n* ARCH & BUILD Sprengwerk *nt*
**trussing** *n* TIMBER Fachwerk *nt*, Unterzug *m*
**trying**: ~ **plane** *n* BUILD MACHIN, TIMBER Langhobel *m*
**tub** *n* WASTE WATER Wanne *f*
**tube** *n* BUILD HARDW Hülse *f*, Rohr *nt*, Röhre *f*, ENVIRON Rohr *nt*, Schlauch *m*, HEAT & VENT, MAT PROP, WASTE WATER Rohr *nt*, Röhre *f*; ~ **clip** *n* BUILD HARDW Rohrschelle *f*; ~ **expander** *n* BUILD HARDW Rohraufweiter *m*; ~**-expanding press** *n* BUILD MACHIN Rohraufweitepresse *f*; ~ **fitting** *n* BUILD HARDW Rohrverschraubung *f*; ~**-jointing sleeve** *n* BUILD HARDW, HEAT & VENT, WASTE WATER Rohrhülse *f*; ~ **nest** *n* BUILD HARDW Rohrbündel *nt*; ~ **sleeve** *n* BUILD HARDW, HEAT & VENT, WASTE WATER Rohrhülse *f*; ~ **socket** *n* BUILD HARDW, HEAT & VENT, WASTE WATER Rohrhülse *f*; ~ **welding** *n* STEEL Rohrschweißung *f*
**tubing** *n* BUILD HARDW Rohrleitung *f*, Verrohrung *f*, ENVIRON Rohrnetz *nt*, Rohrleitung *f*
**tubular** *adj* ARCH & BUILD röhrenförmig; ~ **cylinder** *n* BUILD HARDW *lock* Tubularzylinder *m*; ~ **heating element** *n* HEAT & VENT Röhrenheizkörper *m*; ~ **pile** *n* ARCH & BUILD Hohlpfahl *m*, Rohrpfahl *m*, BUILD MACHIN, INFR & DES Hohlpfahl *m*; ~ **radiator** *n* HEAT & VENT Röhrenheizkörper *m*; ~ **scaffolding** *n* ARCH & BUILD Rohrgerüst *nt*; ~ **steel grating** *n* STEEL Stahlrohrgitter *nt*; ~ **steel scaffolding** *n* ARCH & BUILD, BUILD MACHIN Stahlrohrgerüst *nt*; ~ **well** *n* INFR & DES Rohrbrunnen *m*
**Tudor**: ~ **arch** *n* ARCH & BUILD Tudorbogen *m*
**tuff** *n* INFR & DES Tuff *m*, MAT PROP Traß *m*, STONE Tuff *m*
**tufted**: ~ **floor covering** *n* ARCH & BUILD, BUILD HARDW Nadelvliesteppichboden *m*
**tumbler** *n* BUILD HARDW *lock* Stiftzuhaltung *f*, Zuhaltung *f*, BUILD MACHIN Drehtrommel *f*, Fallmischer *m*, ENVIRON Drehtrommel *f*; ~ **lock** *n* BUILD HARDW Zuhaltungsschloß *nt*
**tundish** *n* BUILD HARDW, WASTE WATER Zwischenbehälter *m*
**tung**: ~ **oil** *n* SURFACE Holzöl *nt*
**tunnel** *n* ENVIRON, INFR & DES Tunnel *m*; ~ **axis** *n* INFR & DES Tunnelachse *f*; ~**-boring machine** *n* BUILD MACHIN Tunnelbohrmaschine *f*; ~ **floor** *n* INFR & DES Tunnelsohle *f*
**tunneling** *AmE see tunnelling BrE*
***tunnel***: ~ **invert** *n* INFR & DES Tunnelsohle *f*
**tunnelling** *n BrE* INFR & DES Tunnelbau *m*, Tunnelvortrieb *m*, Untertunnelung *f*; ~ **machine** *n BrE* BUILD MACHIN Tunnelbaumaschine *f*; ~ **technique** *n BrE* INFR & DES Tunnelbauverfahren *nt*
***tunnel***: ~ **lining** *n* INFR & DES Tunnelauskleidung *f*; ~ **portal** *n* INFR & DES Tunneleinfahrt *f*, Tunneleingang *m*, Tunnelmund *m*; ~ **section** *n* INFR & DES Tunnelabschnitt *m*; ~ **soffit** *n* INFR & DES Tunneldecke *f*; ~ **timbering** *n* INFR & DES, TIMBER Tunnelzimmerung *f*; ~ **vault** *n* INFR & DES Tonnengewölbe *nt*; ~ **wall** *n* INFR & DES Tunnelwandung *f*, Tunnelwand *f*
**tup** *n* BUILD MACHIN Fallgewicht *nt*, Fallbär *m*
**turbidity** *n* CONST LAW, ENVIRON, MAT PROP Trübheit *f*, Trübung *f*; ~ **coefficient** *n* ENVIRON Trübungskoeffizient *m*
**turbine** *n* ELECTR, ENVIRON Turbine *f*; ~ **blade** *n* ENVIRON Turbinenschaufel *f*; ~ **efficiency** *n* ENVIRON Turbinenleistungsvermögen *nt*; ~ **output** *n* ELECTR, ENVIRON Turbinenleistung *f*
**turbulence** *n* ARCH & BUILD, ENVIRON, HEAT & VENT, INFR & DES, WASTE WATER Turbulenz *f*, Wirbelströmung *f*
**turbulent**: ~ **flow** *n* ENVIRON, HEAT & VENT, INFR & DES, WASTE WATER turbulente Strömung *f*, Wirbelströmung *f*
**turf** *n* INFR & DES, MAT PROP Torf *m*
**turn** *n* ARCH & BUILD Biegung *f*, Drehung *f*, Windung *f*,

ELECTR Windung *f*, INFR & DES Biegung *f*; ~ **bridge** *n* INFR & DES Drehbrücke *f*
**turnbuckle** *n* BUILD HARDW *fence* Spannschloß *nt*
**turning** *n* BUILD MACHIN *crane* Drehen *nt*; ~ **bay** *n* INFR & DES Wendefläche *f*, Wendeplatz *m*; ~ **bridge** *n* INFR & DES Drehbrücke *f*; ~ **point** *n* INFR & DES Wendepunkt *m*; ~ **radius** *n* INFR & DES Wenderadius *m*
**turnkey** *adj* ARCH & BUILD schlüsselfertig
***turn***: ~ **knob** *n* BUILD HARDW Drehknopf *m*
**turnlock**: ~ **fastener** *n* BUILD HARDW *half turn* Drehverschluß *m*
***turn***: ~ **pin** *n* BUILD HARDW Drehstift *m*
**turret** *n* ARCH & BUILD Türmchen *nt*, Erkerturm *m*, Dachaufbau *m*
**tusk** *n* ARCH & BUILD Brustzapfenaufwölbung *f*, eingezapfter Mauerstein *m*; ~ **tenon** *n* TIMBER Einlaßzapfen *m*; ~ **tenon joint** *n* ARCH & BUILD Brustzapfenverbindung *f*
**twin**: ~ **bridge** *n* INFR & DES Doppelbrücke *f*; ~ **column** *n* ARCH & BUILD Doppelsäule *f*; ~ **pump** *n* BUILD MACHIN, WASTE WATER Zwillingspumpe *f*; ~ **screw extruder** *n* ENVIRON Doppelschneckenextruder *m*
**twist** *vt* ARCH & BUILD verdrehen, verdrillen
**twist**: ~ **gimlet** *n* BUILD MACHIN, TIMBER Drehbohrer *m*
**two**: ~**-bay** *adj* ARCH & BUILD zweifeldrig, zweischiffig; ~**-bay frame** *n* ARCH & BUILD Zweifeldrahmen *m*; ~**-cell hollow block** *n* STONE Zweikammerstein *m*; ~**-coat work** *n* STONE zweilagiger Putz *m*, SURFACE zweilagiger Anstrich *m*; ~**-dimensional** *adj* ARCH & BUILD, INFR & DES zweidimensional; ~**-flight** *adj* ARCH & BUILD zweiläufig; ~**-hinged arch** *n* ARCH & BUILD Zweigelenkbogen *m*; ~**-leaf door** *n* ARCH & BUILD doppelflügelige Tür *f*; ~**-legged** *adj* TIMBER zweibeinig, zweistielig; ~**-metal connector** *n* STEEL Zweimetallverbinder *m*; ~**-pinned arch** *n* ARCH & BUILD Zweigelenkbogen *m*; ~**-pipe system** *n* INFR & DES, WASTE WATER Trennentwässerung *f*, Trennkanalisation *f*, Trennsystem *nt*; ~ **semidetached houses** *n pl* ARCH & BUILD, INFR & DES Doppelhaus *nt*; ~**-span** *adj* ARCH & BUILD zweischiffig, zweifeldrig; ~**-stage tendering** *n* CONST LAW zweistufige Ausschreibung *f*; ~**-storey** *adj BrE* ARCH & BUILD zweistöckig; ~**-story** *AmE see two-storey BrE*
**tying**: ~**-in of brickwork** *n* STONE Mauerwerkverankerung *f*
**type** *n* ARCH & BUILD Typ *m*, Ausführung *f*, Modell *nt*; ~ **approval** *n* CONST LAW Betriebsgenehmigung *f*, Betriebserlaubnis *f*; ~ **of construction** *n* ARCH & BUILD Bauform *f*; ~ **of glass** *n* MAT PROP Glasart *f*, Glassorte *f*; ~ **of mixture** *n* MAT PROP Mischgutart *f*
**typical**: ~ **cross section** *n* ARCH & BUILD, HEAT & VENT, INFR & DES, WASTE WATER Regelquerschnitt *m*
**Tyton**: ~ **joint** *n* WASTE WATER Tyton-Muffe *f*

# U

**UCV** *abbr* (*upper calorific value*) ENVIRON höhere Energieleistung *f*

**UF** *abbr* (*urea formaldehyde*) ENVIRON, TIMBER UF (*Urea-Formaldehyd*)

**ultimate** *adj* ARCH & BUILD, STONE End-

**ultimate**: **~ compressive strength** *n* MAT PROP Bruchdruckfestigkeit *f*; **~ elongation** *n* MAT PROP Bruchdehnung *f*; **~ limit state** *n* ARCH & BUILD, CONCR, INFR & DES, MAT PROP, STEEL Grenzzustand *m* der Tragfähigkeit; **~ loading** *n* ARCH & BUILD, INFR & DES, MAT PROP Bruchbelastung *f*, Bruchlast *f*; **~ storage** *n* ENVIRON Endlagerung *f*; **~ strength** *n* ARCH & BUILD Bruchfestigkeit *f*, CONCR Endfestigkeit *f*; **~ stress** *n* INFR & DES, MAT PROP Bruchspannung *f*; **~ value** *n* ARCH & BUILD, INFR & DES Endwert *m*; **~ yield strength** *n* MAT PROP, STEEL Bruchstreckgrenze *f*

**ultrasonic**: **~ motion detector** *n* ELECTR Ultraschallbewegungsmelder *m*; **~ probe** *n* ELECTR, INFR & DES, MAT PROP Ultraschallsonde *f*

**ultraviolet**: **~ radiation** *n* ENVIRON, MAT PROP Ultraviolettbestrahlung *f*, Ultraviolettstrahlung *f*

**umbrella**: **~ roof** *n* ARCH & BUILD weit überhängendes Dach *nt*; **~ shell** *n* ARCH & BUILD Regenschirmschale *f*, Pilzschale *f*

**unbreakable** *adj* MAT PROP bruchfest

**uncontrolled**: **~ dumping** *n* ENVIRON ungeordnete Ablagerung *f*; **~ tipping** *n BrE* ENVIRON ungeordnete Deponie *f*, wilde Müllablagerung *f*

**uncouple** *vt* BUILD MACHIN abkuppeln

**uncracked** *adj* ARCH & BUILD, CONCR, MAT PROP, SURFACE ungerissen; **~ flexural stiffness** *n* ARCH & BUILD, CONCR, INFR & DES, MAT PROP Biegesteifigkeit *f* ohne Rißbildung

**under**: **~ construction** *adj* ARCH & BUILD im Bau; **~ public law** *phr* CONST LAW öffentlich-rechtlich

**underbed** *n* STONE Mörtelbett *nt*

**underbridge** *n* INFR & DES Unterführung *f*, Straßenunterführung *f*

**undercoat** *n* STONE Unterputz *m*, SURFACE Voranstrich *m*, Vorstreichfarbe *f*

**undercroft** *n* ARCH & BUILD Gruftgewölbe *nt*

**undercurrent** *n* ENVIRON Unterstrom *m*, Unterströmung *f*

**undercut 1.** *adj* INFR & DES unterhöhlt; **2.** *n* ARCH & BUILD Unterschneidung *f*; **3.** *vt* INFR & DES unterhöhlen

**underdesign** *vt* ARCH & BUILD unterdimensionieren

**underdesigned** *adj* ARCH & BUILD unterdimensioniert

**underfilling** *n* CONCR Unterfüllung *f*

**underfloor** *n* HEAT & VENT Unterboden *m*; **~ duct** *n* ARCH & BUILD, WASTE WATER Unterflurkanal *m*; **~ heating** *n* ARCH & BUILD, HEAT & VENT Fußbodenheizung *f*

**undergrade**: **~ crossing** *n* INFR & DES Straßenunterführung *f*, Wegunterführung *f*

**underground 1.** *adj* ARCH & BUILD Tief-, unter der Erde, unterirdisch, ELECTR, HEAT & VENT, INFR & DES, WASTE WATER erdverlegt, unter der Erde, unterirdisch; **2.** *n BrE* (*cf subway AmE*) INFR & DES U-Bahn *f*, Untergrundbahn *f*

**underground**: **~ cable** *n* INFR & DES Erdkabel *nt*; **~ cabling** *n* INFR & DES Erdverkabelung *f*; **~ depot** *n* ENVIRON Untertagedeponie *f* (*UTD*); **~ distribution chamber** *n* ARCH & BUILD Kabelkeller *m*; **~ garage** *n* ARCH & BUILD, INFR & DES Tiefgarage *f*; **~ hydrant** *n* INFR & DES, WASTE WATER Unterflurhydrant *m*; **~ parking** *n* INFR & DES Kellergarage *f*; **~ railroad** *n AmE* (*cf underground railway BrE*) INFR & DES U-Bahn *f*, Untergrundbahn *f*; **~ railway** *n BrE* (*cf underground railroad AmE*) INFR & DES U-Bahn *f*, Untergrundbahn *f*; **~ station** *n BrE* (*cf subway station AmE*) INFR & DES U-Bahnhof *m*; **~ storage** *n* INFR & DES unterirdische Lagerung *f*; **~ storage tank** *n* HEAT & VENT, WASTE WATER Erdtank *m*; **~ structure** *n* INFR & DES unterirdische Konstruktion *f*; **~ tank** *n* HEAT & VENT, WASTE WATER Erdtank *m*; **~ waste disposal** *n* ENVIRON, INFR & DES unterirdische Abfallbeseitigung *f*; **~ water** *n* ENVIRON, INFR & DES unterirdisches Wasser *nt*, Grundwasser *nt*; **~ water flow** *n* ENVIRON Grundwasserstrom *m*, INFR & DES unterirdischer Wasserfluß *m*

**underlay** *n* INFR & DES Bettung *f*, Bettungsschicht *f*, Trägerschicht *f*, Unterlage *f*

**underlining**: **~ felt** *n* MAT PROP Dachpappenunterlage *f*

**undermine** *vt* INFR & DES unterhöhlen, untergraben

**undermining** *n* INFR & DES Unterspülung *f*, Unterwaschung *f*

**underpass** *n* INFR & DES Unterführung *f*, Straßenunterführung *f*

**underpin** *vt* ARCH & BUILD untermauern, unterfangen, unterbauen

**underpinning** *n* ARCH & BUILD Unterfangung *f*

**underream 1.** *n* ARCH & BUILD *pile foundation*, INFR & DES Fußverbreiterung *f*; **2.** *vt* ARCH & BUILD unterschneiden

**underreamed**: **~ pile** *n* ARCH & BUILD Pfahl *m* mit angeschnittenem Fuß, Knollenfußpfahl *m*, INFR & DES Knollenfußpfahl *m*

**underside** *n* ARCH & BUILD Unterseite *f*, Untersicht *f*

**underview** *n* ARCH & BUILD Untersicht *f*

**underwash** *vt* INFR & DES ausspülen, auswaschen, *erode* unterspülen

**underwater**: **~ atomic explosion** *n* ENVIRON Unterwasseratomexplosion *f*; **~ concrete** *n* CONCR Unterwasserbeton *m*; **~ cutting blowpipe** *n* BUILD MACHIN Unterwasserschneidbrenner *m*; **~ excavation** *n* INFR & DES Unterwasserausgrabung *f*; **~ foundation** *n* INFR & DES Unterwassergründung *f*; **~ tunnel** *n* INFR & DES Unterwassertunnel *m*; **~ welding** *n* STEEL Unterwasserschweißen *nt*, UW-Schweißen *nt*

**undeterminable** *adj* ARCH & BUILD unbestimmbar

**undisturbed**: **~ soil** *n* INFR & DES ungestörter Boden *m*; **~ soil sample** *n* INFR & DES ungestörte Bodenprobe *f*

**uneven** *adj* ARCH & BUILD uneben

**unevenness** *n* ARCH & BUILD Unebenheit *f*

**unfenced** *adj* INFR & DES nicht eingezäunt
**unfinished** *adj* ARCH & BUILD unfertig; ~ **floor** *n* ARCH & BUILD Rohfußboden *m*
**unfixed** *adj* INFR & DES unbefestigt
**unflued**: ~ **heater** *n* HEAT & VENT abzugslose Heizung *f*
**unhardened** *adj* STEEL ungehärtet
**uniaxial** *adj* ARCH & BUILD, BUILD MACHIN, ELECTR einachsig
**uniform** *adj* ARCH & BUILD, INFR & DES einheitlich, gleichförmig; ~ **depth** *n* ARCH & BUILD, CONCR, INFR & DES, STEEL, TIMBER *beam* gleichmäßiger Querschnitt *m*; ~ **grain size** *n* MAT PROP gleichmäßige Korngröße *f*
**uniformly**: ~ **distributed load** *n* INFR & DES Gleichlast *f*, gleichmäßig verteilte Last *f*
**uniform**: ~ **settlement** *n* ARCH & BUILD, INFR & DES gleichmäßige Setzung *f*
**union** *n* ARCH & BUILD Anschlußstück *nt*, *threaded joint* Verschraubung *f*, BUILD HARDW Schraubmuffe *f*, HEAT & VENT *pipe* Rohrverschraubung *f*; ~ **cock** *n* WASTE WATER Anschlußhahn *m*; ~ **elbow** *n* WASTE WATER Überwurfkrümmer *m*
**union-T** *n* HEAT & VENT, WASTE WATER T-Stück *nt*
**unisex** *adj* CONST LAW unisex
**unit** *n* ARCH & BUILD Montageelement *nt*, *boring, milling* Maßeinheit *f*, Bauteil *nt*, Einheit *f*, INFR & DES Montageelement *nt*; ~ **of accommodation** *n* ARCH & BUILD, INFR & DES Wohnungseinheit *f*; ~ **of area** *n* ARCH & BUILD Flächeneinheit *f*; ~ **mass** *n* ARCH & BUILD, CONCR, MAT PROP Einheitsmasse *f*; ~ **price** *n* ARCH & BUILD Einheitspreis *m*; ~ **weight** *n* MAT PROP Gewichtseinheit *f*
**unload** *vt* ARCH & BUILD entlasten, BUILD MACHIN entladen, CONST LAW, INFR & DES entlasten
**unloaded** *adj* ARCH & BUILD entlastet, BUILD MACHIN entladen, CONST LAW, INFR & DES entlastet; ~ **state** *n* ARCH & BUILD, INFR & DES unbelasteter Zustand *m*
**unloading** *n* ARCH & BUILD, CONST LAW, INFR & DES Entlasten *nt*; ~ **area** *n* ARCH & BUILD Entladefläche *f*; ~ **curve** *n* MAT PROP Entlastungskurve *f*; ~ **hopper** *n* ENVIRON Entladebunker *m*; ~ **pump** *n* BUILD MACHIN Entladepumpe *f*
**unmistakable**: ~ **labeling** *AmE*, ~ **labelling** *BrE* *n* ELECTR unverwechselbare Bezeichnung *f*
**unplaned** *adj* TIMBER ungehobelt
**unplastered** *adj* STONE unverputzt
**unpressurized** *adj* WASTE WATER drucklos
**unreinforced** *adj* CONCR, MAT PROP nicht bewehrt, unbewehrt; ~ **concrete** *n* CONCR unbewehrter Beton *m*
**unrust** *vt* SURFACE entrosten
**unrusted** *adj* SURFACE entrostet
**unsaturated** *adj* INFR & DES ungesättigt; ~ **zone** *n* INFR & DES ungesättigter Bereich *m*
**unscrew** *vt* STEEL, TIMBER abschrauben
**unsound** *adj* ARCH & BUILD mangelhaft, *roof, vault* baufällig, CONST LAW fehlerhaft, mangelhaft, schadhaft
**unsoundness** *n* CONST LAW Mangelhaftigkeit *f*
**unspoilt**: ~ **land** *n* ENVIRON gewachsener Boden *m*
**unstable** *adj* INFR & DES instabil; ~ **equilibrium** *n* INFR & DES labiles Gleichgewicht *nt*; ~ **slope** *n* INFR & DES nicht standsichere Böschung *f*; ~ **state** *n* INFR & DES labiler Zustand *m*
**unstrained**: ~ **member** *n* ARCH & BUILD *framework* Nullstab *m*, ungespannter Stab *m*
**unstressed** *adj* ARCH & BUILD, MAT PROP spannungsfrei, schlaff
**unsuitable** *adj* ARCH & BUILD unangemessen, ungeeignet, untauglich
**unsupported**: ~ **beam** *n* ARCH & BUILD strebenloser Balken *m*
**untensioned** *adj* CONCR schlaff bewehrt, schlaff armiert
**untreated** *adj* CONCR, ENVIRON, MAT PROP, SURFACE roh, unbehandelt; ~ **water** *n* INFR & DES Rohwasser *nt*
**untrussed**: ~ **roof** *n* ARCH & BUILD binderlose Dachkonstruktion *f*
**unvented** *adj* HEAT & VENT unbelüftet
**unwanted**: ~ **heat gains** *n pl* HEAT & VENT unerwünschte Wärmezuwächse *m pl*
**unwatering** *n* ARCH & BUILD Grundwasserhaltung *f*, Wasserhaltung *f*, INFR & DES Wasserhaltung *f*
**upgrade** *n* ARCH & BUILD Steigung *f*
**upgrading** *n* ARCH & BUILD Aufwertung *f*
**uphill 1.** *adj* INFR & DES *road* ansteigend; **2.** *adv* INFR & DES bergauf
**uplift** *n* ARCH & BUILD Hebung *f*
**upper**: ~ **boom** *n* ARCH & BUILD, CONCR, INFR & DES, MAT PROP, STEEL Obergurt *m*; ~ **calorific value** *n* (*UCV*) ENVIRON höhere Energieleistung *f*; ~ **edge** *n* ARCH & BUILD Oberkante *f*; ~ **floor** *n* ARCH & BUILD Obergeschoß *nt*, obere Etage *f*; ~ **layer** *n* INFR & DES obere Schicht *f*; ~ **limit** *n* INFR & DES obere Grenze *f*; ~ **storage basin** *n* ENVIRON oberes Speicherbecken *nt*; ~ **storey** *n* *BrE* ARCH & BUILD Obergeschoß *nt*, obere Etage *f*; ~ **story** *AmE* *see upper storey BrE*
**upright 1.** *adj* ARCH & BUILD aufrechtstehend, senkrecht, stehend, INFR & DES senkrecht, stehend; **2.** *n* ARCH & BUILD *stanchion* Stiel *m*, Ständer *m*, Pfosten *m*, *stile* V-Stab *m*, Vertikalstab *m*
***upright***: ~ **course** *n* INFR & DES *masonry bricks* Rollschicht *f*
**uprooting** *n* INFR & DES Rodung *f*, Wurzelbeseitigung *f*
**upstand** *n* ARCH & BUILD Aufkantung *f*; ~ **beam** *n* ARCH & BUILD Überzug *m*; ~ **gutter** *n* WASTE WATER Aufkantungsrinne *f*
**upstream 1.** *adj* ENVIRON, INFR & DES stromaufwärts gelegen; **2.** *adv* ENVIRON, INFR & DES oberhalb, stromaufwärts
***upstream***: ~ **head** *n* ENVIRON Oberwasser *nt*; ~ **water level** *n* INFR & DES Oberwasserstand *m*
**uptake** *n* ARCH & BUILD *chimney* Fuchskanal *m*, HEAT & VENT Entlüftungsschacht *m*
**upward**: ~ **flow** *n* INFR & DES Aufwärtsströmung *f*
**upwards** *adv* INFR & DES aufwärts
**urban** *adj* INFR & DES städtisch; ~ **area** *n* INFR & DES Stadtgebiet *nt*; ~ **renewal** *n* INFR & DES Sanierung *f*; ~ **road** *n* INFR & DES Stadtstraße *f*; ~ **solid waste** *n* ENVIRON Siedlungsabfall *m*; ~ **traffic** *n* INFR & DES Stadtverkehr *m*; ~ **waste** *n* ENVIRON Siedlungsabfall *m*, kommunaler Abfall *m*
**urea**: ~ **formaldehyde** *n* (*UF*) ENVIRON, TIMBER Urea-Formaldehyd *nt* (*UF*); ~ **formaldehyde foam** *n* SOUND & THERMAL Urea-Formaldehydschaum *m*
**urinal** *n* WASTE WATER Urinalbecken *nt*, Wandurinal *nt*
**usable** *adj* MAT PROP verwendbar; ~ **by-products** *n pl* ENVIRON verwendbare Nebenprodukte *nt pl*
**use 1.** *n* ARCH & BUILD, INFR & DES Anwendung *f*, Benutzung *f*, Gebrauch *m*, Verwendung *f*; **2.** *vt* ARCH & BUILD, INFR & DES benutzen, gebrauchen

***use***: ~ **of building** *n* ARCH & BUILD, CONST LAW, INFR & DES Gebäudenutzung *f*
**used**: ~ **car dump** *n* ENVIRON Autofriedhof *m*
**useful** *adj* ARCH & BUILD, INFR & DES brauchbar; ~ **area** *n* ARCH & BUILD, INFR & DES Nutzfläche *f*
**user** *n* CONST LAW Nutzer *m*, Benutzer *m*
**usufruct** *n* CONST LAW Nießbrauch *m*
**usufructuary** *n* CONST LAW Nießbraucher *m*
**utensil** *n* BUILD MACHIN Gerät *nt*, Werkzeug *nt*
**utility** *n* ARCH & BUILD, ELECTR, HEAT & VENT, INFR & DES, WASTE WATER Versorgungseinrichtung *f*; ~ **core** *n* ELECTR, HEAT & VENT, WASTE WATER Installationskern *m*; ~ **line** *n* ARCH & BUILD Versorgungsleitung *f*; ~ **trench** *n* ARCH & BUILD, INFR & DES Leitungsgraben *m*
**utilization** *n* ARCH & BUILD Ausnutzung *f*, Auslastung *f*, Verwertung *f*, Verwendung *f*, CONST LAW, INFR & DES Verwertung *f*; ~ **curve** *n* ENVIRON Ausnutzungskurve *f*; ~ **cycle of materials** *n* ENVIRON Werkstoffnutzungszyklus *m*; ~ **factor** *n* ARCH & BUILD, INFR & DES Ausnutzungsgrad *m*, Belastungsgrad *m*; ~ **of rainwater** *n* ENVIRON Regenwassernutzung *f*
**u-value** *n* SOUND & THERMAL Wärmedurchlaßkoeffizient *m*
**UV**: ~ **radiation** *n* ENVIRON UV-Strahlung *f*

# V

**vacate** *vt* ARCH & BUILD *building* räumen

**vacuum** *n* ARCH & BUILD, MAT PROP Vakuum *nt*; **~ brazing** *n* ARCH & BUILD Vakuumhartlöten *nt*; **~ concrete** *n* CONCR Vakuumbeton *m*; **~ dewatering** *n* CONCR, INFR & DES Vakuumentwässerung *f*; **~ filter** *n* HEAT & VENT Vakuumfilter *m*; **~ lance** *n* INFR & DES Vakuumlanze *f*; **~ pump** *n* HEAT & VENT Vakuumpumpe *f*

**validity** *n* CONST LAW *of a contract* Rechtswirksamkeit *f*; **~ period** *n* CONST LAW Geltungsdauer *f*

**valley** *n* ARCH & BUILD Dachkehle *f*, Kehle *f*, Kehlung *f*; **~ beam** *n* ARCH & BUILD Rinnenbalkenträger *m*, Kehlbalken *m*; **~ board** *n* ARCH & BUILD Kehlbrett *nt*; **~ flashing** *n* STEEL Kehlblech *nt*; **~ girder** *n* ARCH & BUILD Kehlbalken *m*, Rinnenbalkenträger *m*; **~ gutter** *n* ARCH & BUILD Dachkehle *f*, WASTE WATER Kehlrinne *f*; **~ gutter walkway** *n* ARCH & BUILD Kehlrinnenweg *m*; **~ rafter** *n* ARCH & BUILD, TIMBER Kehlgratbalken *m*, Kehlsparren *m*; **~ tile** *n* ARCH & BUILD Kehlziegel *m*, Kehlstein *m*

**value 1.** *n* ARCH & BUILD, CONST LAW, INFR & DES Wert *m*; **2.** *vt* ARCH & BUILD, CONST LAW, INFR & DES bewerten, veranschlagen, MAT PROP bewerten

***value***: **~-added tax** *n* (*VAT*) CONST LAW Mehrwertsteuer *f* (*MwSt*)

**valve** *n* HEAT & VENT, WASTE WATER Absperrorgan *nt*, Ventil *nt*; **~ box** *n* INFR & DES Straßenkappe *f*

**vane** *n* INFR & DES Wetterfahne *f*, Windflügel *m*; **~ pump** *n* BUILD MACHIN Flügelpumpe *f*; **~ shear test** *n* MAT PROP, STEEL Flügelsondenversuch *m*

**vapor** *AmE see vapour BrE*

**vapour** *n BrE* ENVIRON, HEAT & VENT, SOUND & THERMAL Dampf *m*, Dunst *m*, Wrasen *m*; **~ barrier** *n BrE* ARCH & BUILD, INFR & DES Dachhaut *f*, SOUND & THERMAL Feuchtigkeitsabdichtung *f*, Feuchtigkeitssperrschicht *f*, Dachhaut *f*; **~ cooler** *n BrE* HEAT & VENT Brüdenkühler *m*; **~ pipe** *n BrE* HEAT & VENT *vent pipe in kitchens* Wrasenrohr *nt*; **~ plume** *n BrE* ENVIRON Dampffahne *f*; **~ pressure** *n BrE* HEAT & VENT, SOUND & THERMAL Dampfdruck *m*, Dampfspannung *f*; **~ pressure equalizing layer** *n BrE* SOUND & THERMAL Dampfdruckausgleichsschicht *f*; **~-proof** *adj BrE* SOUND & THERMAL dampfdicht; **~-proof barrier** *n BrE* SOUND & THERMAL Dampfsperre *f*; **~-proofing** *n BrE* SOUND & THERMAL Abdichtung *f*; **~-resistant** *adj BrE* SOUND & THERMAL dampfbeständig; **~ seal** *n BrE* SOUND & THERMAL Dampfsperre *f*; **~-tight** *adj BrE* SOUND & THERMAL dampfdicht

**variable 1.** *adj* ARCH & BUILD, CONST LAW variabel, veränderlich; **2.** *n* ARCH & BUILD, CONST LAW Variable *f*

***variable***: **~ head** *n* INFR & DES veränderliches Druckpotential *nt*; **~ size** *n* ARCH & BUILD, CONST LAW veränderliche Größe *f*

**variance** *n* ENVIRON Abweichung *f*

**variant** *n* ARCH & BUILD, CONST LAW Variante *f*

**variation** *n* ARCH & BUILD, CONST LAW, INFR & DES Schwankung *f*, Variation *f*; **~ order** *n* CONST LAW Projektänderung *f*

**varnish 1.** *n* MAT PROP, SURFACE Lack *m*, Lackfarbe *f*, Lackfirnis *m*; **2.** *vt* SURFACE lackieren

**varnishing** *n* SURFACE Lackieren *nt*, Lackierung *f*

**VAT** *abbr* (*value-added tax*) CONST LAW MwSt (*Mehrwertsteuer*)

**vault 1.** *n* ARCH & BUILD Gewölbe *nt*, Wölbung *f*; **2.** *vt* ARCH & BUILD wölben

**vaulted**: **~ ceiling** *n* ARCH & BUILD gewölbte Decke *f*

**vaulting**: **~ rib** *n* ARCH & BUILD Gewölberippe *f*

**V-belt** *n* BUILD MACHIN Keilriemen *m*

**VDE**: **~ test mark** *n* ELECTR VDE-Verbandszeichen *nt*

**vegetation** *n* INFR & DES Bewuchs *m*, Vegetation *f*

**vehicle** *n* BUILD MACHIN Fahrzeug *nt*

**vehicular**: **~ access** *n* INFR & DES Fahrzeugzufahrt *f*

**vein** *n* INFR & DES Ader *f*

**velocity** *n* BUILD MACHIN, HEAT & VENT, INFR & DES, WASTE WATER Geschwindigkeit *f*; **~ coefficient** *n* ENVIRON Geschwindigkeitskoeffizient *m*; **~ diagram** *n* ENVIRON Geschwindigkeitsdiagramm *nt*; **~ distribution** *n* HEAT & VENT, INFR & DES, WASTE WATER Geschwindigkeitsverteilung *f*; **~ of flow** *n* HEAT & VENT, INFR & DES, WASTE WATER Strömungsgeschwindigkeit *f*; **~ head** *n* HEAT & VENT, INFR & DES, WASTE WATER Geschwindigkeitshöhe *f*; **~ index** *n* MAT PROP Geschwindigkeitszahl *f*; **~ potential** *n* ENVIRON Geschwindigkeitspotential *nt*, Potential *nt* der Schallschnelle; **~ pressure** *n* HEAT & VENT, INFR & DES, WASTE WATER Geschwindigkeitsdruck *m*

**velvet**: **~ finish glass** *n* ARCH & BUILD, MAT PROP satiniertes Glas *nt*

**veneer 1.** *n* STONE nicht tragende Mauerverkleidung *f*, SURFACE Furnier *nt*, Furnierholz *nt*, dünne Verblendung *f*, nicht tragende Mauerverkleidung *f*, TIMBER Furnier *nt*; **2.** *vt* ARCH & BUILD, STONE verblenden, TIMBER furnieren

**veneering** *n* SURFACE Beplankung *f*, *facing* Furnieren *nt*

***veneer***: **~ plywood** *n* TIMBER Furnierplatte *f*

**Venetian**: **~ mosaic** *n* STONE Terrazzo *m*

**vent 1.** *n* HEAT & VENT Belüftungsöffnung *f*, Entlüfter *m*, Rauchabzug *m*, Rauchrohr *nt*, WASTE WATER Entlüfter *m*; **2.** *vt* BUILD HARDW belüften, entlüften, lüften, HEAT & VENT *pipes* entlüften, lüften, WASTE WATER *pipes* entlüften

***vent***: **~ cap** *n* BUILD HARDW Entlüftungsklappe *f*; **~ connection** *n* HEAT & VENT Entlüftungsstutzen *m*; **~ hole** *n* HEAT & VENT Entlüftungsloch *nt*, Luftloch *nt*

**ventiduct** *n* BUILD HARDW Lüftungskanal *m*

**ventilate** *vt* BUILD HARDW belüften, lüften, HEAT & VENT belüften, hinterlüften, lüften, ventilieren

**ventilated** *adj* HEAT & VENT belüftet; **~ roof** *n* ARCH & BUILD Kaltdach *nt*

**ventilating**: **~ door** *n* BUILD HARDW Lüftungstür *f*; **~ fan** *n* ARCH & BUILD, BUILD HARDW, HEAT & VENT Gebläse *nt*, Ventilator *m*

**ventilation** *n* ARCH & BUILD Hinterlüftung *f*, HEAT &

VENT Belüftung *f*, Entlüftung *f*, Lüftung *f*, Ventilation *f*; ~ **and air-conditioning ducts** *n pl* HEAT & VENT Lüftungs- und Klimakanäle *m pl*; ~ **and air-conditioning system** *n* HEAT & VENT raumlufttechnische Anlage *f*; ~ **block** *n* ARCH & BUILD, HEAT & VENT, STONE Lüftungsstein *m*; ~ **brick** *n* ARCH & BUILD, HEAT & VENT, STONE Lüftungsstein *m*; ~ **duct** *n* BUILD HARDW Lüftungskanal *m*; ~ **flap** *n* HEAT & VENT Lüftungsklappe *f*; ~ **grille** *n* HEAT & VENT Lüftungsgitter *nt*; ~ **line** *n* HEAT & VENT Lüftungsleitung *f*; ~ **louver** *AmE*, ~ **louvre** *BrE n* HEAT & VENT Lüftungsjalousie *f*; ~ **opening** *n* HEAT & VENT Belüftungsöffnung *f*, Lüftungsöffnung *f*; ~ **pipe** *n* HEAT & VENT Lüftungsrohr *nt*; ~ **rate** *n* HEAT & VENT Ventilationsleistung *f*; ~ **shaft** *n* HEAT & VENT Lüftungsschacht *m*; ~ **slot** *n* HEAT & VENT Belüftungsschlitz *m*; ~ **system** *n* HEAT & VENT Lüftungsanlage *f*; ~ **tile** *n* ARCH & BUILD Siebstein *m*

**ventilator** *n* ARCH & BUILD, BUILD HARDW, HEAT & VENT Gebläse *nt*, Lüfter *m*, Ventilator *m*; ~ **block** *n* ARCH & BUILD, STONE Lüfterstein *m*; ~ **tile** *n* ARCH & BUILD, STONE Lüfterstein *m*

**venting** *n* BUILD HARDW, HEAT & VENT, WASTE WATER Entlüften *nt*

***vent***: ~ **opening** *n* HEAT & VENT Belüftungsöffnung *f*, Lüftungsöffnung *f*; ~ **pipe** *n* BUILD HARDW Abzugsrohr *nt*, Entlüftungsrohr *nt*, HEAT & VENT Abzugsrohr *nt*, Dunstrohr *nt*, WASTE WATER Dunstrohr *nt*; ~ **pipe tile** *n* STONE, WASTE WATER Dunstrohrziegel *m*; ~ **plug** *n* BUILD HARDW Entlüfterstutzen *m*

**venturi**: ~ **scrubber** *n* ENVIRON *gas cleaning* Venturiwäscher *m*

***vent***: ~ **valve** *n* HEAT & VENT Entlüftungsventil *nt*; ~ **window** *n* ARCH & BUILD, BUILD HARDW Lüftungsflügel *m*

**veranda** *n* ARCH & BUILD Veranda *f*

**verdigris** *n* SURFACE Grünspan *m*

**verge** *n* ARCH & BUILD Giebelkante *f*, Saum *m*, *margin* Kante *f*, *edge* Rand *m*, *flashing* Ortgang *m*, INFR & DES Seitenstreifen *m*, *strip on road* Randstreifen *m*; ~ **batten** *n* TIMBER Vordachlattung *f*; ~ **flashing** *n* STEEL Ortblech *nt*, Ortgangverwahrung *f*; ~ **gutter** *n* WASTE WATER Ortgangrinne *f*

**vermiculite** *n* MAT PROP Vermiculite *m*, Blähglimmer *m*

**vermin**: ~-**proof** *adj* INFR & DES, MAT PROP ungezieferbeständig, ungezieferfest, ungeziefersicher; ~-**resistant** *adj* INFR & DES, MAT PROP ungezieferbeständig, ungezieferfest, ungeziefersicher

**vernacular**: ~ **materials** *n pl* MAT PROP einheimische Materialien *nt pl*

**vernier** *n* BUILD MACHIN Nonius *m*

**versatility** *n* INFR & DES Wendigkeit *f*

**vertex** *n* ARCH & BUILD Scheitel *m*, Scheitelpunkt *m*, Spitze *f*, Gipfel *m*

**vertical 1.** *adj* ARCH & BUILD Senkrecht-, senkrecht, lotrecht, INFR & DES senkrecht; **2.** *n* ARCH & BUILD *framework* V-Stab *m*

***vertical***: ~ **alignment** *n* ARCH & BUILD Vertikalanordnung *f*; ~ **bar** *n* ARCH & BUILD Vertikalstab *m*; ~ **bracing** *n* ARCH & BUILD senkrechte Absteifung *f*, Vertikalverband *m*; ~ **cladding rail** *n* BUILD HARDW, STEEL senkrechte Verkleidungsschiene *f*; ~ **curve radius** *n* ARCH & BUILD Ausrundungshalbmesser *m*; ~ **drainage** *n* INFR & DES vertikale Drainage *f*; ~ **joint** *n* STONE Stoßfuge *f*; ~ **load** *n* ARCH & BUILD, TIMBER Vertikallast *f*

**vertically**: ~ **adjustable** *adj* ARCH & BUILD höhenverstellbar; ~ **perforated brick** *n* MAT PROP, STONE Hochlochziegel *m* (*HLZ*)

***vertical***: ~ **member** *n* ARCH & BUILD Senkrechtstab *m*, Ständer *m*, Vertikalstab *m*, senkrechte Stange *f*; ~-**mounted mixer faucet** *n AmE* (*cf vertical-mounted mixer tap BrE*) WASTE WATER Wandbatterie *f*; ~-**mounted mixer tap** *n BrE* (*cf vertical-mounted mixer faucet AmE*) WASTE WATER Wandbatterie *f*; ~ **multimolding** *AmE*, ~ **multimoulding** *n BrE* CONCR *prefabricated concrete elements* Batteriefertigung *f*; ~ **permeability** *n* MAT PROP vertikale Durchlässigkeit *f*; ~ **pipe** *n* WASTE WATER Standrohr *nt*; ~ **plane** *n* ARCH & BUILD Aufriß *m*, INFR & DES Vertikalebene *f*; ~ **section** *n* ARCH & BUILD Senkrechtschnitt *m*; ~ **shear** *n* ARCH & BUILD, CONCR, MAT PROP, STEEL vertikales Abscheren *nt*; ~ **shoring** *n* ARCH & BUILD Unterfangung *f*; ~ **tilework** *n* STONE Wandbelag *m*

**vesica**: ~ **piscis** *n* ARCH & BUILD Fischblasenmotiv *nt*

**vessel** *n* BUILD HARDW, HEAT & VENT Behälter *m*, Hohlkörper *m*

**vestibule** *n* ARCH & BUILD Vestibül *nt*, Vorraum *m*, Eingangshalle *f*; ~ **door** *n* ARCH & BUILD Windfangtür *f*

**viaduct** *n* INFR & DES Viadukt *m*, *on pillars* Hochbrücke *f*

**vibrate** *vt* BUILD MACHIN *concrete* rüttelverdichten

**vibrated**: ~ **concrete** *n* CONCR Rüttelbeton *m*; ~ **crushed rock** *n* INFR & DES Rüttelschotter *m*

**vibrating**: ~ **compactor** *n* BUILD MACHIN Rüttelverdichter *m*, Vibrationsverdichter *m*; ~ **machinery** *n* ARCH & BUILD, BUILD MACHIN vibrierende Maschinen *f pl*; ~ **plate compactor** *n* BUILD MACHIN Plattenrüttler *m*; ~ **roller** *n* BUILD MACHIN Vibrationswalze *f*; ~ **screen** *n* BUILD MACHIN Vibrationssieb *nt*; ~ **sheepsfoot roller** *n* BUILD MACHIN Rüttelschaffußwalze *f*; ~ **table** *n* BUILD MACHIN Vibriertisch *m*; ~ **tamper** *n* BUILD MACHIN Rüttelstampfer *m*, Vibrationsramme *f*

**vibration** *n* ARCH & BUILD, BUILD MACHIN, MAT PROP, SOUND & THERMAL Flattern *nt*, Schwingung *f*, Vibration *f*; ~-**absorbing** *adj* ARCH & BUILD, SOUND & THERMAL schwingungsdämpfend; ~ **driver** *n* BUILD MACHIN Vibrationsramme *f*; ~ **measurement** *n* MAT PROP Schwingungsmessung *f*; ~ **period** *n* MAT PROP Schwingungsperiode *f*; ~ **ram** *n* BUILD MACHIN Vibrationsramme *f*; ~ **sieve** *n* BUILD MACHIN Rüttelsieb *nt*; ~ **test** *n* INFR & DES, MAT PROP Schwingungsversuch *m*

**vibrator** *n* BUILD MACHIN Vibrator *m*, Schwingungserreger *m*, Rüttler *m*, Verdichter *m*; ~ **table** *n* BUILD MACHIN Vibriertisch *m*

**vice** *n BrE* BUILD MACHIN Schraubstock *m*

**Vierendeel**: ~ **column** *n* ARCH & BUILD Vierendeelstütze *f*; ~ **girder** *n* ARCH & BUILD, MAT PROP Vierendeelträger *m*

**viewer** *n* BUILD HARDW Türgucker *m*, Weitwinkelspion *m*

**vinyl**: ~ **lacquer** *n* SURFACE Vinyllack *m*

**visco**: ~-**elasticity** *n* MAT PROP Visko-Elastizität *f*; ~-**plasticity** *n* MAT PROP Visko-Plastizität *f*

**viscosimeter** *n* BUILD MACHIN Viskosimeter *nt*

**viscosity** *n* MAT PROP Viskosität *f*, Zähflüssigkeit *f*

**viscous**: ~ **deformation** *n* MAT PROP viskose Deformation *f*

**vise** *AmE see vice BrE*
**visibility** *n* ARCH & BUILD Sicht *f*
**visible** *adj* ARCH & BUILD sichtbar; **~ area** *n* ARCH & BUILD Sichtfläche *f*; **~ concrete surface** *n* CONCR Betonsichtfläche *f*; **~ face** *n* ARCH & BUILD Sichtseite *f*
**visual**: **~ check** *n* ARCH & BUILD, CONST LAW, INFR & DES Augenscheinkontrolle *f*, Sichtkontrolle *f*, Sichtprüfung *f*; **~ examination** *n* ARCH & BUILD, CONST LAW, INFR & DES Augenscheinkontrolle *f*, Sichtkontrolle *f*, Sichtprüfung *f*
**vitreous**: **~ rock** *n* STONE glasiges Gestein *nt*
**vitrified**: **~ clay pipe** *n* STONE, WASTE WATER Steinzeugrohr *nt*
**void** *n* CONCR Hohlraum *m*, *cavity* Blase *f*, Nest *nt*, Pore *f*, SURFACE Blase *f*; **~ ratio** *n* CONCR Porenzahl *f*, MAT PROP Porengehalt *m*, Porenziffer *f*
**volcanic** *adj* INFR & DES vulkanisch; **~ ash** *n* INFR & DES vulkanische Asche *f*; **~ clay** *n* INFR & DES vulkanischer Ton *m*; **~ earthquake** *n* INFR & DES vulkanisches Erdbeben *nt*; **~ eruption** *n* INFR & DES vulkanischer Ausstoß *m*; **~ rock** *n* INFR & DES vulkanisches Gestein *nt*; **~ slag** *n* INFR & DES vulkanische Schlacke *f*; **~ soil** *n* INFR & DES vulkanischer Boden *m*; **~ zone** *n* INFR & DES vulkanisches Gebiet *nt*
**volcano** *n* INFR & DES Vulkan *m*
**voltage** *n* ELECTR Spannung *f*
**volume** *n* ARCH & BUILD Volumen *nt*, Rauminhalt *m*; **~ change** *n* MAT PROP Volumenänderung *f*; **~ decrease** *n* MAT PROP Volumenabnahme *f*; **~ increase** *n* MAT PROP Volumenerhöhung *f*; **~ level** *n* SOUND & THERMAL Lautstärkepegel *m*; **~ of sewage** *n* ENVIRON Abwasseranfall *m*, Abwassermenge *f*; **~ stability** *n* MAT PROP Raumbeständigkeit *f*, Raumkonstanz *f*
**volumetric**: **~ efficiency** *n* ENVIRON volumetrischer Wirkungsgrad *m*
**volute** *n* ARCH & BUILD Volute *f*, Spirale *f*, Schnecke *f*; **~ compass** *n* ARCH & BUILD Spiralenzirkel *m*
**voussoir** *n* ARCH & BUILD, STONE Bogenziegel *m*, Keilstein *m*
**vulcanized** *adj* STONE vulkanisiert; **~ fiber** *AmE*, **~ fibre** *BrE* *n* MAT PROP Vulkanfiber *f*
**V-unit** *n* ARCH & BUILD Faltwerk *nt*, Falte *f*

**wadding** *n* HEAT & VENT, SOUND & THERMAL Wattierung *f*, Zellstoffwatte *f*
**wafer**: **~-type valve** *n* WASTE WATER Einklemmventil *nt*, Einklemmarmatur *f*
**waffle** *n* ARCH & BUILD *ceiling* Kassette *f*, Kassettenfeld *nt*; **~ panel** *n* ARCH & BUILD Kassettenplatte *f*; **~ plate** *n* ARCH & BUILD Kassettenplatte *f*; **~ slab ceiling** *n* ARCH & BUILD, CONCR, TIMBER Kassettendecke *f*
**wages** *n* ARCH & BUILD, CONST LAW Arbeitslohn *m*
**wagon** *n BrE* (*cf car AmE*) BUILD MACHIN *railway* Lore *f*; **~ vault** *n* INFR & DES Tonnengewölbe *nt*
**wainscoting** *n* ARCH & BUILD Täfelung *f*
**waiting**: **~ hall** *n* ARCH & BUILD Wartehalle *f*
**waler** *n* CONCR, INFR & DES Brustholz *nt*, Gurt *m*, Gurtholz *nt*, Riegel *m*, Verbindungsstück *nt*, MAT PROP Brustholz *nt*, TIMBER Brustholz *nt*, Gurtholz *nt*
**waling** *n* CONCR, INFR & DES Brustholz *nt*, Gurt *m*, Gurtholz *nt*, Riegel *m*, Verbindungsstück *nt*, MAT PROP, TIMBER Brustholz *nt*
**walking**: **~ dragline** *n* BUILD MACHIN Schreitschürfbagger *m*; **~ line** *n* ARCH & BUILD *stairs* Lauflinie *f*, Ganglinie *f*
**walkway** *n* INFR & DES Fußgängerweg *m*, Gehweg *m*
**wall 1.** *n* ARCH & BUILD Wand *f*, Mauer *f*; **2. ~ in** *vt* STONE einmauern, zumauern, vermauern; **~ up** STONE einmauern, vermauern, zumauern
***wall***: **~ arch** *n* ARCH & BUILD Wandbogen *m*; **~ area** *n* ARCH & BUILD Wandfläche *f*; **~ bond** *n* STONE Mauerverband *m*; **~ box** *n* ELECTR Balkenaussparung *f*; **~ bracket** *n* ARCH & BUILD Wandhalterung *f*, Wandkonsole *f*; **~ breakthrough** *n* ARCH & BUILD Wanddurchbruch *m*; **~ bushing** *n* ARCH & BUILD Wanddurchführung *f*; **~ cladding** *n* ARCH & BUILD, SURFACE Wandverkleidung *f*; **~ coat** *n* SURFACE Wandanstrich *m*; **~ coping** *n* ARCH & BUILD Mauerabdeckung *f*; **~ covering** *n* ARCH & BUILD Wandverkleidung *f*, SURFACE Wandverkleidung *f*, *tiling* Wandbelag *m*, *wallpaper* Tapete *f*; **~ crane** *n* BUILD MACHIN Wandkran *m*; **~ crown** *n* ARCH & BUILD Mauerkrone *f*; **~ duct** *n* ARCH & BUILD Wanddurchführung *f*
**walled**: **~ enclosure** *n* ARCH & BUILD ummauerter Raum *m*
***wall***: **~ edge** *n* ARCH & BUILD Mauerkante *f*; **~ effect** *n* INFR & DES Wandeinfluß *m*; **~ footing** *n* ARCH & BUILD, CONCR, INFR & DES Mauergründung *f*; **~ foundation** *n* ARCH & BUILD, CONCR, INFR & DES Mauergründung *f*; **~ friction** *n* INFR & DES Wandreibung *f*; **~ glazing** *n* ARCH & BUILD Wandverglasung *f*; **~ hanging** *n* ARCH & BUILD Wandbehang *m*; **~ holdfast** *n* ARCH & BUILD Wandhalterung *f*; **~ hook** *n* ARCH & BUILD, BUILD HARDW, STONE Maueranker *m*; **~-hung toilet** *n* WASTE WATER wandhängendes WC *nt*; **~-hung urinal** *n* WASTE WATER Wandurinal *nt*
**walling** *n* ARCH & BUILD Mauerwerk *nt*, *action* Mauern *nt*, Gemäuer *nt*, Mauerung *f*, Wandsystem *nt*, Maurerarbeit *f*, *materials* Wandbaustoffe *m pl*, BUILD MACHIN Mauerwerk *nt*, STONE *action* Maurerarbeit *f*, Gemäuer *nt*, Mauerung *f*, *materials* Wandbaustoffe *m pl*, Mauerwerk *nt*
***wall***: **~ insulation** *n* SOUND & THERMAL Wandisolierung *f*; **~ joint** *n* ARCH & BUILD Wandfuge *f*; **~ junction** *n* STONE Wandanschluß *m*; **~ junction profile** *n* STONE Wandanschlußprofil *nt*; **~-mounted door stop** *n* BUILD HARDW Wandtürpuffer *m*; **~-mounted fan heater** *n* HEAT & VENT Wandlufterhitzer *m*; **~-mounted light fixture** *n* BUILD HARDW, ELECTR Wandleuchte *f*; **~-mounted sink** *n* WASTE WATER Wandausgußbecken *nt*; **~-mounted toilet** *n* WASTE WATER wandhängendes WC *nt*; **~-mounted urinal** *n* WASTE WATER Wandurinal *nt*; **~ offset** *n* ARCH & BUILD Mauerabsatz *m*; **~ opening** *n* ARCH & BUILD Wanddurchbruch *m*; **~ paint** *n* SURFACE Wandfarbe *f*, Wandanstrich *m*; **~ panel** *n* ARCH & BUILD Wandplatte *f*, Wandtafel *f*
**wallpaper 1.** *n* MAT PROP, SURFACE Tapete *f*; **2.** *vt* ARCH & BUILD, SURFACE tapezieren
**wallpapering** *n* SURFACE Tapezierarbeiten *f pl*
***wall***: **~ pillar** *n* ARCH & BUILD Wandpfeiler *m*; **~ plaster** *n* STONE Wandputz *m*; **~ plate** *n* ARCH & BUILD Balkenauflagerplatte *f*, Fußpfette *f*, TIMBER Lastverteilungsplatte *f*; **~ post** *n* ARCH & BUILD Wandsäule *f*; **~ pressure** *n* INFR & DES Wanddruck *m*; **~ recess** *n* ARCH & BUILD Wandnische *f*; **~ safe** *n* BUILD HARDW Wandsafe *m*; **~ shoring** *n* ARCH & BUILD Wandabsteifung *f*; **~ slab** *n* ARCH & BUILD Wandplatte *f*; **~ string** *n* ARCH & BUILD Wandwange *f*; **~ thickness** *n* ARCH & BUILD Wanddicke *f*, Wandstärke *f*; **~ tie** *n* ARCH & BUILD, BUILD HARDW, STONE Maueranker *m*, Wandanker *m*; **~ top** *n* ARCH & BUILD Mauerkrone *f*; **~-to-wall carpeting** *n* ARCH & BUILD Teppichboden *m*
**wane** *n* TIMBER Fehlkante *f*
**ward** *n* INFR & DES Schlüsselformblech *nt*
**warded**: **~ lock** *n* BUILD HARDW, INFR & DES Buntbartschloß *nt*
**warden**: **~-assisted** *adj* INFR & DES *for elderly residents* mit Aufsicht
**warehouse** *n* ARCH & BUILD Lagerhaus *nt*, Lager *nt*
**warm** *adj* ARCH & BUILD, INFR & DES, MAT PROP warm; **~-air heating** *n* HEAT & VENT Warmluftheizung *f*; **~ front condensation** *n* SOUND & THERMAL Warmluftfrontkondensation *f*
**warning** *n* BUILD HARDW Alarm *m*, CONST LAW, ELECTR Alarm *m*, Warnmeldung *f*, Warnung *f*; **~ system** *n* BUILD HARDW, CONST LAW, ELECTR Alarmanlage *f*, Warnanlage *f*
**warped**: **~ timber** *n* TIMBER verworfenes Holz *nt*
**warping** *n* MAT PROP Verwerfen *nt*, Verziehen *nt*
**warranty** *n* CONST LAW Garantie *f*, Gewährleistung *f*; **~ inspection** *n* CONST LAW Gewährleistungsabnahme *f*; **~ period** *n* CONST LAW Gewährleistungszeit *f*
**washable** *adj* SURFACE auswaschbar, waschfest, waschbeständig
**wash**: **~ basin** *n* WASTE WATER Waschtisch *m*, Waschbecken *nt*; **~ boring** *n* INFR & DES Spülbohren *nt*; **~ drilling** *n* INFR & DES Spülbohren *nt*

**washed**: ~ **concrete** *n* CONCR Waschbeton *m*
**washer** *n* BUILD HARDW Unterlagscheibe *f*, Beilagescheibe *f*
**washing**: ~ **drum** *n* BUILD MACHIN Waschtrommel *f*; ~ **machine** *n* WASTE WATER Waschmaschine *f*; ~ **screen** *n* BUILD MACHIN Waschsieb *nt*
**washout** *n* ENVIRON, INFR & DES Auswaschung *f*
***wash***: ~ **primer** *n* SURFACE Haftgrundierung *f*, Haftgrund *m*
**washproof** *adj* SURFACE *paint* waschbeständig, waschfest
**wastage** *n* INFR & DES Streu- und Bruchverlust *m*
**waste** *n* ARCH & BUILD, CONST LAW, ENVIRON, INFR & DES Abfall *m*, Müll *m*; ~ **acid** *n* ENVIRON Abfallsäure *f*; ~ **avoidance** *n* ENVIRON Abfallvermeidung *f*
**Waste**: ~ **Avoidance and Management Act** *n* ENVIRON Abfallgesetz *nt* (*AbfG*)
***waste***: ~ **building material** *n* MAT PROP Bauschutt *m*; ~ **chute** *n* ARCH & BUILD, INFR & DES Müllschacht *m*; ~ **collection** *n* CONST LAW, ENVIRON Müllabfuhr *f*; ~ **composition** *n* ENVIRON Hausmüllzusammensetzung *f*; ~ **crusher** *n* ENVIRON Abfallzerkleinerer *m*; ~ **denitrification** *n* ENVIRON Denitrierung *f* des Abfalls; ~ **disinfection** *n* ENVIRON Abfalldesinfektion *f*; ~ **disintegrator** *n* ENVIRON Abfallzerkleinerer *m*; ~ **disposal** *n* CONST LAW, ENVIRON Abfallbeseitigung *f*, Abfallentsorgung *f*, Müllabfuhr *f*, Müllbeseitigung *f*
**Waste**: ~ **Disposal Act** *n* ENVIRON Abfallgesetz *nt* (*AbfG*)
***waste***: ~ **disposal company** *n* ENVIRON Abfallbeseitigungsunternehmen *nt*; ~ **dump** *n* ENVIRON, INFR & DES Deponie *f*, Müllhalde *f*, Müllkippe *f* ; ~**-economical planning** *n* ENVIRON abfallwirtschaftliche Planung *f*; ~ **of energy** *n* ENVIRON Energieverschwendung *f*; ~ **exchange market** *n* ENVIRON Abfallbörse *f*; ~ **formation** *n* ENVIRON Abfallerzeugung *f*, Anfall *m* von Abfällen, Müllanfall *m*; ~ **fuel** *n* ENVIRON Abfallbrennstoff *m*; ~ **gas** *n* ENVIRON, HEAT & VENT Abgas *nt*; ~ **gas cleaning** *n* ENVIRON Abgasreinigung *f*; ~ **gas desulfurization** *AmE*, ~ **gas desulphurization** *BrE* *n* ENVIRON Abgasentschwefelung *f*; ~ **generator** *n* ENVIRON Abfallerzeuger *m*; ~ **glass** *n* ENVIRON Altglas *nt*; ~ **glass container** *n* ENVIRON Altglasbehälter *m*; ~ **heap** *n* INFR & DES Abfallhaufen *m*; ~ **heat** *n* ENVIRON, SOUND & THERMAL Abwärme *f*; ~ **incineration plant** *n* ENVIRON Müllverbrennungsanlage *f* (*MVA*); ~ **injection** *n* ENVIRON Befüllung *f*; ~ **lye** *n* ENVIRON Ablauge *f*; ~ **mass** *n* ENVIRON Deponiegut *nt*; ~ **neutralization** *n* ENVIRON Abfallneutralisation *f*; ~ **oil** *n* ENVIRON Altöl *nt*
**Waste**: ~ **Oil Act** *n* ENVIRON Altölgesetz *nt*
***waste***: ~ **oil preparation** *n* ENVIRON Altölaufbereitung *f*; ~ **oil recovery** *n* ENVIRON Altölrückgewinnung *f*; ~ **oil recycling** *n* ENVIRON Altölwiederverwertung *f*; ~ **paper** *n* ENVIRON Altpapier *nt*; ~ **paper compressor** *n* ENVIRON Altpapierkompressor *m*; ~ **paper preparation** *n* ENVIRON Altpapieraufbereitung *f*; ~ **paper recycling** *n* ENVIRON Altpapierrecycling *nt*; ~ **pipe** *n* ENVIRON, HEAT & VENT Ablauf *m*, INFR & DES Abwasserleitung *f*, Abwasserrohr *nt*, WASTE WATER Abflußrohr *nt*, Ablauf *m*, Abwasserleitung *f*, Abwasserrohr *nt*; ~ **processing** *n* ENVIRON Abfallbehandlung *f*, Abfallverwertung *f*, Umwandlung *f* von Abfallstoffen; ~ **producer** *n* ENVIRON Abfallerzeuger *m*, Abfallverursacher *m*; ~ **product** *n* ENVIRON nicht verwertbarer Rückstand *m*, *non-usable* Abfallprodukt *nt*, *recyclable* Sekundärrohstoff *m*; ~ **production** *n* ENVIRON Abfallerzeugung *f*, Anfall *m* von Abfällen, Müllanfall *m*; ~ **products of civilization** *n pl* ENVIRON Zivilisationsmüll *m*; ~ **pulp** *n* ENVIRON Abfallzellstoff *m*; ~ **recovery** *n* ENVIRON Abfallaufbereitung *f*; ~ **recycling plant** *n* CONST LAW, INFR & DES Müllwiedergewinnungsanlage *f*; ~ **segregation** *n* ENVIRON getrennte Abfallagerung *f*; ~ **site closure** *n* ENVIRON Deponieschluß *m*; ~ **site operation** *n* ENVIRON Deponiebetrieb *m*; ~ **sorting** *n* ENVIRON Sortierung *f* von Abfällen; ~ **sorting plant** *n* ENVIRON Abfallsortieranlage *f*; ~ **storage** *n* ENVIRON Abfallagerung *f*; ~ **tip** *n* *BrE* ENVIRON Deponie *f*, Müllkippe *f*, INFR & DES Deponie *f*, Mülldeponie *f*; ~ **treatment** *n* CONST LAW Müllaufbereitung *f*, ENVIRON Abfallbehandlung *f*, Abfallverwertung *f*, Müllaufbereitung *f*; ~ **treatment plant** *n* CONST LAW Müllaufbereitungsanlage *f*, ENVIRON Abfallverwertungsanlage *f*, INFR & DES Müllaufbereitungsanlage *f*; ~ **utilization plant** *n* ENVIRON Abfallverwertungsanlage *f*; ~ **water** *n* ENVIRON, INFR & DES, WASTE WATER Abwasser *nt*, Schmutzwasser *nt*; ~ **water analysis** *n* ENVIRON, WASTE WATER Abwasseranalyse *f*; ~ **water collection tank** *n* ENVIRON Abwassersammeltank *m*; ~ **water control** *n* ENVIRON Abwasserkontrolle *f*; ~ **water discharge** *n* ENVIRON Abwassereinleitung *f*; ~ **water disposal** *n* ENVIRON, INFR & DES, WASTE WATER Abwasserbeseitigung *f*; ~ **water fishpond** *n* ENVIRON Abwasserfischteich *m*; ~ **water purification plant** *n* ENVIRON, INFR & DES, WASTE WATER Kläranlage *f*; ~ **water renovation** *n* ENVIRON Abwassersanierung *f*; ~ **water sludge** *n* ENVIRON, INFR & DES Klärschlamm *m*; ~ **water stripper** *n* ENVIRON Abwasserstripper *m*; ~ **water treatment** *n* ENVIRON Abwasseraufbereitung *f*; ~ **water treatment plant** *n* ENVIRON, INFR & DES Kläranlage *f*, Klärwerk *nt*, WASTE WATER Kläranlage *f*; ~ **water treatment works** *n* ENVIRON, INFR & DES Klärwerk *nt*
**water 1.** *n* CONST LAW, ENVIRON Gewässer *nt*, WASTE WATER Wasser *nt*; **2.** *vt* WASTE WATER befeuchten, wässern
***water***: ~**-absorbing** *adj* MAT PROP wasserabsorbierend; ~ **absorption** *n* MAT PROP Wasserabsorption *f*, Wasseraufnahme *f*; ~ **balance** *n* CONST LAW Wasserbilanz *f*, Wassermengenwirtschaft *f*; ~ **barrier** *n* INFR & DES, SOUND & THERMAL Wassersperre *f*; ~**-based paint** *n* WASTE WATER Binderfarbe *f*, Dispersionsfarbe *f*, Wasserfarbe *f*; ~ **basin** *n* INFR & DES, WASTE WATER Wasserbecken *nt*; ~**-bearing layers** *n pl* INFR & DES Grundwasserstockwerke *nt pl*; ~**-bearing stratum** *n* INFR & DES wasserführende Schicht *f*; ~**-bound** *adj* INFR & DES *road construction* wassergebunden; ~ **butt** *n* WASTE WATER Wasserfaß *nt*; ~ **carriage** *n* INFR & DES Schiffstransport *m*; ~**-cement ratio** *n* (*w/c ratio*) CONCR, MAT PROP, STONE, WASTE WATER Wasserzementwert *m* (*W/Z Wert*); ~ **channel** *n* INFR & DES offener Wasserkanal *m*; ~ **charge** *n* CONST LAW Wasserabgabe *f*; ~ **circulation** *n* WASTE WATER Wasserzirkulation *f*, Wasserumwälzung *f*; ~ **cistern** *n* BUILD HARDW, WASTE WATER Wasserspülung *f*; ~ **cock** *n* WASTE WATER Wasserhahn *m*; ~ **column** *n* INFR & DES Wassersäule *f*; ~ **condition** *n* CONST LAW,

INFR & DES Wasserbeschaffenheit *f*; ~ **conditioning** *n* CONST LAW, INFR & DES Wasseraufbereitung *f*; ~ **content** *n* INFR & DES, MAT PROP, WASTE WATER Wassergehalt *m*; ~ **course** *n* WASTE WATER Spülkanal *m*, Wasserlauf *m*; ~ **cycle** *n* INFR & DES Wasserkreislauf *m*; ~ **demand** *n* INFR & DES Wasserbedarf *m*; ~ **depth** *n* ENVIRON, INFR & DES Pegelstand *m*, Wassertiefe *f*; ~ **distribution** *n* WASTE WATER Wasserverteilung *f*; ~ **drip** *n* ARCH & BUILD *window, door* Wasserschenkel *m*, Wassernase *f*, *weather* Wetterschenkel *m*, BUILD HARDW Wasserschenkel *m*; ~ **economy** *n* CONST LAW, INFR & DES Wasserwirtschaft *f*; ~ **extraction structure** *n* WASTE WATER Wasserabscheidebauwerk *nt*; ~ **fittings** *n pl* BUILD HARDW Wasserarmatur *f*; ~ **for domestic use** *n* CONST LAW, ENVIRON, INFR & DES, WASTE WATER Brauchwasser *nt*; ~ **for industrial use** *n* CONST LAW, ENVIRON, INFR & DES, WASTE WATER Brauchwasser *nt*; ~ **fountain** *n* INFR & DES Trinkbrunnen *m*

**waterfront** *n* INFR & DES Hafengebiet *nt*; ~ **development** *n* INFR & DES Hafenbauerschließung *f*

***water***: ~ **glass cement** *n* CONCR, MAT PROP Wasserglaskitt *m*; ~-**heating plant** *n* HEAT & VENT Warmwasserbereitungsanlage *f*; ~-**holding capacity** *n* MAT PROP Wasserbindungsvermögen *nt*

**watering** *n* INFR & DES Bewässerung *f*

***water***: ~ **intake** *n* ENVIRON Wassereintritt *m*, Wasserzulauf *m*, WASTE WATER Wasserentnahme *f*, Wasserzulauf *m*; ~ **jet** *n* INFR & DES Wasserstrahl *m*; ~ **layer** *n* INFR & DES Wasserschicht *f*; ~ **level** *n* INFR & DES Wasserspiegel *m*, Wasserstand *m*; ~ **level difference** *n* INFR & DES Wasserstandsdifferenz *f*; ~ **level indicator** *n* BUILD MACHIN, INFR & DES Wasserstandsanzeiger *m*; ~ **lifting** *n* INFR & DES Wasserhebung *f*; ~ **line** *n* WASTE WATER Wasserleitung *f*; ~-**logged** *adj* MAT PROP wassergesättigt; ~ **main** *n* HEAT & VENT, INFR & DES Hauptwasserleitung *f*; ~ **measuring tank** *n* CONCR Wassermeßgefäß *nt*; ~ **meter** *n* WASTE WATER Wasserzähler *m*, Wasseruhr *f*; ~ **overflow pipe** *n* WASTE WATER Wasserüberlauf *m*; ~ **penetration** *n* MAT PROP Wassereindringung *f*; ~ **pollutant** *n* ENVIRON wasserverunreinigender Stoff *m*; ~ **pollution** *n* CONST LAW Wasserverschmutzung *f*, ENVIRON Gewässerbelastung *f*, Wasserverunreinigung *f*, INFR & DES Wasserverschmutzung *f*; ~ **power** *n* ELECTR, ENVIRON, HEAT & VENT, INFR & DES Wasserkraft *f*; ~ **power station** *n* ELECTR, ENVIRON, HEAT & VENT, INFR & DES Wasserkraftwerk *nt*; ~ **pressure** *n* INFR & DES, WASTE WATER Wasserdruck *m*; ~ **pressure test** *n* INFR & DES, WASTE WATER Innendruckprüfung *f*

**waterproof 1.** *adj* MAT PROP wasserdicht, SOUND & THERMAL imprägniert; **2.** *vt* SOUND & THERMAL wasserdicht machen, imprägnieren, absperren

***waterproof***: ~ **concrete** *n* CONCR Sperrbeton *m*

**waterproofer** *n* SOUND & THERMAL Sperrstoff *m*

**waterproofing** *n* SOUND & THERMAL Sperrung *f*, Wasserabdichtung *f*, Bauwerksabdichtung *f*, Abdichtung *f*, WASTE WATER *layer, seal* Wasserabdichtung *f*; ~ **agent** *n* SOUND & THERMAL *concrete, mortar* Dichtungsmittel *nt*; ~ **finish** *n* CONCR, STONE Sperrputz *m*

***waterproof***: ~ **mortar** *n* CONCR, STONE Sperrmörtel *m*; ~ **sheeting** *n* SOUND & THERMAL Dichtungsbahn *f*

***water***: ~ **protection** *n* CONST LAW, ENVIRON Gewässerschutz *m*; ~ **protection area** *n* CONST LAW, INFR & DES Wasserschutzgebiet *nt*; ~ **purification** *n* ENVIRON Wasseraufbereitung *f*; ~ **quality** *n* ENVIRON, WASTE WATER Wassergüte *f*, Wasserqualität *f*; ~-**repellent concrete** *n* CONCR Sperrbeton *m*; ~-**repellent finish** *n* CONCR, STONE Sperrputz *m*; ~-**repellent mortar** *n* CONCR, STONE Sperrmörtel *m*; ~ **repeller** *n* SOUND & THERMAL Dichtungsmittel *nt*; ~ **reservoir** *n* INFR & DES Wasserbehälter *m*; ~ **resource** *n* INFR & DES Wasservorkommen *nt*

**waters** *n pl* CONST LAW, ENVIRON Gewässer *nt*

***water***: ~ **softener** *n* WASTE WATER Wasserenthärter *m*; ~ **softening unit** *n* INFR & DES, WASTE WATER Wasserenthärtungsanlage *f*; ~-**soluble** *adj* WASTE WATER wasserlöslich; ~-**soluble flux** *n* WASTE WATER wasserlösliches Flußmittel *nt*

**waterspout** *n* ARCH & BUILD Wasserspeier *m*

**waterstop** *n* SOUND & THERMAL Fugenband *nt*

***water***: ~ **supply** *n* ENVIRON, INFR & DES, WASTE WATER Wasserversorgung *f*; ~ **surface** *n* INFR & DES Wasseroberfläche *f*; ~ **table** *n* INFR & DES Wasserspiegel *m*; ~ **tanker** *n* WASTE WATER Wassertanker *m*; ~ **tension** *n* MAT PROP Wasserspannung *f*

**watertight** *adj* SOUND & THERMAL, WASTE WATER wasserdicht, wasserundurchlässig

**watertightness** *n* SOUND & THERMAL Wasserdichtigkeit *f*, WASTE WATER Wasserundurchlässigkeit *f*

***water***: ~ **tower** *n* WASTE WATER Wasserturm *m*; ~ **treatment** *n* INFR & DES, WASTE WATER Wasseraufbereitung *f*; ~ **treatment plant** *n* WASTE WATER Wasseraufbereitungsanlage *f*; ~ **truck** *n* BUILD MACHIN Wasserwagen *m*; ~ **turbine** *n* ENVIRON Wasserturbine *f*; ~ **valve** *n* WASTE WATER Wasserventil *nt*; ~ **vapor permeability** *AmE*, ~ **vapour permeability** *BrE n* MAT PROP, SOUND & THERMAL Wasserdampfdurchlässigkeit *f*; ~ **vein** *n* INFR & DES Wasserader *f*

**wave** *n* INFR & DES Welle *f*; ~ **crest** *n* ENVIRON Wellenberg *m*, Wellenkamm *m*; ~ **front** *n* INFR & DES Wellenfront *f*; ~ **momentum per meter of crest** *AmE*, ~ **momentum per metre of crest** *BrE n* ENVIRON Wellenbewegungsenergie *f* pro Meter Woge; ~ **propagation** *n* ENVIRON Wellenausbreitung *f*, Wellenfortpflanzung *f*; ~ **velocity** *n* MAT PROP Wellengeschwindigkeit *f*

**w/c**: ~ **ratio** *n* (*water-cement ratio*) CONCR, MAT PROP, STONE, WASTE WATER W/Z Wert *m* (*Wasserzementwert*)

**weak** *adj* MAT PROP schwach; ~ **soil** *n* INFR & DES weicher Boden *m*

**wear** *n* INFR & DES Verschleiß *m*, Abnutzung *f*

**wearing**: ~ **course** *n* ARCH & BUILD, INFR & DES Deckschicht *f*, Verschleißschicht *f*; ~ **surface** *n* INFR & DES Verschleißschicht *f*

***wear***: ~ **rate** *n* ARCH & BUILD Verschleißwert *m*; ~-**resistant** *adj* MAT PROP verschleißfest

**weather 1.** *n* ENVIRON Wetter *nt*, Witterung *f*; **2.** *vt* MAT PROP abwässern; **3.** *vi* MAT PROP, SURFACE verwittern, auswittern

***weather***: ~ **boarding** *n* ARCH & BUILD Wetterschürze *f*, Stülpschalung *f*, Holzverschalung *f*, TIMBER Holzverschalung *f*; ~ **check** *n* ARCH & BUILD Kehlung *f*, Riß *m*

**weathercoat** *n* SURFACE Außenanstrich *m*

***weather***: ~ **data** *n pl* CONST LAW Wetterdaten *nt pl*; ~-**dependent control** *n* HEAT & VENT *heating* witterungsgeführte Regelung *f*; ~ **drip** *n* ARCH & BUILD Tropfnase *f*, Wasserschenkel *m*, Wassernase *f*, Unterschneidung *f*, Wetterschenkel *m*, BUILD HARDW Wasserschenkel *m*; ~ **groove** *n* ARCH & BUILD Tropfnase *f*, Wassernase *f*, Unterschneidung *f*, Wasserschenkel *m*, Wetterschenkel *m*, BUILD HARDW Wasserschenkel *m*

**weathering** *n* ARCH & BUILD Wasserschräge *f*, BUILD HARDW *window* Wetterschutzabdeckung *f*, MAT PROP Verwitterung *f*, SOUND & THERMAL Feuchtigkeitsabdichtung *f*, SURFACE Verwitterung *f*; ~ **collar** *n* ARCH & BUILD Wettermanschette *f*; ~ **resistance** *n* SURFACE Klimafestigkeit *f*; ~ **test** *n* MAT PROP Verwitterungstest *m*

***weather***: ~ **molding** *AmE*, ~ **moulding** *BrE n* ARCH & BUILD Rinnleiste *f*

**weatherproof** *adj* MAT PROP, SURFACE witterungsbeständig, wetterbeständig

**weatherproofing** *n* SOUND & THERMAL Feuchtigkeitsabdichtung *f*

***weather***: ~-**resistant** *adj* MAT PROP, SURFACE wetterbeständig, witterungsbeständig

**weatherstripping** *n* SOUND & THERMAL Abdichten *nt*

**web** *n* ARCH & BUILD Steg *m*; ~ **member** *n* ARCH & BUILD Fütterungsstab *m*; ~ **reinforcement** *n* ARCH & BUILD, CONCR Schubbewehrung *f*, Stegbewehrung *f*; ~ **splice** *n* TIMBER Stegblechstoß *m*

**wedge** *n* ARCH & BUILD, STONE Keil *m*, Keilstein *m*

**wedged**: ~ **mortice and tenon joint** *n* ARCH & BUILD Keilzapfenverbindung *f*

***wedge***: ~ **finger jointing** *n* TIMBER Keilzinkung *f*

**weep**: ~ **drain** *n* WASTE WATER Sickerdränage *f*; ~ **hole** *n* STONE Entwässerungsrohr *nt*, Entwässerungsöffnung *f*, Leckloch *nt*

**weigh**: ~ **down** *vt* ARCH & BUILD niederdrücken

***weigh***: ~ **office** *n* ENVIRON *depot* Eingangskontrolle *f*

**weight** *n* MAT PROP Gewicht *nt*

**weighted**: ~ **noise level indicator** *n* ENVIRON Geräuschpegelanzeiger *m*

**weighting**: ~ **rail** *n* BUILD HARDW *blind* Beschwerungsschiene *f*

**weighty** *adj* MAT PROP schwer

**weir** *n* ENVIRON, INFR & DES Wehr *nt*; ~ **skimmer** *n* ENVIRON Wehrabschöpfer *m*

**weld** *vt* STEEL schweißen

**weldable** *adj* STEEL einschweißbar

**welded** *adj* BUILD MACHIN, STEEL geschweißt; ~ **asphalt sheeting** *n* SOUND & THERMAL Bitumenbahn *f*, Schweißbahn *f*; ~ **mesh** *n* ARCH & BUILD, CONCR, MAT PROP, STEEL geschweißte Maschen *f pl*; ~ **section** *n* ARCH & BUILD, MAT PROP, STEEL geschweißter Abschnitt *m*; ~ **structural steel** *n* ARCH & BUILD, MAT PROP, STEEL geschweißtes Stahlprofil *nt*; ~ **wire mesh** *n* ARCH & BUILD, CONCR, INFR & DES, MAT PROP, STEEL Betonstahlmatte *f*, Baustahlgewebe *nt*

**welder** *n* STEEL Schweißer *m*

**welding** *n* BUILD MACHIN, STEEL Schweißen *nt*, Schweißung *f*; ~ **blowpipe** *n* BUILD MACHIN Schweißbrenner *m*; ~ **circuit** *n* STEEL Schweißstromkreis *m*; ~ **cycle** *n* STEEL Schweißtakt *m*, Schweißzyklus *m*; ~ **equipment** *n* BUILD MACHIN Schweißapparat *m*, Schweißvorrichtung *f*; ~ **handshield** *n* BUILD MACHIN Schweißerhandschirm *m*, STEEL Handschutzschild *m*; ~ **helmet** *n* BUILD MACHIN Schweißhelm *m*; ~ **neck flange** *n* WASTE WATER Vorschweißflansch *m*; ~ **procedure** *n* STEEL Schweißverfahren *nt*; ~ **process** *n* STEEL Schweißprozeß *m*; ~ **program** *AmE*, ~ **programme** *BrE n* STEEL Schweißprogramm *nt*; ~ **seam** *n* STEEL Schweißnaht *f*; ~ **sequence** *n* STEEL Schweißfolge *f*; ~ **technique** *n* MAT PROP, STEEL Schweißmethode *f*; ~ **wire** *n* STEEL Schweißdraht *m*

**welfare**: ~ **building** *n* CONST LAW Sozialgebäude *nt*

**well** *n* ARCH & BUILD Aufzugsschacht *m*, BUILD HARDW Brunnen *m*, INFR & DES, WASTE WATER Brunnen *m*, Senkbrunnen *m*; ~-**burnt brick** *n* INFR & DES, MAT PROP, STONE Hartbrandziegel *m*, Vormauerziegel *m*; ~ **capacity** *n* INFR & DES Brunnenleistung *f*; ~ **casing** *n* BUILD MACHIN Bohrrohr *nt*, ENVIRON, INFR & DES Brunnenring *m*, Schachtring *m*, WASTE WATER Schachtring *m*; ~ **chamber** *n* INFR & DES, WASTE WATER Sammelkammer *f*; ~ **drain** *n* INFR & DES, WASTE WATER Sickerschacht *m*; ~ **foundation** *n* ARCH & BUILD, INFR & DES Brunnengründung *f*, Senkbrunnengründung *f*

**wellhead** *n* ENVIRON Brunnenkopf *m*; ~ **pressure** *n* ENVIRON Brunnenkopfdruck *m*; ~ **temperature** *n* ENVIRON Brunnenkopftemperatur *f*; ~ **valve** *n* ENVIRON Brunnenkopfventil *nt*

***well***: ~ **hole** *n* ARCH & BUILD Schneckenauge *nt*, Treppenauge *nt*; ~ **house** *n* WASTE WATER Brunnenstube *f*; ~ **logging** *n* ENVIRON radiometrische Bohrlochvermessung *f*; ~ **set** *adj* TIMBER satt aufliegend; ~ **sinking** *n* ENVIRON, INFR & DES Brunnenbau *m*; ~-**swept** *adj* ARCH & BUILD, CONST LAW besenrein

**wet** *vt* ARCH & BUILD anfeuchten, annässen, CONCR benetzen, MAT PROP, STEEL anfeuchten, STONE, SURFACE benetzen

**wet**: ~ **concrete** *n* CONCR, MAT PROP Frischbeton *m*

**wetland** *n* INFR & DES Sumpfland *nt*

***wet***: ~-**on-wet coating** *n* SURFACE Naß-auf-Naßbeschichtung *f*; ~-**on-wet method** *n* CONCR, SURFACE Naß-auf-Naß-Methode *f*; ~ **roof** *n* ARCH & BUILD Naßdach *nt*; ~ **rot** *n* ARCH & BUILD *wood*, MAT PROP, TIMBER Naßfäule *f*; ~ **sand** *n* INFR & DES nasser Sand *m*; ~ **screening** *n* INFR & DES Naßsiebung *f*, Naßsieben *nt*; ~ **soil** *n* INFR & DES nasser Boden *m*

**wetting** *n* ARCH & BUILD Anfeuchten *nt*; ~ **agent** *n* MAT PROP Netzmittel *nt*

**wharf** *n* INFR & DES Hafendamm *m*; ~ **construction** *n* INFR & DES Hafenbau *m*; ~ **development** *n* INFR & DES Hafenbauerschließung *f*

**wheel** *n* ARCH & BUILD, BUILD MACHIN, INFR & DES Rad *nt*

**wheelbarrow** *n* BUILD MACHIN Schubkarre *f*, Schubkarren *m*

**wheelchair** *n* CONST LAW Rollstuhl *m*; ~ **facilities** *n pl* ARCH & BUILD, CONST LAW, INFR & DES Rollstuhlanlagen *f pl*; ~ **ramp** *n* ARCH & BUILD, CONST LAW, INFR & DES Rollstuhlrampe *f*; ~ **space** *n* ARCH & BUILD, CONST LAW Rollstuhlplatz *m*, vorgesehener Raum *m* für Rollstuhl, INFR & DES Rollstuhlplatz *m*

**wheeled**: ~ **vehicle** *n* BUILD MACHIN Radfahrzeug *nt*

***wheel***: ~ **load** *n* INFR & DES Radlast *f*; ~ **pressure** *n* BUILD MACHIN, INFR & DES Raddruck *m*; ~ **window** *n* ARCH & BUILD Rundfenster *nt*, rundes Fenster *nt*, Fensterrosette *f*, gotische Rose *f*

**whirlpool** *n* WASTE WATER *in flow* Wirbel *m*; ~ **basin** *n* INFR & DES, WASTE WATER Tosbecken *nt*

**whiten** *vt* SURFACE weißen, kalken, tünchen

**whitewash** *vt* SURFACE weißen, kalken, tünchen

**whitewashing** *n* SURFACE Kalken *nt*

**wide** *adj* ARCH & BUILD, BUILD MACHIN, MAT PROP breit; **~-angle door viewer** *n* BUILD HARDW Weitwinkelspion *m*; **~-angle judas** *n* BUILD HARDW Weitwinkelspion *m*; **~-flanged beam** *n* ARCH & BUILD, CONCR, STEEL Breitflanschträger *m*; **~-flanged girder** *n* ARCH & BUILD, CONCR, STEEL Breitflanschträger *m*

**width** *n* ARCH & BUILD Breite *f*, Weite *f*

**winch** *n* BUILD HARDW Handkurbel *f*, BUILD MACHIN Winde *f*

**wind 1.** *n* ENVIRON Wind *m*; **2.** *vt* INFR & DES spulen, wickeln

***wind***: **~ brace** *n* ARCH & BUILD Windstrebe *f*, INFR & DES Querträger *m*, Windstrebe *f*, TIMBER Windstrebe *f*, Windverband *m*; **~ bracing** *n* ARCH & BUILD Windaussteifung *f*, INFR & DES Windverband *m*, TIMBER Windaussteifung *f*; **~ energy** *n* ENVIRON Windenergie *f*

**winder** *n* ARCH & BUILD Wendelstufe *f*

***wind***: **~ flow** *n* ARCH & BUILD, INFR & DES Windströmung *f*; **~ force** *n* ENVIRON Windkraft *f*; **~ gage** *AmE*, **~ gauge** *BrE* *n* ENVIRON Windgeschwindigkeitsmesser *m*; **~-induced** *adj* ARCH & BUILD durch Wind erzeugt

**winding** *n* ARCH & BUILD *stairs* Windung *f*, Wendelung *f*, ELECTR Windung *f*

***wind***: **~ load** *n* INFR & DES Windlast *f*; **~ loading** *n* ARCH & BUILD Windbelastung *f*, Windlast *f*

**windmill** *n* ENVIRON Windmühle *f*, Windmühlenrad *nt*, Windrad *nt*; **~ pump** *n* ENVIRON Windmotorpumpe *f*, Windturbinenpumpe *f*; **~ vane** *n* ENVIRON Windmühlenflügel *m*

**window** *n* ARCH & BUILD Fenster *nt*, Sichtfenster *nt*; **~ air-conditioning unit** *n* HEAT & VENT Fensterklimagerät *nt*; **~ area** *n* ARCH & BUILD Fensterfläche *f*; **~ bar** *n* BUILD HARDW Abdichtleiste *f*, Fenstersprosse *f*, Glasleiste *f*, Fenstergitter *nt*, STEEL Fenstergitter *nt*; **~ board** *n* ARCH & BUILD Fensterbrett *nt*, Simsbrett *nt*; **~ catch** *n* BUILD HARDW Fensterriegel *m*; **~-cleaning cradle** *n* ARCH & BUILD Fassadenlift *m*, Fensterputzwagen *m*; **~ column** *n* ARCH & BUILD Fensterpfeiler *m*; **~ drip** *n* ARCH & BUILD Tropfnase *f*; **~ fan** *n* HEAT & VENT Fensterventilator *m*; **~ fastener** *n* BUILD HARDW Fensterschließer *m*; **~ fittings** *n pl* BUILD HARDW Fensterbeschläge *m pl*; **~ frame** *n* ARCH & BUILD Blendrahmen *m*, Blockrahmen *m*, Fensterblendrahmen *m*, Fensterrahmen *m*, BUILD HARDW Fensterblendrahmen *m*; **~ gasket** *n* SOUND & THERMAL Fensterdichtungsprofil *nt*; **~ glass** *n* ARCH & BUILD, BUILD HARDW Fensterglas *nt*; **~ glazing** *n* ARCH & BUILD Fensterverglasung *f*; **~ grille** *n* BUILD HARDW, STEEL Fenstergitter *nt*; **~ guide rail** *n* ARCH & BUILD Fensterführungsschiene *f*; **~ hardware** *n* BUILD HARDW Fensterbeschläge *m pl*; **~ hinge** *n* BUILD HARDW Fensterband *nt*; **~ jamb** *n* ARCH & BUILD, TIMBER Fenstergewände *nt*

**windowless** *adj* ARCH & BUILD fensterlos

***window***: **~ lintel** *n* ARCH & BUILD Fenstersturz *m*; **~ opening** *n* ARCH & BUILD Fensteraussparung *f*, Fensteröffnung *f*, BUILD HARDW Fensteröffnung *f*; **~ pane** *n* ARCH & BUILD, BUILD HARDW Fensterscheibe *f*; **~ parapet** *n* ARCH & BUILD Fensterbrüstung *f*; **~ post** *n* ARCH & BUILD Fensterpfosten *m*, Setzholz *nt*; **~ rabbet** *n* ARCH & BUILD Fensteranschlag *m*; **~ recess arch** *n* ARCH & BUILD Fensternischenbogen *m*; **~ sash** *n* BUILD HARDW Schieberahmen *m*; **~ sealing fillet** *n* SOUND & THERMAL Fensterdichtleiste *f*; **~ sill** *n* ARCH & BUILD Fensterbank *f*, Sohlbank *f*; **~ sill duct** *n* ARCH & BUILD Fensterbankkanal *m*; **~ sill slab** *n* ARCH & BUILD Fensterbankplatte *f*; **~ surround** *n* ARCH & BUILD, TIMBER Fenstergewände *nt*; **~ transom** *n* ARCH & BUILD Fensterkämpfer *m*

***wind***: **~ power** *n* ENVIRON Windkraft *f*; **~ pressure** *n* ARCH & BUILD, INFR & DES Winddruck *m*; **~ protection** *n* ARCH & BUILD, ENVIRON Windschutz *m*; **~ speed data** *n pl* ENVIRON Windgeschwindigkeitsdaten *pl*; **~ suction** *n* ARCH & BUILD, INFR & DES Windsog *m*; **~ sway bracing** *n* ARCH & BUILD Windverband *m*; **~ tunnel** *n* INFR & DES Windkanal *m*; **~ uplift** *n* ARCH & BUILD, INFR & DES Windsog *m*; **~ uplift force** *n* ARCH & BUILD, INFR & DES Windsogkraft *f*; **~ velocity** *n* ENVIRON Windgeschwindigkeit *f*; **~ velocity cubed** *n* ENVIRON Kubikwindgeschwindigkeit *f*

**wing** *n* ARCH & BUILD Gebäudeflügel *m*, BUILD HARDW *of door* Türflügel *m*; **~ bolt** *n* BUILD HARDW Flügelschraube *f*; **~ nut** *n* BUILD HARDW Flügelmutter *f*; **~ screw** *n* BUILD HARDW Flügelschraube *f*; **~ wall** *n* ARCH & BUILD, BUILD HARDW, INFR & DES Flügelmauer *f*

**wipe**: **~-resistant** *adj* SURFACE *paint* wischbeständig

**wire 1.** *n* BUILD HARDW Draht *m*; **2.** *vt* ELECTR verdrahten

***wire***: **~ armoring** *AmE*, **~ armouring** *BrE* *n* ELECTR Drahtbewehrung *f*; **~ core** *n* MAT PROP Drahteinlage *f*; **~ cutters** *n pl* BUILD MACHIN Drahtschere *f*

**wired**: **~ glass** *n* BUILD HARDW Drahtglas *nt*, CONST LAW Welldrahtglas *nt*, MAT PROP Drahtglas *nt*; **~ pattern glass** *n* ARCH & BUILD Drahtornamentglas *nt*

***wire***: **~ fabric** *n* BUILD HARDW, MAT PROP Drahtgeflecht *nt*; **~ fence** *n* INFR & DES Drahtzaun *m*; **~ glass** *n* BUILD HARDW, MAT PROP Drahtglas *nt*; **~ mesh reinforcement** *n* ARCH & BUILD Mattenbewehrung *f*; **~ netting** *n* BUILD HARDW, MAT PROP Drahtgeflecht *nt*; **~ plaster ceiling** *n* ARCH & BUILD, STONE Drahtputzdecke *f*, Rabitzdecke *f*; **~ plaster wall** *n* ARCH & BUILD, STONE Drahtputzwand *f*, Rabitzwand *f*; **~ reinforcement** *n* MAT PROP Drahteinlage *f*; **~ ropeway** *n* INFR & DES, STEEL Drahtseilbahn *f*; **~ strain gage** *AmE*, **~ strain gauge** *BrE* *n* MAT PROP Dehnungsmeßstreifen *m* (*DMS*)

**wiring** *n* ELECTR Verkabelung *f*, Verdrahtung *f*; **~ diagram** *n* ELECTR Schaltbild *nt*, Schaltplan *m*

**withe** *n* ARCH & BUILD Schornsteinzunge *f*

**without**: **~ obligation** *adj* CONST LAW unverbindlich

**wood** *n* TIMBER Holz *nt*; **~-block pavement** *n* TIMBER Holzpflaster *nt*; **~-block paving** *n* TIMBER Holzpflaster *nt*; **~-boring beetle** *n* ENVIRON, MAT PROP, TIMBER Hausbockkäfer *m*; **~ brick** *n* TIMBER Holzziegel *m*; **~ composite** *n* MAT PROP, TIMBER Holzverbundstoff *m*; **~ conditioning** *n* TIMBER Holzkonditionierung *f*; **~ construction** *n* TIMBER Holzbau *m*

**wooden** *adj* TIMBER hölzern, aus Holz; **~ arch** *n* TIMBER Holzbogen *m*; **~ butt strap joint** *n* TIMBER Brettlasche *f*; **~ cleat** *n* TIMBER Holzlasche *f*; **~ crate** *n* TIMBER Holzverschlag *m*; **~ floor** *n* TIMBER Holzfußboden *m*; **~ girder** *n* ARCH & BUILD, TIMBER

Holzträger *m*; **~ joist ceiling** *n* TIMBER Holzbalkendecke *f*; **~ joist floor** *n* TIMBER Holzbalkenboden *m*; **~ lining** *n* TIMBER Holzbekleidung *f*; **~ panel** *n* TIMBER Holztafel *f*; **~ paneling** *AmE*, **~ panelling** *BrE n* TIMBER Holztäfelung *f*; **~ paver** *n* TIMBER Holzpflasterklotz *m*; **~ paving block** *n* TIMBER Holzpflasterklotz *m*; **~ post-and-beam structure** *n* ARCH & BUILD, TIMBER Holzständerkonstruktion *f*; **~ prefabricated construction** *n* ARCH & BUILD, TIMBER Holzmontagebau *m*; **~ skeleton construction** *n* ARCH & BUILD, TIMBER Holzskelettbau *m*; **~ sleeper** *n* TIMBER Holzschwelle *f*; **~ structure** *n* TIMBER Holzkonstruktion *f*; **~ transom** *n* TIMBER *window, door* Kämpferholz *nt*

***wood***: **~ fiber concrete** *AmE*, **~ fibre concrete** *BrE n* CONCR Holzfaserbeton *m*; **~ girder** *n* ARCH & BUILD, TIMBER Holzträger *m*; **~-impregnating plant** *n* TIMBER Holzimprägnieranlage *f*; **~ lacquer** *n* SURFACE, TIMBER Holzlack *m*; **~ moisture** *n* MAT PROP, TIMBER Holzfeuchte *f*, Holzfeuchtigkeit *f*; **~ oil** *n* TIMBER Holzöl *nt*; **~ panel construction** *n* TIMBER Holztafelbauweise *f*; **~ paving block** *n* TIMBER Holzpflasterklotz *m*; **~ preservative** *n* MAT PROP, SURFACE, TIMBER Holzschutzmittel *nt*; **~ primer** *n* SURFACE Holzgrundierung *f*; **~ rasp** *n* BUILD MACHIN Holzraspel *f*; **~ saw** *n* BUILD MACHIN Holzsäge *f*

**woodscrew** *n* TIMBER Holzschraube *f*

***wood***: **~ sealer** *n* SURFACE Holzgrundierung *f*; **~ shingle** *n* TIMBER Holzschindel *f*; **~ skeleton structure** *n* ARCH & BUILD, TIMBER Holzskelettbau *m*; **~ strength** *n* TIMBER Holzfestigkeit *f*; **~-turning lathe** *n* TIMBER Holzdrehbank *f*; **~ waste** *n* ENVIRON Holzabfall *m*; **~ wool** *n* ARCH & BUILD, SOUND & THERMAL Holzwolle *f*; **~ wool building slab** *n* MAT PROP Holzwollebauplatte *f*; **~ wool lightweight building board** *n* BUILD HARDW Holzwolleleichtbauplatte *f*

**woodwork** *n* TIMBER Holzarbeiten *f pl*, *process* Holzbau *m*, *wooden elements* Holzbauteile *m pl*

**woodworm** *n* ENVIRON, TIMBER Holzwurm *m*; **~ treatment** *n* CONST LAW, TIMBER Maßnahmen *f pl* gegen Holzwurmbefall

***wood***: **~ yard** *n* INFR & DES, TIMBER Holzlager *nt*

**wool** *n* SOUND & THERMAL Wolle *f*

**work 1.** *n* ARCH & BUILD Arbeit *f*, Leistung *f*, INFR & DES Werk *nt*; **2. ~ against the grain** *phr* TIMBER gegen die Faser arbeiten

**workability** *n* CONCR, MAT PROP Verarbeitbarkeit *f*

**workable** *adj* ARCH & BUILD, CONCR, MAT PROP, SURFACE bearbeitbar, verarbeitungsfähig, verarbeitbar

**workbench** *n* ARCH & BUILD Werkbank *f*, BUILD MACHIN Hobelbank *f*, Werkbank *f*

**workforce** *n* ARCH & BUILD, CONST LAW Belegschaft *f*, Personal *nt*

**working**: **~ capacity** *n* ENVIRON Arbeitsvermögen *nt*; **~ current** *n* ELECTR Betriebsstrom *m*; **~ drawing** *n* ARCH & BUILD, INFR & DES Ausführungszeichnung *f*, Werkplan *m*; **~ face** *n* ENVIRON Deponieoberfläche *f*; **~ hours** *n* ARCH & BUILD, CONST LAW Arbeitszeit *f*; **~ plan** *n* ARCH & BUILD Bauplan *m*; **~ platform** *n* ARCH & BUILD *scaffold* Arbeitsbühne *f*; **~ pressure** *n* BUILD MACHIN Arbeitsdruck *m*, HEAT & VENT, WASTE WATER Betriebsdruck *m*; **~ safety** *n* BUILD MACHIN, CONST LAW Betriebssicherheit *f*; **~ schedule** *n* ARCH & BUILD, INFR & DES Arbeitsablauf *m*, Arbeitsablaufplan *m*, Bauablaufplan *m*; **~ stress** *n* INFR & DES Gebrauchsspannung *f*; **~ surface** *n* ARCH & BUILD Arbeitsfläche *f*; **~ temperature** *n* BUILD MACHIN Arbeitstemperatur *f*

**workmanlike** *adj* CONST LAW fachgerecht

**workmanship** *n* ARCH & BUILD *execution* Ausführung *f*, Verarbeitung *f*, *skill* Kunstfertigkeit *f*, CONST LAW *skill* Kunstfertigkeit *f*

***work***: **~ noise** *n* SOUND & THERMAL Arbeitslärm *m*; **~ progress** *n* ARCH & BUILD Arbeitsfortgang *m*; **~ schedule** *n* ARCH & BUILD, CONST LAW, INFR & DES Ablaufplan *m*, Bauablaufplan *m*

**workshop**: **~ connection** *n* STEEL, TIMBER Werkstattverbindung *f*

**workspace** *n* ARCH & BUILD, INFR & DES Arbeitsraum *m*

**worm**: **~ feed** *n* BUILD MACHIN Schneckenbeschickung *f*; **~ fence** *n* INFR & DES Scherengitter *nt*

**woven**: **~ filter medium** *n* MAT PROP Filtermatte *f*

**wreath** *n* ARCH & BUILD *wooden stairs* Kranz *m*, Kropfstück *nt*, Krümmling *m*, Kröpfling *m*; **~ piece** *n* ARCH & BUILD *wooden stairs* Kropfstück *nt*, Krümmling *m*, Kröpfling *m*

**wrecking** *n* ARCH & BUILD Abbruch *m*; **~ permit** *n* ARCH & BUILD, CONST LAW, ENVIRON Abbruchgenehmigung *f*

**wrench** *n* BUILD MACHIN Schraubenschlüssel *m*

**wrought**: **~ iron** *n* ARCH & BUILD Schmiedeeisen *nt*

**wye** *n* WASTE WATER Hosenrohr *nt*

# X

**x-axis** *n* ARCH & BUILD Abszissenachse *f*, X-Achse *f*
**X-brace** *n* ARCH & BUILD, INFR & DES Kreuzband *nt*
**X-bracing** *n* ARCH & BUILD, INFR & DES Kreuzverstrebung *f*, Verschwertung *f*
**X-ray** *n* ENVIRON, INFR & DES, MAT PROP *test* Röntgenstrahl *m*
**xylolite**: ~ **slab** *n* STONE Steinholzplatte *f*
**xylophagous** *adj* ENVIRON, TIMBER *pest* holzfressend

# Y

**yard** *n* INFR & DES Hof *m*, *measurement* Yard *nt*
**yardage** *n* ARCH & BUILD Länge *f*
***yard***: **~ gully** *n* WASTE WATER Hofablauf *m*, Hofeinlauf *m*; **~ gully hole** *n* WASTE WATER Hofsinkkasten *m*; **~ inlet** *n* WASTE WATER Hofablauf *m*, Hofsinkkasten *m*, Hoftopf *m*, *pipe* Hofeinlauf *m*; **~ pump** *n* WASTE WATER Hofpumpe *f*
**y-axis** *n* ARCH & BUILD Ordinatenachse *f*, Y-Achse *f*
**year**: **~ of construction** *n* ARCH & BUILD Baujahr *nt*
**yield 1.** *n* ARCH & BUILD Ergeben *nt*, ENVIRON Ertrag *m*, Gewinn *m*, Produktion *f*, MAT PROP plastische Verformung *f*, Fließen *nt*; **2.** *vt* ARCH & BUILD, BUILD HARDW *total* ergeben; **3.** *vi* MAT PROP fließen
***yield***: **~ criterium** *n* MAT PROP Fließkriterium *nt*; **~ law** *n* MAT PROP Fließgesetz *nt*; **~ line** *n* MAT PROP Bruchlinie *f*; **~ line method** *n* INFR & DES Bruchlinientheorie *f*; **~ point** *n* ARCH & BUILD, CONCR, MAT PROP, STEEL Elastizitätsgrenze *f*, Streckgrenze *f*; **~ strength** *n* ARCH & BUILD, MAT PROP, STEEL Streckgrenze *f*
**Y-junction** *n* INFR & DES Gabelung *f*, Straßengabelung *f*
**Y-level** *n* BUILD MACHIN Vermessungsinstrument *nt*
**Young**: **~'s modulus** *n* CONCR, MAT PROP, STEEL Elastizitätsmodul *m* (*E-Modul*)
**Y-tube** *n* WASTE WATER Hosenrohr *nt*

# Z

**zenith**: ~ **angle** *n* ENVIRON Zenitwinkel *m*
**zeolite** *n* CONCR, MAT PROP Zeolith *m*; ~ **cement composite** *n* CONCR Zeolithzementverbundstoff *m*
**zero** *n* ARCH & BUILD Null *f*; ~ **level** *n* INFR & DES Normalnull *nt* (*NN*); ~ **line** *n* ARCH & BUILD, INFR & DES Nullinie *f*; ~ **point of moment** *n* ARCH & BUILD, INFR & DES Momentennullpunkt *m*
**zigzag**: ~ **riveting** *n* STEEL Versatznietung *f*
**zinc** *n* MAT PROP Zink *nt*; ~ **coating** *n* SURFACE Verzinkung *f*; ~ **diecasting** *n* MAT PROP Zinkdruckguß *m*; ~ **paint** *n* SURFACE Zinkfarbe *f*, Zinkschutzfarbe *f*
**zone** *n* ARCH & BUILD Bereich *m*, Feld *nt*, Zone *f*, CONST LAW Bereich *m*, INFR & DES Bereich *m*, Feld *nt*; ~ **of saturation** *n* INFR & DES Sättigungszone *f*
**zoning**: ~ **map** *n* CONST LAW Bebauungsplan *m*, Flächennutzungsplan *m*ÿ